Biotechnology

Second Edition

Volume 1
Biological Fundamentals

Biotechnology

Second Edition

Fundamentals

Volume 1
Biological Fundamentals

Volume 2
Genetic Fundamentals and
Genetic Engineering

Volume 3
Bioprocessing

Volume 4
Measuring, Modelling, and Control

Products

Volume 5
Genetically Engineered Proteins and
Monoclonal Antibodies

Volume 6
Products of Primary Metabolism

Volume 7
Products of Secondary Metabolism

Volume 8
Biotransformations

Special Topics

Volume 9
Enzymes, Biomass, Food and Feed

Volume 10
Special Processes

Volume 11
Environmental Processes

Volume 12
Modern Biotechnology:
Legal, Economic and
Social Dimensions

Distribution:

VCH, P. O. Box 101161, D-6940 Weinheim (Federal Republic of Germany)

Switzerland: VCH, P. O. Box, CH-4020 Basel (Switzerland)

United Kingdom and Ireland: VCH (UK) Ltd., 8 Wellington Court, Cambridge CB1 1HZ (England)

USA and Canada: VCH, 220 East 23rd Street, New York, NY 10010–4606 (USA)

Japan: VCH, Eikow Building, 10-9 Hongo 1-chome, Bunkyo-ku, Tokyo 113 (Japan)

ISBN 3-527-28311-0 (VCH, Weinheim)
Set ISBN 3-527-28310-2 (VCH, Weinheim)

ISBN 1-56081-151-X (VCH, New York)
Set ISBN 1-56081-602-3 (VCH, New York)

A Multi-Volume Comprehensive Treatise

Biotechnology

Second, Completely Revised Edition

Edited by
H.-J. Rehm and G. Reed
in cooperation with
A. Pühler and P. Stadler

Volume 1

Biological Fundamentals

Edited by
H. Sahm

Weinheim · New York
Basel · Cambridge · Tokyo

Series Editors:
Prof. Dr. H.-J. Rehm
Institut für Mikrobiologie
Universität Münster
Corrensstraße 3
D-4400 Münster

Dr. G. Reed
2131 N. Summit Ave.
Apartment #304
Milwaukee, WI 53202-1347
USA

Prof. Dr. A. Pühler
Biologie VI (Genetik)
Universität Bielefeld
P.O. Box 100131
D-4800 Bielefeld 1

Dr. P. J. W. Stadler
Bayer AG
Verfahrensentwicklung Biochemie
Leitung
Friedrich-Ebert-Straße 217
D-5600 Wuppertal 1

Volume Editor:
Prof. Dr. H. Sahm
Institut für
Biotechnologie
Forschungszentrum GmbH
P.O. Box 1913
D-5170 Jülich

Published jointly by
VCH Verlagsgesellschaft mbH, Weinheim (Federal Republic of Germany)
VCH Publishers Inc., New York, NY (USA)

Editorial Director: Dr. Hans-Joachim Kraus
Editorial Manager: Christa Maria Schultz
Copy Editor: Karin Dembowsky
Production Director: Maximilian Montkowski
Production Manager: Dipl. Wirt.-Ing. (FH) Hans-Jochen Schmitt

Library of Congress Card No.: applied for

British Library Cataloguing-in-Publication Data:
A catalogue record for this book is available from the British Library

Die Deutsche Bibliothek – CIP-Einheitsaufnahme
Biotechnology : a multi volume comprehensive treatise / ed. by
H.-J. Rehm and G. Reed. In cooperation with A. Pühler and P.
Stadler. – 2., completely rev. ed. – Weinheim; New York;
Basel; Cambridge; Tokyo: VCH.
NE: Rehm, Hans J. [Hrsg.]

2., completely rev. ed.
Vol. 1. Biological fundamentals / ed. by H. Sahm
– 1993
ISBN 3-527-28311-0 (Weinheim)
ISBN 1-56081-151-X (New York)
NE: Sahm, Hermann [Hrsg.]

Printed on acid-free and low-chlorine paper.

Composition and Printing: Zechnersche Buchdruckerei, D-6720 Speyer.
Bookbinding: Klambt-Druck GmbH, D-6720 Speyer
Printed in the Federal Republic of Germany

Preface

In recognition of the enormous advances in biotechnology in recent years, we are pleased to present this Second Edition of "Biotechnology" relatively soon after the introduction of the First Edition of this multi-volume comprehensive treatise. Since this series was extremely well accepted by the scientific community, we have maintained the overall goal of creating a number of volumes, each devoted to a certain topic, which provide scientists in academia, industry, and public institutions with a well-balanced and comprehensive overview of this growing field. We have fully revised the Second Edition and expanded it from ten to twelve volumes in order to take all recent developments into account.

These twelve volumes are organized into three sections. The first four volumes consider the fundamentals of biotechnology from biological, biochemical, molecular biological, and chemical engineering perspectives. The next four volumes are devoted to products of industrial relevance. Special attention is given here to products derived from genetically engineered microorganisms and mammalian cells. The last four volumes are dedicated to the description of special topics.

The new "Biotechnology" is a reference work, a comprehensive description of the state-of-the-art, and a guide to the original literature. It is specifically directed to microbiologists, biochemists, molecular biologists, bioengineers, chemical engineers, and food and pharmaceutical chemists working in industry, at universities or at public institutions.

A carefully selected and distinguished Scientific Advisory Board stands behind the series. Its members come from key institutions representing scientific input from about twenty countries.

The volume editors and the authors of the individual chapters have been chosen for their recognized expertise and their contributions to the various fields of biotechnology. Their willingness to impart this knowledge to their colleagues forms the basis of "Biotechnology" and is gratefully acknowledged. Moreover, this work could not have been brought to fruition without the foresight and the constant and diligent support of the publisher. We are grateful to VCH for publishing "Biotechnology" with their customary excellence. Special thanks are due Dr. Hans-Joachim Kraus and Christa Schultz, without whose constant efforts the series could not be published. Finally, the editors wish to thank the members of the Scientific Advisory Board for their encouragement, their helpful suggestions, and their constructive criticism.

H.-J. Rehm
G. Reed
A. Pühler
P. Stadler

Scientific Advisory Board

Contents

Introduction 1
H. Sahm

I. General Aspects

1 Cell Structure 5
F. Mayer

2 Metabolism 47
R. Krämer, G. Sprenger

3 Growth of Microorganisms 111
C. H. Posten, C. L. Cooney

4 Overproduction of Metabolites 163
O. M. Neijssel, M. J. Teixeira de Mattos, D. W. Tempest

5 Metabolic Design 189
H. Sahm

6 Special Morphological and Metabolic Behavior of Immobilized Microorganisms 223
H.-J. Rehm, S. H. Omar

II. Biology of Industrial Organisms

7 Methanogens 251
H. König

8 Methylotrophs 265
L. Dijkhuizen

9 Clostridia 285
H. Bahl, P. Dürre

10 Lactic Acid Bacteria 325
M. Teuber

11 *Bacillus* 367
F. G. Priest

12 Pseudomonads 401
G. Auling

13 Streptomycetes and Corynebacteria 433
W. Piepersberg

14 Yeasts 469
J. J. Heinisch, C. P. Hollenberg

15 Filamentous Fungi 515
F. Meinhardt, K. Esser

16 Bacteriophages 543
H. Sandmeier, J. Meyer

17 Plant Cell Cultures 577
M. Petersen, A. W. Alfermann

Index 615

Contributors

Prof. Dr. August Wilhelm Alfermann
Institut für Entwicklungs- und
Molekularbiologie der Pflanzen
Heinrich-Heine-Universität
Universitätsstraße 1
D-4000 Düsseldorf 1
Federal Republic of Germany
Chapter 17

Prof. Dr. Georg Auling
Institut für Mikrobiologie
Universität Hannover
Schneiderberg 50
D-3000 Hannover
Federal Republic of Germany
Chapter 12

Priv.-Doz. Dr. Hubert Bahl
Institut für Mikrobiologie
Universität Göttingen
Grisebachstraße 8
D-3400 Göttingen
Federal Republic of Germany
Chapter 9

Prof. Dr. Charles L. Cooney
Department of Chemical Engineering
Massachusetts Institute of Technology
Cambridge, MA 02139, USA
Chapter 3

Prof. Dr. Lubbert Dijkhuizen
Department of Microbiology
University of Groningen
Kerklaan 30
NL-9751 AA Haren
The Netherlands
Chapter 8

Priv.-Doz. Dr. Peter Dürre
Institut für Mikrobiologie
Universität Göttingen
Grisebachstraße 8
D-3400 Göttingen
Federal Republic of Germany
Chapter 9

Prof. Dr. Karl Esser
Fakultät für Biologie
Lehrstuhl für Allgemeine Botanik
Ruhr-Universität Bochum
Postfach 102148
D-4630 Bochum 1
Federal Republic of Germany
Chapter 15

Dr. Jürgen J. Heinisch
Institut für Mikrobiologie
Heinrich-Heine-Universität
Universitätsstraße 1
D-4000 Düsseldorf 1
Federal Republic of Germany
Chapter 14

Prof. Dr. Cornelis P. Hollenberg
Institut für Mikrobiologie
Heinrich-Heine-Universität
Universitätsstraße 1
D-4000 Düsseldorf 1
Federal Republic of Germany
Chapter 14

Prof. Dr. Helmut König
Abteilung für Angewandte
Mikrobiologie und Mykologie
Universität Ulm
Oberer Eselsberg M 23
D-7900 Ulm
Federal Republic of Germany
Chapter 7

Prof. Dr. Reinhard Krämer
Institut für Biotechnologie
Forschungszentrum GmbH
P.O. Box 1913
D-5170 Jülich
Federal Republic of Germany
Chapter 2

Prof. Dr. Frank Mayer
Institut für Mikrobiologie
Universität Göttingen
Grisebachstraße 8
D-3400 Göttingen
Federal Republic of Germany
Chapter 1

Prof. Dr. Friedhelm Meinhardt
Institut für Mikrobiologie
Universität Münster
Correnstraße 3
D-4400 Münster
Federal Republic of Germany
Chapter 15

Prof. Dr. Jürg Meyer
Zahnärztliches Institut
Universität Basel
Petersplatz 14
CH-4051 Basel
Switzerland
Chapter 16

Prof. Dr. Oense M. Neijssel
Department of Microbiology
University of Amsterdam
Nieuwe Achtergracht 127
NL-1018 WS Amsterdam
The Netherlands
Chapter 4

Prof. Dr. Sanaa Hamdy Omar
Faculty of Science
Department of Botany and Microbiology
Moharram Bey
Alexandria, Egypt
Chapter 6

Dr. Maike Petersen
Institut für Entwicklungs- und
Molekularbiologie der Pflanzen
Heinrich-Heine-Universität
Universitätsstraße 1
D-4000 Düsseldorf 1
Federal Republic of Germany
Chapter 17

Prof. Dr. Wolfgang Piepersberg
Fachbereich 9, Mikrobiologie
Bergische Universität
Gesamthochschule Wuppertal
P.O. Box 100127
D-5600 Wuppertal 1
Federal Republic of Germany
Chapter 13

Dr. Clemens H. Posten
Gesellschaft für Biotechnologische
Forschung mbH
Mascheroder Weg 1
D-3300 Braunschweig
Federal Republic of Germany
Chapter 3

Prof. Dr. Fergus G. Priest
Department of Biological Sciences
Heriot-Watt University
Edinburgh EH1 4AS
Scotland
United Kingdom
Chapter 11

Prof. Dr. Hans-Jürgen Rehm
Institut für Mikrobiologie
Universität Münster
Corrensstraße 3
D-4400 Münster
Federal Republic of Germany
Chapter 6

Prof. Dr. Hermann Sahm
Institut für Biotechnologie
Forschungszentrum GmbH
P.O. Box 1913
D-5170 Jülich
Federal Republic of Germany
Chapter 5

Dr. Heinrich Sandmeier
Zahnärztliches Institut
Universität Basel
Petersplatz 14
CH-4051 Basel
Switzerland
Chapter 16

Dr. Georg Sprenger
Institut für Biotechnologie
Forschungszentrum GmbH
P.O. Box 1913
D-5170 Jülich
Federal Republic of Germany
Chapter 2

Dr. M. Joost Teixeira de Mattos
Department of Microbiology
University of Amsterdam
Nieuwe Achtergracht 127
NL-1080 WS Amsterdam
The Netherlands
Chapter 4

Prof. Dr. David W. Tempest
Department of Molecular Biology
and Biotechnology
University of Sheffield
Sheffield S10 2TN
United Kingdom
Chapter 4

Prof. Dr. Michael Teuber
Institut für Lebensmittelchemie
ETH-Zentrum/LFV
Universitätsgasse 2
CH-8092 Zürich
Switzerland
Chapter 10

Introduction

HERMANN SAHM

Jülich, Federal Republic of Germany

The first volume of the second edition of "Biotechnology" contains the biological background material which is indispensable for the development of biotechnological processes. The rational establishment of a biotechnological project requires some knowledge of the biology of the organism used in the process. This volume offers a unique collection of current information on ecology, taxonomy, nutrition, biochemistry, physiology, and genetics of organisms used in biotechnological projects. It should be pointed out that this volume was not designed as a textbook of classical and molecular biology, but rather the emphasis is on the biological fundamentals of biotechnology. Furthermore, this is not intended to be an advanced treatise, although it is an authoritative introduction written by leading researchers and covering the major aspects of biotechnology. A substantial collection of references has been included after each chapter for those wishing to pursue more specialized studies. Therefore, this book is an excellent reference for advanced students, as well as for researchers in industrial microbiology, biochemical engineering, food science, or medicine.

In *Part I*, the general biological aspects relevant to biotechnology are presented. First, an insight into various aspects of cell structure and tissue organization is provided to obtain a better understanding of their function. As the understanding of cell metabolism is an essential prerequisite for the development of biotechnological processes, the next chapter deals with the metabolic aspects which are important for all living cells. Microbial growth and product formation occur in response to the environment. Thus, the objective is to quantitate microbial growth, characterize growth in response to the environment and develop some fundamental principles for design of the environment, which will elicit the desired response of growth and product formation. Several examples show that metabolic fluxes and overproduction can be manipulated by culture conditions. Furthermore, the cellular activities can also be changed in a purposeful manner with the application of recombinant DNA technology (metabolic design). Finally, effect of immobilization on the morphology and metabolism of microorganisms is described.

Part II deals with a large assemblage of industrially important organisms: Methanogens, Methylotrophs, Clostridia, Lactic Acid Bacteria, Bacillus, Pseudomonads, Streptomycetes, Corynebacteria, Yeasts, Filamentous Fungi, Bacteriophages, and Plant Cell Cultures*). Al-

* A chapter describing the biological fundamentals of Animals Cell Cultures will be published in Volume 10 "Special Processes".

though these chapters vary in the depth of treatment, generally they deal with taxonomic principles, ecology, isolation, morphology, growth, metabolism, and applied aspects. All of this information will be useful for investigators who suddenly find themselves working on a new biotechnological project involving an organism they have never before encountered.

Im am very grateful to the authors, each of whom is an expert on a particular topic or group of organisms. Although they are extremely busy individuals, they agreed to the task because they realized its importance and necessity.

Jülich, January 1993 Hermann Sahm

I. General Aspects

1 Cell Structure

FRANK MAYER

Göttingen, Federal Republic of Germany

1 Introduction 7
2 Methods of Structural Analyses 7
2.1 Preparation of Light Microscopic Samples 7
2.2 Visualization of Light Microscopic Samples 7
2.2.1 Conventional Light Microscopy 7
2.2.2 Laser Scanning Microscopy 8
2.3 Preparation of Electron Microscopic Samples 9
2.3.1 Samples for Scanning Electron Microscopy 9
2.3.2 Samples for Transmission Electron Microscopy 10
2.3.2.1 Conventional Ultrathin Sectioning Techniques 10
2.3.2.2 Low Temperature Preparation Techniques 10
2.3.2.3 Cytochemical Techniques 10
2.3.2.4 Immunocytochemical Techniques 11
2.3.2.5 Preparation of Biological Macromolecules (Nucleic Acids, Proteins) 11
2.4 Visualization of Electron Microscopic Samples 12
2.4.1 Scanning Techniques 12
2.4.2 Conventional and Low Temperature Transmission Electron Microscopy 12
2.4.3 Element Analysis 13
2.5 Image Evaluation 13
3 Structural Organization of the Prokaryotic Cell 13
3.1 Principles 13
3.2 Cell Structures Exposed to the Environment 16
3.2.1 Slime and Capsules 16
3.2.2 Pili and Fimbriae 17
3.3 Cell Wall Structures 19
3.3.1 Bacteria with an Outer Membrane 19
3.3.2 Bacteria without an Outer Membrane 20
3.3.3 Bacterial Surface Layers 22
3.4 Cytoplasmic and Intracellular Membranes 23
3.4.1 Cytoplasmic Membrane 23
3.4.2 Intracellular Membranes 24

3.4.2.1 Phototrophic Bacteria 24
3.4.2.2 Non-Phototrophic Bacteria 25
3.5 Inclusion Bodies 26
3.5.1 PHA Inclusions 26
3.5.2 Protein Inclusions 26
3.5.3 Other Inclusion Bodies 27
3.6 Extracellular Bacterial Enzymes 27
3.6.1 Amylases, Pullulanases, and Related Enzymes 27
3.6.2 Cellulases 28
4 Structural Organization of the Eukaryotic Cell 30
4.1 Principles 30
4.2 Yeasts as Examples for Eukaryotic Microorganisms 32
4.2.1 Cell Envelope 32
4.2.2 Microbodies 33
4.3 The Plant Cell 34
4.3.1 General Plant-Specific Aspects 34
4.3.2 Cell Surface Structures 34
4.3.3 Cell Wall Structures 35
4.3.4 Vacuoles and Microbodies as Compartments in the Plant Cell 36
4.3.5 Plant Tissues 37
4.4 The Animal Cell 38
4.4.1 General Animal-Specific Aspects 38
4.4.2 Cell Envelope 38
4.4.3 Endomembrane Systems Involved in Synthesis and Export 39
4.4.4 Cell–Cell Connections in Animal Tissues 40
4.5 Eukaryotic Cells in Culture 40
5 References 42

1 Introduction

Biotechnology is a technology using biological systems. The basic unit of any biological system is the cell. In the past, at present and in the future, biotechnological approaches involve cells of many kinds, belonging to bacteria, eukaryotic microorganisms, plants and animals (including man). Depending on the specific purposes and needs, wild-type cells, naturally occurring mutants, or genetically manipulated cells and organisms are employed. In many cases, the organisms or cells are not grown under their natural conditions; instead, they are cultivated, under strict control, in an artificial or semi-artificial environment. Cells and organisms grown and maintaind under these conditions might exhibit changes in their morphology and ultrastructure. The aim of this chapter is to provide insight into selected aspects of the structural cell and tissue organization relevant in biotechnology, and to present data suited for a better understanding of functional implications.

2 Methods of Structural Analyses

2.1 Preparation of Light Microscopic Samples

At the level of laboratory experiments and, after scale-up, at the production level, biotechnological systems have to be carefully controlled and regulated. A major prerequisite for success is an intimate knowledge of a number of parameters such as cell density, purity of the sample (i. e., exclusion of contaminating cells), structural preservation of the cell (i.e., no lysis of the cells), and a controlled state of the sample with respect to cell distribution, aggregation or immobilization.

A routine tool in the hands of a scientist, an engineer or a technician should be the light microscope (FRANÇON, 1967). This instrument should be used continuously for checks of cell samples. Often, a fast glance may reveal the reason why a system runs out of control.

Light microscopy can also be used for additional purposes. Immunofluorescence techniques may be applied in studies aimed at the analysis of overproducing cells or the optimization of expression systems, provided well characterized antisera or antibody preparations are available.

Very simple sample preparation techniques may be used for routine light microscopy (MAYER, 1986). An aliquot of culture broth containing the cells is put, as a drop, on the surface of a glass slide, and covered with a cover slip. In most cases, when prokaryotic or eukaryotic cells are to be checked, no staining is needed. A variety of staining and differentiating procedures can be followed in cases where cultured cells have to be investigated.

The preparation of samples used for fluorescence (DODDEMA and VOGELS, 1978) and immunofluorescence microscopy (Fig. 1) is more sophisticated. These techniques are not suitable for routine checks. However, executed by an experienced worker, these approaches may add valuable information.

2.2 Visualization of Light Microscopic Samples

2.2.1 Conventional Light Microscopy

Light micrographs, taken along with routine checks, may be helpful documents when unexpected results are obtained, or when variations are introduced in order to optimize a system.

Large cells may be analyzed by bright field amplitude contrast transmission microscopy, whereas cells of the size of bacteria need application of phase contrast, a technique which can also be used for larger cells (Fig. 2). For all cell sizes, Nomarski interference contrast may be used. It results in micrographs exhibiting a three-dimensional appearance of the cells (Fig. 3). These techniques provide data on the shape, size and integrity of cells, and on the presence of contaminations. Cell surface structures such as slime and bacterial capsules may be visualized by mixing the cell suspension with indian ink. After air drying, the cells appear light on a dark background.

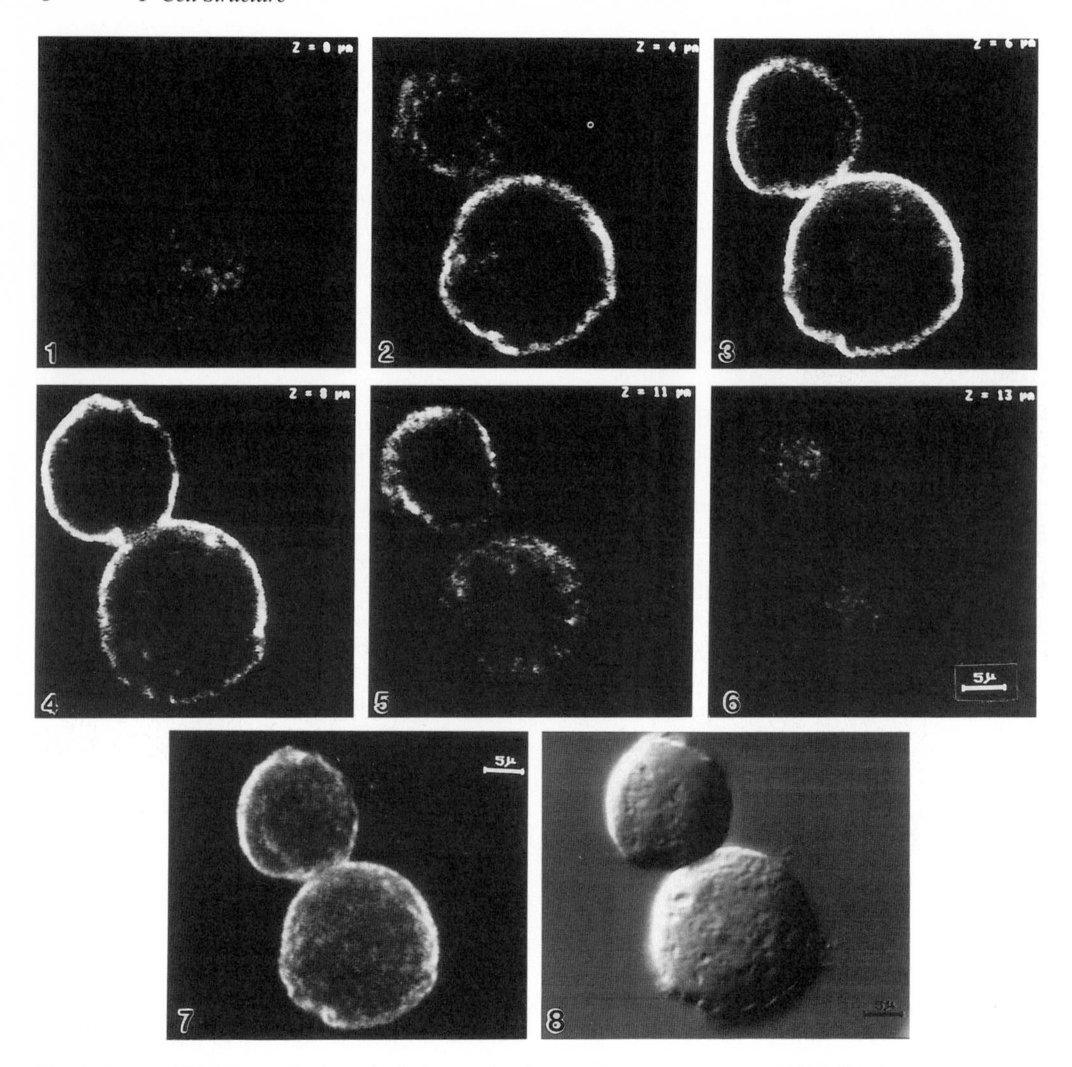

Fig. 1. Immunofluorescence micrograph demonstrating patches of enzyme cathepsin B cell-surface epitopes on human SB-3 tumor cells. Shown are a number of "optical sections" (1 to 6) obtained from a sample which was prepared by applying immunolabeling to living cells. (7) exhibits an "extended focus" reconstruction whereas (8) shows a view obtained by using differential interference contrast. Micrography was performed with a confocal laser scanning microscope.
Original micrograph H. Spring and E. Spiess

Fluorescence light microscopy, i.e., primary or secondary fluorescence in cell samples, can be performed with ultraviolet, blue or green light as excitation irradiation (Fig. 4). To obtain fluorescence, usually an adequate dye has to be introduced, in combination with a system allowing a highly specific staining or labeling of the cell structures of interest (see above, Sect. 2.1).

2.2.2 Laser Scanning Miroscopy

Laser scanning microscopy, i.e., confocal light microscopy (see Fig. 1), is suited for all types of light microscopic samples. All modes of visualization techniques (bright field, phase contrast, Nomarski contrast) are possible. Laser light of various wavelengths can be used. There are major advantages compared to con-

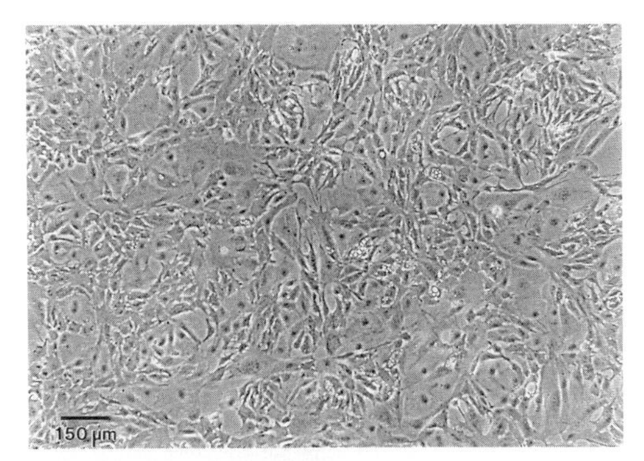

Fig. 2. Phase contrast light micrograph of human SB-3 tumor cells.
Original micrograph CH. SPIESS and E. SPIESS

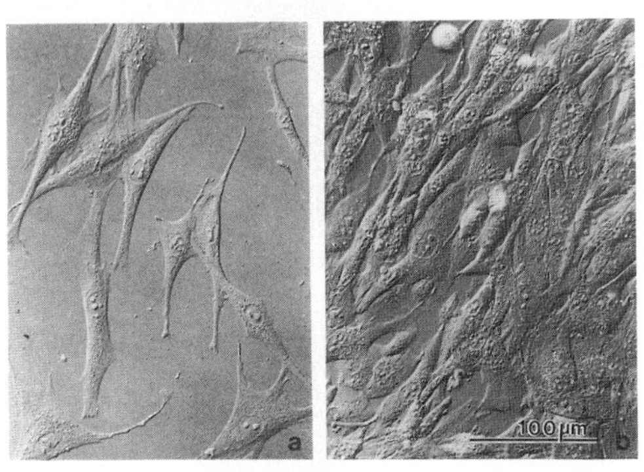

Fig. 3. Interference contrast micrographs of animal cells (baby hamster kidney) grown in small plastic dishes. (a) single cells, (b) beginning confluence and contact inhibition.
From KLEINIG and SITTE (1986), reproduced by permission of Gustav Fischer Verlag

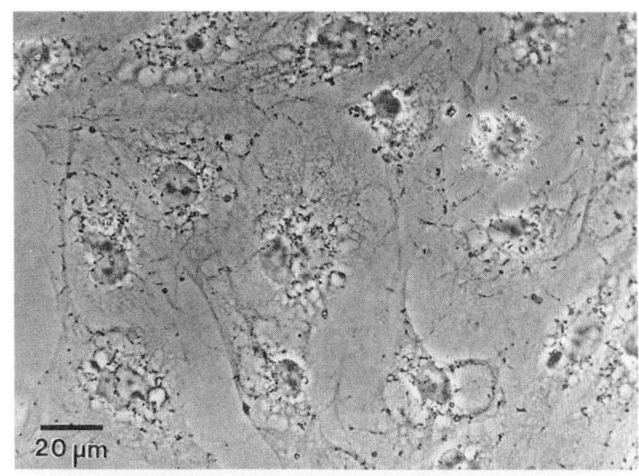

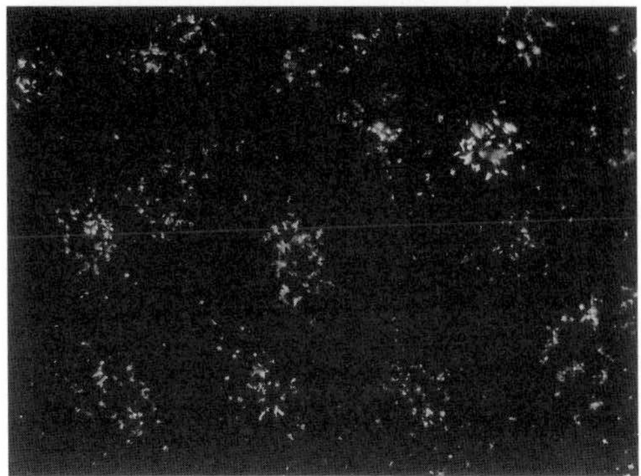

Fig. 4. Fluorescence-microscopical demonstration of the activity of cathepsin B present in lysosomes primarily located close to the cell nucleus in human SB-3 tumor cells.
Top, phase contrast micrograph; bottom, fluorescence micrograph of the identical area.
Original micrograph CH. SPIESS and E. SPIESS

ventional light microscopy: high contrast, high resolution in three dimensions, quantitation and processing of signals.

2.3 Preparation of Electron Microscopic Samples

2.3.1 Samples for Scanning Electron Microscopy

Conventional scanning electron microscopy (HAYAT, 1978; ROBINSON et al., 1985, 1987; ROSENBAUER and KEGEL, 1978) is applied to the documentation of parameters of the cell topology (see Fig. 13). Simple air drying is known to cause severe damage in samples containing high amounts of water. Such an artificial deformation can be minimized if, prior to drying, water is replaced by solvents with low surface tension. Two approaches may be used, critical point drying and freeze drying. Critical point drying is based on the fact that the vapor of the solvent replacing the water (ethanol, acetone) and the solvent itself have the same volume, identical density and are coexistent. This brings the surface tension to zero and considerably reduces shrinking of the sample. However, ethanol or acetone may themselves induce artifacts. Good structural preservation can also be obtained by freeze drying. It is achieved by shock freezing followed by a controlled sublimation of ice. Both for samples prepared by critical point drying and by freeze drying, contrast for electron microscopy is obtained by metal shadowing. Samples for scanning transmission electron microscopy are prepared according to procedures used for transmission electron microscopy (see below).

2.3.2 Samples for Transmission Electron Microscopy

2.3.2.1 Conventional Ultrathin Sectioning Techniques

A prerequisite for routine transmission electron microscopy is the removal of water from the sample. Without precautionary measures, artificially altered cells would be visualized. Therefore, chemical fixation with aldehydes is used to keep the cell components in place also after removal of the water (see Figs. 23 and 40). However, this goal cannot be reached to full satisfaction, as additional steps such as application of organic solvents, staining, embedding in liquid (monomeric) resin and polymerization of the resin at elevated temperatures, ultrathin sectioning (introducing distorsions) and post-staining with metal salt solutions, followed by air drying of the sections, each contribute to artificial alterations (ROBINSON et al., 1987).

2.3.2.2 Low Temperature Preparation Techniques

Drawbacks of chemical treatment of the samples as described above can be avoided by physical fixation, i.e., shock freezing of the sample (MOOR, 1964; MÜLLER, 1988; ROBARDS and SLEYTR, 1985; SLOT et al., 1988). After application of very low concentrations of chemical fixatives (or even without chemical fixation), the sample is shock-frozen (severe damage by ice crystal formation can be avoided); removal of water (in the frozen state) is achieved with organic solvents applied at –35 °C, and the sample is then infiltrated, still at low temperature, with resin. Polymerization of the resin is performed, still in the cold, by ultraviolet light. The ultrathin sections are cut at room temperature. Preparations of this kind are especially suited for immuno electron microscopy. This holds also true for samples prepared by cryo ultramicrotomy (see Fig. 7); there, sectioning is done with the sample kept in the frozen hydrated state. The sections can either be viewed, after stabilization and removal of water, in the conventional transmission electron microscope, or they may be visualized, still frozen, in a cryo electron microscope.

Freeze fracturing and freeze etching (MOOR, 1964; MÜLLER, 1988; ROBARDS and SLEYTR, 1985) are approaches which also start from shock-frozen samples. However, the samples are not sectioned but rather cleaved under high vacuum conditions in the cold; "etching" may be achieved by sublimation of ice from the surfaces exposed by cleavage. The visualization is not done with the sample proper but by production of replica of the surfaces exposed by cleavage and ice sublimation. The replica give an impression of the three-dimensional inner structure of cells and the topology of outer cell surfaces (see Fig. 30).

2.3.2.3 Cytochemical Techniques

Data on the identification, chemical composition, location, and activity (in case of certain enzymes) of cell components in dimensions within and beyond the resolving power of the light microscope can be obtained by various electron cytochemical techniques (MAYER, 1986). Gram staining is a light microscopic procedure used for the analysis of the wall type of bacteria. A variation for electron microscopy uses, instead of KJ-J_2, a platinum compound ("TPt") which produces an anion in aqueous solution which is compatible with crystal violet of the Gram stain. Polyanions and acidic mucosubstances, occurring at cell surfaces, can be visualized by ruthenium red staining (see Fig. 7). Staining is done prior to embedding in resin. The result visible in ultrathin sections is an electron dense appearance of layers containing these types of substances. Identification and localization of specific enzymes can be achieved by procedures resulting in electron dense precipitates which are formed by reactions catalyzed by the enzymes *in situ*. Protocols for the cytochemical demonstration of ATPase (TAUSCHEL, 1988) and a number of other enzymes are available.

2.3.2.4 Immunocytochemical Techniques

Immunocytochemical techniques are employed for the identification and localization of cell components at the outside (ACKER, 1988) or within cells (see Fig. 14). Various approaches can be chosen, depending on what kind of information is needed (ROHDE et al., 1988; SLOT et al., 1988). "Pre-embedding" labeling is used when cell components are to be analyzed which are exclusively located at the cell surface. "Post-embedding" is applied when components within a cell are of interest. "Pre" means that intact cells or isolated organelles, prior to embedding for ultrathin sectioning, are incubated in a solution containing polyclonal antiserum or monoclonal IgG antibodies directed specifically against an "antigen" exposed at the cell surface. "Post" indicates that the labeling is performed after cutting the ultrathin sections. This second approach allows the detection of antigens within a cell, provided they are exposed at the surface of the section. In any case the IgG antibodies which are bound to their specific antigens have to be — indirectly — visualized by antibody-specific markers such as colloidal gold or ferritin. These metal-containing markers can be seen in the electron microscope as small dark dots, and from the procedure used for the experiment it is safe to conclude that these dots are in the immediate neighborhood of the cell components in question. A semi-quantitiative evaluation of the marker pattern is only possible after carefully executed calibration experiments.

2.3.2.5 Preparation of Biological Macromolecules (Nucleic Acids, Proteins)

Nucleic acid molecules such as plasmids used in biotechnological experiments can be characterized by transmission electron microscopy (SPIESS and LURZ, 1988). In solution these molecules are coiled. Therefore, for the measurement of their length (which allows estimation of their molecular weight) and determination of their type (covalently closed or open circular, linear) they have to be transformed into a two-dimensional state without exerting physical stress which would otherwise reduce the reliability of the measurements. A number of preparation techniques are available: conventional cytochrome c spreading, cytochrome c spreading in a version called droplet diffusion technique, carbonate spreading, direct mounting onto mica. In the first two versions cytochrome c is used to complex the nucleic acid which can afterwards, after transfer onto a support film, be contrasted by metal shadowing. In the mica technique, the nucleic acid dissolved in buffer is attached to the clean surface of a piece of mica, stained, metal shadowed (still adhering to the mica), stabilized by a thin layer of sublimated carbon, and finally transferred to a support film by floating off the sample from the mica surface.

Isolated protein molecules such as enzyme complexes can be visualized by "negative staining" (HOLZENBURG, 1988; MAYER and ROHDE, 1988; ROBINSON et al., 1987; SLEYTR et al., 1988). The protein is dissolved in buffer; from a piece of mica which is covered with a thin film of sublimated carbon, the carbon is floated off onto the surface of the buffer solution; within a short period of time, the protein molecules attach to the lower face of the floating carbon film. This film is then transferred, again supported by the original piece of mica, onto the surface of a washing solution and then onto the surface of the aqueous staining solution (uranyl acetate or neutralized phosphotungstic acid solution). Finally, the carbon film with the adhering protein molecules is removed from the staining solution with a copper grid, and most of the staining solution is soaked off with filter paper. The remaining part of the staining solution is allowed to air-dry, thus surrounding the protein molecules attached to the carbon support film. The dried stain appears dark in the electron beam, and the protein molecules are bright due to stain exclusion ("negative" staining) (see Fig. 26).

An alternative approach is the preparation of protein molecules (and particles of larger size such as viruses and membrane vesicles) in the frozen-hydrated state (ROBARDS and SLEYTR, 1985). This is done by dipping a support grid, covered with a support film containing holes, into the enzyme solution. After removal from the solution, the grid is kept in a vertical position, and the liquid is allowed to run down the grid. Excess liquid is soaked off with filter paper. Finally, a very thin layer of liquid is spanning the holes in the support film. Now the sample is shock-frozen and kept cold for inspection in a precooled electron microscope. Application of chemical fixatives and staining salts can be omitted.

2.4 Visualization of Electron Microscopic Samples

2.4.1 Scanning Techniques

Conventional Scanning Electron Microscopy

This technique is used for the study of the topology of cells and tissues (HAYAT, 1978; ROBINSON et al., 1987). It makes use of secondary and backscattered electrons, i.e., signals produced by interaction of primary electrons with the surface layer of the sample. "Scanning" indicates that the primary electron beam is scanned over the sample; the image obtained by the procedure is, thus, composed of lines; it is made visible on a monitor. A major advantage of scanning procedures — both in electron and in light microscopy — over conventional techniques is the fact that the signals can be processed.

Scanning Transmission Electron Microscopy

This mode is applied to the study of samples which are thin enough to allow the electrons to pass through. Thus, the image is formed as a result of interactions of the primary electrons not only with atoms in the surface of the sample (see above); it comprises signals obtained from all layers of the sample. Again the incident electron beam is scanned over the sample, in contrast to the conventional transmission electron microscopy (see below).

2.4.2 Conventional and Low Temperature Transmission Electron Microscopy

Both techniques use a primary electron beam wide enough to illuminate a large area of the sample; the beam is not scanned over the sample, and the image can be seen on a fluorescent screen (HAYAT, 1981; REIMER, 1967). In electron microscopes operated at low temperature, the sample (and its surroundings) is cooled with liquid nitrogen. This allows the analysis of ultrathin cryo sections without thawing, and of samples in the frozen-hydrated state. A necessity is that these samples have to be checked using only very low electron doses; high doses would cause melting of the ice and immediate complete damage of the sample.

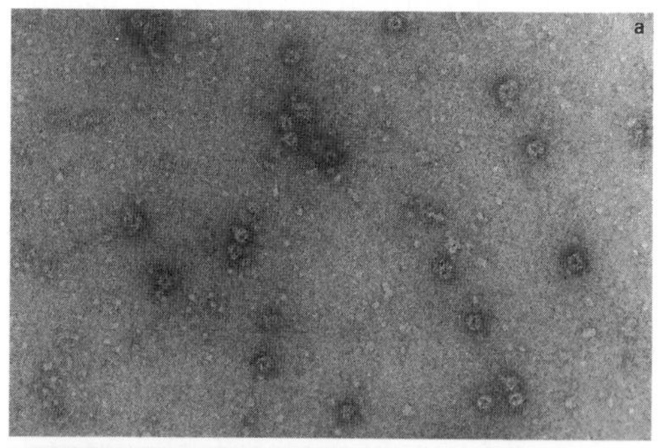

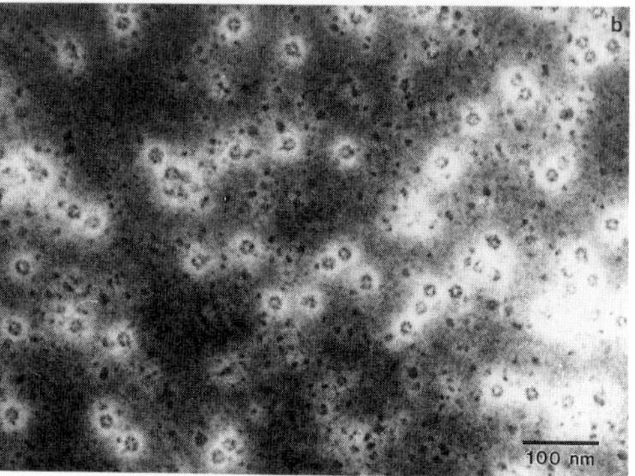

Fig. 5. Hydrogenase isolated from the bacterium *Methanobacterium thermoautotrophicum* negatively stained with uranyl acetate.
(a) Conventional transmission electron microscopy, (b) depicted by electron spectroscopic imaging (ESI). Original micrographs I. BRAKS

2.4.3 Element Analysis

Various approaches can be chosen for qualitative and quantitative analyses of elements in electron microscopic samples. One is the dispersive X-ray microanalysis (ZIEROLD, 1988). Electrons interacting with atoms cause, in addition to emission of secondary electrons, also element-specific X-ray radiation which can be collected, analyzed, and associated with certain cell structures. Electron energy loss spectroscopy and electron spectroscopic imaging are techniques which allow very sensitive detection and localization of elements with low atomic number (BAUER, 1988). Especially interesting is electron spectroscopic imaging, because the respective element distribution is directly superimposed on a high-resolution electron micrograph of the sample. Element-specific imaging of this kind can also be used for the improvement of imaging quality because the contrast can be enhanced, and the background noise present in conventional transmission electron micrographs can be drastically reduced (Fig. 5).

2.5 Image Evaluation

Provided techniques of preparation for electron microscopy have been applied which are designed for good structural preservation of biological samples, electron micrographs can be evaluated with respect to quantitative parameters (ROBINSON et al., 1987). Besides sizes of cells and their organelles, also numbers, e.g., of inclusion bodies per cell, may be determined on the basis of statistical analyses. From these numbers and from the values calculated for the volume of cells and of inclusion bodies, the relative and absolute mass of inclusions in cells may be estimated. Under certain conditions, semi-quantitative evaluation of immunolabeling experiments is possible, in addition to the cellular localization of antigens (at the cell surface, attached to the outside or the inside of the cytoplasmic cell membrane, within organelles, in the cytoplasm) (ROHDE et al., 1988). Size determinations (length, molecular mass) of double- and single-stranded nucleic acids (plasmids, virus nucleic acids) (SPIESS and LURZ, 1988) can be performed with high accuracy, provided standard molecules of known molecular mass are evaluated for comparison, and the imaging systems are carefully calibrated.

3 Structural Organization of the Prokaryotic Cell

3.1 Principles

Biotechnology with prokaryotic cells makes use of an intimate knowledge of the general and the detailed rules governing the life of these

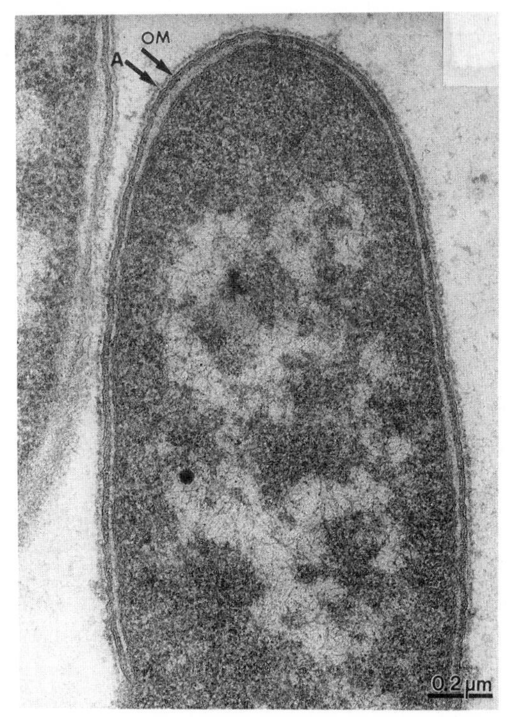

Fig. 6. Conventionally chemically fixed, embedded and ultrathin sectioned cell of the smooth strain Ye75S of *Yersinia enterocolitica*, treated, by the preembedding labeling technique, with anti-Ye75S antiserum. Between the outer leaflet of the outer membrane (arrow OM) and the antiserum depositions (arrow A) a thin layer (light halo) can be seen, formed by O-specific chains of the lipopolysaccharide layer. From ACKER (1988), reproduced by permission of Academic Press

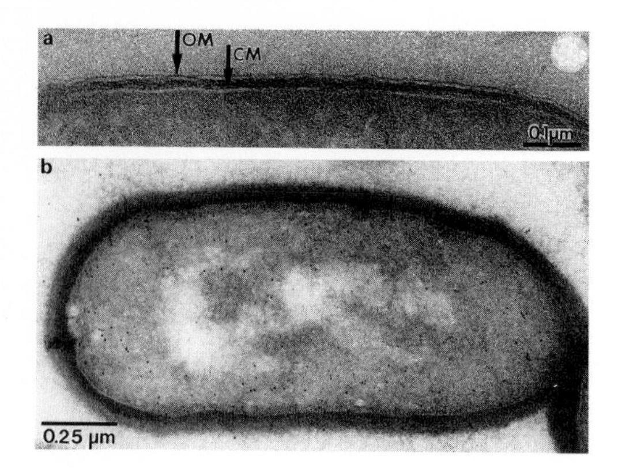

Fig. 7. (a) Cryosection of an *Escherichia coli* cell (ECA-negative mutant F 1283), contrasted with uranyl acetate. OM, outer membrane; CM, cytoplasmic membrane; between these two layers, the electron dense periplasmic gel is visible.
From ACKER (1988), reproduced by permission of Academic Press
(b) Cryosection of a *Bacillus* sp. cell contrasted with ruthenium red and immuno labeled (post-embedding technique) for the localization of cellulase.
Original micrograph D. MILLER

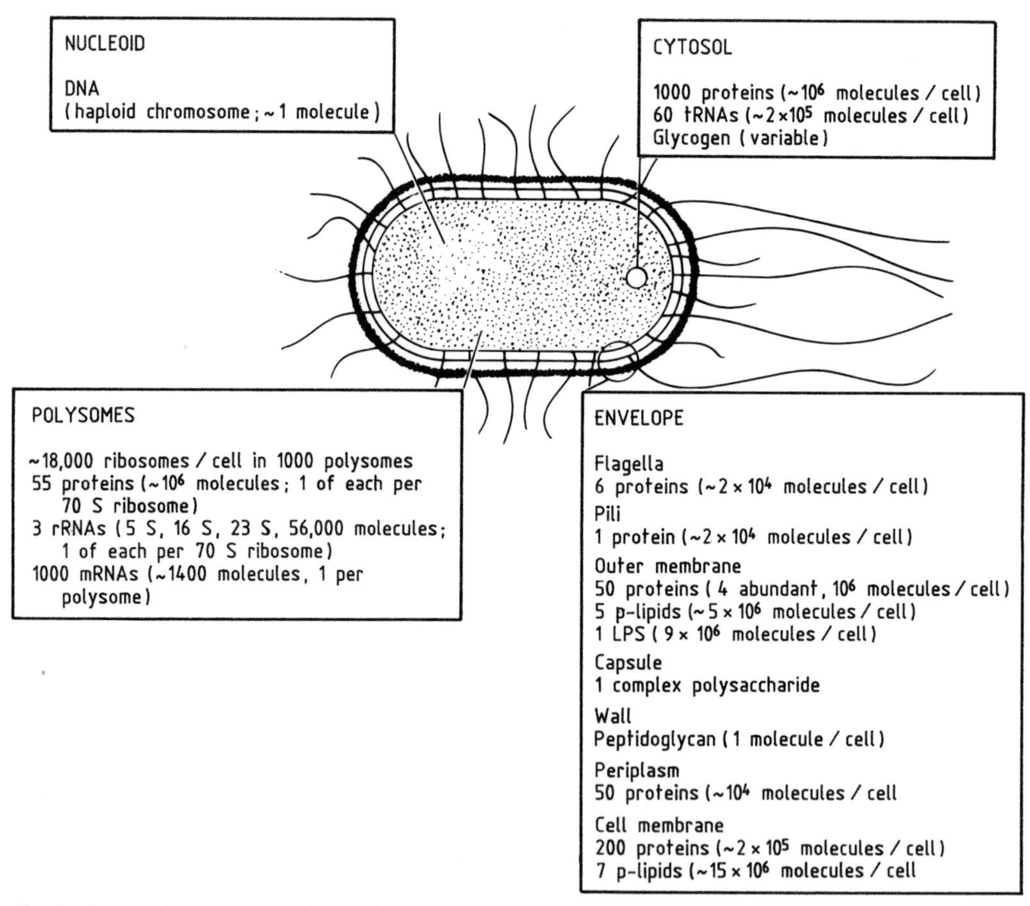

Fig. 8. Macromolecular composition of structures of an average *Escherichia coli* cell.
From INGRAHAM et al. (1983), reproduced by permission of Sinauer

cells. Only this knowledge allows a controlled manipulation of the cell. Cellular reactions are linked to cell structures. The topic of this section is to provide the data necessary for the understanding of bacterial structure–function relationships.

Structural principles common to all bacteria, irrespective of the various degrees of complexity observed in cells belonging to different physiological or taxonomic groups, are that these organisms have an envelope consisting of a cytoplasmic membrane and, in most cases, a wall. The envelope encloses the cytoplasm and interacts, with its exposed layers, with the environment (Figs. 6 and 7). So far, the parallels with the eukaryotic cell are evident. However, the cytoplasm contains, in most cases without obvious compartmentation, nucleic acids, ribosomes, storage materials, enzymes and low-molecular weight components within the cytosol. The lack of obvious compartmentation is characteristic of prokaryotic cells. Intracytoplasmic membranes, formed as differentiations from the cytoplasmic membrane, are characteristic of phototrophic and certain other bacteria. Only in rare cases, these membranes form specific compartments.

Additional features of the prokaryotic cell are its small size and its relatively large surface compared to its volume. This kind of organization has the consequence that the exchange between cell and environment can be high and fast due to the large surface, that the distances are short, and that the absolute masses are low. These features enable the bacterial cell to grow and duplicate very fast, and advantage is taken of these properties for biotechnological purposes.

Figs. 8 and 9 show the macromolecular composition of structures of an average *Escherichia coli* cell, and the types and amounts of macromolecules needed to produce 10^{12} of these cells (1 g, dry weight). Fig. 10 gives additional information. It presents an overall view of the metabolism leading to the chemical synthesis of *E. coli* from glucose. Precursor metabolites

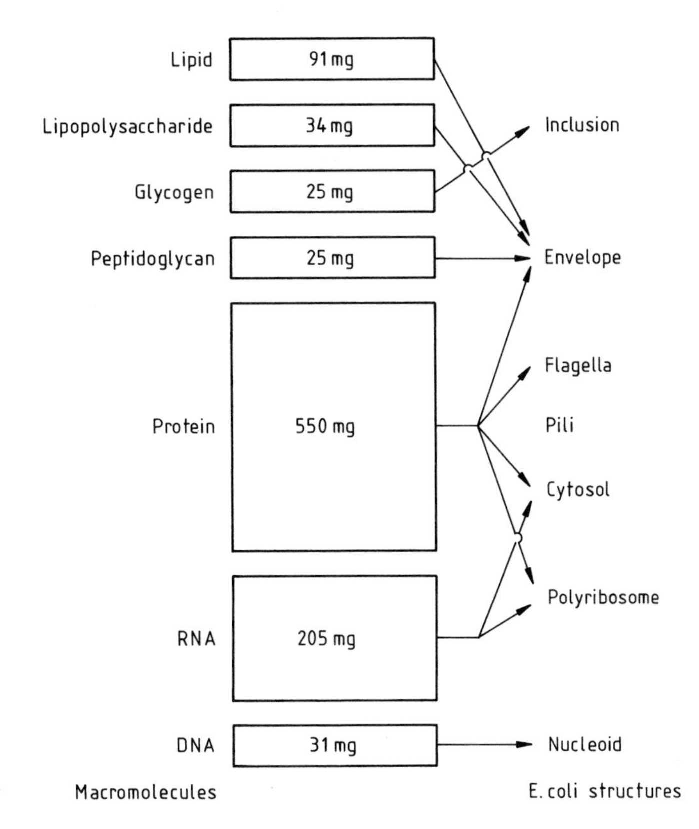

Fig. 9. Macromolecules needed to produce 10^{12} cells (1 g, dry weight) of *Escherichia coli* b/r. The capsule has been omitted.
From INGRAHAM et al. (1983), reproduced by permission of Sinauer

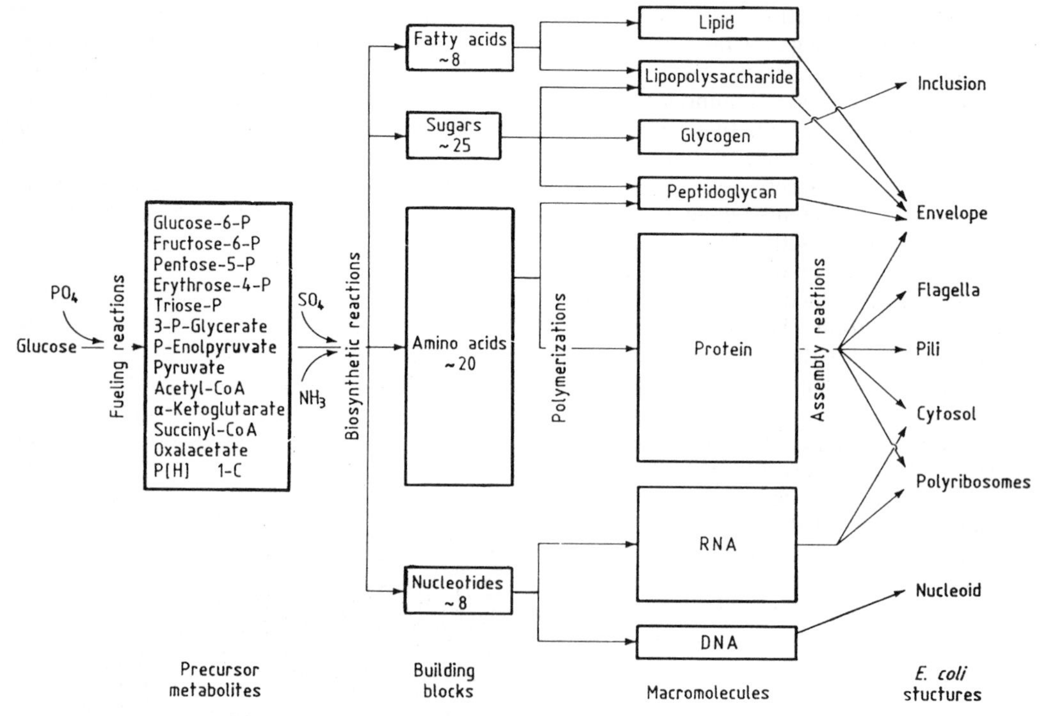

Fig. 10. Overall view of metabolism leading to chemical synthesis of *Escherichia coli* from glucose. Boxes are proportional to need in making *E. coli.*
From INGRAHAM et al. (1983), reproduced by permission of Sinauer

are used, by biochemical reactions, for the formation of building blocks; these form, by polymerization reactions, macromolecules which finally, by assembly reactions, create the cell structures. These diagrams illustrate that an organism is more than the sum of its components; "more" is brought about by the intrinsic order or organization not only of the final structure but also by the way how this final structure is formed.

3.2 Cell Structures Exposed to the Environment

3.2.1 Slime and Capsules

The spectrum of interactions of bacterial cells is large, ranging from contact with partner cells of the same species, with other bacteria, with solid surfaces of biotic or abiotic origin. Active and inactive interactions are observed, and bacterial cells can also be the target of activities of eukaryotic organisms. These interactions involve a diversity of chemical and structural features of components exposed at the cell surface.

Application of indian ink is suited to demonstrate the presence of outermost bacterial surface layers which are called "slime" and "capsule". Depending on the specific bacterium, slime may be of various consistency, and it is loosely organized. It appears as a highly hydrated network of polysaccharide fibrils emanating from the cell wall (COSTERTON et al., 1974; 1978; 1981). These exopolysaccharides are not bound to the cell wall proper by covalent bonds, but by simple entanglement with the outer wall polymers. They are acidic in nature and can be removed by washing at appropriate pH values and ionic strength. Slime may interfere with biotechnological processes; it is of importance to use slime-free mutants in order to avoid severe technical drawbacks caused by large volumes of slime. On the other hand, technical applications of this kind of material are discussed, e.g., for material obtained from certain archaebacteria (ANTÓN et al., 1988). Capsules are less voluminous than slime (COSTERTON et al., 1981). In Gram-negative bacteria, they are composed of an outer part, the O antigens (polysaccharides) which are, as all the wall components, genetically determined, a central part with rather constant composition (also made up of sugars) and an inner part, the lipid A which is integrated into the outer cell membrane. The term "lipopolysaccharide" is used for the complex (Figs. 11 and 15). In many Gram-positive bacteria, neutral polysaccharides are present which are covalently linked to the peptidoglycan (see Fig. 7). However, there are exceptions from the rule. A major side aspect regarding the outermost wall layers described above is that the polysaccharides may play a role in pathogenesis (NEWMAN, 1976).

Difficulties arise when slime and capsules are to be prepared for electron microscopy (COSTERTON, 1984). Any technique implying removal of water without stabilization of these highly hydrated cell components will immediately cause complete loss of structural integrity. Stabilization can be obtained by application of polycationized ferritin (WEISS et al., 1979) or specific antisera (COSTERTON et al., 1981); both approaches prevent severe damage; however, as they add considerable mass to the structure in question, the native structural organization of these outermost wall layers cannot be unequivocally described with these methods. The situation can be improved by physical instead of chemical stabilization, i.e., by application of low temperature preparation procedures. Frozen-hydrated cryo sections, provided sufficient contrast in the electron microscope can be obtained, should be suited for an artifact-free imaging of these delicate cell components. Alternative techniques of preparation (see also Sect. 4) may be freeze substitution, freeze drying or freeze etching (MÜLLER, 1988).

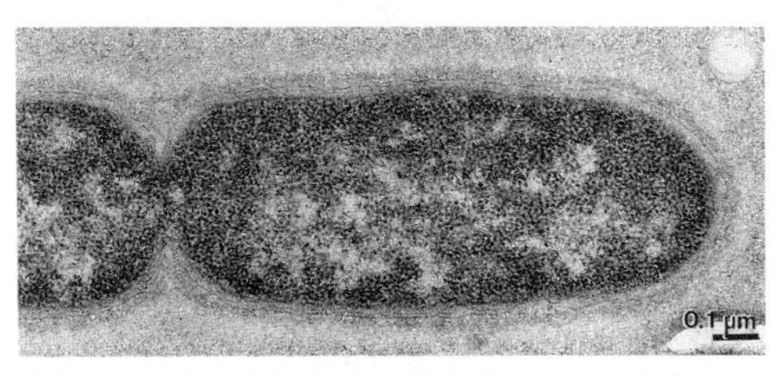

Fig. 11. Low temperature resin (Lowicryl K4M)-embedded sample of gelatin enclosed cells of *Escherichia coli* D 1737 K1. The capsule was stabilzed with anti-K1 polysaccharide antiserum prior to embedding.
From ACKER (1988), reproduced by permission of Academic Press

3.2.2 Pili and Fimbriae

As mentioned above (Sect. 3.2.1), interactions of bacterial cells with their environment are mediated by structural components exposed at the cell surface. Besides fibrillar polysaccharides forming a dense cover totally sur-

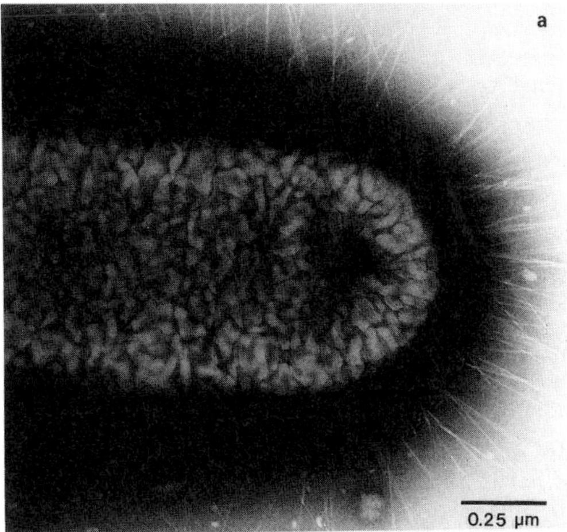

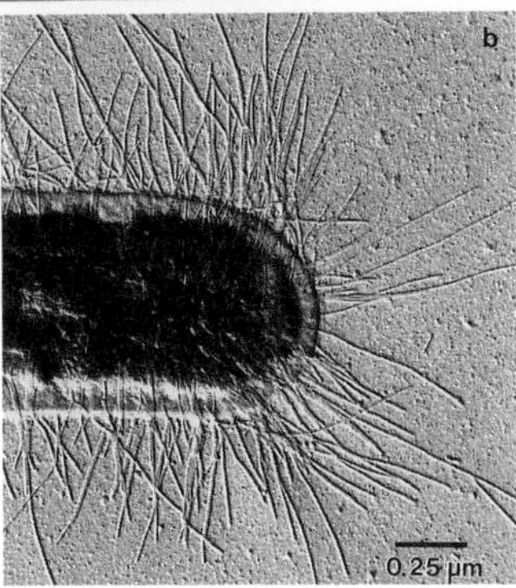

Fig. 12. Cells of *Escherichia coli* exhibiting type I pili (fimbriae).
(a) negative staining, (b) metal shadowing.
Original micrographs B. VOGT

rounding a bacterial cell, nature has invented additional cell structures engaged in interactions, i.e., pili and fimbriae (BRINTON, 1959, 1971; DUGUID et al., 1966) (Fig. 12). These filamentous cell appendages exhibit diameters between 3 and 10 nm and various lengths. They are composed of helically arranged protein subunits forming a tubelike structure. Their number per cell may vary depending on the type of appendage and the state of the cell (growth phase, nutritional conditions). In general, cells having pili and fimbriae have advantages over cells only exhibiting slime and/or capsules when the problem of cell adhesion has to be solved in nature. Cell surfaces and other surfaces involved in interactions often carry surface charges; therefore, interactions between surfaces with identical charge may be hindered by repulsion. Pili and fimbriae can be assumed to be tools of the bacterial cell used to penetrate into and through a "cloud" of surface charges, thus allowing contact with identically surface-charged cells, etc. The pilus retraction observed for the F pilus of *Escherichia coli* is a phenomenon which finally leads to intimate cell–cell contact necessary for gene transfer during conjugation. A second advantage of carrying pili and fimbriae (also exemplified by the F pilus system of *E. coli*) is obvious when the necessity for highly specific interactions is considered, e.g., during conjugation or in pheromone-mediated cell–cell interaction (GALLI et al., 1989; WIRTH et al., 1990). In such a situation a prerequisite is the specific recognition of the partner cell, and this can be mediated by selectively binding cell appendages (Figs. 13 and 14).

The nomenclature of this kind of cell appendages is not yet unequivocally settled. A practicable way would be to call those appendages "pili" which are involved in specific recognition of cells for conjugation. All the other fibrillar appendages of this kind would then be "fimbriae" (MAYER, 1986). However, for *E. coli* this classification is not generally accepted, as the various types of fibrillar appendages of these cells were named "pili", irrespective of their functional role.

In conclusion, phenomena of adhesion observed or intended in biotechnological procedures involving bacteria in many cases are brought about by the cell surface structures described so far. However, additional bacterial

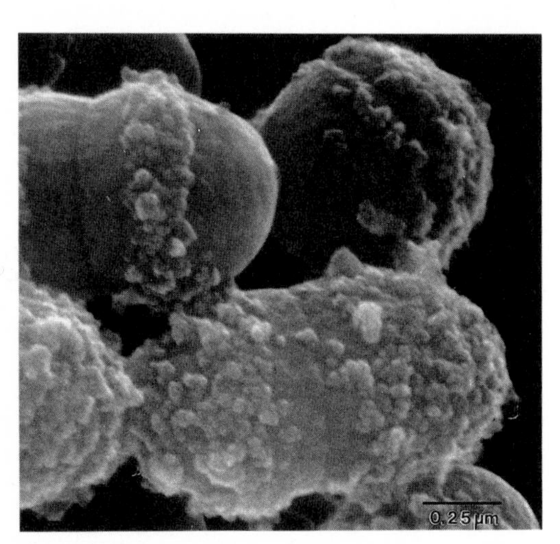

Fig. 13. Demonstration of the presence of adhesin on the cell surface of *Enterococcus faecalis* after induction by sex pheromones. High resolution scanning electron micrograph of cells directly labeled with immunogold. The adhesin is unevenly distributed. Original micrograph R. WIRTH

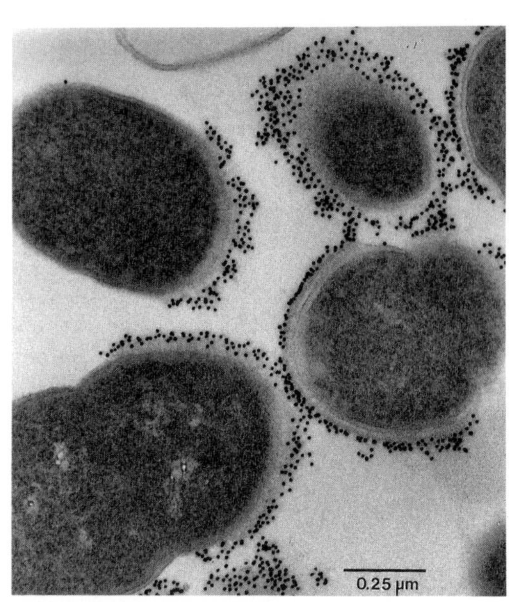

Fig. 14. As Fig. 13; labeling with anti-adhesin antiserum complexed, via a second type of antibody, with 10 nm colloidal gold. The adhesin is present as hairlike structures located between cell wall proper and gold label. Ultrathin section. Original micrograph R. WIRTH

surface components may act in adhesion; these are spinae (BAYER and EASTERBROOK, 1991; EASTERBROOK and ALEXANDER, 1983; MOLL and AHRENS, 1970), surface layers (S-layers) consisting of proteins or glycoproteins (see Sect. 4.3.3), and enzyme complexes, e.g., cellulases, attached to the outside of bacterial walls (see Sect. 3.6).

3.3 Cell Wall Structures

3.3.1 Bacteria with an Outer Membrane

Bacteria with an outer membrane (INGRAHAM et al., 1983; MIZUSHIMA, 1985) exhibit a negative reaction when light microscopic Gram staining is applied. This is caused by the fact that the wall structures, in contrast to Gram-positive bacteria, are so delicate that they do not retain the reaction product of the staining procedure within the cell boundary. As described above (see Sect. 3.2.1), the outermost layer of a typical Gram-negative bacterium is made up of polysaccharides. If they are organized as a capsule, their base, i.e., the lipid A, is integrated into the structural component called outer membrane (the "inner membrane" is the cytoplasmic membrane). A simplified diagram of this layer (INGRAHAM et al., 1983) shows that the lipid A forms the major part of the outer leaflet of the outer membrane (Fig. 15). In electron micrographs of conventionally (by chemical fixation) prepared bacterial cells, this outer leaflet appears dark. The inner leaflet of this membrane, made up of phospholipids, also looks dark, and it is separated from the outer leaflet by an electron translucent zone. So far, outer and cytoplasmic membrane have a very similar appearance (see Fig. 7). The additional components in the outer membrane (porins, lipoprotein, other proteins) cannot be resolved by conventional transmission electron microscopy of ultrathin sections. Occasionally, a monolayer of proteins, the "surface layer" (SLEYTR et al., 1988) (Sect. 3.3.3) is seen attached to the outside of the outer membrane.

Enclosed between the outer membrane and the cytoplasmic membrane is the "periplasmic space". In electron micrographs of conventionally chemically fixed and resin-embedded samples this space is electron transparent, and it contains a fine dark line interpreted as peptidoglycan. However, application of cryo preparation techniques demonstrated that the classical view of the architecture of this part of the cell envelope has to be revised (see Fig. 7). In-

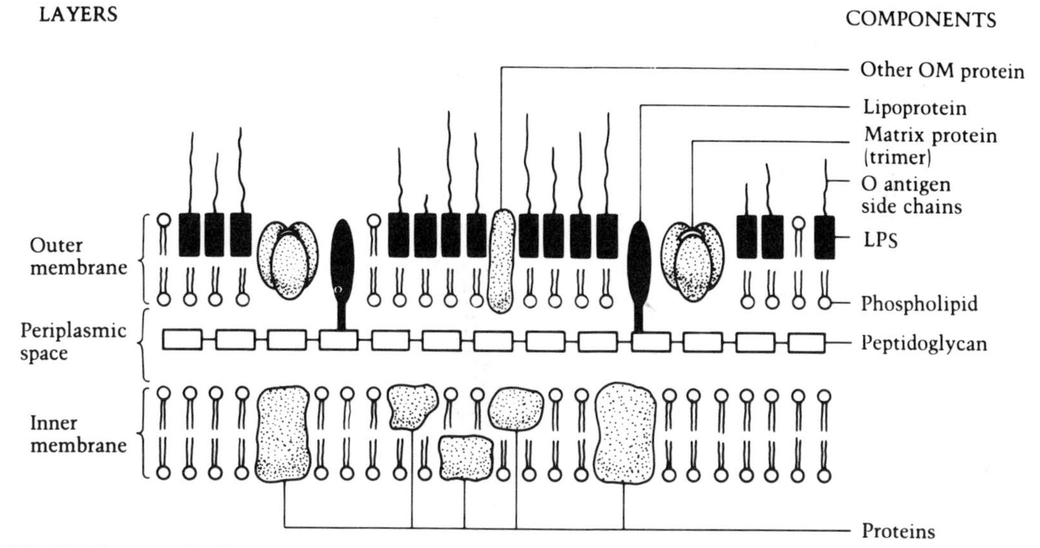

Fig. 15. Diagram of a Gram-negative cell envelope. LPS, lipopolysacharide; OM, outer membrane. From INGRAHAM et al. (1983), reproduced by permission of Sinauer

stead of periplasmic space, the term "periplasmic gel" was coined, indicating that the contents of the space is gel-like (HOBOT et al., 1984). It contains numerous types of proteins. A very loosely organized three-dimensional network of peptidoglycan — instead of only a thin layer — seems to fill the entire space between the outer and the cytoplasmic membrane, and the whole system is highly hydrated. Classical preparation techniques for electron microscopy obviously introduce artifactual changes giving rise to misinterpretations, especially with respect to the three-dimensionality of the peptidoglycan moiety. Techniques for the determination of the volume enclosed between outer and cytoplasmic membrane have been developed (VAN WIELINK and DUINE, 1990). It turned out that this volume may vary, depending on the physiological state of the cell.

Contacts between outer and cytoplasmic membrane were assumed; their existence was made likely by the observation that they were maintained after plasmolysis, and that they could be seen at attachment sites of bacteriophages (ALBERTS et al., 1983) and at sites of synthesis of lipopolysaccharides (BAYER and THUROW, 1977). The periplasm, together with the cytoplasmic membrane is the site within the cell where those components are located which function as receptors of chemotaxis.

3.3.2 Bacteria without an Outer Membrane

As mentioned above, bacteria may be subdivided into the groups of Gram-negative and Gram-positive cells. Gram-positive bacteria do not possess an outer membrane; instead, they typically contain a thick peptidoglycan layer (see Fig. 7) interspersed with additional components such as teichoic acids. The observable thickness of this wall as revealed by conventional electron microscopy depends on the mode of sample preparation, but also on species and growth phase (MAYER, 1986). By chemical and physical analyses insight into the composition and architecture of peptidoglycan was obtained (FREHEL and RYTER, 1979; HOYLE and BEVERIDGE, 1984; SCHLEIFER and KANDLER, 1972). In addition, the synthesis and macromolecular structure of this wall polymer and many of its variations as well as relationships to chitin and cellulose are now well understood. In Gram-positive bacteria, a peptide of up to four different types of amino acids may form bridges giving rise to a more complex nature of this type of peptidoglycan as compared to the peptidoglycan of Gram-negative bacteria. Classification of this wall polymer has been proposed on the basis of its chemical nature. Types of crosslinking and bridging have been used in these systems of classification, and taxonomic implications have become evident which concern both Gram-positive and Gram-negative bacteria.

In contrast to the peptidoglycan of Gram-negative bacteria, dynamics in the assembly and turnover of this wall component in Gram-positive bacteria have been postulated (INGRAHAM et al., 1983). It is assumed that this thick multilayered peptidoglycan is in a dynamic state of synthesis at the surface of the cytoplasmic membrane and sloughing off at the cell periphery (Fig. 16). One could speculate that the outermost bacterial components such as extracellular enzymes might find their way through this thick wall layer together with newly synthesized peptidoglycan moieties which make their way to the cell periphery through the wall from a site directly outside the cytoplasmic membrane, until they are sloughed off (MAYER et al., 1987).

Members of the group of bacteria without an outer membrane can be found as parasitic organisms such as mycoplasmas and mycoplasm-like organisms. They lack not only the outer membrane but also the peptidoglycan, and they are assumed to have developed from typical eubacteria. The lack of a wall may be explained by their specific habitats. They are known to be causative agents for a variety of diseases (KAHANE et al., 1982; NIEHAUS and SIKORA, 1979).

Bacterial L-forms (MADOFF, 1981), a further type of bacteria lacking all wall layers, are also called L-phase variants, indicating that they represent specific growth phases. They can be produced by suitable treatment of normal cells of Gram-positive and Gram-negative bacteria. The existence of "stable" L-forms has been reported. However, under certain modifications

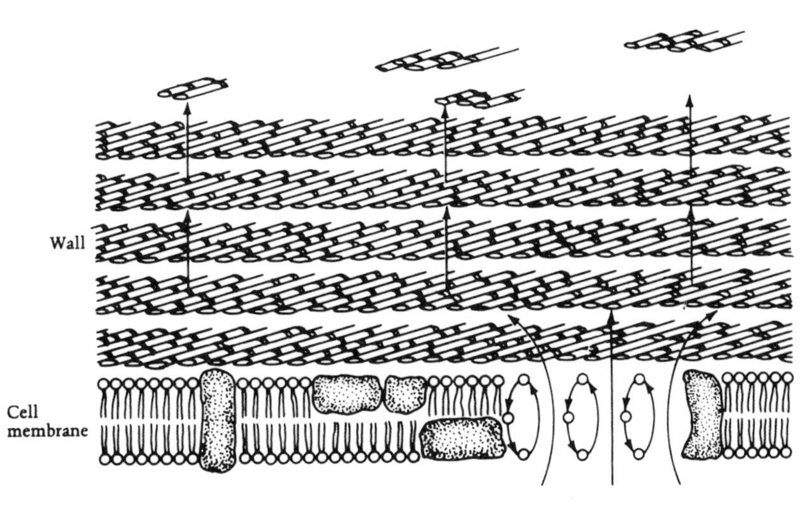

Fig. 16. Wall assembly and turnover in Gram-positive bacteria. The arrows show the progression of newly formed peptidoglycan from its site of synthesis at the cell membrane to the periphery of the cell where it is sloughed off.
From INGRAHAM et al. (1983), reproduced by permission of Sinauer

of growth conditions, even these stable forms have been found to revert to vegetative cells. It has been speculated that a protein, an "autolysine" of the vegetative cell which continues to be produced by L-forms, might be responsible for the inhibition of the re-establishment of peptidoglycan. Anything that destroys this autolysine or prevents its production favors reversion.

A prominent subgroup of bacteria lacking an outer membrane are the archaebacteria (WOESE, 1981). They are separated from the eubacteria by a number of reasons, one being the absence not only of an outer membrane but of typical peptidoglycan as found in eubacteria. The structural and functional role of the cell wall components so far described for typical eubacteria, i.e., peptidoglycan (or murein) and outer membrane, is fulfilled by a variety of other components. Under the electron microscope, wall surfaces of archaebacteria resemble, in many cases, typical surface layers of eubacteria (Sect. 3.3.3). They can be visualized as mosaic patterns of proteinaceous subunits (Fig. 17) (KÖNIG, 1988).

Instead of peptidoglycan, the walls of a variety of methanogenic bacteria consist of polypeptide envelopes or modified polypeptides, polypeptides and polysaccharides in addition, protein sheaths, or heteropolysaccharides containing uronic acid, neutral sugars and amino sugars (KANDLER, 1979). As with peptidoglycan, the polypeptides of methanobacteria probably comprise covalently linked units, each of which has L-glutamic acid, L-lysine and L-alanine in defined ratios and which may have a residue of either N-acetylglucosamine or N-acetyl-galactosamine attached. The term "pseudomurein" was introduced for the wall polymers of the methanogenic bacteria.

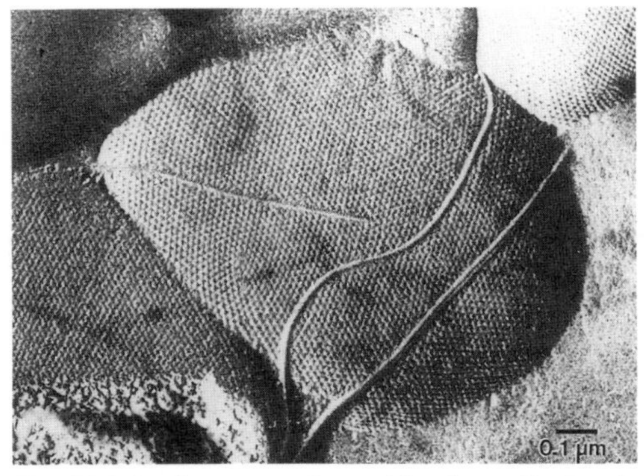

Fig. 17. Mosaic pattern formed by the wall subunits of the methanogenic bacterium *Methanogenium marisnigri*. Freeze-etching preparation.
From ROMESSER et al. (1979)

Data obtained for the composition of the wall of halobacteria (König, 1988) demonstrated that one group of them has, as the structural component of the thick wall, a sulfated heteroglycan containing glucose, mannose, galactose, glucosamine, galactosamine, glucuronic acid and galacturonic acid. It has a branched structure, acidic groups in the form of sulfate and carboxylates that mimic the functions of carboxyl and phosphate groups of peptidoglycan and teichoic acids. Another group of halobacteria does not have peptidoglycan or other carbohydrates in its place. Cells of this group are sensitive to osmotic damage.

3.3.3 Bacterial Surface Layers

Regular arrays of proteins or glycoproteins forming two-dimensional lattices of macromolecules, so-called surface layers (S-layers) have been observed outside of the cell wall proper of many Gram-positive and several Gram-negative bacteria including cyanobacteria (Sleytr and Messner, 1983; Sleytr et al., 1988). Very similar layers have been found in certain archaebacteria that have no rigid pseudomurein wall (see Sect. 3.3.2). Eubacteria often lose their surface layer when cultivated in the laboratory. The S-layers cover the whole bacterium, leaving no gaps. Data have been collected indicating that such a layer may help to protect a cell in a hostile environment. In addition, it was speculated that surface layers might trap enzymes otherwise lost to the environment.

Usually, surface layers exhibit a hexagonal or tetragonal symmetry (Fig. 18). In contrast to the cylindrical part of the cell which is covered with an undistorted lattice of morphological units (consisting of subunits), faults in the lattice can be observed at the cell poles and the sites of cell septation. It was shown that morphological units and small aggregates of units, i.e., crystallites, may rearrange and fuse on the cell surface to form larger areas of regular patterns.

Surface layer molecules can be removed from the cell by application of high concentrations of chaotropic agents such as urea or guanidine hydrochloride. Most isolated S-layer units were shown to disintegrate into their subunits by lowering the pH below a value of 3.

Chemical characterization of S-layer subunits has been performed for eubacteria and selected archaebacteria. For eubacteria subunit species with molecular weights ranging from 100 to 180 kDa, with or without carbohydrate residues, were found. Archaebacteria such as *Sulfolobus acidocaldarius* and *Halobacterium salinarium* exhibit surface layers composed of glycoproteins with molecular weights between 140 and 200 kDa.

Technical applications of isolated surface layers have been developed (Pum et al., 1989; Sara et al., 1988). After chemical modification, S-layers may be used for ultrafiltration purposes, for coating with monolayers of biologically active molecules such as enzymes, or as supports for covalent attachment of macromolecules for specific and rapid determination of substances by enzymic or immunological

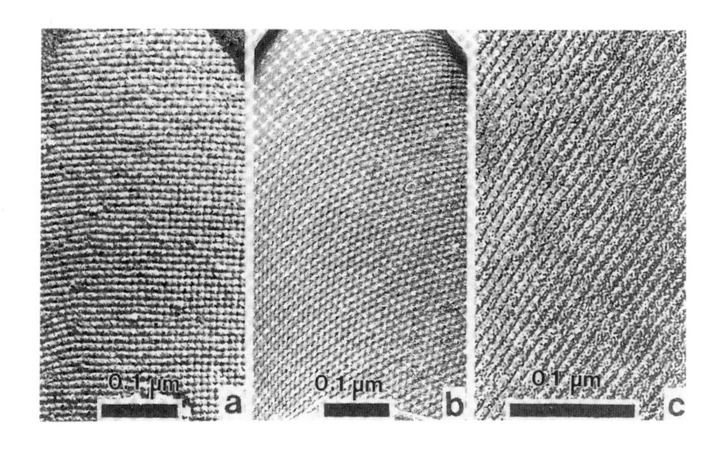

Fig. 18. Surface layers of bacteria.
(a) *Desulfotomaculum nigrificans*, (b) *Clostridium thermohydrosulfuricum*, (c) *Bacillus stearothermophilus.* Freeze-fracture preparations.
From Sleytr and Messner (1983), reproduced by permission of Annual Reviews of Microbiology Inc.

methods. It has been speculated that studies of complex biological systems such as surface layers might influence the design and fabrication of molecular machines and electronic devices.

3.4 Cytoplasmic and Intracellular Membranes

3.4.1 Cytoplasmic Membrane

In electron microscopic samples prepared by application of conventional chemical fixation and staining procedures, a membrane is visible which surrounds the cytoplasm and which exhibits a trilaminar ultrastructure (INGRAHAM et al., 1983) (see Figs. 6 and 7). This membrane acts as a diffusion barrier and as a system regulating the "traffic" of substances from and to the inside of the cell. One major aspect is that of its bioenergetic functions such as ATP synthesis and hydrolysis, oxidoreductions, and cell motility.

The cytoplasmic membrane is composed primarily of phospholipids and proteins. Other components are glycolipids, glycophospholipids, hydrocarbons, quinones and isoprenic alcohols. The proteins may be integral or "intrinsic", or peripheral or "extrinsic" (see Fig. 15). Intrinsic proteins cannot easily be removed from the membrane whereas peripheral proteins may be separated from the membrane by altering the ionic strength of the suspending buffer.

Peripheral proteins associated with a membrane are attached principally by ionic bonds involving cations such as Mg^{2+}. Proteins account for 60 to 70 % of the dry weight of membrane preparations. There are no covalent links between the membrane components. The "fluid mosaic" model (SINGER and NICHOLSON, 1972) has been developed to describe the fact that the membrane components do not form a rigid system, but may diffuse and rearrange laterally. The fluidity of a membrane of a bacterium at a given temperature is dependent on its lipid composition. Bacteria with very high growth temperature optima may exhibit clear differences in lipid composition compared to bacteria growing at room temperature. A designer of artificial membrane systems used for the production of proteoliposomes should keep this situation in mind. There are striking biochemical differences between membranes of eubacteria and archaebacteria with respect to the linkage of residues in the lipid components (WOESE, 1981). Some extreme halophiles have the fatty acids replaced by ether-substituted alkyl residues; lipids containing glycerol ether-linked residues were detected in

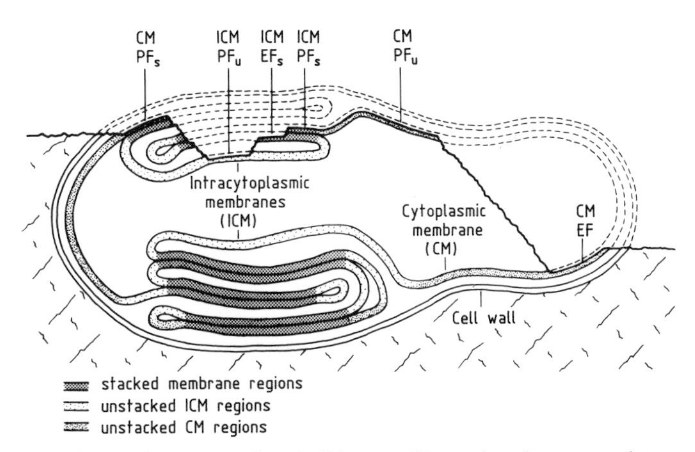

Fig. 19. Photosynthetic membranes in *Rhodopseudomonas palustris.* Diagram illustrating the nomenclature for the freeze-fractured membranes.
CM, cytoplasmic membrane; part of CM (stacked CM regions) participates in photosynthesis; ICM, intracytoplasmic membranes; EF, E-face; PF, P-face; s, stacked; u, unstacked.
From VARGA and STAEHELIN (1983) reproduced by permission of the American Society for Microbiology

Thermoplasma acidophilum; in methanogens the lipids were also found to be branched glycerol ethers.

Cytoplasmic membranes with obvious differentiations are common in bacteria using light as one or the sole energy source (KONDRATIEVA et al., 1992). In *Chlorobium* and *Chloroflexus* this membrane carries the reaction centers and the attachment sites for the chlorosomes. In purple bacteria the physical continuity of cytoplasmic and intracytoplasmic membranes was shown. Parts of the cytoplasmic membrane may function as photosynthetic membranes (Fig. 19). Nature found similar solutions for the location of parts of the photosynthetic system in certain filamentous cyanobacteria. The cytoplasmic membrane of halobacteria (BLAUROCK and STOECKENIUS, 1971; OESTERHELT and STOECKENIUS, 1973) contains patches of membrane-integrated bacteriorhodopsin molecules ("purple membrane"). Technical applications of this kind of photoreceptor systems are considered.

3.4.2 Intracellular Membranes

3.4.2.1 Phototrophic Bacteria

As mentioned above (Sect. 3.4.1), intracytoplasmic membrane systems may differ from the cytoplasmic membrane. A variety of types have been observed (Fig. 20), and groupings of photosynthetic bacteria on the basis of type and arrangement of intracytoplasmic membranes have been proposed (KONDRATIEVA et al., 1992). Complexity and organization of the intracytoplasmic membrane systems of certain phototrophic bacteria have been shown to vary, depending on a number of parameters such as aerobic or anaerobic growth, or variation of incident light intensity (GOLECKI et al., 1980). Dynamic changes in membrane composition and in the composition of membrane-attached complexes harboring the reaction center and the light harvesting components have also been detected (VARGA and STAEHELIN, 1983). The structure of the protein sub-

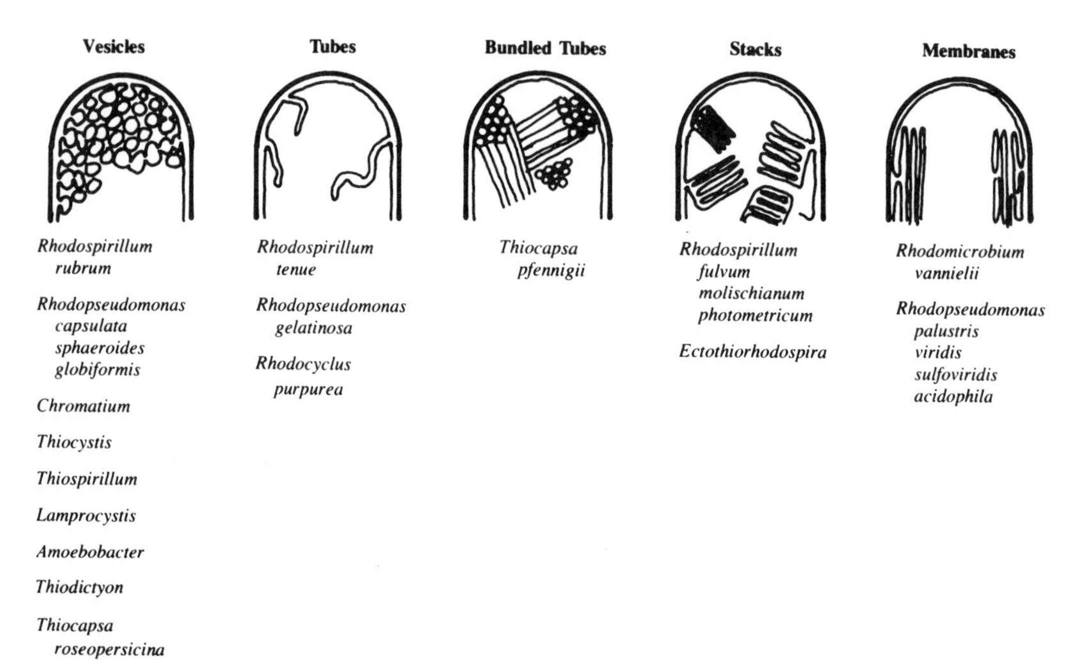

Fig. 20. Types of intracytoplasmic membranes occurring in genera and species of the Rhodospirillaceae and Chromatiaceae.
From TRÜPER and PFENNIG (1981), reproduced by permission of Springer-Verlag

units in a bacterial photosynthetic reaction center has been described in detail (DEISENHOFER et al., 1985).

3.4.2.2 Non-Phototrophic Bacteria

Extensive intracellular membrane systems have been described in bacteria such as dinitrogen fixers (Figs. 21 and 22) (WATSON et al., 1981), methane utilizers and methanogens. These membranes are also derived from the cytoplasmic membrane by invagination. In *Azotobacter vinelandii* morphological and ultrastructural variations depending on aeration were analyzed in continuous culture, and membrane development was followed with either atmospheric nitrogen or ammonium as nitrogen sources. It could be stated that oxygen can mediate changes in the ratio of intracytoplasmic to cytoplasmic membrane surface areas only under conditions of nitrogen fixation. On the basis of extensive experiments it was concluded that intracytoplasmic and cytoplasmic membranes do not differ with respect to respiratory activities and that both types of membranes represent differently located parts of an otherwise identical membrane system (POST et al., 1983). Convoluted internal membrane systems described in *Methanobacterium thermoautotrophicum* (ZEIKUS and WOLFE, 1973) were shown to be preparation artifacts (ALDRICH et al., 1987; KRÄMER and SCHÖNHEIT, 1987). The appearance of "mesosomes" in eubacteria, structures frequently observed in the electron microscope after conventional chemical fixation, was also recognized to be the result of an artifactual change of the membrane structure caused by chemical or physical impairment (NANNINGA et al., 1984). One could assume that the cytoplasmic membrane, at sites where mesosomes are usually formed, might have a composition different from that in other areas, and that this difference makes this specific site of the membrane susceptible to deformation.

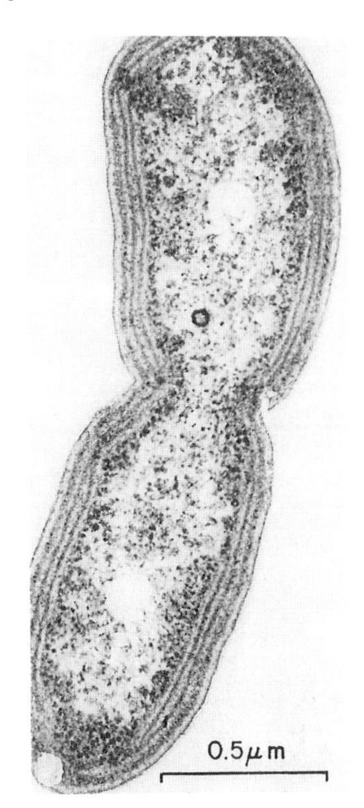

Fig. 21. *Nitrosomonas europaea* exhibiting intracytoplasmic membranes. Ultrathin section.
From WATSON et al. (1981), reproduced by permission of Springer-Verlag

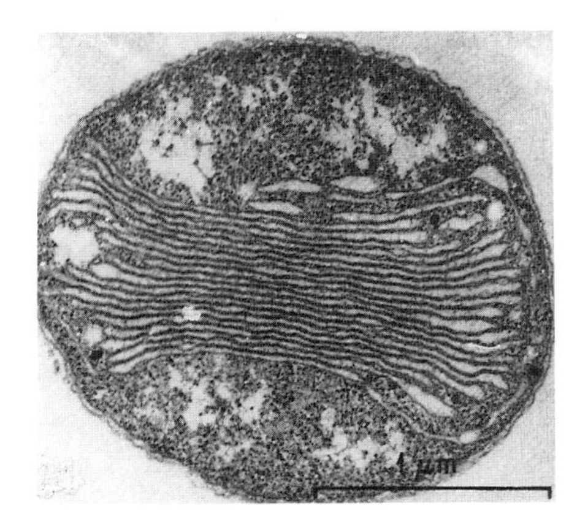

Fig. 22. *Nitrosococcus oceanus* exhibiting intracytoplasmic membranes. Ultrathin section.
From WATSON et al. (1981), reproduced by permission of Springer-Verlag

3.5 Inclusion Bodies

3.5.1 PHA Inclusions

Polyhydroxybutyric acid (PHB) and related polyesters on the basis of hydroxyalkanoic acids (PHA) form inclusion bodies, functioning as storage materials, accumulated in a wide variety of aerobic, anaerobic, heterotrophic and autotrophic bacteria (SCHLEGEL and GOTTSCHALK, 1962; SCHLEGEL et al., 1961). These inclusions usually are more or less spherical; they appear to be surrounded by a faint membrane different in structure from a typical unit membrane. Molecular methods have been applied to analyze the respective biosynthetic pathways of these polymers (HUSTEDE et al., submitted; STEINBÜCHEL and SCHLEGEL, 1991). Considerable progress has been made also in attempts to increase the amount of inclusions per cell (Fig. 23) in order to use them as biodegradable thermoplastic polyesters, as packaging materials, in medicine, and in agriculture. The market for these products is expected to expand. Procedures have been developed for large-scale production of these inclusions; a two-step fed-batch procedure was designed using glucose and propionic acid as substrates. Other approaches avoiding propionic acid are also considered.

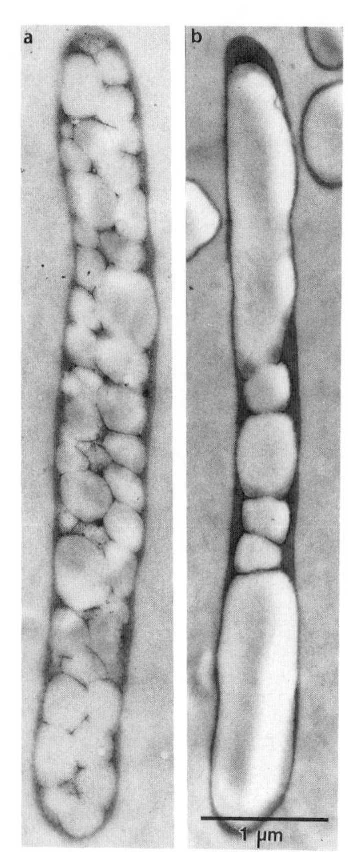

Fig. 23. Results of heterologous expression of PHB-biosynthetic genes of non-sulfur purple bacteria in *Alcaligenes eutrophus*.
(a) Harboring plasmid pEH 140, (b) harboring cosmid pEH 211. Ultrathin sections.
Original micrographs E. HUSTEDE, A. STEINBÜCHEL, H. G. SCHLEGEL

3.5.2 Protein Inclusions

Inclusion bodies containing proteins can be found in a variety of bacteria (MAYER, 1986; SHIVELY, 1974). They are assumed to function as nitrogen storage. Carboxysomes, inclusion bodies observed in many species of chemolithotrophic bacteria and in cyanobacteria, are polyhedral particles containing the enzyme ribulose-1,5-bisphosphate carboxylase-oxygenase. Their shell forms a pentagonal dodecahedron; it might be composed of protein or glycoprotein.

Bacillus thuringiensis produces, during sporulation, a proteinaceous inclusion body which was found to be highly toxic to insects when ingested (STAHLY et al., 1992). Subspecies *israelensis* is especially interesting because it is very toxic to dipterans such as mosquitoes, causing lysis of cells in the midgut epithelium of the larvae. Cell lysis is thought to result primarily from the action of the toxins on membrane phospholipids. Further analyses gave a better insight into interactions between the toxin and its target cells and into cytolytic activity and immunological similarity between toxins of different subspecies of *B. thuringiensis*. They provided evidence for a lectin-like receptor in the larval mosquito, and they showed that the toxin is a glycoprotein. Application of the toxin as a biological control agent against mosquitoes is considered.

Products of heterologous gene expression in bacteria, i.e., polypeptides, often tend to form inclusion bodies (Fig. 24) due to non-physiologically high concentrations (SCHONER et al., 1985). These inclusions contain densely packed aggregates of the products. Sometimes, they even exhibit structural features indicating a paracrystalline order of the constituents.

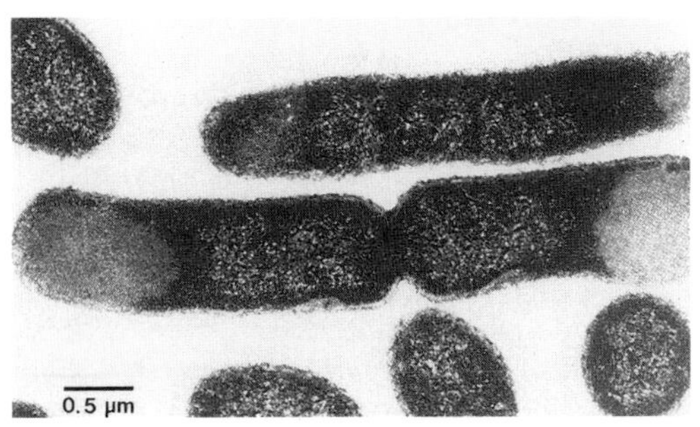

Fig. 24. Ultrathin section through *Escherichia coli* cells producing human proinsulin concentrated close to the cell poles.
From KLEINIG and SITTE (1986) reproduced by permission of Gustav Fischer Verlag

3.5.3 Other Inclusion Bodies

A variety of other inclusions has been detected in bacteria, ranging from well defined structures such as gas vacuoles, magnetosomes, chlorosomes and phycobilisomes (which all are integral components of the respective cells) to sulfur globules, membrane-enclosed polyglucoside (glycogen) granules, glycogen granules without a membrane, polyphosphate (volutin, metachromatic) granules, protein crystals and paracrystalline arrays of various origins, cyanophycin granules, to defective and intact bacteriophages and refractile bodies (MAYER, 1986). The functions of these inclusions are very diverse. In terms of a biotechnological use of bacteria, one should keep in mind that inclusions of many kinds may interfere with procedures used for purification of products manufactured in or with bacteria as soon as the cells have to be opened up. Bacteriophage infections of bacteria used in production may be a major hindrance to high efficiency of a process, and measures should be taken to work with very well defined growth conditions and with bacterial strains or mutants constructed in such a way that certain inclusions are not formed at all.

3.6 Extracellular Bacterial Enzymes

3.6.1 Amylases, Pullulanases, and Related Enzymes

"Extracellular" indicates that these enzymes are exported through the cytoplasmic membrane. There is an obvious necessity for this phenomenon: high-molecular weight substrates (such as starch; but also cellulose, see below) cannot be taken up by bacteria; bacteria lack the potential to perform endocytosis. Bacteria solve this problem by secretion of enzymes, via states where the enzymes are still at tached to the outside of the cytoplasmic membrane or to wall components, into the medium.

Branched glucose polymers with α-1,4- and α-1,6-glycosidic bonds are degraded by enzymes (amylases, pullulanases, and related enzymes) which can be obtained from bacterial sources (AUNSTRUP, 1979; PRIEST, 1984). Major advantages of approaches developed recently (BURCHHARDT et al., 1991; ANTRANIKIAN et al., 1987a) are the following: enzymes produced by thermophilic anaerobic bacteria such as clostridia were shown to be stable up to high temperatures (see also below); overproduction and secretion into the medium could be obtained by variation of the growth parameters pH, temperature, dilution rate, concentration of the substrate, and limitation of phosphate. Enzymes needed in a process should not have different pH optima; this precondition could be fulfilled in the approaches mentioned above. Continuous cultures for enzyme production are superior to batch cultures; the new approaches involve continuous cultures. There may be a physiological explanation for this superiority: enzyme synthesis and secretion are regulated; synthesis is induced by starch, and glucose causes catabolite repression; starch limitation obviously completely derepresses the enzymes and, therefore, causes their secretion into the medium. This secretion was paralleled by a change in the ultrastructure of the cell envelope (ANTRANIKIAN et al., 1987b), i.e., a degradation of the surface layer and the murein, and a deformation (bleb and vesicle formation) of the cytoplasmic membrane (Fig. 25). Also a change in the biochemical composition of the cell wall was observed, depending on the growth phases: peptidoglycan was reduced during the stationary growth phase, whereas galactosamine in the accessory polymers increased. Genetic analyses of the process of enzyme secretion revealed multiple steps of processing.

A search for highly thermostable amylolytic enzymes (ANTRANIKIAN et al., 1990) included a number of hyperthermophiles; the highest temperature at which enzymes from these organisms still degrade starch was 130 °C, and they tolerated a treatment at 120 °C for up to six hours. In addition, metal ions or other intracellular components were not needed for high activity. The implications of these findings for a technical application of these enzymes are obvious.

A further practical improvement for enzyme production has been obtained by immobilization of enzyme-secreting bacteria (KLINGENBERG et al., 1990); thus, cell-free enzyme samples could easily be prepared. Immunoanalytical systems have been designed to perform automatic on-line control of enzyme production (FREITAG et al., 1990), and biosensors have been developed which allow continuous on-line determination of product concentrations (HURDECK et al., 1990).

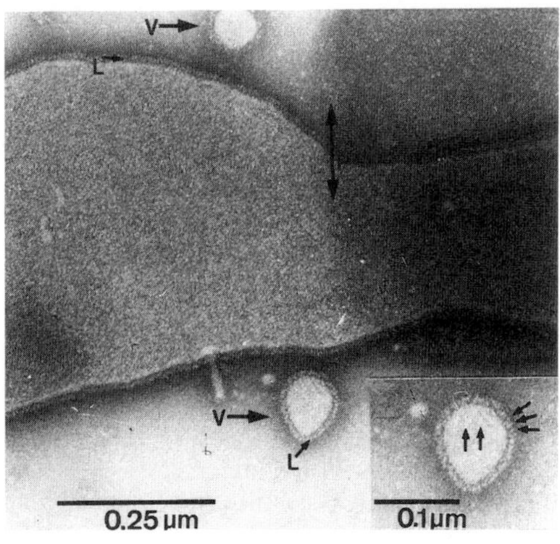

Fig. 25. Cell of *Clostridium* sp. strain EM1 during massive production of α-amylase and pullulanase. The cell wall is partially degraded (double-headed arrow), and vesicles (V) were formed from the cytoplasmic membrane, covered with a faint monolayer (L, and arrows) of particles interpreted as enzyme molecules. Negative staining.
From ANTRANIKIAN et al. (1987b)

3.6.2 Cellulases

Cellulases secreted by bacteria may, at the moment, be of minor practical importance compared to amylolytic enzymes. Nevertheless, a better understanding of their genetics, synthesis, secretion and ultrastructural organization may help to develop ideas for their more widespread application.

Bacillus cellulases (carboxymethylcellulases) (COUGHLAN and MAYER, 1992) appear to be rather simple enzyme systems (Figs. 7 and 26). They are secreted into the medium, and they can be used for the degradation of amorphous cellulose. A possible practical application could be in cotton fiber production where they could remove amorphous cellulose pieces from the otherwise crystalline cellulose fiber which is not attacked by them.

The cellulolytic enzyme system of *Clostridium thermocellum* (Fig. 27) is much more complicated (COUGHLAN and MAYER, 1992). It degrades crystalline (and amorphous) cellulose.

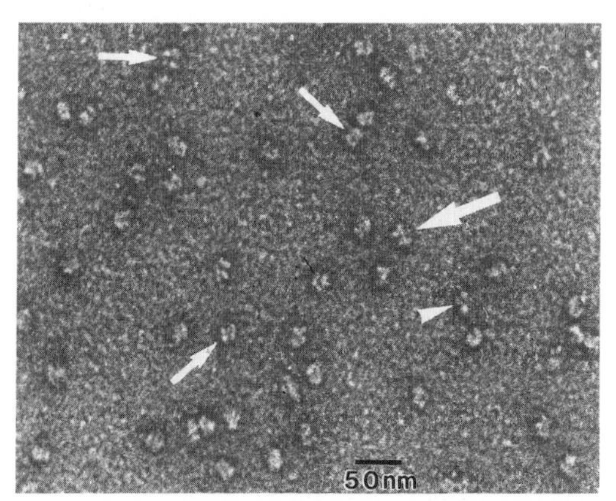

Fig. 26. Isolated *Bacillus* sp. carboxymethylcellulase (CMC) molecules. The arrows mark various projection types of the oligomeric enzyme. Negative staining.
Original micrograph J. KRICKE

It was detected as an adhesion factor attached to the outer surface of the cell wall (LAMED et al., 1983a). Finally, it turned out to be a system organized as "cellulosomes" composed of many types of subunits ranging in their molecular weights between 20 and 200 kDa (LAMED et al., 1983b). More than 20 genes coding for components of the cellulolytic enzyme system

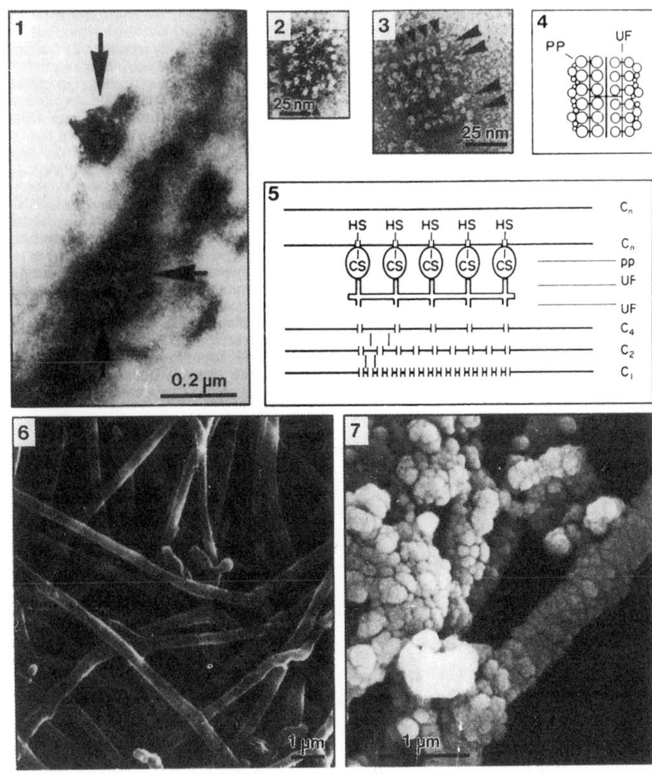

Fig. 27. Cellulolytic enzyme system of *Clostridium thermocellum:* electron microscopic observations and proposed mechanism of action.
(1) Transmission electron micrograph of a negatively stained cell exhibiting clusters ("polycellulosomes") of spherical particles ("cellulosomes") attached to the cell surface.
(2) Structurally well preserved cellulosome isolated from the growth medium. The particle is composed of a number of polypeptides (arrow heads) of varying size.
(3) Artificially flattened cellulosome isolated from the growth medium. The particle has been flattened by the mounting procedure for electron microscopy. Its polypeptides (small arrow heads) are arranged in parallel rows (large arrow heads) held together by additional structural components of unknown composition, forming an ordered network (see MAYER et al., 1987).
(4) Diagrammatic view of the major structural features of a cellulosome as derived from electron micrographs. The cellulosome is comprised of polypeptides (PP), some ordered into groups of rows, and some, at the periphery, non-ordered. The polypeptides are held together by fibrils (UF) of unknown chemical composition.
(5) Diagrammatic view of the structural concept of the cellulosome and of the inherent structure–function relationships. CS, catalytic site; HS, hydrolysis site, i.e., the site on the cellulose chain at which a glycosidic bond is cleaved; UF, fibrils; C_n, C_4, C_2, C_1 are cellodextrins of various lenghts (C_n is cellulose and C_1 is cellobiose); pp, polypeptides.
(6) Scanning electron micrograph of cells of *Thermomonospora curvata*, shown for comparison, grown on cellobiose. No polycellulosomes.
(7) As (6), but grown on protein-free lucerne fibers. Hyperproduction of polycellulosomes.
From COUGHLAN and MAYER (1992)

of clostridia have been found; it appears that there is a considerable degree of sequence homology within these genes. Several of the gene products seem to be glycoproteins. Analyses of the macromolecular organization of the cellulosome (MAYER et al., 1987) revealed that it is constructed out of rows of equidistantly spaced, similarly sized polypeptide subunits, held together by an unknown substance. A working model proposing a concerted action of several (identical) subunits within a cellulosome still attached to the cell, or detached from the cell and bound to the substrate, was developed. Concerted actions of this kind may finally cause hydrolysis of the cellulose elementary fibril into cellobiose, via larger fragments 4 or 2 cellobioses long. Thus, crystalline cellulose may be degraded, by a number of multicutting events, which are made possible by the high order within the enzyme complex. A detailed electron microscopic study of the decomposition of the enzyme complex during storage revealed that, finally, the originally large cellulosomes with particle weights of several millions Da are degraded into single free polypeptides.

4 Structural Organization of the Eukaryotic Cell

4.1 Principles

Obvious similarities and clear differences can be stated when prokaryotic and eukaryotic cells are compared (FAWCETT, 1981; KLEINIG and SITTE, 1986).

A membrane surrounding the protoplast acts as a diffusion barrier and, at the same time, regulates transport from and into the cell; it also contains components which are part of systems for cell–cell interaction and "communication". In plants and prokaryotes, the cells are surrounded by a wall which stabilizes the architecture of the organism, whereas in animals the membrane or a glycocalyx may be the outermost cell layer, and cell turgor, together with microtubules forming the cytoskeleton, preserves and determines cell shape and size.

Microtubules have been detected both in animal and plant cells. In electron micrographs they can often be seen close to the plasma membrane (cortical microtubuli); however, they may also traverse areas of the cytosol, and they occur in the nucleoplasm where they are involved in the rearrangement of the chromosomes during cell division (Fig. 28). Microfilaments, further fibrillar cell strucures, do not form tubules; they are thin fibers. Their function appears to be linked to intracellular transport processes, and bundles of microfilaments can be visualized by light and electron microscopy.

In many fungi and in arthropods, the cell wall or superimposed layers contain chitin, whereas in plants the major polysaccharide in the cell wall is cellulose.

One of the major differences, regarding the structural organization, between a prokaryotic and a eukaryotic cell is the fact that the latter contains "compartments". They are defined as membrane-bound volumes within the protoplast. In general, the membrane surrounding the volume of a compartment is, by definition, part of this compartment. However, this definition is not always strictly valid. Membranes may by themselves be considered to be "compartments", i.e., sites within a cell which harbor components performing specific functions. Fig. 29 shows a drawing specifying the com-

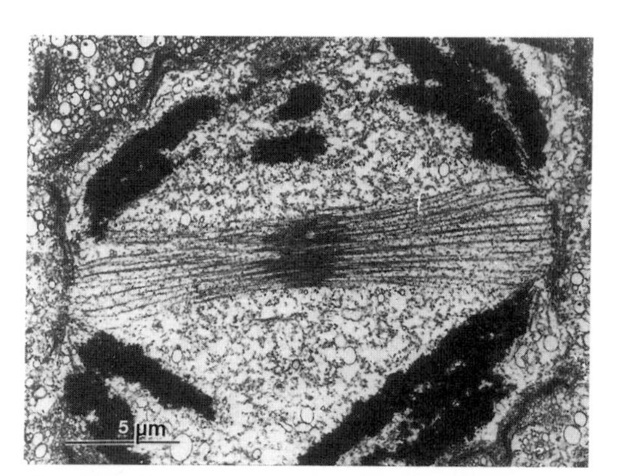

Fig. 28. Mitosis spindle. Mitosis in diatomea. The chromosomes have reached the poles (late anaphase); the microtubules forming the spindle are clearly seen.
From KLEINIG and SITTE (1986) reproduced by permission of Gustav Fischer Verlag

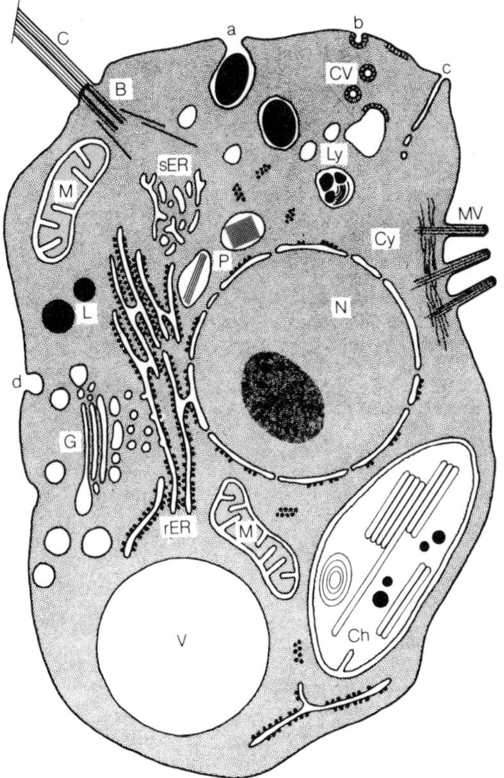

Fig. 29. Compartmentation in a eukaryotic cell. N, nucleus and nucleolus; Cy, cytoplasm with free polysomes; Ch, chloroplast with thylacoids, starch and lipid droplets; M, mitochondria; sER, smooth endoplasmic reticulum; rER, rough endoplasmic reticulum; G, Golgi apparatus (dictyosome); V, vacuole; Ly, lysosomes; CV, coat vesicles; P, peroxisomes with protein crystals; L, lyosomes; C, cilia or flagellum; B, basal body; MV, microvilli; a, b, c, types of endocytosis; d, exocytosis.
From KLEINIG and SITTE (1986), reproduced by permission of Gustav Fischer Verlag

partmentation of a typical eukaryotic cell. As can be deduced from the shading of the various areas within the cell, the volumes within a number of compartments are, at least in a formal sense, interconnected. Examples are the lumen of the endoplasmic reticulum, the dictyosomes, the vacuole, and the volume formed between the two membranes of the nucleus. The compartment "Golgi apparatus" in a cell is composed of all dictyosomes within the cell, the compartment "mitochondrion" comprises all mitochondria within the cell, etc. It was proposed that a biological membrane always separates a "plasmatic phase" from a "non-plasmatic phase". Three plasms in animal and four plasms in plant cells are distinguished: the cytoplasm, the caryoplasm, the mitoplasm, and the pastoplasm. Always two membranes, with a non-plasmatic phase between them, have to be passed in order to go from one plasm to another. In fusion or vesicle formations only plasms with the same denomination can be interacting. Plasmatic phases are mixed or separated during processes of fertilization and multiplication, non-plasmatic phases during processes involved in metabolism and growth.

In many cases, a "compartment" is also an "organelle"; however, in a number of examples, the two terms do not denominate the same structure. For example, a polysome is an organelle, but not a compartment, whereas the nucleus is both an organelle and a compartment.

Based on a huge body of observations, it is known that membranes of various compartments can fuse. The two terms "membrane flow" and "membrane transformation" indicate that a typical cell is, at least formally, in a permanent state of dynamic changes (MORRÉ et al., 1979: ROBINSON and KRISTEN, 1982). Any micrograph only depicts a momentary state of the cell. Mitochondria and chloroplasts do not participate in this change. They form an exception which is explained by their assumed origin, i.e., their formal derivation from endosymbionts. Export and import of material formed within the cell or from outside take place by exocytosis and endocytosis, respectively, i.e., by specific events involving the membrane surrounding the protoplast. Exo- and endocytosis have not been observed in prokaryotic cells.

A property typical for plant and animal cells is that they differentiate to form tissues and to fulfill specialized functions within an organism. One of the consequences is that in differentiated cells often not all organelles potentially present in a eukaryotic cell are detectable, whereas others are very prominent. Examples are xylem cells in plants or muscle cells in animals. It would be beyond the scope of this section to describe in detail the multiple solutions nature has found in this respect to deal with specific requirements. The section will rather be restricted to a description of the basic aspects of organization of eukaryotic microorganisms and plant and animal cells, with occasional remarks on properties relevant in biotechnology.

4.2 Yeasts as Examples for Eukaryotic Microorganisms

4.2.1 Cell Envelope

After conventional preparation including chemical fixation yeast cells exhibit a smooth surface. The true ultrastructure of yeast cell surface structures was only revealed when low-temperature techniques were introduced for electron microscopic preparation (MÜLLER, 1988). It turned out that, typically, the outermost layer of yeast cells exhibits densely packed fibrillar material ("fimbriae") covering the whole surface (Fig. 30). Depending on the growth conditions, especially the substrate, these fibrils were seen to be variable in length. They were removed by protease treatment. These fimbriae are assumed to be involved in cell–cell (agglutination, mating, cell divisions) and cell–substrate interactions (Fig. 31). It was concluded from results of biochemical, enzymological, and immunological investigations that at least some of the fimbriae may contain acid phosphatase. The scattered distribution of this enzyme over the cell surface was demonstrated by application of antisera specific against the enzyme. By comparison with an electron micrograph showing the cell topology, the sites where specific antigen–antibody reaction had occurred could be pinpointed. Scanning electron microscopic immunocytochemistry was used in this analysis, with secondary electrons for visualization of cell topology, and backscattered electrons for localization of colloidal gold particles complexed with antibod-

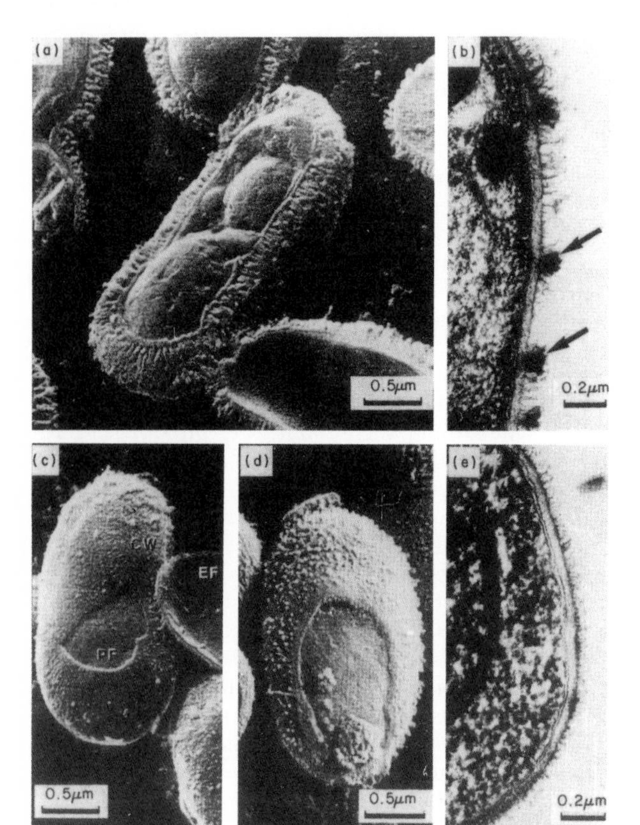

Fig. 30. *Candida tropicalis* grown in continuous culture with hexadecane (a) and (b) and glucose (d) and (e) as carbon sources. Cells were rapidly frozen for physical fixation and freeze-dried for scanning electron microscopy (a), (c), and (d), or freeze-substituted for transmission electron microscopy (b) and (e). In (a) and (b), the fimbriae are very long compared to (d) and (e). They were removed in (c) by protease treatment. (a), (c), and (d), freeze-fractured samples; (b), (e), ultrathin sections.
From MÜLLER (1988), reproduced by permission of Academic Press

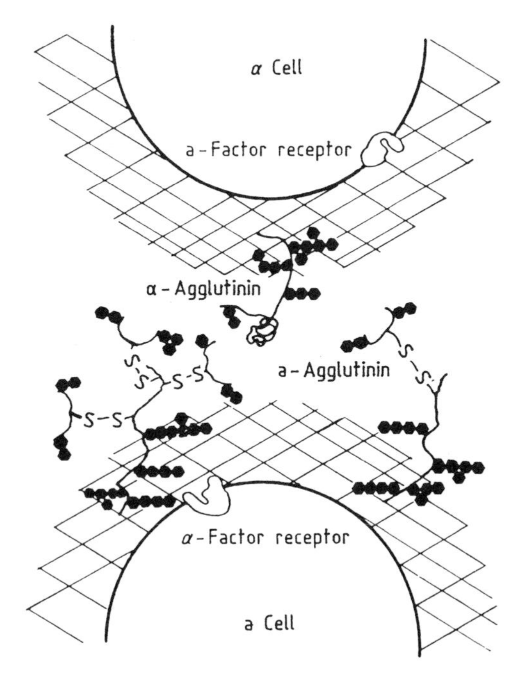

Fig. 31. Model for sexual agglutinins in budding yeasts. Plasma membranes are shown as thick lines; cell walls are crosshatched; hexagons represent agglutinin carbohydrate moieties. α-Agglutinin analogs are composed of a single polypeptide with N- and possibly O-linked carbohydrate. The a-agglutinin analogs consist of a highly O-glycosylated core subunit, which mediates cell surface anchorage, and one or more binding subunits. *Hansenula wingei* and *Pichia amethionina* a-agglutinins analogs are mutivalent (left), and *Saccharomyces cerevisiae* and *S. kluyveri* a-agglutinins are monovalent (right).
From LIPKE and KURJAN (1992), reproduced by permission of the American Society for Microbiology

ies. Part of the fimbriae might correspond to agglutinin-like structures; low-molecular weight proteins and high-molecular weight glycoproteins were shown to be components present in these fimbriae, and a part of about 60 kDa of α-agglutinin, a glycoprotein larger than 200 kDa, was assumed to protrude out of the cell wall (LIPKE and KURJAN, 1992).

4.2.2 Microbodies

Microbodies (peroxisomes, glyoxisomes) (VEENHUIS and HARDER, 1987) are compartments bound by a unit membrane. These organelles harbor enzymic functions required for the metabolism of the carbon and/or nitrogen source (Fig. 32). Enzymes such as catalase, H_2O_2-producing oxidases, enzymes involved in β-oxidation, glyoxylate cycle enzymes, dehydrogenases, an amino transferase and a transketolase have been found in these bodies. It was shown that their functioning often requires metabolic interaction with mitochondria, microsomes or the cytosol. Yeast cells involved in ethanol metabolism exhibited an induction of microbodies, and growth was paralleled by enhanced levels of isocitrate lyase and malate synthase. Fractionation experiments revealed that these enzymes, together with catalase and part of the malate dehydrogenase, were exclusively present in the microbodies. According to this set of enzymes the microbodies were glyoxisomes. Two further glyoxylate cycle enzymes, citrate synthase and aconitase, however, were found to be located in the mitochondria. This situation necessitates interlocking of the enzyme activities of glyoxisomes and mitochondria.

Yeast microbodies exhibit an internal pH around 5.9 whereas the cytosolic pH is approximately 7.1. These values indicate that an electrochemical gradient exists across the microbody membrane which was found to be generated by a membrane-bound proton ATPase. It was suggested that this gradient may play an important role in transport processes across the microbody membrane including uptake of matrix proteins and transport of substrates and/or metabolic intermediates.

It is assumed that microbodies do not arise from the endoplasmic reticulum or *de novo* but develop from already existing organelles. Microbody-matrix (and membrane) protein subunits are synthesized in their mature form on cytosolic polysomes and are imported into the microbody post-translationally.

Massive microbody proliferation was reported in yeasts growing in media containing methanol or *n*-alkanes as the sole carbon source or other unusual carbon and/or nitrogen sources. In general, adaptation of the cell to a new environment involves processes such as synthesis of specific enzyme proteins and inactivation or repression of enzymes. The loss of activity of the majority of microbody matrix

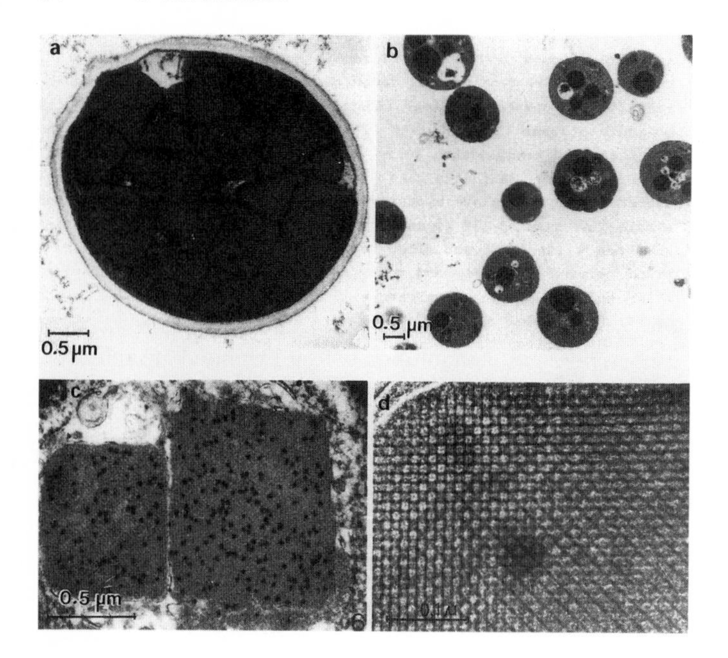

Fig. 32. Peroxisomes in methanol-grown *Hansenula polymorpha.* (a) General view, (b) demonstration of catalase activity, (c), (d) immunocytochemical demonstration of alcohol oxidase; crystallinity of the enzyme is very obvious in (d) (cryosection). From VEENHUIS and HARDER (1987), reproduced by permission of Springer-Verlag

enzymes after a shift can be explained by dilution of existing enzyme protein over newly formed cells. However, examples for selective enzyme inactivation are known. In an additional mechanism, single peroxisomes may be taken up in autophagic vacuoles by means of membrane fusion, followed by proteolysis. At the ultrastructural level, the contents of microbodies often turned out to be crystalline; cells grown under methanol-limited conditions contained a completely crystalline contents in their peroxisomes, due to crystallization of alcohol oxidase molecules.

4.3 The Plant Cell

4.3.1 General Plant-Specific Aspects

Cells of plants, as those of animals, exhibit an enormous variety of specialization depending on their function within a tissue or an organism (GUNNING and STEER, 1975). An obvious plant-specific aspect is their potential to perform photosynthesis (STAEHELIN and ARNTZEN, 1986) (see Fig. 35). Less obvious is the fact that typical plants preserve permanently embryonic cells in certain tissues (the meristemes) in defined areas of the organism. From these regions, differentiation originates, with the consequence that cells of all degrees of differentiation can be seen between the meristematic region and the fully developed areas in the tissue. Experimenters culturing plant cells or tissues (see below) should keep in mind that successful approaches in this field have to consider the respective specific needs. A further obvious plant-specific aspect is that typical plant cells are enclosed by a rigid wall which functions as mechanical stabilizer. Though such a cell wall appears very solid and tight, nature has developed systems which enable a plant cell to perform communication. One of these is the differentiation of plasmodesmata (cell–cell communication, through gaps in the wall, within a tissue). The exposition of lectins at the cell periphery is a way for more general interactions, though specific lectins are usually involved in very specific recognition processes.

4.3.2 Cell Surface Structures

The outermost layer of a cell wall may contain macromolecules which are not involved in the basic function of the wall and which are exposed to the environment (KLEINIG and SITTE, 1986). They are assumed to play a role in cell–cell recognition and cell adhesion. Above (see Sect. 4.2) such a component has already been mentioned. The terms "lectin" and "phytohemagglutinin" have been introduced for these molecules. They are di- or multivalent sugar-binding proteins or glycoproteins without enzymatic activity directed against the bound sugars. Their occurrence is not restricted to plants; they have also been extracted from animal cells. For several of the lectins indications exist that they are not only exposed at the cell periphery but also reside within the cell.

The potential for cell agglutination is based on the fact that lectins, being di-, tetra- or hexamers, contain the respective number of binding sites for sugars (one per subunit). As a rule, the specificity for sugars is very high.

A well-known naturally occurring reaction involving lectins is the recognition of the host root cells by *Rhizobium*. The plant lectin is assumed to interact with the lipopolysaccharide and the fimbriae of the bacterium.

The properties of lectin allow their application in fields such as the detection of sugar-containing macromolecules in membranes, and they can be labeled with markers. This simplifies identification and localization experiments by microscopic and autoradiographic techniques.

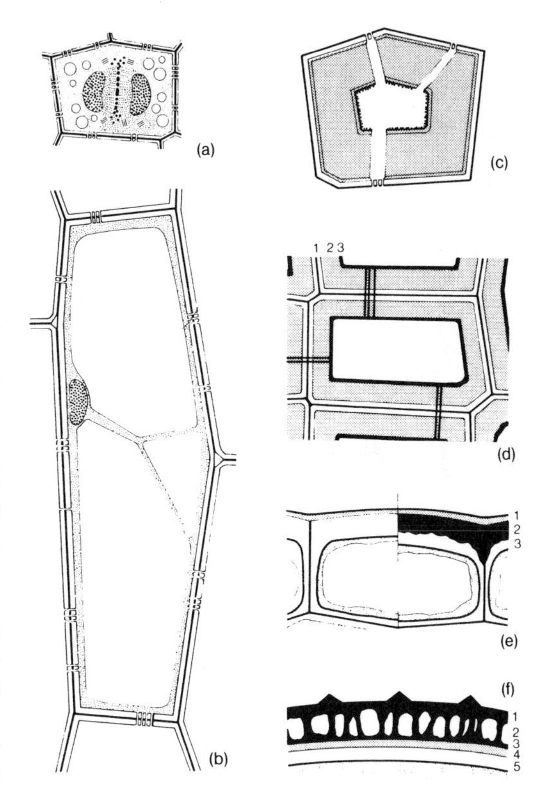

Fig. 33. Walls in plant cells.
(a), (b) Primary wall, (c) primary and secondary wall, (d) primary, secondary and tertiary wall, (e) epidermis with cuticula, (f) sporoderm covering secondary wall.
From KLEINIG and SITTE (1986), reproduced by permission of Gustav Fischer Verlag

4.3.3 Cell Wall Structures

Typical plant cells are surrounded by a cell wall which contains polymerized sugars as the major constituent (FREY-WYSSLING, 1976). Depending on their specific functions, the walls exhibit different thickness, complexity and physical parameters such as rigidity or flexibility (Fig. 33). Often, the outside of a cell wall exposed to the environment is covered by wax which forms a cuticula. This layer reduces evaporation of water.

Cell wall synthesis starts with the formation of a matrix positioned between the two membranes surrounding two adjacent (daughter) cells (MCNEIL et al., 1984). This matrix is very hydrophilic; it is composed of acidic or neutral polysaccharides and contains glycoproteins. They are synthesized within the endoplasmic reticulum (ER) and the dictyosomes, and they are exported by exocytosis. Cellulose, which is embedded later on during the biogenesis of a cell wall, is not synthesized in the ER or the dictyosomes, but by membrane-integrated cellulose synthase complexes (cellulose glucosyl transferase) which form the β-glucan chain of the cellulose macromolecule right at the site where this polymer is needed. The cellulose synthase complex is assumed to make contact, at the inside of the membrane, with components of the cytoskeleton. Cortically positioned microtubuli may direct the lateral movement of the cellulose synthase particle within the membrane, giving rise to the growth

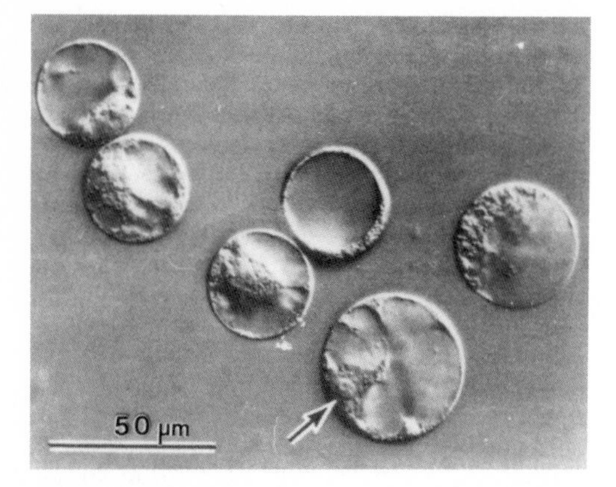

Fig. 34. Light micrograph (interference contrast) of plant protoplasts *(Petroselinum hortense)* obtained by treatment with cellulase and pectinase. Arrow points to a nucleus.
From KLEINIG and SITTE (19886), reproduced by permission of Gustav Fischer Verlag

of oriented cellulose elementary fibrils at the outside of the membrane (KLEINIG and SITTE, 1986).

Cell walls of specialized plant cells contain additional structural components, or the matrix material is even replaced by other compounds. A typical constituent of this kind is the lignin, an amorphous phenol derivative. Together with cellulose, it forms a wall which is most rigid and — at the same time — flexible. Modern building techniques simulate this kind of compound material.

The application of cellulase for the degradation of plant waste materials as sources for biotechnologically manufactured products is very much complicated by the presence of lignin and other constituents. Only pretreatment of the raw material, or the application of mixtures of various polysaccharide-degrading enzyme systems might help to reach reasonable rates of degradation of these waste materials.

Plant cells can be converted to protoplasts (Fig. 34) by a treatment with wall-degrading enzymes, especially cellulases and pectinases. As a consequence, the cells adopt a spherical shape. Regeneration of the wall depends on the specific growth conditions.

4.3.4 Vacuoles and Microbodies as Compartments in the Plant Cell

The basic rules governing the architecture of eukaryotic cells have already been described (see above, Sects. 4.1 and 4.2). There, the existence of compartments was mentioned as one of the prominent structural and functional features typical for cells of higher organisms. A most important compartment or organelle in a fully differentiated plant cell is the vacuole (GUNNING and STEER, 1975; ROBINSON, 1985). It is bound by the tonoplast, a membrane which appears to have an asymmetric ultrastructure. This may be caused by specific proteins integrated into the membrane which are involved in various transport processes. One of them is a proton ATPase necessary for sugar translocation.

A number of functions are ascribed to the vacuoles. Vacuoles contain lytic enzymes, and they may in this respect be considered to be equivalent to the lysosomes in the animal cell. Vacuoles may be the site where secondary plant compounds are stored which might be toxic as long as they are present in the cytoplasm. Very many of these compounds are stored in a glycosilated state. In addition, vacuoles contain high concentrations of dissolved compounds resulting in a high osmotic pressure used by the cell to maintain its turgescence.

One property of vacuoles is especially interesting from the point of view of application. Vacuoles may contain proteins which inhibit proteolytic enzymes (trypsin, chymotrypsin) in the digestive tract of animals. The plant uses them as a biological defense mechanism which protects the plant population. This kind of proteins, after isolation, is considered for use as a biological agent against insect larvae feeding on economically important plants (NEUBERGER and BROCKLEHURST, 1987).

Besides vacuoles, microbodies as described above (see Sect. 4.2) also occur in plants. Two classes are distinguished, peroxisomes and glyoxisomes (TOLBERT, 1981). The peroxisomes convert glycolate produced in the chloroplasts into CO_2 and other compounds. As a rule, these microbodies are positioned in the immediate neighborhood of chloroplasts (Fig.

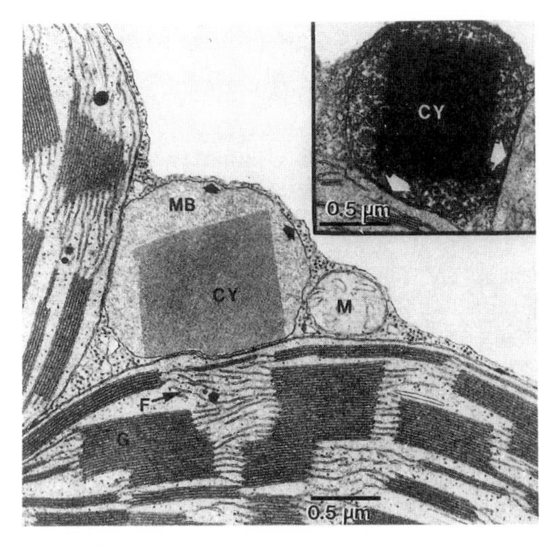

Fig. 35. Microbody (peroxisome) (MB) in a plant cell, in the immediate neighborhood of chloroplasts and a mitochondrion (M).
CY, protein crystal; F, stroma thylakoids; G, grana. Inset: Demonstration of catalase activity in a peroxisome.
From GUNNING and STEER (1986), reproduced by permission of Gustav Fischer Verlag

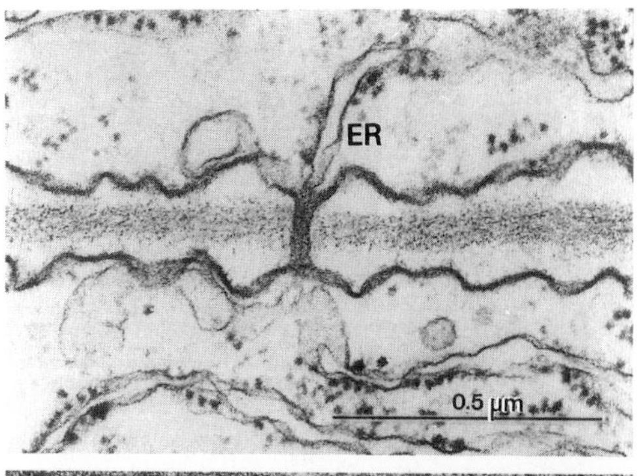

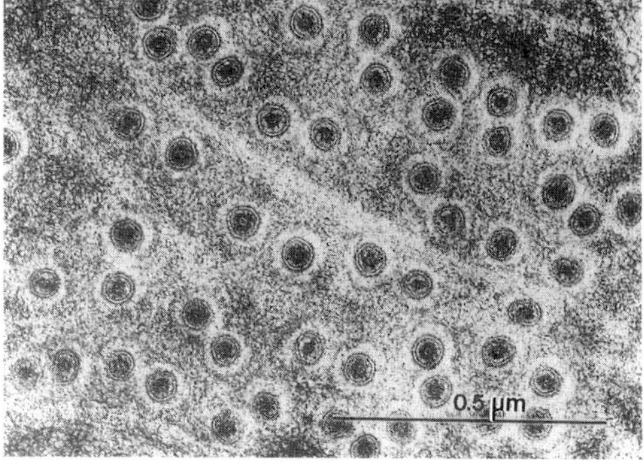

Fig. 36. Plasmodesmata (top: *Cuscuta*, longitudinal section; bottom: *Metasequoia*, cross-section). The ER cisterna traversing the gap in the wall is clearly visible.
From KLEINIG and SITTE (1986), reproduced by permission of Gustav Fischer Verlag

35). Glyoxisomes contain enzymes of the glyoxylic acid cycle; they use lipids stored in lipid droplets, and they are located close to these droplets. During the early stages of the development of plants, microbodies of the glyoxisome type originally present in the cells can be converted into peroxisomes when the stored lipids are used up and fully functioning chloroplasts are differentiated.

4.3.5 Plant Tissues

It is one of the properties of higher pants that they are composed of cells differentiated in very many ways. They have this property in common with higher animals. However, there are obvious differences not only in the pathway of differentiation of cells or formation of tissues but also in the mode which cells use for interaction and communication (STAEHELIN and HULL, 1978). All living cells within a plant have contact – via their neighboring cells – with all the other plant cells. This is brought about by the formation of holes in the wall which are lined by membrane tubes (plasmodesmata) forming a connection between the cell plasms. These tubes are part of the plasma membranes of the adjacent cells. From ultrathin sections it is obvious that, in addition, cisternae of the endoplasmic reticulum pass the plasmodesmata, linking the ER system of the neighboring cells (Fig. 36).

Several functions have been ascribed to the plasmodesmata. One is that they might, in certain aspects, be functionally equivalent to gap junctions of the animal cell (see below), i.e., that they are the route along which information is passed. A major and obvious property of plasmodesmata is that they are responsible for the transport of nutrients. In contrast to animals where nutrient transport takes place in the intercellular space, plants need to involve the lumina within the cells for this purpose.

4.4 The Animal Cell

4.4.1 General Animal-Specific Aspects

Similar to plant cells, typical animal cells (Fig. 37) contain a variety of organelles involved in basic and specialized functions, depending on the degree and type of differentiation (KLEINIG and SITTE, 1986). Compared to plants, animals have an extended range of specialized cells and tissues, with very specific functions: muscles, nerves, sensory systems, blood, the immune system, etc. This section concentrates on the description of selected aspects.

Cell walls comparable to those of prokaryotic or plant cells are missing in animal cells.

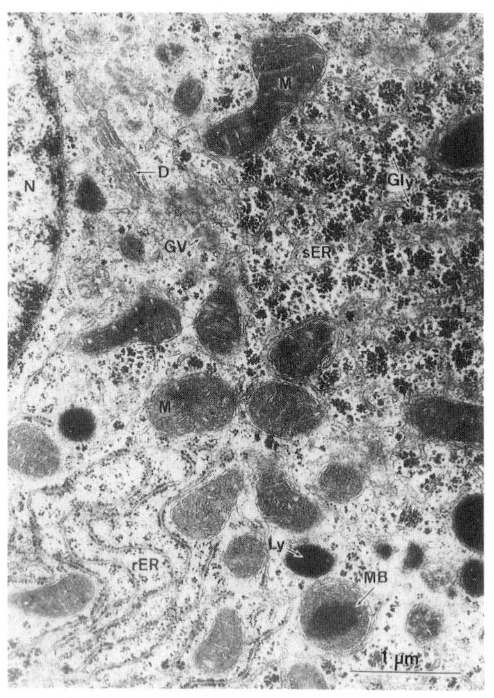

Fig. 37. Section through a typical animal (liver) cell. N, nucleus; M, mitochondrion; D, dictyosome; GV, Golgi vesicles; sER, smooth endoplasmic reticulum; rER, rough endoplasmic reticulum; Ly, lysosome; MB, microbody; Gly, glycogen.
From KLEINIG and SITTE (1986), reproduced by permission of Gustav Fischer Verlag

The shape of the cell is rather determined and maintained by the so-called cytoskeleton. The protein filaments making up the cytoskeleton are actin filaments, cytokeratin filaments, desmin filaments, vimentin filaments, microtubuli, neurofilaments, and glia filaments. The architecture of this system is not static; drastic changes occur during cell growth, division, and differentiation. Interactions with the plasma membrane are evident. The regulation of the processes leading to changes in the ultrastructural order within the cytoskeleton can be disturbed experimentally, for example, by application of cytochalasin which acts as an inhibitor of functions performed by actin. Light microscopic techniques involving immuno methods are especially suited to demonstrate the complexity of the cytoskeleton. Certain components of the cytoskeleton, especially actin in cooperation with myosin, tropomyosin and troponin, and microtubuli together with dynein have additional functions; they are involved in intracellular motility, i.e., streaming of cytoplasm, a phenomenon also observed in plant cells.

In close contact with the cytoskeleton are the components of the "membrane skeleton", the spectrin and the ankyrin. The latter mediates the contact of the spectrin molecules with the inside of the plasma membrane.

4.4.2 Cell Envelope

The outermost layer of typical animal cells (EDELMAN, 1985) exposed to the environment, e.g., of epithelial gut cells with microvilli, is formed by the glycocalyx (Fig. 38). It can be shown by electron microscopy as a network of filaments covering the outside of the plasma membrane. The filaments are in fact bundles of exposed components of membrane-integrated glycoproteins and glycolipids. They are oligosaccharide chains usually containing N-acetylneuraminic acid residues (sialinic acid). They cause a negative charge of the cell surface. Extracellular glycoproteins and polysaccharides have been found to be associated with these oligomers. The exposed components of the envelope are defined as plasma membrane-specific antigens. In experimental approaches involving recognition of specific cells these

Fig. 38. Glycocalyx on microvilli (containing actin filaments) of a gut epithelium cell.
From KLEINIG and SITTE (1986), reproduced by permission of Gustav Fischer Verlag

antigens are of major importance for cell discrimination.

This complex envelope protects the cell and mediates the interactions of the cell with the environment, e.g., by perceiving chemical signals which come from the outside and which are identified by specific receptor molecules located at the cell periphery.

4.4.3 Endomembrane Systems Involved in Synthesis and Export

In electron micrographs of ultrathin sectioned animal (and plant) cells (KLEINIG and SITTE, 1986) complex endomembrane systems can be seen. They can formally be divided into the endoplasmic reticulum (ER), the Golgi apparatus, and vesicles of various kinds such as peroxisomes, glyoxisomes, lysosomes, vacuoles, coated vesicles, and secretion vesicles. The outer membrane of the nucleus also belongs to these systems. The membranes of all the systems can be interconverted (see above, Sect. 4, membrane flow, membrane transformation). Often, these systems are much more evident in animal cells than in plant cells. Therefore, they are described in more detail here and not in Sect. 4.3.

The endoplasmic reticulum consists of interconnected cisternae and tubular and reticular structures. A rough and a smooth ER are distinguished by the attachment or the missing of ribosomes at the plasmatic side of the membranes. Protein synthesis only occurs at the rough ER (WALTER et al., 1984). Both types of ER can be present in a cell at the same time. The ratio of both types depends on the physiological state of the cell.

As a whole, the ER performs a multiplicity of reactions connected to synthesis. It is the site of synthesis of membrane lipids (phospholipids, glycolipids, sterols), storage lipids and other lipophilic compounds such as steroids. Many of the enzymes involved in these reactions are integral or peripheric membrane proteins. They may be divided, according to the properties of their substrates, into enzymes which deal with hydrophilic compounds which orginate from the cytoplasm and are converted to lipophilic membrane-bound compounds, and enzymes which process lipophilic substances. Two functionally important components of the ER are the cytochrome b_5 system, an oxidase, and the cytochrome P450 system which also catalyzes a number of oxidative reactions. It was shown to metabolize lipophilic substances such as pesticides, herbicides, insecticides and fungicides and, thus, acts as a detoxification system.

Furthermore, the ER is the site of storage of Ca^{2+} ions which are involved, via their action on calmodulin, in various kinds of regulation of very many enzyme activities.

The Golgi apparatus (DUMPHY and ROTHMAN, 1985) is constituted by dictyosomes (see Fig. 29), i.e., stacks of flat membrane vesicles. Such a stack exhibits a polarity, with one face (the *cis* face) pointing to the rough ER or the nucleus, and the other face (the *trans* face) to the cell periphery. This polarity is brought about by a preferred occurrence of small vesicles (primary vesicles) at the *cis* face, and large vesicles (Golgi vesicles, secretory vesicles) at the *trans* face, and by differences in the thickness of the membranes. The functional implication is that the dictyosome is an organelle in or by which compounds are synthesized or processed which are finally exported from the cell by exocytosis.

4.4.4 Cell–Cell Connection in Animal Tissues

Functions in cells and tissues of an organism have to be regulated in order to warrant their concerted action. Therefore, the cells in a tissue have to be held together or attached to the basal lamina, and ways for communication between cells have to exist (KLEINIG and SITTE, 1986; STAEHELIN and HULL, 1978). In animal tissues, various kinds of junctions between cells have been found, each with specific functions.

Desmosomes (Fig. 39) fullfil a mechanical function: they keep the cells together and are very prominent in epithelia and other tissues exposed to mechanical stress. From the desmosomes, directed towards the interior of the cell, protein filaments can be seen which belong to the cytoskeleton and could be identified as cytokeratins.

Tight junctions (Fig. 39) keep the internal intercellular milieu of an organ separated from the exterior and prevent diffusion. The analysis of their ultrastructure revealed that within a tight junction the membranes of the cells are in direct contact, and a direct interaction of integral membrane proteins in these regions is assumed. Proteins forming a tight junction cannot float, by lateral diffusion, within the membrane. In plants, specialized structures within the walls of the root endoderms (Casparian strip) fullfil a function similar to that of the tight junctions in animals (GUNNING and STEER, 1975).

Gap junctions are direct plasmatic connections between animal cells. They have a function similar to one of the roles of plant plasmodesmata, i.e., they are the routes for cell–cell communication. Ions and molecules up to a molecular mass of around 1 kDa can move from cell to cell. In contrast to plant cell plasmodesmata, which in fact are "holes", the gap junctions are formed by regularly ordered particles forming pores. Ca^{2+} ions regulate the degree of "porosity" by interaction with the pore-forming protein.

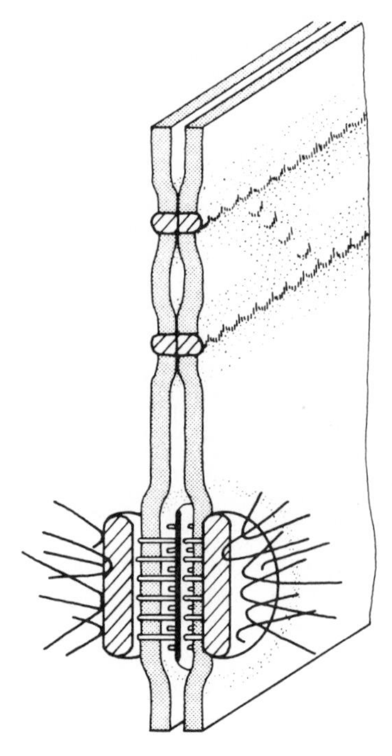

Fig. 39. Diagrammatic view of a tight junction (upper part) and a desmosome (lower part).
From KLEINIG and SITTE (1986), reproduced by permission of Gustav Fischer Verlag

4.5 Eukaryotic Cells in Culture

Plant and animal cells can be grown in culture (FRESHNEY, 1987; LYDERSON, 1987; WEBB and MAVITUNA, 1987). Usually, plant cells are cultured in suspension whereas animal cells are grown, as monolayers or cultures consisting of several layers, on a support. However, under certain conditions animal cells can also be cultured in suspension (see below) (Fig. 40). In all cases, the physiological conditions of the natural environment of the cells are no longer maintained. This has the consequence that the properties of these cells are most probably very different from those of cells of

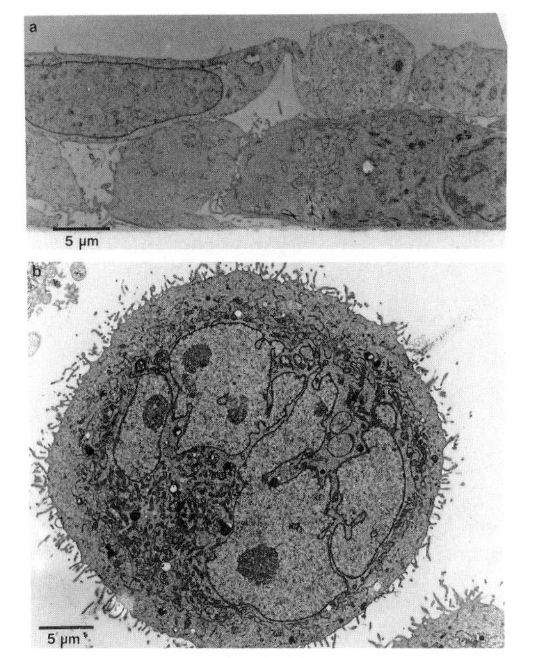

Fig. 40. Human HS-24 tumor cells.
(a) Several layers of adherent cells growing on glass.
(b) As above; however, cells grown in suspension culture, exhibiting a spherical shape and many microvilli.
Original micrographs M. ERDEL and E. SPIESS

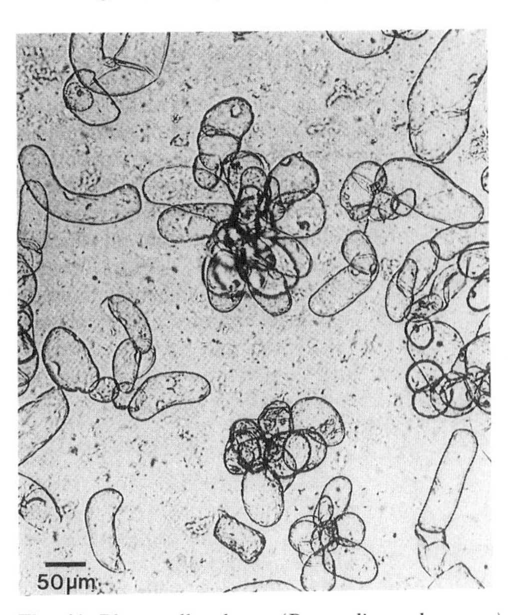

Fig. 41. Plant cell culture *(Petroselinum hortense)*. The cells exhibit heterogeneity in size and shape, and they may form aggregates.
From KLEINIG and SITTE (1986), reproduced by permission of Gustav Fischer Verlag

the same type in the tissue. Therefore, data obtained from cultured cells may not be as reliable as those measured in cells growing in their natural environment.

Recipes and procedures have been developed for the culture of plant and various animal (and human) cells. A first step is to produce single cells; this is usually done by digestion of the walls with enzymes or by gentle application of shearing force.

Photosynthetically active plant cells often only need CO_2 as carbon source and nitrogen-containing salts. Most plant cells kept in culture, however, cannot perform photosynthesis. They are heterotrophic, and they need organic carbon sources, e.g., sucrose. In contrast to cultured animal cells, plant cells of a culture are often heterogeneous in size and shape (Fig. 41). Besides single cells, these cultures usually contain cell aggregates, depending on the type of the original cell. The primary cells often originate from explantats which form a callus, or protoplasts are used at the beginning of a culture. Plant protoplasts regenerate their walls during culture (see Sect. 4.3.3).

Experiments aimed at the culture of vertebrate cells revealed that these cells need organic and inorganic compounds and serum containing growth factors. Most vertebrate cells need surfaces to which they can adhere. Extracellular proteins (adhesion proteins), e.g., fibronectin, mediate the contact to these surfaces. However, procedures have also been designed for continuous culture of animal cells in suspensions; this can be achieved in a carefully controlled airlift reactor equipped with a sedimenter and a cooling device. Cooling reduces metabolism of the cells in the sedimenter; thus insufficiency of oxygen supply can be avoided (HÜLSCHER and ONKEN, 1992).

Differentiated animal cells are difficult to cultivate. Easier to handle are fibroblasts and transformed (tumor) (Fig. 42) or other non-differentiated cells. After adhesion to the substrate, the cells start to flatten, and they can even migrate. Cells in a dense monolayer stop to migrate and to grow ("contact inhibition").

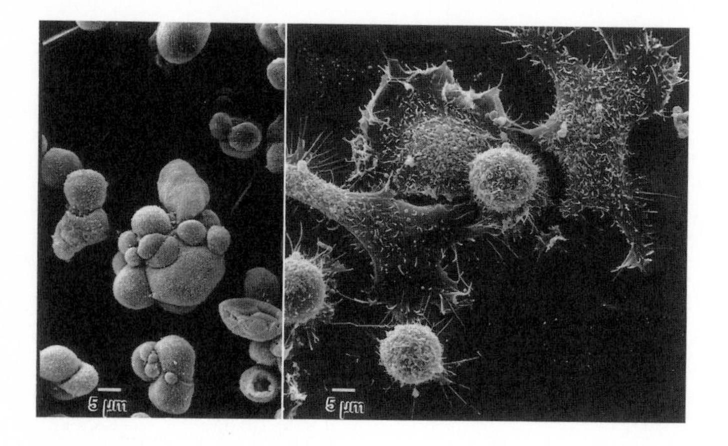

Fig. 42. Scanning electron micrographs of human SB-3 tumor cells.
Left: Cells cultured in suspension; right: adherent cells.
Original micrographs
M. ERDEL and E. SPIESS

Trypsin treatment can be used to remove them from the support, and after dilution and transfer to a new support they regain their original properties.

Cultured cells usually stop growth and division after a certain number of generations due to "ageing", a phenomenon probably caused by a naturally limited life span of cells. Transformed cells are potentially permanent (eternal) cell lines which do not exhibit contact inhibition and are not dependent on extracellular adhesion proteins.

Acknowledgement

The author thanks I. BRAKS, M. ERDEL, J. KRICKE, D. MILLER, A. NOLTE, CH. SPIESS, E. SPIESS, H. SPRING, B. VOGT and R. WIRTH for contributing original micrographs. Supported by the Fonds der Chemischen Industrie.

5 References

ACKER, G. (1988), Immunoelectron microscopy of the surface antigens (polysaccharides) of Gram-negative bacteria using pre- and post-embedding techniques, *Methods Microbiol.* **20,** 147–174.

ALBERTS, B., BRAY, D., LEWIS, J., RAFF, M., ROBERTS, K., WATSON, J. D. (1983), *Molecular Biology of the Cell*, New York: Garland.

ALDRICH, H. C., BEIMBORN, B. D., BOKRANZ, M., SCHÖNHEIT, P. (1987), Immunocytochemical localization of coenzyme M-reductase in *Methanobacterium thermoautotrophicum, Arch. Microbiol.* **174,** 190–194.

ANTÓN, J. J., MESEGUER, J., RODRIGUEZ-VALERA, F. (1988), Production of extracellular polysaccharide by *Haloferax mediterranei, Appl. Environ Microbiol.* **54,** 2381–2386.

ANTRANIKIAN, G., ZABLONSKI, P., GOTTSCHALK, G. (1987a), Conditions for the overproduction and excretion of thermostable α-amylase and pullulanase from *Clostridium thermohydrosulfuricum* DSM 567, *Appl. Microbiol. Biotechnol.* **27,** 75–81.

ANTRANIKIAN, G., HERZBERG, C., MAYER, F., GOTTSCHALK, G. (1987b), Changes in the cell envelope structure of *Clostridium* sp. strain EM1 during massive production of α-amylase and pullulanase, *FEMS Microbiol. Lett.* **41,** 193–197.

ANTRANIKIAN, G., KOCH, R., SPREINAT, A., LEMKE, K. (1990) Hyperthermoactive amylases from the archaebacteria *Pyrococcus woesii* and *Pyrococcus furiosus, Proc. 5th Eur. Congr. Biotechnol.,* pp. 1–4, Copenhagen, Denmark.

AUNSTRUP, K. (1979), Production, isolation, and economics of extracellular enzymes, *Appl. Biochem. Bioeng.* **2,** 27–69.

BAUER, R. (1988), Electron spectroscopic imaging: an advanced technique for imaging and analysis in transmission electron microscopy, *Methods Microbiol.* **20,** 113–146.

BAYER, M. E., EASTERBROOK, K. (1991), Tubular spinae are long-distance connectors between bacteria, *J. Gen. Microbiol.* **137,** 1081–1086.

BAYER, M. E., THUROW, H. (1977), Polysaccharide capsules of *Escherichia coli.* Microscope study of size, structure and sites of synthesis, *J. Bacteriol.* **130,** 911–936.

BLAUROCK, A. E., STOECKENIUS, W (1971), Structure of the purple membrane, *Nature New Biol.* **233,** 152–155.

BRINTON, C. C., (1959), Non-flagellar appendages of bacteria, *Nature* **183,** 782–786.

BRINTON, C. C. (1971), The properties of sex pili, the viral nature of "conjugal" genetic transfer systems, and some possible approaches to the control of bacterial drug resistance, *Crit. Rev. Microbiol.* **1,** 105–160.

BURCHARDT, G., WIENECKE, A., BAHL, H. (1991), Isolation of the pullulanase gene from *Clostridium thermosulfurogenes* (DSM 3896) and its expression in *Escherichia coli*, *Curr. Microbiol.* **22,** 91–95.

COSTERTON, J. W. (1984), Preservation of the dimensions and spatial arrangements of extracellular hydrated polysaccharide structures in scanning and transmission electron microscopy, in: *Abstr. 8th Eur. Congr. Electron Microscopy*, Vol. III, pp. 2375–2376, Budapest.

COSTERTON, J. W., INGRAM, J. M., CHENG, K.-J. (1974), Structure and function of the cell envelope of Gram-negative bacteria, *Bacteriol. Rev.* **38,** 87–110.

COSTERTON, J. W., GEESEY, G. G., CHENG, K.-J. (1978), How bacteria stick, *Sci. Am.* **238,** 86–95.

COSTERTON, J. W., IRVIN, R. T., CHENG, K.-J. (1981), The bacterial glycocalix in nature and disease, *Annu. Rev. Microbiol.* **35,** 000–000.

COUGHLAN, M. P., MAYER, F. (1992), The cellulose-decomposing bacteria and their enzyme systems, in: *The Prokaryotes* (BALOW, A., TRÜPER, H. G., DWORKIN, M., HARDER, W., SCHLEIFER, K.-H., Eds.), Vol. I, pp. 460–516, Berlin: Springer.

DEISENHOFER, J., EPP, O., MIKI, K., HUBER, R., MICHEL, H. (1985), Structure of the protein subunits in the photosynthetic reaction centre of *Rhodopseudomonas viridis* at 3 Å resolution, *Nature* **318,** 618–624.

DODDEMA, J. J., VOGELS, G. D. (1978), Improved identification of methanogenic bacteria by fluorescence microscopy, *Appl. Environ. Microbiol.* **36,** 752–754.

DUGUID, H. P., ANDERSON, E. S., CAMPBELL, J. (1966), Fimbriae and adhesive properties in *Salmonella*, *J. Pathol. Bacteriol.* **92,** 107–138.

DUMPHY, W. G., ROTHMAN, J. E. (1985), Compartmental organization of the Golgi stack, *Cell* **42,** 13–21.

EASTERBROOK, K. B., ALEXANDER, S. A. (1983), The initiation and growth of bacterial spinae, *Can. J. Microbiol.* **29,** 476–487.

EDELMAN, G. M. (1985), Cell adhesion and the molecular process of morphogenesis, *Annu. Rev. Biochem.* **54,** 135–169.

FAWCETT, D. W. (1981), *The Cell,* Philadelphia: Saunders.

FRANÇON, M. (1967), *Einführung in die neueren Methoden der Lichtmikroskopie*, Karlsruhe: Braun.

FREHEL, C., RYTER, A. (1979), Peptidoglycan turnover during growth of a *Bacillus megaterium* Dap Lys Mutant, *J. Bacteriol.* **137,** 947–955.

FREITAG, R., SPREINAT, A., ANTRANIKIAN, G., SCHEPER, T. (1990), Immunoassay in process analysis, *Proc. 5th Eur. Congr. Biotechnol.,* p. 135, Copenhagen, Denmark.

FRESHNEY, R. I. (1987), *Culture of Animal Cells*, 2nd Ed., New York: Liss.

FREY-WYSSLING, A. (1976), *The Plant Cell Wall, Handbuch der Pflanzenanatomie* III/4, 2nd Ed. Berlin: Borntraeger.

GALLI, D., WIRTH, R., WANNER, G. (1989), Identification of aggregation substances of *Enterococcus faecalis* cells after induction by sex pheromones, *Arch. Microbiol.* **151,** 486–490.

GOLECKI, J. R., SCHUMACHER, A., DREWS, G. (1980), The differentiation of the photosynthetic apparatus and the intracytoplasmic membrane in cells of *Rhodopseudomonas capsulata* upon variation of light intensity, *Eur. J. Cell Biol.* **23,** 1–5.

GUNNING, B. E. S., STEER, M. W. (1975), *Ultrastructure and the Biology of the Plant Cells*, London: Arnold.

GUNNING, B. E. S., STEER, M. W. (1986), *Bildatlas zur Biologie der Pflanzenzelle*, Stuttgart: Gustav Fischer.

HAYAT, M. A. (1978), *Introduction to Biological Scanning Electron Microscopy*, Baltimore: University Park Press.

HAYAT, M. A. (1981), *Principles and Techniques of Electron Microscopy — Biological Applications*, London: Arnold.

HOBOT, J. A., CARLEMALM, E., VILLIGER, W., KELLENBERGER, E. (1984), Periplasmic gel: new concept resulting from reinvestigation of bacterial cell envelope ultracstructure by new methods, *J. Bacteriol.* **160,** 143–152.

HOLZENBURG, A. (1988), Preparation of two-dimensional arrays of soluble proteins as demonstrated for bacterial D-ribulose-1,5-bisphosphate carboxylase/oxygenase, *Methods Microbiol.* **20,** 341–356.

HOYLE, B. D., BEVERIDGE, T. J. (1984), Metal binding by the peptidoglycan sacculus of *Escherichia coli* K-12, *Can. J. Microbiol.* **30,** 204–211.

HÜLSCHER, M., ONKEN, U. (1992), Verfahrenstechnische Probleme beim Einsatz von Airlift-Schlaufenreaktoren für die Kultivierung von Tierzellen, *Bio Engineering* **8,** 30–37.

HURDECK, H. G., KOCH, R., ANTRANIKIAN, G., SCHEPER, T. (1990), Use of an enzyme thermistor for process control inl biotechnology, *Proc. 5th Eur. Congr. Biotechnol.*, p. 313, Copenhagen, Denmark.

HUSTEDE, E., STEINBÜCHEL, A., SCHLEGEL, H. G., Cloning of poly(3-hydroxybutyric acid) synthase genes of *Rhodobacter sphaeroides* and *Rhodospi-*

rillum rubrum and heterologous expression in *Alcaligenes eutrophus, FEMS Microbiol. Lett.*, submitted.

INGRAHAM, J. L., MAALOE, O., NEIDHARDT, F. C. (1983), *Growth of the Bacterial Cell*, Sunderland, MA: Sinauer.

KAHANE, I., BANAI, M., RAZIN, S., FELDNER, J. (1982), Attachment of mycoplasmas to host cell membranes, *Rev. Infect. Dis.* **4** (Suppl.), 185–192.

KANDLER, O. (1979), Zellwandstrukturen bei Methan-Bakterien, *Naturwissenschaften* **66**, 95–105.

KLEINIG, H., SITTE, P. (1986), *Zellbiologie.* Stuttgart: Gustav Fischer.

KLINGENBERG, M., VORLOP, K. D., ANTRANIKIAN, G. (1990), Immobilization of anaerobic thermophilic bacteria for the production of cell-free thermostable α-amylases and pullulanases, *Appl. Microbiol. Biotechnol.* **33**, 495–500.

KÖNIG, H. (1988), Archaebacterial cell envelopes, *Can. J. Microbiol.* **34**, 395–406.

KONDRATIEVA, E. N., PFENNIG, N., TRÜPER, H. G. (1992). The phototrophic prokaryotes, in: *The Prokaryotes* (BALOW, A., TRÜPER, H. G., DWORKIN, M., HARDER, W., SCHLEIFER, K. H., Eds.), Vol. I,. pp. 312–330, Berlin: Springer.

KRÄMER, R., SCHÖNHEIT, P. (1987), Testing the "methanochondrion" concept: are nucleotides transported across internal membranes in *Methanobacterium thermoautotrophicum? Arch. Microbiol.* **146**, 370–376.

LAMED, R., SETTER, E., BAYER, E. A. (1983a), Characterization of a cellulose-binding, cellulase-containing complex in *Clostridium thermocellum, J. Bacteriol.* **156**, 828–836.

LAMED, R., SETTER, E., KENIG, R., BAYER, E. A. (1983b), The cellulosome — A discrete cell surface organelle of *Clostridium thermocellum* which exhibits separate antigenic cellulose-binding and various cellulolytic activities, *Biotechnol. Bioeng. Symp.* **13**, 163–181.

LIPKE, P. N., KURJAN, J. (1992), Sexual agglutination in budding yeasts: Structure, function, and regulation of adhesion glycoproteins, *Microbiol. Rev.* **56**, 180–194.

LYDERSON, B. K., (Ed.) (1987), *Large Scale Cell Culture Technology,* München: Hanser.

MADOFF, S. (1981), The L-forms of bacteria, in: *The Prokaryotes* (STARR, M. P., STOLP, H., TRÜPER, H. G., BALOWS, A., SCHLEGEL, H. G., Eds.), Vol. II, pp. 2225–2237, Berlin: Springer.

MAYER, F. (1986), *Cytology and Morphogenesis of Bacteria*, Berlin: Borntraeger.

MAYER, F., ROHDE, M. (1988), Analysis of dimensions and structural organization of proteoliposomes, *Methods Microbiol.* **20**, 283–292.

MAYER, F., COUGHLAN, M. P., MORI, Y., LJUNGDAHL, L. G. (1987), Macromolecular organization of the cellulolytic enzyme complex of *Clostridium thermocellum* as revealed by electron microscopy, *Appl. Environ. Microbiol.* **53**, 2785–2792.

MCNEIL, M., DARVILL, A. G., FRY, S. C., ALBERSHEIM, P. (1984), Structure and function of the primary cell walls of plants, *Annu. Rev. Biochem.* **53**, 625–663.

MIZUSHIMA, S. (1985), Structure and assembly of the outer membrane, in: *Molecular Cytology of Escherichia coli* (NANNINGA, N., Ed.), pp. 39–75, New York: Academic Press.

MOLL, G., AHRENS, R. (1970), Ein neuer Fimbrientyp, *Arch. Microbiol.* **70**, 361–368.

MOOR, H. (1964), Die Gefrierfixation lebender Zellen und ihre Anwendung in der Elektronenmikroskopie, *Z. Zellforsch.* **62**, 546–580.

MORRÉ, J. D., KARTENBECK, J., FRANKE, W. W. (1979), Membrane flow and interconversion among endomembranes, *Biochim. Biophys. Acta* **559**, 71–152.

MÜLLER, M. (1988), Cryopreparation of microorganisms for electron microscopy, *Methods Microbiol.* **20**, 1–28.

NANNINGA, N., BRAKENHOFF, G. J., MEIJER, M., WOLDRINGH, C. L. (1984), Bacterial anatomy in retrospect and prospect, *Antonie van Leeuwenhoek J. Microbiol. Serol.* **50**, 433–460.

NEUBERGER, A., BROCKLEHURST, K. (Eds.) (1987), *Hydrolytic Enzymes,* Amsterdam: Elsevier.

NEWMAN, H. N. (1976), Dental plaque, in: *Microbial Ultrastructure* (FULLER, R., LOVELOCK, D. W., Eds.), pp. 224–263, London: Academic Press.

NIENHAUS, F., SIKORA, R. A. (1979), Mycoplasmas, sprioplasmas, and *Rickettsia*-like organisms as plant pathogens, *Annu. Rev. Phytopathol.* **17**, 37–58.

OESTERHELT, D., STOECKENIUS, W. (1971), Functions of a new photoreceptor membrane, *Proc. Natl. Acad. Sci. USA* **70**, 2853–2857.

POST, E., VAKALOPOULOU, E., OELZE, J. (1983), On the relationship of intracytoplasmic to cytoplasmic membranes in nitrogen-fixing *Azotobacter vinelandii, Arch. Microbiol.* **134**, 265–269.

PRIEST, F. G. (1984), *Extracellular Enzymes. Aspects of Microbiology*, Vol. 9, London: Van Nostrand Reinhold.

PUM, D., SARA, M., SLEYTR, U. B. (1989), Use of two-dimensional protein crystals form bacteria for nonbiological applications, *J. Vac. Sci. Technol.* **B7** (6), 1391–1397.

REIMER, L. (1967), *Elektronenmikroskopische Untersuchungs- und Präparationsmethoden*, Berlin: Springer.

ROBARDS, A. W., SLEYTR, U. B. (1985) *Low Temperature Biological Electron Microscopy*, Amsterdam: Elsevier.

ROBINSON, D. G. (1985), *Plant Membranes — Endo-*

and Plasma Membranes of Plant Cells, New York: Wiley and Sons.

ROBINSON, D. G., KRISTEN, K. (1982), Membrane flow via the Golgi apparatus of higher plant cells, *Int. Rev. Cytol.* **77,** 89–127.

ROBINSON, D. G., EHLERS, U., HERKEN, R., HERMANN, B., MAYER, F., SCHÜRMANN, F. W. (1985), *Präparationsmethodik in der Elektronenmikroskopie — Eine Einführung für Biologen und Mediziner,* Berlin: Springer.

ROBINSON, D. G., EHLERS, U., HERKEN, R., HERRMANN, B., MAYER F., SCHÜRMANN, F.-W. (1987), *Methods of Preparation for Electron Microscopy — an Introduction for the Biomedical Sciences,* New York: Springer.

ROHDE, M., GERBERDING, H., MUND, T., KOHRING, G. W. (1988), Immunoelectron microscopic localization of bacterial enzymes: pre- und postembedding labelling techniques on resin-embedded samples, *Methods Microbiol.* **20,** 175–210.

ROMESSER, J. A., WOLFE, R. S., MAYER, F., SPIESS, E., WALTHER-MAURUSCHAT, A. (1979), *Methanogenium,* a new genus of marine methanogenic bacteria, and characterization of *Methanogenium cariaci* sp. nov. and *Methanogenium marisnigri* sp. nov., *Arch. Microbiol.* **121,** 147–154.

ROSENBAUER, K. A., KEGEL, B. H. (1978), *Rasterelektronenmikroskopische Technik: Präparationsverfahren in Medizin und Biologie,* Stuttgart: Thieme.

SARA, M., WOLF, G., KÜPCÜ, S., PUM, D., SLEYTR, U. B. (1988), Use of crystalline bacterial cell envelope layers as ultrafiltration membranes and supports for immobilization of macromolecules, in: *DECHEMA Biotechnology Conferences,* Vol. 2, pp. 35–51, Weinheim: VCH.

SCHLEGEL, H. G., GOTTSCHALK, G. (1962), Poly-β-hydroxybuttersäure, ihre Verbreitung, Funktion und Biosynthese, *Angew. Chem.* **74,** 342–347.

SCHLEGEL, H. G., GOTTSCHALK, G., VON BARTHA, R. (1961), Formation and utilization of poly-β-hydroxybutyric acid by Knallgas-bacteria *(Hydrogenomonas), Nature* **191,** 463–465.

SCHLEIFER, K. H., KANDLER, O. (1972), Peptidoglycan types of bacterial cell walls and their taxonomic implications, *Bacteriol. Rev.* **36,** 407–477.

SCHONER, R. G., ELLIS, L. F., SCHONER, B. E. (1985), Isolation and purification of protein granules from *Escherichia coli* cells overproducing bovine growth hormone, *Bio/Technology* **3,** 151–154.

SHIVELY, J. M. (1974), Inclusion bodies of prokaryotes, *Annu. Rev. Microbiol.* **28,** 167–187.

SINGER, S. J., NICHOLSON, G. L. (1972), The fluid mosaic model of the structure of cell membranes, *Science* **175,** 720–731.

SLEYTR, U. B., MESSNER, P. (1983), Crystalline surface layers on bacteria, *Annu. Rev. Microbiol.* **37,** 311–339.

SLEYTR, U. B., MESSNER, P., PUM, D. (1988), Analysis of crystalline bacterial surface layers by freeze-etching, metal shadowing, negative staining and ultrathin sectioning, *Methods Microbiol.* **20,** 29–60.

SLOT, J. W., GEUZE, H. J., WEERKAMP, A. J. (1988), Localization of macromolecular components by application of the immunogold technique on cryosectioned bacteria. *Methods Microbiol.* **20,** 211–236.

SPIESS, E., LURZ, R. (1988), Electron microscopic analysis of nucleic acids and nucleic acid-protein complexes, *Methods Microbiol.* **20,** 293–323.

STAEHELIN, L. A., ARNTZEN, C. J. (Eds.) (1986), *Photosynthesis III. Encyclopedia of Plant Physiology,* New Series, Vol. 19, Berlin: Springer.

STAEHELIN, L. A., HULL, B. E. (1978), Junctions between living cells, *Sci. Am.,* May 1978, 141–152.

STAHLY, D. P., ANDREWS, R. E., YOUSTEN, A. A. (1992), The genus *Bacilus* — insect pathogens, in: *The Prokaryotes* (BALOWS, A., TRÜPER, H. G., DWORKIN, M., HARDER, W., SCHLEIFER, K.-H., Eds.), Vol. II, pp. 1697–1745, Berlin: Springer.

STEINBÜCHEL, A., SCHLEGEL, H. G. (1991), Physiology and molecular genetics of poly(β-hydroxyalkanoic acid) synthesis in *Alcaligenes eutrophus, Mol. Microbiol.* **5,** 535–542.

TAUSCHEL, H.-D. (1988), Localization of bacterial enzymes by electron microscopy cytochemistry as demonstrated for the polar organelle, *Methods Microbiol.* **20,** 237–259.

TOLBERT, N. E. (1981), Metabolic pathways in peroxisomes and glyoxisomes, *Annu. Rev. Biochem.* **50,** 133–157.

TRÜPER, H. G., PFENNIG, N. (1981), Characterization and identification of anoxygenic phototrophic bacteria, in: *The Prokaryotes* (STARR, M. P., STOLP, H., TRÜPER, H. G., BALOWS, A., SCHLEGEL, H. G., Eds.), Vol. I, pp. 299–312, Berlin: Springer.

VAN WIELINK, J. E., DUINE, J. A. (1990), How big is the periplasmic space? *TIBS* **15,** 136–137.

VARGA, A. R., STAEHELIN, L. A. (1983), Spatial differentiation in photosynthetic and non-photosynthetic membranes of *Rhodopseudomonas palustris, J. Bacteriol.* **154,** 1414–1430.

VEENHUIS, M., HARDER, W. (1987), Metabolic significance and biogenesis of microbodies in yeasts, in: *Peroxisomes in Biology and Medicine* (FAHIMI, H. D., SIES, H., Eds.), pp. 436–458, Berlin: Springer.

WALTER, P., GILMORE, R., BLOBEL, G. (1984), Protein translocation across the endoplasmic reticulum, *Cell* **38,** 5–8.

WATSON, S. W., VALOIS, F. W., WATERBURY, J. B. (1981), The family nitrobacteraceae, in: *The Pro-*

karyotes (STARR, M. P., STOLP, H., TRÜPER, H. G., BALOWS, A., SCHLEGEL, H. G., Eds.), Vol. I, pp. 1005–1021, Berlin: Springer.

WEBB, C., MAVITUNA, F. (Eds.) (1987), *Plant and Animal Cells: Process Possibilities*, Chichester: Ellis Horwood.

WEISS, E., SCHIEFER, H.-G., KRAUSS, H. (1979), Ultrastructural visualization of *Klebsiella* capsules by polycationic ferritin, *FEMS Microbiol. Lett.* **6,** 435–4y37.

WIRTH, R., WANNER, G., GALLI, D. (1990), Das Sex-Pheromon-System von *Enterococcus faecalis:* ein einzigartiger Mechanismus zum Aufsammeln von Plasmiden, *Forum Microbiol.* **13,** 321–332.

WOESE, C. R. (1981), Archaebacteria, *Sci. Am.* **44,** 94–106.

ZEIKUS, J. G., WOLFE, R. S. (1973), Fine structure of *Methanobacterium thermoautotrophicum:* Effect of growth temperature on morphology and ultrastructure, *J. Bacteriol.* **113,** 461–467.

ZIEROLD, K. (1988), Electron probe microanalysis of cryosections from cell suspensions, *Methods Microbiol.* **20,** 91–111.

2 Metabolism

REINHARD KRÄMER
GEORG SPRENGER
Jülich, Federal Republic of Germany

1 Introduction 50
2 Principles of Energy Metabolism 50
  2.1 Catabolism and Anabolism, Sources of Energy and Carbon 50
  2.2 Thermodynamic Principles of Energy Metabolism 52
    2.2.1 Enthalpy and the First Law of Thermodynamics 52
    2.2.2 Entropy and the Second Law of Thermodynamics 52
    2.2.3 Free Energy 53
    2.2.4 Chemical Equilibria in Solution 53
    2.2.5 Group Transfer Potentials and the Adenylate System 54
    2.2.6 Free Energy and Electrical Work, the Electrochemical Potential 55
  2.3 Supply of Useful Energy for the Cell 56
    2.3.1 Substrate Level Phosphorylation 56
    2.3.2 Electron Transport Phosphorylation 57
  2.4 Energy Coupling 62
3 Enzyme Catalysis 65
  3.1 Classification of Enzymes 66
  3.2 Energetics of Enzyme Catalysis 66
  3.3 Kinetics of Enzyme-Catalyzed Reactions 67
  3.4 Coenzymes, Prosthetic Groups, Cofactors 70
4 Membrane Transport 72
  4.1 The Carrier Concept 73
  4.2 Mechanisms and Energetics of Solute Transport 73
  4.3 Transport Kinetics 77
  4.4 Transport of Macromolecules 79
5 Metabolic Pathways 79
  5.1 Utilization of Carbon Sources 80
    5.1.1 Metabolism of Carbohydrates 81
    5.1.2 Metabolism of Fatty Acids, Aliphatic and Aromatic Hydrocarbons 88
  5.2 Nitrogen Metabolism 90

5.2.1 The Nitrogen Cycle 90
5.2.2 Amino Acid Degradation 91
5.2.3 Biosynthesis of Amino Acids 92
5.3 Sulfur Metabolism 93
5.4 Synthesis of Biopolymers 95
5.4.1 Polymerization of Glycogen in Bacteria 95
5.4.2 Polymerization of DNA (DNA Replication) 95
5.4.3 RNA Polymerization 97
5.4.4 Protein Synthesis 99
6 Regulation and Coordination of Metabolism 100
6.1 Regulation of Enzyme Activity 101
6.1.1 Covalent Modifications of Enyzmes 101
6.1.2 Allosteric Control 101
6.2 Regulation of Enzyme Synthesis 105
6.2.1 Sigma Factors as Regulatory Components 105
6.2.2 DNA Binding Proteins: Activators and Repressors 106
6.2.3 Two-Component Systems of Regulation 106
6.2.4 Regulatory Mechanisms at the RNA Level: The Attenuator Model 106
6.2.5 Eukaryotic Gene Regulation 107
7 Further Reading 109

List of Symbols

E	internal energy
E_O	standard redox potential (V)
H	enthalpy (kJ/mol)
J	flux
K_{eq}	equilibrium constant
K_m	Michaelis constant
S	entropy
V_{max}	maximum velocity
ΔG^O	standard free energy (kJ/mol)
Δp	proton-motive force (V)
Δ pH	pH gradient
$\Delta \tilde{\mu}_{ion}$	electrochemical ion gradient
$\Delta \Psi$	membrane potential (V)

1 Introduction

The detailed understanding of cell metabolism is an essential prerequisite for every biotechnologist. Only with the knowledge of metabolic processes, the extraordinary diversity and versatility of microorganisms in maintaining metabolism and growth even under unfavorable conditions can successfully be used for biotechnological purposes.

In view of the enormous complexity of cellular metabolism, it is somewhat ambiguous to define which metabolic process is "basic" and should, therefore, be discussed here, since this classification depends to a considerable extent on the particular subject of interest. Thus, instead of applying a descriptive approach, we believe that "the nature of metabolic order is better perceived from a teleonomic viewpoint: metabolism is the ensemble of reactions by which cells utilize resources to obtain useful energy and chemical building blocks and to provide the goods and services required for their continued existence, growth and reproduction" (HAROLD, 1986).

The main emphasis is placed on processes which are, on the one hand, essential for all living cells, and, on the other hand, not extensively dealt with in textbooks commonly read by biotechnologists. These are in our view, (1) energy transduction resp. energy coupling mechanisms and, closely connected with this topic, (2) the mechanisms of substrate uptake and product secretion. Furthermore, (3) the various patterns and mechanisms will be discussed, by which the cell regulates carbon flow, thus successfully organizing the network of intermediary metabolism.

All reactions catalyzed by the microbial metabolism, including those which the biotechnologist would like to have carried out additionally, must obey the basic principles of thermodynamics and kinetics. Thus, in order to deal with the above mentioned topics on a solid basis, the first part of this chapter will briefly consider the application of thermodynamics to biochemistry, followed by a short introduction in kinetics.

2 Principles of Energy Metabolism

In this part, we will provide a short review on the basis thermodynamic equations and explain their application to biochemistry. The concept of energy metabolism by coupling exergonic and endergonic, or catabolic and anabolic reactions within the metabolic network shall be outlined.

2.1 Catabolism and Anabolism, Sources of Energy and Carbon

From the standpoint of energy supplementation, organisms fall into two classes, phototrophs which obtain their energy directly from the sun, and chemotrophs which depend on redox reactions for their energy supply. Directly or indirectly, chemotrophs (except chemolithotrophs) use the energy-rich compounds which are synthesized by the phototrophs (Fig. 1). When considering the source of carbon for the essential building blocks, heterotrophic organisms which use organic material as carbon source can be distinguished from autotrophs which obtain the majority of cell carbon from CO_2 fixation. Finally, organisms can be classified by the type of hydrogen donors used: organotrophs use organic compounds and lithotrophs inorganic hydrogen donors such as H_2, H_2S, etc.

Fig. 2 explains the basic concept of metabolism in a heterotrophic cell, applicable to a large number of cells, from *Escherichia coli* to human. In catabolic reactions, organic compounds are degraded, on the one hand, to CO_2 and water for providing reducing power (NADPH, NADH, and $FADH_2$) and energy supply (protonmotive force, ATP, and some other energy-rich compounds). On the other hand, these reactions make available carbon as small precursor metabolites, as well as nitrogen, sulfur and phosphorus. The energy conserved and the metabolic products provided by these fueling reactions are then used for all cellular activities, including biosynthetic work (anabolism), mechanical work (motility), and

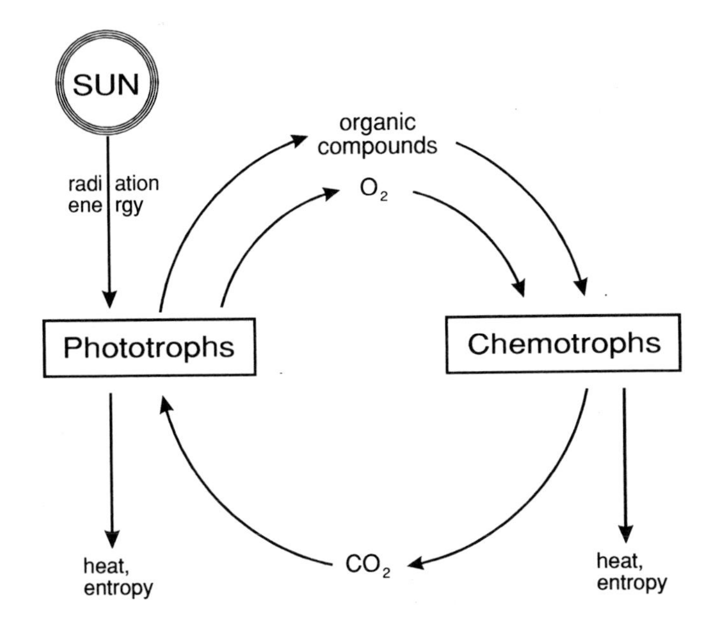

Fig. 1. Flow of energy and carbon in the biosphere.

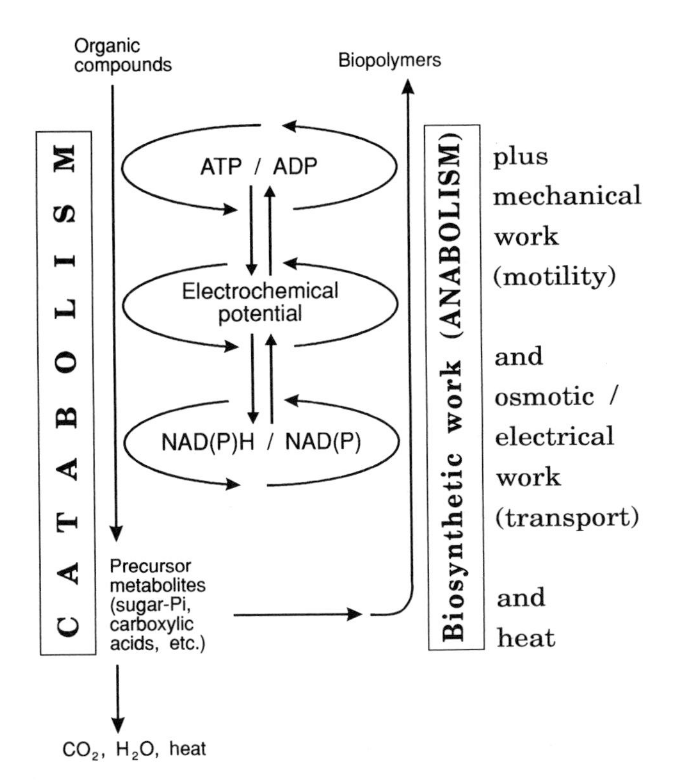

Fig. 2. Flow of energy and carbon in the metabolism of a heterotrophic cell.

osmotic or electrical work (transport). In autotrophs the situation is somewhat different. The fueling reactions are divided into two different sets: one that provides energy and reducing power (photosynthesis) and another that synthesizes the precursor molecules necessary for all further anabolic reactions.

Both catabolism and anabolism include hundreds of different reactions, some common to all cells, some confined to particular cells. The concept of metabolism is expressed in the fact that only a small number of molecules connects the two blocks, catabolism and anabolism. In addition to ATP and pyridine nucleotides which are used in a cyclic manner, only about a dozen metabolites are necessary for most biosynthetic pathways (see Fig. 20). It should be noted, however, that the distinction between catabolic and anabolic pathways is not as strict as drawn in Fig. 2. Particularly glycolysis, the pentose phosphate cycle and the citric acid cycle have both catabolic and anabolic meaning, they are thus better designated as amphibolic pathways.

2.2 Thermodynamic Principles of Energy Metabolism

Classical thermodynamics defines statistical laws for equilibria in closed systems. In contrast, the living cell is an open system which is never in equilibrium. It is essential that life proceeds irreversibly; a reversible reaction cannot perform work on a biological time scale, and reactions in equilibrium cannot be regulated. Can thermodynamics thus be applied to understand the principles of living cells? Some of these restrictions have in fact been overcome by introducing "non-equilibrium" or "irreversible" thermodynamics. Thus, biochemical systems far from equilibrium including flow of material into and out of the system (open systems, e.g., bacterial growth) have successfully been analyzed (WESTERHOFF and VAN DAM, 1987). This interesting approach, however, cannot be discussed here.

The main importance of thermodynamics for the biochemist lies in the fact that it can be used to predict whether or not a reaction is possible under given conditions. In addition, however, it obviously depends on the presence of a suitable pathway and an enzyme, whether a particular reaction will in fact occur in the cell.

2.2.1 Enthalpy and the First Law of Thermodynamics

Every molecule or system has a certain internal energy (E) which depends only on its present state. A change in the state, i.e., an exchange of energy between the system and the surroundings, may occur by gaining or losing heat to the surroundings or by the exchange of work with the surroundings. The first law of thermodynamics is the principle of energy conservation: in any process the total energy of the system plus surroundings remains constant. In other words, the internal energy (E) can be changed only by the flow of energy as heat (Q, positive when absorbed by the system) or work (W, positive when done by the system on the surroundings).

$$\Delta E = E\text{ (products)} - E\text{ (reactants)} = Q - W \quad (1)$$

The quantity normally used for describing the exchange of heat is the enthalpy H, which also includes the work done by volume changes (ΔV) at defined pressure (P).

$$\Delta H_P = \Delta E_P + P\Delta V \quad (P = \text{const.}) \quad (2)$$

A negative sign of ΔH indicates an exothermic reaction. For most biochemical reactions the differences between ΔH and ΔE are insignificant because of the small changes in pressure or volume.

2.2.2 Entropy and the Second Law of Thermodynamics

Enthalpy changes cannot serve as a reliable criterion of spontaneity. Melting of ice, for example, is a spontaneous reaction, although ΔH is positive. For explaining that kind of reactions, entropy has to be introduced, which can be recognized as a measure of "microscopic disorder".

The second law of thermodynamics can be stated in many different ways. It says that systems proceed from a state of low probability (high order) to a state of high probability (low order). The measure for this effect is the entropy (S). In other words, the entropy of the universe always increases. For a reversible reaction at constant temperature and pressure, the change in entropy is expressed by Eq. (3):

$$\Delta H = T\Delta S \tag{3}$$

2.2.3 Free Energy

A system tends toward the lowest enthalpy and the highest entropy. Since neither ΔH or ΔS alone are sufficient to predict whether a reaction will proceed spontaneously, a new function, the free energy G was introduced. The free energy of a reaction is defined by

$$\Delta G = \Delta H - T\Delta S \tag{4}$$

ΔG is negative for any spontaneous process, such a process is called exergonic. An endergonic process is characterized by a positive ΔG and will not proceed spontaneously. The magnitude of the decrease in free energy ($-\Delta G$) is a measure of the maximum work which can be obtained from a certain reaction. In most biochemical reactions, the term $T\,\Delta S$ is relatively small as compared to ΔH, i.e., for catabolic processes in general ΔH is not greatly different from ΔG.

An important consequence of the second law for energy conversion lies in the fact that, although free energy of any kind (chemical, mechanical, electrical, etc.) can be interconverted and transformed into heat, which on the other hand, cannot be transformed into free energy under isothermic conditions. Any heat produced in the metabolism is lost for the cell, thus the metabolism is constructed to minimize heat losses in metabolic reactions. This becomes obvious by the finding that the thermodynamic efficiency of, e.g., glucose respiration to CO_2 reaches 59 %, which is an impressively high value as compared to anabolic reactions (THAUER et al., 1977). Synthesis of cell material, for instance, is characterized by a thermodynamic efficiency of less than 10 %.

The standard free energy of formation (ΔG^O) of many compounds can be found in tables (THAUER et al., 1977). For most compounds ΔG^O is negative. By subtracting the sum of ΔG^O of the reactants from that of the products, the change in free energy of any reaction can be calculated. If two or more chemical equations are involved in a certain reaction, the ΔG^O of the resulting overall reaction is the sum of the changes in free energy for the individual equations. As an example, the standard free energy of ATP hydrolysis can be calculated by adding the ΔG^O values of two reactions:

$$\text{Glucose} + \text{ATP} \rightleftharpoons \text{Glucose 6P} + \text{ADP} \qquad \Delta G^{O\prime} = -21.0 \text{ kJ/mol}$$

$$\text{Glucose 6P} + H_2O \rightleftharpoons \text{Glucose} + P_i \qquad \Delta G^{O\prime} = -12.5 \text{ kJ/mol}$$

$$\text{ATP} + H_2O \rightleftharpoons \text{ADP} + P_i \qquad \Delta G^{O\prime} = -33.5 \text{ kJ/mol}$$

It is of importance to keep in mind that the value of ΔG^O for a particular reaction refers to standard conditions only, when all reactants and products, including protons, are present in 1 M (1 mol/L) concentration. In biochemistry, $\Delta G^{O\prime}$ values are used; the additional superscript indicates that standard conditions are defined at pH 7, 25 °C.

2.2.4 Chemical Equilibria in Solution

Based on the definition of the free energy of reactions in equilibrium it becomes clear that there must be a direct relationship between ΔG of a reaction and its equilibrium constant. For the reversible reaction

$$A + B \rightleftharpoons C + D$$

the equilibrium constant K_{eq} can be calculated from the concentrations of the reactants at equilibrium

$$K_{eq} = \frac{[C]_{eq}[D]_{eq}}{[A]_{eq}[B]_{eq}} \tag{5}$$

The free energy change now is a function of displacement of the reaction from equilibrium

$$\Delta G = RT\left[\ln\frac{[C][D]}{[A][B]} - \ln\frac{[C]_{eq}[D]_{eq}}{[A]_{eq}[B]_{eq}}\right] \quad (6)$$

If all concentrations $[A]=[B]=[C]=[D]=1\,M$, ΔG becomes ΔG^O

$$\Delta G^O = -RT\ln K_{eq} \quad (7)$$

ΔG^O represents the change in free energy under standard conditions and is constant for a given reaction. Eqs. (5) to (7) can be combined to

$$\Delta G = \Delta G^O + RT\ln\frac{[C][D]}{[A][B]} \quad (8)$$

This equation is very important in biochemistry, the value of ΔG (and not ΔG^O) determines whether a reaction may occur spontaneously under certain conditions. This becomes clear in practical terms when rationalizing the relation of ΔG and ΔG^O by the statement that ΔG is a function of displacement of the reaction from equilibrium which is represented by ΔG^O. Thus, for the cell the actual value of ΔG for a certain reaction (and not ΔG^O) is important. If the concentrations in Eq. (8) are equilibrium concentrations (as in Eq. 7), ΔG becomes zero, i.e., the system cannot do any work. Only a system displaced from equilibrium can do work in the cell.

2.2.5 Group Transfer Potentials and the Adenylate System

The importance of these relations is well exemplified by the thermodynamics associated with ATP hydrolysis. ATP is the universal molecular carrier for biological energy. Hydrolysis of ATP is described by the reaction

$$ATP^{4-} + H_2O \rightleftharpoons ADP^{3-} + HPO_4^{2-} + H^+,$$

the corresponding equilibrium constant is

$$K_{eq} = \frac{[ADP^{3-}][HPO_4^{2-}][H^+]}{[ATP^{4-}]} = 0.63 \quad (9)$$

and the standard free energy for ATP at pH 7 and 25 °C ($\Delta G^{O\prime}$) is –34.5 kJ/mol. However, since the reactants are certainly not present at standard concentrations in the cell, and a major fraction of ATP is complexed with Mg^{2+}, the free energy required for ATP synthesis *in vivo* is not equal to the standard free energy, but depends on the actual concentrations of ATP, ADP, P_i, and the cytoplasmic pH (according to Eq. 9). Additionally, the Mg^{2+} concentration plays a role, since the different nucleotide compounds form different complexes with divalent cations. Estimations of the value of the free energy of ATP hydrolysis in the cytoplasm are in the range of 42–50 kJ/mol (Harold, 1986; Thauer et al., 1977).

In energy transfer reactions, ATP transfers its terminal phosphate group(s) to other compounds. It was originally stated that ATP and other activated compounds contain "energy-rich" phosphate bonds, symbolized by a squiggle (~P). This term, however, is a misleading expression, implying that the terminal phosphate bonds in ATP are an exceptional kind of chemical bond. A more appropriate description is the expression "high group transfer potential". Group transfer potential means the tendency of a particular group to be transferred to an appropriate acceptor. In ATP hydrolysis, the acceptor for the transfer of the phosphate group is water, in this case the group transfer potential is the free energy of hydrolysis.

Because of the central role of ATP in energy transfer reactions, the actual free energy of the ATP/ADP system must be effectively regulated by the cell, i.e., kept constant and at a high value (Harold, 1986). Based on the $\Delta G^{O\prime}$ (ΔG^O at pH 7) of ATP hydrolysis, the expected ATP/ADP ratio would be about 10^{-7}–10^{-8}. By coupling the adenylate system to exergonic reactions (see Sect. 2.4), cells maintain this ratio around 10, thus displacing it from equilibrium by a factor of at least 10^8.

When quantifying the energy status of the cell, respectively that of the adenylate system, both the phosphate and proton concentrations are in general not taken into account, because they are relatively constant in the cytosol. On the other hand, adenine nucleotides are not only present in the form of ATP and ADP. Because of the function of the adenylate kinase, AMP has to be also included.

$$ATP + AMP \rightleftharpoons 2\ ADP$$

Consequently, the activity of metabolic enzymes, e.g., in glycolysis, is modulated not only by ATP and ADP, but also by AMP (see Sect. 6.1.2). On the basis of these considerations, the adenylate "energy charge" as a description of the energy status of the adenine nucleotide pool in the cell was introduced.

$$\text{energy charge} = \frac{[ATP]+\frac{1}{2}[ADP]}{[ATP]+[ADP]+[AMP]} \quad (10)$$

The energy charge can range between 1 and 0 and indicates the fraction of the nucleotide pool which is charged with activated phosphoryl groups. Under normal conditions, the regulatory properties of catabolic and anabolic enzymes keep the energy charge in most cells at about 0.8–0.9.

2.2.6 Free Energy and Electrical Work, the Electrochemical Potential

In oxidation and reduction reactions, electrons are transferred between the reactants. The oxidant accepts electrons (electron acceptor) and the reductant donates electrons (electron donor). The potential difference E_h of a given reduction-oxidation system

$$\text{oxidant} + e^- \rightleftharpoons \text{reductant}$$

is expressed by the Nernst equation

$$E_h = E_O + \frac{RT}{nF} \ln \frac{[\text{oxidant}]}{[\text{reductant}]} \quad (11)$$

where E_O is the standard electrode potential (measured in volts), n is the number of electrons transferred, and F is the Faraday constant. In many reactions protons are involved,

$$\text{oxidant} + e^- + H^+ \rightleftharpoons \text{reductant-H}$$

thus E_O becomes pH-dependent and is normally expressed as E'_O which means E_O at pH 7. A strong reducing agent has a negative E'_O (e.g., NADH: $E'_O = -0.32$ V), whereas a strong oxidizing agent has a positive one (e.g., O_2: $E'_O = +0.82$ V). The free energy change of a redox reaction can be derived by combining Eqs. (7) and (11)

$$\Delta G^{O\prime} = -nF\Delta E'_O \quad (12)$$

where $\Delta E'_O$ is the difference in redox potential between the oxidant and the reductant couples. Under standard conditions, a positive sign for E'_O (but a negative one for $\Delta G^{O\prime}$) indicates an exergonic reaction.

In biological systems, the other kind of "energy currency" besides ATP is the electrochemical potential which combines a concentration gradient (chemical potential) with an electrical potential. Analogous to a chemical reaction (Eq. 12), the chemical potential of a solute in a two-compartment system is equivalent to the work required for the transfer of one mole of solute from compartment 1 to compartment 2 which differ in their concentration of S.

$$\Delta G = RT \ln \frac{[S_2]}{[S_1]} \quad (13)$$

The corresponding expression for the transfer of one mole of an ion against an electrical potential ($\Delta\Psi$) is

$$\Delta G = nF\Delta\Psi \quad (14)$$

where n is the valence of the ion. In general, in biological systems both concentration gradient and electrical potential have to be considered, thus the free energy for transfer is equal to the electrochemical potential difference of a particular ion which is correctly expressed as $\Delta\tilde{\mu}_{\text{ion}}$.

$$\Delta G = nF\Delta\Psi + RT \ln \frac{[S_2]}{[S_1]} = \Delta\tilde{\mu}_{\text{ion}} \quad (15)$$

The electrochemical potential difference as driving force for ion movement across membranes is more conveniently expressed in volts, instead of the general expression in kJ/mol. For this purpose, Eq. (20) is divided by F, leading to

$$\frac{\Delta\tilde{\mu}_{\text{ion}}}{F} = n\,\Delta\Psi + \frac{RT}{F}\ln\frac{[S_2]}{[S_1]} \tag{16}$$

Actual values for the components of the electrochemical potential in the cell are discussed in Sect 2.4.

2.3 Supply of Useful Energy for the Cell

A main purpose of catabolism is supplying the cell with useful energy for biosynthesis and work functions. This energy may be provided in different forms, e.g., ATP, reduced pyridine nucleotides, as well as ion gradients. Basically, there exist two different mechanisms by which cells are supplied with energy.

(1) *Substrate level phosphorylation:* This term means group transfer reactions, catalyzed by soluble enzymes, which ultimately lead to production of ATP.

(2a) *Electron transport phosphorylation:* In this class redox reactions are summarized, catalyzed by membrane bound enzymes, which are coupled to ATP production by electrochemical ion (in general proton) potentials.

(2b) *Photophosphorylation:* In this class light energy is converted into the free energy of ATP, the coupling mechanism is again translocation of protons by membrane bound proteins.

2.3.1 Substrate Level Phosphorylation

Substrate level phosphorylation is closely related to the concept of fermentation (see below). The free energy of substrates is conserved by formation of phosphorylated metabolites, e.g., 1,3-diphosphoglycerate during glycolysis.

glyceraldehyde-3-phosphate + P_i + NAD^+ $\rightleftharpoons$
1,3-diphosphoglycerate + NADH + H^+
$\Delta G^{O\prime} = +37$ kJ/mol

1,3-diphosphoglycerate + ADP $\rightleftharpoons$
3-phosphoglycerate + ATP
$\Delta G^{O\prime} = -30.5$ kJ/mol

The activated phosphoryl group is directly transferred to ADP, i.e., at the substrate level. Although fermentation in general includes substrate level phosphorylation, anaerobic metabolism does not necessarily mean energy conservation by this mechanism.

In the course of the years, the term "fermentation" has undergone significant changes. Especially in industrial microbiology, all processes using microorganisms in fermentors for producing metabolites or biomass are called fermentations, regardless whether or not they fall under the correct definition.

A useful biochemical definition of fermentation means utilization of an energy source in the absence of an external electron acceptor, which in most cases means anaerobic conditions. Since also in this case, the energy required for synthesis of ATP is provided by redox processes, oxidation and reduction must be balanced internally. Thus, the electron donors are only partially oxidized. The reducing equivalents obtained (in general NADH) are transferred to some intermediate (electron acceptor) which originates from the original electron donor (Fig. 3). By generation of these fer-

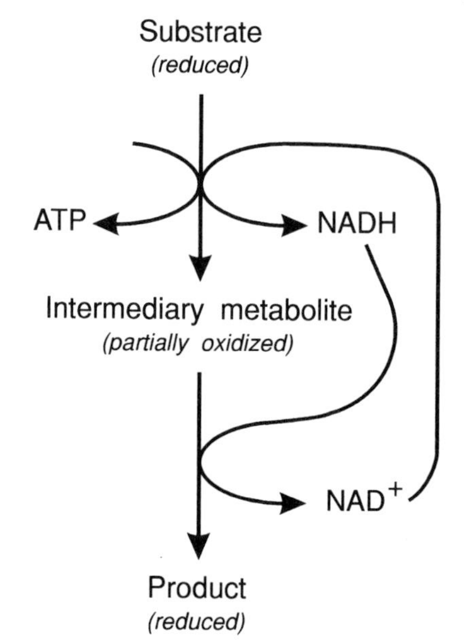

Fig. 3. Scheme of fermentative metabolism: balancing of redox reactions.

mentation products, the cellular redox balance is maintained. Besides the flow of carbon and electrons, energy is provided by coupling the first step (Fig. 3) to ATP synthesis (substrate level phosphorylation, see Sect. 2.3). There are other "energy-rich compounds" besides ATP (compounds with a high group-transfer potential, see Sect. 2.2). They serve for energy conservation in the first step of fermentation and are in energetic equilibrium with the adenylate system. Among these are thioesters (acetyl-CoA, succinyl-CoA) and acid anhydrides (acetylphosphate, carbamoylphosphate).

The need for an appropriate redox balance during fermentation is emphasized by the fact that a number of bacteria produce hydrogen to meet these requirements. Thereby, protons derived from water serve as electron acceptors for the disposal of excess electrons. Transfer of electrons is mediated by ferredoxin, an electron carrier of very low redox potential, and by the use of enzymes called hydrogenases.

The dependence of fermentation on substrate level phosphorylation for supply of ATP leads to the consequence that only exergonic reactions with a change in free energy comparable to that needed for phosphorylation of ADP (about –45 to –50 kJ/mol) can be used in these pathways. By contrast, energy metabolism based on electron transport phosphorylation does not necessarily depend on this "biological energy quantum". Depending on the actual ATP/2e^- stoichiometry of the H^+-ATP synthase, also reactions yielding fractions of this energy quantum can be used, provided they lead to generation of an electrochemical ion gradient. There are certain exceptions where reactions with energy yields definitely below the free energy of ATP are used in fermentative processes. In these cases, however, ATP synthesis proceeds via a chemiosmotic process and not via substrate level phosphorylation. Examples are sodium-coupled membrane energization in some organisms such as *Propiogenium modestum* or *Klebsiella*, as well as generation of a proton potential by electrogenic substrate/product antiport, e.g., in *Oxalobacter formigenes.*

2.3.2 Electron Transport Phosphorylation

The energy for electron transport phosphorylation is provided by oxidation of reduced substrates and transport of electrons along a chain of redox carriers (respiratory chain) to the terminal electron acceptor which may be oxygen (aerobic respiration) or others such as nitrate, sulfate, fumarate, etc. (anaerobic respiration). In contrast to fermentation, the substrate molecules can be oxidized completely, thus a far higher yield of ATP is possible. Whereas in (anaerobic) glycolysis, only 2 mol ATP/mol glucose can be obtained, the maximum ATP yield in aerobic metabolism (theoretically) amounts up to about 38 mol ATP/mol glucose. The respiratory chains in various organisms can be very diverse, however, some central principles are conserved; these may be exemplified by the best known respiratory chain, that of mitochondria and many bacteria (Fig. 4).

Reducing equivalents from various substrates are accepted by flavoproteins either via reduced pyridine nucleotides (NADH: ubiquinone reductase, complex I) or at the flavine coenzyme level (e.g., succinate: ubiquinone reductase, complex II). Electrons are collected in the ubiquinone pool and subsequently transferred to the ubiquinone: cytochrome-c reductase (complex III) which transfers them further on to cytochrome c, the second mobile electron carrier. The cytochrome-c oxidase (complex IV) finally, catalyzes the transfer of electrons to oxygen. The funtional protein complexes of the respiratory chain are, besides the ribosome, the most complex enzyme systems in the cell. They may contain up to 30 different protein subunits each including up to 10 different prosthetic groups (cytochromes, FeS-centers, and molybdenum cofactors in the case of formate dehydrogenase). Some of these protein complexes are sites of proton translocation, i.e., concomitantly with the electron flow, protons are translocated across the membrane from the cytosol into the surroundings. The energy coupling to ATP will be described below.

The respiratory chains of bacteria, which are located in the plasma membrane, are of partic-

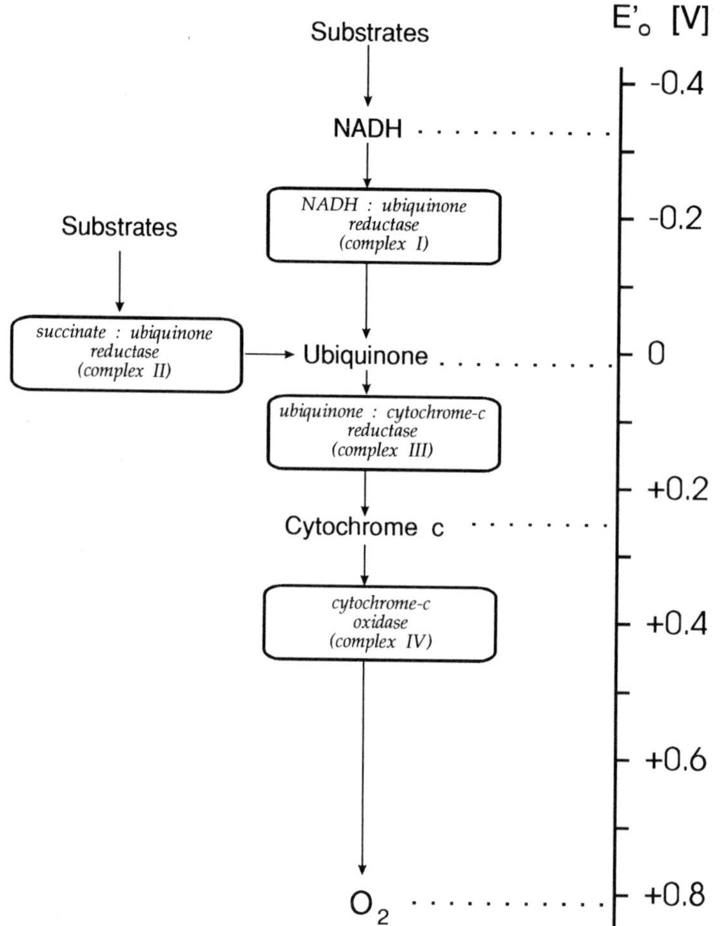

Fig. 4. Scheme of the respiratory chain in mitochondria and some bacteria.

ular diversity. This holds true for the presence of various parts of the chain of redox carriers, especially the type of terminal oxidase. Also the type of quinone may vary (menaquinone or ubiquinone), as well as the particular property of a certain part to translocate protons. Furthermore, bacterial redox chains are generally branched (ANRAKU, 1988), a fact which may be explained by the need of bacteria to respond to changes in environmental conditions by the use of an appropriate respiratory chain. In *Escherichia coli*, for example, the electrons are transferred from ubiquinone either to cytochrome o (low affinity to oxygen) under fully aerobic conditions, or to cytochrome d (high affinity to oxygen) under "micro-aerobic", i.e., low oxygen conditions (Fig. 5). Nevertheless, the general organization into individual complexes and intermediary pools for reducing equivalents is conserved.

Electron flow takes place from the system of low (negative) redox potential to that of high (positive) E_O. In the case of aerobic respiration (Fig. 4) the energy available from substrate oxidation is defined by the span in redox potential between NADH ($E'_O = -0.32$ V) and oxygen ($E'_O = +0.82$ V). According to Eq. (12) this span of 1.23 V is equivalent to $\Delta G^O = -238.5$ kJ/mol. This decrease in free energy is divided up into three parts, each part being used for the extrusion of protons by the three equivalent components of the respiratory chain (complex I, III, and IV). The former explanation of these three parts of the respiratory chain being "coupling sites" for ATP-production originated from the use of "high energy phosphate

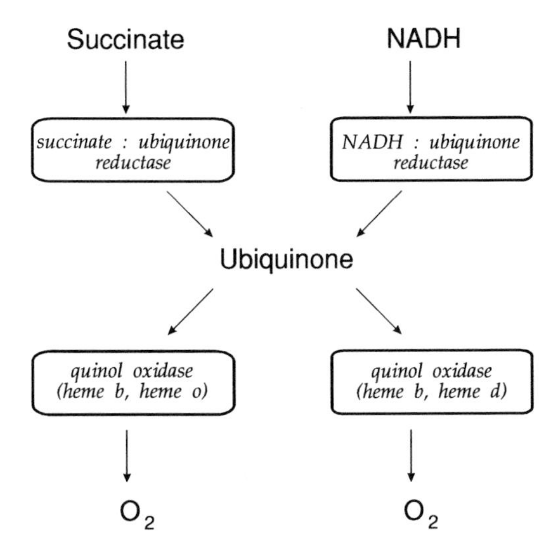

Fig. 5. Scheme of the branched respiratory chain in (aerobic) *Escherichia coli.*

bonds" (~P) and is incorrect. Coupling of the energy supplied by the respiratory chain to the synthesis of ATP proceeds via electrochemical mechanisms (see following section), thus an exact stoichiometry cannot be given.

In *chemo-lithotrophic* bacteria, the oxidation of inorganic compounds is used to generate ATP by electron transport phosphorylation. Electrons are transported from a reduced form of the respective compound to O_2 as the terminal electron acceptor. Hydrogen bacteria oxidize H_2 to 2 protons and 2 electrons, iron bacteria can perform the oxidation of Fe^{2+} ions to Fe^{3+} and e^-. Ammonia oxidizers utilize NH_4^- plus $2\,H_2O$ to NO_2^- plus $8\,H^+\,6\,e^-$. Nitrite oxidizers react nitrite plus water to nitrate plus $2\,H^+$ and $2\,e^-$. Finally, sulfur bacteria can perform three reactions: they either oxidize elemental sulfur (S), sulfite (SO_3^-), or H_2S to elemental sulfur, or sulfite to sulfate (see also Sect. 5.3). As the redox potential of most chemolithotrophic electron donors is lower than that of NADH, NAD reduction cannot be coupled directly to the oxidation of these compounds. In order to reduce NAD, these bacteria have to use reverse electron transport (see Sect. 2.4).

Photosynthesis is the most important biological mechanism of energy transduction. In many respects it resembles that of respiration, including an electron transport chain which translocates protons, the energy of which is finally converted into chemical energy of ATP. The particular aspect of photosynthesis is the use of light energy which is made possible by the presence of certain pigments (chlorophyll). The energy and the reducing equivalents provided by photosynthetic electron transport and photophosphorylation are finally used for assimilation of CO_2 in the Calvin cycle (see Sect. 5.1.1). There exist two different modes of photosynthesis, those in bacteria and in plants, the latter including also cyanobacteria. They differ in the components of light-harvesting complexes and the subsequent electron-transport chains (DREWS, 1986); energy conversion to ATP, however, is identical. The overall reaction of photosynthesis is given by

$$CO_2 + 2\,H_2A \xrightarrow{\text{light}} (CH_2O) + 2\,A + H_2O \quad (17)$$

where H_2A is the donor of reducing equivalents and (CH_2O) means any organic compound at the oxidation level of carbohydrates. Plants and cyanobacteria use water as reductant, thus oxygen is produced during photosynthesis:

$$CO_2 + 2\,H_2O \xrightarrow{\text{light}} (CH_2O) + O_2 + H_2O \quad (18)$$

Photosynthetic bacteria (e.g., purple bacteria) are in general anaerobic bacteria, thus they must substitute water with some other reductant, i.e., H_2S, H_2, or reduced organic compounds. For instance, H_2A in Eq. (17) then becomes H_2S, and sulfur is produced instead of oxygen.

Light energy is absorbed by the photosynthetic pigments in a broad range between 400 and 1100 nm. In order to provide this window of absorption, the photosynthetic apparatus always contains a set of different pigments, including chlorophylls, phycobilins and carotenoids. Chlorophylls are heme-type molecules with Mg^{2+} as the central metal ion, phycobilins are linear tetrapyrrols and carotenoids belong to the class of tetraterpenes. The term "reaction center" designates the structures containing those chlorophyll molecules which directly participate in photosynthetic energy transduction (e.g., P_{700}, see below). The term photosystem includes also other pigments which are involved in energy transfer from light to excited electrons (e.g., PS I, see below). The pigment in the reaction center is always chlorophyll a in the case of oxygenic photosynthesis and bacteriochlorophyll in anaerobic photosynthetic bacteria. The various other chlorophyll forms, as well as the other accessory pigments are mainly used as antenna pigments. They collect photons and deliver them to the reaction center, thereby significantly enhancing the efficiency of the photosynthetic apparatus, especially under situations of weak light.

The photosynthetic energy conversion is initiated when photons are absorbed by the chlorophyll molecule in the reaction center. Excited electrons are produced which means a shift of electrons from the positive redox level of the electron donor (H_2O, H_2S, etc.) to the highly negative one of the primary acceptor of the subsequent electron transport chain. Bacterial photosystems contain only a single reaction center. Fig. 6A schematically shows the pathway of electrons in purple bacteria (e.g., *Rhodopseudomonas viridis*). After light absorption by the reaction center, the excited electron is first transferred to another bacteriochlorophyll, then to bacteriopheophytin, a chlorophyll-type pigment lacking Mg^{2+}, and bound quinone molecules. It enters the quinone pool within the photosynthetic membrane, being finally transferred to a part of the electron transport chain, analogous to complex III of the respiratory chain (see Fig. 4). This cyclic electron flow is closed by transfer of electrons via cytochrome c_2 back to the reaction center.

By contrast, in oxygenic photosynthesis water is oxidized to oxygen and electrons are transferred to $NADP^+$. Thus a redox span significantly larger than in anoxygenic photosynthesis has to be covered which is the reason for an additional photosystem to be used. This results in the so called Z-scheme (Fig. 6B). Photosystem I (with reaction center P_{700}) can again mediate cyclic electron flow via cytochrome b, the plastoquinone pool and two additional redox carriers, cytochrome f and plastocyanine. It can also transfer electrons via a highly negative primary acceptor and ferredoxin to $NADP^+$. At the other end of the scheme, water oxidation is managed by light absorption to photosystem II (with reaction center P_{680}), the

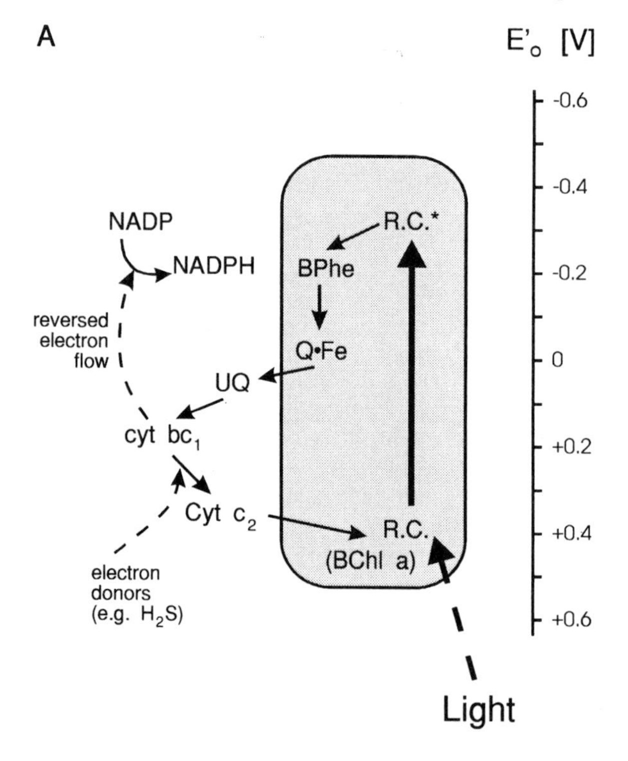

Fig. 6. Electron transport pathways in bacterial (A, anoxygenic photosynthesis in purple bacteria) and plant photosynthesis (B). It should be noted that from the structural point of view, PSII from plants resembles the system from purple bacteria.

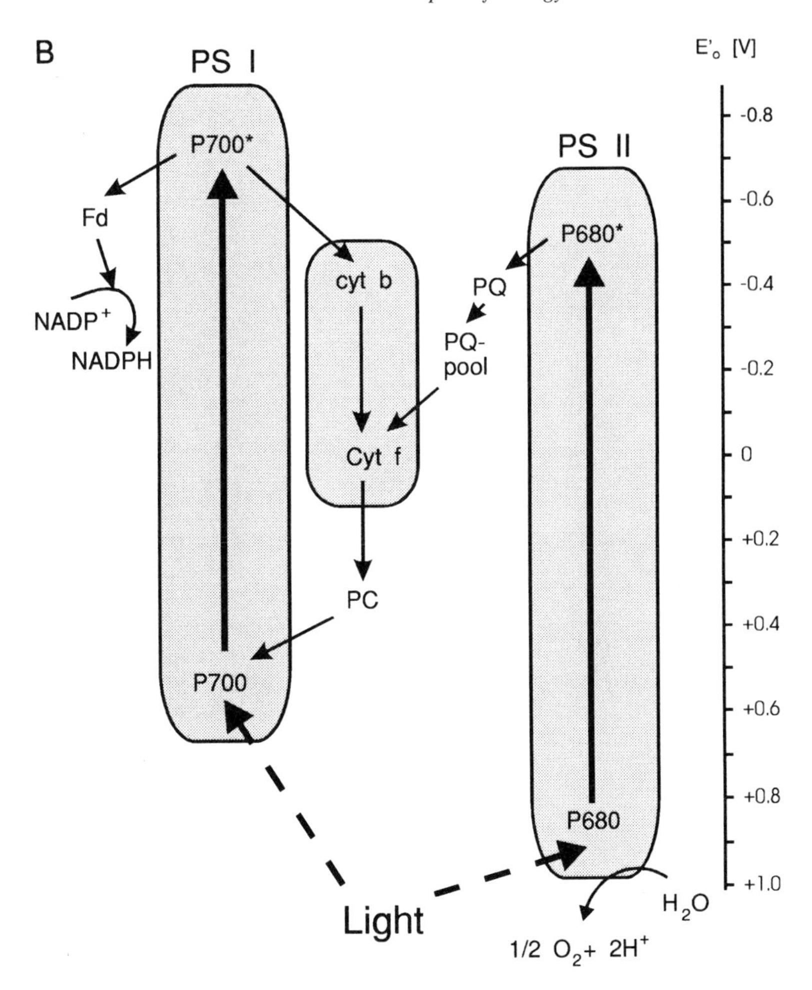

Fig. 6 B

excited electrons are transferred to plastoquinone. PS II and PS I can act in series (non-cyclic electron flow), or PS I may function as the center of cyclic electron flow, similar to that described in bacteria.

Obviously, besides water splitting, the second invention in oxygenic photosynthesis is the direct transfer of electrons to $NADP^+$, thus providing the cell with reducing equivalents for biosynthesis. However, as already stated in the basic reaction of bacterial photosynthesis, also in the anoxygenic process, reducing equivalents can be provided. For this purpose, electrons from appropriate donors (e.g., H_2S) enter at the cytochrome level, are then taken from the quinone pool and transferred to NAD^+ via an ATP-dependent electron flow in the reverse direction of the respiratory chain.

The first step in the energy transduction mechanism is generation of an electric field due to charge separation across the membrane. The reason for this is the vectorial nature of the light-driven redox reaction, i.e. the electron donor and the electron acceptor of the reaction centers are located on opposite sides of the membrane. The charges (electrons) then flow back using the electron transport chains (Fig. 6B) which again cross the respective membrane. Similar to the respiratory chain, electron flow is coupled to translocation of protons across the membrane in which the complexes are embedded. A proton gradient is thus created in all the different systems, by cyclic electron flow in the bacterial system, as well as by cyclic and non-cyclic flow in the systems of oxygenic photosynthesis.

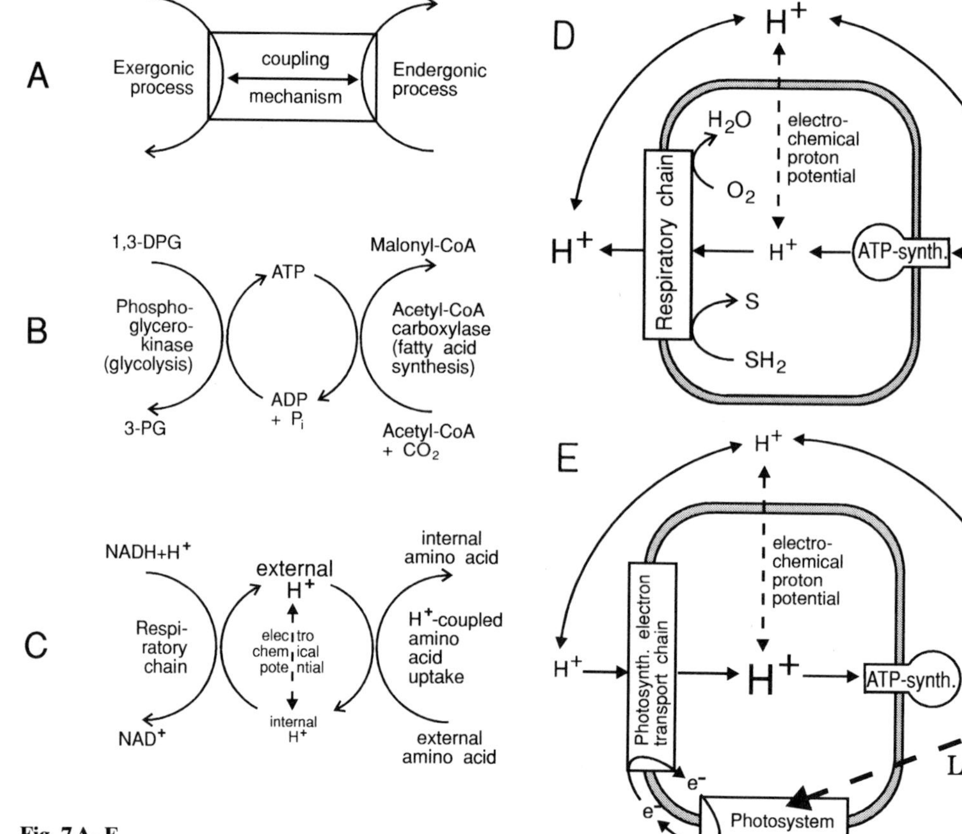

Fig. 7 A–E

2.4 Energy Coupling

The mechanism of energy coupling is the central part in the organization of the metabolic network. As outlined in Fig. 2, it connects the energy fueling reactions with those needing energy for some useful work for the cell. Fig. 7A shows a general "coupling device", exemplifying three main points:

(1) The endergonic reaction at the right side will only take place if it is coupled to the exergonic reaction at the left.
(2) The machinery will work from left to right only if the negative value of ΔG of the exergonic reaction is greater than the positive one of the endergonic reaction. Otherwise it will work the other way round; an example for this is reversed electron flow. Well known examples for reversed electron flow are the reduction of NAD to NADH in lithotrophs or during anoxygenic photosynthesis (see Sect. 2.3).
(3) If the positive ΔG of the reaction on the right side increases (e.g., if some precursors run out or a metabolic pool is filled up), the exergonic reaction will be stopped by the coupling machinery.

For explaining energy coupling in the cell, currently two main mechanisms are known: chemical coupling and coupling by ion currents. Consequently, these two possibilities are connected with two types of general "currency" of energy: ATP (and a few other high-energy compounds) on the one hand, and electrochemical ion potentials on the other.

The principles of chemical coupling have already been mentioned in connection with sub-

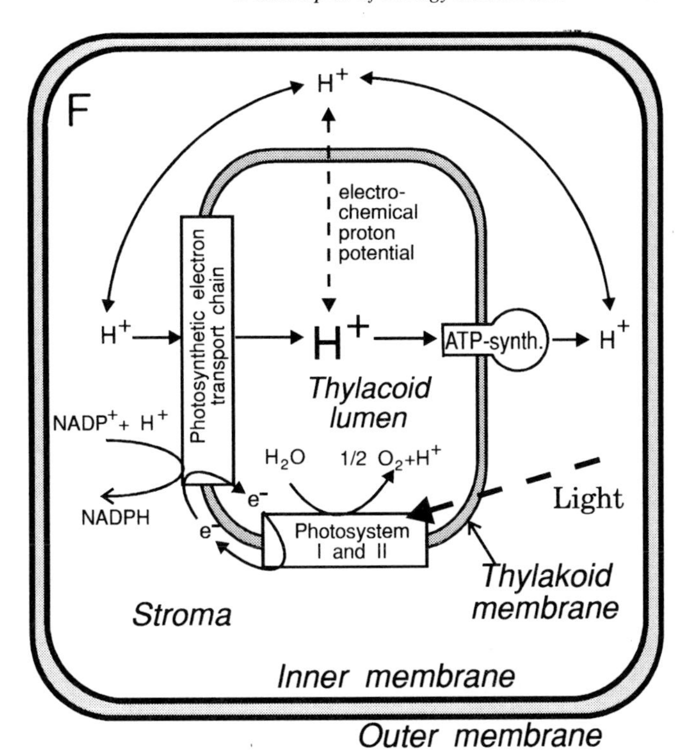

Fig. 7. Energy coupling mechanisms. A, General scheme; B, hypothetic example for direct chemical coupling; C, scheme of electrochemical coupling. Chemiosmotic coupling of respiration and ATP synthesis (D), of photosynthesis and ATP synthesis in bacteria (E), and in plants (F). It has to be mentioned that bacterial chromatophores (E) are formed during isolation; originally the photosynthetic membrane consists of invaginations of the bacterial plasma membrane.

strate level phosphorylation (Sect. 2.2). A general energy pool, represented by substances with a high group transfer potential, i.e., mainly ATP, is filled by catabolic reactions. The high energy compound is used in reactions of biosynthesis, motility or transport to render the overall change in free energy of the particular reaction negative. An example is given in Fig. 7B where, hypothetically, one of the ATP-yielding reactions of glycolytic substrate level phosphorylation is coupled to the ATP-consuming first step in fatty acid synthesis, namely acetyl-CoA carboxylation.

This example shall furthermore be used to emphasize that the formulation, "a reaction is driven by the free energy of ATP hydrolysis" is misleading at best. It is first not the free energy of hydrolysis, but the direct participation of ATP in the chemical reaction which "drives". Hydrolysis of ATP must be prevented in any case, otherwise the free energy would be lost as heat. Secondly, the chemical coupling by ATP does in fact function because the free energy of the left reaction in Fig. 7B is not simply used to synthesize an energy-rich phosphate bond, but is stored in an ATP/ADP ratio which is far from equilibrium (see above).

The second mechanism of energy coupling uses another kind of energy currency, the electrochemical potential (Fig. 7C). An energy-dependent transport system (fueling reaction) pumps ions across the membrane, thus establishing both a chemical gradient of the respective ion and an electrical gradient due to its charge (electrochemical potential). This gradient can then be used as source of free energy for systems whose construction makes possible the conversion of a downhill ion flux into useful work for the cell (chemical, mechanical, electrical, or osmotic energy). It has to be mentioned that both the very first step, i.e., active extrusion of ions and the final step, i.e., conversion of a downhill ion flux in useful work, are still not understood. However, the central part, the role of ion gradients in energy transduction is very well described by Mitchell's chemiosmotic theory. A scheme for energy transduction in an aerobically respiring cell is given in Fig. 7D. The active extrusion of protons is brought about at the expense of free energy provided by oxidation of reduced substrates. The energy stored in the electrochemical potential of protons is used by the ATP synthase for the synthesis of ATP within the cell.

The same mechanism works in transforming energy stored in the proton gradient across the thylakoid membrane or that of chromatophores. Fig. 7E shows the energy transduction in a bacterial chromatophore and Fig. 7F that of a plant chloroplast. Again there is coupling of the electron flow to vectorial movement of protons, and the resulting electrochemical gradient is transformed into chemical energy by the ATP synthase. However, there are several differences between energy transduction in respiration and photosynthesis. Obviously, (1) in one case chemical energy of substrates is used and in the other case light energy and (2) in respiration NADH is consumed, whereas, at least in oxygenic photosynthesis, NADPH is produced. But there are also mechanistic differences. First, the respiratory chain of bacteria and mitochondria expels protons, whereas bacterial chromatophores and chloroplast thylacoids accumulate protons. Second, the main component of the electrochemical driving force in the case of respiration in general is the membrane potential (typically 120–200 mV, exterior positive) whereas in the case of photosynthetic energy transduction it is the pH gradient (typically 3–4 units, interior acid).

It has to be mentioned that there exist further mechanisms of electrochemical energy coupling which cannot be discussed in detail here, for instance, that used by halobacteria. Halobacteria convert light energy directly into vectorial movement of protons as well as other ions across the plasma membrane by the use of small membrane proteins (bacteriorhodopsin and halorhodopsin).

The chemiosmotic mechanism of coupling is based on some fundamental observations. Energy coupling by this kind of mechanism requires closed membrane structures. If a shortcut between the separated spaces is introduced ("uncouplers"), energy coupling is abolished. Many end products of fermentative pathways (see Sect. 5.1), e.g., small organic acids and alcohols, exert uncoupling effects on the plasma membrane of the producing microorganisms. This may be due to transport of protons by the metabolic end product itself (e.g., acetate), or by increasing the passive proton permeability of the bacterial membrane (e.g., butanol).

The three central systems of the cell which are involved in chemiosmotic mechanisms, i.e., respiratory chains, photosynthetic centers, and ATP synthase, all transport protons. In all living cells (and in many organelles) an electrochemical proton potential can be measured, which varies considerably from species to species and depends on particular metabolic conditions (e.g., aerobic/anaerobic state, supply of energy source, external K^+ and NH_3 concentration). Values of 120–200 mV for the electrical part (membrane potential) and pH gradients of up to two units are typical. It should be added here that, although the proton is the general coupling ion of bacteria, sodium ions are also used as coupling ions in many species.

The electrochemical potential was already quantitatively described in Eq. (16). When expressed for protons

$$\frac{\Delta\mu_{H^+}}{F} = \Delta\Psi + \frac{2.3\,RT}{F} \log\frac{[H^+]_{in}}{[H^+]_{ex}} \tag{19}$$

it can be simplified to ($\Delta pH = pH_{in} - pH_{ex}$)

$$\frac{\Delta\mu_{H^+}}{F} = \Delta\Psi - \frac{2.3\,RT}{F}\,\Delta pH = \Delta p \tag{20}$$

Δp is the electrochemical proton potential, in general Mitchell's term proton-motive force is used. At 25 °C, Δp becomes (in millivolts)

$$\Delta p = \Delta\Psi - 59\,\Delta pH \tag{21}$$

It is obvious that the proton motive force can easily be converted into a change in free energy.

$$\Delta G = F\Delta p \tag{22}$$

If a certain reaction (driving or consuming) is in equilibrium with the proton-motive force, the number n of protons transported per reaction cycle has to be taken into account

$$\Delta G = nF\Delta p \tag{23}$$

In general ATP and the proton-motive force are two freely interconvertible energy currencies. That does not mean that the two energy intermediates have the same importance in all cells. In any case, it is clear that all cells need both ATP and proton potential for various processes. Fermentative bacteria in general depend solely on substrate level phosphorylation. The free energy of ATP has thus to be converted into a proton potential by the H^+-ATPase (Fig. 8A). In the case of obligate aerobes, the primary energy form is Δp, thus ATP is provided by the function of the same enzyme (H^+-ATPase), now working in the opposite direction, as an ATP synthase (Fig. 8B). In most cells in the biosphere, the two mechanisms are working together; the two energy forms are then equilibrated by the energy converter ATPase (Fig. 8C).

Photolithotrophic energy metabolism resembles that of obligate aerobes, except that the primary energy form Δp is provided by transformation of light energy. As mentioned above, the two forms of energy are then used differently by the cell, ATP mainly for biosynthesis, whereas mechanical, electrical and osmotic work (motility and transport) uses both kinds of free energy pools in the cell.

It is obvious that "energy coupling" has also a broader meaning in microbiology and biotechnology, i.e., the relation between input of energy and output of biomass. This relation concerns studies on bacterial growth and metabolism, including ATP yield and growth yields, the basis of which should clearly be related to the molecular mechanisms of energy coupling in the cell. This question, however, will be dealt with in Chapters 3 and 4 in this volume.

3 Enzyme Catalysis

Most chemical reactions in the cell do not occur spontaneously because of high kinetic barriers. The cell uses enzymes as catalysts to overcome these barriers, thus making possible energetically favorable reactions to take place at sufficiently high rates. This section will outline the energetic and kinetic background of this catalytic action of enzymes.

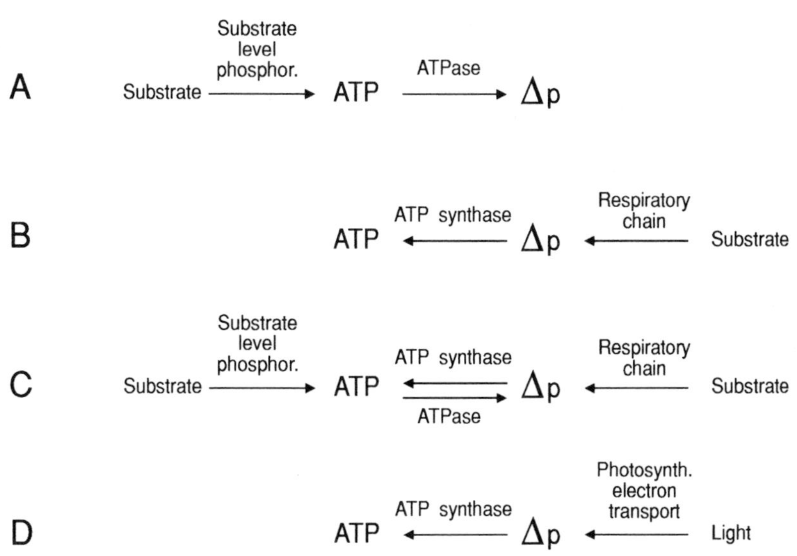

Fig. 8. Energy flow between the ATP/ADP system and the electrochemical proton potential in different organisms. A, fermentative bacteria; B, obligate aerobic organisms; C, facultative bacteria and most eukaryotic cells; D, photolithotrophic bacteria.
Adapted with permission from HAROLD (1986)

3.1 Classification of Enzymes

In view of the vast number of enzymes known, a system of classification is essential to give some rationale of order. According to the system approved by the International Union of Biochemistry (IUB), each enzyme is classified by four numbers. The first number designates the main class (e.g., 2, transferases), the second indicates the subclass (2.6, transfer of N-containing groups), the third number divides subsubclasses (2.6.1, aminotransferases), the last number is the serial number of the particular enzyme (2.6.1.1, aspartate aminotransferase). For most enzymes there exist two names, a common name (1.1.3.4, glucose oxidase) and a systematic one (1.1.3.4, β-D-glucose:oxygen oxidoreductase). Tab. 1 lists the six main classes of enzymes together with some examples.

3.2 Energetics of Enzyme Catalysis

In a reaction of two molecules in homogeneous solution, both the probability of a collision between these molecules and the probability that this collision will successfully lead to conversion of the reactants into products has to be considered. The first factor depends on the concentration of the reactants and on temperature (speed of the molecules), whereas the second factor depends on the height of the activation-energy barrier of this reaction (Fig. 9). The overall free energy of the reaction determines whether a reaction will occur spontaneously, but it does not tell us how long it will take. Since an enzyme as a catalyst does not change the free energy of a reaction, it consequently does not influence the equilibrium, however, by lowering the activation-energy barrier, it dramatically increases the rate.

The common way to rationalize this is illustrated in Fig. 9, which shows a reaction with and without a catalyst (enzyme). In the course of the reaction, the reactants (enzyme and substrate) must undergo conformational changes and constraints (activated complex) which are reflected by a high free energy. The activation energy E_a is the difference between the free energy of the reactants and that of the highest free energy state during the reaction. This state is called the transition state. Since E_a is a free energy it contains both enthalpic and entropic terms (see Sect. 2).

The number of molecules which can cross this barrier is a function of both the temperature and the respective activation energy barrier, the rate constant k for the reaction can be expressed by the following equation

$$k = A\,e^{-E_a/(RT)} \tag{24}$$

where R is the gas constant, T is the absolute temperature, and A is a constant related to the effiency of collision. Whereas chemists can accelerate a given reaction both by changing T (raising the temperature) and by lowering E_a (adding a catalyst), the cell in general depends solely on the presence of a suitable enzyme in order to achieve sufficently high reaction rates. A decrease of the activation energy by 5.8 kJ/mol roughly corresponds to a tenfold increase in the rate. Thus, enzymes can speed up

Tab. 1. Classification of Enzymes

Number	Main Class	Reactions Catalyzed	Example (Common Name)
1.	Oxidoreductases	Oxidation-reduction reactions	1.1.1.1 Alcohol dehydrogenase
2.	Transferases	Transfer of molecular groups	2.6.1.1 Aspartate aminotransferase
3.	Hydrolases	Cleavage of bonds by introduction of water (group transfer to H_2O)	3.1.3.9 Glucose-6-phosphatase
4.	Lyases	Group transfer to double bonds or the reverse	4.1.1.1 Pyruvate decarboxylase
5.	Isomerases	Intramolecular rearrangement	5.1.3.1 Ribulose phosphate epimerase
6.	Ligases (Synthases)	Joining of two molecules with simultaneous cleavage of ATP	6.4.1.1 Pyruvate carboxylase

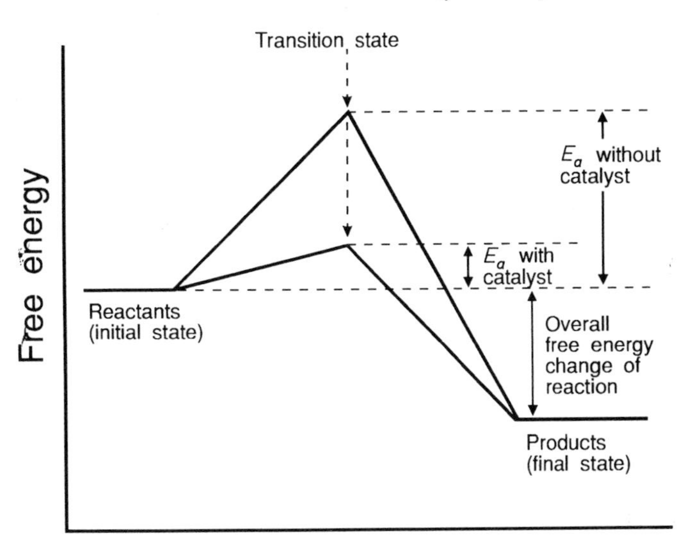

Fig. 9. Overall free energy change and activation energy (E_a) of a chemical reaction.

reactions by factors ranging from 10^8 up to 10^{20}. It should be mentioned here, that in principle all enzymes catalyze reversible reactions, however, highly exergonic reactions are practically unidirectional. That is one of the reasons (the other is regulation) why a strongly exergonic reaction uses a different pathway and an alternative enzyme if it needs to be reversed.

The original imagination of the enzyme-substrate complex was the "lock-and-key" relationship, i.e, an ideal fit of the substrate molecule to the "active site" of the enzyme. We now know that the free enzyme and the substrate do not match, the actual situation is better resembled by an "induced fit". In fact, an important advantage of biochemical catalysts (enzymes) in comparison to chemical catalysts is their flexibility in structure as revealed by recent NMR data. The molecular interaction of enzyme and substrate during catalysis is based on appropriate utilization of binding energy (JENCKS, 1975). The favorable interaction between both reacting and non-reacting parts of enzyme and substrate is referred to as the binding energy. Lowering of the activation energy and thus the rate acceleration brought about by enzymes can be described in terms of tight binding of the substrate to the enzyme exclusively in the transition state. This favorable interaction (binding energy) is not expressed in the initial ES (or the final EP) complex, otherwise the enzyme will be tied up in an energy well. Thus, an "efficient" enzyme has to prevent the formation of a tightly bound substrate in the ES complex ("very high substrate affinity"), since this inevitably will lead to slow reaction rates. Realization of the binding energy in the ES and EP complexes is prevented by conformational and entropic destabilization, i.e., twisting protein side chains of the enzyme and distorting the substrate molecule.

3.3 Kinetics of Enzyme-Catalyzed Reactions

The kinetic analysis of enzyme-catalyzed reactions is an important method to determine the mechanism of enzyme action. The basic kinetic equation, describing a reversible enzyme-catalyzed reaction with a single substrate, includes binding of substrate to the enzyme, thus forming the enzyme-substrate complex and subsequent dissociation into product and enzyme.

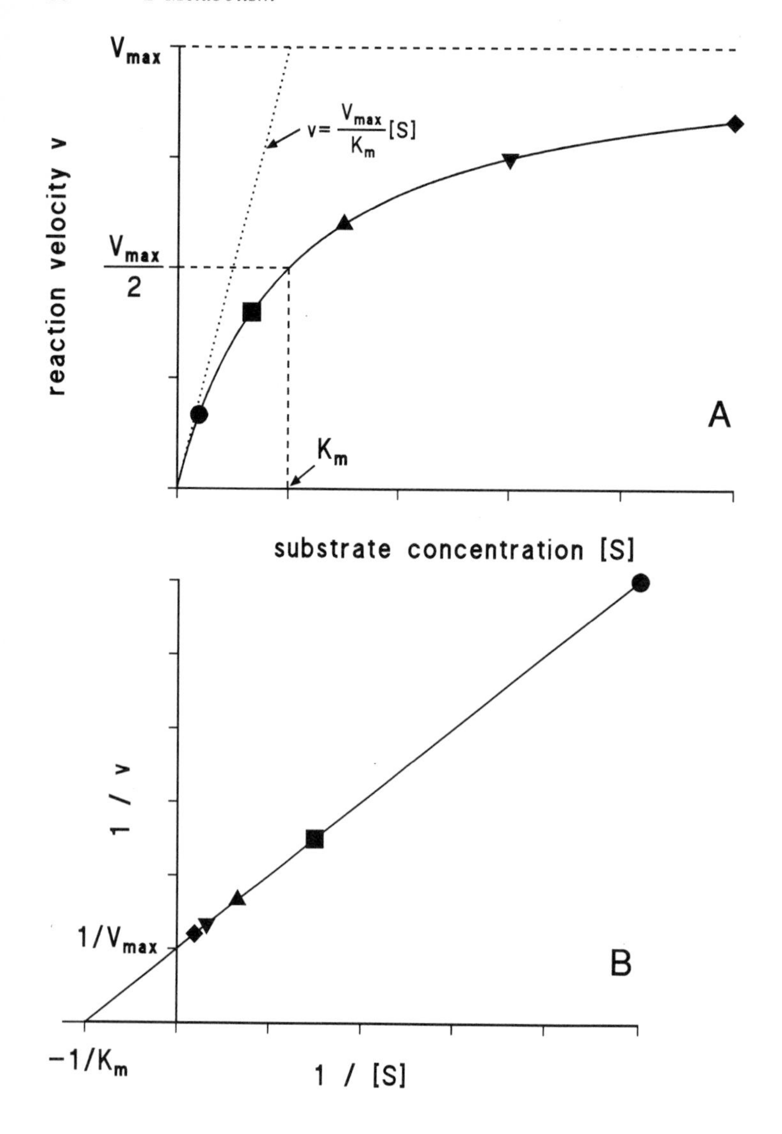

Fig. 10 A and B

$$E + S \underset{k_{-1}}{\overset{k_1}{\rightleftharpoons}} ES \underset{k_{-2}}{\overset{k_2}{\rightleftharpoons}} E + P \qquad (25)$$

Under conditions where the enzyme is in a low, fixed concentration in comparison to the substrate, a steady state is assumed, i.e., formation of ES is balanced by the rate of product formation. If additionally we focus on the initial rate, the back reaction of product and enzyme (k_{-2}) may be ignored. We obtain the Michaelis-Menten equation for the initial rate, the basic equation of enzyme kinetics.

$$v = \frac{V_{max}[S]}{K_m + [S]} = \frac{k_{cat}[E_0][S]}{K_m + [S]} \qquad (26)$$

The initial rate (v) is proportional to the total enzyme concentration [E_0], whereas v follows saturation kinetics with respect to the concentration of substrate [S] (Fig. 10A).

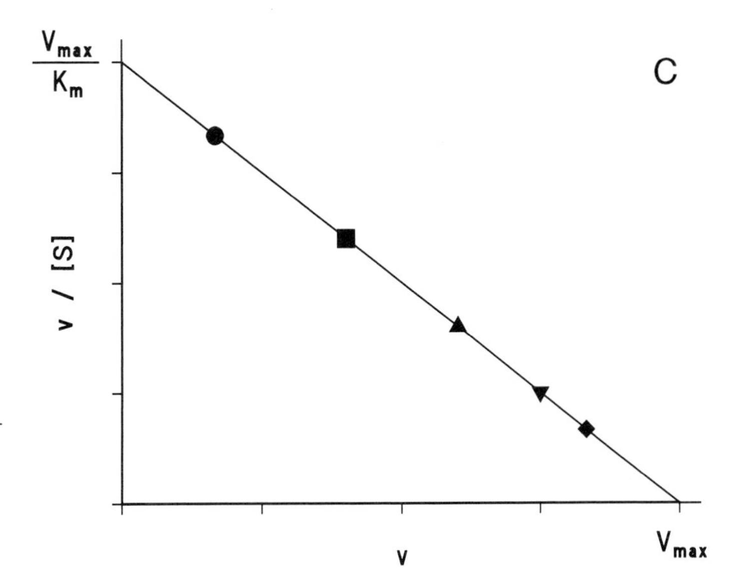

Fig. 10. Graphical analysis of kinetic data. A, Direct Michaelis-Menten plot. Reciprocal plots according to Lineweaver-Burk (B) and Eadie-Hofstee (C).

The significance of the basic parameters in the Michaelis-Menten equation can be described as follows. The constant k_{cat} is in fact identical to k_2 in Eq. (25), it is simply the first order rate constant of conversion of ES to EP. k_{cat} furthermore is called turnover number of the enzyme which correlates to the fact that under conditions when all enzyme is present as ES complex (full saturation, see Fig. 10A), v reaches V_{max}

$$V_{max} = k_{cat}[E_0] \qquad (27)$$

The Michaelis constant $K_m = (k_{-1} + k_2)/k_1$ (Eq. 25) represents the level of S giving $v = V_{max}/2$, and is thus related to the affinity of the enzyme towards its substrate. K_m is in general not identical with the dissociation constant of the enzyme substrate complex but measures the apparent affinity under steady state conditions. Only in cases when $k_{-1} \gg k_2$, i.e., when the breakdown of ES back to E and S is much faster than product formation (equilibrium conditions), K_m becomes k_{-1}/k_1 and is a direct measure of the affinity of the substrate for its binding site at the enzyme. At low [S], where $[S] \ll K_m$, Eq. (26) simplifies to

$$v = \frac{V_{max}[S]}{K_m} = \frac{k_{cat}}{K_m}[E_0][S] \qquad (28)$$

which represents the linear initial part of the Michaelis-Menten curve (Fig. 10A). The parameter k_{cat}/K_m is important in enzyme kinetics since substrate specificity is in fact not directly related to K_m alone, but is determined by the ratio of k_{cat}/K_m for the respective substrates.

Since saturation cannot be fully attained, neither V_{max} nor K_m can be directly deduced from Fig. 10A. This is the reason for transforming Eq. (26) into linear form. There are several possibilities for this procedure, the most widely used are the Lineweaver-Burk plot (which is not the best in terms of distribution of experimental points, Fig. 10B) and the Eadie-Hofstee plot (Fig. 10C). Nowadays it is more convenient to use computer assisted methods for fitting procedures. The main advantage is that the primary plots (like Fig. 10A) are used to obtain the kinetic parameters.

Enzymes frequently catalyze the reaction of two or more different molecules, actually multi-substrate reactions are very common enzyme mechanisms. They fall into two major classes, (1) "sequential" mechanisms, in which all reactants combine with the enzyme before reaction takes place and any products are released, and (2) "ping-pong" mechanisms, in which one or more products are released before all substrates are bound. Typical examples are transaminases for the ping-pong mechanism and NAD(P)-dependent dehydrogenases for (ordered) sequential mechanisms. These types of reactions can also be graphically analyzed by appropriate reciprocal plots.

Another important feature of a particular enzyme, both for understanding its action and for practical use, is the interaction with specific inhibitors. Inhibition may either be reversible or irreversible, the latter being brought about, for instance, by covalent binding of an inhibitor. The analysis of reversible inhibition is an important kinetic tool. Inhibitors which are structurally similar to the respective substrate often cause competitive inhibition, due to competition of the inhibitor with the substrate at the active site of the (free) enzyme. Competitive inhibition is characterized by an unchanged V_{max} and an apparently altered Michaelis constant K'_m which now includes also the terms of the inhibitor concentration [I] and the inhibition constant K_i (Fig. 11). On the other hand, an inhibitor which reacts both with the free and the substrate-loaded enzyme causes non-competitive inhibition. This type of inhibition is kinetically characterized by an unchanged K_m and an apparent decrease of V_{max} to V'_{max}, which in this case includes the terms of [I] and K_i. In contrast to the competetive type, in this kind of inhibition the inhibitor does not compete with the substrate for the same site, but, nevertheless, prevents the formation of the product.

3.4 Coenzymes, Prosthetic Groups, Cofactors

Many enzyme-catalyzed reactions cannot be carried out solely by the use of functional amino acid side chain groups. For particular reactions, especially those which cannot be brought about by acid-base catalysis, special cofactors are used by the cell which are called coenzymes, or prosthetic groups if tightly bound to the enzyme. In general, coenzymes are specifically designed to handle particular groups, as well as hydrogen atoms or electrons, in order to increase the diversity of enzyme reactions. If a coenzyme is removed from the respective enzyme, the remaining protein component is called the apoenzyme. Due to the recent discovery of "new" coenzymes in methanogenic bacteria, the list of coenzymes is especially long in bacteria (Tab. 2). Nucleotide triphosphates such as ATP or GTP are not included, since they are more properly described as substrates of a reaction.

Most microorganisms are able to synthesize all these compounds, however, certain others

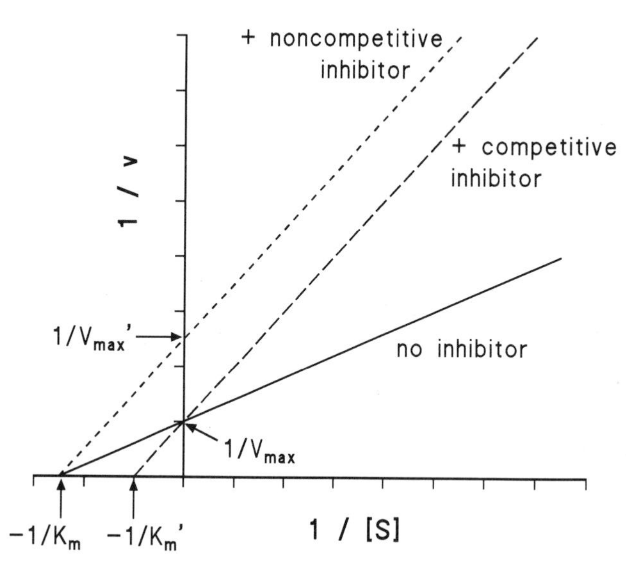

Fig. 11. Graphical analysis of enzyme inhibition. Reciprocal plot of competitive and non-competitive inhibition.

Tab. 2. Coenzymes and Prosthetic Groups: Their Function and Distribution

Coenzyme	Typical reaction	Distribution
NAD, NADP	Hydrogen, e^--transfer	Ubiquitous
FAD	Hydrogen, e^--transfer	Ubiquitous
FMN	Hydrogen, e^--transfer	Ubiquitous
Ubiquinone (benzoquinones)	Hydrogen, e^--transfer	Eukaryotes, some bacteria
Menaquinone (naphthoquinones)	Hydrogen, e^--transfer	Bacteria
PQQ	Hydrogen, e^--transfer	Ubiquitous
F_{420}	Hydrogen, e^--transfer	Methanogenic bacteria
F_{430}	Hydrogen, e^--transfer	Methanogenic bacteria
Heme (in cytochromes)	e^--Transfer	Ubiquitous
Thiamindiphosphate (TPP)	Transfer of active aldehyde	Ubiquitous
Pyridoxalphosphate	Various reactions	Ubiquitous
Coenzyme A (phosphopantetheine)	Acyl group transfer	Ubiquitous
Biotin	Carboxyl transfer	Ubiquitous
Tetrahydrofolic acid (and derivatives)	Formyl-, hydroxymethyl- and methylene transfer	Ubiquitous
Lipoic acid	Oxidative decarboxylation of α-keto-acids	Ubiquitous
Vitamin-B_{12}-coenzymes	Intramolecular shift of C_1-group	Ubiquitous
Coenzyme M	Methyl transfer	Methanogenic bacteria
Methanopterin	Like vitamin-B_{12}-coenzymes	Methanogenic bacteria
Methanofuran	Carboxyl-, formyl transfer	Methanogenic bacteria
HTP	Methyl transfer	Methanogenic bacteria

depend on the presence of coenzymes in the environment. Thus, these compounds are vitamins or growth factors for the respective microorganisms. Well known examples of bacteria with complex vitamin requirements are Streptococci and Lactobacilli.

The pyridine nucleotide coenzymes (nicotinamide adenine dinucleotide, NAD^+, and nicotinamide adenine dinucleotide phosphate, $NADP^+$) function in most reactions as co-substrates rather than as true coenzymes. As carriers of reducing equivalents, they accept two electrons and a proton (equivalent to a hydride ion). The two kinds of pyridine nucleotides are involved in different reactions within different redox systems, one independent from the other. In general, NADH is used in catabolic processes to remove hydrogen from substrates, whereas NADPH serves as reductant in biosynthetic processes. Consequently, the ratios $[NADH]/[NAD^+]$ and $[NADPH]/[NADP^+]$ differ markedly, reflecting a different redox potential of the two systems, that of the NADP-couple being always more negative (reduced).

The flavin coenzymes flavine adenine dinucleotide (FAD) and riboflavine 5′-phosphate (FMN) catalyze a broad variety of redox reactions. Their function differs from that of pyridine nucleotides since formation of a semiquinone (a radical) is involved. Flavines can thus be reduced either by one- or two-electron processes, furthermore they can react directly with O_2 (autoxidation), a reaction which is important for some oxidases and hydroxylases. In 1978 another hydrogen carrier, equivalent to flavines, was discovered in bacteria, namely pyrrolo-quinoline quinone (PQQ, Fig. 12). Most (if not all) PQQ-containing dehydrogenases are located in the periplasm of Gram-negative bacteria. They are involved in the first oxidation step of alcohols (methanol oxidase), amines, and sugars (glucose oxidase). PQQ is also found in some eukaryotic enzymes, e.g., in amine oxidases (dopamine β-hydroxylase).

Whereas higher organisms use only ubiquinone as quinone hydrogen carrier, bacteria contain both naphthoquinones (menaquinone) and benzoquinones (ubiquinone). Naphthoquinones are present in most Gram-positive and anaerobic Gram-negative organisms. Facultative bacteria, like *Escherichia coli* use both quinone types, depending on the metabolic state.

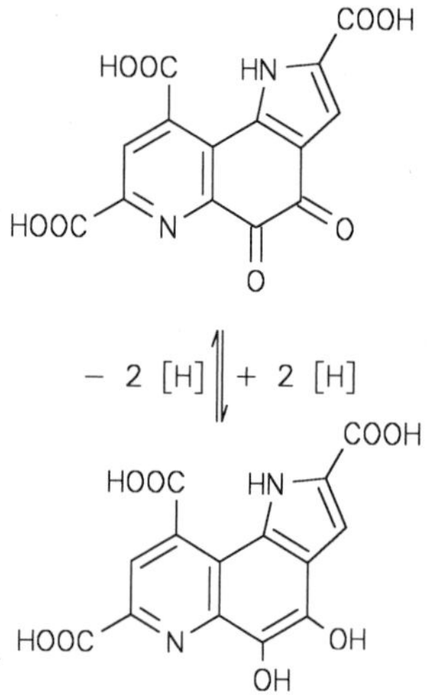

Fig. 12. Structure of pyrrolo-quinoline quinone (PQQ).

Thiamine diphosphate (TPP) typically serves as coenzyme for non-oxidative and oxidative decarboxylation of α-keto acids (example: pyruvate dehydrogenase), as well as for α-ketol formation (transketolase). Pyridoxalphosphate is involved in various amino acid transformation reactions. The primary reaction always is the formation of a Schiff base between amino acid and the carbonyl group of pyridoxal phosphate, then leading to amino group transfer (transaminases), decarboxylation (amino acid decarboxylases), racemization or various eliminiation reactions. Phosphopantetheine groups are carriers for acyl groups both in catabolism (acyl-CoA, acetyl-CoA) and in biosynthesis of fatty acids (acyl carrier protein of the fatty acid synthase). Biotin is covalently bound to lysine residues of the respective enzymes and serves in its active form (carboxybiotin) as a carboxyl group donor in β-carboxylation reactions. The carboxyl group is transferred to biotin either in an ATP-dependent reaction (carboxylases, e.g., pyruvate carboxlyase) or from another carboxylated compound (transcarboxylase). Tetrahydrofolic acid is a member of the widespread family of pteridine compounds. It catalyzes the transfer of various one-carbon fragments, and, according to these functions, occurs in different forms. These forms serve as donors for C1-groups at the oxidation level of formate, formaldehyde and methanol, thus transferring formyl-, hydroxymethyl-, and methyl groups to various substrates.

In recent years a series of coenzymes with surprising chemical structures have been isolated from methanogenic bacteria (cf. Chapter 7 of this volume). They include hydrogen carriers like F420, a deazariboflavine derivative, F430, a Ni^{2+}-tetrapyrrole compound, and the electron donor mercaptoheptanoylthreonine phosphate (HS-HTP), as well as coenzymes for group transfer, such as the methyl-group carrier dithiodiethanesulfonic acid (coenzyme M), the formyl-group carrier methanofuran, and methanopterin, another member of the pteridine family.

4 Membrane Transport

With a few exceptions, such as O_2, CO_2, NH_3, water, and ethanol, molecules that enter or leave the cell, cross the cytoplasmic membrane not by diffusion but via specific transport systems. Diffusion means movement of a solute without direct interaction with membrane proteins, the rate of transfer being directly proportional to the concentration of the solute. Carrier transport involves participation of a proteinaceous component of the membrane, the carrier protein, and shows many properties well known from enzyme kinetics.

All cells need to take up certain compounds from the surroundings in order to survive and to grow. Maintaining the integrity of the cell also requires continuous removal of particular metabolites or ions. Furthermore, excretion of certain metabolites, especially in biotechnology, is of considerable commercial interest. Thus, in order to completely understand these processes, besides the knowledge about the intracellular metabolism, detailed information concerning both uptake of substrates and secretion of products is also essential. The following section describes the mechanisms by which solutes can cross the membrane and the fundamental kinetic and energetic properties of these processes.

4.1 The Carrier Concept

Understanding of transport mediated by carriers was originally based on the imagination of a "ferryboat" moving or rotating as an entity across the membrane. Although this idea is clearly incorrect, the basic concept is still of great validity. In the classical meaning, a "carrier" is a mobile element that, at a given moment, exposes the substrate binding site(s) either to one side of the membrane or to the other, but not to both. From this conception, only the second half, the exclusive exposure of binding sites, is still valid in recent understanding. The alternative concept, that of the "channel" describes a fixed structure in the membrane, classically arranged in form of a "water-filled pore", whose binding sites are accessible from both sides of the membrane at the same time. Fig. 13 exemplifies some hypotheses explaining transport by carriers and channels. The modern concepts now include elements of both ideas. Carriers, even parts of them, are in general not supposed to move, but they may contain channels as intrinsic parts. Obviously, if a channel function is involved, it has to be "closed" by gates, otherwise the free flux of molecules would collapse all gradients across the membrane. Nevertheless, it is clear that examples of the "standard" cases do in fact exist, the antibiotic valinomycin as a classical "carrier" and the antibiotic gramicidin, as well as a variety of ion channels for the "channel" concept.

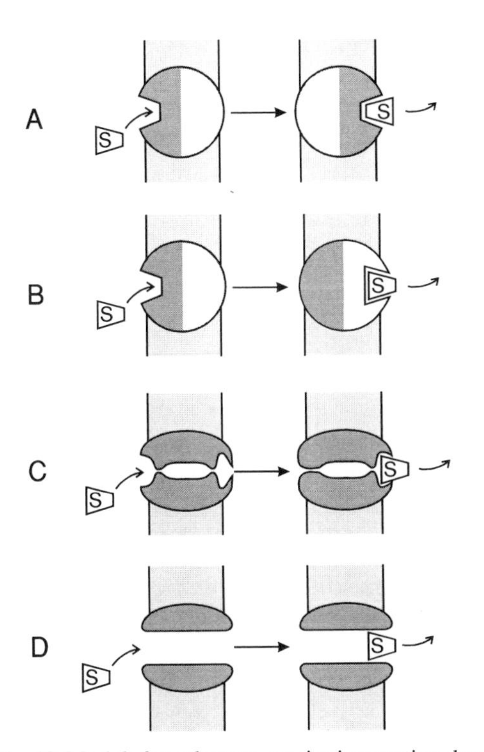

Fig. 13. Models for substrate-carrier interaction during transport catalysis. A, "Mobile carrier", the shaded areas indicate rotation of the carrier molecule; B, fixed carrier with "mobile domains", the substrate binding site is assumed to move from one side to the other; C, carrier concept in recent understanding, "fixed" carrier with alternatively accessible binding sites and an intrinsic channel; D, "channel" concept.

These considerations, however, obscure the fact that, in contrast to the knowledge concerning mechanistical aspects (Sect. 2), as well as kinetic descriptions (Sect. 3), even of extensively studied carriers like the lactose permease in *E. coli*, we have no clear idea whatsoever about the true (molecular) mechanism of any single carrier. This is mainly due to the lack of a three-dimensional structure. But even this structure would not *per se* give us an answer, since, for instance, the mechanism has to be also elucidated by which two ion fluxes are coupled in secondary carriers.

4.2 Mechanisms and Energetics of Solute Transport

According to the usual classification, transport processes are divided into (1) simple diffusion, (2) facilitated diffusion, i.e., carrier mediated, passive transport, and (3) active transport, i.e., movement of solutes against a concentration gradient. This type of classification has serious disadvantages. Substrate symport or antiport, for example, leading to accumulation of a certain solute, thus being "secondary active", is energetically a passive process. Alternatively, transport can be classified according to the energetic properties. Besides category (1) which is again simple diffusion, there are (2) primary transport processes in which a chemical reaction is directly linked to the vectorial solute movement, and (3) secondary transport processes, in which the carrier couples solute movement to the movement of other substances, in any case without direct chemical interaction. Secondary transport systems are further subdivided into uniport, symport, and antiport. Uniport carriers only mediate electrochemical equilibration of a solute, whereas symporters and antiporters, by coupling to ion fluxes in the same or in opposite direction, can catalyze both "downhill" and "uphill" movements of substrates. The classification of carrier mechanisms is exemplified in Tab. 3.

The spectrum of *primary transport systems* is especially diverse in bacteria. The general motif, conversion of chemical or light energy into electrochemical energy, has already been discussed in detail (Sect. 2.4). The first category of primary transport systems comprises the proton translocation mechanisms of respiration and photosynthesis (see Fig. 7). Na^+-ions instead of protons as coupling ions are handled by these systems in other bacteria such as *Vibrio alginolyticus*. A possible primordial case of this type of transport systems is H^+ and Na^+ translocation in halobacteria, in which proton translocation is directly coupled to light absorption (see Sect. 2.4). A second, very heterogeneous class includes primary transport systems coupled to hydrolysis of ATP. These systems are known as ATPases which, however, is an incorrect expression, since the aim of these enzymes is in general either transport or ATP synthesis, but not ATP hydrolysis. They are mainly involved in translocation of small monovalent (H^+, K^+, Na^+) or divalent cations (Ca^{2+}, Mg^{2+}). Like primary transport systems in general, these systems convert free energy of ATP into the electrochemical gradient of a solute. The ATPase systems of respiration and photosynthesis are well known examples (see Fig. 7). ATPases are subdivided into at least three different classes.

(1) F-type ATPases (e.g., F_1F_0-ATPase) are multicomponent systems involved in primary energy conversion, such as respiration and photosynthesis-coupled movement of H^+ (and Na^+) in eubacteria, mitochondria, and chloroplasts
(2) P-type ATPases (e.g., eukaryotic Na^+/K^+-ATPase) are commonly found in the plasma membrane and consist of only one or two subunits. They utilize a phosphorylated high-energy intermediate, and are in general involved in translocation of Na^+, K^+, and Ca^{2+}.
(3) V-type ATPases are present in the vacuolar membrane of eukaryotic cells and are perhaps more closely related to ATPases from archaebacteria.

Tab. 3. Classification of Transport Systems

Class	Subclass	Typical Examples in	
		Prokaryotes	Eukaryotes
Channel		Outer membrane porins (*E. coli*)	Gap junctions
Secondary transport	Uniport	Glucose facilitator (*Zymomonas mobilis*)	Glucose facilitator (erythrocytes)
	Symport	Na^+-coupled proline uptake	Na^+-coupled glucose uptake (intestine)
	Antiport	Sugar-P_i/P_i exchange (*E. coli*)	ADP/ATP-exchange (mitochondria)
Primary transport	Redox-coupled pumps	Respiratory chain (plasma membrane)	Respiratory chain (mitochondria)
	ATP-coupled pumps (ATPases)	H^+-ATPase (*E. coli*)	Na^+/K^+-ATPase (plasma membrane)
	ABC-family	Binding protein dependent systems	Multidrug resistance protein
Group translocation		Phosphotransferase system (sugar uptake)	—

In cells depending on oxidative or photosynthetic phosphorylation for energy generation, in general the F-type ATPase produces ATP and P-type and V-type ATPases consume it. A third class of primary transport systems, also directly coupled to ATP-hydrolysis, are the binding protein dependent transport systems in bacteria (Fig.14). These systems are found in Gram-negative bacteria and comprise a complex ensemble of pores (in the outer membrane), substrate binding proteins (in the bacterial periplasm), components in the plasma membrane for solute translocation, and finally subunits catalyzing the coupling to ATP. Very diverse substrates are transported by these systems: amino acids, peptides, mono- and disaccharides, nucleotides, coenzymes, inorganic ions like sulfate or phosphate, and others.

Another class of transport systems, not mentioned so far, are closely related to primary transport systems and catalyze uptake by a process named group translocation. This means direct coupling of an enzymatic transformation (phosphorylation) of the solute with its translocation across the membrane. The transported solute at the inside thus differs chemically from that at the outside. The only well established systems of this kind are the phosphoenolpyruvate-dependent sugar transport systems (PTS) in eubacteria (Fig. 15). Substrates accepted by PTS-systems include hexoses, sugar alcohols, and β-glycosides. There are recent indications, however, that these systems are in fact not a vectorially oriented metabolic pathway, but fit into the general scheme of the carrier concept. This means that they can be described as a combination of a channel with a phosphorylating enzyme; substrate phosphorylation is used to drive the dissociation of the substrate at the inside.

Secondary transport systems are relatively small single-subunit carrier proteins, showing a common motif of about 12 transmembrane α-helices and a molecular mass of about 45–50 kDa. In principle, they catalyze only facilitated diffusion, but, due to their intrinsic construction, they may couple the flux of one ion to that of another. Thus, all three types, as mentioned above, i.e., uniport, symport and antiport, are observed in this category (Fig. 16). In symport or in antiport, the carrier equilibrates the driving force of the so-called coupling ion (fre-

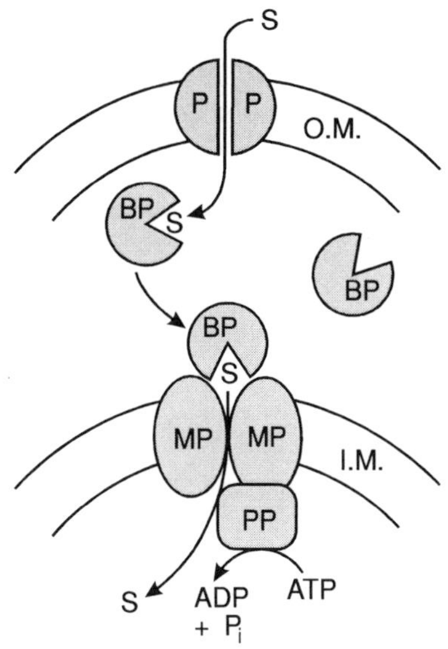

Fig. 14. Binding-protein dependent transport system. The substrate (S) passes the outer membrane (O.M.) through a porin channel (P), binds to the binding protein (BP) which transfers the substrate to the transport components in the inner membrane (I.M.). These comprise membrane proteins (MP) and a peripheral protein (PP) which couples the free energy of ATP to solute translocation.

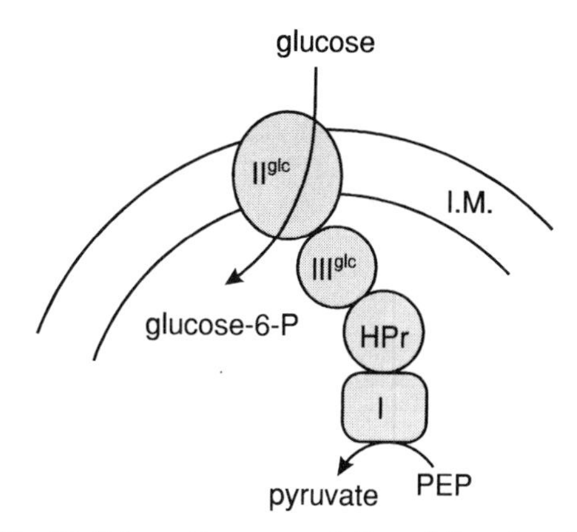

Fig. 15. Schematic representation of the phosphotransferase system (PTS) for glucose transport. A phosphoryl moiety from phosphoenolpyruvate (PEP) is sequentially transferred to the soluble proteins enzyme I (I) and HPr, to enzyme III (III^{glc}) and finally via the integral membrane protein (enzyme II^{glc}) which is specific for glucose (II^{glc}) to the transport substrate glucose. In other PTS-systems the different subunits may be fused in various combinations.

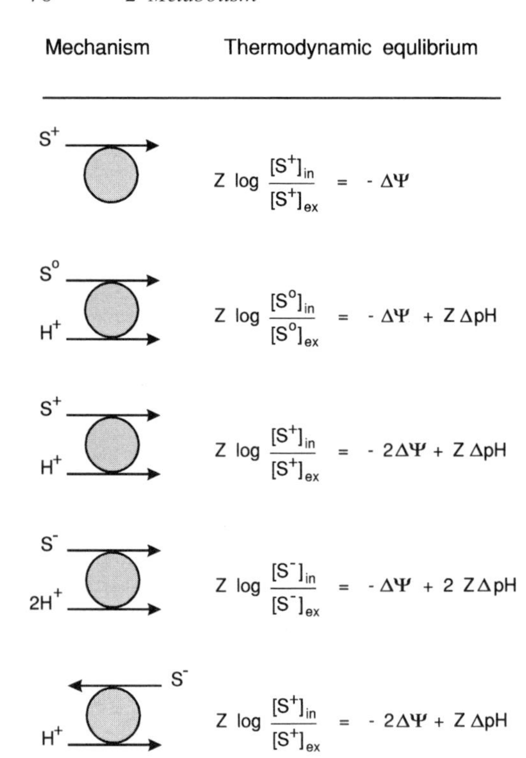

Fig. 16. Examples for proton coupled secondary systems. In thermodynamic equilibrium, the electrochemical potential of the substrate ($Z \log([S]_{in}/[S]_{ex})$, $Z=2.3\ RT/F$) equals the proton-motive force (see Eqs. 2–15 and 2–20). In the figure examples for uniport (top), symport (second to fourth example) and antiport (bottom) are given.

quently H^+- or Na^+-ions) with the driving force of the solute to be translocated. In equilibrium and under the assumption of coupled systems, the free energies (cf. Eq. 15) of the two transported species are equal

$$\Delta G_{\text{coupling ion}} = \Delta G_{\text{solute}} \tag{29}$$

The equilibrium concentration gradient of the solute $[S]_{in}/[S]_{ex}$ will thus depend on the electrochemical gradient of the coupling ion. There are numerous possibilities of secondary coupling, some examples are shown in Fig. 16. Solute symport systems are wide-spread both in bacteria and in eukaryotes. Most amino acids, for example, are taken up in symport with protons or Na^+-ions. Also antiport systems are frequently found. Well known examples are precursor/product antiport systems in bacteria which mediate the coupled exchange of a substrate with the corresponding product of the respective metabolic pathway (e.g., malate and lactate in malolactic fermentation), as well as the large family of exchange carriers for nucleotides, inorganic ions, carboxylates, and amino acids in mitochondria. Uniport systems are common in eukaryotes, e.g., glucose transport in various cells, but rare in bacteria. The only well known systems are the glucose carrier of *Zymomonas mobilis* and presumably the glycerol transport system in *E. coli.* It should be mentioned here, that the "porin" proteins in the outer membrane of Gram-negative bacteria, which function as "size-exclusion pores", do not fit very well into these categories. Porins can be classified into the "channel" type of proteins.

Obviously, so far only uptake transport systems have been discussed. This does not mean that metabolite secretion is unimportant. However, relatively little information is available on efflux systems as compared to uptake. In principle, the same mechanisms, as discussed for solute uptake, may also work in secretion. Hydrophobic metabolic end products, e.g., alcohols (ethanol, butanol), acetone, and some organic acids (in undissociated form) may leave the cell by simple diffusion. Although passive diffusion has been proposed for efflux of many other metabolic products, including amino acids such as glutamate or lysine, evidence is now increasing that most of these compounds are in fact secreted by carrier-mediated processes.

There exist some well known examples for secretion systems functioning in the ion/solute symport mode. In *E. coli* and *Lactococcus lactis* lactate produced in glucose fermentation is excreted in symport with protons, leading to generation of a proton diffusion potential. Other secondary systems for end product secretion function as antiport systems. Some antibiotics, e.g., tetracycline, are secreted in exchange against protons. This leads to resistance against the particular inhibitor. Also precursor/product antiport systems, as mentioned above, are examples of this type of secretion. It should be mentioned that the number of secretion systems which are found to be directly coupled to ATP is increasing. Extrusion of toxic heavy metal ions, for example, is frequently mediated by so called "export ATPases" which are related to the P-type ATPases (see above). Also the family of carrier systems found both in eukaryotes (MDR = multidrug resistance protein) and prokaryotes which show sequence similarity to the energy-coupling subunit of the binding protein-dependent systems (see above) is steadily growing. Because of the common sequence motif of ATP-binding sites, this class of transport ATPases is called the ABC-family (ATP-binding cassette).

The focus in this section clearly lies on transport in bacteria. Transport in eukaryotic cells (and their organelles) is not fundamentally different. Some of the mechanisms discussed above have not been found in eukaryotes, e.g., group translocation or binding protein dependent systems. Primary systems, except the ABC-family, are confined to the transport of inorganic ions. Importance and location of ATPases are different, F_1F_0-ATPase occurs in organelles (of presumably prokaryotic origin), whereas the plasma membrane is energized by the Na^+/K^+-ATPase. Ion fluxes, especially that of Ca^{2+}, are in general more important in eukaryotes than in prokaryotes.

Finally, we may ask what the reason is for the large diversity of systems and why complicated multi-subunit primary carrier devices like binding protein dependent systems or phosphotransferase systems have been developed since alternatively simple single-subunit secondary transport is available. The answer is, at least for bacterial cells, that the different systems are used for different purposes, depending on the availability of a given substrate. Based on their molecular construction, primary transport systems have a high substrate affinity, and they are intrinsically unidirectional, thus making possible very high accumulation ratios. By contrast, secondary systems are in general fully reversible, thus they may lead to leakage of the transport substrate in situations of low substrate concentration or low energy.

4.3 Transport Kinetics

It is not always that simple to uneqivocally prove the involvement of a carrier in a membrane transport process. The main arguments which are commonly taken as indication for the action of a carrier system are mainly of kinetic type. These are, similar to enzymes, (1) saturation kinetics, (2) substrate specificity, (3) inhibition by specific agents, and (4) the particular kinetic observation of counterflow. These properties will be described in the following.

The net flux J of a solute (S) from one compartment (I) to the other (II) by simple diffusion is described by

$$J = PA\,([S]_{I} - [S]_{II}) \tag{30}$$

where A is the area of the membrane and P is the permeability constant with respect to the substrate S. Net diffusion occurs only from the compartment with higher concentration to that of lower concentration and is in general not specific.

For characterizing carrier processes, it is convenient to start with the example of "facilitated diffusion", since it is well suited to describe the basic discrimination between carrier and diffusion. Facilitated diffusion is a passive process, however, since the flux is carrier-mediated, it is in general not simply proportional to the concentration difference of the solute. Many transport systems, in fact not only those of facilitated diffusion, can be described by the well known Michaelis-Menten kinetics

$$J = \frac{J_{\max}\,[S]}{K_m + [S]} \tag{31}$$

K_m is used in analogy to enzyme kinetics; it means the solute concentration at which the transport rate J reaches half its maximum ($J_{\max}$), it is frequently called K_t (for transport).

The approach to carrier kinetics by using exactly the same formalism as established in enzyme kinetics is very useful. The basis for this is the idea that a carrier can be regarded as an enzyme, the reaction of which is not a change of the structure of the substrate but of its location. For kinetic analysis, the same equations, as discussed in Sect 3.3 can be used here, with some precautions. Eq. (31) holds true for the

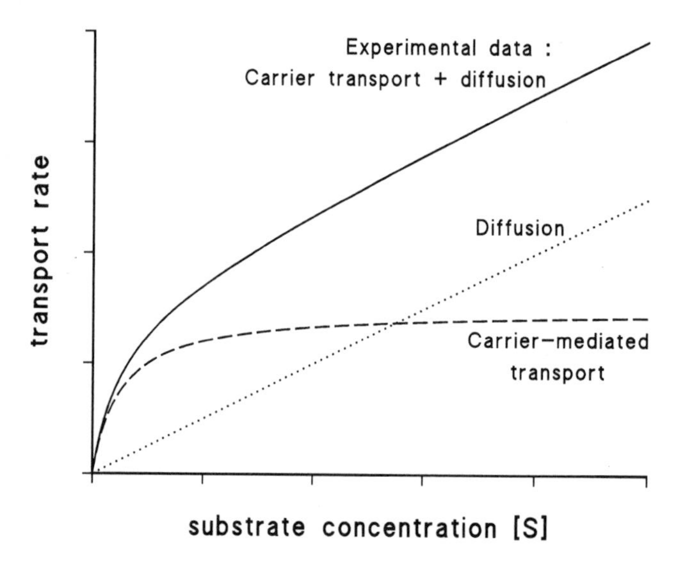

Fig. 17. Dependence of transport rate on substrate concentration in the presence of both a diffusion component and carrier-mediated transport.

dependence of the transport rate on the concentration of the respective substrate in one direction. Net transfer of a solute, when present on both sides of the membrane, by a reversible carrier (i.e., most secondary carriers) must take into account solute translocation in both directions. Graphical analysis of the experimental data can also be performed in the same way as described in Sect. 3.3, this includes analysis of inhibition kinetics.

Two facts have to be considered when using these procedures. (1) Precaution is necessary in interpreting Michaelis-Menten kinetics, since the presence of multiple transport systems may be overlooked in this type of analysis. (2) Depending on the type of solute transported, the presence of a diffusion component, especially when analyzing secretion systems, should always be taken into account (Fig. 17).

A more detailed kinetic description of carrier processes is possible, when single steps within the catalytic cycle of a transport protein are known. This type of interpreting kinetic data is traditionally depicted in schemes as shown in Fig. 18. This kind of models is very useful for understanding carrier mechanisms and, although they are based on the "carrier concept" of transport (see above), the "movement" of the carrier (C) in these schemes is only a formal one and means the conformational changes during transport catalysis. In detailed kinetic analysis a whole set of different binding and

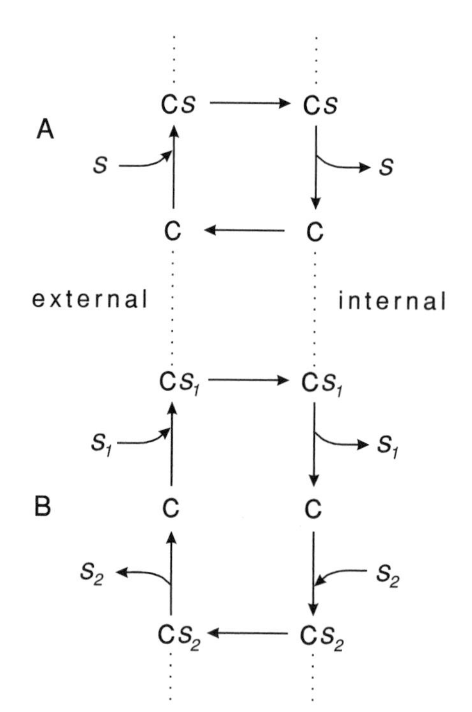

Fig. 18. Kinetic schemes of carrier function for a simple uniport carrier (A) and a coupled antiport carrier (B). Note that translocation of the unloaded carrier species C in (B) is forbidden.

dissociation steps, as well as translocation steps within the catalytic cycle have to be considered and experimentally discriminated.

4.4 Transport of Macromolecules

Although transport is in general discussed with respect to entry and exit of small molecules, it is clear that transport of macromolecules is also essential for the cell. The mechanisms are extremely diverse and include translocation of proteins, complex lipids, as well as nucleic acids (e.g., in transformation). Furthermore, vesicular transport by endocytosis and exocytosis mainly in eukaryotic cells has to be mentioned.

The only example which will be discussed here is the mechanism of protein translocation. In this case we also have at least three different systems, (1) protein secretion into the lumen of the endoplasmic reticulum of eukaryotic cells, (2) protein import into eukaryotic organelles such as mitochondria and chloroplasts, and (3) protein export from bacteria, the most important example in biotechnological terms. Fig. 19 outlines the sequence of events which are necessary for successful transfer of an intracellularly synthesized polypeptide to be transported out of the bacterial cell in a functionally active state. Polypeptides are synthesized with additional sequences in general located at the N-terminal end ("leader sequence") which are essential recognition signals for delivery to the membrane and correct handling by the export machinery. The presence of cytosolic protein factors ("chaperones") guarantees that the polypeptides maintain a loosely-folded, transport-competent conformation. The secretory protein is then recognized by a complex machinery in the plasma membrane, consisting of intrinsic membrane components (channel?) and energy coupling units (ATPase). At the exterior side, the leader sequence is split off by a specific protease ("leader peptidase") and the protein folds into its native conformation, sometimes by help of additional cofactors. In Gram-negative bacteria a second transport step is necessary for crossing the outer membrane present in these organisms.

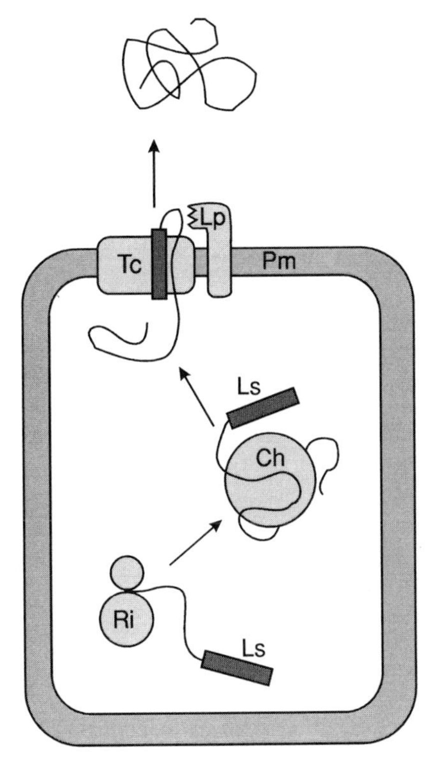

Fig. 19. Scheme of events during synthesis and excretion of a secretory protein in *E. coli*. Ri, ribosome; Ls, leader sequence; Ch, chaperone; Tc, translocation complex (membrane protein components); Pm, plasma membrane; Lp, leader peptidase.

5 Metabolic Pathways

Microorganisms, through action of a plentitude of metabolic pathways, can supply energy for cell maintenance and growth (energy metabolism) and derive the constituents of their cells from a variety of sources. The major cell constituents (proteins, nucleic acids, carbohydrates, lipids) consist of the chemical elements carbon (C), nitrogen (N), oxygen (O), hydrogen (H), sulfur (S), and phosphorus (P). Other elements (trace elements) are needed to maintain enzyme activities and cell homeostasis (potassium, magnesium, sodium, selenium, iron, zinc, and other metal ions).

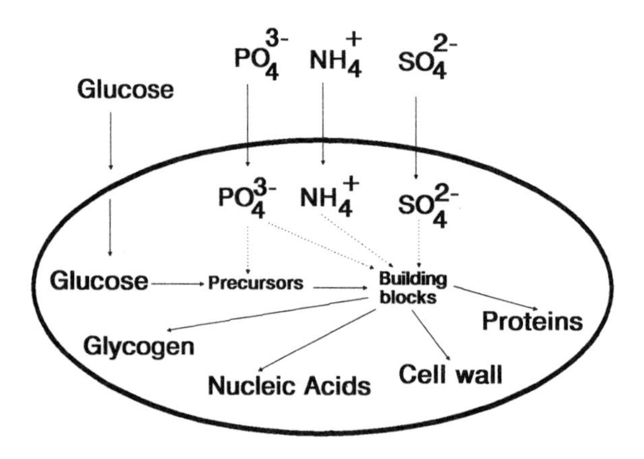

Fig. 20. Scheme of a bacterial cell growing on glucose as sole carbon source. Glucose is transformed to precursors, therefrom building blocks are formed which serve to build up cell constituents. Inorganic compounds (PO_4^{3-}, NH_4^+, and SO_4^{2-}) are assimilated and used for the formation of building blocks.

Microorganisms take up these elements from the surrounding environment (medium) and incorporate them (Fig. 20) into their cell material (assimilation). This can happen either from simple compounds (salts, CO_2, N_2) or from complex media (rich in amino acids, nucleotides, fats, vitamins, salts or sugars). Cells which can derive all their energy and carbon requirements from simple inorganic compounds (including CO_2 or other inorganic carbon) are termed chemoautotrophic. When the energy stems solely from sunlight, organisms are phototrophic. If organic compounds are used for energy and carbon metabolism, cells are called heterotrophic. If additional organic compounds, which they cannot synthesize on their own (vitamins, amino acids, nucleotides, and others), are needed in the growth medium, cells are termed auxotrophic.

5.1 Utilization of Carbon Sources

The potential to utilize carbon sources for metabolism varies enormously among micro-

Tab. 4. Carbon Sources and Their Utilization

Carbon Source	Class	Major Metabolic Pathways	Utilizing Organism(s)
Methanol (C_1)	C_1-compounds	PPP	Methylotrophic bacteria and yeasts
CO_2 (C_1)	C_1-compounds	Reductive PPP	CO_2-fixing organisms (green plants, photosynthetic bacteria)
Acetate (C_2)	Organic acids	Glyoxylate shunt	Bacteria
Lactate (C_3)	Organic acids	Gluconeogenesis,TCA cycle	Bacteria
Succinate (C_4)	Organic acids	TCA cycle, gluconeogenesis	Bacteria
Xylose (C_5)	Sugars	PPP, glycolysis	Bacteria, yeasts
Glucose (C_6)	Sugars	Glycolysis, PPP, KDPG	Pro- and eukaryotes
Sucrose (C_{12})	Disaccharides	Hydrolysis, glycolysis	Pro- and eukaryotes
Raffinose (C_{18})	Trisaccharides	Hydrolysis, glycolysis	Pro- and eukaryotes
Starch (C_n)	Polysaccharides	Hydrolysis, glycolysis	Pro- and eukaryotes
Cellulose (C_n)	Polysaccharides	Hydrolysis, glycolysis	Anaerobic bacteria
Amino acids	Amino acids	TCA cycle, gluconeogenesis	Pro- and eukaryotes
Proteins	Polypeptides	Proteolysis, TCA cycle	Pro- and eukaryotes
Fatty acids	Fatty acids	β-Oxidation, TCA cycle	Pro- and eukaryotes
Aromatic hydrocarbons		*ortho*- or *meta*-cleavage TCA cycle	Bacteria, yeasts

organisms. From so-called C_1-compounds (CO, CO_2, formaldehyde, methanol, methylamine) as smallest molecules up to large and complex macromolecules (glycogen, starch, proteins, cellulose, nucleic acids), nearly all naturally occurring organic compounds can be utilized as carbon and/or energy sources by specialized microbes. A brief survey is presented in Tab. 4 and Fig. 21. Once these complex macromolecules have been depolymerized to smaller subunits (oligomers or monomers), peripheral metabolic pathways are used to transform the C-sources into metabolites which can be further catabolized by central metabolic pathways (see below). Di- and oligosaccharides (sucrose, maltose, and others) can be either split externally by hydrolases to monomers (glucose, fructose) or are first taken up by transport systems and cleaved then. Oligopeptides and oligonucleotides can also be depolymerized extra- or intracellularly by peptidases and endo- or exonucleases, respectively. Fats as C-sources are first split by lipases into glycerol-P and the fatty acids moieties. Glycerol-P enters glycolysis at the triose phosphate step (see Fig. 24); fatty acids are decomposed to acetyl-CoA (and/or propionyl-CoA) and can enter the TCA cycle.

5.1.1 Metabolism of Carbohydrates

Carbohydrates are the major energy and carbon sources for microorganisms. They have a dual function as they supply precursor metabolites for biosynthetic purposes (anabolism) and can be used as energy sources (formation of AcCoA or ATP with a high group transfer potential) when broken down to simple compounds (catabolism) such as CO_2, H_2O, organic acids, or alcohols. Reactions which are involved both in anabolic and in catabolic pathways are termed amphibolic (TCA cycle, pentose phosphate cycle). Central metabolism provides the cells with a dozen of precursor metabolites, which are essential to all cells (Fig. 22). Therefrom, building blocks such as amino acids, nucleotides, vitamins, or fatty acids are derived for macromolecule biosyntheses (Fig. 23). The main pathways of carbohydrate metabolism are glycolysis (glucose breakdown), the pentose phosphate pathway (PPP), the tricarboxylic acid (TCA) cycle/glyoxylate cycle, and gluconeogenesis.

The breakdown of storage polysaccharides such as starch or glycogen and of oligosaccharides, in the peripheral catabolism (Sect. 5.1) in bacteria, leads to the formation of phosphorylated sugars such as Glc 6-P or Fru 6-P. Some

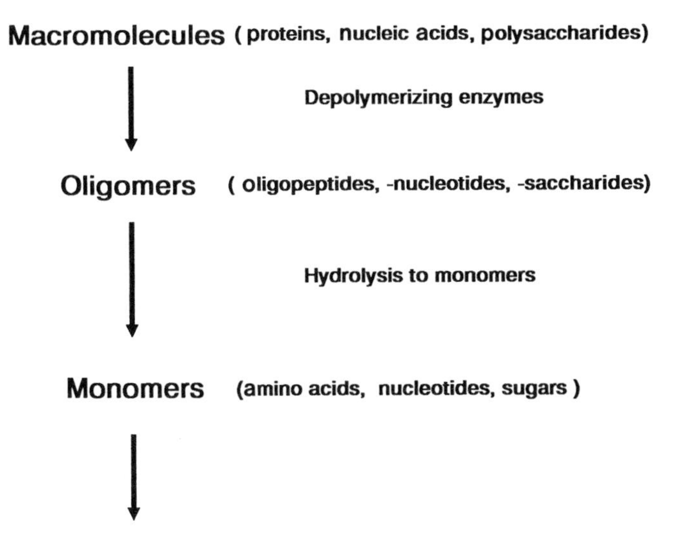

Fig. 21. Flow diagram of major biodegradation pathways leading to monomer carbon sources.

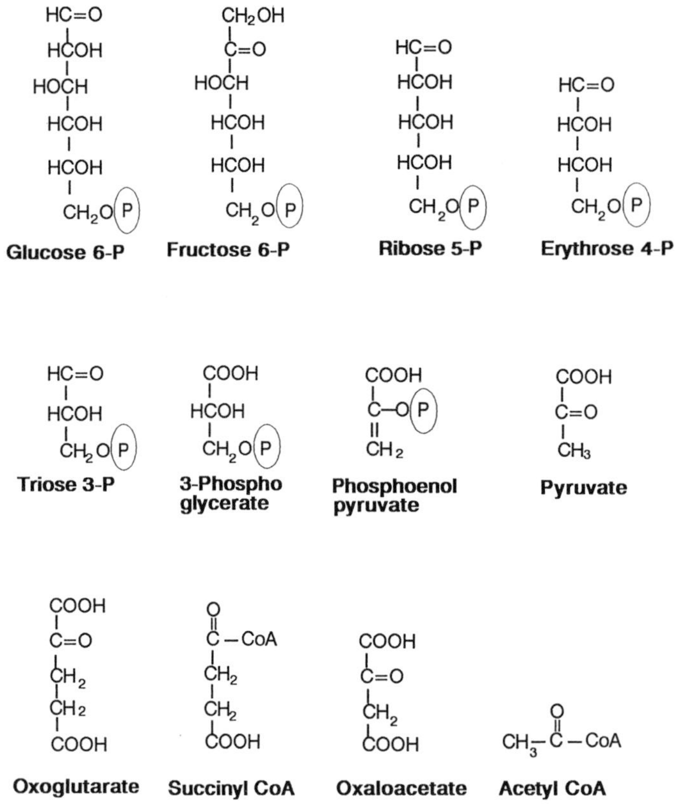

Glucose 6-P Fructose 6-P Ribose 5-P Erythrose 4-P

Triose 3-P 3-Phospho glycerate Phosphoenol pyruvate Pyruvate

Oxoglutarate Succinyl CoA Oxaloacetate Acetyl CoA

Fig. 22. Structural formulas of the twelve major precursor metabolites.

carbohydrates are more reduced or more oxidized than hexose sugars (mannitol, sorbitol, gluconic acid) or are substituted (N-acetylated, methylated) or are isomers/epimers of the central pathway sugars. These first have to be transformed into central metabolites by action of, e.g., dehydrogenases, kinases, deacetylases, or isomerases.

At least three different routes can lead to the breakdown of glucose to small compounds such as pyruvate or finally CO_2 and H_2O. The best known (and occurring in most organisms) is the Embden-Meyerhof-Parnas (EMP) pathway of glycolysis via Fru 1,6-bisphosphate (Fig. 24). Glucose has first to be activated by phosphorylation at the C_6 atom. By action of phosphoglucose-isomerase (PGI) and phosphofructokinase (PFK), Fru 1,6-bisphosphate is formed at the expense of an ATP molecule. Aldolase splits Fru 1,6-BP into two triose phosphate molecules (dihydroxyacetone phosphate or DHAP, and glyceraldehyde 3-P or GA 3-P). Interconversion of these two compounds is accomplished by triose phosphate isomerase (TPI). GA 3-P is the starting metabolite of the lower half of the glycolytic pathway which uses only C3-atom molecules. By a dehydrogenation/phosphorylation step with inorganic phosphate (P_i), 1,3-diphosphoglycerate (1,3-DPGA) is formed. Its subsequent conversion to 3-phosphoglycerate delivers the first ATP molecules of this pathway via substrate level phosphorylation. At the step from PEP to pyruvate, another ATP is gained.

Other glucose degradation pathways (Fig. 24) start by the NADP-dependent oxidation of Glc 6-P to 6-P-gluconolactone, which is then transformed to 6-P-gluconate (6-PG). 6-PG can be decarboxylated (NADP-dependent) to ribulose 5-P (hexose monophosphate shunt) to enter the pentose phosphate cycle. In some organisms such as Pseudomonads or *Zymomonas mobilis*, 6-PG is transformed by the subsequent action of a dehydratase and a specific al-

Precursor metabolite(s)		Biosynthetic block
Glucose 6 - P	→	ADP - Glucose
Fructose 6 - P	→	UDP- NAc - Glucosamine
Ribose 5- P	→	Histidine Nucleotides ATP, GTP, UTP, CTP, TTP Deoxynucleotides
Erythrose 4 - P + Phosphoenolpyruvate	→	Aromatic amino acids (Tryptophan, Tyrosine, Phenylalanine) p-Aminobenzoate, 2,3 Dihydroxybenzoate
Triose 3-P	→	Glycerol 3- P
3 - phosphoglycerate	→	Serine, Glycine, Cysteine, C 1 -compounds (Folate), Heme
Pyruvate	→	Alanine, Valine, Leucine
Oxoglutarate	→	Glutamate, Glutamine Proline, Arginine
Succinyl CoA	→	Heme, Lysine
Oxaloacetate	→	Aspartate, Asparagine Lysine, Methionine, Threonine Isoleucine Pyrimidine nucleotides
Acetyl CoA	→	Fatty acids Isopentenyl-PP

Fig. 23. Precursor metabolites and some of their derivatives (biosynthetic building blocks).

dolase via 2-keto 3-deoxy 6-PG (KDPG) to pyruvate and GA 3-P (Entner-Doudoroff- or KDPG pathway). GA 3-P and pyruvate then can be metabolized by the same reactions as in the EMP pathway. The KDPG pathway is also used in gluconate catabolism of Enterobacteria.

Pentose Phosphate Pathways (PPP)

In the pentose phosphate pathway, sugar phosphates consisting of 3 to 7 C-atoms are interconverted (see Fig. 24). Xylulose 5-P and ribose 5-P are formed by action of ribulose phosphate epimerase and ribose phosphate isomerase. The two pentose phosphates are substrates for the glycolaldehyde-transferring transketolase which forms sedoheptulose 7-P and glyceraldehyde 3-P (GA 3-P). The glyceraldehyde-transferring transaldolase then catalyzes the formation of erythrose 4-P and fructose 6-P. Transketolase can also react with Fru 6-P and GA 3-P (to xylulose 5-P and erythrose 4-P) so that the pathway can be replenished from several sides. No ATP is gained in this pathway and no reducing pyridine nucleotides are involved (non-oxidative). However, the

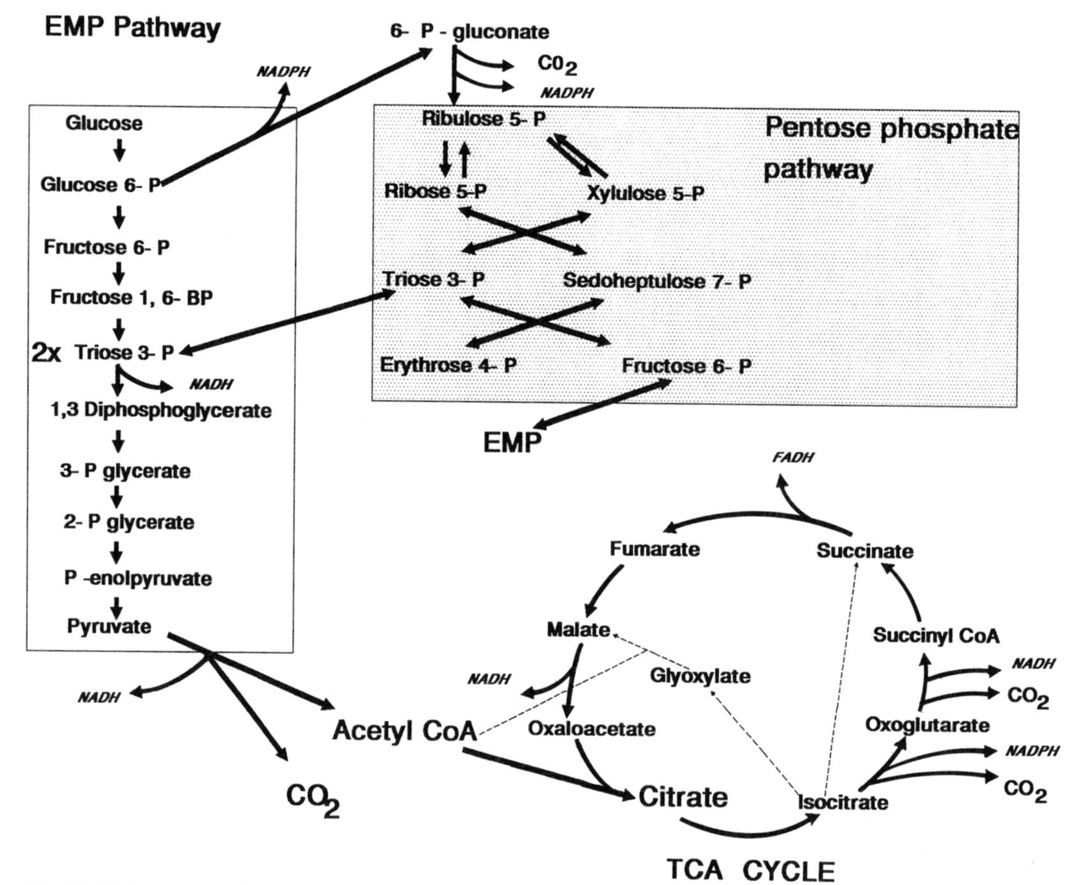

Fig. 24. Major routes of glucose breakdown: Embden-Meyerhof-Parnas (EMP) pathway, pentose phosphate pathway, TCA cycle, glyoxylate shunt (dashed lines) and interconnecting reactions. Adapted with permission from NEIDHARDT et al. (1990)

intermediates can be drawn off this pathway and further metabolized. The C_4- and C_5- sugar phosphates are used for biosynthetic purposes (nucleotides and aromatic amino acids).

The pentose phosphate pathway is also used to fix C_1-compounds like carbon dioxide or formaldehyde (from methanol) by diverse organisms (Fig. 25). Green plants and photosynthetic bacteria use the enzyme ribulose bisphosphate carboxylase/oxygenase (Rubisco) to combine carbon dioxide with ribulose 1,5-bisphosphate resulting in 2 molecules of 3-phosphoglycerate (3-PGA). 3-PGA is converted by phosphoglycerate kinase (ATP) and GA 3-P dehydrogenase (NADH) into GA 3-P (reductive branch of the pentose phosphate cycle or Calvin cycle). Triose phosphates can be used to replenish the pentose phosphate cycle or for anabolic/catabolic purposes. Ribulose 1,5-BP is formed by phosphorylation of ribulose 5-P (action of phosphoribulokinase) (Fig. 25).

In organisms which can grow on methanol, methylamine or similar compounds (methylotrophs), besides the reductive Calvin cycle, two other pathways (XuMP and RuMP cycles) involving pentose phosphates are found (Fig. 25). In the xylulose monophosphate (XuMP) cycle, formaldehyde (formed by the oxidation of methanol) is combined with xylulose 5-P via a transketolase-type enzyme (dihydroxyacetone synthase) to GA 3-P and dihydroxyacetone which is phosphorylated to DHAP. Another route is the RuMP cycle (ribulose mono-

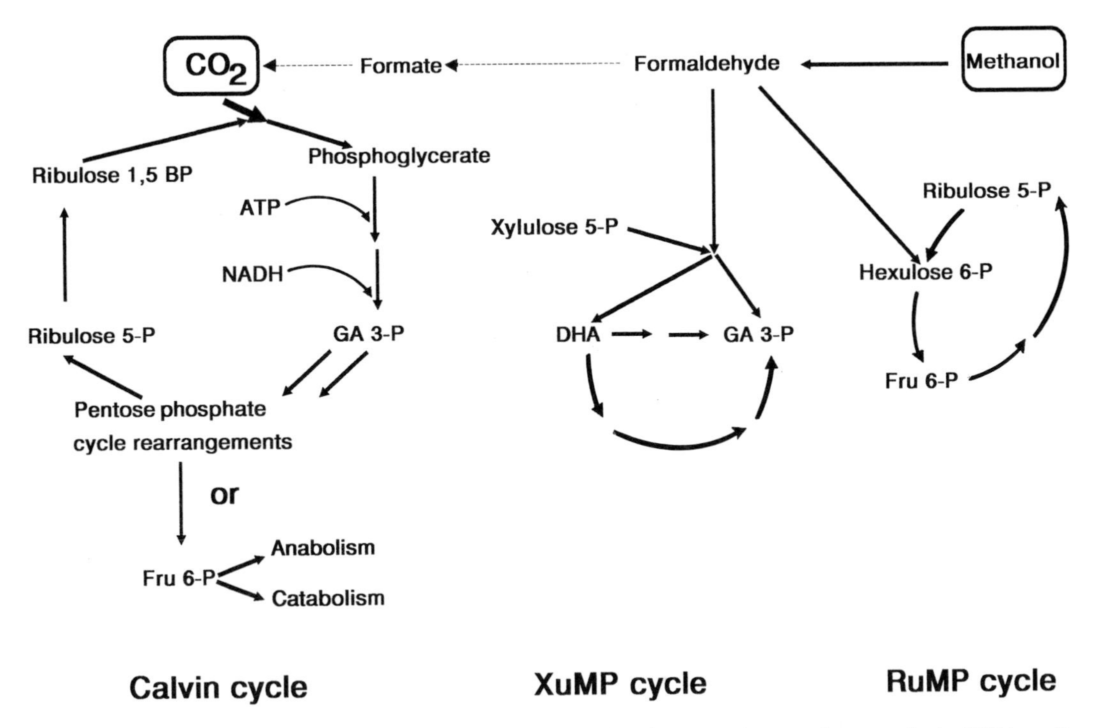

Fig. 25. Three cyclic pathways of C_1-compound fixation in microorganisms and green plants. DHA = dihydroxyacetone. The dotted line between formate and CO_2 indicates that some methylotrophs use the Calvin cycle to utilize methanol via CO_2 as intermediate.

phosphate) wherein ribulose 5-P and formaldehyde give a C_6-compound (hexulose monophosphate) which is then isomerized to Fru 6-P. In both cycles, using the other pentose phosphate cycle enzymes, xylulose 5-P and ribulose 5-P can be replenished.

TCA Cycles and Gluconeogenesis

The tricarboxylic acid (TCA- or Krebs) cycle is used for several purposes. It is able to convert acetyl-CoA (from pyruvate through action of the pyruvate dehydrogenase complex or from fatty acid catabolism) to CO_2 and H_2O with the gain of NADH, NADPH and $FADH_2$ (for energy metabolism or anabolism). The cycle also interconverts several organic acids which are derived from amino acid metabolism (see Fig. 33). It also provides precursor metabolites for biosynthetic purposes (amino acids a.o.) (see Fig. 35) . Therefore, the TCA cycle is an essential central metabolic route acting as a metabolic turn-table.

The glyoxylate shunt by-passes several reactions of the TCA cycle and is a means of interconverting C2- and C4-atom molecules, the prerequisite for microbial growth on compounds as acetate or fatty acids. It starts with the splitting of isocitrate into succinate and glyoxylate. Glyoxylate can react with AcCoA to deliver malate.

Gluconeogenesis and Anaplerotic Reactions

In principle, many reactions of the TCA cycle and the EMP pathway are reversible, i.e., the same enzyme can be recruited for both catabolism and anabolism of sugars and their derivatives. However, several steps are practically irreversible (hexokinase, phosphofructokinase, pyruvate kinase). Thus, in order to "reverse" glycolysis, gluconeogenesis (formation of glucose from smaller compounds) has to by-pass at these steps (see also Sect. 6.1 and Fig. 41). Pyruvate can be converted to PEP via

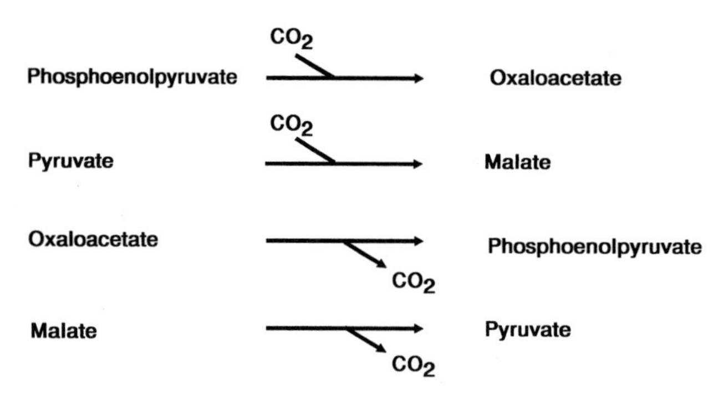

Fig. 26. Anaplerotic reactions to refill the TCA cycle and to enter gluconeogenesis. PEP is transformed to oxaloacetate by a PEP-carboxylase, pyruvate carboxylase leads from pyruvate to malate, PEP-carboxykinase leads from oxaloacetate to PEP, and the reaction malate to pyruvate is catalyzed by the malate enzyme.

a phosphoenolpyruvate synthetase or via the combined actions of a pyruvate carboxylase and phosphoenolpyruvate carboxykinase instead of the catabolic pyruvate kinase. Similarly, the PFK reaction is irreversible, and therefore Fru 1,6-BP has to be cleaved by an enzyme Fru 1,6-BP phosphatase to result in Fru 6-P.

During growth on organic acids such as acetate, succinate or lactate as sole carbon sources or due to metabolite drain from the TCA cycle during biosynthesis, the cycle must be re-

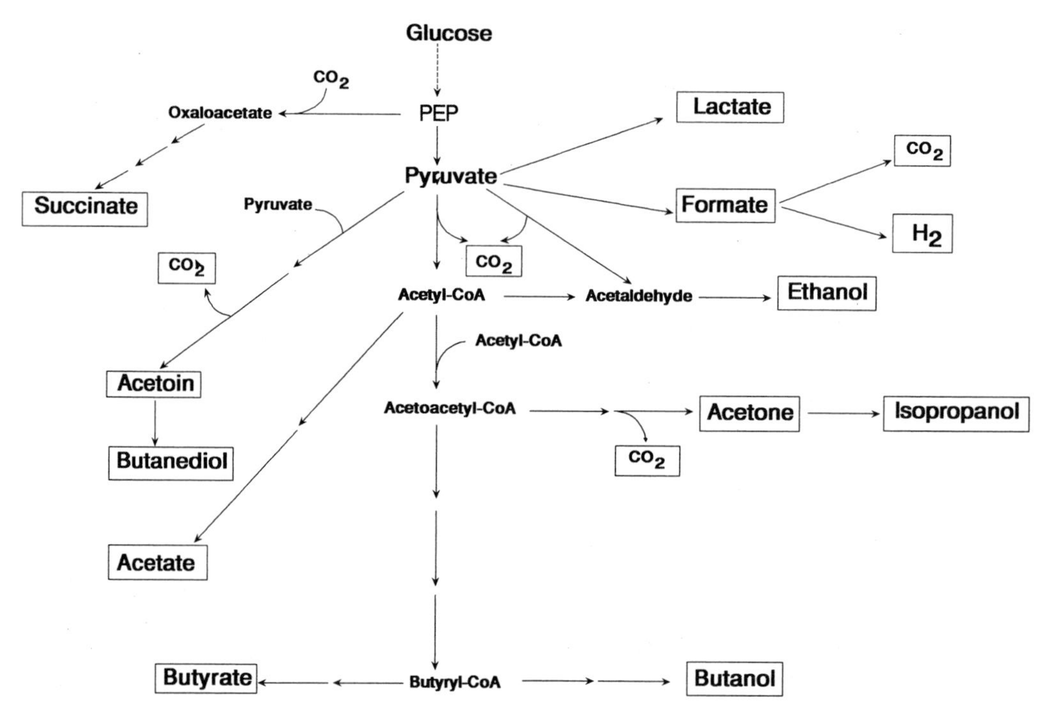

Fig. 27. Fermentation end products from bacterial glucose fermentation. Bacteria (mainly Clostridia, Enterobacteria and lactic acid bacteria) form versatile fermentation products mainly from pyruvate or PEP, although no single species forms all end products shown in the figure.

plenished by so-called anaplerotic reactions. To refill the cycle, interconnecting, reversing or by-passing reactions are necessary (Fig. 26). This is of special interest for biotechnological processes where amino acids or citric acid are produced by microorganisms.

Fermentative Metabolism

Microorganisms not always produce CO_2 and H_2O as final products from the degradation of carbohydrates or organic acids. Instead, a variety of fermentation end products may occur during the course of "fermentations". A survey is given in Fig. 27. By generation of these fermentation products, the cellular redox balance is maintained in the absence of oxygen or other terminal electron acceptors. Clostridial fermentations produce solvents and acids (see Chapter 9 of this volume). Other fermentations are lactic acid fermentations (see Chapter 10), homo-acetate fermentations (Chapter 9), and ethanol fermentations (see Chapter 14).

In the facultatively anaerobic Enterobacteria, a mixture of acidic and ethanolic fermentation products can be formed (Fig. 27). Lactate is derived from pyruvate through lactate de-

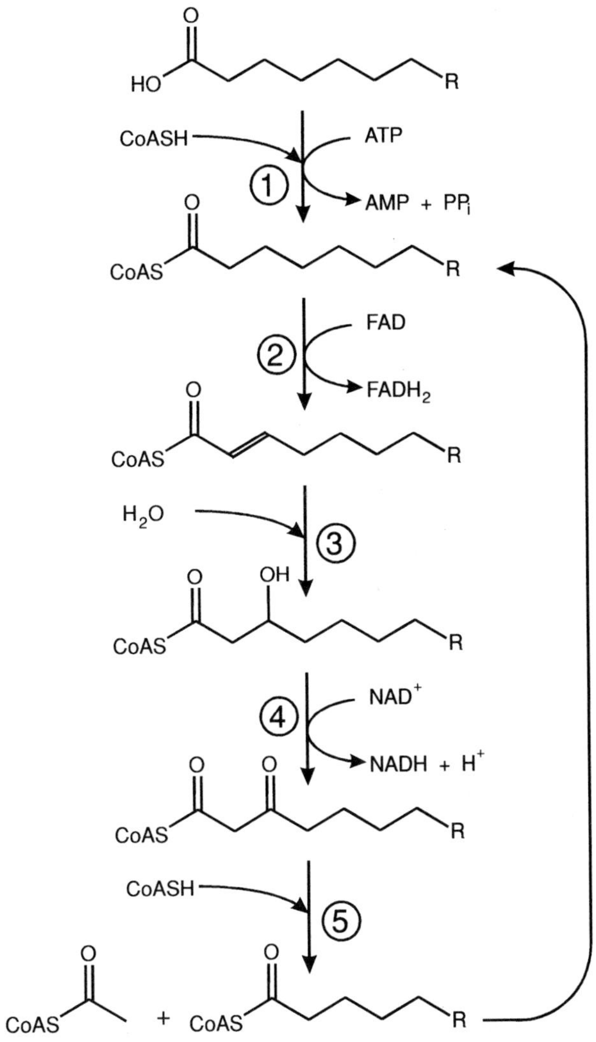

Fig. 28. Fatty acid degradation by β-oxidation. The fatty acid is first activated to the corresponding CoA ester by acyl-CoA synthetase (1). The β-oxidation cycle starts with oxidation by the acyl-CoA dehydrogenase (2), the α,β-unsaturated acyl-CoA compound is hydrated by the 3-hydroxyacyl-CoA hydrolase (3). The next oxidation step is catalyzed by the L-3-hydroxyacyl-CoA dehydrogenase (4). Finally, cleavage of the β-oxoacyl-CoA with CoA by the β-ketothiolase (5) yields acetyl-CoA and an acyl-CoA compound, which again enters the β-oxidation cycle.

hydrogenase. Succinate stems from carboxylation of PEP to oxaloacetate via malate and fumarate. Formate is a product of the pyruvate-formate lyase reaction with pyruvate. The other product (acetyl-CoA) can either be used for ATP generation (via acetyl-P to acetate) or can be reduced to ethanol. Formate is a substrate for formate dehydrogenase delivering H_2 and CO_2. By condensation of two pyruvate molecules (acetolactate synthase), acetolactate is formed and subsequently decarboxylated to acetoin, which can be further reduced to butane-2,3-diol (Fig. 27).

5.1.2 Metabolism of Fatty Acids, Aliphatic and Aromatic Hydrocarbons

Fatty Acid Degradation and Biosynthesis

Whereas fatty acids are a major energy source for mammals, their importance as substrate for bacterial growth is relatively limited. A number of aerobes (pseudomonads, bacilli, *Escherichia, Salmonella typhimurium*) degrade fatty acids by the pathway of β-oxidation. In all

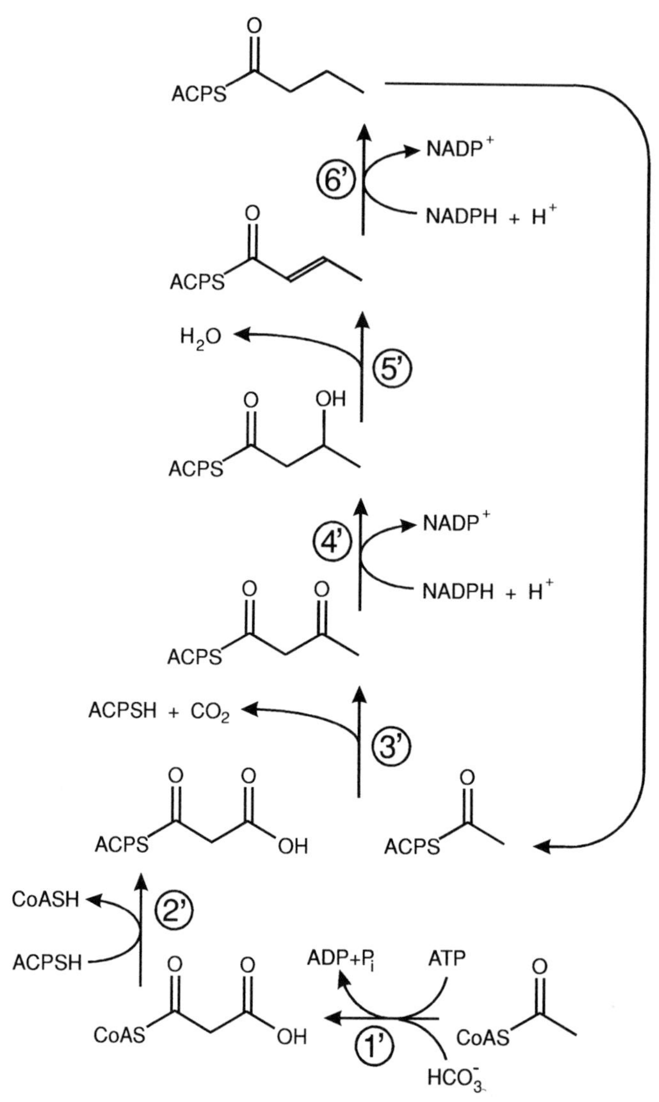

Fig. 29. Fatty acid biosynthesis. The anabolic pathway starts with activation of acetyl-CoA to malonyl-CoA by the acetyl CoA carboxylase (1′); malonyl-CoA is subsequently transferred to the acyl-carrier-protein (ACP). In the following biosynthetic cycle, the reactions 3′ to 6′ are analogous to the corresponding reactions 5 to 2 of fatty acid degradation (Fig. 28), except that NADPH is used as reductant.

bacteria, however, fatty acids can directly be used for synthesis of complex lipids. Furthermore, fatty acid metabolism is important in the degradation of aliphatic hydrocarbons (see below).

After uptake by a specific carrier, long-chain fatty acids (LCFA) are first activated to an acyl-CoA thioester and then enter the cyclic β-oxidation pathway (Fig. 28). This pathway is more or less identical in bacteria and in eukaryotes. In each turn of the cycle, a two-carbon fragment is split off as acetyl-CoA, in addition two reducing equivalents, one $FADH_2$ and one NADH, are generated. After thiolytic cleavage the remaining acyl-CoA compound re-enters the degradation cycle. The acetyl-CoA units can be directly fed into the citric acid cycle.

Bacteria do not use lipids as reserve material, but they contain considerable amounts of lipids (phospholipids and glycolipids) within their membranes. Although the cycle of chemical reactions involved in fatty acid biosynthesis is nearly identical to that used for fatty acid degradation, the two pathways use different enzymes and different mechanisms in some steps (Fig. 29). During biosynthesis the fatty acids are bound to the acyl carrier protein (ACP) which, however, uses the same binding principle as CoA, i.e., a thioester link to a phosphopantetheine group. Since biosynthesis is an energy requiring process, for every elongation step the newly introduced two-carbon unit (acetyl-CoA) must first be activated to malonyl-CoA. For the two reduction steps within the biosynthetic cycle, NADPH is used instead of NADH, as generally found in anabolic pathways. The main product of fatty acid biosynthesis is palmitoyl-ACP (C_{16}). For synthesis of unsaturated fatty acids, i.e., when double bonds are introduced, additional enzyme systems are involved.

Metabolism of Aliphatic and Aromatic Hydrocarbons

Many microorganisms can use aliphatic and aromatic hydrocarbons for growth. Aromatic compounds can be degraded both aerobically (LEAHY and COLWELL, 1990) and anaerobically (FUCHS, 1992). Aliphatic hydrocarbons (alkanes and alkenes) in general are primarily attacked by a monooxygenase at the methyl group. The resulting primary alcohol is further oxidized by alcohol dehydrogenase and aldehyde dehydrogenase to the corresponding fatty acid which then enters the β-oxidation cycle (Fig. 30). Aromatic compounds are important substrates, originating from biosynthesis in plants (e.g., lignin, flavons), or from proteins (aromatic amino acids). In aerobic degradation by bacteria and fungi mainly three general intermediates are used, namely catechol, protocatechuate and gentisate or homogentisate (Fig. 31). Depending on the organism, these "starting substrates" are then further degraded by two types of ring cleavage, either *ortho-*

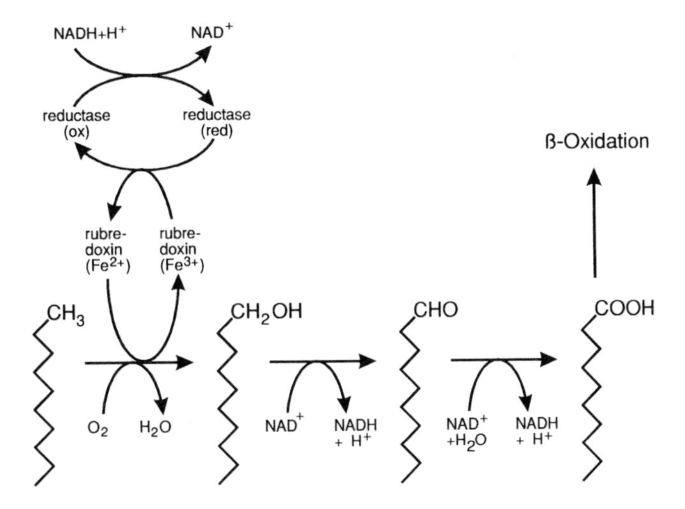

Fig. 30. Terminal oxidation of an *n*-alkane.

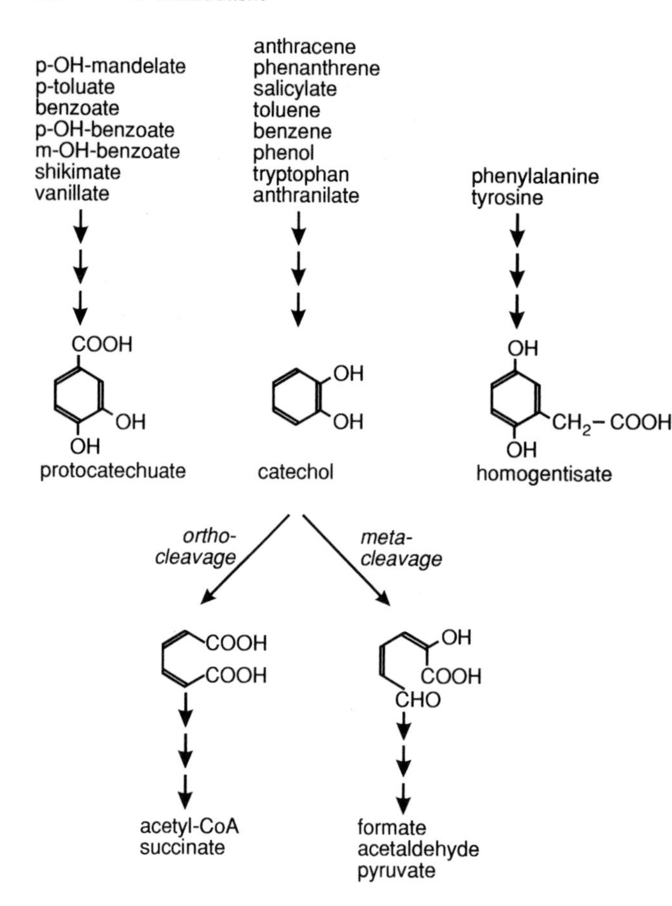

Fig. 31. Schematic representation of degradation pathways for aromatic hydrocarbons. The various aromatic compounds are converted to three main intermediates which are further cleaved by two alternative pathways into central metabolites.

cleavage or *meta*-cleavage. The pathways finally lead to compounds of the central metabolism. Reactions of these pathways are furthermore important in degradation of many xenobiotica.

5.2 Nitrogen Metabolism

5.2.1 The Nitrogen Cycle

Nitrogen (N) is an essential constituent (up to 15% of dry weight) of all cells, it is, for instance, necessary for amino acid and nucleotide formation. To utilize nitrogen, N_2 has to be "activated" and be brought into a biocompatible form (nitrogen fixation), e.g., as ammonium (NH_4^+) or as nitrate/nitrite. Only some prokaryotic organisms (mainly soil bacteria) are able to fix atmospheric N_2 to NH_3 (diazotrophic bacteria). Nitrogen fixation is a highly energy-consuming step (at least 6 mol ATP are hydrolyzed per mol N_2 fixed), and the main enzyme (nitrogenase) is extremely oxygen-sensitive. Both free-living (*Klebsiella, Azotobacter*) and symbiotic (associated with plant root nodules) bacteria (*Rhizobium, Frankia*) are able to fix N_2. The process involves about 20 structural and regulatory proteins and is highly regulated.

NH_3 can then be utilized as N-source for the formation of amino acids (by reductive amination). NH_3 can also be converted via NO_2^- to NO_3^- by nitrifying bacteria such as *Nitrosomonas* or *Nitrobacter*. The N-cycle (Fig. 32) is closed, when denitrifying bacteria (e.g., *Paracoccus denitrificans*) form N_2 (gaseous) from nitrate. Thereby, nitrogen loss from, e.g., soils to the atmosphere occurs which is important

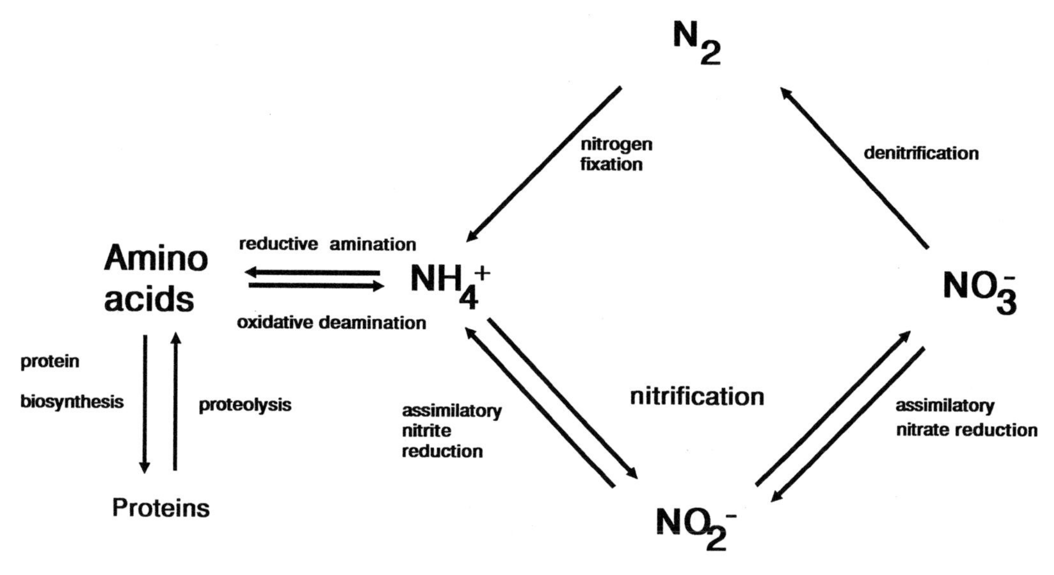

Fig. 32. The nitrogen cycle.

for the loss of soil fertility. Nitrate which is not decomposed, however, can reach the groundwater and may constitute a serious problem for drinking water supply.

Several bacteria are able to reduce nitrate to nitrite by assimilatory nitrate reduction and further to NH_4^+ by assimilatory nitrite reduction. Nitrate can also serve as terminal electron acceptor in anaerobic respiration; this process is called dissimilatory nitrate reduction.

Ammonia assimilation is either by reductive amination using oxo-acids to deliver L-amino acids, mainly glutamate via GDH (Sect. 5.2.2) or by the GS-GOGAT cycle with the enzymes glutamine synthase and glutamine: oxoglutarate aminotransferase. Glutamate/glutamine then serve as amino-group donors for the synthesis of other amino acids.

5.2.2 Amino Acid Degradation

Many microorganisms can utilize amino acids and low-molecular-weight peptides produced by proteases as sole source of carbon, nitrogen, and energy. To enter the pool of central intermediates, most amino acids are first converted to the corresponding keto acids. Four general reactions are used for this purpose.

(1) In oxidative deamination, relatively unspecific cytochrome-linked D- and L-aminooxidases feed the electrons via a flavine coenzyme into the respiratory chain.

$$R-\underset{\displaystyle NH_2}{\underset{|}{CH}}-COOH + {}^1/_2\,O_2 \rightarrow R-\underset{\displaystyle O}{\underset{\|}{C}}-COOH + NH_3$$

(2) The same reaction can also be catalyzed by NAD(P)-linked dehydrogenases, an important example being *glutamate dehydrogenase.* This reaction is also used for anabolic purposes, e.g., the reductive synthesis of amino acids.

$$R-\underset{\displaystyle NH_2}{\underset{|}{CH}}-COOH + NADP^+ \rightleftharpoons R-\underset{\displaystyle O}{\underset{\|}{C}}-COOH + NADPH + NH_4^+$$

(3) Very frequently amino acids are converted by *transamination* mainly with oxoglutarate or pyruvate as acceptor of the amino group. Also this reaction has both catabolic and anabolic meaning.

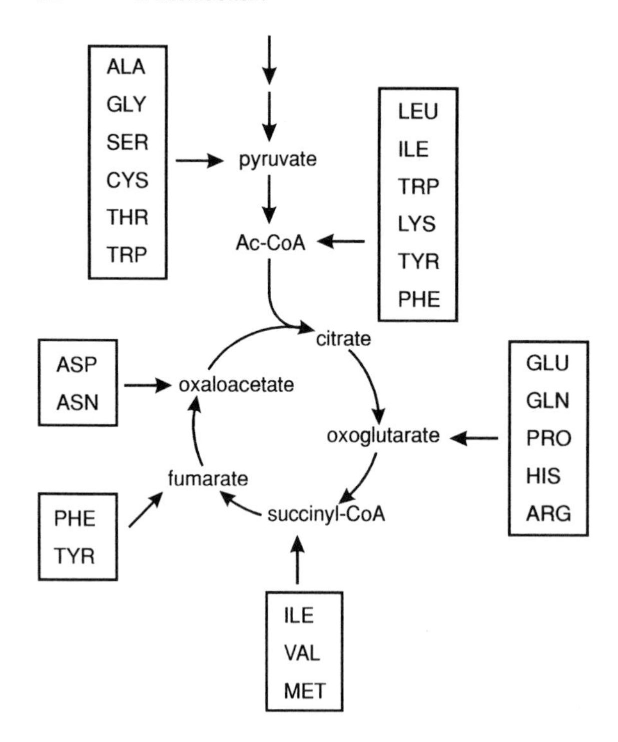

Fig. 33. Amino acid degradation: Classification into families leading to common metabolites.

$$\underset{\underset{NH_2}{|}}{R-CH}-COOH + CH_3-\underset{\underset{O}{\|}}{C}-COOH \rightleftharpoons R-\underset{\underset{O}{\|}}{C}-COOH + CH_3-\underset{\underset{NH_2}{|}}{C}-COOH$$

(4) A fourth reaction for removal of α-amino groups is *deamination* of amino acids which have substituents at the β-carbon atom, such as serine, threonine, aspartate, and histidine. Since the reaction starts with removal of water, these enzymes are called dehydratases.

$$\underset{\underset{OH}{|}}{CH_2}-\underset{\underset{NH_2}{|}}{CH}-COOH \xrightarrow{\text{serine dehydratase}} CH_3-\underset{\underset{O}{\|}}{C}-COOH + NH_3$$

According to the different carbon skeletons, further breakdown of the various keto acids is catalyzed by a variety of enzymic reactions. In Fig. 33, the 20 amino acids are combined in families, according to the main metabolites which are the final products of their degradation. A large number of amino acids has direct connection to intermediates of the tricarboxylic acid cycle (e.g., Asp, AsN, Glu, GlN, and Ala), whereas the degradation of others involves a long series of complicated reactions and is therefore not present in many microorganisms (e.g., Lys, Ile, Val, Leu, Phe, Tyr, and Trp).

5.2.3 Biosynthesis of Amino Acids

The first general step in amino acid biosynthesis is the assimilation of ammonia which is the preferred source of nitrogen for growth. Nitrogen for biosynthetic purposes is derived in the cell from the amino group of glutamate or the amido group of glutamine. Incorporation of ammonia into these compounds is catalyzed by two processes. If the external ammonia concentration is sufficiently high (≥ 1 mM), incorporation proceeds mainly via the glutamate dehydrogenase reaction (see above). The amino group of the glutamate formed can then be transferred to various keto acids (Fig. 34A). If

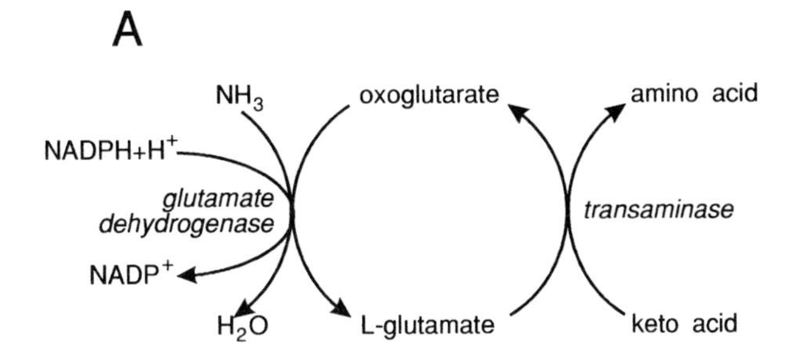

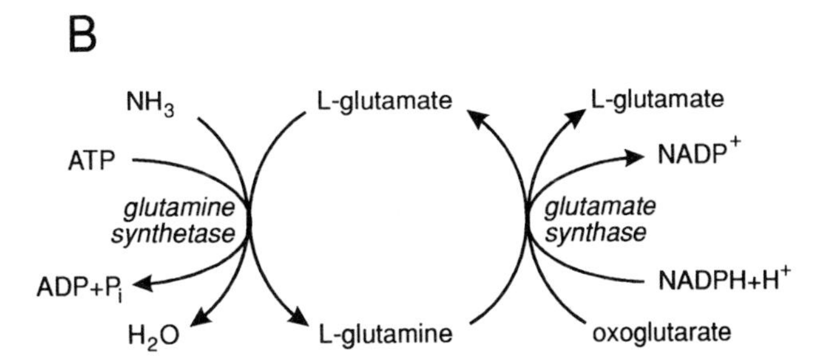

Fig. 34. Incorporation of ammonia into glutamate by the glutamate dehydrogenase (A) and the GS-GOGAT reaction (B).

the external ammonia concentration is low (≤ 0.1 mM), assimilation is catalyzed by a combination of two enzymes, glutamine synthetase and glutamate synthase (Fig. 34B). The latter enzyme is frequently called GOGAT (glutamine:oxoglutarate aminotransferase). The combined reaction proceeds at the expense of ATP. It is essential at low external ammonia because of its high overall negative free energy (equilibrium far on the side of assimilation) and the low K_m-value of the glutamine synthase for ammonia (10^{-4} M).

Only a few precursor compounds serve as substrates in amino acid synthesis, thus the different amino acids are classified in "families", according to the type of starting precursor (Fig. 35). These are: (1) The oxaloacetate (or aspartate) family (Asp, AsN, Lys, Thr, Met, Ile). (2) The pyruvate family (Ala, Val, Leu). Although belonging to different families, the main part of the biosynthetic pathway of Ile, Val, and Leu includes similar reactions and is catalyzed by the same enzymes. (3) The phosphoglycerate family (Ser, Gly, Cys). Conversion of Ser to Cys is a main reaction of assimilatory sulfate reduction (see Sect. 5.3). (4) The oxoglutarate family (Glu, GlN, Pro, Arg). (5) The family of aromatic amino acids (Phe, Tyr, Trp) starting from erythrose 4-P and PEP involves a series of reactions for biosynthesis. Finally, histidine is synthesized in an independent pathway starting from 5-phosphoribosylpyrophosphate. Two additional amino acids are essential constituents of bacterial cell walls: ornithine which is an intermediate in arginine biosynthesis, and ε-diaminopimelate which is the direct precursor of lysine.

5.3 Sulfur Metabolism

Sulfur is essential for the cell because of its role in cysteine, methionine, and some coenzymes. Of the many possible oxidation states of sulfur, three are of practical significance, sulfate (+6), elemental sulfur (0), and sulfide or organic sulfur components (R-SH, −2). These forms of sulfur are transformed in a combination of reactions, both by microbial activity and by chemical reactions. A simplified version of this cycle, emphasizing the involvement of bacteria in these transformations, is outlined in Fig. 36.

The majority of inorganic sulfur is found in sulfate minerals. Sulfate is converted to sulfide and organic sulfur compounds in two different

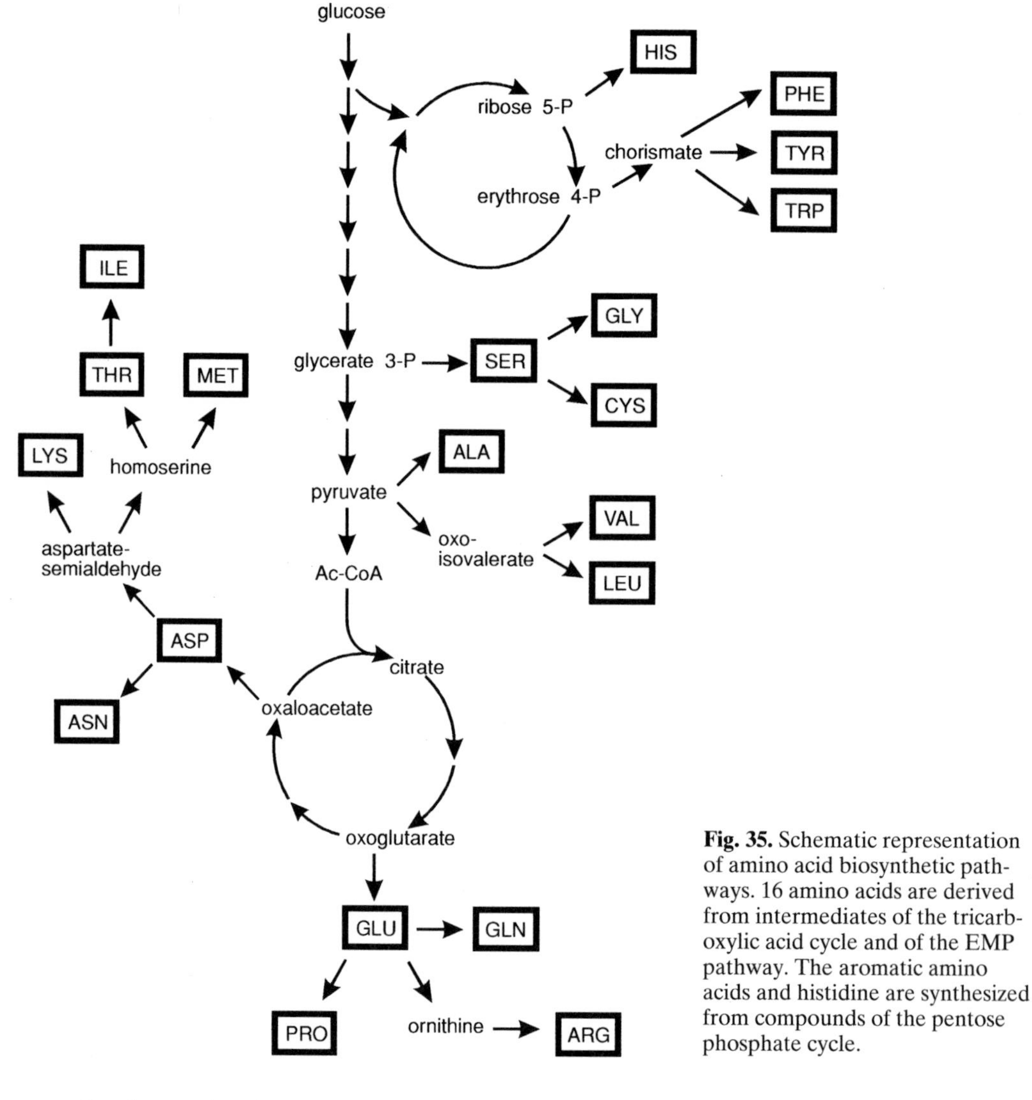

Fig. 35. Schematic representation of amino acid biosynthetic pathways. 16 amino acids are derived from intermediates of the tricarboxylic acid cycle and of the EMP pathway. The aromatic amino acids and histidine are synthesized from compounds of the pentose phosphate cycle.

ways. On the one hand, a great number of organisms, including bacteria, fungi and plants, reduce sulfur for biosynthetic purposes, i.e., using sulfate as sulfur source and converting it to organic sulfur (R-SH, reaction 1 in Fig. 36). This pathway is called assimilatory sulfate reduction. On the other hand, the obligate anaerobic sulfate-reducing bacteria (including various genera, e.g., *Desulfovibrio* or *Desulfotomaculum*) use sulfate as electron acceptor in anaerobic sulfate respiration (see Sect. 2.3.2) and excrete H_2S as end product (Fig. 36: pathway 2). Consequently, this reaction is called dissimilatory sulfate reduction. Besides this, sulfide is also formed by decomposition of organic sulfur compounds during protein degradation (pathway 3) which is called desulfurylation. The electron donors for the reactions 1 and 2 are small organic compounds such as lactate or pyruvate, H_2 is also used by sulfate-reducing bacteria. Since sulfate is rather stable, it must first be activated by ATP, forming adenosine phosphosulfate (APS). APS resembles ADP carrying a sulfate instead of the terminal phosphate group. In dissimilatory sulfate reduction, the activated sulfate is reduced to H_2S

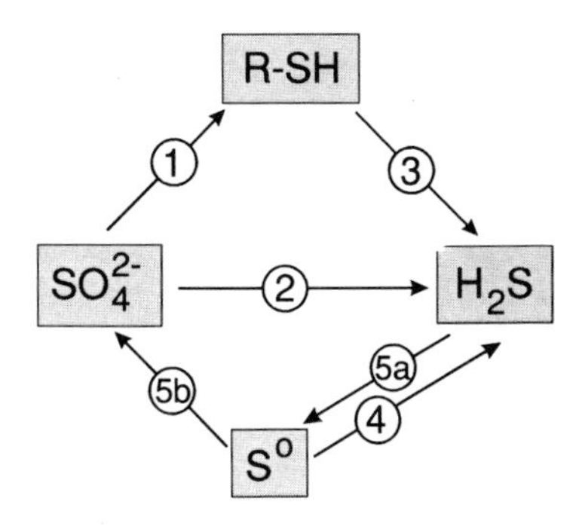

Fig. 36. The sulfur cycle. The individual reactions are (1) assimilative sulfate reduction, (2) dissimilative sulfate reduction (sulfate respiration), (3) desulfurylation, (4) dissimilative sulfur reduction (sulfur respiration), (5) sulfur oxidation.

in several steps by transfer of 8 electrons: H_2S is released. The simple respiratory chain, which provides the electrons for sulfate reduction, generates an electrochemical proton potential which is used for ATP synthesis. The pathway for assimilatory sulfate reduction starts with exactly the same activation step as described above for reaction 2. APS is phosphorylated at the ribose moiety to phosphoadenosine phosphosulfate (PAPS). Reduction of this compound finally leads to organic sulfur compounds. Some bacteria (e.g., *Desulfuromonas acetoxidans*) also use elemental sulfur as electron acceptor in the process of sulfur respiration (Fig. 36, pathway 4).

In contrast to these reactions, bacteria mainly belonging to *Thiobacillus* species use reduced sulfur compounds as electron donors in energy metabolism (sulfur oxidation). H_2S is converted to sulfate via elemental sulfur (pathways 5a and 5b) by the so-called "sulfur bacteria". In addition to sulfur oxidation, *Thiobacillus ferrooxidans* also oxidizes Fe(II) ions and is thus very important in the process of metal leaching. Sulfur oxidation occurs in several stages. The electrons are transferred to an electron transport chain, the transfer to oxygen is coupled to extrusion of protons. Besides ATP-generation by electron transport phosphorylation, some sulfur bacteria can synthesize ATP also by substrate level phosphorylation, making again use of the high-energy compound adenosine phosphosulfate (APS) (see above). The second class of sulfur-oxidizing bacteria are purple and green phototrophic bacteria. These organisms use H_2S in anoxygenic photosynthesis as electron donor, H_2S is reduced to sulfate via elemental sulfur which is frequently deposited inside or outside the cells.

5.4 Synthesis of Biopolymers

Living cells carry out at least four different types of metabolic reactions: Besides the fueling and biosynthetic reactions, these include polymerization reactions linking building blocks into macromolecules (glycogen, RNA, DNA, and proteins) and assembly reactions which put together complex molecules in a sequential order or at specific locations (polysomes, flagella, cell envelopes and enzyme complexes).

5.4.1 Polymerization of Glycogen in Bacteria

Bacterial cells are able to store carbon sources as polymers (e.g., glycogen) mainly when the carbon source is in good supply, whereas other factors are growth-limiting (depletion of N- or S-source). For glycogen formation, glucose has to be "activated" to ADP-glucose (see Fig. 23) or to UDP-glucose in eukaryotes. ADP-glucose is added by a glycogen synthase to the non-reducing end of the growing chain of α-1,4-linked glucose residues, thereby releasing ADP. Another enzyme, transglucosylase, cleaves small oligosaccharides from the end of a chain and attaches them in an α-1,6-linkage on internal glucose residues. This leads to the branching in glycogen formation.

5.4.2 Polymerization of DNA (DNA Replication)

DNA (deoxyribonucleic acid) is organized in cells as a double-stranded antiparallel helix. It is a linear (e.g., unbranched) macromolecule that is formed by template-based polymeriza-

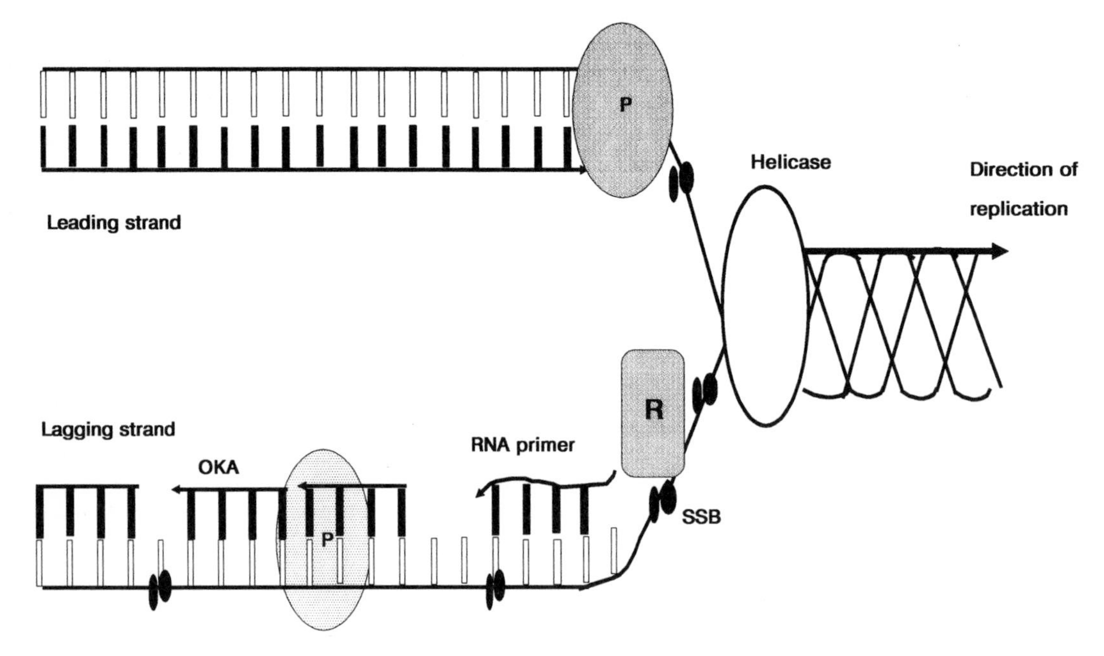

Fig. 37. Model of DNA replication at a replication fork. Helicase unwinds the DNA double strand allowing access for SSB (P, single-strand binding) proteins. DNA polymerase III (shaded oval) adds complementary nucleotides (black rectangle) to the leading strand (upper part). A specific RNA polymerase (R) forms a short RNA primer which is used by DNA polymerase I (blank oval) to form Okazaki (OKA) fragments of 500–2000 bases length on the lagging strand (lower part). The figure is not drawn to scale.
Adapted with permission from NEIDHARDT (1987)

tion (DNA replication) using four activated building blocks (see Fig. 23), namely deoxyribonucleoside triphosphates (NTPs: dATP, dCTP, dGTP, TTP) and one strand of parent DNA as matrix (template). Copying (replication) of this macromolecule starts by opening of the double helix at a number of specific sites (origins of replication) and the two single strands separate. According to the Watson–Crick model, nucleoside triphosphates pair with their complementary bases (dA with dT, dT with dA, dG with dC, dC with dG) of the two exposed single strands (template or parent strands) and are polymerized (Fig. 37), i.e., phosphodiester bonds between adjacent nucleotides are formed. Thus two daughter helices arise which consist of one old and one newly synthesized strand each (semi-conservative mode of replication).

In bacteria, at least twelve proteins form a multienzyme complex (replication apparatus) involved in the steps of:

(1) DNA unwinding and strand separation (ATP-dependent helicase) in 5′ to 3′ direction (designating the orientation of the phosphate diester bonds connecting two adjacent deoxyribonucleotides).
(2) Tight binding of single-stranded DNA by a protein ("SSB") which helps to separate the two DNA strands.
(3) Polymerization: 5′ NTPs pair to their complementary bases on the template strand. They are connected by phosphodiester bonds between the NTP and the terminal 3′ OH group of the growing strand (primer). For driving the process, a pyrophosphate moiety is cleaved from the NTP. Polymerization and proofreading (removal of base mismatches in the newly synthesized DNA) are mediated by the DNA polymerase III complex which consists of at least 7 different subunits. The replica-

tion of one single strand (leading strand) is continuous and proceeds behind the DNA helicase as it moves in the 5′ to 3′ direction. For the replication of the other (lagging strand), it is necessary to make discontinuously short stretches (500 to 2000 nucleotides in length) of newly synthesized DNA (Fig. 37) in opposite direction. For the polymerization of these DNA oligomers (Okazaki fragments), a short RNA primer molecule is necessary.

(4) The formation of the RNA primer is catalyzed by a particular RNA polymerase (primase). DNA polymerase then synthesizes the lagging strand.

(5) When a short stretch of newly synthesized DNA comes in contact with the RNA primer of the next stretch, this primer is hydrolyzed and replaced by DNA through the action of DNA polymerase I.

(6) Gaps between the DNA segments are then sealed by a DNA ligase, and a continuous strand is formed.

(7) Finally, DNA is wound up again and supercoiled by a DNA gyrase.

Overall, DNA replication proceeds (in bacteria) at a constant rate (about 40000 nucleotides polymerized per min at 37 °C). The parental DNA is copied at high fidelity, e.g., less than one mistake (incorrect base pairing) per bacterial genome copy, due to the proofreading (=mismatch removing) activity of DNA polymerase III.

5.4.3 RNA Polymerization

RNA molecules in cells are single-stranded linear nucleic acids formed by template-based polymerization of ribonucleotide triphosphates (ATP, CTP, GTP, UTP) using DNA as template for their formation. RNA-nucleotides may afterwards undergo chemical modifications (e.g., modified bases in tRNA).

Copying of DNA into an RNA molecule (transcription) starts by a partial opening of the DNA double helix and strand separation through the DNA-dependent RNA polymerase complex (RPC). One of the two DNA strands is chosen as template. Ribonucleoside triphosphates which match with their complementary deoxynucleotide bases in the DNA strand (C with dG, G with dC, A with dT, and U with dA) are connected (polymerized) by the RPC without a primer. The events at the initiation, elongation and termination of transcription are described elsewhere (Sect. 6.2). Three forms of RNA are polymerized from their respective genes: messenger RNA (mRNA), ribosomal RNA (rRNA), and transfer RNA (tRNA).

In bacteria, one *mRNA* transcript can be complementary to one or several adjacent genes (mono- or polycistronic RNA), whereas eukaryotic cells normally produce monocistronic transcripts. Eukaryotic mRNAs are modified after synthesis by capping (addition of modified GTP molecules) at their 5′ ends and by addition of several ATPs to their 3′ ends (poly A tail), both modifications help to stabilize the messenger. Moreover, eukaryotic mRNAs may consist of so-called exons (short coding regions) and introns (long non-coding regions). These mRNAs can be processed, for instance the introns are cleaved out and the exons are religated (splicing).

In prokaryotes, mRNA molecules can be directly used as templates for translation into polypeptides. In eukaryotes, the mRNAs must be exported from their sites of origin (normally the nucleus) to the cytosol, where they are translated (after splicing) by the ribosome complexes (polysomes). Bacterial mRNAs are short-lived (half-lives of decay from 30 s to about 20 min), since they are degraded by RNAses, which is also a mode of regulating gene activity. Eukaryotic mRNAs are generally more resistant (half-lives of up to several hours) to degradation.

tRNA and *rRNA* molecules are also transcribed from their respective genes but are not translated into proteins. They are more stable than mRNAs. Transcripts of tRNA genes are processed by nucleases and undergo base modifications to result in the mature tRNA forms. tRNAs have a length of about 75 bases and form L-shaped, three-dimensional structures. Therein, complementary bases of the single strand form pairs to give stem structures, non-pairing intervening bases result in loop structures (Fig. 38). In one of these loops, three se-

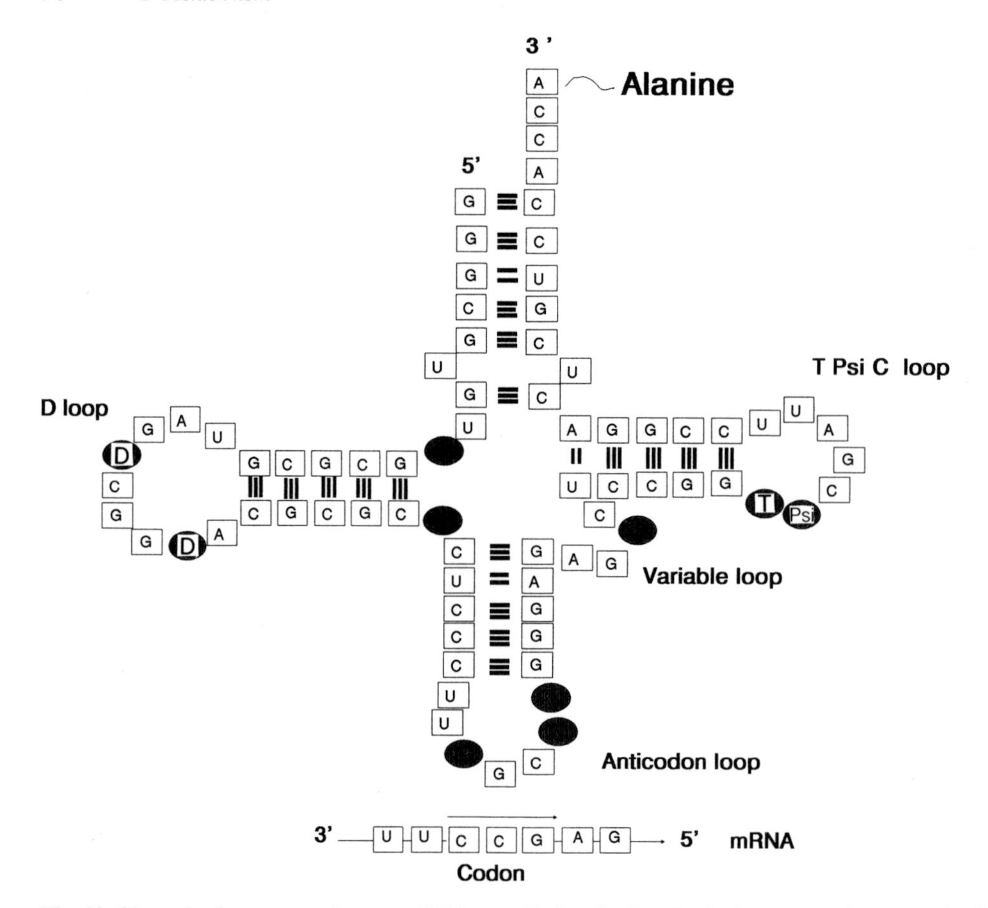

Fig. 38. Clover-leaf structure of a yeast tRNA specific for alanine. At the bottom , a short stretch of mRNA with a codon is shown. The D loop of the tRNA contains two unusual RNA bases (dihydrouridine), the T Psi C loop contains the unusual RNA bases thymidine (T) and pseudo-uridine (Psi) besides a characteristic cytidine (C). Gray circles represent other modified RNA bases.

quential bases (anti-codon) are able to bind their complementary bases (codons) on the mRNA. At the 3′ end of tRNAs, amino acids are attached by specific aminoacyltransferases (see Sect. 5.4.4). For each of the 64 possible codons on the mRNA, one specific tRNA may exist, thus up to 61 tRNA species are possible, taking into account that the three stop codons (UAG, UGA, UAA) normally do not have counterparts at the level of tRNAs.

By mutation, bacterial tRNAs can gain affinity to one of the stop codons, these are called suppressor tRNAs. These can lead to the insertion of an amino acid into a nascent polypeptide chain instead of stopping the respective translation, thus suppressing a mutation that otherwise leads to aborted translation.

Ribosomal RNAs (rRNAs) are involved in the formation of the ribosome structure and in the translation of mRNA into proteins. Prokaryotic rRNAs are classified into three types according to their sedimentation behavior in centrifuges (5S, 16S, and 23S). Eukaryotic rRNAs show more different types (e.g., 5.8S, 18S and 28S). In eubacteria, 16S rRNA forms together with 21 proteins the small subunit of ribosomes (see Sect.5.4.4). Therein, the 3′ end of 16S rRNA binds to the initiation or Shine-Dalgarno (=ribosome binding) site of mRNA, close to the AUG starting codon. 5S rRNA and 23S rRNA form together with 34 proteins the large ribosomal subunit. rRNAs also form stable, three-dimensional structures within ribosomes. Their structures are highly conserved throughout all organisms and are the basis of molecular taxonomy. The functional role of rRNA molecules in ribosomes is still largely unknown, although their involvement in tRNA recognition and in stabilization of the ribosome complex is obvious.

5.4.4 Protein Synthesis

Proteins are polypeptides which consist of 20 different proteinogenic L-amino acids (in eukaryotic proteins some amino acid residues may be modified post-translationally, e.g., hydroxyproline from L-proline). Genes (DNA) are transcribed (copied) into mRNA and therefrom proteins are synthesized at the ribosome. In protein biosynthesis (Fig. 39), nucleic acids are used as (a) templates or blue-prints (mRNA), (b) as amino acid-transferring compounds (tRNA), and (c) as ribosome constituents (rRNA) together with ribosomal proteins. The 20 different amino acids are encoded by up to 61 different mRNA codons (genetic code).

Amino acids have to be "activated" by ATP expenditure to aminoacyl-AMP and are the

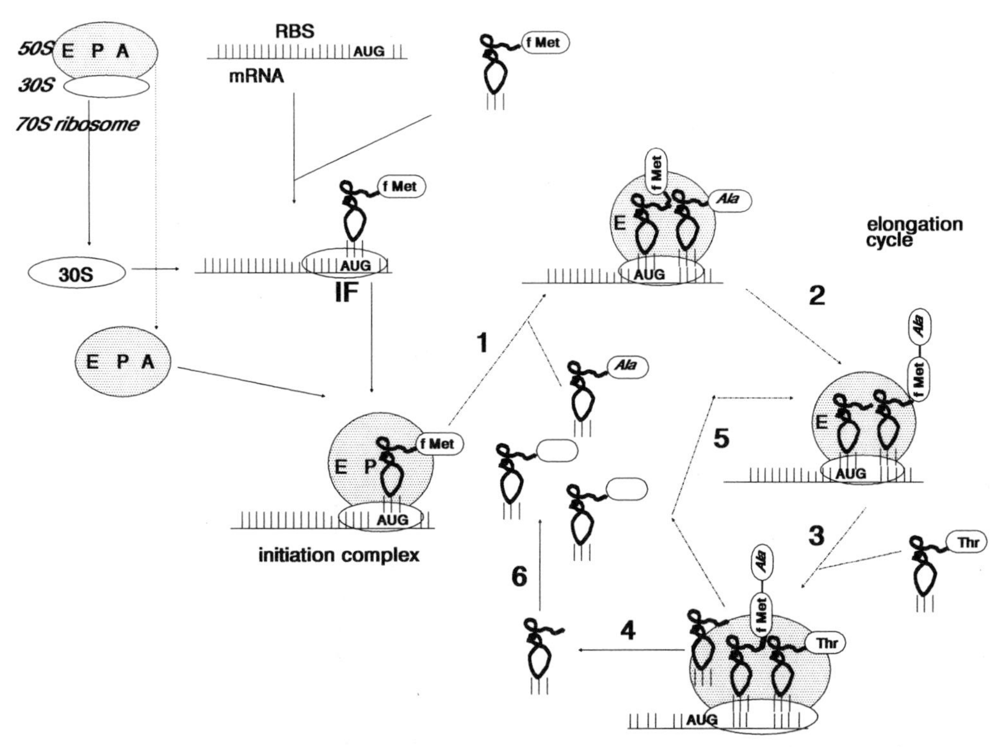

Fig. 39. Scheme of events during translation at bacterial ribosomes. In the upper row, the three main components of translation are shown: The 70S ribosome consisting of two subunits (50S, 30S), mRNA containing a ribosome binding site (RBS) and a start codon, and a charged tRNA molecule. During initiation, the fMet-charged tRNA binds to mRNA and the 30S subunit. Initiation factors (IF) help to form the complex. The 50S ribosomal subunit enters the complex and the fMet tRNA is bound to the acceptor (A) site of it. The elongation cycle starts (1) with the movement of the first bound tRNA to the peptide (P) site while another charged tRNA is bound to the A site. The peptide bond is formed in the next step (2). During translocation (3), the first tRNA moves to the exit (E) site, the peptidyl-charged tRNA moves to the P site and another charged tRNA is bound next to it at the A site. While the first tRNA leaves the complex (4) to be recharged with amino acids (6), other amino acid residues are linked to the growing polypeptide chain (5). When the ribosome complex reaches a stop codon, the complex dissociates to its components.

substrate for amino acyl-tRNA synthetases which connect a given L-amino acid to the 3′ (CCA) end of its specific carrier tRNA. These charged tRNAs can then be utilized by the mRNA/ribosome complex for the elongation of a new polypeptide chain.

In prokaryotes such as *Escherichia coli*, mRNA is synthesized as a copy of its respective gene (transcription) and is utilized directly for protein biosynthesis (translation) at the ribosomes. Three sequential bases (codon) in a given order ("reading frame") encode a specific amino acid in the final protein. Thus, the DNA sequence is translated via an mRNA copy into an amino acid sequence. All bacterial proteins are synthesized with a formyl-substituted methionine (f-met) as the first amino acid residue (the formyl group is removed after the protein is synthesized, sometimes also the methionine may be cleaved off). From several methionine codons (AUG) in a given mRNA, the translation machinery chooses the (first) one that is in a distance of about 10 nucleotides from a site (ribosome binding site) which is complementary to the 3′ end of 16S rRNA in the ribosome (see Sect. 5.4.3). At this site, the 30S ribosome subunit is positioned and starts to form the initiation complex.

Initiation factors are involved in the binding of f-met-charged tRNA to the 30S subunit and the binding to the mRNA. Then the 50S subunit is docked to form the 70S initiation complex, and the first elongation cycle starts. For the elongation of a growing peptide chain, additional protein factors (elongation factors) are necessary.

The 70S ribosome displays three sites (A for acceptor, P for peptide and E for exit of tRNAs). At the A site, a charged tRNA is bound via pairing of its anticodon to a codon on the mRNA mediated by EF-Tu and GTP as well as EF-Ts (removing GDP). At the P site resides a tRNA with the previously formed peptide. By action of the 50S ribosomal subunit, this peptide is transferred to the tRNA at the A site by breakage of the acyl bond and formation of a new peptide bond (linking a C- with an N-atom), resulting in a peptide which is one amino acid longer. The now uncharged tRNA at the P site moves along to the E site, the peptide-loaded tRNA moves to the P site, and the ribosome moves one codon down the mRNA. This process is termed translocation, it requires an accessory protein (EF-G) and completes an elongation cycle.

Termination of protein synthesis and release of the completed polypeptide chain occurs when the ribosome meets one of the three stop codons (or nonsense codons: UAA, UAG, or UGA) on the mRNA. Release factors are involved in the process of hydrolysis of the peptidyl-tRNA to the final peptide.

In eukaryotes, protein synthesis is still more complex: up to ten different initiation factors are used, the ribosomes are more complex (80S ribosomes) and are made up of more proteins. Eukaryotic mRNA must be transported from the nucleus to the cytosol where proteins are synthesized, and may undergo splicing before it can be translated into protein (see Sect. 5.4.3).

6 Regulation and Coordination of Metabolism

Free-living microbial cells face frequent changes in their environments (temperature, osmolarity, pH, supply of nutrients) in contrast to cells of multicellular organisms which dwell in relatively stable environments (e.g., in their respective tissues). Microorganisms can react to changes in the surrounding medium with the rapid onset or shutdown of protein synthesis and of metabolic pathways. Regulation on the level of protein synthesis is generally more economical than the inactivation/degradation of already synthesized proteins (which is faster), as the latter represents a waste of energy and building blocks. Therefore, regulatory steps are mainly found at the transcription and/or translation initiation state. For rapid adjustments, the regulation of enzyme activity at crucial steps of multi-step biosyntheses or catabolic pathways is an additional regulatory tool making use mainly of allosteric control mechanisms.

Microbial cells have to coordinate anabolic and catabolic pathways. This avoids the undesired overproduction of precursor metabolites and/or building blocks, whereas on the other hand the formation of macromolecules (e.g., nucleic acids, proteins, or membranes) must be coordinated (e.g., synchronized) to allow optimal cell growth during changing intracellular and extracellular conditions. The carbon flow through different metabolic pathways must be fine-tuned to avoid metabolic bottlenecks or overshoot reactions as far as possible. For biotechnological purposes, however, the elimination of these regulatory circuits is often desired and may lead to the overproduction of desired compounds (see Sect. 6.1.2, flux control of metabolism).

6.1 Regulation of Enzyme Activity

6.1.1 Covalent Modifications of Enzymes

Enzymes can both be inactivated and activated by covalent modification. Small chemical groups (phosphoryl, methyl, acetyl, adenylyl) can be added or removed from proteins, sometimes drastically influencing their activity. Phosphorylation leads, for instance, to altered enzyme activities of the phosphofructokinase in eukaryotes, or of protein kinases. Certain regulator proteins of the two-component regulatory systems (Sect. 6.2.3) are also activated by phosphorylation. Adenylylation alters glutamine synthetase in *E. coli.*

6.1.2 Allosteric Control

Besides by direct kinetic effects of substrate(s) and product(s) of a particular reaction on the respective enzyme (see Sect. 3.3), many enzymes are controlled by the presence of certain effectors. These effectors may be substrates or products of a given enzyme or of a different enzyme. Other regulatory metabolites may act which are neither substrates nor products and lack any structural similarity to these. Effectors modify either the affinity of the enzyme for its substrate(s) or the reaction rate. Modulation may be stimulatory (positive effector) or inhibitory (negative effector). These regulatory mechanisms have in common that an effector bound to a certain site influences the binding of another ligand (substrate) to a second site not overlapping with the first one. This mechanism of interaction is called allosteric activation or inhibition. Frequently observed mechanisms are precursor activation and feedback inhibition.

Allosterically regulated enzymes are in general oligomeric proteins with multiple binding sites. The conformational interaction of the subunits which depends on the occupancy of the respective effector and substrate binding sites, causes changes in substrate affinity. In allosterically modulated oligomeric enzymes, binding at one substrate binding site influences the binding at other sites which is the reason why the binding sites tend to be either all occupied or all free. The so-called positive cooperativity between the binding sites leads to a sigmoidal dependence of the reaction rate on the substrate concentration (Fig. 40).

Furthermore, the enzyme can exist in two conformations, an active (a) and an inactive conformation (i). Binding of a positive effector will increase the fraction of enzyme which is in state a , whereas binding of the negative effector shifts the equilibrium towards state i. This situation is reflected in Fig. 40. Addition of a positive effector shifts the curve to the left, thereby abolishing the positive cooperativity; the enzyme now behaves according to a Michaelis-Menten-type enzyme. A negative effector, on the other hand, augments the cooperative phenomena.

Allosterically controlled enzymes are frequently found at metabolic branching points or at practically irreversible steps in metabolism (mainly ATP-requiring or ATP-delivering steps), for example (Fig. 41), in glycolysis, gluconeogenesis and the TCA cycle. This allows, for instance, the concurring reactions of gluconeogenesis and glycolysis to proceed in the cell without the need for a compartmentation. Small ligands (e.g., AMP, NAD, ADP, Fru 1,6-BP) act as allosteric effectors. In glucose catabolism, the concentration of the effectors AMP/ADP/ATP or NADH reflects the current state of the cell's energy or redox charge. Thereby, energy metabolism can be slowed down (excess ATP) or enhanced (high AMP and ADP).

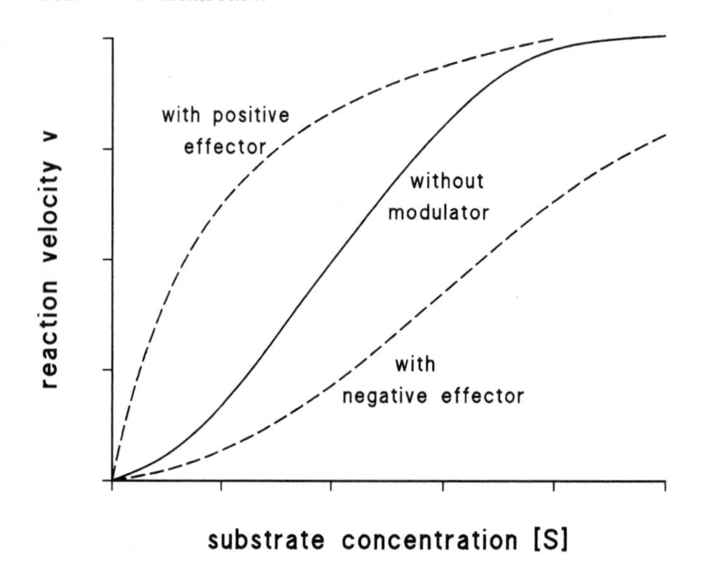

Fig. 40. Allosteric modulation of enzyme activity.

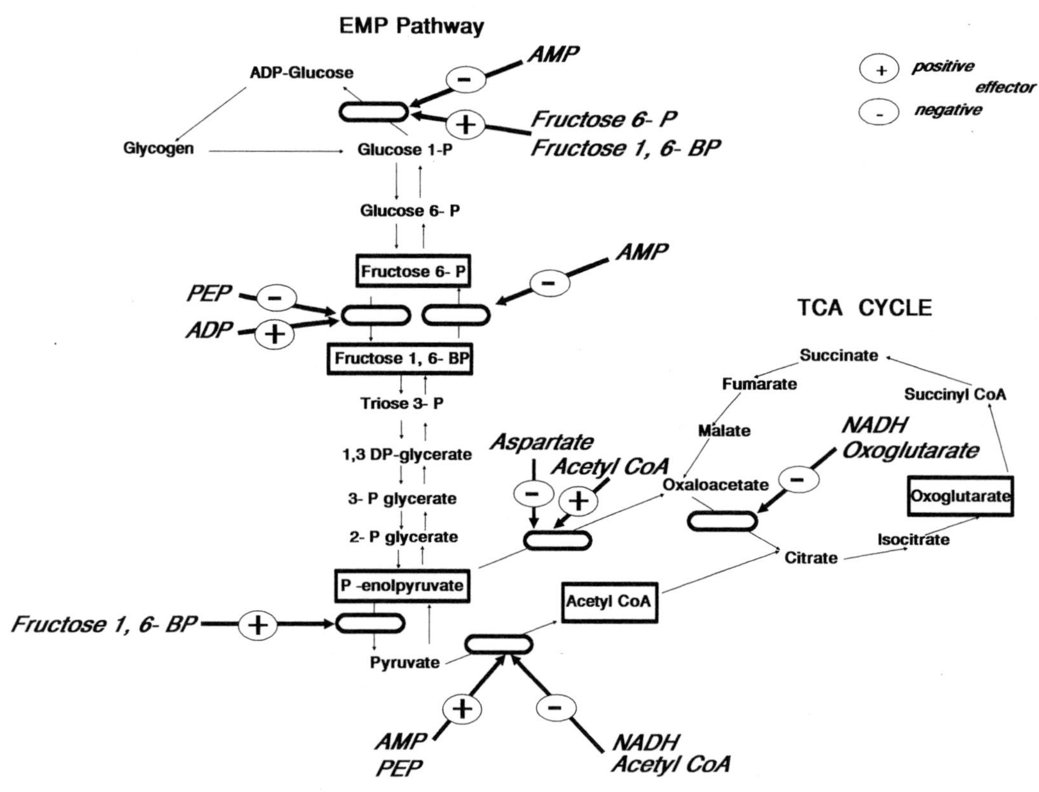

Fig. 41. Allosteric control sites in glycolysis (respectively glycogen formation) and its connection to the TCA cycle in *E. coli.* Effectors which stem from these pathways are boxed; blank ovals represent allosterically regulated targets (enzymes).

Feedback inhibition (negative allosteric control) often occurs in anabolic pathways. For example, tryptophan inhibits one isoenzyme which catalyzes the first committed step in aromatic amino acid biosynthesis (condensation of erythrose 4-P with PEP) of *E. coli.* The other two isoenzymes (Fig. 42A) are regulated separately by the aromatic amino acids phenylalanine and tyrosine, respectively. This control by the end products constitutes a negative feedback loop and enables the cells to keep the concentration of a certain metabolite at a constant level. In fact, most of the 12 precursor metabolites (Fig. 22) of building blocks (Sect. 5.1) stem from pathways which are under allosteric control. This links macromolecule synthesis adjustably to catabolism.

Enzymes at metabolic branching points may be separately regulated by the different end products (e.g., the amino acids threonine and lysine act on the aspartokinase which provides the common precursor aspartyl-P (Fig. 42).

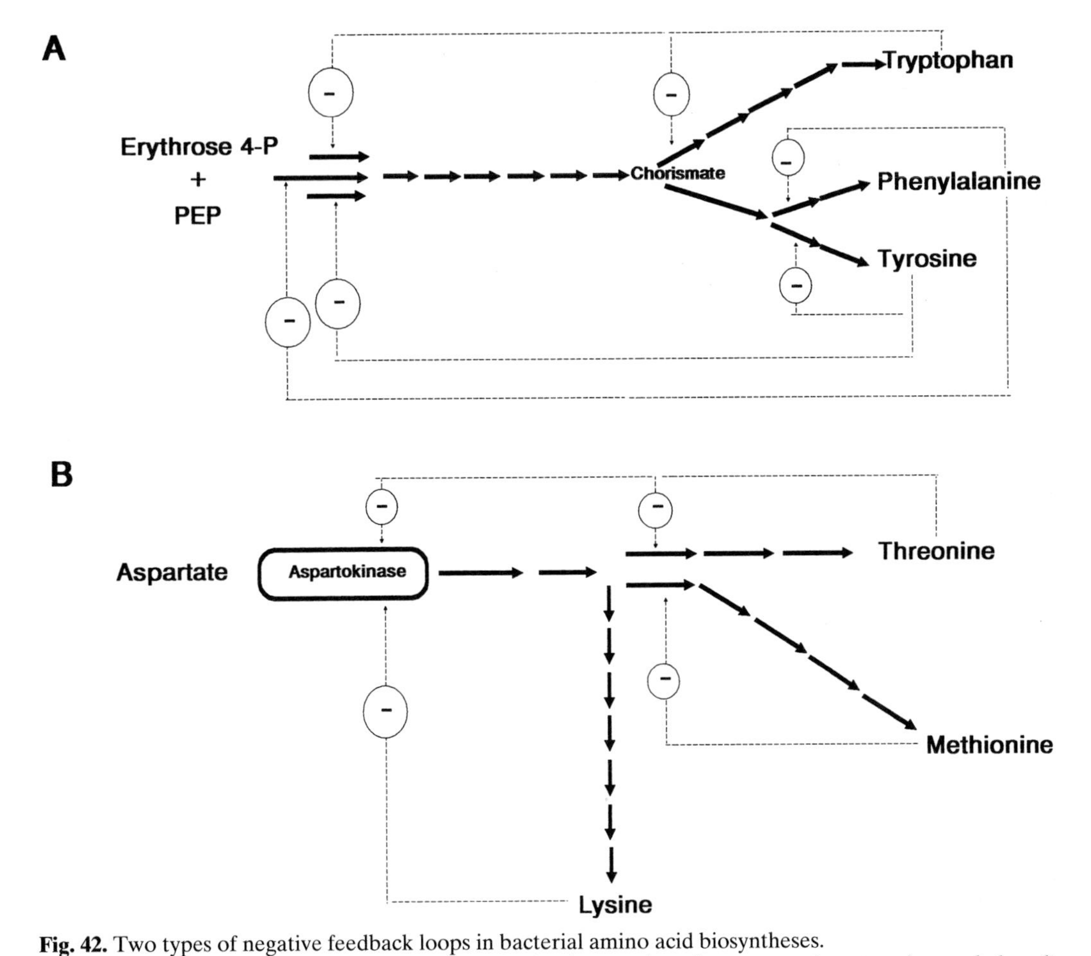

Fig. 42. Two types of negative feedback loops in bacterial amino acid biosyntheses.
(A) In the aromatic amino acid synthesis of *E. coli*, the three end products tryptophan, tyrosine, and phenylalanine inhibit the first committed step in their respective pathways *and* the first enzymatic step of aromatic amino acid synthesis, i.e., connection of erythrose 4-P and phosphoenolpyruvate (PEP). In *E. coli*, three isofunctional enzymes catalyze the first step and are differentially regulated by the end products.
(B) In Gram-positive bacteria, the aspartate-derived amino acids threonine and lysine regulate in a cumulative manner the single common enzyme aspartokinase. Threonine and methionine moreover regulate the first steps to their synthesis.

Flux Control of Metabolism

Metabolic pathways, which are linear and unbranched, result in a steady-state metabolic flux (J) from one substrate (S) to one product (P) through different intermediate metabolites (A, B, C, D, E ...). Substrates or products of an enzyme in a late step of a given pathway may act as feedback inhibitors (I) on an early step enzyme (see above). The intracellular amounts of the different enzymes of a pathway may vary.

Two extreme types of metabolic pathways can be assumed, although in nature they are not realized in their pure forms: In channelled-type pathways, particular intermediates remain protein-associated (localized), sometimes in supramolecular structures made up of enzymes that catalyze succeeding steps in a given metabolism. These intermediates are not freely exchanged with an intermediate pool.

The other extreme is the pool-type pathway where all intermediates are freely exchangeable (diffusible) or delocalized within the cell. Thus, a particular metabolite could act both as substrate for one enzyme and as effector (I) of another enzyme. Metabolic control theory applied to the pool type of metabolism uses

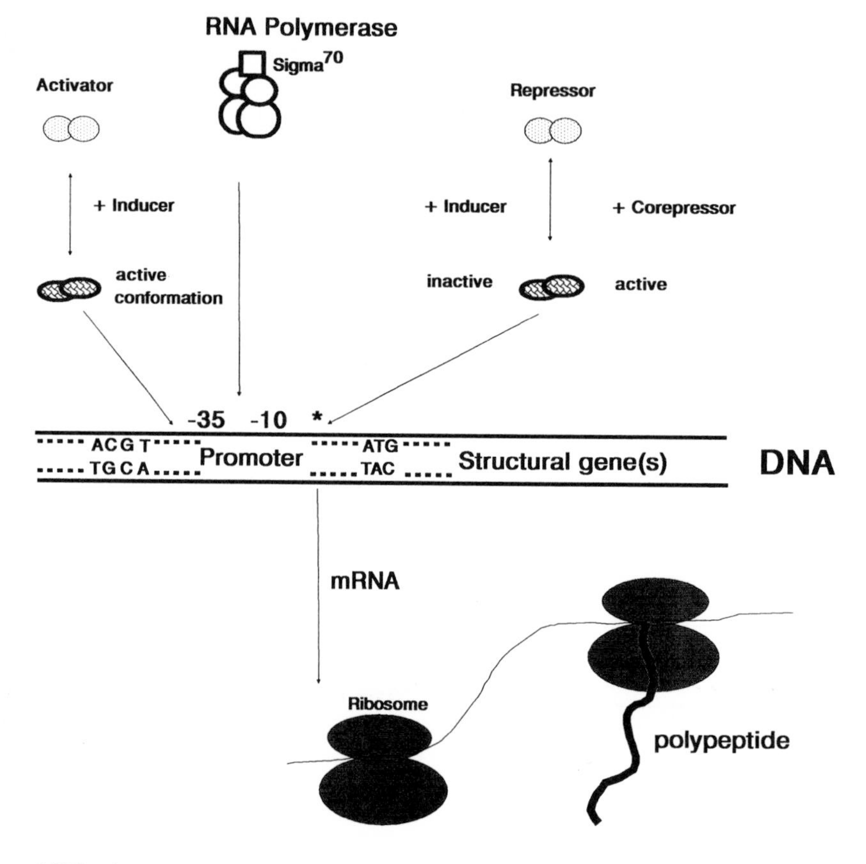

Fig. 43. Possible regulatory sites in transcription and translation of bacterial genes. Adapted with permission from NEIDHARDT et al. (1990)

mathematical terms (differential equations) to describe metabolic fluxes and controlling steps in metabolic pathways (for details see Chapter 4 of this volume).

6.2 Regulation of Enzyme Synthesis

Regulation of enzyme synthesis in microorganisms occurs predominantly at either the transcription initiation step (binding of RNA polymerase to DNA) or at the beginning of translation. In Fig. 43, a scheme of possible regulatory sites during enzyme synthesis in bacteria is given.

Many bacterial genes are transcribed at steady rates under differing growth conditions. Among them are so-called housekeeping enzymes which are involved in central metabolic pathways (e.g., glycolytic enzymes). Their transcription begins in *E. coli* with the binding of the RNA polymerase complex (RPC) to a DNA sequence (which is termed promoter) upstream of a given gene, and a closed complex of DNA and the RPC is formed (Fig. 44). The RPC for housekeeping genes has a subunit, sigma factor 70 or σ^{70} (the superscript indicates the molecular weight of the protein), which determines the promoter region and the start of DNA transcription into mRNA. As the RPC moves along the DNA, the two DNA strands separate (open complex) and thus allow the beginning of transcription. In *E. coli*, the σ^{70} subunit is now exchanged for another protein (NusA) which stays with the RNA-polymerase during transcription until a DNA structure is reached which determines the termination of the transcription (Fig. 44). Then, the RNA polymerase and the NusA protein dissociate and can enter a new round of transcription.

6.2.1 Sigma Factors as Regulatory Components

In *Escherichia coli*, DNA recognition is governed by the RPC sigma subunit σ^{70} which

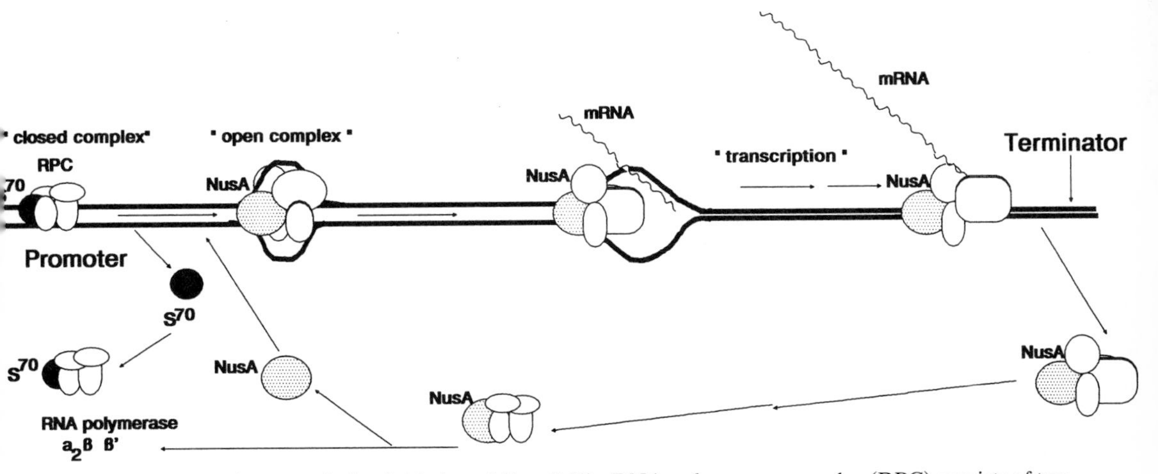

Fig. 44. Early steps in transcription initiation of *E. coli*. The RNA polymerase complex (RPC) consists of two α, one β and one β' subunit together with the sigma factor σ^{70}. The adventive protein NusA replaces the σ^{70} at the open complex and stays at the RPC until a terminator structure in the DNA is reached. Then the RPC dissociates from the DNA and can start a new round of transcription.

leads and binds the RNA polymerase to a promoter region. The σ^{70} subunit is specific for two boxes of 6 to 8 base pairs (bp) of DNA at a distance of 10 or 35 bp upstream of the transcriptional start point (–10 and –35 boxes). A σ^{70} promoter consists mainly of the bases A or T. The –10 and –35 boxes are separated by 16 to 18 base pairs of intervening DNA. Alternative sigma factors (for example, σ^{32}, σ^{43}, σ^{54} with molecular weights of 32 kDa and so on) are known for *E. coli*, *Bacillus subtilis*, *Streptomyces* and others, which are specific for promoter regions with features that are different from the σ^{70} promoter consensus region. These sigma factors govern sets of genes (up to 50 and more each) which are only transcribed under certain environmental conditions (e.g., N-limitation, heat-shock conditions, sporulation, oxygen stress). The formation of these alternative sigma factors itself is strictly regulated and responds to the different environmental conditions.

6.2.2 DNA Binding Proteins: Activators and Repressors

In bacteria, DNA binding proteins are known which interact with DNA regions that lie ahead of genes. These proteins permit or prevent the transcription of the downstream sequences by interference with the RNA polymerase. Proteins which lead to a tighter (more specific) binding of the RNA polymerase with DNA (positive regulation) are called activators, those which halt the transcription at an early stage (negative regulation) are termed repressors (see Fig. 43). These proteins often depend on small effector molecules to gain full DNA binding activity. For example, the compound 3′–5′ cyclic AMP (cAMP) forms a complex with the cAMP-accepting protein (CAP) of Enterobacteria. This allows the binding of the complex to specific sites on the DNA. The complex improves RNA polymerase binding and transcription (=activation). Tryptophan, the end product of a long biosynthetic pathway acts as a co-repressor, changing the TrpR repressor molecule into its active form (=repression of the *trp* gene transcription). On the other hand, repressor molecules in catabolic pathways may be inactivated by inducer molecules which are often the substrate (or a derivative thereof) of this catabolism, e.g., lactose in the *lac* genes. In the case of lactose, the synthesis of the enzymes of lactose catabolism is induced by the presence of lactose.

In some cases, the same protein molecule can even act both as an activator and/or a repressor, depending on the substrate and the DNA region to which it binds. An example is the L-arabinose utilization system of Enterobacteria.

6.2.3 Two-Component Systems of Regulation

In recent years, more and more examples of two-component regulatory systems in bacteria were unravelled. They consist of two proteins, one of which is termed sensor (or transmitter) and the other regulator (or receiver). Sensor molecules are often membrane-spanning proteins and act as protein kinases able to catalyze both ATP-dependent autophosphorylation and phosphoryl group transfer to other proteins (e.g., regulators). When stimulated by external or internal stimuli (e.g., osmotic pressure, chemoattractants, presence or absence of specific molecules) the sensor molecule undergoes autophosphorylation, thus altering a stimulus into a biochemical signal (phosphorylation state of a sensor molecule). This signal can be passed onto the soluble regulator protein by phosphoryl-group transfer (Fig. 45). The phosphorylated regulator then may interact with DNA sequences either as activator or as repressor of transcription. The two-component systems are models for signal transduction in microorganisms; they are, for instance, involved in phosphate and nitrogen regulation.

6.2.4 Regulatory Mechanisms at the RNA Level: The Attenuator Model

Once a transcript (mRNA) has been formed, there are still ways to control translation of this

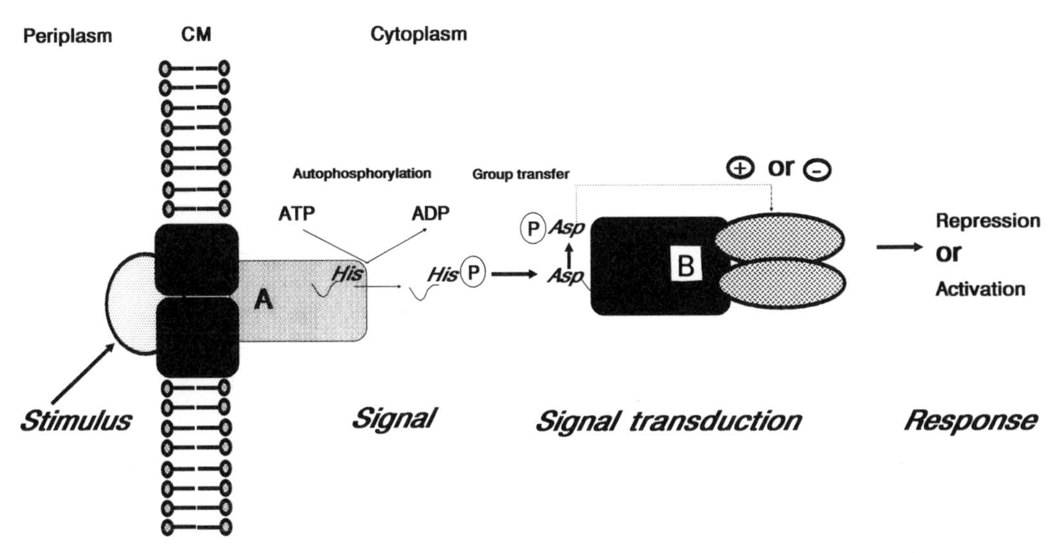

Fig. 45. Scheme of a two-component regulatory system in bacteria. Usually, a membrane-spanning protein (A) serves as sensor (or transmitter), and a soluble protein (B) acts as response regulator (or receiver). A stimulus is altered into a signal by autophosphorylation. The phosphoryl group can be transferred to the regulator protein which then can act as either repressor or activator on certain genes. CM = cytoplasmic membrane.

transcript. A famous paradigm for regulation at the mRNA level is the so-called attenuator model which is found in pyrimidine and several amino acid (histidine, phenylalanine, or tryptophan) biosyntheses of, e.g., *E. coli*. The mRNA contains a leader region upstream of the first structural gene showing an unusual accumulation of codons for the amino acid being formed by the anabolic route. In the famous case of the tryptophan attenuator (Fig. 46), 2 Trp codons lie adjacently in the leader region (Trp codons are normally rare in proteins). The attenuator model discerns between two cases: a) the cells are rich in an amino acid or b) are lacking a certain amino acid.

In the case of tryptophan, if the cell has plenty of Trp-carrying tRNAs, the leader peptide is formed without delay. Then the ribosome complex working on the trp mRNA keeps pace with the RNA polymerase complex which synthesizes the trp mRNA. So, the first configuration of alternatively possible mRNA secondary structures (stem loops) forms a pause loop (1:2) plus a terminator loop (3:4) (Fig. 46). If this happens, chances are high (9 out of 10 cases) that the RNA polymerase complex is stopped at the terminator loop and dissociates. Subsequently, the transcription of the structural genes is aborted ("terminated").

If the translation of the *trp* leader region is halted because of a lack of charged $tRNA^{trp}$ (absence of tryptophan), the ribosome is "stalled" at the Trp codons and prevents physically the formation of the pause loop. As a consequence, the antiterminator loop (2:3) can form, which in turn inhibits the formation of the terminator loop. Thus, the RNA polymerase is not stopped and proceeds transcribing the *trp* genes.

6.2.5 Eukaryotic Gene Regulation

Whereas bacterial gene regulation systems are quite well understood (for example, the *E. coli lac* system), eukaryotic gene regulation appears to be more complex and often proceeds in gradual steps. DNA binding proteins with regulatory functions are known for eukaryotic systems, as well as promoter structures. However, in contrast to most bacterial genes, DNA structures with regulatory functions need not be upstream of a given structural gene. Some structures (UAS for upstream activating se-

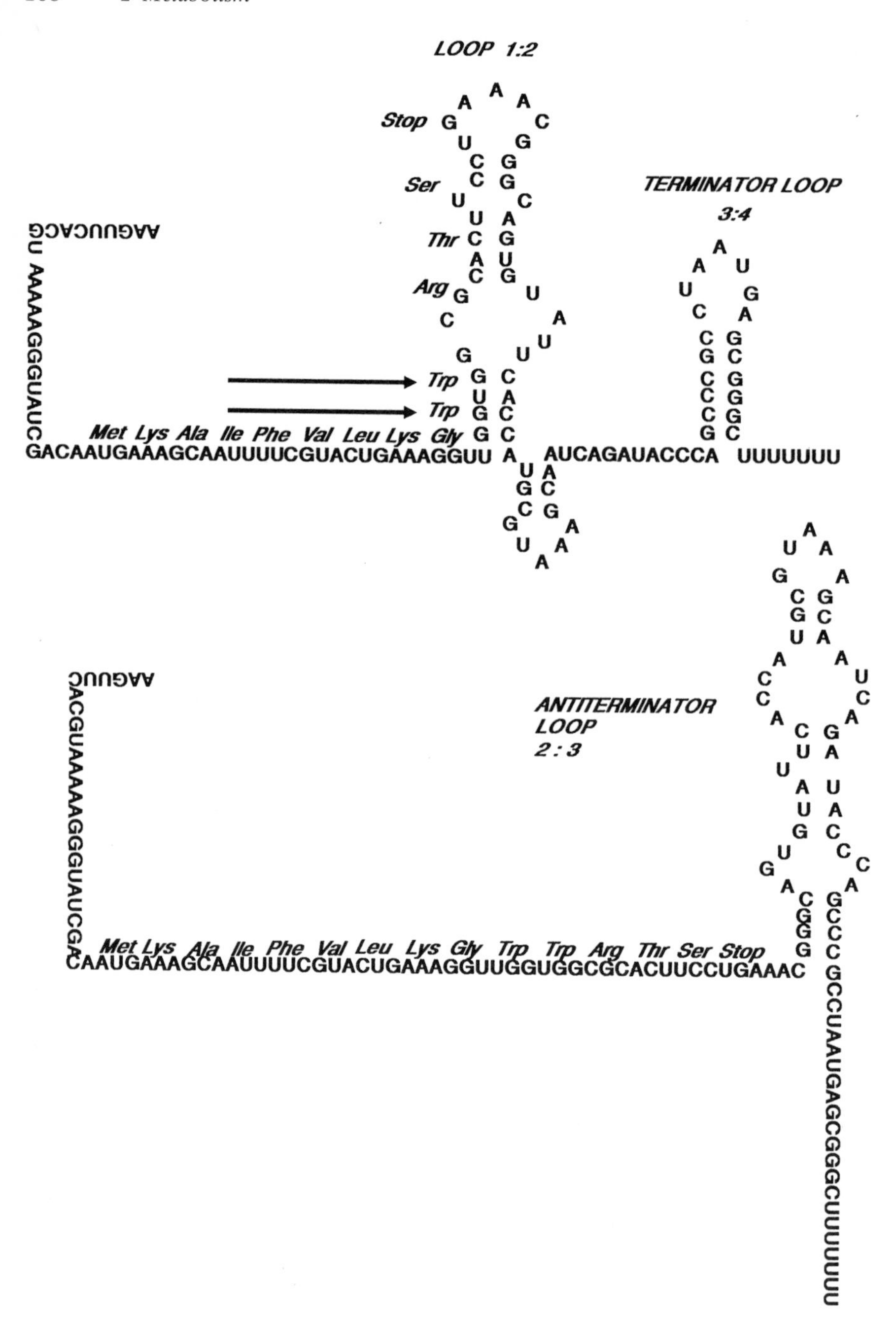

Fig. 46. The attenuator model of tryptophan regulation in *E. coli*. The early part of the single-stranded *trp* mRNA may form two alternative stem-loop structures (1:2 and 3:4 or 2:3). The 1:2 structure involves two codons for tryptophan (marked by arrows).
Adapted with permission from NEIDHARDT (1987)

quences) can function at a distance of several kilobases apart from their target gene and in both directions (enhancers). DNA methylation also plays an important role in determining the functional state of DNA (e.g., in tissues).

7 Further Reading

Textbooks

BROCK, T. D., MADIGAN, M. T. (1991), *Biology of Microorganisms*, 6th Ed., Englewood Cliffs, NJ: Prentice Hall.

GOTTSCHALK, G. (1986), *Bacterial Metabolism*, 2nd Ed., New York: Springer.

MOAT, A. G., FOSTER, J. W. (1988), *Microbial Physiology*, 2nd Ed., New York: John Wiley.

NEIDHARDT, F. C., INGRAHAM, J. L., SCHAECHTER, M. (1990), *Physiology of the Bacterial Cell: A Molecular Approach*, Sunderland, MA: Sinauer Associates Inc.

SCHLEGEL, H. G. (1992), *Allgemeine Mikrobiologie*, 7th Ed., Stuttgart: Thieme.

STRYER, L. (1988), *Biochemistry*, 3rd Ed., San Francisco: W. H . Freeman and Co.

ZUBAY, G. (1988), *Biochemistry*, 2nd Ed., New York: Macmillan.

Section 2

ANTHONY, C. (1988), *Bacterial Energy Transduction*, New York: Academic Press.

DAWES, E. A. (1986), *Microbial Energetics*, Glasgow: Blackie and Sons, Ltd.

DIMROTH, P. (1987), Sodium ion transport decarboxylases and other aspects of sodium ion cycling in bacteria, *Microbiol. Rev.* **51,** 320–340.

DREWS, G. (1986), Structure and functional organization of light-harvesting complexes and photochemical reaction centers in membranes of phototrophic bacteria, *Microbiol. Rev.* **49,** 59–70.

EDSALL, J. T., GUTFREUND, H. (1983), *Biothermodynamics: The Study of Biochemical Processes at Equilibrium*, New York: J. Wiley.

FUTAI, M., NOUMI, T., MAEDA, M. (1989), ATP Synthase (H^+-ATPase): Results by combined biochemical and molecular biological approaches, *Annu. Rev. Biochem.* **58,** 111–136.

HAROLD, F. M. (1986), *The Vital Force: A Study of Bioenergetics*, New York: W. H. Freeman.

REES, D. C., KOMIYA, H., YEATES, T. O., ALLEN, J. P., FEHER, G. (1989), The bacterial photosynthetic reaction center as a model for membrane proteins, *Annu. Rev. Biochem.* **58,** 607–634.

SENIOR, A. E. (1988), ATP synthesis by oxidative phosphorylation, *Physiol. Rev.* **68,** 177–231.

THAUER, R. K., JUNGERMANN, K., DECKER, K. (1977), Energy conservation in chemotrophic anaerobic bacteria, *Bacteriol. Rev.* **41,** 100–180.

WESTERHOFF, H. V., VAN DAM, K. (1987), *Thermodynamics and Control of Biological Free-Energy Transduction*, Amsterdam: Elsevier.

Section 3

DIMARCO, A. A., BOBIK, T. A., WOLFE, R. S. (1990), Unusual coenzymes of methanogenesis, *Annu. Rev. Biochem.* **59,** 355–394.

DIXON, M., WEBB, E. C. (1979), *Enzymes*, 3rd Ed., London: Longman.

DUINE, J. A., JONGEJAN, J. A. (1989), Quinoproteins, enzymes with pyrrolo-quinoline quinone as cofactor, *Annu. Rev. Biochem.* **58,** 403–426.

FERSHT, A. (1983), *Enzyme Structure and Mechanism*, 2nd Ed., San Francisco: W. H. Freeman and Co.

JENCKS, W. P. (1975), Binding energy, specificity, and enzymic catalysis: The Circe effect, *Adv. Enzymol.* **43,** 219–410.

WALSH, C. (1979), *Enzymatic Reaction Mechanisms*, San Francisco: W. H. Freeman and Co.

Section 4

AMES, G. F.-L., MIMURE, D., SHYAMALA, V. (1990), Bacterial periplasmic permeases belong to a family of transport proteins operating from *Escherichia coli* to human: Traffic ATPases, *FEMS Microbiol. Rev.* **75,** 429–446.

ANTHONY, C. (1988), *Bacterial Energy Transduction*, New York: Academic Press.

BALDWIN, S. A., HENDERSON, P. J. F. (1989), Homologies between sugar transporters from eukaryotes and prokaryotes, *Annu. Rev. Physiol.* **51,** 243–256.

HAROLD, F. M. (1986), *The Vital Force: A Study of Bioenergetics*, New York: W. H. Freeman and Co.

KABACK, H. R. (1988), Site-directed mutagenesis and ion-gradient driven active transport: On the path of the proton, *Annu. Rev. Physiol.* **50,** 243–256.

KONINGS, W. N., POOLMAN, B., DRIESSEN, A. J. M. (1992), Can the excretion of metabolites by bacteria be manipulated? *FEMS Microbiol. Rev.* **88,** 93–108.

KRULWICH, T. A. (1990), *Bacterial Energetics*, New York: Academic Press.

MALONEY, P. C., AMBUDKAR, S. V., ANATHARAM, V., SONNA, L. A., VARADHACHARY, A. (1990), Anion-exchange mechanisms in bacteria, *Microbiol. Rev.* **54,** 1–17.

Meadow, N. D., Fox. D. K., Roseman, S. (1990), The bacterial phosphoenolpyruvate : glycose phosphotransferase system, *Annu. Rev. Biochem.* **59,** 497–542.

Nikaido, H. (1992), Porins and specific channels of bacterial outer membranes, *Mol. Microbiol.* **6,** 435–442.

Rosen, B. P., Silver, S. (Eds.) (1987), *Ion Transport in Prokaryotes*, San Diego: Academic Press.

Tanford, C. (1980), *The Hydrophobic Effect: Formation of Micelles and Biological Membranes*, 2nd Ed., New York: Wiley-Interscience.

Wickner, W., Driessen, A. J. M., Hartl, F.-U. (1991), The enzymology of protein translocation across the *Escherichia coli* plasma membrane, *Annu. Rev. Biochem.* **60,** 101–124.

Section 5

Anraku, Y. (1988), Bacterial electron transfer chain, *Annu. Rev. Biochem.* **57,** 101–132.

Barker, H. A. (1981), Amino acid degradation by anaerobic bacteria, *Annu. Rev. Biochem.* **50,** 23–40.

Freifelder, D. (1987), *Microbial Genetics*, Boston, MA: Jones and Bartlett.

Fuchs, G., et al. (1992), Biochemistry of anaerobic biodegradation of aromatic compounds, in: *Biochemistry of Microbial Degradation* (Ratledge, E., Ed.), Dordrecht: Kluwer Acad. Publ.

Gibson, D. T. (Ed.) (1984), *Microbial Degradation of Organic Compounds*, New York: Marcel Dekker Inc.

Gottschalk, G. (1986), *Bacterial Metabolism*, 2nd Ed., New York: Springer.

Leahy J. G., Colwell, R. R. (1990), Microbial degradation of hydrocarbons in the environment, *Microbiol. Rev.* **54,** 305–315.

Lewin, B. (1990), *Genes IV*, Oxford: Oxford University Press.

Lin, E. C. C., Iuchi, S. (1991), Regulation of gene expression in fermentative and respiratory systems in *Escherichia coli* and related bacteria, *Annu. Rev. Genet.* **25,** 361–387.

Neidhardt, F. C. (Ed.) (1987), *Escherichia coli and Salmonella typhimurium. Cellular and Molecular Biology*, Washington, DC: American Society for Microbiology.

Neidhardt, F. C., Ingraham, J. L., Schaechter, M. (1990), *Physiology of the Bacterial Cell: A Molecular Approach*, Sunderland, MA: Sinauer Associates Inc.

Postgate, J. (1989), Trends and perspectives in nitrogen fixation research, *Adv. Microb. Physiol.* **30,** 1–22.

Stewart, V. (1988), Nitrate respiration in relation to facultative metabolism in Enterobacteria, *Microbiol. Rev.* **52,** 190–232.

Tabita, F. R. (1988), Molecular and cellular regulation of autotrophic carbon dioxide fixation in microorganisms, *Microbiol. Rev.* **52,** 155–189.

Watson, J. D., Hopkins, N. H., Roberts, J. W., Steitz, J. A., Weiner, A. M. (1987), *Molecular Biology of the Gene*, 4th Ed., Menlo Park, CA: Benjamin/Cummings.

Section 6

Adhya, S., Garges, S. (1990), Positive control, *J. Biol. Chem.* **265,** 10797–10800.

Beckwith, J., Davies, J., Gallant, J. A. (Eds.) (1983), *Gene Function in Prokaryotes*, Cold Spring Harbor, NY: Cold Spring Harbor Laboratory Press.

Collado-Vides, J., Magasanik, B., Gralla, J. D. (1991), Control site location and transcriptional regulation in *Escherichia coli*, *Microbiol Rev.* **55,** 371–394.

Cornish-Bowden, A. (Ed.) (1990), *Control of Metabolic Processes*, New York: Plenum Press.

Greenblatt, J. (1991), RNA polymerase-associated transcription factors, *Trends Biochem. Sci.* **16,** 408–411.

Kell, D. B., Westerhoff, H. V. (1986), Metabolic control theory: its role in microbiology and biotechnology, *FEMS Microbiol. Lett.* **39,** 305–320.

McCarthy, J. E. G., Gualerzi, C. (1990), Translational control of prokaryotic gene expression, *Trends Genet.* **6,** 78–85.

Merrick, W. C. (1992), Mechanism and regulation of eukaryotic protein synthesis, *Microbiol. Rev.* **56,** 291–315.

Neidhardt, F. C. (Ed.) (1987), *Escherichia coli and Salmonella typhimurium. Cellular and Molecular Biology*, Washington, DC: American Society for Microbiology.

Neidhardt, F. C., Ingraham, J. L., Schaechter, M. (1990), *Physiology of the Bacterial Cell: A Molecular Approach*, Sunderland, MA: Sinauer Associates Inc.

Roeder, R. G. (1991), The complexities of eukaryotic transcription initiation: Regulation of preinitiation complex assembly, *Trends Biochem. Sci.* **16,** 402–408.

Stock, J. B., Ninfa, A. J., Stock, A. M. (1989), Protein phosphorylation and regulation of adaptive responses in bacteria, *Microbiol. Rev.* **53,** 450–490.

Yanofsky, C. (1988), Transcription attenuation, *J. Biol. Chem.* **263,** 609–612.

3 Growth of Microorganisms

CLEMENS H. POSTEN
Braunschweig, Federal Republic of Germany

CHARLES L. COONEY
Cambridge, MA 02139, USA

1 Introduction 113
2 Modes of Growth 113
  2.1 Bacterial Growth 113
  2.2 Yeast Growth 115
  2.3 Mycelial Growth 116
3 Measurement and Characterization of Growth 118
  3.1 Measurement of Cell Number 119
    3.1.1 Direct Microscopic Count 119
    3.1.2 Viable Plate Counts 120
    3.1.3 Coulter Counter 121
    3.1.4 Flow Cytometer 121
  3.2 Measurement of Cell Mass 121
    3.2.1 Dry Cell Weight 121
    3.2.2 Packed Cell Volume 123
    3.2.3 Turbidity 123
    3.2.4 Image Analysis 123
    3.2.5 Model-Based Estimation 124
    3.2.6 Other Measuring Principles 126
4 Growth in Response to the Environment 127
  4.1 Physical Environment 127
    4.1.1 Effect of Temperature 127
    4.1.2 Mechanical Effects 129
    4.1.3 Effect of Water Activity 130
    4.1.4 Effect of pH 131
  4.2 Chemical Environment 131
    4.2.1 Effect of Carbon Source 132

4.2.2 Effect of Nitrogen Source 133
4.2.3 Dissolved Oxygen 134
4.2.4 Effect of Carbon Dioxide 135
4.2.5 Alkanols 135
4.2.6 Organic Acids 136
4.2.7 Other Chemical Nutrients 137
4.3 Variability and Constraints of Growth 138
4.3.1 Growth for Different Groups of Microorganisms 138
4.3.2 Macromolecular Composition of Cells 138
4.3.3 Limiting Steps 139
4.3.4 Energy Demand of Growth 141
4.3.5 Heat Evolution of Microorganisms 141
5 Design of Media for Growth and Production 143
5.1 Nutrient Requirements 143
5.1.1 Carbon 143
5.1.2 Nitrogen 144
5.1.3 Macro-Elements 145
5.1.4 Trace (Micro-)Elements 145
5.1.5 Growth Factors 145
5.2 Medium Optimization 145
5.2.1 Material Balances 145
5.2.2 Chemostat Pulse Method and Factorial Design 146
5.2.3 Other Mathematical Optimization Techniques 147
5.3 Technical Media 148
5.3.1 Molasses 148
5.3.2 Cheese Whey 149
5.3.3 Corn Steep Liquor 149
5.3.4 Cassava 149
5.3.5 Bagasse and Other Cellulosic Materials 150
6 Culture Techniques 150
6.1 Batch Culture 150
6.2 Fed-Batch Culture 154
6.3 Continuous Culture 155
7 References 159

1 Introduction

The capacity to grow, and ultimately to multiply, is one of the most fundamental characteristics of living cells. However, the definitions of growth are as broad as the scientific fields dealing with this topic.

Cell biologists are often interested in morphological changes of single cells especially during cell fission. Biochemistry investigates growth resulting from a few thousands of enzymatically catalyzed biosynthetic steps, organized in biochemical pathways, the kinetics of the growth process being therefore the overall kinetics of integrated enzymatic activity. From a biophysicist's point of view, cells are open systems far from a thermodynamic equilibrium, exchange material and energy with their environment, and especially exhibit a large outflow of entropy. In chemical engineering, growth is referred to as an increasing amount of biocatalyst. Mathematical descriptions of growth are restricted to a couple of equations employing hyperbolic and exponential terms. Reduced to a common denominator, growth is usually considered as an increase of cell material expressed in terms of mass or cell number.

The objective of this chapter is, to give a general view of some aspects of growth of microorganisms. These aspects include modes of growth of single cells and cell aggregates, measurement of growth, limited to concurrent measurement of only some facets of biomass, dependence of growth on environmental factors and finally growth in bioreactors, influenced by man-made and somehow controllable impacts.

2 Modes of Growth

For the purpose of microbial growth quantitation, it is useful to consider some of the general growth properties of different classes of microorganisms and to attempt to establish some principles useful for interpreting measurements of their growth. The discussion of growth will focus on bacteria, yeasts, and mycelial molds. This grouping is chosen to reflect the mechanism of growth rather than usual phylogenic lines. The rationale for this is that problems in quantitation of microbial growth are largely a result of the physical form of the microorganism, i.e., single cell versus mycelia, and the mechanism by which they divide, i.e., fission, budding or chain elongation. The interest here is to provide a basis for quantitative microbiology. In this context, each group of organisms is considered in terms of its physical properties, mechanism of replication, and form of growth.

2.1 Bacterial Growth

Bacteria exhibit a wide diversity in metabolic activities, but all have similar cellular structure and reproduction mechanisms. They are classed as *prokaryotic* organisms (BALOWS et al., 1992). The various genera of bacteria are related back to a common ancestral state (WOESE, 1987), however, through evolution, substantial structural and physiological diversity has developed. In Gram-positive cells the polysaccharide murein forms up to 30 molecular layers, while in Gram-negative cells only a single murein layer is present, and lipopolysaccharides and lipoproteins are the main constituents of the cell wall. Prokaryotes also have no nuclear membrane or other intracellular organelles.

Despite this diversity, there are several features common to many bacteria that are important to quantitative microbiology. Bacteria generally reproduce by the process of binary fission, illustrated in Fig. 1, resulting in two daughter cells of equal size. A cell grows by increasing in size, during which time the amount of each new cell component, e.g., protein, RNA, etc., is doubled and the genome is replicated. Cell division is initiated by ingrowth of the cell wall and eventual formation of a transverse septum. Cell separation proceeds by cleavage of the septum, and two identical daughter cells are formed. There is a difference, however, in how the Gram-positive and Gram-negative bacteria synthesize their cell wall material. Gram-positive bacteria synthesize new cell wall in an equatorial zone along an axis, whereas Gram-negative bacteria

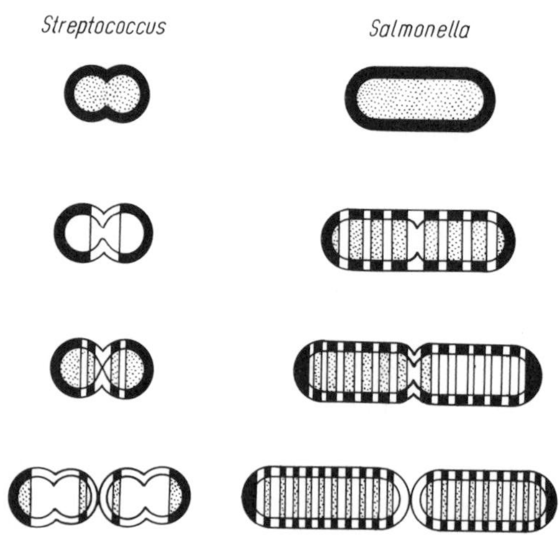

Fig. 1. Process of binary fission and cell wall addition of a Gram-positive species of *Streptococcus* and a Gram-negative species of *Salmonella*; dark areas are old cell wall material, light areas represent newly added compounds.

synthesize cell wall by intercollation along the whole wall. Incomplete cleavage of the septa will result in chains as is the case for streptococci. A microscopic photograph is shown in Fig. 2. Delayed cleavage will result in elongated bacilli structures.

The concentration of cell components, i.e., RNA, enzymes, metabolites, etc., in each daughter will be the same as in the parent. This is true, however, only when the cells are growing in an environment that does not necessitate a change in some cell property with time. If, for instance, an environmental change calling for induction or repression of some enzyme occurs between initiation of a cell cycle and cell division, then the daughter cells will have a different level of one or more enzymes than the original parent. Some bacteria, especially members of the family Bacillaceae and some gliding bacteria (Fig. 3), have the ability to form spores to survive in adverse conditions.

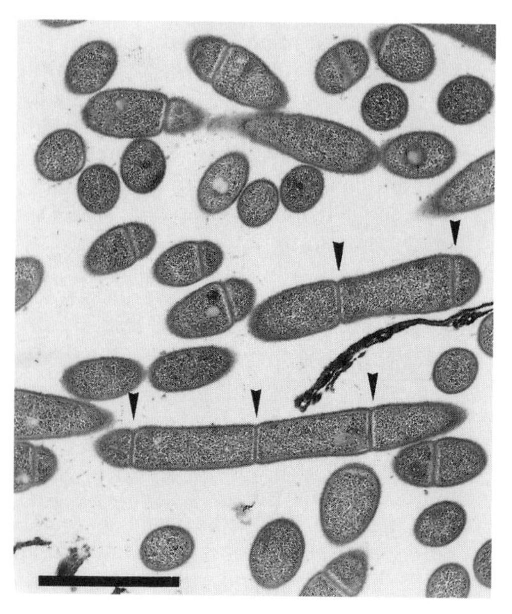

Fig. 2. Electron micrograph of an ultrathin section of filamentous gliding Gram-negative *Herpetosiphon auranticus*. Several bacteria are sectioned longitudinally and show complete septa formation (arrow heads), bar = 2 μm (LÜNSDORF, GBF, 1993).

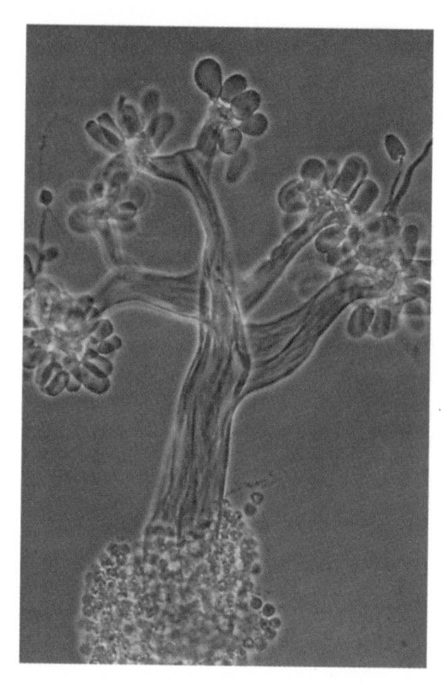

Fig. 3. Several fruiting bodies on a sporangiophore and sporangioles of the gliding myxobacterium *Condromyces crocatus* (H. REICHENBACH, GBF).

From this discussion, it should be apparent that bacteria do not have a wide age distribution. One can talk about "old" or "young" cultures in reference to how long a flask has been incubating or in terms of how long a culture has been left standing, but the true age of a cell relates to the time since its last division.

Bacteria are generally small with a characteristic dimension of about 1 μm. They may exist as spheres or cocci, or as rod or bacilli in shape. The cells may exist as single cells or as groups or chains, e.g., diplococci (pair), streptococci (chains), staphylococci (clusters). Furthermore, many organisms contain flagella and are motile. A typical bacterial cell has a wet density (specific mass) of 1.05 to 1.1 g cm^{-3} and weighs about 10^{-12} g as a dry particle. The density is about 1.25 g cm^{-3}. The actual size of a given cell will depend on its growth rate; faster growing cells are often larger. Bacteria growing on Petri dishes form colonies with a species specific appearance. More details about industrial microorganisms can be found in DEMAIN and SOLOMON (1985).

2.2 Yeast Growth

Yeasts are *eukaryotic* organisms which belong to the fungi. The most prominent members are baker's yeast *Saccharomyces cerevisiae*, brewer's yeast *S. uvarum (carlsbergensis)* and the fodder yeasts *Candida utilis* and *C. tropicalis*. They are fungi that do not form asexual spores or aerial structures and exist as single cells during at least part of their vegetative growth cycle. The most common form of cell division is budding (Fig. 4), however, fission following cross-wall formation, mycelial growth by chain elongation, and branching are also observed in some yeasts (PHAFF et al., 1978). The yeasts are non-motile and non-photosynthetic. They are either oval or spherical in shape. The size of a yeast cell is dependent on the growth rate; the shorter the doubling time, the larger the cell volume. A typical cell may be about 3–7 μm of width, 5–15 μm length and has a dry cell weight (see Sect. 3.2.1) of about 10^{-11} g, its density being 1.05–1.1 g cm^{-3} in the living state.

During growth of yeast by budding, there are several distinct events. Initially, the yeast

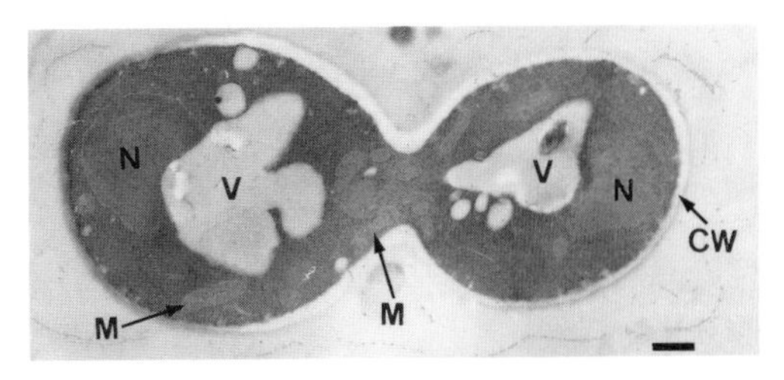

Fig. 4. Electron micrograph of an ultrathin section of *Saccharomyces cerevisiae*; bar represents 0.5 μm (CW, cell wall; M, mitochondria; N, nucleus; V, vacuole); mitochondria accumulate in the zone of highest metabolic activity (RHODE, GBF, 1993).

cell undergoes a period of expansion: its volume increases. Shortly after the cell stops expanding, bud emergence occurs. During bud formation, the total volume of the mother plus daughter bud cell is constant, so that bud growth occurs as a consequence of depletion of the mother cell. The bud separates as a single, but smaller cell from the mother. Once separated, the new daughter cell and the original mother cell grow and reach the same size at the same time; thus, the daughter grows faster than the mother (THOMAS et al., 1980). The mother cell will have a bud scar on its surface for each bud that has separated; these can be seen with fluorescent techniques. Unlike bacteria, mother and daughter yeast cells are different. They possess different growth rates, and their cell surface is different. It is possible to count the number of scars and to establish a cell's age in the broth. Thus, there is a distribution of cells having different ages. While yeasts are typically single-celled, their progeny,or daughter cells, will sometimes not separate, see Fig. 5. When buds do not separate from the mother, the resulting chains of cells are called pseudomycelia.

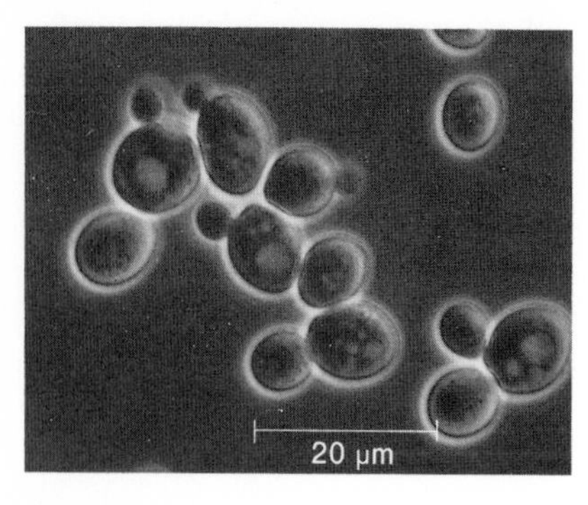

Fig. 5. Budding yeasts cells (*Saccharomyces cerevisiae*) with 1st and 2nd generation daughter cells (NK2, isolated from African palm wine, A. EJIOFOR and M. RHODE, GBF, 1993).

2.3 Mycelial Growth

Molds, actinomycetes, and some yeasts under aerobic conditions, predominantly grow by the process of hypha chain elongation (WEBSTER, 1980) (e.g., elongation at the tip, also called apical growth) and branching, as shown in Fig. 6. The hypha, which is divided into individual cells, is a branching tubular structure of 2–10 μm in diameter. The intertwining strands of hyphae are called mycelium. Many fungi can form asexual spores called conidia (Fig. 7). A comprehensive overview of fungal biotechnology has been given by ELANDER and LOWE (1992).

The length of a hyphal chain depends on the growth environment. If left undisturbed, as on an agar surface, the chain can become quite long. In submerged culture, however, there are shear forces that cause hyphae fragmentation. This results in shorter, but more highly branched mycelia. In submerged cultures, the mycelia may exist along with dispersed, diffuse mycelia or may form pellets; the form of growth has an important effect on growth and product formation. A detailed discussion of mycelial growth kinetics has been presented by RIGHELATO (1979). Model-based simulations (YANG et al., 1992) (Fig. 8) have brought more insight into the process of mycelial growth.

Microorganisms respond to their environment. Cells growing within a pellet will "see" a very different environment than those growing in a more diffuse manner. A measurable bulk concentration of nutrient or product exists in the broth. However, the concentration of *nutrient* on the pellet surface is lower due to diffu-

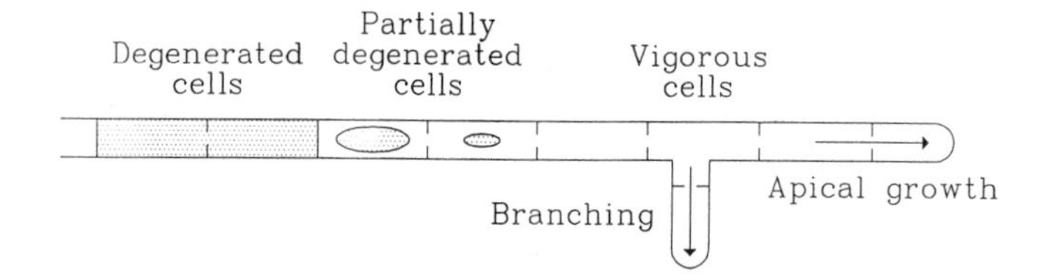

Fig. 6. Schematic diagram of mycelial growth by branching and chain elongation. Shaded areas represent degenerated regions, while unshaded areas are cytoplasmic regions.

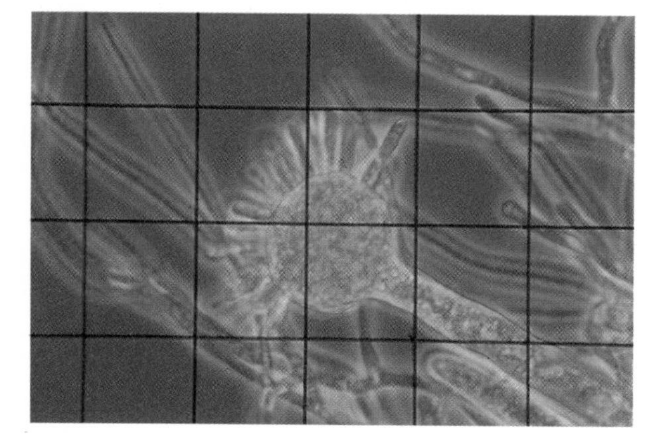

Fig. 7. Hyphae and conidial structure with conidiophore, sterigmata, and conidia (spores) of the fungus *Aspergillus niger*. (FRINKEN and HEIBER, GBF, 1991).

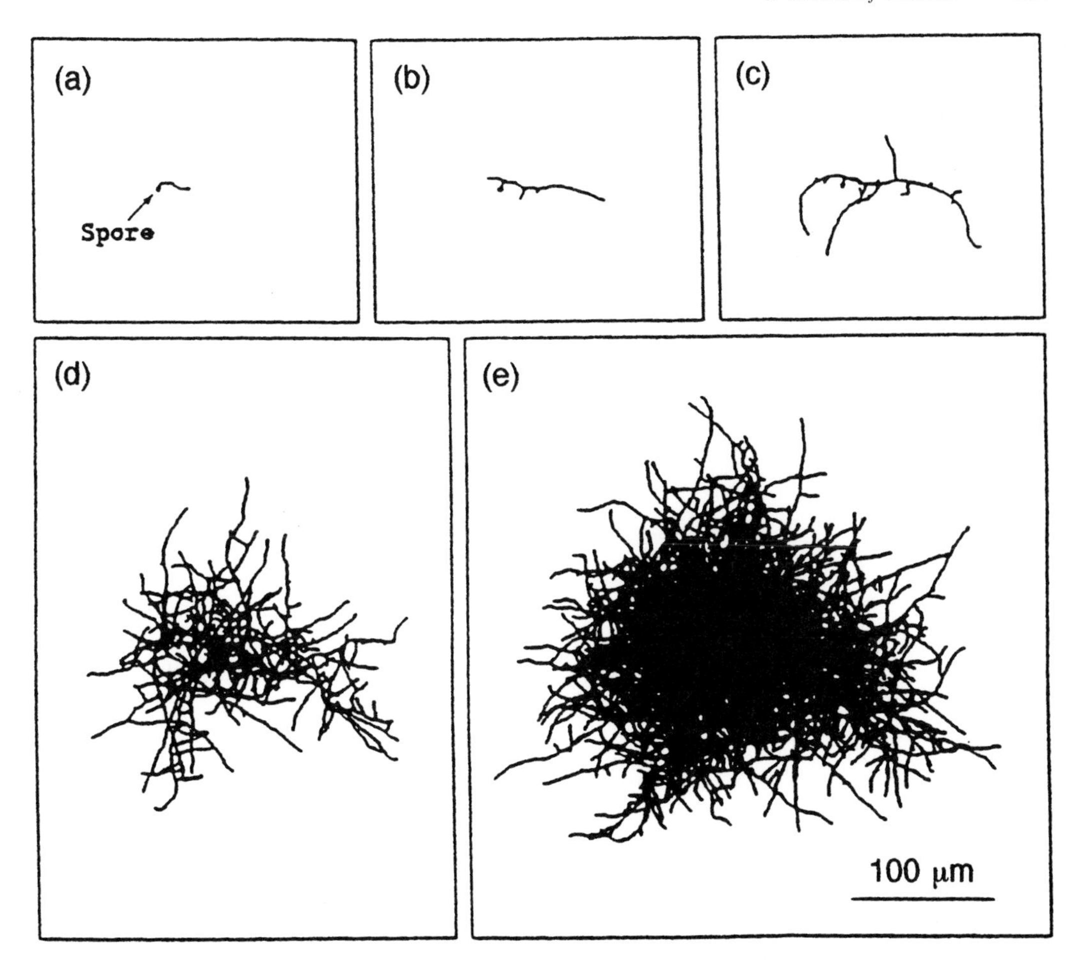

Fig. 8. Simulated morphological development of the growth of a mycelium emerging from a spore: (a) $t=2.5$ h, (b) $t=4$ h, (c) $t=6$ h, (d) $t=10$ h, and (e) $t=14$ h (with kind permission from YANG et al., 1992).

sion through a stagnant liquid boundary layer and cell metabolism. The concentration is reduced further inside in the pellet as a result of metabolism, to such a degree, that, at the center of the pellet, there may be little or no nutrient. This problem is especially important when considering oxygen supply (WITTLER et al., 1986). The *product* concentration, on the other hand, will be higher within the pellet than outside resulting from further diffusional limitation. This is an important issue only if there is product inhibition.

The physical structure of pellets can be seen in Fig. 9. Early in a fermentation, the cells in the pellet see a nutrient-rich environment, later the pellets become more dense, and growth and metabolism occur predominantly on the periphery. As a restult, cells near the pellet center become starved for one or more nutrients.

Mycelial growth leads to an age distribution of cells. Younger cells at the hyphal tip will have different metabolism than cells at or near the origin of growth. This is due to both the aging process inherent in the cell and the differing environments of young and old cells. Ideally, methods of measurement would allow the investigator to distinguish among cells of different ages. However, this is difficult to achieve, and in most cases, the performance of a mycelial population is *normalized* to the total amount of cell mass present.

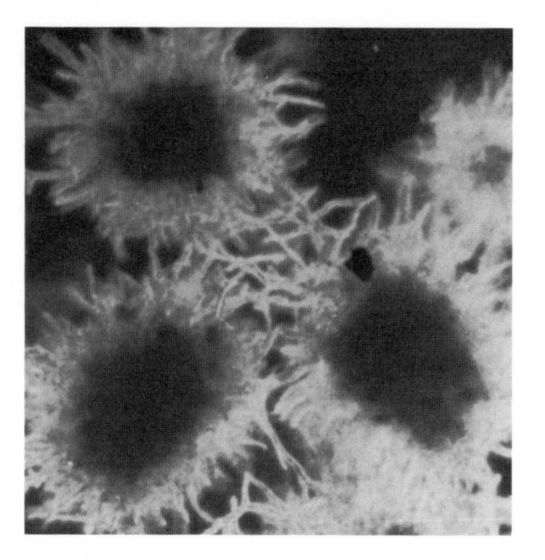

Fig. 9. Photograph showing a pellet during a fermentation of *Penicillium chrysogenum*.

3 Measurement and Characterization of Growth

An essential activity of fermentation technology is the quantitation of both the amount and the rate of change of microbial cell mass. This can often be achieved by direct measurement of either cell mass or number. In many cases, direct methods of measurement are not applicable, and the physiological activity, which is partially related to the amount of biomass, must be measured. Since the main point of interest for many practical applications is the activity of microorganisms, many people make a virtue of necessity. A detailed review is given in SONNLEITNER et al. (1992). Unfortunately, there is no single method that is suitable for all fermentations, and a fermentation technologist needs sets of methods for growth measurement of bacteria, yeast and molds in various fermentation media. The objective of the following discussion is to present the reader with a summary of the most common methods available for microbial growth quantitation. These methods must be examined carefully for each application to be sure that they are appropriate in each case. The methods are categorized as measurement of cell number and measurement of cell mass.

From a practical point of view, it is important to classify measurements as on-line or off-line measurements. The focus lies on the latter ones, but some of the principles (e.g., turbidity, model-based estimation) are applied on-line, too. Excellent reviews on such measurement principles have been presented by REARDON and SCHEPER (1991) and LOCHER et al. (1992). In practical fermentations, sampling is done in constant time intervals. Since bioprocesses are non-linear and time-variable, optimal sampling can reduce the effort and increase the information content of the measurements (MUNACK, 1991).

Apart from the problems to define the amount of biomass, the mathematical definition of the specific growth rate is unequivocal. An increase (or decrease) in the amount of biomass X or cell number N is assumed to be proportional to the total amount actually present, thus yielding

$$\dot{X}(t)=\mu(t)\cdot X(t) \tag{1a}$$

where the (arbitrarily time-dependent) factor μ is called the specific growth rate. The definition makes sense especially if each part and fraction of the biomass has the equal capacity to grow. This lumped definition of growth rate does not imply any assumptions about the status of individual cells, where some may proliferate, while others may show lysis.

The definition of the doubling time t_d of a cell population, where

$$N(t+t_d)=2\cdot N(t) \tag{1b}$$

is commonly used to describe growth in terms of cell number or cell mass. Integration of Eq. (1a) (see Sect. 6.1 and Eq. 12) and application to Eq. (1b) gives

$$t_d=\ln 2/\mu \tag{1c}$$

which relates the number- or mass-based doubling time to the specific growth rate.

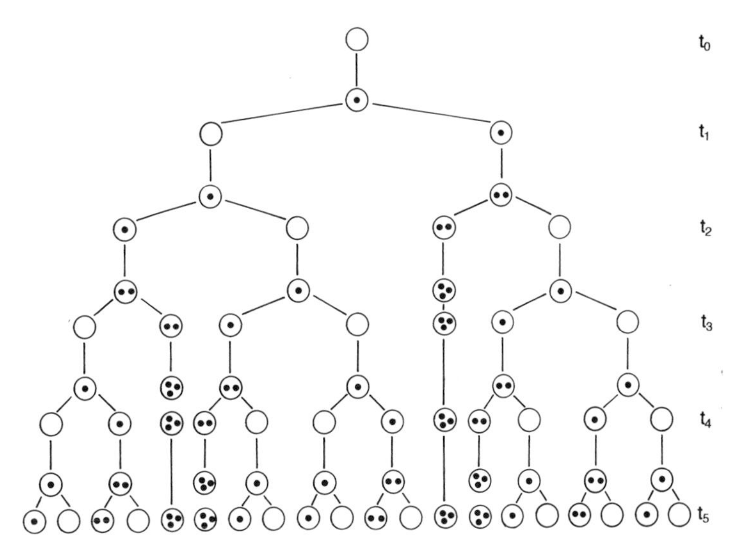

Fig. 10. Simplified model of segregational plasmid instability. The growth rate of plasmid-containing, protein-producing cells (black dots) is decreased compared to the growth rate of plasmid-free cells with the result of a changing cell distribution.

The number-based definition is less useful in cases of segregated biomass by age or activity distribution or even of different organisms. Fig. 10 shows the development of a population of recombinant cells, where different growth rates for different cell types apply. If only a couple of single cells are investigated according to cell number where each cell fission changes the population remarkably, then statistical definitions for growth are implemented.

3.1 Measurement of Cell Number

Cell number is, at least in the case of singly suspended cells, a biologically appropriate defined value to measure an amount of biological material. Conceptual and practical problems arise if formation of cell aggregates occur. These problems include how to define the point where a growing bud of a yeast cell becomes an individual cell, how to distinguish between a dividing cell and two just divided cells in a chain of bacteria or how to count fungal cells in a pellet, especially if fast formation of biological material at the tip of the hyphae and lysis in elder parts is observed. The concept of viability plays an important role. The definition of a single cell to be counted involves not only particles surrounded by a cell membrane, but also qualities representing life directly, such as physiological activity and capability to divide measured as capability of colony formation. The techniques of cell counting presented here provide the reader with an account of the present possibilities. However, for practical application a textbook for microbiological methods should be consulted (e.g., GERHARDT et al., 1981).

3.1.1 Direct Microscopic Count

A rapid method for counting the total number of cells is direct microscopic count utilizing a counting chamber, such as the Petroff–Hausser slide. It consists of a microscopic slide with a depression of known depth and a cover slip marked with a grid of known area. With a phase contrast microscope, a sufficiently high number of cells (typically 500 and more) is counted under an appropriate number of grid squares. The cell density is then calculated from the average cell number per square divided by the corresponding fluid volume beneath the square. The resulting data indicate the total number of cells, but do not quantitate the number of viable cells unless a viable stain such as methylene blue is used. This stain is oxidized to a colorless form by cells capable of respiring, a trait usually associated with viability although it is possible for non-dividing cells to still respire. Dead or non-respiring cells will stain blue.

The use of direct microscopic counts depends on the ability to distinguish and count individual cells and is inapplicable to mycelial or chain-forming organisms. Actively mobile bacteria can be sedimented by adding 0.5% of formalin to the cell suspension; this facilitates counting.

Measurement of cell number in counting chambers is primarily designed for microbial cultures which usually must be substantially diluted prior to counting. In environmental samples, such as river and lake water, the cell number usually is too low for this procedure. The most widely accepted technique to count these microorganisms is membrane filtration combined with epifluorescence microscopy which makes detection of bacteria easier and allows discrimination among living cells, dead cells and detritus to some extent. The preparation of the filter (preferably polycarbonate) involves drying, staining with a fluorochrome dye such as acridine orange, decoloration and embedding in immersion oil.

3.1.2 Viable Plate Counts

The most common method for determining the number of viable cells is to perform a viable plate count. This technique is based on the principle that a population of cells can be spread across a solid nutrient medium such that each cell is separated from the other and, upon division, will form a distinct colony in which all progeny are derived from a single parent forming what is called a clone; this process is called cloning. The most common solid medium is agar. The nutrient medium is prepared by mixing the required growth nutrient with typically 1.5–2.5% agar and, after heating to above 100°C, solidifies when cooled below about 45°C. This medium is commonly poured into Petri dishes. Culture samples need to be diluted such that 30 to 300 cells are applied to one 100 mm diameter Petri dish. More than 300 colonies (about 25 mm^2 of plate area per colony) will lead to overcrowding and difficulty when counting the colonies.

Considerable care is needed when interpreting the results of a viable plate count. This technique measures the number of colony-forming units (CFU) that can form in the environment of the plates. Cells that agglomerate, form chains or hyphae, cannot be counted by this technique, for colonies may evolve from more than a single cell. The solid surface of agar is a different environment than submerged culture, and viability (percent of total organism that will divide and grow) on a plate may be very different than in liquid broth. Motile bacteria will swarm over the plate surface and not form distinct colonies, particularly if the agar surface was not properly dried.

Plate counting requires little capital expense, but significant labor is involved in preparing and counting plates. After plating, the cultures are typically incubated for 24–72 h to allow the colonies to form.

A modified version of the plate count is the slide culture technique. This technique, described by POSTGATE et al. (1969), allows one to place a culture sample onto a small agar medium plate mounted on a microscopic slide. After incubation for 2–4 doubling times, microcolonies will form from viable cells. With the microscope, it is possible to do both a total and viable cell count.

A convenient method to do total bacterial counts in dilute suspensions is to filter a known volume sample with a membrane filter (pore size 0.22 μm) such that the cells are trapped on the filter surface. The filter then is laid on top of a suitable nutrient agar medium and incubated until colonies appear; knowing the volume of the sample filtered, the original concentration of viable organisms can be calculated. With a microscope, both total and viable cell counts can be made; often staining procedures and side lighting aid counting.

All the methods described also apply to viable counts in natural water and soils. However, it must be kept in mind that these samples contain a highly diverse bacterial population of which only a small fraction will grow on the medium and under the conditions provided. Only in highly polluted environments will the viable counts approach the numbers directly counted on membrane filters.

3.1.3 Coulter Counter

Manual methods for cell number estimation are slow and labor-intensive. Thus, there is considerable need for enrichment, selection, and automated techniques which increase both the number of colonies examined and the probability that the organism desired will be among the number examined. Although the classical viable count techniques have been automated to some extent, electronic devices for measuring the total cell number exhibit a much greater potential. One of these is the Coulter counter (COULTER, 1956; HARRIS and KELL, 1985). This device is based on measurement of a current passing through a small, e.g., 30–500 μm, orifice. Every time a microorganism passes through this orifice, the resistance increases and the current falls. The current response can be recorded not only as a pulse or singular event, but also the magnitude of the current decrease is proportional to the cell volume. Thus, the instrument gives both cell number and cell size distribution but no cell shape information. The cell counts provide total and not viable cell numbers. The particle size range that can be counted is 0.5–200 μm, thus, bacteria are near the lower limit of detection and measurement. Using the Coulter counter is preferable in clean cell suspensions with single cells and not with agglomerates, chains or hyphae. When working with populations of cells having different sizes, e.g., budding yeasts, it is possible to follow each population independently by looking at the cell volume distribution.

3.1.4 Flow Cytometer

Cell size and some intracellular and cell surface qualities can be measured by using optical properties of cells. In a flow cytometer a culture sample is diluted in a flowing stream which separates each cell longitudinally from its neighbors. As it passes through a laser beam, absorption, fluorescence or light diffusion is used to register a count. The sensitivity and selectivity of a flow cytometer can be improved by labeling the cells with a fluorescent dye. Thus, many of the methods for vital staining and metabolic activity can be adapted for use in a flow cytometer. Measurements of viability, nucleic acid or protein content, as well as intracellular pH have been reported (REARDON and SCHEPER, 1991).

Cell sorting is done by placing a charge on the drops containing the desired cells; this may be done in response to the fluorescent signal from the cell. When the drops pass charged deflection plates, they are deflected to a collector. With flow cytometry it is possible to count, analyze and collect up to 10000 cells/s.

3.2 Measurement of Cell Mass

As manifold as the qualities and activities of living cells are, are the approaches to measure the amount of the great unknown *X* which is referred to as biomass (SONNLEITNER et al., 1992). To have a mass is only one physical property of cell material. Others are volume, optical appearance, material exchange with environment or response to other (intentionally applied) physical signals. Fig. 11 shows measurement results for different variable characterizing growth as pointed out in the following sections. Different measuring principles focus on different qualities of biological material and, therefore, exhibit non-proportional results during a bioprocess.

3.2.1 Dry Cell Weight

One of the most useful measurements characterizing a fermentation is dry cell weight (DCW, which actually is mass, measured in grams). Although it provides a simple means of quantifying the amount of cell mass in many fermentations, it, in fact, is difficult to interpret. In any case, it differs substantially from the mass of living cells which contain approximately 80% water. Dry cell weight is commonly used as a reference value to calculate specific rates of growth, product synthesis or nutrient consumption.

DCW is measured after a known sample of broth is centrifuged or filtered, washed with water or buffer, and dried typically at 80 °C for 24 h or at 110 °C for 8 h. For cells grown in a water-soluble medium, this technique is generally applicable as a means of quantifying cel-

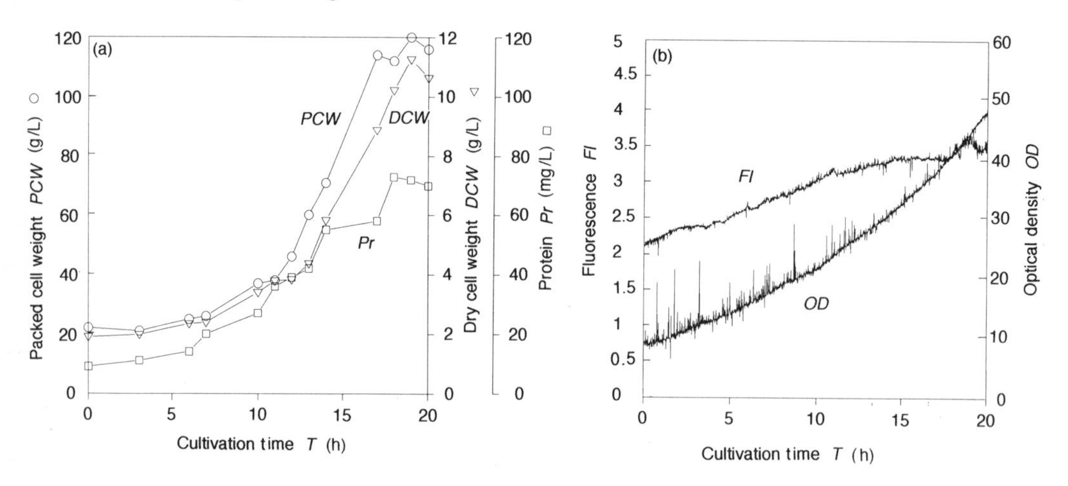

Fig. 11. Real-life measurements with different disturbances; (a) off-line measurements, (b) on-line measurements.
Off-line measurements exhibit obvious out-liers, some of them are correlated (e.g., wrong sampling time or volume). The fluorescence signal shows high offset, while both on-line signals are disturbed by drift and high peaks, resulting from air bubbles and particles. Besides these disturbances, the different measuring methods show different aspects of growth, e.g., during lag phase and stationary phase turbidity increases, while dry cell weight is constant or even decreasing.

lular material. However, to fully interpret the results, it is useful to examine each step of the procedure in the context of the real world of fermentation technology.

First, a broth sample is removed from the fermentor. It must be assumed that this sample reflects the cell concentration found in the whole fermentor. However, many large fermentors are not well mixed and a sample collected from the surface, behind a baffle or at the bottom is not the same as one from the center. This is especially a problem in mycelial fermentations. The volume of the sample must be known. If a mycelial culture with high cell concentration is sampled, the broth may contain gas resulting in a density substantially less than 1.0 g cm^{-3}. Thus, calculation of DCW per unit volume sampled may give a concentration that is as much as 10–25% lower than the actual cell mass per unit of liquid. For this reason, samples are often taken on a weight rather than on a volume basis. If substantial amounts of insoluble nutrients, e.g., soybean meal, cellulose, hydrocarbons, etc., are present, they further reduce the amount of liquid per unit volume, and again samples are best taken on a mass (weight) rather than a volume basis.

Cell separation is usually achieved by centrifugation or filtration. These techniques not only remove cell mass but also other insoluble materials which are commonly found in industrial fermentation media. In some cases, the non-cellular solids can be reduced by washing. For instance, salts such as calcium carbonate, added as a slow release buffer, may be removed by an acid wash. Hydrocarbons or oils can be dissolved in a solvent wash. It is important that these washing procedures do not also remove cell material. Frequent washing may remove low-molecular weight soluble material from the cell; this may account for 5–10% of the cell weight.

After washing, the cell mass must be dried and weighed. Too little drying will result in some retention of water; too much drying will result in charring and volatilization of some cell material. Clearly, the conditions must be optimized.

The net result of DCW analysis may be ±30% or more of the actual cell mass in the fermentor. Furthermore, it reflects total cell mass present, dead or alive, and does not account for cell mass formed and then lost by cell lysis. Nevertheless, DCW is a very useful parameter for normalizing fermentation data, e.g., calculation of specific rates such as mass of product per mass of cells present and specific yields such as mass of product per mass of cells.

3.2.2 Packed Cell Volume

A quick routine method often used in industrial plants to follow fermentation kinetics is packed cell volume (PCV) or "spin down". The measurement employs a conical, graduated centrifuge tube that allows the volume of sediment to be measured after centrifugation. The PCV will depend not only on cell mass concentration, but also on centrifugation speed and time, the amount of non-cellular solids, the culture morphology, and the osmolarity of the medium. Centrifugation conditions are usually set to minimize the effect of these variables. The amount of non-cellular solids, which are usually nutrients, will decrease with time. It is interesting to note that these solids usually settle faster than cells; therefore, it is possible to use this technique to measure, at least semiquantitatively, the residual solids in the broth (this value is useful for correcting dry cell weight measurements to reflect the cell concentration per unit volume or weight of liquid). The morphology of mycelial organisms will alter the packing density in the pellet and is reflected in the volume measurement. Changes in morphology are usually consistent for a given fermentation, thus allowing fermentation runs to be standardized; however, this variable prevents accurate estimation of cell mass from cell volume. Likewise, osmolarity of the broth will affect cell swelling and, hence, volume.

Despite the above problems, the packed cell volume technique is useful to follow kinetics and observe color, morphology, and non-cellular solids variation during the fermentation. A rough estimation of cell mass can be made by dividing the cell volume per liter by 5 mL per g cell. This assumes that a cell is 20% dry solids and that density is about 1.0 g/mL.

3.2.3 Turbidity

Perhaps the most widely used method to follow growth of bacterial or yeast cultures is broth turbidity. The measured turbidity is a sum of absorption and forward scatter of light by medium, cells, and other particles (WYATT, 1973). Turbidity measurements are usually made in the region of 600–700 nm (red filter) where light absorption by cell components is at a minimum. A spectrophotometer is linear from 0.2 to 1.0 absorbance units. Of course, linearity must be confirmed for different culture conditions, because the measured values depend on the color of the broth and the size of the cells. One absorbance unit is about 0.5 g/L of cell dry weight. However, conversion of absorbance values to cell dry weight must be treated with care, for a conversion factor is valid only for the culture conditions for which it was determined and may vary with cell composition and shape as well as with the medium composition. Turbidity readings from different photometers are not exactly comparable, even if the same wavelength is used, for the position of the cuvettes relative to the detector may vary from instrument to instrument influencing the part of the scattered light reaching the detector.

3.2.4 Image Analysis

Especially in the case of filamentous growth and pellet formation it is difficult to measure the growth of individual cells or the total cell mass. In these cases image analysis is a useful tool for the classification and estimation of the biomass (COX and THOMAS, 1992). Samples of fermentation broth are investigated under a microscope with zoom facility (magnification about 10 times). Captured images are then processed by a computer using software for image analysis. However, this approach still needs some heuristical adjustment to recognize thresholds and exclude impurities.

3.2.5 Model-Based Estimation

In many bioprocesses not only conceptual problems defining "biomass", but also practical problems measuring cell number or cell mass may arise. First, reliable or representative samples cannot be obtained. This is due to inhomogeneities caused by wall growth and immobilization, incomplete mixing, flocculation or pellet formation. Second, the growth medium contains substantial non-cellular solids, e.g., from natural complex media or biomass derived particles like cell envelopes. Third, direct methods do not meet the objectives of the measurement. From an engineering point of view, for example, overall activity of the biocatalyst in the reactor may be of more interest than a biomass concentration value which is difficult to interpret. In other cases, direct biomass measurement might be too expensive or too slow for control purposes. Then, the physiological interactions of the cells with their environment can be exploited as a source of information for indirect biomass estimation (STEPHANOPOULOS and SAN, 1984). The relation between measurable material or energy exchange and the amount of cell mass is described by a so-called mathematical model. Such models exhibit a wide range of complexity; for a classification and discussion see BELLGARDT (1991). Model-based biomass estimation procedures are sometimes called, impressively but not quite correctly, "software sensors".

A simple case of a model is to consider the bioprocess as a chemical transformation. A generalized form of stoichiometry, showing major nutrients, for cell growth and product formation is:

$$\text{C-source} + O_2 + \text{N-source} \rightarrow \text{cell mass} + CO_2 + H_2O + \text{products} + \text{heat} \quad (2a)$$

$$a\,C_xH_yO_z + b\,O_2 + c\,N_uH_vO_w \rightarrow d\,C_\alpha H_\beta O_\gamma N_\delta + e\,H_2O + f\,CO_2 + g\,C_nH_mO_pN_r + \Delta H \quad (2b)$$

where a through g are stoichiometric coefficients. From this stoichiometry it is apparent that measurement of substrate (e.g., C-source, N-source, O_2, etc.) consumption, product (e.g., chemical product, CO_2, heat, etc.) formation or some elemental or chemical component of the cell can be related to the cell mass. One way to evaluate this transformation for biomass estimation is to set up elemental and mass balances as described by COONEY et al. (1977) and further developed by ROELS (1983), where, as an example, the carbon balance

$$(a \cdot x) - (d \cdot a) - f(1-g)n = 0 \quad (3a)$$

and the nitrogen balance

$$(c \cdot u) - (d \cdot \delta) = 0 \quad (3b)$$

are given in molar amounts.

The only biological information needed is the elemental biomass composition for all growth conditions of interest (Fig. 12). Even slight errors in these values lead to large errors in the estimation results. In principle, one elemental balance is enough to calculate the biomass concentration. Depending on the number of measurable and the number of non-measurable (H_2O formation!) compounds, the biomass concentration can be estimated with higher accuracy, if more equations are employed. This technique has been applied to yeast in defined medium (WANG et al., 1977) and mold growth in complex medium (MOU and COONEY, 1983b).

The general scheme, Eq. (2), must be modified for different process situations, where different representations of the balance equations apply (see Sect. 6). For on-line use, where many substrates and products are barely measurable, a most common approach is the measurement of the ammonia uptake rate, which can be derived from pH control, and of CO_2 and O_2 in the off-gas (STEPHANOPOULOS and SAN, 1984). Since overall turnover rates are measured, this approach gives an estimation not directly of biomass, but of the overall growth rate $\mu \cdot X \cdot V$.

Although elemental balances must be valid in every case, they often cannot be evaluated. Possible reasons for this are undefined media, lack of measurement devices, such as balances or off-gas analyzers, or a wide product spectrum, where not all compounds are measured. A mathematical pitfall comes from the linear dependencies of the balance equations. The mass balance, for example, is exactly the sum

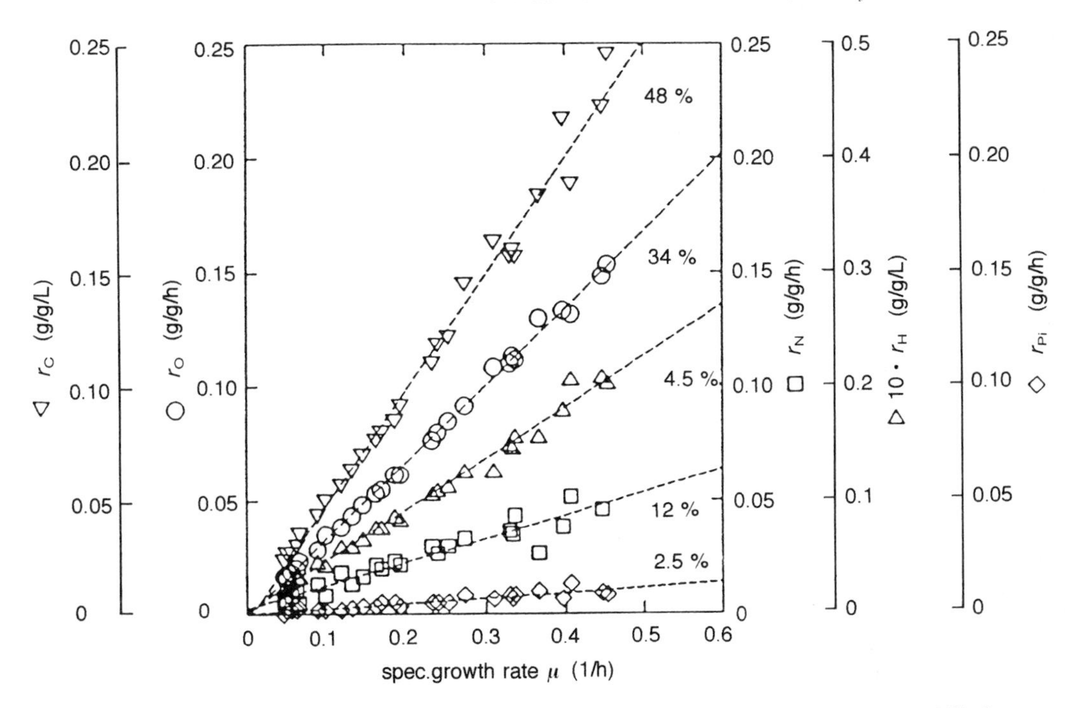

Fig. 12. Evaluation of elemental balances of different continuous cultivations of *Zymomonas mobilis* the sum of elemental fluxes of nutrients and products are shown as a function of specific growth rate; the slope of the regression line is the (unkown) elemental composition of the cells (from POSTEN, 1989).

of all elemental balances and, therefore, contains no additional information. If a product exhibits the same relative composition as the substrate (e.g., glucose and acetate), they cannot be estimated from the balance equations.

If well known yield coefficients in Eq. (2b) are available as additional process-specific information, then additional linear equations can be set up to allow for biomass estimation, even if some compounds in the medium are not measurable (e.g., MOU and COONEY, 1983a). A comprehensive treatise of more complex cases has been given by BASTIN and DOCHAIN (1990). In a simple case (Fig. 13), one easily measurable variable, in this case oxygen consumption, exhibits a strong correlation with biomass (see also Sect. 4.3.3). The reliability of this approach depends on the constancy of the correlation coefficient between the measurement and the cell mass. If, for example, uncoupled growth is possible, then indirect observation via respiration will disguise possible toxic effects.

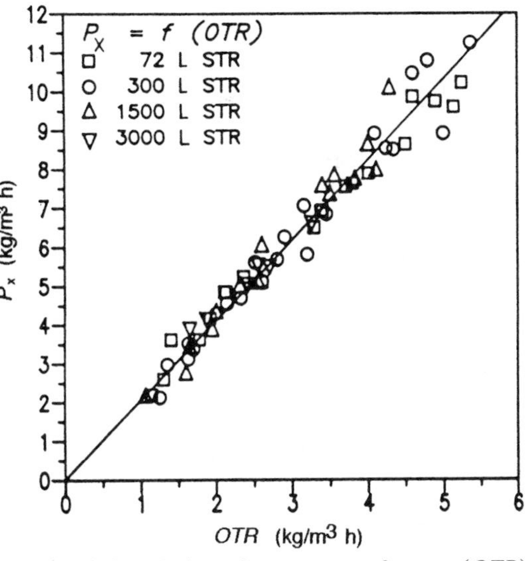

Fig. 13. Correlation of oxygen transfer rate (*OTR*) and biomass formation (P_X) for different bioreactors and different growth conditions; STR, stirred tank reactor (from SCHLÜTER et al., 1992).

More complex models consider, in addition to elemental balances and yields, (non-linear) kinetics, e.g., for substrate uptake or product formation. Kinetic models are very useful for biomass estimation and control purposes. Specific growth rate μ and biomass concentration X are estimated separately as a function of the measurable bioprocess variables, usually using a Kalman filter (STEPHANOPOULOS and PARK, 1991). However, such biomass estimations require the exact measurement of compounds in the fermentation broth. A detailed description of the state of the art method has been given by SCHÜGERL (1991a, b). Analytes of special interest in the context of "software sensors" are listed in Tab. 1. For details on available biosensors see SCHELLER and SCHMID (1992).

If even intracellular structures and dynamics are taken into consideration, a so-called structured model has been established (BELLGARDT, 1991; NIELSON and VILLADSEN, 1992). Such models can provide a deeper understanding of growth processes, but practical applications in biomass estimation are rare, because of the expensive procedure required to construct such a model.

3.2.6 Other Measuring Principles

Since nearly all chemical and physical parameters in a fermentation broth are somehow related to biomass amount or activity, many different proposals for indirect biomass measurements have been made (REARDON and SCHEPER, 1991; SONNLEITNER et al., 1992).

Viscosity measurements have been used to estimate growth, when either substrate (e.g., cellulose) or product (e.g., xanthan) exhibit high viscosity. The *heat of fermentation* is stoichiometrically related to metabolic activity and, therefore, suitable for measurement of bacterial growth. However, quantitative calorimetric measurements are very expensive and are especially useful for studying changes in the stoichiometry of bioprocesses (see Sect. 4.3.5). In large bioreactors, the amount and temperature of cooling water can be employed as an approximation to measure microbial activity.

It is of vital interest for controlling and optimizing bioprocesses to be able to measure the

Tab. 1. Methods for Measurement of Analytes Relevant for Indirect Biomass Estimation

Elements	Analytes	Methods
Carbon	Alcohol (e.g., ethanol)	Enzymatic assay Flow injection analysis Gas chromatography Gas membrane sensor High pressure liquid chromatography Infrared analysis Mass spectrometry Nuclear magnetic resonance spectroscopy
	Carbohydrates (e.g., glucose)	Enzymatic assay Flow injection analysis Glucose electrode High pressure liquid chromatography Thick film biosensor
	Hydrocarbons	Gas chromatography High pressure liquid chromatography
Nitrogen	Ammonia	Ammonia gas electrode Ammonia ion electrode Chemical analysis Enzymatic assay Mass spectrometry
	Nitrate	Chemical analysis Enzymatic assay Nitrate electrode
	Urea	Enzymatic assay Flow injection analysis Biosensors
Oxygen		Flow injection analysis Galvanic electrode Mass spectrometry Paramagnetic analyzer Polarographic electrode
Phosphorus	Phosphate	Chemical analysis Enzymatic assay Flow injection analysis High pressure liquid chromatography Nuclear magnetic resonance spectroscopy
Others	Amino acids	Biosensors Chemical analysis Enzymatic assay High pressure liquid chromatography
	Minerals	Atomic absorption Specific ion electrodes

activity and the amount of biomass on-line, especially with simple and sterilizable sensors. *Capacitance* measurement aims directly to the dielectric properties of the cells (FEHRENBACH et al., 1992), while *conductivity* is a measure for the amount of solutes in the medium and, therefore, a more indirect measurement. The most popular measurements are *turbidity* and *fluorescence* (SCHEPER et al., 1987), which are reviewed in REARDON and SCHEPER (1992). These are different from the other methods mentioned in the previous sections because they are not passive, but rather apply an active physical stimulus to the cells. Along with problems with non-linearities and sensitivities to other medium components, it must be mentioned that a fluorescence sensor, e.g., is not a "biomass sensor", because not only the biomass, but also the physiological state of the cells influence the measured signal. Nevertheless, if a careful interpretation is assumed, such measurements are reliable to a certain extent to monitor and evaluate bioprocesses with respect to growth and even more to characterize the metabolic state of the cells.

4 Growth in Response to the Environment

Microbial growth and product formation occur in response to the environment. Therefore, to understand growth and product formation, it is essential to understand the relationship between regulation of microbial metabolism and the physical and chemical environment. Each microorganism responds uniquely with its "own personality" to the environment, and it is this "personality" which provides the microorganisms with a selective and competitive advantage in their usual ecological niche. Microbial response to the environment is often interactive; while growing in or adapting to an environment, microorganisms will alter the environment as a consequence of their growth activities and, in some cases, as a means to improve their competitive advantage against other organisms. Microorganisms respond with both physical and chemical mechanisms which provide them with a selective advantage. The objective here is to characterize growth in response to the environment and as a result of generally applicable biological and physical laws.

4.1 Physical Environment

Most microorganisms are separated from their environment by cell membranes or even cell envelopes. Thus, they can control material fluxes into the cells, often very accurately. For some parameters, like pH or water activity, only a rough moderation by the membrane itself is achieved, while the cytoplasm, when exposed to energy inputs, like heat, is totally unprotected. In such cases, cell internal activities can be observed to compensate for the eventually harmful effects. The following sections give some impressions of the influence of different physical stimuli on growth.

4.1.1 Effect of Temperature

Microbial growth rate, as all chemical reactions, is a function of temperature. In general, microorganisms will grow over a temperature range of 25 to 30 °C. However, it is important to note that there are, in the environment, microorganisms that grow at temperatures below 0 °C and above 90 °C — the primary requirement is liquid water. The effect of temperature on growth rate may be seen in Fig. 14. Shown here are four groups of microorganisms, the psychrophilic, mesophilic, thermophilic, and extreme thermophilic. There are three generalizations that can be made concerning the effect of temperature on growth rate. First, growth only occurs over a 30 °C range for any microorganism; second, growth rate increases slowly from a minimum temperature below which growth no longer occurs, with increasing temperature until the maximum growth rate is reached; and third, above the temperature for the maximum, growth rate falls rapidly with further increase to the maximum temperature above which growth is not possible. Temperatures for minimum and maximum growth rates are called the cardinal temperatures.

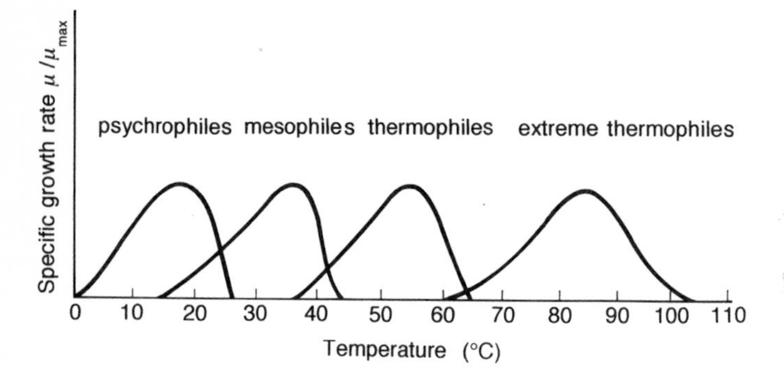

Fig. 14. The effect of temperature on specific growth rates of psychrophiles, mesophiles, thermophiles, and extreme thermophiles.

The effect of temperature on the specific growth rate can be described by the Arrhenius equation reflecting growth as an overall chemical process,

$$\mu = A\,e^{-E_a/RT} \tag{4a}$$

where A is the Arrhenius constant, E_a is the activation energy, R is the universal gas constant, and T is absolute temperature. A plot of the growth rate versus reciprocal of absolute temperature according to

$$\ln \mu = \ln A - \frac{E_a}{R} \tag{4b}$$

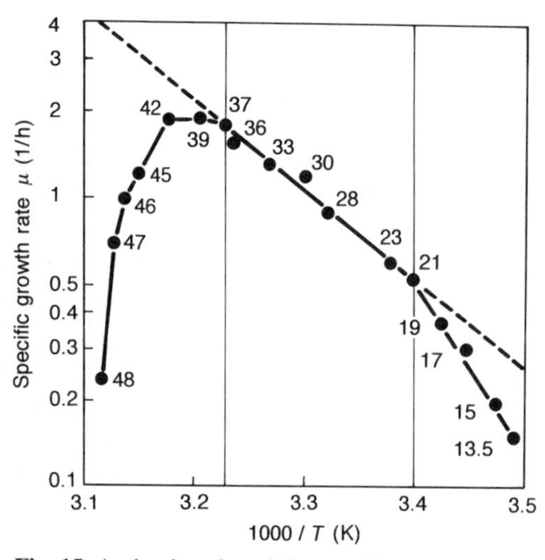

Fig. 15. Arrhenius plot of the specific growth rate of *Escherichia coli* B/r on glucose-rich medium. Individual data points are marked with corresponding degrees Celsius (after HERENDEEN et al., 1979).

is shown in Fig. 15. Typical values of activation energy for microbial growth are in the range of 50 to 70 kJ/mol. Above the optimum temperature for growth the overall growth rate, Eq. (4a), begins to fall. This results from the increased rate of microbial death caused by protein denaturation and finally thermal lysis. Death rate is a strong function of temperature and is characterized by activation energies of 300–380 kJ/mol. This high value of E_a for microbial death means that the rate of death increases much faster with temperature than the rate of growth (low value of E_a). Hence, overall growth rates rapidly decline above the optimum. For temperatures below the minimum, membrane gelling occurs with a subsequent diminished transport process.

Temperature also affects the efficiency of the carbon-energy substrate conversion to cell mass. The maximum conversion yield occurs at a temperature that is *less* than the temperature for maximum growth rate. This point is particularly important in process optimization, when it is desired to maximize yield, but *not* growth rate. It is not surprising that temperature affects a variety of metabolic processes differently. Macromolecular composition, especially RNA content, is a strong function of temperature as well as of growth rate. The effect of temperature on RNA content reflects its influence on the rate of protein synthesis and the need to maintain a given rate of protein synthesis to support a given growth rate. For the production of recombinant proteins, a temperature shift from 30 to 42 °C is used to induce product protein formation. In process optimization, it is important to realize that temperature may affect growth rate and product synthesis rate differently. Thus, these two objective functions must be optimized independently.

Thermal deactivation plays an important role in inactivation techniques (Wallhäusser, 1988). The vegetative cells of most bacteria and fungi are killed at temperatures around 60 °C within 15–20 min; yeasts and fungal spores are killed only above 80 °C, while bacterial spores need about 15–20 min at 121 °C. Sterilization above the boiling point of water is usually done in an autoclave under pressure of about 2 bar. Microbial growth follows first-order kinetics. The reader is referred to the chapter on Sterilization in Volume 3 of this Series by Raju and Cooney.

4.1.2 Mechanical Effects

Most microorganisms require an aqueous environment, because they depend on the transport of soluble substrates and nutrients across their plasma membranes. In natural habitats, being heterogeneous systems, they often accumulate at phase interfaces or in biofilms. This has an distinct effect on microbial behavior (Marshall, 1991). One example is the differentiation of *Vibrio parahaemolyticus* swimmer cells with one polar flagellum, occurring in free water, into elongated swarmer cells with numerous lateral flagella when they attach on surfaces.

In bioreactors, however, growth in response to interfacial phenomena also plays a role. Immobilization, either by entrapment into polymers or by adsorption onto a porous support, is used to prevent wash-out of cells even for high dilution rates (Fig. 16). The major effect on growth in these cases is limitation of mass transfer diffusion of substrates and oxygen. For reviews on these aspects see Karel et al. (1985) and Fletcher (1991). Products (possibly toxic) can accumulate in a local area around the cells and influence the micro-environment. Other physico-chemical effects, like shifting of surface charges or pH gradients, from direct contact between the cell surface and the support seem to have less importance than suspected (Rouxhet and Mozes, 1990).

Some other surface-related phenomena are relevant to biochemical processes. In stirred tank reactors, zones of high energy input are located near the impeller tips. Here the problem of shear stress arises especially for filamentous microorganisms (Prokop and Bajpai, 1992). To prevent cell damage, only low agitator speed and, therefore, low oxygen input is possible. Flotation is the interaction between microorganisms and gas–liquid interfaces. Some sensitive cells can be destroyed by the high surface tension of air bubbles in bio-

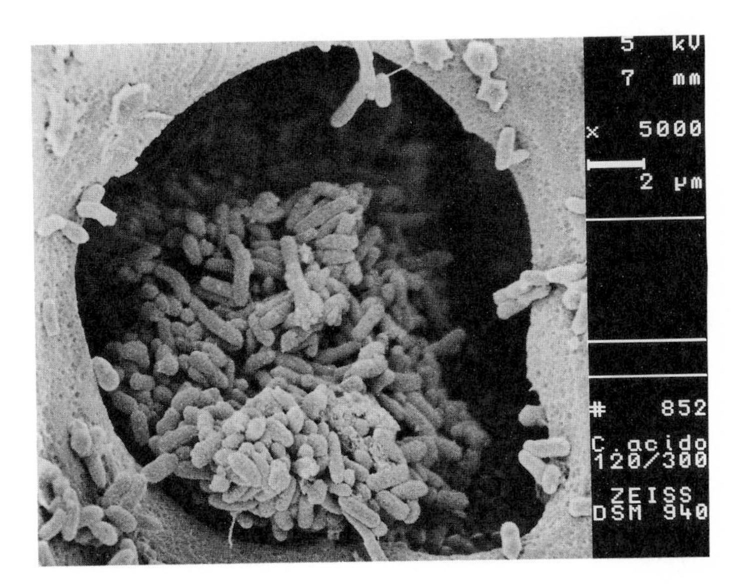

Fig. 16. Immobilized bacterial cells *Comamonas acidovorans* in the internal porous structure of sintered glass spheres (diameter 2 mm) (Lünsdorf and Ulonska, GBF, 1993).

reactors. Flocculation is a result of an interaction between cell surfaces among each other. This process is used to enhance separation of cells and medium either for product recovery or for cell recycle. Another means of cell recycle and dialysis of products are membrane modules. Here again shear stress is an important factor influencing overall growth rate and viability of the cells.

4.1.3 Effect of Water Activity

Water is important for the cell not only because it represents, on the average, about 80% of a cell's weight, but also as a reactant (as in hydrolysis reactions) and a product (as from reduction of oxygen in the electron transport chain). In addition, water provides the most common environment for microbial growth. Not surprisingly, therefore, water has substantial influence on cell growth. The effect of water is most often described by the water activity, which is defined as $a_W = P_S/P_W$, where P_S and P_W are the vapor pressure of water in a solution and that of pure water ($a_W = 1$) at the same temperature. In low solute strength solutions osmolality (mole solubles per kg solute, pure water $Os = 0$) is the more correct measurement. Examples of natural habitats with low water activity are fruit syrup (e.g, $a_W = 0.9$, $Os = 6.0$) which exhibits high sugar concentrations and seawater (3.5% salt, $a_W = 0.98$, $Os = 1$) with relatively high salt content. Even higher values ($a_W = 0.7$, $Os = 20$) can be found in dried fruit and in salt lakes.

Microorganisms respond differently to water activity as shown in Fig. 17. The minimum water activity for growth of Gram-negative bacteria is generally higher than that for Gram-positive bacteria. Many yeasts tolerate minimum water activities similar to those of Gram-positive bacteria. The filamentous fungi range from osmosensitive to the extremely osmophilic species (BLOMBERG and ADLER, 1992). This is, in fact, an important method to prevent microbial contamination of liquids. By reducing the water activity to below 0.95, bacteria will not grow; it must be lowered to below 0.7 to prevent mold growth. Halophiles as a special case of osmophilic organisms grow optimally at high salt concentrations, while halotolerant organisms can stand high salt concentrations, but grow optimally under lower values (Fig. 18).

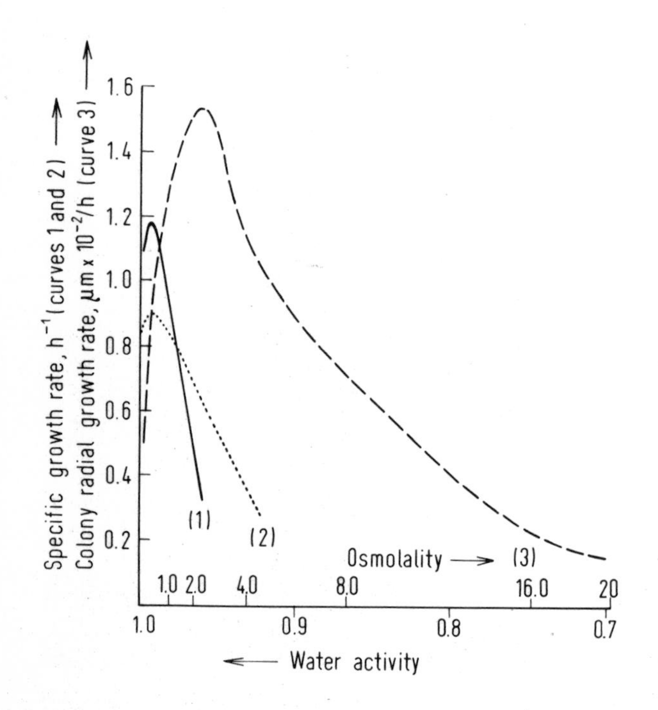

Fig. 17. Specific growth rates as functions of water activity and tonicity of the medium. (1) *Salmonella newport*, (2) *Staphylococcus aureus*, (3) *Aspergillus amstelodami* (from PIRT, 1975).

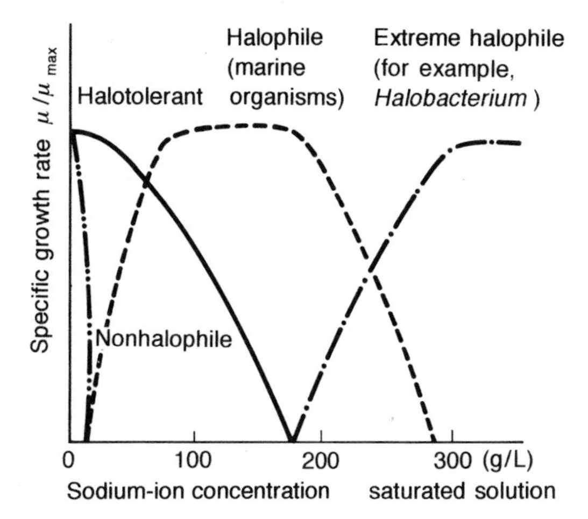

Fig. 18. Effect of sodium ion concentration on growth of microorganisms having different salt tolerances (after BROCK and MADIGAN, 1991).

4.1.4 Effect of pH

The pH, a measure of the hydrogen ion (H^+) activity, is particularly important as a parameter of microbial growth. Most microorganisms grow over a pH range of 3 to 4 pH units; this represents a 1000- to 10000-fold range of hydrogen ion concentration. It is possible to make some generalizations that are useful in defining a chemical environment for the cell. Most bacteria have a pH optimum (i.e., a pH at which growth rate is maximum) around pH 6.5–7.5. Below 5.0 and above 8.5, they do not grow well. Exceptions are the acetic acid bacteria (i.e., *Acetobacter suboxydans*) and the sulfur oxidizing bacteria. Yeasts, on the other hand, prefer pH values of 4–5 and will grow over the range of 3.5 to 7.5. Their preference for low pH provides an important advantage for yeast over bacteria. It is common to grow yeast at low pH as a means of preventing bacterial contamination. Molds often have pH optima between 5 and 7 but, like yeasts, will tolerate a wide range of pH value, e.g., 3 to 8.5. Since pH has an influence on membrane processes (transmembrane pH gradient), it is not surprising that there is a relationship between pH and temperature. As the temperature for growth is increased, the observed pH optimum increases. This suggests a useful strategy for design of microbial environments and conversely for design of environments in which microbial growth is undesired.

While control of pH is important to maintain an optimal environment for growth and/or product formation, it is also important to examine the means of pH control. The pH changes in response to microbial activities: H^+ is generated during NH_4^+ uptake, H^+ is consumed during NO_3^- metabolism, H^+ is consumed when amino acids are used as a carbon source. Thus, by monitoring and controlling pH, it is possible to gain both qualitative and quantitative insight into the fermentation. Titration of pH changes, however, will also alter the chemical environment of the cell. If ammonia is used to balance H^+ generation from ammonia metabolism, then the NH_4^+ concentration will be maintained constant (cf. Sects. 4.2.2 and 6.2). However, if it is used to titrate an acid that is being formed, ammonia toxicity may be developed. When titrating acid production, it is more common to use NaOH or $Ca(OH)_2$, however, care is needed here, because they affect the ionic strength of the medium and also the product solubility. Thus, in industrial fermentations, the maintenance and control of pH is a critical variable.

4.2 Chemical Environment

Chemical compounds influence growth in different manners. If chemical compounds are used as *nutrients*, growth and uptake rate depend on the concentration in the micro-environment of the microorganism. In other cases substances may constitute a stimulus and trigger metabolic events without being directly involved in metabolic pathways. In ecological systems exchange of chemical compounds between different kinds of microorganisms is a prerequisite for survival. In natural habitats and in many bioprocesses, growth is not only determined by the concentration and availability of nutrients but also by the influence of growth *inhibitors*, in fact, this plays an enormous economic role. Especially in anaerobic fermentations growth inhibition by fermentation products determines productivity and yield (BAJPAI and IANNOTTI, 1988). The mechanisms reported to be involved, include:

modulation of different key enzymes, interference with metabolic control and especially energy metabolism, and last but not least spatial effects on the cell membrane. The net result may be reduced growth, reduced substrate to product conversion, and/or reduced yield.

4.2.1 Effect of Carbon Source

An adequate supply of the carbon-energy source is critical for optimal growth and product formation. Calculation of the need for the carbon source will be discussed later. Of importance here is how the concentration of the carbon source affects microbial growth rate.

It is to be expected that growth rate, as any chemical reaction rate, will depend on the concentration of chemical nutrients. The most common expression used to describe the dependence is the Monod relationship

$$\mu = \mu_{max} \frac{S}{k_S + S} \tag{5a}$$

where μ and μ_{max} are the specific and maximum growth rate (h^{-1}), respectively, S is substrate concentration (g/L), and k_S the value of the substrate concentration at $\mu = 0.5\mu_{max}$, as shown in Fig. 19.

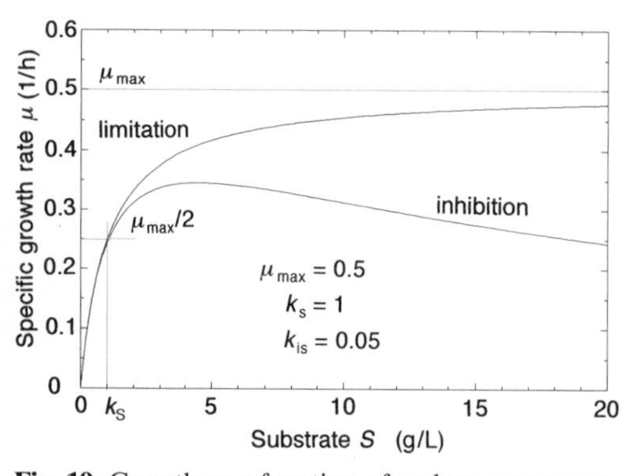

Fig. 19. Growth as a function of carbon source concentration. For non-inhibited Monod kinetics 90% μ_{max} is reached at about 10 k_S; the quadratic substrate inhibition term causes a remarkable inhibition even for low substrate concentrations, what is (luckily) not valid in most bioprocesses.

This expression was proposed by MONOD (1942 and 1949; Nobel prize 1965) to describe experimental data. Its formulation, based on the analogy with saturation kinetics in monomolecular adsorption, is quite convenient. The model suggests that the dependence of growth on chemical concentration can be described by two constants, the maximum growth rate and the limitation constant, k_S. Some typical values are listed in Tab. 2, comprehensive lists were provided by PIRT (1975). Therefore, a cell will grow at near μ_{max} when the carbon source concentration is in excess of about 10 times k_S.

The description of substrate-dependent growth rate by two parameters does not automatically imply reliability of Monod kinetics (BELLGARDT, 1991). Growth dependence on substrate concentrations reflects, in many cases, the mechanism of substrate uptake. While passive or even facilitated diffusion results in k_S values of about 100–300 mg/L, carrier-mediated and active transport allows for lower values in the range of 1–10 mg/L. In practice, growth kinetics are often too similar to achieve a clear distinction between the mechanisms of transport; for an example see Fig. 20.

All nutrients have an upper concentration limit, above which further increase will cause a

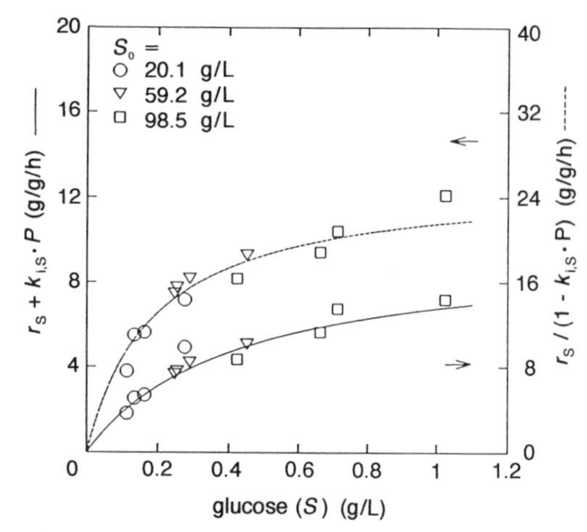

Fig. 20. Measured values for specific uptake rates corrected with additive inhibition (left axis) and linear inhibition (right axis); straight line is the simulation of Michaelis–Menten kinetics after parameter estimation; one set of measurements can coincide with two different kinetics.

Tab. 2. Values for Kinetic Parameters for Different Organisms and Different Substrates (all items are from sources where defined substrates and explicit modeling have been employed)

Organism	Substrate (g/L)	Max. Growth Rate μ_{max} (1/h)	Yield $y_{X,S}$ (g/g)	Limitation k_S (g/L)	Maintenance μ_m (1/h)	Inhibitor (g/L) ($\mu=0$)
Bacteria						
Escherichia coli	Glucose	0.5	0.50	0.005	0.1	
E. coli	Glycerol	0.3	0.50	0.60	0.1	
Acetogenium kivui	Glucose	0.79	0.28	0.18	0.12	95.0
Zymomonas mobilis	Glucose	0.19	0.020	0.19	0.10	85.0
Clostridium butyricum	Glycerol	0.58	0.045	–	0.03	55.0
Pseudomonas capacia	Phenol	0.32	0.74	0.001	0.02	1.3
P. capacia	Oxygen	0.32	0.52	–	0.03	–
Yeasts						
Saccharomyces cerevisiae	Glucose (aer.)	0.25 (no Crab.)	0.50	0.4–0.5	0.0	80.0–90.0
S. cerevisiae	Glucose (aer.)	0.5 (Crab.)	0.15	0.4–0.5	0.0	80.0–90.0
S. cerevisiae	Ethanol (aer.)	0.15–0.2	0.7–0.8	0.1–0.2	0.0	
S. cerevisiae	Glucose (anaer.)	0.0–0.10	0.05–0.1	0.4–0.5	0.0	
Trichosporon cutaneum	Glucose	0.37	0.87	0.2	0.045	
T. cutaneum	Oxygen (gluc.)	0.37	2.8	$4.6 \cdot 10^{-5}$	0.023	
Fungi						
Trichoderma reesei	Glucose	0.12	0.5	0.1–0.2	0.005	
T. reesei	Cellulose	0.04	0.28	–	–	
Penicillium chrysogenum	Glucose	0.12	0.48	0.1–0.2	0.015	

Values for kinetic parameters found in the literature have to be treated with care, because they strongly depend on the strain used, on the cultivation conditions, e.g., temperature, pH, or trace elements. The concentration of the limiting substrate in a sample changes remarkably within a few seconds what makes determination of k_S difficult. Anaerobic growth sometimes (e.g., yeasts) depends on traces of oxygen available, which are usually not controlled or monitored in real fermentations.

decrease in growth rate. This effect is generally called substrate inhibition. Of importance here is the concentration at which it occurs. In the case of carbohydrates, substrate inhibition may begin above 50 g/L, although often not until 100–150 g/L; the reason is usually osmotic pressure. The increasing solute concentration causes partial dehydration of the cell, and the result is reduced growth rate (see Sect. 4.1.3). Substrate inhibition is usually described by

$$\mu = \mu_{max} \frac{S}{S + k_S + k_{i,S} S^2} \tag{5b}$$

as formal kinetics (Fig. 19).

The carbon source' concentration may also affect product formation. An important example is the Crabtree effect in yeasts. When yeasts are grown in the presence of high sugar concentration, even when fully aerobic conditions are maintained, they carry out anaerobic metabolism and produce alcohol from glucose. This is illustrated in Fig. 21. When the glucose concentration is above 0.15 g/L, ethanol production occurs. To prevent ethanol synthesis, the growth rate and glucose concentration must be below 0.22 h^{-1} and 0.15 g/L, respectively. In cases such as this, the techniques of fed-batch and continuous culture (see Sects. 6.2 and 6.3) are particularly useful for growing cells at high rates while maintaining low substrate concentration.

4.2.2 Effect of Nitrogen Source

Ammonia is the most common source for nitrogen in most microorganisms. Many bac-

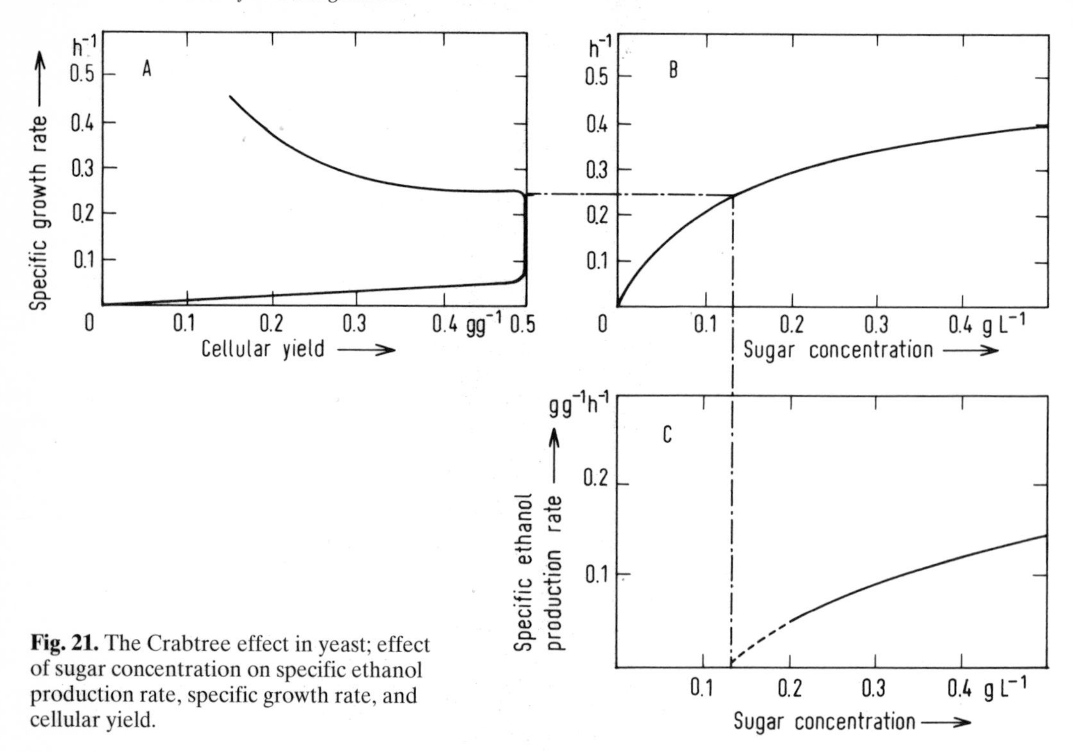

Fig. 21. The Crabtree effect in yeast; effect of sugar concentration on specific ethanol production rate, specific growth rate, and cellular yield.

teria can utilize nitrate via reduction to nitrite and finally ammonia to cover their need for the synthesis of nitrogen containing cell compounds, such as amino acids. In anaerobic respiration nitrate is used as the terminal electron acceptor. Many microorganisms prefer ammonia. Complex media often contain organic compounds like amino acids which are also consumed by the microorganisms.

Values for the k_S for ammonia are typically 0.1–1.0 mg/L. The low k_S values reflect the effectiveness of the active transport systems of cells for these materials. The typical value for k_S for amino acids is about 0.003 to 0.2 mg/L. Clearly, microorganisms are able to grow at their maximum growth rate in very low nitrogen concentrations. In general, the concentration of nitrogen compounds in the environment is low and as a consequence, cells that can grow rapidly in low concentrations have an important competitive advantage.

Satisfying the low concentration limit for microbial growth does not usually cause a problem for the fermentation technologist. More often, problems of substrate inhibition develop from providing too much nitrogen. Ammonia, above 3–5 g/L, will often cause inhibition. If a culture becomes starved of carbon source in the presence of amino acids, the carbon skeleton of the amino acids is consumed and ammonia accumulates. If NO_3^- is used as nitrogen source, it can be partially reduced to NO_2^- which may be toxic to the cell. Thus, the effect of a nitrogen source may be derived from its metabolic products as well as from its original form. Such a phenomenon can occur with other metabolites as well.

4.2.3 Dissolved Oxygen

Oxygen plays various roles for different classes of microorganisms. Obligate aerobes can obtain their energy only via aerobic respiration and are dependent on O_2. In this case, oxygen acts as a terminal electron acceptor (KRÄMER and SPRENGER, this volume, Chapter 2). This enables the microorganisms to oxidize carbohydrates and other substrates completely to water and carbon dioxide. Incorporation of molecular oxygen into the metabolic compounds only occurs, if highly reduced substrates are utilized.

Facultative aerobes do not require oxygen, but if it is available they grow better, because the energy yield of respiration is about 18 times higher than that of fermentation. Microaerophilic organisms require low levels of oxygen. Among the anaerobes the aerotorant cells can tolerate oxygen in their environment, but do not need it, while it is harmful or lethal for the obligate anaerobes.

For all organisms, including obligate aerobes, oxygen may be toxic at any concentrations. The mechanism of oxygen toxicity is through the formation of singlet oxygen, superoxide radicals O_2^-, peroxide O_2^{2-} or hydroxyl free radical $OH^{\cdot}$ which are destructive to many cell components (GREGORY and FRIDOVICH, 1973; HARRISON, 1972). These radicals are generated in some physical and biochemical processes including respiration. Aerobic organisms have developed mechanisms to protect them against oxygen toxicity by activity of catalase, peroxidase or superoxide dismutase (CLARKSON et al., 1991). Obligate anaerobes protect themselves against oxygen mainly by the latter enzyme, but some even contain cytochromes (ISHIKAWA et al., 1990).

In biotechnology, oxygen supply for aerobic processes is often a limiting factor on the way to high growth rates and high biomass concentrations. On the shake-flask scale, oxygen for aerobic cultures is supplied via the headspace and adequate mixing. For anaerobic cultures reducing agents like thioglycolate are used. Anaerobic surface cultures are grown in jars, where oxygen is replaced by hydrogen and remaining traces are removed by a catalyst (see, e.g., BROCK and MADIGAN, 1991). On bioreactor- and pilot scale oxygen is provided by a gas sparger at the bottom of the reactor. The concentration, under which oxygen limitation occurs, depends on culture conditions and in particular on the growth rate, because oxygen uptake is directly coupled to energy metabolism and is taken up only as required. Typical values for limiting pO_2 values are in the range of 1–20% of air saturation. For facultative aerobes oxygen supply can be used as a means to control growth and product composition. Recent interest has focused on the influence of oxygen partial pressure on product formation, especially in the area of antibiotics formation (ZHOU et al., 1992). For many aerobic organisms toxic effects on growth up to 100% of air saturation at 1 bar (partial pressure pO_2, about 7 mg O_2/L) are negligibly small (for a review see ONKEN and LIEFKE, 1989).

4.2.4 Effect of Carbon Dioxide

Carbon dioxide is an end product of respiration, as well as of fermentation. It is produced in large amounts in nearly all bioprocesses, but it also plays an important role as a nutrient. Phototrophic bacteria like purple bacteria and green bacteria (Rhodospirillales) and the cyanobacteria, which all depend on light as energy source, and the chemolithotrophic bacteria can utilize CO_2 as their sole carbon source. Heterotrophic microorganisms, which are of more biotechnological interest, need carbon dioxide fixation at several metabolic steps for anaplerotic sequences, thus depending on CO_2 in their environment. Most microorganisms are adapted to low CO_2 concentrations (0.02 to 0.04% v/v). Usually this demand is met by the metabolic activity of the cells themselves, but CO_2 limitation conditions may arise at the beginning of a fermentation resulting in a long lag phase, especially if the gas is stripped out by aeration (DIJKHUIZEN and SCHLEGEL, 1987). At the end of fermentation during fast product formation, dissolved CO_2 can accumulate in the broth, possibly with physiological effects (JONES and GREENFIELD, 1982).

4.2.5 Alkanols

In most anaerobic and some aerobic bioprocesses alkanols are formed as the desired product or as by-product. Among these are ethanol, butanol, 2,3-butanediol, propanol, isopropanol, or 1,3-propanediol. The involved microorganisms (e.g., the yeasts *Saccharomyces* spp., the bacteria *Zymomonas* spp. and *Clostridium* spp.) can utilize different kinds of carbon sources, but sugars are generally preferred. Ethanol production is one of the most important biotechnological processes. Besides the traditional yeast process, ethanol fermentation with the bacterium *Zymomonas mobilis* (ROGERS et al., 1982; SAHM and BRINGER-MEYER, 1987) is of interest. Volumetric productivity and final product concentration are limited by the inhibitory effects of ethanol and other alkanols on growth and product formation of yeasts and of bacteria. Most bacteria exhibit a dose-dependent inhibition of growth over the range from 1 g/L to 100 g/L ethanol. Some yeasts can tolerate ethanol in excess of 120 g/L. "It is likely that many "targets" will be discovered for ethanol-related phenomena with new "targets" becoming relevant at different concentrations of alcohol" (INGRAM and BUTTKE, 1984).

Many different kinetics for ethanol dependent growth rates of yeasts have been described by VAN UDEN (1985). Enzyme assays show remarkable effects only for high ethanol concentrations and cannot explain all the observed effects even if intracellular accumulation of ethanol is assumed (DASARI et al., 1950). Inhibition of glucose uptake as an example of electroneutral carrier-mediated transport (DIMARCO and ROMANO, 1985) and ammonia uptake as an example of an electrogenic transmembrane process (LEAO and VAN UDEN, 1983; POSTEN, 1989) have been reported. The major effect seems to be the interaction of ethanol with the cell membrane which increases the membrane fluidity. Thus, the lipid environment for membrane-located enzymes is disturbed, with intelligible effect on their function. Finally, membrane leakage of ions and metabolites occurs, especially the passive influx of protons is enhanced resulting in the breakdown of the transmembrane potential and the related ion transport and energy transforming processes (Fig. 22). Other environmental parameters which have an influence on membrane processes, such as temperature and pH, show strong interference with elevated ethanol concentration (LAWFORD et al., 1988; MICHALCAKOVA and REPOVA, 1992). Yeasts and some bacteria are able to alter their membrane lipid composition as an adaptive response to such stress factors (BRINGER et al., 1985).

4.2.6 Organic Acids

A large number of organic acids are produced by microorganisms via incomplete oxidation of the carbon source. Among others, the low-molecular weight organic acids are the most common and are used in industrial processes (CHERRINGTON et al., 1991). Although these acids are products of a fermentative metabolism, they are often excreted by microorganisms under aerobic conditions as well, par-

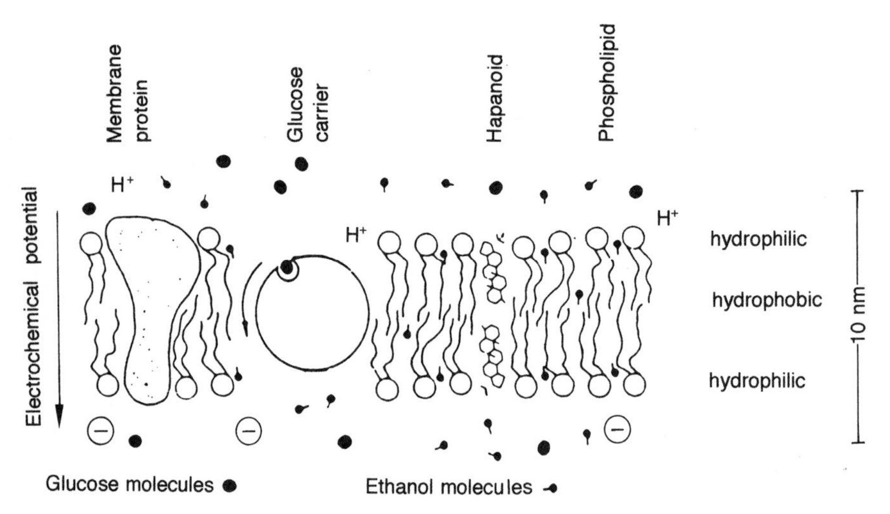

Fig. 22. Structure of a microbial cell membrane and the inhibitory effect of ethanol; after INGRAM and BUTTKE (1984), modified after BRINGER et al. (1985). Many inhibitors disturb membrane organization and reduce membrane potential.

ticularly under restricted oxygen supply and/or at high growth rates (LULI and STRAHL, 1990). Organic acids are well known to inhibit growth of microorganisms, particularly at low pH values. For a number of bacteria and yeasts it was found that the undissociated forms of organic acids are primarily responsible for the toxicity (TANG et al., 1989; ZENG et al., 1991). The inhibition of growth appears to be non-competitive. A survey has been given by BAJPAI and IANNOTTI (1988) along with growth inhibition of other fermentation products. It was reported that organic acids interfere with membrane functions, and their toxicity appears to be related to their solubility in the membrane lipids (DÜRRE et al., 1988). Consequently, an interference with ethanol was observed (PAMPULHA and LOUREIRO, 1989). The critical concentration of undissociated organic acids, which is the concentration above which cells cease to grow, is found to be in the range of 0.4 to 0.5 g/L and is nearly the same for a number of microorganisms and different acids (BIEBL, 1991). It is not clear at present, if the critical concentration of an organic acid is a biological constant. However, undissociated organic acids can traverse the cytoplasmic membrane of microbes and decrease the internal pH value, thus acidifying the cytoplasm (BOOTH, 1985). They interfere with energy-coupling processes, resulting in lower efficiency of energy production and higher energy demand for cell growth. The net results of inhibition are changed metabolic pathways and a reduced rate of product formation and cell growth.

4.2.7 Other Chemical Nutrients

In general, it is essential to maintain a minimum level of any required nutrient. This is the concentration that will allow for the transmembrane processes to support a desired growth rate. However, there is also a maximum level,

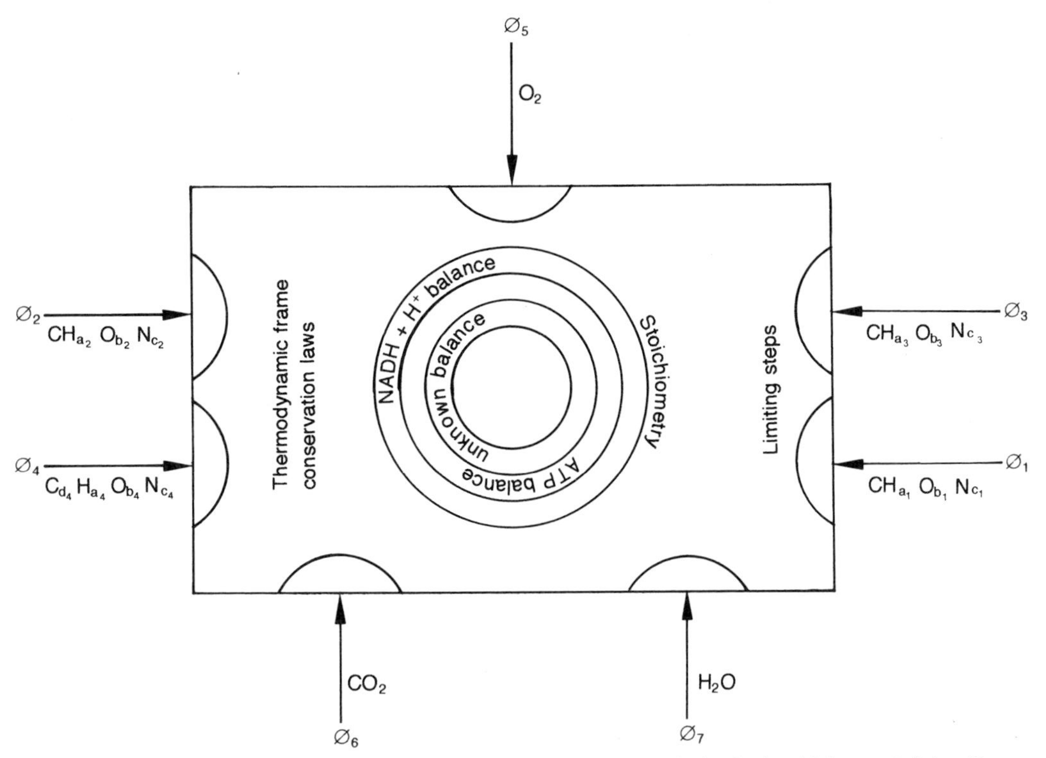

Fig. 23. System, flows, and their interconnections for macroscopic analysis of microbial growth (after ROELS, 1983).

above which substrate inhibition occurs. This inhibition can manifest itself in reduced growth rates or product synthesis rates or both. Clearly, care is needed in defining these boundaries. There are few literature references defining what these boundaries are for materials such as minerals, vitamins, and other growth factors. Quantification of mineral needs is especially difficult because of the occurrence of metal complexes, some cellular and others, non-cellular in origin. Materials such as phosphate or sulfate also can exert specific catabolite repression (KRÄMER and SPRENGER, Chapter 2, this volume) that affects specific metabolic pathways while not affecting growth. For example, many microorganisms take up phosphate very rapidly and store it for later use. In such cases it is impossible to define an optimal concentration in the broth. Such phenomena are often overlooked in process optimization.

4.3 Variability and Constraints of Growth

Just as microorganisms show manifold responses to their environment, growth rate is influenced by endogenous factors, including morphological organization, metabolic patterns represented by rate-limiting steps, stoichiometry, and intracellular control, and by physical constraints resulting from conservation laws for materials and energy. Such interactions have been studied in terms of macroscopic analysis (ROELS, 1983), where cells are described as open systems (Fig. 23) and thermodynamic laws apply.

4.3.1 Growth for Different Groups of Microorganisms

Since growth can be regarded as the sum of complex biochemical reactions on the one hand, and as a result of changing environmental conditions on the other, it is possible to make some generalizations about growth rates for microorganisms. Growth is compared to the relaxation times of different chemical and biological procedures in Fig. 24. Growth can also be compared amongst the diversity of microbial life, as seen in Fig. 25. The frequency with which a given doubling time is observed for bacteria, yeasts, and molds falls into a general pattern. These relationships are meant to provide guidelines for estimation of typical doubling times for each group of organisms.

4.3.2 Macromolecular Composition of Cells

Under the given conditions for supply of nutrients, growth factors and physical stimuli the cells try to "make the best of it". Depending on their ecological niche this may be achievement

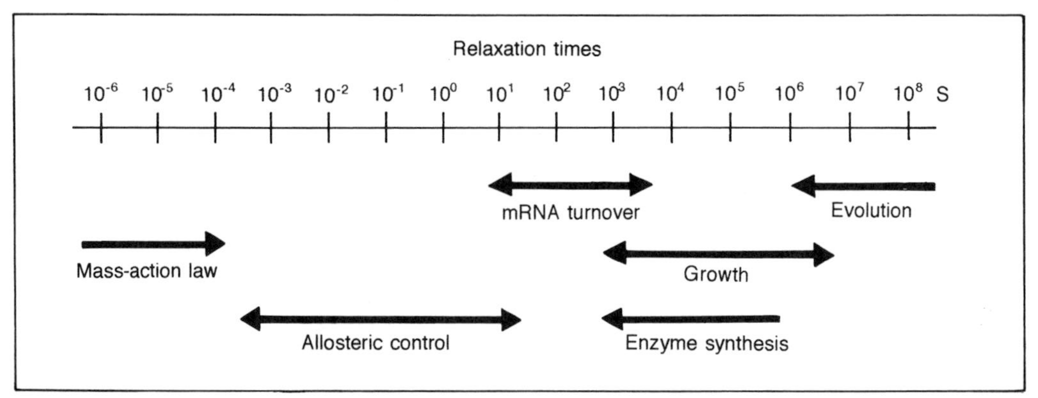

Fig. 24. Relaxation times of biochemical reactions, growth, and proceedings in the environment influencing growth (after ROELS, 1983).

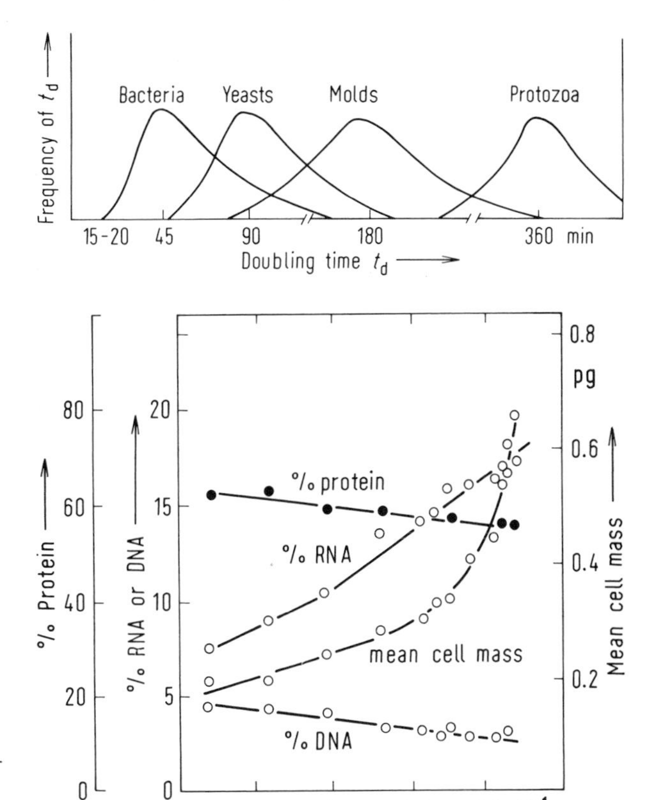

Fig. 25. Frequency of doubling times for various groups of microorganisms.

Fig. 26. Variation of the macromolecular composition of *Kluyveromyces aerogenes* as a function of growth rate in a nitrogen-limited chemostat.

of high yields or of high specific growth rates. However, many microorganisms can change their intracellular composition to adapt to the environment. Balanced growth is obtained if all components of the proliferating cells are reproduced identically to the past generation. Under such conditions the increase of cell number is proportional to the increase of cell mass, and biomass composition does not change. In bioreactors, balanced growth is obtained usually only in continuous culture (see Sect. 6.3) for one dilution rate. Remarkable changes of cell composition can be observed for different growth rates (Fig. 26). Especially, the RNA content increases with increasing growth rate corresponding to the high rate of protein syntheses.

4.3.3 Limiting Steps

Of similar importance as Monod's famous equation, Eq. (5a), for growth rate dependence on substrate concentration was the discovery of a linear relationship

$$r_S = \frac{1}{y_{X,S}} \cdot \mu + m \tag{6a}$$

between substrate uptake (and/or product formation) and growth (NOORMAN et al., 1991, and more historical references therein). The constant m represents the "maintenance" part of substrate uptake (PIRT, 1965) (see Sect. 4.3.4). Eqs. (5a) and (6a) make up the nucleus of a kinetic model, which is frequently used in the literature.

In fact, in many cases substrate uptake is controlled by extracellular substrate concentration via an enzymatically mediated step in membrane transport, which can be described by Michaelis–Menten kinetics

$$r_S = r_{S,\max} \cdot \frac{S}{k_s + S} \tag{6b}$$

Growth rate itself is then controlled by this substrate uptake rate like

$$\mu = y_{X,S} - y_m \tag{6c}$$

Eqs. (6b) and (6c) again form the core of a kinetic model which is formally similar (but not equivalent) to the first one, but having a different biological interpretation.

In most heterotrophic organisms the organic carbon source also acts as an energy source. While one part of the substrate is incorporated into biomass, the remainder is oxidized for energy generation and finally excreted, but giving different yields for different substrates (see Tab. 2). These results have been discussed in terms of thermodynamics of growth by ROELS (1983) and SANDLER and ORBEY (1991), where also more values for yield can be found. In many aerobic and anaerobic bioprocesses uptake rate and energy generation from the organic substrate is the limiting factor for faster growth or higher yields (see Sect. 4.3.4). This is illustrated in Fig. 27 in a block diagram representation. A growth condition, in which growth is controlled via limited substrate uptake, is usually called energy limited growth.

Also for other nutrients, e.g. ammonia, often a linear relationship is observed. This stoichiometry is forced by special stoichiometric properties of the biochemical pathways employed (HERBERT, 1976) or generally by conservation laws of chemical elements. If a microorganism maintains a given intracellular composition, the degree of freedom for the spectrum of nutrient uptake rates and product formation rates is reduced. This finding has been exploited for the design of "software sensors" (Sect. 3.2.5). For energy-limited growth, e.g., ammonia is taken up only as the growth rate demands, even if it is in excess in the medium. On the other hand, if ammonia concentration is low, ammonia uptake controls

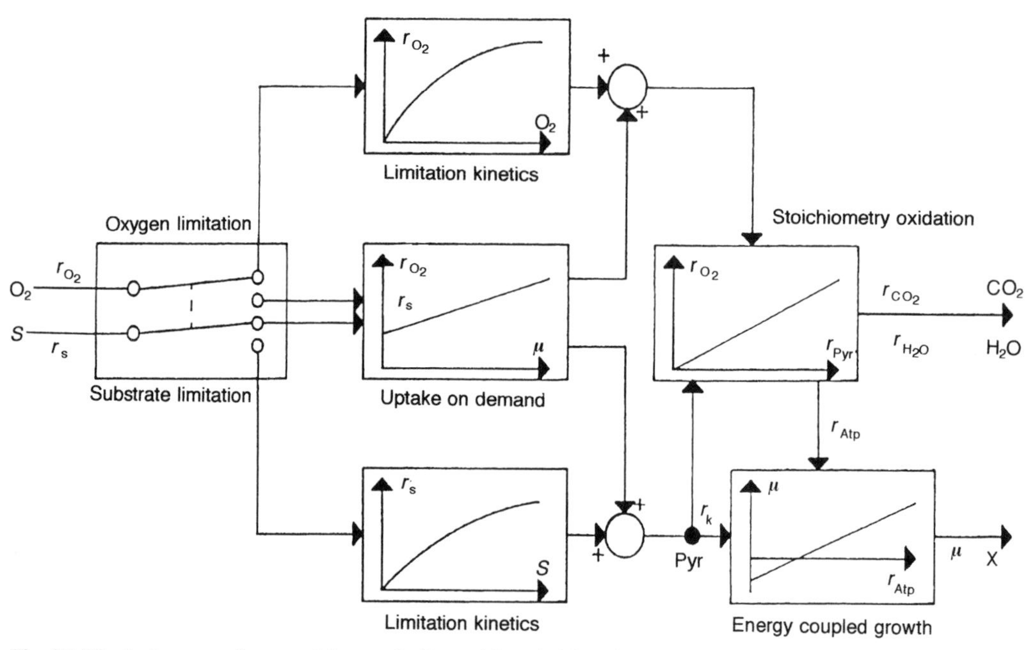

Fig. 27. Block diagram of an aerobic metabolism with switching characteristics for two limiting rates (POSTEN and MUNACK, 1990).

growth, which then is referred to as ammonia limited growth, and other uptake rates are adjusted by the cells according to stoichiometry. Thus, only a few limiting steps in nutrient uptake or in intracellular metabolism can determine the overall growth rate and the whole fermentation pattern. A practical application is found in yeast production to avoid oxygen limitation. At the end of a fermentation, substrate feeding is reduced which leads to reduced growth rate and finally to reduced oxygen demand of the cells (see also Sect. 6.2). The alternative strategies have been described by O'CONNOR et al. (1992).

4.3.4 Energy Demand of Growth

From the analysis of growth dependence on different substrates it became clear that growth could be stoichiometrically related to the amount of ATP produced. A review has been given by TEMPEST and NEIJSSEL (1984). The energy yielding processes in microorganisms are known quite well. For details of biochemical pathways see Chapter 2 of this volume. In aerobic respiration, theoretically, 2–3 mol ATP are formed per atom equivalent of oxygen consumed (P/O ratio). The energy yield of the substrate depends on the degree of reduction. For glucose it is up to 38 ATP per mol glucose catabolized for respiration. In anaerobic fermentation processes oxygen is not available as an electron acceptor and electrons are transferred from the organic substrate to catabolites which are finally excreted as products. The energy yield of this process can be calculated from the available electrons of the substrate (ONER et al., 1984) and is 1–3 mol ATP per mol glucose consumed depending on the product and the biochemical pathway employed. This represents only about 2% of the combustion energy of glucose.

The mechanisms for energy partition are much more complex. Generally a constant and a growth-dependent energy consumption can be observed. PIRT (1965) introduced the concept of maintenance energy for the non-growth-associated energy demand. This includes turnover of cellular constituents, intracellular control purposes, maintenance of membrane potential and cell motility. Without energy expenditure the biochemical system "the cell" would disintegrate towards thermodynamic equilibrium which finally means death. Some bacteria have developed a process called starvation survival (KJELLEBERG et al., 1987). In other concepts of maintenance this is no longer restricted to a formal definition of the non-growth-associated part of energy utilization but focuses on the (occasionally growth-dependent) underlying physiological events (TEMPEST and NEIJSSEL, 1984).

For many different microorganisms the ATP yield Y_{ATP} for growth is about 10.5 g cells formed per mol ATP generated by substrate catabolism. Calculations of values for synthesis of cell compounds from monomers or glucose (STOUTHAMER, 1977) gave theoretical yields of about 32 g cells per mol ATP. Similar values (29 g/mol) have been reported by ONER et al. (1984). This large discrepancy between theoretical and experimental values makes it clear that a highly organized cell is more than a mixture of enzymes and metabolites. The apparent low energy efficiency of microbial growth is correlated with a high entropy change (see Sect. 4.3.5). Besides synthesis of chemical compounds it is also associated with cellular organization, as can be expected from the second law of thermodynamics (HAROLD, 1986). This is manifested in transport processes or high turnover rates for proteins or nucleic acids. Furthermore, most of all enzymatic reactions are maintained far from their thermodynamic equilibrium and have to be calculated on the basis of linear non-equilibrium thermodynamics, where uncoupling, e.g., of oxidative phosphorylation can to be expected. A comprehensive review of thermodynamic aspects of growth has been given by WESTERHOFF et al. (1982) and WESTERHOFF and VAN DAM (1987).

4.3.5 Heat Evolution of Microorganisms

Microbial heat evolution has been recognized for many years as a problem in self-heating hay-stacks leading to their spontaneous combustion and also in heat removal from large fermentors where the surface area to volume ratio and the difference between cooling water and fermentor temperature both are low. Here, however, it is useful to take advantage of heat evolution as an indirect means of assessing microbial growth.

In each single metabolic step, as in any chemical reaction, there is an enthalpy change, which in living cells normally is accompanied with heat evolution. During growth, cells capture "useful" energy in high energy bonds of ATP. This provides common energy currency that can be used to drive the wide variety of biosynthesis reactions. In formation of ATP, only 40–75% of the energy associated with the reaction is conserved in a usable form as ATP; the rest is evolved as heat. When the ATP is used, additional heat is generated. The net difference between heat evolved and energy retained by the cell mass synthesized is seen in Fig. 28. It is also apparent that heat evolution continues after cell mass accumulation ceases. This is due to non-growth related metabolism of the cell.

COONEY et al. (1968) showed a correlation between heat evolution and oxygen consumption for many different aerobic fermentations; from this work and more recent results (Fig. 29), the heat evolution was estimated to be about 440 kJ per mole oxygen consumed. As the heat of combustion for glucose is about 460 kJ/mol oxygen, calorimetric measurements must be very accurate to allow for a more detailed interpretation of growth processes, e.g., changes in efficiency of energy metabolism, than is allowed by measurement of oxygen consumption only.

For anaerobic processes the heat of cultivation depends on the fermentation products and is generally lower (about 20 kJ per mole glucose, assuming two moles of ethanol and two moles of ATP) than that of aerobic processes. Changes in metabolism can be seen in the calorimetric measurement. For organisms with a mixed anaerobic and aerobic metabolism, the relative part can be quantified by calorimetric measurements (VON STOCKAR and BIROU, 1989).

If substrates and products are completely and quantitatively known, the heat of a bioprocess can be calculated with the aid of a thermodynamic cycle. As combustion temperatures for nearly all chemical compounds are known, it is useful to construct a cycle for the combustion products. Thus, the heat of a reaction can be calculated as

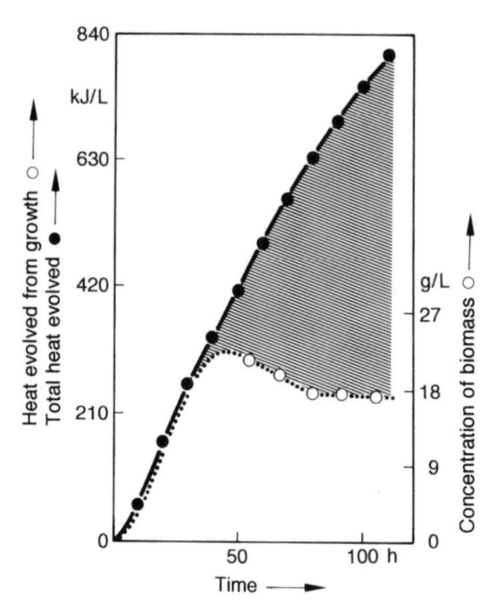

Fig. 28. Heat evolution during a *Streptomyces* fermentation. The shaded area represents the heat evolved for culture maintenance and product formation, and the clear area represents the enthalpy conserved in cell mass.

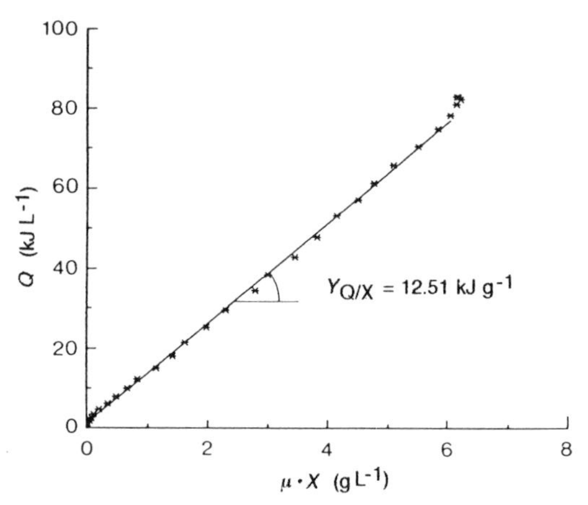

Fig. 29. Determination of $Y_{Q/X}$ for *Kluyeromyces fragilis* grown aerobically on a glucose-limited minimal salt medium from measurements of the total heat evolved and biomass formed (from VON STOCKAR and BIROU, 1989).

$$\Delta H(\text{reaction}) = \sum \Delta H_{\text{com}}(\text{products}) - \sum \Delta H_{\text{com}}(\text{substrates}) \quad (7)$$

Unfortunately, in practice there are some difficulties. As the whole growth process is run in aqueous solution, solution temperatures must be taken into account. Another point is the elemental composition and the heat of combustion of the biomass. The values given in the literature are often not very accurate and the contents of phosphorus, sulfur, and trace elements are in most cases not exactly known. Also the measurement of substrate, product, and biomass are of limited accuracy. This leads to the situation that cultivation temperatures cannot, as yet, be predicted accurately. Nevertheless, such thermodynamic calculations are very worthwhile when used to help estimate the influence of single factors on the heat evolution.

Besides these conceptual problems, there are many practical difficulties to measure heat evolution in a bioreactor (for a review see VAN STOCKAR and MARISON, 1989). For anaerobic fermentations it is possible to work with an external microcalorimeter, where the cultivation liquid is pumped through the measuring cell. For aerobically grown organisms it is recommended to use a fermentor calorimeter, where many precautions to measure heat of agitation, evaporation loss, conduction losses and loss to cooling system raise the experimental effort. However, in many cases calorimetric investigations are very useful to follow changes in the stoichiometry of bioprocesses on-line.

5 Design of Media for Growth and Production

One of the most important factors in optimization of a fermentation process is the design of the growth and production medium. The medium must meet the needs for synthesis of cell materials and for biosynthetic processes, as well as for environmental requirements of the microorganism. The medium must, in addition, take into account several economic and technical constraints. The ultimate objective function is the development of a fermentation with minimum cost per unit product. This usually requires a trade-off between minimizing medium cost and maximizing product yield and titer. In the following discussion, the general principles of medium design are presented. They represent an integration of the knowledge of cell composition and physiology with engineering and economic considerations. The discussion will attempt to summarize the general principles of medium design taking into account the nutrient requirements and the nature of commonly used technical media.

5.1 Nutrient Requirements

Qualitative and quantitative nutritional requirements vary with the microorganisms. In any case, as a consequence of the laws of conservation of mass, it is essential to balance the elemental requirements for growth and product formation. This implies that the culture media must contain all the essential growth substances in proportions similar to those of the cells and products (ANTIER et al., 1990). Tabs. 3 and 4 show the variations in quality and quantity for supplements of basal media used by a cross-section of researchers in recent years for the cultivation of yeasts and *Escherichia coli*, respectively (FIECHTER, 1984).

Initiation and maintenance of metabolic activities by microorganisms are dependent upon the availability of the necessary nutrients in a form suitable for uptake and utilization. The necessary components of growth media are listed in the following sections.

5.1.1 Carbon

Carbon (C) is the major constituent of the microbial cell, making up approximately 50% of the dry weight. Carbohydrate and non-carbohydrate (methanol, alkanes) compounds are used as carbon and energy sources and usually are the main substrates in growth media. Sources of carbon in fermentation media include starch, saccharose, molasses, and vegetable oils.

Tab. 3. Synopsis of Media Used for Yeast Growth Experiments (from FIECHTER, 1984)

Components	Form	Amount per Liter
Ca	$CaCl_2$	50 μg–500 mg
Cu	$CuSO_4$	80 μg
Fe	$FeCl_3$, $FeCl_4$, $FeSO_4$	1 μg–70 mg
K	KCl, KI, K_2HPO_4, KH_2PO_4	200 μg–7 g
Mg	$MgCl_2$, $MgSO_4$	0.2 g–10 g
Mn	$MnCl_2$, $MnSO_4$	30 μg–100 mg
Na	NaCl	50–100 mg
Mo	Na_2MoO_4	10 μg
NH_4	$(NH_4)Cl$, $(NH_4)_2SO_4$, $(NH_4)_2HPO_4$	0.2 g–8 g
Zn	$ZnSO_4$	80 μg–6.99 mg
B	H_3BO_3	200 μg
Biotin		23 μg
Ca-pantothenate		23.3 mg
m-Inositol		46.6 mg
Phenol-red		1 mg
Pyridoxin		1.16 mg
Thiamine		3 μg–4.66 mg
Yeast extract		100 mg
Substrate	*n*-Dodecane, *n*-tetradecane, *n*-hexadecane, gas oil	10–15 g

Tab. 4. Synopsis of Media Used in Experimentation Including Mass Culture of *Escherichia coli* (from FIECHTER, 1984)

Components	Form	Amount per Liter
Carbon source	Glucose, glycerol, casein	2–33 g
Amino acids	Casein	6.7–30 g
Growth factors	YEPD	0.5–31 g
K	K_2HPO_4, KH_2PO_4	1 g–13.6 g
Na	$NaHPO_4$, NaCl, Na_2SO_4	11 mg–7 g
NH_4	$(NH_4)_2HPO_4$, $(NH_4)_2SO_4$, NH_4Cl, NH_4NO_3	0.8 g–8.0 g
Mg	$MgCl_2$, $MgSO_4$	2.4 mg–2 g
Na	NaCl, Na_2SO_4	11 mg–3 g
Ca	$CaCl_2$	1 mg–40 mg
Fe	$FeSO_4$, $FeCl_3$	0.25 mg–40 mg
Cl	NH_4Cl, $FeCl_3$, $CaCl_2$, NaCl, $MgCl_2$	0.5 mg–3 g
Trace elements		
Antifoam		

5.1.2 Nitrogen

Nitrogen (N) is second to carbon in terms of quantity and economic importance. It consists approximately 10% of the dry weight. It is available to microorganisms in the form of ammonium salts, nitrates or urea, amino acids, peptides, purines and pyrimidines. It serves as the building block for the syntheses of proteins and other cellular macro-molecules. Ammonia gas and solution (NH_4^+) is popular in the industry as a source of nitrogen because it is inexpensive and easy to use. Urea has a considerable buffering capacity and is also used.

Other sources of nitrogen in fermentation media include corn steep liquor, soy bean meal and flour, fish meal, distillers solubles, yeast extract, and protein hydrolysate.

5.1.3 Macro-Elements

Macro-elements include phosphorus (P), potassium (K), magnesium (Mg) and sulfur (S). Phosphorus constitutes approximately 1.5% of the dry cell weight and is present in the microbial cell as phosphate sugars, nucleic acids and nucleotides. The actual amount varies in proportion to the nucleic acid content of the cell which is dependent on growth rate. It is made available to the microorganisms as dehydrogeno-phosphate or orthophosphate ions, $H_2PO_4^-$. K is required for maintenance of ionic balance across the cell membranes, and as stabilizing component of RNA. Mg is an essential enzyme activator and component of the cell membrane. S is an important component of amino acids and coenzymes.

5.1.4 Trace (Micro-)Elements

Trace (micro-)elements include Fe, Mn, Zn, Cu, Co, Ni, B, Cl, Na, and Si. They are required in extremely small amounts, for example, in the scale of mM or μM. They are important as effectors of enzymes and coenzymes. They normally are present in complex media and also in tap water in varying amounts.

5.1.5 Growth Factors

Growth factors are organic compounds normally required in small amounts for growth and for obtaining optimum yield. They act as precursors or constituents of growth factors. They include vitamins, amino acids, unsaturated fatty acids and sterols. The most important vitamins are thiamine, riboflavin, pantothenic acid, pyridoxine, nicotinic acid, biotin, *p*-aminobenzoic acid, and folic acid. They are available either in pure form or as components of certain complex media.

The macro- and micro-elements, as well as the growth factors, are also found in the complex media that serve as carbon and nitrogen sources. Although it is essential that all the necessary nutrients be provided in aqueous solution, some water-insoluble substrates such as hydrocarbons and vegetable oils may also serve as carbon and energy sources and require special treatment. In any case the interaction between the various elements must be taken into account as it could lead to inhibitory actions or chelation at high concentrations (JONES and GREENFIELD, 1984).

5.2 Medium Optimization

Medium optimization is becoming an increasingly complex and painstaking operation. This complexity follows an increasing recognition of the need to provide the required nutrients in the exact amounts. If successfully implemented, the benefits are two-fold. First, it enhances the growth of the microorganisms under conditions optimal for product formation and release. Second, it protects against wastage of materials and, as a result, facilitates cost effectiveness.

Shake flask experiments tend to be highly sophisticated and disproportionately expensive, especially when the objective is merely scientific with little economic consideration. Besides, growth and product formation are investigated only after a given time without any information on the time course. This may be accepted at the beginning of a process development, but much effort has been carried out to give medium a rational basis as pointed out below.

5.2.1 Material Balances

The first step in medium optimization is the formulation of the overall stoichiometry

$$\begin{aligned}&\text{C-source} + \text{N-source} + O_2 + \text{Minerals} +\\&+ \text{Specific nutrients} \rightarrow\\&\text{Cell mass} + \text{Product} + CO_2 + H_2O\end{aligned} \tag{8}$$

similar to that pointed out in Sect. 3.2.5, Eq. (2).

In the past, it was logical to follow this stoichiometry with calculations of the minimum nutrient requirements. For example, to grow 20 g/L of cell mass containing 0.5% magnesium will require at least 0.1 g/L of Mg or 0.5 g/L of $MgSO_4$ or 1.0 g/L of $MgSO_4 \cdot 7H_2O$, since the relative molecular masses of the chemical species indicated are approximately in the ratio of 1:5:10. In the absence of cell composition data, it is possible to experimentally determine the requirement in a classical approach by carrying out a series of batch fermentations in which the nutrient of interest is available in limited supply, while all other nutrients are available in excess. The results usually resemble those presented in Fig. 30. The slope of the curve is the cell yield coefficient ($Y_{x/s}$) on nutrient. Eventually, as the nutrient concentration is increased, it becomes available in excess and some other nutrient becomes growth-limiting. Occasionally, when such an experiment is done, the intercept is not at the origin. If there is carry-over of nutrients from the seed culture, there will be growth at zero (added) nutrient concentration as shown in curve A. A second type of anomalous result occurs, if there is a critical concentration requirement of the nutrient. This can occur, if the nutrient binds to some non-cellular material, forms a precipitate or complex, is destroyed or otherwise lost at low concentration during sterilization. The major drawbacks of this single-dimensional approach are that it is cumbersome, time-consuming and inefficient particularly when a large number of variables (which include nutrients as well as physical factors like temperature, pH, O_2 etc.) are considered. In addition and regrettably, it still does not guarantee the achievement of optimal conditions. For example, the amount of a nutrient such as NH_4^+, which in this kind of estimation is considered adequate, may actually be toxic to the organism at high initial concentration. On the other hand, an amount of a nutrient, such as yeast extract, which might be considered sufficient for the batch fermentation, is usually used up quite rapidly and would be lacking in the later stages of the fermentation. When such estimates are extrapolated in industrial fermentations, the results are bound to be spurious and unreliable. Consequently, other approaches were developed including an elegant modification of the single-dimensional approach, the chemostat pulse method and the full-factorial method.

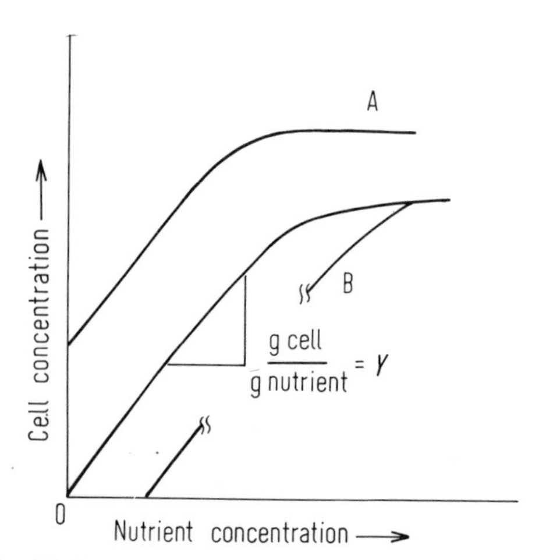

Fig. 30. Dependence of final cell concentration on nutrient concentration. Curve A results when some nutrient is carried over from the seed culture. Curve B results from the need for a critical concentration in the medium.

5.2.2 Chemostat Pulse Method and Factorial Design

The chemostat pulse method (FIECHTER, 1984) involves the sequential pulsing of single components of a medium during a steady- or pseudo-steady state, as shown in Fig. 31. A

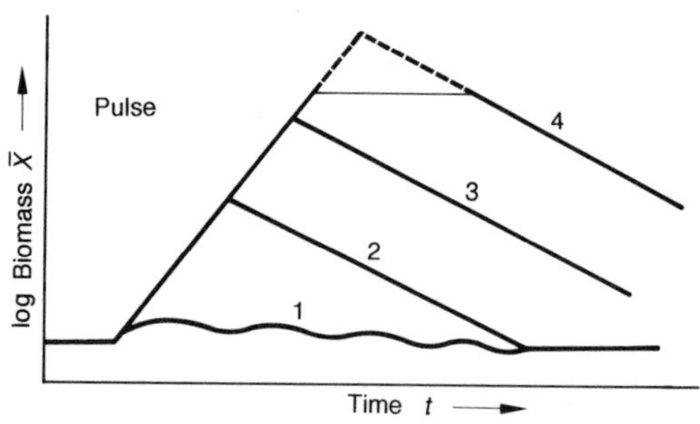

Fig. 31. Chemostat pulse method, types of response to injections of nutrient components during steady state. (1) Injection of a non-essential component: unspecific response; (2), (3) essential medium component; peak height is a function of the injected amount; (4) injection of an essential component in excess; some other component becomes limiting (after FIECHTER, 1984).

suitable amount of the substrate of interest is injected at, e.g., $D=\mu_{max}/2$, where the actual value of S is very small. A rapid response takes place after injection. One major requirement for the success of this technique is that the limiting nutrients (C, N, P, K, Mg) must be individually identified and their dosage determined in advance. This is a problem, for a large number of experiments will still be required to meet this condition. This problem seems to have been taken care of by the application of an automatic process control to the chemostat pulse procedure. The task can be performed unattended by a repetitive batch cultivation in a chemostat under sterile conditions. In an experiment described by LOCHER et al. (1991), 14 pulses of three different nutrients in varying amounts were injected, triggered by the automatic detection of the relevant batch phase. The result of this application is presented in Fig. 32.

The full-factorial method is a statistical approach that examines all possible combinations of independent variables at appropriate levels. Although this approach guarantees faster achievement of optimal conditions with a few nutrients at, say, two concentrations, it is certainly impractical for a large number of variables. This is due to the incredibly large number of experimental runs necessary to complete the procedure. For example, the combination of eight variables at three levels will require 6561 (3^8) experimental runs. This is impractical and, therefore, called for the development of more practical techniques. An extension of the full-factorial approach of interest is the Plackett–Burman two-factorial design (GREASHAM and INAMINE, 1986). This is a fractional factorial design which allows the combination of up to $N-1$ variables in N experiments and is therefore appropriate, when more than five independent variables are being investigated. Unfortunately, the fractional factorial technique largely depends on the ability to guess from theoretical considerations that only some of the variables are likely to be important. By taking this risk, the entire design is reduced to fewer combinations (MORGAN, 1991), which has the disadvantage of providing less information.

5.2.3 Other Mathematical Optimization Techniques

Besides the application of automatic process control to the chemostat pulse system, another very recent development in media design technology is the application of neural networks which facilitate a 63% savings in the number of experimental runs that need to be conducted in the factorial approach (KENNEDY et al., 1992). They can generate a model of a system by mimicking the learning process of the brain. Neural networks could be trained on a few ex-

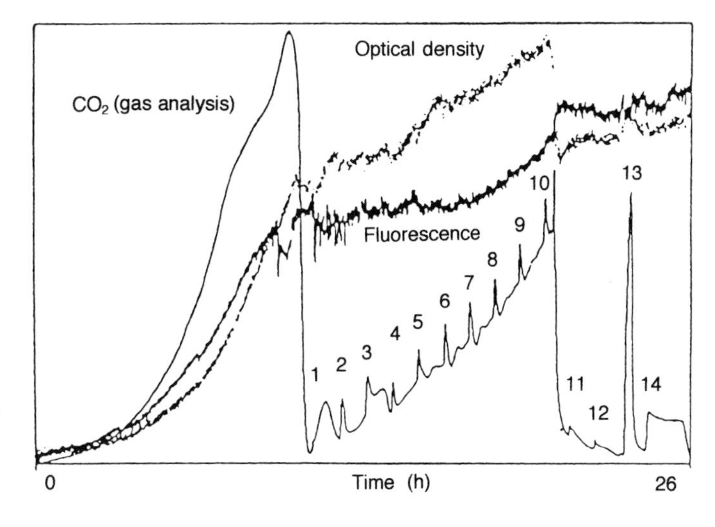

Fig. 32. Automated medium design with 14 nutrient pulses of three substances, which have been applied to a batch culture in different amounts (from LOCHER et al., 1991).

periments after which they predict the rest of test experiments with considerable precision. In addition to fewer experiments, neural networks provide the possibility of capturing data in a form that is easily accessible. KENNEDY et al. (1992) have presented comparative data obtained by neural networks and factorial designs.

Another computer-assisted approach to process control through medium optimization has been reported by FREYER et al. (1992). In this approach, optimization is made feasible both in continuous (on-line) and batch (off-line) processes. The on-line procedure utilizes two dosing units, one for feeding and the other for new medium preparation. Both units are controlled by an optimization algorithm. Care, however, must be taken to avoid precipitation of salts. The off-line procedure utilizes a genetic algorithm which is an iterative procedure based on the principles of natural selection and natural genetic phenomena over a number of generations. Although this procedure is also based on batch fermentations, many fewer trials are required.

5.3 Technical Media

Apart from capital investments, raw materials for medium composition constitute the principal costs, sometimes comprising about 70% of the total cost of even low-value products. It is therefore important to relate medium design data to the nature, availability, and cost of such media. Complex natural media often work better than synthetic media and they are generally less costly. This is because they contain other essential nutrients such as vitamins and mineral elements thereby making supplementation, where necessary, minimal. However, they also contain some substances which are either unusable by the microorganisms or may even be toxic to them. Such substances may cause problems during downstream processes for product recovery and waste management. They may also affect product quality and interfere with process control. Synthetic media do not have these short-comings. In any case, the advantages of complex media which will be enumerated far outweigh these short-comings.

5.3.1 Molasses

Molasses is a by-product of sugar produced from sugar cane or sugar beet. It is obtained from the mother liquor when sugar can no longer be crystallized by conventional methods. In the past, molasses was regarded as a waste product until it found wide-spread use as a sugar source for many types of fermentation including ethanol, amino acids, organic acids (e.g., citric acid), antibiotics and microbial biomass production. Molasses differ from each other depending on their sources, the stage at which they were obtained, and the batch. The differences and similarities between cane or blackstrap "mother liquor" molasses and beet

Tab. 5. Average Values for Constituents in Beet and Cane Molasses at 75% Dry Matter (from BOZE et al., 1992)

Constituent	Dimension	Beet Molasses	Cane Molasses
Total sugars	%	48–52	48–56
Non sugar organic matter	%	12–17	9–12
Protein (N X 6.25)	%	6–10	2–4
Potassium	%	2.0–7.0	1.5–5.0
Calcium	%	0.1–0.5	0.4–0.8
Magnesium	%	0.09	0.06
Phosphorus	%	0.02–0.07	0.6–2.0
Biotin	$\mu g \cdot g^{-1}$	0.02–0.15	1.0–3.0
Pantothenic acid	$\mu g \cdot g^{-1}$	50–110	15–55
Inositol	$\mu g \cdot g^{-1}$	5000–8000	2500–6000
Thiamine	$\mu g \cdot g^{-1}$	1.3	1.8

molasses are shown in Tab. 5. In general, both contain an average of 50% sugar (BOZE et al., 1992). Cane molasses contain more essential mineral elements than beet molasses and will therefore, require less supplementation. The nitrogen content is quite high except, perhaps, the biotin content of cane molasses which will require considerable supplementation in some fermentation processes.

The so-called high-test or invert molasses is actually a product obtained from raw cane juice in which crystallization has been inhibited and the purity reduced by hydrolysis with yeast invertase. It contains 70–85% of sugar due to the high solubility of invert sugar (MINODA, 1986). The world production of molasses is approximately one-third of the total sugar production. Cane molasses account for 21 million of the 33 million tons produced annually, while beet molasses accounts for 12 million (HACKING, 1986). Cane molasses comes from tropical countries where sugar cane is grown, while beet molasses comes from the temperate regions of the world. Some 80% of the world's molasses production is consumed within the producing country, while the remainder is traded around the world. Molasses is inexpensive, because it is a by-product, but its price is greatly affected by the cost of transportation which fluctuates greatly. The large demand for molasses, which is not matched by an increase in production, since it is not an end product *per se*, is a cause for concern. It raises the possibility of scarcity in the near future and the need for new substrates as substitutes or as supplements. Furthermore, improved sugar technology has caused a decrease in the quality of molasses now produced.

5.3.2 Cheese Whey

The most plentiful alternative to molasses as a technical-grade medium is cheese whey which is a by-product of cheese production after fat and casein have been removed. Approximately $36–45 \times 10^9$ kg of whey is produced in the world *per annum* (OURA, 1986) in the process of producing $4–5 \times 10^9$ kg of cheese. The composition of the whey varies with the type of cheese produced, but, generally, contains 6.5–7.0% solids of which lactose is one of the major components and 4.5–5.2% of the whey. Whey also contains 0.3% fat, 0.5–6.0% minerals, 0.7% proteins and 93% water. It has been used as fermentation substrate in the production of ethanol and microbial biomass. Its use, however, is limited by the inability of many micoorganisms to metabolize lactose. In addition, its solubility is so poor that it crystallizes out of solutions having more than 20% concentration. Although whey is extremely inexpensive, due to its potential for environmental pollution, the cost of its concentration is an additional problem.

5.3.3 Corn Steep Liquor

Corn steep liquor (CSL) is a waste product of the wet milling of corn during the manufacture of corn starch, various sweeteners, and other by-products. In the process, corn is steeped in hot water (45–55 °C) and acidified with sulfur dioxide for 30–50 h to loosen its constituents. Acidification encourages the growth of thermophilic lactic acid bacteria including *Lactobacillus bulgaricus* which play a significant role in the loosening process. In the later stages of steeping, some carbohydrates and low-molecular weight nitrogenous compounds are leached into the liquor. The amount of these constituents varies according to the manufacturing process, as well as the quality of the corn (age, origin, germination potency). Corn steep liquor is used as nitrogen source in the production of many industrial products such as enzymes, penicillin G, amino acids, and microbial biomass including *Bacillus sphaericus* as bioinsecticide. Production and availability of corn steep liquor is dependent on the world production and processing of corn which is above 352×10^9 tons annually (HACKING, 1986). Corn steep liquor is relatively inexpensive.

5.3.4 Cassava

Cassava (*Manihot esculenta* Crantz), also known as manioc or tapioca is a root crop of tropical and subtropical regions of the world, especially Brazil, Zaire, Indonesia, Thailand, and Nigeria. It contains 20–40% starch,

0.9–2.3% protein, 0.1–0.7% fat, 0.7–2.0% ash, 0.3–0.8% crude fiber, and 50–70% water (MINODA, 1986). This product is gaining increasing attention as a bulk raw material for ethanol production. It is also being used in Japan for fermentations leading to amino acid production. It compares favorably with sugar cane to the extent that it can grow in poorer soil and can be stored as dehydrated chips or extracted starch for use during the non-harvest season. Under these conditions it keeps much better than sugar cane. However, the cost of dehydration from 50–70% to 20% moisture, the cost of handling and treatment of raw cassava and residue, respectively, imposes serious economic constraints.

5.3.5 Bagasse and Other Cellulosic Materials

Another raw material of interest as technical medium is bagasse. Bagasse is the fibrous material remaining after extraction of sugar from sugar cane. Ordinarily, this waste is burnt to provide heat and electrical energy for distillation during ethanol production and for steam production in vegetable and orange juice industries. It is also used for furfural, paper and cellulose production. Autohydrolyzed bagasse is used as a constituent of animal feed. Being cellulosic in nature, hydrolysis of bagasse helps to produce oligosaccharides and subsequently fermentable sugars. Hydrolysis can be achieved either by chemical means, enzymatically or by direct microbial digestion. The sugars so obtained are available for fermentation. Although bagasse is a waste product, its use is highly competed for by the various industries already mentioned and thus reducing its availability for industrial fermentations.

6 Culture Techniques

The interaction between medium and organisms can be influenced by time-dependent feeding of new medium into the vessel and removal of fermentation broth. Thus, different operation modes can be obtained with different advantages for research and production. For details on some aspects of this section see BAILEY and OLLIS (1986). Some of the basic strategies are outlined here.

Reactor equations for a one-stage continuous stirred tank reactor (CSTR) (Fig. 33) can be set up by material balances considering terms for

input − output +/− conversion = accumulation

$$\begin{aligned} &-(q_{out}/V \cdot X) + (\mu \cdot X) = \dot{X} \text{ biomass,} \\ &(q_{in}/V \cdot S_0) - (q_{out}/V \cdot S) - (r_S \cdot X) = \dot{S} \text{ substrate,} \\ &-(q_{out}/V \cdot P) + (r_P \cdot X) = \dot{P} \text{ product and} \\ &q_{in} - q_{out} = \dot{V} \text{ working volume,} \end{aligned} \tag{9}$$

where the dots indicate the derivatives, e.g., $\dot{X} = \mathrm{d}X/\mathrm{d}t$, which represent the changes in time.

The different operating modes can be obtained as special cases of this system of equations. In the following sections, performance of typical aerobic and anaerobic fermentations in these operating modes are discussed. For simulation studies a simple kinetic model with

$$r_S = r_{S,max} \cdot \frac{S}{k_S + S} \cdot (1 - k_{i,P} \cdot P) \tag{10a}$$

for Michaelis–Menten type substrate uptake kinetics and linear product inhibition

$$\mu = y_{X,S} \cdot r_S - \mu_m \tag{10b}$$

to describe a linear relationship between substrate uptake and growth, motivated by energy limitation (see Sect. 4.3.3, Eqs. 5c, 6b, and BELLGARDT (1991). This model has been preferred to the one consisting of Eqs. (5a and 6a) because its chain of causation is reasonable and because it does not predict a substrate uptake if the substrate concentration equals zero. The parameter values are listed in Tab. 6.

6.1 Batch Culture

Batch culture is a process where the reactants (nutrients in the medium, cells via inoculum, gases by permanent aeration) are brought

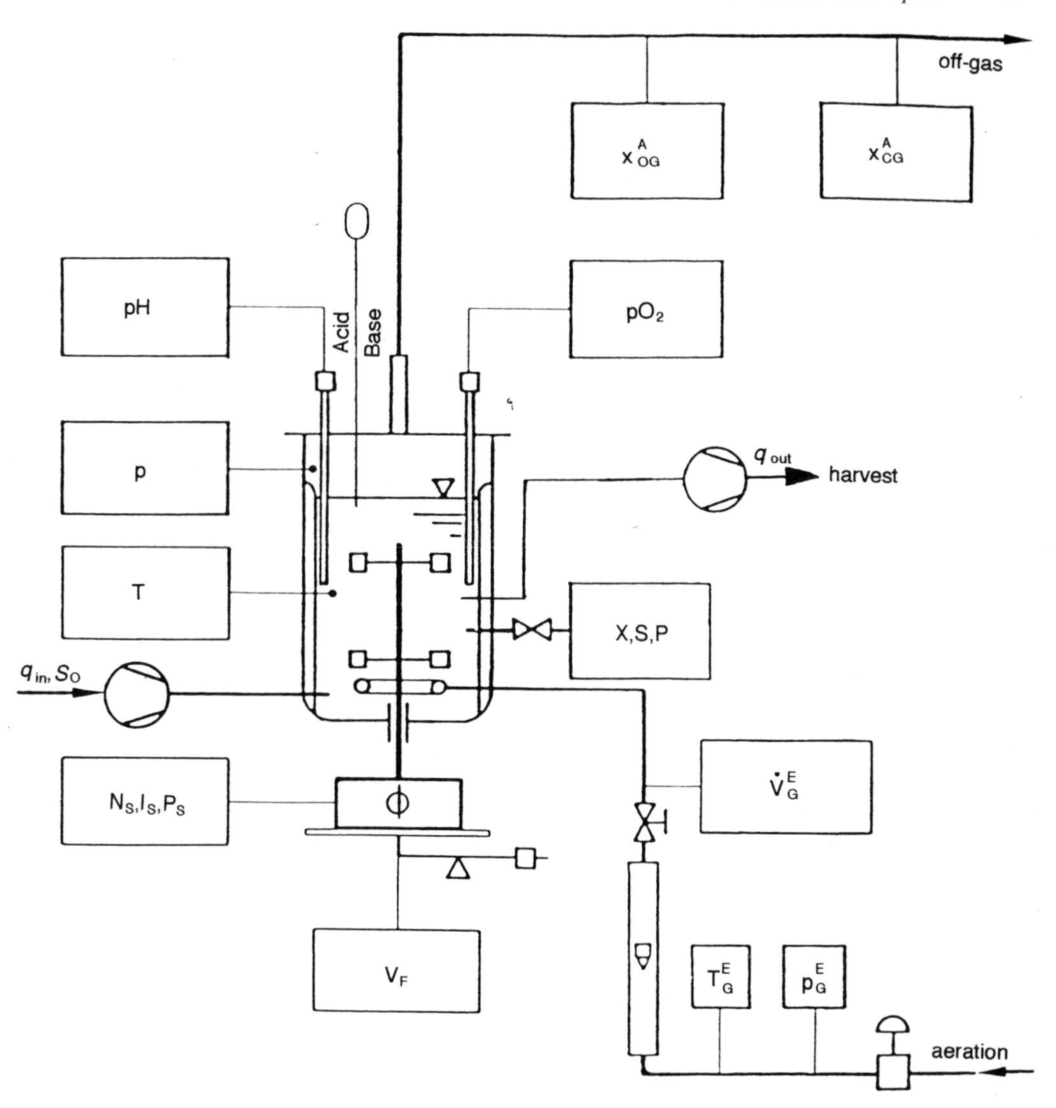

Fig. 33. Typically equipped bioreactor with different measuring and control facilities.

Tab. 6. Parameter Values for Simulations

	$r_{S,max}$	$y_{X,S}$	$y_{P,S}$	μ_m	k_S	$k_{i,P}$	S_0	q_0	V
Aerobic	2.0	0.5	–	0.05	1.0	–	100.0	0.05	1.0
Anaerobic	10.0	0.05	0.5	0.05	1.0	0.01	100.0	0.05	1.0

Data of real-life fermentations are from different authors, where simulations are also available.

together to induce a growth process. It represents in principle a closed system, except for possible gas flow. All mass flows of fluids (q_{in} and q_{out} in Eq. 9) are zero, yielding the simple system of differential equations

$$\begin{aligned} \dot{X} &= \mu \cdot X \\ \dot{S} &= -r_S \cdot X \quad \text{and} \\ \dot{P} &= r_P \cdot X \end{aligned} \tag{11}$$

for the material balances of a batch process, while the volume keeps constant.

Despite the variability of microbial growth, different phases for a typical batch culture are common (BULL, 1974). The different phases are shown in Fig. 34.

During the *lag phase* (I) there is no net increase in microbial mass, both biomass and cell number are constant, but the cells adapt to the new growth conditions, as is obvious from RNA synthesis. Duration of lag phase depends on the quality of inoculum, which should be derived from an exponentially growing culture, and also on physiological tolerance to the medium, which may contain high concentration of osmotically active compounds or exhibit low CO_2 partial pressure.

The *acceleration phase* (II) is usually of short duration. It is neither mathematically nor biologically well described, but, at the end, most of the cells reach their maximum growth rate possible under provided conditions.

Exponential growth (III) behavior occurs, if all cells are adapted to the medium and are able to divide, nutrients are in excess, and inhibitors are absent. Here the differential equation for biomass has the simple solution

$$X(t) = X_0 \cdot e^{\mu_{max} \cdot t} \tag{12}$$

which is the reason for the name of this phase. In terms of control theory, the bioreactor exhibits unstable dynamics. Duration of a batch fermentation depends not only on lag phase and maximum specific growth rate, but also strongly on biomass concentration at the beginning of this growth phase X_0.

During *deceleration or retardation phase* (IV) "unlimited growth" is no longer possible, due to the exhaustion of nutrients and accumulation of inhibitors. Then the specific growth rate becomes a function of nutrient and product concentrations and time. The cell population is still increasing, but the specific growth rate at any particular process time tends to become smaller and smaller. The cells may undergo metabolic and morphological changes.

After most of the nutrients have been used up, overall growth is no longer possible and X becomes constant with time, corresponding to zero net growth rate; this is referred to as *stationary phase* (V). Products of lysis or intracellular material may still allow proliferation of some cells while others already die.

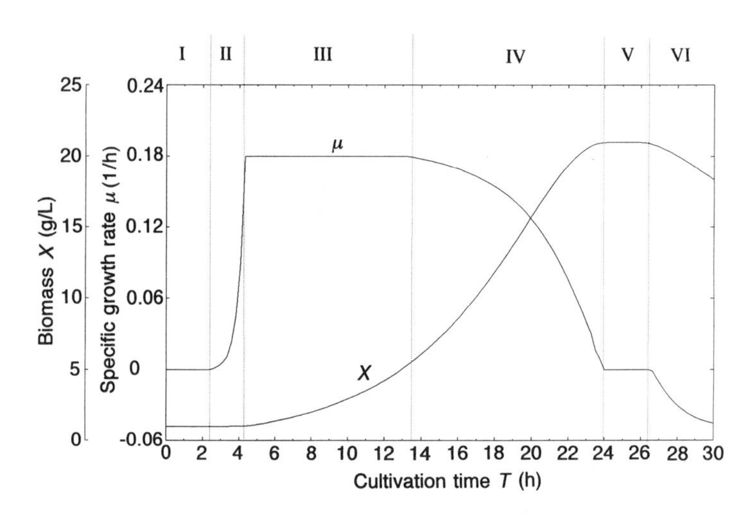

Fig. 34. Idealized characteristics of growth in a batch culture of microorganisms; for details see text.

Finally, the overall specific growth rate becomes negative in the *death phase* (VI). Cells cannot maintain their physiological activities and autolysis can occur.

Batch culture is still the most important process operation used in industry. There are technical and biological reasons for the choice of this strategy. Operation is very simple and can be carried out without much effort in chemical engineering, measurement, and control. Upstream and downstream processes, such as medium preparation, sterilization, or cell harvesting by centrifugation, are usually carried out in batchwise operation, too. The product is of uniform quality in one production batch. Some products cause strong growth inhibition (e.g., ethanol) or are produced only at low growth rates (e.g., antibiotics). Here the batch operation offers the advantage that a high growth rate at the beginning of the fermentation is followed by high product formation at the end. Despite the increasing product inhibition, a high overall volumetric productivity can be achieved combined with high product concentration (Fig. 35c, d). Batch culture is less suitable, if substrate inhibition or growth dependent (undesired) inhibitory by-product formation occurs.

Also for scientific examination of growth related phenomena batch culture is of limited usefulness. Permanent adaptation of the cells to changing medium conditions does not allow for a careful evaluation of kinetics, especially if complex media are used. But even in defined media, some important parameters describing microbial growth such as the Monod constant k_S cannot be accurately estimated in batch culture (HOLMBERG, 1982), because substrate limitation occurs only during a short time peri-

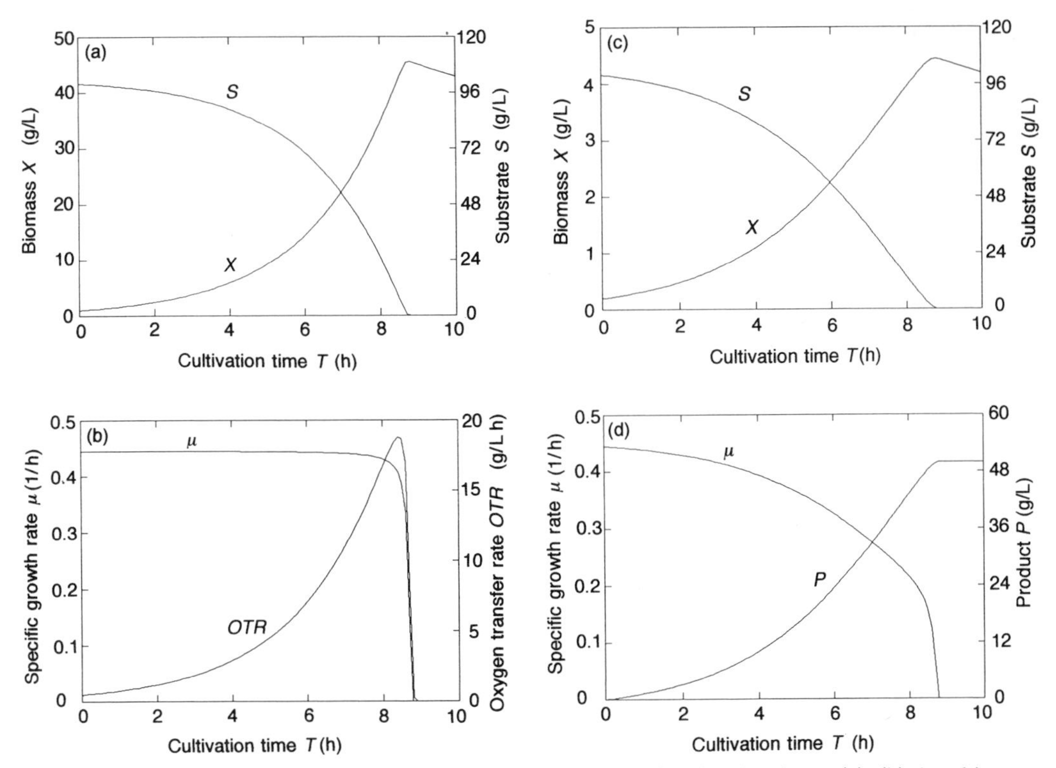

Fig. 35. Model simulations of some important process parameters in a batch culture. (a), (b) Aerobic case: specific growth rate is constantly high until substrate concentration decreases rapidly to the level of k_S and finally to zero. (c), (d) Anaerobic case: specific growth rate decreases with increasing ethanol inhibition; typically final biomass concentrations are lower in this case due to lower yield.

od as can be seen in Fig. 35a,b. In general, there are few possibilities for influencing the process with regard to kinetic investigations. The special case of diauxic growth is shown in Fig. 36. As long as glucose concentration is high, yeasts only take up this preferred substrate and even produce ethanol, which is utilized later when the carbohydrate is no longer available.

6.2 Fed-Batch Culture

Some of the problems of batch culture can be overcome with fed-batch operation. Here the nutrients are permanently fed to the reactor with $q_{in}=q_{in}(t)$, while the volume of the broth increases. Because no removal of fermentation broth occurs ($q_{out}=0$), the reactor equations read

$$\begin{aligned}\dot{X}&=q_{in}/(V\cdot X)-(\mu\cdot X)\\ \dot{S}&=q_{in}/V\cdot(S_0-S)-(r_S\cdot X)\\ \dot{P}&=-q_{in}/(V\cdot P)+(r_P\cdot X)\quad\text{and}\\ \dot{V}&=q_{in}\end{aligned}\tag{13}$$

Of special interest is an exponentially increasing feeding strategy

$$q(t)=q_0\cdot e^{\mu_{max}\cdot t}\tag{14}$$

which results in a constant growth rate $\mu(<\mu_{max})=\mu_{set}$, at least as long as yield remains constant, which is, e.g., not the case during the Crabtree effect. However, this is a very convenient way to control the specific growth rate and related phenomena in a bioprocess. Extensive simulation studies for fed-batch processes can be found in YAMANE and SHIMIZU (1984) and LIM et al. (1986).

Although fed-batch culture needs additional vessels, mass flow control, and balances, it is the operation of choice for many bioprocesses. One main reason is that unpleasant effects of inoculation into pure medium such as substrate inhibition or ammonium toxicity (even of concentrations that are necessary for final yield) are avoided. For production of yeast biomass, growth has to be controlled to avoid ethanol formation through the Crabtree effect, which would cause a longer fermentation duration and lower yield. In the later process phases, where high biomass concentration would cause oxygen limitation, a constant feeding rate corresponds to constant oxygen transfer, proper respiration of the cells assumed, but causes a decreasing growth rate. This is also of advantage for yeast quality (DECKWER et al., 1991), because the fraction of budding cells is low at low growth rates and elder cells are allowed to mature.

In the so-called high density cultivation (RIESENBERG et al., 1991), high biomass con-

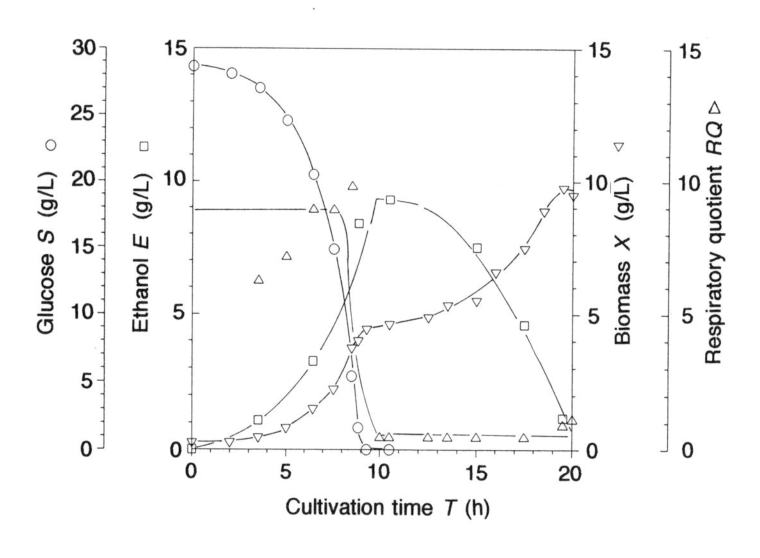

Fig. 36. Diauxic growth of the yeast *Saccharomyces cerevisiae* in batch culture. Symbols represent measurements, straight lines are model simulations. The yeast cells utilize glucose and produce ethanol (Crabtree effect), which is taken up after a short diauxic lag phase.

centrations (>120 g/L) are achieved with *Escherichia coli* cells by maintaining low growth rate, where no inhibiting acetate formation occurs. However, same or better productivities and product concentrations than in batch culture can be achieved. While growth rate (and for substrate limitation also substrate concentration) can be kept constant, products and some medium compounds may accumulate in the broth. Fed-batch cultivation is also a tool for investigation of microbial kinetics with, compared to continuous culture, low experimental effort and high accuracy of results (POSTEN and MUNACK, 1990).

The simulation study in Fig. 37 compares the effects of exponential feeding for the aerobic and the anaerobic models. After an initial batch phase, caused by substrate in the inoculum and possibly not proper adjusted initial feeding rate q_0, the specific growth rate is maintained constant in both cases. But, in the system with product inhibition, substrate increases to compensate for the increasing ethanol inhibition. So, limitation is increasingly superseded by inhibition, which may change microbial physiology, even if the specific growth rate is constant. With increasing time initial volume and initial concentrations become less important and a quasi-steady state is reached. In Fig. 38 measurements of a real-life penicillin production process are compared with simulations.

6.3 Continuous Culture

Continuous culture is an open system. Addition of nutrients and removal of an equal fraction of the reaction mixture ($q_{in}= q_{out}=q$ in

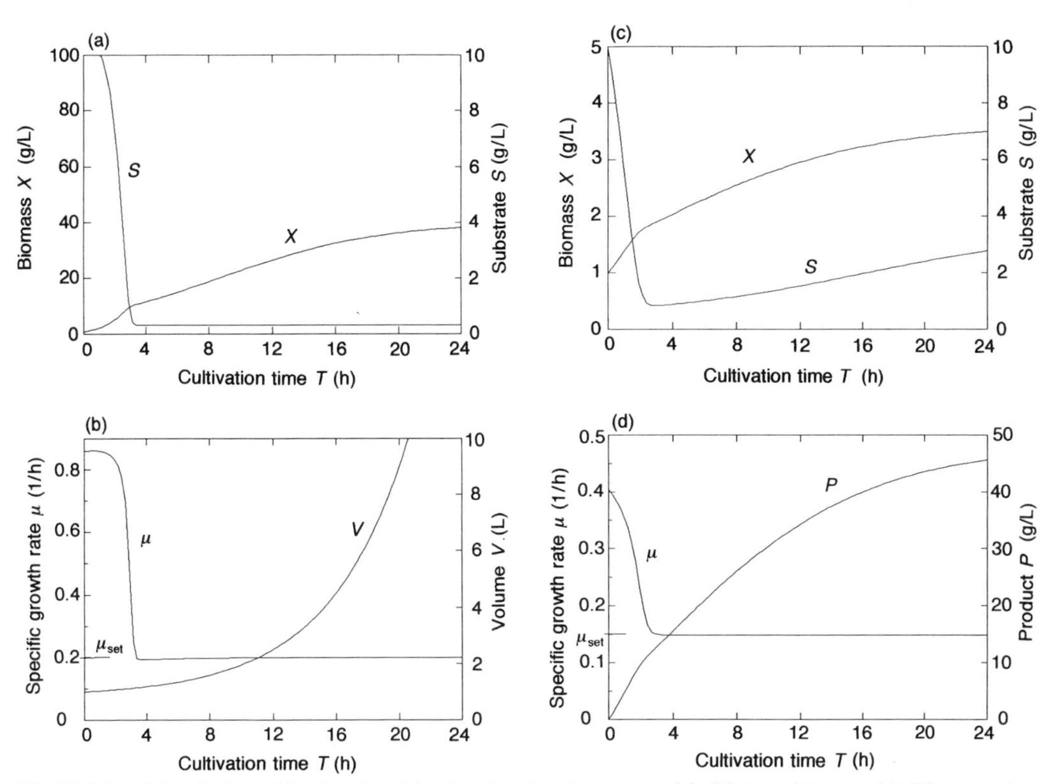

Fig. 37. Model simulation of fed-batch cultivation; for details see text. (a), (b) Aerobic case; (c), (d) anaerobic case.

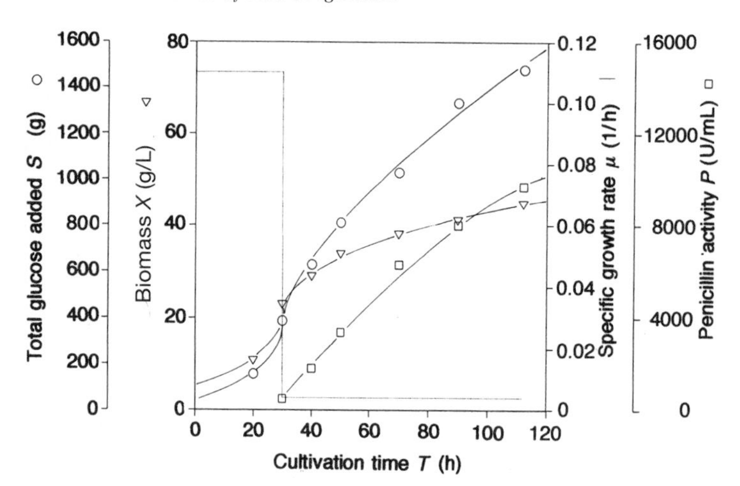

Fig. 38. Penicillin production with the fungus *Penicillium* by fed-batch cultivation. A high growth rate (0.15 L/h) optimal for biomass formation is followed by a low growth rate (0.006 L/h) optimal for production, thus yielding an optimal overall volumetric productivity. Symbols represent measurements, straight lines are simulations (after GUTHKE and KNORRE, 1987).

Eq. 9) are maintained continuously. If the flow rates, the working volume in the reactor vessel, and all other reaction parameters are kept constant, a *steady state* of population characteristics will be reached, where no net accumulation occurs. Consequently, such a culture is often referred to as chemostat. The corresponding reactor balances are no longer differential, but algebraic equations (derivatives in Eq. 9 are zero) and read

$$(D \cdot X) - (\mu \cdot X) = 0$$
$$D \cdot (S_0 - S) - (r_S \cdot X) = 0 \quad \text{and} \qquad (15)$$
$$-(D \cdot P) + (r_P \cdot X) = 0$$

with $D = q/V$ called the dilution rate. Behind the term "dilution rate" (h^{-1}) stands the idea of diluting the fermentation broth continuously with fresh medium. Growth and conversion processes in the reactor, indeed, counterbalance this effect with a steady state growth as a result. From a practical point of view, the relation of the feeding rate q to the actual volume V allows for a discussion of simulations and experimental results independent of the size of the fermentation vessel or the physical magnitude of the feeding stream.

Due to the fact that growth rates solely depend on the rate of feed, a strict control over population dynamics is set up. The investigation of continuous culture dates back to HERBERT (1961); for a recent summary of different kinetic aspects see, e.g., DAWSON (1989). The steady-state condition, especially $\mu = D$, can be achieved for various values of μ and D ranging from zero to the maximum specific growth rate μ_{max} of the organism in the medium used. The corresponding dilution rate is referred to as critical dilution rate D_{crit}. Higher values will result in washout of the fermenter.

The situation is illustrated in Fig. 39. At low dilution rates, biomass concentration (and yield) is low, due to the high ratio of maintenance to the total metabolic energy turnover. At higher dilution rates, substrate concentration increases to allow higher growth rates according to substrate uptake kinetics. Finally, maximum substrate uptake rate is reached, a further increase of dilution rate and substrate concentration no longer supports increasing growth rate and washout occurs. In the anaerobic case, the transition from substrate limitation to washout is prolonged, caused by the slowly decreasing product inhibition. It should be stated that the cells do not "see" the dilution rate directly, but only the actual substrate concentration.

In terms of control theory a steady state is usually stable. This is due to the fact that the presence of substrate S promotes cell proliferation, e.g., increase of biomass X. In turn, an increase of X leads to a decrease of S, so that constancy of feed rate effects constancy of both S and X and keeps growth rate and the cell population in equilibrium independent of time. The situation is, however, more complicated. Besides the stable operating point

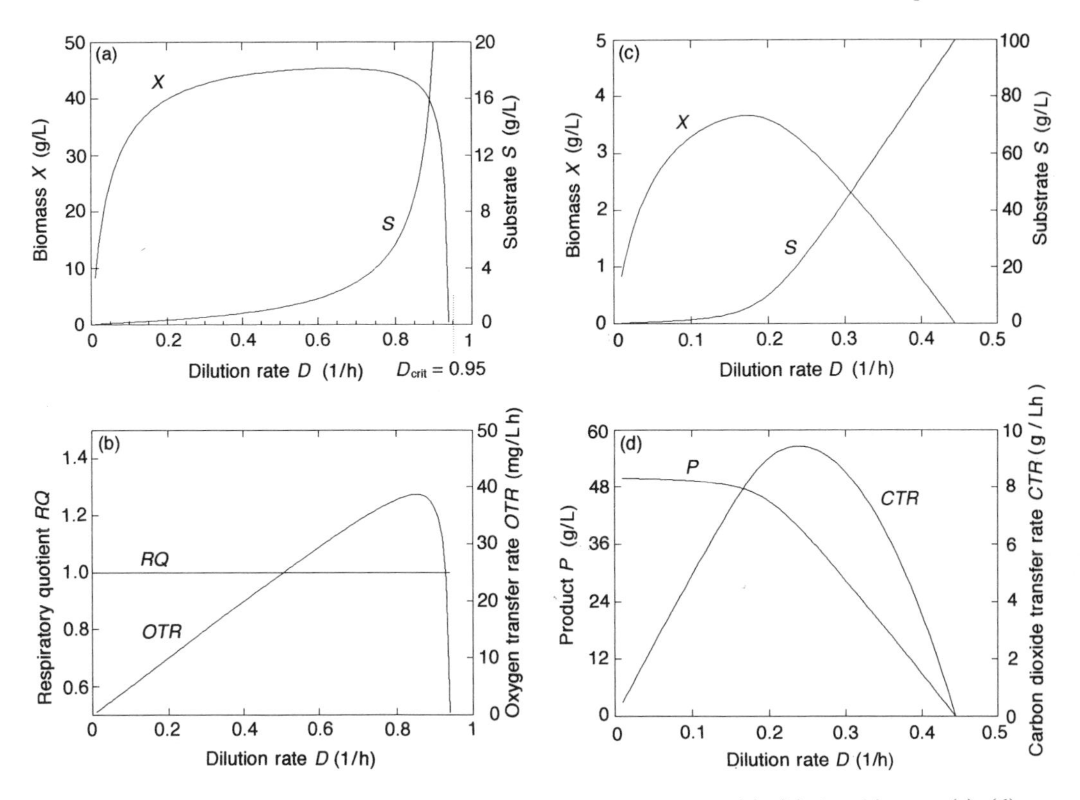

Fig. 39. Model simulation of continuous cultivation; for details see text. (a), (b) Aerobic case; (c), (d) anaerobic case.

characterized by substrate limitation, other conditions are also solutions of Eq. (15) and, therefore, also possible steady states. One of these solutions is always the washout case with $S=S_0$, $P=0$, and $X=0$. If μ as a function of substrate concentration is not monotonic, but reveals a maximum like in substrate inhibition kinetics, substrate-inhibited culture may, for a range of dilution rates, also be a physiologically feasible steady state, as shown in Fig. 40. If, in addition, product inhibition has to be taken into account, instable solutions can occur,

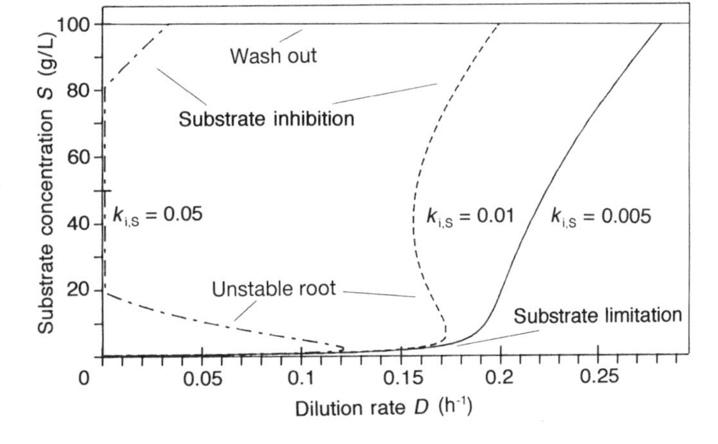

Fig. 40. Substrate concentration as a function of dilution rate and degree of substrate inhibition (for $k_{i,S}$, see Eq. 5b). Substrate-inhibited steady states (usually not desired) can occur for high $k_{i,S}$ values and low dilution rates; "unstable roots" cause oscillatory behavior of a culture, as occasionally observed.

which may lead to prolonged oscillations, as have often been observed in such cultures. Which of the steady states, especially if they are adjoining, are obtained during practical fermentation, depends on the starting conditions, which, therefore, require careful attention (SCHÜGERL, 1987).

Experiments are usually carried out by using stepwise shifts from low dilution rates to higher ones. For practical use, a state is considered as a steady state after five mean residence times $\tau = 1/D$ or if changes are no longer visible. This procedure has to be considered with care, because actual time constants especially at high dilution rates may be higher and changes, e.g., in biomass concentration, may be small but spread over a long duration. A simulated experiment is shown in Fig. 41. Especially for high dilution rates, steady state is not achieved after five residence times. This dynamic behavior can be accelerated by use of a turbidostat, where turbidity is controlled via dilution rate.

Single-stage chemostat methods have found increasing application in basic research in recent years, for experimental advantages of this a growth system are enormous. Cell populations of typical growth characteristics are produced experimentally and independently of time. Kinetics can be investigated under balanced growth, excluding the combined effects of reactions to the environment and cell internal adaptation dynamics; for an example with

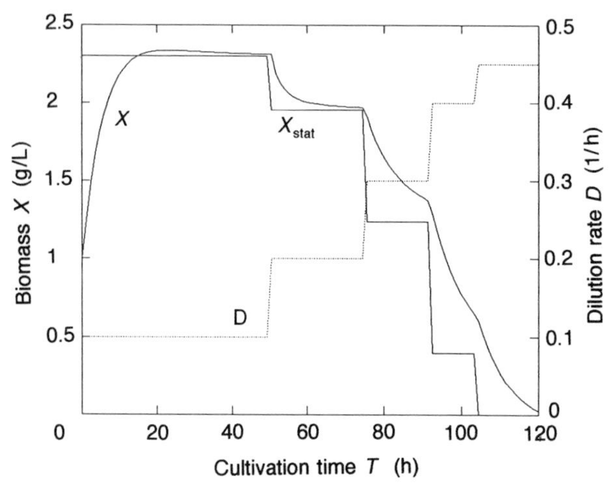

Fig. 41. Simulation of an experiment, where each particular dilution rate was maintained for five residence times. The theoretical steady state is achieved only for low dilution rates. The situation would be even worse, if dynamic adaptation of the cells had to be considered.

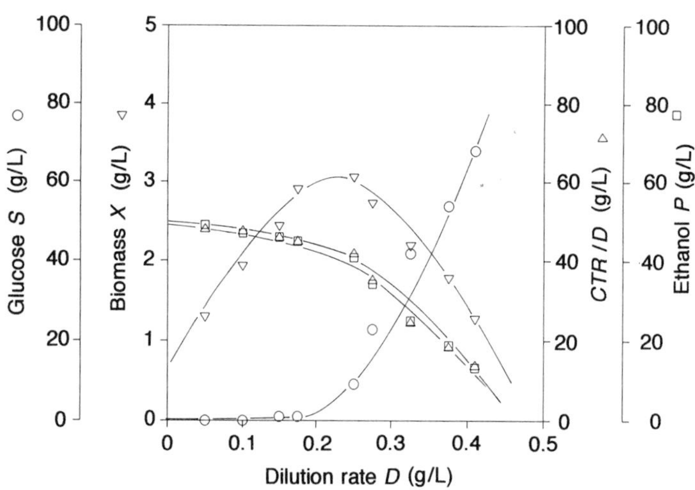

Fig. 42. Real-life measurements of continuous cultivation of the bacterium *Zymomonas mobilis*. Symbols represent measurements, straight lines are simulations; $S_0 = 100$ g/L; CTR/D is shown to make ethanol and CO_2 formation comparable (after POSTEN, 1989).

measured data see Fig. 42. One drawback of a continuous culture is the large quantity of medium required to achieve and maintain different steady states.

Besides the high effort to perform continuous culture on an industrial scale, and technical problems, e.g., with viscous media, some other less profitable features must be mentioned. The permanent loss of biocatalyst often leads to low productivities for cell mass or product. Especially when growth is product-inhibited, only low yield and low product concentration can be achieved. Cascades of several reactors, immobilization of biomass, or dialysis of inhibitors by membranes can diminish these problems. If this is true, the concentrations in the reactor will differ from the concentrations in the outlet and the related terms in Eq. (9) must be modified respectively.

7 References

ANTIER, P., MOULLIN, G., GALZY, P. (1990), Infuence of composition of the culture behaviour of *Kluyveromyces fragilis* in chemostat culture, *Process Bioechem.* **25,** 9–13.

BAILEY, J. E., OLLIS, D. F. (1986), *Biochemical Engineering Fundamentals*, 2nd Ed., New York: McGraw-Hill.

BAJPAI, R. K., IANNOTTI, E. L. (1988), Product inhibition, *Handbook of Anaerobic Fermentation.* (ERICKSON, L. E., FUNG, Y. C., Eds.), pp. 207–241, New York: Marcel Dekker.

BALOWS, A., TRÜPER, H. G., DWORKIN, M., HARDER, W., SCHLEIFER, K.-H. (1992), *The Prokaryotes, A Handbook on the Biology of Bacteria*, New York–Berlin: Springer.

BASTIN, G., DOCHAIN, D. (1990), *On-line Estimation and Adaptive Contol of Bioreactors*, Amsterdam: Elsevier.

BELLGARDT, K.-H. (1991), Cell models, in: *Biotechnology*, 2nd Ed. (REHM, H.-J., REED, G., Eds.), Vol. 4, pp. 267–298, Weinheim–New York–Basel–Cambridge: VCH.

BIEBL, H. (1991), Glycerol fermentation to 1,3-propanediol by *Clostridium butyricum*. Measurement of product inhibition by use of a pH-auxostat, *Appl. Mirobiol. Biotechnol.* **35,** 701–705.

BLOMBERG, A., ADLER, L. (1992), Physiology of osmotolerance in fungi, *Adv. Microb. Physiol.* **33,** 145–208.

BOOTH, I. R. (1985), Regulation of cytoplasmatic pH in bacteria, *Microbiol. Rev.* **49,** 359–378.

BOZE, H., MOULLIN, G., GALZY, P. (1992), Production of food and fodder yeasts, *Crit. Rev. Biotechnol.* **12,** 65–86.

BRINGER, S., HÄRTNER, T., PORALLA, K., SAHM, H. (1985), Influence of ethanol on the hopanoid content and the fatty acid pattern in batch and continuous cultures of *Zymomonas mobilis*, *Arch. Microbiol.* **140,** 312–316.

BROCK, T. D., MADIGAN, M. T. (1991), *Biology of Microorganisms*, 6th Ed., Englewood Cliffs: Prentice Hall.

BULL, A. T. (1974), Microbial growth, in: *Companion to Biochemistry* (BULL, A. T., LAGNADO, J. R., THOMAS, J. O., TIPTON, K. F., Eds.), London: Longman.

CHERRINGTON, C. A., HINTON, M., MEAD, G. C., CHOPRA, I. (1991), Organic acids: chemistry, antibacterial activity and practical applications, *Adv. Microb. Physiol.* **32,** 87–00.

CLARKSON, S. P., LARGE, P. L., BOULTON, C. A., BAMFORTH, C. W. (1991), Synthesis of superoxide dismutase, catalase and other enzymes on oxygen and superoxide toxicity during changes in oxygen concentration in cultures of brewing yeast, *Yeast* **7,** 91–103.

COONEY, C. L., WANG, D. I. C., MATELES, R. I. (1968), Measurement of heat evolution and correlation with oxygen consumption during microbial growth, *Biothechnol. Bioeng.* **11,** 269–281.

COONEY, C. L., WANG, H., WANG, D. I. C. (1977), Computer-aided material balancing for predicting biological parameters, *Biotechnol. Bioeng.* **19,** 55–66.

COULTER, W. H. (1956), *Proc. Natl. Electron. Conf.* **12,** 1034.

COX, P. W., THOMAS, C. R. (1992), Classification and measurement of fungal pellets by automated image analysis, *Biotechnol. Bioeng.* **39,** 945–952.

DASARI, G., WORTH, M. A., CONNER, M. A., PAMMENT, N. B. (1990), Reasons for the apparent difference in the effect of produced and added ethanol on culture viability during rapid fermentations by *Saccharamyces cerevisiae*, *Biotechnol. Bioeng.* **35,** 109–122.

DAWSON, P. S. S. (1984), Continuous cultivation of microorganisms, *CRC Crit. Rev. Biotechnol.* **2,** 315–372.

DECKWER, W.-D., YUAN, J.-Q., BELLGARDT, K.-H., JIANG, W.-S. (1991), A dynamic cell cycling model for growth of baker's yeast and its application in profit optimization, *Bioprocess Eng.* **6,** 265–272.

DEMAIN, A. L., SOLOMON, N. A. (1985), *Biology of Industrial Microorganisms*, London: The Benjamin/Cummings Publishing Company.

DIJKHUIZEN, L., SCHLEGEL, H. G. (1987), The biochemical basis of carbon dioxide requirement, *Proc. 4th Eur. Biotechnology*, Vol. 4 (NEIJSSEL, R. R., VAN DER MEER, R. R., LUYBEN, K. CH. A. M., Eds.), Amsterdam: Elsevier Science Publishers.

DIMARCO, A. A., ROMANO, A. H. (1985), D-Glucose transport system of *Zymomonas mobilis*, *Appl. Environ. Microbiol.* **49,** 151–157.

DOELLE, H. W., MCGREGOR, A. N. (1983), The inhibitory effect of ethanol on ethanol production by *Zymomonas mobilis. Biotechnol. Lett.* **5,** 423–428.

DÜRRE, P., BAHL, H., GOTTSCHALK, G. (1988), Membrane processes and product formation in aerobes, in: *Handbook on Anaerobic Fermentation*, (ERICKSON, L. E., FUNG, Y. C., Eds.), pp. 187–205, New York: Marcel Dekker.

ELANDER, R. P., LOWE, D. A. (1992), Fungal biotechnology: An overview, in: *Handbook of Applied Mycology*, (ARORA, D. K., ELANDER, R. P., MUKERJI, K. G., Eds.), Vol. 4: *Fungal Biology*, New York: Marcel Dekker.

FEHRENBACH, R., COMBERBACH, M., PETRE, J. O. (1992), On-line biomass monitoring by capitance measurement, *J. Biotechnol.* **23,** 303–314.

FIECHTER, A. (1984), Physical and chemical parameters of microbial growth, *Adv. Biochem. Eng. Biotechnol.* **30,** 7–60.

FLETCHER, M. (1991), The physiological activity of bacteria attached to solid surfaces, *Adv. Microb. Physiol.* **32,** 53–82.

FREYER, O., ROTHE, D., WENSTER-BOTZ, C. W. (1992), Computer-aided medium optimization, in: *Microbial Principles on Bioprocess, Cell Culture Technology, Downstream Processing and Recovery* (KREYSA, G., DRIESEL, A. J., Eds.), *DECHEMA Biotechnol. Conf.* **5,** 387–392.

GERHARDT, P. (Ed.) (1981), *Manual of Methods for General Bacteriology*, Washington, DC: American Society for Mirobiology.

GREASHAM, R., INAMINE, E. (1986), Nutritional improvement processes, in: *Manual of Industrial Microbiology and Biotechnology* (DEMAIN, A. L., SOLOMON, A. A., Eds.), pp. 41–48, Washington, DC: American Society for Microbiology.

GREGORY, E. M., FRIDOVICH, I. (1973), Oxygen toxicity and the superoxide dismutase, *J. Bacteriol.* **114,** 1193–1197.

GUTHKE, R., KNORRE, W. A. (1987), Model aided design of repeated fed-batch penicillin fermentation, *Bioprocess Eng.* **2,** 169–173.

HACKING, A. J. (1986), *Economic Aspects of Biotechnology*, Cambridge: Cambridge University Press.

HAROLD, F. M. (1986), *The Vital Force: A Study of Bioenergetics*, New York: W. H. Freeman & Co.

HARRIS, C. M., KELL, D. B. (1985), The estimation of microbial biomass, *Biosensors* **1,** 17–84.

HARRISON, D. E. F. (1972), Physiological effects of dissolved oxygen tension and redox potential on growing populations of microorganisms, *J. Appl. Chem.* **22,** 417–440

HERBERT, D. (1961), The chemical composition of microorganisms as a function of their environment, in: *Microbial Reactions to the Environment* (MEYNELL, G. G., GOODER, H., Eds), pp. 341, Cambridge: Cambridge University Press.

HERBERT, D. (1976), Stoichiometric aspects of microbial growth, in: *Continuous Culture 6: Applications and New Fields* (DEAN, A. C. R., ELLWOOD, D. C., EVANS, C. G. T., MELLING, J., Eds.), pp. 1–30, Chichester: Ellis Horwood.

HERENDEEN, S. L., VAN BOGELEN, R. A., NEIDHARDT, F. C. (1979), Levels of major proteins of *Escherichia coli* during growth at different temperatures, *J. Bacteriol.* **139,** 185–194.

HOLMBERG, A. (1982), On the practical identifiability of microbial growth models incorporating Michaelis–Menten type non-linearities. *Math. Biosci.* **62,** 23–43.

INGRAM, L. O., BUTTKE, T. M. (1984), Effects of alcohol on micro-organisms, *Adv. Microb. Physiol.* **25,** 253–300.

ISHIKAWA, H., NOBAYASHI, H., TANAKA, H. (1990), Fermentation performane of *Zymomonas mobilis* against oxygen supply, *J. Ferment. Bioeng.* **70,** 34–40.

JEFFERSON, C. L., LIM, H. C. (1982), The growth and dynamics of *Saccharomyces cerevisiae*, *Annu. Rep. Ferment. Processes* **5,** 211–262.

JONES, R. P., GREENFIELD, P. F. (1984), A review of yeast ionic nutrition. I: Growth and fermentation requirements, *Process Biochem.* **19,** 48–60.

KAREL, S. F., LIBICKI, S. B., CHANNING, R. R. (1985), The immobilization of whole cells: Engineering principles, *Chem. Eng. Sci.* **40,** 1321–1354.

KENNEDY, M. J., PRAPULLA, S. G., THAKUR, M. S. (1992), Designing fermentation media: A comparison of neural networks to factorial design, *Biotechnol. Tech.* **6,** 293–298.

KJELLEBERG, S., HERMANSSON, M., MARDEN, P., JONES, G. W. (1987), The transient phase between growth and nongrowth of heterotrophic bacteria with emphasis on marine environment, *Annu. Rev. Microbiol.* **41,** 25–49.

LAWFORD, H., HOLLOWAY, P., RUGGIERO, A. (1988), Effect of pH on growth and ethanol production by *Zymomonas*, *Biotechnol. Lett.* **10,** 809–814.

LEAO, C., VAN UDEN, N. (1983), Effect of ethanol and other alcanols on the ammonium transport system of *Saccharomyces cerevisiae*, *Biotechnol. Bioeng.* **25,** 2085–2090.

LIM, H. C., TAYEB, Y. J., MODAK, J. M., BONTE, P. (1986), Computational algorithms for optimal feed rates for a class of fed-batch fermentations:

Numerical results for penicillin and cell mass production, *Biotechnol. Bioeng.* **28,** 1408–1420.

LOCHER, G., SONNLEITNER, B., FIECHTER, A. (1991), Automatic bioprocess control 3: Impacts on process perception, *J. Biotechnol.* **19,** 173–192.

LOCHER, G., SONNLEITNER, B., FIECHTER, A. (1992), On-line measurement in biotechnology: Techniques, *J. Biotechnol.* **25,** 23–53.

LULI, G. W., STROHL, W. R. (1990), Comparison of growth, acetate production, and acetate inhibition of *Escherichia coli* strains in batch and fed-batch fermentations, *Appl. Environ. Microbiol.* **56,** 1004–1011.

MARSHALL, K. C. (1991), The importance of studying microbial cell surfaces, in: *Microbial Cell Surface Analysis — Structural and Physicochemical Methods* (MOZES, N., HANDLEY, P. S., BUSSCHER, H. J., ROUXHET, P. G., Eds.), pp. 3–19, Weinheim–NewYork–Basel–Cambridge: VCH.

MICHALCAKOVA, S., REPOVA, L. (1992), Effect of ethanol, temperature and pH on the stability of killer yeast strains, *Acta Biotechnol.* **12,** 163–168.

MINODA, Y. (1986), Raw materials for amino acid fermentation, *Prog. Ind. Microbiol.* **24,** 51–65.

MONOD, J. (1942), *Recherches sur la Croissance des Cultures Bacteriennes*, Paris: Herman et Cie.

MONOD, J. (1949), The growth of bacterial cultures, *Annu. Rev. Microbiol.* **3,** 371–394.

MORGAN, E. (1991), *Chemometrics: Experimental Design*, New York: John Wiley & Sons.

MOU, D. G., COONEY, C. L. (1983a), Growth monitoring and control through computer-aided online mass balancing in a fed-batch penicillin fermentation, *Biotechnol. Bioeng.* **25,** 225–255.

MOU, D. G, COONEY, C. L. (1983b), Growth monitoring and control in complex medium — a case study employing fed-batch penicillin fermentation and computer-aided on-line mass balancing, *Biotechnol. Bioeng.* **25,** 257–269.

MUNACK, A. (1991), Optimization of sampling, in: *Biotechnology*, 2nd Ed. (REHM, H.-J., REED, G., Eds.), Vol. 4, pp. 251–266, Weinheim–New York–Basel–Cambridge: VCH.

NIELSON, J., VILLADSEN, J. (1992), Modelling of microbial kinetics, *Chem. Eng. Sci.* **47,** 4225– 4270.

NOORMAN, H. J., HEIJNEN, J. J., LUYBEN, K. CH. A. M. (1991), Linear relations in microbial reaction systems: A general overview of their origin, form and use, *Biotechnol. Bioeng.* **38,** 603–618.

O'CONNOR, G. M., SANCHEZ-RIERA, F., COONEY, C. L. (1992), Design and evaluation of control strategies for high cell density fermentations, *Biotechnol. Bioeng.* **39,** 293–304.

ONER, M. D., ERICKSON, L. E., YANG, S. S. (1984), Estimation of yield, maintenance, and product formation kinetic parameters in anaerobic fermentations, *Biotechnol. Bioeng.* **26,** 1436–1444.

ONKEN, U., LIEFKE, E. (1989), Effect of total and partial pressure (oxygen and carbon dioxide) on aerobic microbial processes, *Adv. Biochem. Eng. Biotechnol.* **40,** 137–169.

OURA, E. (1986), Biomass from carbohydrates, in: *Biotechnology* (REHM, H.-J., REED, G., Eds.), Vol. 3, pp. 1–41, Weinheim–Deerfield Beach/Florida–Basel: Verlag Chemie.

PAMPULHA, M. E., LOUUREIRO, V. (1989), Interaction of the effect of acetic acid and ethanol on the inhibition of fermentation in *Saccharamyces cerevisiae*, *Biotechnol. Lett.* **11,** 269–274.

PHAFF, H. J., MILLER, M. W., MRAK, E. M. (1978), *The Life of Yeasts*, Cambridge, MA: Harvard University Press.

PIRT, S. J. (1965), The maintenance energy of bacteria in growing cultures, *Proc. R. Soc. London Ser. B* **163,** 224–231.

PIRT, S. J. (1975), *Principles of Microbe and Cell Cultivation*, London: Blackwell Scientific Publishers.

POSTEN, C. (1989), Modelling of the metabolism of *Zymomonas mobilis* growing on a defined medium, *Bioprocess Eng.* **4,** 217–222.

POSTEN, C., MUNACK, A. (1990), Improved modelling of plant cell suspension cultures by optimum experiment design, *Proc. 11th IFAC World Congress*, Tallinn, pp. 268–273.

POSTGATE, J. R. (1969), Viable counts and viability, *Methods Microbiol.* **1,** 611–628.

PROKOP, A., BAJPAI, K. (1992), The sensitivity of biocatalysts to the hydrodynamic shear stress, *Adv. Appl. Microbiol.* **37,** 165–232.

REARDON K. F., SCHEPER, T. H. (1991), Determination of cell concentration and characterization of cells, in: *Biotechnology*, 2nd Ed. (REHM, H.-J., REED, G., Eds.), Vol. 4, pp. 179–223, Weinheim–New York–Basel–Cambridge: VCH.

RIESENBERG, D., SCHULZ, V., KNORRE, W. A., POHL, H.-D., KORZ, D., SANDERS, E. A., ROSS, A., DECKWER, W.-D. (1991), High cell density cultivation of *Escherichia coli* at controlled specific growth rate, *J . Biotechnol.* **20,** 17–28.

RIGHELATO, R. C. (1979), in: *Fungal Walls and Hyphal Growth* (BURNETT, J. H., TRINCI, A. P. J., Eds.), Cambridge: Cambridge University Press.

ROELS, J. A. (1983), *Energetics and Kinetics in Biotechnology*, Amsterdam: Elsevier Biomedical Press.

ROGERS, P. L., LEE, K. J., SKOTNICKI, M. L., TRIBE, D. E. (1982), Ethanol production by *Zymomonas mobilis*, *Adv. Biochem. Eng.* **23,** 37–84.

ROUXHET, P. G., MOZES, N. (1990), The Micro-environment of immobilized cells: critical assessment of the influence of surfaces and local concentrations, in: *Physiology of Immobilized Cells* (DE BONT, J. A. M., VISSER, J., MATTIASSON, B., TRAMPER, J., Eds.), Amsterdam: Elsevier.

SAHM, H., BRINGER-MEYER, S. (1987), Continuous ethanol production by *Zymomonas mobilis* on an industrial scale, *Acta Biotechnol.* **7,** 307–313.

SANDLER, S. I., ORBEY, H. (1991), On the thermodynamics of microbial growth processes, *Biotechnol. Bioeng.* **38,** 697–718.

SCHELLER, F., SCHMID, R. D. (1992), *Biosensors: Fundamentals, Technologies and Applications, GBF Monographs* **17,** Weinheim–New York–Basel–Cambridge: VCH.

SCHEPER, T., LORENZ, T., SCHMIDT, W., SCHÜGERL, K. (1987), On-line measurement of culture fluorescence of process monitoring and control of biotechnological processes, *Ann. N.Y. Acad. Sci.* **506,** 431–445.

SCHÜGERL, K. (1987), *Bioreaction Engineering*, Vol. 1: *Fundamentals, Thermodynamics, Formal Kinetics, Idealized Reactor Types and Operation Modes*, New York: J. Wiley & Sons, Inc.

SCHÜGERL, K. (1991 a). Common instruments of process analysis and control, in: *Biotechnology*, 2nd Ed. (REHM, H.-J., REED, G., Eds.), Vol. 4, pp. 5–25, Weinheim–New York–Basel–Cambridge: VCH.

SCHÜGERL, K. (1991 b), On-line analysis of broth, in: *Biotechnology*, 2nd Ed. (REHM, H.-J., REED, G., Eds.), Vol. 4, pp. 149–178, Weinheim–New York–Basel–Cambridge: VCH.

SONNLEITNER, B., LOCHER, G., FIECHTER, A. (1992), Biomass determination, *J. Biotechnol.* **25,** 5–22.

STEIN, W. D. (1981), Concepts of mediated transport, in: *Membrane Transport* (BONTING, DE PONT, Eds.), pp. 123–157, Amsterdam: Elsevier.

STEPHANOPOULOS, G., PARK, S. (1991), Bioreactor state estimation, in: *Biotechnology*, 2nd Ed. (REHM, H.-J., REED, G., Eds.), Vol. 4, pp. 225–249, Weinheim–New York–Basel–Cambridge: VCH.

STOUTHAMER, A. H. (1977), Energetic aspects of the growth of micro-organisms, *Symp. Soc. Gen. Microbiol.* **27,** 285–315.

TANG, I. C., OKOS, M. R., YANG, S. T. (1989), Effect of pH and acetic acid on homoacetic fermentation of lactate by *Clostridium formicoaceticum*, *Biotechnol. Bioeng.* **34,** 1063–1074.

TEMPEST, D. W., NEIJSSEL, O. M. (1984), The status of Y_{ATP} and maintenance energy as biologically interpretable phenomena, *Annu. Rev. Microbiol.* **38,** 459–486.

THOMAS, K. C., DAWSON, P. S. S., GAMBORG, B. L. (1980), Differential growth rates of *Candida utilis* mother and daughter cells under phased cultivation, *J. Bacteriol.* **141,** 1–9.

VAN UDEN, N. (1985), Ethanol toxicity and ethanol tolerance in yeasts, *Annu. Rep. Ferment. Processes* **8,** 11–58.

VON STOCKAR, U., BIROU, B. (1989), The heat generated by yeast cultures with a mixed metabolism in the transition between respiration and fermentation, *Biotechnol. Bioeng.* **34,** 86–101.

VON STOCKAR, U., MARISON, W. (1989), The use of calorimetry in biotechnology, *Adv. Biochem. Eng./Biotechnol.* **40,** 93–136.

WALLHÄUSSER, K. H. (1988), *Praxis der Sterilisation, Desinfektion, Konservierung, Keimidentifizierung, Betriebshygiene*, 4th Ed., Stuttgart–New York: Thieme.

WANG, H., COONEY, C. L., WANG, D. I. C. (1977), Computer-aided baker's yeast fermentations, *Biotechnol. Bioeng.* **19,** 69–86.

WEBSTER, J. (1980), *Introduction to Fungi*, 2nd Ed., Cambridge: Cambridge University Press.

WESTERHOFF, H. V., VAN DAM, K. (1987), *Thermodynamics and Control of Biological Free-Energy Transduction*, Amsterdam: Elsevier.

WESTERHOFF, H. V., LOLKEMA, J. S., OTTO, R., HELLINGWERF, K. J. (1982), Thermodynamics of growth — non-equilibrium thermodynamics of bacterial growth — the phenomenological and the mosaic approach, *Biochim. Biophys. Acta* **683,** 181–220.

WITTLER, R., BAUMGARTL, H., LÜBBERS, D. W., SCHÜGERL, K. (1986), Investigations of oxygen transfer into *Penicillium chrysogenum* pellets by microprobe measurements, *Biotechnol. Bioeng.* **28,** 1024–1036.

WOESE, C. R. (1987), Bacterial evolution, *Microbiol. Rev.* **51,** 221–272.

WYATT, P. J. (1973), Differential light scattering techniques for microbiology, *Methods Microbiol.* **8,** 183–263.

YAMANE, T., SHIMIZU, S. (1984), Fed-batch techniques in microbial processes, *Adv. Biochem. Eng. Biotechnol.* **30,** 147–194.

YANG, H., KING, R., REICHL, U., GILLES, E. D. (1992), Mathematical model for apical growth, septation, and branching of mycelial microorganisms, *Biotechnol. Bioeng.* **39,** 49–58.

ZENG, A.-P., BIEBL, H., DECKWER, W.-D. (1990), Effect of pH and acetic acid on growth and 2,3-butanediol production by *Enterobacter aerogenes* in continuous culture, *Appl. Microbiol. Biotechnol.* **33,** 485–489.

ZHOU, W., HOLZHAUER-RIEGER, K., DORS, M., SCHÜGERL, K. (1992), Influence of dissolved oxygen concentration on the biosynthesis of cephalosporin C, *Enzyme Microb. Technol.* **14,** 848–854.

4 Overproduction of Metabolites

OENSE M. NEIJSSEL
M. JOOST TEIXEIRA DE MATTOS
Amsterdam, The Netherlands

DAVID W. TEMPEST
Sheffield, United Kingdom

1 Introduction 164
2 Flux Control Analysis 165
3 Relation between the Energetics of Microbial Growth and Product Formation 167
3.1 General Considerations 167
3.2 Energetics of Microbial Growth 168
3.3 Effect of Growth Energetics on Product Formation 169
4 Effect of Medium Composition 170
4.1 Carbon and Energy Limitation 170
4.2 Effects of a Nitrogen or Sulfur Limitation on Product Formation 172
4.3 Effects of a Potassium Limitation on Product Formation 176
4.4 Effects of a Phosphate, Magnesium, or Iron Limitation on Product Formation 179
5 Futile Cycles and Effect of Culture pH Value 180
6 Concluding Remarks 183
7 References 183

1 Introduction

Microorganisms manifest themselves to us by their products. Sometimes the product is microbial biomass, such as algal blooms or baker's yeast, sometimes the product is of a different nature, such as ethanol, exoenzymes, toxins, antibiotics, etc. Even before he knew about the existence of microorganisms Man had already acquired the skill to use microbes for the production of foodstuffs and chemicals. Until relatively recently the optimization of these production processes was a matter of trial and error. Nevertheless, one can only be impressed by the extent to which this approach has led to the development of processes for the production of compounds such as ethanol, lactic acid, or penicillin, and by the efficiency of these processes.

This chapter deals with metabolite overproduction. In the context of this contribution this means: production in quantities higher than those required in intracellular metabolism, i.e. excreted into the culture fluids (therefore this process is called *over*production). Although the word "metabolite" is generally used for low molecular compounds, in some cases the production of polymers such as polysaccharides will also be briefly discussed.

The emphasis will be put on a discussion of the physiology of metabolite overproduction. Microbial physiology can be defined as the study of the relation between cell structure and function, and how they respond to changes in the environment. It follows that much emphasis will be put on the functional significance of product formation. The reason for this approach is the following: if by a careful choice of the growth conditions an organism can be forced to excrete metabolites in order to increase its survival potential, such a production process will be stable and reversion to a low productivity will not occur. If, on the other hand, production of a compound creates a growth disadvantage, there will be a strong selection pressure on organisms that have lost this property. This means that in this latter case special precautions have to be taken (usually via genetic engineering) in order to avoid the loss of the desired property.

Some scientists treat the microbe as a black box, assuming one particular type of metabolic behavior, but it is the purpose of this chapter to show that this approach ignores the metabolic versatility of microbes and that metabolic fluxes are highly dependent on the growth environment, and therefore can be manipulated by a careful choice of the culture conditions. An example will prove this point: a wild type strain of *Escherichia coli*, grown under fully aerobic conditions in a batch culture on a medium containing glucose as the main carbon and energy source, will produce new cells and usually some acetate. It comes as a surprise to many scientists that the same strain can be a reasonably good gluconate producer (63% of glucose converted into gluconate) when it is grown under special environmental conditions (see below). The crucial point about this example is that this productivity has been achieved with a wild-type organism and without any genetic engineering. It is clear that application of this latter technique could quickly lead to even higher productivities.

A radically different approach is also possible: one tries to reorganize the metabolic pathways of a particular organism by introducing new genes coding for enzymes the host organism did not possess, thereby creating new pathways. This metabolic engineering has yielded already some very interesting results (for a recent review, e.g., see: BAILEY, 1991).

Similarly, production of amino acids by bacteria, which has been initiated and subsequently improved very considerably by Japanese scientists, is based on deregulation of metabolism by selecting mutants that lack the usual tight regulation of the synthesis of these compounds (for review, e.g., see: KINOSHITA, 1987). Cometabolism (conversion of a compound into a product without concomitant assimilation) has also been used in this context. Thus, GROEGER and SAHM (1987) fed the precursor α-ketoisocaproate to cultures of *Corynebacterium glutamicum* growing on glucose. The organism converted this compound into L-leucine (yield 91%). When α-ketobutyrate was added to a culture of this organism isoleucine was formed (EGGELING et al., 1987). One is forced to conclude, however, that the physiological significance of amino acid overproduction by microorganisms is uncertain, and one could argue perhaps that it is a pathological rather than a physiological process taking place in the producer organisms.

The functional significance of the production of antibiotics and other secondary metabolites is also open to dispute. The biosynthetic pathways of these compounds are complex, and the genes necessary for the biosynthesis of secondary metabolites are clustered (together with regulatory and resistance genes). This has led STONE and WILLIAMS (1992) to conclude that organisms producing these compounds must have obtained a selectional advantage, and they hypothesize that this advantage is protection against predation by other organisms during periods of depletion of the supply of one or more nutrients. A similar view had already been expressed by VINING (1990), who stated: "The biosynthesis and regulation of secondary metabolites is a complex, highly coordinated activity that is lost in mutants defective in any of a considerable number of genes. To disbelieve that such pathways are retained because they confer selective value would be truly heretical". On the other hand, the same author also suggests that secondary metabolites do not have any single function that is all inclusive. Again, this lack of knowledge about the function(s) of secondary metabolites is a severe obstacle in the development of a process of antibiotic overproduction that is based on physiological considerations. In such a case a program of selective, or even random mutagenesis to improve product formation is much more rewarding and straightforward.

In spite of the problems discussed above, the most important factor in all production processes is the control of metabolic fluxes, and the main problem to be solved in any production process is how to maximize the flux leading to the product and how to minimize fluxes leading to unwanted side reactions. It is therefore necessary to discuss first the factors that play a role in the control of every metabolic flux.

2 Flux Control Analysis

An important contribution to the quantitative description of metabolic fluxes was made by the work of KACSER and BURNS (1973), and HEINRICH and RAPOPORT (1974). It is outside the scope of this chapter to deal extensively with this theory, but its implications for metabolite overproduction are so great that one cannot ignore it (for a more detailed review, e.g., see: KELL and WESTERHOFF, 1986).

$$S \xrightarrow{E_1} I_1 \xrightarrow{E_2} I_2 \xrightarrow{E_3} I_3 \xrightarrow{E_4} P$$

Fig. 1. Hypothetical metabolic pathway from substrate S, via 3 intermediates (I_1, I_2 and I_3) to product P, catalyzed by 4 enzymes.

Fig. 1 shows a hypothetical pathway from substrate S to product P catalyzed by four enzymes. Flux control analysis enables the researcher to quantify the influence of the properties of each enzyme (K_m, V_{max}, etc.) and the concentrations of all compounds involved in the pathway on the flux through the total pathway. A measure for the influence of one particular enzyme on the flux is its control coefficient. By axiom the sum of the control coefficients of all enzymes taking part in the pathway is set at unity (this is called the summation theorem). Thus, in this hypothetical example, the four enzymes can have many different control coefficients, as is shown in Tab. 1.

Tab. 1. Hypothetical Control Coefficients of Enzymes Involved in the Pathway Shown in Fig. 1

Example	*S*	Control Coefficient of			
		E_1	E_2	E_3	E_4
1	1	0.25	0.25	0.25	0.25
2	1	0.01	0.01	0.97	0.01
3	1	0.00	0.00	0.00	1.0
4	1	0.10	0.01	0.10	0.79

S, hypothetical substrate concentration

The first example shows that the influence of each enzyme of the pathway shown in Fig. 1 on the flux through the pathway is equal. The second case shows that one enzyme (in this case E_3) has a great influence, whereas that of the others is almost (but not completely!) negligible. The importance of this example lies in two consequences that have a great impact on the optimization of product formation.

The first is that this theory predicts that all enzymes that are involved in a pathway will have an influence on the flux through this pathway. It is therefore usually wrong to consider a single particular reaction in a pathway as the "rate-limiting" step. This is only the case when the control coefficients of all other enzymes involved in the pathway are equal to 0, a very exceptional situation. The third case in Tab. 1 shows this case: enzyme E_3 has a control coefficient of 1 and therefore controls the flux for 100%. One would be tempted to state that enzyme E_3 is catalyzing the rate-limiting step in the pathway from S to P, but one should not forget that the concentrations of S and P also play a role. Other values of these concentrations could give rise to a different situation.

Let us assume that in an industrial process the concentration of S can be kept constant at a value that causes enzyme E_3 to have a control coefficient of 1. Let us further assume that the scientists of the company have cloned the gene coding for enzyme E_3 and re-introduced it into the production strain. Let us finally assume that the genetically modified producer organism contains, under the process conditions, twice the amount of enzyme E_3. What would be the effect on the flux through the hypothetical pathway? It could be that the flux is doubled, but this would be extremely unlikely. Much more likely is that the output of the process is only slightly increased, because the control coefficients of all enzymes in the pathway have changed and that, for instance, the new (hypothetical) situation is as is shown in Tab. 1 line 4. This, at first sight curious, result is caused by the fact that the higher V_{max} of E_3 has led to different concentrations of all intermediates in the pathway and of the product.

Literature data on the application of flux control analysis on biotechnologically relevant processes are, to the best of our knowledge, not available. This is probably caused by a major disadvantage of the use of this type of analysis: one needs to know all relevant properties (K_m, V_{max}, effects of inhibitors, allosteric effects, etc.) of all enzymes involved in the pathway from the substrate to the product. This necessitates the purification of all these enzymes and an extensive study of their properties *in vitro* and *in vivo*. Moreover, such a study has to be carried out with the organism used in the process, because other organisms (even those that belong to the same species) may possess different enzyme levels and properties. Nevertheless, the relatively few reports that have been published show the importance of flux control analysis. Thus, RUYTER et al. (1991) have studied the quantitative effects of variations in the cellular amount of Enzyme II^{Glc} of the phosphoenolpyruvate:glucose phosphotransferase system on glucose metabolism in *Escherichia coli*. These authors found that at a cellular concentration of Enzyme II^{Glc} that was the same as in the wild-type organism, the control coefficient of this enzyme on the growth rate on glucose and on the rate of glucose oxidation was low (i.e., a slight variation of Enzyme II^{Glc} activity had no effects on the rate of growth and of glucose oxidation).

The second consequence of this theory is that the longer the pathway (i.e., the more enzymes involved) the smaller the influence of a single enzyme is on the flux through the pathway. This is a consequence of the summation theorem: the sum of all control coefficients of the enzymes of a pathway is 1, independent of the "length" (number of enzymes involved) of the pathway. Thus, if the number of enzymes in Tab. 1 is increased by one, the inevitable consequence will be that the control coefficients of enzymes E_1–E_4 will be lowered, unless the last enzyme has a control coefficient of 0 (see above). This is particularly relevant in the context of antibiotic production. Here the pathway for the synthesis of these compounds are often complex and consist of numerous steps and intermediates. This means that any modification of the properties of one single enzyme has usually only a limited effect on product formation.

In this connection, it has to be mentioned that a control coefficient can also be negative. This occurs at branchpoints in metabolism (Fig. 2). Let us assume that the pathway from

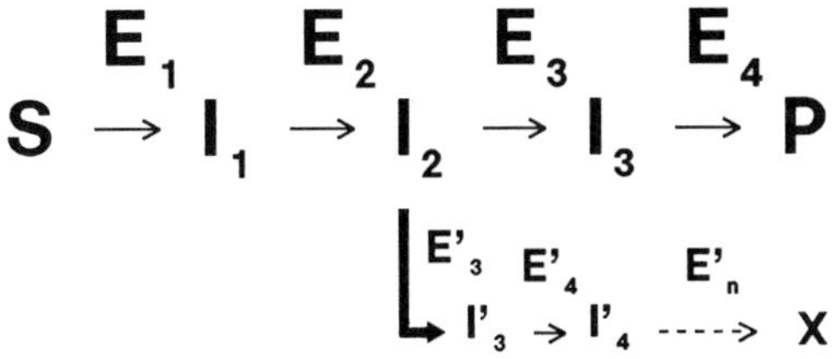

$$S \xrightarrow{E_1} I_1 \xrightarrow{E_2} I_2 \xrightarrow{E_3} I_3 \xrightarrow{E_4} P$$

$$\hookrightarrow \xrightarrow{E'_3} I'_3 \xrightarrow{E'_4} I'_4 \xrightarrow{E'_n} X$$

Fig. 2. Hypothetical metabolic pathway from substrate S, via 3 intermediates to product P, catalyzed by 4 enzymes, and a branch point with a pathway leading to product X.

substrate S to (the desired) product P has one branch which leads to product X (unwanted by-product). Clearly the flux through the pathway leading to P will be influenced by the fluxes through the branch leading to X. The enzymes in the branch leading to X will have negative control coefficients for the flux to P. On the other hand, this example shows clearly that manipulation of the properties of enzymes involved in unwanted side-reactions does have biotechnological relevance.

After this discussion of the implications of flux control analysis one may wonder how microbes are able to produce any substances at all, since one would get the impression that a metabolic network as present in an organism such as *E. coli* inevitably is (in view of the summation theorem) a rigid structure (e.g., see: STEPHANOPOULOS and VALLINO, 1991). This is, as we know, not true and the reason is that during evolution the selection pressure has acted in such a way that those organisms that acquired a high maximal growth rate have become dominant. And for growth to proceed at a high rate the fluxes through all pathways that are involved in what is usually called primary metabolism must necessarily be commensurate with the growth rate. Moreover, it must be realized that the flux control theory does not provide information on the magnitude of fluxes, but on their control.

3 Relation between the Energetics of Microbial Growth and Product Formation

3.1 General Considerations

Of course, the quantitatively most important products (apart from new cells) from any chemotrophic organism are those that are produced by their energy generating reactions. Thus, water and carbon dioxide are the major products of an aerobically growing culture of *Escherichia coli*. Although these compounds are usually not relevant in the context of biotechnology (but think about baking bread!), it is often overlooked that many important products (ethanol, acetic acid, gluconic acid, etc.) are related to energy metabolism. The chemical pathways that lead to the synthesis of these compounds have all been elucidated a long time ago. The remaining problem is therefore how one can maximize the flux through these pathways so as to maximize the rate of product formation and to minimize the flux to pathways leading to undesired side-products (including the synthesis of biomass!); this is discussed in more detail below.

The first thought might be that an increase in the concentration of the substrate, in other words "to put more pressure on the system", might be the answer. This may provide sometimes a solution, but usually other phenomena also occur. This question is dealt with in Sect. 4. Moreover, this approach ignores the fact that metabolic pathways are regulated, and that the activity of the participating enzymes have a great influence on the flux. In Sect. 2 this problem has been dealt with in more detail.

A more rewarding approach is to study the energetic effect of the production of a particular compound. Here, three possibilities exist: a biologically exergonic reaction (i.e., yielding ATP or a proton-motive force), a biologically endergonic reaction, and an energetically neutral reaction. In order to understand the effect of energy metabolism on product formation it is necessary to discuss the energetics of microbial growth.

3.2 Energetics of Microbial Growth

For what purposes is the energy generated in a microbial cell used? This is not a trivial question. Of course, the main energy consuming process is growth, and growth means here the building of new cell material (i.e., a net increase in biomass), but there are other energy-consuming processes as well. Some of these processes are known. For instance, every living cell needs energy to maintain its structural and functional integrity. This is known as the maintenance energy requirement. MONOD (1942) already postulated that there had to be a maintenance energy requirement, but he was unable to measure this consumption of energy in the batch cultures he used for his growth experiments. When *Klebsiella pneumoniae* (now called *Aerobacter aerogenes*) was grown in an aerobic carbon-limited chemostat culture, HERBERT (1958) noted that there was a linear relationship between the rate of oxygen consumption (which is proportional to the rate of energy generation) and the growth rate, but that this line did not pass through the origin, i.e., extrapolation indicated that at zero growth rate the cells consumed oxygen. HERBERT (1958) considered this to be the maintenance energy requirement. Similar data obtained with *E. coli*, grown under the same conditions as those of HERBERT (1958) are shown in Fig. 3.

In spite of the fact that one can get an indication of the magnitude of the consumption of energy by non-growing cells, the problem remains how much energy is minimally needed for the maintenance of a functional cell. This became evident from two types of observations. First, NEIJSSEL and TEMPEST (1976a), using the same organism as HERBERT and colleagues, showed that in chemostat cultures in which glucose was present in excess of the growth requirement of the cells (nitrogen-, sulfate-, or phosphate-limited cultures) there was again a linear relationship between the specific rate of oxygen consumption and the growth rate (Fig. 4). However, in these cultures the specific rates of oxygen consumption at $\mu=0$ (obtained by extrapolation) were invariably higher than those obtained with carbon-limit-

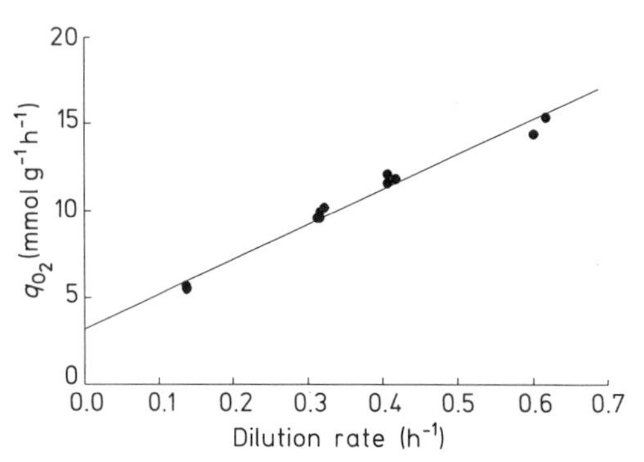

Fig. 3. Relationship between the specific rate of oxygen consumption (q_{O_2}) and the dilution rate of *Escherichia coli* GR70N, grown in a glucose-limited, aerobic chemostat culture on a mineral medium (EVANS et al., 1970), pH=7.0, 37 °C.

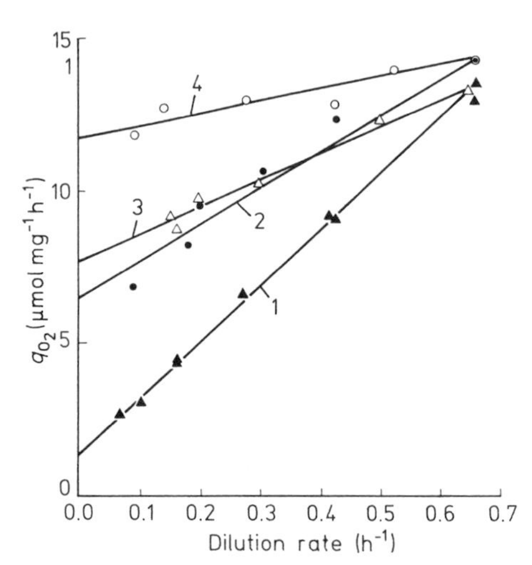

Fig. 4. Relationship between the specific growth rate and the specific rate of oxygen consumption in variously limited chemostat cultures of *Klebsiella aerogenes* growing in a glucose-containing medium. Cultures were carbon-limited (1), ammonium-limited (2), sulfate-limited (3) and phosphate-limited (4), respectively. Data of NEIJSSEL and TEMPEST (1976a), with permission of Springer-Verlag.

ed cultures. Since it is extremely unlikely that the maintenance energy requirement of cells grown in the presence of excess glucose is higher than those of glucose-limited cells, one must conclude that other types of energy consuming reactions were active in these cells. Based on physiological considerations it was proposed that some form of energy spilling must have occurred in glucose-excess cultures, but the specific reactions that were responsible for this wastage were not yet identified (NEIJSSEL and TEMPEST, 1976b).

Another observation points to the same problem. The following experiment makes this clear: when an organism such as *E. coli* or *Klebsiella pneumoniae* is grown in an aerobic glucose-limited chemostat culture at a low rate (for instance, dilution rate, $D = 0.1\ h^{-1}$) all glucose is consumed and converted into carbon dioxide and new cells. Such a culture is fully viable and stable, which suggests that the energetic costs of maintaining cell integrity are fully covered. When the growth limitation is relieved by the sudden addition of an extra amount of glucose to the growth vessel, one observes an immediate and sharp increase in the oxygen consumption and carbon dioxide production rates (NEIJSSEL and TEMPEST, 1976b; NEIJSSEL et al., 1977). A more detailed analysis will show that the culture is not growing immediately at a faster rate. The conclusion is again that a major part of the energy generated after the addition of the extra glucose must have been wasted. The identity of these energy-spilling reactions is again largely unknown, but some ideas about their physiological significance have been put forward. Another important conclusion is that the enzyme(s) catalyzing these reactions is (are) expressed constitutively.

From a physiological point of view one can therefore define three types of energy consuming reactions: the net synthesis of new cell material, maintenance of cell integrity and viability, and finally energy-spilling reactions (whatever their function may be). Some authors divide the energy consuming reactions in only two classes: growth (increase in mass) and maintenance (no increase in mass). In the context of this contribution it is not necessary to discuss the merits of the one or the other classification, as will be made clear below.

3.3 Effect of Growth Energetics on Product Formation

Quite a number of useful compounds are produced by energy generating reactions. Examples are gluconic, lactic, and acetic acid, ethanol, acetone, butanol, iso-propanol, etc. Although most of these products are formed in a fermentation process (*sensu stricto:* no involvement of a respiratory chain) and others in a process in which respiration is involved, there is no principal difference between these two processes in one important respect: their production is biologically exergonic.

In view of the discussion in the previous section it will now be clear that an increased rate of energy consumption will lead to an increased rate of production of these compounds. This may seem like a simple solution, but there is a problem: the main energy consuming process is growth and, with the exception of baker's yeast and food or feed yeast from *Candida utilis*, microbial biomass is waste product. Thus, one should aim at increasing the other energy consuming reactions and minimize the production of new cells. It will now be clear why maintenance of energy and energy-spilling reaction can play an important, positive role in product formation: they consume energy, but do not lead to the production of waste material, although the generation of extra heat (the inevitable consequence of energy-spilling reactions) could have negative economic consequences. The next question is how one can manipulate this energy drain.

Many observations have shown that different strategies can be used. Firstly, as indicated above, some types of growth environment lead to this wasteful energy dissipation (Sect. 4.3). Secondly, some microorganisms excrete products that induce energy spilling reactions (Sect. 5). Thirdly, one could add to the growth medium compounds that stimulate the wastage of energy (provided, of course, that such compounds do not interfere with other parts of the process, such as downstream processing, or have a negative effect on the quality of the final product) (Sect. 5). Finally, one could use oscillating culture conditions. The experiment described above (the addition of an excess of glucose to a glucose-limited culture) indicates that it is, at least in principle, possible to induce energy-spilling reactions in a culture that was previously growing with a maximal energetic efficiency. These strategies will be discussed in more detail below.

4 Effect of Medium Composition

4.1 Carbon and Energy Limitation

The composition of the medium has a great effect on product formation by microorganisms. This applies equally to primary products, such as ethanol or citrate, and to so-called secondary metabolites, such as antibiotics. Although the important role of the concentrations of some macronutrients such as the nitrogen source or phosphate in the production of antibiotics is well documented, a clear overall view is lacking. Particularly in the production of antibiotics it seems that almost every antibiotic requires its own type of medium composition and growth conditions (e.g., see DEMAIN, 1982; HÜTTER, 1986). A discussion of this type of metabolite overproduction would only lead to a phenomenological treatise. The physiological significance of the production of primary metabolites is much better understood and therefore this topic will be discussed.

It is to be expected that when one particular nutrient element is present in the growth medium at such low concentrations that it limits growth, the microorganism will contain the minimum amount of this element. In this context "minimum" means the amount that is just sufficient to allow the cell to survive – any surplus will not be present in the cell. Hence, one would not expect accumulation of storage polymers, such as glycogen or poly-β-hydroxybutyrate, to occur under carbon-limited conditions, unless the organism shows this behavior constitutively. But even if the latter is the case, one will observe that certain carbon-excess conditions (notably, a nitrogen limitation, see below) will stimulate the production of these storage polymers in such an organism even further.

When heterotrophic organisms are grown with a growth limiting supply of the carbon and energy source they will only produce extracellular products other than carbon dioxide and water when they carry out a fermentation. Under these circumstances the organisms grow with the maximum energetic efficiency that they can obtain through their metabolism. This can be illustrated by the following example. ROSENBERGER and ELSDEN (1960) have grown *Streptococcus faecalis* in chemostat culture under two different limitations; a glucose limitation and a tryptophan limitation. Since this organism uses glucose exclusively for energy generation (organic nutrients in the medium supply the building blocks for the synthesis of new cell material) these experiments are easy to interpret from an energetic point of view.

The data in Tab. 2 show that when the organism grew faster the specific rates of glucose consumption and of lactate production increased. This can be explained by a greater energy demand for cell synthesis. It is also evident that the organism used more glucose when it was grown under a tryptophan limitation. This is the same phenomenon as discussed in Sect. 3.2: under conditions of energy excess, more of the energy source is consumed. However, there is an interesting problem with these data. When glucose is metabolized completely to lactate one would expect an overall reaction as shown in Fig. 5 for a homofermentative pathway. The specific rate of ATP production is therefore equal to the specific rate of lactate production, and this latter rate should be twice the rate of glucose consumption. It can be observed that this stoichiometry has not been obtained. The reason is that the organism has an alternative pathway for glucose fermentation (Fig. 5, heterofermentative pathway; Tab. 2). The important point is that this organism, which according to the textbooks produces predominantly lactic acid (e.g., HARDY, 1986), also can produce substantial quantities of ethanol and acetate. As is evident from the scheme shown in Fig. 5, production of acetate is an energetically favorable process. The use of this pathway leads to an overall ATP/glucose ratio of 3. One can conclude therefore that the nature of the growth limitation has at least three effects on this organism: (1) it uses a more efficient pathway for energy generation

Tab. 2. Specific Rates of Glucose Consumption and of Product Formation in Chemostat Cultures of *Enterococcus (Streptococcus) faecalis*

Limitation	D	q_{Glc}	q_{Lac}	q_{Ac}	q_{EtOH}	q_{For}	q_{CO_2}	q_{ATP}
Glucose[a]	0.22	4.9	3.8	ND	ND	ND	ND	> 9.8
	0.31	7.5	6.8	ND	ND	ND	ND	>15
	0.43	9.6	17.6	ND	ND	ND	ND	>19.2
Tryptophan[a]	0.22	12.9	13.4	ND	ND	ND	ND	>25.8
	0.31	16.6	20.2	ND	ND	ND	ND	>33.2
	0.43	16.1	30.6	ND	ND	ND	ND	>32.2
Glucose[b]	0.1	2.5	1.2	1.9	1.8	3.7	0.5	6.8
	0.3	7.4	4.1	4.0	6.7	8.1	1.4	18.8
	0.5	12.6	15.2	3.6	5.7	7.8	1.5	28.1

Abbreviations: ND, not determined; Glc, glucose; Lac, lactate; Ac, acetate; EtOH, ethanol; For, formate
[a] Data of ROSENBERGER and ELSDEN (1960); q_{ATP} has been calculated as $>2\ q_{Glc}$
[b] Data of L. SNOEP, unpublished results; q_{ATP} has been calculated according to $q_{ATP}=q_{Lac}+q_{EtOH}+2\ q_{Ac}$

when the energy source is limiting growth; (2) even though a less efficient pathway of energy generation is used, it still generates more energy when the energy source is in excess of its growth requirement; (3) the production of lactate occurs preferentially at higher growth rates.

The same behavior has been observed in many other streptococci (CARLSSON and GRIFFITH, 1974; HAMILTON et al., 1979; THOMAS et al., 1974; YAMADA and CARLSSON, 1975). In addition, it has to be emphasized that one of the most important rules dominating a fermentation (*sensu stricto*) is that the degree of reduction of the substrate(s) and of the product(s) must be equal. In streptococci, this can be checked, because these organisms use organic compounds (e.g., yeast extract) for cell synthesis and need a separate energy source (e.g., glucose). It can easily be seen that a stoichiometric conversion of 1 mol glucose into 2 mol lactate obeys this rule. This means that an energy source more reduced than glucose (e.g., sorbitol) cannot be converted solely into lactate, whereas compounds more oxidized than glucose (e.g., pyruvate and citrate) will gener-

glucose + 2 NAD^+ + 2 ADP + 2 P_i ⟶ 2 pyruvate + 2 NADH + 2 H^+ + 2 ATP

Homofermentative pathway:

2 pyruvate + 2 NADH + 2 H^+ ⟶ 2 lactate + 2 NAD^+

Overall reaction:

glucose + 2 ADP + 2 P_i ⟶ 2 lactate + 2 ATP

Heterofermentative pathway:

2 pyruvate + 2 CoA ⟶ 2 formate + 2 acetyl-CoA

acetyl-CoA + P_i ⟶ acetylphosphate + CoA

acetylphosphate + ADP ⟶ acetate + ATP

acetyl-CoA + 2 NADH + 2 H^+ ⟶ ethanol + 2 NAD^+ + CoA

Overall reactions:

glucose + 3 ADP + 3 P_i ⟶ 2 formate + acetate + ethanol + 3 ATP

Fig. 5. Scheme of the different pathways of glucose fermentation in streptococci.

ate less lactate and more oxidized products of metabolism. This has been observed indeed under many different culture conditions (ABBE et al., 1982; BROWN and PATTERSON, 1973; SNOEP et al., 1990). Thus, the concept of a homolactic fermentation (i.e., lactate as the sole fermentation product) as a fundamental property of a *Streptococcus* species is in its generality incorrect.

Organisms carrying out a respiration show a similar type of behavior. This was first observed with aerobic cultures of the facultative aerobe *Klebsiella pneumoniae* (previously called *Aerobacter aerogenes* and *K. aerogenes*): the most efficient metabolism of the carbon and energy source was observed when this organism was grown in a chemostat under carbon-limited conditions, the sole products of glucose metabolism being carbon dioxide and new cell material (Tab. 3). Under carbon-excess conditions the specific rates of glucose and oxygen consumption were invariably higher. Interestingly, oxidized products of glucose metabolism were excreted into the culture fluids (Tab. 3). The accumulation of these compounds was not caused by cell lysis, because the nature and quantity of these products were dependent upon the nature of the growth limitation (NEIJSSEL and TEMPEST, 1975; 1976a). It has to be emphasized that these cultures were fully aerobic; under an oxygen limitation this organism produces considerable quantities of 2,3-butanediol (PIRT and CALLOW, 1958). The same type of experiment with other organisms, such as *Agrobacterium radiobacter* (LINTON et al., 1987b), *Candida utilis, Bacillus subtilis, Clostridium butyricum* (CRABBENDAM et al., 1985), *Escherichia coli* (Tab. 4), *Pseudomonas* species (HARDY, 1992) have shown that all these organisms behave similarly in that carbon-excess conditions led to a lowered energetic efficiency (higher specific rates of consumption of the carbon and energy source and, where applicable, of oxygen) and overproduction of metabolites. Although it is not possible to review all these experiments in sufficient detail, general trends have emerged. These and the effects of some specific nutrient limitations will now be discussed.

Tab. 3. Conversion of Glucose into Products by Variously Limited Chemostat Cultures of *Klebsiella pneumoniae* ($D=0.17\ h^{-1}$, pH 6.8, 35 °C)

Carbon limitation
12 glucose + 26 O_2 + 9 NH_3 →
1 kg cells + 31 CO_2 + 50 H_2O

Nitrogen limitation
36 glucose + 44 O_2 + 9 NH_3 →
1 kg cells + 40 CO_2 + 70 H_2O + 4 acetate + 3.5 pyruvate + 11 2-ketoglutarate + 1.9 kg extracellular polysaccharide

Nitrogen limitation plus 1 mM DNP
84 glucose + 76 O_2 + 9 NH_3 →
1 kg cells + 35 CO_2 + 123 H_2O + 8 acetate + 19 pyruvate + 7-ketoglutarate + 4 gluconate + 48 2-ketogluconate

Potassium limitation
58 glucose + 106 O_2 + 9 NH_3 →
1 kg cells + 113 CO_2 + 165 H_2O + 13 acetate + 7 pyruvate + 1 2-ketoglutarate + 11 gluconate + 7 2-ketogluconate + 0.4 kg extracellular protein

All coefficients were adjusted to a production rate of 1 kg cells per hour. Substrates and products are given in moles. Data of NEIJSSEL and TEMPEST (1975, 1979)

4.2 Effects of a Nitrogen or Sulfur Limitation on Product Formation

Bacteria contain approximately 13% (w/w) nitrogen, the content of this element is slightly lower in fungi (about 11% (w/w)), because their cell wall contains less nitrogen. When *Klebsiella pneumoniae* and *Candida utilis* were grown under a nitrogen limitation it was observed that the nitrogen content of the biomass decreased (HERBERT, 1976). This was caused by the accumulation of polysaccharide (glycogen). This observation was first made by HOLME (1957) in a nitrogen-limited chemostat culture of *Escherichia coli*, and since then it has become clear that it applies to a multitude of microorganisms; that is why the main feature of the production process of biotechnologically relevant polysaccharides, such as xanthan and others, is the use of nitrogen-limited culture conditions.

For example, LINTON and coworkers have studied the production of an exopolysaccha-

Tab. 4. Specific Rates of Glucose Utilization, Oxygen Utilization and of Production Formation (expressed in mmol g^{-1} h^{-1}) Observed in Chemostat Cultures of *Escherichia coli* B/r, Grown under Different Limitations ($D=0.16\pm0.02$ h^{-1}, pH 5.5, 35 °C), in the Absence and Presence of PQQ (final concentration 0.2 μM)

Lim.	[PQQ]	Glc	GA	2-OG	Pyr	HAc	CO_2	O_2	GA/Glc
C	0	2.0	0	0	0	0	5.2	5.2	0
	0.2	1.9	0	0	0	0	5.6	5.6	0
N	0	2.9	0	0.5	0	0.1	8.8	9.3	0
	0.2	3.6	0.3	0.8	0	0.2	9.6	10.7	8
S	0	2.7	0	0	0.5	0.7	7.7	8.1	0
	0.2	7.3	4.4	0	0.7	0.6	7.9	10.5	60
P	0	3.2	0	0	0	0.1	12.9	12.5	0
	0.2	10.0	6.3	0	0	0.5	13.1	15.2	63
K	0	4.6	0	0	0.6	0.9	16.3	15.4	0
	0.2	6.0	1.1	0	0.3	1.0	16.2	16.2	18

Abbreviations: Glc, glucose; GA, gluconate; 2-OG, 2-oxoglutarate; Pyr, pyruvate; HAc, acetate; GA/Glc, percentage of gluconate formed per glucose consumed. Data of HOMMES et al. (1991)

ride (succinoglycan) by *Agrobacterium radiobacter* NCIB 11883. This polymer contains the constituents glucose, galactose, pyruvate, succinate and acetate in the molar ratio 7:1:1:1: 0.1. It was found that this polymer was synthesized optimally in a nitrogen-limited chemostat, whereas cells grown in a carbon-limited chemostat culture did not produce this compound. The production of this polysaccharide was shown to be an energy-demanding process (LINTON et al., 1987a).

Further studies indicated that this extra energy demand did not lead to a change in the energetic efficiency of the respiratory chain of this organism (CORNISH et al., 1987). This organism possesses a PQQ-linked glucose dehydrogenase, but can only synthesize the apo-glucose dehydrogenase and not pyrroloquinoline quinone (PQQ) (for a more detailed discussion of this phenomenon, see Sect. 4.3). One could imagine that addition of PQQ to a nitrogen-limited culture (grown in the presence of excess glucose) would lead to extra energy generation and therefore be beneficial for the production of the polysaccharide. It was shown, however, that addition of 1–4 μM PQQ led to an immediate production of gluconate (with a specific rate of production of 2.8 mmol g^{-1} h^{-1}) but no increase in the specific rate of polysaccharide production could be observed (Tab. 5). Although it could be argued that the oxidation of PQQH2 by the respiratory chain of this organism did not lead to the generation of extra energy (which implies that the seg-

Tab. 5. Specific Rates of Glucose Consumption, Oxygen Consumption and Product Formation in Chemostat Cultures of *Agrobacterium radiobacter* Grown under Nutrient Limitation

Limitation	D (h^{-1})	Glucose[a]	Gluconate[a]	Polysacch.[b]	O_2[a]	CO_2[a]
C	0.043	0.557	0	0	1.27	1.37
N	0.041	1.83	0	0.18	1.77	2.0
N + PQQ[c]	0.03	4.77	2.8	0.15	3.04	1.96

[a] The rates are expressed in mmol g^{-1} h^{-1}.
[b] The specific rates of production of extracellular polysaccharide are given in g g^{-1} h^{-1}.
[c] 4 μM PQQ
Data of LINTON et al. (1987b)

ment between the quinone pool and oxygen did not contribute to energy generation), one could also argue that the extra energy generated by the activity of this enzyme could not be used for polysaccharide synthesis, since the metabolic flux could not be increased because of other factors.

In this connection, it has to be remembered that the uptake systems for the respective carbon sources (succinate and glucose) will be saturated with their substrates when the organism is growing under nitrogen-limited conditions. The conversion of glucose to gluconate, which is carried out in the periplasm, does not involve transport, but, at least in principle, gluconate could be used for the extra supply of carbon skeletons. However, it has been shown in *E. coli* and *K. pneumoniae* that under glucose-excess conditions gluconate, formed by the activity of the glucose dehydrogenase, is not taken up by the cells (unpublished results). From the data of LINTON et al. (1987b) it is clear that in this respect *A. radiobacter* behaves similarly.

Interestingly, the synthesis of other compounds such as poly-β-hydroxybutyrate and polyphosphate is also stimulated under nitrogen-limited conditions (or in media with a high C/N ratio). Together these compounds have been called "storage polymers", and the regulation of their production was elucidated already some time ago (DAWES and SENIOR, 1973). The underlying mechanism of the stimulating effect of a nitrogen limitation on storage polymer production seems to be that the ample availability of the carbon source leads to an increase of the intracellular NADH/NAD and ATP/ ADP ratios (see Fig. 6). In other words, the cells face the problem of energy surplus. This is for instance shown by experiments with *K. pneumoniae*. This organism produces large amounts of an extracellular polysaccharide when it is grown in a chemostat under an ammonia limitation with glucose present in excess. Addition of 2,4-dinitrophenol (a compound causing energy dissipation, see Sect. 5) to the medium abolished polysaccharide synthesis and led to a drastic re-organization of intermediary metabolism (NEIJSSEL, 1977) (see Tab. 3).

One may wonder why during the course of evolution organisms have acquired the capacity to store polymers. One explanation is that the polymerization of glucose moieties to glycogen consumes energy and therefore solves the problem of energy surplus, while at the same time the carbon skeletons remain available for future growth and energy generation. However, it is incorrect to conclude that all polysaccharides are produced by endergonic reactions. JARMAN and PACE (1984) have performed a systematic study into the energy requirement of microbial polysaccharide synthesis, and they have shown that polysaccharides with a high amount of acidic substituents are actually produced via exergonic reactions.

Another phenomenon that can be observed frequently in nitrogen-limited cultures with ammonium ions as the sole nitrogen source is the production (although not in great quantities) of 2-oxoglutarate (Tabs. 3, 4). This points to another facet of the effect of nutrient limitation on metabolism. The assimilation of ammonia/ammonium occurs via two pathways:

I. 2-oxoglutarate + NH_3 + NADPH $\rightarrow$ glutamate + $NADP^+$
catalyzed by glutamate dehydrogenase

II. glutamate + NH_3 + ATP $\rightarrow$ glutamine + ADP + P_i
glutamine + 2-oxoglutarate + NADPH $\rightarrow$ 2 glutamate + $NADP^+$

catalyzed by glutamine synthetase and glutamate synthase.

Pathway I is used when the concentration of ammonium ions is high (>1 mM), pathway II is

glucose-6-phosphate $\longrightarrow$ glucose-1-phosphate (1)

glucose-1-phosphate + ATP $\longrightarrow$ ADP-glucose + PP_i (2)

ADP-glucose + glycogen primer $\longrightarrow$ glycogen + ADP (3)

Fig. 6. Scheme of the synthesis of glycogen in *Escherichia coli*. Reaction 2, catalyzed by ADP-glucose pyrophosphorylase, is activated by fructose-1,6-bisphosphate and by NADPH, and inhibited by AMP.

used when the concentration of ammonium ions is limiting growth. When one studies the reactants of pathway II, it will become clear that the concentration of ammonia (or ammonium ions) is limiting the rate of reaction 1 of pathway II. Glutamate is present in all bacteria at rather high concentrations (above 100–200 mM), since potassium glutamate is responsible for the turgor pressure of the cells (EPSTEIN and SCHULZ, 1965; TEMPEST et al., 1970; DINNBIER et al., 1988). As argued above, the ATP and NADPH levels will be relatively high, which leaves 2-oxoglutarate as the other important factor in the control of the reaction rate. From the fact that 2-oxoglutarate is excreted, one may conclude that the intracellular levels are also high and it is easy to see that this will have a beneficial effect on the reaction rate.

The reader will have noticed that in this discussion no clear distinction has been made between the ammonia molecule (NH_3) and the ammonium ion (NH_4^+). The reason is that it is not clear which species (ammonia or the ammonium ion) takes part in these reactions. The relevance of this distinction has become clear only recently. Experiments with different bacterial species and with yeast have indicated that the cell membrane of these organisms is exceedingly permeable to ammonia. This means that in these species the intracellular and extracellular concentrations of ammonia molecules will be almost equal under most culture conditions. Since the equilibrium between ammonia and the ammonium ion is dependent on the pH value ($pK_a=9.1$), this has far-reaching consequences. In organisms such as *Escherichia coli*, the intracellular pH value is thought to remain relatively constant at different extracellular pH values (it varies between 7.4 and 7.8 at extracellular pH values between 5 and 9) (BOOTH, 1985). In fact, it is generally assumed that whereas the intracellular pH value of different species may vary from one to another, in general the cytoplasm of one particular species is only varying within a relatively small pH band. When the extracellular pH value is different from the cytoplasmic pH value, one of the consequences of the equilibration of the ammonia concentration across the membrane will be that the concentration of ammonium ions in the two compartments will be different.

This prediction was tested with glucose-limited cultures of *Klebsiella pneumoniae* growing in the presence of an excess of ammonium chloride (75–85 mM in the culture fluid). The organism was grown at pH values between 4.2 and 8.2; when the culture was in a steady state, the intracellular levels of glutamate dehydrogenase and glutamate synthase were measured. At culture pH values above 6, glutamate dehydrogenase levels were high (up to 1 μmol per min per mg protein) and glutamate synthase levels were low (<20 nmol per min per mg protein) – this was no surprise since the culture was carbon-limited. However, at lower culture pH values there was a marked decline in the glutamate dehydrogenase content of the cells (to 0.4 μmol per min per mg protein) and an increase of glutamate synthase activity (to 30–40 nmol per min per mg protein), showing that the cells were now facing a nitrogen limitation even though the extracellular ammonium ion concentration was high (BUURMAN et al., 1989b). The conclusion is that the choice of the type of nitrogen source, even under carbon-limited conditions, should be considered carefully.

The major role of the element sulfur is that it is necessary for the synthesis of the amino acids methionine and cysteine, and is present in some cofactors: e.g., thiamine pyrophosphate, coenzyme A, lipoic acid. These cofactors are for instance essential for the enzymes pyruvate dehydrogenase and 2-oxoglutarate dehydrogenase which both play a central role in metabolism. It is to be expected that some properties of sulfur-limited cultures are similar to those of nitrogen-limited cultures, because both elements are essential for protein synthesis and are not present in some major cell constituents, such as carbohydrates and lipids. Yet there is a major difference, whereas nitrogen is present in proteins, nucleic acids, and cofactors, the presence of sulfur is more restricted: it is not present in nucleic acids.

When *Klebsiella pneumoniae* was grown under a sulfate limitation the specific rate of glucose consumption was higher than that of a glucose-limited culture growing at the same rate, a now familiar phenomenon. In addition, extracellular polysaccharide was produced and in particular pyruvate, but also acetate, 2-oxoglutarate and succinate were excreted (NEIJS-

SEL and TEMPEST, 1975). The production of polysaccharide is again an indication of the cells' condition of energy surplus. The excretion of pyruvate may indicate that the limitation of sulfate is causing a metabolic bottleneck at the level of pyruvate dehydrogenase, because the cofactors of this enzyme (thiamin pyrophosphate, lipoic acid) and a substrate (coenzyme A) all contain sulfur and it is to be expected that a sulfate limitation will lead to low levels of these compounds. On the other hand, acetate and 2-oxoglutarate were also produced indicating that the carbon flux after the pyruvate dehydrogenase was still substantial. In this connection it has to be mentioned that under glucose-excess conditions the citric acid cycle does not function as a cycle, because the synthesis of 2-oxoglutarate dehydrogenase is repressed (HOLLYWOOD and DOELLE, 1976). This could be an explanation for the production of minor amounts of 2-oxoglutarate under limitations other than ammonia. The production of increased amounts of pyruvate under sulfate-limited conditions has also been found with *Escherichia coli* (Tab. 4) and *Bacillus subtilis* (NEIJSSEL and TEMPEST, 1979), indicating that this is a rather specific effect.

Tab. 4 shows that when *E. coli* was grown under a sulfate limitation in the presence of the cofactor pyrroloquinoline quinone (PQQ), large amounts of gluconate were produced, whereas *K. pneumoniae* did not produce this compound when it was grown under the same growth limitation. In Sect. 4.3 the physiological significance of the production of gluconate from glucose via the PQQ-linked glucose dehydrogenase will be discussed extensively, but it can be mentioned here that there was an important difference between the two cultures: *E. coli* was grown at pH 5.5, whereas the data on *K. pneumoniae* were obtained at pH 6.8. At this culture pH value the latter organism only produced gluconate when it was grown under a potassium, phosphate, or magnesium limitation. At lower culture pH values, however, gluconate (and 2-ketogluconate) production could also be observed in nitrogen- or sulfur-limited chemostat cultures of *K. pneumoniae* (HOMMES et al., 1989). Nevertheless, the regulation of the synthesis of the apo-glucose dehydrogenase in *E. coli* seems to be different.

4.3 Effects of a Potassium Limitation on Product Formation

A very interesting growth condition is a potassium limitation. Potassium is an essential element for many cell functions; in bacteria its quantitatively most important function is the maintenance of turgor pressure. In Gram-negative organisms the intracellular potassium concentration is between 150–250 mM, in Gram-positives minimally 400 mM (ROUF, 1964; TEMPEST, 1969). It has been observed that organisms growing in a potassium-limited culture show very high rates of energy consumption and product formation (the nature of the product(s) being, of course, dependent on the type of organism). It is therefore necessary to analyze the effects of this type of growth environment on the physiology of the microorganism in more detail.

Not much information is available about the extracellular potassium concentration in potassium-limited cultures. The only data that have been published are those of chemostat cultures of *E. coli*; MULDER (1988) measured concentrations as low as 0.1 μM at $D=0.28\ h^{-1}$. The implications of these observations are clear: potassium-limited growth conditions lead to a very steep concentration gradient of these ions across the cell membrane, and the question arises as to how these cells are able to build up and maintain such an enormous gradient.

E. coli possesses three uptake systems for potassium: they are called Trk (mnemonic for transport of K^+), Kup (mnemonic for K^+ uptake), and Kdp (mnemonic for K^+-dependent). Not surprisingly, all need energy to carry out their function. The Trk and Kup systems are expressed constitutively, and their K_m values for potassium uptake are 0.9–1.5 mM and 0.37±0.13 mM, respectively (BOSSEMEYER et al., 1989). The Kdp system is synthesized only under potassium-limited growth conditions and has a high affinity for potassium ions ($K_m=$ 2 μM). From this it follows that *E. coli* organisms growing in a potassium-limited medium possess all three systems in their cell membrane. MULDER et al. (1986) have proposed that this must have great consequences for the energy budget of the cells.

Since the extracellular potassium concentration in potassium-limited cultures is so low, the Kdp would be the only system able to take up this ion with a sufficiently fast rate. The Trk system, on the other hand, was assumed to be unable to withstand such a steep potassium gradient and assumed to start to work in reverse: i.e., it would leak potassium ions from the cytoplasm to the extracellular fluids. As long as the rate of uptake via the Kdp would be faster than the rate of export (leak) via the Trk a net accumulation of potassium ions would be possible. From the fact that the organisms are able to grow at such low potassium concentrations one can conclude that they are able to achieve this. The export of a potassium ion via the Trk may deliver up energy (because it is thermodynamically a down-hill process), but MULDER et al. assumed that this amount of energy was less than that invested in the uptake of a potassium ion via the Kdp. The net result of the simultaneous activity of Kdp (uptake) and Trk (leakage) would be that energy is spilled (Fig. 7). The rate at which this would occur can be described as:

$$q_{\text{waste}} = q_{\text{Kdp}} \cdot (\text{X ATP}) - q_{\text{Trk}} \cdot (\text{Y ATP})$$

where X>Y.

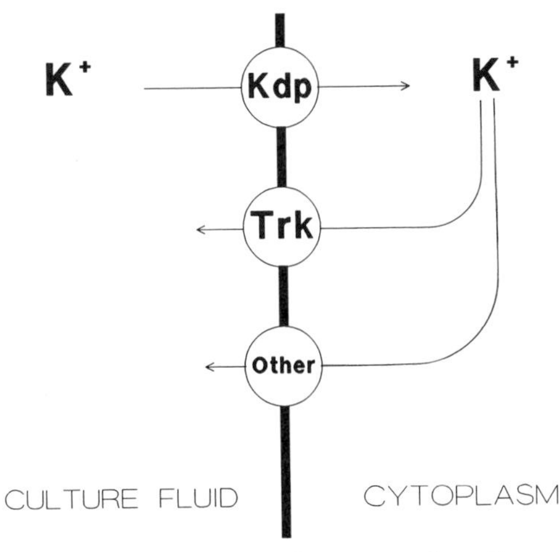

Fig. 7. Futile cycling of K^+ ions across the cell membrane, due to energy-linked uptake via the high-affinity uptake system Kdp, and leakage via the low-affinity uptake system Trk, or other systems (details, see text).

It has to be emphasized that although it was proposed originally that the Trk system would be responsible for leakage of potassium ions, this does not necessarily have to be the case, since a possible involvement of the Kup system has not been considered. Moreover, *Escherichia coli* possesses at least three potassium efflux systems (called KefA, KefB, and KefC, BAKKER et al., 1987), which are also thought to play a role in the regulation of turgor pressure. In spite of these uncertainties, the basic mechanism of the cycling of potassium ions could certainly be correct. The only modification would have to be that one or more of the other transport systems (Kup, KefA, KefB, KefC) is also involved in potassium leakage, and that potassium efflux via one or more of these systems generates less energy than invested in potassium ion uptake. One may wonder whether *E. coli* is unique in possessing so many systems involved in potassium transport and efflux. A clear answer to this question cannot be given, but recent evidence suggests that a system similar to KefC exists in some other Gram-negative bacteria (DOUGLAS et al. 1991).

The nature of the nitrogen source in a potassium-limited culture also plays a role in the energetics of growth. It was observed that potassium-limited cultures grown with ammonium chloride as the nitrogen source, showed an increased steady state dry weight at more alkaline pH values (BUURMAN et al., 1989a; MULDER, 1988). It was subsequently shown that this was due to the accumulation of ammonium ions, and that these ions took over the role of potassium ions. In fact, it was possible to grow *Bacillus stearothermophilus* in medium without added potassium ions, but with a large concentration of ammonium ions (BUURMAN et al., 1989a). The accumulation of ammonium ions under these circumstances is thought to occur via the same mechanism as described above for the depletion of intracellular ammonium ions at acid culture pH values (see Sect. 4.2), but now the cytoplasmic pH is *lower* than the external pH, which leads to the movement of ammonia molecules *into* the cell and the intracellular build-up of ammonium ions. It has to be emphasized that this effect of ammonium ions can only be observed in potassium-limited cultures. On the other hand, the pH-dependent increase in steady state dry weight due to the presence of ammonia has been observed in many different microorganisms: *Klebsiella pneumoniae* (BUURMAN et al., 1989a), *Escherichia coli* (MULDER, 1988), *Bacillus stearothermophilus* (BUURMAN et al., 1989a), *Pseudomonas putida* (HARDY, 1992), *Pseudomonas fluorescens* and *Pseudomonas putida* (HARDY, 1992), *Enterococcus faecalis* (SNOEP, unpublished results), and even *Candida utilis* (BUURMAN, 1991).

The substitution of the potassium ion by a different ion is not unique: it was already shown previously that the rubidium ion functions as such (AIKING and TEMPEST, 1977). The ammonium ion is smaller than the rubidium ion and slightly bigger than the potassium ion. In view of this similarity between potassium and ammonium ions BUURMAN et al. (1991) investigated whether the potassium-transport system Kdp would also transport ammonium ions, and there is now strong evidence that this is indeed the case. When *E. coli* FRAG-1 (wild type with respect to potassium transport systems) and *E. coli* FRAG-5 (containing a deletion in the Kdp gene coding for the high-affinity K^+ transport system) were cultured in the presence of D,L-alanine as the sole nitrogen source, both strains behaved similarly. The use of ammonium chloride as the sole nitrogen source led to a higher specific rate of oxygen consumption of *E. coli* FRAG-1. This observation is explained by a futile cycle generated by ammonium ions (Fig. 8). Thus, the presence of ammonium ions in potassium-limited cultures leads to extra energy consumption.

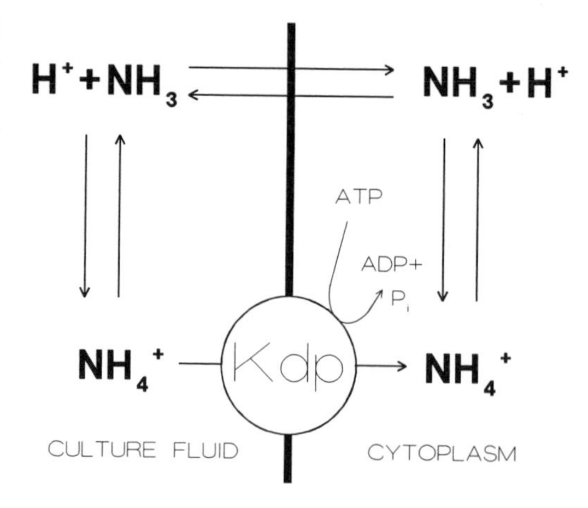

Fig. 8. Futile cycling of NH_4^+ and NH_3 across the cell membrane, due to energy-linked uptake of the NH_4^+ ion via the high-affinity K^+ uptake system Kdp and diffusion of NH_3 through the cell membrane. Pumping of protons by the respiratory chain is not shown.

After this survey of the energetic consequences of potassium-limited growth, the physiology of product formation under these growth conditions will be discussed. As shown in Tab. 3, potassium-limited cultures of *Klebsiella pneumoniae* excrete a number of compounds into the culture fluids. In contrast to nitrogen-limited cultures, potassium-limited cultures do not produce extracellular polysaccharides, nor do they show storage of intracellular glycogen or other polymers, indicating that these cells do not contain a high ATP/ADP ratio. Secondly, potassium-limited cultures of *K. pneumoniae* produce large quantities of gluconate and 2-ketogluconate. It has been found that the oxidation of glucose to gluconate is catalyzed by a glucose dehydrogenase with pyrroloquinoline quinone (PQQ) as its cofactor. The active site of this enzyme is facing the periplasm, so oxidation of glucose does not require transport into the cell. The reduced form of PQQ reacts with ubiquinone of the respiratory chain. Gluconate is oxidized in the periplasm via a flavoprotein gluconate dehydrogenase, which also uses ubiquinone as acceptor. The oxidation of glucose and gluconate is therefore solely contributing to energy generation and not to carbon metabolism. It was therefore proposed that the PQQ-linked glucose dehydrogenase functions as an ancillary energy-generating pathway that is derepressed (c.f. the effect of DNP on nitrogen-limited cultures, Tab. 3) when the cells' energy demand is high (HOMMES et al., 1985).

Potassium-limited cultures of *Escherichia coli*, on the other hand, did not produce gluconate or 2-ketogluconate. It was found, however, that this organism synthesizes the glucose dehydrogenase protein, but is seemingly unable to synthesize the cofactor PQQ (HOMMES et al., 1984), although a recent report suggests that PQQ-producing mutants of *E. coli* can be isolated relatively easily (BIVILLE et al., 1991). When PQQ was added to the medium of potassium-limited cultures of *E. coli* an immediate production of gluconate could be observed and addition of PQQ to the growth medium led to steady state production of gluconate (see Tab. 4). This latter compound was not oxidized to 2-ketogluconate, because the organism does not possess gluconate dehydrogenase activity.

Many other bacterial species possess the glucose dehydrogenase protein (apo-enzyme), but are seemingly unable to synthesize the cofactor PQQ (AMEYAMA et al., 1985; LINTON et al., 1987b; VAN SCHIE et al., 1984, 1987). This is a curious phenomenon, because without the presence of PQQ the apo-enzyme is not functional and could therefore be considered as a nonsense protein. One may wonder therefore why the gene coding for this enzyme has not been lost by these organisms, because production of a non-functional enzyme represents an energetic burden. Nevertheless, all strains of *E. coli* that have been checked until now contain the glucose dehydrogenase apo-enzyme, and some of these strains have been isolated from nature or other sources in the early years of this century! Free PQQ has been found in many natural environments and it could well be that when these organisms are growing in their natural habitat they will frequently be able to assemble the holo-enzyme. In this respect, PQQ could be considered a vitamin (DUINE et al., 1986).

Data on product formation in potassium-limited cultures of other organisms are extremely scarce, but WÜMPELMAN et al. (1984) have shown that in aerobic potassium-limited chemostat cultures of *Saccharomyces cerevisiae* the production of ethanol rose when the input concentration of potassium was lowered. These authors could also demonstrate a linear relationship between the amount of biomass produced per mol ATP generated in energy metabolism (YATP) and the magnitude of the potassium gradient across the cell membrane. This finding is in agreement with the considerations put forward above.

4.4 Effects of a Phosphate, Magnesium, or Iron Limitation on Product Formation

Interestingly, phosphate-limited cultures of *Klebsiella pneumoniae* excrete substantial amounts of gluconate and 2-ketogluconate. *Escherichia coli* shows a similar high productivity of gluconate in a phosphate-limited chemostat culture: at $D=0.15\ h^{-1}$ this organism was shown to convert 63% of the glucose that was consumed into gluconate when 0.2 μM PQQ was present in the growth medium (Tab. 4).

Since it was argued above that this pathway of glucose oxidation is used when the energy demand is high, one may wonder why these culture conditions lead to the derepression of this pathway. A clear answer cannot yet be provided, but it is possible that again a futile cycle is involved. In *E. coli* at least two uptake systems for phosphate have been described, and their properties and regulation are similar to those of the two main potassium uptake systems: a low-affinity system that is constitutively expressed and a high-affinity system that is derepressed in phosphate-limited environments (ROSENBERG et al., 1977). And although the cytoplasmic phosphate concentration is certainly not as high as the intracellular potassium concentration, futile cycling of phosphate ions is a distinct possibility. Phosphate-limited environments could therefore also be used to stimulate energy consumption and product formation. An example is the production of acetone and butanol by *Clostridium acetobutylicum*, which is thought to be triggered by the presence of weak acids in the culture fluids (see: Sect. 5). A phosphate-limited chemostat culture (BAHL et al., 1981) produced even higher amounts of these solvents, although for commercial application this process would need a further improvement of the productivity.

Magnesium-limited cultures of *Klebsiella pneumoniae* again produced gluconate and 2-ketogluconate, provided the medium contained enough calcium ions (NEIJSSEL and TEMPEST, 1979; BUURMAN et al., 1990). When no calcium was added to the growth medium, no gluconate and/or 2-ketogluconate production could be observed. This is caused by the fact that the binding of the cofactor PQQ to the apo-glucose dehydrogenase is dependent upon the presence of magnesium or calcium ions. The steady state extracellular magnesium ion concentration in magnesium-limited cultures was apparently too low to allow this binding to take place. This simple observation points again to the necessity of a careful consideration of the composition of the medium used in a production process. It is not possible to grow *K. pneumoniae* calcium-limited, which indicates that the organism needs only very small amounts (if any) of this ion.

In analogy with a potassium or a phosphate limitation it could be speculated that a futile cycle of magnesium ions across the membrane plays a role. In *E. coli* magnesium ions are taken up by at least two transport systems: one is constitutive and has a relatively low affinity towards the ion, whereas the other, a high affinity system, is derepressed when the availability of magnesium ions is low (NELSON and KENNEDY, 1972). It has to be emphasized, however, that experimental evidence for this proposal is as yet not available.

A low availability of iron has a profound effect on the physiology of microbes. Iron has been shown to be essential for almost all organisms (for a review, see: NEILANDS, 1981; NEILANDS, 1984). It has been found that many microorganisms excrete compounds called siderophores when they are grown under iron-limited conditions. These siderophores are capable of solubilizing and transporting Fe^{3+} into the cell. They are compounds with a low molecular weight (500–1000 Da) and have a high affinity for ferric ions (formation constant for ferri-complex is 10^{20} and higher) (NEILANDS, 1982). Their affinity towards Fe^{2+} ions is very low.

The regulation of their synthesis is similar to that of the uptake systems for potassium, phosphate, or magnesium ions. The siderophore and its membrane-bound receptor protein (called ton A protein in *E. coli*) is the high-affinity system, which is derepressed when the iron concentration is below 1 μM. The low-affinity system is only used when sufficient iron is available.

It has been found that production of siderophores by *Pseudomonas* species has a growth-promoting effect on plants, because they inhibit the growth of fungi (BECKER and COOK, 1988; KLOEPPER et al., 1980). An iron limitation plays a role in the production of citrate by *Aspergillus niger* (KUBICEK and RÖHR, 1977), although a limited availability of other metal ions (zinc and manganese) also stimulates production of this compound. BAHL et al. (1986) noted that during the fermentation of whey by *Clostridium acetobutylicum* an iron limitation (in addition to a phosphate limitation) had a profound influence on the ratio butanol/acetone produced. In a synthetic medium this organism produced butanol and acetone in a ratio of only 2:1, but in whey this ratio was 100:1. The authors hypothesized that the hydrogenase reaction (necessary for the production of hydrogen) is not functioning properly under iron-limited conditions and that, therefore, more butanol (the more reduced substance) is formed.

5 Futile Cycles and Effect of Culture pH Value

When two or more reactions carry out the following conversion:

substrate → intermediate 1 (→ intermediate 2 ... intermediate *n*) → substrate

and in this conversion energy is used, one calls such a reaction sequence a futile cycle. A well known example is the combined action of phosphofructokinase and fructose 1,6-bisphosphate phosphatase (NEWSHOLME and CRABTREE, 1973), which also takes place in *Escherichia coli* (CHAMBOST and FRAENKEL, 1980; DALDAL and FRAENKEL, 1983). Whereas the use of the term "cycle" is obvious, the use of the word "futile" can be criticized. It is clear that such a reaction imposes an energy drain on the organism possessing this cycle, but this brings us to the question of the physiological function of such an energy drain. As long as we do not know why this cycle is operative one should be careful with calling it "futile".

From the perspective of metabolite overproduction futile cycles are very interesting: they belong to the energy-spilling reactions discussed previously (see Sect. 3.2). Thus, their activity could stimulate the production of compounds that have been formed via energy yielding reactions. An interesting question is therefore whether one can manipulate the cycling rate of futile cycles, and, if so, to what extent this is possible.

An important futile cycle has already been discussed: the cycling of potassium and ammonium ions across the cell membrane (see Sect. 4.3), and it has been proposed that phosphate- or magnesium-limited growth conditions may lead to a similar phenomenon. Yet there are other futile cycles that may have importance for metabolite overproduction.

In Sect. 4.2 the effect of the presence of 2,4-dinitrophenol (DNP) in the growth medium was described. It will now be obvious that this compound also induces a futile cycle: the protons pumped out by the respiratory chain (=energy input) are carried back into the cytoplasm by DNP. It has become clear that some fermentation products have a similar effect: weak organic acids such as acetic, propionic, and butyric acid share the same property with DNP (Fig. 9). The undissociated acids are also sufficiently lipophilic and this means that when

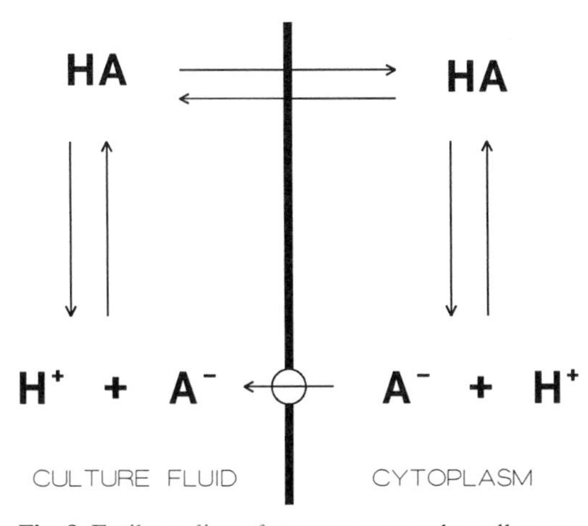

Fig. 9. Futile cycling of protons across the cell membrane due to the presence of a weak acid (HA) in the culture medium. Pumping of protons by the respiratory chain is not shown.

these products accumulate in culture fluids (particularly at pH values near their pK_a values) they will induce a leakage current of protons through the membrane (HERRERO et al., 1985). The microorganism can only survive when its rate of energy generation is greater than the rate of energy dissipation by the acid. In some fermentations this increased rate of energy generation leads to an increased rate of production of the toxic acid, and eventually growth will stop.

GOTTSCHAL and MORRIS (1981) proposed that the intracellular accumulation of the weak acid butyrate is the trigger for solvent production by *Clostridium acetobutylicum;* TERRACCIANO and KASHKET (1986) arrived at the same conclusion. In addition, studies by ZOUTBERG et al. (1989a, b) have shown that *Clostridium butyricum* (which produces butyrate and acetate from glucose) is extremely sensitive towards butyrate at low culture pH values. When a wild-type strain was cultured for a prolonged period of time at low culture pH values in the presence of elevated concentrations of butyrate and acetate (produced by the organisms themselves) selection of variant invariably occurred. This variant was shown to be more resistant to the toxic effect of butyrate and, interestingly, produced the solvents butanol and iso-propanol at a high rate: 2.1 g L^{-1} h^{-1} (ZOUTBERG et al., 1989a, b).

In view of this activity of weak acids it comes as no surprise that the culture pH value can have a profound effect on the production of compounds by microbes. The earliest observation of this effect was made by GALE and EPPS (1942). These authors showed that *E. coli* and *Micrococcus lysodeikticus* possessed a higher activity of decarboxylases when the culture fluids were acidic (thereby avoiding the accumulation of acidic products of metabolism). This so-called compensation mechanism has been observed also in other organisms (Tab. 6). An important factor is that the pyruvate formate lyase activity sharply decreases when the organism is grown at more acid culture pH values. The obvious advantage of this behavior is that an accumulation of weak acids at low culture pH values is diminished.

Recently, it has been found that the culture pH value also has a more indirect effect on microbial metabolism. When *Enterococcus fae-*

Tab. 6. Effect of the Medium pH Value on the Fermentation of Glucose by Different Organisms

Organism	*Streptococcus liquefaciens*[a]			*Escherichia coli*[b]		*Aerobacter cloacae*[c]	
pH Value	5.0	7.0	9.0	6.0	7.8	5.0	7.2
Acetate	12.2	18.8	31.2	36.5	38.7	5.0	69.5
Acetoin	–	–	–	0.1	0.2	0	1.5
2,3-Butanediol	–	–	–	0.3	0.3	38.8	2.3
Butyrate	–	–	–	0	7.1	–	–
Carbon dioxide	–	–	–	88.0	1.8	–	–
Ethanol	7.0	14.6	22.4	49.8	50.5	61.8	67.5
Formate	15.4	33.6	52.8	2.4	86.0	–	–
Glycerol	–	–	–	1.4	0.3	–	–
Hydrogen	–	–	–	75.0	0.3	–	–
Lactate	174	146	122	79.5	70.0	0	3.6
Succinate	–	–	–	10.7	14.8	3.3	6.4
Total acid	201.6	198.4	206.0	129.1	216.6	8.3	79.5
Total neutral incl. CO_2	7.0	14.6	22.4	139.6	53.1	100.6	71.3

[a] GUNSALUS and NIVEN (1942)
[b] BLACKWOOD et al. (1956)
[c] HERNANDEZ and JOHNSON (1967)
– not determined
Products are expressed in mmol produced per 100 mmol of fermented glucose

Tab. 7. Specific Rates (mmol g^{-1} h^{-1}) of Pyruvate Consumption and Product Formation of Anaerobic Pyruvate-Limited Cultures of *Enterococcus faecalis*, Grown at Various Culture pH Values (D=0.1 h^{-1}, 37 °C)

pH	Pyruvate	Acetate	Lactate	Formate	CO_2
5.5	21.2	9.3	10.1	0.0	10.7
6.0	13.5	7.3	5.5	2.2	6.7
6.5	12.5	7.4	3.8	3.9	4.7
7.0	11.6	8.4	2.2	6.2	2.4
7.5	9.5	7.8	1.1	7.0	1.3
8.0	9.1	7.9	0.7	7.2	0.6
8.5	9.2	8.3	0.8	7.5	0.7

Carbon recoveries and redox balances were 91–98 %. The *in vivo* pyruvate dehydrogenase activity can be calculated from the specific rate of acetate production minus the specific rate of formate production. Data of SNOEP et al. (1990)

calis (formerly *Streptococcus faecalis*) was grown anaerobically in a pyruvate-limited chemostat culture, it was found that the culture pH value had a marked influence on the specific rates of production of the fermentation products (Tab. 7). From these data it could be calculated that under these anaerobic conditions the pyruvate dehydrogenase complex activity *in vivo* increased with decreasing culture pH values (0.8 mmol per g dry weight per hour at pH 7 to 9.3 mmol per g dry weight per hour at pH 5.5). A crucial factor in the regulation of the synthesis and activity of the pyruvate dehydrogenase complex was proposed to be the intracellular redox potential which was influenced by the culture pH value (SNOEP et al., 1990) .

It has already been mentioned (Sect. 4.2) that both the synthesis and activity of the PQQ-linked glucose dehydrogenase in *Kleb-*

siella pneumoniae are strongly influenced by the culture pH value. At pH 8.0 no gluconate production could be observed in potassium-limited cultures, whereas at pH 5.0–5.5 maximal production of gluconate and 2-ketogluconate took place (HOMMES et al., 1989). In addition, ammonia- or sulfate-limited cultures of this organism produced gluconate and 2-ketogluconate when grown at pH 5.5, but no production of these compounds could be observed at pH 6.8. This production of acids at low culture pH values seems to contradict the hypothesis put forward by GALE and EPPS (1942), but gluconate and 2-ketogluconate are not as lipophilic as acids, such as acetate or butyrate and are therefore not so toxic.

6 Concluding Remarks

In this chapter we have tried to present the reader with an overview of the effect of different culture conditions on product formation by a variety of microorganisms. Much emphasis has been put on data obtained with cultures of *Klebsiella pneumoniae* or *Escherichia coli*. This does not mean that we wish to argue that these organisms are the best choice for many biotechnological processes, they clearly are not. However, experiments with these model organisms have shown that metabolic fluxes can be manipulated by a careful choice of medium composition and other culture parameters, such as pH value.

The theories that have been put forward could form the basis for a rational design of biotechnological processes, although it should be remembered that other factors, such as economy, nature of the feedstock, efficiency of downstream processing, play a decisive role in the success of such processes. Thus, in the production of ethanol molasses, instead of glucose, is commonly used as the starting material. This implies that one is not free in the choice of a particular nutrient limitation, such as a potassium limitation, because molasses contains a considerable amount of potassium and other nutrients (unpublished observation).

In addition, it should be stressed that, whereas the use of a chemostat culture played a dominant role in the experiments that have been discussed in this contribution, one should not conclude that other culture methods are biotechnologically irrelevant. This is certainly not the case, because high concentrations of biomass (leading to high volumetric productivities) have been achieved with alternative culture methods (immobilized cells, growth in pellets or aggregates, etc.). The advantage of using a chemostat is that it allows the researcher to investigate the effect of a limitation of a specific nutrient, or of a specific environmental parameter (pH, temperature, osmolarity, etc.) on the physiology of microbes that are growing under otherwise equal conditions. This emphasis on physiology does not mean that this approach is the only way forward in the improvement of microbial processes for a biotechnological application. The many successes of genetic engineering prove otherwise. However, only if one combines microbial genetics and physiology will one be able to develop the best possible culture conditions for the overproduction of metabolites.

7 References

ABBE, K., TAKAHASHI, S., YAMADA, T. (1982), Involvement of oxygen sensitive pyruvate formate-lyase in mixed-acid fermentation by *Streptococcus mutans* under strictly anaerobic conditions, *J. Bacteriol.* **152,** 175–182.

AIKING, H., TEMPEST, D.W. (1977), Rubidium as a probe for function and transport in the yeast *Candida utilis* NCYC 321, growing in chemostat culture, *Arch. Microbiol.* **115,** 215–221.

AMEYAMA, M., SHINAGAWA, E., MATSUSHITA, K., ADACHI, O. (1985), Growth stimulating activity for microorganisms in naturally occurring substances and partial characterization of the substance for the activity as PQQ, *Agric. Biol. Chem.* **49,** 699–709.

BAHL, H., ANDERSCH, W., GOTTSCHALK, G. (1981), Continuous production of acetone and butanol by *Clostridium acetobutylicum* in a two-stage phosphate-limited chemostat, *Eur. J. Appl. Microbiol. Biotechnol.* **15,** 201–205.

BAHL, H., GOOTWALD, M., KUHN, A., RALE, V., ANDERSCH, W., GOTTSCHALK, G. (1986), Nutritional factors affecting the ratio of solvents produced by *Clostridium acetobutylicum*, *Appl. Environ. Microbiol.* **52,** 169–172.

BAILEY, J. E. (1991), Toward a science of metabolic engineering, *Science* **252,** 1668–1675.

BAKKER, E. P., BOOTH, I. R., DINNBIER, U., EPSTEIN, W., GAJEWSKA, A. (1987), Evidence for multiple potassium export systems in *Escherichia coli, J. Bacteriol.* **169,** 3743–3749.

BECKER, J. O., COOK, R. J. (1988), Role of siderophores in suppression of *Pythium* species and production of increased/growth response of wheat by fluorescent pseudomonas, *Phytopathology* **78,** 778–782.

BIVILLE, F., TURLIN, E., GASSER, F. (1991), Mutants of *Escherichia coli* producing pyrroloquinoline quinone, *J. Gen. Microbiol.* **137,** 1775–1782.

BLACKWOOD, A. C., NEISH, A. C., LEDINGHAM, G. A. (1956), Dissimilation of glucose at controlled pH values by pigmented and non-pigmented strains of *Escherichia coli, J. Bacteriol.* **72,** 497–499.

BOOTH, I. R. (1985), Regulation of cytoplasmic pH in bacteria, *Microbiol. Rev.* **49,** 359–378.

BOSSEMEYER, D., SCHLÖSSER, A., BAKKER, E. P. (1989), Specific cesium transport via the *Escherichia coli* Kup (TrkD), K^+ uptake system, *J. Bacteriol.* **171,** 2219–2221.

BROWN, A. T., PATTERSON, C. E. (1973), Ethanol production and alcohol dehydrogenase activity in *Streptococcus mutans, Arch. Oral. Biol.* **18,** 127–131.

BUURMAN, E. T. (1991), The effect of cations on microbial metabolism and growth energetics, *PhD Thesis*, University of Amsterdam.

BUURMAN, E. T., PENNOCK, J., TEMPEST, D. W., TEIXEIRA DE MATTOS, M. J., NEIJSSEL, O. M. (1989a), Ammonium ions take over the role of potassium ions in potassium-limited chemostat cultures of different organisms, *Arch. Microbiol.* **152,** 58–63.

BUURMAN, E. T., TEIXEIRA DE MATTOS, M. J., NEIJSSEL, O. M. (1989b), Nitrogen-limited behaviour of microorganisms growing in the presence of large concentrations of ammonium ions, *FEMS Microbiol. Lett.* **58,** 229–232.

BUURMAN, E. T., BOIARDI, J. L., TEIXEIRA DE MATTOS, M. J., NEIJSSEL, O. M. (1990), The role of magnesium and calcium ions in the glucose dehydrogenase activity of *Klebsiella pneumoniae* NCTC 418, *Arch. Microbiol.* **153,** 502–505.

BUURMAN, E. T., TEIXEIRA DE MATTOS, M. J., NEIJSSEL, O. M. (1991), Futile cycling of ammonium ions via the high affinity potassium uptake system (Kdp) of *Escherichia coli, Arch. Microbiol.* **155,** 391–395.

CARLSSON, J., GRIFFITH, C. J. (1974), Fermentation products and bacterial yields in glucose-limited and nitrogen-limited cultures of streptococci, *Arch. Oral Biol.* **19,** 1105–1109.

CHAMBOST, J. P., FRAENKEL, D. G. (1980), The use of C6-labelled-glucose to assess futile cycling in *Escherichia coli, J. Biol. Chem.* **255,** 2867–2869.

CORNISH, A., LINTON, J. D., JONES, C. W. (1987), The effect of growth conditions on the respiratory system of a succinoglucan-producing strain of *Agrobacterium radiobacter, J. Gen. Microbiol.* **133,** 2971–2978.

CRABBENDAM, P. M., NEIJSSEL, O. M., TEMPEST, D. W. (1985), Metabolic and energetic aspects of the growth of *Clostridium butyricum* on glucose in chemostat culture, *Arch. Microbiol.* **142,** 375–382.

.DALDAL, F., FRAENKEL, D. G. (1983), Assessment of a futile cycle involving reconversion of fructose-6-phosphate to fructose-1,6-bisphosphate during gluconeogenic growth of *Escherichia coli, J. Bacteriol.* **153,** 390–394.

DAWES, E. A., SENIOR, P. J. (1973), The role and regulation of energy reserve polymers in microorganisms, *Adv. Microb. Physiol.* **10,** 135–266.

DEMAIN, A. L. (1982), Catabolite regulation in industrial microbiology, in: *Overproduction of Microbial Products* (KRUMPHANZL, V., SIKYTA, B., VANĚK, Z., Eds.), pp. 3–20, London: Academic Press.

DINNBIER, U., LIMPINSEL, E., SCHMID, R., BAKKER, E. P. (1988), Transient accumulation of potassium glutamate and its replacement by trehalose during adaptation of growing cells of *Escherichia coli* K12 to elevated sodium chloride concentrations, *Arch. Microbiol.* **150,** 348–357.

DOUGLAS, R. M., ROBERTS, J. A., MUNRO, A. W., RITCHIE, G. Y., LAMB, A. J., BOOTH, I. R. (1991), The distribution of homologues of the *Escherichia coli* CefC K^+-efflux system in other bacterial species, *J. Gen. Microbiol.* **137,** 1999–2005.

DUINE, J. A., FRANK JZN, J., JONGEJAN, J. A. (1986), PQQ and quinoprotein enzymes in microbial oxidations, *FEMS Microbiol. Rev.* **32,** 165–178.

EGGELING, I., CORDES, C., EGGELING, L., SAHM, H. (1987), Regulation of acetohydroxy acid synthase in *Corynebacterium glutamicum* during fermentation of α-ketobutyrate to isoleucine, *Appl. Microbiol. Biotechnol.* **25,** 346–351.

EPSTEIN, W., SCHULZ, S. G. (1965), Cation transport in *Escherichia coli*. V. Regulation of cation content, *J. Gen. Physiol.* **49,** 221–234.

EVANS, C. G. T., HERBERT, D., TEMPEST, D. W. (1970), The continuous cultivation of microorganisms. II. Construction of a chemostat, in: *Methods in Microbiology* (NORRIS, J. R., RIBBONS, D. W., Eds.), Vol. 2, pp. 277–327, London: Academic Press.

GALE, E. F., EPPS, H. M. R. (1942), The effect of pH of the medium during growth on the enzymic activities of bacteria (*Escherichia coli* and *Micrococcus lysodeikticus*) and the biological significance of the changes produced, *Biochem. J.* **36,** 600–618.

GOTTSCHAL, J. C., MORRIS, J. G. (1981), The induction of acetone and butanol production in cultures of *Clostridium acetobutylicum* by elevated concentrations of acetate and butyrate, *FEMS Microbiol. Lett.* **12,** 385–389.

GROEGER, U., SAHM, H. (1987), Microbial production of L-leucine from α-keto-isocaproate by *Corynebacterium glutamicum*, *Appl. Microbiol. Biotechnol.* **25,** 352–356.

GUNSALUS, I. C., NIVEN, C. F. (1942), The effect of pH on the lactic acid fermentation, *J. Biol. Chem.* **145,** 131–136.

HAMILTON, I. R., PHIPPS, P. J., ELLWOOD, D. C. (1979), Effect of growth rate and glucose concentration on biochemical properties of *Streptococcus mutans* Ingbritt in continuous culture, *Infect. Immun.* **26,** 861–869.

HARDY, G. P. M. A. (1992), Dual glucose metabolism of *Pseudomonas* species in chemostat culture, *PhD Thesis,* University of Amsterdam.

HARDY, J. M. (1986), Genus *Streptococcus* Rosenbach 1884, in: *Bergey's Manual of Systematic Bacteriology* (SNEATH, P. H. A., MAIR, N. S., SHARPE, M. E., HOLT, J. G., Eds.), Vol. 2, pp. 1043–1047, Baltimore: Williams & Wilkins.

HEINRICH, R., RAPOPORT, T. A. (1974), A linear steady-state treatment of enzymatic chains: general properties, control and effector strength, *Eur. J. Biochem.* **42,** 89–95.

HERBERT, D. (1958), Some principles of continuous culture in: *Recent Progress in Microbiology,* pp. 381–396. Symposia held at the VII. Int. Congr. for Microbiology, 1958.

HERBERT, D. (1976), Stoichiometric aspects of microbial growth, in: *Continuous Culture 6, Applications and New Fields* (DEAN, A. C. R., ELLWOOD, D. C., EVANS, C. G. T., MELLING, J., Eds.), pp. 1–30, Chichester: Ellis Horwood.

HERNANDEZ, E., JOHNSON, M. J. (1967), Anaerobic growth yields of *Aerobacter cloacae* and *Escherichia coli*, *J. Bacteriol.* **94,** 991-995.

HERRERO, A. J., GOMEZ, R. F., SNEDECOR, B., TOLMAN, C. J., ROBERTS, M. J. (1985), Growth inhibition of *Clostridium thermocellum* by carboxylic acids: a mechanism based on uncoupling by weak acids, *Appl. Microbiol. Biotechnol.* **22,** 53–62.

HOLLYWOOD, N., DOELLE, H. W. (1976), Effect of specific growth rate and glucose concentration on growth and metabolism of *Escherichia coli* K12, *Microbios* **17,** 23–33.

HOLME, T. (1957), Continuous culture studies on glycogen synthesis in *Escherichia coli* B, *Acta Chem. Scand.* **11,** 763–775.

HOMMES, R. W. J., POSTMA, P. W., NEIJSSEL, O. M., TEMPEST, D. W. (1984), Evidence of a quinoprotein glucose dehydrogenase apo-enzyme in several strains of *Escherichia coli*, *FEMS Microbiol. Lett.* **24,** 329–333.

HOMMES, R. W. J., VAN HELL, B., POSTMA, P. W., NEIJSSEL, O. M., TEMPEST, D. W. (1985), The functional significance of glucose dehydrogenase in *Klebsiella aerogenes*, *Arch. Microbiol.* **143,** 163–168.

HOMMES, R. W. J., POSTMA, P. W., TEMPEST, D. W., NEIJSSEL, O. M. (1989), The influence of the culture pH value on the direct glucose oxidative pathway in *Klebsiella pneumoniae* NCTC 418, *Arch. Microbiol.* **151,** 261–267.

HOMMES, R. W. J., SIMONS, J. A., SNOEP, J. L., POSTMA, P. W., TEMPEST, D. W., NEIJSSEL, O. M. (1991), Quantitative aspects of glucose metabolism by *Escherichia coli* B/r, grown in the presence of pyrroloquinoline quinone, *Antonie van Leeuwenhoek* **60,** 373–382.

HÜTTER, R. (1986), Overproduction of microbial metabolites, in: *Biotechnology* (REHM, H.-J., REED, G., Eds.) Vol. 4 (PAPE, H., REHM, H.-J., Eds.), pp. 3–17, Weinheim: VCH.

JARMAN, T. R., PACE, G. W. (1984), Energy requirements for microbial exopolysaccharide synthesis, *Arch. Microbiol.* **137,** 231–235.

KACSER, H., BURNS, J. A. (1973), The control of flux, in: *Rate Control of Biological Processes* , (DAVIES, D. D., Ed.), *Symp. Soc. Exp. Biol.* **27,** 65–104.

KELL, D. B., WESTERHOFF, H. V. (1986), Metabolic control theory: its role in microbiology and biotechnology, *FEMS Microbiol. Rev.* **39,** 305–320.

KINOSHITA, S. (1987), Thirty years of amino acid fermentation, in: *Proc. ICEAM Int. Symp. on Amino Acid Fermentations* (DEMAIN, A. L., Ed.), *Proc. 4th Eur. Cong. Biotechnol.* (NEIJSSEL, O. M., VAN DER MEER, R. R., LUIJBEN, K. CH. A. M., Eds.), Vol. 4, pp. 679–688, Amsterdam: Elsevier.

KLOEPPER, J. W., LEONG, J., TEINTZE, M., SCHROTH, M. N. (1980), *Pseudomonas* siderophores: a mechanism explaining disease-suppressive soils, *Curr. Microbiol.* **4,** 317–320.

KUBICEK, C. P., RÖHR, M. (1977), Influence of manganese on enzyme synthesis and citric acid production in *Aspergillus niger*, *Eur. J. Appl. Microbiol. Biotechnol.* **4,** 167–175.

LINTON, J. D., EVANS, M. W., JONES, D. S., GOULDNEY, D.G. (1987a), Exocellular succinoglycan production by *Agrobacterium radiobacter* NCIB 11883, *J. Gen. Microbiol.* **133,** 2961–2969.

LINTON, J. D., WOODARD, S., GOULDNEY, D. G. (1987b), The consequences of stimulating glucose dehydrogenase activity by addition of PQQ on metabolite production by *Agrobacterium radiobacter* NCIB 11883, *Appl. Microbiol. Biotechnol.* **25,** 351–361.

MONOD, J. (1942), *Recherches sur la croissance des cultures bactériennes,* Paris: Hermann, Editeurs des Sciences et des Arts.

MULDER, M. M. (1988), Energetic aspects of bacterial growth: a mosaic non-equilibrium thermodynamic approach, *PhD Thesis*, University of Amsterdam.

MULDER, M. M., TEIXEIRA DE MATTOS, M. J., POSTMA, P. W., VAN DAM, K. (1986), Energetic consequences of multiple potassium-uptake systems in *Escherichia coli*, *Biochim. Biophys. Acta* **851,** 223–228.

NEIJSSEL, O. M. (1977), The effect of 2,4-dinitrophenol on the growth of *Klebsiella aerogenes* NCTC 418 in aerobic chemostat cultures, *FEMS Microbiol. Lett.* **1,** 47–50.

NEIJSSEL, O. M., TEMPEST, D. W. (1975), The regulation of carbohydrate metabolism in *Klebsiella aerogenes* NCTC 418 organisms, growing in chemostat culture, *Arch. Microbiol.* **106,** 251–258.

NEIJSSEL, O. M., TEMPEST, D. W. (1976a), Bioenergetic aspects of aerobic growth of *Klebsiella aerogenes* NCTC 418 in carbon-limited and carbon-sufficient chemostat culture, *Arch. Microbiol.* **107,** 215–221.

NEIJSSEL, O.M., TEMPEST, D.W. (1976b), The role of energy-spilling reactions in the growth of *Klebsiella aerogenes* NCTC 418 in aerobic chemostat culture, *Arch. Microbiol.* **110,** 305–311.

NEIJSSEL, O.M., TEMPEST, D.W. (1979), The physiology of metabolite overproduction, in: *Microbial Technology: Current State, Future Prospects, 29th Symp. Soc. Gen. Microbiol.* (BULL, A. T., ELLWOOD, D. C., RATLEDGE, C., Eds.), pp. 53–82, Cambridge: University Press.

NEIJSSEL, O. M., HUETING, S., TEMPEST, D. W. (1977), Glucose transport capacity is not the rate-limiting step in the growth of some wild-type strains of *Escherichia coli* and *Klebsiella aerogenes* in chemostat culture, *FEMS Microbiol. Lett.* **2,** 1–3.

NEILANDS, J. B. (1981), Microbial iron compounds, *Annu. Rev. Biochem.* **50,** 715–731.

NEILANDS, J. B. (1982), Microbial envelope proteins related to iron, *Annu. Rev. Microbiol.* **36,** 285–309.

NEILANDS, J. B. (1984), Siderophores of bacteria and fungi, *Microbiol. Sci.* **1,** 9–14.

NELSON, D. L., KENNEDY, E. P. (1972), Transport of magnesium by a repressible and a nonrepressible system in *Escherichia coli*, *Proc. Natl. Acad. Sci. USA* **69,** 1091–1093.

NEWSHOLME, E.A., CRABTREE, B. (1973), Metabolic aspects of enzyme activity regulation, in: *Rate Control of Biological Processes*, *Symp. Soc. Exp. Biol.* pp. 429–460, Cambridge: University Press.

PIRT, S. J., CALLOW, D. S. (1958), Exocellular product formation by microorganisms in continuous culture. I. Production of 2,3-butanediol by *Aerobacter aerogenes* in a single-stage process, *J. Appl. Bacteriol.* **21,** 188–205.

ROSENBERG, H., GERDES, R. G., CHEGWIDDEN, K. (1977), Two systems for the uptake of phosphate in *Escherichia coli*, *J. Bacteriol.* **131,** 505–511.

ROSENBERGER, R. F., ELSDEN, S. R. (1960), The yields of *Streptococcus faecalis* grown in continuous culture, *J. Gen. Microbiol.* **22,** 726–739.

ROUF, M. A. (1964), Spectrochemical analysis of inorganic elements in bacteria, *J. Bacteriol.* **88,** 1545–1549.

RUYTER, G. J. G., POSTMA, P. W., VAN DAM, K. (1991), Control of glucose metabolism by Enzyme II^{Glc} of the phosphoenolpyruvate-dependent phosphotransferase system in *Escherichia coli*, *J. Bacteriol.* **173,** 6184–6191.

SNOEP, J. L., TEIXEIRA DE MATTOS, M. J., POSTMA, P. W., NEIJSSEL, O. M. (1990), Involvement of pyruvate dehydrogenase in product formation in pyruvate-limited anaerobic chemostat cultures of *Enterococcus faecalis* NCTC 775, *Arch. Microbiol.* **154,** 50–55.

STEPHANOPOULOS, G., VALLINO, J. J. (1991), Network rigidity and metabolic engineering in metabolite overproduction, *Science* **252,** 1675–1681.

STONE, M. J., WILLIAMS, D. H. (1992), On the evolution of functional secondary metabolites (natural products), *Mol. Microbiol.* **6,** 29–34.

Tempest, D.W. (1969), Quantitative relationships between inorganic cations and anionic polymers in growing bacteria, *Symp. Soc. Gen. Microbiol.* **19,** 87–111.

TEMPEST, D. W., MEERS, J. L., BROWN, C. M. (1970), Influence of environment on the content and composition of microbial free amino acid pools, *J. Gen. Microbiol.* **64,** 171–185.

TERRACCIANO, J. S., KASHKET, E. R. (1986), Intracellular conditions required for initiation of solvent production by *Clostridium acetobutylicum*, *Appl. Environ. Microbiol.* **52,** 86–91.

THOMAS, T. D., ELLWOOD, D. C., LONGYEAR, V. M. C. (1974), Change from homo- to heterolactic fermentation by *Streptococcus lactis* resulting from glucose limitation in anaerobic chemostat cultures, *J. Bacteriol.* **138,** 109–117.

VAN SCHIE, B. J., VAN DIJKEN, J. P., KUENEN, J. G. (1984), Non-coordinated synthesis of glucose dehydrogenase and its prosthetic group PQQ in *Acinetobacter* and *Pseudomonas* species, *FEMS Microbiol. Lett.* **24,** 133–138.

VAN SCHIE, B. J., DE MOOY, O. H., LINTON, J. D., VAN DIJKEN, J. P., KUENEN, J. G. (1987), PQQ-dependent production of gluconic acid by *Acinetobacter*, *Agrobacterium*, and *Rhizobium* species, *J. Gen. Microbiol.* **133,** 867–875.

VINING, L. C. (1990), Functions of secondary metabolites, *Annu. Rev. Microbiol.* **44,** 385–427.

WÜMPELMANN, M., KJAERGAARD, L., JOERGENSEN, B. B. (1984), Ethanol production with *Saccharomyces cerevisiae* under aerobic conditions at different potassium concentrations, *Biotechnol. Bioeng.* **26,** 301–307.

YAMADA, T., CARLSSON, J. (1975), Regulation of lactate dehydrogenase and change of fermentation products in streptococci, *J. Bacteriol.* **124,** 55–61.

ZOUTBERG, G. R., WILLEMSBERG, R., SMIT, G., TEIXEIRA DE MATTOS, M. J., NEIJSSEL, O. M. (1989a), Aggregate formation by *Clostridium butyricum, Appl. Microbiol. Biotechnol.* **32,** 17–21.

ZOUTBERG, G. R., WILLEMSBERG, R., SMIT, G., TEIXEIRA DE MATTOS, M. J., NEIJSSEL, O. M. (1989b), Solvent production by an aggregate-forming variant of *Clostridium butyricum*, *Appl. Microbiol. Biotechnol.* **32,** 22–26.

5 Metabolic Design

Hermann Sahm

Jülich, Federal Republic of Germany

1 Introduction 190
2 Measurement of Intracellular Metabolites 190
  2.1 Silicone Oil Centrifugation 191
  2.2 Nuclear Magnetic Resonance (NMR) Spectroscopy 192
3 Expansion of the Substrate Spectrum 195
  3.1 Construction of Lactose-Utilizing Strains 195
  3.2 Development of Xylose-Metabolizing Microorganisms 196
  3.3 Construction of Starch-Degrading Microorganisms 200
  3.4 Novel Metabolic Pathways for the Degradation of Xenobiotics 201
4 Enhancement of Product Yield and Spectrum 204
  4.1 Genetic Improvement of Ethanol Production from Pentoses 204
  4.2 Production of 1,3-Propanediol by a Transformed *Escherichia coli* Strain 206
  4.3 Construction of L-Amino Acid-Producing Bacteria 206
  4.4 Synthesis of Precursors for Vitamins 211
  4.5 Improvement of the Production of Biopolymers 212
  4.6 Redesign of Secondary Metabolic Pathways 214
5 Concluding Remarks and Future Aspects 216
6 References 217

1 Introduction

Strain improvement is an essential part of process development for biotechnological products as a means of reducing costs by developing strains with increased productivity and yield, ability to use cheaper raw materials, or more specialized desirable characteristics such as improved tolerance to high substrate and/or product concentrations. The current practice for the development of primary and secondary metabolites overproducing microorganisms by mutagenesis and selection is a very well established technique (ROWLANDS, 1984). Mutagenic procedures can be optimized in terms of type of mutagen and dose. The understanding of mutagenesis and DNA repair allows mutation procedures to be optimized for the production of desirable mutant types. Screens can be designed to allow maximum expression and detection of the desirable mutant types. Furthermore, automated procedures can be developed using robotics and microprocessors to increase the numbers of isolated mutants that can be processed per unit time. Thus the development of many of the highly productive industrial strains with this procedure has been largely an empirical process. The precise genetic and physiological changes resulting in increased overproduction of metabolites in many of these organisms have remained unknown. Success in attempts to further increase the productivities and yields of already highly productive strains will depend on the availability of detailed information on the metabolic pathways and the mutations.

During the last years genetic engineering and amplification of relevant structural genes have become a fascinating alternative to mutagenesis and random screening procedures (IMANAKA, 1986). Introduction of genes into organisms via recombinant DNA techniques is a most powerful method for the construction of strains with desired genotypes. The opportunity to introduce heterologous genes and regulatory elements permits construction of metabolic configurations with novel and beneficial characteristics. Furthermore, this approach avoids the complication of uncharacterized mutations that are often obtained with classical whole cell mutagenesis. The improvement of cellular activities by manipulation of enzymatic, transport, and regulatory functions of the cell with the application of recombinant DNA technology is called metabolic engineering (BAILEY, 1991); I prefer the term metabolic design since in this way the metabolism can be changed in a purposeful manner.

An early successful use of recombinant bacteria involved altering the nitrogen metabolism of the parent culture to obtain a higher yield of single cell protein (SCP). The process modification was developed by ICI, Great Britain, for making SCP from methanol (WINDASS et al., 1980). The pathway used by the methylotrophic bacterium *Methylophilus methylotrophus* for assimilating ammonia involves an aminotransferase and glutamine synthase; it consumes one mole of ATP for every mole of ammonia assimilated. In *Escherichia coli*, e.g., an alternative pathway uses glutamate dehydrogenase (GDH) for ammonia assimilation and does not consume any ATP. Therefore, the *gdh* gene from *E. coli* was isolated and integrated into a vector that effected the gene expression in *M. methylotrophus*. The expression of this gene in a glutamate synthase negative mutant gave a cell yield 5% higher than that of the parent strain because of the saving in ATP. Recent applications of recombinant DNA technology to restructure metabolic networks and improve production of metabolites will be highlighted in this chapter.

2 Measurement of Intracellular Metabolites

As the metabolic activities of living cells are accomplished by a regulated, coupled network of more than 1000 enzyme-catalyzed reactions and selective membrane transport systems for doing effective metabolic design, it is important to know the concentrations and regulations of corresponding enzymes and their substrates. In principle, the activities of many enzymes can be measured in the crude extracts of the organisms by using specific enzyme assays.

The concentrations of the proteins can be determined from two-dimensional gel electrophoresis, but references are necessary to identify the various enzymes (O'FARRELL, 1975). To determine metabolite concentrations in cells various analytical methods can be used.

2.1 Silicone Oil Centrifugation

For the measurements of intracellular compounds the cells should be separated from the culture medium as some metabolites may also be in the medium. There are basically two types of separation procedures: centrifugation and filtration. A very useful technique is the centrifugation combined with a filtration through silicone oil, a procedure known as centrifugal filtration (PALMIERI and KLINGENBERG, 1979). During centrifugation, the cells pass the silicone layer and become stripped of their surrounding medium. As the metabolic state will change further in the sediment at the bottom of the centrifuge tubes the reactions were stopped in an acid layer. After homogenization of the sedimented cells, the extracts were neutralized and used for analysis of the various compounds. For the determination of internal metabolite concentrations the cytoplasmic volume can be investigated from the distribution of solutes in cell suspensions using ^{14}C taurine as an impermeable marker for the extracellular space and ^{3}H water for the total volume (ROTTENBERG, 1979). Measurement of the volumes occupied by the two compounds allows the calculation of the total cytoplasmic volume of cells (Fig. 1).

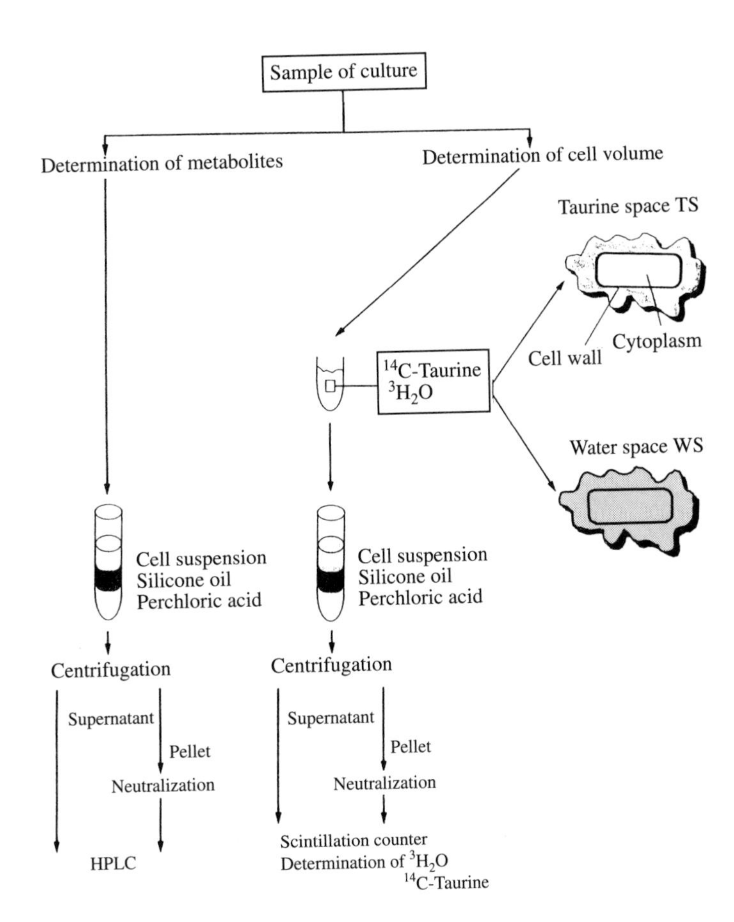

Fig. 1. Scheme of the determination of intracellular metabolite concentrations.

2.2 Nuclear Magnetic Resonance (NMR) Spectroscopy

During the last decade nuclear magnetic resonance methods have become firmly established as a powerful tool for analyzing metabolite concentrations in cell extracts but also in whole cells. One of the great advantages of the NMR applications to biology and medicine is the non-invasive and non-destructive nature of the technique. This allows details of cellular metabolism to be studied *in vivo*. The NMR provides a means for measuring reaction rates and enzymatic activities in bacteria, fungi, algae, plant cells, and animal cells (FERNANDEZ and CLARK, 1987). By relating these measurements to metabolic levels and physiological function it should be possible to get a more detailed understanding of the control and integration of metabolic pathways and physiological activities.

^{31}P NMR was first applied to yeast cells *in vivo* by SALHANY et al. in 1975. It was demonstrated that NMR could detect the presence of various intracellular compounds, e.g., sugar phosphates, ATP, and NAD. Furthermore, it was found that the chemical shift of the phosphate resonance was dependent on the pH of the environment; this behavior can be used to measure the intracellular pH. In the meantime, many ^{31}P NMR investigations of metabolic processes in yeast have been carried out, for example, the effect of oxygen on glycolytic metabolism has been studied (DEN HOLLANDER et al., 1981; GALAZZO et al., 1990). Recently, the metabolism in the ethanol-producing bacterium *Zymomonas mobilis* has been studied by *in vivo* NMR technique (DE GRAAF et al., 1992). As shown in Fig. 2 peaks of various intracellular phosphorylated metabolites were obtained. The sugar phosphate peaks consist of several unresolved resonances. The internal and external inorganic phosphate peaks are well separated, reflecting a pH difference between the inside cells and the culture medium. In order to compensate for the inherent relatively low sensitivity of the NMR method, the measurements were carried out with cell suspensions of high concentration (1.6×10^{11} cells/mL). With the aid of perchloric acid extracts, the various intermediates of glucose metabolism in the ENTNER-DOUDOROFF pathway could be detected (DE GRAAF et al., 1991). In this way it

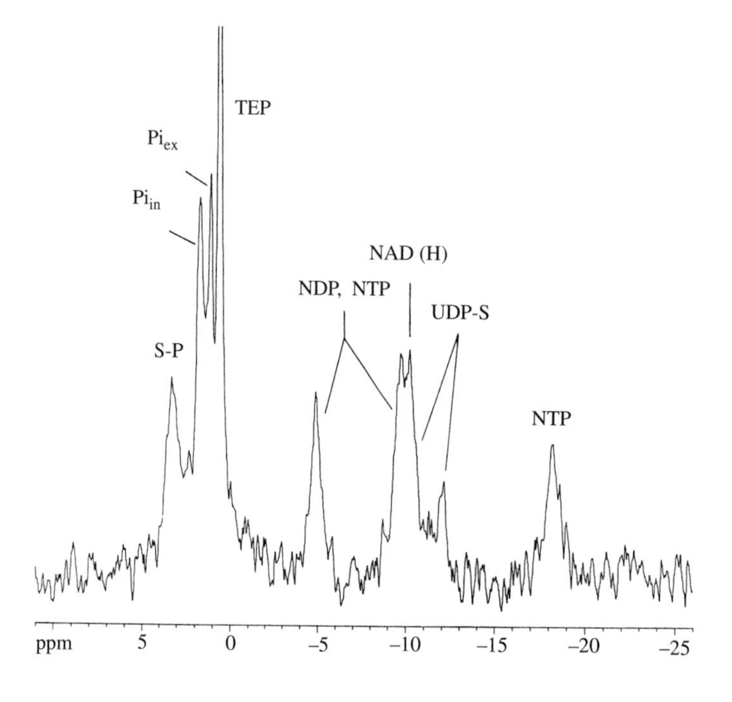

Fig. 2. *In vivo* ^{31}P (162 MHz) spectrum of a highly concentrated suspension of *Zymomonas mobilis* (123 mg dw/mL) metabolizing glucose. Assignments of intracellular resonances are as follows: S-P sugar phosphates, Pi$_{in}$ inorganic phosphate, NDP, NTP nucleoside di- and triphosphates, NAD(H) nicotine adenine dinucleotides, UDP-S uridine diphosphosugars. Extracellular resonances are as follows: Pi$_{ex}$ inorganic phosphate, TEP triethyl phosphate.

could be found out that at high ethanol concentrations (10%) in the culture medium 3-phosphoglycerate was accumulated in the cells.

Although the ^{31}P NMR technique provides an elegant non-invasive means of monitoring the energetic state of intact cells, it has limited potential for elucidating the details of metabolic pathways. A more general possibility in the study of metabolism and regulation is represented by ^{13}C NMR. Because of the low natural abundance of the ^{13}C nucleus and its low NMR sensitivity it is necessary to use ^{13}C enriched compounds in order to get good spectra at short measurement times. Many ^{13}C NMR studies have been carried out to assess details of a variety of metabolic pathways (NICOLAY, 1987). For example, using ^{13}C glucose it could be detected that in *Corynebacterium glutamicum* lysine is synthesized by at least two different pathways (ISHINO et al., 1984). It is in particular this potential of ^{13}C NMR to measure the flows through competing pathways which makes it a very powerful tool in detailed metabolic investigations (GALAZZO and BAILEY, 1990).

Another important application of NMR is the study of transmembrane transport of ions and other solutes into the cell (FERNANDEZ and CLARK, 1987). Recently, *in vivo* measurements of intra- and extracellular concentrations of Na^+ and K^+, and of the transport of these ions across cell membranes have been carried out using ^{23}Na and ^{39}K NMR, respectively. Intra- and extracellular pools of these ions give peaks at the same chemical shift position. The use of impermeable shift reagents, like dysprosiumtripolyphosphate, shifts the peak of the external pool thereby making it possible to measure the intra- and extracellular peak intensities separately (OGINO et al., 1985). This method seems also very promising for measuring the uptake of various sugars by microorganisms (SCHOBERTH and DE GRAAF, 1992).

NMR studies of suspended cells require that the cells be present in very high densities (up to 10^{11} cells/mL) so that spectra with an adequate signal-to-noise ratio can be obtained. However, the maintenance of dense cultures in a well defined state presents a major problem due to limited oxygenation and rapidly changing substrate and product concentrations. To solve

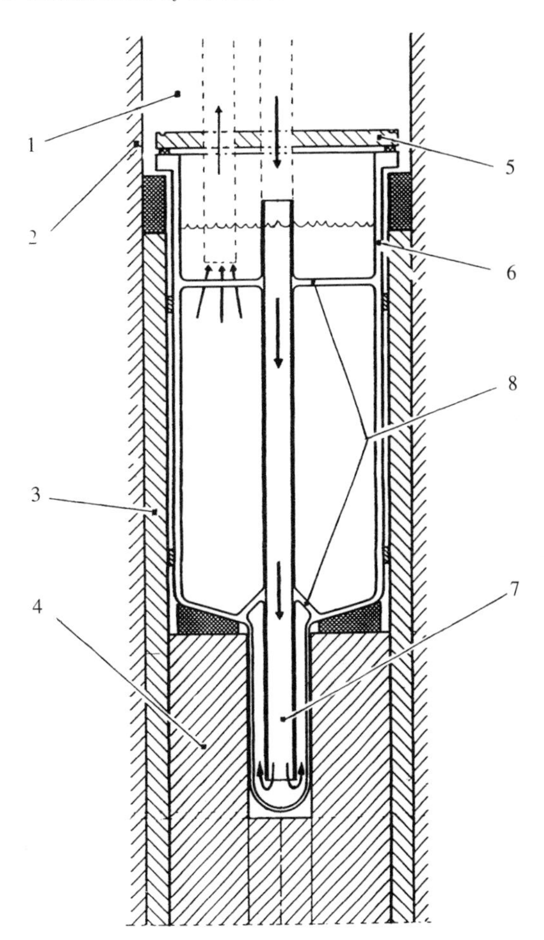

Fig. 3. Schematic drawing of the continuous-flow NMR bioreactor. A small centrifugal pump is positioned 11 cm above the bioreactor lid. The lower part of the bioreactor is positioned inside the sensitive region of the NMR receiver coil. Arrows indicate the flow direction. With 375 mL contents, the suspension level is 16 cm above the magnet isocenter. **1** magnet bore (89 mm), **2** magnet inner wall, **3** lower section of central magnet insert holding the room temperature shim coils and with the turbine part removed, **4** upper part of the NMR probehead with detection chamber, **5** bioreactor lid, **6** bioreactor main section used as prepolarizing chamber, **7** central 10 mm tube, **8** glass supports for central tube.

these problems, cells were cultivated in perfusion systems like hollow fiber bioreactors. All these systems suffer from a non-uniform substrate, product, and oxygen distribution over the culture. Thus, in none of the reported methods can the cells be kept in a well defined physiological state, i.e., a state that is reproducible, uniform over the complete culture volume, stable during long periods of time (many doubling times of the cells), and easily adjustable.

Therefore, a new NMR bioreactor was constructed that fits into the 89 mm vertical bore of the NMR magnet after removal of the upper section of the 50 mm i.d. central insert plus the spinner turbine from the lower section of the central insert (DE GRAAF et al., 1992). A schematic drawing of the bioreactor is shown in Fig. 3. It consists of three main parts: a reactor vessel, a measuring chamber, and a pump (not shown in Fig. 3). The reactor vessel acts as a prepolarizing volume. To ensure complete relaxation before the next excitation, i.e., the maximal NMR signal, the mean residence time of the spins in the vessel should be at least $5*T_1$ under all experimental conditions. Therefore, the vessel volume was chosen as large as possible yet small enough to ensure that everywhere in the vessel the magnetic field strength is more than 90% of the magnetic field strength in the magnet isocenter. The vessel is made of glass, with a diameter of 7 cm, a height of 14 cm and a volume of 366 mL. The measuring chamber consists of the lower 5.5 cm part of a standard 20 mm o.d. NMR sample tube connected with the reactor vessel. The bioreactor is used in combination with a special $^{31}P/^{13}C$ dual tunable 20 mm probe head (Bruker Spectrospin, Faellanden/Switzerland). It was shortened by 2 cm compared to standard probeheads in order to obtain a larger prepolarizing volume. The sensitive volume of the radio frequence coil is about 9 mL. The capacity of the centrifugal pump is adjustable between 150 and 600 L/h. Thereby, the mean residence time in the sensitive volume of the RF coil, t_M, can be varied between 54 ms (expected line broadening, 20 Hz) and 216 ms (expected line broadening, 5 Hz), and the mean residence time in the reactor vessel, t_P, between 2.2 s and 8.8 s. Thus, this reactor allows optimal detection of nuclei with T_1 values ranging from 440

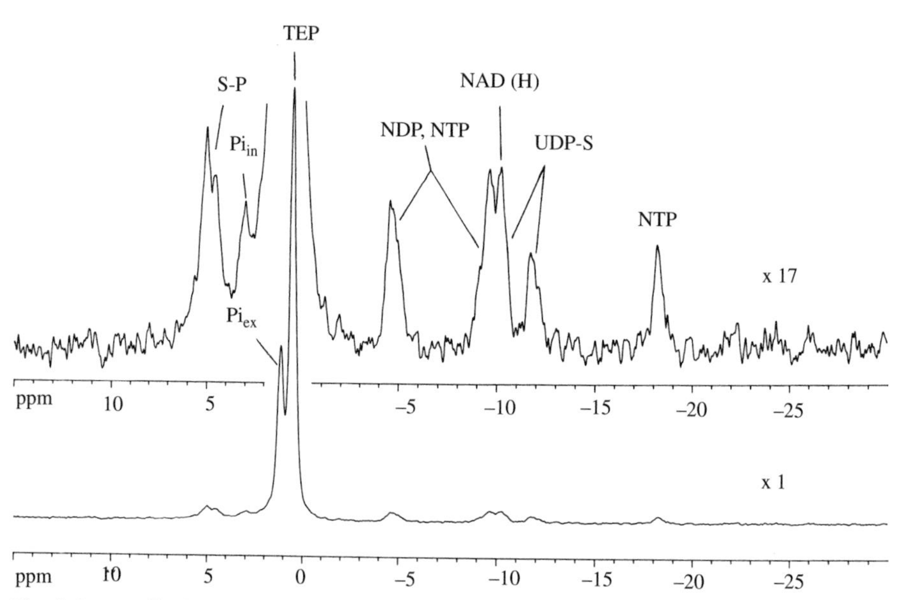

Fig. 4. *In vivo* ^{31}P (162 MHz) spectrum of a dilute suspension of *Zymomonas mobilis* (2 mg dw/mL) metabolizing glucose, using the continuous-flow NMR bioreactor. Assignments are as given in the legend to Fig. 2. Peaks of Pi_{ex} and TEP appear strongly enhanced over intracellular peaks due to the low relative intracellular volume (0.5%).

ms to 1.8 s, and T_2^* values ranging from 18 to 72 ms. This makes it ideal for *in vivo* studies. The bioreactor can be equipped with various internal measuring devices and external control apparatus necessary for cell cultivation (e.g., control of substrate influx, culture efflux, pH, temperature, oxygen tension). Thus it can be operated *in situ* in the NMR magnet as a laboratory bioreactor, for batch as well as for continuous cultivation of microorganisms. With this NMR bioreactor *in vivo* ^{31}P measurements using a bacteria suspension with biomass concentrations as low as 2 mg dw/mL yield spectra with an acceptable S/N ratio within half an hour (Fig. 4). The described experimental setup allows the study of microorganisms by *in vivo* NMR under well defined physiological conditions.

All these studies demonstrate that NMR spectroscopy is a versatile and powerful tool for analyzing cellular metabolism and measuring important biochemical parameters *in vivo*. The limitations with respect to the relative insensitivity and the high cell concentrations may be overcome by the special NMR bioreactor described above. Thus, this technique will certainly make further significant contributions to the study of cell physiology and will support rational metabolic design.

3 Expansion of the Substrate Spectrum

Cloning and expression of heterologous genes can be used for extending the catabolic pathway in an organism; in this manner the substrate spectrum can be enlarged. However, the synthesis of a heterologous protein does not guarantee appearance of the desired activity. The protein must avoid degradation by proteases, it has to fold correctly and accomplish any necessary assembly and prosthetic group acquisition, and the localization must be suitable. Despite these potential barriers several positive experiments for the extension of the substrate spectrum have been published in the last few years. In the following a few examples will be presented.

3.1 Construction of Lactose-Utilizing Strains

Currently there is a great deal of interest in using whey as a cheap nutrient source in biotechnological processes. Whey is a nutrient-rich by-product of the dairy industry, and its disposal is a major pollution problem. The composition of whey generally includes high amounts of lactose – 75% of dry matter –, 12–14% of protein and smaller amounts of organic acids, minerals, and vitamins.

Previously, the lactose transposon, Tn 951, was introduced into the ethanol-producing bacterium *Zymomonas mobilis* (CAREY et al., 1983; GOODMAN et al., 1984). Although the induction of β-galactosidase and the production of ethanol from lactose by *Z. mobilis* strains containing the *E. coli* lactose operon was reported, the ethanol yield was much lower than the theoretical yield. Furthermore, these strains were not able to grow on lactose as a single carbon and energy source. There are two factors which may be responsible for this poor ethanol production on lactose. Firstly, lactose is cleaved to glucose and galactose by β-galactosidase, but only glucose can be fermented to ethanol by *lac*$^+$ *Z. mobilis* strains, as this bacterium is not able to degrade galactose. Therefore, galactose is accumulated during lactose metabolism and this may cause an inhibition (YANASE et al., 1988). Secondly, the poor utilization of lactose by these *Z. mobilis* strains may be due to a very slow uptake of lactose from the medium. To achieve complete utilization of lactose, *Z. mobilis* strains should be constructed that contain both the lactose and the galactose operons.

The amino acid-producing bacterium *Corynebacterium glutamicum* is also not able to metabolize lactose. Recently, the entire *E. coli* lactose operon was inserted into an *E. coli/C. glutamicum* shuttle vector and introduced into the Gram-positive host organism *C. glutamicum* (BRABETZ et al., 1991). Recombinant strains carrying the *lac* genes downstream of an efficient promotor displayed rapid growth with lactose as the sole source of carbon. Two prerequisites were necessary for growth of *C. glutamicum* on lactose: (1) presence of the *lacY* gene (lactose permease) in addition to the *lacZ* gene (β-galactosidase), (2) an appropriate promotor for efficient transcription of these genes. Although the correct membrane insertion of the *E. coli* lactose carrier in *C. glutamicum* is not clear, it is very interesting and important to see that this membrane protein from *E. coli* – a Gram-negative strain – also functions in a Gram-positive bacterium. As the galactose moiety of the lactose was not utilized, further supplementation of these recombinant strains with genes encoding enzymes for galactose metabolism seems necessary.

The use of *Alcaligenes eutrophus* for the production of polyhydroxyalkanoic acids is impaired by its limited capability to grow on cheap carbon and energy sources. Fructose is the only sugar used by *A. eutrophus*, whereas glucose and disaccharides such as lactose, maltose, or sucrose are not metabolized (WILDE, 1962). Many years ago, SCHLEGEL and GOTTSCHALK (1965) were able to isolate glucose-utilizing mutants of this organism. Evidence was obtained that the use of glucose is not possible because of the lack of a transport system allowing passage of the cytoplasmic membrane, and that glucose is taken up by facilitated diffusion by the mutant strain. Glucose-utilizing mutants are used by the industry for the production of polyhydroxybutyric acid (STEINBÜCHEL, 1991). Recently, lactose and glucose-utilizing strains of *A. eutrophus* were obtained by using the recombinant DNA technology (PRIES et al., 1990). The lactose operon of *E. coli* could be expressed in *A. eutrophus* after the promotor of the polyhydroxybutyric acid synthetic gene was ligated to these structural genes. To get stable strains the promotor-*lac* fusion was inserted into the genome of *A. eutrophus;* these clones were able to grow on lactose with a doubling time of 16–23 h. In this case also only glucose was metabolized and galactose was excreted. Therefore, the *E. coli* galactose operon was also transferred. Thereafter the organism could utilize lactose completely, no galactose was secreted, the growth yields increased twofold and the doubling time on lactose was only 6 h.

The *E. coli* lactose operon has been previously used also for the construction of lactose-utilizing strains of *Xanthomonas campestris* (WALSH et al., 1984) and of *Pseudomonas aeruginosa* (KOCH et al., 1988). The *Pseudomonas* strain was able to grow on lactose and produced rhamnolipids which are valuable biosurfactants. A few years ago, the β-galactosidase gene from *Kluyveromyces lactis* was also introduced together with the cloned lactose permease gene into *Saccharomyces cerevisiae*, transformants leading to the ability to ferment lactose (SREEKRISHNA and DICKSON, 1985).

3.2 Development of Xylose-Metabolizing Microorganisms

Xylose is an abundant pentose in plant biomass where it occurs mainly in polymeric structures, e.g., the hemicellulose xylan. It can be used by many microorganisms as the sole

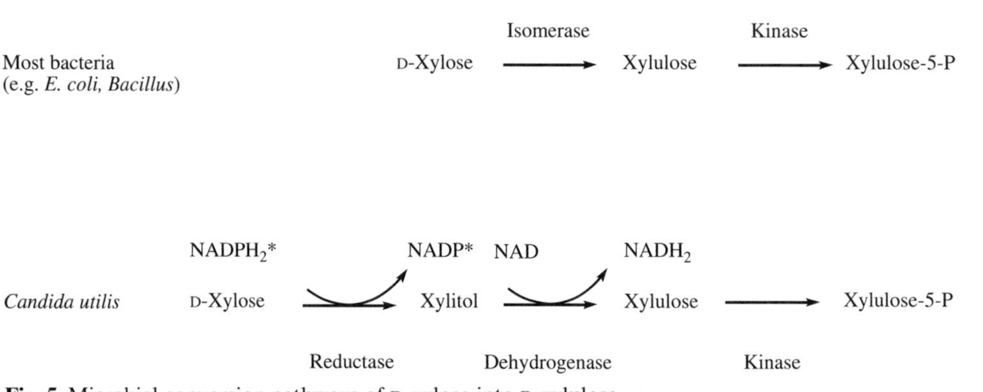

Fig. 5. Microbial conversion pathways of D-xylose into D-xylulose.
* In *Pichia stipitis* the xylose reductase can react with both $NADPH_2$ and $NADH_2$.

source of carbon and energy. In the past, processes were developed to use xylose in the sulfite spent liquor as a substrate for single cell protein production. Metabolic studies demonstrated that in bacteria xylose is converted into xylulose by the enzyme xylose isomerase. As in yeast strains, this reaction is catalyzed by two enzymes (Fig. 5). In the first step xylose is reduced to xylitol by an NADPH-linked xylose reductase, the second reaction is the oxidation of xylitol to xylulose catalyzed by an NAD-linked xylitol dehydrogenase. In bacteria as well as in yeasts xylulose is subsequently phosphorylated by xylulose kinase to form xylulose-5-phosphate which is then channelled into the pentose phosphate pathway (JEFFRIES, 1983). During the last few years the genes encoding enzymes for xylose utilization have been well characterized in *E. coli* and some other bacteria, the expression of these genes appears to be tightly regulated (LAWLIS et al., 1984; RYGUS et al., 1991).

Previously several projects were started to develop microbial strains that are able to use xylose as a carbon source for ethanol fermentation. Although a wide spectrum of microorganisms utilize xylose, none of them can perform an efficient production of ethanol under anaerobic conditions. For example, the yeast *Candida utilis* which can grow on xylose very well can convert xylose into xylulose only under aerobic conditions (BRUINENBERG et al., 1983). Other yeasts, like *Pachysolen tannophilus*, can ferment xylose slowly under anaerobic conditions, but produce considerable amounts of side products (DEBUS et al., 1983). This is also the case with several facultative or obligate anaerobic bacteria that form various organic acids besides alcohol (GOTTSCHALK, 1986). Recently, BRUINENBERG et al. (1984) have reported that the yeast *Pichia stipitis* can perform a significant anaerobic ethanol fermentation on xylose. Although the data for ethanol formation from xylose with this yeast strain are better than those with other strains, the ethanol production rates and yields are far lower than those of *S. cerevisiae* and *Z. mobilis* on glucose (Tab. 1).

Because of the many positive properties of *S. cerevisiae* with regard to ethanol fermentation, the feasibility of constructing strains able to utilize xylose was studied. As this yeast seems able to metabolize xylulose, the construction of a xylose utilizing *S. cerevisiae* strain should be feasible by introducing the xylose isomerization ability from a xylose utilizing organism into the yeast. As the first step towards this goal the xylose isomerase gene from *Bacillus subtilis* was cloned and transferred to the yeast. Although the transformants synthesized the xylose isomerase proteins in high amounts (approx. 5% of the cellular protein), no enzyme activity could be measured (HOLLENBERG and WILHELM, 1987). So far, approaches to establish xylose isomerase from various bacteria in *S. cerevisiae* were not successful (AMORE et al., 1989). Therefore, recently the genes for the xylose–xylulose pathway from *P. stipitis* were transferred to *S. cerevisiae*. Strains transformed with the two genes are able to grow aerobically on xylose as a sole carbon source (KÖTTER et al., 1990). Comparative studies of xylose utilization in these *S. cerevisiae* transformants with the prototypic xylose-utilizing yeast *P. stipitis* showed that in the absence of respiration the *S. cerevisiae*

Tab. 1. Comparison of Xylose- and Glucose-Specific Fermentation Rates of Various Yeasts and *Zymomonas mobilis*[a]

Organism	Substrate	Fermentation Rate (ethanol formation g/g/h)	Yield (g ethanol/ g substrate)	Max. Concentration (g ethanol/L)
Pachysolen thannophilus	Xylose	0.3	0.3	38
Pichia stipitis	Xylose	0.92	0.45	40
P. thannophilus	Glucose	0.22	0.47	56
Saccharomyces cerevisiae	Glucose	1.78	0.50	87
Zymomonas mobilis	Glucose	2.5	0.50	102

[a] modified and completed after JEFFRIES (1983)

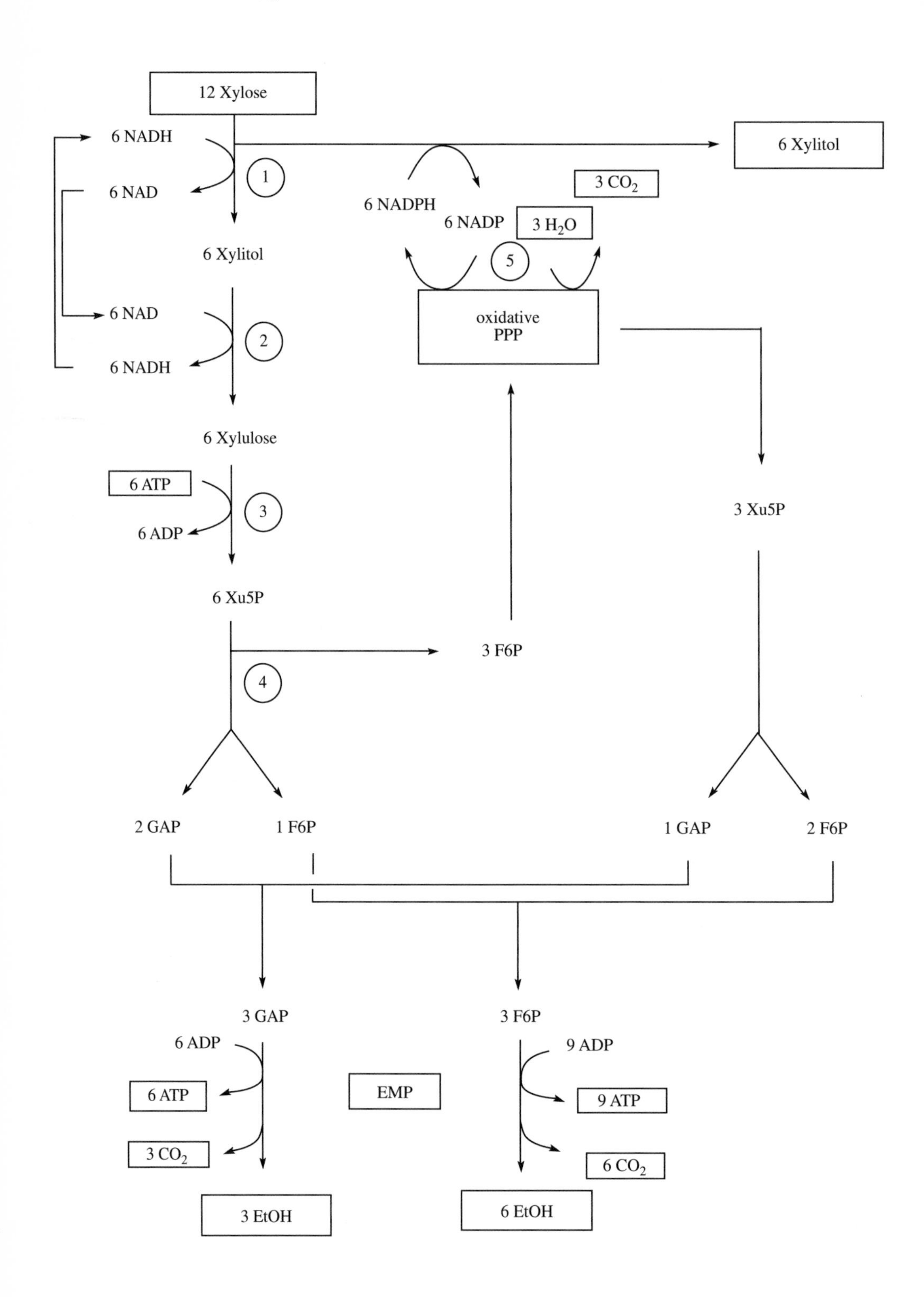

12 Xylose
6 NADH
6 NAD
1
6 Xylitol
6 Xylitol
6 NAD
6 NADH
2
6 Xylulose
6 ATP
6 ADP
3
6 Xu5P
4
3 F6P
2 GAP
1 F6P
6 NADPH
6 NADP
$3\ CO_2$
$3\ H_2O$
5
oxidative
PPP
3 Xu5P
1 GAP
2 F6P
3 GAP
6 ADP
6 ATP
$3\ CO_2$
3 EtOH
EMP
3 F6P
9 ADP
9 ATP
$6\ CO_2$
6 EtOH

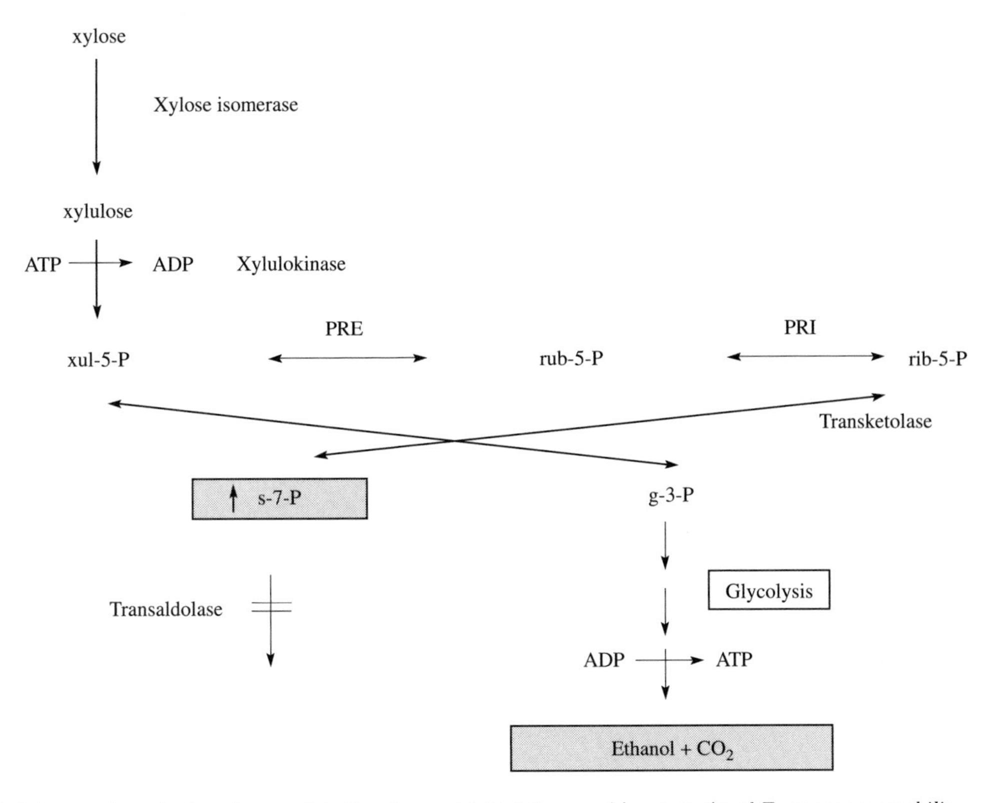

Fig. 7. Scheme of catabolic xylose metabolism in a *xylAB*[+] *tkt*[+] recombinant strain of *Zymomonas mobilis*.

clone converted half of the xylose to xylitol and ethanol, respectively, while *P. stipitis* produced mainly ethanol from xylose (KÖTTER and CIRIACY, 1992). Xylitol production is inter- preted as a result of the dual cofactor dependence of the xylose reductase and the generation of NADPH by the pentose phosphate pathway (Fig. 6). Further limitations of xylose utilization in *S. cerevisiae* cells may be caused by an insufficient capacity of the non-oxidative pentose phosphate pathway as indicated by accumulation of sedoheptulose-7-phosphate.

◀ **Fig. 6.** Scheme of xylose utilization and mechanism of cofactor regeneration in *Saccharomyces cerevisiae* transformants in the absence of respiration. EMP Embden–Meyerhof–Parnas pathway, **1** xylose reductase, **2** xylitol dehydrogenase, **3** xylulose kinase, **4** ribulose phosphate epimerase, ribose phosphate isomerase, transaldolase, and transketolase, **5** glucose-6-phosphate dehydrogenase, 6-phosphogluconate dehydrogenase (KÖTTER and CIRIACY, 1992).

Furthermore, xylose could be an interesting substrate for industrial ethanol production with *Z. mobilis*. This bacterium can take up xylose into the cells via the glucose transport system (DIMARCO and ROMANO, 1985; STRUCH et al., 1991), but no growth on the pentose occurs. Introduction of the genes (*xylAB*) for xylose isomerase and xylulokinase from *Xanthomonas campestris*, led to the successful expression of these enzymes in *Z. mobilis* cells. However, the recombinant strain was also not able to grow on xylose as sole carbon source (LIU et al., 1988). Recently, similar results have been obtained by FELDMANN et al. (1992), but after the gene for transketolase from *E. coli* was also cloned in *Z. mobilis*, a conversion of small amounts of xylose to CO_2 and ethanol occurred. However, no growth on xylose as sole carbon source was detected, instead sedoheptulose-7-phosphate accumulated intracellularly. As shown in Fig. 7, it seems that in this strain finally the transaldolase reaction is limited for an efficient conversion of xylose into ethanol.

3.3 Construction of Starch-Degrading Microorganisms

The use of starch as a carbon and energy source in biotechnological processes depends on the ability of the organisms to degrade this biopolymer. As most microorganisms lack starch-hydrolyzing enzymes, in the last few years some research work has been started to insert genes for enzymatic starch utilization into various strains (KENNEDY et al., 1988). Replacing glucose with starch as feedstock may not only reduce fermentation costs but could also minimize catabolite (glucose) repression and achieve higher yields and concentrations of the desired products. At present processes are in operation that convert starch enzymatically into glucose and some oligosaccharides. These hydrolytic products are then fermented by yeast into fuel-grade ethanol used as an octane booster in unleaded gasoline (FINN, 1987). Recently, a *Saccharomyces cerevisiae* strain has been constructed that contains a glucoamylase gene from *Aspergillus* (INNIS et al., 1985). The enzyme was secreted into the medium and had a specific activity similar to the orginal enzyme isolated from the culture medium of the *Aspergillus* strain. The fermentation of amylodextrins was possible with this recombinant *S. cerevisiae* strain, but the fermentation rate was considerably lower than when the glucoamylases were added into the culture medium.

In brewing the fermentable sugars are provided from barley by partial hydrolysis of starch during the malting process. This process, however, results in a considerable amount of dextrins that cannot be fermented by the yeast *S. cerevisiae*. These dextrins contribute to the high calorie content of beer and thus have to be removed to get light beer. For this purpose the present production process depends on the addition of glucoamylase. An *S. cerevisiae* strain with amylolytic properties would provide a tool to produce beer, especially light beer, which does not depend on adding enzymes. The application of an amylolytic baker's yeast in baking could end the dependence on α-amylase-enriched flour in certain types of bread manufacturing. A prerequisite

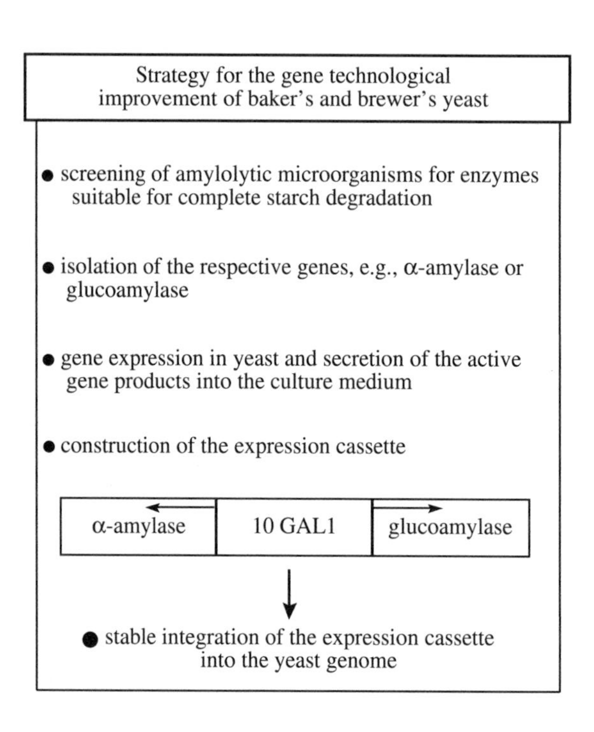

Fig. 8. Strategy for the construction of an amylolytic *Saccharomyces cerevisiae* yeast (HOLLENBERG and STRASSER, 1990).

for the construction of amylolytic *S. cerevisiae* strains is the availability of suitable genes coding for α-amylase and glucoamylase (Fig. 8). As the yeast *Schwanniomyces occidentalis* is able to hydrolyze starch entirely into glucose (SILLS et al., 1984), the α-amylase gene (*AMY*1) of this yeast was isolated and transferred into *S. cerevisiae* cells (HOLLENBERG and STRASSER, 1990). The *AMY*1 gene was found to be expressed in *S. cerevisiae* from its native promotor, leading to an actively secreted gene product. This yeast strain was able to grow on starch. High expression of the *AMY*1 gene in *S. cerevisiae* was achieved after fusion of the GAL10 promotor to this structural gene. In the second step the glucoamylase gene (*GAM*1) was isolated also from *S. occidentalis*, fused to the GAL1 promotor and expressed in *S. cerevisiae*. The resultant transformants secreted active glucoamylase and grew on starch as a sole carbon source quite well. Finally, an expression cassette containing both the *AMY*1 and *GAM*1 gene under control of the GAL10 and GAL1 promotor, respectively (Fig. 8), was integrated into the yeast genome in a stable form. The comparative enzymatic measurements demonstrated that the amylolytic system is as efficient in the genetically engineered *S. cerevisiae* as in the original *S. occidentalis* strain. Stable integration of the expression cassette into an industrial *S. cerevisiae* strain resulted in a clone that excels by high fermentation rates, ethanol tolerance, and the newly acquired ability to degrade starch entirely (DOHMEN et al., 1990). During fermentation of ground liquified wheat, this recombinant yeast strain showed the same ethanol production rate without the addition of saccharifying enzymes as a conventional distillary yeast with malt enzymes added prior to fermentation (*Eur. Patent Appl.* 87110 370.1).

3.4 Novel Metabolic Pathways for the Degradation of Xenobiotics

A large number of industrial organic compounds, particularly those that are structurally related to natural compounds, are readily degraded by soil and water microorganisms. Chemicals having novel structural elements or substituents rarely found in nature (xenobiotics) are slowly decomposed and thus tend to accumulate in the environment. Fortunately, the microorganisms have not only remarkable capacities to degrade a very wide range of various organic compounds but they also can evolve activities for the degradation of new components. However, in nature this evolution of new catabolic pathways proceeds very slowly where multiple genetic changes are required and where the selection pressures may only be effective in selecting the last genetic change to occur. Under laboratory conditions effective selection procedures can be designed for each of the genetic changes required. For example, a few years ago CHAKRABARTY and coworkers (KILBANE et al., 1983) were successful in isolating a *Pseudomonas* strain capable of degrading halogenated compounds such as the herbicide 2,4,5-trichlorophenoxyacetic acid, penta-, tetra-, and 2,4,5-trichlorophenol, etc. They inoculated microbial samples from various toxic waste dump sites into a chemostat, and provided a pool of plasmid genes that allow degradation of chlorinated and non-chlorinated aromatics. At the beginning the mixed cultures were fed with the various components for the effective dissemination of the plasmids into the surrounding bacteria. Over a period of several months 2,4,5-trichlorophenoxyacetic acid degrading bacteria were selected by increasing the concentration of this herbicide in the culture medium and decreasing the concentration of the other aromatic substrates. Finally, a pure culture of *Pseudomonas cepacia* could be isolated that can use 2,4,5-trichlorophenoxyacetic acid as sole source of carbon and energy (KILBANE et al., 1982). Using this strain, CHATTERJEE et al. (1982) showed removal of more than 98% of the 2,4,5-trichlorophenoxyacetic acid present at 1000 ppm from soil within one week. KILBANE et al. (1983) also found that once the bulk of this herbicide was gone, this strain could not compete with the indigenous microorganisms and died rapidly within a few weeks.

Besides the *in vivo* genetic transfer and the powerful chemostat selection, a rational restructuring of catabolic pathways is also possible by sequential mutation as demonstrated in the following example. *Pseudomonas putida* containing the TOL plasmid pWWO can degrade and grow on a variety of substituted benzoates including the 3-methyl-, 4-methyl-, and 3,4-dimethyl derivatives, but not on 4-ethylbenzoate. TIMMIS and coworkers found out that the TOL encoded *meta*-cleavage pathway cannot metabolize 4-ethylbenzoate, because this compound does not activate the positive regulatory protein xylS, and, therefore, *TOL* genes are not expressed (Fig. 9) (RAMOS et al., 1987). Therefore, a procedure was developed for selection of xylS regulator protein mutants that are activated by 4-ethylbenzoate. Although in such a xylS mutant the catabolic enzymes were synthesized, the compound 4-ethylbenzoate was only converted to 4-ethylcatechol. Catechol-2,3-dioxygenase could not attack this intermediate because this compound is a suicide inhibitor of this enzyme. After mutagenesis it was possible to select mutants that synthesized a modified catechol-2,3-dioxygenase capable of also oxidizing 4-ethylcatechol. Thus the TOL plasmid encoded pathway for the degradation of alkylbenzoates has been restructured to permit the degradation of 4-ethylbenzoate. By specific selection of mutants the two elements were changed: the effector specificity of the xylS protein, and the resistance of catechol-2,3-dioxygenase to substrate inactivation (Fig. 9).

Finally, by using the gene cloning techniques a redesigning of an existing pathway is a relatively straightforward approach to obtain microorganisms able to degrade xenobiotics. Industrial wastes frequently contain mixtures of chloro- and methyl aromatics that are partially converted into dead end products by the microorganisms. Therefore, ROJO et al. (1987) constructed a *Pseudomonas* strain able to degrade such a mixture of aromatic compounds. *Pseudomonas* sp. B13 used for this study, can degrade 3-chlorobenzoate via the *ortho*-pathway (Fig. 10). In the first step the substrate spectrum of this bacterium could be expanded to include 4-chlorobenzoate by cloning and expressing the toluate-1,2-dioxygenase of the TOL plasmid. This recombinant strain grew on 3- and 4-chlorobenzoate but not on 4-methylbenzoate because the metabolism of 4-methylcatechol via the modified *ortho*-cleavage pathway leads to the formation of the dead end

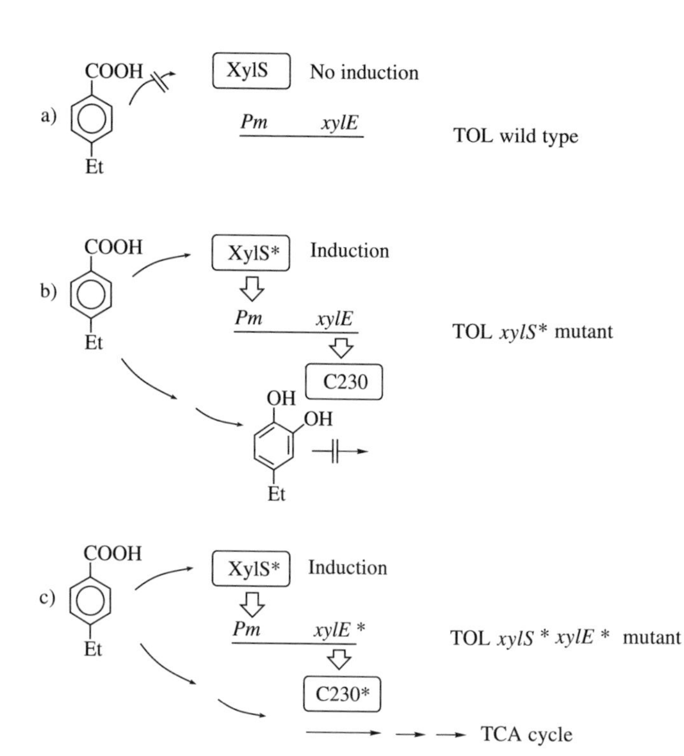

Fig. 9. Stepwise improvements of the TOL plasmid-encoded *m*-cleavage pathway to permit mineralization of 4-ethylbenzoate.

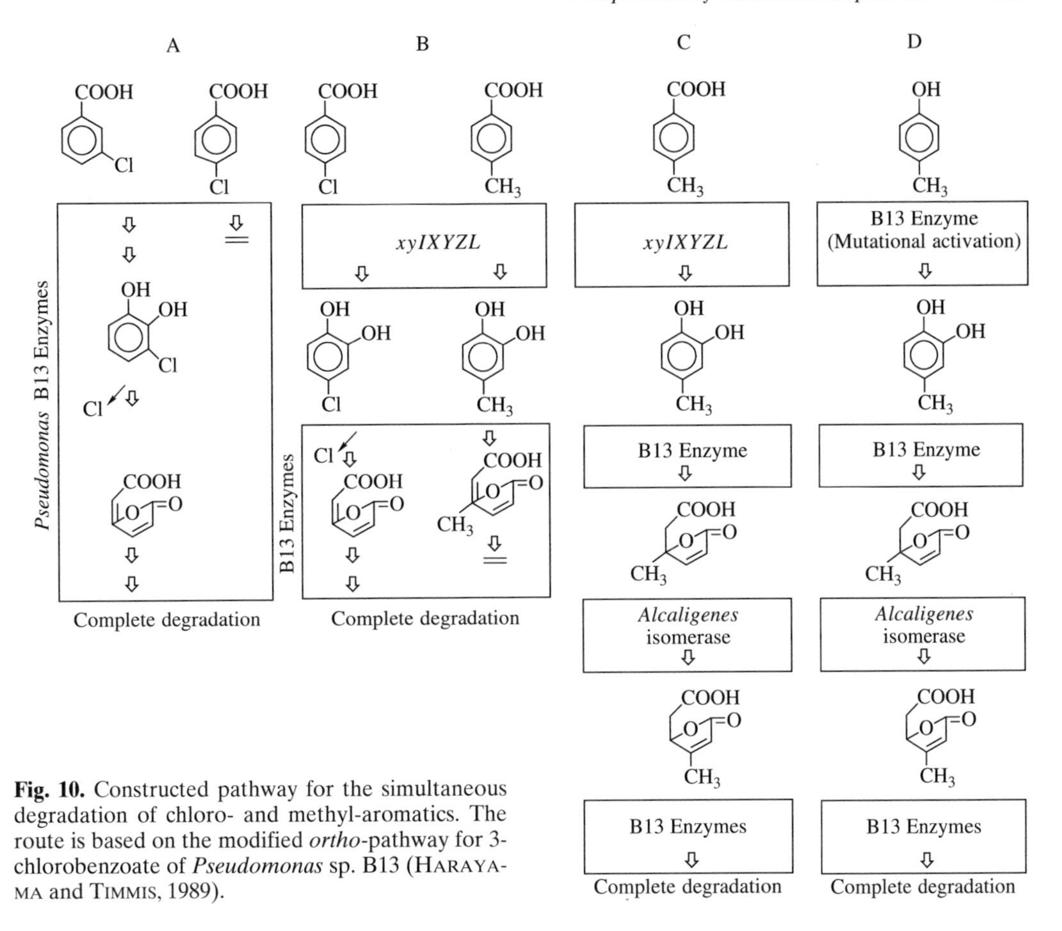

Fig. 10. Constructed pathway for the simultaneous degradation of chloro- and methyl-aromatics. The route is based on the modified *ortho*-pathway for 3-chlorobenzoate of *Pseudomonas* sp. B13 (HARAYAMA and TIMMIS, 1989).

product 4-methyl-2-ene-lactone. Therefore, in the second step the gene for the enzyme 4-methyl-2-ene-lactone isomerase, which converts 4-methyl-2-ene-lactone to 3-methyl-2-ene-lactone, was isolated from *Alcaligenes eutrophus* and expressed in the *Pseudomonas* strain. This new construct could degrade 4-methylbenzoate completely, for example, it could grow on mixtures of 3-chlorobenzoate and 4-methylbenzoate and simultaneously degrade both compounds. Finally, this pathway was further expanded through mutational activation of the cryptic gene for phenol hydroxylase which allowed metabolism of 4-methylphenol exclusively via *ortho*-cleavage (Fig. 10). This newly evolved strain metabolizes chloro- and methyl-substituted phenols and benzoates via *ortho*-cleavage routes, tolerates shock loads of one type of substituted benzoate or phenol, and can degrade both types simultaneously (HARAYAMA and TIMMIS, 1989). Another interesting enzyme system for the degradation of various aromatic compounds is the ligninase of the white rot fungus *Phanerochaete chrysosporium* that can decompose a broad range of xenobiotics, including even such diverse structures as benzopyrene and triphenylmethane dyes (BUMPUS et al., 1985; BUMPUS and BROCK, 1988). Thus, the cloning of especially active biodegradative genes into microorganisms more suited to a particular treatment process affords a challenge for the future.

4 Enhancement of Product Yield and Spectrum

Application of recombinant DNA techniques to restructure metabolic networks can improve production of metabolites by redirecting metabolite flows. In a cell the pathway to a desired product usually has several forks where intermediates can be metabolized via alternative routes. For getting high product yields it is therefore necessary that at each fork the desired pathway has a strong priority and only small amounts of the metabolites are used by the alternative steps. Overexpression of certain enzymes in the desired route is a useful strategy to reach this aim. In the following a few examples will be presented. Furthermore, cloning and expression of heterologous genes can extend an existing pathway to broaden the product spectrum as demonstrated in the last few years.

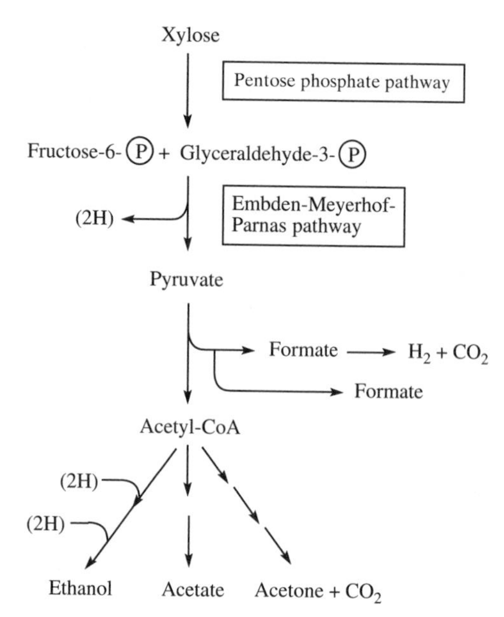

Fig. 11. Pentose fermentation pathways in bacteria with pyruvate-formate lyase activity.

4.1 Genetic Improvement of Ethanol Production from Pentoses

As already described neither the commonly used fermentative yeast strains nor the very potent ethanol-forming bacterium *Zymomonas mobilis* can ferment pentoses, as for example xylose, to ethanol. Attempts are being made to overcome such shortcomings genetically, however, until now with rather limited success. Therefore, in recent years several groups have been exploring the redesign of metabolic pathways of some bacteria that are vigorous pentose fermenting organisms.

For example, enteric bacteria dissimilate xylose by the pentose phosphate pathway to yield fructose-6-phosphate and glyceraldehyde-phosphate (Fig. 11). These are then metabolized to pyruvate by the Embden–Meyerhof pathway. Under anaerobic conditions pyruvate is degraded by the pyruvate–formate lyase; thus acetate, ethanol, and formate are formed as the main fermentation products in the ratio 1:1:2. In order to reduce the spectrum of fermentation products to mainly ethanol, the genes of pyruvate decarboxylase from *Z. mobilis* were transferred to various bacterial strains.

Klebsiella planticola is a Gram-negative facultative anaerobic enteric bacterium that can ferment a broad range of sugars, including all of the natural pentoses. Hexoses and pentoses are mainly fermented to acetate, ethanol, and formate; lactate and 2,3-butanediol are formed in small amounts (Tolan and Finn, 1987). After transferring the pyruvate decarboxylase (*pdc*) gene of *Z. mobilis* into *K. planticola* wild-type cells, the yield of ethanol could be raised from 0.6–0.7 to 1.3 M per mole of xylose and other catabolic end products at low pH (acetate, formate) were produced in smaller amounts (Tolan and Finn, 1987). Such a strong diversion of pyruvate is explained not only by the level of expression of the *pdc* gene but also by the strong affinity of the pyruvate decarboxylase for pyruvate (Bringer-Meyer et al., 1986), as compared to the affinity of competing enzymes such as pyruvate–formate lyase or lactate dehydrogenase. However, slow feeding of nutrients was necessary to attain the high yields cited and in fact the mixed-substrate fermentation lasted almost 100 h, much too long to be economically feasible. If growth and sugar uptake are too fast, organic acids and butanediol accumulate instead of ethanol.

Therefore, further genetic modifications were made to delete undesired metabolic pathways. FELDMANN et al. (1989) isolated a mutant of *K. planticola* deficient in pyruvate–formate lyase. It showed among the fermentation products more than 70% lactate with residual acetate, 2,3-butanediol, and traces of ethanol, formate, and CO_2. After the introduction of a plasmid carrying the *pdc* gene from *Z. mobilis* this mutant became an efficient ethanol producer (Fig. 12). The recombinant strain produced 387 mM ethanol from 275 mM xylose in 80 h, about 83% of the theoretical maximal yield. Furthermore, this mutant consumed more than double the amount of xylose (41 g/L) compared to the wild type, due to reduced production of inhibiting acids during growth. However, the low ethanol tolerance of this organism (2–3%) is a serious problem for practical use.

Recombinants that only contain the pyruvate decarboxylase gene are dependent on endogenous levels of alcohol dehydrogenase activity to couple the reduction of acetaldehyde to the oxidation of NADH. Since ethanol is only one of several abundant fermentation products normally produced by these enteric bacteria, it seemed possible that a deficiency in alcohol dehydrogenase activity and accumulation of NADH could contribute to the formation of various side products. Therefore, recently INGRAM and coworkers (INGRAM and CONWAY, 1988; BEALL et al., 1991) have developed a novel strain for the fermentation of xylose and other sugars to ethanol by expressing the genes for pyruvate decarboxylase and alcohol dehydrogenase from *Z. mobilis* in *E. coli.* The two *Z. mobilis* genes were organized into an artificial operon for ethanol production. In recombinants having both genes, pyruvate was diverted from the native pathways for organic acids and converted to ethanol as the dominant product of fermentation. For example, relatively high concentrations of ethanol (56 g/L) were produced from xylose with excellent efficiencies. Volumetric productivities of up to 1.4 g ethanol/L/h were obtained. Thus this ethanologenic *E. coli* has a number of advantages over previous systems for the conversion of xylose into ethanol. It also has the ability to efficiently ferment – in addition to xylose – all other sugar constituents of lignocellulosic material (glucose, mannose, arabinose, and galactose). By using *Pinus* sp. hemicellulose hydrolysate as a substrate, the ethanol yield reached 91% of the maximum theoretical value in 48 h (DE BARBOSA et al., 1992). The high yield results from an efficient conversion of various sugars to ethanol with minimal side product formation and the small synthesis of bacterial cell mass.

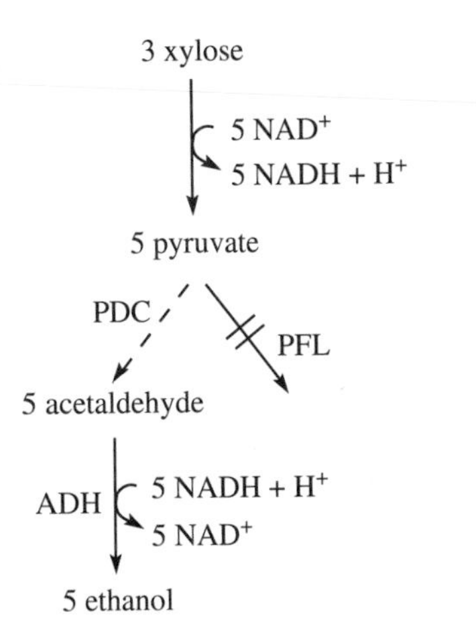

Fig. 12. Scheme for the fermentation of xylose to ethanol by pyruvate-formate lyase-negative mutants of *Klebsiella planticola* carrying the pyruvate decarboxylase gene from *Zymomonas mobilis.* Abbreviations: PFL, pyruvate-formate lyase; PDC, pyruvate decarboxylase; ADH, alcohol dehydrogenase.

Recently, OHTA et al. (1991) have also investigated the expression of the pyruvate decarboxylase (*pdc*) gene and the alcohol dehydrogenase (*adh*) gene from *Z. mobilis* on ethanol production by a related enteric organism, *Klebsiella oxytoca.* Recombinants containing only the *pdc* gene produced more than twice the parental level of ethanol. Recombinants having both *pdc* and *adh* genes produced ethanol very rapidly and with high efficiency. The maximum volumetric productivity was in the range of 2.1 g/L/h for both glucose and xylose. With either sugar, this strain produced approximately 37 g ethanol/L after 30 h. Fermentation of these sugars was essentially completed after 48 h with 45 g ethanol/L. Without the addition of base to control pH, recombinants containg *pdc* and *adh* genes grew to more than twice the density of the parent organism as a result of a reduced rate of acid production. Finally, the plasmid carrying the two genes from *Z. mobilis* was stably maintained in *K. oxytoca* in the absence of antibiotic selection.

4.2 Production of 1,3-Propanediol by a Transformed *Escherichia coli* Strain

1,3-Propanediol is a useful chemical intermediate, e.g., in the synthesis of polyurethanes and polyesters. It is currently derived from acrolein, a petroleum derivative, and is expensive to produce relative to other diols. Recently, TONG et al. (1991) have described 1,3-propanediol production by *E. coli* expressing genes from the *Klebsiella pneumoniae dha* regulon. The *dha* regulon in *K. pneumoniae* enables the organism to grow anaerobically on glycerol and produce 1,3-propanediol. Studies on the pathway have shown that in the first step glycerol is converted into 3-hydroxypropionaldehyde by a coenzyme B_{12}-dependent dehydratase. This compound then is reduced to 1,3-propanediol by an NAD-dependent 1,3-propanediol oxidoreductase and excreted into the culture medium. The genes for these two enzymes are part of the *dha* regulon in *K. pneumoniae.* This *dha* regulon is induced by dihydroxyacetone in the absence of an exogenous electron acceptor, such as oxygen, fumarate, or nitrate (FORAGE and FOSTER, 1982). *E. coli* does not have a *dha* regulon, therefore this organism cannot grow anaerobically on glycerol or dihydroxyacetone without an additional electron acceptor such as nitrate or fumarate. Recently, SPRENGER et al. (1989) have cloned genes of the *dha* regulon in *E. coli*, but they did not detect dehydratase activity, and no 1,3-propanediol was formed.

Now TONG et al. (1991) have used a genomic library of *K. pneumoniae* in *E. coli* and enriched strains with the ability to grow anaerobically on glycerol and dihydroxyacetone. A transformed *E. coli* strain producing 1,3-propanediol had all enzymatic activities associated with the genes of the *dha* regulon. When this *E. coli* strain was grown on a complex medium plus glycerol, the yield of 1,3-propanediol from glycerol was 0.46 mol/mol, other by-products were formate, acetate, and lactate. Further investigations will permit the development of methods to improve the yield and productivity of 1,3-propanediol from glycerol and to extend the substrate range of the pathway to more abundant renewable compounds such as sugars and starch.

4.3 Construction of L-Amino Acid-Producing Bacteria

L-Amino acids have a wide spectrum of commercial use as food additives, feed supplements, infusion compounds, therapeutic agents, and precursors for the synthesis of peptides or agrochemicals. In the mid-fifties, in Japan a bacterium was isolated which excreted large quantities of L-glutamic acid into the culture medium (KINOSHITA and NAKAYAMA, 1978). This bacterium, *Corynebacterium glutamicum*, is a short, aerobic, Gram-positive rod capable of growing on a simple medium (see Chapter 13). Under optimal conditions this organism converts about 100 g/L glucose into 50 g/L glutamic acid within a few days; about 350000 t of L-glutamate are produced annually with this bacterium. In the last 30 years many various mutants of *C. glutamicum* were isolated which are also able to produce significant amounts of other L-amino acids. During the past decade a new generation of strain improvements was developed using the gene cloning technique, as demonstrated by the following few examples.

At present, L-lysine is produced in an amount of about 130000 t/a with strains of *C. glutamicum* or subspecies. The wild types of these organisms do not secrete lysine, but strains producing this amino acid have been obtained by classical screening programs. During these procedures, mutants were selected primarily as resistant or sensitive to a variety of chemicals by assuming as targets enzymes of the amino acid pathway (TOSAKA et al., 1978) or of central metabolism (SHIIO et al., 1984). At each screening step the mutant producing the highest titer of lysine was selected for subsequent mutagenesis. This empirical procedure resulted in strains with high productivity.

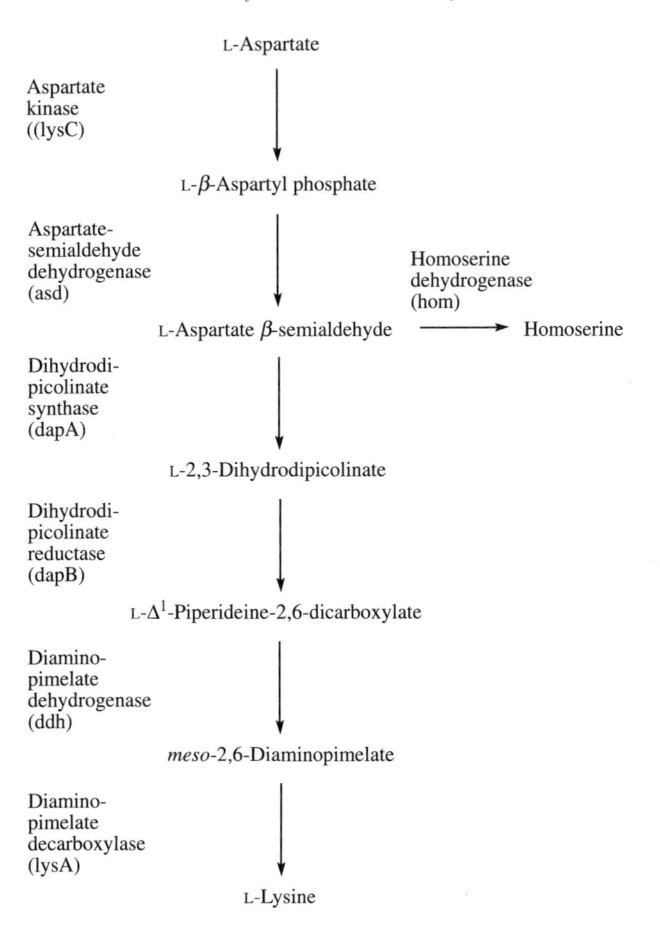

Fig. 13. The lysine biosynthetic sequence via the diaminopimelate dehydrogenase reaction, starting from the central metabolite, L-aspartate.

However, their metabolic changes can be described neither qualitatively nor quantitatively. This also agrees with the fact that it is impossible to reconstruct a good lysine producer by using most of the published strain development schemes. The only exception is the use of the lysine analog S-2-aminoethyl-L-cysteine. Some of the mutants resistant to this analog secrete lysine, which can be attributed to an altered feedback response of the aspartate kinase (SANO and SHIIO, 1970).

Recently, CREMER et al. (1991) have described a rational approach to analysis of lysine production with *C. glutamicum* by overexpression of the various biosynthetic enzymes. Each of the six genes that are involved in the pathway of aspartate to lysine (Fig. 13) was overexpressed individually in the wild type and a mutant with a feedback-resistant aspartate kinase. Analysis of lysine formation revealed that overexpression of the gene for the feedback-resistant aspartate kinase alone suffices to achieve lysine secretion in the wild type. Neither aspartate semialdehyde dehydrogenase, dihydrodipicolinate reductase, diaminopimelate dehydrogenase, nor diaminopimelate decarboxylase overexpression had any effect on lysine overproduction. Most surprisingly, however, overexpression of dihydrodipicolinate synthase also converted the wild type into a lysine secreting strain, although not to the same extent as the deregulated aspartate kinase did. This is a strong indication that dihydrodipicolinate synthase is also involved in the flow control of this pathway. By combined overexpression of aspartate kinase and dihydrodipicolinate synthase lysine secretion could be further increased (10–20%). These results show that of the six enzymes that convert aspartate to lysine, aspartate kinase and dihydrodipicolinate synthase are responsible for metabolic flow control. For the construction of strong lysine-producing strains the activities of these enzymes must be increased.

As shown in Fig. 14 three different pathways of D,L-diaminopimelate and L-lysine synthesis are known in prokaryotes. All bacteria investigated in detail so far appear to utilize only one of the three pathways (WEINBERGER and GILVARY, 1970; WHITE, 1983). However, recently, SCHRUMPF et al. (1991) have detected that in *C. glutamicum* the dehydrogenase variant and the succinylase variant exist side by side, both allowing D,L-aminopimelate and L-lysine synthesis. Mutants with an inactive dehydrogenase pathway are still prototrophic but in lysine

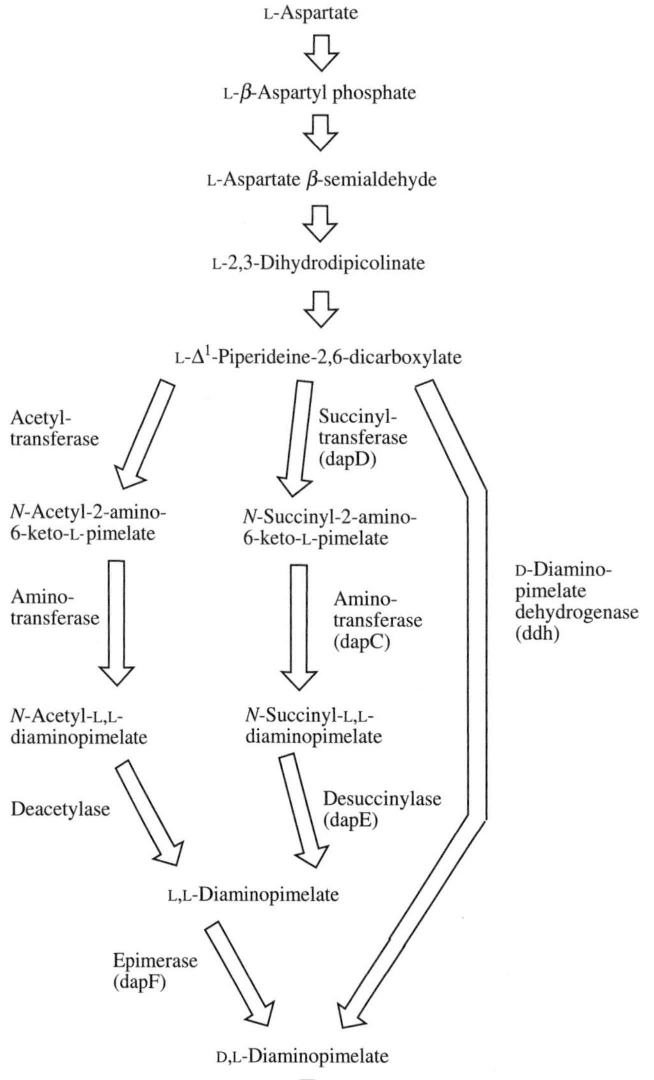

Fig. 14. The three pathways of D,L-diaminopimelate and L-lysine synthesis in prokaryotes.

overproducers lysine secretion is reduced to 50–70%. Thus, although the dehydrogenase pathway is not essential for growth on mineral salt medium, it is a prerequisite for handling an increased flow of metabolites to diaminopimelate and finally to lysine; for further strain improvement this might be very important.

In addition to all steps considered so far to be important for lysine overproduction, the secretion of lysine into the culture medium has also to be noted. Although it is still widely accepted that excretion of amino acids may occur as a result of a physically changed (leaky) membrane (DEMAIN and BIRNBAUM, 1968), recently KRÄMER and coworkers provided evidence for the presence of specific secretion carriers (EBBIGHAUSEN et al., 1989; HOISCHEN and KRÄMER, 1990; BRÖER and KRÄMER, 1991). While in the cells of *C. glutamicum* high concentrations of glutamate (~200 mM) and lysine (~50 mM) could be detected, only lysine was secreted into the culture medium. Thus the secretion of lysine is not the consequence of unspecific permeability of the plasma membrane but is mediated by a secretion carrier which is specific for lysine. The results of all experimental data are summarized in the following model (Fig. 15). In *C. glutamicum* lysine is excreted in symport with two OH^- ions. The substrate-loaded carrier is uncharged. After lysine is transported through the membrane, the carrier changes its orientation back to the cytosol. The pH gradient and the lysine gradient are important for the translocation of the loaded carrier. The membrane potential is important for the reorientation of the positively charged unloaded carrier. All data clearly show that this carrier is a well designed secretion carrier as

- it has a rather high K_m value for lysine (20 mM);
- it is an OH^- symport system in contrast to the uptake systems, that are H^+ antiport systems;
- it is positively charged, thus the membrane potential stimulates secretion.

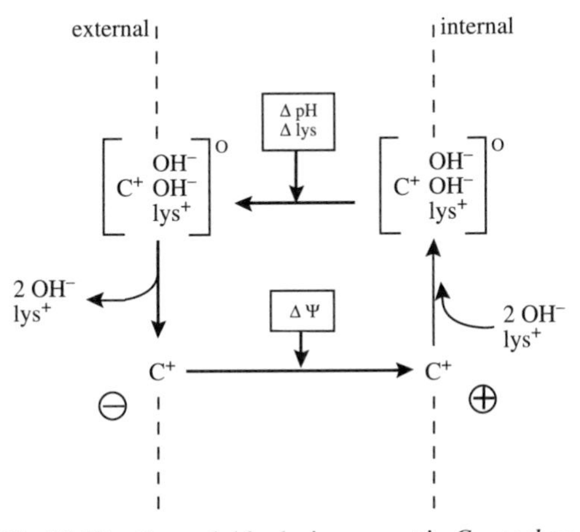

Fig. 15. Kinetic model for lysine export in *Corynebacterium glutamicum* (BRÖER and KRÄMER, 1991).

Recent analysis of lysine hyperproducing *C. glutamicum* strains indicates that this secretion carrier has a strong influence on the overproduction of this amino acid (SCHRUMPF et al., 1992). Thus for the construction of strong lysine-overproducing strains by using the gene cloning technique, the overexpression of the gene(s) for this export system seems also necessary.

Another example of strain improvement by recombinant DNA techniques is the amplification of the threonine biosynthetic genes in *C. glutamicum*. Amplification of the feedback-inhibition insensitive homoserine dehydrogenase and homeserine kinase in a high lysine-overproducing strain permitted channeling of the carbon flow from the intermediate aspartate semialdehyde toward homoserine, resulting in a high accumulation of L-threonine. The final lysine concentration was shifted from 65 g/L to 4 g/L and the final threonine concentration was increased from 0 g/L to 52 g/L (Fig. 16) (KATSUMATA et al., 1987). Overexpression of the third gene for threonine biosynthesis, threonine synthase alone or in combination with the other two genes had no further effect on threonine or lysine overproduction (EIKMANNS et al., 1991). However, threonine production could be increased by 12% by the expression of cloned phosphoenolpyruvate carboxylase (SANO et al., 1987). This improvement can be explained by the fact, that in this strain the formation of the precursor oxaloacetate and thus the carbon flow into the pathway of the biosynthesis of the aspartate family of amino acids is increased.

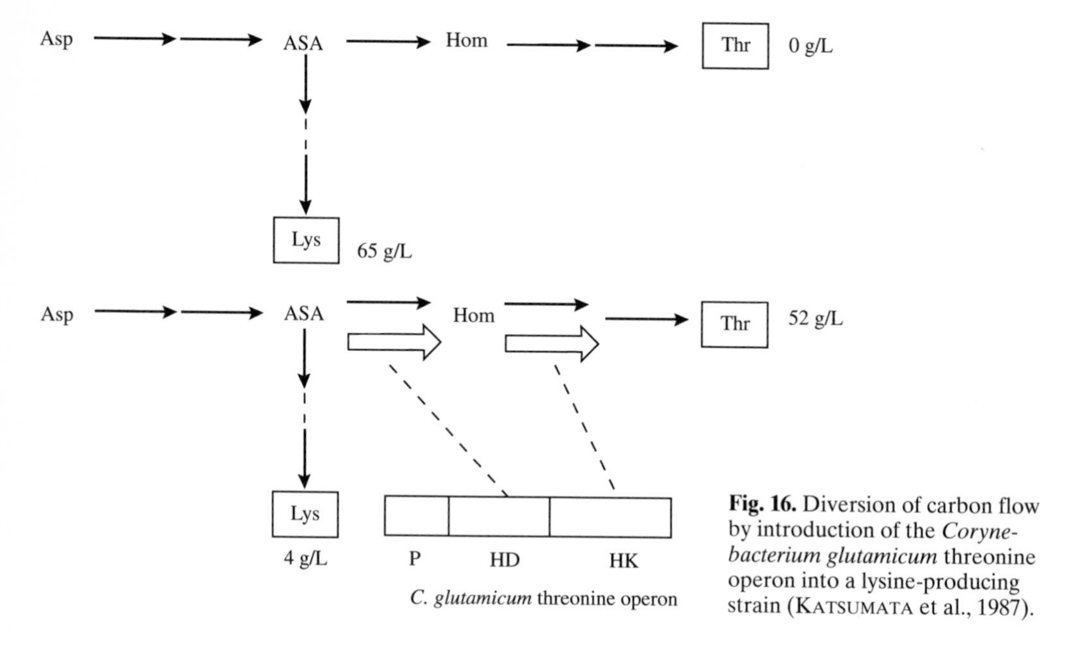

Fig. 16. Diversion of carbon flow by introduction of the *Corynebacterium glutamicum* threonine operon into a lysine-producing strain (KATSUMATA et al., 1987).

Recently, a strain of *C. glutamicum* with the ability to produce 18 g/L of tryptophan has been altered by overexpressing the genes for the deregulated 3-deoxy-D-arabinoheptulosonate-7-phosphate synthase and chorismate mutase to produce a large amount of tyrosine (26 g/L) (IKEDA and KATSUMATA, 1992). A clone overexpressing besides these two genes the gene for prephenate dehydratase produced mainly phenylalanine (28 g/L), because the accelerated carbon flow through the common pathway was redirected to phenylalanine (see Chapter 13).

Furthermore, UHLENBUSCH et al. (1991) were able to construct a *Z. mobilis* strain which excretes L-alanine. The gene *alaD* for L-alanine dehydrogenase from *Bacillus sphaericus* was cloned and introduced into *Z. mobilis*. Under the control of the strong promotor of the pyruvate decarboxylase (*pdc*) gene, the enzyme was expressed up to a specific activity of nearly 1 μmol/min/mg of protein in recombinant cells. As a result of this high L-alanine dehydrogenase activity, growing cells excreted up to 10 mmol of alanine per 280 mmol of glucose utilized. By the addition of 85 mM NH_4^+ to the medium, growth of the recombinant cells stopped, and up to 41 mmol of alanine were secreted. As alanine dehydrogenase competed

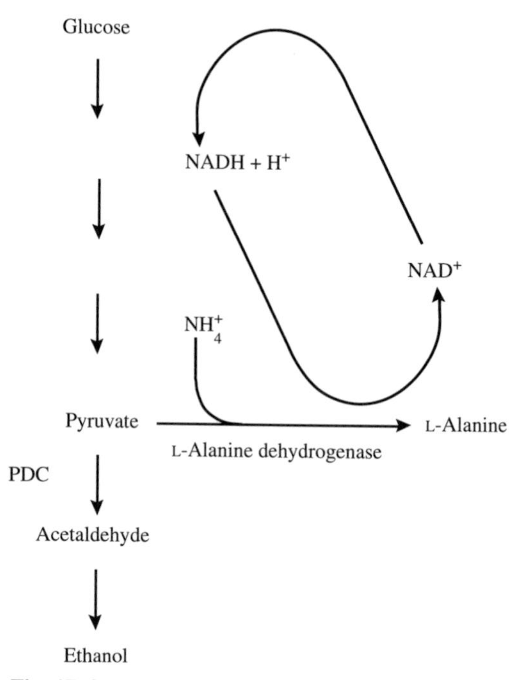

Fig. 17. Scheme of redirection of the carbon flux to alanine formation in *Zymomonas mobilis.*

with pyruvate decarboxylase (PDC) for the same substrate (pyruvate) (Fig. 17), PDC activity was reduced by starvation for the essential PDC cofactor thiamine-PP. A thiamine auxotrophy mutant of *Z. mobilis* which carried the *alaD* gene was starved for 40 h in glucose-supplemented mineral salt medium and then shifted to a medium with 85 mM NH_4^+ and 280 mmol of glucose. Under these conditions the recombinants excreted up to 84 mmol of alanine (7.5 g/L) over 25 h. Alanine excretion proceeded at an initial velocity of 238 nmol/min/mg (dry weight). Despite this high activity, the excretion rate seemed to be a limiting factor, as the intracellular concentration of alanine was as high as 260 mM at the beginning of the excretion phase and decreased to 80–90 mM over 24 h. A further increase in alanine formation seems possible if the pyruvate decarboxylase activity can be diminished more drastically.

4.4 Synthesis of Precursors for Vitamins

L-Ascorbate (vitamin C) can be readily produced from 2-keto-L-gulonate by chemical procedures which are part of the Reichstein synthesis. Several years ago, SONOYAMA et al. (1982) described a process for producing 2-keto-L-gulonate by two successive fermentations. In the first step glucose is oxidized to 2,5-diketo-D-gluconate by an *Erwinia* strain; the intermediate products are D-gluconate and 2-keto-D-gluconate. In the second fermentation a species of *Corynebacterium* converts 2,5-diketo-D-gluconate to 2-keto-L-gulonate. This stereospecific reduction at the C-5 position is catalyzed by on NADPH-requiring 2,5-diketo-D-gluconate reductase (Fig. 18). In order to develop a one-step microbial bioconversion of D-glucose into 2-keto-L-gluconate, the pathway of *Erwinia* was enhanced by gene cloning techniques. The gene specifying the 2,5-diketo-D-gluconate reductase was cloned from *Corynebacterium* and expressed in *Erwinia* (ANDERSON et al., 1985). After optimizing the culture conditions these recombinant strains of *Erwinia* produced about 120 g/L of 2-keto-L-gulonate within 120 h. The molar yield from glucose was in the range of 60%. As this process is fundamentally much simpler than either the current multistep manufacturing process or the two-stage fermentation method, the conversion of glucose to 2-keto-L-gulonate, by a recombinant strain of *Erwinia* may lead to an economical process for vitamin C production (GRINDLEY et al., 1988).

Another example that native metabolites can be converted to preferred end products by the genetic installation of some specific enzymes, is the production of β-carotene – a precursor of vitamin A. Recently, MISAWA et al. (1990) have succeeded in cloning the genes for the biosynthesis of cyclic carotenoids containing β-carotene from *Erwinia uredovora*. As the

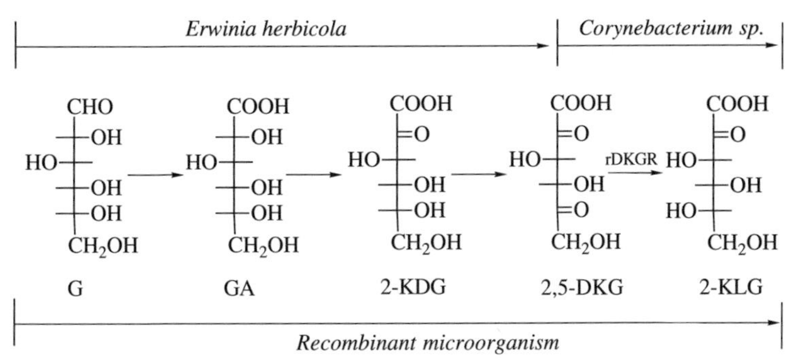

Fig. 18. Carbohydrate metabolites involved in the bioconversion of D-glucose to 2-keto-L-gulonic acid. The top route depicts the two-step tandem fermentation process; the bottom route shows the one-step recombinant process (ANDERSON et al., 1985).
Abbreviations: G, glucose; GA, gluconic acid; 2-KDG, 2-keto-D-gluconic acid; 2,5-DKG, 2,5-diketo-D-gluconic acid; 2-KLG, 2-keto-L-gluconic acid; rDKGR, diketo-D-gluconate reductase.

precursor geranyl-geranyl PP exists in many organisms for the synthesis of sterols, hopanoids, and terpenes, these carotenoid biosynthesis genes should be capable of producing carotenoids in many cells which are not able to form carotenoids. After four genes of the β-carotene biosynthesis were transferred into *Z. mobilis* and *Agrobacterium tumefaciens*, yellow colonies were obtained on agar plates. The transconjugants produced 220–350 μg of β-carotene per gram of dry weight in the stationary phase in liquid culture (MISAWA et al., 1991). In the wild types of *Z. mobilis* and *A. tumefaciens* no carotenoids are synthesized. The *Z. mobilis* cells accumulating β-carotene may be able to be utilized as nutrients for farm animals after carrying out ethanol production.

4.5 Improvement of the Production of Biopolymers

Polyhydroxyalkanoates are a class of carbon and energy storage polymers produced by numerous bacteria in response to several types of environmental limitation (oxygen or nitrogen deprivation, sulfate or magnesium limitation). Under such limiting conditions these polymers may constitute as much as 80–90% of the dry cell weight. Polyhydroxyalkanoates are renewable resources of biodegradable thermoplastic materials that have already small-scale applications. For many bacteria poly-β-hydroxybutyrate (PHB), a homopolymer of D(–)3-hydroxybutyrate is the principal storage compound. The PHB biosynthetic pathway has been studied extensively in the Gram-negative, facultative chemolithoautotrophic bacterium *Alcaligenes eutrophus* (STEINBÜCHEL and SCHLEGEL, 1991). The pathway for the synthesis of PHB consists of three steps (Fig. 19). The first step is catalyzed by the enzyme β-ketothiolase which condenses two acetyl-CoA molecules to acetoacetyl-CoA. This intermediate is reduced to D-β-hydroxybutyryl-CoA by an NADPH-dependent acetoacetyl-CoA reductase. In the last step the enzyme PHB synthase catalyzes the head-to-tail polymerization of the monomer to PHB. Molecular studies revealed that the genes for these three enzymes are organized in a single operon.

Recently, the *A. eutrophus* PHB-biosynthetic genes have been transformed and expressed in *E. coli* (SLATER et al., 1988; SCHUBERT et al., 1988) and in various species of *Pseudomonas* (TIMM and STEINBÜCHEL, 1990). As all recombinant strains accumulated PHB to a significant portion of the cellular dry matter (up to 50%) and deposited the polyester in granules when the cells were cultivated in the presence of an excess carbon source under nitrogen limitation, the *A. eutrophus* pathway is functionally active in these bacteria. It is very interesting that recombinant strains of *Pseudomonas aeruginosa* possess three different pathways for the synthesis of polyhydroxyalkanoates. With gluconate as carbon source, these cells accumulated a polymer consisting of β-hydroxybutyrate, β-hydroxydecanoate, and β-hydroxydodecanoate as the main constituents and of β-hydroxyoctanoate and β-hydroxyhexanoate as minor constituents (TIMM and STEINBÜCHEL, 1990). *A. eutrophus* is also able to synthesize a large variety of polyesters of hydroxyalkanoic

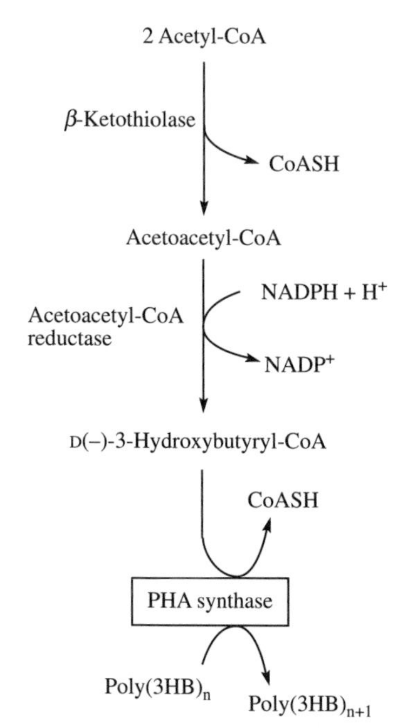

Fig. 19. Biosynthetic pathway of poly-β-hydroxybutyrate (poly-HB) in *Alcaligenes eutrophus*.

acids. The composition of these copolyesters depends on the carbon source and precursors provided to the cells. The copolymer poly-(3-hydroxybutyrate-co-3-hydroxyvalerate) is of particular interest because it is more flexible than the PHB homopolymer. By altering the intermediary metabolism SLATER et al. (1992) have constructed an *E. coli* strain that produces this copolymer quite well. The strategy centered on genetically eliminating the regulation of *E. coli* genes required for propionate metabolism (constitutive expression), and it resulted in the very efficient uptake and incorporation (greater than 95%) into the copolymer poly-(3-hydroxybutyrate-co-3-hydroxyvalerate). 3-Hydroxybutyrate-3-hydroxyvalerate ratios in the copolymer could be manipulated by altering the propionate concentration and/or the glucose concentration in the culture medium. This example demonstrates that metabolic design of the synthesis of this biopolymer can provide greater control over the composition and the quantity of the polymer produced.

Xanthan is an extracellular polysaccharide produced by the Gram-negative bacterium *Xanthomonas campestris*. It has unique rheological properties (high viscosity, pseudoplastic behavior) and is therefore used in a variety of food and industrial applications. The chemical structure of xanthan gum has been extensively studied and shown to consist of a cellulosic (1–4)-β-glucose backbone with trisaccharide side chains composed of two mannose residues and one glucuronic acid residue attached to alternate glucose residues in the backbone (Fig. 20) (JANSSON et al., 1975). The mannose sugars are acetylated and pyruvylated at specific sites, but to various degrees. Recent studies have demonstrated the utility of recombinant DNA technology for cloning of xanthan biosynthetic genes and the potential use of the cloned genes to improve the production and alter the structure (pyruvate content) of xanthan gum (HARDING et al., 1987; MARZOCCA et al., 1991). Many of the genes involved in exopolysaccharide synthesis are of-

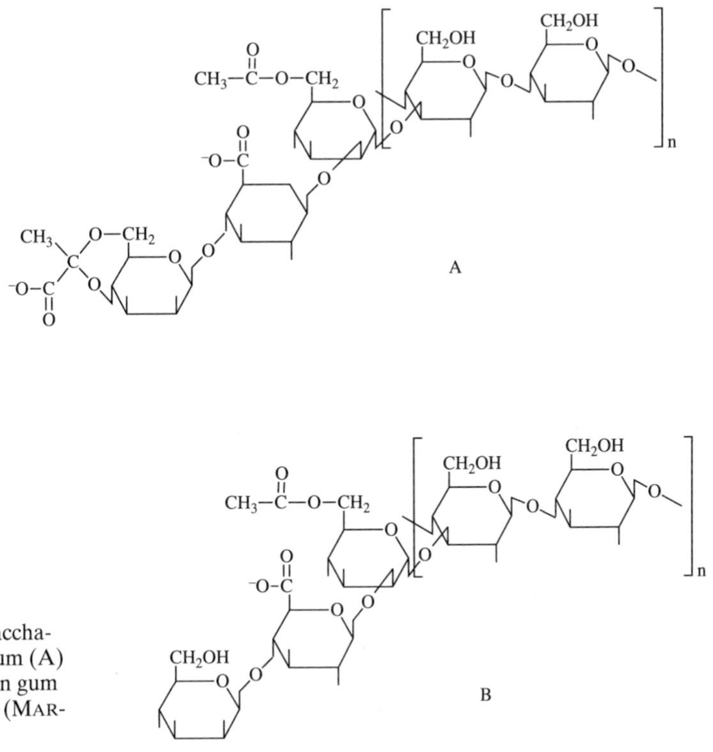

A

B

Fig. 20. Structures of the pentasaccharide-repeating unit of xanthan gum (A) and of the depyruvylated xanthan gum produced by a mutant strain (B) (MARZOCCA et al., 1991).

ten clustered. A cluster of genes essential for xanthan synthesis has also been isolated in *X. campestris* (BARRERE et al., 1986). A plasmid containing several xanthan biosynthetic genes increased the production of xanthan by 10% and the extent of pyruvylation of the xanthan side chains by about 45% (HARDING et al., 1987). By cloning and overexpressing the gene for the enzyme ketal pyruvate transferase, the pyruvate content of xanthan could be modified. Furthermore, using transposon mutagenesis a strain could be constructed which formed xanthan with a severely reduced pyruvate content (Fig. 20) (MARZOCCA et al., 1991). These first results indicate that a successful genetic manipulation of xanthan synthesis and structure will be possible in the near future. These materials should provide insight into the relationship between xanthan structure and rheological behavior.

4.6 Redesign of Secondary Metabolic Pathways

In the last few years it could be demonstrated that for secondary metabolites such as antibiotics the yields can also be increased by overcoming rate-limiting steps in the biosynthetic pathways using gene amplification techniques. For example, the production of cephalosporin C by *Cephalosporium acremonium* could be increased to 15% by overexpressing the *cef EF* gene (SKATRUD et al., 1989). This gene codes for a bifunctional protein that exhibits two sequentially acting cephalosporin biosynthetic enzyme activities: deacetoxycephalosporin C synthetase and deacetylcephalosporin C synthetase. The parent strain of *C. acremonium* used in this study excreted a substantial quantity of penicillin N, a precursor of cephalosporin C, into the culture medium. The recombinant strain with a two-fold increase in the specific activity of deacetoxycephalosporin C synthetase was able to convert penicillin N

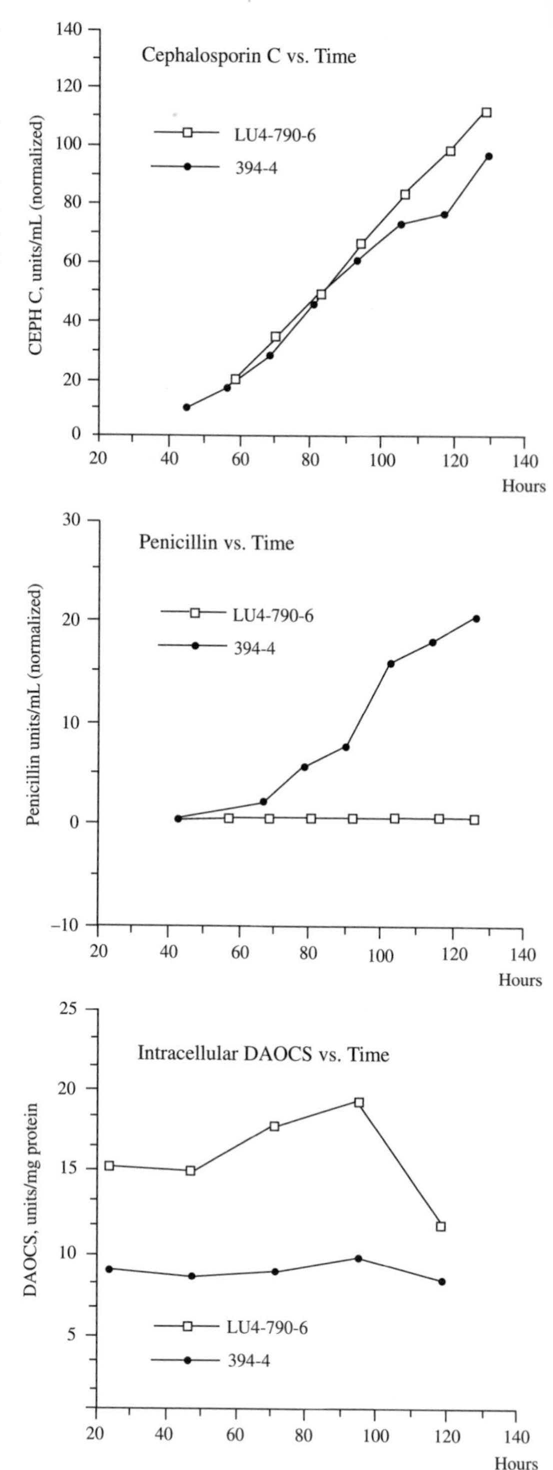

Fig. 21. Cephalosporin C production and specific activity of deacetoxycephalosporin (DAOCS) C synthetase in pilot plant fermenters (SKATRUID et al., 1989).

completely into the final product cephalosporin C (Fig. 21).

Furthermore, there will be great industrial benefits from an understanding of how genes for antibiotic production are regulated. Recent studies indicate that the genes for the biosynthesis of antibiotics are generally clustered together with genes for regulation. CHATER and BRUTON (1985) could increase the production of methylenomycin by disrupting the regulatory region, as in this case a negative control system is involved. However, normally positive control genes have been found in the biosynthetic gene clusters for various secondary metabolites. Therefore, overexpression of these regulatory genes caused, e.g., overproduction of streptomycin, undecylprodigiosin, and actinorhodin, respectively, by the wild-type strains (CHATER, 1990).

Recently, SMITH et al. (1990) have used a cosmid clone containing the penicillin biosynthetic gene cluster from *Penicillium chrysogenum* to transform the filamentous fungi *Neurospora crassa* and *Aspergillus niger*, which do not produce β-lactam antibiotics. Both of the transformed hosts produced penicillin. These data indicate that also in *P. chrysogenum* genes encoding all enzymes necessary for biosynthesis of penicillin are forming a gene cluster which can very well be expressed in heterologous fungal hosts.

As also described in Chapter 13, recombinant DNA techniques can be used for the production of hybrid or even novel antibiotics. Genes for biosynthetic steps in different organisms can be combined in the same organism, thus leading to the production of novel metabolites. HOPWOOD et al. (1985) described the first experiment testing this idea. Part of the cloned pathway for actinorhodin from *Streptomyces coelicolor* was transformed into a *Streptomyces* strain which produces the compound medermycin (Fig. 22). The recombinant produced an additional antibiotic, identified as mederrhodin. The recombinant plasmid used contained the gene coding for the enzyme which catalyzes the β-hydroxylation of actinorhodin. Thus in the recombinant the broad substrate specificity of the enzyme allowed to also hydroxylate medermycin at the analogous position to produce mederrhodin. MCALPINE et al. (1987) have used a similar strategy. They transformed a mutant of *Saccharopolyspora erythraea*, which was blocked in an early step of erythromycin biosynthesis, with a DNA library from the oleandomycin producer *Streptomyces antibioticus*. One recombinant formed an antibiotic active compound which was identified as 2-norerythromycin. In this case a novel structure lacking a functional group ($-CH_3$) was generated.

In each of these examples described novel structures altered by a substitution at one carbon atom were made. A greater challenge for getting novel antibiotics is to alter the backbone structure of a metabolite. *Streptomyces galilaeus* normally produces aclacinomycin A and B. After transformation with the genes for polyketide synthase involved in the synthesis of actinorhodin, clones were obtained which produced an anthraquinone (BARTEL et al., 1990). This is a very interesting result in the use of this approach to produce antibiotics with novel structures. In the near future we may be able to change structures of antibiotics in a rational way. To reach this goal it is also important to change the specificity of the biosynthet-

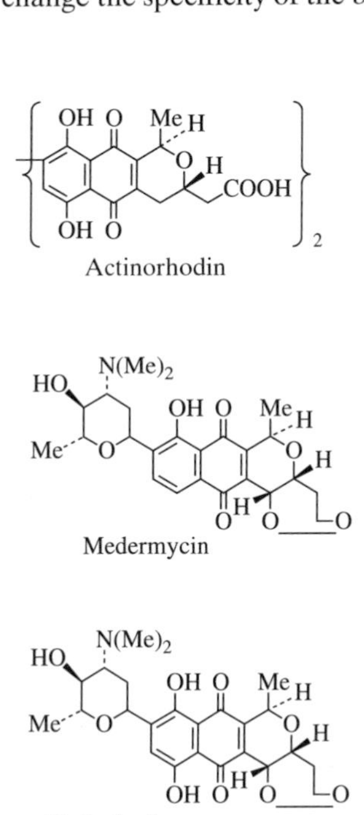

Actinorhodin

Medermycin

Mederrhodin

Fig. 22. Structures of actinorhodin, medermycin, and mederrhodin.

ic enzymes by site-directed mutagenesis and/or genetic engineering. First experiments in this direction were started with the key enzyme in the synthesis of β-lactam antibiotics, the isopenicillin N synthase (BALDWIN and ABRAHAM, 1988).

5 Concluding Remarks and Future Aspects

In the past several years, important strides have been made towards the development of effective tools for the genetic manipulation of important industrial organisms. The possibility to modify the metabolism of cells in a purposeful manner has very important implications for biotechnological applications. In this chapter, examples of the manipulation of catabolic and anabolic pathways were presented. At present, in most experiments a single gene or gene cluster is changed. In some cases an iterative cycle is used; after the genetic modification and analysis of the metabolic change the next genetic modification is made. In order to enhance the yield and productivity of various compounds, the main activities have so far been carried out on enzyme amplification and deregulation of the biosynthetic pathways. However, overproduction of many metabolites also requires a redirection of flux distribution in the metabolism.

Flux control analysis is discussed in detail by NEIJSSEL et al. in Chapter 4. Data on the control strength determined by the sensitivity coefficients, respectively, the elasticity coefficients of individual components in response to metabolite and effector concentrations are useful for rational metabolic design. Recently, NIEDERBERGER et al. (1992) have described a strategy for increasing an *in vivo* flux by genetic manipulation of the tryptophan pathway in yeast. When the five enzymes of the tryptophan biosynthesis were simultaneously increased by a multi-copy vector carrying all five genes, a substantial elevation of the flux to the amino acid tryptophan was obtained. If complex pathways are to be understood, one must not only work out these parameters and the metabolic concentrations of various metabolites and enzymes in the pathway, but one must solve the network interactions to interpret feedback and feedforward effects. As the metabolic activities of cells are accomplished by a network of more than 1000 enzymatic reactions and selective membrane transport systems, it is obvious that computer models and simulations are very useful for solving these problems. A thorough understanding of the elements and mechanisms controlling the biosynthesis of a metabolite should make it possible to influence its rate of overproduction in a predictable way.

Concrete alterations in the metabolism of an organism may also be possible by exchanging functional domains between various enzyme systems. Recently, biochemical and genetic analyses have indicated that during evolution several genes were split into the equivalent of functional enzyme domains and these gene fragments were constantly shuffled and reshuffled while passing through new organisms (LENGELER, 1990). Thus using the DNA technique, for example, one could generate a carbamoyl phosphate synthase from a carbamate kinase, an ATPase domain and a glutamine transaminase (NYUNOYA et al., 1985). In a similar manner the substrate specificity of an amino acid transcarbamoylase could be changed by joining a common carbamoyl phosphate binding domain with distinct amino acid binding domains (HOUGHTON et al., 1989). The formation of chimeric enzymes by exchanging defined genetic cassettes responsible for specific protein domains is an excellent opportunity to modify the specificity or the regulation of an enzyme in a purposeful manner. Thus genetic engineering of protein domains may also be a very important method for rational metabolic design in the near future. A clear understanding of the biological systems will enable biotechnology to make the anticipated important contribution to mankind.

6 References

AMORE, R., WILHELM, M., HOLLENBERG, C. P. (1989), The fermentation of xylose – an analysis of the expression of *Bacillus* and *Actinoplanes* xylose isomerase genes in yeasts, *Appl Microbiol. Biotechnol.* **30,** 351–357.

ANDERSON, S., MARKS, C. B., LAZARUS, R., MILLER, J., STAFFORD, K., SEYMOUR, J., LIGHT, D., RASTETTER, W., ESTELL, D. (1985), Production of 2-keto-L-gulonate, an intermediate in L-ascorbate synthesis by a genetically modified *Erwinia herbicola*, *Science* **230,** 144–149.

BAILEY, J. E. (1991), Toward a science of metabolic engineering, *Science* **252,** 1668.

BALDWIN, J. E., ABRAHAM, E. (1988), The biosynthesis of penicillins and cephalosporins, *Nat. Prod. Rep.* **5,** 129–145.

BARRERE, G. C., BARBER, C. E., DANIELS, M. J. (1986), Molecular cloning of genes involved in the production of the extracellular polysaccharide xanthan by *Xanthomonas campestris* pv. *campestris*, *Int. J. Biol. Macromol.* **8,** 372–374.

BARTEL, P. L., ZHU, C-B., LAMPEL, J. S., DOSCH, D. C., CONNORS, N., STROHL, W. R., BEALE, J. M., FLOSS, H. G. (1990), Biosynthesis of anthraquinones by interspecies cloning of actinorhodin biosynthesis genes in streptomycetes: Clarification of actinorhodin gene functions, *J. Bacteriol.* **172,** 4816–4826.

BEALL, D. S., OHTA, K., INGRAM, L. O. (1991), Parametric studies of ethanol production from xylose and other sugars by recombinant *Escherichia coli*, *Biotechnol. Bioeng.* **38,** 296–303.

BRABETZ, W., LIEBL, W., SCHLEIFER, K.-H. (1991), Studies on the utilization of lactose by *Corynebacterium glutamicum* bearing the lactose operon of *Escherichia coli*, *Arch. Microbiol.* **155,** 607–612.

BRINGER-MEYER, S., SCHIMZ, K. L., SAHM, H. (1986), Pyruvate decarboxylase from *Zymomonas mobilis*. Isolation and partial characterization, *Arch. Microbiol.* **146,** 105–110.

BRÖER, S., KRÄMER, R. (1991), Lysine excretion by *Corynebacterium glutamicum:* 2. Energetics and mechanism of the transport system, *Eur. J. Biochem.* **202,** 137–143.

BRUINENBERG, P. M., DE BOT, P. H. M., VAN DYKEN, J. P., SCHEFFERS, W. A. (1983), The role of redox balances in the anaerobic fermentation of xylose by yeast. *Eur. J. Appl. Microbiol. Biotechnol.* **18,** 287–292.

BRUINENBERG, P. M., DE BOT, P. H. M., VAN DYKEN, J. P., SCHEFFERS, W. A. (1984), NADH-linked aldose reductase: the key to anaerobic alcoholic fermentation of xylose by yeasts. *Eur. J. Appl. Microbiol. Biotechnol.* **19,** 256–269.

BUMPUS, J. A., BROCK, B. J. (1988), Biodegradation of crystal violet by the white rot fungus *Phanerochaete chrysosporium*, *Appl. Environ. Microbiol.* **54,** 1143–1150.

BUMPUS, J. A., TIEN, M., WRIGHT, D., AUST, S. D. (1985), Oxidation of persistent environmental pollutants by a white rot fungus, *Science* **228,** 1434–1436.

CAREY, V. C., WALIA, S. K., INGRAM, L. O. (1983), Expression of a lactose transposon (Tn951) in *Zymomonas mobilis*. *Appl. Environ. Microbiol.* **46,** 1163–1168.

CHATER, K. F. (1990), The improving prospects for yield increase by genetic engineering in antibiotic-producing streptomycetes, *Bio/Technology* **8,** 115–121.

CHATER, K. F., BRUTON, C. J. (1985), Resistance, regulatory and production genes for the antibiotic methylenomycin are clustered, *EMBO J.* **4,** 1893–1897.

CHATTERJEE, D. K., KILBANE, J. J., CHAKRABARTY, A. M. (1982), Biodegradation of 2,4,5-trichlorophenoxyacetic acid in soil by a pure culture of *Pseudomonas cepacia*, *Appl. Environ. Microbiol.* **44,** 514–516.

CREMER, J., EGGELING, L., SAHM, J. (1991), Control of the lysine biosynthesis sequence in *Corynebacterium glutamicum* as analyzed by overexpression of the individual corresponding genes, *Appl. Environ. Microbiol.* **57,** 1746–1752.

DE BARBOSA, M. F. S., BECK, M. J., FEIN, J. E., POTTS, D., INGRAM, L. O. (1992), Efficient fermentation of *Pinus* sp. acid hydrolysates by an ethanologenic strain of *Escherichia coli*, *Appl. Environ. Microbiol.* **58,** 1382–1384.

DE GRAAF, A. A., SCHOBERTH, S. M., PROBST, U., WITTIG, R. M., STROHHÄCKER, J., SAHM, H. (1991), Charakterisierung des Zuckerstoffwechsels im Bakterium *Zymomonas mobilis* mit Hilfe der *in-vivo*-^{31}P-NMR-Spektroskopie. *Chem. Ing. Tech.* **63,** 628–631.

DE GRAAF, A. A., WITTIG, R. M., PROBST, U., STROHHÄCKER, J., SCHOBERTH, S. M., SAHM, H. (1992), Continuous flow NMR bioreactor for *in vivo* studies of microbial cell suspensions with low biomass concentrations, *J. Magn. Reson.* **98,** 654–659.

DEBUS, D., METHNER, H., SCHULZE, D., DELLWEG, H. (1983), Fermentation of xylose with the yeast *Pachysolen tannophilus*. *Eur. J. Appl. Microbiol. Biotechnol.* **17.** 287–291.

DEMAIN, A. L., BIRNBAUM, J. (1968), Alteration of permeability for the release of metabolites from the microbial cell, *Curr. Top. Microbiol. Immunol.* **46,** 1–25.

DEN HOLLANDER, J. A., UGURBIL, K., BROWN, R. T., SHULMAN, R. G. (1981), Phosphorus-31 nuclear magnetic resonance studies of the effect of oxygen upon glycolysis in yeast, *Biochemistry* **20,** 5871–5880.

DiMarco, A. A., Romano, A. H. (1985), D-Glucose transport system of *Zymomonas mobilis. Appl. Environ. Microbiol.* **49,** 151–157.

Dohmen, R. J., Strasser, A. W. M., Dahlems, U. M., Hollenberg, C. P. (1990), Cloning of the *Schwanniomyces occidentalis* glucoamylase gene (*GAM1*) and its expression in *Saccharomyces cerevisiae*, *Gene* **95,** 111–121.

Ebbighausen, H., Weil, B., Krämer, R. (1989), Transport of branched-chain amino acids in *Corynebacterium glutamicum*, *Arch. Microbiol. Biotechnol.* **151,** 238–244.

Eikmanns, B. J., Metzger, M., Reinscheid, D., Kircher, M., Sahm, H. (1991), Amplification of three threonine biosynthesis genes in *Corynebacterium glutamicum* and its influence on carbon flux in different strains, *Appl. Microbiol. Biotechnol.* **34,** 617–622.

Feldmann, S., Sprenger, G. A., Sahm, H. (1989), Ethanol production from xylose with a pyruvate-formate-lyase mutant of *Klebsiella planticola* carrying a pyruvate decarboxylase gene from *Zymomonas mobilis*, *Appl. Microbiol. Biotechnol.* **31,** 152–157.

Feldmann, S. D., Sahm, H., Sprenger, G. A. (1992), Pentose metabolism in *Zymomonas mobilis* wild type and recombinant strains, *Appl. Microbiol. Biotechnol.*, in press.

Fernandez, E. J., Clark, D. S. (1987), NMR spectroscopy: a non-invasive tool for studying intracellular processes, *Enzyme Microb. Technol.* **9,** 259.

Finn, R. K. (1987), Conversion of starch to liquid sugar and ethanol, in: *Biotec*, Vol. 1, pp. 101–107, Stuttgart: Gustav-Fischer-Verlag.

Forage, R. G., Foster, M. A. (1982), Glycerol fermentation in *Klebsiella pneumoniae:* functions of the coenzyme B_{12}-dependent glycerol and diol dehydratases, *J. Bacteriol.* **149,** 413–419.

Galazzo, J. L., Bailey, J. E. (1990), Fermentation pathway kinetics and metabolic flux control in suspended and immobilized *Saccharomyces cerevisiae*, *Enzyme Microb. Technol.* **12,** 162.

Galazzo, J. L., Shanks, J. V., Bailey, J. E. (1990), Comparison of intracellular sugar-phosphate levels from ^{31}P NMR spectroscopy of intact cells and cell-free extracts, *Biotechnol. Bioeng.* **35,** 1164–1168.

Goodman, A. E., Strzelecki, A. T., Rogers, P. L. (1984), Formation of ethanol from lactose by *Zymomonas mobilis*, *J. Biotechnol.* **1,** 219–228.

Gottschalk, G. (1986), *Bacterial Metabolism*, 2nd Ed. New York–Berlin–Heidelberg–Tokyo: Springer-Verlag.

Grindley, J. F., Payton, M. A., van de Pol, H., Hardy, K. G. (1988), Conversion of glucose to 2-keto-L-gulonate, an intermediate in L-ascorbate synthesis, by a recombinant strain of *Erwinia citreus*, *Appl. Environ. Microbiol.* **54,** 1770–1775.

Harayama, S., Timmis, K. N. (1989), Catabolism of aromatic hydrocarbons by *Pseudomonas*, in: *Genetics of Bacterial Diversity* (Hopwood, D. A., Chater, K. F., Eds.), pp. 151–174, London: Academic Press.

Harding, N. E., Cleary, J. M., Cabanas, D. K., Rosen, I. G., Kang, K. S. (1987), Genetic and physical analyses of a cluster of genes essential for xanthan gum biosynthesis in *Xanthomonas campestris*, *J. Bacteriol.* **169,** 2854–2861.

Hoischen, C., Krämer, R. (1990), Membrane alteration is necessary but not sufficient for effective glutamate secretion in *Corynebacterium glutamicum*, *J. Bacteriol.* **172,** 3409–3416.

Hollenberg, C. P., Strasser, A.W. M. (1990), Improvement of baker's and brewer's yeast by gene technology, *Food Biotechnol.* **4,** 527–534.

Hollenberg, C. P., Wilhelm, M. (1987), New substrates for old organisms, in: *Biotec*, Vol. 1, pp. 21–31, Stuttgart: Gustav-Fischer-Verlag.

Hopwood, D. A., Malpartida, F., Kieser, H. M., Ikeda, H., Duncan, J., Fujii, I., Rudd, B. A. M., Floss, H. G., Omura, S. (1985), Production of hybrid antibiotics by genetic engineering, *Nature* **314,** 642–646.

Houghton, J. E., O'Donovan, G. A., Wild, J. R. (1989), Reconstruction of an enzyme by domain substitution effectively switches substrate specificity, *Nature* **338,** 172–174.

Ikeda, M., Katsumata, R. (1992), Metabolic engineering to produce tyrosine or phenylalanine in a tryptophan-producing *Corynebacterium glutamicum* strain, *Appl. Environ. Microbiol.* **58,** 781–785.

Imanaka, T. (1986), Application of recombinant DNA technology to the production of useful biomaterials, *Adv. Biochem. Eng. Biotechnol.* **33,** 1–23.

Ingram, L. O., Conway, T. (1988), Expression of different levels of ethanologenic enzymes from *Zymomonas mobilis* in recombinant strains of *Escherichia coli*, *Appl. Environ. Microbiol.* **54,** 397–404.

Innis, M. A., Holland, M. J., McCabe, P. C., Cole, G. E., Wittman, V. P., Tal, R., Watt, K. W. K., Gelfand, D. H., Holland, J. P., Meade, J. H. (1985), Expression, glycosylation and secretion of an *Aspergillus* glucoamylase by *Saccharomyces cerevisiae*, *Science* **228,** 21–26.

Ishino, S., Yamaguchi, K., Shirahata, K., Araki, K. (1984), Involvement of *meso*-2,6-diaminopimelate-D-dehydrogenase in lysine biosynthesis in *Corynebacterium glutamicum*, *Agric. Biol. Chem.* **48,** 2557–2560.

JANSSON, P. E., KEENE, L., LINDBERG, B. (1975), Structure of the extracellular polysaccharide from *Xanthomonas campestris*, *Carbohydr. Res.* **45,** 275–282.

JEFFRIES, T. W. (1983), Utilization of xylose by bacteria, yeast and fungi, *Adv. Biochem. Eng. Biotechnol.* **27,** 1–32.

KATSUMATA, R., MIZUKAMI, T., OZAKI, A., KIKUCHI, Y., KINO, K., OKA, T., FURUYA, A. (1987), Gene cloning in glutamic acid bacteria: the system and its applications, in: *Proc. 4th Eur. Congr. Biotechnology*, Vol. 4, pp. 767–776 (NEIJSSEL, O. M., VAN DER MEER, R. R., LUYBEN, K. C. A. M., Eds.), Amsterdam: Elsevier.

KENNEDY, J. F., CABALDA, V. M., WHITE, C. A. (1988), Enzymic starch utilization and genetic engineering, *TIBTECH* **6,** 184–189.

KILBANE, J. J., CHATTERJEE, D. K., KARNS, J. S., KELLOGG, S. T., CHAKRABARTY, A. M. (1982), Biodegradation of 2,4,5–trichlorophenoxyacetic acid by a pure culture of *Pseudomonas cepacia*, *Appl. Environ. Microbiol.* **44,** 72–78.

KILBANE, J. J., CHATTERJEE, D. K., CHAKRABARTY, A. M. (1983), Detoxification of 2,4,5-trichlorophenoxyacetic acid from contaminated soil by *Pseudomonas cepacia*, *Appl. Environ. Microbiol.* **45,** 1697–1700.

KINOSHITA, S., NAKAYAMA, K. (1978), Amino acids, in: *Primary Products of Metabolism* (ROSE, A.H., Ed.), pp. 209–261, London: Academic Press.

KOCH, A. K., REISER, J., KÄPPELI, O., FIECHTER, A. (1988), Genetic construction of lactose-utilizing strains of *Pseudomonas aeruginosa* and their application in biosurfactant production, *Bio/Technology* **6,** 1335–1339.

KÖTTER, P., CIRIACY, M. (1992), Xylose fermentation by *Saccharomyces cerevisiae*, *Appl. Microbiol. Biotechnol.*, in press.

KÖTTER, A., AMORE, R., HOLLENBERG, C. P., CIRIACY, M. (1990), Isolation and characterization of the *Pichia stipitis* xylitol dehydrogenase gene, *XYL2*, and construction of a xylose-utilizing *Saccharomyces cerevisiae* transformant, *Curr. Genet.* **18,** 493–500.

LAWLIS, V. B., DENNIS, M. S., CHEN, E. Y., SMITH, D. H., HENNER, D. J. (1984), Cloning and sequencing of the xylose isomerase and the xylulose kinase genes of *Escherichia coli*, *Appl. Environ. Microbiol.* **47,** 15–21.

LENGELER, J. W. (1990), Molecular analysis of the enzymeII–complexes of the bacterial phosphotransferase system (PTS) as carbohydrate transport systems, *Biochim. Biophys. Acta* **1018,** 155–159.

LIU, C.-Q., GOODMAN, A. E., DUNN, N. W. (1988), Expression of the cloned *Xanthomonas* D-xylose catabolic genes in *Zymomonas mobilis*, *J. Biotechnol.* **7,** 61–70.

MARZOCCA, M. P., HARDING, N. E., PETRONI, E. A., CLEARY, J. M., IELPI, L. (1991), Location and cloning of the ketal pyruvate transferase gene of *Xanthomonas campestris*, *J. Bacteriol.* **173,** 7519–7524.

MCALPINE, J. B., TUAN, J. S., BROWN, D. P., GREBNER, K. D., WHITERN, D. N., BAKO, A., KATZ, L. (1987), New antibiotics from genetically engineered actinomycetes. I. 2-norerythromycins, isolation and structural determinations, *J. Antibiot.* **40,** 1115–1122.

MISAWA, N., NAKAGAWA, M., KOBAYASHI, K., YAMANO, S., IZAWA, Y., NAKAMURA, K., HARASHIMA, K. (1990), Elucidation of the *Erwinia uredovora* carotenoid biosynthesis pathway by functional analysis of gene products expressed in *Escherichia coli*, *J. Bacteriol.* **172,** 6704–6712.

MISAWA, N., YAMANO, S., IKENAGA, H. (1991), Production of β-carotene in *Zymomonas mobilis* and *Agrobacterium tumefaciens* by introduction of the biosynthesis genes from *Erwinia uredovora*, *Appl. Environ. Microbiol.* **57,** 1847–1849.

NICOLAY, K. (1987), Applications of NMR in microbial physiology, in: *"Physiological and Genetic Modulation of Product Formation"* (BEHRENS, D., Ed.), *DECHEMA-Monographs* Vol. 105, pp. 97–116, Weinheim: VCH Verlagsgesellschaft.

NIEDERBERGER, P., PRASAD, R., MIOZZARI, G., KACSER, H. (1992), A strategy for increasing an *in vivo* flux by genetic manipulation: the tryptophan system of yeast, *Biochem. J.*, in press.

NYUNOYA, H., BROGLIE, K. E., LUSTY, C. J. (1985), The gene coding for carbamoylphosphate synthetase I was formed by fusion of an ancestral glutaminase gene and a synthetase gene, *J. Biol. Chem.* **259,** 9790–9798.

O'FARRELL, P. H. (1975), High resolution two-dimensional electrophoresis of proteins, *J. Biol. Chem.* **250,** 4007–4021.

OGINO, T., SHULMAN, G. I., AVISON, M. J., GULLANS, S. R., DEN HOLLANDER, J. A., SHULMAN, R. G. (1985), ^{23}Na and ^{39}K NMR studies of ion transport in human erythrocytes, *Proc. Natl. Acad. Sci. USA* **82,** 1099–1104.

OHTA, K., BEALL, D. S., MEJIA, J. P., SHANMUGAM, K. T., INGRAM, L. O. (1991), Metabolic engineering of *Klebsiella oxytoca* M5A1 for ethanol production from xylose and glucose, *Appl. Environ. Microbiol.* **57,** 2810–2815.

PALMIERI, F., KLINGENBERG, M. (1979) Direct methods for measuring metabolite transport and distribution in mitochondria, *Methods Enzymol.* **56,** 279–297.

PRIES, A., STEINBÜCHEL, A., SCHLEGEL, H. G. (1990), Lactose- and galactose-utilizing strains of poly(hydroxyalkanoic acid)-accumulating *Alcaligenes eutrophus* and *Pseudomonas saccharophila* obtained by recombinant DNA technology, *Appl. Microbiol. Biotechnol.* **33,** 410–417.

RAMOS, J. L., WASSERFALLEN, A., ROSE, K., TIMMIS, K. N. (1987), Redesigning metabolic routes: manipulation of TOL plasmid pathway for catabolism of alkylbenzoates, *Science* **235,** 593–596.

ROJO, F., PIEPER, D. H., ENGESSER, K. H., KNACKMUSS, H. J., TIMMIS, K. N. (1987), Assemblage of *ortho*-cleavage route for simultaneous degradation of chloro- and methylaromatics, *Science* **235,** 1395–1398.

ROTTENBERG, H. (1979), The measurement of membrane potential and pH cells, organelles, and vesicles, *Methods Enzymol.* **55,** 547–569.

ROWLANDS, R. T. (1984), Industrial strain improvement: mutagenesis and random screening procedures, *Enzyme Microb. Technol.* **6,** 3–10.

RYGUS, T., SCHELER, A., ALLMANSBERGER, R., HILLEN, W. (1991), Molecular cloning, structure, promoters and regulatory elements for transcription of the *Bacillus megaterium* encoded regulon for xylose utilization, *Arch. Microbiol.* **155,** 535–542.

SALHANY, J. M., YAMANE, T., SHULMAN, R. G., OGAWA, S. (1975), High resolution ^{31}P NMR studies of intact yeast cells. *Proc. Natl. Acad. Sci. USA* **72,** 4966–4970.

SANO, K., SHIIO, I. (1970), Microbial production of L-lysine. III. Production by mutants resistant to S-(2-aminoethyl)-L-cysteine, *J. Gen. Appl. Microbiol.* **16,** 373–391.

SANO, K., ITO, K., MIWA, K., NAKAMORI, S. (1987), Amplification of the phosphoenol pyruvate carboxylase gene of *Brevibacterium lactofermentum* to improve amino acid production, *Agric. Biol. Chem.* **51,** 597–599.

SCHLEGEL, H. G., GOTTSCHALK, G. (1965), Verwertung von Glucose durch eine Mutante von *Hydrogenomonas* H16, *Biochem. Z.* **342,** 249–259.

SCHOBERTH, S. M., DE GRAAF, A. A. (1992), Use of ^{13}C *in vivo* nuclear magnetic resonance spectroscopy to follow uptake of glucose and xylose in *Zymomonas mobilis*, *Anal. Biochem.*, in press.

SCHRUMPF, B., SCHWARZER, A., KALINOWSKI, J., PÜHLER, A., EGGELING, L., SAHM, H. (1991), A functionally split pathway for lysine synthesis in *Corynebacterium glutamicum*, *J. Bacteriol.* **173,** 4510–4516.

SCHRUMPF, B., EGGELING, L., SAHM, H. (1992), Isolation and prominent characteristics of an L-lysine hyperproducing strain of *Corynebacterium glutamicum*, *Appl. Microbiol. Biotechnol.* **37,** 566–571.

SCHUBERT, P., STEINBÜCHEL, A., SCHLEGEL, H. G. (1988), Cloning of the *Alcaligenes eutrophus* genes for synthesis of poly-β-hydroxybutyric acid (PHB) and synthesis of PHB in *Escherichia coli*, *J. Bacteriol.* **170,** 5837–5847.

SHIIO, I., SUGIMOTO, S., TORIDE, Y. (1984), Studies on mechanisms for lysine production by pyruvate kinase-deficient mutants of *Brevibacterium flavum*, *Agric. Biol. Chem.* **48,** 1551–1558.

SILLS, A. M., SAUDER, M. E., STEWARD, G. G. (1984), Isolation and characterization of the amylolytic system of *Schwanniomyces castellii*, *J. Inst. Brew.* **90,** 311–314.

SKATRUD, P. L., TIETZ, A. J., INGOLIA, T. D., CANTWELL, C. A., FISHER, D. L., CHAPMAN, J. L., QUEENER, S. W. (1989), Use of recombinant DNA to improve production of cephalosporin C by *Cephalosporium acremonium*, *Bio/Technology* **7,** 477–485.

SLATER, S. C., VOIGE, W. H., DENNIS, D. E. (1988), Cloning and expression in *Escherichia coli* of the *Alcaligenes eutrophus* H16 poly-β-hydroxybutyrate biosynthetic pathway, *J. Bacteriol.* **170,** 4431–4436.

SLATER, S. C., GALLAHER, T., DENNIS, D. E. (1992), Production of poly-(3-hydroxybutyrate-co-3-hydroxyvalerate) in a recombinant *Escherichia coli* strain, *Appl. Environ. Microbiol.* **58,** 1089–1094.

SMITH, D. J., BURNHAM, M. K. R., EDWARDS, J., EARL, A. J., TURNER, G. (1990), Cloning and heterologous expression of the penicillin biosynthetic gene cluster from *Penicillium chrysogenum*, *Bio/Technology* **8,** 39–41.

SONOYAMA, T., TANI, H., MATSUDA, K., KAGEYAMA, B., TANIMOTO, M., KOBAYASHI, K., YAGI, S., KYOTANI, H., MITSUSHIMA, K. (1982), Production of 2-keto-L-gulonic acid from D-glucose by two-stage fermentation, *Appl. Environ. Microbiol.* **43,** 1064–1069.

SPRENGER, G. A., HAMMER, B. M., JOHNSON, E. A., LIN, E. C. C. (1989), Anaerobic growth of *Escherichia coli* on glycerol by importing genes of the *dha* regulon from *Klebsiella pneumoniae*, *J. Gen. Microbiol.* **135,** 1255–1262.

SREEKRISHNA, K., DICKSON, R. C. (1985), Construction of strains of *Saccharomyces cerevisiae* that grow on lactose, *Proc. Natl. Acad. Sci. USA* **82,** 7909–7913.

STEINBÜCHEL, A. (1991), Polyhydroxyalkanoic acids, in: *Biomaterials, Novel Materials from Biological Sources* (BYROM, D., Ed.), pp. 123–213, Basingstoke: Macmillan Publishers Ltd.

STEINBÜCHEL, A., SCHLEGEL, H. G. (1991), Physiology and molecular genetics of poly(β-hydroxyalkanoic acid) synthesis in *Alcaligenes eutrophus*, *Mol. Microbiol.* **5,** 535–542.

STRUCH, T., NEUSS, B., BRINGER-MEYER, S., SAHM, H. (1991), Osmotic adjustment of *Zymomonas mobilis* to concentrated glucose solutions. *Appl. Microbiol. Biotechnol.* **34,** 518–523.

TIMM, A., STEINBÜCHEL, A. (1990), Formation of polyesters consisting of medium-chain-length

3-hydroxyalkanoic acids from gluconate by *Pseudomonas aeruginosa* and other fluorescent pseudomonads, *Appl. Environ. Microbiol.* **56,** 3360–3367.

TOLAN, J. S., FINN, R. K. (1987), Fermentation of D-xylose to ethanol by modified *Klebsiella planticola*, *Appl. Environ. Microbiol.* **53,** 2039–2044.

TONG, I., LIAO, H. H., CAMERON, D. C. (1991), 1,3-Propanediol production by *Escherichia coli* expressing genes from the *Klebsiella pneumoniae dha* regulon, *Appl. Environ. Microbiol.* **57,** 3541–3546.

TOSAKA, O., TAKINAMI, K., HIROSE, Y. (1978), L-Lysine production by S-(2-aminoethyl)-L-cysteine and α-amino-β-hydroxyvaleric acid resistant mutants of *Brevibacterium lactofermentum*, *Agric. Biol. Chem.* **42,** 745–752.

UHLENBUSCH, I., SAHM, H., SPRENGER, G. A. (1991), Expression of an L-alanine dehydrogenase gene in *Zymomonas mobilis* and excretion of L-alanine, *Appl. Environ. Microbiol.* **57,** 1360–1366.

WALSH, P. M., HAAS, M. J., SOMKUTI, G. A. (1984), Genetic construction of lactose-utilizing *Xanthomonas campestris*, *Appl. Environ. Microbiol.* **47,** 253–257.

WEINBERGER, S., GILVARG, C. (1970), Bacterial distribution of the use of succinyl and acetyl blocking groups in diaminopimelate acid biosynthesis, *J. Bacteriol.* **101,** 323–324.

WHITE, P. J. (1983), The essential role of diaminopimelate dehydrogenase in the biosynthesis of lysine by *Bacillus sphaericus*, *J. Gen. Microbiol.* **129,** 739–749.

WILDE, E. (1962), Untersuchungen über Wachstum und Speicherstoffsynthese von *Hydrogenomonas*, *Arch. Microbiol.* **43,** 109–137.

WINDASS, J. D., WOORSEY, M. J., PIOLI, E. M., BIOLI, D., BARTH, K. T., ATHERTON, K. T., DART, E. C., BYROM, D., POWELL, K., SENIOR, P. J. (1980), Improved conversion of methanol to single-cell protein by *Methylophilus methylotrophus*, *Nature* **287,** 396–205.

YANASE, H., KURII, J., TONOMURA, K. (1988), Fermentation of lactose by *Zymomonas mobilis* carrying a Lac^+ recombinant plasmid, *J. Ferment. Technol.* **66,** 409–415.

6 Special Morphological and Metabolic Behavior of Immobilized Microorganisms

HANS-JÜRGEN REHM
Münster, Federal Republic of Germany

SANAA HAMDY OMAR
Alexandria, Egypt

1 Introduction 224
2 Methods of Immobilization 224
2.1 Immobilization of Microorganisms by Adsorption 224
2.2 Immobilization of Microorganisms by Aggregate Formation 225
2.3 Immobilization of Microorganisms by Covalent Coupling 225
2.4 Immobilization of Microorganisms by Entrapment 225
3 Improvement of Processes by Immobilized Microorganisms 226
4 Parameters Influencing the Development of Immobilized Microorganisms 226
4.1 Oxygen Transfer to Immobilized Microorganisms 226
4.1.1 Oxygen Transfer into Microfilms 227
4.1.2 Oxygen Transfer in Entrapped Microorganisms 227
5 Growth and Colony Formation of Immobilized Microorganisms 230
6 Morphological Alterations of Immobilized Microorganisms 233
7 Altered Physiology of Immobilized Microorganisms 234
7.1 Common Observations on Physiological Alterations 234
7.2 Alterations of Enzyme Activities in Immobilized Miroorganisms 236
7.3 Alteration of Product Formation by Immobilized Microorganisms 239
7.4 Prolongation of Metabolic Activities of Immobilized Miroorganisms 242
7.5 Alteration of Resistance against Poisons 242
8 Conclusion and Future Aspects 244
9 References 245

1 Introduction

It is well known that pure enzymes change their behavior, e.g., their stability, when they are immobilized. In the past two decades the immobilization of microorganisms, cells and parts of cells was gradually introduced into microbiology and biotechnology. Therefore, it is of interest to get information on the behavior of microorganisms under the conditions of immobilization. Many observations concerning morphological and physiological changes of immobilized microorganisms have been described. Investigations dealing with these changes compared to free microorganisms have been published, for example by DE BONT et al. (1990). For a general overview of the physiological aspects of immobilized cells and numerous literature citations see HAHN-HÄGERDAL (1990).

In this chapter some effects of the immobilization on microorganisms and one protozoon (*Tetrahymena thermophila*) shall be described, and some explanations will be given, as far as it is possible to explain these effects. Especially the following questions are of great interest:

- What are the main changes in the behavior of microorganisms when they are immobilized?
- Is it possible to characterize these effects?
- Are these behavioral changes caused by the new environment, which is set up by the immobilization?
- What are the reasons for these effects?
- Are these effects stable enough to be applied in biotechnology?
- How can we observe effects of changing the behavior of microorganisms in nature?

No information will be given on the effects of immobilization on plant cells. Concerning this subject reviews by ROSEVEAR and LAMBE (1985) and by FURASAKI and SEKI (1992) have been published. Some literature on the behavior of immobilized animal cells was reviewed by DE BONT et al. (1990).

2 Methods of Immobilization

Methods of immobilization are different and, therefore, the behavior of microorganisms may also be different. Many reviews of this topic are available, for instance, CHIBATA and WINGARD (1983), MATTIASSON (1983a,b), HARTMEIER (1988), BRODELIUS and VANDAMME (1987), FURUSAKI and SEKI (1992), REHM and REED (in preparation). Therefore, we will summarize here only some important methods, without detailed descriptions.

2.1 Immobilization of Microorganisms by Adsorption

To fix microorganisms for practical purposes, adsorption has been used since a long time, e.g., adsorption of *Acetobacter* for the production of vinegar or adsorption of waste water-degrading bacteria in trickling filters. The use of membrane filters and the growing of the microorganisms, filtered by these membranes, are adsorption immobilization methods, which have been developed since a long time (HAAS, 1956). Recently many other applications of adsorbed microorganisms have been described, e.g., for the production of citric acid, lactic acid, ethanol, butanol, glycerol, antibiotics, ergot alkaloids, and also in environmental biotechnology, e.g., for degradation of xenobiotics or recalcitrant compounds.

Besides others, the following materials for the adsorption of microorganisms have been described (see MATTIASSON, 1983a,b):

- natural materials, e.g., wood shavings, cellulose, etc.
- resins
- glass, sintered glass, etc.
- granular clay, lava and similar products
- activated carbon with varying porosity and surface behavior.

Numerous publications describe the behavior of microorganisms on the surface of the adsorbents, especially the ability to form films and layers and the conditions for the development of microorganisms in such micro- and macrofilms. Changes of physical and chemical conditions, which influence the development of microorganisms, are presented here, e.g., mass transfer, oxygen transfer and others.

2.2 Immobilization of Microorganisms by Aggregate Formation

Immobilization of miroorganisms by conglomeration – a "self-immobilization" – is frequently used for biotechnological purposes, e.g., for anaerobic waste water treatment or ethanol production with *Saccharomyces cerevisiae*. In these cases a special physico-chemical environment influences the development of microorganisms which is different from that for the development of free cells. Sometimes it is similar to that of adsorbed microorganisms, e.g., relating to the oxygen transfer or other mass transfer (SAHM, 1984; MACLEOD et al., 1990).

2.3 Immobilization of Microorganisms by Covalent Coupling

By these methods microorganisms are crosslinked by chemical substances, e.g., by glutardialdehyde. The surfaces (especially the proteins) of microorganisms are linked with the surfaces of other microorganisms by aldehyde groups of glutardialdehyde. Yeast cells, for instance, react with free ε-amino groups or N-terminal amino groups to form imines (Schiff's bases) as shown in Fig. 1. Another reaction mechanism was proposed concerning a conjugated addition of amino groups to double linkages of ε-,β-unsaturated oligomers, which are present in commercial aqueous solutions of glutardialdehyde. This mechanism may explain the stability of the linkages.

Reactions by covalent coupling with glutardialdehyde have been investigated by GHOSE and BANDYOPADHYAY (1980), KENNEDY and CABRAL (1983), VOJTISEK and JIRKU (1983); see also BRODELIUS and VANDAMME (1987). By this chemical linking, growth inhibition and toxic influences on the microorganisms are very intensive. These reactions are only partly understood and can lead to decay or death of the microorganisms.

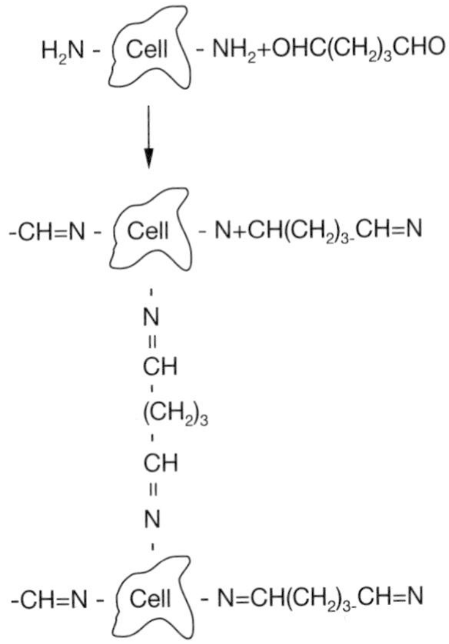

H_2N - Cell - $NH_2+OHC(CH_2)_3CHO$

↓

-CH=N - Cell - $N+CH(CH_2)_3$-CH=N

N = CH – $(CH_2)_3$ – CH = N

-CH=N - Cell - $N=CH(CH_2)_3$-CH=N

Fig. 1. Cross-linking of cells by glutardialdehyde. Reaction of glutardialdehyde with free amino acid groups of the cell wall (REHM and GREVE, 1993).

2.4 Immobilization of Microorganisms by Entrapment

The methods of immobilization by entrapment are diverse. The microorganisms can be directly entrapped in synthetic polymers, most commonly acrylic polymers, e.g., polyacrylamide, and polyurethans. Here diffusion through the hardened resin material has to take place. Most of the corresponding monomers are toxic for the microorganisms. After polymerization the resins mostly do not exhibit toxicity, but during polymerization dangerous effects, e.g., high temperature, can damage the microorganisms. By means of a prepolymerization of the resins these negative effects could be diminished, e.g., for polyacrylamide hydrazide (FREEMANN and AHARONOWITZ, 1981). An excellent review of immobilization of microorganisms after prepolymerization has been written by FUKUI and TANAKA (1984).

Some methods describe immobilization in synthetic polymers or in epoxides together with other substances, such as alginates, which can be dissolved, e.g., by phosphates, after hardening of the synthetic polymer. In this way *cavities* are produced, in which the microorganisms are entrapped. In these cavities, mass transfer goes through the small channels from the outside into the interior parts where the microorganisms are entrapped.

During entrapment of cells in synthetic polymers or epoxide resins many of the microorganisms can be killed, and the biocatalytic functions are then only performed by the remaining active enzymes or by parts of the dead microorganisms. Most of the metabolic potential is lost. Only when mild polymerization methods are applied, viable microorganisms can be expected.

For better entrapment of living microorganisms natural polymers such as alginate, kappa-carrageenan, agar or agarose can be used: The miroorganisms are located in non-toxic gels by non-toxic methods. In these gels they can grow and form microcultures or mycelial colonies. Substrate, oxygen and other substances must be transported to the entrapped microorganisms by diffusion. Often the microorganisms grow up to the surface of the beads in order to be in direct contact with the substrate.

Sometimes the microorganisms have to tolerate high concentrations of the hardening substances, e.g., high calcium concentrations in alginate.

3 Improvement of Processes by Immobilized Microorganisms

Biotechnological processes can be essentially improved by using immobilized microorganisms in comparison to processes with free microorganisms. These improvements have been described as follows:

- prolonged metabolic activities
- prolonged life cycles
- greater tolerance against poisonous substances
- decreased growth and biomass formation
- formation of special products
  and
  many other improvements

Sect. 7.3 of this chapter deals with process improvement by immobilized microorganisms, and literature references are cited there.

It can be concluded from the above that some of the improved abilities of the immobilized microorganisms originate from a change of their metabolism caused by the changed environment during immobilization; but some changes cannot be explained up to now by changed parameters.

4 Parameters Influencing the Development of Immobilized Microorganisms

Some influences of the immobilization techniques on the development of the microorganisms have already been described in Sect. 2. Some others shall be discussed now. These parameters may affect and alter development and behavior of immobilized microorganisms compared to free microorganisms.

4.1 Oxygen Transfer to Immobilized Microorganisms

Availability of oxygen is one of the most important parameters which are different for immobilized and free microorganisms. Free organisms can get oxygen directly from the surrounding air or, in most technical processes, from the liquid, especially water, which contains dissolved oxygen. The transfer dn/dt can be described by the following equation:

$$\frac{dn}{dt} = k_L a (c_g - c_x)$$

$k_L a$ is the volumetric O_2 mass transfer coefficient, due to the oxygen transfer from the gas phase or air, c_g, the surface of the cells, c_x, or to the transfer of oxygen dissolved in water to the surface of the cells. These transfers have been described in numerous papers and examples.

In principle, the same formula can be applied to immobilized microorganisms, but here the conditions are quite different. Two different types can be observed:

1. The adsorbed cells form microbial films of varying thickness during a short incubation time. Oxygen has to be transferred into these microfilms, and due to this, zones of different oxygen concentrations in the films exist, in which the growth of the microorganisms varies in direct relation to the oxygen concentration. Details see GOSMANN and REHM (1988).

2. Entrapped microorganisms depend on the diffusion of oxygen through the gel substance used for their entrapment. They also depend on the diffusion of oxygen into and through the microcolonies which have been formed by their growth. The growth can strongly influence these parameters. For determination of the oxygen diffusion into alginate see GOSMANN and REHM (1988) and BEUNINK et al. (1989).

4.1.1 Oxygen Transfer into Microfilms

A great deal of literature on oxygen availibity in microfilms has been published (e.g., MARSHALL, 1984; BISHOP and KINNER, 1986; CHARACKLIS and WILDERER, 1989). In all investigations high oxygen concentrations were observed on the surface of the microfilms. The quantity of oxygen measured in the film decreases proportionally with the thickness of the film formed. In the interior of thick films microaerophilic and sometimes anaerobic conditions exist (see Fig. 2).

These conditions highly influence the morphological and physiological behavior of the adsorbed microorganisms. Furthermore, they are responsible for a very different mass transfer, e.g., for nutrients, to the microorganisms found in the microfilms and for the transport of metabolic substances or gases (e.g., CO_2) out of the microfilms. The microbial population has been described by BISHOP and KINNER (1986). Components and processes characterizing a biofilm system have been reviewed by WILDERER and CHARACKLIS (1989) and CHARACKLIS and WILDERER (1989). For details on the gradients, especially O_2 gradients in microfilms, see REVSBECH (1989).

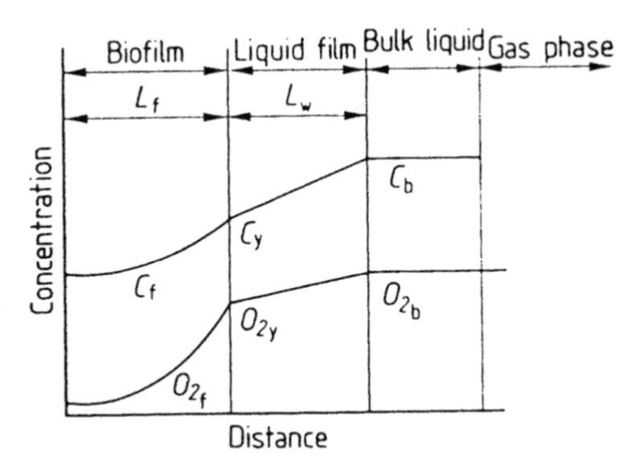

Fig. 2. Concentration profiles of organic substrate (C) and oxygen (O_2) in an idealized biofilm and stagnant liquid layer, with L_f as thickness of the biofilm and L_w thickness of the stagnant liquid film (BISHOP and KINNER, 1986).

4.1.2 Oxygen Transfer in Entrapped Microorganisms

The growth of entrapped microorganisms in the gel affects the transfer of oxygen to the inside. TANAKA et al. (1984) found no reduction of the diffusion coefficients in calcium alginate beads without microorganisms for solutions with substances of a molecular weight of $2 \cdot 10^4$ D in comparison with the free diffusion in water. HIEMSTRA et al. (1983) reported a diffusion coefficient for oxygen in barium alginate gels that was only 25% of that in water. MARTINSEN et al. (1992) found relationships for the diffusion of serum albumin (molecular weight $6.9 \cdot 10^4$ D) in Ca-alginate beads with pH, temperature, the conditions under which the beads are produced, and the concentration and the uronic acid composition of the alginate.

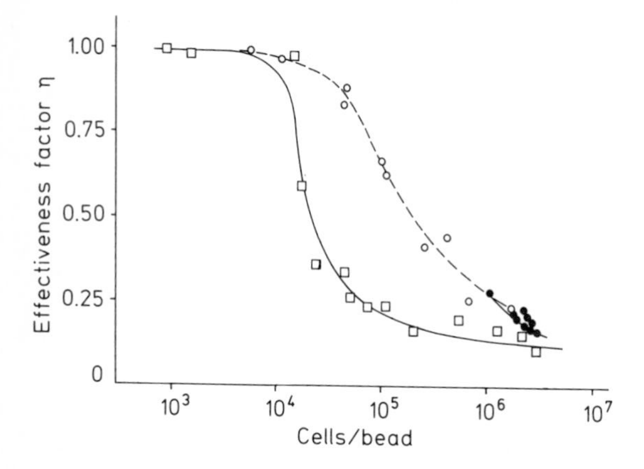

Fig. 3. Effectiveness factor η of oxygen consumption of immobilized *Saccharomyces cerevisiae* cells in relation to the cell concentration. □ homogeneously distributed cells; ○ homogeneously distributed microcolonies; ● dense cell layer at the gel surface (GOSMANN and REHM, 1988, reproduced with permission).

The diffusion of oxygen in alginate decreases with the increase of growth of the miroorganisms in the biocatalyst (see also TRAMPER et al., 1983). A so-called effectiveness factor, η, was introduced by RADOVICH (1985). It permits the determination of the O_2 availability of entrapped cells.

$$\eta = \frac{\text{specific } O_2 \text{ uptake rate of immobilized cells}}{\text{specific } O_2 \text{ uptake rate of free cells}}$$

If the diffusion barrier does not affect the O_2 uptake rate, the effectiveness factor $\eta = 1$.

SATO and TODA (1983) determined the effectiveness factor with a strictly aerophilic yeast, *Candida lipolytica*, entrapped in spherical agar particles. After one day's incubation at 30 °C, the effectiveness factor of the immobilized growing yeast decreased, if the initial effectiveness factor was greater than 0.5. The factor increased, if the initial effectiveness factor was no greater than 0.5. In both cases, the effectiveness factor of the immobilized growing yeast increased by more than the theoretical value calculated according to the reaction–diffusion equation.

Our investigations with *Saccharomyces cerevisiae*, *Pseudomonas putida* P8 and *Aspergillus niger* have demonstrated that the effectiveness factor η increases, when the biomass concentrations of aerobic microorganisms increase in the marginal regions of alginate polymers (GOSMANN and REHM, 1986, 1988). The effectiveness factors of oxygen consumption of immobilized cells of *Saccharomyces cerevisiae* and *Aspergillus niger* are shown in Fig. 3 and Fig. 4, respectively.

Microorganisms in the outer region of alginate beads (and this can also be seen with mi-

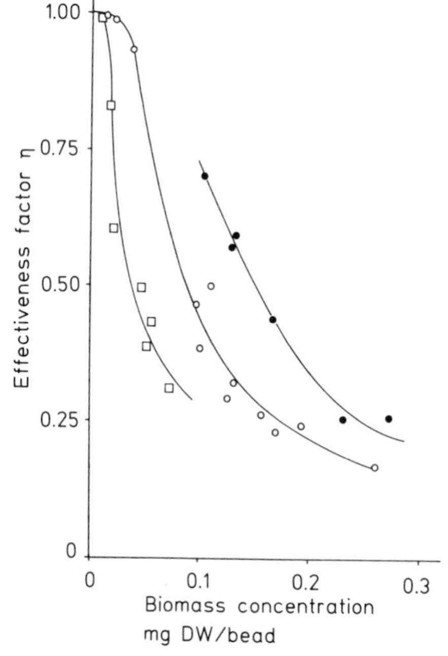

Fig. 4. Effectiveness factor η of oxygen consumption of *Aspergillus niger* in alginate in relation to the biomass content in gel. □ homogeneous mycelia growth; ○ biomass concentration at the gel surface; ● fur-like mycelia coat outside the gel (GOSMANN and REHM, 1988, reproduced with permission).

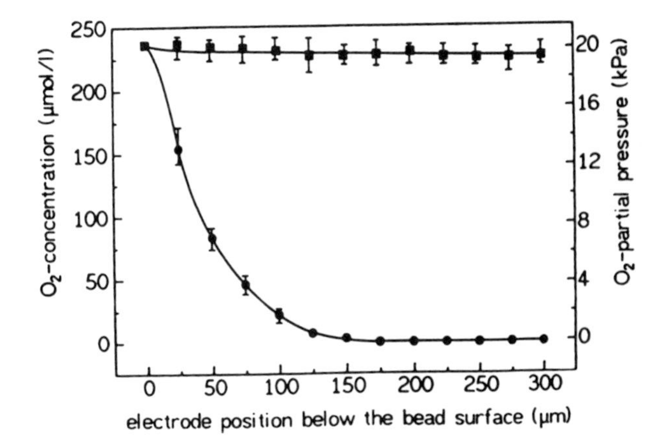

Fig. 5. Oxygen concentration in cell-free ■ and cell-loaded ● alginate beads. Beads loaded with cells of *Enterobacter cloacae* were incubated for 6 h in the fermentation medium. The cell concentration was 1.93×10^8 cells per bead. Bars show the deviation between the maximum and minimum data (from BEUNINK et al., 1989, with permission).

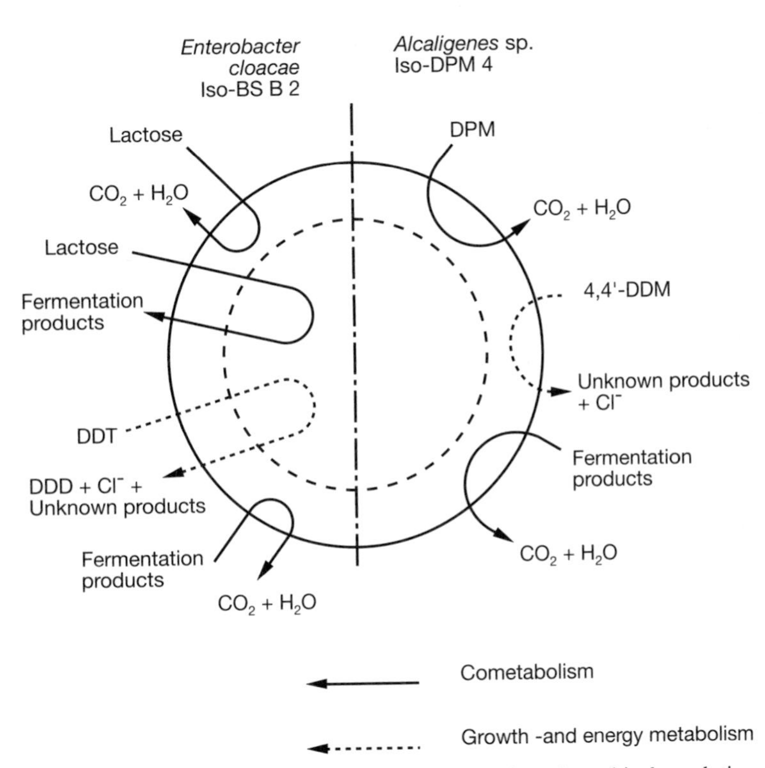

Fig. 6. Metabolic processes in a Ca-alginate bead during the synchronous anaerobic and aerobic degradation of DDT and DDM (BEUNINK and REHM, 1988).

croorganisms adsorbed on the outer regions of films) obtain much oxygen. Consequently, less oxygen diffuses into the interior parts of the beads or of the microorganism films. This is of great importance for the growth and the oxidative metabolism of the microorganisms. New investigations with O_2 microelectrodes could confirm these results (BEUNINK et al., 1989; BEUNINK and REHM, 1990). In the interior parts of cell-loaded beads or films microaerophilic or anaerobic conditions can be observed. Free Ca-ions may have an influence on oxygen diffusion, which can be eliminated by washing. This was practiced with Ca-alginate beads loaded with cells of *Enterobacter cloacae* (BEUNINK et al., 1989; Fig. 5).

An interesting application of an oxygen microsensor in combination with mathematical modelling to elucidate the behavior of immobilized *Thiosphaera pantatropha* has been described by HOOIJMANS et al. (1990).

These findings confirm the results of HANNOUN and STEPHANOPOULOS (1986) and TANAKA et al. (1984) who found that Ca-alginate gel without living cells shows a diffusion coefficient for low-molecular weight substances similar to that in water. For further details about the influence of microbial growth on the diffusion and the availability of oxygen in Ca-alginate for microorganisms see GOSMANN and REHM (1986, 1988). It is not easy to fully explain this influence on immobilized microorganisms. The fact that the concentration of oxygen in the interior of Ca-alginate beads was reduced nearly to anaerobic conditions, was used tp construct aerobic/anaerobic systems to degrade DDT and 4-chloro-2-nitrophenol by co-immobilized mixed culture systems (BEUNINK and REHM, 1988, 1990) (Fig. 6).

MATTIASSON and HAHN-HÄGERDAL (1982) were the first, who published results on the influence of the water activity on immobilized microorganisms. Microorganisms stop growing at low water activity (SCOTT, 1987). Reduced water activity was also observed in immobilized cell preparations by HAHN-HÄGERDAL (1990).

The activity of water is

$$a_w = \frac{P \text{ (water vapor pressure of substrate)}}{P_0 \text{ (water vapor pressure of water alone)}}$$

Altered conditions will influence the metabolism of the microorganisms (see also HOLCBERG and MARGALITH, 1981). MATTIASSON (1983a) demonstrated that a low a_w increases the pool of polyols and enriched ethanol and lactic acid. An example for the influence of a_w is the formation of polyols which are produced by immobilized *Pichia farinosa* in larger amounts to stabilize the osmotic pressure in the cells (BISPING et al., 1990b).

5 Growth and Colony Formation of Immobilized Microorganisms

Immobilized microorganisms differ in their growth rates and show altered morphological forms of colonies. Only some examples shall be described here. A review of such changes is being prepared by OMAR (1993).

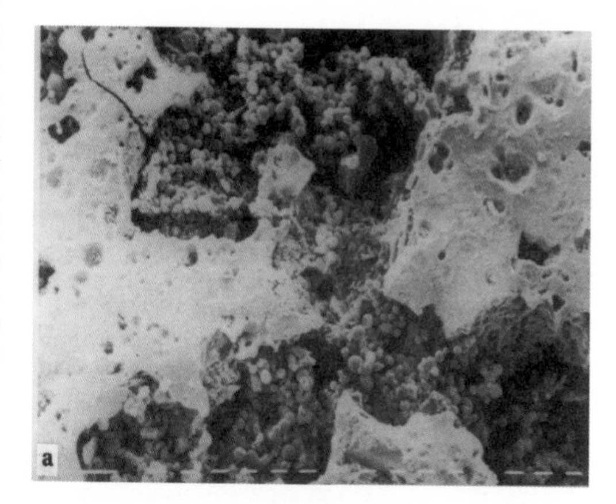

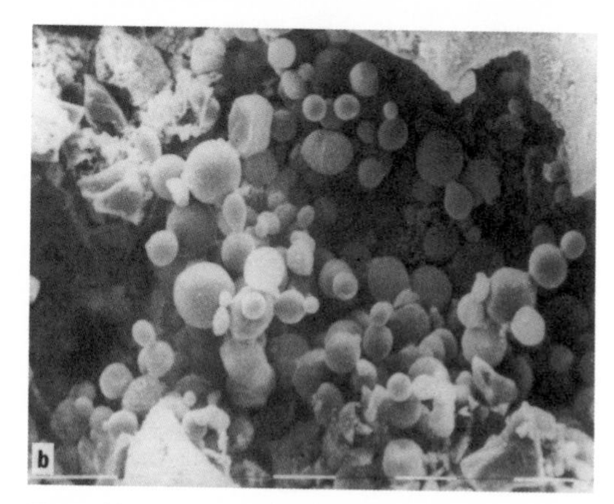

Fig. 7. (a) Cells of *Saccharomyces cerevisiae* immobilized in a pit of a sintered glass Raschig ring after a fermentation time of 400 h, – corresponds to 10 μm. (b) Enlarged part of the same section. Besides some dead cells it is clear that most yeast cells are still budding (BISPING and REHM, 1986).

As described above, adsorbed cells form micro- and macrofilms in which the microorganisms in the outer region have another morphology than in the inner region. Many investigations about microfilms of adsorbed microorganisms have been made (see CHARACKLIS and WILDERER, 1989). These microfilms often show different colony forms in relation to their density. Thick films have a slimy character, thin films often show the presence of individual microorganisms. These characteristics can be observed with bacteria, yeasts, and also with molds. They are caused by differing oxygen concentrations and limited concentrations of nutrients. Figs. 7 and 8 show two species of adsorbed microorganisms.

Fig. 7 shows cells of *Saccharomyces cerevisiae* immobilized in a cave of sintered glass Raschig rings after a glycerol fermentation time of 400 hours. These cells form dense colonies in the caves of the glass rings. The budding activity is preserved, but the separation of the new cells is strongly restricted. The reasons for this restriction are not known (BISPING and REHM, 1986).

Penicillium frequentans forms a dense network on the surface of granular clay (Fig. 8, OMAR and REHM, 1988).

Entrapped bacteria and yeasts can form microcolonies with altered growth, which may also be caused by diffusion limitations of oxygen, nutrients and other parameters. Such colonies can be very dense with large numbers of microorganisms. In other cases the microorganisms are distributed equally along the border of the beads. Fig. 9 demonstrates the formation of microcolonies of *Streptomyces aureofaciens* ATCC 10762 in Ca-alginate. These microcolonies (or micropellets) are very different from pellets of free cells formed by this bacterium in submerged culture (MAHMOUD and REHM, 1986). The application of nitrogen, phosphate, oxygen or other substrates can regulate the formation of these colony forms (see EIKMEIER et al., 1984; DIERKES et al., 1993).

Molds can also be regulated to a fine, even distribution through the beads or to a dense layer at the margin of the beads or at the outer part of the beads, e.g., *Aspergillus niger* after precultivation with varying ammonium nitrate concentrations, see Fig. 10 (EIKMEIER et al., 1984). The influence of this changed morphology on the metabolic behavior has not been investigated (see also OMAR et al., 1992).

Similar observations with *Claviceps purpurea* were published by LOHMEYER et al. (1990) and with *Penicillum roqueforti* by KUSCH and REHM (1986).

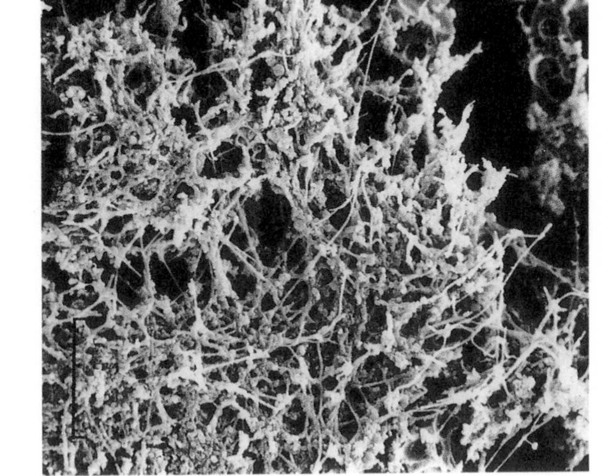

Fig. 8. Scanning electron micrograph of *Penicillium frequentans* (×510), immobilized on granular clay; the intensity of growth over the surface of the clay, like a network, is demonstrated (OMAR and REHM, 1988).

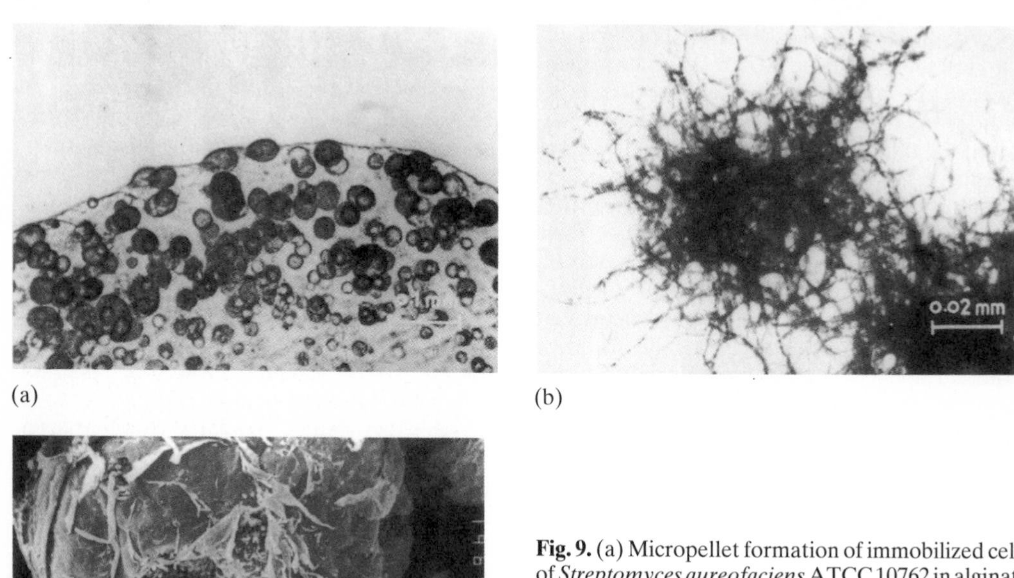

(a) (b)

(c)

Fig. 9. (a) Micropellet formation of immobilized cells of *Streptomyces aureofaciens* ATCC 10762 in alginate after 24 h in growth medium. (b) Pellet formation of free cells of *S. aureofaciens* in submerged fermentation. (c) Electron scanning micrograph of the surface of one micropellet covered with a membrane-like structure of calcium alginate (see the rupture on the surface and growth of mycelia in the interior of the micropellet), – corresponds to 10 μm (MAHMOUD and REHM, 1986).

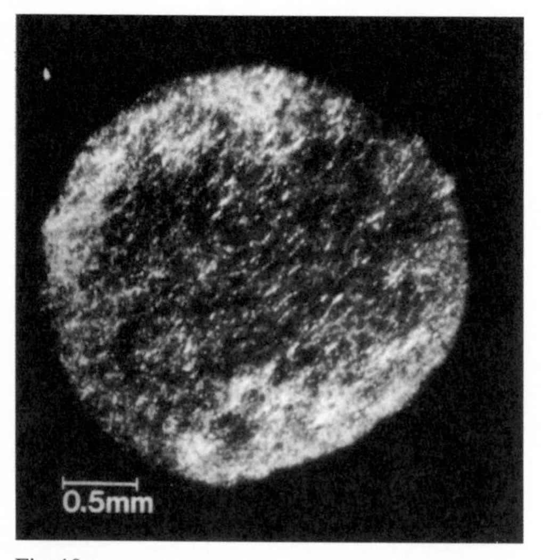

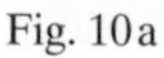

Fig. 10a

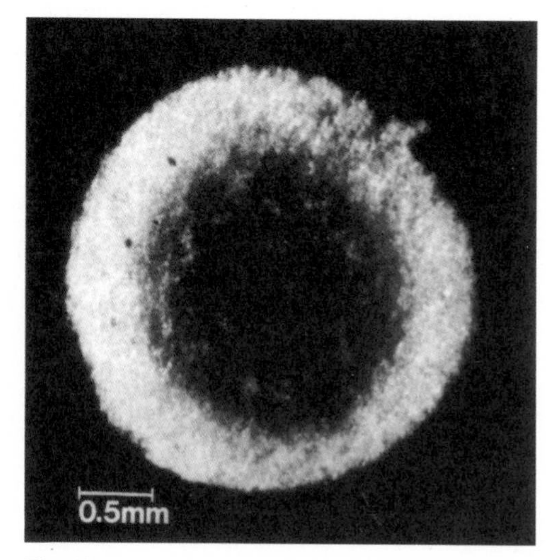

Fig. 10b

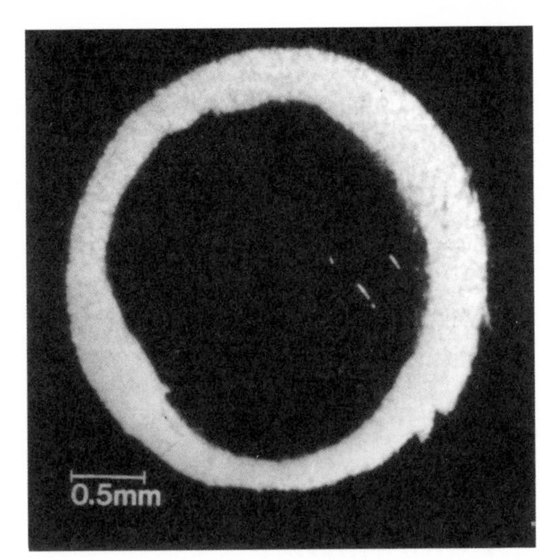

Fig. 10. Growth behavior of Ca-alginate immobilized *Aspergillus niger* after precultivation with various ammonium nitrate concentrations. (a) 0.05 g/L NH_4NO_3; (b) 0.10 g/L NH_4NO_3; (c) 0.20 g/L NH_4NO_3 (EIKMEIER et al., 1984).

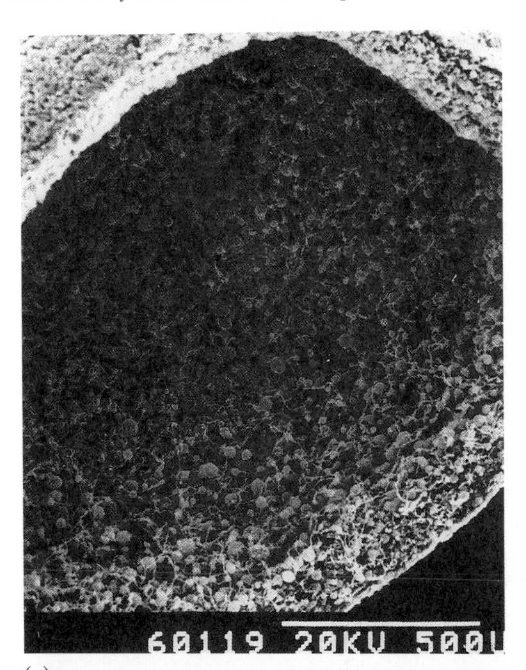

(a)

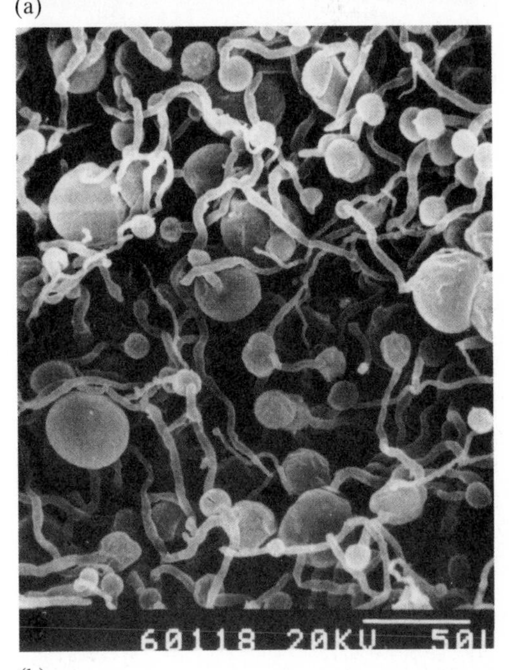

(b)

Fig. 11. Scanning electron micrographs of phosphate-limited immobilized cells of *Aspergillus niger* after 30 days of continuous citric acid production. (a) Interior of a Ca-alginate bead; (b) detail from (a) (HONECKER et al., 1989).

6 Morphological Alterations of Immobilized Microorganisms

As described above, the colonial forms of immobilized microorganisms often differ from those of free microorganisms. Furthermore, many morphological changes of immobilized microorganisms, in comparison to free growing microorganisms, can be observed. *Bacteria* are very often linked together by a net of polysaccharides. The cells often lengthen after immobilization. The morphological changes of *molds* can be better observed than those of bacteria. A strain of *Aspergillus niger* formed enlarged, bulbous cells in the interior of Ca-alginate (Fig. 11). The intensity of bulbous cell formation differs from strain to strain in *Aspergillus niger* (OMAR et al., 1992).

It is well known that free mycelia of *Claviceps pupurea* are very thin and light in color. Ca-alginate entrapped mycelia, on the other hand, form an athrosporoid-like mycelium, which is very similar to the cell morphology found in a sclerotium of this mold (Fig. 12, KOPP and REHM, 1984).

The alginate-immobilized mycelia of *Penicillium roqueforti* showed special structures consisting of spiral-shaped hyphae which, in later stages, looked like balls or clusters. These balls had a diameter of about 40 μm and were not observed with free cells of this strain (KUSCH and REHM, 1986). Many other morphological alterations could be described; but a review on these changes is in preparation and will give much more details (OMAR, 1993).

The causes of these alterations are not yet known.

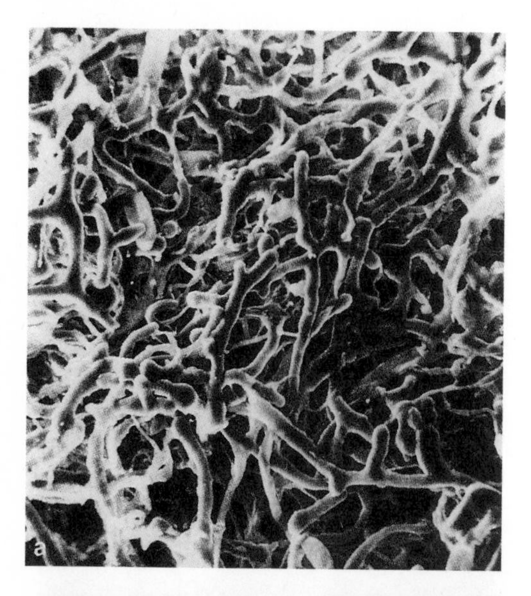

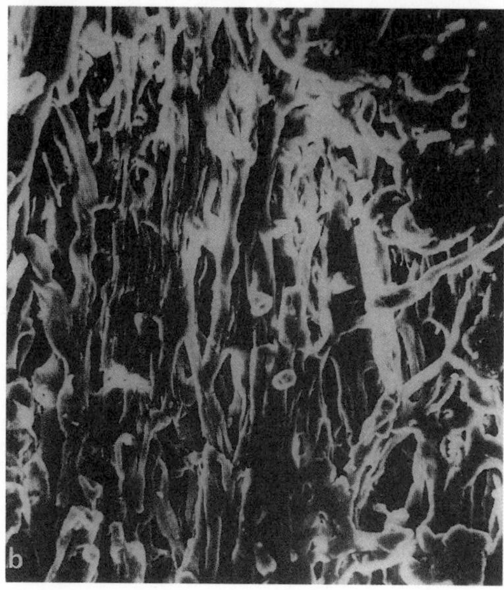

Fig. 12. (a) Surface of a 4% calcium alginate bead covered with sclerotia-like cells of *Claviceps purpurea* (×800); (b) cell morphology of a parasitically formed sclerotium of *Claviceps* sp., longitudinal section (×800) (KOPP and REHM, 1984).

7 Altered Physiology of Immobilized Microorganisms

7.1 Common Observations on Physiological Alterations

The first important review of physiological alterations of immobilized microorganisms was written by MATTIASSON (1983a,b). He described some observations on the changed metabolism of immobilized microorganisms and gave explanation, e.g., the influence of oxygen transfer on the immobilized cells and the water activity. In the meantime many other papers on physiological effects caused by immobilization have been published (see, for example, DE BONT et al., 1990).

As described in Sect. 4, the oxygen transfer plays an important role for the metabolic activity of microorganisms. In dense microcolonies and in the interior part of gels microorganisms are often poorly supplied with oxygen. TANAKA et al. (1986) reported the direct production of ethanol from starch by the synergistic action of co-immobilized aerobic *Aspergillus awamori* and anaerobic *Zymomonas mobilis.*

Furthermore, the fact of limited oxygen concentration in the interior part of Ca-alginate beads, by capturing the oxygen in the peripherally growing microorganisms of an aerobic *Alcaligenes* sp., could be demonstrated by BEUNINK and REHM (1988). They constructed a combined aerobic/anaerobic system with *En-*

terobacter cloacae, which was regulated to the anaerobic metabolism in the interior of the beads, in order to degrade the insecticides DDT and 4-chloro-2-nitrophenol (BEUNINK and REHM, 1990).

These examples show that coupled reductive and oxidative reaction systems can be constructed with immobilized microorganisms by regulation of the oxygen transfer. The mass transfer in the beads from the interior parts to the exterior parts was also used in these systems. In the first system, 1,1-dichloro-2,2-bis(4-chlorophenyl)ethane (DDD), and 4,4′-dichlorodiphenylmethane (DDM), which are anaerobic degradation products of DDT, diffused to the marginal oxidation zone of the beads. In the second system, 4-chloro-2-aminophenol and 4-chloro-2-acetaminophenol, which are anaerobic degradation products of 4-chloro-2-nitrophenol produced by one bacterium, diffused to the marginal zones of the alginate beads to be aerobically degraded by an aerobic bacterium. In these examples the regulation of the facultative aerobic bacteria to an anaerobic metabolism was realized. MATTIASSON (1983 a, b) postulated the Pasteur effect of immobilized *Saccharomyces cerevisiae* at low O_2 concentrations.

SHINMYO et al. (1982) observed that the respiratory activity and the growth rate of *Bacillus amyloliquefaciens* were repressed to 1/2 and 1/6, respectively, when immobilized in kappa-carrageenan beads, but the alpha-amylase productivity was almost the same or higher than that of the cells, released from the beads into the medium.

As described before, the oxygen uptake and, therefore, also the respiratory activity is greatly dependent on the growth and the number of microorganisms in gels or in the biological films (see GOSMANN and REHM, 1986, 1988).

BRIGHT and FLETCHER (1983) reported an increased assimilation of amino acids and decreased respiration rates in attached populations of a marine *Pseudomonas* sp. MORISAKI (1983) found increased respiration rates of *Escherichia coli* at solid–liquid interfaces and explained this with changes in membrane processes.

Adsorbed or covalently attached cells of *Saccharomyces cerevisiae* on porous glass showed a change in metabolic behavior compared to free cells. With adsorbed cells an increase of about 7% in ethanol fermentation was observed. This increase was accompanied by a decrease in CO_2 productivity (NAVARRO and DURAND, 1977). The metabolic mechanisms could not be explained by the authors, but they assumed a change in metabolic patterns in *Saccharomyces cerevisiae.*

VAN LOOSDRECHT et al. (1990) have reviewed the influence of interfaces on microbial activity and concluded that there is no conclusive evidence that adhesion directly influences bacterial metabolism, in the sense that the bacteria undergo a structural change due to the adhesion. This means that all differences between adsorbed and free cells can be attributed to an indirect mechanism, i. e., a mechanism by which the surroundings of the cells are modified due to the presence of surfaces, but not due to the cell itself. They cited many results, which describe increased growth rates of adsorbed microorganisms. These increased growth rates were explained by

- increased substrate concentrations at the interface
- more efficient use of proton native forces
- detoxification of substrate and inhibitor
- pH buffering by ion exchange.

Some observed decreased growth rates were explained by

- less cell surface available for substrate uptake
- higher maintenance coefficient
- substrate limitation.

An observed decreased substrate utilization was explained by

- desorption limitation
- diffusion limitation
- lower substrate concentration.

For more detailed literature references see VAN LOOSDRECHT et al. (1990).

The macromolecular composition of a Ca-alginate entrapped haploid strain of *Saccharomyces cerevisiae* (Tab. 1) differs greatly from that of free cell cultures (SIMON et al., 1990).

Tab. 1. Macromolecular Composition of a Ca-alginate entrapped Haploid Strain of *Saccharomyces cerevisiae in* Comparison to Free Cells (μg per 10^8 cells)

	Free Cell Culture		Entrapped Cell Culture
	Exponential Phase	Stationary Phase	Pseudo-stationary Phase
Sugars			
Trehalose	18	26	18
Glycogen	131	131	357
Mannans	70	69	256
β-Glucans	213	72	223
Total	431	298	864
Proteins			
Cytosol	1160	100	830
Membrane	30	40	170
Total	1190	140	1000

The formation of glycogen, mannans, β-glucans was very much increased. HILGE-ROTMANN and REHM (1991) previously reported on the increase of glycogen in Ca-alginate entrapped cells of *Saccharomyces cerevisiae.*

7.2 Alterations of Enzyme Activities in Immobilized Microorganisms

DORAN and BAILEY (1986a,b) observed a decreased growth of *Saccharomyces cerevisiae*, entrapped in cross-linked gelatin in comparison to free cells. They reported that the specific rate of ethanol production by the immobilized cells was 40–50% greater than that of suspended yeast. The immobilized cells consumed glucose twice as fast as the suspended yeast, but their specific growth rate was reduced to 45%. Yields of biomass from the immobilized cell population were one-third lower than the values for the suspended cells. They assumed a suppressed budding of cells caused by the cell contact with the support material. Furthermore, they observed a stimulating reaction within glycolysis.

GALAZZO et al. (1987) assumed that a lower intracellular pH in Ca-alginate entrapped *S. cerevisiae* cells could be the reason for enhanced glucose turnover, because hexokinase and phosphofructokinase can be stimulated by a lower pH *in vitro.*

Detailed investigations of the key enzymes of glycolysis in *Saccharomyces cerevisiae* en-

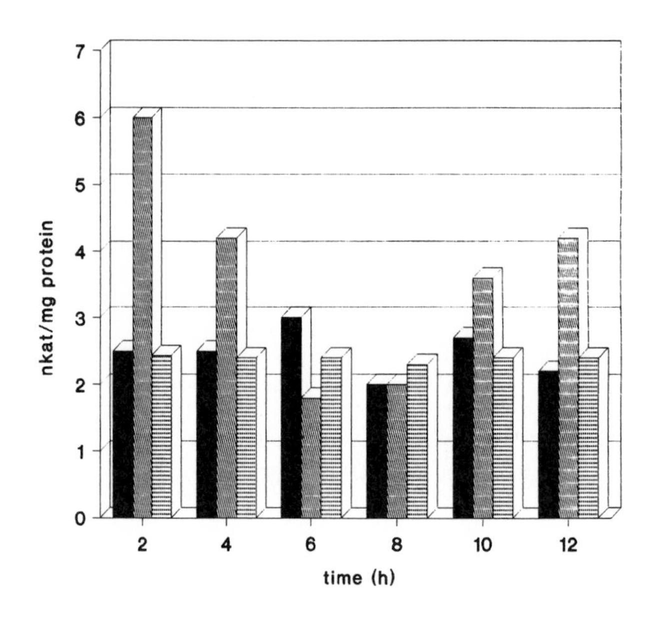

Fig. 13. Specific hexokinase activity of suspended cells, ■, immobilized microcolonies, ▧, and immobilized single cells ⊟, of *Saccharomyces cerevisiae* during fermentation (HILGE-ROTMANN and REHM, 1990).

trapped in Ca-alginate showed that these key enzymes had a faster glucose uptake and ethanol productivity with simultaneously decreased cell yields (HILGE-ROTMANN and REHM, 1990). Increased specific hexokinase and phosphofructokinase activities could be determined in these cells (Figs. 13 and 14) when they grew in microcolonies. Immobilized single cells showed only slightly enhanced glucose turnover and, therefore, higher specific hexokinase activity than free cells, but also an increased and prolonged activity of phosphofructokinase. The aldolase had the same activities in suspended cells, immobilized microcolonies and immobilized single cells.

One well-known regulating factor of phosphofructokinase activity is oxygen. As described above, the O_2 supply can be diminished to 25% or less in gel matrix. The more than tenfold higher specific phosphofructokinase activities in entrapped *Saccharomyces cerevisiae* cells, compared to suspended cells, are apparently not caused by their better protection by the alginate against oxygen, since the fermentations were carried out partly under anaerobic conditions. The dense cell-to-cell or cell-to-carrier contact may play a role in the change of cell behavior, when they are immobilized by directly affecting the cell envelope. ELLWOOD et al. (1982) postulated an altered metabolism of microorganisms growing at surfaces with changes in their membrane permeability for protons.

Heterocystous, free living cyanobacteria, *Anabaena azollae*, were immobilized in polyvinyl and polyurethan foam matrices in order to mimic the *in vivo* environment of the co-sym-

Tab. 2. Hydrogen Production and Nitrogenase and Hydrogenase Mediated by Free Living and Immobilized *Anabaena azollae* (cells were 4 days old and incubation time was 5 hours)

Sample	Hydrogen Production (μmol H_2 per mg Chl)		
	Nitrogenase mediated	Hydrogenase mediated	Ratio
Free living	1.6	6.2	0.26
Immobilized in PV foam (PR 22/60)	3.1	9.4	0.33
Immobilized in PU foam (3300A)	2.2	14.5	0.15

Abbreviations: Chl, chlorophyll; PV, polyvinyl; PU, polyurethan

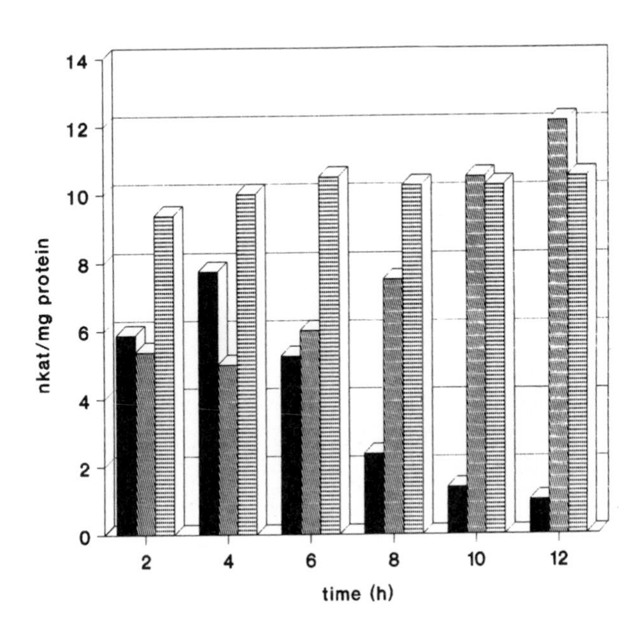

Fig. 14. Specific phosphofructokinase activity of suspended cells ■, immobilized microcolonies, ▧, and immobilized single cells ⊟, of *Saccharomyces cerevisiae* during fermentation (HILGE-ROTMANN and REHM, 1990).

biotic cyanobacteria in the leaf cavity (BROUERS et al., 1990). They were compared with free living and symbiotic *Anabaena azollae*. The immobilized cyanobacteria showed increases in heterocyst frequency and nitrogenase activity (see Tab. 2) with excretion of NH_4^+ and development of a thick mucilage layer.

A new type of peptidase, which cleaved Leu-Lys bonds of α-mating factor, was excreted specifically by *Saccharomyces cerevisiae*, entrapped with a neutral hydrophobic photocrosslinkable resin prepolymer. The enzyme was supposed to be induced to a certain extent due to a kind of stress caused by the entrapment (SONOMOTO et al., 1990).

The covalent binding has a very strong influence on the metabolism. *Saccharomyces cerevisiae* cells immobilized covalently with 0.01% glutardialdehyde (GDA) showed the same growth rate as free cells. Concentrations of 0.05% and 0.1% GDA inhibited the growth by prolongation of the lag phase. 0.01% did not influence glucose consumption, ethanol production and glycerol formation markedly, but concentrations of 0.025% GDA increased glucose consumption and decreased ethanol and glycerol production. On the other hand, the activity of the phosphofructokinase was distinctly reduced at 0.025% GDA. This activity and also that of the alcohol dehydrogenase were

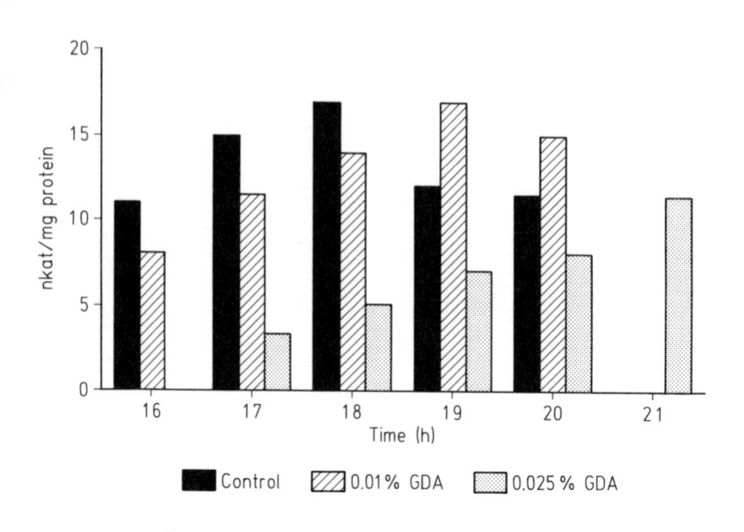

Fig. 15. Specific activity of phosphofructokinase of free and cross-linked cells of *Saccharomyces cerevisiae* (24 h dried) (REHM and GREVE, 1993). GDA, glutardialdehyde.

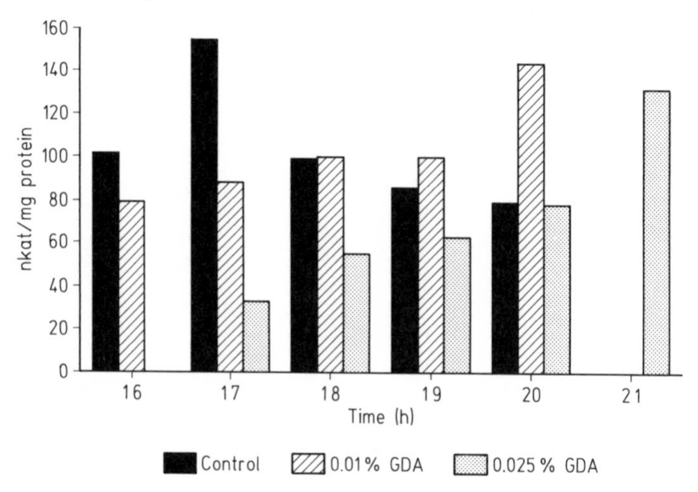

Fig. 16. Specific activity of alcohol dehydrogenase of free and cross-linked cells of *Saccharomyces cerevisiae* (24 h dried) (REHM and GREVE, 1993).

regenerated after some time (see Figs. 15 and 16, REHM and GREVE, 1993, in preparation).

These results show that the intensive linkage of alcohol dehydrogenase in high concentrations reduces the activity of the *Saccharomyces* cells and acts like a poison, but the yeast can regenerate after some time, if the concentrations of GDA are not too high.

Poisonous effects could be observed in many cases, when microorganisms were immobilized in epoxide resins or other polymers. As a result, many microorganisms died. The reactions, which then take place, were catalyzed by enzymes and not by the microorganisms.

KIY and TIEDTKE (1991) entrapped *Tetrahymena thermophila* in hollow-Ca-alginate spheres. Under these conditions the protozoa produced very large amounts of α-glucosidase, β-glucosidase, β-hexosaminidase and acid phosphatase in semicontinuous culture for at least 4 weeks (see Fig. 17).

No literature is available on the mechanism of this intensive production of enzymes.

7.3 Alteration of Product Formation by Immobilized Microorganisms

Many improvements of product formation with immobilized cells have been described (see, e.g., MATTIASSON, 1983a, b; FUKUI and TANAKA, 1984; BRODELIUS and VANDAMME, 1987; HARTMEIER, 1988; DE BONT et al., 1990; FURUSAKI and SEKI 1992). The main goal of these investigations was a prolonged and higher activity of microorganisms in producing the desired substances. Mostly, not much interest was shown in observations of changes of the pattern of substances produced by the immobilized microorganisms.

The prolongation of the metabolic activities has been well documented in many examples of semicontinuous and continuous processes for product formation with immobilized microorganisms, e.g., ethanol, glycerol (BISPING et al., 1989; HECKER et al., 1990), citric acid (EIK-

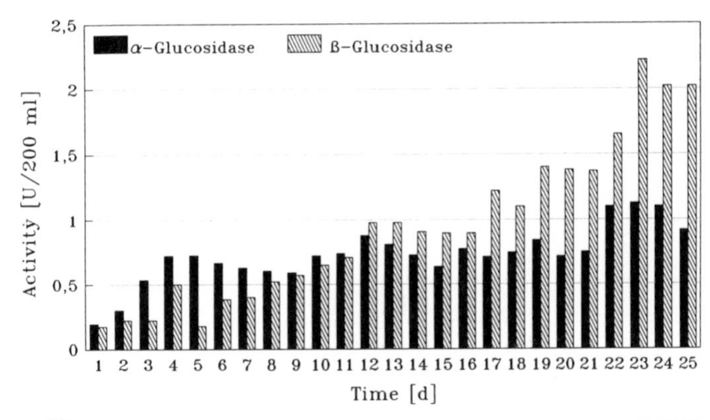

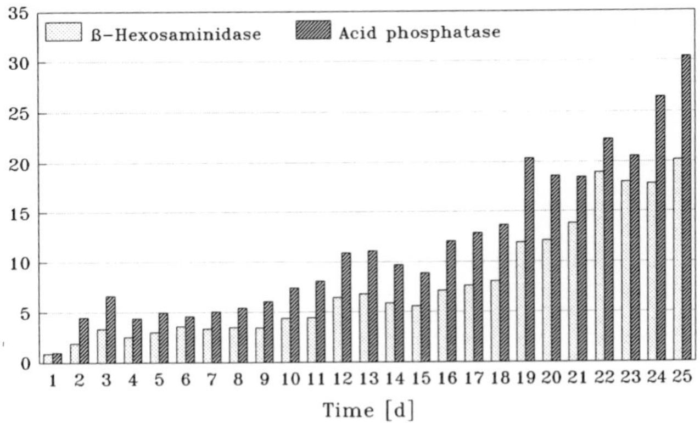

Fig. 17. Secretion of α-glucosidase, β-glucosidase, β-hexosaminidase, and acid phosphatase by encapsulated *Tetrahymena thermophila* during semicontinuous cultivation in a conical bubble column reactor. The supernatant was harvested and replaced by fresh medium every 24 h (KIY and TIEDTKE, 1991, with permission).

MEIER and REHM, 1987), antibiotics (SUZUKI and KARUBE, 1979; VEELKEN and PAPE, 1984), *Claviceps* alkaloids (KOPP and REHM, 1984; DIERKES et al., 1993) and many other products (see MATTIASSON, 1983b; HARTMEIER, 1988; BRODELIUS and VANDAMME, 1987; DE BONT et al., 1990). It was possible to obtain ergot peptides in yields of about 25 mg/L/day or more, in continuous culture of immobilized *Claviceps* for about 30 days. All these results show a distinct prolongation of the metabolic activities of microorganisms. Free cells have only a short metabolic activity, when they are used in batch culture for product formation. In continuous culture free cells grow continuously in contrast to immobilized cells, which are mostly not growing cells or very closely growing cells (see also KOPP and REHM, 1984).

The influence of the Ca-alginate concentration on the formation of clavine alkaloids by a *Claviceps purpurea* strain has been reported by KOPP and REHM (1983) and is illustrated in Fig. 18.

It is obvious that the concentration of Ca-alginate, in which this mold is immobilized, influences the pattern of alkaloids very much and cannot be explained only as a result of a changed mass tranfer into the beads. It may be that the high concentration of agroclavine (with 8% Ca-alginate) is caused by the diffusion barrier for oxygen. On the other hand, cells immobilized in 3% and 4% Ca-alginate produced as much agroclavine as free cells. The conversion of agroclavine into elymoclavine occurred by the action of a peroxidase or oxygen transferase, and the hydroxyl group of elymoclavine originated from molecular oxygen (JINDRA et al., 1968). The assumed barrier for oxygen was, in this case, not strong enough to decrease the oxidation reaction in comparison to that of free cells. Perhaps the enzyme activity was increased.

In semicontinuous culture, free cells of *Claviceps purpurea* produced alkaloids only for five reincubations in new substrate, while cells immobilized in 4% Ca-alginate produced elymoclavine and ergometrine in much higher concentrations for more than 15 reincubations (KOPP and REHM, 1984). The reason for this metabolic change is not known.

With another strain of *Claviceps purpurea* (1029/N5) (a mutant made by KELLER, 1983,

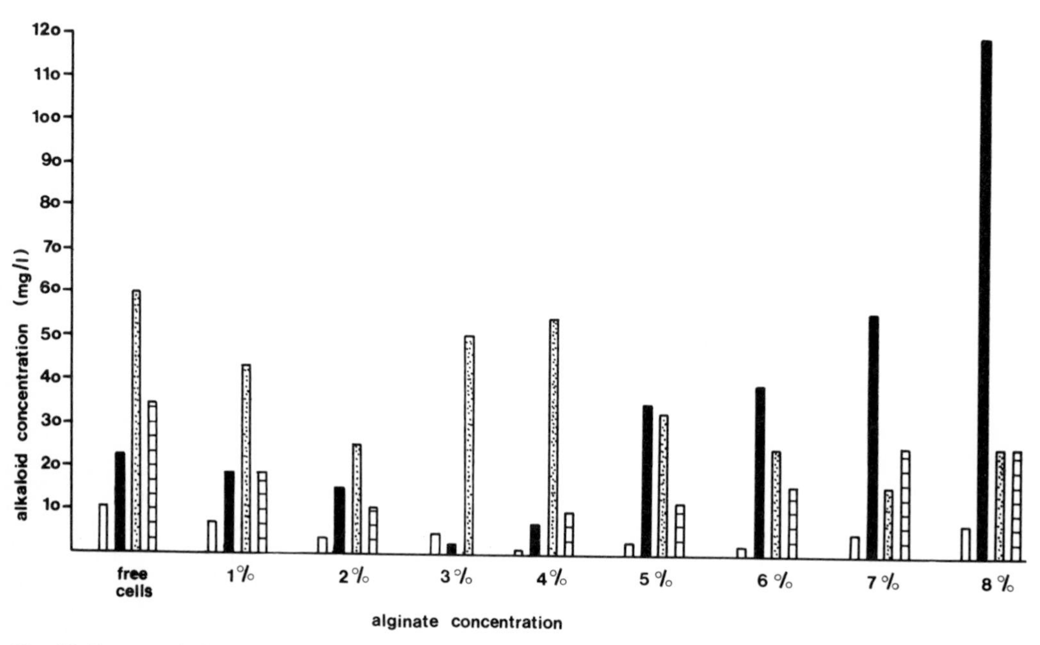

Fig. 18. Pattern of alkaloids produced in different calcium alginate concentrations: chanoclavine □, agroclavine ■, elymoclavine ⊡, ergometrine ⊟, (KOPP and REHM, 1983, reproduced with permission).

which produced and excreted ergot alkaloids, mainly ergotamine and ergocryptine), it could be observed that this strain was regulated to an optimum of ergot alkaloid production with 0.46 mM phosphate when immobilized in Ca-alginate and with 1.84 mM phosphate when the strain was not immobilized (LOHMEYER et al., 1990). This phosphate regulation also influences the amounts of mycelia in the interior of the beads and the morphological behavior of the mycelia. A similar regulatory effect could be observed with *Aspergillus niger* in Ca-alginate and kappa-carrageenan (EIKMEIER et al., 1984).

The entrapment of *Claviceps purpurea* in Ca-alginate beads leads to a microaerophilic metabolism and to the formation of ethanol. Generally immobilization resulted in increased specific activities of phosphofructokinase, but decreased glucose-phosphate dehydrogenase activities. Malate-dehydrogenase and fumarase as representatives of the tricarboxylic cycle are not influenced significantly by immobilization (LOHMEYER et al., 1993).

Investigations with free and immobilized cells adsorbed on sintered glass Raschig rings of an osmotolerant *Pichia farinosa* showed a changed polyol productivity of the adsorbed yeast in comparison with the free cells. At high glucose concentrations (up to 50% glucose), the glycerol-arabitol yield ratio of the immobilized cells was shifted to glycerol (BISPING et al., 1990b). The enhanced osmotic stress is probably responsible for this change of cell metabolism and the increased production of polyols (see also MATTIASSON and HAHN-HÄGERDAL, 1982). The key enzymes of this yeast showed some differences in their activity, whether adsorbed or not. The free cells seemed to have a higher activity of glycerol-3-phosphate dehydrogenase than the adsorbed cells (HÖÖTMANN et al., 1991). For a detailed review of the glycerol and polyol formation see REHM (1988).

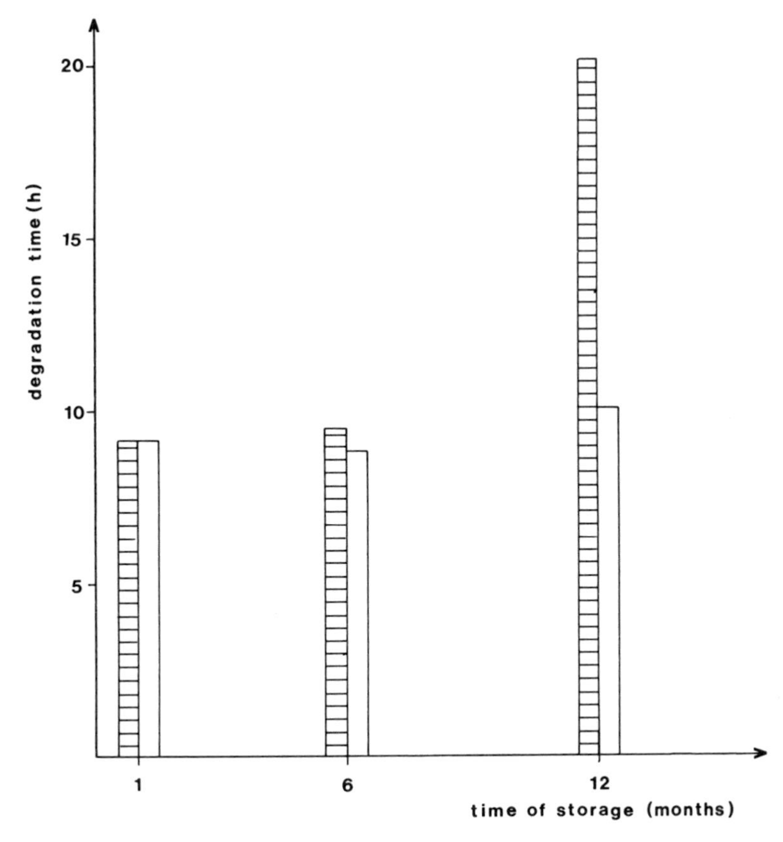

Fig. 19. Storage stability of the mixed culture of *Cryptococcus elinovii* H1 and *Pseudomonas putida* P8 adsorbed on activated carbon. ⊟ first fermentation after storage; □ second fermentation after storage (MÖRSEN and REHM, 1987, reproduced with permission).

A comparison of polyol- and acid formation by different *Aspergillus niger* strains entrapped in Ca-alginate showed that one strain did not produce any polyols, when entrapped, in contrast to free cells. Two other strains produced partly higher concentrations of polyols than free cells (OMAR et al., 1992).

7.4 Prolongation of Metabolic Activities of Immobilized Microorganisms

In many papers (see DE BONT et al., 1990; FURUSAKI and SEKI, 1992) a prolongation of the metabolic activities has been reported. Possible explanations for this prolongation are the use of semicontinuous and continuous processes and also long-time degradation processes, both with immobilized microorganisms.

Furthermore, the metabolic stability and the prolonged life of immobilized microorganisms, which are able to degrade phenol, have been studied. The microorganisms were adsorbed on activated carbon (MÖRSEN and REHM, 1987) or entrapped in Ca-alginate (ZACHE and REHM, 1989). The original degrading capacity for phenol could be observed even after a 6 month storage at 5 °C of Ca-alginate entrapped mixed cultures of *Pseudomonas putida* P8 and *Cryptococcus elinovii* H1. After 4 months at room temperature the phenol degradation capacity decreased to about 20%, although many viable cells could be isolated from the beads (ZACHE and REHM, 1989).

The same two microorganisms were adsorbed on activated carbon. Fig. 19 shows the phenol degradation after 1, 6 and 10 month storage of the viable cells (MÖRSEN and REHM, 1987). The immobilized microorganisms had, after 6 month storage, the same activity as not stored free cells and as cells freshly immobilized by adsorption. Also in this experiment, many viable immobilized cells with degradation capacity could be detected after 12 month storage at 4 °C.

The reasons for a longer life and prolonged metabolic activity of immobilized microorganisms are not known. Many explanations are possible, but will not be discussed here.

7.5 Alteration of Resistance against Poisons

Many observations have been made on increased resistance of immobilized microorganisms against poisons. As an example Fig. 20 shows increased resistance of a *Pseudomonas putida* strain, immobilized in Ca-alginate and in polyacrylamide-hydrazide (PAAH), against phenol compared to free cells. The immobilized bacterium was able to degrade phenol concentrations up to 2 g/L in less than 2 days,

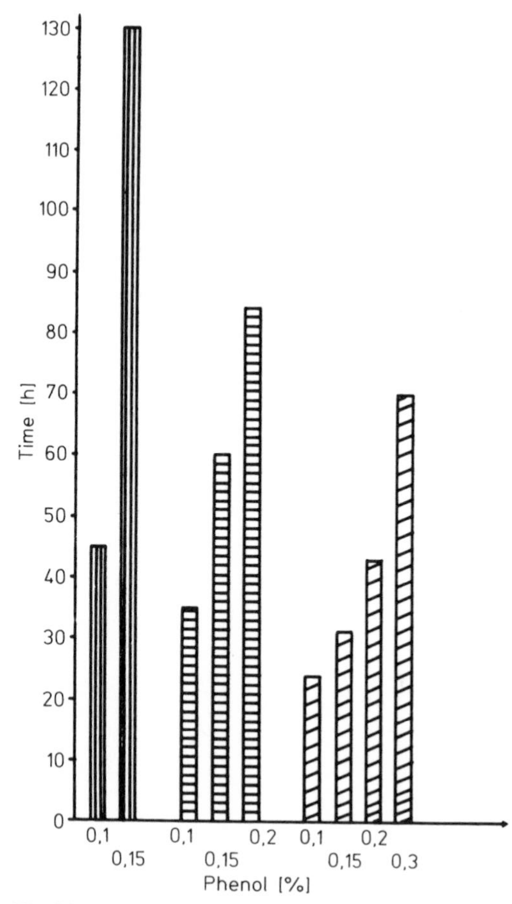

Fig. 20. Duration of the degradation of different phenol concentrations by free (◫) Ca-alginate- (⊟) and PAAH- (▨) immobilized *Pseudomonas* cells (BETTMANN and REHM, 1984, reproduced with permission). PAAH, polyacrylamide-hydrazide.

whereas the free cells did not grow at this concentration (BETTMANN and REHM, 1984).

Although the reason for this increased resistance of entrapped bacteria is not yet known, it can be assumed that the behavior of the microorganisms in the microcolonies found in the beads protects the cells. By forming many colonies at the border of the beads, a diffusion barrier for phenol can be built up. 4-Chlorophenol is also better tolerated by entrapped *Alcaligenes* sp. than by free *Alcaligenes* cells (WESTMEIER and REHM, 1985).

Microorganisms adsorbed on activated carbon, e.g., *Pseudomonas putida* P8 and *Cryptococcus elinovii* H1, showed increased resistance against phenol. While the free cells of *Pseudomonas putida* did not tolerate more than 1.5 g/L phenol, the bacteria adsorbed on activated carbon survived at temporary high phenol concentrations of up to 15 g/L, and they degraded about 80% of the adsorbed phenol (EHRHARDT and REHM, 1985). Here also, the reasons for phenol resistance are unknown. The activated carbon operates like a "depot", the adsorbed phenol can diffuse out of the activated carbon and can be degraded by the microorganisms, but this does not explain the bacterial resistance against the very high phenol concentrations around the cells.

High phenol concentrations (e.g., 17 g/L) are also tolerated by a mixed culture of *Pseudomonas putida* and *Cryptococcus elinovii* (MÖRSEN and REHM, 1987) and also with these microorganisms as a mixed culture entrapped in Ca-alginate (ZACHE and REHM, 1989).

These and many other results suggest metabolic changes in immobilized microorganisms regarding their resistance against poisons. Basic investigations demonstrated that also *Escherichia coli* K12, *Pseudomonas putida* and *Staphylococcus aureus* immobilized in Ca-alginate showed a reduction of the growth inhibition caused by bacteriostatic concentrations of phenol (KEWELOH et al., 1989). In other investigations it was found that growth inhibition by *p*-cresol and 4-chlorophenol of immobilized *Escherichia coli* K12 was equal to that of free cells (HEIPIEPER et al., 1991).

In intensive investigations it was demonstrated that *Escherichia coli* cells, grown entrapped in Ca-alginate, showed low lipid to protein ratios with or without addition of phenol compared to freely grown cells (KEWELOH et al., 1990). This is shown in Tab. 3.

In the presence of sublethal concentrations of phenol, 4-chlorophenol and *p*-cresol, cells of *Escherichia coli* modified the fatty acid composition of their membrane lipids. The result of these changes, induced by the phenols, was an increase in the degree of saturation of lipids in order to compensate for the increase of fluidity of the membrane (KEWELOH et al., 1991).

By entrapment, the formation of microcolonies leads to a connection of membrane gradients of the cells, because of the sharing of a common external milieu. After the loss of metabolites, due to an increase in permeability, the cells are faced with relatively high concentrations of these compounds inside the space of the colonies. In contrast to free cells the instantaneous high dilution of the lost cellular material into the gel-free medium is prevented. Therefore, the efflux from immobilized cells is small and retarded compared with that of free cells. The advantage of the reduced permeability changes for the cell can consist in a saving of cytoplasmic material and of cellular energy for the restoration of a low membrane permeability needed for survival (HEIPIEPER et al., 1991). Furthermore, a change of unsaturated fatty acids from a *cis*-configuration to a *trans*-configuration in the membrane may prevent the efflux of substances in immobilized cells under the influence of phenols (HEIPIEPER et al., 1992).

These results let us suppose that also other membrane-active compounds, at sublethal concentrations, are less inhibitory to immobilized microorganisms, than they are to free

Tab. 3. Lipid–Protein Ratios of Membranes of *Escherichia coli* Cells Grown in Suspension or Immobilized in Ca-alginate

	Cytoplasmic Membrane	Outer Membrane
Freely grown cells	0.794 ± 0.072	0.479 ± 0.089
Immobilized cells	0.578 (72.8)	0.373 (77.9)
Immobilized cells with phenol (0.5 g/L)	0.468 (58.9)	0.308 (64.3)

Values are given in milligrams of lipid per milligrams of protein. The numbers in parentheses denote percentages of the values found for free cells.

organisms. This may be confirmed by a great number of observations that cells in biofilms are effectively protected against biocidal agents and antibiotics (LeChevallier et al., 1988; Nichols et al., 1989).

Also surface-associated bacteria in poly-aggregates (e.g., biofilms) have been reported to be less sensitive to antibacterial agents, such as antibiotics, biocides, disinfectants. Nichols (1989) investigated the reduced toxicity of tobramycin (an aminoglycoside antibioticum) and cefsulodin (α,β-lactam compound) and their toxicity and penetration into biofilms of *Pseudomonas aeruginosa* with a mathematical model. Nichols stated that the restricted penetration in biofilms is probably not a generally applicable mechanism responsible for the lowered sensitivity of bacteria in biofilms, although a significant barrier can exist, if there is a detoxifying enzyme present within the biofilm. He assumed that an alternative reason for the lower sensitivity of cells, deeply embedded in a biofilm, may be a different physiology.

A similar observation was made with *Staphylococcus aureus* cells which colonized the surface of a prosthetic device. These bacteria have often a reduced sensitivity against antibiotics (Peters, 1988). For more literature references on these effects see Brown et al. (1988) and Nichols (1989).

8 Conclusion and Future Aspects

In this chapter effects of the immobilization on the metabolism of microorganisms and other cells have been discussed. Many of these effects are caused by the conditions of immobilization: especially the environmental conditions have changed in comparison to those of free cells. However, besides these changes, other explanations must be found for these effects. Real physiological alterations occur during immobilization the cause of which should be further investigated.

Robert Koch was the first who developed a method to study microorganisms in pure cultures. These pure cultures have brought a very great progress to our knowledge on the physiology of microorganisms. We should always keep in mind that all our results with pure cultures are results with miroorganisms which are not found in their natural environment: "The microorganisms are cultivated in an artificial manner".

In their natural environment many – perhaps most – microorganisms are fully or partly immobilized, and their metabolism is influenced by this fact. Immobilization can be observed with soil microorganisms, where the immobilization takes place by adsorption, e.g., on silicate (comparable with glass adsorption), on clay, or by entrapment, e.g., in slimy polysaccharides, cellulose and others. Often biofilms can be observed in the soil (Burns, 1986, 1989; Fletcher, 1990). Fig. 21 shows a diagrammatic representation of a soil microen-

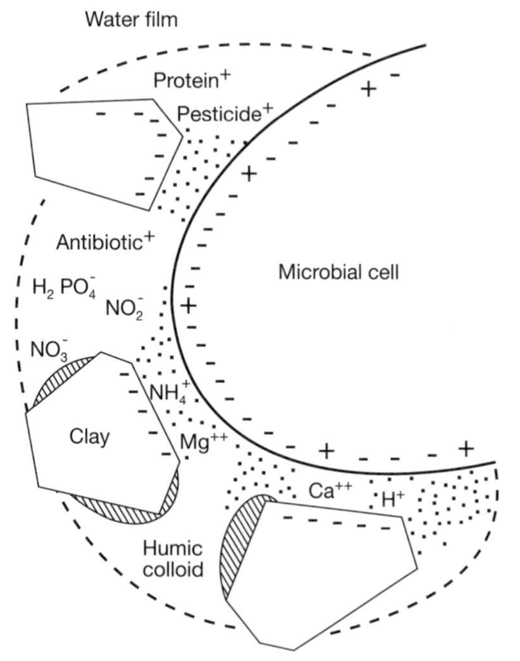

Fig. 21. Diagrammatic representation of a soil microbial miroenvironment containing clay and organic colloids, adsorbed organic and inorganic cations and repulsed anions, and polysaccharide linkages (·:·) aiding the retention of the cell at colloidal surfaces. The diagram represents an area of approximately 8 μm^2 (from Burns, 1989, with permission).

vironment containing clay, organic colloids, inorganic cations and repulsed anions (BURNS, 1989).

In the next decade, many interesting results on the properties of immobilized microorganisms in soil using modern methods can be expected. Combining these results with those we get with immobilized microorganisms in laboratories, we will be able to obtain much more information about the microbial life in nature.

Acknowledgement

We are greatly indepted to Mrs. BARBARA STECKEL for the enormous help in typing and correcting this manuscript.

9 References

BETTMANN, H., REHM, H. J. (1984), Degradation of phenol by polymer entrapped microorganisms, *Appl. Microbiol. Biotechnol.* **20,** 285–290.

BEUNINK, J., REHM, H. J. (1988), Synchronous anaerobic and aerobic degradation of DDT by an immobilized mixed culture system, *Appl. Microbiol. Biotechnol.* **29,** 72–80.

BEUNINK, J., REHM, H. J. (1990), Coupled reductive and oxidative degradation of 4-chloro-2-nitrophenol by a co-immobilized mixed culture system, *Appl. Microbiol. Biotechnol.* **34,** 108–115.

BEUNINK, J., BAUMGAERTL, H., ZIMELKA, W., REHM, H. J. (1989), Determination of oxygen gradients in single Ca-alginate beads by means of oxygen-microelectrodes, *Experientia* **45,** 1041–1047.

BISHOP, P. L., KINNER, N. E. (1986), Aerobic fixed-film processes, in : *Biotechnology* (REHM, H. J., REED, G., Eds.), Vol. 8, pp. 113–176, Weinheim–Basel–New York: VCH.

BISPING, B., REHM, H. J. (1986), Glycerol production by cells of *Saccharomyces cerevisiae* immobilized in sintered glass, *Appl. Microbiol. Biotechnol.* **23,** 174–179.

BISPING, B., HECKER, D., REHM, H. J. (1989), Glycerol production by semicontinuous fed-batch fermentation with immobilized cells of *Saccharomyces cerevisiae, Appl. Microbiol. Biotechnol.* **32,** 119–123.

BISPING, B., BAUMANN, U., REHM, H. J. (1990a), Production of glycerol by immobilized *Pichia farinosa, Appl. Mirobiol. Biotechnol.* **32,** 380–386.

BISPING, B., HELLFORS, H., HONECKER, S., REHM, H. J. (1990b), Formation of citric acid and polyols by immobilized cells of *Aspergillus niger, Food Biotechnol.* **4** (1), 17–23.

BRIGHT, J. J., FLETCHER, M. (1983), Amino acid assimilation and electron transport system activity in attached and free-living marine bacteria, *Appl. Environ. Microbiol.* **45,** 818–825.

BRODELIUS, P., VANDAMME, E. J. (1987), Immobilized cell systems, in: *Biotechnology* (REHM, H. J., REED, G., Eds.), Vol. 7a, pp. 405–464, Weinheim–New York–Cambridge–Basel: VCH.

BROUERS, N., SHI, D. J., HALL, D. O. (1990), Comparative physiology of naturally and artificially immobilized *Anabaena azollae*, in: *Physiology of Immobilized Cells* (DE BONT, J. A. M., VISSER, B., MATTIASSON, B., TRAMPER, J., Eds.), pp. 75–86, *Proc. Int. Symp.* held in Wageningen, The Netherlands, 10–13 December 1989, Amsterdam: Elsevier Science Publishers B.V.

BROWN, M. R. W., ALLISON, D. G., GILBERT, T. (1988), Resistance of bacterial biofilms to antibiotics: a growth-rate related effect? *J. Antimicrob. Chemother.* **22,** 777–780.

BURNS, R. G. (1986), Interactions of enzymes with soil mineral and organic colloids, in: *Interactions of Soil Minerals with Natural Organics and Microbes* (HUANG, P. M., SCHNITZER, M., Eds.), pp. 429–451, Madison, WI: Soil Science Society of America.

BURNS, R. G. (1989), Microbial and enzymatic activities in soil biofilms, in: *Structure and Function of Biofilms* (CHARACKLIS, W. G., WILDERER, P. A., Eds.), pp. 333–349, Chichester: John Wiley & Sons, *Dahlem Workshop Reports.*

CHARACKLIS, W. G., WILDERER, P. A. (1989), *Structure and Function of Biofilms*, Chichester–New York–Brisbane–Toronto–Singapore: John Wiley & Sons: *Dahlem Workshop Reports.*

CHIBATA, I., WINGARD, JR., L. B. (1983), *Applied Biochemistry and Bioengineering*, Vol. 4: *Immobilized Microbial Cells*, New York–London–Paris–San Diego–Sao Paulo–Sydney–Tokyo–Toronto: Academic Press.

DE BONT, J. A. M., VISSER, J., MATTIASSON, B., TRAMPER, J. (1990), *Physiology of Immobilized Cells*, Amsterdam–Oxford–New York–Tokyo: Elsevier.

DIERKES, W., LOHMEYER, M., REHM, H. J. (1993), Long-term production of ergot peptides by immobilized *Claviceps purpurea* in semicontinuous and continuous cultures, submitted for publication.

DORAN, P. M., BAILEY, J. E. (1986a), Effects of immobilization on growth, fermentation properties, and macromolecular composition of *Saccharo-*

myces cerevisiae attached to gelatin, *Biotechnol. Bioeng.* **28,** 73–87.

Doran, P. M., Bailey, J. E. (1986b), Effects of hydroxyurea on immobilized and suspended yeast fermentation rates and cell cycle operation, *Biotechnol. Bioeng.* **28,** 1814–1831.

Ehrhardt, H. M., Rehm, H. J. (1985), Phenol degradation by microorganisms adsorbed on activated carbon, *Appl. Microbiol. Biotechnol.* **21,** 32–36.

Eikmeier, H., Rehm, H. J. (1987), Semicontinuous and continuous production of citric acid with immobilized cells of *Aspergillus niger*, *Z. Naturforsch.* **42c,** 408–413.

Eikmeier, H., Westmeier, F., Rehm, H. J. (1984), Morphological development of *Aspergillus niger* immobilized in Ca-alginate and kappa-carrageenan, *Appl. Microbiol. Biotechnol.* **19,** 53–57.

Ellwood, D. C., Keevel, C. W., Marsh, P. D., Brown, C. M., Wardell, J. N. (1982), Surface associated growth, *Phil. Trans. Soc. Lond. Biol.* **297,** 517–532.

Fletcher, M. (1990), Surfaces and bacteria in natural environments – Nature's banquet table? in: *Physiology of Immobilized Cells* (De Bont, J. A. M., Visser, B., Mattiasson, B., Tramper, J., Eds.), pp. 25–35, *Proc. Int. Symp.* held in Wageningen, The Netherlands, 10–13 December 1989, Amsterdam: Elsevier Science Publishers B.V.

Freeman, A., Aharonowitz, Y. (1981), Immobilization of microbial cells in crosslinked, prepolymerized linear polyacrylamide gels: antibiotic production by immobilized *Streptomyces clavuligerus*, *Biotechnol. Bioeng.* **23,** 2747–2759.

Fukui, S., Tanaka, A. (1984), Application of biocatalysts immobilized by prepolymer methods, *Adv. Biochem. Eng. Biotechnol.* **29,** 1–33.

Furusaki, S., Seki, M. (1992), Use and engineering aspects of immobilized cells in biotechnology, *Adv. Biochem. Eng. Biotechnol.* **46,** 161–185.

Galazzo, J. L., Shanks, J. V., Bailey, J. E. (1987), Comparison of suspended and immobilized yeast metabolism using ^{31}P Nuclear Magnetic Resonance Spectroscopy, *Biotechnol. Tech.* **1,** 1–6.

Ghose, T. K., Bandyopadhyay, K. K. (1980), Rapid ethanol fermentation in immobilized yeast cell reactor, *Biotechnol. Bioeng.* **22,** 1489–1496.

Gosmann, B., Rehm, H. J. (1986), Oxygen uptake by microorganisms entrapped in Ca-alginate, *Appl. Microbiol. Biotechnol.* **23,** 162–167.

Gosmann, B., Rehm, H. J. (1988), Influence of growth behaviour and physiology of alginate-entrapped microorganisms on the oxyen consumption, *Appl. Microbiol. Biotechnol.* **29,** 554–559.

Haas, G. J. (1956), Use of the membrane filter in the brewing laboratory, *Wallerstein Lab. Commun.* **19,** 7–20.

Hahn-Hägerdal, B. (1990), Physiological aspects of immobilized cells: A general overview, in: *Physiology of Immobilized Cells* (De Bont, J. A. M., Visser, B., Mattiasson, B., Tramper, J., Eds.), pp. 481–486, *Proc. Int. Symp.* held in Wageningen, The Netherlands, 10–13 December 1989, Amsterdam: Elsevier Science Publishers B.V.

Hannoun, B. J. M., Stephanopoulos, G. (1986), Diffusion coefficients of glucose and ethanol in cell-free and cell-occupied calcium-alginate membranes, *Biotechnol. Bioeng.* **28,** 829–835.

Hartmeier, W. (1988), *Immobilized Biocatalysts*, Berlin–Heidelberg: Springer-Verlag.

Hecker, D., Bisping, B., Rehm, H. J. (1990), Continuous glycerol production by the sulphite process with immobilized cells of *Saccharomyces cerevisiae, Appl. Microbiol. Biotechnol.* **32,** 627–632.

Heipieper, H. J., Keweloh, H., Rehm, H. J. (1991), Influence of phenols on growth and membrane permeability of free and immobilized *Escherichia coli*, *Appl. Environ. Microbiol.* **57,** 1213–1217.

Heipieper, H. J., Diefenbach, R., Keweloh, H. (1992), Conversion of *cis*-unsaturated fatty acids to *trans,* a possible mechanism for the protection of phenol-degrading *Pseudomonas putida* P8 from substrate toxicity, *Appl. Environ. Microbiol.* **58,** 1847–1852.

Hiemstra, H., Dijhuizen, L., Harder, W. (1983), Diffusion of oxygen in alginate gels related to the kinetics of methanol oxidation by immobilized *Hansenula polymorpha* cells, *Eur. J. Appl. Microbiol. Biotechnol.* **18,** 189–196.

Hilge-Rotmann, B., Rehm, H. J. (1990), Comparison of fermentation properties and specific enzyme activities of free and calcium-alginate-entrapped *Saccharomyces cerevisiae*, *Appl. Microbiol. Biotechnol.* **33,** 54–58.

Hilge-Rotmann, B., Rehm, H. J. (1991), Influence of grwoth behavior relationship between fermentation capability and fatty acid composition of free and immobilized *Saccharomyces cerevisiae*, *Appl. Miocrobiol. Biotechnol.* **34,** 502–508.

Holcberg, I. B., Margalith, P. (1981), Alcoholic fermentation by immobilized yeast at high sugar concentrations, *Eur. J. Appl. Microbiol. Biotechnol.* **13,** 133–140.

Honecker, S., Bisping, B, Yang, Z., Rehm, H. J. (1989), Influence of sucrose concentration and phosphate limitation on citric acid production by immobilized cells of *Aspergillus niger*, *Appl. Microbiol. Biotechnol.* **31,** 17–24.

Hooijmans, C. M., Geraats, S. G. M., van Niel, E. W. J., Robertson, L. A., Heijnen, J. J., Luyben, K. Ch. A. M. (1990), Nitrification/denitrification by immobilized *Thiosphaera pantotropha* in relation to oxygen profiles, in: *Physiology of Immobilized Cells* (De Bont, J. A. M., Visser, B., Mattiasson, B., Tramper, J., Eds.), pp. 361–368,

Proc. Int. Symp. held in Wageningen, The Netherlands, 10–13 December 1989, Amsterdam: Elsevier Science Publishers B.V.

HÖÖTMANN, U., BISPING, B., REHM, H. J. (1991), Physiology of polyol formation by free and immobilized cells of the osmotolerant yeast *Pichia farinosa, Appl. Microbiol. Biotechnol.* **35,** 258–263.

JINDRA, A., RAMSTAD, E., FLOSS, H. G. (1968), Biosynthesis of ergot alkaloids. Peroxidase and the conversion of agroclavine to elymoclavine, *Lloydia* **31,** 190–196.

KELLER, U. (1983), Highly efficient mutagenesis of *Claviceps purpurea* by using protoplasts, *Appl. Environ. Microbiol.* **46,** 580–584.

KENNEDY, J. F., CABRAL, M. S. (1988), Immobilized living cells and their applications, in: *Applied Biochemistry and Bioengineering*, Vol. 4: *Immobilized Microbial Cells* (CHIBATA, I., WINGARD, L. B., Eds.), pp. 189–280, New York–London–San Francisco: Academic Press.

KEWELOH, H., HEIPIEPER, H. J., REHM, H. J. (1989), Protection of bacteria against toxicity of phenol by immobilization in calcium alginate, *Appl. Microbiol. Biotechnol.* **31,** 383–389.

KEWELOH, H., WEYRAUCH, G., REHM, H. J. (1990), Phenol-induced membrane changes in free and immobilized *Escherichia coli, Appl. Microbiol. Biotechnol.* **33,** 66–71.

KEWELOH, H., DEIFENBACH, R., REHM, H. J. (1991), Increase of phenol tolerance of *Escherichia coli* by alterations of the fatty acid composition of the membrane lipids, *Arch. Microbiol.* **157,** 49–53.

KIY, T., TIEDTKE, A. (1991), Lysosomal enzymes produced by immobilized *Tetrahymena thermophila, Appl. Microbiol. Biotechnol.* **35,** 14–18.

KOPP, B., REHM, H. J. (1983), Alkaloid production by immobilized mycelia of *Claviceps purpurea, Eur. J. Appl. Microbiol. Biotechnol.* **18,** 257–263.

KOPP, B., REHM, H. J. (1984), Semicontinuous cultivation of immobilized *Claviceps purpurea, Appl. Microbiol. Biotechnol.* **19,** 141–145.

KUSCH, J., REHM, H. J. (1986), Regulation aspects of roquefortine production by free and Ca-alginate immobilized mycelia of *Penicillium roqueforti, Appl. Microbiol. Biotechnol.* **23,** 394–399.

LECHEVALLIER, M. W., CAWTHON, C. D., LEE, R. G. (1988), Inactivation of biofilm bacteria, *Appl. Environ. Microbiol.* **54,** 2492–2499.

LOHMEYER, M., DIERKES, W., REHM, H. J. (1990), Influence of inorganic phosphate and immobilization on *Claviceps purpurea, Appl. Microbiol. Biotechnol.* **33,** 196–201.

LOHMEYER, M., KÖRNER, M., REHM, H. J. (1993), Primary metabolism and ergot alkaloid production by free and immobilized cells of *Claviceps purpurea*, in preparation.

MACLEOD, F. A., GUIOT, S. R., COSTERTON, J. W. (1990), Layered structure of bacterial aggregates produced in an upflow anaerobic sludge bed and filter reactor, *Appl. Environ. Microbiol.* **56,** 1598–1607.

MAHMOUD, W., REHM, H. J. (1986), Morphological examination of immobilized *Streptomyces aureofaciens* during chlorotetracycline fermentation, *Appl. Microbiol. Biotechnol.* **23,** 305–310.

MARSHALL, K. C. (1984), *Microbial Adhesion and Aggregation*, Berlin–Heidelberg–New York–Tokyo: Springer-Verlag.

MARTINSEN, A., STORRE, I., SKJAK-BREAK, G. (1992), Alginate as immobilization material: III. Diffusional properties, *Biotechnol. Bioeng.* **39,** 186–194.

MATTIASSON, B. (1983a), *Immobilized Cells and Organelles*, Vol. I, Boca Raton, Florida: CRC Press.

MATTIASSON, B. (1983b), *Immobilized Cells and Organelles*, Vol. II, Boca Raton, Florida: CRC Press.

MATTIASSON, B., HAHN-HÄGERDAL, B. (1982), Microenvironmental effects on metabolic behavior of immobilized cells: a hypothesis, *Eur. J. Appl. Microbiol. Biotechnol.* **16,** 52–55.

MORISAKI, H. (1983), Effect of solid–liquid interface on metabolic activity of *Escherichia coli, J. Gen. Appl. Microbiol.* **29,** 195–204.

MÖRSEN, A., REHM, H. J. (1987), Degradation of phenol by a mixed culture of *Pseudomonas putida* and *Cryptococcus elinovii* adsorbed on activated carbon, *Appl. Microbiol. Biotechnol.* **26,** 283–288.

NAVARRO, H. M., DURAND, G. (1977), Modification of yeast metabolism by immobilization onto porous glass, *Eur. J. Appl. Microbiol. Biotechnol.* **4,** 243–254.

NICHOLS, W. W. (1989), Susceptibility of biofilms to toxic compounds, in: *Structure and Function of Biofilms* (CHARACKLIS, W. G., WILDERER, P. A., Eds.), pp. 321–331, Chichester: John Wiley & Sons, *Dahlem Workshop Reports.*

NICHOLS, W. W., EVANS, M. J., SLACK, M. P. E., WALMSLEY, H. L. (1989), The penetration of antibiotics into aggregates of mucoid and non-mucoid *Pseudomonas aeruginosa, J. Gen. Microbiol.* **135,** 1291–1303.

OMAR, S. H. (1993), *Morphological Changes in Immobilized Microorganisms*, an atlas, in preparation.

OMAR, S. H., REHM, H. J. (1988), Degradation of *n*-alkanes by *Candida parapsilosis* and *Penicillium frequentans* immobilized on granular clay and aquifer sand, *Appl. Mirobiol. Biotechnol.* **28,** 103–108.

OMAR, S. H., HONECKER, S., REHM, H. J. (1992), A comparative study on the formation of citric acid and polyols and on morphological changes of three strains of free and immobilized *Aspergillus niger, Appl. Microbiol. Biotechnol.* **36,** 518–524.

PETERS, G. (1988), New considerations of the pathogenesis of coagulase-negative staphylococcal foreign bodies infections, *J. Antimicrob. Chemother.* **21,** Suppl. C, 139–148.

RADOVICH, J. N. (1985), Mass transfer effects in fermentations using immobilized whole cells, *Enzyme Microb. Technol.* **7,** 2–10.

REHM, H. J. (1988), Microbial production of glycerols and other polyols, in: *Biotechnology* (REHM, H. J., REED, G., Eds.), Vol. 6b, pp. 51–69, Weinheim–Basel–Cambridge–New York: VCH.

REHM, H. J., GREVE, R. (1993), Influence of glutaric di-aldehyde on the metabolism of *Saccharomyces cerevisiae*, in preparation.

REHM, H. J., REED, G. (Eds.), *Biotechnology, 2nd. Edition*, Vol. 9, Weinheim–Basel–Cambridge–New York: VCH, in preparation.

REVSBECH, N. P. (1989), Microsensors: Spatial gradients in biofilms, in: *Structure and Function of Biofilms* (CHARACKLIS, W. G., WILDERER, P. A., Eds.), pp. 129–144, Chichester: John Wiley & Sons, *Dahlem Workshop Reports.*

ROSEVEAR, A., LAMBE, C. A. (1985), Immobilized plant cells, *Adv. Biochem. Eng. Biotechnol.* **31,** 37–58.

SAHM, H. (1984), Anaerobic wastewater treatment, *Adv. Biochem. Eng. Biotechnol.* **29,** 83–115.

SATO, K., TODA, K. (1983), Oxygen uptake rate of immobilized growing *Candida lipolytica*, *J. Ferment. Technol.* **61,** 239–245.

SCOTT, C. D. (1987), Immobilized cells: a review of recent literature, *Enzyme Microb. Technol.* **9,** 66–73.

SHINMYO, A., KIMURA, H., OKADA, H. (1982), Physiology of *alpha*-amylase production by immobilized *Bacillus amyloliquefaciens*, *Eur. J. Appl. Microbiol. Biotechnol.* **14,** 7–12.

SIMON, J. P., BENOOT, T., DEFROYENNES, J. P., DECKERTS, B., DEKEGEL, D., VANDEGANS, J. (1990), Physiology and morphology of Ca-alginate entrapped *Saccharomyces cerevisiae*, in: *Physiology of Immobilized Cells* (DE BONT, J. A. M., VISSER, B., MATTIASSON, B., TRAMPER, J., Eds.), pp. 583–590, *Proc. Int. Symp.* held in Wageningen, The Netherlands, 10–13 December 1989, Amsterdam: Elsevier Science Publishers B.V.

SONOMOTO, K., OKADA, T., TANAKA, A. (1990), Unexpected discovery and production of novel site-specific peptidase with immobilized yeast cells, in: *Physiology of Immobilized Cells* (DE BONT, J. A. M., VISSER, B., MATTIASSON, B., TRAMPER, J., Eds.), pp. 311–316, *Proc. Int. Symp.* held in Wageningen, The Netherlands, 10–13 December 1989, Amsterdam: Elsevier Science Publishers B.V.

SUZUKI, S., KARUBE, I. (1979), Production of antibiotics and enzymes by immobilized whole cells, *ACS Symp. Ser.* **106,** 59–72.

TANAKA, H., MATSUMURA, M., VELIKY, I. A. (1984), Diffusion characteristics of substrates in Ca-alginate gel beads, *Biotechnol. Bioeng.* **26,** 53–58.

TANAKA, H., KUROSOWA, H., MURAKAMI, H. (1986), Ethanol production from starch by a coimmobilized mixed culture system of *Aspergillus awamori* and *Zymomonas mobilis*, *Biotechnol. Bioeng.* **28,** 1761–1768.

TRAMPER, J., LUYBEN, K. CH. A. M., VAN DEN TWEEL, W. J. J. (1983), Kinetic aspects of glucose oxidation by *Gluconobacter oxydans* cells immobilized in Ca-alginate, *Eur. J. Appl. Microbiol.* **17,** 13–18.

VAN LOOSDRECHT, M. C. M., LYKLEMA, J., NORDE, W., ZEHNDER, A. J. B. (1990), Influence of interfaces on microbial activity, *Microbiol. Rev.* **54,** 75–87.

VEELKEN, M., PAPE, H. (1984), Production of nikkomycin by immobilized *Streptomyces* cells – Physiological properties, *Appl. Microbiol. Biotechnol.* **19,** 146–152.

VOJTISEK, V., JIRKU, V. (1983), Immobilized cells, *Folia Microbiol.* **28,** 309–340.

WESTMEIER, F., REHM, H. J. (1985), Biodegradation of 4-chlorophenol by entrapped *Alcaligenes* sp. A 7–2, *Appl. Microbiol. Biotechnol.* **22,** 301–305.

WILDERER, P. A., CHARACKLIS, W. G. (1989), Structure and function of biofilms, in: *Structure and Function of Biofilms* (CHARACKLIS, W. G., WILDERER, P. A., Eds.), pp. 5–17, Chichester: John Wiley & Sons, *Dahlem Workshop Reports.*

ZACHE, G., REHM, H. J. (1989), Degradation of phenol by a coimmobilized entrapped mixed culture, *Appl. Microbiol. Biotechnol.* **30,** 426–432.

II. Biology of Industrial Organisms

7 Methanogens

HELMUT KÖNIG

Ulm, Federal Republic of Germany

1 General Aspects 252
2 Taxonomy 252
3 Ecology 255
4 Morphology 256
5 Cell Components 257
5.1 Envelopes 257
5.2 Membranes 259
5.3 Low-Molecular Weight Compounds 259
6 Culture Technique 259
7 Energy Metabolism 260
8 Carbon Metabolism 261
9 Molecular Biology 262
10 Applied Aspects 263
10.1 Biogas Production 263
10.2 Restriction Enzymes 263
11 References 263

1 General Aspects

The first report on methane emanation from aquatic muds was given by VOLTA in 1776. About a century later it was recognized that methane production is linked to anaerobic microbial action, namely cellulose degradation. The first methanogen in pure culture was not obtained before 1947. After overcoming the difficulties in growing the strict anaerobic methanogens by developing special culture techniques (BALCH et al., 1979) the interest in methanogens increased in the last decade. The distinct phylogenetic position — the methanogens were the first archaea detected — as well as their unique biochemical and genetic properties stimulated basic investigations of this microbial group.

Besides basic science, economic and geochemical aspects also led to studies with methanogens. Since methanogens are involved in anaerobic degradation of organic material to methane, they play an important role in the global carbon cycle. The oil crisis provoked the search for alternative energy sources such as natural gas. Problems with environmental pollution, waste treatment, the involvement of methane in atmospheric reaction sequences leading to ozone depletion and the greenhouse effect are further applied factors. Furthermore, the cattle industry is interested in rumen fermentation, because methane production means a loss of energy.

2 Taxonomy

The methanogenic bacteria are distinguished from all other prokaryotes by the production of methane (CH_4), which is a common physiological feature of these otherwise heterogeneous microorganisms (BALCH et al., 1979; JONES et al., 1987; JAIN et al., 1988; BOONE and MAH, 1989). Sequence analysis of their 16S rRNA led to the detection of the archaea, which in addition to the bacteria and the eucarya represent the third evolutionary domain of life (Fig. 1; WOESE, 1987; WOESE et al., 1990). The analysis of pairs of duplicated genes revealed with most probability that the archaea are more closely related to the eucarya than to the bacteria (IWABE et al., 1989). The archaea include several diverse groups with distinct features; they are divided into two main lineages: the methanogenic branch (euryarchaeota) and the branch of extreme thermophilic sulfur metabolizers (crenarchaeota). Methanogens are among the first microbial groups, whose taxonomy is based on natural phylogeny. So far the methanogenic bacteria are grouped in 3 orders, 7 families and 21 genera.

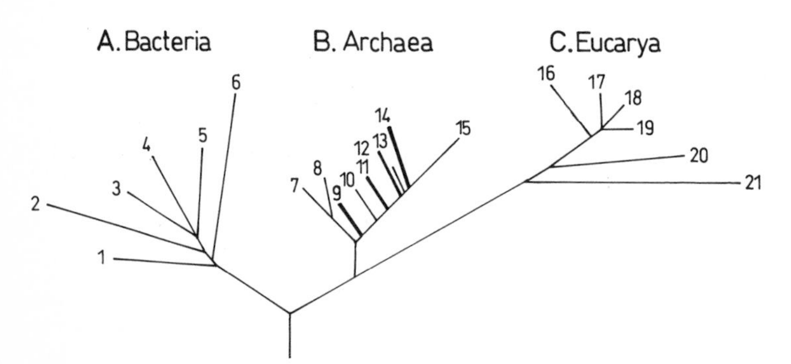

Fig. 1. Phylogenetic tree of organisms (from WOESE et al., 1990, modified; bold lines, methanogens).
Domain A: Bacteria. 1. Thermotogales, 2. Flavobacteria, 3. Cyanobacteria, 4. Purple bacteria, 5. Gram-positive bacteria, 6. Green non-sulfur bacteria.
Domain B: Archaea. Kingdom I: Crenarchaeota. 7. *Pyrodictium,* 8. *Thermoproteus.* Kingdom II: Euryarchaeota. 9. *Methanopyrus*, 10. Thermococcales, 11. Methanococcales, 12. Methanobacteriales, 13. *Archaeoglobus*, 14. Methanomicrobiales, 15. Extreme halophiles.
Domain C: Eucarya. 16. Animals, 17. Ciliates, 18. Plants, 19. Fungi, 20. Flagellates, 21. Microsporidia.

Up to now 65 species are included in the Approved List of Bacterial Names (Tab. 1).

Many isolates exhibit a similar morphology and lack easily determinable physiological features. For the identification of the species it is often necessary to examine also biochemical and molecular markers which are more difficult to determine. Minimal standards for the description of new taxa have been described (BOONE and WHITMAN, 1988). Characteristic features for the determination of methanogens up to the genus level are included in Tab. 1.

Tab. 1. Systematics and Determinative Features of Methanogens
(n.d. not described so far; * type species)

Domain: Archaea
The archaea are distinguished from the other two domains by the sequence signature of the 16S rRNA. The cytoplasmic membranes contain phytanyl ether lipids. The multisubunit RNA polymerases are of the AB′B″C type, (A + C)B′B″ type or BAC type. Elongation factor EF-2 is ADP-ribosylated by diphtheria toxin. Cell walls lack muramic acid. Gram-positive species possess pseudomurein, methanochondroitin or heteropolysaccharide cell walls. Gram-negative organisms have only (glyco-)protein surface layers. Ribothymidine is replaced by 1-methylpseudouridine or pseudouridine in the "common" arm of the tRNAs.

Subdomain: Euryarchaeota
Methanogens
Strict anaerobes. Greenish fluorescence in the fluorescent microscope. Cells are rods, short lancet-shaped rods, cocci, irregular plates or spirilli. Cells produce methane from methyl-CoM catalyzed by the enzyme methyl-CoM reductase. Cofactors F_{420}, F_{430}, methanopterin and CoM are present. RNA polymerase is of the AB′B″C type. H_2/CO_2, H_2/CO, formate, acetate, methanol, methanol/H_2, methylamines or dimethyl sulfide serve as substrates. Some species can use primary or secondary alcohols as electron donors. Obligate nickel requirement.

Order I: Methanobacteriales
Gram-positive organisms. Rods, short, lancet-shaped rods, cocci. Cells occur singly, in chains or in packets. Pseudomurein cell walls. Cell division by binary fission. C_{20} isopranyl diethers and C_{40} isopranyl tetraethers are found. Mesophiles to extreme thermophiles.

Family I: Methanobacteriaceae
Growth optimum between 37 °C and 70 °C. Energy source is H_2 or formate. Cell walls contain no S-layer.

Genus 1: *Methanobrevibacter*
Cells are short, lancet-shaped rods. DNA base composition is 27–32 mol% G + C. Two species contain Thr or Orn in the peptide moiety, one species has exclusively GalN instead of GlcN in the glycan strand of the pseudomurein. Mesophilic growth. Some species have one flagellum.
Species: *Mbr. arboriphilus, Mbr. ruminatium*, Mbr. smithii*

Genus 2: *Methanobacterium*
Cell are straight, long, sometimes irregular rods.DNA base composition is 33–61 mol% G + C. Non-motile. Mesophilic to thermophilic growth.
Species: *Mb. alcaliphilum, Mb. bryantii, Mb. espanolae, Mb. formicicum*, Mb. ivanovii, Mb. palustre, Mb. thermoaggregans, Mb. thermoalcaliphilum, Mb. thermoautotrophicum, Mb. thermoformicicum, Mb. uliginosum, Mb. wolfei*

Genus 3: *Methanosphaera*
Coccoid cells. Single and in packets. Methane produced from methanol and H_2. CO_2 and acetate serve as carbon sources. Pseudomurein contains Ser. The DNA base composition is 23–26 mol% G + C.
Species: *Msp. cuniculi, Msp. stadtmanae**

Family II: Methanothermaceae
Genus 1: *Methanothermus*
Long rods growing optimally above 80 °C. Maximal growth temperature 97 °C. Double-layered cell envelope composed of pseudomurein and a glycoprotein layer. Bipolar polytrichous flagellation. DNA base composition is around 33 mol% G + C.
Species: *Mt. fervidus*, Mt. sociabilis*

Order II: Methanococcales
Family I: Methanococcaceae
Genus 1: *Methanococcus*
Cells are regular to irregular Gram-negative cocci. Energy sources are H_2/CO_2 and formate. Surface layer composed of non-glycosylated protein subunits. Except for one species only C_{20} isopranyl glycerol ethers are present. DNA base composition is 31–41 mol% G + C. Mesophilic to extremly thermophilic growth.
Species: *Mc. deltae, Mc. igneus, Mc. jannaschii, Mc. maripaludis, Mc. thermolithothrophicus, Mc. vannielii*, Mc. voltae*

Tab. 1. Continued

Order III: Methanomicrobiales
Cells are rods, spirilli, filaments, cocci and plate-shaped cells. Gram-negative or Gram-positive. No growth above 60 °C, H_2/CO_2, acetate, formate, methanol, methylamines and dimethyl sulfide can serve as substrates. Genera growing on acetate or methanol possess cytochromes.

Family I: Methanomicrobiaceae
Gram-negative, irregular cocci, rods and spirilli. H_2 and formate serve as electron donor. Proteinaceous cell envelopes. DNA base composition is 45–62 mol% G + C.

Genus 1: *Methanomicrobium*
Short, rod-shaped cells with a subunit cell envelope. Monotrichous flagellation. Acetate is required as carbon source. H_2 and formate serve as energy source. Complex growth requirements. DNA base composition is 49 mol% G + C.
Species: *Mm. mobile**

Genus 2: *Methanolacinia*
Short, irregular rods or coccoid to lobe-shaped cells. Flagellated. Methane produced from H_2/CO_2, 2-propanol/CO_2, 2-butanol/CO_2 and cyclopentanol/CO_2. Acetate required. Polyamine and lipid pattern is different from *Methanomicrobium mobile.* DNA base composition is 44 mol% G + C.
Species: *Ml. paynteri**

Genus 3: *Methanogenium*
Highly irregular cocci. DNA base composition is 47–61 mol% G + C. Substrates are H_2/CO_2, formate, and sometimes alcohols. Polytrichous or monotrichous flagellation. Most strains require acetate as carbon source. Higher halotolerance than *Methanoculleus*. Best growth around 1 M Na^+.
Species: *Mg. cariaci*, Mg. liminatans, Mg. organophilum, Mg. tationis*

Genus 4: *Methanoculleus*
Irregular coccoids. H_2/CO_2, formate, and sometimes secondary alcohols function as substrates. Organic growth factors required. DNA base composition is 54–62 mol% G + C. Best growth at a Na^+ concentration of 0.1–0.4 M. Distinguished from *Methanogenium* by low DNA–DNA hybridization.
Species: *Mcl. bourgense*, Mcl. marisnigri, Mcl. olentangyi, Mcl. thermophilicus*

Genus 5: *Methanospirillum*
Rods forming long filaments. Several cells are surrounded by an SDS-resistant protein sheath, and single cells are separated by spacers. C_{20} isopranyl diether and C_{40} isopranyl tetraether present. Polar flagellation. DNA base composition is 46–50 mol% G + C.
Species; *Msp. hungatei**

Family II: Methanoplanaceae
Genus 1: *Methanoplanus*
Gram-negative, plate-shaped cells with sharp edges. Glycoprotein S-layer. Acetate strictly required as carbon source. Polar tuft of flagella. H_2 and formate serve as electron donors. DNA base composition is 38–48 mol% G + C.
Species: *Mp. endosymbiosus, Mp. limicola**

Family III: Methanosarcinaceae
Cells are cocci occurring singly or in large clumps. Gram-positive or Gram-negative. Methanol and methylamines serve as carbon and energy sources. Some species use also H_2/CO_2 or acetate. Two genera containing rods use only acetate as substrate.

Genus 1: *Methanosarcina*
Cells can grow on acetate, methanol, methylamines and most species also on H_2/CO_2. Formate is not used. Most species are Gram-positive and contain methanochondroitin cell walls. Cytochromes present. DNA base composition is 40–51 mol% G + C. Some species exhibit a life cycle.
Species: *Ms. acetivorans, Ms. barkeri*, Ms. frisia, Ms. mazei, Ms. methanica, Ms. thermophila, Ms. vacuolata.*

Genus 2: *Methanolobus*
Obligate methylotrophs. Cells can only use methanol and methylamines as energy and carbon source. Monotrichous flagellation or flagella absent. Cytochromes present. DNA base composition is around 39–46 mol% G + C.
Species: *Mlb. sicilae, Mlb. tindarius*, Mlb. vulcani*

Genus 3: *Methanococcoides*
Description as for *Methanolobus,* DNA–DNA homology is 2–4% with *Methanolobus*. Cytoplasmic membrane contains only C_{20} isopranyl diethers. DNA base composition is 42 mol% G + C.
Species: *Mcc. methylutens**

Genus 4: *Methanothrix*
Thick rods forming long filaments. Several cells are surrounded by an SDS-resistant protein sheath. Individual cells are separated by spacer plugs. Flat ends. Acetate serves solely as substrate. Formate is oxidized. C_{20} isopranyl diether found. DNA base composition is 52 mol% G + C. No species exists in pure culture.
Species: *Mtx. soehngenii**

Tab. 1. Continued

Genus 5: *Methanosaeta*
Description as for *Methanothrix*. DNA base composition is 61 mol% G + C.
Species: *Mst. concilii**, *Mst. thermoacetophila*

Genus 6: *Methanohalophilus*
Irregular cocci. Moderately halophilic. Optimal salinity is 0.6–2.5 M NaCl. Lysis with 0.05% sodium dodecyl sulfate. Substrates are methylamines. Most strains use also methanol.
Species: *Mh. halophilus**, *Mh. mahi*, *Mh. oregonense*, *Mh. zhilinae*

Genus 7: *Methanohalobium*
Irregular plate-shaped cells. Extremely halophilic. Optimal salinity is 2.5–4.3 M NaCl. Substrates are methylamines. Methanol is not used. No lysis with sodium dodecyl sulfate (0.05%). Low DNA–DNA hybridization with *Methanohalophilus*.
Species: *Mhm. evestigatus**

Genus 8: *Halomethanococcus*
Pleomorphic cocci. Minimum salt requirement is 1.8 M NaCl, salt optimum is 3.0 M NaCl. Substrates are methylamines and methanol. Acetate and rumen fluid are required for growth. Temperature range is 5–45 °C with an optimum at 35 °C. DNA base composition is 43 mol% G + C.
Species: *Hmc. doii**

Family IV: Methanocorpusculaceae
Genus 1: *Methanocorpusculum*
Small mesophilic cocci (<1 μm). Gram-negative. Monotrichous flagellation. H_2/CO_2, formate and 2-propanol/CO_2 serve as substrates. Acetate, yeast extract and tungstate required. Cytochromes present. DNA base composition is 48–52 mol% G + C.
Species: *Mcp. aggregans*, *Mcp. bavaricum*, *Mcp. labreanum*, *Mcp. parvum**, *Mcp. sinense*

Order IV: n. d.
Family 1: n. d.
Genus: *Methanopyrus*
Gram-positive, motile rods. Double-layered cell envelope consisting of pseudomurein and S-layer. Pseudomurein lacks N-acetylglucosamine, but contains ornithine. Growth at 4% NaCl. Maximal growth temperature 110 °C.
Species: *Mpr. kandleri**

3 Ecology

The methanogenic bacteria are widely distributed in anoxic environments (Balch et al., 1979: Jones et al., 1987; Jain et al., 1988; Oremland, 1988). Organic material is anaerobically degraded consecutively during a (1) hydrolytic, (2) fermentative, (3) acetogenic/hydrogenic and a final methanogenic (4) phase. Interspecies hydrogen transfer enables the degradation of longer-chain fatty acids to acetate, propionate and H_2. Acetate, CO_2 and H_2 are the products of the breakdown process, which function as substrates for methanogens. Under light, sulfate or nitrate limitation methanogens play an important role in anaerobic acetate metabolism. Less competitive substrates in the presence of sulfate or nitrate are methanol, methylamines and dimethyl sulfide, which are derived from pectin or compatible solutes. Examples of such terminal niches of an anaerobic food chain are freshwater and marine sediments, sewage sludge and soil, especially paddy fields and landfills.

Furthermore, methanogens occur in living beings, where they are involved in the digestion of food. Methanogens occur in the rumen and cecum of herbivorous mammals, in the digestive tract of fishes, the forestomach of whales and the hindgut of xylophagous insects such as termites. In humans they have been found in the large bowel and dental plaques. In contrast to sediments, not acetate but H_2/CO_2 and formate are the main methanogenic substrates. In the rumen methanogens dominate over acetogens, but in the gut of certain animals acetogens are preferentially found, even if the affinity for H_2 is lower. Methanogens have been isolated from the heartwood of trees, where cellulose and pectin are degraded by soil bacteria. Methanogens live in syntrophy with hydrogen producing bacteria or as endosymbionts of anaerobic protozoa (interspecies

H_2-transfer), which is also the case with termites.

Geothermally heated areas, alkalophilic and hypersaline environments are other, but more extreme habitats of methanogens. In continental solfataras and marine hydrothermal systems the substrates (H_2/CO_2) are, however, mainly geochemically produced. Geothermally heated areas, where methanogens have been detected are, e.g., Iceland, Yellowstone National Park or Volcano. The Great Salt Lake is an example of a hypersaline environment inhabited by methanogenic bacteria.

4 Morphology

The strict anaerobic methanogens contain diverse morphological forms (Figs. 2 and 3; Tab. 1). Members of the Gram-positive Methanobacteriales are short to long rods, lancet-shaped cocci or cocci occurring singly or in packets. *Methanosarcina*, the only Gram-positive genus in the order Methanomicrobiales, occurs as single cells or can form large tissue-like aggregates (Fig. 2). The other methanogens belonging to the Methanomicrobiales and

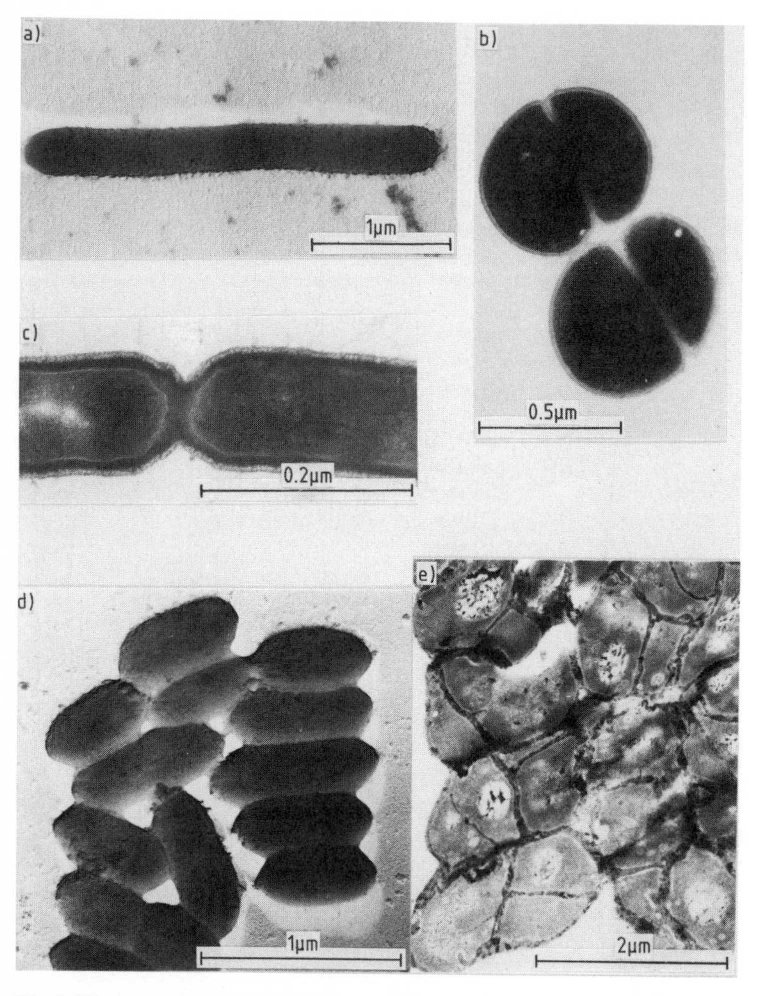

Fig. 2. Electron micrographs of Gram-positive methanogens (a, d, platinum shadowing; b, c, e, thin sections). (a) *Methanobacterium uliginosum*, (b) *Methanosphaera stadtmanae*, (c) *Methanothermus fervidus*, (d) *Methanobrevibacter smithii*, (e) *Methanosarcina* sp.

Methanococcales react Gram-negative and are irregular cocci and irregular plates or rods forming long filaments (Fig. 3) occurring as single cells or in aggregates.

5 Cell Components

5.1 Envelopes

Methanogens possess no common cell wall polymer, but rather diverse cell envelope types are found (KÖNIG, 1988). All methanogens lack murein.

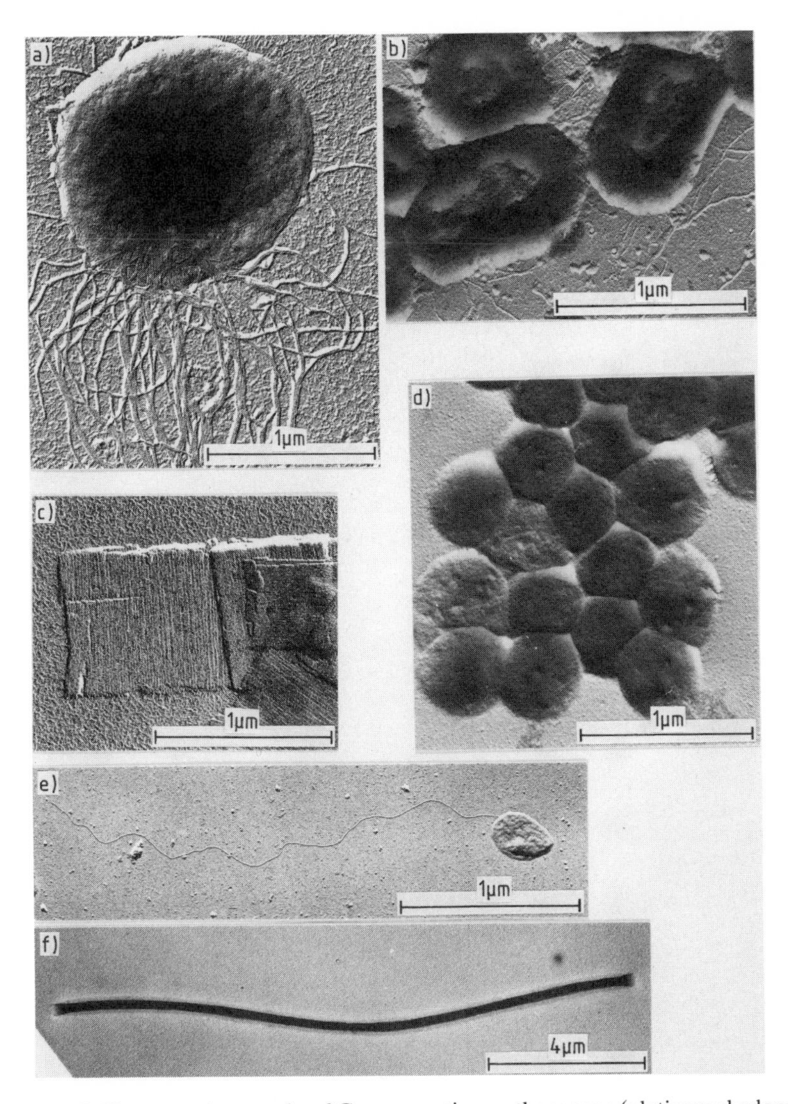

Fig. 3. Electron micrographs of Gram-negative methanogens (platinum shadowing). (a) *Methanococcus jannaschii*, (b) *Methanoplanus limicola*, (c) *Methanospirillum hungatei* (protein sheath), (d) *Methanolobus vulcani*, (e) *Methanomicrobium mobile*, (f) *Methanospirillum hungatei* (filament).

The cell walls of the Gram-positive methanogens consists of pseudomurein (Methanobacteriales; Figs. 2 and 4) or of methanochondroitin (*Methanosarcina;* Figs. 2 and 4; Tab. 1). *Methanothermus* and *Methanopyrus* have a double-layered cell envelope consisting of pseudomurein and an outer proteinaceous surface layer (Fig. 2). Some species of *Methanosarcina* exhibit a life cycle, where the cells occur singly or in aggregates. During disintegration of the aggregates the methanochondroitin matrix is lost.

The Gram-negative methanogenic bacteria have cell envelopes composed of single-layered crystalline protein (e.g., *Methanococcus*) or glycoprotein (e.g., *Methanoplanus*) subunits. In the case of *Methanospirillum* and *Methanothrix* the individual cells occur in long chains, which are surrounded by a layer of unknown chemical nature. The cells are separated by spacer plugs, and the outer envelope of the filaments is formed by a tubular proteinaceous sheath. Cells with different cell envelopes have distinct modes of cell division such as binary fission by septum formation (pseudomurein, *Methanobacterium*), constriction (S-layer, *Methanococcus*) or complicated modes of fragmentation (sheath, *Methanothrix*). Due

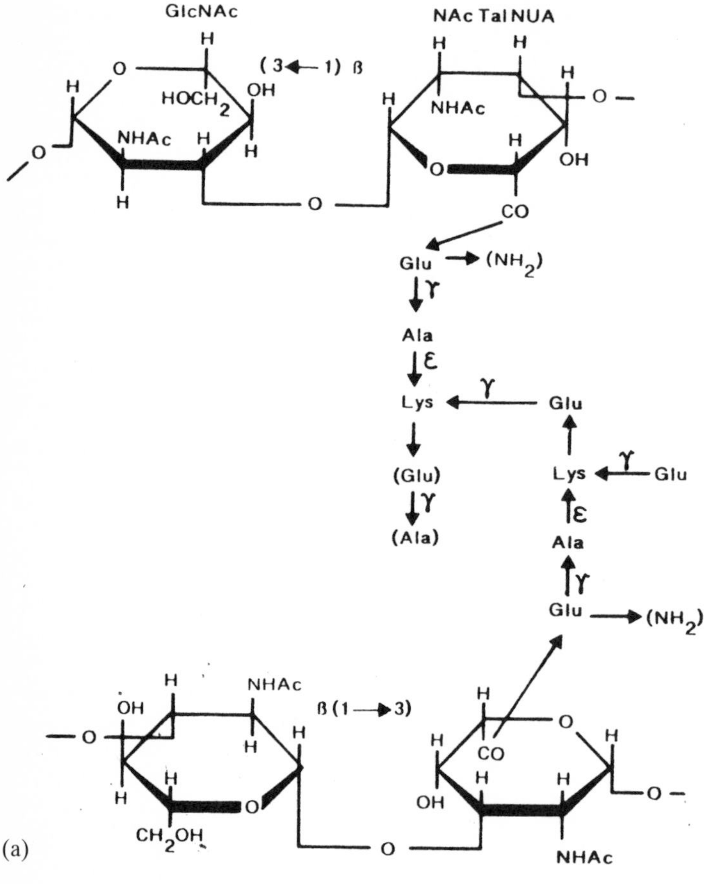

(a)

(b) [—> 4)-ß-D-GlcA-(1——> 3)-ß-D-GalNAc-(1——> 4)-ß-D-GalNAc-(1——> $]_n$

Fig. 4. Primary structure of (a) pseudomurein and (b) methanochondroitin.

to their unusual cell wall structures methanogens are not sensitive to many antibiotics affecting cell wall biosynthesis or to lytic agents.

5.2 Membranes

Fatty acid ester-linked glycerol lipids are the predominant constituents of the hydrophobic core of the bacterial and eukaryotic membranes. Typical fatty acid glycerolipids are absent in methanogens, except *Methanosphaera*, where fatty acids have been found. Phytanyl diether and diphytanyl tetraether lipids with a chain length of 20, 25 or 40 carbon atoms form the lipophilic part of the membranes (LANGWORTHY and POND, 1986). In contrast to the bacteria the glycerol residues possess the sn-2,3 stereoconfiguraton. The diethers allow for the formation of a normal bilayer, but the carbon chains of the tetraether extend the membrane forming a monolayer (Fig. 5).

5.3 Low-Molecular Weight Compounds

In methanogens a series of unique cofactors have been found (FRIEDMANN et al., 1990; WOLFE, 1991; JONES et al., 1987; VOGELS et al., 1988). The blue-green fluorescence of the methanogens in the epifluorescence microscope is caused by coenzyme F_{420}. It is a deazariboflavin derivative substituted with ribitol, phosphate and glutamyl residues possessing an absorption maximum at 420 nm in the oxidized state. The reduced form of coenzyme F_{420} is colorless. Coenzyme F_{420} functions as low potential electron carrier in two electron transfer reactions (hydride transfer). F_{420} is a coenzyme of, e.g., one out of two hydrogenases or the formate dehydrogenase.

Methanofuran, a second cofactor, is the first coenzyme involved in the reduction of CO_2 to methane. It functions as formyl carrier in the first step of CO_2 reduction. The formyl groups are then transferred to methanopterin (F_{432}), a 7-methylpterin derivative.

Coenzyme F_{430} is a nickel tetrapyrrole, which plays a role in the terminal step of methane reduction. It is a prosthetic group of the methyl coenzyme M reductase. It contains Ni in a divalent state and is the most reduced bond system known for a cyclic tetrapyrrole. Additional cofactors of the methyl CoM reductase system are 7-mercapto-N-heptanoyl-O-phospho-L-threonine (= component B, HS-HTP) and 2-mercaptoethanesulfonic acid (coenzyme M).

Corrinoid-dependent (factor III) methyltransfer reactions are involved in the acetate formation (acetyl-CoA pathway) and methylation of coenzyme M (methanogenesis).

Cyclo-2,3-diphosphoglycerate is supposed to be involved in the thermostabilization of enzymes of thermophilic methanogens.

6 Culture Technique

Methanogenic bacteria are strict anaerobes. All procedures such as the preparation of the media or the transfer of bacteria have to be performed in an oxygen-free atmosphere (BALCH et al., 1979; HIPPE, 1991). During sam-

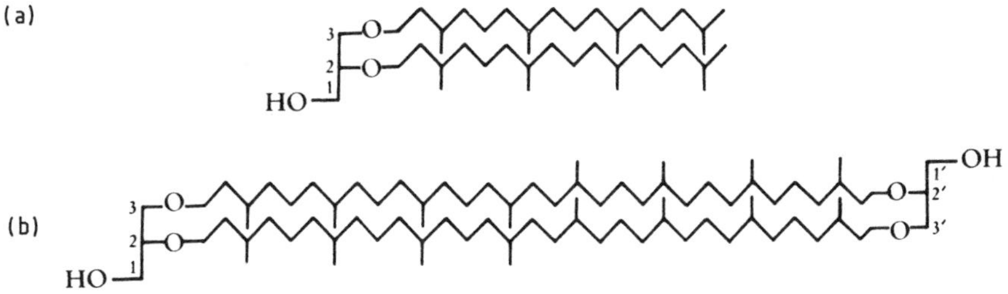

Fig. 5. Phytanylether lipids of methanogens. (a) Phytanyl diether, (b) biphytanyl tetraether.

pling the exposure to oxygen has to be minimized. Samples are transferred to rubber stoppered glass vials, which should contain a reducing agent, e.g., dithionite. The gases (H_2/CO_2, N_2) used are freed of oxygen by passing through reducing columns filled with copper at 350 °C or with a platinum catalyst at room temperature. Oxygen is removed from the medium by boiling and/or outgassing with N_2. Reducing agents such as cysteine and sulfide are added, and the pH is adjusted to the desired value with diluted sulfuric acid. The medium is then transferred in an anaerobic glove box and dispensed in serum bottles (Bellco Glass Inc.). The bottles are stoppered with butyl rubber septa and crimped with aluminum seals. The tubes are pressurized to 100–200 kPa with H_2/CO_2. After sterilization volatile ingredients like methanol or fatty acids are added with a syringe from sterile concentrated anaerobic stock solutions. Solid media in petri dishes are prepared in the anaerobic glove boxes and incubated in pressurized oxygen-free metal jars. The bacteria are transferred with syringes.

The nutritional requirements range from simple in chemolithotrophs to complex in some heterotrophs (JARREL and KALMOKOFF, 1987). Substrates and energy sources for methanogens are H_2/CO_2, methanol, methylamines, formate, acetate and dimethyl sulfide. Many methanogens are stimulated by complex media containing yeast extract, peptone or rumen fluid. Ammonium and sulfide are the main nitrogen and sulfur sources, respectively. A diverse spectrum of trace elements and vitamins is required by the different species. Nickel is obligately required. The uptake of monosaccharides has not been demonstrated. The growth temperatures range from mesophilic (25 °C) to extremely thermophilic (110 °C). The temperature optima are 30°–40 °C for the mesophiles, 60°–70 °C for the thermophiles and 80°–98 °C for the hyperthermophiles. The methanogens grow best under neutral conditions (pH 6.5–7.5), but some alkalophilic organisms have also been isolated. Acidophilic methanogens have not yet been described. Extremely halophilic methanogens need between 2.5 and 4.3 M NaCl for optimal growth. The combination of a newly developed anaerobic culture technique (serum bottle technique; Wolin-Miller tube), which has been derived from the roll tube technique of Hungate, and the anaerobic glove box method, led to the isolation of a series of pure cultures (Tab. 1; BALCH et al., 1979; Catalog of Strains 1989, Deutsche Sammlung von Mikroorganismen, Braunschweig). Most methanogens can be grown in one of three media described by BALCH et al. (1979).

Maintenance techniques of methanogens have been described by HIPPE (1991). The viability of methanogens is prolonged for some weeks, when growing cultures are placed at 4 °C. Several methanogens have been preserved for years by freezing with glycerol (10%) or dimethyl sulfoxide (5%) at –18 °C, –70 °C or in liquid nitrogen. Anaerobic heat-sealed glass ampoules or glass capillary tubes are used for storage.

7 Energy Metabolism

Methanogenic bacteria gain energy by methane formation from different methanogenic substrates such as H_2/CO_2, methanol, methylamines, dimethyl sulfide, formate and acetate (Tab. 2). Secondary alcohols (e.g., 2-propanol) can also be utilized as electron donors.

The pathway of CO_2 reduction with molecular hydrogen to methane has largely been elucidated part (Fig. 6, WOLFE, 1991; FRIEDMANN et al., 1990; VOGELS et al., 1988). The cofactors methanofuran, methanopterin, coenzyme M, component B, F_{430}, and corrinoids are involved in methanogenesis. CO_2 is sequentially reduced to CH_4. The first stable compound in-

Tab. 2. Energy Yielding Reactions in Methanogenic Bacteria

$4H_2 + CO_2$	$\rightarrow CH_4 + 2H_2O$
$4HCOOH$	$\rightarrow 3CO_2 + CH_4 + 2H_2O$
CH_3COOH	$\rightarrow CH_4 + CO_2$
$4CH_3OH$	$\rightarrow 3CH_4 + CO_2 + 2H_2O$
$4CH_3NH_3^+ + 2H_2O$	$\rightarrow 3CH_4 + CO_2 + 4NH_4^+$
$4CH_3CHOHCH_3 + CO_2$	$\rightarrow 4CH_3COCH_3 + CH_4 + 2H_2O$
$(CH_3)_2S + H_2O$	$\rightarrow 0.5CO_2 + 1.5CH_4 + H_2S$

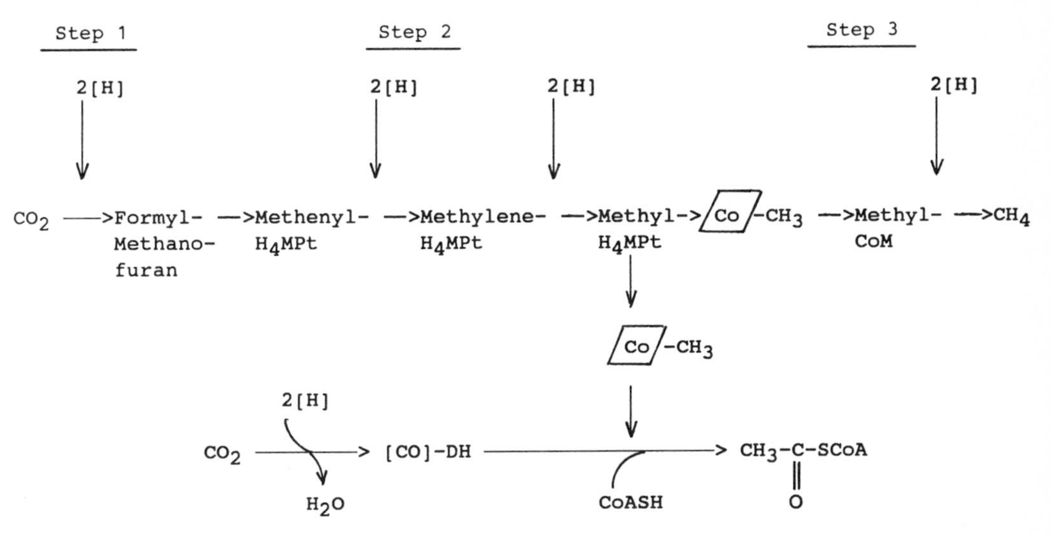

Fig. 6. Pathway of methane formation from CO_2 and H_2 and autotrophic CO_2 fixation in *Methanobacterium thermoautotrophicum*.
Co corrinoid protein; [CO]-DH, CO-dehydrogenase; CoM, coenzyme M; H_4MPt, tetrahydromethanopterin.

volved in CO_2 reduction is formylmethanofuran. The transfer of the formyl group results in the formation of 5-formyl tetrahydromethanopterin (5-formyl H_4MPT). Successively 5-formyl H_4MPT is reduced to 5–10-methenyl H_4MPT, 5–10-methylene H_4MPT and 5-methyl H_4MPT. As in tetrahydrofolic acid the N^5 and N^{10} positions are involved in formaldehyde reduction. Most probably a cobamide accepts the methyl group, which is then transferred to HS-CoM and CH_3-CoM is formed. In the last step methane and a heterodisulfide are formed, which is regenerated to CoM and components B (Fig. 6).

In methylotrophic methanogens methanol is disproportionated to carbon dioxide and methane. Methanol and the methyl groups of methylamines are transferred to coenzyme M. Acetoclastic methanogens dismutate acetate to methane and carbon dioxide. Methylotrophic and acetoclastic methanogens have cytochromes of the b and c type.

The reduction of methyl CoM to methane is an exergonic process and is coupled to ATP synthesis. Experiments with *Methanosarcina barkeri, Methanobacterium thermoautotrophicum* and other methanogens suggest that energy is gained by a chemiosmotic mechanism and an electrochemical Na^+ potential is involved. No evidence has been found for a substrate level phosphorylation (Kaesler and Schönheit, 1989; Blaut er al., 1990; Müller et al., 1990).

8 Carbon Metabolism

Many methanogenic bacteria can grow autotrophically using CO_2 as sole carbon source for biosynthesis. A pentose phosphate cycle (Calvin cycle) is lacking. In *Methanobacterium thermoautotrophicum* about 90% of the fixed CO_2 is used for energy generation and only 10% is used for biosynthesis of cell constituents. In methanogens acetyl-CoA plays a central role in CO_2 fixation (acetyl-CoA pathway; Fuchs, 1989). Acetate is synthesized from 2 carbon dioxide molecules. The methyl carbon is successively provided via N^5-methyltetrahydromethanopterin, an intermediate of the CO_2 reduction pathway to CH_4, and a methyl cobamide derivative. The carboxyl carbon is provided by reduction of CO_2 to carbon monoxide. The reduction of CO_2 to a nickel-bound carbon monoxide is catalyzed by a CO-dehydrogenase. In *Methanobacterium, Methanospirillum* and *Methanococcus* an incomplete reductive carboxylic acid pathway is operating. An incomplete oxidative carboxylic acid pathway is found in *Methanosarcina*. Biosynthesis of hexoses occurs via classical gluconeogenesis. Methanogens do not take up sugars, but can form glycogen granules. Details of the glycogen metabolism have not been studied.

9 Molecular Biology

Methanobacterium thermoautotrophicum shows a genome size of 1.1×10^9 dalton, which is smaller than that of *Escherichia coli*. The G + C content of the DNA of different methanogens exhibits a broad range from 23 to 62 mol%. The high G + C content of some species does not correlate with the growth temperature. Histone-like proteins, DNA-binding proteins and high salt concentrations are believed to stabilize the DNA in different methanogens.

During the last years the genome structure of methanogens has attracted remarkable interest (KLEIN, 1988; BROWN et al., 1989; HAUSNER et al., 1991). A series of genes has been sequenced, and information has been obtained on codon usage, processing signals, replication and transcription. From several gene sequences the corresponding gene products are known such as EF-Tu and EF-G in *Methanococcus vannielii*, nifH in *Methanococcus voltae*, methyl coenzyme M reductase in *Methanosarcina barkeri* and other species, glyceraldehyde-3-phosphate dehydrogenase and the S-layer glycoprotein of *Methanothermus fervidus*. It was shown that methanobacterial genes have a similar complexity as eubacterial genes.

Cloned methanogenic genes can complement auxotrophic mutants of *E.coli*. As shown for multi-subunit proteins (e.g., methyl CoM reductase) the genes of the subunits are transcribed in a polycistronic way. Shine–Dalgarno type sequences have been found. Transcription starts in an AT-rich region, and an AT-rich region contains the termination signal. The main stop signal appears to be a stem/loop structure followed by an oligo T sequence. Consensus sequences involved in transcription initiation and termination have been found in tRNA genes. By cell-free transcription experiments the sequence (box A)

$$^{TT}_{AA}TATATA$$

has been found to be the most important part of the promotor of stable tRNA genes. It is centered at –25 from the transcription start. The transcription starts at a G within a second conserved motif (box B). Highly A + T-rich regions are dispersed throughout the genome of methanogens separating G + C-rich genes. Insertion elements have also been found.

Plasmids have been identified in *Methanococci, Methanolobus* and *Methanobacterium*. A phage has been found in *Methanobacterium thermoautotrophicum*, and virus-like particles have been characterized in *Methanolobus vulcanii*.

A series of DNA-associated proteins have been characterized. The DNA replicating polymerases of methanogens are sensitive to aphidicoline and butyl-phenyl-GTP, which are inhibitors of the eukaryotic DNA polymerase α. The DNA-dependent RNA polymerases are multi-subunit enzymes and resemble the eukaryotic DNA-dependent α-polymerase. The largest components of the DNA polymerases of the three domains exhibited immunological cross-reaction suggesting a monophyletic origin of these proteins. A gyrase activity is also present, and a number of restriction enzymes have been purified.

Many of the structural details of the t-RNAs are unique. Ribothymidine has been replaced by pseudouridine or 1-methylpseudouridine. The initiator tRNA is not formylated. T-RNA genes occur in operons or as individual transcription units.

The overall features of the methanogenic ribosome resemble those of eubacteria. They sediment with 70S and dissociate into 30S and 50S subunits. The electron microscopic analysis of the fine structure revealed, however, some specific structural features. The total number of proteins is 56 in the case of *Methanobacterium bryantii*. The ribosomal RNAs have the coding sequence 16S-23S-5S. The rRNAs have been extensively sequenced. Elongation factor II contains diphthamide and is ADP-ribosylated.

10 Applied Aspects

10.1 Biogas Production

Methanogens play a significant role in the anaerobic recycling of carbon fixed by photosynthesis (OREMLAND, 1988; CONRAD, 1987). In this process about 5% of the photosynthetically fixed carbon is converted to atmospheric methane.

Atmospheric methane is formed mainly from surface areas, while methane produced in deeper sediments is mainly oxidized to CO_2. 65% of the total output of methane originates from biogenic sources like paddy fields, swamps, marshes, freshwater and marine sediments and volcanic areas. Apart from these habitats a considerable part of the biogenic methane originates from the gastrointestinal tract of animals, especially from the rumen of cattle. A cow produces about 200 liters, a man about 3 L methane per day. Atmospheric methane from termites is estimated to be $2–5 \times 10^{12}$ g per year. The total output of methane on earth is about 430×10^{12} g, and about 280×10^{12} g are biogenic. The atmospheric methane increases 2% per year, and the tropospheric concentration has reached 1.6 ppm.

The biological methanogenesis is practically applied in the anaerobic treatment of sewage sludge, agricultural, municipal and industrial wastes, manure of cattle, pigs and chicken. Immunological methods have been developed for the identification and maintenance of a desired methanogenic flora in waste treatment plants. Several waste treatment systems and reactor designs have been built. The anaerobic process has several advantages over the aerobic process. It stabilizes the sludge with formation of methane and offsets the disposal costs of wastes. Some methane processes are economically not viable compared to cheaper energy sources like fuel.

10.2 Restriction Enzymes

Site-specific endonucleases have become indispensable tools in the analysis and manipulation of nucleic acids, and for some specific purposes new enzymes are required. Three type II restriction endonucleases with novel site specificities have been isolated from *Methanococcus "aeolicus"*. The recognition sites of the enzymes MaeI, MaeII and MaeIII are C↓TAG, A↓CGT and ↓GTNAC, respectively.

11 References

BALCH, W. E., FOX, G. E., MAGRUM, L. J., WOESE, C. R., WOLFE, R. S. (1979), Methanogens: reevaluation of a unique biological group, *Microbiol. Rev.* **43,** 260–296.

BLAUT, M., PEINEMANN, S., DEPPENMEIER, U., GOTTSCHALK, G. (1990), Energy transduction in vesicles of the methanogenic strain Gö1, *FEMS Microbiol. Rev.* **87,** 367–372.

BOONE, D. R., MAH, R. A. (1989), Methanogenic archaebacteria, in: *Bergey's Manual of Systematic Bacteriology* (STALEY, J. T., BRYANT, M. P., PFENNIG N., HOLT, J. G., Eds.), Vol. 3, pp. 2173–2216, Baltimore: Williams & Wilkins.

BOONE, D. R., WHITMAN, W. B. (1988), Proposal of minimal standards for describing new taxa of methanogenic bacteria, *Int. J. Syst. Bacteriol.* **38,** 212–219.

BROWN, J. W., DANIELS, C. J., REEVE, J. N. (1989), Gene structure, organisation, and expression in archaebacteria, *CRC Crit. Rev. Microbiol.* **16,** 287–338.

CONRAD, R. (1987), The methane cycle. fundamentals and importance. *Forum Mikrobiol.* **10,** 320–329.

FRIEDMANN, H. C., KLEIN, A., THAUER, R. K. (1990), Structure and function of the nickel porphinoid, coenzyme F_{430}, and of its enzyme, methyl coenzyme M reductase, *FEMS Microbiol. Rev.* **87,** 339–348.

FUCHS, G. (1989), Alternative pathways of autotrophic carbon dioxide fixation in autotrophic bacteria, in: *Biology of Autotrophic Bacteria* (SCHLEGEL, H. G., BOWIEN, B., Eds.), pp. 365–382. Madison: Science Technical Publishers.

HAUSNER, W., FREY, G., THOMM, M. (1991), Control regions of an archaeal gene. A TATA box and an initiator element promote cell-free transcription of the $tRNA^{Val}$ Gene of *Methanococcus vannielii*, *J. Mol. Biol.* **222,** 495–508.

HIPPE, H. (1991), Maintenance of methanogenic bacteria, in: *Maintenance of Microorganisms. A Manual of Laboratory Methods*, pp. 101–113, New York: Academic Press.

IWABE, N., KUMA, K., HASEGAWA, M., OSAWA, S., MIYATA, T. (1989), Evolutionary relationship of archaebacteria, eubacteria and eukaryotes inferred from phylogenetic trees of duplicated genes, *Proc. Natl. Acad. Sci. USA* **86,** 9355–9359.

JAIN, M. K., BHATNAGER, L., ZEIKUS, J. G. (1988), A taxonomic overview of methanogens, *Ind. J. Microbiol.* **28,** 143–177.

JARRELL, K. F., KALMOKOFF, M. L. (1987), Nutritional requirements of the methanogenic archaebacteria, *Can. J . Microbiol.* **34,** 557–576.

JONES, W. J., NAGLE, J. R., WHITMAN, W. B. (1987), Methanogens and the diversity of archaebacteria, *Microbiol. Rev.* **51** 135–177.

KAESLER, B., SCHÖNHEIT, P. (1989), The sodium cycle in methanogenesis. CO_2 reduction to the formaldehyde level in methanogenic bacteria is driven by a primary electrochemical potential of Na^+ generated by formaldehyde reduction to CH_4, *Eur. J. Biochem.* **186,** 309–316.

KLEIN, A. (1988), Genome organization and gene structure in methanogenic archaebacteria, *Forum Mikrobiol.* **11,** 82–86.

KÖNIG, H. (1988), Archaeobacterial cell envelopes, *Can. J. Microbiol.* **34,** 395–406.

LANGWORTHY, T. A., POND, J. L. (1986), Archaebacterial ether lipids and chemotaxonomy, *Syst. Appl. Microbiol.* **7,** 253–257.

MÜLLER, V., BLAUT, M., HEISE, R., WINNER, C., GOTTSCHALK, G. (1990), Sodium bioenergetics in methanogens and acetogens, *FEMS Microbiol. Rev.* **87,** 373–376.

OREMLAND, R. S. (1988), Biogeochemistry of methanogenic bacteria, in: *Biology of Anaerobic Microorganisms* (ZEHNDER, A. J. B., Ed.), pp. 641–705, New York: J. Wiley & Sons.

VOGELS, G. D., KELTJENS, J. T., VAN DER DRIFT, C. (1988), Biochemistry of methane production, in: *Biology of Anaerobic Microorganisms* (ZEHNDER, A. J. B., Ed.), pp. 707–770, New York: J. Wiley & Sons.

WOESE, C. R. (1987), Bacterial evolution, *Microbiol. Rev.* **51,** 221–271.

WOESE, C. R., KANDLER, O., WHEELIS, M. L. (1990), Towards a natural system of organisms: Proposal for the domains Archaea, Bacteria, and Eucarya, *Proc. Natl. Acad. Sci. USA* **87,** 4576–4579.

WOLFE, R. S. (1991), Novel coenzymes of archaebacteria, in: *The Molecular Basis of Bacterial Metabolism* (HAUSKA, G., THAUER, R., Eds.), pp. 1–12, Heidelberg: Springer-Verlag.

8 Methylotrophs

LUBBERT DIJKHUIZEN
Groningen, The Netherlands

1 Introduction 266
2 Taxonomy and Ecology 268
  2.1 Occurrence of One-Carbon Compounds 268
  2.2 Isolation of Methylotrophs 268
  2.3 Methylotrophic Yeasts 269
  2.4 Methylotrophic Bacteria 269
    2.4.1 Methanotrophic Bacteria 269
    2.4.2 Gram-Negative Non-Methane-Utilizing Obligate Methylotrophs 269
    2.4.3 Gram-Negative Non-Methane-Utilizing Facultative Methylotrophs 270
    2.4.4 Gram-Positive Non-Methane-Utilizing Methylotrophs 270
  2.5 Further Outlook 270
3 Metabolism and Physiology 271
  3.1 General 271
  3.2 Methane Oxidation 271
  3.3 Methanol Oxidation 271
  3.4 Methylamine Oxidation 274
  3.5 Dimethylsulfide Oxidation 274
  3.6 Formaldehyde and Formate Oxidation 275
  3.7 Assimilation of C_1 Units 275
  3.8 Bioenergetics of Growth 278
4 Industrial Applications and Products 278
  4.1 Synthesis of Single Cell Protein 278
  4.2 Synthesis of Amino Acids 279
  4.3 Exploitation of Unique Metabolic Features of Methylotrophs 280
5 References 280

1 Introduction

Studies on the microbial metabolism of C_1 compounds were initiated in 1892, just over a century ago, by LOEW who reported the isolation and characterization of a pink-pigmented bacterium, *Bacillus methylicus*, growing in a mineral salts medium with methanol, methylamine, formaldehyde, formate, and on a variety of multi-carbon substrates (QUAYLE, 1987). The number of microorganisms isolated from natural habitats on the basis of their ability to convert one-carbon compounds has increased rapidly in the second half of this century. Subsequent studies on the physiology and biochemistry of C_1 compound utilization have resulted in the elucidation of a fascinating set of special metabolic features employed by these organisms (ANTHONY, 1982, 1986; HEIJTHUIJSEN and HANSEN, 1990; DE KONING and HARDER, 1992). Much of this work has been stimulated strongly by the (initially) high expectations about potential applications of these organisms (single cell protein, enzymes, metabolites) in industrial processes (LIDSTROM and STIRLING, 1990; LEAK, 1992). Progress in other areas has been relatively slow, and especially studies on the taxonomy and molecular genetics of C_1 compound utilizing microbes have been hampered by an initial general lack of suitable techniques (GREEN, 1992; LIDSTROM, 1992).

Only aerobic bacteria and yeasts utilizing C_1 compounds as sole carbon- and energy source for growth are under consideration in this chapter (Tab. 1). The oxidation of C_1 compounds provides these microorganisms with the metabolic energy and carbon building blocks (formaldehyde and carbon dioxide) required for the biosynthetic processes yielding new cell material. Microorganisms growing on C_1 compounds possess a special set of enzymes for the above-mentioned activities. The pathways in which these enzymes function are shown schematically in Fig. 1 and discussed in more detail in Sect. 3 on metabolism and physiology. The combined action of these dissimilatory and assimilatory pathways results in the synthesis of a metabolite containing a C_3 skeleton from C_1 units (ANTHONY, 1982; DIJKHUIZEN et al., 1992). Once such a compound has been produced, the synthesis of all cell constituents (polysaccharides, proteins, nucleic acids and lipids) further proceeds via the general pathways of intermediary metabolism, as found in other organisms. Detailed studies by many research groups over the last 30 years have generated a great deal of knowledge on the enzymology of C_1 metabolism. To date this has led to the discovery and biochemical description of four different pathways which effect the synthesis of compounds containing carbon–carbon bonds from C_1 units. These pathways are the Calvin cycle or ribulose bisphosphate (RuBP) pathway of carbon dioxide

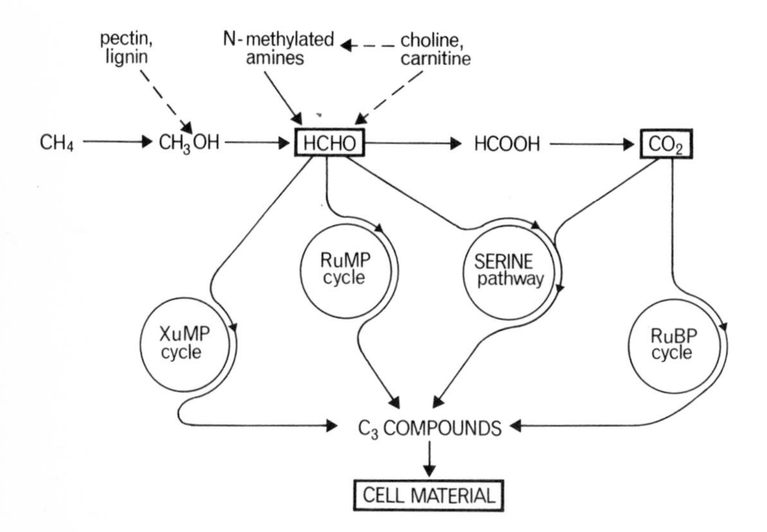

Fig. 1. Dissimilatory and assimilatory pathways in microorganisms utilizing C_1 compounds.

Tab. 1. Characteristics of Aerobic Methylotrophic Bacteria (Modified from LIDSTROM, 1990a)

Group	Growth Substrates	Major Assimilation Pathway	N_2 Fixing	GC Content (mol%)	Gram Reaction [a]	Reference [b]
Obligate Methylotrophs						
Type I Methanotrophs						
Methylomonas	CH_4, CH_3OH [c]	RuMP	No	50–54	Neg	1
Methylobacter	CH_4, CH_3OH [c]	RuMP	No	50–54	Neg	1
Methylococcus	CH_4, CH_3OH [c]	RuMP	Yes	62–64	Neg	1
Type II Methanotrophs						
Methylosinus	CH_4, CH_3OH [c]	Serine	Yes	62.5	Neg	1
Methylocystis	CH_4, CH_3OH [c]	Serine	Yes	62.5	Neg	1
Methanol Utilizers						
Methylophilus	CH_3OH, CH_3NH_2, Glucose	RuMP	No	50–55	Neg	2
Methylobacillus	CH_3OH, CH_3NH_2	RuMP	No	50–55	Neg	3
Methylovorus	CH_3OH, Glucose	RuMP	No	56–57	Neg	4
Methylophaga	CH_3OH, CH_3NH_2, Fructose	RuMP	No	38–46	Neg	5
Facultative Methylotrophs						
Methylobacterium	CH_3OH, CH_3NH_2, [C—C]	Serine	No	65–68	Neg	6
Hyphomicrobium	CH_3OH, CH_3NH_2 [c], DMSO [c], DMS [c], Some C_2 and C_4 compounds [c]	Serine	No	60–66	Neg	7, 8
Acidomonas	CH_3OH [c], [C—C]	RuMP	ND	63–65	Neg	9
Paracoccus	CH_3OH, CH_3NH_2, [C—C]	RuBP	No	66	Neg	10
Xanthobacter	CH_3OH, [C—C]	RuBP	Yes	67–69	Neg	11
Microcyclus	CH_3OH, [C—C]	RuBP	Yes	65–67	Neg	3
Thiobacillus	CH_3OH [c], CH_3NH_2 [c], [C—C]	RuBP	No	65–69	Neg	1
Rhodospeudomonas	CH_3OH [c], [C—C]	RuBP	No	62–72	Neg	1
Mycobacterium	CH_4 [c], CH_3OH [c], [C—C]	RuMP	No	65–69	Pos	1, 10, 13
Bacillus	CH_3OH, [C—C]	RuMP	No	48–50	Pos	14
Arthrobacter	CH_3NH_2, [C—C]	RuMP	No	ND	Pos	15
Amycolatopsis	CH_3OH, [C—C]	RuMP	No	ND	Pos	16

Abbreviations: [C—C], multi-carbon compounds; RuMP, ribulose monophosphate; RuBP, ribulose bisphosphate; ND, no data

[a] Neg, Gram-negative; Pos, Gram-positive

[b] References; 1) ANTHONY, 1982; 2) JENKINS et al., 1987; 3) URAKAMI and KOMAGATA, 1986; 4) GOVORUKHINA and TROTSENKO, 1991; 5) JANVIER et al., 1985; 6) GREEN et al., 1988; 7) HARDER and ATTWOOD, 1978; 8) STACKEBRANDT et al., 1988; 9) URAKAMI et al., 1989; 10) COX and QUAYLE, 1975; 11) MEIJER et al., 1990; 12) REED and DUGAN, 1987; 13) KATO et al., 1988; 14) ARFMAN et al., 1992; 15) LEVERING et al., 1981; 16) DE BOER et al., 1990

[c] Some strains

fixation, the ribulose monophosphate (RuMP) cycle of formaldehyde fixation, the xylulose monophosphate (XuMP) cycle of formaldehyde fixation and the serine pathway, in which both formaldehyde and carbon dioxide are fixed (Fig. 1). One of these pathways, the XuMP cycle, is only present in yeasts (VAN DIJKEN et al., 1978).

Methylotrophs are characterized by their ability to utilize carbon compounds containing one or more carbon atoms (but no carbon–carbon bonds) as sole carbon- and energy source for growth, assimilating carbon as formaldehyde or as a mixture of formaldehyde and carbon dioxide (Fig. 1). Those methylotrophs able to utilize methane as sole carbon- and energy source generally are referred to as methanotrophs. Autotrophic organisms, on the other hand, synthesize cell material via assimilation of carbon dioxide (Fig. 1). ANTHONY (1982) defined methylotrophs more generally as those microbes able to grow at the expense of C_1 compounds, irrespective of the way C_1 units are assimilated. Methanol-utilizing bacteria which assimilate carbon dioxide via the Calvin cycle (Tab. 1), such as *Xanthobacter flavus* (MEIJER et al., 1990) and *Paracoccus denitrificans* (COX and QUAYLE, 1975), are included in this classification.

2 Taxonomy and Ecology

2.1 Occurrence of One-Carbon Compounds

One-carbon compounds at all oxidation levels between methane and carbon dioxide can be found in nature. Methane, the most abundant of these compounds and a significant greenhouse gas, is present in fossil deposits, in soils, ponds, marshes, and the digestive tracts of ruminants and other animals, and is formed by methanogenic bacteria. Methanol is formed by the hydrolysis of plant methyl esters and ethers such as pectin and lignin. Methylated amines (mono-, di-, and trimethylamine) are produced by microbial degradation of choline derivatives present in plant membrane material and animal tissue. Dimethylsulfide arises from the cleavage of dimethylsulfoniopropionate, a product of algal sulfur metabolism (ANDREAE, 1990). Formate is a major end product of mixed acid fermentation, and carbon dioxide is present in the atmosphere and, as carbonates, in natural waters and soil. Finally, chlorinated methanes are frequently used as solvents and are produced in enormous quantities in industry.

2.2 Isolation of Methylotrophs

Microorganisms able to use the above-mentioned C_1 compounds as carbon- and energy sources for growth have been isolated from many aquatic and terrestrial habitats. At the moment only limited knowledge is available about their distribution and role in nature. Methanotrophs form the major biological sink for methane in the biosphere, and although a large variety of methylotrophs has been isolated and characterized, it is clear that many naturally occurring strains have not yet been identified (HANSON and WATTENBERG, 1991). Also the question whether the dominant methylotrophs in any environment in fact have been identified cannot yet be answered with certainty. At the moment rapid progress is made in the development of methods for the estimation of the diversity of methylotrophic bacteria in natural populations, using gene probes and signature probes that are complementary to consensus sequences uniquely found in 16S rRNA of the various groups of organisms under study (HANSON et al., 1992). Symbiotic associations between methanotrophs and mussels or tubeworms, invertebrates found at methane seeps in deep-sea reducing sediments and hydrothermal vents, have also been described (CAVANAUGH, et al., 1987; CHILDRESS et al., 1986; SCHMALJOHANN and FLÜGEL, 1987). This type of symbiosis is viewed as an adaptation by methane-utilizing bacteria insuring simultaneous access to methane and oxygen, typically available from anaerobic and aerobic environments, respectively (CAVANAUGH et al., 1987). A detailed characterization of the methanotrophic bacteria involved in these consortia has not yet been possible. A further search for methylotroph–invertebrate associations in marine and freshwater habitats undoubtedly will be most rewarding.

2.3 Methylotrophic Yeasts

Yeast species isolated on the basis of their ability to use methanol as carbon- and energy source belong to the genera *Pichia* (*Hansenula*) and *Candida* (*Torulopsis*) (DE KONING and HARDER, 1992). These are versatile organisms predominantly isolated from soils, rotting fruits and vegetables, exudates of trees, or from the frass of insects living on trees. The ecological position of these organisms is still largely unknown, but it has been suggested that in the habitats mentioned above methanol can be derived from the methoxy groups present in wood lignin or fruit pectin (DE KONING and HARDER, 1992).

2.4 Methylotrophic Bacteria

Many strains of methylotrophic bacteria are typical obligate methylotrophs. Organisms that in addition to their methylotrophic character also grow (usually poorly) only on glucose or fructose will be considered here as obligate methylotrophs as well. Most of the methanotrophs (employing either the serine or the RuMP pathway) and the Gram-negative methanol-utilizing RuMP cycle bacteria are obligate methylotrophs. The methanol-utilizing serine pathway and RuBP cycle organisms are generally more versatile methylotrophs (see Tab. 1). Most of the methylotrophic species studied are Gram-negative bacteria, and only in recent years an increasing number of Gram-positive methylotrophic bacteria have been isolated and characterized. It is becoming increasingly clear that these Gram-positive RuMP cycle bacteria are almost exclusively facultative methylotrophs. Classification of methylotrophs was hampered initially by a general lack of suitable taxonomic criteria, especially with respect to the restricted nutritional abilities of many of the new isolates. Based on insufficiently detailed phenotypic studies a bewildering variety of new names has been proposed in the literature. Progress in the classification of some groups of methylotrophic bacteria, therefore, is best described as a semichaotic flux (GREEN et al., 1984; GREEN, 1992).

2.4.1 Methanotrophic Bacteria

With the exception of *Mycobacterium* species ID-Y (REED and DUGAN, 1987), all methane-utilizing bacteria described are Gram-negative and obligate methylotrophs (LIDSTROM, 1990a). Methanotrophs share the ability to grow aerobically on methane; and based on morphology, resting stages formed, intracytoplasmic membrane structures, and other physiological characteristics, five genera have been proposed (*Methylomonas*, *Methylobacter*, *Methylococcus*, *Methylosinus*, *Methylocystis*). The methanotrophs are broadly split into two groups (type I and type II) on the basis of the arrangement of their intracytoplasmic membranes (bundles of disk-shaped vesicles stacked throughout the center of the cells versus rings at the periphery of the cells, respectively), the major carbon assimilation pathway used (RuMP cycle versus serine pathway; Fig. 1), and ability to fix atmospheric nitrogen. *Methylococcus* strains are often classified as type X because of their somewhat intermediate position (Tab. 1; WHITTENBURY and KRIEG, 1984).

2.4.2 Gram-Negative Non-Methane-Utilizing Obligate Methylotrophs

The group of non-methane-utilizing obligate methylotrophs consists of Gram-negative, rod-shaped bacteria that superficially resemble pseudomonads (Tab. 1; JENKINS et al., 1987; GREEN, 1992). These organisms lack the intracytoplasmic membrane structures characteristic of methanotrophs and all assimilate formaldehyde via the RuMP pathway. Four different genera of Gram-negative bacteria are currently recognized (*Methylobacillus*, *Methylophilus*, *Methylophaga*, *Methylovorus*). Key differentiating criteria between the genera *Methylobacillus* and *Methylophilus* are the low (ca. 22%) DNA–DNA homology levels, (in)ability to use glucose as sole carbon- and energy source (Tab. 1), and polar lipid composition (JENKINS and JONES, 1987; GREEN, 1992). The genera *Methylophaga* and *Methylovorus* have been proposed more recently and show very low levels (5–15%) of DNA–DNA homology with the type strains of *Methylobacillus* and *Methylophilus. Methylovorus* strains were shown to utilize glucose (GOVORUKHINA and TROTSENKO, 1991), whereas representatives of the genus *Methylophaga* are marine bacteria (JANVIER et al., 1985, 1992), and fructose was identified as the sole non-C_1 substrate tested that supported growth of these organisms.

2.4.3 Gram-Negative Non-Methane-Utilizing Facultative Methylotrophs

The methanol-utilizing facultative methylotrophs display a much stronger variation in many properties. The largest group is represented by members of the genus *Methylobacterium* (e.g., *Methylobacterium extorquens* AM1, previously known as *Pseudomonas* AM1), pink-pigmented facultative methylotrophs (PPFMs) that use the serine pathway for formaldehyde assimilation and grow on a range of multi-carbon substrates (GREEN et al., 1988). The budding methylotrophs are classified as *Hyphomicrobium* (HARDER and ATTWOOD, 1978; STACKEBRANDT et al., 1988). Members of this genus are serine pathway methylotrophs which use a limited range of multi-carbon C_2 and C_4 compounds that can be metabolized by means of acetyl-CoA. The inability of these organisms to grow, for instance, on pyruvate and succinate is most likely due to the absence of a functional pyruvate dehydrogenase (PDH) complex. This was confirmed in what may be described as early gene therapy experiments (DIJKHUIZEN et al., 1984). Introduction of the *Escherichia coli pdh* genes in *Hyphomicrobium* X resulted in appearance of all three PDH complex proteins and PDH complex enzyme activity, conferring the ability to grow in mineral medium with pyruvate or succinate. Hyphomicrobia are selectively enriched in methanol mineral salts media using nitrate as the terminal electron acceptor for anaerobic respiration (SPERL and HOARE, 1971; ATTWOOD and HARDER, 1972). Some strains utilize (di)chloromethane (KOHLER-STAUB and LEISINGER, 1985; GALLI and LEISINGER, 1985), dimethylsulfoxide and dimethylsulfide (DE BONT et al., 1981; SUYLEN and KUENEN, 1986), or monomethylsulfate (GHISALBA and KÜENZI, 1983a). Gram-negative acidophilic bacteria belonging to the newly created genus *Acidomonas* (formerly *Acetobacter methanolicus*; STEUDEL et al., 1980; URAKAMI et al., 1989) are exceptional in combining formaldehyde assimilation via the RuMP pathway with the clear-cut phenotype of facultative methylotrophs. Finally, a number of methanol-utilizing autotrophic bacteria are listed in Tab. 1, all of which are facultative methylotrophs, able to grow on multi-carbon substrates as well.

2.4.4 Gram-Positive Non-Methane-Utilizing Methylotrophs

As mentioned above, only a limited number of Gram-positive methylotrophic bacteria have been identified. Those isolates that have been characterized in some detail are all versatile organisms, belonging either to the coryneform bacteria, the actinomycetes (e.g., *Amycolatopsis methanolica*; DE BOER et al., 1990), or the genus *Bacillus* (e.g., *Bacillus methanolicus*; ARFMAN et al., 1992), and assimilate formaldehyde via the RuMP pathway (see Tab. 1). *B. methanolicus* strains are able to grow in methanol mineral medium over a temperature range of 35–60 °C. At the optimum growth temperatures (50–55 °C) they display doubling times between 40 and 80 min. The metabolism of these strains is strictly respiratory. Further interesting features are their high molar growth yield with methanol and their marked resistance to methanol concentrations up to 1.5–2.0 M.

2.5 Further Outlook

The taxonomic status of methylotrophic bacteria thus far has been based largely on phenotypic and chemotaxonomic data; only recently ribosomal RNA sequence data have become available which provide a more phylogenetic basis for taxonomy. Currently, 16S and 5S ribosomal RNAs from over 30 and 80 strains, respectively, of methylotrophic bacteria have been sequenced, allowing sequence comparisons to determine phylogenetic relationships. The current data indicate that type I (RuMP pathway) and type II (serine pathway) methanotrophs cluster within the gamma and alpha subdivision of the proteobacteria, respectively. The non-methane-utilizing methylotrophic bacteria belong to the alpha subdivision (serine pathway methylotrophs) or constitute a separate branch within the beta subdivision (RuMP pathway methylotrophs) of the proteobacteria. The phylogenetic data also suggest that the methylotrophic bacteria do not form a single line of descent; methylotrophy probably has a polyphyletic origin (WOLFRUM and STOLP, 1987; ANDO et al., 1989; BULYGINA et al., 1990, 1992; TSUJI et al., 1990; HANSON et al., 1992).

3 Metabolism and Physiology

3.1 General

The physiology and biochemistry of the utilization of C_1 compounds by aerobic methylotrophic bacteria and yeasts has been dealt with in various reviews (ANTHONY, 1986; DE VRIES et al., 1990) and monographs (ANTHONY, 1982; LARGE and BAMFORTH, 1988; LIDSTROM, 1990a, b). Various aspects of the regulation of C_1-specific enzymes in microorganisms employing the Calvin cycle (TABITA, 1988), the serine pathway (GOODWIN, 1990), the RuMP cycle (DIJKHUIZEN and SOKOLOV, 1991; DIJKHUIZEN et al., 1992) and the XuMP cycle (DE KONING and HARDER, 1992) also have been reviewed in recent years. A detailed description of the properties of enzymes involved in the utilization of C_1 compounds is outside the scope of this chapter. Instead, the metabolism of the C_1 compounds methane, methanol, methylamine, dimethylsulfide as elucidated in various methylotrophs will be outlined in this section with emphasis on the physiological role of the enzymes constituting the special pathways involved, and recent developments.

3.2 Methane Oxidation

Methane oxidation in methanotrophs is catalyzed by either a soluble (sMMO) or membrane-bound (pMMO) mixed-function methane monooxygenase (MMO; EC 1.14.13.25) enzyme complex, as shown below:

$$CH_4 + NADH + H^+ + O_2 \rightarrow H_2O + NAD^+ + CH_3OH$$

The activity levels of the two forms of MMO in *Methylosinus trichosporium* OB3b and *Methylococcus capsulatus* (Bath) strongly depend on the level of copper ions in the environment, with excess copper resulting in predominantly pMMO (STANLEY, et al., 1983). The sMMO enzymes purified from both type I and type II methanotrophs are very similar and characterized by an extremely broad substrate specificity (COLBY et al., 1977; STIRLING et al., 1979). The sMMO enzyme has been characterized extensively, both at the biochemical and the genetic level (DALTON, 1992; MURRELL, 1992). Only limited information is currently available about the pMMO enzyme, mainly because purification of active protein is difficult. Compared to sMMO, pMMO allows a higher specific growth rate, growth efficiency, and affinity towards methane. Recent studies suggest that pMMO in fact is a copper-containing protein, able to oxidize a limited number of alkanes and alkenes (DALTON, 1992; CHAN et al., 1992).

3.3 Methanol Oxidation

The methanol-oxidizing enzymes in yeast and Gram-negative or Gram-positive bacteria are rather different. Methylotrophic yeasts employ an alcohol oxidase (EC 1.1.3.13) which has been purified from various sources and studied extensively (WOODWARD, 1990; MÜLLER et al., 1992). This enzyme is a homo-octameric

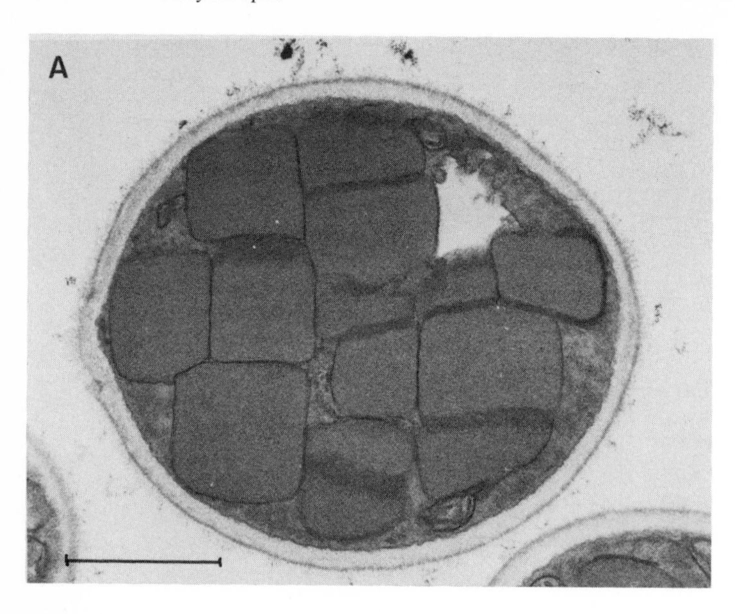

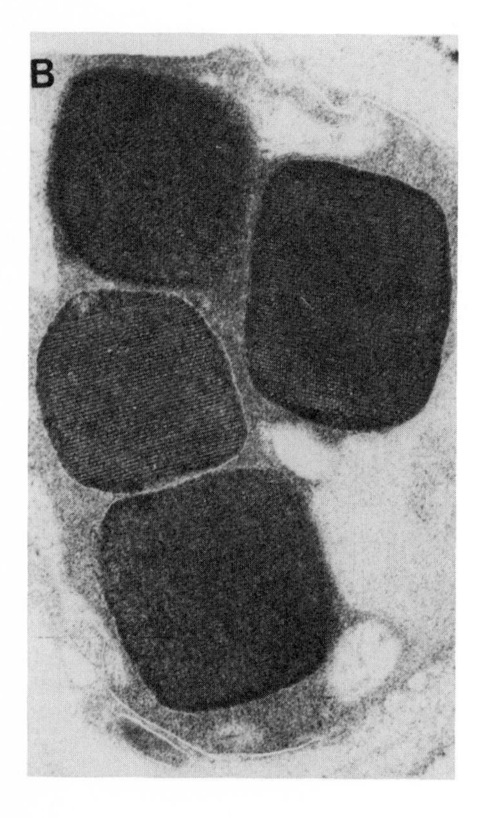

Fig. 2. Cells of the yeast *Hansenula polymorpha*, grown in a methanol-limited chemostat at a dilution rate of $D = 0.05\ h^{-1}$, closely packed with several cubically shaped peroxisomes. (A) Fixation $KMnO_4$. Alcohol oxidase activity in the organelles is demonstrated after incubation of glutaraldehyde-fixed spheroplasts of these cells with $CeCl_3$ and methanol. Note that the crystalline substructure of the peroxisomal matrix is well preserved under these conditions. (B) Fixation glutaraldehyde-OsO_4. The bar represents 1 μm. Courtesy of M. VEENHUIS.

protein with a molecular mass around 600 kDa, oxidizing primary aliphatic alcohols and producing hydrogen peroxide. Alcohol oxidase contains FAD as a prosthetic group and is localized in microbodies (peroxisomes), together with catalase (WOODWARD, 1990; DE KONING and HARDER, 1992). Due to the crystalline nature of the alcohol oxidase in these organelles, yeasts grown on methanol possess a unique ultrastructure (Fig. 2). As evident from Fig. 2, yeast cells grown in methanol-limited chemostats synthesize the enzyme up to very high levels. Under these conditions the main intracellular role of FAD is, therefore, that of a prosthetic group of alcohol oxidase. This necessitates an accurate coordinate induction of the FAD biosynthetic enzymes (BROOKE et al., 1986). Depending upon growth conditions the enzyme displays a variable affinity for methanol. The latter phenomenon can be explained at least partly by the observation that alcohol oxidase possesses two chemically distinct forms of FAD, the ratio of which varies with growth conditions (SHERRY and ABELES, 1985; BYSTRYKH et al., 1991; KELLOGG et al., 1992).

An entirely different enzyme, methanol dehydrogenase (MDH; EC 1.1.99.8), is involved in the oxidation of methanol to formaldehyde in the periplasm of Gram-negative bacteria (ANTHONY, 1982, 1986). The enzyme has an $\alpha_2\beta_2$ configuration with subunits of approximately 66 and 9 kDa, respectively. It contains two moles of pyrroloquinoline quinone (PQQ) as a prosthetic group (DUINE, 1991; JONGEJAN and DUINE, 1989) and a single atom of calcium which is apparently involved in binding PQQ to the active site (RICHARDSON et al., 1992). The β-subunit has an unusually high lysine content (20%). In the periplasm MDH is embedded in a pool of cytochrome c proteins which accept electrons from PQQ and donate these reducing equivalents directly to a terminal oxidase (ANTHONY, 1986). Most MDH enzymes oxidize a range of primary alcohols; the affinity for these substrates decreases with increasing carbon chain length. Formaldehyde may also serve as a substrate and is converted into formate. The MDH enzymes from various methylotrophic bacteria have been characterized in detail at the biochemical and genetical level and are highly conserved at the amino acid sequence level (ANTHONY, 1986; LIDSTROM, 1992). MDH activity *in vivo* is most likely influenced by additional factors. A heat-stable, oxygen-labile, low-molecular weight factor X isolated from *Hyphomicrobium* X was found to stimulate electron transport from MDH to cytochrome c when tested in an assay system under physiological conditions (DIJKSTRA et al., 1988). Moreover, a so-called modifier protein of approximately 130 kDa (most likely a homo-tetramer with 45 kDa subunits), present in the periplasm of several Gram-negative methylotrophic bacteria, was reported to act on the MDH protein, effectively decreasing the V_{max} and the affinity of MDH for the substrate formaldehyde (LONG and ANTHONY, 1991).

In recent years a limited number of Gram-positive methylotrophic bacteria have been isolated in pure culture (Tab. 1) and studied with respect to the biochemistry of methanol oxidation. The results of these studies with thermotolerant *Bacillus methanolicus* strains (ARFMAN, 1991; ARFMAN et al., 1992) and the actinomycetes *Amycolatopsis methanolica* (DE BOER et al., 1990) and *Mycobacterium gastri* (KATO et al., 1988) indicate that these organisms employ a novel type (type III) of cytoplasmic alcohol oxidoreductase for methanol oxidation (ARFMAN, 1991; DE VRIES et al., 1992; BYSTRYKH et al., 1993). Under methanol-limiting conditions *B. methanolicus* strain C_1 may synthesize up to 30% of total soluble protein in the form of this methanol dehydrogenase. Characterization of purified MDH from *B. methanolicus* strain C_1 revealed that it is composed of 10 identical subunits of 43 kDa, each of which contains 1 zinc and 1–2 magnesium ions and a tightly (but non-covalently) bound NAD(P)H cofactor which participates in the alcohol oxidation reaction (ARFMAN, 1991; VONCK et al., 1991). The enzyme subunits are arranged in two pentagonal rings of five subunits each. The structural gene encoding MDH of *B. methanolicus* strain C_1 has been cloned and sequenced recently (DE VRIES et al., 1992) and found to possess extensive amino acid sequence homology with the N-termini and internal peptide fragments of the methanol oxidizing enzymes purified from the two above-mentioned actinomycetes (BYSTRYKH et al., 1993). In each of these organisms additional proteins appear to participate in the transfer of reducing equivalents from the NAD(P)H cofactors to free NAD coenzyme and/or the electron transport chain (ARFMAN, 1991; ARFMAN et al., 1991; BYSTRYKH et al., 1993).

3.4 Methylamine Oxidation

Methylamine oxidation to formaldehyde occurs either via a methylamine dehydrogenase, an amine oxidase, or a methylglutamate dehydrogenase. Whereas methylamine dehydrogenase is a periplasmic enzyme, the latter two enzymes are located in the cytoplasm of the cells. Rapid progress has been made especially in studies of the three-dimensional structure of the methylamine dehydrogenase and the identification of the cofactors involved in some of these enzymes. During growth on methylamine, Gram-negative bacteria such as *Methylobacterium extorquens* AM1 employ a methylamine dehydrogenase (EC 1.4.99.3) with an $\alpha_2\beta_2$ configuration and subunits of approximately 40 and 14 kDa, respectively. A blue copper periplasmic protein, amicyanin, is the physiological electron acceptor of the enzyme. The small subunit of the methylamine dehydrogenase contains the covalently bound cofactor tryptophan tryptophylquinone (TTQ) which is derived from two cross-linked, modified tryptophan residues (VELLIEUX et al., 1989; MCINTIRE et al., 1991). This cofactor is apparently synthesized by posttranslational modification of two tryptophan residues in each of the small subunit polypeptides, a process involving both oxidative and cross-linking reactions. The genetic organization of methylamine utilization genes in *Methylobacterium extorquens* AM1 has been reported by CHISTOSERDOV et al. (1991). The Gram-positive bacterium *Arthrobacter* P1 employs a copper-containing enzyme, amine oxidase (EC 1.4.3.6), during growth on methylamine. It now appears likely that copper-containing amine oxidases ranging from microbes to man also contain a novel quinoid cofactor, topaquinone (TPQ), which is covalently bound to the enzyme and in fact is part of the primary protein structure (DUINE, 1991). The structures of these quinoid cofactors are shown in Fig. 3.

3.5 Dimethylsulfide Oxidation

Only a limited number of aerobic bacteria able to grow on dimethylsulfide (DMS) have been characterized thus far. These strains are *Hyphomicrobium* (DE BONT et al., 1981; SUYLEN and KUENEN, 1986; ZHANG et al., 1991) or *Thiobacillus* (KANAGAWA and KELLY, 1986; VISSCHER et al., 1991). No information thus far has been reported about the identity and properties of enzymes involved in the conversion of DMS into methanethiol. Methanethiol itself is oxidized by an oxidase, yielding formaldehyde, hydrogen sulfide and hydrogen peroxide (SUYLEN et al., 1987). GOULD and KANAGAWA (1992) have purified and characterized this enzyme from *Thiobacillus thioparus*.

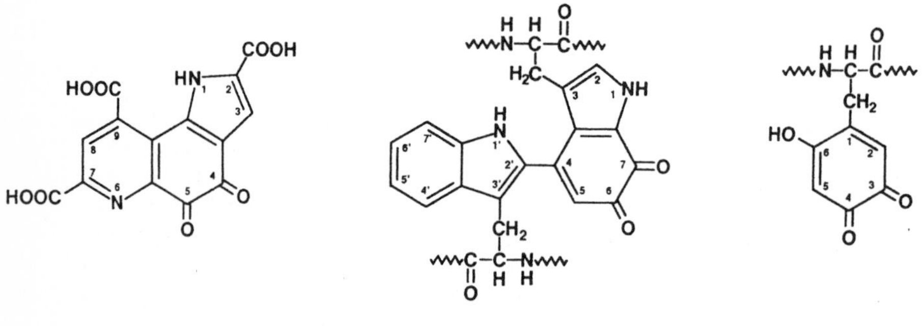

PQQ TTQ TPQ

Fig. 3. Structure of pyrroloquinoline quinone (PQQ), tryptophan tryptophylquinone (TTQ) and topaquinone (TPQ).

3.6 Formaldehyde and Formate Oxidation

In many methylotrophs the TCA cycle plays a minor role in energy generation during growth on C_1 compounds. In these organisms formaldehyde is oxidized by NAD-dependent (EC 1.2.1.46) or dye-linked (EC 1.2.99.3) formaldehyde and formate dehydrogenases to carbon dioxide (see Fig. 1). Although the route from formaldehyde to CO_2 is straightforward, a confusingly large number of formaldehyde oxidizing enzymes has been described in the literature (DIJKHUIZEN and SOKOLOV, 1991; DIJKHUIZEN et al., 1992). The NAD-linked formaldehyde dehydrogenases can be further subdivided according to the nature of the additionally required cofactors, such as reduced glutathione (EC 1.2.1.1) in several Gram-negative bacteria and all yeast species tested, or heat-stable, low-molecular weight factors in (non)methylotrophic Gram-positive bacteria, e.g., the actinomycete *Rhodococcus erythropolis* (EGGELING and SAHM, 1984, 1985). In yeasts, an NAD- and glutathione-dependent formaldehyde dehydrogenase (EC 1.2.1.1) is exclusively involved in formaldehyde oxidation. The product of this reaction, S-formylglutathione, is further metabolized to CO_2 via either a hydrolase and an NAD-dependent formate dehydrogenase (SCHÜTTE et al., 1976), or oxidized directly via the latter enzyme (see DE KONING and HARDER, 1992). Finally, formate oxidation is carried out by soluble NAD-dependent formate dehydrogenases in most organisms, and rarely is a dye-linked, membrane-associated enzyme involved (DIJKHUIZEN et al., 1992).

RuMP cycle methylotrophic bacteria may employ either of two different routes for formaldehyde oxidation (Fig. 4). The direct linear route proceeds via formate as described above (Fig. 4A). Another possible mechanism for complete oxidation of formaldehyde to CO_2 is based on a cyclic sequence of reactions in which enzymes of the oxidative pentose phosphate pathway and the RuMP cycle play an important role (Fig. 4B; see below). Various studies, including mutant evidence for *Methylobacillus flagellatum* KT (KLETSOVA et al., 1988), indicate that the dissimilatory RuMP cycle is the major pathway for the generation of reducing equivalents in the (non-methane-utilizing) obligate RuMP cycle methylotrophs.

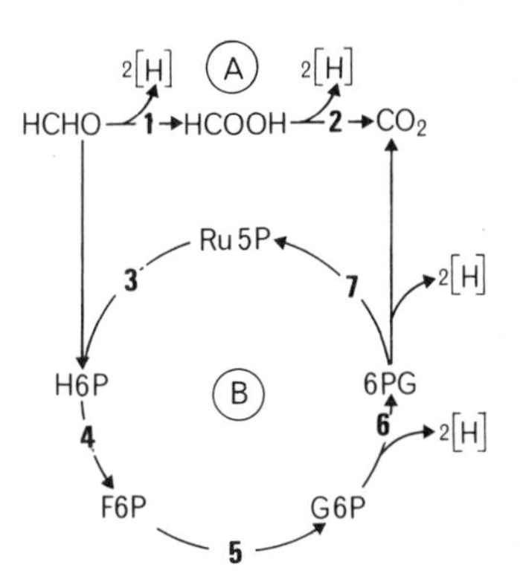

Fig. 4. Mechanism of formaldehyde fixation in RuMP cycle methylotrophic bacteria. (A) Linear oxidation pathway; (B) dissimilatory RuMP cycle. The numbers refer to the following enzymes: **1,** formaldehyde dehydrogenase; **2,** formate dehydrogenase; **3,** hexulose-6-phosphate (H6P) synthase; **4,** hexulose-6-phosphate isomerase; **5,** phosphoglucose isomerase; **6,** glucose-6-phosphate (G6P) dehydrogenase; **7,** 6-phosphogluconate (6PG) dehydrogenase.

3.7 Assimilation of C_1 Units

Three of the pathways specifically employed for conversion of C_1 units in intermediates of central metabolism involve various sugar phosphate molecules. From their schematic representation in Fig. 5 it is clear that these pathways are very similar in design. They differ in the identity of the enzymes catalyzing the actual C_1 assimilatory reactions. These enzymes are ribulose-1,5-bisphosphate carboxylase-oxygenase/phosphoribulokinase (Calvin cycle), hexulose-6-phosphate synthase/hexulose-6-phosphate isomerase (RuMP cycle) and dihydroxyacetone synthase/dihydroxyacetone kinase (XuMP cycle). Further steps in all three cycles are catalyzed by the general enzymes from the glycolytic or Entner–Doudoroff pathways (i.e., fructose bisphosphate aldolase, FBPA, or 2-keto-3-desoxy-6-phosphogluconate aldolase, KDPGA, respectively) and the oxidative pentose phosphate pathway (involving either sedoheptulose bisphosphatase, SBPase, or transaldolase, TA). Based on the var-

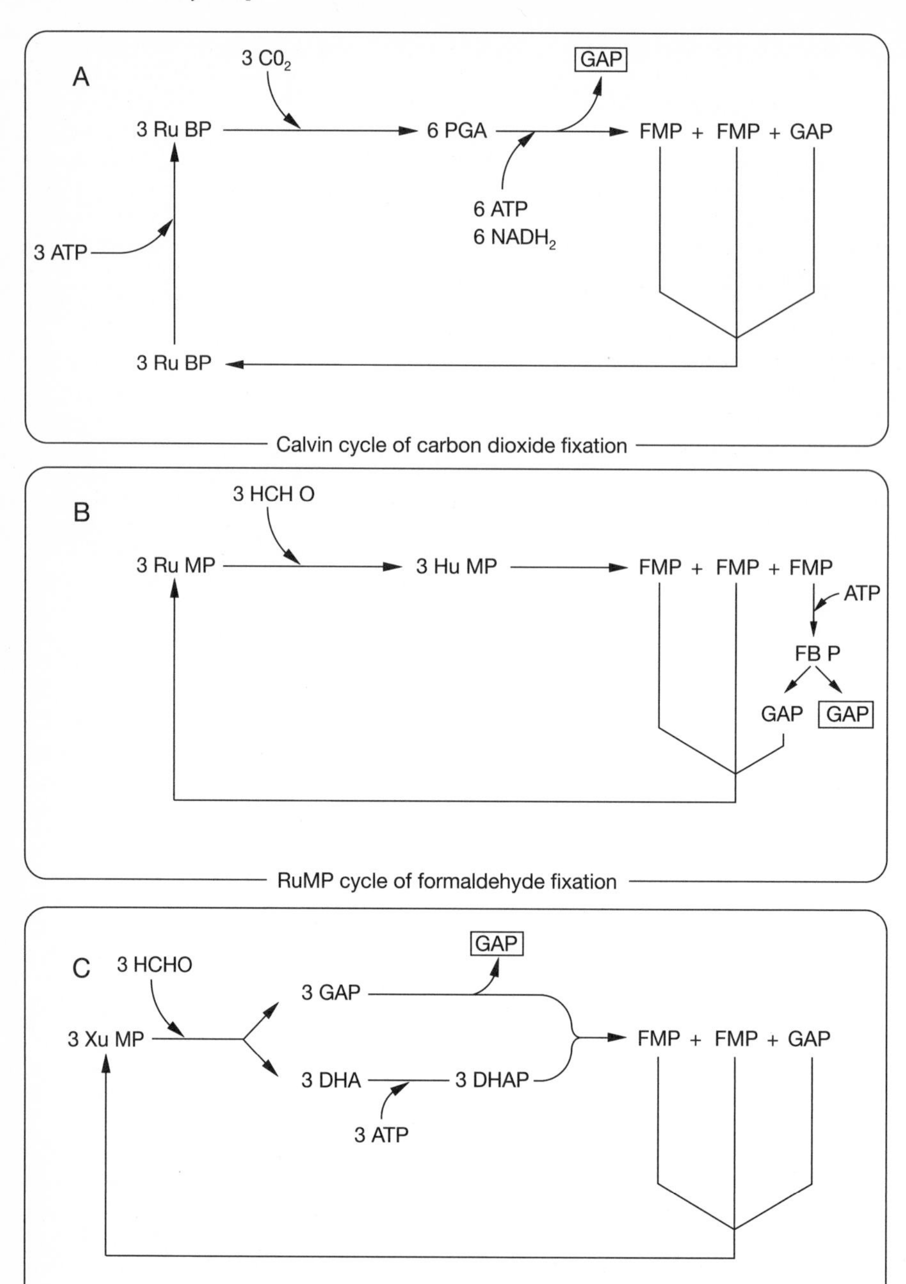

Fig. 5. Schematic representation of the Calvin cycle (A), the ribulose monophosphate (RuMP) cycle (B) and the xylulose monophosphate (XuMP) cycle (C). RuBP, ribulose-1,5-bisphosphate; RuMP, ribulose-5-phosphate; PGA, 3-phosphoglycerate; GAP, glyceraldehyde-3-phosphate; FMP, fructose-6-phosphate; HuMP, 3-hexulose-6-phosphate; FBP, fructose-1,6-bisphosphate; XuMP, xylulose-5-phosphate; DHA, dihydroxyacetone; DHAP, dihydroxyacetone-phosphate.

ious possible combinations of these enzyme systems, three variants of the RuMP cycle actually have been encountered in different methylotrophic bacteria (see Tab. 2; DIJKHUIZEN et al., 1985, 1992).

Intermediates of the serine pathway are mainly organic acids and amino acids (Fig. 6). Enzymes specifically involved in this pathway are serine transhydroxymethylase, serine:glyoxylate aminotransferase and hydroxypyruvate reductase. Further steps in this pathway are catalyzed by enzymes of the citric acid cycle. In most serine pathway organisms the glyoxylate cyle enzyme, isocitrate lyase, is absent, however, and the identity of the enzymes involved in the regeneration of glyoxylate remains to be established (GOODWIN, 1990). Efforts in this direction with *Methylobacterium extorquens* AM1 now concentrate on the characterization of cloned DNA fragments complementing glyoxylate-requiring mutants of the organism, blocked in the conversion of acetyl CoA to glyoxylate (SMITH and GOODWIN, 1992).

Tab. 2. Metabolic Energy Required for the Conversion of C_1 Units into C_3 Compounds

Cycle	C_1 Units Fixed	Primary Product	Variant	ATP Required	$NAD(P)H_2$ Required
Calvin	$3\,CO_2$	PGA	SBPase	8	5
RuMP	3HCHO	GAP	FBPA/TA	1	0
		GAP	FBPA/SBPase	2	0
		pyruvate	KDPGA/TA	0	–1
Serine	2HCHO + $1\,CO_2$	PGA	icl$^+$	3	2
XuMP	3HCHO	GAP	SBPase	3	0

SBPase, sedoheptulose bisphosphatase; FBPA, fructose bisphosphate aldolase; TA, transaldolase; KDPGA, 2-keto-3-desoxy-6-phosphogluconate aldolase; icl$^+$, isocitrate lyase positive. For other abbreviations, see legend to Fig. 5.

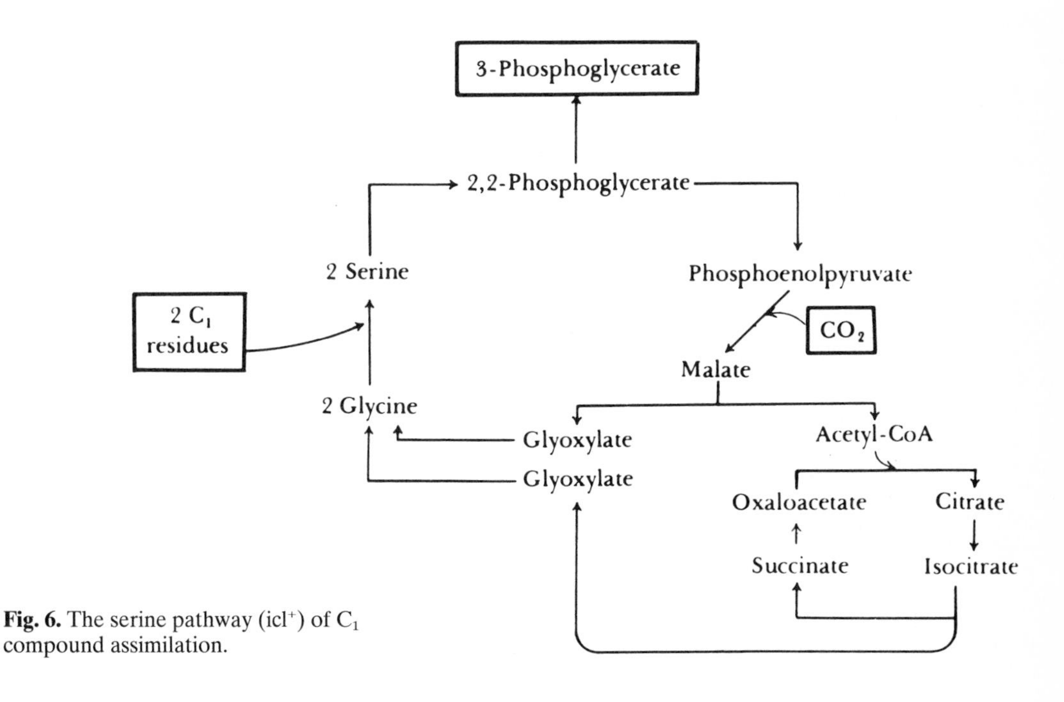

Fig. 6. The serine pathway (icl$^+$) of C_1 compound assimilation.

3.8 Bioenergetics of Growth

From Fig. 5 it is already apparent that the metabolic energy required for the synthesis of the primary C_1 output molecule strongly varies with the type of assimilation pathway. More detailed information is provided in Tab. 2, and the data show that the RuMP cycle in fact is energetically the most efficient, followed by the XuMP pathway in yeasts (DIJKHUIZEN et al., 1985). As expected, these different metabolic energy requirements are reflected to a large extent in the experimental growth yields obtained with methanol. In RuMP cycle organisms these yield values range from 0.55 to 0.6 g dry weight per g methanol, the highest values being observed with the Gram-positive bacteria employing an NAD-dependent methanol dehydrogenase (DIJKHUIZEN et al., 1988). In serine pathway organisms and yeasts the yield values range from 0.37 to 0.45 g g^{-1}. The relatively low growth yields observed with yeasts are probably due to the involvement of an alcohol oxidase in the initial conversion of methanol to formaldehyde.

4 Industrial Applications and Products

It is evident from the above that a great diversity of aerobic methylotrophs exists in nature, employing specific sets of enzymes for the conversion of one-carbon compounds. Over the years there has been a strong interest in possible industrial applications for these methylotrophic organisms, either aiming for the production of (components of) cell material, biocatalytic conversion of (non-)growth substrates into value-added products, bioremediation of environmental pollutants, or as a vehicle for overexpression of heterologous proteins. Much of this interest is stimulated by the fact that methane and methanol are attractive feedstocks for industrial fermentation processes, mainly because of their relatively low cost, ease of handling, and abundant availability. Compared to other fermentation feedstocks, contamination of concentrated stocks of methanol is unlikely to occur. Drawbacks of methane are its low solubility, the high oxygen demand of methane-based fermentations, and the explosive nature of methane/air gas mixtures. By comparison, methanol has several distinctive additional advantages; it is completely miscible with water, is easy to store and to transport, and available in high purity. The methanol-utilizing bacteria generally combine a relatively low oxygen demand with higher rates and yields of product formation. Potential biotechnological applications for methylotrophs are numerous and have been reviewed frequently (LIDSTROM and STIRLING, 1990; DUINE and VAN DIJKEN, 1991; DE KONING and HARDER, 1992; LEAK, 1992). These include various commodity chemicals (citrate, dihydroxyacetone, formaldehyde, glycerol, sorbitol), fine chemicals (ATP, FAD, glutathione, NAD, PQQ, riboflavin, vitamin B12) and specialty chemicals (polyhydroxybutyrate, polysaccharides). Today no major commercial success has been achieved, however. Only a limited number of potential applications, which are still actively investigated, are described in the following sections, with emphasis on recent developments where appropriate.

4.1 Synthesis of Single Cell Protein

Single cell protein (SCP) production processes have been developed with methanotrophs, with methanol-utilizing bacteria as well as yeasts (LARGE and BAMFORTH, 1988; EKEROTH and VILLADSEN, 1991; LEAK, 1992). Natural gas, however, contains several higher hydrocarbons in significant amounts. Several of these impurities are likely to be converted by the broad specificity methane monooxygenase present in methanotrophs, posing the danger of accumulation of toxic compounds in further (co)oxidation steps. SCP processes based on methane nevertheless are still being investigated. Methanol has been the substrate of choice for the ICI process to produce a SCP product known as Pruteen, using the obligate methylotroph *Methylophilus methylotrophus* AS1. This RuMP cycle organism combines a relatively high growth temperature (approx. 40 °C) with a high growth rate. Methanol metabolism in yeasts is energetically less favorable than in bacteria employing the RuMP cycle (see Tab. 2) and results in a higher heat production, thereby increasing the demand for cooling. Phillips Petroleum successfully developed the Provesteen process, based on the yeast *Pichia pastoris*, which is grown in a large-scale continuous fermentor up to a density of more than 125 g dry weight of cells per liter. The economics of SCP production processes, however, are extremely sensitive to changes in world market prices of other possible feedstocks; when issued as a fodder protein, the price of interchangeable protein sources like soybeans also has to be considered. At the moment SCP from C_1 compounds fails to compete with these other protein products, which have remained relatively inexpensive. The available expertise with high cell density yeast fermentations serves to produce large amounts of the enzyme alcohol oxidase (see Sect. 3.3). This enzyme may be used for the production of hydrogen peroxide as a bleaching agent (GIUSEPPIN, 1988), and for the determination of alcohol (WOODWARD, 1990).

4.2 Synthesis of Amino Acids

Amino acid fermentative processes using methanol media are of interest, because they will allow a reduction in feedstock prices and relieve the high burden of waste water treatment of currently employed large-scale fermentations with molasses. Studies on amino acid overproduction by methylotrophic bacteria mainly concentrate on serine and aromatic amino acids, in view of the special metabolic features of these organisms. One example is the stoichiometric conversion of glycine plus C_1 units (derived from methanol) into serine, based on high activity levels of serine hydroxymethylase in serine pathway (Fig. 6) methylotrophs. Serine is used primarily as a precursor for tryptophan biosynthesis. BEHRENDT et al. (1984) further optimized serine synthesis in *Pseudomonas* 3ab by allowing direct conversion of the serine produced by a tryptophan-producing strain of *E. coli.* This resulted in production of 7 g L^{-1} serine from 10 g L^{-1} glycine and 20 g L^{-1} methanol.

The most efficient carbon assimilation pathway in methylotrophic bacteria is the RuMP cycle (see Tab. 2). Unfortunately, most RuMP cycle organisms are obligate methylotrophs and difficult to manipulate (see Tab. 1). Gram-positive non-methylotrophic bacteria have found wide application in the production of fine chemicals such as amino acids. It is only recently that a number of versatile RuMP cycle Gram-positive methylotrophic bacteria have become available; these organisms are of considerable interest from a fundamental as well as an applied point of view (see Tab. 1). In our experience, the thermotolerant *B. methanolicus* (ARFMAN, 1991; ARFMAN et al., 1992) and *Amycolatopsis methanolica* strains (DE BOER et al., 1990) are relatively insensitive to the growth inhibition normally exerted by relatively high methanol concentrations (1.5–2 M). Both organisms also are readily amenable to the (extensive) physiological and genetical manipulations required for strain development. SCHENDEL et al., (1990) reported the construction of analog-resistant, auxotrophic mutants of their thermotolerant *Bacillus* isolate, excreting the amino acid lysine (20 g L^{-1}). The aromatic amino acids are derived from erythrose-4-phosphate, a central intermediate in the RuMP cycle, which will offer at least some advantages when attempting to construct overproducing strains of these methylotrophs. Analog-resistant, auxotrophic mutants of *A. methanolica* accumulate 1–5 g L^{-1} of L-phenylalanine. In further work genetic techniques for this organism have been developed, based on an efficient whole-cell transformation system (DIJKHUIZEN et al., 1993). In addition, *A. methanolica* and closely related actinomycetes (DE BOER et al., 1990) also synthesize a number of secondary metabolites which are derived from the shikimate pathway. Attempts to increase amino acid and secondary metabolite yields through genetic manipulation of *A. methanolica* are currently in progress.

Increasing success is also achieved in the manipulation of obligate RuMP cycle methylotrophs for amino acid production, e.g., glutamate, threonine and lysine synthesis by *Methylobacillus glycogenes* (MOTOYAMA et al., 1992).

4.3 Exploitation of Unique Metabolic Features of Methylotrophs

Various enzymes specifically synthesized by methylotrophs are studied with respect to possible applications as biocatalysts or as biosensors (DUINE and VAN DIJKEN, 1991). The soluble methane monooxygenase (sMMO) enzyme employed by methanotrophs possesses an extraordinarily broad substrate specificity (COLBY et al., 1977; STIRLING et al., 1979). The possible application of the enzyme as a biocatalyst for synthesis of, for instance, epoxides from alkenes (e.g., propene oxide from propene) subsequently has received strong interest (DALTON, 1992). Today no commercial process for this biocatalyst has been realized, however. Recent studies indicate that a further application for the sMMO enzyme may be found in the treatment of potential organic environmental pollutants. Methanotrophs and sMMO are able to degrade trichloroethylene and low-molecular weight halogenated hydrocarbons (OLDENHUIS et al., 1989; TSIEN et al., 1989). Various other chemical waste components also may be degraded by specialized methylotrophs, e.g., dichloromethane (GALLI and LEISINGER, 1985), methylsulfides (DE BONT et al., 1981; SUYLEN et al., 1987; ZHANG et al., 1991), methylsulfate (GHISALBA and KÜENZI, 1983a) and methylated amines (GHISALBA and KÜENZI, 1983b; GHISALBA et al., 1985). Methylotrophs capable of degrading those compounds have already been isolated and could form the basis of waste (water) treatment processes.

The methylotrophic yeasts *Pichia pastoris* and *Hansenula polymorpha* express very high levels of alcohol oxidase. This has been exploited by using the regulatory sequences involved for the production of various heterologous proteins (e.g., hepatitis B surface antigen, tumor necrosis factor) from a chromosomally integrated cassette (TSCHOPP and CREGG, 1991; BUCKHOLZ and GLEESON, 1991). The yeast alcohol oxidase promoter can be switched off by using glucose or glycerol as growth substrates for these yeasts. The metabolic burden of the presence of heterologous genes thus is minimal, further reducing the risk that cells lose the heterologous gene. Other advantages of the use of *P. pastoris* are the very high cell densities that can be achieved. Attempts to utilize methylotrophs for the production of heterologous proteins have not yet resulted in commercial processes, however.

5 References

ANDO, S., KATO, S., KOMAGATA, K. (1989), Phylogenetic diversity of methanol-utilizing bacteria deduced from their 5S ribosomal RNA sequences, *J. Gen. Appl. Microbiol.* **35,** 351–361.

ANDREAE, M. O. (1990), Ocean–atmosphere interactions in the global biogeochemical sulfur cycle, *Mar. Chem.* **30,** 1–29.

ANTHONY, C. (1982), *The Biochemistry of Methylotrophs*, London: Academic Press.

ANTHONY, C. (1986), Bacterial oxidation of methane and methanol, *Adv. Microb. Physiol.* **27,** 113–210.

ARFMAN, N. (1991), Methanol metabolism in thermotolerant bacilli, *Ph. D. Thesis*, University of Groningen, The Netherlands.

ARFMAN, N., VAN BEEUMEN, J., DE VRIES, G.E., HARDER, W., DIJKHUIZEN, L. (1991), Purification and characterization of an activator protein for methanol dehydrogenase from thermotolerant *Bacillus* spp., *J. Biol. Chem.* **266,** 3955–3960.

ARFMAN, N., DIJKHUIZEN, L., KIRCHHOF, G., LUDWIG, W., SCHLEIFER, K. H., BULYGINA, E. S., CHUMAKOV, K. M., GOVORUKHINA, N. I., TROTSENKO, Y. A., WHITE, D., SHARP, R. J. (1992), *Bacillus methanolicus* sp. nov., a new species of thermotolerant, methanol-utilizing, endospore-forming bacteria, *Int. J. Syst. Bacteriol.* **42,** 439–445.

ATTWOOD, M. M., HARDER, W. (1972), A rapid and specific enrichment procedure for *Hyphomicrobium* spp., *Antonie van Leeuwenhoek* **38,** 369–378.

BEHRENDT, U., BANG, W. G., WAGNER, F. (1984), The production of L-serine with a methylotrophic microorganism using the L-serine pathway and coupling with an L-tryptophan-producing process, *Biotechnol. Bioeng.* **26,** 308–314.

BROOKE, A. G., DIJKHUIZEN, L., HARDER, W. (1986), Regulation of flavin biosynthesis in the methylotrophic yeast *Hansenula polymorpha, Arch. Microbiol.* **145,** 62–70.

BUCKHOLZ, R. G., GLEESON, M. A. G. (1991), Yeast systems for the commercial production of heterologous proteins, *Bio/Technology* **9,** 1067–1072.

BULYGINA, E. S., GALCHENKO, V. F., GOVORUKHINA, N. I., NETRUSOV, A. I., NIKITIN, D. I., TROTSENKO, YU. A., CHUMAKOV, K. M. (1990), Taxonomic studies on methylotrophic bacteria by 5S ribosomal RNA sequencing, *J. Gen. Microbiol.* **136,** 441–446.

BULYGINA, E., CHUMAKOV, K., NETRUSOV, A. (1992), Systematics of Gram negative methylotrophic bacteria based on 5S rRNA sequences, *Abstr. 7th Int. Symp. Microbial Growth on C_1 Compounds,* L22.

BYSTRYKH, L. V., DIJKHUIZEN, L., HARDER, W. (1991), Modification of flavin adenine dinucleotide in alcohol oxidase of the yeast *Hansenula polymorpha, J. Gen. Microbiol.* **137,** 2381–2386.

BYSTRYKH, L. V., ARFMAN, N., DIJKHUIZEN, L. (1993), Methanol oxidizing enzyme systems in Gram-positive methylotrophic bacteria, in: *Microbial Growth on C_1 Compounds* (MURRELL, J. C., KELLY, D. P., Eds.), in press.

CAVANAUGH, C. M., LEVERING, P.R., MAKI, J. S., MITCHELL, R., LIDSTROM, M. (1987), Symbiosis of methylotrophic bacteria and deep-sea mussels, *Nature* **325,** 346–348.

CHAN, S. I., NGUYEN, H-H. T., SHIEMKE, A. K., LIDSTROM, M. E. (1992), Biochemical and biophysical studies towards characterization of the particulate methane monooxygenase, *Abstr. 7th Int. Symp. Microbial Growth on C_1 Compounds,* L7.

CHILDRESS, J. J., FISHER, C. R., BROOKS, J. M., KENNICUTT, M. C., II, BRIDIGARE, R., ANDERSON, A. E. (1986), A methanotrophic marine molluscan (bivalvia, mytilidae) symbiosis: methane-oxidizing mussels, *Science* **233,** 1306–1308.

CHISTOSERDOV, A. Y., TSYGANKOV, Y. D., LIDSTROM, M. E. (1991), Genetic organization of methylamine utilization genes from *Methylobacterium extorquens* AM1, *J. Bacteriol.* **173,** 5901–5908.

COLBY, J., STIRLING, D. I., DALTON, H. (1977), The soluble methane monooxygenase of *Methylococcus capsulatus* (Bath). Its ability to oxygenate *n*-alkanes, *n*-alkenes, ethers, alicyclic, aromatic and heterocyclic compounds, *Biochem. J.* **165,** 395–403.

COX, R. B., QUAYLE, J. R. (1975), The autotrophic growth of *Micrococcus denitrificans* on methanol, *Biochem. J.* **150,** 569–571.

DALTON, H. (1992), Methane oxidation by methanotrophs, in: *Biotechnology Handbooks,* Vol. 5, *Methane and Methanol Utilizers,* (MURRELL, J. C, DALTON, H., Eds.), pp. 85–114, New York: Plenum Press.

DE BOER, L., DIJKHUIZEN, L., GROBBEN, G., GOODFELLOW, M., STACKEBRANDT, E., PARLETT, J. H., WHITEHEAD, D, WITT, D. (1990), *Amycolatopsis methanolica* sp. nov., a facultatively methylotrophic actinomycete, *Int. J. Syst. Bacteriol.* **40,** 194–204.

DE BONT, J.A.M., VAN DIJKEN, J.P., HARDER, W. (1981), Dimethylsulphoxide and dimethylsulphide as a carbon, sulphur and energy source for growth of *Hyphomicrobium S, J. Gen. Microbiol.* **127,** 315–323.

DE KONING, W., HARDER, W. (1992), Methanol-utilizing yeasts, in: *Biotechnology Handbooks,* Vol. 5, *Methane and Methanol Utilizers* (MURRELL, J. C, DALTON, H., Eds.), pp. 207–244, New York: Plenum Press.

DE VRIES, G. E., KÜES, U., STAHL, U. (1990), Physiology and genetics of methylotrophic bacteria, *FEMS Microbiol. Rev.* **75,** 57–101.

DE VRIES, G. E., ARFMAN, N., TERPSTRA, P., DIJKHUIZEN, L. (1992), Cloning, expression and sequence analysis of the *Bacillus methanolicus* C_1 methanol dehydrogenase gene, *J. Bacteriol.* **174,** 5346–5353.

DIJKHUIZEN, L., SOKOLOV, I. G. (1991), Regulation of oxidation and assimilation of one-carbon compounds in methylotrophic bacteria, in: *Biology of Methylotrophs* (GOLDBERG, I., ROKEM, J. S., Eds.), pp. 127–148, Boston: Butterworth-Heinemann.

DIJKHUIZEN, L., HARDER W., DE BOER, L., VAN BOVEN, A., CLEMENT, W., BRON, S., VENEMA, G. (1984), Genetic manipulation of the restricted facultative methylotroph *Hyphomicrobium* X by the R-plasmid-mediated introduction of the *Escherichia coli pdh* genes, *Arch. Microbiol.* **139,** 311–318.

DIJKHUIZEN, L., HANSEN, T. A., HARDER, W. (1985), Methanol: a potential feedstock for biotechnological processes, *Trends Biotechnol.* **3,** 262–267.

DIJKHUIZEN, L., ARFMAN, N., ATTWOOD, M. M., BROOKE, A. G., HARDER, W., WATLING, E. M. (1988), Isolation and initial characterization of thermotolerant methylotrophic *Bacillus* strains, *FEMS Microbiol. Lett.* **52,** 209–214.

DIJKHUIZEN, L., LEVERING, P. R., DE VRIES, G. E. (1992), The physiology and biochemistry of aerobic methanol-utilizing Gram-negative and Gram-positive bacteria, in: *Biotechnology Handbooks,* Vol. 5, *Methane and Methanol Utilizers* (MURRELL, J. C., DALTON, H., Eds.), pp. 149–181, New York: Plenum Press.

DIJKHUIZEN, L., EUVERINK, G. J. W., HESSELS, G. I., VRIJBLOED, J. W. (1993), L-Phenylalanine synthesis by the facultative RuMP cycle methylotroph *Amycolatopsis methanolica,* in: *Microbial Growth on C_1 Compounds* (MURRELL, J. C., KELLY, D. P., Eds.), in press.

DIJKSTRA, M., FRANK, J., DUINE, J. A. (1988), Methanol oxidation under physiological conditions using methanol dehydrogenase and a factor isolated from *Hyphomicrobium* X, *FEBS Lett.* **227,** 198–202.

DUINE, J. A. (1991), Quinoproteins: enzymes containing the quinoid cofactor pyrroloquinoline quinone, topaquinone or tryptophan-tryptophan quinone, *Eur. J. Biochem.* **200,** 271–284.

DUINE, J. A., VAN DIJKEN, J. P. (1991), Enzymes of

industrial potential from methylotrophs, in: *Biology of Methylotrophs* (GOLDBERG, I., ROKEM, J. S., Eds.), pp. 233–252, Boston: Butterworth-Heinemann.

EGGELING, L., SAHM, H. (1984), An unusual formaldehyde oxidizing enzyme in *Rhodococcus erythropolis* grown on compounds containing methyl groups, *FEMS Microbiol. Lett.* **25,** 253–257.

EGGELING, L., SAHM, H. (1985), The formaldehyde dehydrogenase of *Rhodococcus erythropolis*, a trimeric enzyme requiring a cofactor and active with alcohols, *Eur. J. Biochem.* **150,** 129–134.

EKEROTH, L., VILLADSEN, J. (1991), Single cell protein production from C_1 compounds, in: *Biology of Methylotrophs* (GOLDBERG, I., ROKEM, J. S., Eds.), pp. 205–231, Boston: Butterworth-Heinemann.

GALLI, R., LEISINGER, T. (1985), Specialized bacterial strains for the removal of dichloromethane from industrial waste, *Conserv. Recycl.* **8,** 91–100.

GHISALBA, O., KÜENZI, M. (1983a), Biodegradation and utilization of monomethyl sulfate by specialized methylotrophs, *Experientia* **39,** 1257–1263.

GHISALBA, O., KÜENZI, M. (1983b), Biodegradation and utilization of quaternary alkylammonium compounds by specialized methylotrophs, *Experientia* **39,** 1264–1271.

GHISALBA, O., CEVEY, P., KÜENZI, M., SCHÄR, H-P. (1985), Biodegradation of chemical waste by specialized methylotrophs, an alternative to physical methods of waste disposal, *Conserv. Recycl.* **8,** 47–71.

GIUSEPPIN, M. L. F. (1988), Optimization of methanol oxidase production by *Hansenula polymorpha*: an applied study on physiology and fermentation, *Ph. D. Thesis*, Technical University of Delft, The Netherlands.

GOODWIN, P. M. (1990), The biochemistry and genetics of C_1 metabolism in the pink pigmented facultative methylotrophs, in: *Advances in Autotrophic Microbiology and One-Carbon Metabolism* (CODD, G. A., DIJKHUIZEN, L., TABITA, F. R., Eds.), pp. 143–162, Dordrecht: Kluwer Academic Publishers.

GOULD, W. D., KANAGAWA, T. (1992), Purification and properties of methyl mercaptan oxidase from *Thiobacillus thioparus* TK-m, *J. Gen. Microbiol.* **138,** 217–221.

GOVORUKHINA, N. I., TROTSENKO, YU. A. (1991), *Methylovorus* – a new genus of restricted facultatively methylotrophic bacteria, *Int. J. Syst. Bacteriol.* **41,** 158–162.

GREEN, P. N. (1992), Taxonomy of methylotrophic bacteria, in: *Biotechnology Handbooks*, Vol. 5, *Methane and Methanol Utilizers* (MURRELL, J. C., DALTON, H., Eds.), pp. 23–84, New York: Plenum Press.

GREEN, P. N., HOOD, D., DOW, C. S. (1984), Taxonomic status of some methylotrophic bacteria, in: *Microbial Growth on C_1 Compounds* (CRAWFORD, R. L., HANSON, R. S., Eds.), pp. 251–254, Washington, DC: American Society for Microbiology.

GREEN, P. N., BOUSFIELD, I. J., HOOD, D. (1988), Three new *Methylobacterium* species: M. *rhodesianum* sp. nov., *M. zatmanii* sp. nov., and *M. fujisawaense* sp. nov., *Int. J. Syst. Bacteriol.* **38,** 124–127.

HANSON, R. S., WATTENBERG, E. V. (1991), Ecology of methylotrophic bacteria, in: *Biology of Methylotrophs* (GOLDBERG, I., ROKEM, J. S., Eds.), pp. 325–348, Boston: Butterworth-Heinemann.

HANSON, R. S., BRATINA, B. J., BRUSSEAU, G. A. (1992), Phylogenetic and ecological studies of methylotrophs, *Abstr. 7th Int. Symp. Microbial Growth on C_1 Compounds*, L23.

HARDER, W., ATTWOOD, M. M. (1978), Biology, physiology and biochemistry of hyphomicrobia, *Adv. Microb. Physiol.* **17,** 303–359.

HEIJTHUIJSEN, J. H. F. G., HANSEN, T. A. (1990), C_1-Metabolism in anaerobic non-methanogenic bacteria, in: *Advances in Autotrophic Microbiology and One-Carbon Metabolism* (CODD, G. A., DIJKHUIZEN, L., TABITA, F. R., Eds.), pp. 163–191, Dordrecht: Kluwer Academic Publishers.

JANVIER, M., FREHEL, C., GRIMONT, F., GASSER, F. (1985), *Methylophaga marina* gen. nov., sp. nov. and *Methylophaga thalassica* sp. nov., marine methylotrophs, *Int. J. Syst. Bacteriol.* **35,** 131–139.

JANVIER, M., FRANK, J., LUTTIK, M., GASSER, F. (1992), Isolation and phenotypic characterization of methanol oxidation mutants of the restricted facultative methylotroph *Methylophaga marina*, *J. Gen. Microbiol.* **138,** 2113–2123.

JENKINS, O., JONES, D. (1987), Taxonomic studies on some Gram-negative methylotrophic bacteria, *J. Gen. Microbiol.* **133,** 453–473.

JENKINS, O., BYROM, D., JONES, D. (1987), *Methylophilus:* a new genus of methanol utilizing bacteria, *Int. J. Syst. Bacteriol.* **37,** 446–448.

JONGEJAN, J. A., DUINE, J. A. (Eds.) (1989), Special issue: PQQ and quinoproteins, *Antonie van Leeuwenhoek* **56** (1).

KANAGAWA, T., KELLY, D. P. (1986), Breakdown of DMS by mixed cultures and by *Thiobacillus thioparus*, *FEMS Microbiol. Lett.* **34,** 13–19.

KATO, N., MIYAMOTO, N., SHIMAO, M., SAKAZAWA, C. (1988), 3-Hexulose phosphate synthase from a new facultative methylotroph, *Mycobacterium gastri* MB19, *Agric. Biol. Chem.* **52,** 2659–2661.

KELLOGG, R. M., KRUIZINGA, W., BYSTRYKH, L. V., DIJKHUIZEN, L., HARDER, W. (1992), Structural analysis of a stereochemical modification of flavin adenine dinucleotide in alcohol oxidase from methylotrophic yeasts, *Tetrahedron* **48,** 4147–4162.

KLETSOVA, L. V., CHIBISOVA, E. S., TSYGANKOV, Y. D. (1988), Mutants of the obligate methylotroph *Methylobacillus flagellatum* KT defective in genes of the ribulose monophosphate cycle of formaldehyde fixation, *Arch. Microbiol.* **149,** 441–446.

KOHLER-STAUB, D., LEISINGER, T. (1985), Dichloromethane dehalogenase of *Hyphomicrobium* sp. strain DMZ, *J. Bacteriol.* **162,** 676–681.

LARGE, P. J., BAMFORTH, C. W. (1988), *Methylotrophy and Biotechnology*. Harlow: Longman.

LEAK, D. J. (1992), Biotechnological and applied aspects of methane and methanol utilizers, in: *Biotechnology Handbooks*, Vol. 5, *Methane and Methanol Utilizers* (MURRELL, J. C, DALTON, H., Eds.), pp. 245–279, New York: Plenum Press.

LEVERING, P .R., VAN DIJKEN, J. P., VEENHUIS, M., HARDER, W. (1981), *Arthrobacter* P1, a fast growing versatile methylotroph with amine oxidase as a key enzyme in the metabolism of methylated amines, *Arch. Microbiol.* **129,** 72–80.

LIDSTROM, M. E. (1990a), The aerobic methylotrophic bacteria, in: *The Prokaryotes* (BALOWS, A., TRÜPER, H. G., DWORKIN, M., HARDER, W., SCHLEIFER, K. H., Eds.), pp. 431–445, New York: Springer-Verlag.

LIDSTROM, M. E. (Ed.) (1990b), Hydrocarbons and methylotrophy. *Methods in Enzymology*, Vol. 188, San Diego: Academic Press.

LIDSTROM, M. E. (1992), The genetics and molecular biology of methanol-utilizing bacteria, in: *Biotechnology Handbooks*, Vol. 5, *Methane and Methanol Utilizers* (MURRELL, J. C., DALTON, H., Eds.), pp. 183–206, New York: Plenum Press.

LIDSTROM, M. E., STIRLING, D. I. (1990), Methylotrophs: Genetics and commercial applications, *Annu. Rev. Microbiol.* **44,** 27–58.

LONG, A. R., ANTHONY, C. (1991), The periplasmic modifier protein for methanol dehydrogenase in the methylotrophs *Methylophilus methylotrophus* and *Paracoccus denitrificans*, *J. Gen. Microbiol.* **137,** 2353–2360.

MCINTIRE, W. S., WEMMER, D. E., CHISTOSERDOV, A. Y., LIDSTROM, M. E. (1991), A new cofactor in a prokaryotic enzyme: tryptophan tryptophylquinone as the redox prosthetic group in methylamine dehydrogenase, *Science* **252,** 817–824.

MEIJER, W. G., CROES, L.M., JENNI, B., LEHMICKE, L. G., LIDSTROM, M. E., DIJKHUIZEN, L. (1990), Characterization of *Xanthobacter* strains H4–14 and enzyme profiles after growth under autotrophic and heterotrophic conditions, *Arch. Microbiol.* **153,** 360–367.

MOTOYAMA, H., ANAZAWA, H., MAKI, K., TESHIBA, S. (1992), Amino acid production by *Methylobacillus glycogenes*: Isolation and characterization of amino acid producing mutants from *M. glycogenes* ATCC 21276 and ATCC 21371, *Abstr. 7th Int. Symp. Microbial Growth on C_1 Compounds*, C112.

MÜLLER, F., HOPKINS, T. R., LEE, J., BASTIAENS, P. I. H. (1992), Methanol oxidase, in: *Chemistry and Biochemistry of Flavoenzymes*, Vol. III (MÜLLER, F., Ed.), pp. 95–119, Boca Raton: CRC Press.

MURRELL, J. C. (1992), The genetics and molecular biology of obligate methane-oxidizing bacteria, in: *Biotechnology Handbooks*, Vol. 5, *Methane and Methanol Utilizers* (MURRELL, J. C, DALTON, H., Eds.), pp. 115–148, New York: Plenum Press.

OLDENHUIS, R., VINK, R., JANSSEN, D. B., WITHOLT, B. (1989), Degradation of chlorinated aliphatic hydrocarbons by *Methylosinus trichosporium* OB3b expressing soluble methane monooxygenase, *Appl. Environ. Microbiol.* **55,** 2819–2826.

QUAYLE, J. R. (1987), An eightieth anniversary of the study of microbial C_1 metabolism, in: *Microbial Growth on C_1 Compounds* (VAN VERSEVELD, H. W., DUINE, J. A., Eds.), pp. 1–5, Dordrecht: Martinus Nijhoff Publishers.

REED, W. M., DUGAN, P. R. (1987), Isolation and characterization of the facultative methylotroph *Mycobacterium* ID-Y, *J. Gen. Microbiol.* **133,** 1389–1395.

RICHARDSON, I. W., AVEZOUX, A., ANTHONY, C. (1992), Methanol dehydrogenase lacking an essential calcium ion, *Abstr. 7th Int. Symp. Microbial Growth on C_1 Compounds*, B55.

SCHENDEL, F. J., BREMMON, C. E., FLICKINGER, M. C., GUETTLER, M., HANSON, R. S. (1990), L-Lysine production at 50 °C by mutants of a newly isolated and characterized methylotrophic *Bacillus* sp., *Appl. Environ. Microbiol.* **56,** 963–970.

SCHMALJOHANN, R., FLÜGEL, H. (1987), Methane-oxidizing bacteria in Pogonophora, *Sarsia* **72,** 91–98.

SCHÜTTE, H., FLOSSDORF, J., SAHM, H., KULA, M. R. (1976), Purification and properties of formaldehyde dehydrogenase and formate dehydrogenase from *Candida boidinii*, *Eur. J. Biochem.* **62,** 151–160.

SHERRY, B., ABELES, R. H. (1985), Mechanism of action of methanol oxidase, reconstitution of methanol oxidase with 5-deazaflavin, and inactivation of methanol oxidase by cyclopropanol, *Biochemistry* **24,** 2594–2605.

SMITH, L. M., GOODWIN, P. M. (1992), Complementation analysis of mutants defective in the conversion of acetyl CoA to glyoxylate in *Methylobacterium extorquens* AM1, *Abstr. 7th Int. Symp. Microbial Growth on C_1 Compounds*, B54.

SPERL, G. T., HOARE D. S. (1971), Denitrification with methanol: a selective enrichment for *Hyphomicrobium* species, *J. Bacteriol.* **108,** 733–736.

STACKEBRANDT, E., FISCHER, A., ROGGENTIN, T.,

WEHMEYER, U., BOMAR, D., SMIDA, J. (1988), A phylogenetic survey of budding, and/or prosthecate, non-phototrophic eubacteria: membership of *Hyphomicrobium*, *Hyphomonas*, *Pedomicrobium*, *Filomicrobium*, *Caulobacter* and "*Dichotomicrobium*" to the alpha-subdivision of purple non-sulfur bacteria, *Arch. Microbiol.* **149,** 547–556.

STANLEY, S. H., PRIOR, S. D., LEAK, D. J., DALTON, H. (1983), Copper stress underlies the fundamental change in intracellular location of the methane monooxygenase in methanotrophs, *Biotechnol. Lett.* **5,** 487–492.

STEUDEL, A., MIETHE, D., BABEL, W. (1980), Bakterium MB58, ein methylotrophes „Essigsäurebakterium", *Z. Allg. Mikrobiol.* **20,** 663–672.

STIRLING, D. L., COLBY, J., DALTON, H. (1979), A comparison of the substrate and electron-donor specificities of the methane mono-oxygenases from three strains of methane-oxidizing bacteria, *Biochem. J.* **177,** 361–364.

SUYLEN, G. M.H., KUENEN, J. G. (1986), Chemostat enrichment and isolation of *Hyphomicrobium* EG, a dimethyl sulphide oxidizing methylotroph and reevaluation of *Thiobacillus* MS1, *Antonie van Leeuwenhoek* **52,** 281–293.

SUYLEN, G. M. H., LARGE, P. J., VAN DIJKEN, J. P., KUENEN, J. G. (1987), Methylmercaptan oxidase, a key enzyme in the metabolism of methylated sulphur compounds by *Hyphomicrobium* EG, *J. Gen. Microbiol.* **133,** 2989–2997.

TABITA, F. R. (1988), Molecular and cellular regulation of autotrophic carbon dioxide fixation in microorganisms, *Microbiol. Rev.* **52,** 155–189.

TSCHOPP, J. F., CREGG, J. M. (1991), Heterologous gene expression in methylotrophic yeast, in: *Biology of Methylotrophs* (GOLDBERG, I., ROKEM, J. S., Eds.), pp. 305–322, Boston: Butterworth-Heinemann.

TSIEN, H. C., BRUSSEAU, G. A., HANSON, R. S., WACKETT, L. P. (1989), Biodegradation of trichloroethylene by *Methylosinus trichosporium* OB3b, *Appl. Environ. Microbiol.* **55,** 3155–3161.

TSUJI, K., TSIEN, H. C., HANSON, R. S., DEPALMA, S. R., SCHOLTZ, R., LAROCHE, S. (1990), 16S ribosomal RNA sequence analysis for determination of phylogenetic relationship among methylotrophs, *J. Gen. Microbiol.* **136,** 1–10.

URAKAMI, T., KOMAGATA K. (1986), Emendation of *Methylobacillus* Yordy and Weaver 1977, a genus for methanol utilizing bacteria, *Int. J. Syst. Bacteriol.* **36,** 502–511.

URAKAMI, T., TAMAOKA, J., SUZUKI, K.-I., KOMAGATA, K. (1989), *Acidomonas* gen. nov., incorporating *Acetobacter methanolicus* as *Acidomonas methanolica* comb. nov., *Int. J. Syst. Bacteriol.* **39,** 50–55.

VAN DIJKEN, J. P., HARDER, W., BEARDSMORE, A. J., QUAYLE, J. R. (1978), Dihydroxyacetone: an intermediate in the assimilation of methanol by yeasts? *FEMS Microbiol. Lett.* **4,** 97–102.

VELLIEUX, F. M. D., HUITEMA, F., GROENDIJK, H., KALK, K. H., FRANK, J. JZN., JONGEJAN, J. A., DUINE, J. A., PETRATOS, K., DRENTH, J., HOL, W. G. J. (1989), Structure of quinoprotein methylamine dehydrogenase at 2.25 Å resolution, *EMBO J.* **8,** 2171–2178.

VISSCHER, P. T., QUIST, P, VAN GEMERDEN, H. (1991), Methylated sulfur compounds in microbial mats: *In situ*, concentrations and metabolism by a colorless sulfur bacterium, *Appl. Environ. Microbiol.* **57,** 1758–1763.

VONCK, J., ARFMAN, N., DE VRIES, G. E., VAN BEEUMEN, J., VAN BRUGGEN, E. F. J., DIJKHUIZEN, L. (1991), Electron microscopic analysis and biochemical characterization of a novel methanol dehydrogenase from the thermotolerant *Bacillus* sp. C_1, *J. Biol. Chem.* **266,** 3949–3954.

WHITTENBURY, R., KRIEG, N. R. (1984), Family IV. *Methylococcaceae*, in: *Bergey's Manual of Systematic Bacteriology* (KRIEG, N. R., HOLT, J. G., Eds.), pp. 256–261, Baltimore: Williams & Wilkins.

WOLFRUM, T., STOLP, H. (1987), Comparative studies on 5S RNA sequences of RuMP-type methylotrophic bacteria, *Syst. Appl. Microbiol.* **9,** 273–276.

WOODWARD, J. R. (1990), Biochemistry and applications of alcohol oxidase from methylotrophic yeasts, in: *Advances in Autotrophic Microbiology and One-Carbon Metabolism* (CODD, G. A., DIJKHUIZEN, L., TABITA, F. R., Eds.), pp. 193–225, Dordrecht: Kluwer Academic Publishers.

ZHANG, L., HIRAI, M., SHODA, M. (1991), Removal characteristics of dimethyl sulfide, methanethiol and hydrogen sulfide by *Hyphomicrobium* sp. 155 isolated from peat biofilter, *J. Ferment. Bioeng.* **5,** 392–396.

9 Clostridia

HUBERT BAHL
PETER DÜRRE
Göttingen, Federal Republic of Germany

1 Introduction 286
2 General Properties of Clostridia 286
3 Growth 293
3.1 Oxygen Sensitivity and Redox Potential 293
3.2 pH and Temperature Range 294
3.3 Nutritional Requirements 294
3.4 Isolation, Handling, and Storage 295
4 Metabolism and Physiology 297
4.1 Degradation of Polysaccharides 297
4.2 Growth on Mono- and Disaccharides 298
4.3 Growth on Alcohols, Aromatic Compounds, and Organic Acids 300
4.4 Fermentation of Nitrogenous Compounds 306
4.5 Gaseous Substrates 308
5 Industrially Important Products 308
5.1 Low-Molecular Weight Compounds 308
5.2 Proteins 310
5.3 Stereospecific Compounds 311
5.3.1 Reduction Reactions 311
5.3.2 Bile Acids and Steroid Conversions 311
5.3.3 Dechlorination 313
6 References 314

1 Introduction

Clostridia are becoming more and more recognized as organisms of biotechnological potential both in the production of chemical feedstock and as biotransformation agents. The acetone-butanol fermentation using *Clostridium acetobutylicum* strains was one of the largest industrial feedstock production processes involving anaerobic bacteria. It was the major route for the synthesis of these solvents during the first part of this century and became uneconomical only after cheap raw oil was available for the petrochemical industry (BAHL and GOTTSCHALK, 1988; JONES and WOODS, 1986). The prospect of converting agricultural and industrial waste products of little or no economic value into raw materials for chemical transformations is the focus of renewed industrial interest. Enormous progress has been achieved not only in handling of the strictly anaerobic bacteria and in strain improvement by conventional procedures such as mutagenesis and screening, but also in the development of recombinant DNA techniques for several species of this genus which opens the possibility of defined construction of new production strains (YOUNG et al., 1989). Furthermore, specific enzymes from these bacteria have been found to be useful in chemical transformation reactions (MORRIS, 1989; SAHA et al., 1989). Even the toxins produced by the few pathogenic species are now becoming indispensable for medical treatment of special diseases (SCHANTZ and JOHNSON, 1992; SHONE and HAMBLETON, 1989). Undoubtedly, the list of potential biotechnological applications of clostridia will grow at fast rate in the near future.

2 General Properties of Clostridia

So far, more than 120 species of clostridia have been validly described (Tab. 1). Thus, this genus is one of the largest within the prokaryotic kingdom. A bacterium will be assigned to this group if four conditions are fulfilled: It must possess a Gram-positive cell wall structure, its metabolism must be obligatorily fermentative (i.e., anaerobic), it must be able to form heat-resistant endospores, and it must be unable to perform dissimilatory sulfate reduction. Since these criteria are relatively broad, it is not surprising that the genus *Clostridium* is very heterogeneous. The G+C content ranges from 21 to 55 mol% (see Tab. 1), already indicating large phylogenetic differences. Recent attempts at classification in defined groups have been based on 16S rRNA cataloging and comparison of complete 5S rRNA sequences (CATO and STACKEBRANDT, 1989). The results show that the genus *Clostridium* is a phylogenetically incoherent taxon and that many species are more closely related to various other anaerobic eubacteria, even to those with Gram-negative staining behavior.

In more than 20 species only one strain is available for respective studies. This represents a problem for taxonomists. Correct nomenclature of quite a few clostridia seems to be a matter of debate (CATO et al., 1986; HIPPE et al., 1992). Names given in Tab. 1 are mostly those now in common use and are in accordance with the last extensive review of this genus (HIPPE et al., 1992). Clostridia are ubiquitous due to their ability to form heat-resistant endospores (Fig. 1). Successful isolation therefore depends on choosing appropriate growth conditions (e.g., oxygen content, energy and carbon source, temperature) that will be dealt with in the following section. Species of this genus are usually oxygen-sensitive (thus, vegetative cells live and survive only under strictly anaerobic conditions), rod-shaped, and motile by means of peritrichously-arranged flagella, although several exceptions from these general rules have been observed (cp. HIPPE et al., 1992). The energy metabolism is anaerobic and relies on fermentation of suitable carbon sources. Membrane-associated redox carriers such as menaquinone and cytochromes have only been found in homoacetogens such as *C. aceticum*, *C. formicoaceticum*, *C. thermoaceticum*, and *C. thermoautotrophicum* (BRAUN et al., 1981; GOTTWALD et al., 1975; DAS et al., 1989; LJUNGDAHL et al., 1989). Only few clostridia are pathogenic due to the excretion of proteinaceous exotoxins that necrotize body tissues or interfere with nerve transmission. Examples are *C. botulinum*, causing the food-

Tab. 1. Clostridial Species[a]

Species	G+C Content (mol%)	Optimal Growth Temperature (°C)	Saccharolytic	Proteolytic	Special Substrates	Pathogenic	Biotechnologically Interesting Characteristics
C. absonum	n.r.[b]	30–45	+				Gonane (sterane) derivative conversions
C. aceticum	33	30	+		CO, H_2-CO_2, methoxylated aromatic compounds		Homoacetogen
C. acetobutylicum	28–29	37	+	+			Produces acetone, butanol, 1,3-propanediol
C. acidiurici	27–30	40–45			Glycine, purines		
C. aerotolerans	40	38	+		Xylan		
C. aldrichii	40	35			Cellulose, xylan		
C. aminovalericum	33	37			5-Aminovalerate		
C. arcticum	n.r.	22–25	+				
C. argentinense	26–28	37		+		+	
C. aurantibutyricum	27–28	37	+		Pectin		Produces acetone, butanol, isopropanol
C. baratii	28	35–40	+				
C. barkeri	45	30–37	+		Nicotinic acid		
C. beijerinckii	26–28	30–37	+		Pectin		Produces acetone, butanol, isopropanol
C. bifermentans	27	30–37	+	+			Gonane (sterane) derivative conversions, specific reduction reactions
C. botulinum	26–28	30–40	+	+		+	Produces toxins used in medical treatment
C. butyricum	27–28	25-37	+		Pectin		Produces 1,3-propanediol, dechlorination reactions
C. cadaveris	27	30–37	+	+			Gonane (sterane) derivative conversions
C. carnis	25–28	37	+			+	
C. celatum	n.r.	37	+				
C. celerecrescens	38	35	+		Cellulose		
C. cellobioparum	28	30–37	+		Cellulose		
C. cellulofermentans	34	37–40	+		Cellulose		
C. cellulolyticum	41	32–35	+		Cellulose		
C. cellulosi	35	55–60	+		Cellulose		
C. cellulovorans	26–27	32–35	+		Cellulose, pectin, xylan		
C. chartatabidum	31	39	+		Cellulose		
C. chauvoei	27	37	+			+	
C. clostridiiforme	47–50	30–37	+				
C. coccoides	43–45	37	+				

Tab. 1. (Continued)

Species	G+C Content (mol %)	Optimal Growth Temperature (°C)	Saccharolytic	Proteolytic	Special Substrates	Pathogenic	Biotechnologically Interesting Characteristics
C. cochlearium	27–28	37–45			Glutamate, glutamine, histidine		
C. cocleatum	28–29	37–45	+				
C. colinum	n.r.	37	+			+	
C. collagenovorans	24	30–37		+			
C. cylindrosporum	27–30	40–45			Glycine, purines		
C. difficile	28	30–37	+			+	Gonane (sterane) derivative conversions
C. disporicum	40	25–26, 39–44	+				
C. durum	50	30	+				
C. fallax	26	37–45	+			+	
C. felsineum	26	25–37	+		Pectin		Produces butanol
C. fervidum	39	68	+		Xylan		
C. formicoaceticum	34	30–37	+		Methoxylated aromatic compounds, CO, H_2-CO_2, pectin		Homoacetogen, specific reduction reactions
C. ghoni	27	37		+			
C. glycolicum	29	25–37	+		Cellulose, 1,2-ethanediol, 1,2-propanediol, uracil		
C. haemolyticum	21, 26–27	37	+	+		+	
C. halophilium	27	41	+		Amino acids, betaine		
C. hastiforme	n.r.	30–37		+			
C. histolyticum	n.r.	37		+		+	
C. homopropionicum	32	28–37	+		Hydroxybutyrate		
C. indolis	44	37	+		Pectin		
C. innocuum	43–44	37	+				Gonane (sterane) derivative conversions
C. intestinale	26–28	37	+				
C. irregulare	n.r.	30–37			Gelatin		
C. josui	40	45	+		cellulose, xylan		
C. kluyveri	30	35–37			Ethanol/propanol-acetate/succinate		Stereospecific biohydrogenation reactions
C. lentocellum	36	40	+		Cellulose, xylan		
C. leptum	51–52	37	+				Gonane (sterane) derivative conversions
C. limosum	24	37		+	Glutamate	+	Gonane (sterane) derivative conversions

Tab. 1. (Continued)

Species	G+C Content (mol%)	Optimal Growth Temperature (°C)	Saccharolytic	Proteolytic	Special Substrates	Pathogenic	Biotechnologically Interesting Characteristics
C. litorale	26	28			Betaine/sarcosine-amino acids		
C. lituseburense	27	30–37	+	+			
C. longisporum	33	35–42	+	+	Cellulose,pectin		
C. magnum	29	30–32	+				Homoacetogen
C. malenominatum	28	37			Glutamate, peptone, uric acid		
C. mangenotii	n.r.	30–37		+			
C. mayombei	26	33	+		Amino acids,H_2-CO_2		Homoacetogen
C. methylpentosum	46	45	+				
C. neopropionicum	35	30	+		Amino acids, propanol/acetate		Can produce propanol
C. nexile	40–41	30–37	+				
C. novyi	23–29	37–45	+	+		+	
C. oceanicum	26–28	30–37	+	+			
C. orbiscindens	56–57	37			Complex media		Gonane (sterane) derivative conversions
C. oroticum	44	30–37	+		Orotic acid		
C. oxalicum	36	28–30			Acetate-oxalate		
C. papyrosolvens	30	20–30	+		Cellulose		
C. paraputrificum	26–27	30–37	+				Gonane (sterane) derivative conversions
C. pasteurianum	26–28	37	+				Produces butanol, 1,3-propanediol; dechlorination and specific reduction reactions
C. perfringens	24–27	37–45	+	+		+	Gonane (sterane) derivative conversions
C. pfennigii	38	36–38			CO, pyruvate, methoxylated monobenzoids		
C. polysaccharolyticum	42	30–38	+		Cellulose, xylan		
C. populeti	28	35	+		Cellulose, pectin, xylan		
C. propionicum	35	30–37			Acrylate, amino acids, lactate		Can produce acrylic acid
C. proteolyticum	30	30–37		+			
C. puniceum	28–29	23–33	+		Pectin		Produces butanol
C. purinolyticum	29	36			Glycine, purines		Gonane (sterane) derivative conversions
C. putrefaciens	22–25	20–25	+	+			
C. putrificum	27	37	+	+			

Tab. 1. (Continued)

Species	G+C Content (mol %)	Optimal Growth Tempe-rature (°C)	Sac-charo-lytic	Proteo-lytic	Special Substrates	Patho-genic	Biotechnologically Interesting Characteristics
C. quercicolum	52–54	25–30	+				
C. quinii	28	40–45	+				
C. ramosum	26–27	37	+				
C. rectum	26	37–45	+				Dechlorination reactions
C. roseum	n.r.	37	+		Pectin		Produces butanol
C. saccharolyticum	28	37	+				
C. sardiniense	n.r.	25–37	+				
C. sartagoforme	28	30–37	+				
C. scatologenes	27	30–37	+				
C. scindens	45	45	+				Gonane (sterane) derivative conversions
C. septicum	24	35–40	+			+	
C. sordellii	24–26	30–37	+	+		+	Gonane (sterane) derivative conversions, specific reduction reactions
C. sphenoides	41–42	30–37	+				Dechlorination reactions, produces 1,2-propanediol
C. spiroforme	27	30–40	+			+	
C. sporogenes	26	30–40	+	+	Amino acids, chitin		Specific reduction reactions
C. sporosphaeroides	27	37–45			Citrate, glutamate, lactate, pyruvate		
C. stercorarium	39	65	+		Cellulose, xylan		
C. sticklandii	31	30–37	+		Amino acids		
C. subterminale	28	37		+	Amino acids		
C. symbiosum	46	37	+		Glutamate		
C. termitidis	39	37	+		Cellulose		
C. tertium	24–26	37	+				Gonane (sterane) derivative conversions
C. tetani	25–26	37	+	+	Amino acids	+	
C. tetanomorphum	25–28	37	+		Glutamate, histidine, threonine		Produces butanol
C. thermoaceticum	54	55–60	+		CO, H_2—CO_2, oxalate		Homoacetogen, dechlorination and specific reduction reactions
C. thermoautotrophicum	53–55	55–60	+		CO, H_2—CO_2, oxalate		Homoacetogen
C. thermobutyricum	37	55	+		Cellulose		
C. thermocellum	38–40	55–69			Cellobiose, cellulose, pectin		

Tab. 1. (Continued)

Species	G+C Content (mol%)	Optimal Growth Temperature (°C)	Saccharolytic	Proteolytic	Special Substrates	Pathogenic	Biotechnologically Interesting Characteristics
C. thermocopriae	37–38	60–65	+		Cellulose, xylan		
C. thermohydrosulfuricum	30–32, 34–38	65–70	+		Pectin		Produces thermostable amylolytic enzymes, specific reduction reactions
C. thermolacticum	41–42	60–65	+		Cellulose, xylan, pectin		
C. thermopalmarium	36	50–55	+				
C. thermopapyrolyticum	34	59	+		Cellulose		
C. thermosaccharolyticum	29–32	55–62	+		Pectin		Produces butanol, 1,2-propanediol, thermostable pullulanase
C. thermosuccinogenes	36	58	+				
C. thermosulfurogenes	33	60	+		Pectin		Produces thermostable amylolytic enzymes
C. tyrobutyricum	28	30–37	+				Stereospecific biohydrogenation reactions
C. uzonii	34	65	+		Pectin		
C. villosum	n.r.	37			Complex media		
C. xylanolyticum	40	35	+		Xylan		

[a] Data taken from ANDREESEN et al. (1970, 1989), BRYSON and DRAKE (1988), CATO et al. (1986), DANIEL and DRAKE (1991), DÖRNER and SCHINK (1990), DRENT et al. (1991), EGLI et al. (1988), FENDRICH et al. (1990), HETHENER et al. (1992), HIPPE et al. (1992), KANE et al. (1991), KRIVENKO et al. (1990), LE RUYET et al. (1985), LUX et al. (1989), MCCOY and MCCLUNG (1935), MÉNDEZ et al. (1991), MORRIS (1989), OHISA et al. (1980), ROGERS and BAECKER (1991), SAVAGE and DRAKE (1986), SMITH (1992), SOH et al. (1991), SPINNLER et al. (1986), SVENSSON et al. (1992), THOLOZON et al. (1992), VAREL (1989), WINTER et al. (1991), YANG et al. (1990), and YANLING et al. (1991). Some features are specific for only one or few strains of a species.

[b] n.r., not reported

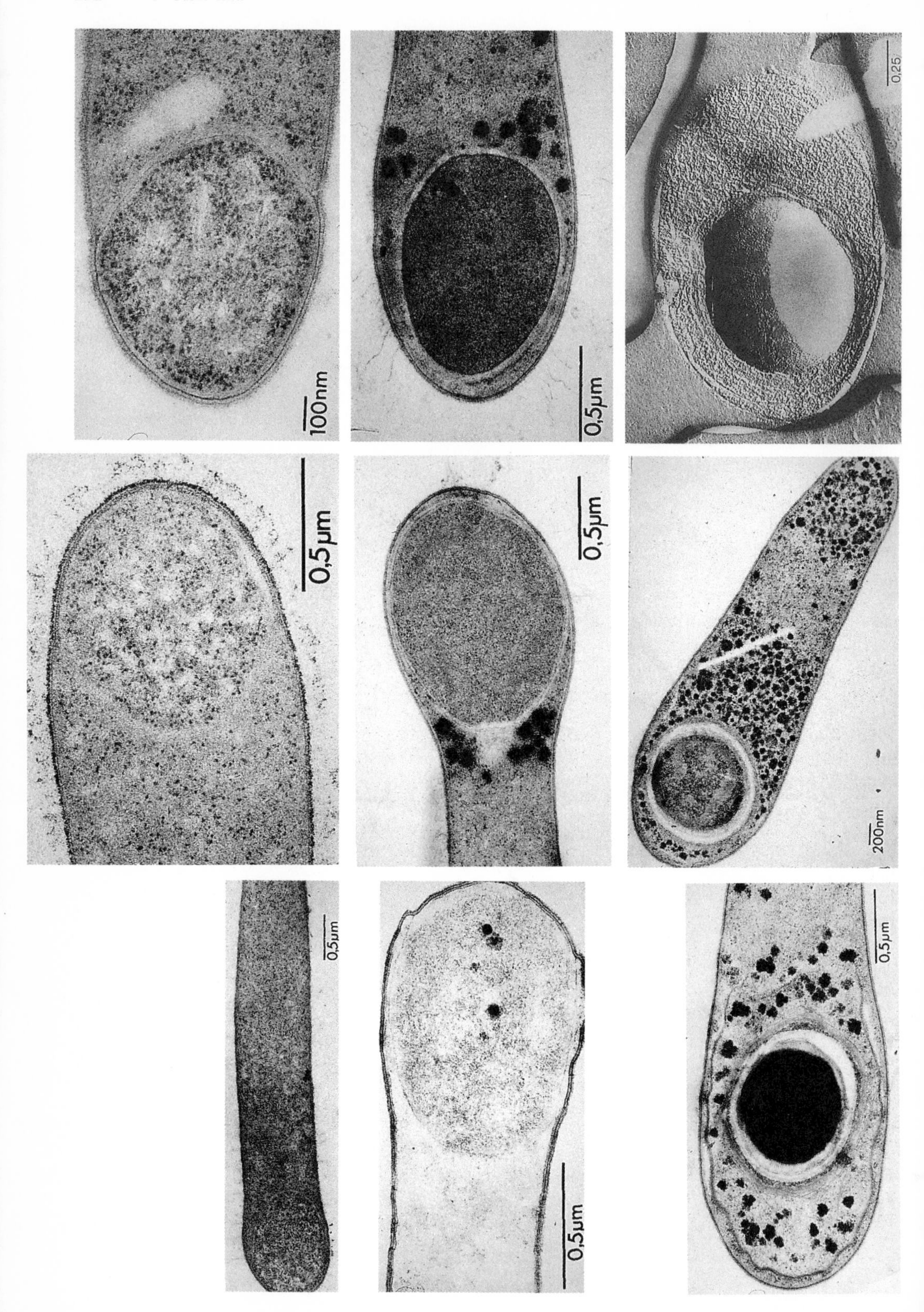
100nm
0,5µm
0,25
0,5µm
0,5µm
200nm
0,5µm
0,5µm
0,5µm

borne disease botulism by production of a neurotoxin (the most potent microbial toxin known so far), *C. perfringens* and *C. histolyticum*, causing gas gangrene in humans (during warfare and under poor hygienic conditions wounds are places of infections), and *C. tetani* causing tetanus by production of a toxin that leads to uncontrolled spasm of voluntary muscles. Unfortunately, these species are responsible for the dangerous image of clostridia still sometimes expressed in public which somewhat suppresses the enormous biotechnological potential of this bacterial group. However, even the very powerful clostridial toxins become now a valuable tool in medical treatment as will be discussed in a later section.

Development of genetic systems has mainly focused on *C. acetobutylicum* and *C. perfringens* as preferred model organisms for apathogenic and pathogenic species, respectively (YOUNG et al., 1989). The molecular genetics of *C. perfringens* have been summarized in a recent review (ROOD and COLE, 1991). For *C. acetobutylicum* transformation of whole cells by electroporation with reasonable frequencies (MERMELSTEIN et al., 1992; OULTRAM et al., 1988), stable expression of shuttle plasmids (MINTON and OULTRAM, 1988; MERMELSTEIN et al., 1992; OULTRAM et al., 1988; TRUFFAUT et al., 1989; YOON et al., 1991; YOSHINO et al., 1990), and transposon mutagenesis (BERTRAM and DÜRRE, 1989; BERTRAM et al., 1990; WOOLLEY et al., 1989) have been achieved. A number of genes, mostly involved in fermentative metabolism, solvent production, ammonia assimilation, and the heat-shock response have been cloned, partially sequenced, and their regulation has been investigated. For further information the reader is referred to a recent compilation of several reviews covering most aspects of the molecular genetics of *C. acetobutylicum* and other clostridia (WOODS, 1992). The enormous progress made in the fields of recombinant DNA techniques and studies of regulatory mechanisms raises expectations that in the near future the construction of strains will be possible that overproduce, e.g., solvents or valuable proteins.

3 Growth

3.1 Oxygen Sensitivity and Redox Potential

By reacting with reduced cellular constituents (e.g., ferredoxin, thiols, or flavoproteins) potentially lethal substances are formed from oxygen. These include hydrogen peroxide, singlet oxygen (only under simultaneous illumination), the superoxide anion, and the hydroxyl radical. The latter two are most probably responsible for the damaging effects on anaerobic bacteria observed after exposure towards oxygen (MORRIS, 1980). Targets of these highly reactive products could probably be lipid cell membranes or sensitive enzymes, particularly those catalyzing dehydrogenation reactions or possessing autoxidizable groups at their active center(s). As defense mechanisms living cells use catalase (or peroxidases) to convert hydrogen peroxide (or peroxide groups) to water (or hydroxyl groups) and oxygen, and superoxide dismutase to destroy the superoxide anion. However, the hypothesis that aerotolerant organisms do not contain catalase and strictly anaerobic bacteria lack both enzymes proved to be wrong, since superoxide dismutase activity could be demonstrated in *C. acetobutylicum, C. beijerinckii, C. bifermentans, C. butyricum, C. pasteurianum, C. perfringens*, and *C. sporogenes* (HEWITT and MORRIS, 1975). Furthermore, the presence of this enzyme did not correlate with the degree of oxygen tolerance several clostridia showed. Only *C. aerotolerans, C. carnis, C. durum, C. histolyticum, C. intestinale*, and *C. tertium* have been described as growing on solid media or in liquid culture in the presence of air (CATO et al., 1986; HIPPE et al., 1992). Reports on oxygen tolerance of *C. haemolyticum* and *C. novyi*, however, are somewhat contradictory (CATO et al., 1986; HIPPE et al., 1992). Some clostridia (e.g., *C. acetobutylicum*) survive even exposure to air for several hours (O'BRIEN and MORRIS, 1971). However, this refers only to relatively dense suspensions of actively grow-

◀ **Fig. 1.** Spore formation in *Clostridium formicoaceticum*. Thin sections of different stages of development and a freeze-etching sample of a mature spore are shown. Original micrographs from MAYER, 1986, reproduced with permission.

ing cells, whereby a certain percentage is killed and lysed. Plating under air usually leads to a drastic or even complete loss of viable organisms.

The presence of oxygen in the culture broth causes a positive redox potential (E_h), whereas growing clostridia need and indeed establish a very low E_h, depending on the type of medium used. For *C. sporogenes* values of −450 to −138 mV have been reported, and with *C. acetobutylicum* −400 to −250 mV have been determined. Growth was still possible up to +150 and −50 mV, respectively (MORRIS and O'BRIEN, 1971; O'BRIEN and MORRIS, 1971). In case of *C. acetobutylicum* experiments have been performed in which the medium was kept anaerobically while the redox potential was artificially raised by addition of potassium ferricyanide to a value of +370 mV. Growth rate, substrate consumption, and acid production were virtually unaffected. However, if the culture was made aerobic and the potential was artificially lowered to −50 mV by addition of dithiothreitol, the intracellular ATP content dropped dramatically and growth stopped immediately (O'BRIEN and MORRIS, 1971). These results again demonstrate that oxygen or its products are toxic, not the redox potential.

3.2 pH and Temperature Range

The optimal pH for the onset of growth for most clostridia is in the range of 6.5 to 7.5 (ANDREESEN et al., 1989). Exceptions are, e.g., *C. aceticum* that has a pH optimum of 8.3 for autotrophic growth (BRAUN et al., 1981) and strains of purinolytic clostridia that grow best on hypoxanthine if the initial pH is 8.0 (SCHIEFER-ULLRICH et al., 1984). The latter organisms are adapted to higher pH values since degradation of the nitrogenous heteroaromatic compounds liberates considerable amounts of ammonia as will be discussed in a later section. Acid-producing clostridia usually lower the pH of the medium to a value of around 4.5 during growth. Examples are the homoacetogen *C. thermoaceticum*, and the acetate- and butyrate-forming *C. acetobutylicum* and *C. butyricum*. A newly isolated strain, *"C. ljungdahlii"* (not yet validly described), is even able to produce acetate at a pH as low as 4 which makes it interesting for a commercial acetogenic fermentation (TANNER and YANG, 1990). The low pH is responsible for increasing amounts of the undissociated forms of these acids that bring about dissipation of the proton gradient across bacterial membranes and eventually lead to cell death (DÜRRE et al., 1988). It is advisable to try several different pH values for testing utilization of a substrate (even sugars), since the range of the optimal initial pH for growth is relatively broad and standardized test media for identification purposes (HOLDEMAN et al., 1977) only show a very narrow window of pH 6.7–7.0. Thus, several examples of misclassified carbohydrate fermentation patterns have been described (ANDREESEN et al., 1989).

The range of optimal growth temperatures in the genus *Clostridium* is even larger and spans the whole spectrum from psychrophilic to thermophilic species (see Tab. 1). Two clostridia with a low temperature optimum of around 22 °C (*C. arcticum* and *C. putrefaciens*) have been described so far, one of which (*C. putrefaciens*) is truly psychrophilic, i.e., unable to grow at more than 30 °C (HIPPE et al., 1992). The majority of members of this genus is mesophilic with optima between 30 and 40 °C. However, many thermophilic species have been described, especially in the last few years, which reflects the increased interest in fermentations at high temperatures (low danger of contamination, no cost and equipment for cooling) and in thermostable enzymes.

3.3 Nutritional Requirements

In general, clostridia are either saccharolytic, proteolytic, a combination of both, or specialists with respect to their preferred or characteristic substrate (see Tab. 1; HIPPE et al., 1992). Thus, the substrate spectrum of clostridia covers a wide range of organic compounds, reflecting the assembly of quite diverse microorganisms in this genus and the ubiquitous distribution of its members. Although the nutrient requirements for most of the many species have not been determined in detail, synthetic media are known for several members of each of the four nutritional groups.

In most cases vitamins and amino acids have to be added to carbon and nitrogen source,

Tab. 2. Composition of Synthetic Media for *C. bifermentans and C. acetobutylicum*

Ingredient	*C. bifermentans*[a] (g L^{-1})	*C. acetobutylicum*[b] (g L^{-1})
Glucose	20.00	20.00
$(NH_4)_2SO_4$	–	2.00
$Fe(NO_3)_3 \cdot 9H_2O$	$0.10 \cdot 10^{-3}$	–
$CaCl_2 \cdot 2H_2O$	0.22	0.01
KCl	0.40	–
$MgSO_4 \cdot 7H_2O$	0.20	0.10
$ZnSO_4$	$0.10 \cdot 10^{-4}$	–
NaCl	5.20	0.01
$NaH_2PO_4 \cdot H_2O$	1.95	1.00
$Na_2HPO_4 \cdot H_2O$	1.53	1.00
$NaHCO_3$	2.00	–
$Na_2MoO_4 \cdot 2H_2O$	–	0.01
$MnSO_4 \cdot H_2O$	–	$15.00 \cdot 10^{-3}$
$FeSO_4 \cdot 7H_2O$	–	$15.00 \cdot 10^{-3}$
Biotin	$0.10 \cdot 10^{-5}$	$0.10 \cdot 10^{-3}$
p-Aminobenzoate	–	$2.00 \cdot 10^{-3}$
D-Ca-pantothenate	$0.10 \cdot 10^{-3}$	–
Nicotinamide	$0.20 \cdot 10^{-3}$	–
Pyridoxal-HCl	$0.20 \cdot 10^{-3}$	–
L-Alanine	0.30	–
L-Arginine	0.40	–
L-Aspartic acid	0.50	–
L-Cysteine	$0.10 \cdot 10^{-2}$	–
L-Glutamic acid	3.00	–
Glycine	0.30	–
L-Histidine	0.70	–
L-Leucine	1.10	–
L-Methionine	0.70	–
L-Phenylalanine	0.60	–
L-Threonine	0.40	–
L-Trytophan	0.20	–
L-Tyrosine	0.30	–
L-Valine	0.40	–
Sodium thioglycolate	0.10	0.10

[a] according to HOLLAND and COX (1975)
[b] according to BAHL et al. (1982b), LAMPEN and PETERSON (1943)

common salt, and trace element solutions in order to prepare a synthetic medium for a certain clostridial strain. Thus, *p*-aminobenzoate and biotin are required by several clostridia, e.g., *C. acetobutylicum* (LAMPEN and PETERSON, 1943), *C. kluyveri* (BORNSTEIN and BARKER, 1948), and *C. pasteurianum* (SERGEANT et al., 1968). Other species need thiamine (e.g., *C. acidiurici*, SCHIEFER-ULLRICH et al., 1984; *C. purinolyticum*, DÜRRE et al., 1981), pyridoxine (e.g., *C. formicoaceticum*, LEONHARDT and ANDREESEN, 1977; *C. sporogenes*, CHAIGNEAU et al., 1974), pantothenate (e.g., *C. bifermentans*, HOLLAND and COX, 1975; *C. glycolicum*, GASTON and STADTMAN, 1963; *C. thermocellum*, FLEMMING and QUINN, 1971), and/or nicotinate (*C. thermoaceticum*, LUNDIE and DRAKE, 1984). Some species have a pronounced requirement for amino acids. In Tab. 2 the components of a synthetic medium for such an organism, *C. bifermentans*, are listed. The ingredients of a synthetic medium for *C. acetobutylicum*, a representative of the genus with comparatively moderate requirements, are also shown. For detailed information on nutritional requirements of a specific member of the genus *Clostridium*, especially of a specialist, the reader is referred to the review of HIPPE et al. (1992) and the references given therein.

Tab. 3. Clostridial Complex Medium (GIBBS and HIRSCH, 1956)

Ingredient	Concentration (g L^{-1})
Yeast extract	3.0
Meat extract	10.0
Peptone	10.0
Glucose	5.0
Starch	1.0
Sodium acetate	5.0
Cysteine-HCl	0.5

Since growth of most clostridia is stimulated if the medium contains complex nutrients, such media are routinely used in the laboratory for cell propagation. Examples are the PYG (peptone-yeast extract-glucose) or the CMC (chopped meat-carbohydrate) media as recommended by HOLDEMAN et al. (1977). These types of complex media for clostridia are to some extent commercially available, and the composition of a typical clostridial complex medium is given in Tab. 3.

3.4 Isolation, Handling, and Storage

Due to their ability to form endospores, which are resistant to several kinds of environmental hazards, the clostridia are potentially ubiquitously distributed. Therefore, if a natural sample (soil, mud sewage, feces) is put into an environment where the physical parameters such as pO_2, E_h, pH, and temperature and the presence of an organic compound allow growth of these microorganisms, spores will germinate. Which clostridial group or even which species will develop, then depends on the medium selected for the enrichment culture.

In enrichment cultures, non-sporeformers can be eliminated by pasteurization or by ethanol treatment. For pasteurization 1 g of material chosen as a possible habitat of the species to be isolated is placed in a sterile test tube and suspended in 5 mL of 0.9% NaCl. After flushing with oxygen-free nitrogen gas it is closed with a rubber stopper and incubated in a water bath. When the temperature in the test tube has reached 80 °C, as monitored in a control tube, the tube is incubated at this temperature for ten minutes and quickly cooled to room temperature. Alternatively, the suspended samples can be mixed with an equal volume of filter-sterilized absolute ethanol (JOHNSTON et al., 1964; KORANSKY et al., 1978). Incubation of this mixture at room temperature for 60 min will eliminate vegetative cells. In either case, after the treatment the suspension can be used to inoculate enrichment cultures.

Common saccharolytic and proteolytic clostridia can be enriched using the complex media mentioned above. They can be supplemented with carbohydrates, certain protein extracts, or amino acids to make them more selective. Special media have been designed for certain groups of clostridia or even a certain species, e.g., for the isolation of saccharolytic clostridia (VELDKAMP, 1965), saccharolytic nitrogen-fixing clostridia (SKINNER, 1971), lactate and acetate fermenting clostridia (BHAT and BARKER, 1947), glutamate fermenting clostridia (BARKER, 1939), purine fermenting clostridia (BARKER and BECK, 1942; CHAMPION and RABINOWITZ, 1977), cellulolytic clostridia (OMELIANSKI, 1902; SKINNER, 1960, 1971), and for the isolation of *C. sphenoides* (WALTHER et al., 1977) and of *C. kluyveri* (BORNSTEIN and BARKER, 1948).

Due to the oxygen intolerance of the clostridia, oxygen has to be excluded from the culture medium. Usually, prereduced media are used in order to ensure a quick onset of growth and to avoid long lag phases. Boiling of the heat-stable ingredients of the medium, purging with oxygen-free nitrogen or argon and addition of reducing agents (e.g., sodium thioglycolate, cysteine-HCl + Na_2S, titanium(III) citrate) leads to redox potentials of at least as low as –110 mV. To remove all oxygen from gases they are passed through a column containing heated copper, which is oxidized by traces of oxygen present in the gases and regenerated to the reduced form by purging with hydrogen. The redox potential is easily controlled by the presence of a non-toxic E_h indicator (resorufin/dihydroresorufin) that is colorless at –110 mV and turns pink at higher redox potentials (COSTILOW, 1981). Before autoclaving, the medium is dispensed into tubes or bottles, purged with nitrogen and equipped with flanged butyl rubber stopper and screw caps with an opening in the middle. Thus, the stoppers are held in position during autoclaving and they prevent oxygen to enter the medium.

Inoculation of the medium or transfer of cultures is easily achievable with disposable syringes and hypodermic needles flushed with oxygen-free gas. This procedure minimizes the risk that medium or cells are exposed to air. Clostridia are often able to reduce the E_h themselves, probably by action of NADH-oxidizing activities that are enhanced after exposure to air (MORRIS and O'BRIEN, 1971). The use of relatively large inoculums of actively growing cells (in the range of 10%) thus could help to overcome the normally lethal conditions of oxygen-contaminated media.

The roll-tube technique originally described by HUNGATE (1969) has meanwhile been replaced by anaerobic chambers and jars in order to grow clostridia on agar plates and to isolate single colonies, a prerequisite for genetic work with these microorganisms. Detailed descriptions of the technique have been given by COSTILOW (1981).

Clostridial strains can be stored as spores over a long period of time. Spores can be kept in dried sand or soil (BEESCH, 1952; SPIVEY, 1978). Alternatives are to lyophilize the spores (LAPAGE et al., 1970), to keep them in milk medium (BAHL et al., 1982b) or in fermentation broth (MONOT et al. 1982; LIN and BLASCHEK, 1982). However, a general procedure for the initiation of spore formation in clostridia cannot be given, and this abilitiy is in fact difficult to demonstrate for a number of species. Growth on the surface of agar media, incubation at suboptimal temperature, excess or absence of carbohydrates are parameters favoring sporulation in some cases (NASUNO and ASAI, 1960; PERKINS, 1965; ROBERTS, 1967; BÉRGERE and HERMIER, 1970). Chopped meat medium (HOLDEMAN et al., 1977) is a reliable sporulation medium for several species whereas specific media have been described, e.g., for *C. pasteurianum* (EMTSEV, 1963; MACKEY and MORRIS, 1971), *C. perfringens* (DUNCAN and STRONG, 1968), and *C. thermosaccharolyticum* (HSU and ORDAL, 1969). In our laboratory vegetative cells of some saccharolytic clostridia, e.g., *C. pasteurianum*, could be stored at −70 °C in the presence of 10% (vol/vol) dimethylsulfoxide (DMSO) for at least one year without a significant loss in cell viability.

4 Metabolism and Physiology

4.1 Degradation of Polysaccharides

Cellulose, xylan, starch, and pectin are abundant carbohydrate sources, available from plant biomass. These polymers are important substrates for clostridia in nature.

Cellulose consisting of glucose units joined together in β-1,4-linkages occurs in several crystalline forms in which amorphic regions might occur (ATALLA and VANDERHART, 1984; BLACKWELL, 1982; COWLING, 1963). Hydrolysis of the crystalline part is slower than that of the amorphous region. Among clostridia, cellulase has been extensively studied in *C. thermocellum*. In this organism the various cellulase components are found in an extracellular tightly associated multienzyme complex, designated the cellulosome (LAMED and BAYER, 1988). This structure, which contains about 20 polypeptides, mediates binding to the cellulose fibers and is subsequently released from the cell surface (BAYER and LAMED, 1986; LAMED et al., 1983; MAYER, 1988; NOLTE and MAYER, 1989). The cellulolytic enzymes of *C. thermocellum* degrade crystalline cellulose very efficiently with cellobiose as main product. The corresponding system of *C. stercorarium*, another cellulolytic thermophilic *Clostridium* species, is markedly different. Five components were isolated: avicelase I (endo- and exoglucanase activities), avicelase II (endoglucanase activity), β-cellobiase I and II, and a β-glucosidase (BRONNENMEIER and STAUDENBAUER, 1988, 1990; BRONNENMEIER et al., 1991). Most of the cellulolytic clostridia produce ethanol and acetate from cellulose. Only *C. populeti* and *C. cellulovorans* form butyrate as the predominant product (SLEAT and MAH, 1985; SLEAT et al., 1984).

In most cases natural cellulose is closely associated with xylans, the major component of hemicellulose. Xylan is a xylose polymer (β-1,4-linkages), which is branched and contains arabinose, glucose, galactose, and glucuronate. Although xylan degradation has been described for several clostridia (see Tab. 1), the enzymology of clostridial xylanases is not well defined. Xylanases from *C. acetobutylicum* (LEE et al., 1987; ZAPPE et al., 1987), *C. stercorarium* (BÉRENGER et al., 1985), and *C. thermocellum* (GRÉPINET et al., 1988) have been characterized, and the kinetics of xylan degradation by *C. thermosaccharolyticum*, *C. thermohydrosulfuricum*, and *C. thermocellum* have been compared by WIEGEL et al. (1985).

Starch can be degraded by quite a number of saccharolytic clostridia. This polysaccharide is composed of amylopectin (branched glucose polymer, α-1,6- in addition to α-1,4-linkages) and amylose (linear, only α-1,4-linkages). Several enzymes such as α-amylase, β-amylase, glucoamylase, α-glucosidase, and cyclodextrin glycosyltransferase are involved in the breakdown of this polymer (VIHINEN and MÄNTSÄLÄ, 1989), which all have been identified in several clostridia (ANTRANIKIAN, 1990; ANTRANIKIAN et al., 1987; ENSLEY et al., 1975; HYUN and ZEIKUS, 1985a, b; SPECKA et al., 1991). Major products of starch hydrolysis are glucose, maltose, maltotriose and other variously sized oligomers, and limit dextrin.

Pectins are complex carbohydrates composed of polygalacturonic acid (α-1,4-linkages) esterified with methyl alcohol. They have a helical structure, which is interrupted by rhamnose bound by α-1,2-linkages. In addition the

C2 and C3 positions of galacturonic acid and rhamnose as well are substituted with acetate, arabinose, fucose, galactose, xylose, or glucose. Pectin-degrading enzymes include esterases and several kinds of depolymerases (SAHA et al., 1989). Pectin esterase has been identified in *C. aurantibutyricum*, *C. butyricum*, *C. thermosulfurogenes*, and *C. roseum*; polygalacturonase in *C. felsineum*, *C. thermosulfurogenes*, and *C. roseum*; and polygalacturonate lyase in *C. aurantibutyricum*, *C. felsineum*, and *C. butyricum* (FOGARTY and KELLY, 1983; LUND and BROCKLEHURST, 1978; MACMILLAN et al., 1964; SCHINK and ZEIKUS, 1983).

4.2 Growth on Mono- and Disaccharides

A wide range of mono- and disaccharides and related compounds can be used by clostridia as carbon and energy source. This includes arabinose, cellobiose, fructose, galactose, galacturonate, gluconate, glucose, glucosamine, glycerol, lactose, maltose, mannose, rhamnose, ribose, sucrose, and xylose. Only little is known about how these substrates are transported in the clostridial cell. Several species including *C. acetobutylicum* are known to possess a phosphotransferase system (BOOTH and MORRIS, 1982; HUTKINS and KASHKET, 1986; MITCHELL and BOOTH, 1984; PATNI and ALEXANDER, 1971; VON HUGO and GOTTSCHALK, 1974). However, only the glucose phosphotransferase system of *C. acetobutylicum* has been described in detail (MITCHELL et al., 1991). It is likely that in addition other transport systems such as symport mechanisms driven by the transmembrane proton gradient or binding protein-dependent transport systems energized by direct hydrolysis of ATP are active to translocate some of the substrates listed above. In fact, two genes encoding proteins similar to the maltose transport proteins MalF and MalG of *Escherichia coli* (binding protein-dependent system) were identified in *C. thermosulfurogenes* (BAHL et al., 1991 a). Disaccharides such as maltose and lactose might then be cleaved intracellularly by an appropriate phosphorylase.

Hexose phosphates are metabolized by clostridia via the Embden-Meyerhof-Parnas (EMP) pathway (GOTTSCHALK, 1986; ROGERS, 1986; THAUER et al., 1977). The conversion of 1 mol hexose to 2 mol of pyruvate results in the net production of 2 mol of ATP and 2 mol of NADH.

Breakdown of pentoses involves phosphorylation in position 5, isomerization, and epimerization yielding ribose-5-phosphate and xylulose-5-phosphate. By means of transketolase and transaldolase, these compounds are converted to fructose-6-phosphate and glyceraldehyde-3-phosphate, intermediates of the EMP pathway. The corresponding enzymes have been identified in *C. perfringens* (CYNKIN and DELWICHE, 1958), *C. beijerinckii*, *C. butyricum* (CYNKIN and GIBBS, 1958), and *C. thermoaceticum* (GOTTWALD, 1973). The oxidation of 3 mol of pentose to pyruvate yields 5 mol of ATP and 5 mol of NADH.

Sugar acids such as gluconate are degraded via a modified Entner-Doudoroff pathway (ANDREESEN and GOTTSCHALK, 1969). Gluconate is first dehydrated and subsequently phosphorylated yielding 2-keto-3-deoxy-6-phosphogluconate, which is cleaved by an aldolase into pyruvate and glyceraldehyde-3-phosphate. Thus, the conversion of gluconate to 2 mol of pyruvate results in the net production of 1 mol ATP and 1 mol NADH.

The fermentation of glycerol can be accomplished by several solventogenic clostridia (FORSBERG, 1987). Part of the glycerol is oxidized to dihydroxyacetone by glycerol dehydrogenase and phosphorylated by dihydroxyacetone kinase. Further catabolism proceeds through glycolysis. The remaining part of the glycerol is reduced to 1,3-propanediol after dehydration to 3-hydroxypropionaldehyde. The existence of a specific 1,3-propanediol dehydrogenase has been demonstrated for *C. pasteurianum* (HEYNDRICKX et al., 1991). Two moles of ATP and two moles of NADH are generated per mole glycerol converted to pyruvate.

The degradation pathways of all these compounds make it clear that pyruvate is a central intermediate in the formation of fermentation products from carbohydrates by clostridia. Pyruvate is cleaved by pyruvate:ferredoxin oxidoreductase to yield acetyl-CoA, carbon dioxide, and reduced ferredoxin, an iron sulfur protein with a low redox potential. Electrons from the reduced form can be transferred to protons via hydrogenase or to NAD via NADH:ferredoxin oxidoreductase. When the further breakdown of pyruvate is impaired by inhibition of hydrogenase by carbon monoxide (DABROCK et al., 1992; KIM et al., 1984) or by iron depletion (BAHL et al., 1986; DABROCK et al., 1992), some clostridia like *C. acetobutylicum* and *C. pasteurianum* produce considerable amounts of lactate at pH values above 5.0. Under these conditions the intracellular concentration of fructose-1,6-bisphosphate reaches values at which a NADH-specific lactate dehydrogenase is activated (FREIER and GOTTSCHALK, 1987).

The fate of acetyl-CoA depends very much on the clostridial species and on the conditions under which it is grown. It can be converted to acetate, butyrate, ethanol, butanol, acetone, and/or 2-propanol. In Fig. 2 the carbon and

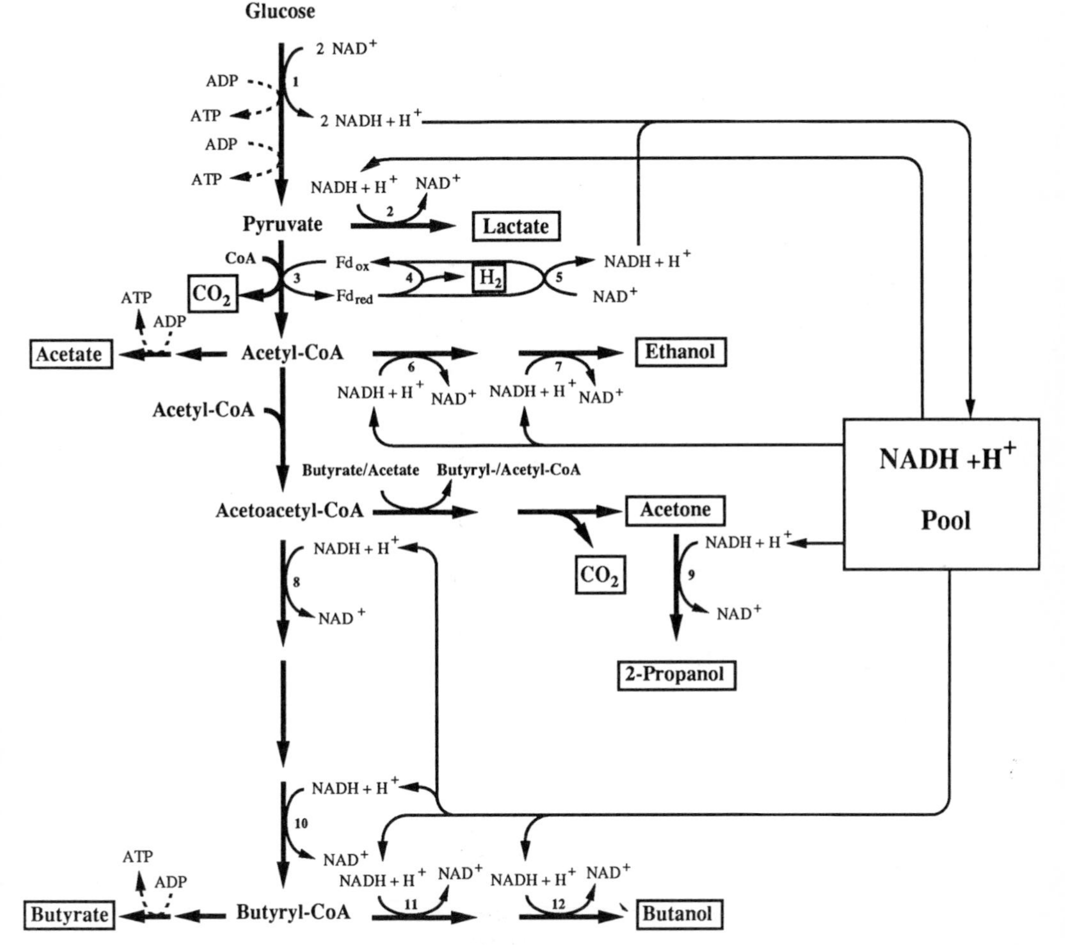

Fig. 2. Carbon (thick arrows) and electron (thin arrows) flow during glucose fermentation in clostridia. Enzymes involved in oxidation-reduction reactions are numbered as follows: **1**, glyceraldehyde-3-phosphate dehydrogenase; **2**, lactate dehydrogenase; **3**, pyruvate:ferredoxin oxidoreductase; **4**, hydrogenase; **5**, NADH: ferredoxin oxidoreductase; **6**, acetaldehyde dehydrogenase; **7**, ethanol dehydrogenase; **8**, β-hydroxybutyryl-CoA dehydrogenase; **9**, 2-propanol dehydrogenase; **10**, butyryl-CoA dehydrogenase; **11**, butyraldehyde dehydrogenase; **12**, butanol dehydrogenase. In certain reactions some species use $NADP^+$/NADPH + H^+ as electron acceptor/donor instead of NAD^+/NADH + H^+, which are generally shown.

electron flow through the branched metabolic pathway leading to the formation of acids and solvents is shown. Key intermediates at branch points are acetyl-CoA, acetoacetyl-CoA, and butyryl-CoA. The formation of acetate and butyrate is coupled to the generation of ATP.

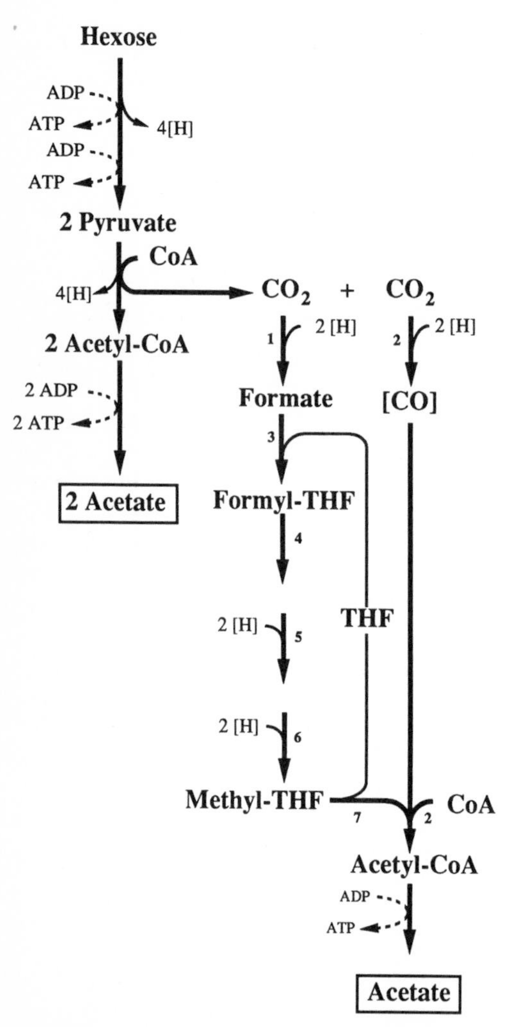

Fig. 3. Fermentation of hexose to 3 moles of acetate by homoacetogenic clostridia (LJUNGDAHL, 1986). The reactions using CO_2 as electron acceptor and as a source of the third mol acetate are numbered as follows: **1**, formate dehydrogenase, **2**, carbon monoxide dehydrogenase, **3**, formyl-tetrahydrofolate (THF) synthetase; **4**, methenyl-THF cyclohydrolase; **5**, methylene-THF dehydrogenase; **6**, methylene-THF reductase; **7**, transmethylase. Other reactions are identical to those in Fig. 2.

However, in some cases parameters such as redox balance or inhibitory effects of the products formed are more important for the cell than a maximum ATP yield and the more reduced neutral compounds are formed. Thus, solvents can be found among the products of many clostridial species, although often in low amounts. However, *C. acetobutylicum* is known as a potent solvent producer, and the formation of acetone, butanol, and ethanol by this microorganism has been well studied (BAHL and GOTTSCHALK, 1988; JONES and WOODS, 1986).

The homoacetogenic clostridia, e.g., *C. aceticum*, have developed a special way of disposing reducing equivalents. These are transferred to CO_2 to yield acetate (Fig. 3). Thus, these bacteria produce almost 3 mol acetate from 1 mol glucose or fructose (FUCHS, 1986; LJUNGDAHL, 1986).

Another example that certain growth conditions initiate the formation of special products is the fermentation of glucose to 1,2-propanediol by *C. sphenoides* (TRAN-DINH and GOTTSCHALK, 1985). Under phosphate limitation the so-called methylglyoxal pathway is active for the formation of this solvent (Fig. 4), which is also produced by *C. thermosaccharolyticum* (CAMERON and COONEY, 1986).

4.3 Growth on Alcohols, Aromatic Compounds, and Organic Acids

The primary alcohols methanol and ethanol can be metabolized by some homoacetogenic clostridia. However, there are clear differences within this group. *C. formicoaceticum*, *C. thermoaceticum*, and *C. thermoautotrophicum* do ferment methanol, while *C. aceticum* cannot utilize this compound. On the other hand, *C. aceticum* and *C. formicoaceticum* grow on ethanol as sole carbon source, whereas *C. thermoautotrophicum* is unable to ferment this substance (BRAUN et al., 1981; WIEGEL et al., 1981; WIEGEL and GARRISON, 1985). The only product, even from methanol, is acetate (BERESTOVSKAYA et al., 1987). It might be possible that utilization of these alcohols in some strains is strictly dependent on substrate concentration and addition of sufficient amounts of bicarbonate that serves as an electron accep-

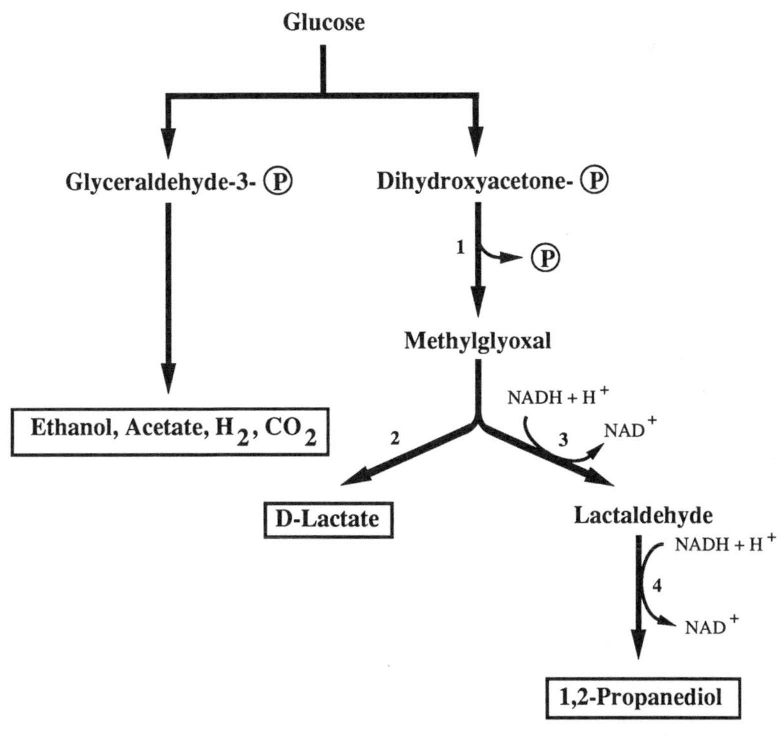

Fig. 4. Formation of 1,2-propanediol from glucose by *Clostridium sphenoides* (TRAN-DINH and GOTTSCHALK, 1985). **1**, methylglyoxal synthase; **2**, glyoxalase; **3**, methylglyoxal reductase; **4**, 1,2-propanediol dehydrogenase. Other reactions are identical to those in Fig. 2.

tor. Variation of these conditions allowed the detection of ethanol utilization by *Acetobacterium woodii,* a non-sporulating anaerobic homoacetogen (BUSCHHORN et al., 1989). Ethanol is also degraded by *C. kluyveri* in a peculiar fermentation in which acetate is additionally required as electron acceptor (summarized in GOTTSCHALK, 1986). Products are butyrate, caproate, and hydrogen. Propanol and succinate can replace ethanol and acetate in substrate combinations (KENEALY and WASELEFSKY, 1985). *Clostridium neopropionicum* ferments ethanol to acetate, propionate, and some propanol. In combination with acetate, propanol can be used as a substrate, too (THOLOZAN et al., 1992).

Several different diols also serve as carbon and energy sources for clostridia. Quite common is the utilization of glycerol, which has already been mentioned. Besides some homoacetogens such as *C. formicoaceticum* and *C. mayombei* and other saccharolytic strains such as *C. papyrosolvens* and *C. quercicolum* many butyrate-forming species ferment this compound (HIPPE et al., 1992). It is remarkable, however, that during growth on glycerol *C. acetobutylicum*, *C. beijerinckii*, *C. butyricum*, and *C. pasteurianum* form large amounts of the solvent 1,3-propanediol as an end product (BIEBL, 1991; DABROCK et al., 1992; FORSBERG, 1987; HEYNDRICKX et al., 1991). This is an important intermediate for the synthesis of heterocyclic compounds and the production of polymers and thus, due to its high price from chemical production, it is of considerable biotechnological interest. 1,2-Ethanediol (ethylene glycol) is fermented by *C. aceticum* and *C. glycolicum*. The initial step of degradation in *C. glycolicum* yielding acetaldehyde is catalyzed by an integral membrane protein and seems to employ a radical mechanism (HARTMANIS and STADTMAN, 1986). The organism can also grow on 1,2-propanediol. Alcohols with 4 carbons such as 2,3-butanediol and acetoin are utilized by the homoacetogenic *C. aceticum* and *C. magnum* (SCHINK, 1984). Butanol (as well as methanol, ethanol, and propanol) can also be metabolized by *C. thermoaceticum* if either dimethylsulfoxide or thiosulfate are present that serve as electron acceptors. Reduction products are dimethylsulfide or sulfide, respectively. The alcohols are oxidized to the corresponding acids (BEATY and LJUNGDAHL, 1991).

Methoxy groups from aromatic monobenzoids can be used for growth by *C. aceticum*, *C. formicoaceticum*, *C. mayombei*, *C. thermoaceticum*, and *C. thermoautotrophicum* (DANIEL et al., 1988; KANE et al., 1991; LUX et al., 1989). No further degradation of the aromatic compounds occurs, except for *C. thermoaceticum* that has been reported to decarboxylate, e.g., vanillate in addition to the O-demethylation that is catalyzed by an inducible enzyme system (Fig. 5) (HSU et al., 1990; WU et al., 1988). It is remarkable that *C. aceticum* is able to grow at the expense of methoxy groups although it cannot utilize methanol in contrast to the other three homoacetogenic clostridia mentioned. An identical phenomenon has been found with *C. pfennigii* that produces butyrate from methoxy groups of compounds

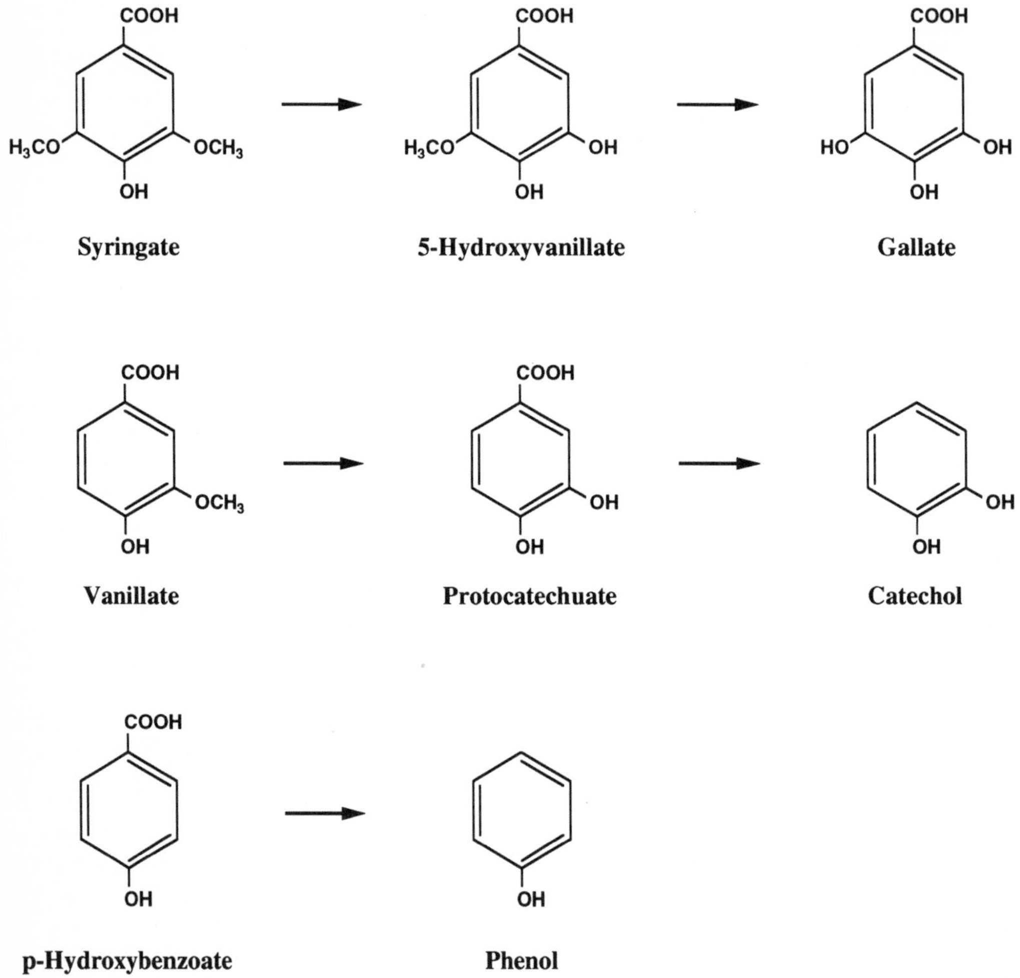

COOH
H_3CO OCH$_3$
OH
Syringate → COOH H_3CO OH OH **5-Hydroxyvanillate** → COOH HO OH OH **Gallate**

COOH OCH$_3$ OH **Vanillate** → COOH OH OH **Protocatechuate** → OH OH **Catechol**

COOH OH **p-Hydroxybenzoate** → OH **Phenol**

Fig. 5. Examples of O-demethylation and decarboxylation of aromatic compounds by clostridia.

such as ferulate, but cannot grow on methanol (KRUMHOLZ and BRYANT, 1985). These reactions might be useful in the anaerobic breakdown of lignin to yield an alternative supply of aromatic carbon to the plastics and synthetic fiber industry (GRIFFIN and MAGOR, 1987). Reports on cleavage of aromatic ring systems by clostridia are scarce. Flavonoids are attacked by *C. scindens* that cleaves the C-ring of naringenin as well as by *C. orbiscindens* that degrades quercitin to phloroglucinol and 3,4-dihydroxyphenylacetic acid and also cleaves kaempherol and naringenin (WINTER et al., 1989, 1991). A total fermentation of dihydroxybenzene (resorcinol) and two different dihydroxybenzoates to acetate and butyrate has been reported for two as yet unclassified clostridia. However, this decomposition was only performed in co-culture with *Campylobacter*-like organisms (TSCHECH and SCHINK, 1985).

Several *monocarboxylic acids* can be used for growth by clostridia. Formate represents a substrate for the homoacetogens such as *C. aceticum*, *C. formicoaceticum*, *C. mayombei*, and *C. thermoautotrophicum* (BRAUN et al., 1981; KANE et al., 1991; LUX et al., 1989; WIEGEL et al., 1981). Although *C. thermoaceticum* originally was reported to convert formate and carbon dioxide to acetate only in resting cell suspensions (LENTZ and WOOD, 1955), mutant strains have recently been isolated that use formate as the sole energy source (PAREKH and CHERYAN, 1991). Acetate serves as a cosubstrate in fermentations of *C. kluyveri* (in combination with ethanol), *C. oxalicum* (in combination with oxalate or oxamate), and *C. tyrobutyricum* (in combination with lactate). *C. kluyveri* can also use propionate, butyrate, or (to a lesser extent) valerate as well as the unsaturated 2-butenoate (crotonate) and 3-butenoate (vinylacetate) instead of acetate (BORNSTEIN and BARKER, 1948). Another unsaturated acid (acrylate) represents an intermediate in the fermentation of lactate to propionate by *C. propionicum* and can serve as substrate for this organism (CARDON and BARKER, 1947). 2-Butenoate, 3-butenoate, and acrylate are also metabolized by *C. homopropionicum* (DÖRNER and SCHINK, 1990). Monocarboxylic acids containing additional functional groups such as glycerate, lactate, and pyruvate are fermented by many clostridia, including acetate-forming (e.g., *C. aceticum*, *C. formicoaceticum*, *C. mayombei*, *C. thermoautotrophicum*), propionate-forming (e.g., *C. homopropionicum*, *C. propionicum*), butyrate-forming (e.g., *C. acetobutylicum*, *C. thermopalmarium*), and ethanol/acetate-forming (e.g., *C. thermohydrosulfuricum*) species (ANDREESEN et al., 1970; BAHL et al., 1986; BRAUN et al., 1981; DÖRNER and SCHINK, 1990; JANSSEN, 1991; JUNELLES et al., 1987; KANE et al., 1991; SOH et al., 1991; WIEGEL et al., 1979, 1981). However, as the carbon chain becomes longer, only few fermenters are known. 2-, 3-, 4-Hydroxybutyrate, and 4-chlorobutyrate represent good substrates for *C. homopropionicum*, and 4-hydroxybutyrate is metabolized by *C. kluyveri* (DÖRNER and SCHINK, 1990; KENEALY and WASELEFSKY, 1985).

The simplest *dicarboxylic acid* oxalate and some of its derivatives are fermented by *C. oxalicum* (which can also use oxamate; fermentation products are formate and CO_2) as well as by *C. thermoaceticum* and *C. thermoautotrophicum* (to acetate) which are in addition able to utilize glyoxylate for growth (DANIEL and DRAKE, 1991; DEHNING and SCHINK, 1989). Tartrate is a substrate for a clostridial strain formerly called *"C. tartarivorum"*, but now assigned to the species *C. thermosaccharolyticum* (MATTEUZI et al., 1978). Several other dicarboxylic acids are also fermented by clostridia. Examples are fumarate (by *C. aceticum* and *C. formicoaceticum*), malate (*C. aceticum*, *C. formicoaceticum*, *C. magnum*, *C. mayombei*), and oxaloacetate (*C. propionicum*) (BRAUN et al., 1981; JANSSEN, 1991; KANE et al., 1991; SCHINK, 1984). Succinate is of special interest, because a fermentation of this compound to propionate and CO_2 allows only a free energy change ($\Delta G^{o\prime}$) of –20 kJ/mol, not enough for either substrate-level phosphorylation or electron-transport phosphorylation. In *Propionigenium modestum* such a fermentation was found to proceed by a membrane-bound decarboxylase pumping sodium ions to the outside during the reaction. This Na^+ gradient could be used for ATP synthesis by a Na^+-translocating ATPase (for a recent review see DIMROTH, 1991). So far, within the genus *Clostridium* only *C. mayombei* has been reported to convert succinate as sole carbon and energy source into propionate and CO_2 (KANE et al., 1991). Presumably,

Tab. 4. Amino Acid Fermentations by Clostridia [a]

Key Reaction(s)	Amino Acid(s) Fermented	Reaction Products	Organisms	Remarks
Reductive processes, exchange of amino group by hydroxyl group, activation by coenzyme A, dehydration to enoyl CoA, reduction to saturated fatty acid	Alanine, cysteine, glutamate, histidine, leucine, methionine, serine, threonine	Acetate, butyrate, CO_2, H_2, 3-methylbutyrate, NH_4^+, propionate	*C. neopropionicum*, *C. propionicum*, *C. sporogenes*, *C. sporosphaeroides*, *C. symbiosum*	Most common pathway of amino acid degradation
Reaction sequence: $R{-}CH(NH_2){-}COOH \rightarrow R{-}C(=O){-}COOH \rightarrow R{-}CH(OH){-}COOH \rightarrow R{-}CH(OH){-}C(=O){-}SCoA \rightarrow R{=}CH{-}C(=O){-}SCoA \rightarrow R{-}CH_2{-}C(=O){-}SCoA$				
Coenzyme B_{12}-dependent carbon-carbon rearrangement to yield β-amino acid, elimination of NH_4^+, addition of water, cleavage into pyruvate and acetate	Glutamate	→ 3-Methylaspartate → acetate, butyrate, CO_2, NH_4^+	*C. cochlearium*, *C. limosum*, *C. malenominatum*, *C. tetanomorphum*, *C. tetani*	This reaction led to the discovery of the biological function of coenzyme B_{12}
Reaction sequence: $HOOC{-}CH_2{-}CH_2{-}CH(NH_2){-}COOH \rightarrow HOOC{-}CH(CH_3){-}CH(NH_2){-}COOH \rightarrow HOOC{-}C(CH_3){=}CH{-}COOH \rightarrow HOOC{-}C(CH_3)(OH){-}CH_2{-}COOH \rightarrow HOOC{-}C(=O){-}CH_3 + CH_3{-}COOH$				
α,β-Elimination of ammonium, formation of unsaturated acid	Aspartate, histidine	Furmarate, urocanate	*C. tetanomorphum*	Several clostridia use aspartate, but the respective degradation pathways have not been elucidated
Reaction sequence: $R{-}CH(NH_2){-}COOH \rightarrow R{=}CH{-}COOH + NH_4^+$				

Reductive cleavage of C_α-N-bond	Glycine, proline	Acetate, 5-aminovalerate, CO_2, NH_4^+	*C. acidiurici*, *C. cylindrosporum*, *C. histolyticum*, *C. litorale*, *C. purinolyticum*, *C. sporogenes*, *C. sticklandii*	Selenocysteine is an essential component of the glycine reductase that catalyzes an ATP-yielding reaction

Reaction sequence:

$$\underset{\displaystyle CH_2}{\overset{\displaystyle NH_2}{|}}-COOH \rightarrow CH_3-COOH + NH_4^+$$

2,3-Shift of the amino group	Leucine, lysine	Acetate, butyrate, 4-methylvalerate, NH_4^+	*C. cochlearium*, *C. sporogenes*, *C. subterminale*	Lysine 2,3-amino mutase contains S-adenosylmethionine instead of adenosylcobalamine

Reaction sequence:

$$H_2N-[CH_2]_3-CH_2-\underset{}{\overset{NH_2}{\overset{|}{CH}}}-COOH \rightarrow H_2N-[CH_2]_3-\overset{NH_2}{\overset{|}{CH}}-CH_2-COOH \rightarrow CH_3-\overset{NH_2}{\overset{|}{CH}}-CH_2-\overset{NH_2}{\overset{|}{CH}}-CH_2-COOH \rightarrow$$

$$CH_3-\overset{NH_2}{\overset{|}{CH}}-CH_2-\overset{O}{\overset{\|}{C}}-CH_2-COOH \xrightarrow{+Acetyl\text{-}CoA} CH_3-\overset{O}{\overset{\|}{C}}-CH_2-COOH + CH_3-\overset{NH_2}{\overset{|}{CH}}-CH_2-C-SCoA \rightarrow Butyrate$$

Stickland reaction (coupled oxidation and reduction of amino acid pairs). *Oxidation:* conversion to the corresponding 2-oxoacids by oxidative deamination, transamination, or α,β-elimination, followed by oxidative decarboxylation. *Reduction:* involves the reaction mechanisms desribed above	Oxidized: alanine, arginine, glycine, isoleucine, (leucine, phenylalanine)[b], ornithine, serine, tyrosine, valine. Reduced: glycine, leucine, phenylalanine, proline, tryptophan, tyrosine	CO_2, fatty acids, NH_4^+	*C. halophilium*, *C. litorale*, *C. mayombei*, *C. sporogenes*, *C. sticklandii*	Enoate reductases are required for oxidation of, e.g., phenylalanine to phenylpropionate

Reaction sequence:

$$R'-\overset{NH_2}{\overset{|}{CH}}-COOH + 2R''-\overset{NH_2}{\overset{|}{CH}}-COOH + H_2O \rightarrow R'-COOH + 2R''-CH_2-COOH + CO_2 + 3NH_3$$

[a] Data taken from ANDREESEN et al. (1989), BARKER (1981), BUCKEL (1991), BADER et al. (1982), FENDRICH et al. (1990), KANE et al. (1991), THOLOZAN et al. (1992)

[b] only in the absence of isoleucine or valine

a similar mechanism is operating in this organism. As already mentioned, *C. kluyveri* is also able to use succinate for growth. However, ethanol is needed in this case as a cosubstrate.

The *tricarboxylic acid* citrate is fermented by a variety of clostridial species. *C. indolis*, *C. magnum*, *C. rectum*, *C. sphenoides*, *C. sporogenes*, *C. sporosphaeroides*, *C. subterminale*, and *C. symbiosum* probably all use citrate lyase as the key enzyme for the anaerobic degradation (ANTRANIKIAN et al., 1984; SCHINK, 1984). The different and complex regulation mechanisms of the anaerobic citrate metabolism have been reviewed (ANTRANIKIAN and GIFFHORN, 1987).

4.4 Fermentation of Nitrogenous Compounds

Many clostridia are able to utilize amino acids for growth. Tab. 4 summarizes the reaction mechanisms so far known for such fermentations (BUCKEL, 1991). Both, oxidative and reductive processes are involved. The oxidation reactions such as oxidative deaminations, transaminations, or α,β-eliminations are usually similar or identical to corresponding reactions catalyzed by aerobic bacteria. However, due to the absence of molecular oxygen, oxygenation reactions and fatty acid oxidations do not occur (BARKER, 1981; BUCKEL, 1991). It is obvious that all essential amino acids, the building blocks of proteins, can be degraded by clostridia, thus stressing the enormous importance of this genus in the remineralization of organic material under anaerobic conditions. Most amino acids are converted into short-chain fatty acids and finally into acetate and CO_2, the precursors of methane formation. However, branched-chain and aromatic amino acids are only partially degraded to isobutyrate, isovalerate, isocaproate, and various aromatic compounds. As mentioned earlier, no decomposition of the benzene or indole ring systems has been reported for this genus. Detailed reviews on amino acid fermentation by clostridia have been presented by ANDREESEN et al. (1989) and BUCKEL (1991).

Although homoaromatic systems are usually not attacked, several clostridial species are able to degrade heterocyclic compounds. Purines are decomposed by strains of *C. malenominatum*, *C. sartagoforme*, *C. sporogenes*, *C. sticklandii*, and *C. tertium* (BARNES and IMPEY, 1974; MEAD and ADAMS, 1975; SCHÄFER and SCHWARTZ, 1976). All these species can also use sugars and/or amino acids for growth (see Tab. 1). However, three clostridia (*C. acidiurici*, *C. cylindrosporum*, *C. purinolyticum*) are obligately purinolytic and ferment only purines and some of the degradation products (BARKER and BECK, 1941, 1942; DÜRRE et al., 1981). By action of xanthine dehydrogenase and deaminases the heterocyclic compounds are converted into xanthine, the central intermediate from which ring cleavage starts (DÜRRE and ANDREESEN, 1982a, 1983; RABINOWITZ, 1963). Degradation proceeds by a series of hydrolytic steps that are shown in Fig. 6. Formate is further oxidized to CO_2, if an oxidized compound such as uric acid serves as a substrate. Energy-yielding reactions are the conversion of formyltetrahydrofolate into formate and tetrahydrofolate and the reduction of glycine to acetate and ammonia (DÜRRE and ANDREESEN, 1982b, 1983). The latter process is also used by many strains performing a Stickland reaction with glycine. The catalyzing enzyme (glycine reductase) contains selenocysteine as an essential component and forms acetylphosphate as a direct product, possibly via a ketene intermediate (ARKOWITZ and ABELES, 1989; BUCKEL, 1991; SLIWKOWSKI and STADTMAN, 1988). The required reducing equivalents are probably transferred to glycine reductase via dihydrolipoamide dehydrogenase and/or a thioredoxin reductase-like flavoprotein (DIETRICHS et al., 1991).

The purinolytic clostridia might soon become subject of biotechnological application. Purines such as uric acid are a main part of poultry excrements. These represent a major environmental problem for large poultry breeding farms, since due to the bad odor the removal of the excrements is subject to strict regulation. The obligately purinolytic clostridia have been shown to completely ferment the uric acid content of chicken feces (SCHIEFER-ULLRICH, 1984). Such a fermentation could be helpful in the conversion of problematic waste products (which require expensive removal) into compounds to be used directly for fertil-

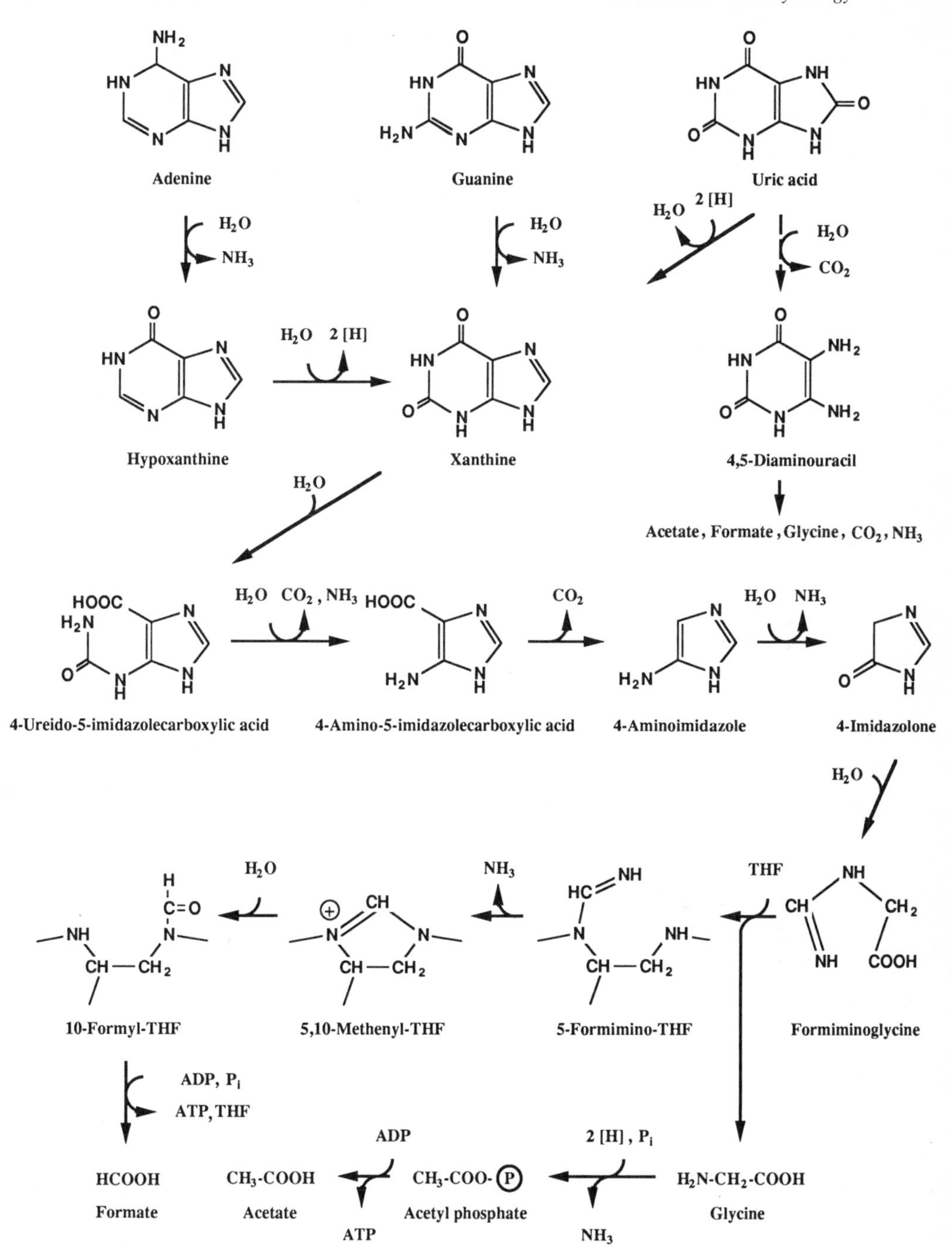

Fig. 6. Purine degradation by obligately purinolytic clostridia. P_i, inorganic phosphate; THF, tetrahydrofolate.

ization. A degradation of uric acid under selenium deficiency via pyrimidine derivatives by *C. purinolyticum* has also been described. This pathway probably is used as metabolic escape reaction, but might have implications for alloxan-induced diabetes in humans (DÜRRE and ANDREESEN, 1982c)

Pyrimidines such as uracil and orotic acid can be completely degraded by *C. glycolicum* and *C. oroticum*, respectively. Products are β-alanine, aspartate, ammonia, and CO_2. Again, hydrolytic reactions are involved in ring cleavage (for reviews see ANDREESEN et al., 1989; VOGELS and VAN DER DRIFT, 1976). Complete decomposition of pyridine derivatives has been described for *C. barkeri* (which ferments nicotinic acid to acetate, propionate, ammonia, and CO_2) and an as yet unclassified *Clostridium* that degrades mimosine, 3(OH)-4-(1H)-pyridone, 2,3-dihydroxypyridine, and dihydroxyphenylalanine to predominantly acetate, butyrate, and ethanol (DOMINGUEZ-BELLO and STEWART, 1991; STADTMAN et al., 1972). The unidentified strain thus represents the first example of breakdown of a benzoid system by a *Clostridium*.

Several other amine compounds are also degraded by clostridia. 5-Aminovalerate and 4-aminobutyrate are fermented to acetate, butyrate, propionate, valerate, and ammonium by *C. aminovalericum* and "*C. aminobutyricum*", the latter representing a not validly described species (BUCKEL, 1991). Betaine, choline, creatine, creatinine, and ethanolamine represent single or co-substrates for a variety of mesophilic clostridia, whereby betaine is used in the Stickland reaction and is always cleaved into trimethylamine and acetate (FENDRICH et al., 1990; MÖLLER et al., 1986; NAUMANN et al., 1983).

4.5 Gaseous Substrates

The homoacetogens *C. aceticum*, *C. formicoaceticum*, *C. mayombei*, *C. thermoaceticum*, and *C. thermoautotrophicum* are all able to grow at the expense of hydrogen and carbon dioxide that are converted to acetate. With the exception of *C. mayombei* that has not been tested in this respect, they also use carbon monoxide as carbon and energy source (BRAUN et al., 1981; KANE et al., 1991; LJUNGDAHL et al., 1989; LUX et al., 1989; SAVAGE and DRAKE, 1986). Carbon fixation occurs by the reductive acetyl-CoA pathway or Wood pathway (cp. Fig. 2). The respective reactions and the enzymes involved have been dealt with in a number of recent reviews (FUCHS, 1986; LJUNGDAHL, 1986; WOOD, 1991; RAGSDALE, 1991). However, *C. pfennigii* only uses CO (or methoxyl groups from monobenzoids, see Sect. 4.2) and converts it into acetate and butyrate (KRUMHOLZ and BRYANT, 1985). The respective reactions have not yet been elucidated.

Biological fixation of dinitrogen is a common feature of many clostridial species, including mesophilic and thermophilic strains (BOGDAHN et al., 1983; BOGDAHN and KLEINER, 1986; ROSENBLUM and WILSON, 1949). The N_2-fixing system of *C. pasteurianum* that was purified and extensively characterized with respect to structural properties and the mechanism of catalysis has received most attention (MORTENSON and THORNELEY, 1979). Genetic investigations confirm the distinct structural features of the *C. pasteurianum* nitrogenase that prevent the formation of active heterologous complexes with components of the enzymes from other organisms (CHEN et al., 1986).

5 Industrially Important Products

5.1 Low-Molecular Weight Compounds

In the previous sections the abundance of biochemical reactions carried out by clostridia has been outlined. Several of the many fermentation products such as acetate, 1-butanol, ethanol, and 1,3-propanediol are of potential importance. In addition, clostridial fermentations have the advantage that problems connected with the need of high aeration rates of aerobic cultures (agitation, cooling) do not exist. On the other hand, there are some limitations which represent major drawbacks for the production of fuels and chemicals by clostridial fermentation processes.

First, in most cases the desired product is obtained in a mixture with other fermentation products resulting in low yields. The clostridia take advantage of their branched fermentation pathways to control product formation and, depending on the growth conditions, they choose between different routes (GOTTSCHALK, 1986; ANDREESEN et al., 1989). This flexibility enables them to balance their oxidation-reduction state, a crucial problem in clostridia, and to avoid the accumulation of a single product, which may become toxic in high concentrations (DÜRRE et al., 1988).

Second, the tolerance of clostridia towards high concentrations of such products as acetic acid or butanol is low, leading to low productivities and low final concentration of the product. This again results in high costs for product recovery from the dilute solution. The deleterious effects of butanol include abolition of the pH gradient across the membrane, decrease of the intracellular ATP level, release of intracellular metabolites, and inhibition of sugar uptake (BOWLES and ELLEFSON, 1985; GOTTWALD and GOTTSCHALK, 1985; HUTKINS and KASHKET, 1986; MOREIRA et al., 1981). The production of acids lowers the pH of the growth medium and results in increased amounts of the undissociated form, which is membrane-permeable and is able to bring about dissipation of the transmembrane proton gradient (DÜRRE et al., 1988; PADAN et al., 1981).

In this section the future prospects for biotechnological applications of some clostridial fermentations will be briefly considered. The only fermentation of a member of the genus *Clostridium* that was operated commercially on a large scale is the production of acetone and butanol by *C. acetobutylicum*. Fermentation of 100 kg of sugar at 6% initial concentration yielded 18 kg butanol, 9 kg acetone, 3 kg ethanol, 49.5 kg CO_2, 2 kg H_2, and biomass (SPIVEY, 1978). This traditional batch process is no longer in use, at least in Western countries. The main reasons for the unfavorable economics were high raw material (starch, molasses) and transportation costs and the above mentioned intrinsic limitations such as low solvent yield, low final concentrations, and undesirable solvent ratios. This resulted in high processing costs and waste disposal problems.

As a result of the oil crisis, there is a renewed interest in the acetone-butanol fermentation, and recent studies were devoted to the development of a modern fermentation process. This work includes development of a continuous process (BAHL and GOTTSCHALK, 1984; BAHL et al., 1982a, 1982b; MONOT and ENGASSER, 1983) with cell recycling (AFSCHAR et al., 1985; SCHLOTE and GOTTSCHALK, 1986) and cell immobilization (FORBERG and HÄGGSTRÖM, 1985; LARGIER et al., 1985). In addition, alternate product recovery techniques such as reverse osmosis, pervaporation, liquid-liquid extraction, and other chemical recovery methods have been applied to the acetone-butanol fermentation (ENNIS et al., 1986; ROFFLER et al., 1987; WAYMAN and PAREKH, 1987). Furthermore, the products could be shifted towards butanol (BAHL et al., 1986; KIM et al., 1984; MADDOX, 1980; MEYER et al., 1986), and mutants resistant to higher solvent concentrations could be isolated (HERMANN et al., 1985; LIN and BLASCHEK, 1982). Thus, significant progress was made regarding the biology of *C. acetobutylicum* and the fermentation process in recent years. However, as long as cheap crude oil sources are available, the acetone-butanol fermentation at its present state cannot compete with the existing petroleum-based production of these solvents.

Other solvents which have a potential of commercial exploitation are ethanol produced by some thermophilic clostridia (*C. thermohydrosulfuricum*, *C. thermocellum*; KUROSE et al., 1988; LOVITT et al., 1984; PARKKINEN, 1986; WIEGEL, 1980) and 1,3-propanediol formed by several species, e.g., *C. butyricum* and *C. pasteurianum*, from glycerol (BIEBL, 1991; DABROCK et al., 1992; FORSBERG, 1987; HEYNDRICKX et al., 1991).

Organic acids like acetate are produced from petrochemicals. However, the high yield of homoacetogenic clostridia (e.g., 3 mol acetate per mol glucose) has stimulated research to develop an economically feasible fermentation process for the synthesis of acetate. Acetate production using *C. thermoaceticum* and *C. thermoautotrophicum* has been studied by LJUNGDAHL et al. (1985), REED and BOGDAN (1985), and WANG and WANG (1984). In a batch process up to 45 g L^{-1} acetate were obtained with a production rate of 0.8 g L^{-1} h^{-1}.

In a system with cell recycling together with cell adsorption higher productivities (14.3 g L^{-1} h^{-1}) were reached, however, at decreased acetate concentrations (7.1 g L^{-1}). As in the ace- tone-butanol fermentation, product recovery accounts for a major part of the production costs. In addition to the most obvious way of distillation after acidification of the fermentation broth, several processes have been examined for the recovery of acetate from fermentation media (BUSCHE, 1985). The production of acetate by the acetogenic clostridia has been considered a source of calcium magnesium acetate, which could be used as a highway de-icer (LJUNGDAHL et al., 1989).

5.2 Proteins

Clostridia are able to use a variety of polysaccharides and proteins as substrates. Therefore, they produce a number of extracellular hydrolytic enzymes for the breakdown of these polymers. Recently, much attention has been paid to the stable starch- and cellulose-degrading enzymes of thermophilic clostridia, which are of commercial importance (reviewed by ANTRANIKIAN, 1990; SAHA et al., 1989).

The starch-degrading enzymes of *C. thermosulfurogenes* (HYUN and ZEIKUS, 1985a; MADI et al., 1987), *C. thermosaccharolyticum* (KOCH et al., 1987), and *C. thermohydrosulfuricum* (HYUN and ZEIKUS, 1985b; MELASNIEMI, 1987; SAHA et al., 1988) have been extensively studied. The properties of some of these enzymes are listed in Tab. 5. In addition, as a prerequisite for massive enzyme production by recombinant DNA technology, some of the corresponding genes have been cloned and expressed in *Escherichia coli* or *Bacillus subtilis* (BAHL et al., 1991b; BURCHHARDT et al., 1991; HAECKEL and BAHL, 1989; KITAMOTO et al., 1988; MELASNIEMI and PALOHEIMO, 1989).

The cellulolytic activity of *C. thermocellum* on crystalline cellulose is superior to that of the well studied fungus *Trichoderma reesei* (JOHNSON et al., 1982). Unfortunately, glucanohydrolases and cellobiohydrolases are inhibited by their end products. Since an appropiate process to overcome this problem has not yet been developed, the conversion of cellulose to glucose on an industrial scale by the cellulolytic enzymes of *C. thermocellum* does not seem to be feasible at the moment. On the other hand, cellulases of *C. thermocellum* have been considered for utilization in direct ethanol fermentation from cellulose and for use in laundry detergents. The cellulase-encoding genes of *C. thermocellum* have been cloned and expressed in *E. coli* (BÉGUIN, 1990; CORNET et al., 1983;

Tab. 5. Some Properties of Starch-Degrading Enzymes from Thermophilic Clostridia[a]

Organism	Enzymes	Main Products	Temperature Optimum (°C)	pH Optimum
C. thermohydrosulfuricum				
Strain E101-69	α-Amylase, pullulanase, α-glucosidase	Glucose, maltose	85–90	5.6
Strain 39E	Glucoamylase, pullulanase	Glucose	85–90	5.5
C. thermosaccharolyticum	α-Amylase, pullulanase, α-glucosidase, glucoamylase	Glucose	70–75	5.5
C. thermosulfurogenes 4B	β-Amylase	Maltose, limit dextrin	70–75	5.5
C. thermosulfurogenes EM1	α-Amylase, pullulanase, α-glucosidase	Glucose, maltose	65–70	5.5

[a] Data taken from ANTRANIKIAN (1990) and SPECKA et al. (1991)

ROMANIEC et al., 1987; SCHWARZ et al., 1987). However, due to the complex structure of the cellulolytic system of *C. thermocellum* (cellulosome), the hydrolysis of cellulose does not only depend on large amounts of these enzymes, but requires synergistic effects of the several polypeptides of the cellulosome, which are not known in detail.

Clostridial botulinum toxin became the first microbial protein to be used by injection for the treatment of human disease. Its mode of action involves recognition of and binding to a receptor on the nerve ending, internalization of a portion of the protein into the nerve cell, and finally action of this internal toxin fragment to prevent acetylcholine release from the nerve ending. Weakening of the skeletal muscle is the result, and death may come about through paralysis of the muscles of respiration (HATHEWAY, 1990). Paralytic action on survivors lasted for many weeks as determined in animal studies. These properties led to the rationale to use the toxin in medical treatment of hyperactive muscles. Successful examples are correction of strabismus, a disorder of vision caused by hyperactivity of muscles controlling eye position, and blepharospasm, in which patients suffer from involuntary eyelid closure (SCHANTZ and JOHNSON, 1992; SCOTT, 1989). Thus, a simple injection relieves symptoms formerly to be corrected only by surgery. Adverse side effects of this therapy do not seem to occur, due to the low dose of toxin applied. Positive results have also been achieved in treatment of a number of other involuntary muscle movements (e.g., writer's and musician's cramp, hemifacial spasm, spasmodic torticollis; for a recent review see SCHANTZ and JOHNSON, 1992). The clostridial binary toxins (possessing ADP-ribosylating activity) might also become accepted as valuable drugs and research tools due to their ability to bind to a wide variety of cells (CONSIDINE and SIMPSON, 1991).

5.3 Stereospecific Compounds

5.3.1 Reduction Reactions

The investigation of enoate reductases, enzymes required, e.g., for conversion of aromatic amino acids in the Stickland reaction (see Sect. 4.3), led to the discovery that they are able to perform stereospecific reductions on a variety of unsaturated chemical substances (SIMON et al., 1985). This opened the way for producing chiral substrates in high yield and at comparatively low cost. In addition, various clostridia are able to reduce ketones to secondary alcohols, carboxylic acids to primary alcohols, N-allylhydroxylamines to N-allylamines, and aliphatic nitro compounds to the respective amino compounds. A summary of these reactions is presented in Tab. 6. In the course of 2-nitroethanol reduction 1,4-butanediol and ethylene glycol are formed as side products. Nitroaryl compounds are also reduced (ANGERMAIER and SIMON, 1983). An interesting feature of many reduction reactions is that electrons can be provided by reduced methyl viologen, which acts as an artificial mediator and uses in turn the cathode of an electrochemical cell as electron donor. This has been shown for a variety of clostridial species and proved to be by far superior to chemical synthesis of several stereospecific compounds (SIMON et al., 1985). Finally, this ability can be used to transfer electrons to methyl viologen-dependent NAD(P) reductases that are quite common in the genus *Clostridium* and mediate the regeneration of the pyridine nucleotides in preparative selective reactions (SIMON, 1988; SIMON et al., 1984).

5.3.2 Bile Acids and Steroid Conversions

Like many other intestinal bacteria several clostridia are able to act on compounds containing the gonane (often incorrectly called sterane) ring system. Side chain cleavage of the peptide bond in conjugated bile acids (desmolase activity) has been reported for *C. cadaveris*, *C. paraputrificum*, *C. perfringens*, and *C. scindens* (MORRIS, 1989; WINTER et al., 1989). Hydroxyl groups at various positions of the gonane ring system can be epimerized by a number of clostridial species. Dehydrogenases for the $3\alpha,\beta$- (*C. perfringens*, *C. purinolyticum*), $3\alpha,5\beta$- (*C. innocuum*, *C. paraputrificum*), 6β- (*Clostridium* sp. R6X76), $7\alpha,\beta$- (*C.*

Tab. 6. Reduction Reactions Performed by Clostridia[a]

Substrate	Product	Electron Donor	Organism
$R_2R_3C{=}C(X)R_1$ [b]	$R_2R_3HC{-}CHR_1X$	H_2, reduced methyl viologen [c]	*C. kluyveri* *C. tyrobutyricum*
$R_2R_3C{=}C(CH_2OH)R_1$	$R_2R_3HC{-}CHR_1CH_2OH$	Ethanol, H_2	*C. kluyveri*, *C. tyrobutyricum*
$R_2R_3C{=}C{=}C(COOH)R_1$	$R_2R_3C{=}CH{-}CH(COOH)R_1$	H_2, reduced methyl viologen [c]	*C. kluyveri* *C. tyrobutyricum*
$R_2{-}C({=}O){-}R_1$	$R_2{-}CHOH{-}R_1$	CO, formate, H_2	*C. kluyveri*, *C. sporogenes*, *C. thermohydrosulfuricum*, *C. tyrobutyricum*
$R{-}COOH$	$R{-}CH_2OH$	CO, formate	*C. formicoaceticum*, *C. thermoaceticum*
Linoleic acid	Transvaccenic acid	n.d. [d]	*C. bifermentans*, *C. sordellii*, *C. sporogenes*
$R_2HC{=}C(R_3){-}CH(R_1)NHOH$	$R_2HC{=}C(R_3){-}CH(R_1)NH_2$	CO, formate, H_2	*C. kluyveri*, *C. thermoaceticum*, *C. tyrobutyricum*
$O_3N{-}CH_2{-}CH_2OH$	$H_2N{-}CH_2{-}CH_2OH$	H_2	*C. kluyveri*, *C. pasteurianum*, *C. sporogenes*, *C. tyrobutyricum*

[a] Data taken from ANGERMAIER and SIMON (1983), BADER and SIMON (1980), BRAUN et al. (1991 a), SIMON (1988), SIMON et al. (1985), VERHULST et al. (1985)
[b] X = −COOH, −CHO
[c] The reduction can also be performed by purified enoate reductase using reduced methyl viologen as an electron donor
[d] n.d., not determined

absonum, C. bifermentans, C. limosum), 12α- (*C. leptum, Clostridium* sp. ATCC 29733, *Clostridium* sp. strain C 48-50), 12β- (*C. difficile, C. paraputrificum, C. tertium*), and 20α-position (*C. scindens*) have been described and in some cases purified (BRAUN et al., 1991 b; DÜRRE, 1981, HYLEMON, 1985; KRAFFT et al., 1987; MORRIS, 1989; WINTER et al., 1989). Dehydroxylation can be performed at position 7α,β by *C. bifermentans, C. leptum, C. perfringens*, and *C. sordellii* and at position 21 by *C. orbiscindens* (HYLEMON, 1985; MORRIS, 1989; WINTER et al., 1989, 1991).

All these reactions are not required for growth, and their physiological role is unknown. Maybe they represent some sort of a detoxification mechanism or unspecific side reactions of special enzymes (e.g., xanthine dehydrogenase in case of *C. purinolyticum*). *C. innocuum* and *C. paraputrificum* might be responsible for the undesired inactivation of contraceptive steroid hormones in the human intestine. Synthetic progestins proved to be more resistant (BOKKENHEUSER et al., 1983). The site- and stereospecific steroid conversion reactions of clostridia might prove useful for the preparation of new oral contraceptives or other hormones by the pharmaceutical industry.

5.3.3 Dechlorination

Halogenated compounds represent a growing environmental problem. The insecticide lindane (γ-hexachlorocyclohexane, γ-HCH) is attacked by several clostridial species. Growing cells of *C. butyricum* and *C. pasteurianum* completely dechlorinate this compound (Fig. 7). The α-isomer is similarly degraded by *C. butyricum*, whereas *C. pasteurianum* only achieves partial dehalogenation. β- and δ-Isomers are decomposed more slowly and to a lesser extent (JAGNOW et al., 1977). Cell suspensions and cell-free extracts of *C. rectum* convert lindane into monochlorobenzene and γ-tetrachlorocyclohexene (γ-TCH; OHISA et al., 1980). A conversion of γ-HCH to γ-tetrachloro-1-cyclohexene was also reported for cell-free extracts and membrane fractions of *C. sphenoides* (HERITAGE and MACRAE, 1977).

Tetrachloromethane can be completely degraded by growing cells of *C. thermoaceticum*. The homoacetogen *Acetobacterium woodii* performs a similar reaction (EGLI et al., 1988). For the latter organism it has been demonstrated that components of autoclaved cells still perform an oxidative dehalogenation of CCl_4 (EGLI et al., 1990).

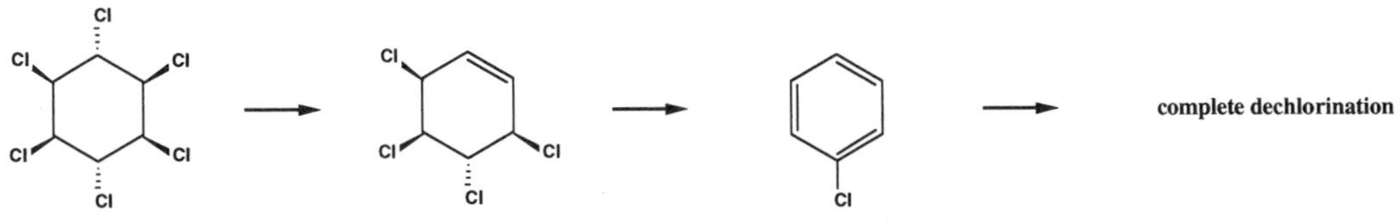

Fig. 7. Dehalogenation of cyclic compounds by clostridia. Identified intermediates of the dechlorination of lindane by various clostridia are shown (see text for details).

CCl_4 ⟶ $CHCl_3$ ⟶ **CH_2Cl_2, unidentified products (including totally dechlorinated compounds)**

CCl_3-CH_3 ⟶ **$CHCl_2$-CH_3, acetate, and unidentified products**

Fig. 8. Identified intermediates of the dechlorination of halogenated hydrocarbons by clostridia.

An as yet uncharacterized *Clostridium* (strain TCAIIB) is able to reductively dehalogenate 1,1,1-trichloroethane (TCA), tetrachloromethane, and trichloromethane. Degradation of TCA leeds to 1,1-dichloroethane, acetic acid, and unidentified products. Tetrachloromethane is dechlorinated via the intermediate trichloromethane to dichloromethane and unidentified products (Fig. 8). Dense cell suspensions of the *Clostridium* might be able to degrade as much as 75 μmol L^{-1} TCA within one day. Tetrachloroethene, trichloroethene, 1,1-dichloroethene, chloroethane, 1,1-dichloroethane, and dichloromethane are not significantly biotransformed (GÄLLI and McCarty, 1989a, 1989b).

Hence, clostridial strains might prove useful in the microbiological clean-up of contaminated soils and ground water. Such biotreatment has a high potential and is becoming a fast developing technology (MORGAN and WATKINSON, 1989).

Acknowledgement
Work in the authors' laboratories has been supported by grants from the Deutsche Forschungsgemeinschaft and the Bundesminister für Forschung und Technologie.

6 References

AFSCHAR, A. S., BIEBL, H., SCHALLER, K., SCHÜGERL, K. (1985), Production of acetone and butanol by *Clostridium acetobutylicum* in continuous culture with cell recycle, *Appl. Microbiol. Biotechnol.* **22,** 394–398.

ANDREESEN, J. R., GOTTSCHALK, G. (1969), The occurrence of a modified Entner-Doudoroff pathway in *Clostridium aceticum*, *Arch. Mikrobiol.* **69,** 160–170.

ANDREESEN, J. R., GOTTSCHALK, G., SCHLEGEL, H. G. (1970), *Clostridium formicoaceticum* nov. spec. Isolation, description and distinction from *C. aceticum* and *C. thermoaceticum*, *Arch. Mikrobiol.* **72,** 154–174.

ANDREESEN, J. R., BAHL, H., GOTTSCHALK, G. (1989), Introduction to the physiology and biochemistry of the genus *Clostridium*, in: *Biotechnology Handbooks*, (MINTON, N. P., CLARKE, D. J., Eds.), Vol. 3: *Clostridia*, pp. 27–62, New York: Plenum Press.

ANGERMAIER, L., SIMON, H. (1983), On the reduction of aliphatic and aromatic nitro compounds by clostridia, the role of ferredoxin and its stabilization, *Hoppe-Seyler's Z. Physiol. Chem.* **364,** 961–975.

ANTRANIKIAN, G. (1990), Physiology and enzymology of thermophilic anaerobic bacteria degrading starch, *FEMS Microbiol. Rev.* **75,** 201–218.

ANTRANIKIAN, G., GIFFHORN, F. (1987), Citrate metabolism in anaerobic bacteria, *FEMS Microbiol. Rev.* **46,** 175–198.

ANTRANIKIAN, G., FRIESE, C., QUENTMEIER, A., HIPPE, H., GOTTSCHALK, G. (1984), Distribution of the ability for citrate utilization amongst clostridia, *Arch. Microbiol.* **138,** 179–182.

ANTRANIKIAN, G., ZABLOWSKI, P., GOTTSCHALK, G. (1987), Conditions for the overproduction and excretion of thermostable α-amylase and pullulánase from *Clostridium thermohydrosulfuricum* DSM 567, *Appl. Microbiol. Biotechnol.* **27,** 75–81.

ARKOWITZ, R. A., ABELES, R. H. (1989), Identification of acetyl phosphate as the product of clostridial glycine reductase: Evidence for an acyl enzyme intermediate, *Biochemistry* **28,** 4639–4644.

ATALLA, R. H., VANDERHART, D. L. (1984), Native cellulose: A composite of two distinct crystalline forms, *Science* **223,** 283–285.

BADER, J., SIMON, H. (1980), The activities of hydrogenase and enoate reductase in two *Clostridium* species, their interrelationship and dependence on growth conditions, *Arch. Microbiol.* **127,** 279–287.

BADER, J., RAUSCHENBACH, P., SIMON, H. (1982), On a hitherto unknown fermentation path of several amino acids by proteolytic clostridia, *FEBS Lett.* **140,** 67–72.

BAHL, H., GOTTSCHALK, G. (1984), Parameters affecting solvent production by *Clostridium acetobutylicum* in continuous culture, *Biotechnol. Bioeng. Symp.* **14,** 215–223.

BAHL, H., GOTTSCHALK, G. (1988), Microbial production of butanol/acetone, in: *Biotechnology* (REHM, H.-J., REED, G., Eds.), Vol. 6b, pp. 1–30, Weinheim: VCH Verlagsgesellschaft.

BAHL, H., ANDERSCH, W., BRAUN, K., GOTTSCHALK, G. (1982a), Effect of pH and butyrate concentration on the production of acetone and butanol by *Clostridium acetobutylicum* grown in continuous culture, *Eur. J. Appl. Microbiol. Biotechnol.* **14,** 17–20.

BAHL, H., ANDERSCH, W., GOTTSCHALK, G. (1982b), Continuous production of acetone and butanol by *Clostridium acetobutylicum* in a two-stage phosphate limited chemostat, *Eur. J. Appl. Microbiol. Biotechnol.* **15,** 201–205.

BAHL, H., GOTTWALD, M., KUHN, A., RALE, V., ANDERSCH, W., GOTTSCHALK, G. (1986), Nutritional factors affecting the ratio of solvents produced by

Clostridium acetobutylicum, *Appl. Environ. Microbiol.* **52,** 169–172.

Bahl, H., Burchhardt, G., Wienecke, A. (1991 a), Nucleotide sequence of two *Clostridium thermosulfurogenes* EM1 genes homologous to *Escherichia coli* genes encoding integral membrane components of binding protein-dependent transport systems, *FEMS Microbiol. Lett.* **81,** 83–88.

Bahl, H., Burchhardt, G., Spreinat, A., Haeckel, K., Wienecke, A., Schmidt, B., Antranikian, G. (1991 b), α-Amylase of *Clostridium thermosulfurogenes* EM1: Nucleotide sequence of the gene, processing of the enzyme, and comparison to other α-amylases, *Appl. Microbiol. Biotechnol.* **57,** 1554- 1559.

Barker, H. A. (1939), The use of glutamic acid for the isolation and identification of *Clostridium cochlearium* and *Cl. tetanomorphum*, *Arch. Mikrobiol.* **10,** 376-384.

Barker, H. A. (1981), Amino acid degradation by anaerobic bacteria, *Annu. Rev. Biochem.* **50,** 23–40.

Barker, H. A., Beck, J. V. (1941), The fermentative decomposition of purines by *Clostridium acidi-urici* and *Clostridium cylindrosporum*, *J. Biol. Chem.* **141,** 3–27.

Barker, H. A., Beck, J. V. (1942), *Clostridium acidi-urici* and *Clostridium cylindrosporum*, organisms fermenting uric acid and some other purines, *J. Bacteriol.* **43,** 291–304.

Barnes, E. M., Impey, C. S. (1974), The occurrence and properties of uric acid decomposing anaerobic bacteria in the avian caecum, *J. Appl. Bacteriol.* **37,** 393–409.

Bayer, E. A., Lamed, R. (1986), Ultrastructure of the cell surface cellulosome of *Clostridium thermocellum* and its interaction with cellulose, *J. Bacteriol.* **167,** 828–836.

Beaty, P. S., Ljungdahl, L. G. (1991), Growth of *Clostridium thermoaceticum* on methanol, ethanol, propanol and butanol in medium containing either thiosulfate or dimethylsulfoxide, *Abstr. Annu. Meet. Am. Soc. Microbiol.* **1991,** p. 236, K-131.

Beesch, S.C. (1952), Acetone-butanol fermentation of sugars, *Eng. Proc. Dev.* **44,** 1677–1682.

Béguin, P. (1990), Molecular biology of cellulose degradation, *Annu. Rev. Microbiol.* **44,** 219–248.

Bérenger, J.-F., Frixon, C., Bigliardi, J., Creuzet, N. (1985), Production, purification, and properties of thermostable xylanase from *Clostridium stercorarium*, *Can. J. Microbiol.* **31,** 635–643.

Berestovskaya, Y. Y., Kryukov, V. R., Bodnar, I. V., Pusheva, M. A. (1987), Growth of the homoacetic bacterium *Clostridium thermoautotrophicum*, *Microbiologia* **56,** 642–647.

Bergère, J.-L., Hermier, J. (1970), Spore properties of clostridia occurring in cheese, *J. Appl. Bacteriol.* **33,** 167–179.

Bertram, J., Dürre, P. (1989), Conjugal transfer and expression of streptococcal transposons in *Clostridium acetobutylicum*, *Arch. Microbiol.* **151,** 551–557.

Bertram, J., Kuhn, A., Dürre, P. (1990), Tn*916*-induced mutants of *Clostridium acetobutylicum* defective in regulation of solvent formation, *Arch. Microbiol.* **153,** 373–377.

Bhat, J. V., Barker, H. A.(1947), *Clostridium lactoacidophilum* nov. spec. and the role of acetic acid in the butyric acid fermentation of lactate, *J. Bacteriol.* **54,** 381–391.

Biebl, H. (1991), Glycerol fermentation of 1,3-propanediol by *Clostridium butyricum*. Measurement of product inhibition by use of a pH-auxostat, *Appl. Microbiol. Biotechnol.* **35,** 701–705.

Blackwell, J. (1982), The macromolecular organization of cellulose and chitin, in: *Cellulose and Other Natural Polymer Systems: Biogenesis, Structure and Degradation* (Brown, R.M. Jr., Ed.), pp. 403–428 New York: Plenum Press.

Bogdahn, M., Kleiner, D. (1986), N_2 fixation and NH_4^+ assimilation in the thermophilic anaerobes *Clostridium thermosaccharolyticum* and *Clostridium thermoautotrophicum*, *Arch. Microbiol.* **144,** 102–104.

Bogdahn, M., Andreesen, J. R., Kleiner, D. (1983), Pathways and regulation of N_2, ammonium and glutamate assimilation by *Clostridium formicoaceticum*, *Arch. Microbiol.* **134,** 167–169.

Bokkenheuser, V. D., Winter, J., Cohen, B. I., O'Rourke, S., Mosbach, E. H. (1983), Inactivation of contraceptive steroid hormones by human intestinal clostridia, *J. Clin. Microbiol.* **18,** 500–504.

Booth, I. R., Morris, J. G. (1982), Carbohydrate transport in *Clostridium pasteurianum*, *Biosci. Rep.* **2,** 47–53.

Bornstein, B. T., Barker, H. A. (1948), The nutrition of *Clostridium kluyveri*, *J. Bacteriol.* **55,** 223–230.

Bowles, L. K., Ellefson, W. L. (1985), Effects of butanol on *Clostridium acetobutylicum*, *Appl. Environ. Microbiol.* **50,** 1165–1170.

Braun, M., Mayer, F., Gottschalk, G. (1981), *Clostridium aceticum* (Wieringa), a microorganism producing acetic acid from molecular hydrogen and carbon dioxide, *Arch. Microbiol.* **128,** 288–293.

Braun, H., Schmidtchen, F. P., Schneider, A., Simon, H. (1991 a), Microbial reduction of allylhydroxylamines to N-allylamines using clostridia, *Tetrahedron* **47,** 3329–3334.

Braun, M., Lünsdorf, H., Bückmann, A. F. (1991b), 12α-Hydroxysteroid dehydrogenase from

Clostridium group P, strain C 48-50. Production, purification and characterization, *Eur. J. Biochem.* **196,** 439–450.

BRONNENMEIER, K., STAUDENBAUER, W. L. (1988), Resolution of *Clostridium stercorarium* cellulase by fast protein liquid chromatography, *Appl. Microbiol. Biotechnol.* **27,** 432–436.

BRONNENMEIER, K., STAUDENBAUER, W. L. (1990), Cellulose hydrolysis by a highly thermostable endo-1,4-β-glucanase (Avicelase I) from *Clostridium stercorarium, Enzyme Microb. Technol.* **12,** 431–436.

BRONNENMEIER, K., RÜCKNAGEL, K. P., STAUDENBAUER, W. L. (1991), Purification and properties of a novel type of exo-1,4-β-glucanase (Avicelase II) from the cellulolytic thermophile *Clostridium stercorarium, Eur. J. Biochem.* **200,** 379–385.

BRYSON, M. F., DRAKE, H. L. (1988), A reevaluation of the metabolic potential of *Clostridium formicoaceticum, Abstr. Annu. Meet. Am. Soc. Microbiol.* **1988,** p. 198, I-107.

BUCKEL, W. (1991), Ungewöhnliche Chemie bei der Fermentation von Aminosäuren durch anaerobe Bakterien, *Bioforum* **14,** 7–19.

BURCHHARDT, G., WIENECKE, A., BAHL, H. (1991), Isolation of the pullulanase gene from *Clostridium thermosulfurogenes* (DSM 3896) and its expression in *Escherichia coli, Curr. Microbiol.* **22,** 91–95.

BUSCHE, R. M. (1985), Acetic acid manufacture-fermentation alternatives, in: *Biotechnology Applications and Research* (CHEREMISINOFF, P. N., OULETTE, P., Eds.), pp. 88–102, Lancaster: Technicon.

BUSCHHORN, H., DÜRRE, P., GOTTSCHALK, G. (1989), Production and utilization of ethanol by the homoacetogen *Acetobacterium woodii, Appl. Environ. Microbiol.* **55,** 1835–1840.

CAMERON, D. C., COONEY, C. L. (1986), A novel fermentation: The production of R(-)-1,2-propanediol and acetol by *Clostridium thermosaccharolyticum, Biotechnology* **4,** 651–654.

CARDON, B. P., BARKER, H. A. (1947), Amino acid fermentations by *Clostridium propionicum* and *Diplococcus glycinophilus, Arch. Biochem.* **12,** 165–180.

CATO, E. P., STACKEBRANDT, E. (1989), Taxonomy and phylogeny, in: *Biotechnology Handbooks,* Vol. **3**, *Clostridia*, pp. 1–26 (MINTON, N. P., CLARKE, D. J., Eds.), New York: Plenum Press.

CATO, E. P., GEORGE, W. L., FINEGOLD, S. M. (1986), Genus *Clostridium* Prazmowski 1880, 23^{AL}, in: *Bergey's Manual of Systematic Bacteriology*, Vol. 2, pp. 1141–1200 (SNEATH, P. H. A., MAIR, N. S., SHARPE, M. E., HOLT, J. G., Eds.), Baltimore: Williams & Wilkins.

CHAIGNEAU, A. B., BORY, J., LABARRE, C., DESCROZAILLES, J. (1974), Composition des gazes dégagés par *Welchia perfringens* et *Clostridium sporogenes* cultivés en différents milieux synthétiques, *Ann. Pharm. Fr.* **32,** 619–622.

CHAMPION, A. B., RABINOWITZ, J. C. (1977), Ferredoxin and formyltetrahydrofolate synthetase: Comparative studies with *Clostridium acidiurici, Clostridium cylindrosporum*, and newly isolated anaerobic uric acid-fermenting strains, *J. Bacteriol.* **132,** 1003–1020.

CHEN, K. C.-K., CHEN, J.-S., JOHNSON, J. L. (1986), Structural features of multiple *nifH*-like sequences and very biased codon usage in nitrogenase genes of *Clostridium pasteurianum, J. Bacteriol.* **166,** 162–172.

CONSIDINE, R. V., SIMPSON, L. L. (1991), Cellular and molecular actions of binary toxins possessing ADP-ribosyltransferase activity, *Toxicon* **29,** 913–936.

CORNET, P., MILLET, J., BÉGUIN, P., AUBERT, J.-P. (1983), Characterization of two *cel* (cellulose degradation) genes of *Clostridium thermocellum* coding for endoglucanases, *Biotechnology* **1,** 589–594.

COSTILOW, R. N. (1981), Biophysical factors in growth, in: *Manual of Methods for General Bacteriology* (GERHARDT, P., MURRAY, R. G. E., COSTILOW, R. N., NESTER, E. W., WOOD, W. A., KRIEG, N. R., PHILLIPS, G. B., Eds.), pp. 66–78, Washington: American Society for Microbiology.

COWLING, E. B. (1963), Structural features of cellulose that influence its susceptibility to enzymatic hydrolysis, in: *Advances in Enzymic Hydrolysis of Cellulose and Related Materials* (RESE, E. T., Ed.), pp. 1–32, New York: Macmillan.

CYNKIN, M. A., DELWICHE, E. A. (1958), Metabolism of pentoses by clostridia. I. Enzymes of ribose dissimilation in extracts of *Clostridium perfringens, J. Bacteriol.* **75,** 331–334.

CYNKIN, M. A., GIBBS, M. (1958), Metabolism of pentoses by clostridia. II. The fermentation of C^{14}-labeled pentoses by *Clostridium perfringens, C. beijerinckii*, and *C. butylicum, J. Bacteriol.* **75,** 335–338.

DABROCK, B., BAHL, H., GOTTSCHALK, G. (1992), Parameters affecting solvent production by *Clostridium pasteurianum, Appl. Environ. Microbiol.* **58,** 1233–1239.

DANIEL, S. L., DRAKE, H. L. (1991), Acetogenesis from two-carbon compounds by *Clostridium thermoaceticum, Abstr. Annu. Meet. Am. Soc. Microbiol.* **1991,** p. 237, K-137.

DANIEL, S. L., WU, Z., DRAKE, H. L. (1988), Growth of thermophilic acetogenic bacteria on methoxylated aromatic acids, *FEMS Microbiol. Lett.* **52,** 25–28.

DAS, A., HUGENHOLTZ, J., VAN HALBEEK, H., LJUNGDAHL, L. G. (1989), Structure and function of a menaquinone involved in electron transport in membranes of *Clostridium thermoaceticum* and *Clostridium thermoautotrophicum*, *J. Bacteriol.* **171,** 5823–5829.

DEHNING, I., SCHINK, B. (1989), Two new species of anaerobic oxalate-fermenting bacteria, *Oxalobacter vibrioformis* sp. nov. and *Clostridium oxalicum* sp. nov., from sediment samples, *Arch. Microbiol.* **153,** 79–84.

DIETRICHS, D., MEYER, M., RIETH, M., ANDREESEN, J. R. (1991), Interaction of selenoprotein P_A and the thioredoxin system, components of the NADPH-dependent reduction of glycine in *Eubacterium acidaminophilum* and *Clostridium litoralis*, *J. Bacteriol.* **173,** 5983–5991.

DIMROTH, P. (1991), Na^+-coupled alternative to H^+-coupled primary transport systems in bacteria, *Bioessays* **13,** 463–468.

DOMINGUEZ-BELLO, M. G., STEWART, C.S. (1991), Characteristics of a rumen *Clostridium* capable of degrading mimosine, 3(OH)-4-(1 H)-pyridone and 2,3-dihydroxypyridine, *Syst. Appl. Microbiol.* **14,** 67–71.

DÖRNER, C., SCHINK, B. (1990), *Clostridium homopropionicum* sp. nov., a new strict anaerobe growing with 2-, 3-, or 4-hydroxybutyrate, *Arch. Microbiol.* **154,** 342–348.

DRENT, W. J., LAHPOR, G. A., WIEGANT, W. M., GOTTSCHAL, J. C. (1991), Fermentation of inulin by *Clostridium thermosuccinogenes* sp. nov., a thermophilic anaerobic bacterium isolated from various habitats, *Appl. Environ. Microbiol.* **57,** 455–462.

DUNCAN, C. L., STRONG, D. H. (1968), Improved medium for sporulation of *Clostridum perfringens*, *Appl. Microbiol.* **16,** 82–89.

DÜRRE, P. (1981), Selenabhängige Vergärung von Purinen und Glycin durch *Clostridium purinolyticum* und *Peptococcus glycinophilus*, *PhD Thesis*, University of Göttingen, Germany.

DÜRRE, P., ANDREESEN, J. R. (1982a), Separation and quantitation of purines and their anaerobic and aerobic degradation products by high-pressure liquid chromatography, *Anal. Biochem.* **123,** 32–40.

DÜRRE, P., ANDREESEN, J. R. (1982b), Selenium-dependent growth and glycine fermentation by *Clostridium purinolyticum*, *J. Gen. Microbiol.* **128,** 1457-1466.

DÜRRE, P., ANDREESEN, J. R. (1982c), Anaerobic degradation of uric acid via pyrimidine derivatives by selenium-starved cells of *Clostridium purinolyticum*, *Arch. Microbiol.* **131,** 255–260.

DÜRRE, P., ANDREESEN, J. R. (1983), Purine and glycine metabolism by purinolytic clostridia, *J. Bacteriol.* **154,** 192–199.

DÜRRE, P., ANDERSCH, W., ANDREESEN, J. R. (1981), Isolation and characterization of an adenine-utilizing, anaerobic sporeformer, *Clostridium purinolyticum* sp. nov., *Int. J. Syst. Bacteriol.* **31,** 184–194.

DÜRRE, P., BAHL, H., GOTTSCHALK, G. (1988), Membrane processes and product formation in anaerobes, in: *Handbook on Anaerobic Fermentations* (ERICKSON, L. E., FUNG, D. Y-C., Eds.), pp. 187–206, New York: Marcel Dekker.

EGLI, C., TSCHAN, T., SCHOLTZ, R., COOK, A. M., LEISINGER, T. (1988), Transformation of tetrachloromethane to dichloromethane and carbon dioxide by *Acetobacterium woodii*, *Appl. Environ. Microbiol.* **54,** 2819 -2824.

EGLI, C., STROMEYER, S., COOK, A. M., LEISINGER, T. (1990), Transformation of tetra- and trichloromethane to CO_2 by anaerobic bacteria is a non-enzymic process, *FEMS Microbiol. Lett.* **68,** 207–212.

EMTSEV, V. T. (1963), Sporulation in *Clostridium pasteurianum*, *Microbiologia* **32,** 434–438.

ENNIS, B. M., GUTIERREZ, N. A., MADDOX, I. S. (1986), The acetone-butanol-ethanol fermentation: A current assessment, *Process Biochem.* **21,** 131–147.

ENSLEY, B., MCHUGH, J. J., BARTON, L. L. (1975), Effect of carbon sources on formation of α-amylase and glucoamylase by *Clostridium acetobutylicum*, *J. Gen. Appl. Microbiol.* **21,** 51–59.

FENDRICH, C., HIPPE, H., GOTTSCHALK, G. (1990), *Clostridium halophilium* sp. nov. and *C. litorale* sp. nov., an obligate halophilic and a marine species degrading betaine in the Stickland reaction, *Arch. Microbiol.* **154,** 127–132.

FLEMMING, R. W., QUINN, L. Y. (1971), Chemically defined medium for growth of *Clostridium thermocellum*, a cellulolytic thermophilic anaerobe, *Appl. Microbiol.* **21,** 967.

FOGARTY, W. M., KELLY, C. T. (1983), Pectic enzymes, in: *Microbial Enzymes and Biotechnology* (FOGARTY, W. M., Ed.), pp. 131–182, London: Applied Science Publishers.

FORBERG, C., HÄGGSTRÖM, L. (1985), Control of cell adhesion and activity during continuous production of acetone and butanol with adsorbed cells, *Enzyme Microbial Technol.* **7,** 230–234.

FORSBERG, C. W. (1987), Production of 1,3-propanediol from glycerol by *Clostridium acetobutylicum* and other *Clostridium* species, *Appl. Environ. Microbiol.* **53,** 639–643.

FREIER, D., GOTTSCHALK, G. (1987), L(+)-lactate dehydrogenase of *Clostridium acetobutylicum* is activated by fructose-1,6-bisphosphate, *FEMS Microbiol. Lett.* **43,** 229–233.

FUCHS, G. (1986), CO_2 fixation in acetogenic bacteria: variations on a theme, *FEMS Microbiol. Rev.* **39,** 181–213.

GÄLLI, R., MCCARTY, P. L. (1989a), Biotransformation of 1,1,1-trichloroethane, trichloromethane, and tetrachloromethane by a *Clostridium* sp., *Appl. Environ. Microbiol.* **55,** 837–844.

GÄLLI, R., MCCARTY, P. L. (1989 b), Kinetics of biotransformation of 1,1,1-trichloroethane by *Clostridium* sp. strain TCAIIB, *Appl. Environ. Microbiol.* **55,** 845–851.

GASTON, L. W., STADTMAN, E. R. (1963), Fermentation of ethylene glycol by *Clostridium glycolicum* sp. n., *J. Bacteriol.* **85,** 356–362.

GIBBS, B. M., HIRSCH, A. (1956), Spore formation by *Clostridium* species in an artificial medium, *J. Appl. Bacteriol.* **19,** 129–141.

GOTTSCHALK, G. (1986), *Bacterial metabolism*, 2nd Ed., New York: Springer Verlag.

GOTTWALD, M. (1973), Untersuchungen zum Pentosestoffwechsel bei Homoacetatgärern, *Diploma Thesis*, University of Göttingen, Germany.

GOTTWALD, M., GOTTSCHALK, G. (1985), The internal pH of *Clostridium acetobutylicum* and its effect on the shift from acid to solvent formation, *Arch. Microbiol.* **143,** 42–46.

GOTTWALD, M., ANDREESEN, J. R., LEGALL, J., LJUNGDAHL, L. G. (1975), Presence of cytochrome and menaquinone in *Clostridium formicoaceticum* and *Clostridium thermoaceticum*, *J. Bacteriol.* **122,** 325–328.

GRÉPINET, O., CHEBROU, M.-C., BÉGUIN, P. (1988), Nucleotide sequence and deletion analysis of the xylanase gene *xynZ* of *Clostridium thermocellum*, *J. Bacteriol.* **170,** 4582–4588.

GRIFFIN, M., MAGOR, A. M. (1987), Plastics and synthetic fibres from microorganisms: a dream or a potential reality?, *Microbiol. Sci.* **4,** 357–361.

HAECKEL, K., BAHL, H. (1989), Cloning and expression of the thermostable α-amylase gene from *Clostridium thermosulfurogenes* (DSM 3896) in *Escherichia coli*, *FEMS Microbiol. Lett.* **60,** 333–338.

HARTMANIS, M. G. N., STADTMAN, T. C. (1986), Diol metabolism and diol dehydratase in *Clostridium glycolicum*, *Arch. Biochem. Biophys.* **245,** 144–152.

HATHEWAY, C. L. (1990), Toxigenic clostridia, *Clin. Microbiol. Rev.* **3,** 66–98.

HERITAGE, A. D., MACRAE, I. C. (1977), Degradation of lindane by cell-free preparations of *Clostridium sphenoides*, *Appl. Environ. Microbiol.* **34,** 222–224.

HERMANN, M., FAYOLLE, F., MARCHAL, R., PODVIN, L., SEBALD, M., VANDECASTEELE, J. P. (1985), Isolation and characterization of butanol-resistant mutants of *Clostridium acetobutylicum*, *Appl. Environ. Microbiol.* **50,** 1238–1243.

HETHENER, P., BRAUMAN, A., GARCIA, J.-L. (1992), *Clostridium termitidis* sp. nov., a cellulolytic bacterium from the gut of the wood-feeding termite, *Nasutitermes lujae*, *Syst. Appl. Microbiol.* **15,** 52–58.

Hewitt, J., Morris, J. G. (1975), Superoxide dismutase in some obligately anaerobic bacteria, *FEBS Lett.* **50,** 315–318.

HEYNDRICKX, M., DE VOS, P., VANCANNEYT, M., DE LEY, J. (1991), The fermentation of glycerol by *Clostridium butyricum* LMG 1212t_2 and 1213t_1 and *C. pasteurianum* LMG 3285, *Appl. Microbiol. Biotechnol.* **34,** 637–642.

HIPPE, H., ANDREESEN, J. R., GOTTSCHALK, G. (1992), The genus *Clostridium*-Nonmedical, in: *The Prokaryotes* (BALOWS, A., TRÜPER, H. G., DWORKIN, M., HARDER, W., SCHLEIFER, K.-H., Eds.), 2nd Ed., pp. 1800–1866, New York: Springer-Verlag.

HOLDEMAN, L. V., CATO, E. P., MOORE, W. E. C. (1977), *Anaerobe Laboratory Manual*, 4th Ed., Blacksburg: Anaerobe Laboratory, Virginia Polytechnic Institute and State University.

HOLLAND, K. T., COX, D. J. (1975), A synthetic medium for the growth of *Clostridium bifermentans*, *J. Appl. Bacteriol.* **38,** 193–198.

HSU, E. J., ORDAL, Z. J. (1969), Sporulation of *Clostridium thermosaccharolyticum*. *Appl. Microbiol.* **18,** 958–960.

HSU, T., DANIEL, S. L., LUX, M. F., DRAKE, H. L. (1990), Biotransformations of carboxylated aromatic compounds by the acetogen *Clostridium thermoaceticum*: Generation of growth-supportive CO_2 equivalents under CO_2-limited conditions, *J. Bacteriol.* **172,** 212–217.

HUNGATE, R. E. (1969), A roll tube technique for cultivation of strict anaerobes, in: *Methods in Microbiology*, Vol. 3B (NORRIS, J. R., RIBBONS, D. W., Eds.), pp. 92-123, New York: Academic Press.

HUTKINS, R. W., KASHKET, E. R. (1986), Phosphotransferase activity in *Clostridium acetobutylicum* from acidogenic to solventogenic phases of growth, *Appl. Environ. Microbiol.* **51,** 1121–1123.

HYLEMON, P. B. (1985), Metabolism of bile acids in intestinal microflora, in: *Sterols and Bile Acids* (DANIELSSON, H., SJÖVALL, J., Eds.), pp. 331–343, Amsterdam: Elsevier Science Publ.

HYUN, H. H., ZEIKUS, J. G. (1985 a), General biochemical characterization of thermostable β-amylase from *Clostridium thermosulfurogenes*, *Appl. Environ. Microbiol.* **49,** 1162–1167.

HYUN, H. H., ZEIKUS, J. G. (1985 b), General biochemical characterization of thermostable pullulanase and glucoamylase from *Clostridium thermohydrosulfuricum*, *Appl. Environ. Microbiol.* **49,** 1168–1173.

JAGNOW, G., HAIDER, K., ELLWARDT, P.-C. (1977), Anaerobic dechlorination and degradation of hexachlorocyclohexane isomers by anaerobic and facultative anaerobic bacteria, *Arch. Microbiol.* **115,** 285–292.

JANSSEN, P. H. (1991), Isolation of *Clostridium pro-*

pionicum strain 19acry3 and further characteristics of the species, *Arch. Microbiol.* **155,** 566–571.

JOHNSON, E. A., SAKAJOH, M., HALLIWELL, G., MADIA, A., DEMAIN, A. L. (1982), Saccharification of complex cellulosic substrates by the cellulase from *Clostridium thermocellum*, *Appl. Environ. Microbiol.* **43,** 1125–1132.

JOHNSTON, R., HARMON, S., KAUTTER, D. (1964), Method to facilitate the isolation of *Clostridium botulinum* type E, *J. Bacteriol.* **88,** 1521–1522.

JONES, D. T., WOODS, D. R. (1986), Acetone-butanol fermentation revisited, *Microbiol. Rev.* **50,** 484–524.

JUNELLES, A.-M., JANATI-IDRISSI, R., PETITDEMANGE, H., GAY, R. (1987), Effect of pyruvate on glucose metabolism in *Clostridium acetobutylicum*, *Biochimie* **69,** 1183–1190.

KANE, M. D., BRAUMAN, A., BREZNAK, J.A. (1991), *Clostridium mayombei* sp. nov., an H_2/CO_2 acetogenic bacterium from the gut of the African soil-feeding termite, *Cubitermes speciosus*, *Arch. Microbiol.* **156,** 99–104.

KENEALY, W. R., WASELEFSKY, D. M. (1985), Studies on the substrate range of *Clostridium kluyveri:* the use of propanol and succinate, *Arch. Microbiol.* **141,** 187–194.

KIM, B. H., BELLOWS, P., DATTA, R., ZEIKUS, J. G. (1984), Control of carbon and electron flow in *Clostridium acetobutylicum* fermentations: Utilization of carbon monoxide to inhibit hydrogen production and to enhance butanol yield, *Appl. Environ. Microbiol.* **48,** 764–770.

KITAMOTO, N., YAMAGATA, H., KATO, T, TSUKAGOSHI, N., UDAKA, S. (1988),Cloning and sequencing of the gene encoding thermophilic β-amylase of *Clostridium thermosulfurogenes*, *J. Bacteriol.* **170,** 5848–5854.

KOCH, R., ZABLOWSKI, P., ANTRANIKIAN, G. (1987), Highly active and thermostable amylases and pullulanases from various anaerobic thermophiles, *Appl. Microbiol. Biotechnol.* **27,** 192–198.

KORANSKY, J. R., ALLEN, S. D., DOWELL, V. R., Jr. (1978), Use of ethanol for selective isolation of sporeforming microorganisms, *Appl. Environ. Microbiol.* **35,** 762–765.

KRAFFT, A. E., WINTER, J., BOKKENHEUSER, V. D., HYLEMON, P.B. (1987), Cofactor requirement of steroid-17-20-desmolase and 20α-hydroxysteroid dehydrogenase activities in cell extracts of *Clostridium scindens*, *J. Steroid Biochem.* **28,** 49–54.

KRIVENKO, V. V., VADACHLORIYA, R. M., CHERMYKH, N. A., MITYUSHINA, L. L., KRASIL'NIKOVA, E. N. (1990), *Clostridium uzonii* sp. nov., an anaerobic thermophilic glycolytic bacterium isolated from hot springs in the Kamchatka peninsula, *Microbiology USSR, Engl. Transl.* **59,** 741–748.

KRUMHOLZ, L. R., BRYANT, M. P. (1985), *Clostridium pfennigii* sp. nov. uses methoxyl groups of monobenzoids and produces butyrate, *Int. J. Syst. Bacteriol.* **35,** 454–456.

KUROSE, N., KINOSHITA, S., YAGYU, J., UCHIDA, M., HANAI, S., OBAYASHI, A. (1988), Improvement of ethanol production of thermophilic *Clostridium* sp. by mutation, *J. Ferment. Technol.* **66,** 467–472.

LAMED, R., BAYER, E. A. (1988), The cellulosome of *Clostridium thermocellum*, *Adv. Appl. Microbiol.* **33,** 1–46.

LAMED, R., SETTER, E., BAYER, E. A. (1983), Characterization of a cellulose-binding, cellulase-containing complex in *Clostridium thermocellum*, *J. Bacteriol.* **156,** 828–836.

LAMPEN, J. H., PETERSON, E. H. (1943), Growth factor requirements of clostridia, *Arch. Biochem.* **2,** 443–449.

LAPAGE, S. P., SHELTON, J. E., MITCHELL, T. G., (1970), Media for the maintenance and preservation of bacteria, in: *Methods in Microbiology*, Vol. 3A (NORRIS, J. R., RIBBONS, D. W., Eds.), pp. 1–228, London: Academic Press.

LARGIER, S. T., LONG, S., SANTANGELO, J. D., JONES, D. T., WOODS, D. R. (1985), Immobilized *Clostridium acetobutylicum* P262 mutants for solvent production, *Appl. Environ. Microbiol.* **50,** 477–481.

LEE, S. F., FORSBERG, C. W., RATTRAY, J. B. (1987), Purification and characterization of two endoxylanases from *Clostridium acetobutylicum* ATCC 824, *Appl. Environ. Microbiol.* **53,** 644–650.

LENTZ, K., WOOD, H. G. (1955), Synthesis of acetate from formate and carbon dioxide by *Clostridium thermoaceticum*, *J. Biol. Chem.* **215,** 645–654.

LEONHARDT, U., ANDREESEN, J. R. (1977), Some properties of formate dehydrogenase, accumulation and incorporation of ^{185}W-tungsten into proteins of *Clostridium formicoaceticum*, *Arch. Microbiol.* **115,** 277–284.

LE RUYET, P., DUBOURGUIER, H. C., ALBAGNAC, G., PRENSIER, G. (1985), Characterization of *Clostridium thermolacticum* sp. nov., a hydrolytic thermophilic anaerobe producing high amounts of lactate, *Syst. Appl. Microbiol.* **6,** 196–202.

LIN, Y., BLASCHEK, H. P. (1982), Butanol production by a butanol-tolerant strain of *Clostridium acetobutylicum* in extruded corn broth, *Appl. Environ. Microbiol.* **45,** 966–973.

LJUNGDAHL, L. G. (1986), The autotrophic pathway of acetate synthesis in acetogenic bacteria, *Annu. Rev. Microbiol.* **40,** 415–450.

LJUNGDAHL, L. G., CARREIRA, L. H., GARRISON, R. J., RABEK, N. E., WIEGEL, J. (1985), Comparison of three thermophilic acetogenic bacteria for the production of calcium magnesium acetate. *Biotechnol. Bioeng. Symp.* **15,** 207–223.

LJUNGDAHL, L. G., HUGENHOLTZ, J., WIEGEL, J. (1989), Acetogenic and acid-producing clostridia, in: *Biotechnology Handbooks*, Vol. 3: *Clostridia* (MINTON, N. P., CLARKE, D. J., Eds.), pp. 145–191, New York: Plenum Press.

LOVITT, R. W., LONGIN, R., ZEIKUS, J. G. (1984), Ethanol production by thermophilic bacteria: Physiological comparison of solvent effects on parent and alcohol-tolerant strains of *Clostridium thermohydrosulfuricum*, *Appl. Environ. Microbiol.* **48,** 171–177.

LUND, B. M., BROCKLEHURST, T. F. (1978), Pectic enzymes of pigmented strains of *Clostridium*, *J. Gen. Microbiol.* **104,** 59–66.

LUNDIE, L. L. Jr., DRAKE, H. L.(1984), Development of a minimally defined medium for the acetogen *Clostridium thermoaceticum*, *J. Bacteriol.* **159,** 700–703.

LUX, M. F., HSU, T., WU, Z., DRAKE, H. L. (1989), Growth of *Clostridium formicoaceticum* with CO, formate, and aromatic compounds: Comparative study with *Clostridium aceticum*, *Abstr. 6th Int. Symp. Microbial Growth on C_1 Compounds*, Göttingen 1989, p. 429.

MACKEY, B. M., MORRIS, J. G. (1971), Sporulation in *Clostridium pasteurianum*, in: *Spore Research* (BARKER, A. N., GOULD, G. W., WOLF, J., Eds.), p. 343, London: Academic Press.

MACMILLAN, J. D., PHAFF, H. J., VAUGHN, R. H. (1964), The pattern of action of an exopolygalacturonic acid trans-eliminase from *Clostridium multifermentans*, *Biochemistry* **3,** 572–578.

MADDOX, I. S. (1980), Production of *n*-butanol from whey filtrate using *Clostridium acetobutylicum* NCIB 2951, *Biotechnol. Lett.* **2,** 493–498.

MADI, E., ANTRANIKIAN, G., OHMIYA, K., GOTTSCHALK, G. (1987), Thermostable amylolytic enzymes from a new *Clostridium* isolate, *Appl. Environ. Microbiol.* **53,** 1661–1667.

MATTEUZZI, D., HOLLAUS, F., BIAVATI, B. (1978), Proposal of neotype for *Clostridium thermohydrosulfuricum* and the merging of *Clostridium tartarivorum* with *Clostridium thermosaccharolyticum*, *Int. J. Syst. Bacteriol.* **28,** 528–531.

MAYER, F. (1986), *Cytology and Morphogenesis of Bacteria*, Berlin–Stuttgart: Gebrüder Borntraeger.

MAYER, F. (1988), Cellulolysis: Ultrastructural aspects of bacterial systems, *Electron Microsc. Rev.* **1,** 69–85.

McCOY, E., McCLUNG, L. S. (1935), Studies on anaerobic bacteria. VI. The nature and systematic position of a new chromogenic *Clostridium*, *Arch. Mikrobiol.* **6,** 230–238.

MEAD, G. C., ADAMS, B. W. (1975), Some observations on the caecal microflora of the chick during the first two weeks of life, *Br. Poult. Sci.* **16,** 169–176.

MELASNIEMI, H. (1987), Characterization of α-amylase and pullulanase activities of *Clostridium thermohydrosulfuricum*, *Biochem. J.* **246,** 193–197.

MELASNIEMI, H., PALOHEIMO, M. (1989), Cloning and expression of the *Clostridium thermohydrosulfuricum α*-amylase-pullulanase gene in *Escherichia coli*, *J. Gen Microbiol.* **135,** 1755–1762.

MÉNDEZ, B., PETTINARI, M. J., IVANIER, S. E., RAMOS, C. A., SIÑERIZ, F. (1991), *Clostridium thermopapyrolyticum* sp. nov., a cellulolytic thermophile. *Int. J. Syst. Bacteriol.* **41,** 281–283.

MERMELSTEIN, L. D., WELKER, N. E., BENNETT, G. N., PAPOUTSAKIS, E. T. (1992), Expression of cloned homologous fermentative genes in *Clostridium acetobutylicum* ATCC 824, *Biotechnology* **10,** 190–195.

MEYER, C. L., ROOS, J. W., PAPOUTSAKIS, E. T. (1986), Carbon monoxide gassing leads to alcohol production and butyrate uptake without acetone formation in continuous cultures of *Clostridum acetobutylicum*, *Appl. Microbiol. Biotechnol.* **24,** 159–167.

MINTON, N. P., OULTRAM, J. D. (1988), Host: vector systems for gene cloning in *Clostridium*, *Microbiol. Sci.* **5,** 310–315.

MITCHELL, W. J., BOOTH, I. R. (1984), Characterization of the *Clostridium pasteurianum* phosphotransferase system, *J. Gen. Microbiol.* **130,** 2193–2200.

MITCHELL, W. J., SHAW, J. E., ANDREWS, L. (1991), Properties of the glucose phosphotransferase system of *Clostridium acetobutylicum* NCIB 8052, *Appl. Environ. Microbiol.* **57,** 2534–2539.

MÖLLER, B., HIPPE, H., GOTTSCHALK, G. (1986), Degradation of various amine compounds by mesophilic clostridia, *Arch. Microbiol.* **145,** 85–90.

MONOT, F., ENGASSER, J. M. (1983), Production of acetone and butanol by batch and continuous culture of *Clostridium acetobutylicum* under nitrogen limitation, *Biotechnol. Lett.* **5,** 213–218.

MONOT, F., MARTIN, J.-R., PETITDEMANGE, H., GAY, R. (1982), Acetone and butanol production by *Clostridium acetobutylicum* in a synthetic medium, *Appl. Environ. Microbiol.* **44,** 1318–1324.

MOREIRA, A. R., ULMER, D. C., LINDEN, J. C. (1981), Butanol toxicity in the butylic fermentation, *Biotechnol. Bioeng. Symp.* **11,** 567–579.

MORGAN, P., WATKINSON, R. J. (1989), Microbiological methods for the cleanup of soil and ground water contaminated with halogenated organic compounds, *FEMS Microbiol. Rev.* **63,** 277–300.

MORRIS, J. G. (1980), Oxygen tolerance/intolerance of anaerobic bacteria, in: *Anaerobes and Anaerobic Infections* (GOTTSCHALK, G., PFENNIG, N., WERNER, H., Eds.), pp. 7–15, Stuttgart: Gustav Fischer Verlag.

MORRIS, J. G. (1989), Bioconversions, in: *Biotechnology Handbooks,* Vol. 3: *Clostridia* (MINTON, N. P.,

CLARKE, D.J., Eds.), pp. 193–225, New York: Plenum Press.

MORRIS, J. G., O'BRIEN, R. W. (1971), Oxygen and clostridia: A review, in: *Spore Research 1971* (BARKER, A. N., GOULD, G.W., WOLF, J., Eds.), pp. 1–37, New York: Academic Press.

MORTENSEN, L. E., THORNELEY, R. N. F. (1979), Structure and function of nitrogenase. *Annu. Rev. Biochem.* **48,** 387–418.

NASUNO, S., ASAI, T. (1960), Some environmental factors affecting sporulation in butanol and butyric acid bacteria, *J. Gen. Appl. Microbiol.* **6,** 71–82.

NAUMANN, E., HIPPE, H., GOTTSCHALK, G. (1983), Betaine: New oxidant in the Stickland reaction and methanogenesis from betaine and L-alanine by a *Clostridium sporogenes-Methanosarcina barkeri* coculture, *Appl. Environ. Microbiol.* **45,** 474–483.

NOLTE, A., MAYER, F. (1989), Localization and immunological characterization of the cellulolytic enzyme system in *Clostridium thermocellum*, *FEMS Microbiol. Lett.* **61,** 65–72.

O'BRIEN, R. W., MORRIS, J. G. (1971), Oxygen and the growth and metabolism of *Clostridium acetobutylicum*, *J. Gen. Microbiol.* **68,** 307–318.

OHISA, N., YAMAGUCHI, M., KURIHARA, N. (1980), Lindane degradation by cell-free extracts of *Clostridium rectum*, *Arch. Microbiol.* **125,** 221–225.

OMELIANSKI, W. (1902), Ueber die Gärung der Cellulose, *Centralbl. Bakteriol. Parasitenkd. Infektionskrankh., II. Abt. Orig.* **8,** 225–231, 257–263, 289–294, 321–326, 353–361, 385–391.

OULTRAM, J. D., LOUGHLIN, M., SWINFIELD, T.-J., BREHM, J. K., THOMPSON, D. E., MINTON, N. P. (1988), Introduction of plasmids into whole cells of *Clostridium acetobutylicum* by electroporation, *FEMS Microbiol. Lett.* **56,** 83–88.

PADAN, E., ZILBERSTEIN, D., SCHULDINER, S. (1981), pH homeostasis in bacteria, *Biochim. Biophys. Acta* **650,** 151–166.

PAREKH, S. R., CHERYAN, M. (1991), Production of acetate by mutant strains of *Clostridium thermoaceticum*, *Appl. Microbiol. Biotechnol.* **36,** 384–387.

PARKKINEN, E. (1986), Conversion of starch into ethanol by *Clostridium thermohydrosulfuricum*, *Appl. Microbiol. Biotechnol.* **25,** 213–219.

PATNI, N. J., ALEXANDER, J. K.(1971), Catabolism of fructose and mannitol in *Clostridium thermocellum:* Presence of phosphoenolpyruvate:fructose phosphotransferase, fructose 1-phosphate kinase, phosphoenolpyruvate : mannitol phosphotransferase, and mannitol 1-phosphate dehydrogenase in cell extracts, *J. Bacteriol.* **105,** 226–231.

PERKINS, W. E. (1965), Production of clostridial spores, *J. Appl. Bacteriol.* **28,** 1–16.

RABINOWITZ, J. C. (1963), Intermediates in purine breakdown, *Methods Enzymol.* **6,** 703–713.

RAGSDALE, S. W. (1991), Enzymology of the acetyl-CoA pathway of CO_2 fixation, *Crit. Rev. Biochem. Mol. Biol.* **26,** 261–300.

REED, W. M., BOGDAN, M. E. (1985), Application of cell recycling to continuous fermentative acetic acid production, *Biotechnol. Bioeng. Symp.* **15,** 641–647.

ROBERTS, T. A. (1967), Sporulation of mesophilic clostridia, *J. Appl. Bacteriol.* **30,** 430–443.

ROFFLER, S. R., BLANCH, H. W., WILKE, C. R. (1987), Extractive fermentation of acetone and butanol: Process design and economic evaluation, *Biotechnol. Prog.* **3,** 131–140.

ROGERS, P. (1986), Genetics and biochemistry of *Clostridium* relevant to development of fermentation processes, *Adv. Appl. Microbiol.* **31,** 1–60.

ROGERS, G. M., BAECKER, A. A. W. (1991), *Clostridium xylanolyticum* sp. nov., an anaerobic xylanolytic bacterium from decayed *Pinus patula* wood chips, *Int. J. Syst. Bacteriol.* **41,** 140–143.

ROMANIEC, M. P. M., CLARKE, N. G., HAZLEWOOD, G. P. (1987), Molecular cloning of *Clostridium thermocellum* DNA and the expression of further novel β-glucanase genes in *Escherichia coli, J. Gen. Microbiol.* **133,** 1297–1307.

ROOD, J. I., COLE, S. T. (1991), Molecular genetics and pathogenesis of *Clostridium perfringens*, *Microbiol. Rev.* **55,** 621–648.

ROSENBLUM, E. D., WILSON, P. W. (1949), Fixation of isotopic nitrogen by *Clostridium*, *J. Bacteriol.* **57,** 413–414.

SAHA, B. C., MATHUPALA, S. P., ZEIKUS, J. G. (1988), Purification and characterization of a highly thermostable, novel pullulanase from *Clostridium thermohydrosulfuricum*, *Biochem. J.* **247,** 343–348.

SAHA, B. C., LAMED, R., ZEIKUS, J. G. (1989), Clostridial enzymes, in: *Biotechnology Handbooks*, Vol. 3: *Clostridia* (MINTON, N. P., CLARKE, D. J., Eds.), pp. 227–263, New York: Plenum Press.

SAVAGE, M. D., DRAKE, H. L. (1986), Carbon monoxide-dependent growth of *Clostridium thermoautotrophicum:* Role of carbon dioxide, *Abstr. Annu. Meet. Am. Soc. Microbiol.* **1986,** p. 168, I-23.

SCHÄFER, R., SCHWARTZ, A. C. (1976), Catabolism of purines in *Clostridium sticklandii*, *Zentralbl. Bakteriol. Hyg., I. Abt. Orig. A* **235,** 165–172.

SCHANTZ, E. J., JOHNSON, E. A. (1992), Properties and use of botulinum toxin and other microbial neurotoxins in medicine, *Microbiol. Rev.* **56,** 80–99.

SCHIEFER-ULLRICH, H. (1984), Harnsäureabbau im Hühnerdarm durch anaerobe Bakterien, *PhD Thesis*, University of Göttingen, Germany.

SCHIEFER-ULLRICH, H., WAGNER, R., DÜRRE, P., ANDREESEN, J. R. (1984), Comparative studies on physiology and taxonomy of obligately purinolytic clostridia, *Arch. Microbiol.* **138,** 345–353.

SCHINK, B. (1984), *Clostridium magnum* sp. nov., a

non-autotrophic homoacetogenic bacterium, *Arch. Microbiol.* **137,** 250–255.

SCHINK, B., ZEIKUS, J. G. (1983), Characterization of pectinolytic enzymes from *Clostridium thermosulfurogenes*, *FEMS Microbiol. Lett.* **17,** 295–298.

SCHLOTE, D., GOTTSCHALK, G. (1986), Effect of cell recycle on continuous butanol-acetone fermentation with *Clostridium acetobutylicum* under phosphate limitation, *Appl. Microbiol. Biotechnol.* **24,** 1–6.

SCHWARZ, W. H., SCHIMMING, K. S., STAUDENBAUER, W. L. (1987), High-level expression of *Clostridium thermocellum* cellulase genes in *Escherichia coli*, *Appl. Microbiol. Biotechnol.* **27,** 50–57.

SCOTT, A. B. (1989), Clostridial toxins as therapeutic agents, in: *Botulinum Neurotoxin and Tetanus Toxin* (SIMPSON, L. L., Ed.), pp. 399–412, San Diego: Academic Press.

SERGEANT, K., FORD, J. W. S, LONGYEAR, V. M. C. (1968), Production of *Clostridium pasteurianum* in a defined medium, *Appl. Microbiol.* **16,** 296–300.

SHONE, C. C., HAMBLETON, P. (1989), Toxigenic clostridia, in: *Biotechnology Handbooks,* Vol. 3: *Clostridia* (MINTON, N. P., CLARKE, D. J., Eds.), pp. 265–292, New York: Plenum Press.

SIMON, H. (1988), Elektro-mikrobielle und andere unkonventionelle biologische Oxidoreduktionen zur Herstellung chiraler Verbindungen, *GIT Fachz. Lab.* **5/88,** 458–465.

SIMON, H., BADER, J., GÜNTHER, H., NEUMANN, S., THANOS, J. (1984), Biohydrogenation and electromicrobial and electroenzymatic reduction methods for the preparation of chiral compounds, *Ann. N. Y. Acad. Sci.* **434,** 171–185.

SIMON, H., BADER, J., GÜNTHER, H., NEUMANN, S., THANOS, J. (1985), Chirale Verbindungen durch biokatalytische Reduktionen, *Angew. Chem.* **97,** 541–555.

SKINNER, F. A. (1960), The isolation of anaerobic cellulose-decomposing bacteria from soil, *J. Gen. Microbiol.* **22,** 539–554.

SKINNER, F. A. (1971), The isolation of soil bacteria, in: *Isolation of Anaerobes* (SHAPTON, D. A., BOARD, R. G., Eds.), pp. 57–78, London: Academic Press.

SLEAT, R., MAH, R. A. (1985), *Clostridium populeti* sp. nov., a cellulolytic species from a wood-biomass digestor, *Int. J. Syst. Bacteriol.* **35,** 160–163.

SLEAT, R., MAH, R. A., ROBINSON, R. (1984), Isolation and characterization of an anaerobic, cellulolytic bacterium, *Clostridium cellulovorans* sp. nov., *Appl. Environ. Microbiol.* **48,** 88–93.

SLIWKOWSKI, M. X., STADTMAN, T. C. (1988), Selenium-dependent glycine reductase: differences in physicochemical properties and biological activities of selenoprotein A components isolated from *Clostridium sticklandii* and *Clostridium purinolyticum*, *Biofactors* **1,** 293–296.

SMITH, L. D. S. (1992), The genus *Clostridium*-Medical, in: *The Prokaryotes* (BALOWS, A., TRÜPER, H. G., DWORKIN, M., HARDER, W., SCHLEIFER, K.-H., Eds.), 2nd Ed., pp. 1867–1878, New York: Springer-Verlag.

SOH, A. L. A., RALAMBOTIANA, H., OLLIVIER, B., PRENSIER, G., TINE, E., GARCIA, J.-L. (1991), *Clostridium thermopalmarium* sp. nov., a moderately thermophilic butyrate-producing bacterium isolated from palm wine in Senegal, *Syst. Appl. Microbiol.* **14,** 135–139.

SPECKA, U., MAYER, F., ANTRANIKIAN, G. (1991), Purification and properties of a thermoactive glucoamylase from *Clostridium thermosaccharolyticum*, *Appl. Environ. Microbiol.* **57,** 2317–2323.

SPINNLER, H. E., LAVIGNE, B., BLACHERE, H. (1986), Pectinolytic activity of *Clostridium thermocellum:* Its use for anaerobic fermentation of sugar beet pulp, *Appl. Microbiol. Biotechnol.* **23,** 434–437.

SPIVEY, M. J. (1978), The acetone/butanol fermentation, *Process Biochem.* **13,** 2–5.

STADTMAN, E. R., STADTMAN, T. C., PASTAN, I., SMITH, L. D. (1972), *Clostridium barkeri* sp. n., *J. Bacteriol.* **110,** 758–760.

SVENSSON, B. H., DUBOURGUIER, H.-C., PRENSIER, G., ZEHNDER, A. J. B. (1992), *Clostridium quinii* sp. nov., a new saccharolytic anaerobic bacterium isolated from granular sludge, *Arch. Microbiol.* **157,** 97–103.

TANNER, R. S., YANG, D. (1990), *Clostridium ljungdahlii* PETC sp. nov., a new, acetogenic, Gram-positive, anaerobic bacterium, *Abstr. Annu. Meet. Am. Soc. Microbiol.* **1990,** p. 249, R-21.

THAUER, R. K., JUNGERMANN, K., DECKER, K. (1977), Energy conservation in chemotrophic anaerobic bacteria, *Bacteriol. Rev.* **41,** 100–180.

THOLOZAN, J. L., TOUZEL, J. P., SAMAIN, E., GRIVET, J. P., PRENSIER, G., ALBAGNAC, G. (1992), *Clostridium neopropionicum* sp. nov., a strict anaerobe fermenting ethanol to propionate through acrylate pathway, *Arch. Microbiol.* **157,** 249–257.

TRAN-DINH, K., GOTTSCHALK, G. (1985), Formation of D(–)-1,2-propanediol and D(–)-lactate from glucose by *Clostridium sphenoides* under phosphate limitation, *Arch. Microbiol.* **142,** 87–92.

TRUFFAUT, N., HUBERT, J., REYSSET, G. (1989), Construction of shuttle vectors useful for transforming *Clostridium acetobutylicum*, *FEMS Microbiol. Lett.* **58,** 15–20.

TSCHECH, A., SCHINK, B. (1985), Fermentative degradation of resorcinol and resorcylic acids, *Arch. Microbiol.* **143,** 52–59.

VAREL, V. H. (1989), Reisolation and characterization of *Clostridium longisporum*, a ruminal sporeforming cellulolytic anaerobe, *Arch. Microbiol.* **152,** 209–214.

VELDKAMP, H. (1965), Enrichment cultures of prokaryotic organisms, in: *Methods in Microbiology* (NORRIS, J. R., RIBBONS, D. W., Eds.), Vol. 3a, pp. 305–361, London: Academic Press.

VERHULST, A., SEMJEN, G., MEERTS, U., JANSSEN, G., PARMENTIER, G., ASSELBERGHS, S., VAN HESPEN, H., EYSSEN, H. (1985), Biohydrogenation of linoleic acid by *Clostridium sporogenes, Clostridium bifermentans, Clostridium sordellii* and *Bacteroides* sp., *FEMS Microbiol. Ecol.* **31,** 255–259.

VIHINEN, M., MÄNTSÄLÄ, P. (1989), Microbial amylolytic enzymes, *Crit. Rev. Biochem. Mol. Biol.* **24,** 329–418.

VOGELS, G. D., VAN DER DRIFT, C. (1976), Degradation of purines and pyrimidines by microorganisms, *Bacteriol. Rev.* **40,** 403–468.

VON HUGO, H., GOTTSCHALK, G. (1974), Distribution of 1-phosphofructokinase and PEP:fructose phosphotransferase activity in clostridia, *FEBS Lett.* **46,** 106–108

WALTHER, R., HIPPE, H., GOTTSCHALK, G. (1977), Citrate, a specific substrate for the isolation of *Clostridium sphenoides, Appl. Environ. Microbiol.* **33,** 955–962.

WANG, G., WANG, D. (1984), Elucidation of growth inhibition and acetic acid production by *Clostridium thermoaceticum, Appl. Environ. Microbiol.* **47,** 294–298.

WAYMAN, M., PAREKH, R. (1987), Production of acetone-butanol by extractive fermentation using dibutylphthalate as extractant, *J. Ferment. Technol.* **65,** 295–300.

WIEGEL, J. (1980), Formation of ethanol by bacteria. A pledge for the use of extreme thermophilic anaerobic bacteria in industrial ethanol fermentation processes, *Experientia* **36,** 1431–1446.

WIEGEL, J., GARRISON, R. (1985), Utilization of methanol by *Clostridium thermoaceticum, Abstr. Annu. Meet. Am. Soc. Microbiol.* **1985,** p. 165, I–115.

WIEGEL, J., LJUNGDAHL, L. G., RAWSON, J. R. (1979), Isolation from soil and properties of the extreme thermophile *Clostridium thermohydrosulfuricum, J. Bacteriol.* **139,** 800–810.

WIEGEL, J., BRAUN, M., GOTTSCHALK, G. (1981), *Clostridium thermoautotrophicum* species novum, a thermophile producing acetate from molecular hydrogen and carbon dioxide, *Curr. Microbiol.* **5,** 255–260.

WIEGEL, J., MOTHERSHED, C. P., PULS, J. (1985), Differences in xylan degradation by various noncellulolytic thermophilic anaerobes and *Clostridium thermocellum, Appl. Environ. Microbiol.* **49,** 656–659.

WINTER, J., MOORE, L. H., DOWELL, V. R., Jr., BOKKENHEUSER, V.D. (1989), C-ring cleavage of flavonoids by human intestinal bacteria, *Appl. Environ. Microbiol.* **55,** 1203–1208.

WINTER, J., POPOFF, M. R., GRIMONT, P., BOKKENHEUSER, V. D. (1991), *Clostridium orbiscindens* sp. nov., a human intestinal bacterium capable of cleaving the flavonoid C-ring, *Int. J. Syst. Bacteriol.* **41,** 355–357.

WOOD, H. G. (1991), Life with CO or CO_2 and H_2 as a source of carbon and energy, *FASEB J.* **5,** 156–163.

WOODS, D. R. (1992), *The Clostridia and Biotechnology*, Stoneham: Butterworth Publishers, in press.

WOOLLEY, R. C., PENNOCK, A., ASHTON, R. J., DAVIES, A., YOUNG, M. (1989), Transfer of Tn*1545* and Tn*916* to *Clostridium acetobutylicum, Plasmid* **22,** 169–174.

WU, Z., DANIEL, S. L., DRAKE, H. L. (1988), Characterization of a CO-dependent O-demethylating enzyme system from the acetogen *Clostridium thermoaceticum, J. Bacteriol.* **170,** 5747–5750.

YANG, J. C., CHYNOWETH, D. P., WILLIAMS, D. S., LI, A. (1990), *Clostridium aldrichii* sp. nov., a cellulolytic mesophile inhabiting a wood-fermenting anaerobic digester, *Int. J. Syst. Bacteriol.* **40,** 268–272.

YANLING, H., YOUFANG, D., YANQUAN, L. (1991), Two cellulolytic *Clostridium* species: *Clostridium cellulosi* sp. nov. and *Clostridium cellulofermentans* sp. nov., *Int. J. Syst. Bacteriol.* **41,** 306–309.

YOON, K.-H., LEE, J.-K., KIM, B. H. (1991), Construction of a *Clostridium acetobutylicum-Escherichia coli* shuttle vector, *Biotechnol. Lett.* **13,** 1–6.

YOSHINO, S., YOSHINO, T., HARA, S., OGATA, S., HAYASHIDA, S. (1990), Construction of shuttle vector plasmid between *Clostridium acetobutylicum* and *Escherichia coli, Agric. Biol. Chem.* **54,** 437–441.

YOUNG, M., MINTON, N. P., STAUDENBAUER, W. L. (1989), Recent advances in the genetics of the clostridia, *FEMS Microbiol. Rev.* **63,** 301–326.

ZAPPE, H., JONES, D. T., WOODS, D. R. (1987), Cloning and expression of a xylanase gene from *Clostridium acetobutylicum* P262 in *Escherichia coli, Appl. Microbiol. Biotechnol.* **27,** 57–63.

10 Lactic Acid Bacteria

MICHAEL TEUBER
Zurich, Switzerland

1 Introduction 326
2 Taxonomy – Occurrence – Biotopes 328
 2.1 Evolutionary Positions of Lactic Acid Bacteria 328
 2.2 Animals and Plants as Biotopes 330
 2.3 The Genus *Lactobacillus* 330
 2.4 The Genus *Pediococcus* 339
 2.5 The Genus *Leuconostoc* 340
 2.6 The Genus *Lactococcus* 343
 2.7 The Genus *Streptococcus* 344
 2.8 The Genus *Enterococcus* 346
 2.9 The Genus *Bifidobacterium* 346
3 Isolation and Cultivation 347
4 Metabolism 350
 4.1 Carbohydrate Metabolism and Metabolic Energy 350
  4.1.1 Bioenergetics and Solute Transport 350
  4.1.2 Energetic Aspects of Carbohydrate Metabolism 351
  4.1.3 Energetic Aspects of Oxygen Metabolism and Oxygen Detoxification 352
 4.2 Biochemistry of Carbohydrate Metabolism 352
 4.3 Protein, Peptide and Amino Acid Metabolism 354
  4.3.1 The Cell Wall Protease of *Lactococcus lactis* 354
  4.3.2 The Peptidases of Lactococci 355
  4.3.3 Amino Acid Metabolism 358
  4.3.4 Amino Acid and Peptide Transport 358
  4.3.5 Bacteriocins 358
5 Genetics and Genetic Engineering 360
6 Industrial Applications 362
7 References 364

1 Introduction

Lactic acid bacteria (LAB) are a group of Gram-positive, non-spore forming, anaerobic bacteria which excrete lactic acid as the main fermentation product into the medium if supplied with suitable carbohydrates. This biochemical definition is reflected by the fact that lactic acid producing bacteria belong to different genera including *Bifidobacterium*, *Enterococcus*, *Lactobacillus*, *Lactococcus*, *Leuconostoc*, *Pediococcus* and *Streptococcus*. In phylogenetic terms, they may be as far apart as in the *Actinomyces* branch (*Bifidobacterium*) and the *Clostridium* branch (all others) of bacterial evolution (STACKEBRANDT and TEUBER, 1988).

Lactic acid bacteria are important microorganisms in the body and environment of human beings. Some colonize the mouth and the nasopharyngal mucosa (oral streptococci), the gut and intestine (bifidobacteria, enterococci, some lactobacilli) and the mucosa of the vagina (specific lactobacilli). In these biotopes, they are apathogenic, eventually useful, however, their exact functions in the human body are still unclear. A few species of the streptococci are pathogenic, especially the hemolytic streptococci, others like the enterococci may be pathogens if they penetrate to unusual places in the body causing for example endocarditis or urinary tract infections. Other lactic acid bacteria have not been with certainty identified as pathogens including the fulfillment of Koch's postulates although certain lactobacilli (e.g., *L. rhamnosus*) are occasionally observed in septicaemias in immunocompromised patients.

In the environment, lactic acid bacteria are prominent in spontaneous fermentations of organic matter of animal and plant origin which contain sufficient levels of mono- and disaccharides. Such spontaneous fermentation processes which were experienced by mankind as soon as moist organic matter was stored at ambient temperatures, developed into a series of

Tab. 1. Food and Feed Prepared With the Aid of Lactic Acid Bacteria (LAB)

Product	Raw Material	Role of LAB
Plant Origin		
Silage	Grass, maize	Homo- and heterolactic fermentation
Sauerkraut	Cabbage	Homo- and heterolactic fermentation
Pickles	Cucumber	
Olives	Olives	
Beans	Beans	
Sour dough	Rye flour	Lactic fermentation
"Weissbier"	Beer wort	Lactic fermentation
"Berliner Weisse"	Beer wort	
Wine	Grape must	Malo-lactic fermentation
Whisky	Sour mash	Lactic fermentation
Soy sauce (shoyu)	Soy beans	Lactic fermentation
Soy paste (miso)	Rice, wheat	
Animal Origin		
Sour milk	Milk	Homolactic fermentation
Sour cream		Aroma formation
Cheese		Cheese ripening
Kefir	Milk	Homo- and heterolactic fermentation
Raw sausage	Minced and	Homolactic fermentation
Raw ham	whole meat	
Fermented fish	Fish meat	Homolactic fermentation

food and feed preparation biotechnologies still in use in our days (Tab. 1; TEUBER et al., 1987). Although the use of spontaneous fermentations for food making in ancient times is mostly speculative, there is some clear archaeological evidence as early as in the 4th millenium B.C. To my knowledge, the earliest physical remains of an intact, though carbonized loaf of bread were excavated in 1975 from a Cortaillod period settlement close to Twann at lake Biel in Switzerland. An exact dating of its manufacture was possible due to the availability of unequivocal tree ring dendrochronological records to the years between 3560 and 3530 B.C. (WÄHREN, 1990). On the basis of pore structure, sizes and distribution, it was proposed that a sour dough had been used. Physical remains from bread of that time are also known from Mesopotamia and Egypt.

Even more exciting, however, was the discovery of H. J. NISSEN and his coworkers (1990) that the archaic texts from Uruk/Warka (Irak) dated from about 3200 B.C. already contain references to fermented dairy products: cheese (GA′AR), butter and yoghurt (KISIM). These texts are written on small clay tablets in archaic signs which later developed into the Cuneiform Writing. These earliest written documents of mankind which can be interpreted are documents of book keeping. In the case of the dairy products, they list the name of the shepherd, the number and kind of animals (cows, sheep, or goats), and the kind and number/ amount of products which had to be delivered by the shepherd. The fact that in some of these documents 6480 and 18120 pieces of GA′AR are mentioned, shows that these products were manufactured in huge amounts and were probably an important part of the daily diet (Fig. 1). Of course, we have no direct evidence that lactic acid bacteria have been used. However, from products still made by nomads from milk in our days in Irak, Kurdistan and Yemen (FOX, 1989), we may safely conclude that spontaneous lactic fermentations were also occurring in ancient Uruk. All later written antique documents from the Middle East and Mediterranean (Mesopotamia, Egypt, Palestine, Greece, and Rome) contain ample documents of fermented milk, meat, fish, vegetable and cereals (e.g., sour dough). Application of spontaneous fermentation of

Fig. 1. Clay tablet from Uruk/Warka (about 3200 B.C.) showing the first known sign for cheese (GA′AR in Sumerian language; NISSEN et al., 1990). The arrow on top points to the sign KISIM (for yoghurt, or butterfat?), the arrow at the right hand side to GA′AR (cheese), the arrow to the left to the amount of cheese (2×120 units in the bisexagesinal system B) and the arrow at the bottom to the name of the sheppard. The three signs between KISIM and GA′AR denote textile(s) products (made of wool?), the sign below GA′AR is at the moment discussed to stand for dung. In summary, the tablet lists the quantity of different products the named sheppard had to deliver from his herd (courtesy of Deutsches Archäologisches Institut, Bagdad).

organic matter of animal and plant origin with lactic acid bacteria for the production of particular food items is part of the cultural heritage of mankind. The exponential growth of sour dough (due to its content of microorganisms) was so well known to everybody because of daily experience in every household that it became a parable – understood by everybody – for a powerful development with either a positive or negative image, for example in the new testament of the bible: "The kingdom of heaven is like unto leaven, which a woman took, and hid in three measures of meal, till the whole was leavened" (Mat. 13,33), or "Know ye not that a little leaven leaveneth the whole lump?" (1st Cor. 5,6), and "... beware of the leaven of the Pharasees ..." (Mat. 16,6).

One indication that cheesemakers formed a guild as early as during the time of the second temple (about 150 B.C.) is the first reference to the Tyropoeon Valley ("cheesemaker" valley) in Jerusalem (in front of the Western Wall) by FLAVIUS JOSEPHUS (De Bello Judaico 5,140).

The scientific exploration of lactic fermentations started with the isolation and chemical characterization of lactic acid from fermented milk by CARL WILHELM SCHEELE (1780) in Sweden. The famous paper by LOUIS PASTEUR (1857) on the lactic fermentation not only destroyed the theory of spontaneous generation but also proved bacteria to be the cause of fermentation. In the course of experiments with heated milk to prove PASTEUR's germ hypothesis, JOSEPH LISTER (1873) obtained by chance the first bacterial pure culture. He described it as *Bacterium lactis.* This species was renamed by LÖHNIS (1909) as *Streptococcus lactis*, its accepted name is now *Lactococcus lactis* (SCHLEIFER, 1987). On the basis of PASTEUR's and LISTER's experiments, WILHELM STORCH in Copenhagen and HERMANN WEIGMANN in Kiel were the first (around 1890) to isolate the lactic acid bacteria from spontaneously fermented milk and cream that are responsible for sour milk and cheese fermentations, with the aim to produce pure cultures for the dairy industry (WEIGMANN, 1905–1908). The starter culture (WEIGMANN used the term "Säurewecker") was born.

Lactic acid bacteria also contributed to the field of genetics, biochemistry and molecular biology. GRIFFITH's work (1928) on the virulence of smooth forms and the avirulence of rough variants of *Diplococcus* (now *Streptococcus*) *pneumoniae*, lead to the discovery of genetic transformation of living bacteria with extracts from heat killed ones. The identification of the chemical nature of the transforming agent as DNA by AVERY, MACLEOD, and MCCARTHY (1944) was the beginning of molecular biology.

In biochemistry and physiology, lactic acid bacteria were important tools – starting in 1939 – to perform quantitative determinations of vitamins as exemplified by the pioneering work of SNELL (1952).

2 Taxonomy – Occurrence – Biotopes

The classical term "lactic acid bacteria" merely describes this group of microorganisms as those bacteria which produce lactic acid as one of their main fermentation products. This property alone is not sufficient for a clear taxonomic positioning as we now know. The pioneer of lactic acid bacteria taxonomy was ORLA-JENSEN (1919) who proposed a very clear cut differentiation on the basis of sugar fermentation, gas formation, (hetero- or homolactic fermentation), morphology (cocci or rods), behavior against oxygen (catalase positive or negative, microaerophilic or strictly anaerobic), and optimum growth temperature (meso- or thermophilic). His systematic survived until modern biochemistry, cell wall analysis (SCHLEIFER and KANDLER, 1972) and molecular biology (sequencing and comparison of ribosomal RNA) led to the proposals of "natural" evolutionary trees during the last 20 years.

On a species and strain level, the recent construction of DNA oligonucleotides with strain and species-specific nucleotide sequences allows the rapid identification of specific strains even in complex mixtures with other microbes, e.g., by colony hybridization (LICK and TEUBER, 1992).

However, for the sake of practical application, ORLA-JENSEN's scheme is still very valuable because it uses properties which can be easily determined in a routine bacteriological laboratory. Using the feature pairs coccus/rod and homofermentative/heterofermentative fermentation, four squares in a double cross are obtained harboring all important genera of lactic acid bacteria from the *Clostridium* branch of the Gram-positive bacteria (Fig. 2).

2.1 Evolutionary Positions of Lactic Acid Bacteria

With the exception of the bifidobacteria placed in the *Actinomyces*-branch of the Gram-positive bacteria, all other LAB are members of the *Clostridium* branch of the bac-

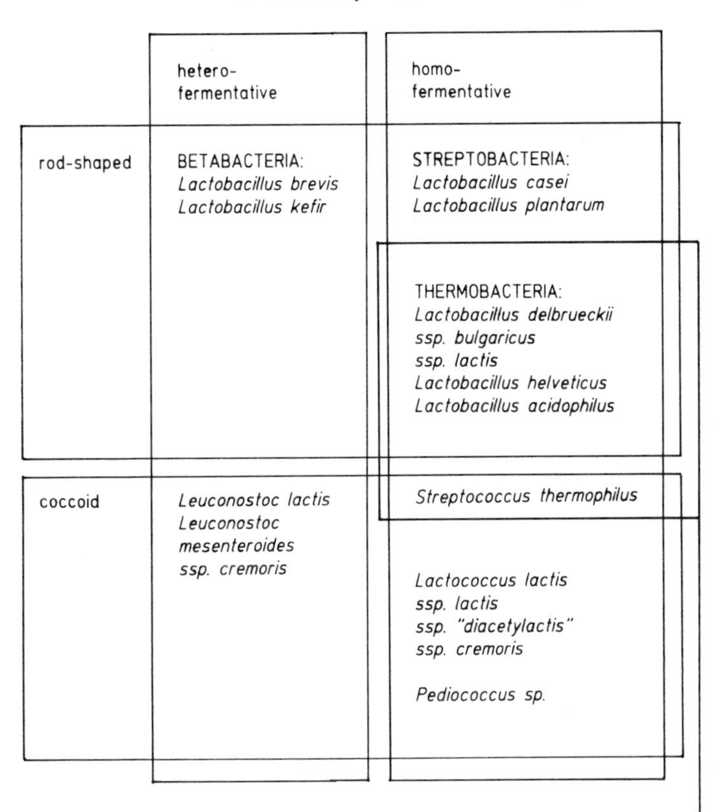

Fig. 2. Functional classification of lactic acid bacteria on the basis of morphology and basic carbohydrate catabolic pathways. Thermophilic bacteria are grouped in addition. The terms Betabacteria, Streptobacteria and Thermobacteria were devised by ORLA-JENSEN (1919).

terial evolution tree. On the basis of oligonucleotide similarities of 16S ribosomal RNA, a schematic tree was constructed (STACKEBRANDT and TEUBER, 1988; Fig. 3). Whereas the genus *Clostridium*, being strictly anaerobic, is the oldest part of the branch, the genera *Streptococcus*, *Lactococcus*, *Lactobacillus*, *Leuconostoc* and *Pediococcus* branch off at about the same S_{AB}-value (similarity), the youngest part being the *Enterococcus–Listeria–Bacillus–Staphylococcus* cluster.

Out of these genera, only the genus *Enterococcus* is regarded to belong to the LAB. This part comprises organisms possessing true respiratory chains like *Bacillus*, or organisms containing rudimentary electron transfer chains like *Enterococcus.* Anyhow, it is tempting to speculate that this "aerobic" part of the branch has developed after the accumulation of oxygen in the earth's atmosphere starting about $2 \cdot 10^9$ years ago.

In contrast to *Clostridium*, most leuconostocs, lactobacilli, pediococci, lactococci and streptococci are not strictly anaerobic, but can grow more or less abundantly in the presence of oxygen at low concentrations (see below). What is most noteworthy from this evolutionary tree is the fact that the species used for the biotechnology of food and fodder are very clearly separated from the pathogenic species. On the other hand, the enterococci which are not really universally accepted as starter culture, are the "nearest" relatives of the pathogenic listerias. The intimate relationship of mankind with lactic acid bacteria consumed with food and growing on and in the human body speaks for a co-evolution guaranteeing a useful coexistence for both sides. By aquiring certain pathogenicity determinants, however, certain streptococci have become powerful, life-threatening pathogens.

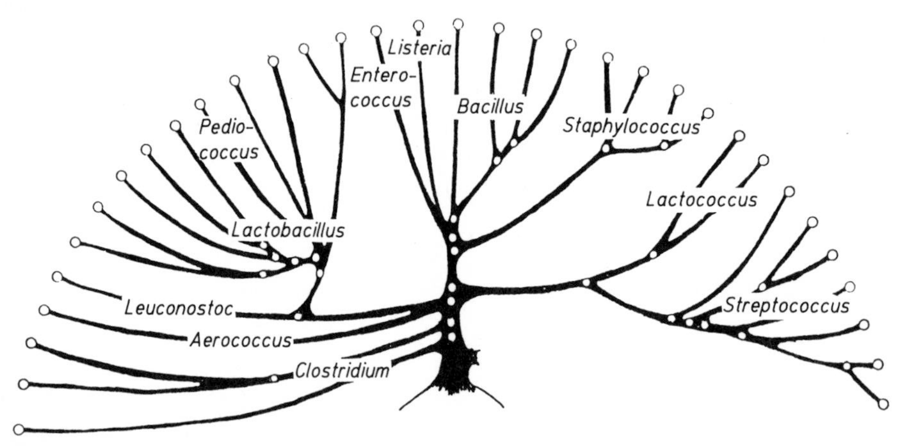

Fig. 3. Evolutionary tree of the *Clostridium* branch of Gram-positive bacteria. The tree has been drawn on the basis of 16s rRNA oligonucleotide sequence similarities (STACKEBRANDT and TEUBER, 1988).

2.2 Animals and Plants as Biotopes

All lactic acid bacteria are typical inhabitants of plants and animals. Accordingly, they are nutritionally fastidious. They are adapted to be supplied by their host environments with carbohydrates, amino acids, vitamins, nucleic acid derivatives, fatty acids and others. They can survive and grow at low pH values (down to about 3). In Sect. 3 on isolation and cultivation, it will become evident that many of the just named substrates have to be added to the growth medium in order to keep lactic acid bacteria in the laboratory. The same is true for substrates applied in the biotechnology of LAB. The dependence of LAB on many different growth factors has for a long time severely inhibited their genetic analysis since defined minimal media are difficult to devise, and auxotrophic mutants are hard to get and maintain. The following sections on habitats and biotopes of LAB are ordered on the basis of the different genera.

2.3 The Genus *Lactobacillus*

(HAMMES et al., 1991)

Lactobacilli demand carbohydrates, protein breakdown products, vitamins, and usually a low oxygen tension. Their main product of fermentation is lactic acid. A broad diversity of habitats is evident from the following discussions.

Oral cavity

Lactobacilli constitute only about 1% of the total microbes in a healthy human mouth. In children, the predominant species are *Lactobacillus casei*, *L. rhamnosus*, *L. fermentum*, occasionally *L. acidophilus*, *L. brevis*, *L. buchneri*, and *L. plantarum*. *L. salivarius*, although representing only 2% of all oral lactobacilli seems to have its specific habitat in the mouth. The prevalent *Lactobacillus* in plaques is *L. casei*, but it is not able to form plaques alone (i.e., without *Streptococcus mutans* or *S. sanguis*). It is believed that the high number of *L. casei* found in carious cavities may be the result, but not the cause of caries.

Intestinal tract

The gastro-intestine of new born children and animals is free of microbes. During birth and in the following days the new-born is contaminated with bacteria from the vagina, the rectum, the hands, the mouth and the breast of the mother with a diversity of microorganisms. After an initial growth of Gram-negative bacteria and clostridia, lactic acid bacteria start to dominate and to displace the initial flora. Bifidobacteria constitute the dominating flora, especially in breast fed infants, however *L. ac-*

idophilus may be equally dominant, e.g., in bottle fed infants (both species form up to 10^{10} viable counts per gram of feces). In adults, lactobacilli may not be more than 1% of the total fecal flora. In the stomach (about 10^3–10^5/mL), and the small intestine (about 10^5 viable counts/mL), however, lactobacilli are predominant including in addition to *Lactobacillus acidophilus* also the species *L. fermentum*, *L. reuteri*, *L. lactis* and *L. salivarius*. *L. reuteri*, *L. salivarius*, *L. delbrueckii* and *L. fermentum* are also dominant in the rumen of young calves, the oesophagus and stomach of piglets, and the crop of chicken and other birds. These lactobacilli thickly colonize the epithelial cells of the named organs, and constitute a constant reservoir from which the farther gastro-intestinal tract is colonized. Many of these lactobacilli produce potent bacteriocins (e.g., reuterin, acidophilin) providing a selective advantage. In addition, they possess lectin-like surface structures (adhesins) which enable them to adhere to the surface of epithelial cells. Feeding people with lactobacilli in the diet does lead to a transient but not a permanent colonization of the intestine.

For this reason, health claims regarding lactic acid bacteria in the diet (bifidobacteria, *L. acidophilus*, *L. casei*) have still to be treated with care and reservation (GURR, 1983). A role in the ecology of the stomach and small intestine is nevertheless quite reasonable. All the other claims of the market (e.g., lowering of the cholesterol level of blood, stimulation of the immunoregulatory system, antitumor activities, prevention of diarrhoe by enteropathogenic enteric bacteria) have never been proved.

Human vagina

The human vagina has a pH value of <4.5. This low value is important to keep out undesirable microorganisms. The low pH value is clearly attributed to the presence of lactobacilli (10^5–10^7 per mL vaginal fluid), the dominating species being *L. acidophilus*, *L. reuteri* and *L. casei* ssp. *rhamnosus*. It seems that most *L. acidophilus* isolates, however, have been inadequately characterized. They have been reclassified as *L. gasseri* or *L. crispatus*. Most women only have one species in the vaginal flora, some have more than one. The substrate for the lactic fermentation is glycogen which can be split by some *L. acidophilus* and *L. crispatus* strains. Part of glycogen is probably also degraded by tissue cells and other microbes. The protective action of the lactobacilli may be further enhanced by the likely production of hydrogen peroxide (see Sect. 4.1.2).

Pathogenicity of lactobacilli

There only are 20 to 30 reports that lactobacilli are potentially pathogenic for man. Occasionally, *L. rhamnosus* has been reported from persons with bacterial endocarditis, systemic septicemia (sometimes in HIV positive individuals) and abscesses. Several other lactobacilli have been isolated from opportunistic infections: *L. gasseri*, *L. acidophilus*, *L. plantarum* and *L. salivarius*. However, KOCH's postulates have not been fulfilled. Therefore, the genus *Lactobacillus* is generally recognized to be non-pathogenic.

Substrates of animal origin

Two main substrates of animal origin (besides eggs) are used in enormous quantities for human consumption: meat and milk. If animal bodies contain lactobacilli as just described, they can be expected to contaminate products made from them. It then depends on the further handling and storage conditions whether lactobacilli will finally be the dominating majority of the developing microflora.

Meat and meat products

The numbers of lactobacilli in fresh meat and meat products are hardly above 10^3 per g or 10^2 per cm^2 of surface area. Modern packaging technologies use air tight plastic sheets which allow wrapping of meat under vacuum or with a modified atmosphere (N_2/CO_2). If the product is stored at low temperatures, the growth of the usual putrefactive meat flora (Gram-negative enterobacteria and pseudomonads) is inhibited. However, the development of psychrotrophic lactobacilli, leuconostocs, corynebacteria and *Brochothrix thermosphacta* is induced, leading to slime formation, souring, off-odor, and greening. A thorough analysis of the lactobacilli of meat and meat products by REUTER (1975) established *Lactobacillus curvatus* and *L. sake* as the main species, complemented by *L. plantarum*, *L. casei*,

L. farciminis, *L. alimentarius*, *L. brevis* and *L. halotolerans*. A typical *Lactobacillus* defect is greening. It is caused by two mechanisms:

(1) The lactobacilli form H_2O_2 which converts hemoglobin into the green choleglobin.
(2) The lactobacilli especially in vacuum packed products form H_2S which yields the green sulfomyoglobin. Lactobacilli in spoiled meat products may be as numerous as 10^7–10^8 viable counts per gram, e.g., if sugar has been added to the product.

For fermented sausages (salami) or hams made from raw meat with the addition of salt and nitrate/nitrite, lactobacilli play an indispensable part in the ripening process together with certain micrococci, staphylococci, and pediococci. The dominating species are *L. curvatus* and *L. sake*, with *L. plantarum*, *L. alimentarius* and *L. farciminis* as minor components. Typically, sugar is added to the minced meat as a substrate for the lactobacilli. The resulting lactic fermentation lowers the pH to values inhibitory to spoilage bacteria. Starter cultures containing proper lactobacilli are commercially available (see below).

Milk and fermented milk products (TEUBER, 1987)

Freshly collected milk contains lactobacilli in the order of 10^2 to 10^3 per mL. The species identified are *Lactobacillus casei*, *L. plantarum*, *L. brevis*, *L. coryneformis*, *L. curvatus*, *L. buchneri*, *L. lactis*, and *L. fermentum*. All are killed by pasteurization and therefore do not cause spoilage in products made from pasteurized milk. All our fermented dairy products have evolved over thousands of years from spontaneous fermentations (see introduction) which means that these are important biotopes for lactic acid bacteria. Lactobacilli are prominent in yoghurt where *L. delbrueckii* ssp. *bulgaricus* ferments the milk in collaboration with *Streptococcus thermophilus* at temperatures of 42 to 45 °C. The synergism of the two species is explained by the formation of peptides by the proteolytically active *L. bulgaricus* which aids the proteolytically inactive *Streptococcus*. This species in turn provides formic acid (around 10 ppm) which stimulates the *Lactobacillus*. The dominant aroma compound of yoghurt is acetaldehyde which is formed from threonine by a specific aldolase by *L. bulgaricus*. Yoghurt is a typical spontaneously fermented milk product obtained after the milk has been left at temperatures around 40 °C.

Another fermented milk containing lactobacilli is kefir. The kefir grain contains yeasts (*Candida kefir*), the lactobacilli *Lactobacillus kefir* and *L. kefiranofaciens*, and mesophilic lactococci and leuconostocs, eventually acetic acid bacteria. *L. kefiranofaciens* produces the glycocalyx which keeps the kefir grain together. Fig. 4 shows a scanning electron micrograph of the surface of a kefir grain with all components of its complex microflora.

In cheese making, lactobacilli are important for the classical hard cheeses (Emmental, Gruyère, Grana, Parmesan). In the production

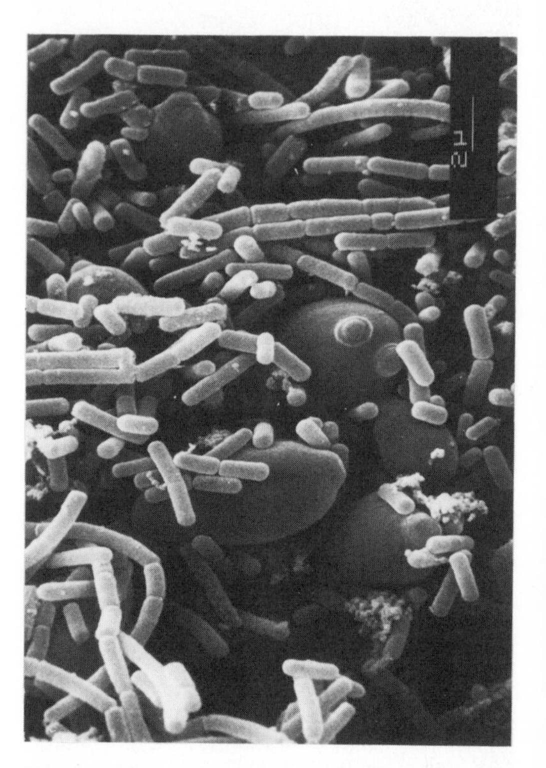

Fig. 4. Surface of kefir grain as seen by a scanning electron microscope. The typical members of the mixed microflora are easily identified: yeasts, lactobacilli and lactococci (courtesy of Dr. H. NEVE, Kiel).

of these cheeses, the cheese curd is heated up to 50–55 °C during processing with the consequence that thermophilic lactobacilli survive and perform the lactic fermentation during cooling down (Fig. 5). Together with *Streptococcus thermophilus*, this fermentation is due to *L. helveticus*, *L. delbrueckii* ssp. *lactis* and ssp. *bulgaricus*. During ripening, mesophilic lactobacilli like *L. casei*, *L. plantarum*, *L. brevis* and *L. buchneri* may dominate with numbers of 10^6–10^8 per gram.

L. acidophilus has recently been used to augment classical starter cultures with the aim to provide the consumer with new bacteria for their intestinal microflora. The corresponding health claims are still questionable (see above).

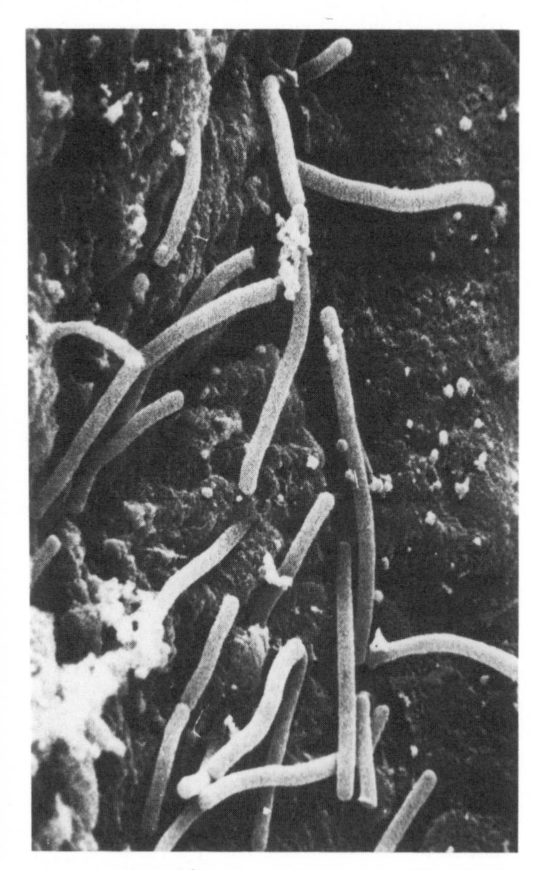

Fig. 5. Scanning electron micrograph of lactobacilli (*Lactobacillus helveticus*) on the surface of a cheese hole (courtesy of Dr. H. NEVE, Kiel).

Substrates of plant origin

Lactic acid bacteria (LAB) are found on plant leaves in numbers of 10^1 to 10^3 per gram accounting for not more than 0.01 to 1% of the total microflora. Leuconostocs predominate (80%). Lactobacilli of the following species make up 10% of the recovered LAB: *Lactobacillus plantarum*, *L. fermentum*, occasionally *L. brevis*, *L. casei*, *L. vividescens*, *L. cellobiosus*, and *L. salivarius*. In the plant rhizosphere, again *L. plantarum*, *L. brevis* and *L. fermentum* are most frequently isolated LAB.

Fermented plant materials as food and fodder

Silage made from grass and maize is the most abundant product made by spontaneous fermentation with LAB. In cut grass filled into silos or piled under air tight plastic foils, a rapid lactic fermentation takes place. Such fermentations are characterized by a typical succession of different types of microorganisms: within one day or two, aerobic bacteria like coliforms use up the available oxygen. The lactic fermentation is started by leuconostocs and enterococci, followed by pediococci and lactobacilli including *L. plantarum*, *L. casei*, *L. brevis*, *L. buchneri*, *L. coryneformis*, *L. curvatus*, *L. fermentum*, *L. acidophilus*, *L. salvarius* and *L. graminis*. pH values depending on available carbohydrates may be as low as 2.8 to 3, lactic acid concentrations around 2% (WOOLFORD, 1984). In sauerkraut made from shredded cabbage with the addition of sodium chloride, a very similar succession of microbial species during fermentation is evident, the lactobacilli including *L. brevis*, followed by *L. plantarum*, *L. curvatus*, and *L. sake*, minor components being *L. paracasei* and *L. bavaricus*. In pickle fermentations, the microbial succession is *Leuconostoc mesenteroides*, *Enterococcus faecalis*, *Pediococcus cerevisiae* (now *damnosus*), *L. brevis*, and *L. plantarum*. The enterococci probably have their origin from manure.

Another important biotope of lactobacilli is sour dough for breadmaking, a spontaneous process in use for thousands of years. Besides yeasts, homo- and heterofermentative lactobacilli are predominant. *Lactobacillus delbrueckii*, *L. acidophilus*, *L. plantarum*, *L. casei*, *L. farciminis*, *L. homohiochii*, *L. brevis*, *L. buchneri*, *L. fermentum*, *L. hilgardii*, *L. sanfrancisco*, and

L. viridescens. There is no doubt that *L. sanfrancisco* and certain *L. brevis* strains are the species contributing mainly to the sour dough fermentation (SPICHER and STEPHAN, 1987).

In soil, water, sewage and manure, lactobacilli are present if decaying organic material is available. The flora is then determined by the original flora of the decaying material. In sewage, lactobacilli numbers between 10^4 and 10^5 per mL. *L. vaccinostercus* have only been found in cow's dung.

Lactobacilli play definite roles in the fermentation and spoilage of fermented beverages (wine, beer, cider, whisky, etc.).

Wine

Although lactobacilli are rarely detected on grapes and in grape must, they are common in wine despite high ethanol, organic acid (pH 3.2–3.8) and SO_2 concentrations. As a matter of fact, in spontaneous fermentations of grape must relying on the autochthonous must microflora there is always a danger that lactic fermentation by lactic acid bacteria could proceed faster than the alcoholic fermentation by "wild" yeasts. One of the reasons for the traditional addition of SO_2 (since Roman times) is the initial inhibition of LAB, in addition to control of acetic acid bacteria. With the use of modern pure culture technology, this problem has more or less vanished. The winery and its equipment are probably the source of the main lactobacilli in wine. The substrates available for an energy yielding fermentation are organic acids, especially malic, citric, and tartaric acid in wine, and glucose and other fermentable sugars in must before the alcoholic fermentation. The malo-lactic fermentation trans- forming the strong or hard tasting malic acid into the soft tasting lactic acid is of prime importance for many top quality wines. In addition to leuconostocs and pediococci, the following lactobacilli may participate in malo-lactic fermentations: *Lactobacillus plantarum*, *L. casei*, *L. brevis*, *L. buchneri*, *L. hilgardii*, *L. trichodes*, *L. fructivorans*, *L. desidiosus*, and *L. mali.* The degradation of tartaric acid by *L. plantarum* or *L. brevis* in low acid wine is as detrimental as the transformation of citric acid into diacetyl by a variety of lactobacilli. Excess formation of mannitol from fructose may cause bitterness.

Apple cider

As in wineries, the apple cider plants harbor an indigenous *Lactobacillus* flora. Again, the malo-lactic fermentation will improve the flavor of cider. In contrast, the metabolization of fructose and quinic acid which is abundant in apple must, yields acetate, CO_2 and dihydroshikimate. Especially acetate is detrimental to the flavor of cider. The main lactobacilli isolated from cider are *L. brevis*, *L. plantarum* and *L. mali.* Strains producing extracellular polysaccharides from glucose, fructose or maltose may cause ropiness.

Beer

Maltose fermenting lactobacilli are used in breweries in sour wort fermentations run at 48–50 °C. 1–2 % of this product is added to normal wort, yielding a pH drop of 0.2– 0.4 units, higher enzyme activities, better separation of proteins during wort boiling, an improved beer quality, more mellow in taste, lighter in color, and a higher foam stability. Strains used include *Lactobacillus delbrueckii* ssp. *delbrueckii*, *L. delbrueckii* ssp. *lactis*, *L. amylovorus*, *L. fermentum*, *L. rhamnosus,* and *L. helveticus.* For the mixed fermentation in the preparation of "Berliner Weisse", *L. brevis* is applied as starter culture in addition to *Saccharomyces cerevisiae.*

A broad range of LAB are able to grow in finished beer causing turbidity and flavor defects: *L. brevis*, *L. casei*, *L. coryneformis*, *L. curvatus*, *L. plantarum*, *L. buchneri* besides *Pediococcus damnosus* (previously *cerevisiae*). These LAB are adapted to low pH values around 3.8– 4.3, and especially resistant to isohumulones, the hop residues in wort, which normally have an appreciable antibiotic potency against Gram-positive bacteria (SCHMALRECK et al., 1975).

Grain and fruit mashes

In these mashes undergoing an alcoholic fermentation to yield the raw material for distilled alcoholic beverages (e.g., Whisky, "Geist"), lactobacilli can cause fermentation losses and flavor problems if they grow up to high numbers. Commonly, *L. plantarum*, *L. suebicus*, *L. brevis*, *L. hilgardii*, *L. fermentum*, *L. casei,* and *L. delbrueckii* are isolated.

Sugar processing

Sugar cane and cane syrup (about 15% sucrose) are very sensitive to lactic fermentation caused by specific osmotolerant lactobacilli comprising *Lactobacillus confusus*, *L. plantarum* and *L. casei*. In addition to souring, these strains produce large amounts of dextran. A similar spoilage may occur in sugar beet juice by strains of *L. casei*, *L. plantarum*, *L. cellobiosus* and *L. fermentum*.

Spoilage of sour (acetic acid) food preserves

A species which can tolerate 4–5% acetic acid at pH 3.5 has been detected in fermenting rice vinegar and named *Lactobacillus acetotolerans*. In mayonnaise, dressings and salads preserved by acetic acid at pH values around 3.7 to 3.8, the following six species have been recorded in spoiled samples: *L. plantarum*, *L. buchneri*, *L. brevis*, *L. delbrueckii*, *L. casei*, and *L. fructivorans*. These strains are commonly also resistant to added benzoic and sorbic acid. In marinated fish, lactobacilli produce gas by fermentation of carbohydrate and decarboxylation of amino acids with biogenic amines as byproducts: γ-aminobutyric acid, cadaverine, tyramine, and histamine. The responsible species are *L. brevis*, *L. buchneri*, *L. fermentum*, *L. delbrueckii*, *L. plantarum* and *L. casei*.

The known species of lactobacilli are listed in Tabs. 2 to 4, put into the physiological categories:

- Obligately homofermentative: more than 85% lactic acid from glucose via the Embden–Meyerhof pathway; pentoses and gluconate are not fermented (lack of phosphoketolase).
- Facultatively heterofermentative: 85% lactic acid from glucose by Embden–Meyerhof pathway; pentoses and gluconate are fermented via pentose–phosphate pathway.
- Obligately heterofermentative: formation of equimolar amounts of CO_2, lactic acid, and acetic acid and/or ethanol.

As explained in Sect. 4, this ranking is valid for metabolism under anaerobic conditions. The morphology of some typical lactobacilli is shown in Fig. 6.

Some atypical lactobacilli isolated from refrigerated meat and chicken have been put into the new genus *Carnobacterium*, comprising the species *C. divergens*, *C. piscicola*, *C. gallinarum*, and *C. mobile* (HAMMES et al., 1991). The most important habitats seem to be unprocessed, refrigerated red meat. In evolutionary terms, they branch off the *Enterococcus–Bacillus* cluster of the *Clostridium* branch of Gram-positive bacteria. Isolation, cultivation, and identification are described in detail by HAMMES et al. (1991).

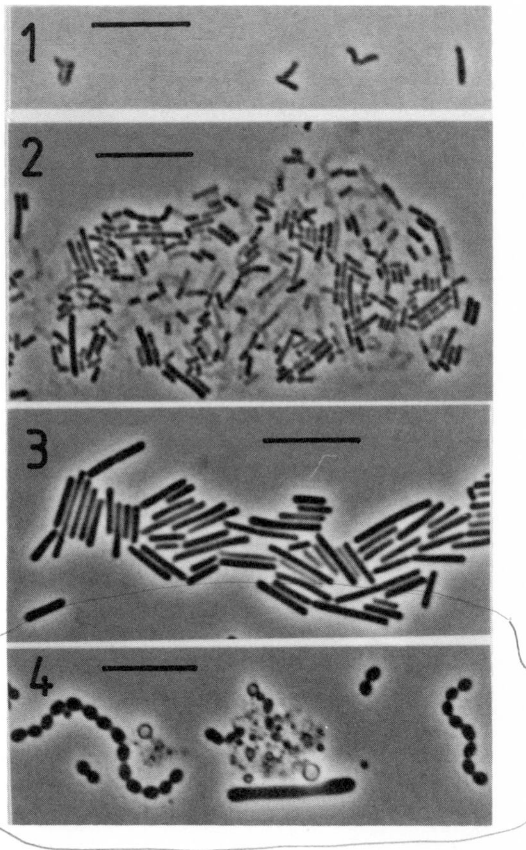

Fig. 6. Phase contrast microscopy of some important lactic acid bacteria. **1** *Bifidobacterium* sp., **2** *L. brevis*, **3** *L. helveticus*, **4** *L. delbrueckii* ssp. *bulgaricus* and *Streptococcus thermophilus* in yoghurt. The bar represents 10 μm.

Tab. 2. Key Characteristics of the Obligately Homofermentative *Lactobacillus* Species (modified from HAMMES et al., 1991)

Species	Lactic Acid Isomer(s)	Growth at 15 °C	NH_3 from Arginine	Carbohydrate Fermented											
				Amygdalin	Cellobiose	Galactose	Lactose	Maltose	Mannitol	Mannose	Melibiose	Raffinose	Salicin	Sucrose	Trehalose
L. delbrueckii ssp. *delbrueckii*	D	–	d	–	d	–	–	d	–	+	–	–	–	+	d
L. delbrueckii ssp. *lactis*	D	–	d	+	d	d	+	+	–	+	–	–	+	+	+
L. delbrueckii ssp. *bulgaricus*	D	–	–	–	–	–	+	–	–	–	–	–	–	–	–
L acidophilus	DL	–	–	+	+	+	+	+	–	+	d	d	+	+	d
L. amylophilus	L	+	ND	–	–	+	–	+	–	+	–	–	–	–	–
L. amylovorus	DL	–	ND	+	+	+	–	+	–	+	–	–	+	+	+
L. animalis	L	–	–	d	+	+	+	+	–	+	+	+	+	+	–
L. aviarius ssp. *araffinosus*	L(D)	–	ND	d	d	–	–	+	–	+	–	–	d	+	+
L. aviarius ssp. *aviarius*	DL	–	ND	d	+	d	d	+	–	+	d	+	+	+	+
L. crispatus	DL	–	–	+	+	+	+	+	–	+	–	–	+	+	–
L. farciminis	L(D)	+	+	+	+	+	+	+	–	+	–	–	+	+	+
L. gasseri	DL	–	–	+	+	+	d	d	–	+	d	d	+	+	d
L. hamsteri	DL	–	ND	+	+	+	+	+	–	+	+	+	+	+	d
L. helveticus	DL	–	–	–	–	+	+	d	–	d	–	–	–	–	d
L. jensenii	D	–	+	+	+	+	–	d	d	+	–	–	+	+	+
L. kefiranofaciens	D(L)	–	ND	–	–	+	+	+	–	ND	+	+	–	+	–
L. mali	L	+	–	+	d	d	–	–	–	+	–	–	+	+	+
L. ruminis	L	–	–	+	+	+	d	+	–	+	+	+	+	+	–
L. salivarius	L	–	–	–	–	+	+	+	+	–	+	+	d	+	+
L. sharpeae	L	+	–	+	+	+	+	+	–	+	–	–	+	–	–
L. vitulinus	D	–	–	+	+	+	+	+	–	+	+	+	+	+	d

Symbols: +, 90% or more of strains are positive; –, 90% or more of strains are negative; d, 11–89% of strains are positive; ND, no data available; parenthesized isomers indicate <15% of total lactic acid

Tab. 3. Key Characteristics of the Facultative Heterofermentative *Lactobacillus* Species (modified from HAMMES et al., 1991)

Species	Lactic Acid Isomer(s)	Growth at 15 °C	Carbohydrate Fermented												
			Amygdalin	Arabinose	Cellobiose	Esculin	Gluconate	Mannitol	Melezitose	Melibiose	Raffinose	Ribose	Sorbitol	Sucrose	Xylose
L. acetotolerans I	DL	–	–	–	–	w	–	+	–	–	–	+	–	–	–
L. acetotolerans II	D(L)	–	–	–	+	+	–	–	–	–	–	–	–	–	–
L. agilis	L	–	+	–	+	+	–	+	+	+	+	+	d	+	–
L. alimentarius	L(D)	+	ND	d	+	+	+	–	–	–	–	+	–	+	–
L. bavaricus	L	+	–	–	+	+	+	–	–	+	–	+	–	+	–
L. casei	L	+	+	–	+	+	+	+	+	–	–	+	+	+	–
L. coryniformis ssp. *coryniformis*	D(L)	+	–	–	–	d	+	+	–	d	d	–	d	+	–
L. coryniformis ssp. *torquens*	D	+	–	–	–	–	+	+	–	–	–	–	–	+	–
L. curvatus	DL	+	–	–	+	+	+	–	–	–	–	+	–	d	–
L. graminis	DL	+	+	–	+	+	–	–	–	–	–	–	–	+	+
L. homohiochii	DL	+	–	–	d	ND	–	d	–	–	–	d	–	–	–
L. maltaromicus [a]	L	+	+	–	+	ND	ND	+	+	+	–	+	+	+	–
L. murinus	L	–	d	+	+	+	–	d	–	+	+	+	–	+	–
L. paracasei ssp. *paracasei*	L	+	+	–	+	+	+	+	+	–	–	+	d	+	–
L. paracasei ssp. *tolerans*	L	+	–	–	–	–	w	–	–	–	–	–	–	–	–
L. pentosus	DL	+	+	+	+	ND	+	+	d	+	+	+	+	+	+
L. plantarum	DL	+	+	d	+	+	+	+	+	+	+	+	+	+	d
L. rhamnosus	L	+	+	d	+	+	+	+	+	–	–	+	+	+	–
L. sake	DL	+	+	+	+	+	+	+	–	+	–	+	–	+	–

For symbols see Tab. 2
[a] is now grouped into the new genus *Carnobacterium*

Tab. 4. Key Physiological Characteristics of the Obligately Heterofermentative *Lactobacillus* Species (modified from HAMMES et al., 1991)

Species	Growth at 15 °C	NH_3 from Arginine	Carbohydrate Fermented[a]												
			Arabinose	Cellobiose	Esculin	Galactose	Maltose	Mannose	Melezitose	Melibiose	Raffinose	Ribose	Sucrose	Terhalose	Xylose
L. bifermentans	+	–	–	–	–	+	+	+	–	–	–	+	–	–	–
L. brevis	+	+	+	–	d	d	+	–	–	+	d	+	d	–	d
L. buchneri	+	+	+	–	d	d	+	–	+	+	d	+	d	–	d
L. collinoides	+	+	+	–	+	+	+	–	+	+	–	+	–	–	+
L. confusus	+	+	–	+	+	+	+	+	–	–	–	+	+	–	+
L. fermentum	–	+	d	d	–	+	+	w	–	+	+	+	+	d	d
L. fructivorans	+	+	–	–	–	–	d	–	–	–	–	w	d	–	–
L. fructosus	+	–	–	–	ND	–	–	–	–	–	–	–	–	–	–
L. halotolerans	+	+	–	–	–	–	+	+	–	–	–	+	–	+	–
L. hilgardii	+	+	–	–	–	d	+	–	d	–	–	+	d	–	+
L. kandleri	+	+	–	–	–	+	–	–	–	–	–	+	–	–	–
L. kefir	+	+	d	–	–	–	+	–	–	+	–	+	–	–	–
L. malefermentans	+	+	–	–	–	–	+	–	–	–	–	+	–	–	–
L. minor	+	+	–	+	+	–	+	+	+	–	–	+	+	+	–
L. oris	–	–	+	d	d	+	+	d	–	+	+	+	+	d	+
L. parabuchneri	+	+	+	–	–	+	+	ND	+	+	+	+	+	–	–
L. reuteri	–	+	+	–	ND	+	+	–	–	+	+	+	+	–	–
L. sanfrancisco	+	–	–	–	ND	+	+	–	–	–	–	–	–	–	–
L. suebicus	+	ND	+	d	–	+	+	ND	–	d	–	+	d	–	+
L. vaginalis	–	ND	–	–	d	+	+	+	–	+	+	d	+	–	–
L. vaccinostercus	–	–	+	w	–	w	+	–	–	–	–	+	–	–	+
L. viridescens	+	–	–	–	–	–	+	+	–	–	–	–	d	d	–

For symbols, see Tab. 2

[a] The following sugars are generally fermented: fructose (exceptions: *L. malefermentans*, *L. sanfrancisco*, and *L. vaccinostercus*) and glucose. The following sugars are generally not fermented: amygdalin (exceptions: *L. confusus* and *L. oris*), mannitol (exceptions: *L. bifermentans* and *L. kandleri*), rhamnose (exception: *L. bifermentans*), and sorbitol.

2.4 The Genus *Pediococcus*
(WEISS, 1991)

This genus is characterized by a typical morphological appearance, the formation of tetrads in one plane, see Fig. 7 (2) and (3). Taxonomically it is positioned within the *Lactobacillus* branch. Is it a genus which has lost the ability to form rod-shaped cells, but otherwise a *Lactobacillus*? For historical reasons, the existing nomenclature must be kept. Further characteristics are their homofermentative metabolism and tolerance to high osmotic values, combined with tolerance to low pH values. Their main characteristics are listed in Tab. 5.

Habitats

Organisms described as pediococci ("beer sarcina") have been studied for a long time mainly in relation to problems in the brewing industry. These microorganisms are potentially dangerous beer spoilers in that they are microaerophilic and relatively tolerant of alcohol and antibiotic hop resins.

Spoilage of beer by *Pediococcus damnosus* (previously *cerevisiae*) is apparent by the secretion of voluminous capsular material which may turn beer viscous and ropy. Pediococcal contamination as a cause of diacetyl formation in beer has also been reported. In modern brewing, the *Pediococcus* problem has vanished due to advanced food technology, especially cleaning in place (CIP) techniques for the fermentation equipment, postfermentation-pasteurization, and aseptic filling.

Slime-forming strains of *P. damnosus* have been isolated from outbreaks of ropiness in cider plants. Pediococci are commonly found in a great variety of fermenting plant materials such as silage, sauerkraut, cucumbers and olives. These microorganisms have also been detected in proteinaceous foods such as fresh and cured meat, raw sausages, fresh and marinated fish and poultry as well as in cheese.

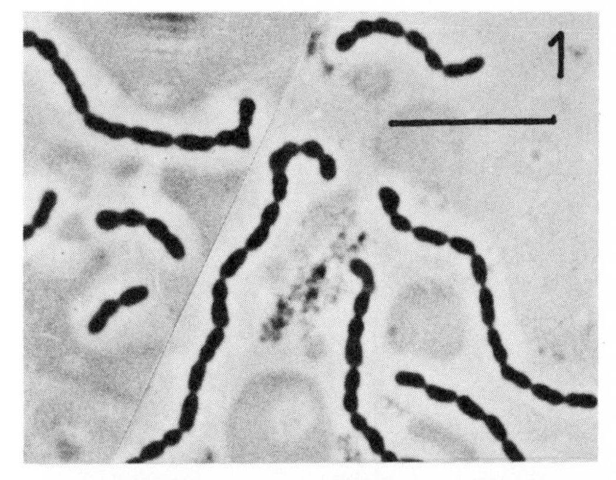

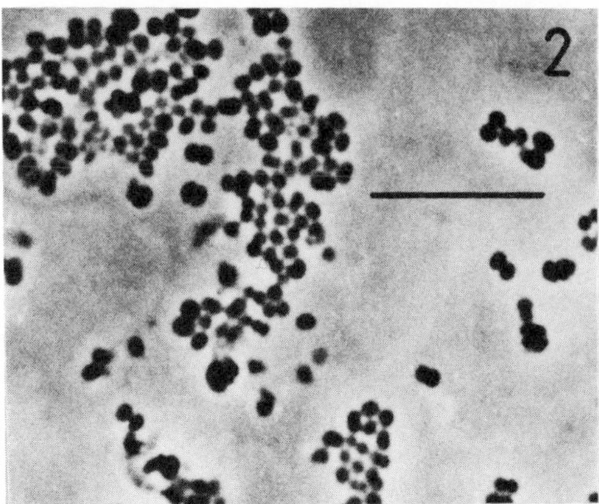

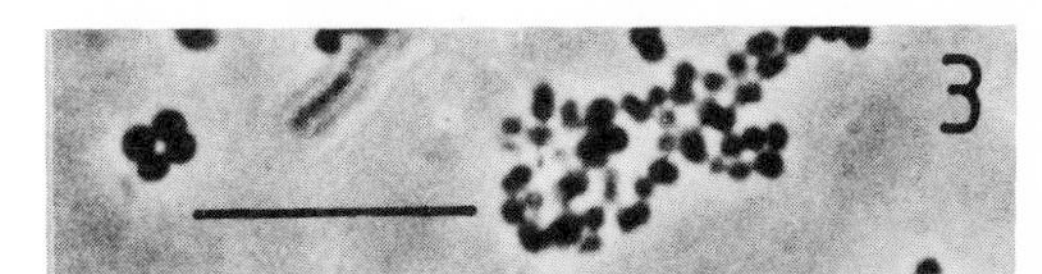

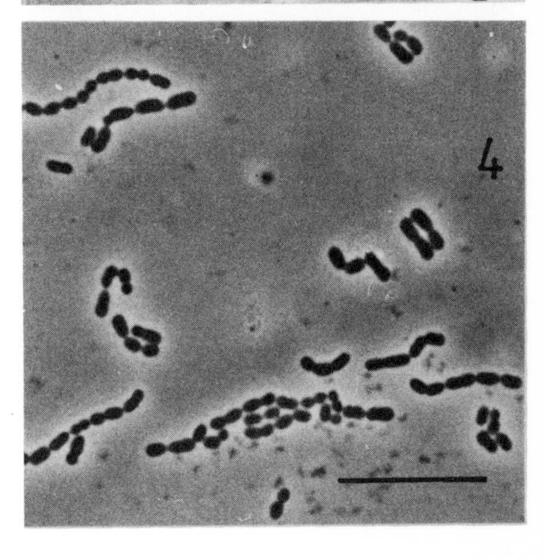

Fig. 7. Phase contrast microscopy of coccoid lactic acid bacteria. **1** *Leuconostoc mesenteroides* with visible, extracellular slime matrix, **2** *Pediococcus damnosus* (*cerevisiae*), **3** *P. dextrinicus*, **4** *Lactococcus lactis*. The bar represents 10 μm.

Tab. 5. Characteristics of *Pediococcus* Species (modified from WEISS, 1991)

Species	Lactate Isomer	Growth at pH 4.5	Growth at pH 7	Growth at 45 °C	NH_3 from Arginine	Carbohydrate Fermented										
						Ribose	Arabinose	Xylose	Mannitol	Lactose	Maltose	Sucrose	Trehalose	Melezitose	Dextrin	Starch
P. damnosus[a]	DL	+	–	–	–	–	–	–	–	–	d	d	+	d	–	–
P. parvulus	DL	+	+	–	–	–	–	–	–	–	+	–	d	–	–	–
P. inopinatus	DL	ND	+	–	–	–	–	–	–	+	+	d	+	–	d	–
P. dextrinicus	L(+)	–	+	–	–	–	–	–	–	d	+	d	–	–	+	+
P. pentosaceus	DL	+	+	d	+	+	+	d	–	d	+	–	+	–	–	–
P. acidilactici	DL	+	+	+	+	+	d	+	–	d	–	–	d	–	–	–
P. halophilus	L(+)	–	+	–	–	+	+	–	–	–	+	+	+	+	–	–
P. urinaeequi	L(+)	–	+	–	–	ND	d	d	d	d	+	+	+	–	+	–

[a] previously *P. cerevisiae*; +, more than 90% of strains positive; d, 11–89% of strains positive; ND, no data available; *P. pentosaceus* and *P. acidilactici* possess pseudocatalase; most strains ferment fructose, galactose, glucose, mannose, cellobiose, amygdalin, esculin and salicin; most strains do not ferment sorbitol, glycerol, sorbose, inulin, dulcitol, and inositol.

The isolation of pediococci from the rumen of cows and from feces of turkey has been reported. In the production of various kinds of raw sausages and bacon, starter cultures of *P. damnosus* (previously *cerevisiae*) are used in some European countries and the USA. As a result of the growth of the starter culture, the pH value of the raw material decreases rapidly. Due to this low pH value, contamination by potentially pathogenic bacteria like *Staphylococcus aureus* and members of the family Enterobacteriaceae is reduced. In addition, the use of starter cultures in the production of meat products shortens the production time and improves the quality of the product and the bacteriological safety of the production process. *P. halophilus* ssp. *soyae* is common in soy sauce fermentation. It may not be a true *Pediococcus* since it does not tolerate low pH values. Its reclassification into the new genus *Tetragenococcus* has been proposed.

2.5 The Genus *Leuconostoc*

(HOLZAPFEL and SCHILLINGER, 1991)

Leuconostoc together with *Lactobacillus* and *Pediococcus* forms a common cluster in the evolutionary tree of the Gram-positive bacteria (see Fig. 3). Leuconostocs are coccoid bacteria (see Fig. 7) which are heterofermentative, fastidious and widely distributed in nature with plant materials as the main habitats. Like lactobacilli, they need carbohydrates, amino acids and peptides, fatty acids, nucleic acids and especially vitamins, e.g., biotin, nicotin, thiamin, and panthotenic acid. With the exception of *L. oenos* adapted to the low pH of wine, leuconostocs prefer neutral or slightly acidic media. They are therefore less aciduric than most lactobacilli. Although their preferred growth temperature is between 20 and 30 °C, psychrotrophic strains have recently been isolated from spoiled vacuum packed, cold stored meat with a high selective advantage at 1–2 °C. Leuconostocs are aerotolerant, but usually grow better in the absence of oxygen. Many leuconostocs are known for their slime production.

Animal products as habitats

Together with *Brochothrix thermosphacta*, *Lactobacillus curvatus*, and *L. sake*, leuconostocs have been recognized as main components of the spoilage microflora of vacuum packed, or modified atmosphere covered meat and meat products (including cured sausages of the "Vienna"-type) when stored under refrigeration. A main species is *Leuconostoc mesenteroides* ssp. *mesenteroides*. New species which grow down to 1 °C, but not at 37 °C or 45 °C, have been isolated and characterized from such products as *L. carnorum* and *L. gelidum*. Whereas at 5 °C, leuconostocs account for about half of the present bacteria, they may account for 100% at 1–2 °C reaching cell numbers of up to more than 10^8 per cm^2 meat surface. Spoilage leads to discoloration, off-odor and off-taste as well as slime formation. Leuconostocs are not suited as starter cultures for fermented meat products.

Only a minor part (around 1%) of the spoilage or contamination microflora of raw milk is made up of leuconostocs. In contrast to the situation in meat, refrigerated (2–5 °C) raw or pasteurized milk is no suitable substrate for the proliferation of leuconostocs. Contaminants identified include *L. mesenteroides* ssp. *mesenteroides* and ssp. *cremoris*, *L. lactis*, *L. amelibiosum* and *L. paramesenteroides*. Of a certain importance, however, is the use of *L. mesenteroides* ssp. *cremoris* and *L. lactis* as components of starter cultures for sour milk, cream and cheeses like Gouda and Edam, since they contribute diacetyl via their heterofermentative carbohydrate fermentation and some CO_2 which determines eye formation in cheese. Citrate-lyase positive strains provide additional diacetyl. Leuconostocs alone do not grow well in milk, they prefer to be grown together, e.g., with mesophilic lactococci, because leuconostocs lack an active proteolytic system necessary to break down caseins for their nitrogen supply.

Plant products as habitats

As already mentioned in the section on lactobacilli, leuconostocs can be isolated in small numbers (10^3 per gram) from leaf material constituting, however, 80% of the prevailing lactic acid bacteria, followed by pediococci and lactobacilli with about 10% each. The dominating species is *L. mesenteroides* ssp. *mesenteroides*.

In the early stages of silage fermentation, *L. mesenteroides* ssp. *mesenteroides* together with *Enterococcus faecalis* displaces the aerobic Gram-negative microflora dominating on the plant surface at the moment of harvest. Leuconostocs and enterococci are soon followed by heterofermentative and homofermentative lactobacilli and pediococci as mentioned above. A similar role is played by *L. mesenteroides* ssp. *mesenteroides* in the fermentations of sauerkraut, olives and pickles. This general succession is also observed in food fermentations in tropic climates like the Indonesian Sayur-Asin prepared from mustard cabbage, or the African sour product from *Ensete ventricosa*. *Leuconostoc mesenteroides* ssp. *mesenteroides* has been found to degrade the cyanogenic glucoside linamarin in cassava, and to sour and leaven doughs for production of the Indian idli or dona bread, or the Ethiopian tef bread. The same species is participating in the not very well defined spontaneous fermentation steps during processing of vanilla, coffee, and cocoa beans.

The most important biotechnological function of leuconostocs is the so-called malo-lactic fermentation in red wine (RENAULT et al., 1988). The species *L. oenos* responsible for this conversion has taxonomically its own position, being also the only really aciduric species, and the only one ever isolated from wine at pH values below 3.5. *L. oenos* is also well adapted to high ethanol concentrations and the presence of SO_2. It is commercially available and also applied to malo-lactic fermentation in cider. *L. mesenteroides* ssp. *mesenteroides* is present in sour mashes for grain and malt whisky distillery fermentations and is probably contributing to the final flavor of the product. In pulque, a Mexican beverage prepared from agave juice, leuconostocs produce the polysaccharides providing the typical viscosity of this beverage. Acid-tolerant strains of *L. mesenteroides* ssp. *mesenteroides* have been isolated from sour fruit mashes. They may also cause spoilage in fruit juices (citrus, orange, mango) by generation of off-flavors through increased levels of acetoin and diacetyl.

Leuconostoc and pathogenicity? Vancomycin-resistant leuconostocs have recently been

Tab. 6. Differentiation Within the Genus *Leuconostoc* (modified from HOLZAPFEL and SCHILLINGER, 1991)

Leuconostoc Species	Growth at 1 °C	Growth at 37 °C	Growth at pH 4.8	Growth in 10% Ethanol	Esculin Hydrolysis	Acid Formed from:															Dextran from Sucrose	Lemon-yellow Pigment
						L-Arabinose	Cellobiose	Fructose	Galactose	Lactose	Maltose	Mannitol	Mannose	Melibiose	Raffinose	Ribose	Salicin	Sucrose	Trehalose	D-Xylose		
L. amelibiosum	+	ND	–	ND	+	+	+	+	ND	–	+	–	ND	–	–	–	+	+	+	–	+	–
L. carnosum	+	∓	ND	ND	d	–	d	+	–	–	+	–	d	d	–	d	d	+	+	–	±	–
L. citreum	ND	d	ND	ND	±	±	±	+	∓	–	+	d	+	∓	–	–	±	+	+	∓	ND	±
L. gelidum	+	∓	ND	ND	+	+	+	+	–	–	d	–	+	+	+	d	+	+	+	+	±	–
L. lactis	ND	+	–	–	–	–	–	+	+	+	+	–	d	d	d	∓	d	+	–	–	–	–
L. mesenteroides ssp. *cremoris*	ND	–	–	–	–	–	–	–	+	+	–	–	–	–	–	–	–	–	–	–	–	∓
L. mesenteroides ssp. *dextranicum*	ND	+	–	–	d	–	∓	±	d	d	±	∓	d	d	d	ND	∓	±	+	d	+	–
L. mesenteroides ssp. *mesenteroides*	ND	d	–	–	±	+	d	+	±	d	±	d	±	±	d	±	±	+	+	d	+	–
L. oenos	ND	d	+	+	+	±	±	+	d	–	–	–	d	d	–	d	d	–	+	d	–	–
L. paramesenteroides	ND	d	∓	–	d	±	d	+	+	d	+	∓	+	+	d	+	–	±	+	∓	–	–
L. pseudomesenteroides	ND	+	ND	ND	d	±	±	+	±	d	+	∓	+	±	±	+	d	±	+	+	ND	∓

+, positive reaction; –, negative reaction; ±, most strains positive; ∓, most strains negative; d, variable reaction; ND, no data isolated from clinical sources like purulent meningitis and odontogenic infections. The isolates have been described as the new species *L. citreum* and *L. pseudomesenteroides*. However, whether these vancomycin-resistant leuconostocs are true pathogens is still doubtful, since KOCH's postulates have still to be demonstrated.

The differential properties of *Leuconostoc* are compiled in Tab. 6.

2.6 The Genus *Lactococcus*

(TEUBER et al., 1991)

The genus *Lactococcus* is the earliest offspring of the *Lactococcus–Streptococcus* cluster within the *Clostridium* branch of the Gram-positive bacteria (see Fig. 3). Lactococci perform a homolactic fermentation with L(+)lactate as the main fermentation product. In dairy products, pH values of about 4.5 can be obtained which is sufficient for a conservation and inhibition of putrefactive bacteria. Their morphology is shown in Fig. 7.

Habitats

The lactococci comprise the species *Lactococcus lactis*, *L. garviae*, *L. plantarum*, and *L. raffinolactis* (Tab. 7). *L. lactis* ssp. *lactis* and *L. lactis* ssp. *lactis* biovar *diacetylactis* have commonly been detected directly or following enrichment in plant material, including fresh and frozen corn, corn silks, navy beans, cabbage, lettuce, peas, wheat middlings, grass, clover, potatoes, cucumbers, and cantaloupe. Lactococci are usually not found in fecal material or soil. Only small numbers occur on the surface of the cow and in its saliva. Since raw cow's milk consistently contains *L. lactis* ssp. *lactis*, and to a much lesser extent *L. lactis* ssp. *"diacetylactis"* and *L. lactis* ssp. *cremoris*, it is tempting to suggest that lactococci enter the milk from the exterior of the udder during milking and from the feed, which may be the primary source of inoculation. *L. lactis* ssp. *cremoris* has hitherto not been isolated with certainty from habitats other than milk, fermented milk, cheese, and starter cultures.

Information on the habitats of the species *L. garviae*, *L. plantarum*, and *L. raffinolactis* is

Tab. 7. Characteristics of *Lactococcus* Species (modified from TEUBER et al., 1991)

Species and Subspecies	Source	Hydrolysis of Arginine	Acid Production from						
			Galactose[a]	Lactose	Maltose	Melibiose	Melezitose	Raffinose	Ribose
L. lactis ssp. *lactis*	Raw milk and dairy products	+	+	+	+	–	–	–	+
L. lactis ssp. *cremoris*	Raw milk and dairy products	–	+	+	–	–	–	–	–
L. lactis ssp. *hordniae*	Leaf hopper	+	–	–	–	–	–	–	–
L. garviae	Bovine samples	+	+	+	V	V	–	–	+
L. plantarum	Frozen peas	–	–	–	+	–	+	–	–
L. raffinolactis	Raw milk	V	+	+	+	+	V	+	V

For symbols see Tab. 6; V, variable

scarce. Only a few strains have been isolated from samples of cows with mastitis. KOCH's postulates for *L. garviae* as infective agent of mastitis have not been fulfilled. In the horse stomach, where an active lactic fermentation takes place, $3 \cdot 10^8$ lactococci have been counted per gram of content. The most important habitats for lactococci, however, are in the dairy industry. They are applied as pure or mixed starter cultures in the production of cheese like Cheddar, Camembert, Tilsit, Edam, Gouda, of sour milk and cream, and of Scandinavian ropy milks (Taette, Viili). Most technologically important functions in these bacteria are encoded on a great variety of plasmids (see below).

2.7 The Genus *Streptococcus*

(HARDIE and WHILEY, 1991; RUOFF, 1991)

Previously, the genera *Enterococcus* and *Lactococcus* were included in the genus *Streptococcus.* It was due to modern techniques of molecular taxonomy that the former genera could be elevated to their own genus rank diminishing the confusion which existed for almost 100 years and which was based on too much emphasis on the morphology of these bacteria (partly due to lack of other criteria). The genus *Streptococcus* contains two functionally different groups: the oral streptococci and the hemolytic streptococci (Tabs. 8 and 9). These bacteria are most intimately connected with the human body, many of them are extremely potent pathogens. One noteworthy exception is S*treptococcus thermophilus* – a member of the oral group – which is commonly used for yoghurt and cheese production. All streptococci are homofermentative and produce mainly L(+)lactic acid. Streptococci are generally much less aciduric than lactobacilli. Many are killed in unbuffered media by their own lactic acid and grow only on very complex media such as fresh or boiled blood agar.

Habitats

The oral streptococci (see Tab. 8) have their habitat in the different parts of the mouth and pharynx: teeth, tongue, mucosa and saliva. They have to withstand a constant flow of saliva and swallowing. To avoid a washout, oral streptococci have invented mechanisms to adhere to the walls of their biotope by the building of extracellular polysaccharides and adhesins (fimbriae and pili). This behavior is the basis of dental caries. From experiments with gnotobiotic animals it became clear that monocultures of *S. mutans*, certain strains of *S. milleri*, *S. salivarius*, and *S. oralis* do induce dental caries. The disease was transmissible to other animals. *S. sanguis* and *S. mitis* had a much lower or no cariogenic potential. Human caries is also predominantly initiated and caused by *S. mutans*. People with more than 10^5–10^6 colony forming units of *S. mutans* per mL of saliva are likely to develop new carious lesions. The cariogenicity of the *mutans* streptococci is due to their potential to colonize the tooth surface, to become part of the dental plaque and to produce acid. Sucrose is an excellent source of acid production, but also a substrate for glucosyl- and fructosyl-transferases synthesizing water-soluble and -insoluble polysaccharides (e.g., $\alpha(1\rightarrow3)$-glucans). Mutants with decreased acid and polymer production also have a diminished cariogenicity. Other diseases caused by mutant streptococci include purulent infections (especially by the *S. milleri* group) and infective endocarditis (about 50% of all cases). Infection occurs via lesions, for example during extraction of a tooth.

Streptococcus thermophilus has its habitats only in such fermented dairy products which are made at elevated temperatures, e.g., yoghurt at 42 °C, or pass through processing steps at elevated temperatures (e.g., heating of cheese curd in Emmental production at 50 to 55 °C). They are usually associated with thermophilic lactobacilli (*Lactobacillus helveticus, L. delbrueckii* ssp. *bulgaricus*; see Fig. 7). *S. thermophilus* is non-pathogenic. Some strains produce extracellular polysaccharides which are important in the texture of yoghurt (CERNING, 1990).

In the context of this treatment, it must suffice to list the pathogenic potential and the habitats of the hemolytic streptococci in Tab. 9 without further discussion. However, it should be mentioned that raw milk may contain hemolytic streptoccci due to mastitis of the dairy cow. In the latent early period of such an infection, the streptococci (and other causative

Tab. 8. Species Groups and Subgroups Within the Oral Streptococci (modified from HARDIE and WHILEY, 1991)

Main Group	Subgroup	Species
S. mutans	A	*S. mutans*
		S. rattus
	B	*S. sobrinus*
		S. downei
		S. cricetus
	Ungrouped	*S. ferus*
	Not known	*S. macacae*
S. oralis	A	*S. oralis*
		S. mitis
		(S. pneumoniae)
	B	*S. sanguis*
		S. gordonii
		S. parasanguis
		"Tufted fibril group"
	S. milleri group	*S. anginosus*
		S. constellatus
		S. intermedius
S. salivarius		*S. salivarius*
		S. vestibularis
		S. thermophilus
Nutritionally variant streptococci (NVS)	Not known	*S. adjacens*
	Not known	*S. defectivus*

Tab. 9. Habitats and Pathogenicity of Clinically Important Streptococci (modified from RUOFF, 1991)

Species or Group	Habitat	Pathogenicity
S. pyogenes	Throat	Pharyngitis
	Skin	Respiratory tract infection
	Rectum	Skin and soft tissue infection
		Rheumatic fever
		Glomerulonephritis
		Endocarditis
S. agalactiae	Genital tract	Neonatal infection
	Gastrointestinal tract	Urogenital infection
	Throat	Skin and soft tissue infection
	Skin	Endocarditis
		Respiratory tract infection
Group C and G streptococci (large-colony forms)	Throat	Respiratory tract infection
	Genital tract	Pharyngitis
	Gastrointestinal tract	Endocarditis
	Skin	Skin and soft tissue infection
		Meningitis
		Bone and joint infection
		Nephritis
S. pneumoniae	Throat	Respiratory tract infection
		Ear and eye infection
		Septicemia
S. bovis	Gastrointestinal tract	Endocarditis
		Bacteremia associated with colonic cancer

agents, like staphylococci) may already be shed into the milk. Since the farmer does not yet recognize this infection, he is not able to single out this milk from distribution.

2.8 The Genus *Enterococcus*

(DEVRIESE et al., 1991)

The name indicates that this genus is living in the enteric system of animals, the intestine. Enterococci colonize and dominate the intestine of newborns during their first 2 or 3 days. *Enterococcus faecalis* and *E. faecium* are always found in the feces of adult human beings and animals including insects. In some countries, enterococci have therefore the legal function as indicator organisms for fecal contamination of water and food. In certain food, however, enterococci may also act as index of improper hygienic handling since enterococci may multiply during food processing from low to high numbers. The enterococci are characterized by tolerance to high salt concentrations (growth at >6.5% sodium chloride) and optimum growth temperatures around 40 °C and more. *E. faecalis* and *E. faecium* are found on plants. Strains from plants and insects, as far as investigated, tend to be more amylolytic than strains from the human intestine. Enterococci disappear from the plants in winter and reappear during spring and summer, possibly due to inoculation by insects. It is not clear whether "plant" enterococci are genetically different from "enteric" enterococci. *E. faecium* has recently been propagated as a "probiotic" to be added to animal fodder in order to protect the intestine of young animals (piglets, chicken, etc.) from infections with enteropathogenic enteric bacteria (*Salmonella*, *Escherichia coli*, *Shigella*). The results are still quite controversial and contradicting (STAVRIC et al., 1992; FULLER, 1992). The ecology of the enterococci is interesting on a genetic basis, as they possess plasmids and transposons which can migrate to all other Gram-positive species of the *Clostridium* branch and even into *E. coli* (see below). The use of enterococci as components of starter cultures in the dairy industry is not generally accepted due to the above mentioned functions as index and indicator organisms. In addition, *E. faecium* and *E. faecalis* are pathogenic under certain conditions. In the human body, enterococci can cause endocarditis and urinary infections (as sole organisms involved). In intraabdominal and pelvic infections, they may be part of microbial associations leading to these infections. Recently, their role in nosocomial infections has been increasing, probably due to their broad resistance patterns agains many commonly used antibiotics. Antibiotic-resistant enterococci are also found in fermented food made from raw milk and meat (TEUBER et al., unpublished data). Under these circumstances, it may be wise not to use enterococci as starter cultures and probiotics, unless their lack of antibiotic resistance genes and the corresponding transfer systems has been proved.

2.9 The Genus *Bifidobacterium*

(BIAVATI et al., 1991)

Bifidobacteria do not belong to the *Clostridium* branch of Gram-positive bacteria like all the other lactic acid bacteria (LAB). Nevertheless, they are regarded as LAB due to the presence of lactic acid as a main product of fermentation. Bifidobacteria possess a fructose-6-phosphoketolase forming acetyl-phosphate and erythrose-4-phosphate which by the action of transketolase and transaldolase is finally converted into pentose-phosphate, being split by the pentose-phosphate phosphoketolase into triosephosphate (the source of lactate). The habitats of the strictly anaerobic bacteria are the feces of human and animal infants and adults, of the human vagina, the intestine of insects, the rumen of cattle, the oral cavity, and sewage. Of the 25 acknowledged species, *B. bifidum*, *B. longum*, *B. breve*, *B. thermophilum* and *B. pseudolongum* should be mentioned, since they are used as starter cultures for products of human consumption or therapeutic preparations for animals. This application is based on the "belief" that bifidobacteria are helpful in maintaining a proper balance in the human intestinal flora. However, the scientific evidence for this is inversely proportional to the large number of papers on the subject. In milk, bifidobacteria can only be propagated if "bifidofactors" such as lactulose, N-acetylglucosamine, and peptide hydrolysates are added.

The bifidobacteria containing fermented milk products (mainly yoghurt, recently soft cheese) contain *Streptococcus thermophilus* and *Lactobacillus acidophilus* in addition. Very often, the bifidobacteria claimed on the label can not be isolated from the product. Under these circumstances, it is extremely doubtful whether bifidobacteria consumed with these specific products reach the intestine of the consumer at all. Whether a few bacteria surviving the passage of the stomach will be able to establish themselves in an intestine and its contents where already 10^{10} to 10^{12} bacteria per gram are present, has to be questioned (GURR, 1983).

3 Isolation and Cultivation

Isolation of lactic acid bacteria (LAB) from a certain habitat or medium may not be difficult if they represent the main part of the microflora. If not, enrichment procedures have to be applied. As already mentioned, all LAB are nutritionally fastidious requiring rich organic media. Usually, the pH value of the media is lowered to take advantage of the acidophilic properties of many LAB. Upon first isolation, it may be advantageous to incubate agar plates under anaerobic conditions, e.g., in an atmosphere of 10% CO_2 and 90% N_2. For strictly anaerobic species, e.g., bifidobacteria, the medium may have to be prereduced by overnight incubation in 10% CO_2 and 90% H_2. Incubation temperatures depend on the biotope investigated, e.g., 22 °C for low temperature isolates, 30 °C for isolates from plants and food, 37 °C for isolates from warmblooded animals. In attempts to isolate thermophilic species out of a mixture with mesophilic species, an incubation temperature of 40 to 45 °C may be useful. LAB form rather small colonies even after 2 to 4 days of incubation (1–2 mm in size). The first step in isolation often is incubation of the analytical sample in broth which, after growth, is plated onto the surface of the corresponding agar medium. This allows differentiation on the basis of colony morphology, and to take samples directly for microscopic examination. This is very important because many of the classical media support growth of more than one species, and microscopy is the only way to come to an immediate decision on what has been isolated. An excellent collection of many different media is given in the 2nd edition of *Prokaryotes* which should be consulted for details (see references to the different genera). Here, one or two examples of media for the main species are given.

Lactobacilli

When lactobacilli are expected to constitute the majority of the population, direct plating onto the classical MRS agar is first choice:

MRS-agar for isolating and propagating lactobacilli (DE MAN et al., 1960). For 1 liter of medium, dissolve 15 g agar in distilled water by steaming. Add:

oxoid peptone	10.00 g
meat extract	10.00 g
yeast extract	5.00 g
K_2HPO_4	2.00 g
diammonium citrate	2.00 g
glucose	20.00 g
tween 80	1 mL
Na-acetate	5.00 g
$MgSO_4 \cdot 7H_2O$	0.58 g
$MnSO_4 \cdot 4H_2O$	0.25 g

Adjust pH to 6.2 or 6.4 and sterilize at 121 °C for 15 min. When lactobacilli are expected to be only a (minor) part of the investigated sample, the selective SL-medium is of advantage, although it also provides growth of pediococci, leuconostocs, some enterococci, bifidobacteria, and yeasts. Yeasts may be inhibited by the addition of cycloheximide at 10 mg per liter medium.

Selective SL medium for isolating lactobacilli (ROGOSA et al., 1951). For 1 liter of medium:

trypticase	10.00 g
yeast extract	10.00 g
KH_2PO_4	6.00 g
diammonium citrate	2.00 g
$MgSO_4 \cdot 7H_2O$	0.58 g
$MnSO_4 \cdot 4H_2O$	0.15 g
$FeSO_4 \cdot 7H_2O$	0.03 g
glucose	20.00 g
tween 80	1 mL
Na-acetate $\cdot$ $2H_2O$	25.00 g
agar	15.00 g

Dissolve agar in 500 mL water by boiling. Dissolve all other ingredients in 500 mL water, adjust pH to 5.4 with glacial acetic acid, mix with the melted agar, boil another 5 min, and pour plates.

The very fastidious obligately heterofermentative lactobacilli are isolated and propagated on a modified Homohiochii medium at its best (KLEYNMANS et al., 1989). For 1 liter of medium, dissolve 15 g agar in distilled water by steaming and add:

tryptone	10.00 g
yeast extract	7.00 g
meat extract	2.00 g
glucose	5.00 g
fructose	5.00 g
maltose	2.00 g
Na-gluconate	5.00 g
diammonium citrate	2.00 g
Na-acetate	5.00 g
tween 80	1 mL
$MgSO_4 \cdot 7H_2O$	0.20 g
$MnSO_4 \cdot 4H_2O$	0.05 g
$FeSO_4 \cdot 7H_2O$	0.01 g
mevalonic acid lactone	0.03 g
cysteine hydrochloride	0.50 g

Adjust pH to 5.4, sterilize at 121 °C for 15 min, add 40 mL of ethanol per liter.

Pediococci

Pediococci from plant material can be isolated on MRS agar or SL-medium. For the isolation from beer, MRS agar is adjusted to pH 5.5. It may also be mixed for this purpose 1:1 with beer. Incubation should be effected in an atmosphere of 90% N_2 + 10% CO_2 at 22 °C. Halophilic pediococci from soy sauce fermentation are isolated on MRS-agar pH 7.0 supplemented with 4 to 6% NaCl and cycloheximide to suppress the yeasts.

Leuconostocs

If leuconostocs predominate like as in early stages of silage fermentations, MRS-agar is the medium of choice. For the isolation from meat, MRS-agar is adjusted to pH 5.7 and supplemented with 0.2% potassium sorbate. For the enumeration of leuconostocs in dairy starter cultures, the HP medium of PEARCE and HALLIGAN (1978) is the medium of choice. Composition per liter:

phytone	20.00 g
yeast extract	6.00 g
beef extract	10.00 g
tween 80	0.50 g
ammonium citrate	5.00 g
$FeSO_4 \cdot 7H_2O$	0.04 g
$MgSO_4 \cdot 7H_2O$	0.20 g
glucose	10.00 g (sterilized separately)

The addition of tetracycline (0.12 mg/mL) to the medium selectively inhibits the growth of lactococci.

For the isolation of *Leuconostoc oenos* from wine, the acidic tomato broth (ATB) is a classical and useful medium (GARVIE, 1967). The medium contains per liter:

peptone	10.00 g
Yeastrel	5.00 g
glucose	10.00 g
$MgSO_4 \cdot 7H_2O$	0.20 g
$MnSO_4 \cdot 4H_2O$	0.05 g
tomato juice	250.00 g

The pH is 4.8. Sterilization is for 15 min at 121 °C. Before use, a solution of cysteine-hydrochloride sterilized by filtration is added to a final concentration of 0.05%.

Lactococci and *Streptococcus thermophilus*

The classical medium has been the unbuffered Elliker-agar-medium (ELLIKER et al., 1956). However, since lactococci rapidly die at pH values below 5, the medium of choice is the excellent M17 medium developed by TERZAGHI and SANDINE (1975). It is buffered with β-disodium glycerophosphate which inhibits lactobacilli, pediococci and leuconostocs almost completely. Per liter of medium add:

phytone peptone	5.00 g
polypeptone	5.00 g
yeast extract	5.00 g
beef extract	5.00 g
lactose	5.00 g
ascorbic acid	0.50 g
β-disodium glycerophosphate	19.00 g
1.0 M $MgSO_4 \cdot 7H_2O$	1 mL
glass-distilled water	

The medium is sterilized at 121 °C for 15 min, pH is 7.1. For solid media, 10 g agar is added. If incubated at 42 °C, the medium is selective for *S. thermophilus*, at 25 to 30 °C for *Lactococcus.*

Oral streptococci

The most commonly used selective agar for streptococci contains sucrose which leads to characteristic colonies due to extracellular polysaccharide formation.

The trypticase-yeast-extract-cystine (TYC) agar has the following composition (DE STOPPELAAR et al., 1967). One liter contains:

trypticase (BBL)	15.00 g
yeast extract (Difco)	5.00 g
L-cystine	0.20 g
Na_2SO_3	0.10 g
NaCl	1.00 g
$Na_2HPO_4 \cdot 12H_2O$	2.00 g
$NaHCO_3$	2.00 g
Na-acetate $\cdot$ $3H_2O$	20.00 g
sucrose	50.00 g
agar	12.00 g

pH is 7.3, autoclaved at 121 °C for 15 min. For the description of the typical colonies the specific literature should be consulted (HARDIE and WHILEY, 1991)

Hemolytic streptococci

Rich agar-containing media with little or no reducing sugar (tryptic soy, heart infusion) and supplemented with 5% animal blood (sheep or horse) are excellent for the cultivation of streptococci and determination of hemolysis (RUOFF, 1991).

Enterococcus

More than 60 different selective media have been described for the isolation of enterococci. They are all based on the fact that enterococci can withstand many adverse conditions, especially many antibiotics active against other Gram-positive bacteria.

One medium with an excellent selectivity due to its contents of sodium azide and kanamycin has been developed by MOSSEL et al. (1978). It can be used as broth and agar. On agar, enterococci grow as large shiny black colonies due to hydrolysis of esculin and resulting formation of black iron complexes. The medium has the following composition (per liter):

tryptone	20.00 g
yeast extract	5.00 g
NaCl	5.00 g
Na-citrate	1.00 g
esculin	1.00 g
ferric ammonium citrate	0.50 g
Na-azide	0.15 g
kanamycin sulfate	0.02 g

Agar 10 to 12 g, sterilization at 121 °C for 15 min.

Propagation and preservation of starter cultures (TEUBER et al., 1991)

Starter cultures are needed and applied by the industry producing food and feed as indicated in Tab. 1. Whereas silage is mainly fermented by the spontaneously developing autochthonous microflora, dairy products are almost always made with commercial cultures, although silage inoculation is gaining more and more importance. Because of its advanced state, the manufacture of lactococcal dairy cultures is described, but the principles are the same for all lactic acid bacteria.

Liquid, dried, and frozen starter cultures are in use. Starter cultures must have a high survival rate of microorganisms coupled with optimum activity for the desired technological performance: e.g., the fermentation of lactose to lactate, controlled proteolysis of casein, and production of aroma compounds like diacetyl. Since the genes for lactose and citrate fermentation as well as for certain proteases are located on plasmids, continuous culture has not been successful because fermentation-defective variants easily develop. In most instances, pasteurized or sterilized skim milk is the basic nutrient medium for the large-scale production of starter cultures because it ensures that only lactococci fully adapted to the complex substrate milk will develop. For liquid starter cultures, the basic milk medium may be supplemented with yeast extract, glucose, lactose, and calcium carbonate. To obtain optimum activity and survival, it may be necessary to neutralize the lactic acid that is produced by addition of sodium or ammonium hydroxide. Since many strains of lactic streptococci produce hydrogen peroxide during growth under microaerophilic conditions, it has been beneficial to add catalase to the growth media, thus leading to cell densities of more than 10^{10} viable units per mL of culture.

For the preparation of concentrated starters, the media are clarified by proteolytic digestion of skim milk with papain or bacterial enzymes to avoid precipitation of casein in the separator used to collect the lactococcal biomass. The available self-cleaning clarifiers, e.g., bactofuges, concentrate the fermentation broth to cell concentrations of about 10^{12} viable bacteria per mL. These concentrates can either be lyophilized or are preferentially transferred dropwise into liquid nitrogen. The formed pellets are packed in metal cans or cartons and are kept and shipped at −70 °C (dry ice). These modern starter preparations allow the direct inoculation of cheese vats, since the bacteria immediately resume logarithmic growth, if harvested in the late logarithmic growth phase. This modern technology shifts most of the microbiological work and responsibility from the cheese factory to the starter producer.

The classical liquid starter culture with about 10^9 viable cells per mL must be further propagated in the factory. The same refers to the traditional lyophilized cultures containing about 10^9 viable bacteria per gram. However, they can be easily shipped and kept at ambient temperatures for several months. Concentrated lyophilized starters are also suited for direct vat inoculation having a short lag phase, however, before growth is resumed. Direct vat inoculation may have a certain advantage regarding protection against bacteriophages which are common in the open dairy fermentation systems (for details see Chapter 14 of this volume, and TEUBER, 1987; BRAUN et al., 1989).

4 Metabolism

In the carbon remineralization cycle of organic matter in this world, the lactic acid bacteria (LAB) are involved in the very first steps including depolymerization of biopolymers, and fermentation of carbohydrates and certain amino acids. Since LAB are devoid of respiratory chains, they use the "archaic" mechanisms of substrate level phosphorylation by intramolecular and intermolecular oxidation/reduction to gain enough energy for their life. As a consequence, they have developed efficient transport and metabolic systems to achieve a rapid turnover of fermentable carbohydrates for a maximum energy yield from the available substrates. Part of this efficient evolutionary adaptation is the resistance of LAB to their own end product of fermentation, i.e., lactic acid at pH levels as low as pH 3 and the development of ion pumps to provide additional metabolic energy. The very clearcut definition of LAB as bacteria forming lactic acid as the main product of fermentation of carbohydrates fades, if one looks at fermentation under aerobic conditions. Nevertheless, biochemical investigations have solved many confusing problems and shown exciting alternatives to classical pathways in LAB growing under stress conditions.

4.1 Carbohydrate Metabolism and Metabolic Energy

LAB are able to utilize a broad range of carbohydrates as evident from Tabs. 2 to 7. A preliminary to fermentation is transport into the microbial cells. For more detail see KANDLER, 1983; KASHKET, 1987; KONINGS and OTTO, 1983; LONDON, 1990.

4.1.1 Bioenergetics and Solute Transport

Lactococcus lactis ssp. *cremoris* and *Enterococcus faecalis* have been most intensively studied as general models, since they lack cytochromes, do not store glycogen, and do not possess an outer membrane facilitating the experimental use of ionophores like valinomycin, nigericin and protonophores. KONINGS and OTTO (1983) have measured in *Lactococcus lactis* ssp. *cremoris* the lapse of the proton motive force (Δp), membrane potential ($\Delta\psi$), pH gradient and lactate gradient during growth with lactose (Fig. 8). From these data they developed a model shown in Fig. 9. At pH 6.7, internal (cytoplasmic) lactate is around 200 mM. Lactate is exported in symport with 2 protons. By this event, the burden is taken from mem-

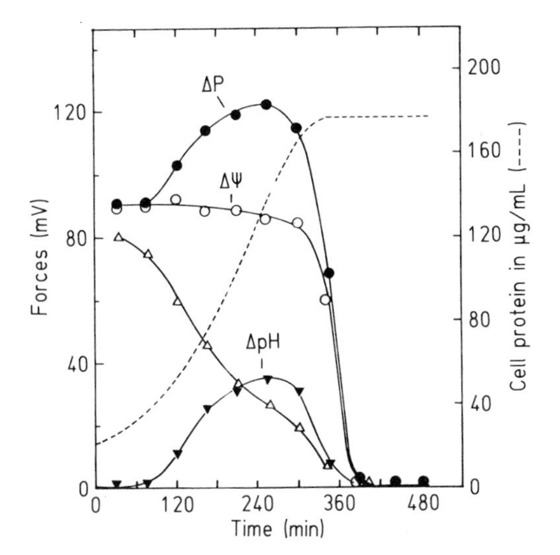

Fig. 8. The proton motive force and the lactate gradient in *Lactococcus lactis* ssp. *cremoris* during growth in batch culture with 2 g lactose per liter complex medium. The different gradients are expressed in mV: ΔP, proton motive force; $\Delta\psi$, membrane potential; △ lactate gradient; --- cell protein (courtesy of KONINGS and OTTO, 1983).

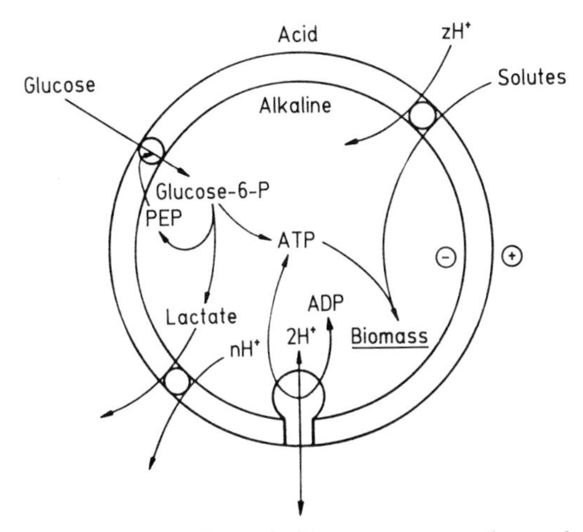

Fig. 9. Scheme of metabolic energy-generating and energy-consuming processes in lactococci (courtesy of KONINGS and OTTO, 1983).

brane ATPase F_0F_1 to split ATP in order to generate a proton motive force. In this way the reverse of the uptake process occurs: whereas in the uptake process the energy of the proton motive force is converted into the energy of a solute gradient, in the efflux process, the energy of the lactate gradient is converted into a proton motive force. If in *Lactococcus* the internal pH has dropped from pH 7.6 to 5.7 (at outside pH 4.5), growth and fermentation cease, and the cells start to die. At the beginning of a fermentation when the lactate gradient and proton motive forces are high, the proton motive force may contribute substantially (about 50%) to the metabolic energy supply. Lactococci growing in mixed culture with a lactate scavenging organism like *Pseudomonas stutzeri* exhibited an increase in the molar growth yield of *Lactococcus* by 60 to 70%.

Three different solute transport systems are present in the membranes of lactococci and enterococci:

(1) a phosphoenolpyruvate (PEP)-dependent sugar transport system characterized by vectorial phosphorylation and release of the phosphorylated sugar into the cytoplasm (PTS system),
(2) a secondary transport system energized by the electrochemical potential,
(3) an ATP-dependent transport system driven by the high energy phosphate bond directly.

Mechanism (1) is realized for the transport of lactose, glucose, sucrose and gluconate, mechanism (2) for galactose, histidine, alanine, glycine, leucine, serine, threonine and lactate, and mechanism (3) for K^+, Na^+, Ca^{2+} (extrusion), glutamate, aspartate and phosphate. Some of the secondary transport systems translocate protons together with the substrate (=symport) (in enterococci). For sophisticated recent models including uniport, proton–substrate symport, proton–sodium antiport, sodium– substrate symport, and neutral anion exchange, the work of MALONEY (1990) should be consulted.

4.1.2 Energetic Aspects of Carbohydrate Metabolism

The biochemical pathways of glucose metabolism in homo- and heterofermentative lactic acid bacteria are well known (see Figs. 10 and 11, and KANDLER, 1983). Under anaerobic conditions, homofermentative LAB gain 2 ATP per mole of glucose fermented into 2 moles of lactate, heterofermentative LAB gain only 1 ATP per mole of glucose since acetylphosphate is completely converted into ethanol in order to reoxidize enough NADH to NAD^+, as described by CONDON (1987) for leuconostocs. In the presence of oxygen, mainly acetate is formed instead of ethanol and CO_2. ATP yield under these conditions is 2 ATP per mole of glucose which is also evident by an increase of the molar growth yield by a factor of 2.

4.1.3 Energetic Aspects of Oxygen Metabolism and Oxygen Detoxification

An important contribution of biochemistry to the understanding of the growth and metabolism of lactic acid bacteria (LAB) in the presence of oxygen was the discovery of a number of enzymes directly metabolizing oxygen. The enzymatic reactions that have been detected in LAB or are suggested to be present are shown in Tab. 10. Superoxide dismutase has not been detected in the true LAB, instead, Mn^{2+} is responsible for a catalytic scavenging of O_2^-. This explains the absolute requirement of LAB for Mn^{2+} in the growth medium in addition to the Mn^{2+}-dependence of the RNA polymerases. If the H_2O_2 producing systems are more active than the H_2O_2 destroying ones under aerobic conditions, the bacteria are inhibited by their own products of oxygen metabolism, which may accumulate up to 10 mg H_2O_2 per liter growth medium. With regard to metabolic energy, the growth of LAB in the presence of O_2 very often results in higher growth yields and shorter generation times, and may also allow the use of substrates not metabolized under anaerobic conditions, e.g., glycerol, mannitol, sorbitol and lactate. Lactate, for example, is oxidized by *Lactobacillus plantarum*, when glucose is depleted in the medium. The mechanism is probably the oxidation of lactate to pyruvate by either a NAD-independent and/or a NAD-dependent lactate dehydrogenase followed by pyruvate oxidase. The formed acetylphosphate yields another ATP.

4.2 Biochemistry of Carbohydrate Metabolism

The classical pathways for the fermentation of glucose are given in Fig. 10. Glycolytic rates, e.g., in *Lactococcus* are around 50 mmol per min per liter cell water.

This means that these bacteria metabolize their own body weight in glucose in about 100 minutes.

The obligately homofermentative lactobacilli (see Tab. 2) lack the enzyme phosphoket-

Tab. 10. Reactions Involving Molecular Oxygen or Oxygen Metabolites Catalyzed by Enzymes of Lactic Acid Bacteria (modified from CONDON, 1987)

1. $NADH + H^+ + O_2 \xrightarrow{\text{NADH:H}_2\text{O}_2\text{ oxidase}} NAD^+ + H_2O_2$

2. $2\,NADH + 2H^+ + O_2 \xrightarrow{\text{NADH:H}_2\text{O oxidase}} 2\,NAD^+ + 2H_2O$

3. $\text{pyruvate} + \text{phosphate} + O_2 \xrightarrow[\text{TPP, FAD}]{\text{pyruvate oxidase}} \text{acetylphosphate} + CO_2 + H_2O_2$

4. $\alpha\text{-glycerophosphate} + O_2 \xrightarrow{\text{oxidase }\alpha\text{-glycerophosphate}} \text{dihydroxyacetone phosphate} + H_2O_2$

5. $2O_2^- + 2H^+ \xrightarrow{\text{superoxide dismutase}} H_2O_2 + O_2$

6. $NADH + H^+ + H_2O_2 \xrightarrow{\text{NADH peroxidase}} 2H_2O$

TPP, thiamine pyrophosphate; FAD, flavin adenine dinucleotide

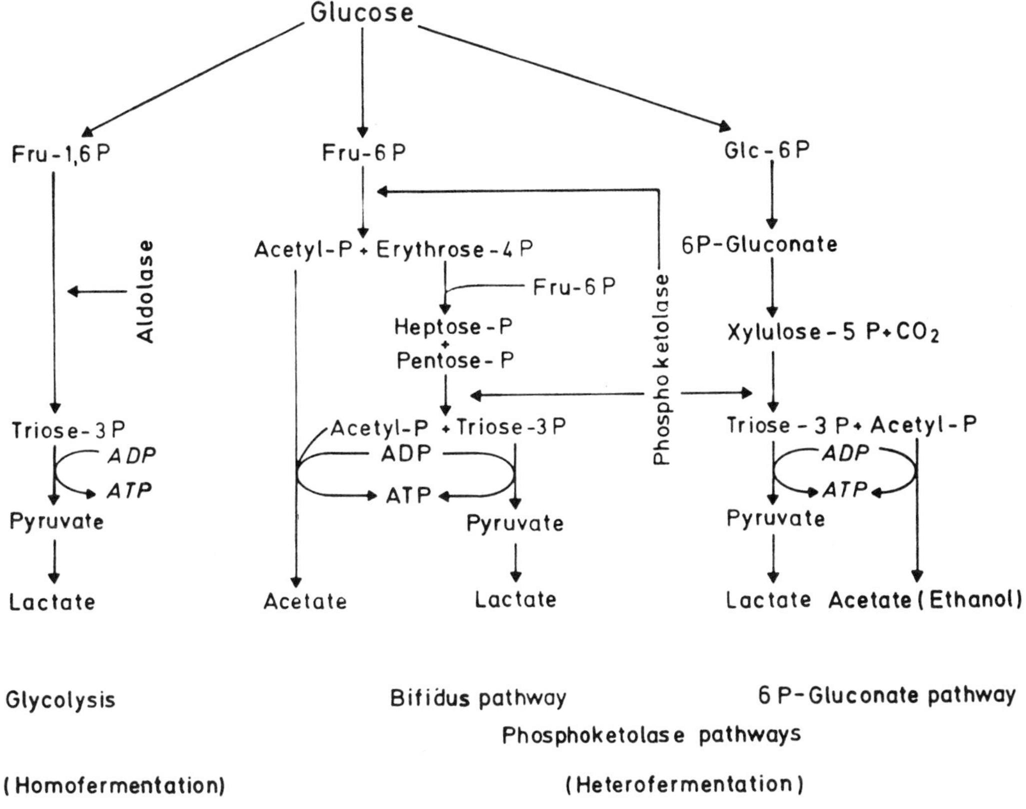

Fig. 10. Scheme of the main pathways of hexose fermentation in lactic acid bacteria (courtesy of KANDLER, 1983).

olase, which is present in the facultative and obligate heterofermenters. The obligate heterofermenters lack the enzyme aldolase which splits fructose-1,6-diphosphate in the homofermenters.

In all lactococci, the investigated enterococci and some lactobacilli lactose is transported through the cytoplasmic membrane by the PTS system leading to the accumulation of lactose-6-phosphate which is split by a phospho-β-galactosidase. Galactose-6-phosphate is further metabolized by the tagatose-6-P pathway (Fig. 11). THOMPSON (1988) has proposed a sophisticated model how lactose uptake is regulated by the PEP-dependent phosphorylating system: FDP is a positive effector for pyruvate kinase and HPr-kinase, inorganic phosphate a negative effector of pyruvate kinase but an activator of HPr(ser)P-phosphatase. This explains the build-up of the PEP potential in starved cells which allows immediate uptake of fermentable substrate if it should become available. Fermentation rates of bacteria using the PTS system are much higher than those of bacteria lacking it. Initial uptake rates for glucose in *Lactococcus lactis* are about 400 μmol sugar/g dry weight cells per min compared to about 30 μmol in PTS-defective mutants.

Diacetyl giving sour cream, sour milk and lactic butter its characteristic butter taste, is produced by lactic acid bacteria from pyruvate as outlined in Fig. 12. In case of *Lactococcus lactis* ssp. *lactis* biovar *diacetylactis* citrate from milk is transported into the cells by a plasmid coded citrate permease. Inside the cells, citrate is cleaved by citrate lyase into acetate and oxaloacetate which is further decarboxylated to pyruvate.

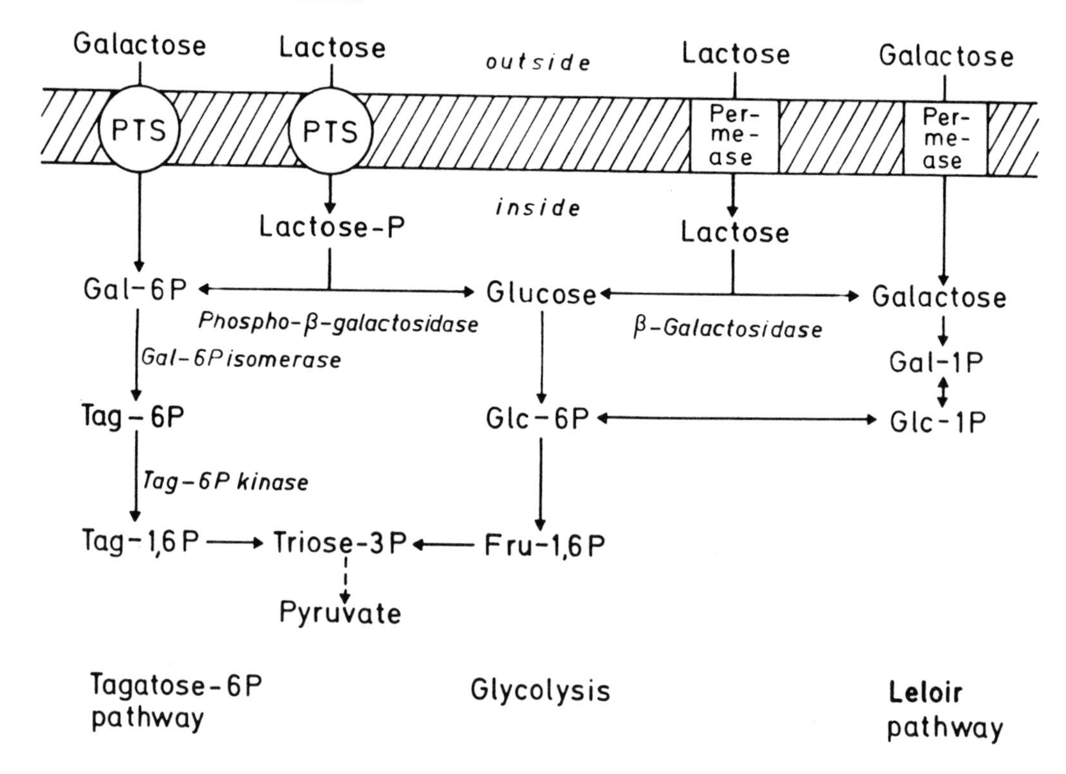

Fig. 11. Scheme of lactose and galactose uptake and dissimilation in some lactic acid bacteria (courtesy of KANDLER, 1983).

4.3 Protein, Peptide and Amino Acid Metabolism

(KOK, 1990; SMID et al., 1991)

The pathways of protein metabolism are important in fermented milk, since milk does not contain enough free amino acids and low-molecular weight peptides to allow proliferation of lactic acid bacteria to more than about 10^7 viable cells per liter. For a sufficient lactic fermentation, at least 10^9 cells per mL are necessary. For that reason, the proteolytic system has been intensively investigated in recent years with emphasis on *Lactococcus lactis*, since the genetic information for the main cell wall protease is encoded on plasmids in these bacteria which greatly facilitated elucidation of its genetics and biochemistry.

4.3.1 The Cell Wall Protease of *Lactococcus lactis*

Purification was difficult since without proper precaution the enzyme digests itself. It was quite clear from the beginning that it is a cell wall-bound enzyme because intact cells are able to utilize specifically *β*-casein in most instances (proteinase type PI). Some strains use *α*- and *κ*-caseins in addition (proteinase type PIII). The enzyme, which can be released from intact cells by washing in the absence of Ca^{2+}, has an apparent molecular weight of about 135000 Da which rapidly disintegrates into smaller polypeptides. The specificity for a PI-type enzyme from *L. lactis* ssp. *cremoris* AC1 is shown in Fig. 13 (MONNET et al., 1989). In the meantime, a proteinase model has been developed which demonstrates that the purified so-

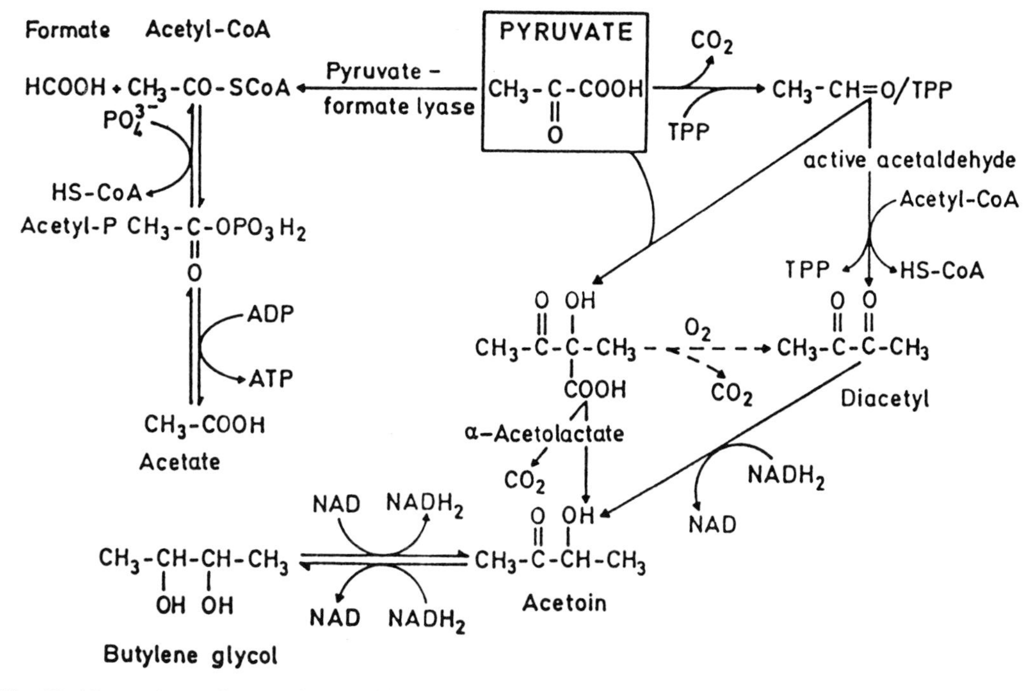

Fig. 12. Alternative pathways of anaerobic dissimilation of pyruvate leading to the butter aroma diacetyl in lactic acid bacteria (courtesy of KANDLER, 1983).

luble enzymes are self-digestion products, although still active with an identical specificity as cell wall-bound enzymes in intact cells. This model takes into account the genetic organization of proteinase plasmids (see below).

The proteinase genes code for a maturation protein (prt M) and the proteinase protein. The maturation protein is anchored with a lipid-modified cysteine in the cytoplasmic membrane, the pre-pro-proteinase (corresponding to a calculated molecular weight of about 220000 Da) transported through the membrane into the cell wall with the loss of the signal peptide. It is anchored into the membrane with its carboxy terminal end. In the cell wall region, the maturation protein splits off the N-terminal pro-region of proteinase generating the active cell wall proteinase.

In Ca^{2+}-free buffer, self-digestion of the proteinase leads to the release of an enzymatically active fraction (MW 135000 Da) into the medium. The amino acid sequences of PI and PIII enzymes have been deduced from the nucleotide sequences of the cloned genes. It showed the enzymes to be serine proteinases homologous to subtilisin, which was substantiated by inhibition of enzyme activity by phenyl methylsulfonyl fluoride and diisopropyl fluorophosphate. The amino acid differences, although few in number, do divide the enzymes into the two classes PI and PIII. Enzymes with chimeric properties are found in some strains and have been constructed by genetic engineering. Similar cell-bound proteinase activities have been detected in lactobacilli known to be proteolytically active in milk, e.g., *Lactobacillus helveticus*, *L. acidophilus* and *L. delbrueckii* ssp. *bulgaricus*.

4.3.2 The Peptidases of Lactococci

Since lactococci can only transport di- to hexapeptides they must possess peptidases to split larger peptides generated by the cell wall proteinase. The following activities have been described, most of them have been purified,

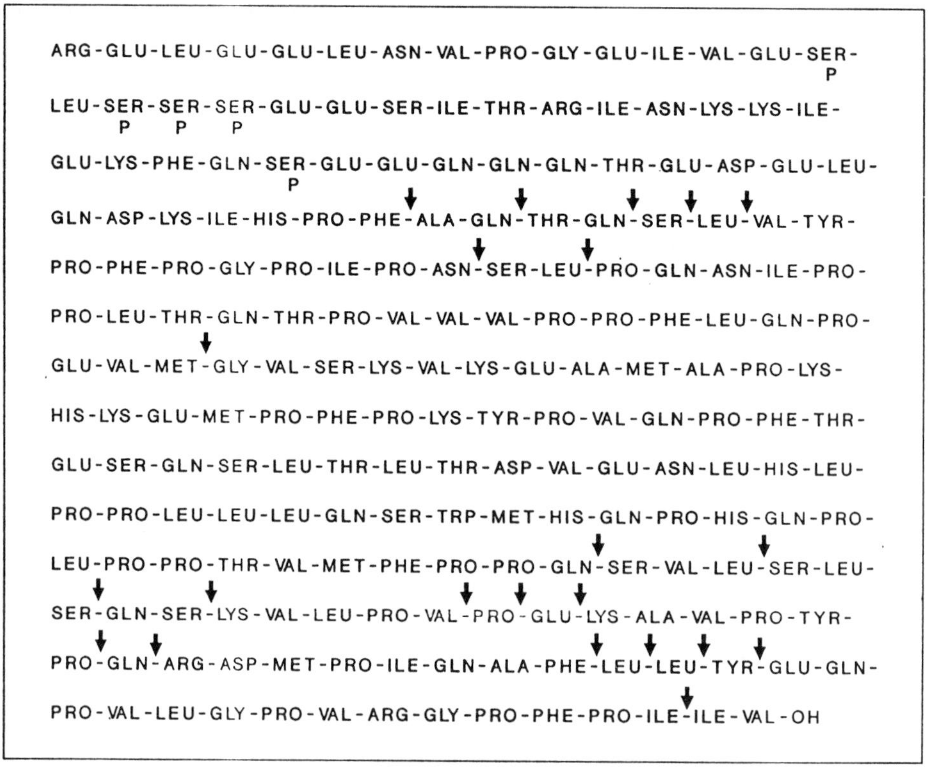

ARG-GLU-LEU-GLU-GLU-LEU-ASN-VAL-PRO-GLY-GLU-ILE-VAL-GLU-SER-
P
LEU-SER-SER-SER-GLU-GLU-SER-ILE-THR-ARG-ILE-ASN-LYS-LYS-ILE-
P P P
GLU-LYS-PHE-GLN-SER-GLU-GLU-GLN-GLN-GLN-THR-GLU-ASP-GLU-LEU-
P
GLN-ASP-LYS-ILE-HIS-PRO-PHE-ALA-GLN-THR-GLN-SER-LEU-VAL-TYR-
PRO-PHE-PRO-GLY-PRO-ILE-PRO-ASN-SER-LEU-PRO-GLN-ASN-ILE-PRO-
PRO-LEU-THR-GLN-THR-PRO-VAL-VAL-VAL-PRO-PRO-PHE-LEU-GLN-PRO-
GLU-VAL-MET-GLY-VAL-SER-LYS-VAL-LYS-GLU-ALA-MET-ALA-PRO-LYS-
HIS-LYS-GLU-MET-PRO-PHE-PRO-LYS-TYR-PRO-VAL-GLN-PRO-PHE-THR-
GLU-SER-GLN-SER-LEU-THR-LEU-THR-ASP-VAL-GLU-ASN-LEU-HIS-LEU-
PRO-PRO-LEU-LEU-LEU-GLN-SER-TRP-MET-HIS-GLN-PRO-HIS-GLN-PRO-
LEU-PRO-PRO-THR-VAL-MET-PHE-PRO-PRO-GLN-SER-VAL-LEU-SER-LEU-
SER-GLN-SER-LYS-VAL-LEU-PRO-VAL-PRO-GLU-LYS-ALA-VAL-PRO-TYR-
PRO-GLN-ARG-ASP-MET-PRO-ILE-GLN-ALA-PHE-LEU-LEU-TYR-GLU-GLN-
PRO-VAL-LEU-GLY-PRO-VAL-ARG-GLY-PRO-PHE-PRO-ILE-ILE-VAL-OH

Fig. 13. Specificity of the cell wall proteinase of lactococci for β-casein (modified from MONNET et al., 1989).

some have been cloned and sequenced (e.g., the X-prolyl dipetidyl aminopeptidase; MAYO et al., 1990).

The following peptidases have been purified and characterized from lactic acid bacteria:

(1) a general aminopeptidase (metallo type) from lactococci, *Lactobacillus acidophilus*, *L. delbrueckii* ssp. *lactis* and *bulgaricus*,
(2) an X-prolyl dipeptidyl aminopeptidase (serine type) from lactococci, *Streptococcus thermophilus*, *Lactobacillus helveticus,* and *L. delbrueckii* ssp. *lactis* and ssp. *bulgaricus*,
(3) a prolidase (metallo type) from *Lactococcus lactis* ssp. *cremoris*,
(4) a proline iminopeptidase (metallo type) from *L. lactis* ssp. *cremoris*,
(5) dipeptidases from lactococci (metallo type),
(6) tripeptidases from lactococci (metallo type),
(7) endopeptidases from lactococci (metallo type).

Aryl peptidylamidases have not been detected in the above mentioned LAB.

A schematic representation of the breakdown of β-casein by the proteolytic system of *Lactococcus lactis* is given in Fig. 14. This concept includes the postulation that some of the peptidases must also be localized at the cell surface, although genetic analysis does not support this because signal peptide sequences are missing in cloned genes.

However, recently an unconventional complex transport system has been suggested for the export of bacteriocins in lactococci (see below) which could also function with peptidases.

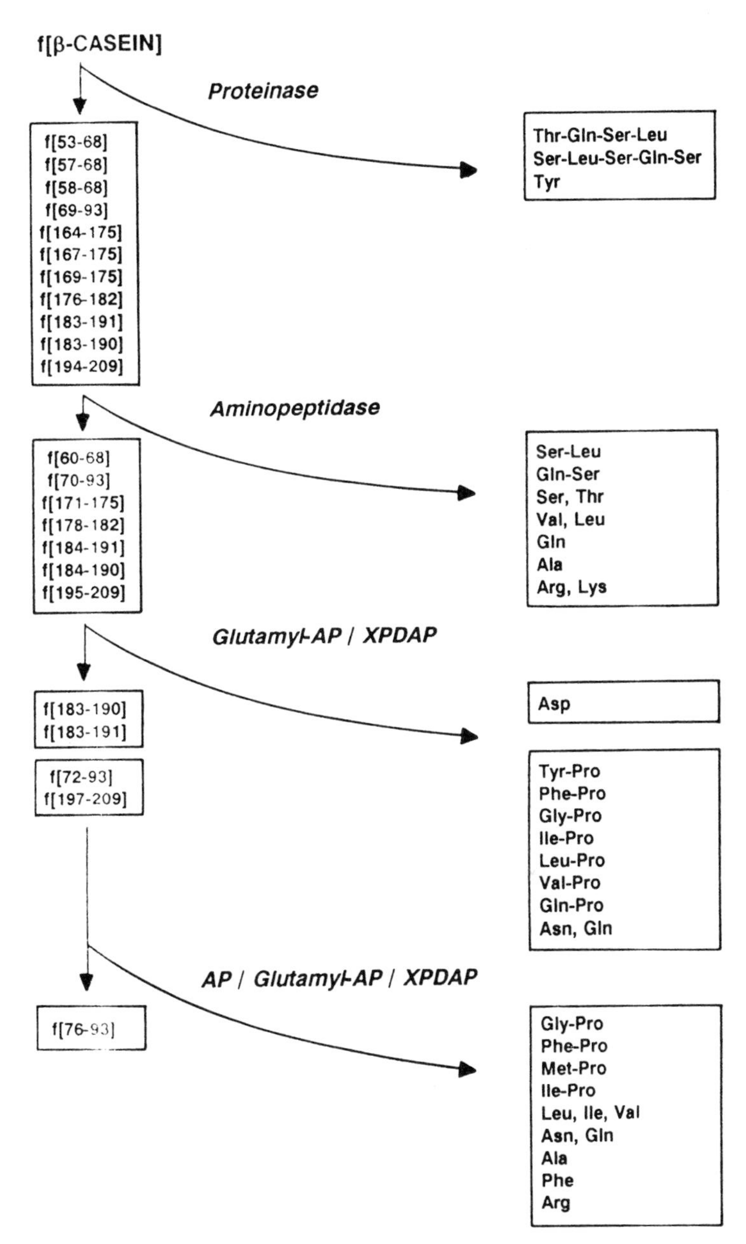

Fig. 14. Scheme of the degradation of β-casein into non-transportable and transportable peptides by the proteolytic system of lactococci: cell wall proteinase, aminopeptidase (AP), X-prolyl dipeptidyl aminopeptidase (XPDAP), glutamyl aminopeptidase (glutamyl AP) (courtesy of SMID et al., 1991).

4.3.3 Amino Acid Metabolism

One important feature is the decarboxylation of amino acids which leads to the accumulation of biogenic amines in food, wine and beer. Cadaverine, putrecine, histamine, tryptamine and tyramine have been identified. Species known to have active decarboxylases include *Lactobacillus buchneri*, pediococci, enterococci, and leuconostocs.

An interesting pathway supplying metabolic energy is the arginine dihydrolase way in *L. buchneri* and a few other lactic acid bacteria (MANCA DE NADRA et al., 1988): Arginine is deaminated into citrulline by arginine deiminase. Citrulline is transformed in the presence of inorganic phosphate into ornithine and carbamyl phosphate. Kinase produces ATP from ADP and carbamyl phosphate. For the heterofermentative *L. buchneri* gaining only one ATP by the heterolactic fermentation under anaerobic conditions, the arginine dihydrolase pathway doubles the available metabolic energy.

Recently, two unusual amino acid derivatives were detected in rather high concentrations in the amino acid pool of lactococci: N^5-(carboxyethyl)-ornithine and N^6-(carboxyethyl)-lysine. Such compounds are rare in nature, they constitute the opines produced in crown galls of plants infected with *Agrobacterium tumefaciens*. N^5-(carboxyethyl)-ornithine is produced by an NADPH-dependent reductive condensation between ornithine and pyruvate by an $NADP^+$-specific N^5-(L-1-carboxyethyl)-L-ornithine oxidoreductase (EC 1.5.1.24).

4.3.4 Amino Acid and Peptide Transport

In lactococci, amino acids are taken up by three mechanisms:

(1) Proton motive force-driven transport: L-methionine, L-leucine, L-isoleucine, L-valine, L-serine, L-threonine, L-alanine, glycine, and L-lysine, possibly also L-histidine, L-cysteine, L-tyrosine, and L-phenylalanine.
(2) Phosphate bond-linked transport driven by ATP or ATP-derived metabolite: glutamate, glutamine, asparagine, and possibly aspartate.
(3) Exchange-transport: arginine/ornithine antiport, driven by the concentration gradient of the two amino acids resulting from arginine uptake and ornithine excretion. Arginine is metabolized by the arginine deiminase pathway (see above).

In lactococci, a dipeptide transport system with a broad specificity is present. It is driven by proton motive forces, in part generated by the splitting of the transported peptides by peptidases in the cytoplasm.

4.3.5 Bacteriocins

There are innumerable papers on inhibitory substances produced by lactic acid bacteria which claim inhibitory activities against putrefactive and pathogenic microorganisms including *Salmonella* and other Gram-negative bacteria. However, only a few active principles have been identified including lactic acid, diacetyl and H_2O_2. A class of substances is gaining increasing interest: bacteriocins which are defined as proteins produced by bacteria and usually inhibiting closely related strains and species only.

The best known compound is nisin, a polypeptide produced by some strains of *Lactococcus lactis*. It has been discovered in 1928 and been in use as food preservative even before its chemical structure and mechanism of action were known. Nisin is inhibitory for most Gram-positive bacteria including some staphylococci, enterococci, pediococci, lactobacilli, leuconostocs, listerias, corynebacteria, *Mycobacterium tuberculosis*, and especially germinating spores of bacilli and clostridia, e.g., *Clostridium botulinum*. The mechanism of action involves the breakdown of the electrochemical potential of bacterial membranes. The chemical structure has been elucidated and is shown in Fig. 15.

Recently, the gene for nisin has been cloned and sequenced. The pentacyclic peptide of 34 amino acid residues contains the three unusual amino acids dehydroalanine, lanthionine and β-methyllanthionine. Nisin is ribosomally synthesized as a 57 amino acids containing prepeptide which is post-translationally processed. The *meso*-lanthionine and the 3-methyllanthionine bridges are formed by post-translational modifications at a cysteine, serine and threonine residue, respectively. The nucleotide sequence of the structural gene is shown in Fig. 16. Nisin is a classical "lantibiotic" being named due to their lanthionine content. Although other bacteriocins from lactococci have been known for a while, e.g., diplococcins, lactacins, and lactococcins (GEIS et al., 1983), their structures have only recently been elucidated due to modern techniques of molecular genetics. VENEMA's group in Groningen (VAN BELKUM et al., 1992) and MCKAY's group in Minneapolis (STODDARD et al., 1992) have cloned and analyzed the plasmid-coded bacteriocin opera from two strains of *Lactococcus lactis* ssp. *lactis* and *cremoris*, respectively. The *"cremoris"* bacteriocin plasmid p9B4-6 (NEVE et al., 1984) contained structural genes for three (!) different polypeptides with bacteriocin activities and immunities explaining the unusually severe problems to purify the active peptides. The nucleotide sequence of the plasmid DNA responsible for lactococcin A revealed several striking features:

(1) the Lcn-operon (of *L. lactis* ssp. *lactis*) contains 4 open reading frames LcnD, LcnC, LcnA, and LciA under the direction of one promoter.
(2) LcnA is the structural gene for the active bacteriocin. Although it has an N-

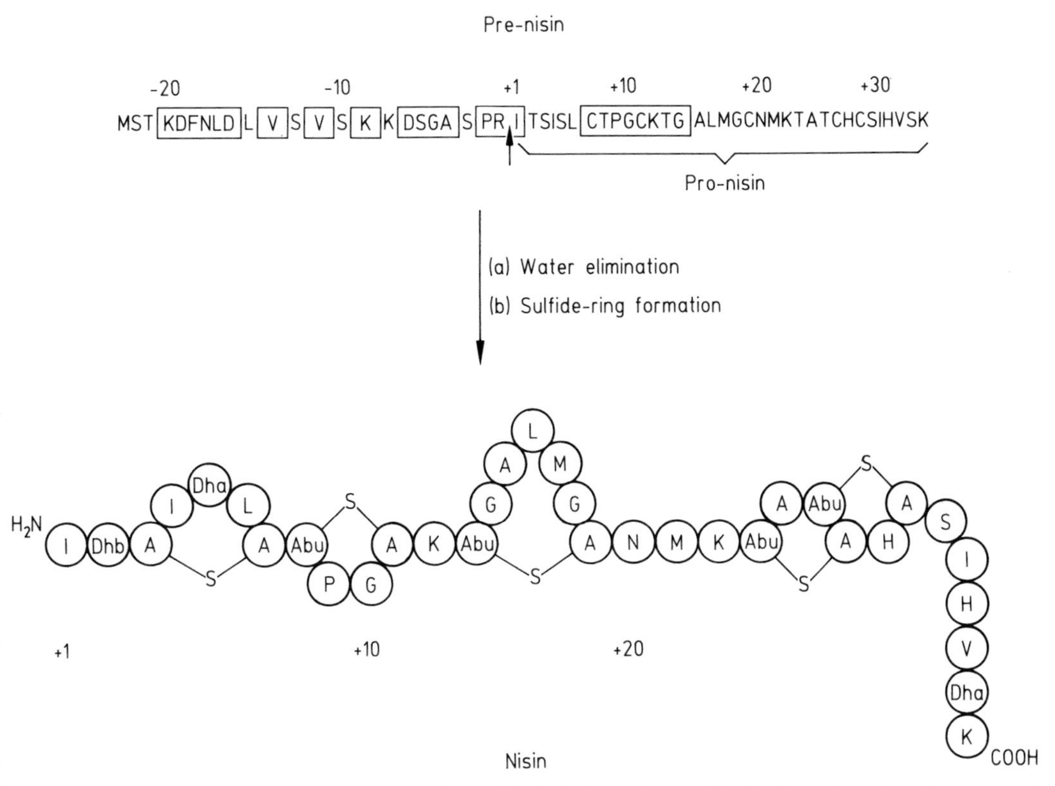

Fig. 15. Processing of the pre-pro-nisin amino acid chain into the antibiotically active nisin (courtesy of KALETTA and ENTIAN, 1989).

terminal stretch of 23 amino acids which is missing in the final polypeptide, this is not a typical signal peptide. LciA is the gene determining lactococcin immunity.

(3) LcnC contains 474 codons and is proposedly a protein of 52.5 kDa. Regarding its presumptive amino acid sequence, it has a 28.7% identity and a 18.7% similarity with gene HlyB, an ATP-dependent membrane translocator protein for a hemolysin in *Escherichia coli.*

(4) LcnD contains 716 codons for a 79.9 kDa protein. It has less than 20% identity and 13% similarity with HlyD.

This is the first experimental report that protein export in Gram-positive bacteria may use a similar signal peptide-independent export system like Gram-negative bacteria.

Bacteriocins have been detected in all investigated lactic acid bacteria including lactobacilli, pediococci, leuconostocs, and especially enterococci (KLAENHAMMER, 1988). However, their state of analysis is usually far from the sophistication reached in lactococci. It is thought that production of bacteriocins provides a selective advantage in an environment where many different, but closely related organisms fight for survival.

5 Genetics and Genetic Engineering

Despite the early breakthrough of GRIFFITH (1928) and AVERY, MACLEOD and MCCARTHY (1944) with *Streptococcus pneumoniae*, molecular genetics of lactic acid bacteria (LAB) started rather late in 1972, when MCKAY's group obtained the first evidence for the presence of plasmids in lactococci (MCKAY et al., 1972). By 1980, the presence of plasmids in all investigated lactococcal strains was established (PECHMANN and TEUBER, 1980). In 1979, GASSON and DAVIES reported the first transfer of lactose genes in lactococci by natural conjugation, and in 1980 the conjugal transfer of the MLS resistance plasmid pAMβ from *Enterococcus* into *Lactococcus.* Since then and with these basic phenomena as tools, the molecular genetics of the lactococci and other LAB have enormously expanded. For the most advanced lactococci, the present knowledge can be briefly summarized as follows (TEUBER et al., 1991; GASSON, 1990).

Lactococcus lactis

The great excitement over the plasmids of lactococci (see Fig. 17) has its explanation in

```
AAAATAAATTATAAGGAGGCACTCAAA ATG AGT ACA AAA GAT TTT AAC TTG GAT
                             M   S   T   K   D   F   N   L   D
                                        -20

TTG GTA TCT GTT TCG AAG AAA GAT TCA GGT GCA TCA CCA CGC ATT ACA
 L   V   S   V   S   K   K   D   S   G   A   S   P   R ↑ I   T
                -10                                 -1   +1

AGT ATT TCG CTA TGT ACA CCC GGT TGT AAA ACA GGA GCT CTG ATG GGT
 S   I   S   L   C   T   P   G   C   K   T   G   A   L   M   G
                            +10

TGT AAC ATG AAA ACA GCA ACT TGT CAT TGT AGT ATT CAC GTA AGT AAA
 C   N   M   K   T   A   T   C   H   C   S   I   H   V   S   K
    +20                                     +30

TAACCAAATCAAAGGATAGTATTTT
OC
```

Fig. 16. Nucleotide sequence of the nisin structural gene (*nisA*) from *Lactococcus lactis* (courtesy of KALETTA and ENTIAN, 1989).

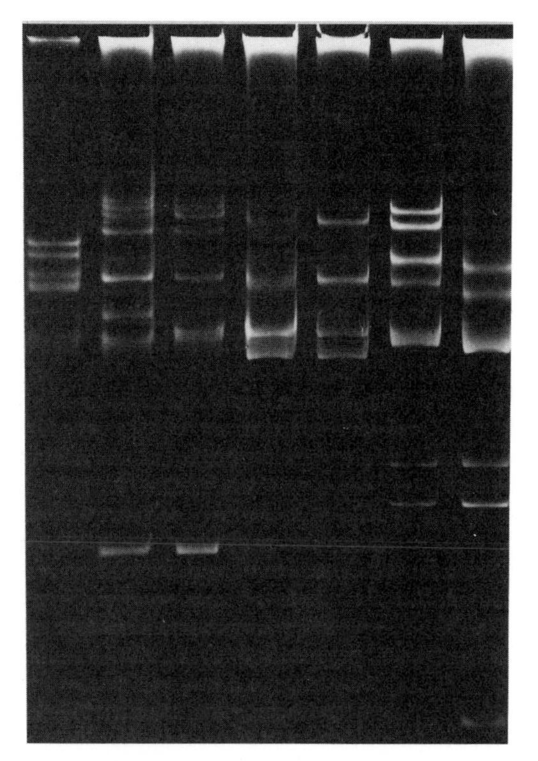

Fig. 17. Plasmid profiles from different lactococcal strains isolated from one starter culture.

the fact that many technologically necessary functions for dairy fermentations are coded on plasmids (Tab. 11). Many plasmids are self-transmissible by natural conjugation. Non-self-transmissible plasmids may be mobilized by conjugative plasmids. Unfortunately, the biochemical and ultrastructural details of this conjugation are almost completely unknown. In addition to conjugation, DNA transfer into lactococci can be accomplished by

- transduction since many strains carry prophages,
- protoplast transformation and transfection,
- protoplast fusion, and
- transformation by electroporation of intact cells.

These methods of gene transfer have allowed the allocation of specific biochemical traits to specific plasmids when transfer into plasmid cured derivatives was obtained. Lactococci harbor at least two different IS elements, ISS1 and IS904, which both can be found in plasmids and in the chromosome (SCHÄFER et al., 1991). IS904 shows some homology with the transposases of IS3 and IS600 from *E. coli* and *Shigella dysenteriae*, respectively. During conjugation, conjugative and non-conjugative plasmids frequently form co-integrates with the aid of IS elements which may be the basic mechanism of co-transfer of plasmids in lactococci (and of recombination). Conjugation is especially effective in strains exhibiting a clumping phenomenon, a favorable condition for intimate cell to cell contact between donor and recipient. Conjugation efficiency may be 10^{-4} and higher. During conjugation, plasmids carrying IS elements may also be integrated into the chromosome of the recipient cell. After excision from the chromosome, they may carry chromosomal DNA sequences flanked by the corresponding IS elements. Obviously, the presence of plasmids and their exchangeability by conjugation between different strains provide the dairy lactococci with an almost unlimited genetic flexibility regarding plasmid coded functions. Such a flexibility may be of great advantage in a microbiologically open system as it is represented in a dairy plant. Since milk for cheese production cannot be sterilized, lactococci have to compete constantly with microbes of the endogenous flora. The permanent mixing of genetic material allows for the selection of the most adapted under the prevailing conditions.

Gene expression signals in *Lactococcus lactis*, such as promoters and terminators, tran-

Tab. 11. Plasmid Coded Functions in Lactococci (TEUBER, 1990)

(1)	Lactose metabolism: enzyme II and factor III of lactose phosphotransferase system, P-β-galactosidase, tagatose-6-P isomerase, tagatose-6-P kinase, tagatose-1-6-P aldolase
(2)	Citrate metabolism: citrate permease
(3)	Proteolytic system: β-casein specific protease, maturation protein for the protease
(4)	Sucrose-6-P-hydrolase
(5)	Slime production in ropy strains
(6)	Bacteriocin production and immunity
(7)	Bacteriophage insensitivity phenomena

scription initiation sites, and translation initiation signals show considerable similarities to those from *E. coli* and *Bacillus subtilis* (VAN DE GUCHTE et al., 1992). Regarding codon usage, a high within-species diversity is observed.

Homologous and heterologous gene expression has been achieved in *Lactococcus* by the following strategies:

(a) Use of heterologous plasmids replicating in *Lactococcus.* The best known example is plasmid pAMβ1 originating from *Enterococcus faecalis* which carries MLS resistance genes (macrolides, lincosamides, and streptogramin B) and is self-transmissible.

(b) Construction of low (6–9; pIL277) and high copy (45–85; pIL253) number vectors on the basis of pAMβ1, by insertion of a poly-restriction cloning site, and shortening of the remaining pAMβ1 part to the MLS region (coding for the convenient erythromycin resistance) and the replication region. These plasmids carry and stably maintain large DNA-fragments up to 30 kilobases (Fig. 18; SIMON and CHOPIN, 1988).

(c) Construction of cloning vectors on the basis of small cryptic plasmids of *Lactococcus lactis* (DE VOS, 1987). One example are the vectors developed from the smallest cryptic plasmid (2059 base pairs) of *L. lactis* ssp. *cremoris* Wg_2 by incorporation of multiple cloning sites and proper antibiotic resistance markers (e.g., chloramphenicol transferase from *Staphylococcus aureus*, MLS from *Enterococcus*, TEM-*β*-lactamase from *Escherichia coli*). Most remarkably, such vectors are able to cross the lines between *Lactococcus*, *Bacillus subtilis* and even *E. coli.* At least two genetically different replication systems are present in lactococci (JAHNS et al., 1991).

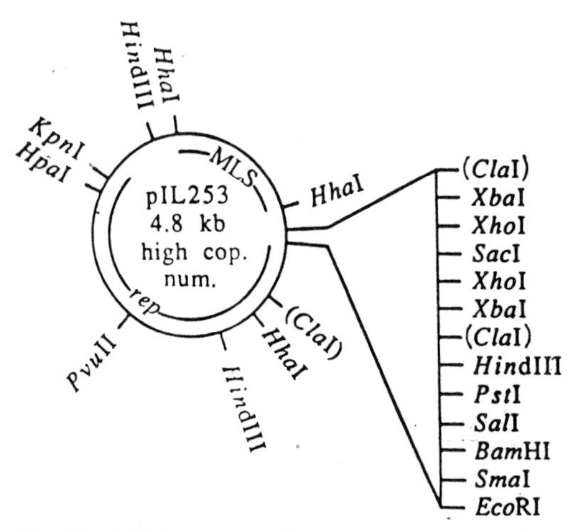

Fig. 18. A high copy number vector constructed on the basis of a small cryptic plasmid from *Lactococcus* supplemented with a multi-restriction site, and the MLS-region from an *Enterococcus* plasmid (courtesy of SIMON and CHOPIN, 1988).

Heterologous genes expressed include the mentioned antibiotic resistance genes, *E. coli lacZ*, *Bacillus licheniformis* α-amylase, *Bacillus subtilis* neutral protease, *Cyamopsis tetragonoloba* α-galactosidase, bovine chymosin, chicken lysozyme, and *Thaumatococcus daniellii* thaumatin (VAN DE GUCHTE et al., 1992). Noteworthy is also the recent conjugal transfer of transposons such as Tn916, Tn919, and Tn1545 from *Streptococcus sanguis*, *Enterococcus faecalis*, and *S. pneumoniae*, respectively (for a review see TEUBER et al., 1991).

The scientific progress in the field of genetic engineering of lactic acid bacteria has been greatly advanced by three large research projects within the BEP, BAP and BRIDGE programs of the European Community starting with four laboratories in the mid nineteen eighties, and comprising now about 25 laboratories in Europe. In the current BRIDGE program, the genetically less developed genera, e.g., *Leuconostoc*, *Lactobacillus*, *Pediococcus* and *Streptococcus thermophilus* will be given proper attention.

Acceptance of genetically modified LAB in fermented food by the consumer and legislation is still a problem (TEUBER, 1990, 1992).

6 Industrial Applications

Without doubt, fermented food and feed, and the corresponding starter cultures are the most important industrial applications of lactic acid bacteria (LAB), though still an appreciable amount is made by spontaneous fermentations. These products will be described in detail in Volume 9 of *Biotechnology.*

Tab. 12 lists the main products and the LAB used for their manufacture.

A minor specialty product is streptokinase isolated from *Streptococcus pyogenes* which activates plasminogen for conversion into plasmin. Streptokinase is an indispensable drug for the dissolution of blood clots in cases of clogging of blood vessels. Freeze-dried preparations of *Bifidobacterium bifidum* and other species of this type, *Lactobacillus acidophilus* and *L. casei*, are given to patients with disturbances of the intestinal microflora caused by antibiotic treatment. Similarly, *L. acidophilus* incorporated into suppositories is applied to restore the vaginal microflora after antibiotic treatment of inflammatory processes of the human vagina. Nisin (see Sect. 4.3.4) is legal and used in some countries as preservative for, e.g., processed cheese to protect it against outgrowth of *Clostridium botulinum*.

Many vitamins can be conveniently analyzed with a high precision at low concentrations by taking advantage of the requirements of many LAB for vitamins for growth. This elegant method has been invented by E. E. SNELL in 1939 (SNELL, 1952) and is the basis of a number of official AOAC methods (see Tab. 13).

Tab. 12. Examples for Main Lactic Acid Bacteria Used in Spontaneous and Controlled Industrial Fermentations of Food and Feed (TEUBER et al., 1987)

Product	Spontaneous Flora	Starter Cultures
Silage	*Leuconostoc* sp., *Enterococcus* sp., *Pediococcus* sp., *Lactobacillus plantarum*	*Lactobacillus plantarum*, *Enterococcus* sp.
Sauerkraut, pickles, olives, etc.	Same as silage	*Lactobacillus bavaricus*
Sour dough		*Lactobacillus sanfrancisco*, *Lactobacillus brevis*, *Lactobacillus plantarum*
Wine, cider		*Leuconostoc oenos*
Yoghurt		*Lactobacillus delbrueckii*, ssp. *bulgaricus*[a], *Streptococcus thermophilus*
Emmental cheese		*Lactobacillus helveticus*
Sour cream/milk		*Lactococcus lactis* ssp. *lactis*, *Lactococcus lactis* ssp. *lactis* biovar *diacetylactis*, *Lactococcus lactis* ssp. *cremoris*, *Leuconostoc lactis*, *Leuconostoc mesenteroides* ssp. *cremoris*
Gouda, Edam		Same as sour cream
Kefir	Kefir grains *Lactobacillus kefir* *Lactobacillus kefiranofaciens* *Lactococcus lactis*	
Salami, cured ham		*Lactobacillus curvatus*, *Lactobacillus sake*, *Pediococcus* sp.
Soy sauce/paste	*Pediococcus* sp.	*Pediococcus* sp.

[a] in modern mild tasting products, *Lactobacillus delbrueckii* ssp. *bulgaricus* is frequently exchanged for *Lactobacillus acidophilus* and/or *Bifidobacterium* sp.

Tab. 13. Lactic Acid Bacteria Used For the Microbiological Determination of Vitamins (SNELL, 1952; and Official AOAC Methods)

Vitamin	Used Species	
Thiamine	*Lactobacillus viridescens*	ATCC 12706
Riboflavine	*Lactobacillus casei*	ATCC 7649
Niacin	*Lactobacillus plantarum*	ATCC 8014
Folic acid	*Lactobacillus casei*	ATCC 7469
B12	*Lactobacillus leichmannii*	ATCC 7830
Panthotenate	*Lactobacillus plantarum*	ATCC 8014
Biotin	*Lactobacillus plantarum*	ATCC 8014

7 References

AVERY, O. T., MACLEOD, C. M., MCCARTHY, M. (1944), Studies on the chemical nature of the substance inducing transformation of pneumococcal types. I. Induction of transformation by a deoxyribonucleic acid fraction isolated from *Pneumococcus* type III, *J. Exp. Med.* **79,** 137–158.

BIAVATI, B., SGORBATI, B., SCARDOVI, V. (1991), The genus *Bifidobacterium*, in: *The Prokaryotes* (BALOWS, A., TRÜPER, H. G., DWORKIN, M., HARDER, W., SCHLEIFER, K. H., Eds.), 2nd Ed., Vol. 1, pp. 816–833, New York: Springer-Verlag.

BRAUN, V., Jr., HERTWIG, S., NEVE, H., GEIS, A., TEUBER, M. (1989), Taxonomic differentiation of bacteriophages of *Lactococcus lactis* by electron microscopy, DNA–DNA hybridization, and protein profiles, *J. Gen. Microbiol.* **135,** 2551–2560.

CERNING, J. (1990), Exocellular polysaccharides produced by lactic acid bacteria, *FEMS Microbiol. Rev.* **87,** 113–130.

CONDON, S. (1987), Responses of lactic acid bacteria to oxygen, *FEMS Microbiol. Rev.* **46,** 269–280.

DE MAN, J. C., ROGOSA, M., SHARPE, E. (1960), A medium for the cultivation of lactobacilli, *J. Appl. Bacteriol.* **23,** 130–135.

DE STOPPELAAR, J. D., VAN HOUTE, J., DE MOOR, C. E. (1967), The presence of dextran forming bacteria resembling *Streptococcus bovis* and *Streptococcus sanguis* in human dental plaque, *Arch. Oral Biol.* **12,** 1199–1201.

DE VOS, W. M. (1987), Gene cloning and expression in lactic streptococci, *FEMS Microbiol. Rev.* **46,** 281–295.

DEVRIESE, L. A., COLLINS, M. D., WIRTH, R. (1991), The genus *Enterococcus*, in: *The Prokaryotes* (BALOWS, A., TRÜPER, H. G., DWORKIN, M., HARDER, W., SCHLEIFER, K. H., Eds.), 2nd Ed., Vol. 2, pp. 1465–1481, New York: Springer-Verlag.

ELLIKER, P. R., ANDERSON, A. W., HANNESSON, G. (1956), An agar medium for lactic acid streptococci and lactobacilli, *J. Dairy Sci.* **39,** 1611–1612.

FOX, P. F. (1989), Some non-European cheese varieties, in: *Cheese: Chemistry, Physics and Microbiology*, Vol. 2: *Major Cheese Groups* (FOX, P. F., Ed.), pp. 311–337, London: Elsevier Applied Science.

FULLER, R. (Ed.) (1992), *Probiotics, the Scientific Basis*, London: Chapman & Hall.

GARVIE, E. I. (1967), *Leuconostoc oenos* sp. nov., *J. Gen. Microbiol.* **48,** 431–438.

GASSON, M. J. (1990), *In vivo* genetic systems in lactic acid bacteria, *FEMS Microbiol. Rev.* **87,** 43–60.

GASSON, M. J., DAVIES, L. B. (1979), Conjugal transfer of lactose genes in group N streptococci, *Soc. Gen. Microbiol. Quart.* **6,** 86.

GASSON, M. J., DAVIES, F. L. (1980), Conjugal transfer of the drug resistance plasmid pAMβ in the lactic streptococci, *FEMS Microbiol. Lett.* **7,** 51–53.

GEIS, A., SINGH, J., TEUBER, M. (1983), Potential of lactic streptococci to produce bacteriocins, *Appl. Environ. Microbiol.* **45,** 205–211.

GRIFFITH, F. (1928), Significance of pneumococcal types, *J. Hyg.* (London) **27,** 113–159.

GURR, M. (Ed.) (1983), Cultured dairy foods in human nutrition, *Int. Dairy Fed. Bull.* Document **159,** Brussels: IDF.

HAMMES, W. P., WEISS, N., HOLZAPFEL, W. (1991), The genera *Lactobacillus* and *Carnobacterium*, in: *The Prokaryotes* (BALOWS. A., TRÜPER, H. G., DWORKIN, M., HARDER, W., SCHLEIFER, K. H., Eds.), 2nd Ed., Vol. 2, pp. 1535–1594, New York: Springer-Verlag.

HARDIE, J. M., WHILEY, R. A. (1991), The genus *Streptococcus* – oral, in: *The Prokaryotes* (BALOWS, A., TRÜPER, H. G., DWORKIN, M., HARDER, W., SCHLEIFER, K. H., Eds.), 2nd Ed., Vol. 2, pp. 1421–1449, New York: Springer-Verlag.

HOLZAPFEL, W. H., SCHILLINGER, U. (1991), The genus *Leuconostoc*, in: *The Prokaryotes* (BALOWS, A., TRÜPER, H. G., DWORKIN, M., HARDER, W.,

SCHLEIFER, K. H., Eds.), 2nd Ed., Vol. 2, pp. 1508–1534, New York: Springer-Verlag.

JAHNS, A., SCHÄFER, A., GEIS, A., TEUBER, M. (1991), Identification, cloning and sequencing of the replication region of *Lactococcus lactis* ssp. *lactis* biovar *diacetylactis* Bu2 citrate plasmid pSL2, *FEMS Microbiol. Lett.* **80,** 253–258.

KALETTA, C., ENTIAN, K. D. (1989), Nisin, a peptide antibiotic: cloning and sequencing of the *nisA* gene and posttranslational processing of its peptide product, *J. Bacteriol.* **171,** 1597–1601.

KANDLER, O. (1983), Carbohydrate metabolism in lactic acid bacteria, *Antonie van Leeuwenhoek* **49,** 209–224.

KASHKET, E. R. (1987), Bioenergetics of lactic acid bacteria: cytoplasmic pH and osmotolerance, *FEMS Microbiol. Rev.* **46,** 233–244.

KLAENHAMMER, T. R. (1988), Bacteriocins of lactic acid bacteria, *Biochimie* **70,** 337–349.

KLEYNMANS, U., HEINZEL, H., HAMMES, W. (1989), *Lactobacillus suebicus* spec. nov., an obligately heterofermentative *Lactobacillus* species isolated from fruit mashes, *Syst. Appl. Microbiol.* **11,** 267–271.

KOK, J. (1990), Genetics of the proteolytic system of lactic acid bacteria, *FEMS Microbiol. Rev.* **87,** 15–42.

KONINGS, W. N., OTTO, R. (1983), Energy transduction and solute transport in streptococci, *Antonie van Leeuwenhoek* **49,** 247–257.

LICK, S., TEUBER, M. (1992), Construction of a species-specific DNA oligonucleotide probe for *Streptococcus thermophilus* on the basis of a chromosomal *lacZ* gene, *Syst. Appl. Microbiol.* **15,** 456–459.

LISTER, J. (1873), A further contribution to the natural history of bacteria and the germ theory of fermentative changes, *Quart. Microbiol. Sci.* **13,** 380–408.

LÖHNIS, F. (1909), Die Benennung der Milchsäurebakterien, *Zentralbl. Bakteriol. Parasitenkd. Infektionskr. Hyg.,* Abt. 2, **22,** 553–555.

LONDON, J. (1990), Uncommon pathways of metabolism among lactic acid bacteria, *FEMS Microbiol. Rev.* **87,** 103–112.

MALONEY, P. C. (1990), Microbes and membrane biology, *FEMS Microbiol. Rev.* **87,** 91–102.

MANCA DE NADRA, M. C., PESCE DE RUIZ HOLGADO, A. A., OLIVER, G. (1988) Arginine dihydrolase pathway in *Lactobacillus buchneri:* a review, *Biochimie* **70,** 367–374.

MAYO, B., KOK, J., VENEMA, K., BOCKELMANN, W., TEUBER, M., REINKE, H., VENEMA, G. (1991), Molecular cloning and sequence analysis of the X-prolyl dipeptidyl aminopeptidase gene from *Lactococcus lactis* ssp. *cremoris*, *Appl. Environ. Microbiol.* **57,** 38–44.

MCKAY, L. L., BALDWIN, K. A., ZOTTOLA, E. A. (1972), Loss of lactose metabolism in lactic streptococci, *Appl. Microbiol.* **23,** 1090–1096.

MONNET, V., BOCKELMANN, W., GRIPON, J. C., TEUBER, M. (1989), Comparison of cell wall proteinases from *Lactococcus lactis* ssp. *cremoris* AC1 and *Lactococcus lactis* ssp. *lactis* NCDO 763, *Appl. Microbiol. Biotechnol.* **31,** 112–118.

MOSSEL, D. A. A., BIJKER, P. G. H., EELDERINK, I. (1978), Streptococci of Lancefield groups A, B and D and those of buccal origin in foods: their public health significance, monitoring and control, in: *Streptococci* (SKINNER, F. A., QUESNEL, L. B., Eds.), pp. 315–333, London: Academic Press.

NEVE, H., GEIS, A., TEUBER, M. (1984), Conjugal transfer and characterization of bacteriocin plasmids in group N (lactic) streptococci, *J. Bacteriol.* **157,** 833–838.

NISSEN, H. J., DAMEROW, P., EGLUND, R. K. (1990), *Frühe Schrift und Techniken der Wirtschaftsverwaltung im alten Vorderen Orient – Informationsspeicherung und -verarbeitung vor 5000 Jahren*, Berlin: Verlag Franzbecker.

ORLA-JENSEN, S. (1919), *The lactic acid bacteria*, Copenhagen: Host & Son.

PASTEUR, L. (1857), Mémoire sur la fermentation appelée lactique, *C. R. Séances 'Acad. Sci*, Tome **45,** 913–916.

PEARCE, L. E., HALLIGAN, A. C. (1978), Cultural characteristics of *Leuconostoc* strains from cheese starters, in: *Proc. 20th Int. Dairy Congr.*, pp. 520–521, Paris: Congrilait.

PECHMANN, H., TEUBER, M. (1980), Plasmid pattern of group N (lactic) streptococci, *Zentralbl. Bakteriol. Hyg., I. Abt. Orig.* **C 1,** 133–136.

RENAULT, P., GAILLARDIN, C., HESLOT, H. (1988), Role of malolactic fermentation in lactic acid bacteria, *Biochimie* **70,** 375–379.

REUTER, G. (1975), Classification problems, ecology and some biochemical activities of lactobacilli of meat products, in: *Lactic acid bacteria in beverages and food* (CARR, G. J., CUTTING, C. V., WHITING, G. C., Eds.), pp. 221–229, London: Academic Press.

ROGOSA, M., MITCHELL, J. A., WISEMAN, R. F. (1951), A selective medium for the isolation and enumeration of oral and fecal lactobacilli, *J. Bacteriol.* **62,** 132–133.

RUOFF, K. L. (1991), The genus *Streptococcus* – medical, in: *The Prokaryotes* (BALOWS, A., TRÜPER, H. G., DWORKIN, M., HARDER, W., SCHLEIFER, K. H., Eds.), 2nd Ed., Vol. 2, pp. 1450–1464, New York: Springer-Verlag.

SCHÄFER, A., JAHNS, A., GEIS, A., TEUBER, M. (1991), Distribution of the IS elements *ISS1* and *IS904* in lactococci, *FEMS Microbiol. Lett.* **80,** 311–318.

SCHEELE, C. W. (1780), Von der Milch und ihrer Säure, in: *Carl Wilhelm Scheele, Sämmtliche Physische und Chemische Werke* (HERMBSTÄDT, D. S. F., Ed.), Vol. 1, 1793, pp. 249–260. Berlin: Heinrich August Rottmann.

SCHLEIFER, K. H. (1987), Recent changes in the taxonomy of lactic acid bacteria, *FEMS Microbiol. Rev.* **46,** 201–203.

SCHLEIFER, K. H., KANDLER, O. (1972), Peptidoglycan types of bacterial cell walls and their taxonomic implications, *Bacteriol. Rev.* **36,** 407–477.

SCHMALRECK, A. F., TEUBER, M., REININGER, W., HARTL, A. (1975), Structural features determining the antibiotic potencies of natural and synthetic hop bitter resins, their precursors and derivatives, *Can. J. Microbiol.* **21,** 205–212.

SIMON, D., CHOPIN, A. (1988), Construction of a vector plasmid family and its use for molecular cloning in *Streptococcus lactis*, *Biochimie* **70,** 559–566.

SMID, E. J., POOLMAN, B., KONINGS, W. N. (1991), Casein utilization by lactococci, *Appl. Environ. Microbiol.* **57,** 2447–2452.

SNELL, E. E. (1952), The nutrition of lactic acid bacteria, *Bacteriol. Rev.* **16,** 235–241.

SPICHER, G. STEPHAN, H. (1987), *Handbuch Sauerteig: Biologie, Biochemie, Technologie*, Hamburg: Behr's Verlag.

STACKEBRANDT, E., TEUBER, M. (1988), Molecular taxonomy and phylogenetic position of lactic acid bacteria, *Biochimie* **70,** 317–324.

STAVRIC, S., GLEESON, T.M., BUCHANAN, B., BLANCHFIELD, B. (1992), Experience of the use of probiotics for Salmonellae control in poultry, *Lett. Appl. Microbiol.* **14,** 69–71.

STODDARD, G. W., PETZEL, J. P., VAN BELKUM, M. J., KOK, J., MCKAY, L. L. (1992), Molecular analyses of the lactococcin A gene cluster from *Lactococcus lactis* subsp. *lactis* biovar *diacetylactis* WM4, *Appl. Environ. Microbiol.* **58,** 1952–1961.

TERZAGHI, B. E., SANDINE, W. E. (1975), Improved medium for lactic streptococci and their bacteriophages, *Appl. Microbiol.* **29,** 807–813.

TEUBER, M. (1987), *Grundriß der praktischen Mikrobiologie für das Molkereifach*, 2nd Ed., Gelsenkirchen-Buer: Th. Mann.

TEUBER, M. (1990), Strategies for genetic modification of lactococci, *Food Biotechnol.* **4,** 537–546.

TEUBER, M. (1992), Exploitation of genetically-modified microorganisms in the food industry, in: *The release of genetically modified microorganisms – Regem 2* (STEWART-TULL, D. E. S., SUSSMAN, M., Eds.), pp. 59–69, New York: Plenum Press.

TEUBER, M., GEIS, A. (1981), The family Streptococcaceae (nonmedical aspects), in: *The Prokaryotes* (STARR, M. P., STOLP, H., TRÜPER, H. G., BALOWS, A., SCHLEGEL, H. G., Eds.), Vol. 2, pp. 1614–1630, New York: Springer-Verlag.

TEUBER, M., GEIS, A., KRUSCH, U., LEMBKE, J., MOEBUS, O. (1987), Biotechnologische Verfahren zur Herstellung von Lebensmitteln und Futtermitteln, in: *Handbuch der Biotechnologie* (PRÄVE, P., FAUST, U., SITTIG, W., SUKATSCH, D. A., Eds.), 3rd Ed., pp. 269–318, München: R. Oldenbourg.

TEUBER, M., GEIS, A., NEVE, H. (1991), The genus *Lactococcus*, in: *The Prokaryotes* (BALOWS, A., TRÜPER, H. G., DWORKIN, M., HARDER, W., SCHLEIFER, K. H., Eds.), 2nd Ed., Vol. 2, pp. 1482–1501, New York: Springer-Verlag.

THOMPSON, J. (1988), Lactic acid bacteria: model systems for *in vivo* studies of sugar transport and metabolism in Gram-positive organisms, *Biochimie* **70,** 325–336.

VAN BELKUM, M. J., KOK, J., VENEMA, G. (1992), Cloning, sequencing, and expression in *Escherichia coli* of *lncB*, a third bacteriocin determinant from the lactococcal bacteriocin plasmid p9B4-6, *Appl. Environ. Microbiol.* **58,** 572–577.

VAN DE GUCHTE, M., KOK, J., VENEMA, G. (1992), Gene expression in *Lactococcus lactis*, *FEMS Microbiol. Rev.* **88,** 73–92.

WÄHREN, M. (1990), Brot und Getreide in der Urgeschichte, in: *Die ersten Bauern*, Vol. 1, pp. 117–118, Zürich: Schweizerisches Landesmuseum.

WEIGMANN, H. (1905–1908), Das Reinzuchtsystem in der Butterbereitung und in der Käserei, in: *Handbuch der Technischen Mykologie*, Vol. 2, *Mykologie der Nahrungsmittelgewerbe* (LAFAR, F., Ed.), pp. 2293–2309, Jena: Gustav Fischer Verlag.

WEISS, N. (1991), The genera *Pediococcus* and *Aerococcus*, in: *The Prokaryotes* (BALOWS, A., TRÜPER, H. G., DWORKIN, M., HARDER, W., SCHLEIFER, K. H., Eds.), 2nd Ed., Vol. 2, pp. 1502–1507, New York: Springer-Verlag.

WOOLFORD, M. K. (1984), *The Silage Fermentation*, New York: Marcel Dekker.

11 *Bacillus*

FERGUS G. PRIEST
Edinburgh, Scotland, U.K.

1 Introduction 368
2 Taxonomy 368
 2.1 Traditional Approaches 368
 2.2 Numerical Phenetics and Chemotaxonomy 369
 2.3 Phylogenetic Analyses and Evolution 372
 2.4 Identification of *Bacillus* Species 372
3 Ecology 373
 3.1 Distribution and Habitats 374
 3.2 Associations with Plants 375
 3.3 Associations with Animals 376
  3.3.1 Pathogens of Man and Mammals 376
  3.3.2 Pathogens of Insects 376
4 Extremophiles 378
 4.1 Thermophiles 378
 4.2 Psychrophiles 379
 4.3 Alkaliphiles 379
 4.4 Acidophiles 381
5 Physiology 381
 5.1 Relations to Oxygen 381
 5.2 Carbon Metabolism 383
  5.2.1 Catabolite Repression 386
 5.3 Nitrogen Metabolism 387
6 The Endospore 388
 6.1 Endospore Structure and Properties 388
 6.2 Sporulation 391
 6.3 Germination 393
7 Genetics and Molecular Biology 394
 7.1 Transformation 395
 7.2 Plasmids 395
 7.3 Recombinant Gene Products from *Bacillus* 396
8 Concluding Remarks 396
9 References 397

1 Introduction

Bacteria belonging to the genus *Bacillus* have a long and distinguished history in the realms of biotechnology. They were probably first used by the Japanese in the preparation of a traditional fermented food from rice straw and soybean, itohiki-natto. This derives from the action of *"Bacillus natto"* (a derivative of *B. subtilis*) on steamed soybean and results in a viscous, sticky polymer (primarily polyglutamic acid) that forms long, thin threads when touched. Natto has been prepared in Japan for at least four hundred years, and currently consumption is about 10^8 kg per annum (UEDA, 1989).

Exploitation of *Bacillus* in the west is more recent. Manufacture of extracellular amylases and proteases for industrial applications began early this century, but significant production and usage was delayed until after the 1950s when the advantages of including the alkaline protease (subtilisin Carlsberg) of *Bacillus licheniformis* in washing detergents was realized. This was followed by developments in the starch processing industry based on the α-amylase from *B. licheniformis*, particularly the conversion of starch to high-fructose corn syrups as sucrose replacements in foods and beverages (PRIEST, 1989b).

The bacilli include many versatile bacteria and the most effective bacterial control agents for various insect pests. *Bacillus thuringiensis* was isolated early this century in Japan from diseased silkworms and in Germany from the Mediterranean mealmoth. Its value as a control agent for lepidoptera was realized soon thereafter, but only developed commercially in the last 40 years. It now has an annual market value of US $ 107 million which is forecast to reach $ 300 million by the year 1999 (see FEITELSON et al., 1992). Bacilli are also sources of numerous antibiotics, flavor enhancers such as purine nucleosides, surfactants, and various other products (PRIEST, 1989b). These biotechnological attributes, together with sporulation and germination, the molecular biology of which provides a model system for differentiation, have led to great interest in these bacteria and resulted in *B. subtilis* becoming the most advanced genetic system available in any Gram-positive bacterium (SONENSHEIN et al., 1993). But as with *Escherichia coli*, this concentration on one species has resulted in many others being ignored and the great diversity within the genus *Bacillus* is seldom appreciated. In this chapter, I shall redress the balance and review current knowledge of the biology of these fascinating bacteria.

2 Taxonomy

2.1 Traditional Approaches

Gram-positive, rod-shaped bacteria that differentiate into heat-resistant endospores under aerobic conditions are placed in the genus *Bacillus*. Of the other endospore-forming bacteria, strict anaerobes are allocated to *Clostridium*, cocci to *Sporosarcina* and branching filamentous forms to *Thermoactinomyces* (PRIEST and GRIGOROVA, 1990). It is therefore quite simple to identify bacilli microscopically, and early this century this led to many new "species" being described. It was much simpler to give a new isolate a new species name than to attempt to identify it! By the 1940s some 150 species had been proposed, many synonymous, and most with poor descriptions. The basis of *Bacillus* classification and identification was established at that time by TOM GIBSON working in Edinburgh and RUTH GORDON, FRANK CLARK and NATHAN SMITH at the Northern Regional Research Laboratory, Peoria, Illinois. SMITH and his colleagues examined 1134 strains representing over 150 species and allocated them to just 19 species. In a later, comprehensive monograph (GORDON et al., 1973), strain and species descriptions were provided complete with full methodology, and this became the primary reference for isolation and identification of bacilli.

GIBSON and GORDON also promoted the concept of morphological groups of bacilli based on the shape of the endospore (oval or spherical) and its position in the mother cell or sporangium. Thus, group I species, including *B. subtilis* and many other common bacilli, differentiate into oval spores of the same diameter as the mother cell; group II species (*B. polymyxa*, etc.) possess oval spores that distend the mother cell, and group III species (notably *B. sphaericus*) produce spherical spores. This division of the genus has been very useful for identification purposes because it reduces the taxon into groups of more manageable size. However, it is sometimes difficult to distinguish the different classes, and this can lead to serious misidentification (GORDON, 1981). Interestingly, the morphological groups of bacilli have recently been shown to have some phylogenetic validity (see Sect. 2.3).

2.2 Numerical Phenetics and Chemotaxonomy

With the introduction of modern taxonomic techniques such as numerical phenetics, DNA base composition determinations, and DNA reassociation experiments which allow DNA sequence homology between strains to be estimated, it became apparent that the bacilli were more heterogeneous than hitherto suspected. The range of DNA base composition among strains is a good indicator of genetic diversity; indeed it is generally agreed that species in a genus should vary by no more than 10–12 mol% G+C (JOHNSON, 1989). In the case of *Bacillus*, the range is about 33 to 65% although strains of most species cluster between 40 and 50% (PRIEST, 1981). This indicates considerable genetic diversity among species and suggests that the genus should perhaps be split into several, more homogeneous taxa.

Bacilli are also physiologically diverse (see Sect. 5), and this can best be appreciated by numerical classification in which strains are examined for numerous physiological, biochemical and morphological characters and grouped together on the basis of character similarities. Clusters of phenotypically similar strains are revealed by this process and these can usually be equated with species (AUSTIN and PRIEST, 1986). Three comprehensive numerical studies of *Bacillus* strains (LOGAN and BERKELEY, 1981; PRIEST et al., 1988; KAMPFER, 1991) have produced essentially similar results. The bacteria have been recovered in six large groups or aggregates of clusters which in many ways equate with genera. Within these groups are numerous clusters or species (Tab. 1). The following description is based on the numerical classification of PRIEST et al. (1988).

Group I includes *B. polymyxa* as a reference organism and comprises species such as *B. alvei*, *B. circulans* and *B. macerans* which produce oval spores that distend the mother cell. These bacteria are facultative anaerobes that ferment a variety of sugars and have reasonably fastidious growth requirements in the form of vitamins and amino acids. They secrete numerous extracellular carbohydrases such as amylases, β-glucanases including cellulases, pectinases and pullulanases.

B. subtilis and its relatives, *B. amyloliquefaciens*, *B. licheniformis* and *B. pumilus*, are included in *group II*. These bacteria differentiate into oval endospores that do not distend the mother cell. Most of these bacteria are regarded as strict aerobes but many, such as *B. subtilis*, have a limited ability to ferment sugars and will grow readily anaerobically in the presence of glucose and nitrate as a terminal electron acceptor. Some species, such as *B. anthracis*, *B. cereus*, *B. licheniformis* and *B. thuringiensis*, are true facultative anaerobes. These bacteria secrete numerous extracellular enzymes including many commercially important amylases, β-glucanases and proteases (PRIEST, 1977).

Group III species are perhaps taxonomically the least satisfactory and are rather physiologically heterogeneous. The group is based on *B. brevis* which is a strict aerobe that does not produce appreciable acid from sugars and differentiates into an oval endospore that distends the sporangium. Other species in this group might include *B. badius* and *"B. freudenreichii"*.

Bacilli which differentiate into spherical endospores are allocated to *group IV*. This is a phylogenetically homogeneous group of species including *B. sphaericus*, the psychrophiles *B. insolitus* and *B. psychrophilus* and some other species. These bacteria are also distinguished from virtually all other bacilli by the replacement of *meso*-diaminopimelic acid in the peptidoglycan of their cell walls by lysine or ornithine. These bacteria are all strict aerobes and, in the case of *B. sphaericus*, do not use sugars for growth. Acetate is a pre-

Tab. 1. Allocation of Some Bacillus Species[a] to Groups Based on Phenotypic Similarities

Species	mol% G+C[b]	RNA Group[c]	Group Characteristics
Group I. The *B. polymyxa* group			
B. alvei	46	3	All species are facultative anaerobes and grow strongly in the absence of oxygen. Acid is produced from a variety of sugars. Endospores are ellipsoidal and swell the mother cell.
B. amylolyticus	53	3	
"B. apiarius"	–	–	
B. azotofixans	52	3	
B. circulans	39	1	
B. glucanolyticus	48	–	
B. larvae	38	3	
B. lautus	51	1	
B. lentimorbus	38	1	
B. macerans	52	3	
B. macquariensis	40	3	
B. pabuli	49	3	
B. polymyxa	44	3	
B. popilliae	41	1	
B. psychrosaccharolyticus	44	1	
B. pulvifaciens	44	3	
B. thiaminolyticus	53	–	
B. validus	54	–	
Group II. The *B. subtilis* group			
B. alcalophilus	37	UG	All species produce acid from a variety of sugars including glucose. Most are able to grow at least weakly in the absence of oxygen, particularly if nitrate is present. Spores are ellipsoidal and do not swell the mother cell.
B. amyloliquefaciens	43	1	
B. anthracis	33	1	
B. atrophaeus	42	1	
B. carotarum	–	–	
B. firmus	41	1	
B. flexus	38	–	
B. laterosporus	40	5	
B. lentus	36	1	
B. licheniformis	45	1	
B. megaterium	37	1	
B. mycoides	34	1	
B. niacini	38	–	
B. pantothenticus	37	1	
B. pumilus	41	1	
B. simplex	41	1	
B. subtilis	43	1	
B. thuringiensis	34	1	
Group III. The *B. brevis* group			
(B. alginolyticus)[d]	48	–	Strict aerobes that do not produce acid from sugars; names in parentheses are exceptions. They produce ellipsoidal spores that swell the mother cell.
"B. aneurinolyticus"	42	UG	
B. azotoformans	39	1	
B. badius	44	1	
B. brevis	47	4	
(B. chondroitinus)	47	–	
"B. freudenreichii"	44	–	
B. gordonae	55	3	
Group IV. The *B. sphaericus* group			
("B. aminovorans")	40	–	All species produce spherical spores which may
B. fusiformis	36	2	

Tab. 1. Continued

Species	mol% G+C[b]	RNA Group[c]	Group Characteristics
B. globisporus	40	2	swell the mother cell. They
B. insolitus	36	2	contain L-lysine or ornithine
B. marinus	39	–	in their cell wall. All species
B. pasteurii	38	2	are strictly aerobic but some
(B. psychrophilus)	42	2	have limited ability to
B. sphaericus[e]	37	2	produce acid from sugars.
Group V. The thermophiles			
B. coagulans	44	1	All these bacteria grow
"B. flavothermus"	61	–	optimally at 50 °C or above.
B. kaustophilus	53	5	Physiologically and
B. pallidus	40	–	morphologically they are
B. schlegelii	64	–	heterogeneous but most
B. smithii	39	1	produce oval spores that
B. stearothermophilus	52	5	swell the mother cell.
B. thermocatenulatus	69	–	
B. thermocloacae	42	–	
B. thermodenitrificans	52	–	
B. thermoglucosidasius	45	5	
B. thermooleovorans	55	–	
B. thermoruber	57	–	
B. tusciae	58	–	
Group VI. *Alicyclobacillus*			
A. acidocaldarius	60	6	Thermophilic, acidophilic
A. acidoterrestris	52	6	species with membraneous
A. cycloheptanicus	56	6	ω-alicyclic fatty acids.
Unassigned species			
B. benzoevorans	41	1	
B. fastidiosus	35	1	
B. naganoensis	45	–	

[a] Names in quotation marks refer to taxa which do not appear in the Approved Lists of Bacterial Names or its supplements and therefore have not been validly published.
[b] Base composition is given either as the figure for the type strain or as the mean of a range for several strains.
[c] RNA groups are based on the work of ASH et al. (1991) and WISOTZKEY et al. (1992). UG, ungrouped; –, no data available.
[d] Names in parentheses refer to species that are atypical of the general description.
[e] *B. sphaericus* includes at least five "species" of round-spored strains.
Table reproduced from PRIEST (1993) with permission.

ferred carbon and energy source, although amino acids such as arginine, glutamate and histidine can also be metabolized.

Finally, most numerical classification studies have recovered the thermophilic bacilli as a separate group (*group V*, Tab. 1). This includes a physiologically and morphologically heterogeneous collection of species with various forms of energy metabolism ranging from strict aerobes to microaerophilic types. Indeed some species such as *B. schlegelii* are chemolithoautotrophs which can grow with carbon dioxide or carbon monoxide as sole carbon source. The thermophilic bacilli are phylogenetically diverse (ASH et al., 1991), and acidophilic thermophiles have recently been allocated to a new genus *Alicyclobacillus* (WISOTZKEY et al., 1992) (*group VI*). It appears that thermophily has evolved independently in many lineages.

Numerical classification has also helped clarify relationships between bacilli at the species level, although in most cases this is better done by DNA reassociation studies. It is reassuring that, in general, numerical classification and DNA homology have given concordant results. In many areas, for example *B. circulans*, *B. megaterium*, *B. sphaericus*, *B. stearothermophilus* and *B. subtilis*, examination of strains by these techniques has revealed that GORDON et al. (1973) "lumped" strains into species rather too enthusiastically and that each of these species probably represents several taxa. *B. subtilis sensu lato*, for example, is now known to include *B. amyloliquefaciens* and *B. atropheus* as well as *B. subtilis* itself and *B. circulans sensu lato* encompasses numerous species including *B. alginolyticus*, *B. amylolyticus*, *B. chondroitinus*, *B. glucanolyticus*, *B. lautus*, *B.pabuli*, and *B. validus* as well as some unnamed DNA homology groups. These revisions of several taxa, together with the isolation and naming of new strains, has led to the expansion of *Bacillus*, and the genus now includes at least 67 validly described species (Tab. 1).

2.3 Phylogenetic Analyses and Evolution

There are essentially two types of relationships on which classifications of organisms can be based. Phenetic relationships are those which encompass the complete organism and describe the genotypes and/or phenotypes of the organisms under study, and phylogenetic relationships are intended to represent the evolutionary branching patterns of the bacteria under study. Classifications based on phylogenetic relationships are termed cladistic. Such classifications can be derived from comparisons of gene sequences. In particular ribosomal (r)RNA sequences are amenable to this type of analysis (WOESE, 1987). Such studies reveal possible evolutionary patterns among the bacilli and are used to group *Bacillus* species into phylogenetically-based taxa (ASH et al., 1991; RÖSSLER et al., 1991).

It is becoming apparent that many strains cluster around *B. subtilis*, and indeed most of these are members of the phenetic group II. There are anomalies, however, including *B. circulans* and *B. coagulans* which apparently diverged from *B. subtilis* relatively recently and yet are phenetically dissimilar. One of the more robust groups is that based on *B. sphaericus* which is distinct both phenetically and phylogenetically indicating that these bacteria diverged from the main *Bacillus* lineage at an early date and have subsequently evolved at a constant rate. Similarly, the *B. polymyxa* group diverged from *B. subtilis* at a very early stage and is the equivalent of a separate genus. The thermophilic bacilli are placed in several areas of the tree including two nuclei based on *B. stearothermophilus* and *Alicyclobacillus acidocaldarius.* The latter is a new genus established to accommodate *"B. acidocaldarius"* and some other acidophilic thermophiles which show an early divergence from the main *Bacillus* lineage and are considered sufficiently different to warrant separate genus status (WISOTZKEY et al., 1992).

2.4 Identification of *Bacillus* Species

An important goal of the taxonomist is to produce an accurate and simple identification system from the classification. Such schemes should be multi-purpose and enable the identification of strains of, for example, medical, ecological or biotechnological relevance, in an unambiguous manner.

The traditional approach to the identification of bacilli was based on the three morphological groups mentioned in Sect. 2.1. Having assigned an isolate to one of these groups, the bacterium was identified to the species level using a panel of physiological and biochemical tests. This system was workable, but familiarity with these bacteria was often necessary in order to distinguish spore morphologies. Largely because of this, later schemes disregarded spore morphology (CLAUS and BERKELEY, 1986), but then the number of tests to effect an identification had to be increased. There was also the problem of the ever-increasing number of new species to be included in the tables.

The expansion of the genus and growing interest in these bacteria prompted the development of computerized identification systems. BERKELEY et al. (1984) used API 50 CH trays (miniaturized kits each enabling the examination of 50 phenotypic tests for one bacterium) to characterize a large number of strains and from the results developed a computerized identification matrix. An alternative computer-assisted identification matrix is available from this laboratory. This is based on 30 tests and 44 species. The tests include common physiological and biochemical features such as starch and casein hydrolysis, acid production from sugars, etc., and the 44 taxa represent all of the common species. With this system we can routinely identify about 50% of environmental isolates of *Bacillus*. Unidentified strains presumably represent undescribed taxa which we do not have in the matrix or variants of established species.

Of the various chemotaxonomic approaches available for identification of *Bacillus* strains, pyrolysis mass spectrometry seems to hold particular promise (BERKELEY et al.,1984). Pyrolysis involves the combustion of a sample in an inert atmosphere. The fragments produced are then separated on the basis of their mass/charge ratio using a mass spectrometer. A library of pyrograms or mass spectra is held in a computer and when a new isolate is to be identified its pyrogram is compared with the library using multivariate statistics. This procedure suffers from lack of reproducibility and drift over a period; pyrograms prepared at six-month intervals often give different results. The procedure is very quick, however, a few minutes per sample for colonies from a Petri plate, so standards can be readily examined alongside the unknown to overcome the drift of the machine. This approach lends itself to rapid and highly specific typing of bacteria including *Bacillus* spp.

A second chemotaxonomic approach of promise for routine identification of bacilli is based on fatty acid composition. A microbial identification system using gas chromatographic analysis of methyl esters of cellular fatty acids is marketed by Hewlett-Packard. The production of a profile takes about 60–90 minutes (including preparation time which can be reduced when samples are prepared in batches) and the profile is compared with a library of profiles for *Bacillus* species. Matching of the profile with one from the library effects an identification. This approach has recently been successfully used for the identification of *B. sphaericus* strains including types pathogenic for mosquitoes (FRACHON et al., 1991).

The few reports of DNA probes and related technology for the identification and typing of bacilli concentrate on specific groups of particular environmental or biotechnological importance although a probe for *B. subtilis* based on the 23S rRNA has been prepared. The insect pathogen *B. thuringiensis* has received most attention in this respect. Oligonucleotide probes directed at the toxin genes have been successful for the detection and identification of *B. thuringiensis* (reviewed in PRIEST and GRIGOROVA, 1990), and a typing procedure using polymerase chain reaction (PCR) products of toxin genes has also been reported (CAROZZI et al., 1991).

3 Ecology

The aerobic endospore-forming bacteria are widespread and can be recovered from almost every environment in the biosphere. Bacilli have been isolated from dry antarctic valleys and from thermal sites. They are prevalent in marine and other aquatic sites and, of course, they comprise a major proportion of the soil microflora. One of the most interesting aspects of the ecology of these bacteria is their role in these varied locations. Bacilli are almost invariably isolated from environmental samples by heat destruction of vegetative cells (usually by incubating samples at 80°C for about 10 min) or some other procedures such as ethanol inactivation. The spores are then germinated and colonies grown on suitable media in air. This simple process is absolutely specific for aerobic, endospore-forming bacteria; hence its attraction (see PRIEST and GRIGOROVA, 1990).

However, this process only recovers spores from the environment and no indication is provided as to the contribution of these microorganisms to the habitat from which they were isolated. For example, the isolation of strict thermophiles from garden soils, or alkaliphilic bacilli from acid soils, suggests that the spores had simply accumulated but were unable to germinate and were unlikely to contribute to the ecology of that area. Nevertheless, this approach has its uses, and spores of thermoactinomycetes are used to determine runoff of soils into estuarine waters, since the spores do not germinate in the cold waters. But it also provides problems in considering spore-formers in terms of habitats (SLEPECKY and LEADBETTER, 1983). A high number of spores of a certain type in a given habitat may indicate continued or previous growth of the bacterium, but in other circumstances it may reflect conditions conducive to spore dormancy (i.e, darkness, dryness and constant cool temperature). In other studies, bacilli have been recovered almost exclusively as vegetative cells, and it is then easier to draw conclusions about their metabolic contributions to the environment.

A second problem has been the identification of environmental isolates. In the absence of good classification and identification systems it is impossible to make sensible statements about the ecology of microorganisms. Fortunately, such classifications are now becoming available (see Sects. 2.2 and 2.3).

3.1 Distribution and Habitats

The primary habitat of most bacilli is the soil. These are saprophytic bacteria that grow at the expense of dead plant matter and other nutrients. They secrete a range of extracellular carbohydrases and proteases that permits hydrolysis of cellulose (although few hydrolyze crystalline cellulose), starch and other glucans, pectin and proteins. The major soil bacilli are the members of the *B. subtilis* group and include *B. cereus*, *B. licheniformis*, *B. megaterium*, *B. pumilus* and *B. subtilis* itself, and members of the *B. sphaericus* group. Numbers are normally high in cultivated soils (10^7 or 10^8 per gram) and the bacteria become scarce (around 10^3 per gram) in low-nutrient soils such as moorland, desert or tundra soils. Similarly, the diversity of types varies with the nutrient status of the soil; the more nutritionally demanding species of group I such as *B. alvei*, *B. macerans* and *B. polymyxa* can be isolated from composts and soils rich in organic matter in addition to the above-mentioned species.

Certain environments enrich for particular types of endospore-formers indicating a role for the bacterium in nutrient turnover. Poultry litter, for example, has high numbers of *B. fastidiosus* which is suited to the metabolism of the uric acid it contains. Large numbers of alkaliphilic bacilli inhabit alkaline soils, although they are by no means limited to such environments and may be isolated from normal soils. Similarly, thermophilic bacilli are more prevalent in heated soils and composts.

Estuarine waters and muds are rich in *Bacillus* spores and cells (BONDE, 1981). Bacteria of phenotypic group II predominate in these areas and in particular *B. firmus* and *B. lentus* are commonly isolated from samples high in sodium chloride. Many of these bacteria are of uncertain taxonomic position and have been simply labeled as *B. firmus–B. lentus* intermediates. They are often pigmented yellow or red, possibly providing some resistance to sunlight on exposure of muds at low tide. Bacilli are rare in more concentrated saline habitats such as salterns or saline lakes.

Bacilli in the oceans have often been considered as "runoff" spores that have been washed into the sea from coastal and river soils. In some cases, e.g. with the thermophiles, this is undoubtedly the case, and probably explains the higher numbers of bacilli in coastal waters (8–20% of total bacterial population) compared with oceanic sites (0–7%). Nevertheless, some species are present in such high numbers that they can be considered as primary marine organisms. *B. firmus*, *B. licheniformis* and *B. subtilis* are found in both polluted and non-polluted sites, whereas *B. cereus* is restricted to non-polluted sites (BONDE, 1981). The frequency of bacilli in sediments is generally greater than in the water column, and at lower levels *Bacillus* spores predominate and may account for the bulk of the total heterotrophic flora. It seems that most marine strains of *Bacillus* conform to the established species and have no special characteristics. However, some species may be restricted to this environment such as *B. marinus*, and some species of uncertain rank including *"B. cirroflagellosus"*, *"B. epiphytus"* and *"B. filicolonicus"*.

Most bacilli in freshwater probably represent bacteria of soil origin, the vast majority of aquatic organisms being Gram-negative. Bacilli are more plentiful in the sediments and include the common species of *B. licheniformis*, *B. megaterium*, and *B. subtilis*.

3.2 Associations with Plants

It is becoming apparent that bacilli may often be closely associated with plants, in particular the facultative anaerobes of phenotypic group I. Several species such as *B. azotofixans*, *B. macerans* and *B. polymyxa* fix nitrogen under anaerobic conditions; indeed in one study *B. polymyxa* was found to be the predominant nitrogen-fixing bacterium in temperate forest and tundra soils (reviewed in PRIEST, 1993). It has long been suspected that nitrogen-fixing bacteria might feed on exudates from plant roots and in return provide fixed nitrogen. In an examination of Marram grass colonization of nutrient-poor sand dunes in Wales, such a symbiotic relationship with *B. polymyxa* and *B. macerans*-like bacteria was strongly indicated (RHODES-ROBERTS, 1981). *B. polymyxa* and related bacteria are also important in the healthy development of Canadian wheat. Inoculation of spring wheat plants with *B. polymyxa* encouraged strong development of the plants which received up to 10% of their total nitrogen from the bacteria (KUCEY, 1988).

Bacteria may also be beneficial to plant growth by helping to prevent disease. *B. subtilis* has been used to control several types of fungal disease associated with field crops, fruit and vegetables. It is thought that the protection is derived from secretion of antifungal antibiotics by the bacteria. Take-all disease of wheat (caused by *Gaeumannomyces graminis*) is suppressed by *B. polymyxa* and *B. pumilus* and damping-off of alpha-alpha seedlings by a strain of *B. cereus* has been noted.

Bacteria of phenotypic group II are commonly encountered on foods and vegetables. They presumably are derived from the soil as spores are blown onto the crops in the field. Some colonization of crops, particularly cereal crops and rice, seems to occur. Since the spore is resistant to heat and desiccation, bacilli are more common on desiccated and dried foods of plant origin than on animal products. Most of these associations are superficial and of little consequence, since the bacteria are harmless, but there are a few instances in which the presence of the bacteria is important or relevant to the food.

The most common agent of *Bacillus* food poisoning is *B. cereus*. The spores of this bacterium are common on rice, pulses, lentils and other dried seeds and will cause no harm. If, however, rice is cooked and left at room temperature for several hours, the heat-activated spores germinate and the bacterium multiplies rapidly. During growth a peptide toxin is released that is highly heat-stable (survives 126 °C for 1.5 h) and causes vomitting when ingested. This emetic type of food poisoning is unpleasant but not particularly serious. A second type of food poisoning from *B. cereus* is the diarrheal form. This is also an intoxication derived from eating contaminated foodstuffs, including meats, vegetables, puddings and sauces. The diarrheal enterotoxin is a thermolabile protein that again accumulates during growth of the *B. cereus* (KRAMER and GILBERT, 1989).

Other bacilli implicated in food poisoning include *B. licheniformis* and *B. subtilis* but rarely *B. pumilus*. In these cases, vomitting and diarrhea have been associated with ingestion of large numbers of the bacteria in the absence of demonstrable established food-borne pathogens (reviewed in KRAMER and GILBERT, 1989).

The traditional cocoa fermentation is largely dependent on bacilli. To produce cocoa, the beans are fermented in their pods in a sterile sugary pulp. During this fermentation bacilli originating from the air become the dominant flora (about 10^6 per gram), and the spores survive the subsequent processing of the beans. After roasting, (150 °C, 40 min) only bacilli remain (notably *B. coagulans*, *B. megaterium*, *B. pumilus*, *B. subtilis* and *B. stearothermophilus*), but the heat reduces the numbers to about 100 to 1000 per gram in a hygienically manufactured product (CARR, 1981).

Spices also have a *Bacillus* flora derived from the original plants or air contamination and subsequently enriched during drying. Since these spices are often to be added to foods, such as sausages or sauces, this flora is a consideration with regard to storage of the items. Typical spice floras include *B. cereus* and *B. subtilis* as well as thermophilic types (PIVNICK, 1980).

Sugar is extracted from the beet by steeping in a hot-water diffuser which usually has a high population of facultatively anaerobic, thermophilic bacilli. In one of the most comprehensive surveys of the bacteria present, numerous thermophiles were described by numerical taxonomy including *B. coagulans*, *B. kaustophilus*, *B. stearothermophilus* and *B. thermodenitrificans* (KLAUSHOFER et al., 1971).

At the other extreme, psychrotrophic bacilli have often been isolated from frozen and chilled foods including vegetables and fruits. *B. globisporus*, *B. insolitus* and *B. psychrosaccharolyticus* have been detected in these foods on several occasions (SCHOFIELD, 1992).

3.3 Associations with Animals

3.3.1 Pathogens of Man and Mammals

The only severe human pathogen in the genus *Bacillus* is *B. anthracis*, the causative agent of anthrax in man and animals. Despite its scarcity in the west, this remains an important pathogen in tropical countries which lack an effective vaccination program, and few developed countries are entirely free of the disease (TURNBULL, 1990). The virulence factors of *B. anthracis* (capsule and toxin) are encoded on separate plasmids (pX02 and pX01, respectively). Curing strains of pX01 results in non-toxic derivatives and similarly non-capsulated strains devoid of pX02 are weakly pathogenic.

B. cereus is also capable of invasive infection in man and animals and has been implicated in septicemia, endocarditis, meningitis, osteomyelitis, wound infections and urinary tract infections (TURNBULL et al., 1979). Although often associated with debilitated hosts, many infections are primary infections in persons with normal immunity. Other bacilli involved in septicemia, abcesses and infected wounds in man include: *B. alvei*, *B. brevis*, *B. cereus*, *B. licheniformis*, *B. macerans*, *B. sphaericus* and *B. subtilis*, but virtually all cases involved immunodeficient or severely compromised patients. This potential pathogenicity has obvious implications with regard to safety of biotechnological products from bacilli. However, in a recent review of infections caused by *B. amyloliquefaciens* and *B. subtilis*, DE BOER and DIDERICHSEN (1991) concluded that no case of invasive infection with these bacteria has been described, supporting the safety of strains of these two species in general biotechnological practice.

3.3.2 Pathogens of Insects

Many bacilli are associated with insects either as commensals or as pathogens. Facultative anaerobes of phenotypic group I, such as *B. circulans*, may be isolated from the gut or feces of many insects, but they are not considered pathogenic. Similarly, *B. alvei*, *B. apiarius*, *B. pulvifaciens* and *B. thiaminolyticus* have been isolated from living and dead honeybees but are not thought to be pathogenic. *B. alvei* is associated with European foulbrood of honeybees but is not the primary pathogen.

B. lentimorbus and *B. popilliae* are described as obligate insect pathogens, because they only sporulate efficiently when growing in larval hosts. These bacteria have complex growth requirements and will not grow on normal laboratory media such as nutrient agar. These bacteria cause the "milky" disease of scarbaeid beetle larvae in which foraging larvae eat the spores of the bacterium. The spores germinate in the guts of the larvae and the growing bacteria invade the hemolymph. Massive growth and sporulation of the bacteria causes opacity of the virtually transparent larvae. Eventually the larvae die releasing spores into the soil and foliage for future infection. In this way a persistent infection of larvae can be established for long lasting control of insect pests. *B. popilliae* has been used for control of Japanese beetle and some other pests for many years in the USA (BULLA et al., 1978).

Some strains of *B. sphaericus* are pathogenic towards mosquito larvae, particularly species of *Culex* and some *Anopheles* and *Mansonia* species. These bacteria have low toxicity towards most *Aedes* species (BAUMANN et al., 1991). The pathogens differ from *B. sphaericus sensu stricto* both genotypically (from DNA sequence homology studies) and phenotypically and really belong to a separate species although this division has not been formally adopted. Strains with high and low toxicity have been recognized, but only the former produce a protein toxin during sporulation which accumulates as a small crystal. This parasporal crystal comprises equal amounts of two proteins, the 41.9 and 51.4 kD toxins. Both proteins are required for toxicity and appear to act as a binary toxin. Upon ingestion by susceptible larvae, the crystal dissolves in the high pH of the larval gut, and the proteins are processed by larval proteases to highly toxic polypeptides. The precise mode of action is not known (BAUMANN et al., 1991).

Toxin genes are chromosomal rather than plasmid located in *B. sphaericus*. Examples from several strains have been sequenced, and they show remarkable conservation, although the strains originate from geographically diverse areas. The genes and toxins are very different from those of *B. thuringiensis*. Low toxicity strains lack the genes for the 42 and 51 kD proteins and do not make a crystal but contain a separate toxin. This ADP-ribosylating-type toxin is also present in the crystal-forming high-toxicity strains (THANABALU et al., 1991).

B. sphaericus has several attractive attributes as a biocontrol agent, hence its adoption by the WHO for the control of mosquitoes in tropical countries (YOUSTEN, 1984). The bacterium and its toxins are reasonably persistent, and control of mosquito larval populations may last for up to 3 weeks after spraying with *B. sphaericus* formulations. There is also some evidence that the bacterium may grow in the larval cadaver, resporulate and thus be recycled in the environment (reviewed in PRIEST, 1992).

The most successful of all microbial control agents of insects is *B. thuringiensis* (FEITELSON et al., 1992). This close relative of *B. cereus* and *B. anthracis* produces a crystal parasporal body or delta endotoxin which is toxic for various insect larvae. *B. thuringiensis* strains have been allocated to named serotypes based on flagella antigens. In most cases there is little correlation between serotype and toxicity, but the system has been invaluable as a basis for typing these bacteria. An exception is serotype 13 (*B. thuringiensis* serovar *israelensis*), strains of which are almost invariably associated with toxicity for mosquito (dipteran) larvae (DE BARJAC, 1990).

Toxin genes in *B. thuringiensis* are generally located on large conjugative plasmids, although in a minority of cases the genes are chromosomal. The genes have been allocated to five classes based on sequence homologies (HOFTE and WHITELEY, 1989; FEITELSON et al., 1992). These classes correlate with toxicity against different insects. Crystals of the *cryI* type are bipyrimidal in shape and toxic for lepidopteran (moth and butterfly) larvae. Most isolated strains of *B. thuringiensis* are in this category. The toxin proteins are large (130–140 kD) but contain a toxic core of about 60–70 kD comprising the N-terminal portion of the protein. *CryII* proteins are smaller (comprise the toxic core only), relatively rare and toxic for both lepidopteran and dipteran larvae. The *cryIII* encoded crystals are relatively recent discoveries. These flat, wafer-shaped crystals are toxic for coleopteran (beetle) larvae including an important pest of potato crops, the Colorado potato beetle. *CryIV* genes code for the proteins of the amorphous crystal found in mosquito pathogenic varieties of *B. thuringiensis*. These crystals are different from the others in comprising four proteins derived from the *cryIV A*, *B*, *C*, and *D* genes. An additional protein from the *cytA* gene is not needed for toxicity. Finally, *cryV* and *cryVI* genes, which code for proteins toxic toward nematodes, have been described recently and extend the pathogenicity of *B. thuringiensis* beyond the insect kingdom (FEITELSON et al., 1992).

Upon ingestion, the crystal protein is processed by larval proteases to a toxic core of 60–70 kD. The protein has three domains (LI et al., 1991). One area interacts with receptors on the epithelial cells of the gut membrane. Binding results in the insertion of a hydrophobic portion into the membrane and in the establishment of small pores in the membrane. Ions enter via the pores, and the cells then accumulate water by osmosis in an attempt to equilibrate the ionic balance. This "colloid osmotic lysis" leads to disruption of the cell and overall destruction of the gut lining. The larvae quickly stop feeding and subsequently die. It was originally thought that resistance to *B. thuringiensis* toxins would be improbable because of the "complexity" of the interaction between toxin and insect. However, *B. thuringiensis* resistant insects can be bred in the laboratory in as few as 12 generations which may indicate that natural resistant populations might appear relatively rapidly. Indeed, heavy usage of *B. thuringiensis* products in selected applications has already resulted in field resistance in populations of diamond back moths, and resistant strains of several other insects have been generated in the laboratory. There is therefore an urgent need to monitor closely the usage of this important bioinsecticide (Fox, 1991).

4 Extremophiles

One of the fascinating and biotechnologically attractive aspects of the genus *Bacillus* is the great variety of biochemistry represented by the various species and the occurrence of adaption to physiological extremes. Several of these extremophiles, notably the alkaliphiles and thermophiles have been exploited industrially, particularly for the manufacture of enzymes with desirable industrial properties.

4.1 Thermophiles

Many bacilli conform to a generally accepted definition of thermophily: maximum growth temperature above 60 °C and an optimum above 50 °C with a minimum above 30 °C; although few could be described as extreme thermophiles with growth temperatures above 90 °C. Earlier this century, many thermophilic bacilli had been described and given species status. The taxonomic revisions of GORDON and SMITH (see Sect. 2.1) suggested that most were synonymous with either *B. coagulans* or *B. stearothermophilus* and others were thermotolerant variants of mesophilic species such as *B. brevis*, *B. sphaericus* or *B. subtilis*. This view predominated until biotechnological interests in the past two decades prompted the isolation and description of diverse thermophilic bacilli. It is now apparent that some of the earlier species, such as *B. kaustophilus* and *B. thermodenitrificans* were indeed valid species and several new species have been added. The classification of the thermophilic bacilli remains confused but numerous valid species have been recognized (Tab. 1). Most of these have been described in a recent review (SHARP et al., 1992).

Of particular interest are the acidophilic species. *B. coagulans* is slightly acidophilic, preferring growth at about pH 6, but highly acidophilic species were encountered with the isolation of *"B. acidocaldarius"*, now *Alicyclobacillus acidocaldarius*, (WISOTZKEY et al., 1992) from pools and hot springs in Yellowstone national park. Generally, strains of *A. acidocaldarius* grow at 45–70 °C and pH 2 to 6, and the bacterium is common in acid soils and pools in most geothermal areas. Subsequently, *A. acidoterrestris* and *A. cycloheptanicus*, two acidophilic thermotolerant species, were isolated from soils.

The ecology of the thermophilic bacilli is heavily influenced by the endospore. Thus, other thermophilic bacteria such as *Thermus* strains tend to be restricted to certain geothermal areas, and isolates from geographically widespread areas differ. Thermophilic bacilli, on the other hand, can be isolated from cool environments where the spores are presumably lying dormant. Moreover, thermophilic bacilli do not seem to be peculiar to certain geographical locations; isolates of *B. stearothermophilus* from hot springs in Iceland or New Zealand are essentially the same. This emphasizes the importance of the spore in dissemination of these bacteria across great distances.

Explanations for the molecular basis of thermophily have centered on increased temperature tolerance of proteins, protein synthesizing machinery and other cell constituents. An interesting, but poorly documented aspect of thermophily is the demonstration that *B. subtilis* can be readily transformed to thermophily (growth at 70 °C) using chromosomal DNA from the thermophile *B. caldolyticus* (reviewed in SHARP et al., 1992).

With increasing demand from the biotechnology industry for robust enzymes for use in high temperature applications, thermophilic bacilli were screened for suitable products. In general, thermophiles produce enzymes with greater intrinsic thermostability than do mesophiles (although there are exceptions such as the highly thermostable α-amylase from the thermotolerant mesophile *B. licheniformis*), and this also provides other attractive features such as prolonged shelf life of the enzymes and increased resistance to denaturants such as detergents and organic solvents. However, the observed slower reaction rates at high temperature are disadvantageous. When thermostable mesophilic enzymes, which may display an optimum reaction rate at 40 °C, are used at high temperature (e.g., 70 °C) there is a corresponding increase in reaction rate. This would not be the case with an enzyme from a thermophile, the optimum for which may be 70 °C. Nevertheless, several enzymes from thermophiles have reached the market and many more have been evaluated (KRISTJANSSON, 1989). In particular, several thermostable extracellular enzymes such as neutral protease "thermolysin", lipases and pullulanase are manufactured from thermophilic bacilli. The intracellular enzyme fructose (glucose) isomerase, used for the preparation of high-fructose corn syrups from starch hydroylysates, is manufactured in immobilized form from *B. coagulans* by Novo Nordisk. In the laboratory, the development of the polymerase chain reaction has created a demand for thermostable DNA polymerases and other nucleic acid modifying enzymes which is being met, in part, from thermophilic bacilli.

Numerous genes from thermophiles such as the maltogenic α-amylase of *B. stearothermophilus* have been cloned and expressed in mesophiles. This approach circumvents many of the problems associated with the large-scale fermentation of thermophiles (poor growth yields, new technology, non-GRAS clearance etc.). On the other hand, thermophilic bacilli provide inherent advantages for industrial practice such as reduction in cooling costs, reduced contamination (not generally borne out in practice!) and reduced medium viscosity. Thermophilic bacilli may therefore be attractive as producers of primary products, or for the industrial preparation of cloned gene products, as well as a source of useful genetic material.

Exploitation of the thermophiles is hampered by difficulties with genetic engineering in these bacteria. Several plasmid cloning vectors are available for *B. stearothermophilus* which were originally discovered as antibiotic resistance plasmids. Shuttle vectors for the transfer of cloned genes from *B. subtilis* to *B. stearothermophilus* were later constructed (ZHANG et al., 1988), but transformation of *B. stearothermophilus* remains a formidable barrier. Protoplast transformation of *B. stearothermophilus* is strain-dependent and reproducibility remains elusive (SHARP et al., 1992). Nevertheless, several genes have been successfully cloned and expressed in *B. stearothermophilus* including the penicillinase from *B. licheniformis* and the neutral protease and α-amylase genes from *B. subtilis* (reviewed by SHARP et al., 1992). Other important applications of thermophilic bacilli include waste management, in particular aerobic sewage treatment and composting.

4.2 Psychrophiles

Bacteria with an optimum growth temperature of 15 °C or lower, a maximum about 20 °C and a minimum of 0 °C or lower are generally regarded as psychrophilic. Several bacilli conform to this description, but unlike true psychrophiles, have a maximum growth temperature of around 25 °C and are best described as psychrotrophic. Many were thought to be variants of mesophilic bacilli, but several taxonomic studies have shown that they are in fact different taxa (Tab. 1). Little is known about these bacteria, they are widely distributed in the marine and terrestrial environments and they commonly occur in refrigerated and frozen foods and as spoilage agents in milk. Their ability to germinate and grow at normal refrigeration temperatures indicates that they could be important spoilage agents of refrigerated foods (SCHOFIELD, 1992).

4.3 Alkaliphiles

The genus *Bacillus* is a rich source of bacteria able to grow at high pH. Those with a pH optimum above pH 8 and usually above pH 9 are termed alkaliphiles. As with other extremophiles, we recognize obligate alkaliphiles, which are unable to grow at neutrality and have pH optima around pH 10 or above, and alkalitolerant strains which can grow at pH 7. The classification of these bacteria is inadequate. The first alkaliphilic *Bacillus* was isolated by VEDDER in 1934 and was named *B. alcalophilus*. Subsequently, many alkalitolerant and alkaliphilic bacilli were isolated, and it was suggested that these were variants of neutrophilic bacilli, particularly *B. firmus* and *B. lentus*. Recent taxonomic studies have shown that this is an oversimplification and that most of these strains are in fact unrelated to *B. alcalophilus* or *B. firmus*. It seems likely from the distribution of extracellular enzymes and other properties in these bacteria that they represent several new species (FRITZE et al., 1990).

Alkaliphilic bacilli are widespread. Their numbers are greater in alkaline soils, but they can be readily recovered from neutrophilic and even acid soils where they may occur at up to 10^2 per gram (HORIKOSHI and AKIBA, 1982). Isolation of alkaliphiles requires media that are buffered to high pH (at least pH 10) with a plentiful supply of sodium, which is a requirement for growth, sporulation and germination of these bacteria. Normally, a sodium carbonate buffer is used. Ammonium ion is lost as ammonia at high pH, so peptone or some other organic nitrogen source is generally used in these media. A final problem is the maintenance of high pH during incubation; absorption of carbon dioxide and the ability of the bacteria to modify the pH of their environment significantly, can lead to a dramatic reduction in alkalinity (GRANT and TINDALL, 1986).

The bioenergetics of alkaliphilic growth has received considerable attention (GRANT and HORIKOSHI, 1989; KRULWICH and GUFFANTI, 1989) and will not be reviewed here. Suffice to say, the bacteria maintain an internal pH within the range 7–9 and do not have a modified protein synthesizing machinery. The basic problem is a "reversed pH gradient" arising from an internal pH which is lower than the surrounding pH; and alkaliphiles seem to have generated at least two different strategies for coping with this. One, derived from studies of a Gram-negative bacterium, is based on a sodium motive force (SMF) for generating ATP rather than the proton motive force (PMF) as used by neutrophilic bacteria. However, KRULWICH and GUFFANTI (1989) have been unable to find evidence for an SMF in a range of alkaliphilic bacilli. From studies of *B. firmus* strain RAB, they concluded that a normal PMF was used for ATP synthesis and that the protons follow some localized pathway that avoids equilibration with the protons in the highly alkaline external medium. Such cells, however, still have a requirement for sodium ions in pH homeostasis.

The main biotechnological application for alkaliphilic bacilli is in the enzyme industry, particularly the laundry detergent market. The idea of including proteases in washing detergents is not new. At the turn of the century, OTTO ROHM added pancreatic extracts to washing powders, but the properties of the enzymes were such that they were best suited for presoaking. The first suitable enzyme for use in a laundry detergent was the serine protease from *B. licheniformis* ("subtilisin Carlsberg"). The high pH optimum of this enzyme, its temperature tolerance, resistance to chelating agents and reasonable resistance to oxidizing agents made it ideal for this purpose. More recently, protein engineering has been used to improve the thermostability and resistance of subtilisin Carlsberg to oxidizing agents (WELLS and ESTELL, 1988).

The need for proteases with exceptionally high pH optima prompted screening of alkaliphilic bacilli. As a result, several proteases were introduced onto the market by Novo Nordisk such as Esperase, Maxatase and Savinase. These enzymes are used in heavy duty detergents and for the dehairing and bating of hides in the leather industry.

Other alkaliphilic enzymes for inclusion in laundry detergents include α-amylase and lipase which, like protease, remove organic molecules which bind the dirt to the fabric and cause stains. An interesting new approach to fabric conditioning is the inclusion of cellulase (an alkaline endo-cellulase) in detergents. This enzyme removes fine fibrils from denatured cotton fibers in fabrics. The fibrils reflect light and give the garment a bleached "matted" appearance and make the cotton fabric feel rough. Their removal recovers the original color and texture of the garment.

Numerous other enzymes from alkaliphiles have been investigated including amylases, cellulases, penicillinases and xylanases, but few have commercial potential (HORIKOSHI and AKIBA, 1982; SHARP and MUNSTER, 1986). There is one notable exception, however, cyclodextrin glucanotransferase (CGT). CGT was originally identified in *B. macerans* and hydrolyzes starch with the production of 6-, 7-, or 8-membered cyclic 1,4-α-linked oligosaccharides, the Schardinger dextrins. Cyclodextrins have an interesting, and practically very important, asymmetry in which the central cavity is apolar due to a lining of hydrogen and ether-oxygen atoms. Adding any hydrophobic substances to a solution of cyclodextrin in water results in the inclusion of the molecules into the cyclodextrin cavity to form a complex (SZEJTLI, 1991).

Cyclodextrins have found numerous industrial and pharmaceutical uses. Volatile liquids can be stabilized and flavors and flavor enhancers can be complexed to avoid volatilization. Unstable, poorly soluble drugs can be solubilized, and even injectable, aqueous solutions can be prepared from insoluble drugs. Light-sensitive molecules can be protected, toxic molecules can be detoxified; the list of new applications and patents continues to grow monthly. Initially, cyclodextrins were manufactured from starch using the CGT from *B. macerans* but the thermal stability of this enzyme is poor, and the conversion of amylose to cyclodextrin is rarely in excess of 55%. The enzyme has also been identified in thermophilic bacilli, but the CGT from alkaliphilic *Bacillus* No. 38-2 has formed the backbone of the cyclodextrin industry in Japan. This enzyme has a broad pH activity profile, is more thermostable than the enzyme from *B. macerans* and cyclodextrin yields in excess of 70% from potato starch are obtained. This example indicates the importance of considering the alkaliphilic bacilli as a source of novel biotechnological products which need not necessarily be relevant to processes which are performed at high pH.

4.4 Acidophiles

Both thermophilic and mesophilic acidophilic bacilli have been described. The thermophilic species *Alicyclobacillus acidocaldarius*, *A. acidoterrestris* and *A. cycloheptanicus* have interesting membrane compositions and other features which set them apart from other bacilli to the extent that they were placed in a separate genus. Like other bacilli, they possess menaquinone (MK) with 7 isoprene units (MK-7) as a major component but they also possess MK-8 and MK-9. *A. acidocaldarius* and *A. acidoterrestris* also contain the unique fatty acid ω-cyclohexane and *A. cycloheptanicus* possesses ω-cycloheptane. Moreover, *A. acidocaldarius* and *A. acidoterrestris* contain hopanoids in their membranes (WISOTZKEY et al., 1992).

Mesophilic acidophiles include *"Bacillus acidopullulyticus"*, which is used for the commercial production of the starch debranching enzyme pullulanase, and *B. naganoensis*, another pullulanase-secreting bacterium; Again these bacteria have unusual fatty acid profiles, the latter comprising major amounts of 14-methylpentadecanoic acid (*iso*-C_{16}).

5 Physiology

Most bacilli are chemoorganotrophs and gain energy from the oxidation of organic compounds, although some thermophiles are facultative autotrophs, notably *B. schlegelii*, and are able to grow with carbon dioxide or carbon monoxide as sole carbon and energy source. With regard to nutrition, many species such as *B. subtilis* are prototrophic, while others, including *B. cereus* and *B. sphaericus* have minimal organic requirements such as the odd vitamin or amino acid. In general, the members of phenotypic group I have more demanding requirements than other groups, but only the obligate insect pathogens *B. lentimorbus* and *B. popilliae* have such complex nutritional requirements that they cannot be grown on general media such as nutrient agar or broth (BULLA et al., 1978).

5.1 Relations to Oxygen

Bacilli are, by definition, able to grow and sporulate in the presence of air, thus distinguishing them from clostridia. Most are catalase positive. It is important at this point to distinguish fermentation (anaerobic growth with the use of indigenous electron acceptors for the generation of NAD^+ from NADH) from anaerobic respiration in which some exogenous compound such as nitrate acts as an electron acceptor. Both forms of energy metabolism operate in bacilli.

The members of phenotypic group I are true facultative anaerobes and ferment sugars in the absence of exogenous electron acceptors. *B. polymyxa* conducts a butanediol fermentation and *B. macerans* an ethanolic fermentation (Fig. 1). In both cases, the ratio of end products is pH-dependent. At low pH, ethanol is the major fermentation product of *B. macerans*, but at high pH acetate predominates. Similarly, in *B. polymyxa*, ethanol and formate replace butanediol as electron sinks when the bacterium is grown at a higher pH (MAGEE and KOSARIC, 1987). Both of these species release gas when growing anaerobically. Other members of this group, such as *B. circulans*, show a mixed acid fermentation pattern.

Members of phenotypic group II are less well equipped to grow anaerobically in the absence of exogenous electron acceptors. Fermentation by *B. licheniformis* is restricted to glucose catabolism, and the bacterium is incapable of adjusting to fermentation of more reduced or oxidized substrates such as sorbitol or gluconate. The major products from fermentation of glucose are butanediol, glycerol, ethanol, formate, acetate, succinate, pyruvate and carbon dioxide with lactate and butanediol as the major electron sinks at high and low pH, respectively (RASPOET et al., 1991). *B. cereus* and *B. thuringiensis* are also capable of fermentation of glucose with the formation of butanediol and acids; indeed, the common formation of acetoin and butanediol by these bacteria is revealed by the distribution of the Voges Proskauer test. *B. subtilis* is something of an enigma. It is generally regarded as a strict aerobe but this is misleading, since the bacterium grows weakly in the absence of oxygen with the production of acetoin and 2,3-butanediol from pyruvate via acetolactate. Acetate and other acids are also produced. Since the bacterium can regenerate NAD^+ from the re-

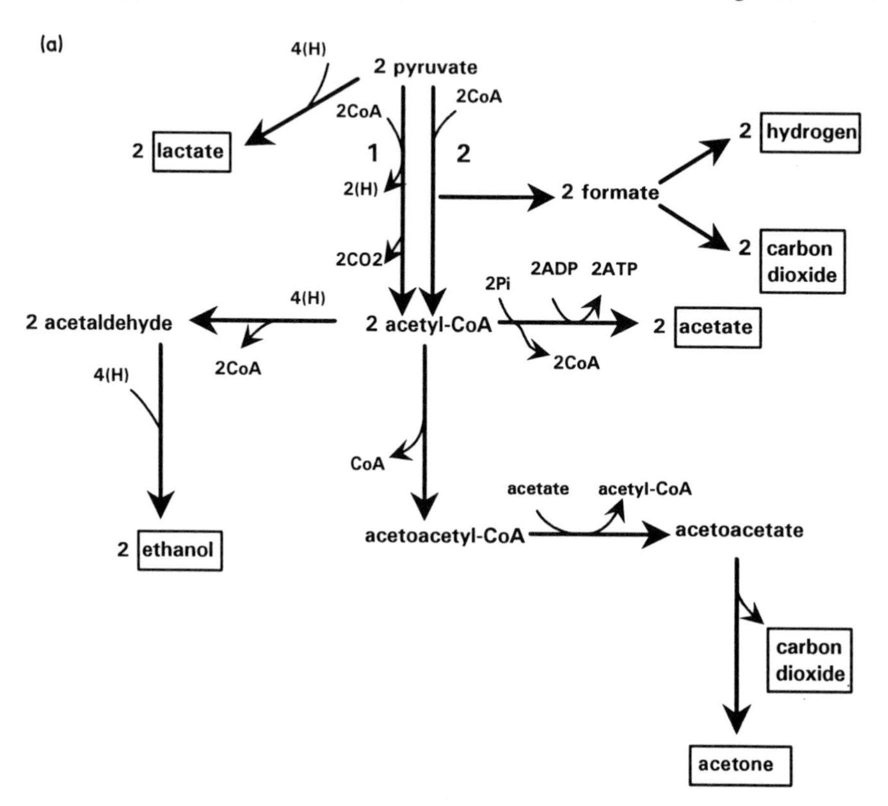

duction of acetoin to butanediol, it is not clear why it cannot grow more strongly anaerobically and emulate *B. cereus* or *B. licheniformis.*

Bacilli of groups III and IV are generally strict aerobes that do not grow in the absence of air, although several reduce nitrate in anaerobic respiration reactions. Most are characterized by a rise in the pH of the medium when grown in nutrient broth (GORDON et al., 1973) which arises from the oxidation of amino and organic acids.

Many bacilli of groups I and II and some thermophiles can use nitrate as a terminal electron acceptor and are positive in the traditional nitrate reduction test. *B. licheniformis*, *B. stearothermophilus* and *B. subtilis* all grow strongly on various sugars in the presence of nitrate under anaerobic conditions. Nitrate reductase has been purified from these three bacteria, and the enzyme is induced by nitrate and anaerobic conditions. Nitrate has a profound effect on the metabolic products from *B. licheniformis* grown anaerobically on glucose. The formation of all reduced compounds such as acetoin and butanediol is totally inhibited as is formate. Instead, acetate appears in large amounts. It seems likely that the fermentative pyruvate formate lyase is repressed under anaerobic conditions and replaced by a respiratory formate dehydrogenase which channels electrons via the quinone pool to nitrate reductase as it does in *Escherichia coli* (SHARIATI, 1992).

Several bacilli, including *B. azotoformans*, *B. licheniformis* and *B. thermodenitrificans*, reduce nitrate to molecular nitrogen, and unusual strains have been isolated by providing N_2O and NO as electron acceptors under anaerobic conditions (reviewed in PRIEST, 1989 a).

5.2 Carbon Metabolism

Members of phenotypic groups I, II and III have similar carbon metabolic pathways as is known so far. Therefore, comments will be restricted to *B. subtilis* as the best studied example.

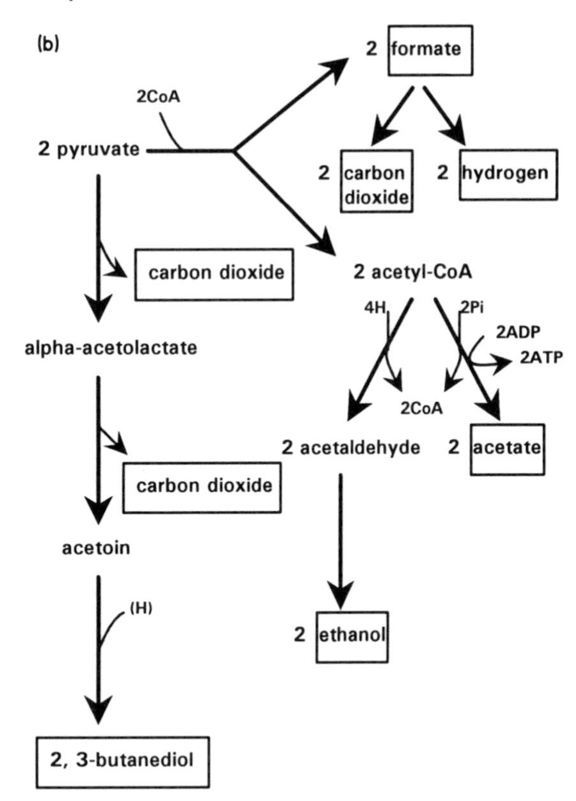

Fig. 1. Fermentation pathways of some solvent-producing bacilli.
(a) Dissimilation of pyruvate by *Bacillus macerans.* Products are shown in boxes, the primary metabolite is ethanol. **1,** pyruvate ferredoxin oxidoreductase; **2,** pyruvate formate lyase.
(b) Dissimilation of pyruvate by *B. polymyxa.* The primary metabolite is 2,3-butanediol.

Many bacilli secrete extracellular enzymes. With the exception of cell-wall metabolizing enzymes, these are scavenger enzymes responsible for the degradation of polymeric molecules in the environment which are too large to enter the cell. The low-molecular weight products can then be transported into the cell as carbon and energy sources.

Many bacilli hydrolyze starch through the action of extracellular α-amylases. These enzymes have been allocated to three major types. *B. amyloliquefaciens, B. licheniformis* and *B. stearothermophilus* secrete the liquefying-type of α-amylase that hydrolyzes starch to high-molecular weight oligosaccharides predominantly of five or six glucose residues. These are the enzymes that have found application in the starch processing industry, particularly the thermostable enzyme from *B. licheniformis. B. subtilis* secretes a saccharifying α-amylase which hydrolyzes starch to lower-molecular weight products, principally glucose and maltose. This enzyme is less thermostable than its counterpart from *B. licheniformis* and has limited industrial application. There is little sequence homology between the two types of amylase (SVENSSON et al., 1991). A third group of α-amylases comprises the cyclodextrin-forming enzymes from *B. macerans* which produce six- or seven-membered cyclic (Schardinger) dextrins from starch (see Sect. 4.3). Moreover, several species, including *B. megaterium* and *B. polymyxa*, secrete β-amylases which are similar to the plant enzymes. None of these enzymes hydrolyze the 1,6-α-branch points of amylopectin, but numerous bacilli secrete pullulanase which accomplishes this reaction (PRIEST and STARK, 1991).

Several other polysaccharide-degrading enzymes are produced from bacilli. Barley β-glucan is a mixed linkage β-1,3-1,4-linked polysaccharide and is hydrolyzed by the β-glucanases of *B. circulans* and *B. subtilis*. Although bacilli are not noted for their ability to hydrolyze cellulose, endoglucanases that hydrolyze soluble derivatives of cellulose such as carboxymethyl cellulose are prevalent and some species such as *B. lautus* hydrolyze crystalline cellulose (BÉGUIN, 1990; GILKES et al., 1991). Xylanases, pectinases and other extracellular enzymes have been described from *Bacillus* species (PRIEST, 1977).

It is important that the bacterium uses extracellular enzymes sparingly if it is to compete effectively in the environment. Uncontrolled degradation of polysaccharides will permit the

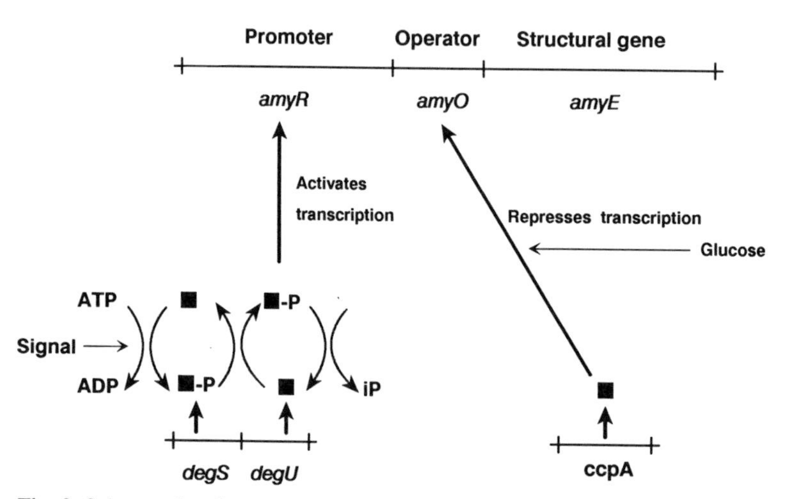

Fig. 2. Schemes for the regulation of α-amylase synthesis in *Bacillus subtilis* by catabolite repression and DegS/DegU. For catabolite repression, glucose interacts, perhaps via a small molecule, with the CcpA protein which represses transcription of the structural gene by binding to the operator region. The general control of degradative enzyme biosynthesis is effected by a two-component system in which the DegS protein autophosphorylates at a conserved histidine residue in response to a signal. DegS passes the phosphate on to the DegU protein which in its phosphorylated state activates transcription of the amylase structural gene.

proliferation of competing microorganisms. This may be the reason why extracellular enzyme synthesis is governed by complex control mechanisms generally restricting their synthesis to early stationary phase growth. The DegS/DegU system is a two-component system that controls the synthesis of numerous extracellular enzymes including amylase, β-glucanase, cellulase, xylanase and protease in *B. subtilis* and other species (STOCK et al., 1989; BOURRET et al., 1991). DegS is a histidine protein kinase sensor molecule that autophosphorylates at a conserved His residue upon receiving a particular signal. This protein then transfers the phosphate to a response regulator, in this case DegU, which, when phosphorylated, acts at sites about 100 bases upstream of extracellular enzyme promoters to activate transcription. Thus in response to some (still unknown) signal associated with entry into stationary phase, DegS phosphorylates DegU, and certain extracellular enzyme genes are activated (Fig. 2). Several other regulatory systems also interact in the regulation of expression of these genes (PRIEST, 1991).

Low-molecular weight sugars in the environment, perhaps the products of extracellular enzyme activity, are transported into *B. subtilis*

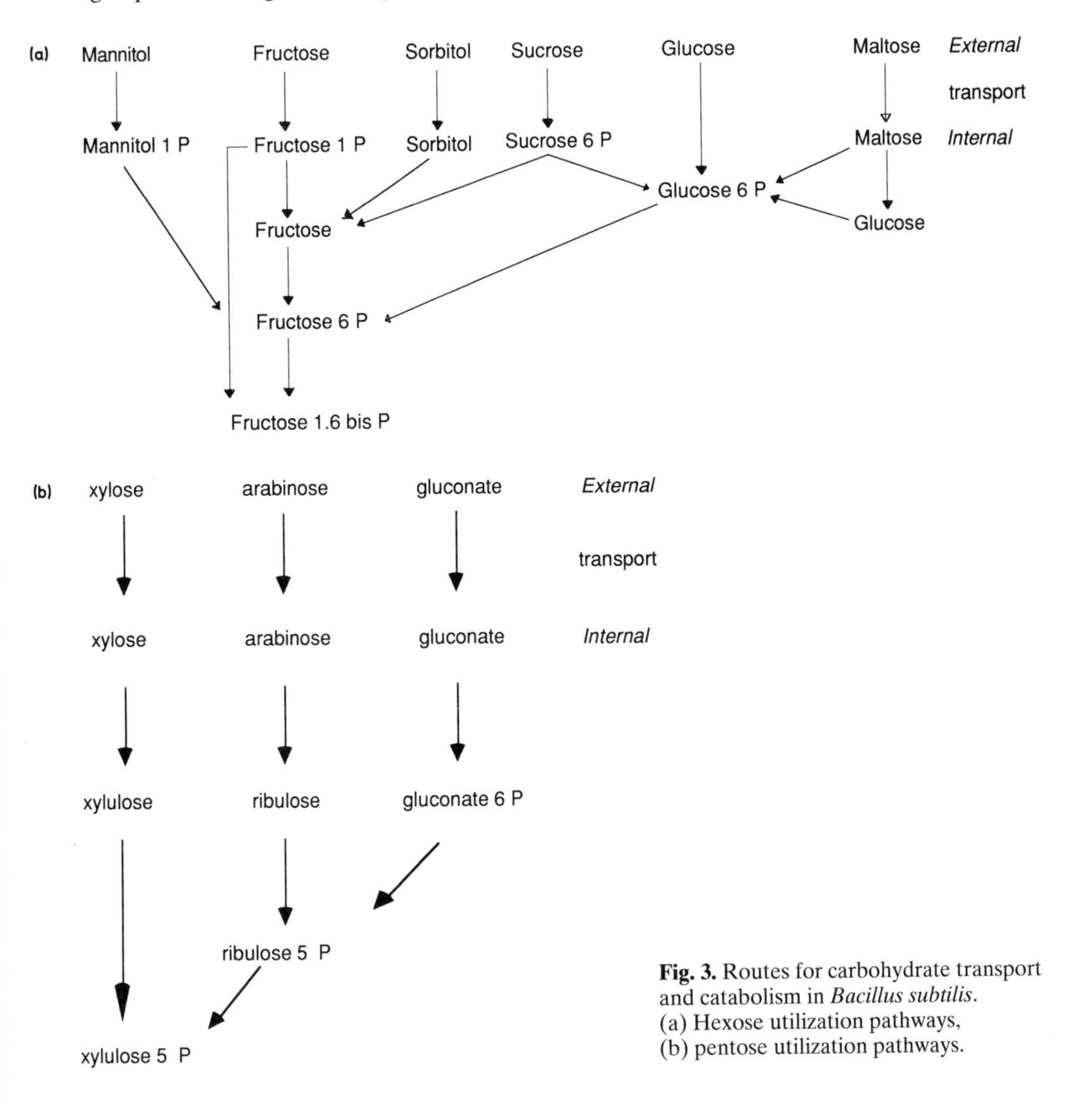

Fig. 3. Routes for carbohydrate transport and catabolism in *Bacillus subtilis*.
(a) Hexose utilization pathways,
(b) pentose utilization pathways.

by various routes. The phosphoenolpyruvate (PEP):carbohydrate phosphotransferase system (PTS) is a major system for sugar transport into the cell. The PTS comprises two classes of components, the general cytoplasmic proteins (enzyme I and HPr) and the sugar-specific proteins (enzymes II and III) located in or on the membrane. Enzyme I transfers the phosphate from PEP to a histidine residue in HPr forming HPr-P. HPr-P may either donate its phosphate direct to enzyme II in the membrane, concomitant with the transport of the sugar, or this reaction may be mediated by a specific enzyme III. In either case the sugar enters the cell as a sugar phosphate which permits sequestration of the sugar (MEADOW et al., 1990). In *B. subtilis,* enzyme I and HPr are encoded by adjacent genes which form an operon, and glucose is transported by a combined enzyme II/III^{Glc} rather than the separate proteins which operate in *E. coli.* Other sugars transported by the PTS in *B. subtilis* are fructose, sucrose and mannitol. Sugars which enter by non-PTS routes are arabinose, gluconate, glycerol, maltose, sorbitol and xylose (Fig. 3). Further metabolism of these carbon sources has been examined in varying detail (KLIER and RAPOPORT, 1988). Other sugars, which are metabolized by *B. subtilis*, but have not been examined in any detail, are mannose, raffinose and salicin. Hexose metabolism is through the Embden–Meyerhof pathway to pyruvate, and pentose catabolism is via the pentose phosphate route. There are no Entner–Doudoroff reactions, so gluconate is metabolized by gluconate-6-phosphate into the pentose phosphate pathway.

Growing aerobically in the presence of glucose, several bacilli, including *B. cereus, B. subtilis*, and *B. thuringiensis*, exhibit "metabolite overflow". Pyruvate is converted to acetate and partially oxidized products such as acetoin which are excreted into the environment; but as the cells enter the stationary phase, the acetoin is oxidized to acetate and the acetate metabolized via acetyl phosphate and acetyl-CoA through the Krebs cycle. In aerobic, carbon-limited cultures, however, pyruvate is metabolized exclusively through the Krebs cycle.

B. sphaericus and other members of phenotypic groups III and IV are strict aerobes that do not metabolize carbohydrates. *B. sphaericus* lacks many of the enzymes of the Embden–Meyerhof pathway. For example, this organism shows a total inability to phosphorylate hexoses (RUSSELL et al., 1989). Indeed, *B. sphaericus* cannot transport or catabolize hexoses or pentose sugars, although it grows slowly on gluconate by way of the pentose phosphate pathway. *B. sphaericus* prefers acetate and other organic acids as carbon and energy source.

5.2.1 Catabolite Repression

Catabolite repression is the inhibition of synthesis of enzymes of peripheral carbon metabolism by the catabolism of a rapidly metabolized carbon source. This provides a hierarchy of carbon sources which are used by the bacterium in turn, starting with the most readily utilized and ending with the most slowly metabolized. This consecutive usage of carbon sources has obvious energetic advantages over the concerted induction of operons and enzymes to permit the simultaneous catabolism of numerous carbon sources in the environment.

Although diauxie can be demonstrated in bacilli, and glucose is a strong repressor of sporulation, the molecular mechanism(s) must be different from those in *Escherichia coli*, because cAMP is not present in these bacteria in physiologically meaningful concentrations. Indeed, it is remarkable that such an important regulatory system in an industrial bacterium is still so poorly understood. Nevertheless, advances in this area are being made, particularly from studies of α-amylase synthesis. This enzyme is secreted by *B. subtilis* as the bacterium enters the stationary phase, but it is repressed by the presence of glucose. Mutants that retain the temporal regulation (i.e., show derepressed synthesis of amylase in stationary phase) but which secrete amylase irrespective of the presence of glucose, distinguish temporal regulation from catabolite repression. Such mutants of the amylase operon have been isolated which have lesions in a region between –3 and +11 (where +1 is the start point of mRNA synthesis). This *amyO* locus is similar in sequence to the *lac* operator, the binding site of the *lac* repressor protein responsible for induction of the *lac* operon (WEICKERT and CHAMBLISS, 1990). Subsequently, transposon mutagenesis was used to identify the *ccpA* (catabolite control protein) gene which encodes a protein with striking homology to the *lac* repressor and has the features of a DNA binding protein (HENKIN et al., 1991). It therefore seems likely that, in the presence of glucose, this protein is modified or interacts with a small molecule such that it binds to the *amyO* locus and inhibits expression of the gene (Fig. 2). CcpA protein does not explain all aspects of catabolite repression, however, since the amylase gene is repressed by glycerol in *ccpA* mutants and other binding proteins have been identified. Other genes subject to catabolite repression and details of possible regulatory systems have been reviewed by FISHER and SONENSHEIN (1991).

It is important to note that acetate acts as a catabolite repressing carbon source in *B. sphaericus* just as glucose does in *B. subtilis*. The atypical metabolism of *B. sphaericus* emphasizes the dangers of extrapolating aspects of carbon and energy metabolism from *B. subtilis* into all types of bacilli, just as it is important to distinguish *B. subtilis* from *E. coli*.

5.3 Nitrogen Metabolism

Reduced forms of nitrogen such as ammonium, glutamate and glutamine are usually the preferred nitrogen sources for assimilation into cellular materials. The principal routes for ammonium assimilation are shown in the following reactions;

NH_3 + glutamate + ATP →
glutamine + ADP + Pi (1)

glutamine + 2-ketoglutarate + NADPH →
2-glutamate + $NADP^+$ (2)

NH_3 + 2-ketoglutarate + NADPH →
glutamate + $NADP^+$ (3)

In (1), glutamine synthetase catalyzes the incorporation of ammonium into glutamine. This is the only route for nitrogen assimilation in *B. subtilis* and is the route adopted by *E. coli* when the external ammonium concentration is below 1 mM. This is linked to the glutamate synthase reaction shown in (2) which effects the synthesis of glutamate, the direct precursor of many biosynthetic routes. Direct reductive amination of 2-ketoglutarate by glutamate dehydrogenase (reaction 3) is induced in *E. coli* when the external ammonium concentration is high (>1 mM) but does not operate as an assimilatory pathway in *B. subtilis*. Consequently, ammonium ions can only be assimilated via glutamine synthetase in *B. subtilis*. Unlike *E. coli*, but similar to *Klebsiella aerogenes*, *B. subtilis* can use oxidized forms of nitrogen such as nitrate as a nitrogen source.

Since most of the cells' nitrogen-containing compounds are synthesized from glutamine, glutamine production is potentially a very important control point. Glutamine synthetase is closely regulated in *B. subtilis*, and cells grown under nitrogen limitation or with a poor nitrogen source have higher levels of the enzyme than those grown in nitrogen sufficiency or with a preferred nitrogen source such as glutamine. The enzyme is not regulated by adenylylation, as it is in *E. coli*, but transcriptional control is important. The structural gene for glutamine synthetase (*glnA*) is located adjacent to a regulatory gene *glnR* in a dicistronic operon transcribed by the vegetative (σ^A) form of RNA polymerase. Mutations in *glnR* give rise to constitutive synthesis of glutamine synthetase, indicating that the GlnR protein is a repressor of *glnA* expression. Indeed GlnR binds to two sites situated within and adjacent to the *glnRA* promoter and inhibits transcriptional initiation (GUTOWSKI and SCHREIER, 1992). However, the simplest model that GlnR may sense the relative concentrations of glutamine and 2-ketoglutarate in some way and modulate the transcription of *glnA* accordingly does not account for all aspects of the observed regulation (FISHER and SONENSHEIN, 1991).

Asparaginase, urease and other peripheral enzymes of nitrogen metabolism are derepressed about 20-fold when *B. licheniformis* and *B. subtilis* are grown in media of low nitrogen status. These enzymes seem to be subject to a global nitrogen catabolite repression control (equivalent of the Ntr system of enteric bacteria) and are expressed constitutively in *glnA* mutants, although they are not affected by GlnR. This suggests a role for glutamine synthetase in the regulation of these genes, but the details are unknown.

6 The Endospore

The endospore of the Bacillaceae is a truly differentiated structure and for this reason has attracted much attention as a "simple" differentiation process (see Figs. 4 and 5). The spores of bacilli, clostridia and thermoactinomycetes are structurally similar and have components such as the small acid-soluble spore proteins (SASP) that are highly conserved. It therefore seems probable that sporulation evolved only once and within the clostridia which is thought to be the ancestral member of the low G+C content Gram-positive bacteria. Non-sporing relatives such as the lactobacilli and staphylococci must have lost the ability to differentiate at some stage.

The elaborate morphology, remarkable resistance properties and inordinate longevity (some thermoactinomycete spores may have survived for almost 2000 years) have fascinated scientists since the endospore was discovered about 100 years ago. Recent reviews of sporulation include those by MURRELL (1988), DOI (1989) and several articles in SMITH et al. (1990).

6.1 Endospore Structure and Properties

In outlining the structure of the endospore, it will be useful to also explain, so far as is possible, its resistance properties. The major resistance properties are towards heat, irradiation (particularly UV light), and harsh chemicals such as organic solvents. It is important to note that the mechanisms of resistance towards these various insults to the integrity of the spore will differ, and that several determinants or processes may be responsible for each resistance. For example, thermotolerance may result from dehydration and mineralization both acting in a different way. Recent reviews on resistance mechanisms tend to concentrate on thermotolerance and UV resistance and include those by GERHARDT and MARQUIS (1990) and SETLOW (1992b).

The typical endospore comprises an inner protoplast in which the DNA, ribosomes and other cellular components reside (Fig. 4). The DNA is associated with SASPs. These proteins of 12–15 kD may be of the α- or β-type and are found in all endospores. They have been highly conserved throughout evolution; indeed the SASPs from *B. subtilis*, *Thermoactinomyces vulgaris* and clostridia show considerable homology (SETLOW, 1992b). The SASPs exist in multiple forms (at least seven in *B. megaterium*) and comprise 5–12% of the total spore protein. All α/β-type SASPs are DNA-binding proteins with a stoichiometry of 1 protein per 5 base pairs. The binding of the SASPs to DNA has two effects on the DNA *in vitro* which are also thought to occur *in vivo*. Negative supertwists are introduced into the DNA, and the molecule alters its conformation. DNA can assume a variety of helical shapes, notably the right-handed helices or A- and B-forms and the left-handed helix or Z-form. A- and B-forms differ in the number of base pairs per turn (higher in the A-form) which results in a wider minor groove in the molecule. DNA generally assumes the B-conformation in living cells but there is now overwhelming evidence that in the endospore, SASP binding results in an A-like conformation of the chromosome (SETLOW, 1992a). An important implication of this conformational switch is the reaction of the DNA to UV irradiation. Rather than the typical photoproducts of DNA, the cyclobutane thymine dimers, it has long been known that spores produce the so-called "spore photoproduct" after UV irradiation. This is 5-thymimyl-5,6-dihydrothymine which is the *in vitro* product of UV irradiation of A-DNA. This change in UV photochemistry is the reason for the resistance of the spore to UV irradiation and the reason why spores do not accumulate thymine dimers in their dormant state. Mutants lacking α/β-SASP produce normal thymine dimers upon irradiation and are UV-sensitive. SASPs have no role in heat resistance.

Low water content has long been connected with the thermotolerance of the endospore.

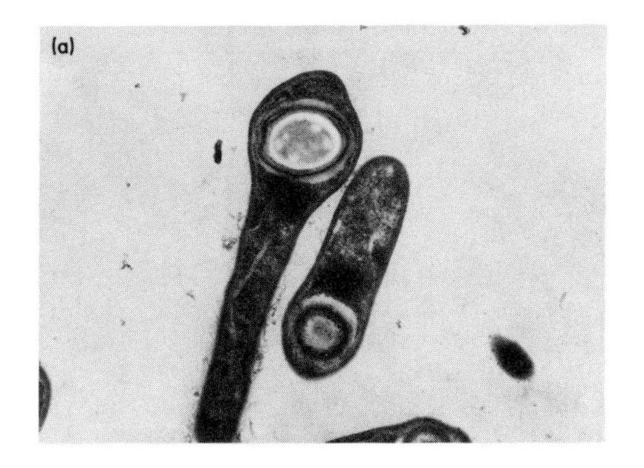

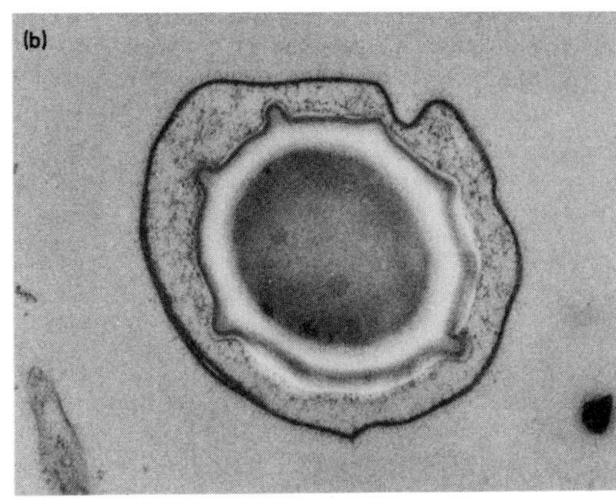

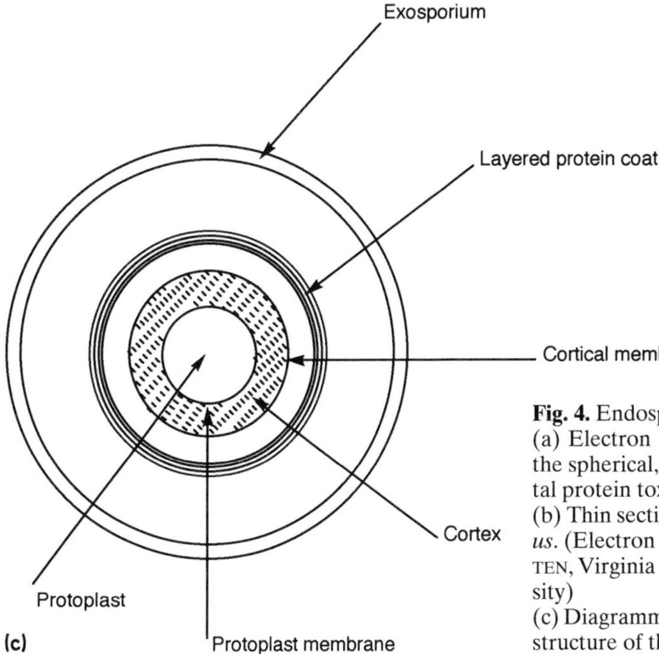

Fig. 4. Endospores of *Bacillus.*
(a) Electron micrographs of *B. sphaericus* showing the spherical, terminal spore and an associated crystal protein toxic for mosquito larvae.
(b) Thin section through an endospore from *B. cereus*. (Electron micrographs courtesy of . A. A. YOUSTEN, Virginia Polytechnic Institute and State University)
(c) Diagrammatic version of (b) showing the internal structure of the endospore.

Current estimates suggest that the fully hydrated spore protoplast of *B. megaterium* has a water content of about 30% although some estimates place it higher at up to 50% wet weight (reviewed by GERHARDT and MARQUIS, 1990). There is close correlation between protoplast dehydration and moist heat resistance. As dehydration increases so does thermoresistance. Indeed, dehydration is the only determinant both sufficient and necessary for elevated heat resistance. There are various suggestions as to the mechanism by which the spore protoplast becomes dehydrated, including contraction of the cortex and osmoregulation. A recent suggestion is potassium-coupled osmotic egress of water in which loss of potassium from the forespore is associated with osmotic loss of water (GERHARDT and MARQUIS, 1990).

The protoplast contains extraordinarily high levels of dipicolinic acid (DPA) chelated with Ca and other ions including Mn and Mg. DPA is unique to endospores, and the older literature suggests that it may be responsible for thermoresistance of the endospore. Isolation of mutants that produced endospores virtually devoid of DPA but which retained thermotolerance, and several other lines of evidence have since disproved this theory. Nevertheless, increased Ca is associated with increased heat resistance, on average a 30-fold increase in D_{100} with each percentage increase in Ca content over the range of 1.2–3.2% (MURRELL, 1988). Mineralization is accompanied by reduction in water content, so this observation may be partly explained by dehydration of the protoplast. Nevertheless at the lower and upper limits of protoplast water content, mineralization may act independently. Mineralization is probably achieved by active transport systems, although these have been studied only in a preliminary way.

At this point, it is appropriate to mention a third aspect of thermoresistance, that is associated with thermal adaptation. In general, thermophilic bacilli produce spores that have higher heat resistance than mesophilic bacilli and psychrophiles produce spores with lower heat resistance. Indeed, in each species the spore is more resistant than the vegetative cell by about 40 °C. Superimposed on this genetic adaptation, spores of a single strain will show increased thermal resistance when the bacterium is grown at a higher temperature. Thermal tolerance therefore comprises at least three facets, dehydration of the protoplast, mineralization and thermal adaptation.

The protoplast is surrounded by the cytoplasmic membrane which provides the normal

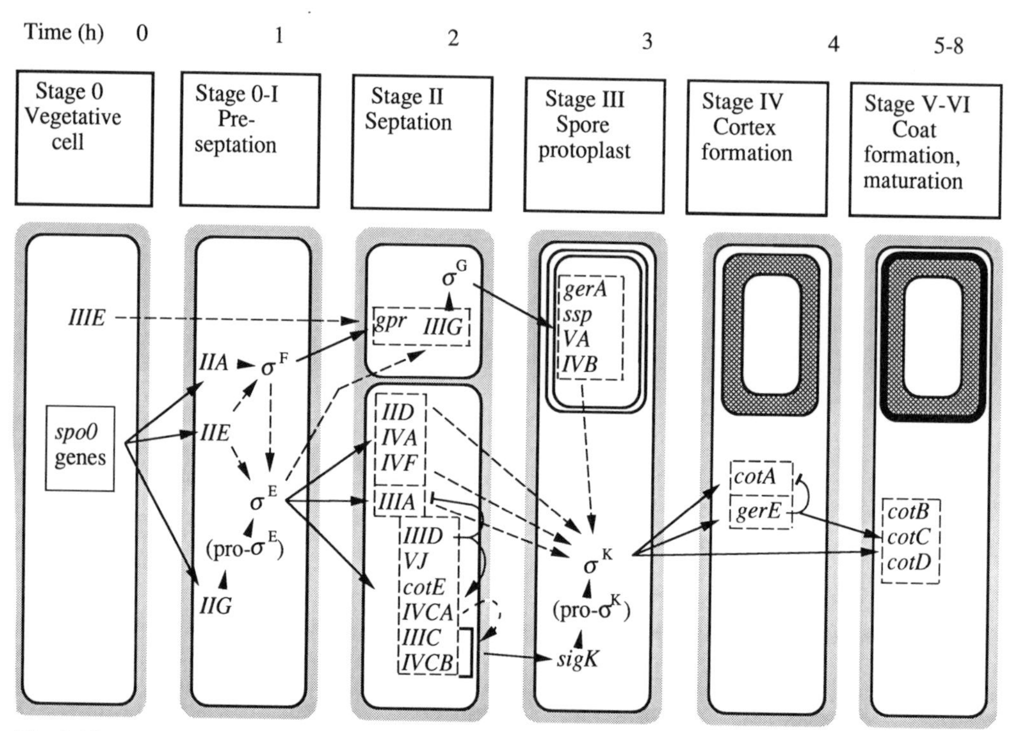

Fig. 5. The stages of sporulation and the known dependence relationships between sporulation genes. The stages of sporulation are indicated with an approximate time scale. Sporulation genes (for clarity the *spo* gene designation has been omitted) are placed on the figure according to the time at which they are expressed. Genes with promoters recognized by the same sigma-factor are grouped together in dotted boxes. The solid arrows represent direct effects on transcription, either activation (arrows) or repression (bars). Dotted arrows represent other, generally unknown effects on transcription. The bracket indicates a recombination event which brings together the two separate segments of the hybrid *sigK* gene encoding the precursor of σ^K. No attempt has been made to represent the complex series of interactions between the *spo0* genes (see Fig. 6). (Reproduced with permission from ERRINGTON, 1992.)

attributes of the lipid bilayer. The endospore, however, possesses a second "cytoplasmic" membrane which is recruited during stage III of sporulation as the prespore is engulfed by the mother cell (Fig. 5). This membrane is "inside out" and between the two membranes lies the spore cortex. The cortex comprises peptidoglycan. Cortical peptidoglycans isolated by mechanical means tend to have a high net negative charge, a low degree of cross-linking and many muramyl residues substituted with L-alanine or with a *delta*-lactam structure instead of a peptide. However, extraction of the peptidoglycan by chemical processes results in a more highly cross-linked peptidoglycan structure. This more closely represents the *in situ* material and the low cross-linked structures isolated by mechanical procedures represent damage from the isolation process. The probable function of cortical peptidoglycan is to act as a mechanical restraining structure to maintain the dehydration of the protoplast (GERHARDT and MARQUIS, 1990).

Spore coat proteins surround the outer cortical membrane. These tough, layered proteins are major components of the spore and may comprise 40–80% of the total spore protein. They are rather difficult to study because of their insolubility and the requirement of harsh extraction procedures. At least 12 major polypeptides have been characterized and genes encoding 65 kD (*cotA*), 59 kD (*cotB*), 12 kD (*cotC*), 11 kD (*cotD*), 24 kD (*cotE*), and 12.5 processed to 7.8 kD (*cotT*) have been cloned (reviewed by MOIR and SMITH, 1990). Mutants with altered coat proteins often show atypical germination characteristics (hence their early confusion with *ger* genes) and lowered heat resistance and sensitivity to lysozyme. Resistance to solvents and other harsh chemicals is also associated with the coat proteins.

The spore may be surrounded by an exosporium, a loose, protein-lipid membraneous covering of unknown function.

6.2 Sporulation

Electron microscopy of thin sections of sporulating cells permitted division of the process into seven morphological stages (Fig. 5). Mutants have been isolated which are blocked at most of these stages (stage I is the exception) and are described as Spo0 or SpoII mutants etc. depending on the stage at which the block on sporulation becomes apparent. Each class of mutants comprises several loci designated by a letter. For example, *spoIII* mutants are allocated to six loci; *spoIIA* through *F*. In this way, many hundreds of mutations have been mapped to more than 60 loci classed as *spo* (sporulation), *ger* (germination), *out* (outgrowth), *cot* (spore coat protein) and *ssp* (SASP) mutations (PIGGOT and HOCH, 1989). It has become apparent that many of these loci are in fact operons and comprise several genes, so the total number of genes involved is very large and probably about 200. The genes are scattered on the circular genetic map of *B. subtilis*, although some clusters are apparent.

Complete differentiation in *B. subtilis* takes 6–8 h growing at 37 °C. Cells at the end of exponential growth are at stage 0. Mutants unable to enter the sporulation cycle are termed Spo0 mutants. As nutrients become limiting, the cells enter sporulation and the chromosome thickens, taking the form of an axial filament. These stage I cells are not committed to sporulation because if placed in a plentiful environment they will return to vegetative cells. There are no mutants blocked at stage I and it is of dubious significance. The first true morphological event is the formation of an asymmetric cell division septum at stage II resulting in two unequal compartments. The two compartments have identical chromosomes, and the smaller will become the endospore while the larger, the mother cell, will participate in the formation of the spore but then lyse. During stage III the mother cell membrane engulfs the forespore surrounding the potential spore with a second membrane in the opposite orientation to the original forespore membrane. A layer of peptidoglycan is deposited between the two membranes forming the cortex (stage IV). A tough protein coat is then layered around the prespore by the mother cell during stage V, and maturation proceeds with the development of the typical resistance properties of the spore during stage IV. Finally in stage VII the mother cell lyses and releases the mature spore (see Fig. 5; reviewed by DOI, 1989).

Biochemically, sporulation can be viewed in three phases; initiation, a period of programmed differential gene expression terminating around stage IV, and construction and maturation of the spore which can proceed in the absence of further protein synthesis. Most studies have attempted to understand the first two of these phases.

The signal responding to nutrient deprivation (carbon, nitrogen or phosphate) and triggering the initiation of sporulation remains obscure. There is, however, a significant reduction of the intracellular concentration of GTP and GDP when cells are induced to sporulate. Indeed, decoyinine, which artificially reduces the level of GTP by inhibiting GMP synthetase, can induce sporulation. Perhaps the GTP interacts with a GTP-dependent regulatory cascade as found in eukaryotic cells. All Spo0 products are present during vegetative growth and seem to be involved in gathering and transducing signals, thus prompting the initiation of sporulation. Of the seven genes, *spo0A* is central to the process, since mutations in this gene are pleiotropically affected in many ways and totally blocked in sporulation. SpoA is controlled by a phosphorelay (BURBULYS et al., 1991). Several histidine protein kinases are involved with the initiation of sporulation, in particular the product of the *kinA* gene (also known as *spoIIJ*). These kinases autophosphorylate in response to certain signals and pass the phosphate on to Spo0F (Fig. 6). SpoF passes the phosphate on to a phosphotransferase, SpoB, which in turn passes the phosphate on to Spo0A which becomes phosphorylated. SpoA is a response regulator receiving signals from the "sensors" or histidine protein kinases (see Sect. 4.2). Spo0A in its phosphorylated form is a severe repressor of another regulatory gene, *abrB*. AbrB is evident in vegetative cells and represses the expression of many genes involved in stationary phase physiology such as extracellular enzyme synthesis, competence for DNA mediated transformation and sporulation. Inhibition of *abrB* by phosporylated Spo0A reduces the AbrB repressor and enhances transcription of stationary phase genes. Phosphorylated Spo0A also enhances transcription of *spoIIA*, the gene for a sporulation-specific sigma factor. It is thought that Spo0B is a critical point in the phosphorelay and that spo0B might be influenced by the product of a second gene in the *spo0B* operon, the Obg protein. Obg is a GTPase and links the reduction of GTP levels associated with the onset of sporulation with the phosphorelay (BURBULYS et al., 1991; GROSSMAN, 1991).

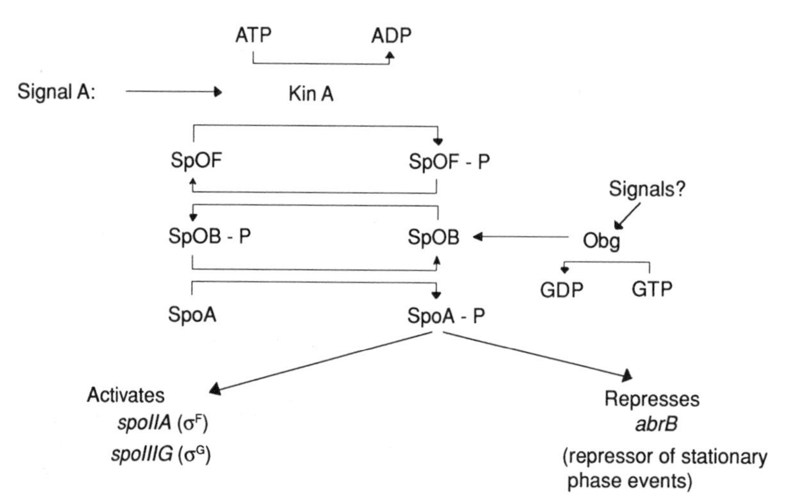

Fig. 6. The reactions of the phosphorelay leading to the phosphorylation of Spo0A and the initiation of sporulation (see BURBULYS et al., 1991; GROSSMAN, 1991). Spo0F is phosphorylated by several kinases, in particular KinA. The phosphate is passed on to Spo0A via Spo0F. A second route to phosphorylated Spo0A might be via the Obg protein interacting through Spo0B. Phosphorylated Spo0A is a response regulator that activates at least two key sporulation genes, those encoding σ^F and σ^G, and represses the synthesis of abrB, an important repressor of stationary phase events.

Following initiation, sporulation proceeds as a programmed series of transcriptional events. The timing of expression of the genes can be investigated by examining the expression of sporulation genes as β-galactosidase fusions in which the promoter of the gene is fused to the reporter gene *lacZ*. The expression of the gene is indicated by β-galactosidase synthesis. In this way the *timing of expression* of many genes has been determined. The *dependence of transcription* on previous events can be assessed by examining *spo-lacZ* fusion gene expression in a mutant strain. If a *spoIIA-lacZ* fusion is expressed in a *spoIVA* host, but not in a *spo0H* strain, then the expression of *spoIIA* is dependent on the SpoH protein but not on the SpoIVA protein. Large numbers of such experiments have determined a dependent series of events or a pathway of gene expression during sporulation (LOSICK et al., 1989). Interestingly, this is a branched pathway, beginning in the undifferentiated cell and then diverging with one line of events taking place in the mother cell and the other in the spore (see Fig. 5).

A second major form of transcriptional control operates during sporulation. The core RNA polymerase enzyme is competent for transcription but lacks specificity. The sigma-factor is a polypeptide that acts as an adaptor that provides the specificity of binding to the core enzyme and ensures that the correct promoter site is recognized. It follows that changing the sigma-factor will change the specificity of RNA polymerase and direct the enzyme to alternative promoters. By progressive changes of the sigma-factor, the transcription of different sets of genes can be programmed into the cell (MORAN, 1989). There are at least nine sigma-factors in *B. subtilis*, each denoted by a superscript letter (the earlier notation used the molecular mass of the polypetide). RNA polymerase containing σ^A is the vegetative form of the enzyme and it transcribes growth phase genes. Three other sigma-factors are also synthesized during growth but need not concern us here. Five important sigma-factors are involved in the pathways of sporulation. One of the first genes to be expressed is *spo0H* which is necessary for the initiation of sporulation and encodes σ^H. RNA polymerase containing σ^H transcribes several sporulation-specific genes and is also involved in the induction of competence for DNA-mediated transformation. σ^E and σ^F are encoded by *spoIIGB* and *spoIIAC*, respectively, and synthesized early in sporulation in the predivisional cell. It seems that σ^E and σ^F function in the undifferentiated cell but are responsible for events that will take place later in the mother cell and forespore, respectively. For example, mutations in σ^F that abolish sporulation prevent expression of prespore-specific genes and in a complementary way σ^E direct expression of mother cell-specific genes (ERRINGTON, 1991). In the forespore transcription is directed by RNA polymerase containing σ^G, the product of the *spoIIIG* gene. The genes for the SASPs are transcribed by this enzyme as are presumably other spore-specific proteins. Similarly, in the mother cell, transcription is effected by RNA polymerase containing σ^K (the product of the *spoIVCB* and *spoIIIC* genes). Spore coat proteins are synthesized exclusively in the mother cell by this form of RNA polymerase.

The fine-timing of sporulation gene expression may include a third aspect of transcriptional regulation, that is effected by DNA binding proteins of the activator or repressor class. Timing is of course very important and has been evaluated at length elsewhere (LOSICK and KROOS, 1989). Another aspect is compartmentalization and the regulation of morphogenesis. The mounting evidence for coupling between morphogenesis and the molecular pathways has also been discussed by ERRINGTON (1992).

6.3 Germination

Conversion of the spore back into a vegetative cell can be divided into two stages. Germination itself is an irreversible reaction in which metabolism is reactivated. Various degradative biochemical reactions take place resulting in the breaking of dormancy and the loss of typical spore properties such as refractility and heat- and chemical resistance. Germination is followed by outgrowth in which RNA, DNA and protein syntheses lead to normal vegetative metabolism and finally cell division.

For efficient germination the dormant spore must be activated. Sublethal heat treatment or extreme pH are effective activation steps but the process is unknown. The exquisite properties of the spore are reflected in the paradox that it is resistant to various extreme conditions and yet sensitive to low concentrations of molecules which will trigger germination (MOIR and SMITH, 1990). L-Alanine is a powerful germinant in *B. subtilis* but other species respond to sugars plus inorganic ions or ions alone. A mixture of glucose, fructose, asparagine and potassium ions (AGFK) also triggers germination in *B. subtilis*. The amino acid germinants act allosterically, D-alanine, for example, will inhibit the effect of L-alanine. The identity and localization of the germinant receptor(s) are still obscure although there is mounting evidence for a protein receptor in the inner membrane (MOIR and SMITH, 1990). Most physiological experiments have used *B. megaterium* as a model system, whereas genetic studies have obviously centered on *B. subtilis*.

In *B. megaterium*, heat shock is thought to activate the receptor(s) and exposure to L-alanine triggers spore germination rapidly. Within three minutes exposure, 50% of spores are committed to germination. Following interaction of the alanine with the receptor, proteolytic activity becomes detectable and activation of cortex lytic enzymes ensues. DPA is lost, and selective cortex hydrolysis begins with the concomitant uptake of water. The core swells to occupy the space previously occupied by the cortex. This is a critical period, the spores lose refractility and heat resistance, and it is thought that a cortex-lytic enzyme regulates the rehydration of the spore core. Water uptake leads to rehydration of spore proteins and metabolic activity. In *B. megaterium* these early germination events take about six minutes post commitment and are followed by the late events including protein turnover, SASPs providing amino acids for protein synthesis, ATP synthesis and the onset of general metabolism.

Genetic analysis of germination is reaching a very exciting stage (MOIR et al., 1991). Mutants affected in the L-Ala triggered pathway have been mapped to *gerA* which is the locus encoding the L-Ala receptor. *GerA* is now known to encode three membrane-located proteins. Interestingly, *gerA* is transcribed in the forespore by RNA polymerase associated with σ^G, so the likely fate of, these proteins is to be retained in, or at the surface of, the inner spore membrane. Mutants defective in the AGFK triggered germination pathway have been mapped to *gerB*, indicating that this is a separate system. Indeed, it is now thought that there is a family of GerA-like membrane-located proteins, each member of which responds to different stimuli. Two groups of mutants (*gerD* and *gerF*) that interfere with both germination pathways have been identified. *GerD* is also transcribed by the σ^G form of RNA polymerase and is thought to encode a protein that is exported from the forespore into the cortex. The products of these two genes are essential for the early stages of germination prior to loss of heat resistance. Finally, there are several types of germination mutants (*gerE*, *J*, and *M*) which block at a later stage after loss of heat resistance and in which cortex hydrolysis is incomplete. These pleiotropic mutants are also affected in the resistance properties of the spore, and may reflect defects in spore structure. These mutants demonstrate the interplay between sporulation and germination and the dangers of considering the two processes in isolation.

7 Genetics and Molecular Biology

Our understanding of sporulation and germination and the exploitation of *Bacillus* spp. by industry could not have come about without the availability of efficient genetic manipulation. Following the discovery of DNA-mediated transformation of certain *B. subtilis* strains in the late 1950s, *B. subtilis* became a favorite Gram-positive genetic system and is now second only to *Escherichia coli* in its genetic and molecular systems (SONENSHEIN et al., 1993). It is pertinent to cover briefly some of the molecular biology of *Bacillus* that is relevant to biotechnology; several comprehensive books on the subject are available (HARWOOD, 1989; HARWOOD and CUTTING, 1990; SMITH et al., 1990; HOCH et al., 1992).

7.1 Transformation

Bacillus genetics began with DNA-mediated transformation, and this process is still a vital issue since all recombinant DNA technology depends on efficient return of the *in vitro* modified DNA into the host cell. Competent cell transformation is well known in the *B. subtilis* Marburg strain and its derivative 168 and also occurs in *B. licheniformis* and perhaps *B. thuringiensis*. *B. subtilis* can be induced to enter the competent state at the end of exponential growth by a nutritional "shift down". A fraction (about 10–20%) of the cell population becomes competent and will bind exogenous DNA, whether chromosomal or plasmid in nature. The genetic control of competence and its involvement with the initiation of sporulation and other stationary phase phenomena have been reviewed recently (DUBNAU, 1991). The double-stranded DNA is cleaved into fragments of about 20–30 kb and enters the cell as a single-stranded molecule. In the case of chromosomal DNA, this single strand will replace the homologous strand in the chromosome by recombination. With plasmid DNA, the fragments of plasmid that enter the cell will recombine with other fragments to result in complete plasmids if sufficient areas of overlapping homologous sequence are available. If such DNA is not available, the transformed DNA will be lost. It is for this reason that only plasmid multimers (2 or more covalently joined copies of the plasmid) are effective in competent cell transformation of *B. subtilis*. Plasmid monomers or single, ligated recombinant forms are not transformable. To circumvent this problem, plasmid "rescue" systems were developed in which a homologous resident plasmid provides DNA for recombination with the incoming plasmid in much the same way as the chromosome rescues incoming chromosomal DNA. Alternatively, vectors with internal repeated sequences were constructed to provide the areas of homology for plasmid circularization in the cell (BRON, 1990).

Protoplast transformation, in which cells are stripped of their walls, transformed with plasmid DNA and cultured on osmotically stabilized media that encourage regeneration of the cell wall, avoids many of the problems associated with plasmid transformation of competent cells. Protoplasts can be transformed with ligated DNA or plasmid monomers, and with recent improvements in host strains and vectors, high transformation frequencies can be achieved (BRON, 1990). Electroporation of bacilli with plasmid DNA has been reported for *B. subtilis* and several other species (BRON, 1990).

7.2 Plasmids

The first plasmid vectors for *Bacillus* were derived from *Staphylococcus aureus* and comprised small antibiotic resistance plasmids such as pUB110 (kanamycin resistance), pE194 (erythromycin resistance) and pC194 (chloramphenicol resistance). Similar, if not identical plasmids have since been identified in natural *Bacillus* isolates, so these are now considered to be general Gram-positive plasmids. Although these should be ideal cloning vectors, severe instability problems were encountered. It was later realized that this instability derived largely from the mode of replication of these plasmids, which was based on "rolling circle" replication rather like that found in the single-stranded DNA phages of *Escherichia coli* and involved a single-stranded intermediate. This form of replication gives rise to extensive recombination and, under certain circumstances, the accumulation of high-molecular weight multimers that lead to segregational instability (GRUSS and ERLICH, 1989). These plasmids are still used for recombinant DNA studies, but are being replaced by plasmid vectors such as those based on pAMß1 which replicate in a normal, semi-conservative fashion (BRON, 1990).

The great range of plasmid vectors with sophisticated features that is available for *E. coli* is more limited in *B. subtilis*. However, one area in which the bacilli lead, is in insertional mutagenesis. YOUNGMAN et al. (1989) and others have exploited recombination in *B. subtilis* with a splendid range of vectors for mutagenesis and the construction of LacZ fusions. The transposon Tn*917*, which encodes erythromycin resistance, has been incorporated into derivatives of pE194 which are temperature-sensitive for replication. Growth of transformed cells at high temperature in the presence of erythromycin allows for selection of transposon insertions into the chromosome. The transposon integrates randomly giving rise to mutants. After initial development in *B. subtilis*, this technique has been used successfully in several other *Bacillus* species and other Gram-positive genera such as *Listeria*. Later versions of the Tn*917* containing vectors included the *E. coli lacZ* gene to provide β-galactosidase fusions when the transposon integrated upstream of a promoter. Additional features include an *E. coli* origin of replication to allow simple cloning of integrants into *E. coli*.

Insertional mutagenesis can also be achieved by cloning random (or specific) fragments of chromosomal DNA into an *E. coli* plasmid that contains a *cat* gene or a similar selectable marker that is funtional in *B. subtilis*. The recombinant plasmids are transformed into *B. subtilis* followed by chloramphenicol selection. *E. coli* plasmids are unable to replicate in *B. subtilis*, so only cells in which the *cat* gene is inserted into the chromosome at the sites of homology provided by the cloned fragments are able to grow on the chloramphenicol medium. In addition to these useful plasmid systems, ERRINGTON (1990) has developed sophisticated phage vectors for cloning and the construction of fusion proteins in *B. subtilis*.

Expression vectors employing powerful, regulated promoters for industrial production of cloned gene products are not readily available, but there is little doubt that companies manufacturing products from bacilli have such systems.

7.3 Recombinant Gene Products from *Bacillus*

The safety of *B. subtilis* and its ability to secrete large amounts of protein initially attracted many biotechnology companies to this bacterium as a host for expression of heterologous gene products. Seemingly insoluble problems with vector instability and proteolytic degradation of products disillusioned many, and the inability to produce correctly glycosylated eukaryotic proteins led numerous groups into the development of eukaryotic systems such as those based on animal cells or yeast. But *Bacillus* retained its attraction to those companies traditionally involved in manufacturing products from these bacteria, such as enzymes and insect toxins, and familiar with the shortcomings as well as the benefits of these organisms. Problems with vector instability were overcome, either by producing new plasmid vectors (see Sect. 7.2), or by amplifying the foreign gene integrated in the chromosome (PETIT et al., 1990). Strains almost totally deficient in protease have been prepared (WU et al., 1991) which give greatly enhanced accumulation of extracellular products. As a result, *Bacillus* enzymes are manufactured extensively from recombinant *Bacillus* strains (VEHMAANPERÄ and KORHOLA, 1986; MOUNTAIN, 1989) although the details are usually kept secret.

Some companies have persevered with *Bacillus* for the manufacture of mammalian-derived products and have developed highly efficient systems. Gist Brocades, for example, investigated the use of *E. coli*, *B. licheniformis*, *Saccharomyces* and animal cells for the manufacture of human interleukin 3. The yeast glycosylated the product, but incorrectly, and therefore offered no advantages. Animal cells produced the correct product but were declined on the basis of cost and low yield. *E. coli* could not efficiently secrete the product, and there was the possibility of endotoxin contamination. *B. licheniformis*, however, produced large quantities of secreted IL-3 (4 g could be purified from 50 liters of culture) inexpensively, and although it was not glycosylated this was found to be unimportant (VAN LEEN et al., 1991). Clinical trials are now underway with the material.

8 Concluding Remarks

Bacillus is at the crossroads in its biotechnological exploitation. The main route leads to ever increasing sophistication of current products and systems. The molecular biology of *B. subtilis* has developed dramatically in the past five years, many of the initial problems have been at least explained, if not overcome, and interest in these bacteria as hosts for cloned gene products on a commercial scale has been resurrected. This will no doubt lead to improved and simplified genetic engineering in these bacteria. On the product side, protein engineering of proteases has been, and will continue to be, successful in providing new enzymes with useful properties. But insect toxins for biocontrol present complex challenges to protein engineers. Insects will increasingly become resistant to *B. sphaericus* and *B. thuringiensis* toxins as the products are used on a greater scale. Can we modify existing toxins such that their spectrum of activity is changed in a predictable way? In the future, most probably yes, but this is still a long way ahead. So in the meantime the other route will be followed in which the ecologist and systematist will have an important input. Biotechnology has driven much of the recent research in diversity and ecology of endospore-forming bacteria and much remains to be learned. There are at least as many (and probably twice as many) *Bacillus* species waiting to be discovered as there are described in the literature. New enzymes, insect toxins, antibiotics and other valuable products await discovery and with this will come a greater appreciation of the biology of this remarkable group of microorganisms.

Acknowledgements

I am very grateful to J. ERRINGTON, R.W. LOVITT and A.A. YOUSTEN for providing material for inclusion in this review and to MAS RINA WATI for assistance with preparing the figures. Work in the author's laboratory was supported by grants from the SERC and the German Collection of Microorganisms and Cell Cultures.

9 References

ASH, C., FARROW, A. E., WALLBANKS, S., COLLINS, M. D. (1991), Phylogenetic heterogeneity of the genus *Bacillus* revealed by small-subunit-ribosomal RNA sequences, *Lett. Appl. Microbiol.* **13,** 202–206.

AUSTIN, B. A., PRIEST, F. G. (1986). *Modern Bacterial Taxonomy*, Wokingham, England: Van Nostrand Reinhold.

BAUMANN, N. P., CLARK, M. A., BAUMANN, L., BROADWELL, A. H. (1991), *Bacillus sphaericus* as a mosquito pathogen: properties of the organism and its toxins, *Microbiol. Rev.* **55,** 425–436.

BÉGUIN, P. (1990), Molecular biology of cellulose degradation. *Annu. Rev. Microbiol.* **44,** 219–248.

BERKELEY, R. C. W., LOGAN, N. A., SHUTE, L. A., CAPEY, A. G. (1984), Identification of *Bacillus* species, *Methods Microbiol.* **16,** 291–328.

BONDE, G. J. (1981), *Bacillus* from marine habitats: allocation to phena established by numerical techniques, in: *The Aerobic Endospore-forming Bacteria, Classification and Identification* (BERKELEY, R. C. W., GOODFELLOW, M., Eds.), pp. 181–216, London: Academic Press.

BOURRET, R. B., BORKOVICH, K. A., SIMON, M. I. (1991), Signal transduction pathways involving protein phosphorylation in prokaryotes, *Annu. Rev. Biochem.* **60,** 401–442.

BRON, S. (1990), Plasmids, in: *Molecular Biological Methods for Bacillus* (HARWOOD, C. R., CUTTING, S. M., Eds.), pp. 75–174, Chichester: John Wiley & Sons.

BULLA, Jr., L. A., COSTILOW, R., SHARPE, E. S. (1978), Biology of *Bacillus popilliae*, *Adv. Appl. Microbiol.* **23,** 1–18.

BURBULYS, D., TRACH, K. A., HOCH, J. A. (1991). Initiation of sporulation in *B. subtilis* is controlled by a multicomponent phosphorelay, *Cell* **64,** 545–552.

CAROZZI, N. B., KRAMER, V. C., WARREN, G. W., EVOLA, S., KOZIEL, M. G. (1991), Prediction of insecticidal activity of *Bacillus thuringiensis* strains by polymerase chain reaction product profiles, *Appl. Environ. Microbiol.* **57,** 3057–3061.

CARR, J. G. (1981), Microbes I have known, *J. Appl. Bacteriol.* **55,** 383–402.

CLAUS, D., BERKELEY, R. C. W. (1986), Genus *Bacillus* Cohn 1872, in: *Bergey's Manual of Systematic Bacteriology*, Vol. 2, (SNEATH, P. H. A., Ed.), pp. 1105–1139, Baltimore: Williams & Wilkins.

DE BARJAC, H. (1990), Characterization and prospective view of *Bacillus thuringiensis israelensis*, in: *Bacterial Control of Mosquitoes and Blackflies* (DE BARJAC, H., SUTHERLAND, D. J., Eds.), pp. 10–15, London: Unwin Hyman.

DE BOER, A. S., DIDERICHSEN, B. (1991), On the safety of *Bacillus subtilis* and *Bacillus amyloliquefaciens*, *Appl. Microbiol. Biotechnol.* **36,** 1–4.

DOI, R. H. (1989), Sporulation and germination, in: *Biotechnology Handbooks*, Vol. 2: *Bacillus* (HARWOOD, C. R., Ed.), pp. 169–215, New York: Plenum Press.

DUBNAU, D. (1991), Genetic competence in *Bacillus subtilis*, *Microbiol. Rev.* **55,** 395–424.

ERRINGTON, K. (1990), Gene cloning techniques, in: *Molecular Biological Methods for Bacillus* (HARWOOD, C. R., CUTTING, S. M., Eds.), pp. 175–220, Chichester: John Wiley.

ERRINGTON, J. (1991), Possible intermediate steps in the evolution of a prokaryotic developmental system, *Proc. R. Soc. London B* **224,** 117–121.

FISHER, S. H., SONENSHEIN, A. L. (1991), Control of carbon and nitrogen metabolism in *Bacillus*, *Annu. Rev. Microbiol.* **45,** 107–135.

FEITELSON, J. S., PAYNE, J., KIM, L. (1992), *Bacillus thuringiensis:* insects and beyond, *Bio/Technology* **10,** 271–275.

FOX, J. L. (1991), Bt resistance prompts early planning, *Bio/Technology* **9,** 1319.

FRITZE, D., FLOSSDORF, J., CLAUS, D. (1990), Taxonomy of alkaliphilic *Bacillus* strains, *Int. J. Syst. Bacteriol.* **40,** 92–97.

FRACHON, E., HAMON, S., NICOLAS, L., DE BARJAC, H. (1991), Cellular fatty acid analysis as a potential tool for predicting mosquitocidal activity of *Bacillus sphaericus* strains, *Appl. Environ. Microbiol.* **57,** 3394–3398.

GERHARDT, P., MARQUIS, R. E. (1990), Spore thermoresistance mechanisms, in: *Regulation of Procaryotic Development* (SMITH, I., SLEPECKY, R. A., SETLOW, P. Eds.), pp. 43–63, Washington, DC: American Society for Microbiology.

GILKES, N. R., HENRISSAT, B., KILBURN, D. T., MILLER, Jr., R. C. , WARREN, R. A. J. (1991), Domains in microbial β-1,4-glucanases: sequence conservation, function and enzyme families, *Microbiol. Rev.* **55,** 303–315.

Gordon, R. E. (1981), One hundred and seven years of the genus *Bacillus,* in: *The Aerobic Endospore-forming Bacteria: Classification and Identification*, (BERKELEY, R. C. W., GOODFELLOW. M., Eds.), pp. 1–15, London: Academic Press.

GORDON, R. E., HAYNES, W. C., PANG, C. H.-N. (1973), The Genus *Bacillus, Agriculture Handbook No. 427.* Washington, DC: United States Department of Agriculture.

GRANT, W. D., HORIKOSHI, K. (1989), Alkaliphiles, in: *Microbiology of Extreme Environments and its Potential for Biotechnology*, (DA COSTA, M. S., DUARTE, J. C., WILLIAMS, R. A. D., Eds.), pp. 346–366, London–New York: Elsevier Applied Science.

GRANT, W. D., TINDALL, B. J. (1986), The alkaline saline environment, in: *Microbes in Extreme Environments* (HERBERT, R. A., CODD, G. A., Eds.), pp. 25–54, London: Academic Press.

GROSSMAN, A. D. (1991), Integration of developmental signals and the initiation of sporulation in *B. subtilis*, *Cell* **65,** 5–8.

GRUSS, A., ERLICH, S. D. (1989), The family of highly interrelated single-stranded deoxyribonucleic acid plasmids. *Microbiol. Rev.* **53,** 231–241.

GUTOWSKI, J. C., SCHREIER, H. J. (1992), Interaction of the *Bacillus subtilis glnRA* repressor with operator and promoter sequences *in vivo*, *J. Bacteriol.* **174,** 671–681.

HARWOOD, C. R. (Ed.) (1989), *Biotechnology Handbooks*, Vol. 2, *Bacillus*, New York: Plenum Press.

HARWOOD, C. R., CUTTING, S. (Eds.) (1990), *Molecular Biological Methods for Bacillus*, Chichester: John Wiley & Sons.

HENKIN, T. M., GRUNDY, F. J., NICHOLSON, W. L., CHAMBLISS, G. H. (1991), Catabolite repression of α-amylase gene expression in *Bacillus subtilis* involves a *trans*-acting gene product homologous to the *Escherichia coli lacI* and *galR* repressors. *Mol. Microbiol.* **5,** 575–584.

HOFTE, H., WHITELEY, H. R. (1989), Insecticidal crystal proteins of *Bacillus thuringiensis*, *Microbiol. Rev.* **53,** 242–255.

HORIKOSHI, K., AKIBA, T. (1982), *Alkalophilic Microorganisms – A New Microbial World*, Berlin: Springer-Verlag.

JOHNSON, J. L. (1989), Nucleic acids in bacterial identification, in: *Bergey's Manual of Systematic Bacteriology,* Vol. 4; (WILLIAMS, S T., SHARPE, M. E., HOLT, J. G., Eds.), pp. 2305–2306, Baltimore : Williams & Wilkins.

KAMPFER, P. (1991), Application of miniaturized physiological tests in numerical classification and identification of some bacilli, *J. Gen. Appl. Microbiol.* **37,** 225–247.

KLAUSHOFER, H., HOLLAUS, F., POLLACH, G. (1971), Microbiology of beet sugar manufacture, *Process Biochem.* **6,** 39–41.

KLIER, A. F., RAPOPORT, G. (1988), Genetics and regulation of carbohydrate catabolism in *Bacillus*, *Annu. Rev. Microbiol.* **42,** 65–95.

KRAMER, J. M., GILBERT, R J. (1989), *Bacillus cereus* and other *Bacillus* species, in: *Foodborne Bacterial Pathogens* (DOYLE, M. P., Ed.), pp. 21–70, New York: Marcel Dekker.

KRISTJANNSON, J. K. (1989), Thermophilic organisms as sources of thermostable enzymes, *Trends Biotechnol.* **7,** 349–353.

KRULWICH, T. A., GUFFANTI, A. A. (1989), Alkalophilic bacteria, *Annu. Rev. Microbiol.* **43,** 435–463.

KUCEY, R. M. N. (1988), Alteration of wheat root systems and nitrogen fixation by associative nitrogen-fixing bacteria measured under field conditions, *Can. J. Microbiol.* **34,** 735–739.

Li, J., Carroll, J, Ellar, D. J. (1991). Crystal structure of insecticidal δ-endotoxin from *Bacillus thuringiensis* at 2.5 Å resolution, *Nature* **353,** 815–821.

LOGAN, N., BERKELEY, R. C. W. (1981), Classification and identification of members of the genus *Bacillus* using API tests, in: *The Aerobic Endospore-forming Bacteria: Classification and Identifi-*

cation (BERKELEY, R. C. W., GOODFELLOW, M., Eds.), pp. 105–140, London: Academic Press.

LOSICK, R., KROOS, L. (1989), Dependence pathways for the expression of genes involved in endospore formation in *Bacillus subtilis*, in: *Regulation of Procaryotic Development* (SMITH, I., SLEPECKY, R. A., SETLOW, P., Eds.), pp. 223–242, Washington, DC: American Society for Microbiology.

LOSICK, R., KROOS, L., ERRINGTON, J., YOUNGMAN, P. (1989), Pathways of developmentally regulated gene expression in *Bacillus subtilis*, in: *Genetics of Bacterial Diversity* (HOPWOOD, D. A., CHATER, K. E., Eds.), pp. 221–242, London: Academic Press.

MAGEE, R. J., KOSARIC, N. (1987), The microbial production of 2,3-butanediol, *Adv. Appl. Microbiol.* **32,** 89–159.

MEADOW, N. D., FOX, D. K., ROSEMAN, S. (1990), The bacterial phosphoenolpyruvate: glycose phosphotransferase system, *Annu. Rev. Biochem.* **59,** 497–542.

MOIR, A., SMITH, D. A. (1990), The genetics of bacterial spore germination, *Annu. Rev. Microbiol.* **544,** 531–553.

MOIR, A., YAZDI, M. A., KEMP, E. H. (1991), Spore germination genes of *Bacillus subtilis* 168, *Res. Microbiol.* **142,** 847–850.

MORAN, Jr., C. P. (1989), Sigma factors and the regulation of transcription, in: *Regulation of Procaryotic Development* (SMITH, I., SLEPECKY, R. A., SETLOW, P., Eds.), pp. 167–184, Washington, DC: American Society for Microbiology.

MOUNTAIN, A. (1989), Gene expression systems for *Bacillus subtilis*, in: *Handbooks of Biotechnology*, Vol. 2: *Bacillus* (Harwood, C. R., Ed.), pp. 13–114, New York: Plenum Press..

MURRELL, W. G. (1988), Bacterial spores – nature's ultimate survival package, in: *Mirobiology in Action* (MURRELL, W. G., KENNEDY, I. R., Eds.), New York: John Wiley & Sons.

PETIT, M. A., JOLIFF, G., MESAS, J. M., KLIER, A., RAPOPORT, G., ERLICH, S. D. (1990), Hypersecretion of a cellulase from *Clostridium thermocellum* in *Bacillus subtilis* by induction of chromosomal DNA amplification, *Bio/Technology* **6,** 559–563.

PIGGOT, P,. J., HOCH, J. A. (1989), Updated linkage map of *B. subtilis*, in: *Handbooks of Biotechnology,* Vol. 2: *Bacillus* (HARWOOD, C. R., Ed.), pp. 363–406, New York: Plenum Press.

PIVNICK, H. (1980), Spices, in: *Microbial Ecology of Foods*, Vol. 2: *Food Commodities,* pp. 731–751, The International Commission of Microbiological Specifications for Foods, New York: Academic Press.

PRIEST, F. G. (1977), Extracellular enzyme synthesis in the genus *Bacillus, Bacteriol. Rev.* **41,** 711–753.

PRIEST, F. G. (1981), DNA homology in the genus *Bacillus*, in: *The Aerobic, Endospore-forming Bacteria: Classification and Identification* (BERKELEY, R. C. W., GOODFELLOW, M., Eds.), pp. 33–57, London: Academic Press.

PRIEST, F. G. (1989a), Isolation and identification, in: *Biotechnology Handbooks*, Vol. 2: *Bacillus* (HARWOOD, C. R., Ed.), pp. 27–56, New York: Plenum Press.

PRIEST, F. G. (1989b), Products and applications, in: *Biotechnology Handbooks*, Vol. 2: *Bacillus* (HARWOOD, C. R., Ed.), pp. 293–320, New York: Plenum Press.

PRIEST, F. G. (1991), Synthesis and secretion of extracellular enzymes in bacteria, in: *Microbial Degradation of Natural Products* (WINKELMANN, G., Ed.), pp. 1–26, Weinheim: VCH.

PRIEST, F. G. (1992), Bacterial control of mosquitoes and other biting flies using *Bacillus sphaericus* and *Bacillus thuringiensis*, a review, *J. Appl. Bacteriol.* **72,** 357–369.

PRIEST, F. G. (1993), Systematics and ecology, in: *Bacillus subtilis and Other Gram-Positive Bacteria: Biochemistry, Physiology, and Molecular Genetics* (SONENSHEIN, A. L., LOSICK, R., HOCH, J. A., Eds.), Washington, DC: American Society for Microbiology.

PRIEST, F. G., GRIGOROVA, R. (1990), Methods for studying the ecology of endospore-forming bacteria, *Methods Microbiol.* **22,** 565–591.

PRIEST, F. G., STARK, J. R. (1991), Starch-hydrolyzing enzymes with novel properties, in: *Biotechnology of Amylodextrin Oligosaccharides* (FRIEDMAN, R. B., Ed.), pp. 72–85, Washington, DC: American Chemical Society.

PRIEST, F. G., GOODFELLOW, M., TODD, C. (1988), A numerical classification of the genus *Bacillus*, *J. Gen. Microbiol.* **134,** 1847–1882.

RASPOET, D., POT, B., DE DEYN, D., DE VOS, P., KERSTERS, K., DE LEY, J. (1991), Differentiation between 2,3-butanediol producing *Bacillus licheniformis* and *B. polymyxa* strains by fermentation product profiles and whole-cell electrophoretic patterns, *Syst. Appl. Microbiol.* **14,** 1–7.

RHODES-ROBERTS, M. E. (1981), The taxonomy of some nitrogen-fixing *Bacillus* species from Ynyslas sand dunes, in: *The Aerobic Endospore-forming Bacteria: Classification and Identification* (BERKELEY, R. C. W., GOODFELLOW, M., Eds.), pp. 315–336, London: Academic Press.

RÖSSLER, D., LUDWIG, W., SCHLEIFER, K. H., LIN, C., MCGILL, T. J., WISOTZKEY, J. D., JURTSHUK, Jr., P., FOX, G. E. (1991), Phylogenetic diversity in the genus *Bacillus* as seen by 16S RNA sequencing studies, *Syst. Appl. Microbiol.* **14,** 266–269.

RUSSEL, B. L., JELLEY, S. A., YOUSTEN, A. A. (1989), Carbohydrate metabolism in the mosquito pathogen *Bacillus sphaericus* 2362, *Appl. Environ. Microbiol.* **55,** 294–297.

SCHOFIELD, G. M. (1992), Emerging food-borne pathogens and their significance in chilled foods, *J. Appl. Bacteriol.* **72,** 267–273.

SETLOW, P. (1992a), DNA in dormant spores of *Bacillus* species is in an A-like conformation. *Mol. Microbiol.* **6,** 563–567.

SETLOW, P. (1992b), I will survive: protecting and repairing spore DNA, *J. Bacteriol.* **174,** 2737-2741.

SHARIATI, P. (1992), Anaerobic Metabolism in *Bacillus licheniformis, MSc Thesis*, Heriot-Watt University, Edinburgh.

SHARP, R. J., MUNSTER, M. J. (1986), Biotechnological implications for microorganisms from extreme environments, in: *Microbes in Extreme Environments* (HERBERT, R. A., CODD, G. A., (Eds.), pp. 215–295, London: Academic Press.

SHARP, R. J., RILEY, P. W., WHITE, D. (1992), Heterotrophic thermophilic bacilli, in: *Thermophilic Eubacteria* (KRISTJANSSON, J. K., Ed.), pp. 19–50, Boca Raton: CRC Press.

SLEPECKY, R. A., LEADBETTER, E. R. (1983), On the prevalence and roles of spore-forming bacteria and their spores in nature, in: *The Bacterial Spore*, Vol. 2, (HURST, A., GOULD, G. W., Eds.), pp. 79–101, London: Academic Press.

SMITH, I., SLEPECKY, R. A., SETLOW, P. (Eds.) (1990), *Regulation of Procaryotic Development*, Washington, DC: American Society for Microbiology.

SONENSHEIN, A. L., LOSICK, R., HOCH, J. A. (Eds.) (1993), *Bacillus subtilis and Other Gram-Positive Bacteria: Biochemistry, Physiology, and Molecular Genetics*, Washington, DC: American Society for Microbiology.

STOCK, J. B., NINJA, A. J., STOCK, A. M. (1989), Protein phosphorylation and regulation of adaptive responses in bacteria, *Microbiol. Rev.* **53,** 450–490.

SVENSSON, B., SIERKS, M. R., JESPERSEN, H., SOGAARD, M. (1991), Structure-function relationships in amylases, in : *Biotechnology of Amylodextrin Oligosaccharides* (FRIEDMAN, R. B., Ed.), pp. 28–43, Washington, DC: American Chemical Society.

SZEJTLI, J. (1991), Helical and cyclic structures in starch chemistry, in: *Biotechnology of Amylodextrin Oligosaccharides* (FRIEDMAN, R. B., Ed.), pp. 2–10, Washington, DC: American Chemical Society.

THANABALU, T., HINDLEY, J., JACKSON-YAP, BERRY, C. (1991), Cloning, sequencing, and expression of a gene encoding a 100-kilodalton mosquitocidal toxin from *Bacillus sphaericus* SSII-1, *J. Bacteriol.* **173,** 2276–2285.

TURNBULL, P. C. B. (Ed.) (1990), *Proceedings of the International Workshop on Anthrax, Salisbury Medical Bulletin*, Special Supplement **68.**

TURNBULL, P. C. B., JORGENSEN, K., KRAMER, J. M., GILBERT, R. J., PARRY, J. M. (1979), Severe clinical conditions associated with *Bacillus cereus* and the apparent involvement of exotoxins, *J. Clin. Pathol.* **32,** 289–293.

UEDA, S. (1989), Utilization of soybean as natto, a traditional Japanese food, in: *Bacillus subtilis: Molecular Biology and Industrial Application* (MARUO, B., YOSHIKAWA, H., Eds.), pp. 143–161, Amsterdam: Elsevier.

VAN LEEN, R. W., BAKHUIS, J. G., VAN BECKHOVEN, R. F. W. C., BURGER, H., DORSSERS, L. C. J., HOMMER, R. W. J., LEMSON, P J., NOORDAM, B., PERSOON, N. L. M., WAGEMAKER, G. (1991), Production of human interleukin-3 using industrial microorganisms, *Bio/Technology* **9,** 47–52.

VEHMAANPERÄ, J. O., KORHOLA, M. P. (1986), Stability of the recombinant plasmid carrying the *Bacillus amyloliquefaciens* α-amylase gene in *B. subtilis. Appl. Microbiol. Biotechnol.* **23,** 456–461.

WEICKERT, M., CHAMBLISS, G. H. (1990), Site directed mutagenesis of a catabolite repression operator sequence in *Bacillus subtilis*, *Proc. Natl. Acad. Sci. USA* **87,** 6238–6242.

WELLS, J. A., ESTELL, D. A. (1988), Subtilisin – an enzyme designed to be engineered, *Trends Biochem. Sci.* **13,** 291–297.

WISOTZKEY, J. D., JURTSHUK, Jr., P., FOX, G. E., REINHARD, G., PORALLA, K. (1992), Comparative sequence analyses on the 16S rRNA (rDNA) of *Bacillus acidocaldarius*, and *Bacillus cycloheptanicus* and proposal for creation of a new genus, *Alicyclobacillus* gen. nov., *Int. J. Syst. Bacteriol.* **42,** 263–269.

WOESE, C. R. (1987), Bacterial evolution, *Microbiol. Rev.* **51,** 221–271.

WU, X.-C., LEE, W., TRAN, L., WONG, S.-L. (1991), Engineering a *Bacillus subtilis* expression-secretion system with a strain deficient in six extracellular proteases *J. Bacteriol.* **173,** 4952–4958.

YOUNGMAN, P., POTH, M., GREEN, B., YORK, K., OLINEDO, G., SMITH, K. (1989), Methods for genetic manipulation, cloning and functional analysis of sporulation genes in *Bacillus subtilis*, in: *Regulation of Prokaryotic Development* (SMITH, I., SLEPECKY, R. A., SETLOW, P., Eds.), pp. 65–88, Washington, DC: American Society for Microbiology.

YOUSTEN, A. A. (1984), *Bacillus sphaericus:* microbiological factors related to its potential as a mosquito larvicide, *Adv. Biotechnol. Proc.* **3,** 315–343.

ZHANG, M., NAKAI, H., IMANAKA, T. (1988), Useful host-vector systems in *Bacillus stearothermophilus*, *Appl. Environ. Microbiol.* **54,** 3162–3167.

12 Pseudomonads

GEORG AULING

Hannover, Federal Republic of Germany

1 Definition of the Family Pseudomonadaceae and the Genus *Pseudomonas* 402
2 Phylogenetic Status of the Family Pseudomonadaceae and Internal Structure of the Genus *Pseudomonas* 402
3 Rapid Identification of Authentic versus Misclassified and/or "Honorary" Pseudomonads by Chemotaxonomy 406
4 Genome Organization and Plasmids 409
5 Ecology of Pseudomonads 411
6 Metabolic Properties of Pseudomonads 414
7 Applications of Pseudomonads 420
8 References 424

1 Definition of the Family Pseudomonadaceae and the Genus *Pseudomonas*

The large and important family of Pseudomonadaceae suffers from inconsistent definition. The present physiological characteristics of the Pseudomonadaceae, such as heterotrophic nutrition, respiratory metabolism, absence of fermentation, nutritional versatility except growth on one-carbon compounds, absence of photosynthesis or nitrogen fixation (PALLERONI, 1984), does not allow clear-cut differentiation from other groups of aerobic polarly flagellated Gram-negative bacteria. The circumscription of the genus *Pseudomonas* is unsatisfactory as well, and two thirds of the *Pseudomonas* species are generically misnamed. The basic morphological characteristics common to all pseudomonads are a negative reaction in the Gram-stain, the rod shape of cells, motility by polar flagellae, and the absence of spores. These criteria of low diagnostic value have led to a huge extension of both the genus and the family. STANIER et al. (1966) analyzed a large collection of pseudomonads by phenotypic characterization. PALLERONI et al. (1973), in the same laboratory, first detected the heterogeneity of what was called *Pseudomonas* by competitive ribosomal RNA hybridization. Ironically this did not cause a revision of the genus, rather its inappropriate status was conserved (DOUDOROFF and PALLERONI, 1974; PALLERONI, 1984, 1992a, b). Finally "honorary pseudomonads" were created (SILVER et al., 1990). While this reluctance to change a merely determinative system conferred on the genus *Pseudomonas* a sacrosanct character, it nevertheless teaches the microbiologists a lesson on the importance of nomenclature in bacterial classification (WOESE et al., 1985). Placing any new bacterial isolate in such a determinative bacterial classification scheme will not provide preexisting knowledge from related species but mislead further investigation on the isolate and work against its understanding unless the taxon changes to have phylogenetic validity.

The confusion arising from grouping together phenotypically disparate taxa into the Pseudomonadaceae has been a major obstacle to general and applied microbiology of the genus *Pseudomonas* within the last decades. Recent research on pseudomonads has been narrowly based with greatest emphasis given to such areas as molecular biology, where funds have been more plentiful. Thus, pseudomonad taxonomy and nomenclature remained perplexing since systematics received little attention. Abundant reports exist in the literature concerning environmental or biodegradation studies of strains allocated to *Pseudomonas* without any sound taxonomic data. This is unfortunate, since the defining of taxonomic relationships provides valuable information for those who are searching for clues concerning bacterial properties that relate to catabolic pathways or environmental fitness.

2 Phylogenetic Status of the Family Pseudomonadaceae and Internal Structure of the Genus *Pseudomonas*

When it became clear in the early seventies that methods such as numerical taxonomy, DNA base ratio, and DNA-DNA hybridization were merely able to elucidate the upper ramifications of relatedness of bacteria but not the major branches of the phylogenetic stem, two groups started comparing ribosomal RNA of bacteria. The 16S rRNA cataloguing method, followed by total sequencing, generated the present tripartite phylogenetic system (WOESE, 1992). Within this phylogenetic system the majority of Gram-negative bacteria were placed into the alpha, beta, gamma, and delta subdivision of "purple bacteria and relatives" (WOESE, 1987) and later given class status as the Proteobacteria (STACKEBRANDT et al., 1988). However, frequently only one strain of a species has been included in these studies although taxonomic conclusions should not be drawn from one single strain. Taxonomic experience tells us that even the type strain may be rather frequently atypical for its taxon. The incredibly large numbers of DNA-rRNA hybridizations carried out on the Gram-negative bacteria in the laboratory of J. DE LEY (Ghent, Belgium) firmly established the details of the phylogenetic structure of the alpha to gamma subclasses of the Proteobacteria at the generic and suprageneric levels (DE LEY, 1992). Fortunately, the data from both the 16S rRNA cataloguing and the DNA-rRNA hybridization are in excellent agreement. Because the hybridization approach was the major breakthrough for the taxonomy of the genus *Pseudomonas* it is briefly explained here. The most significant parameter of a DNA-rRNA hybrid is $T_{m(e)}$, the temperature at which half of the DNA-rRNA hybrid is denatured in a solution with defined salt composition and concentration. Because this parameter is a measure of the base sequence similarities between rRNA cistrons, it has been used for taxonomic differentiation displayed in $T_{m(e)}$ dendrograms (Fig. 1). The type species *Pseudomonas aeruginosa* and the phylogenetic nucleus of the genus *Pseudomonas*, designated *Pseudomonas fluorescens* complex in the terminology of DE VOS and DE LEY (1983) are located in the rRNA superfamily II (Fig. 1). Therefore, the *Pseudomonas fluorescens* complex is referred to as authentic genus *Pseudomonas* and authentic pseudomonads, respectively.

According to DE LEY (1992), a decisive criterion in the redefinition of the authentic genus *Pseudomonas* is that all members have a difference in $T_{m(e)}$ of $\Delta 6\,°C$ from the type strain of the species *Pseudomonas fluorescens*. Using only this criterion, the dinitrogen-fixing bacteria of the genera *Azotobacter* and *Azomonas* grouped together in the family Azotobacteriaceae as cyst-forming organisms (TCHAN, 1984) would clearly be members of the authentic genus *Pseudomonas* (DE LEY, 1992). Such a redefinition of the authentic genus *Pseudomonas* is supported by conservation of the four alginate gene sequences (FIALHO et al., 1990) in all members of this enlarged taxon. Similar amino acid sequences of cytochrome c_{551} of both *Azotobacter vinelandii* and five species of the authentic pseudomonads (AMBLER, 1973) also suggest a close relatedness. A cross-reacting monoclonal antibody indicates conservation of the outer membrane protein H2 in both the authentic genus *Pseudomonas* and in *Azotobact-*

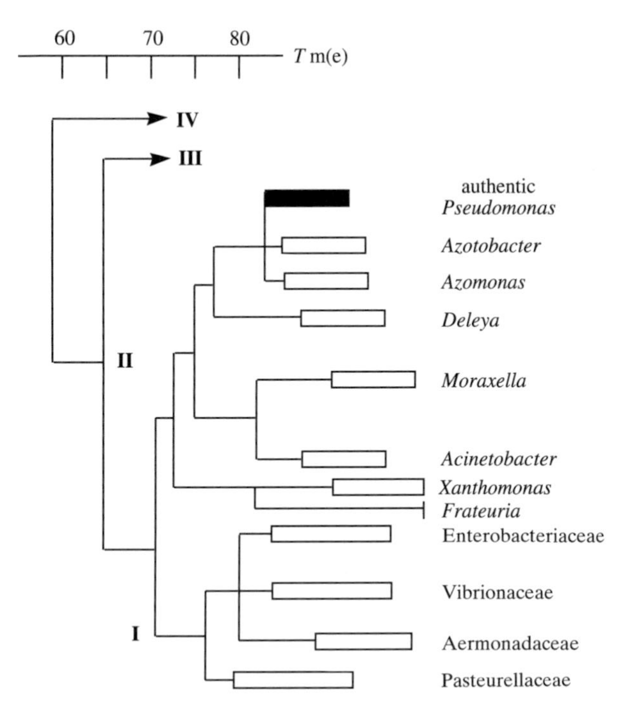

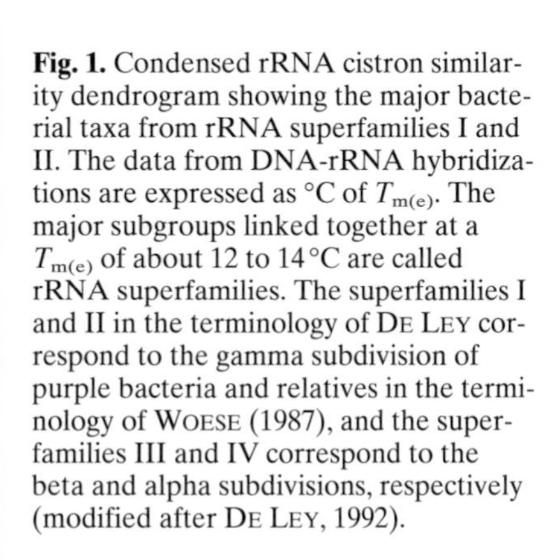

Fig. 1. Condensed rRNA cistron similarity dendrogram showing the major bacterial taxa from rRNA superfamilies I and II. The data from DNA-rRNA hybridizations are expressed as °C of $T_{m(e)}$. The major subgroups linked together at a $T_{m(e)}$ of about 12 to 14 °C are called rRNA superfamilies. The superfamilies I and II in the terminology of DE LEY correspond to the gamma subdivision of purple bacteria and relatives in the terminology of WOESE (1987), and the superfamilies III and IV correspond to the beta and alpha subdivisions, respectively (modified after DE LEY, 1992).

er vinelandii (MUTHARIA et al., 1985) In addition, members of the genera *Azotobacter* and *Azomonas* have chemotaxonomic properties like the authentic pseudomonads (cf. Sect. 3), i.e., they possess a ubiquinone with nine isoprenoid units in the side chain (Q-9), and putrescine and spermidine as their characteristic polyamines (BUSSE and AULING, unpublished results). In addition, all members of the authentic pseudomonads are very sensitive to EDTA and barium salts (SMIRNOV and KIPRIANOVA, 1990) and it would be of interest whether *Azotobacter* displays the same sensitivity. At the least, a revised family Pseudomonadaceae containing the authentic genus *Pseudomonas* and the genera *Azotobacter* and *Azomonas* has been envisaged (DE LEY, 1992). Phylogenetically, the genus *Serpens* is also highly related to the authentic genus *Pseudomonas* (WOESE et al., 1982). Phenotypically, it appears to be rather diverse. On the other hand, the chemotaxonomic marker rhodoquinone (cf. Fig. 2 and Sect. 3) allows an easy discrimination (HIRAISHI et al., 1992) between the inadequately described genus *Zoogloea* (DUGAN et al., 1992) and a revised family Pseudomonadaceae. Due to their low $T_{m(e)}$ value to members of the authentic genus *Pseudomonas* (Fig. 1), the genera *Xanthomonas* and *Frateuria* should be excluded from the Pseudomonadaceae and may constitute another bacterial family to be established in a future classification system (DE LEY, 1992), integrating phylogenetic, genotypic, and phenotypic information. Tab. 1 displays the possible revision of the family Pseudomonadaceae.

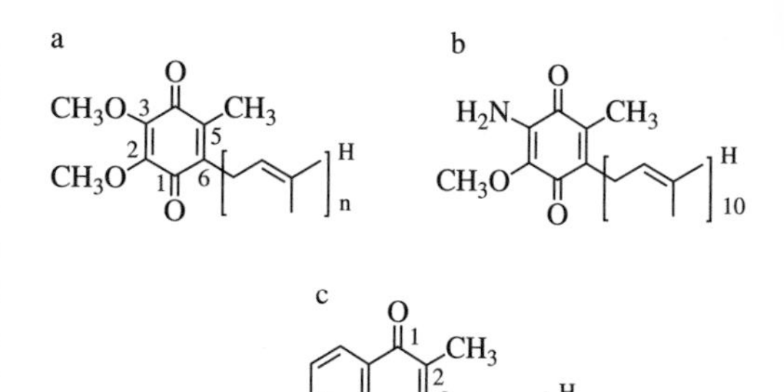

Fig. 2. Structure of prokaryotic respiratory quinones: (a) ubiquinone, (b) rhodoquinone, and (c) menaquinone. The multiprenyl side chain of menaquinones may be partially saturated or demethylated.

The internal structure of the genus *Pseudomonas* is perplexing in spite of the enormous amount of previous work. The pseudomonads used to be divided into fluorescent and nonfluorescent species (PALLERONI, 1984). However, the title role-trait of production of water-soluble fluorescent pigments (cf. Sect. 7) is not reliable for assignment (for discussion of de-

Tab. 1. Proposal for a Revision of the Family Pseudomonadaceae

Present Status	Emerging Natural Classification	
Family		
Pseudomonadaceae	Pseudomonadaceae [a]	Xanthomonadaceae?
Genera		
Pseudomonas	*Pseudomonas* (authentic)	*Xanthomonas*
Xanthomonas	*Azotobacter*	(*X. maltophilia*) [b]
Frateuria	*Azomonas*	*Frateuria*
Zoogloea	*Serpens*?	

[a] The exclusion of *Zoogloea* from and the inclusion of the Azotobacteriaceae into a phylogenetically defined family Pseudomonadaceae is supported by chemotaxonomic data (HIRAISHI et al., 1992), by conservation of cytochrome c (AMBLER, 1973), outer membrane protein H2 (MUTHARIA et al., 1985), and alginate genes (FIALHO et al., 1990).

[b] Although strains of *Pseudomonas maltophilia* were clearly excluded from the genus *Pseudomonas* by DNA-rRNA hybridization data which justify the transfer to *Xanthomonas* (SWINGS et al., 1983), they may ultimately give rise to a new genus when the boundary of the genus *Xanthomonas* will have been redefined (J. SWINGS, personal communication; van ZYL and STEYN, 1992).

tails see SCHROTH et al., 1992). STANIER et al. (1966) took a broad view in defining species, and placed most of the fluorescent strains into the two species *P. fluorescens* and *P. putida*, which were further subdivided into biotypes or biovars. Liquefaction is the main property which separates *P. putida* from *P. fluorescens*. However, in retrospective with the present-day request for arbitration of species by high DNA-DNA homology (WAYNE et al., 1987), it is evident that the range of DNA homology values within the two species and even within their individual biovars is extremely heterogeneous (PALLERONI et al., 1972). A recently introduced approach is restriction length polymorphism fingerprinting. However, as discussed by SCHROTH et al. (1992), application of this technique for taxonomic conclusions requires arbitration by DNA:DNA hybridization. We have to be aware that the authentic genus *Pseudomonas* consists of more than one hundred species (SCHROTH et al., 1992), considering the phylogenetic definition of modern taxonomy. The problem will not be solved unless a large polyphasic study collecting chemotaxonomic, phenotypic, and genotypic data on saprophytic and plant-pathogenic pseudomonads is carried out to provide the background for a sound reclassification of species within the authentic genus *Pseudomonas*.

The completion of this study should permit both an understanding of the relationships among the strains of *Pseudomonas* and the development of more useful identification schemes. For the moment the most important of the present species of the authentic genus *Pseudomonas* have been compiled in Tab. 2. Their chemotaxonomic characteristics have been included in this table, because the chemotaxonomic approach allows an unequivocal allocation to the authentic genus *Pseudomonas* as discussed in the following section.

Tab. 2. Present[a] Species of the Authentic[b] Genus *Pseudomonas* and Their Chemotaxonomic Characteristics[c]

Designation	Quinone	Polyamines	Diagnostic Fatty Acids
Fluorescent			
P. aeruginosa	Q-9	PUT, SPD	3-OH $C_{10:0}$, 3-OH $C_{12:0}$
P. fluorescens[d]	Q-9	PUT, SPD	3-OH $C_{10:0}$, 3-OH $C_{12:0}$
P. chlororaphis	Q-9	PUT, SPD	3-OH $C_{10:0}$, 3-OH $C_{12:0}$
P. aureofaciens	n.d.	n.d.	3-OH $C_{10:0}$, 3-OH $C_{12:0}$
P. putida	Q-9	PUT, SPD	3-OH $C_{10:0}$, 3-OH $C_{12:0}$
P. fragi	n.d.	PUT, SPD	n.d.
Non-fluorescent			
P. stutzeri[e]	Q-9	PUT, SPD	3-OH $C_{10:0}$, 3-OH $C_{12:0}$
P. mendocina	Q-9	PUT, SPD	3-OH $C_{10:0}$, 3-OH $C_{12:0}$
P. alcaligenes	Q-9	PUT, SPD	3-OH $C_{10:0}$, 3-OH $C_{12:0}$
P. pseudoalcaligenes	n.d.	PUT, SPD	3-OH $C_{10:0}$, 3-OH $C_{12:0}$
P. oleovorans	n.d.	PUT, SPD	n.d.
P. citronellolis	n.d.	PUT, SPD	n.d.
P. syringae[f]	Q-9	PUT, SPD	3-OH $C_{10:0}$, 3-OH $C_{12:0}$

[a] Compiled according to PALLERONI (1984, 1992 a, b),
[b] The allocation of species corresponds to RNA homology group I of PALLERONI extended by the work of the Ghent laboratory as compiled by DE LEY (1992).
[c] The data for chemotaxonomic markers are from different sources; quinones: (YAMADA et al., 1982), fatty acids: OYAIZU and KOMAGATA, 1983; STEAD, 1992), polyamines: (BUSSE and AULING, 1988; AULING et al., 1991 a).
[d] Several biovars exist, some of them have species character, e.g ., *P. lundensis*.
[e] Seven genomovars have been described (ROSSELLO et al., 1991).
[f] The special problems related to plant-pathogenic *Pseudomonas* species which can be easily identified by analysis of fatty acids (STEAD, 1992) or polyamines (AULING et al., 1991 a) have been discussed elsewhere (KLEMENT et al., 1990; SCHROTH et al., 1992).
n.d., not determined; PUT, putrescine; SPD, spermidine

3 Rapid Identification of Authentic versus Misclassified and/or "Honorary" Pseudomonads by Chemotaxonomy

It should be clear from the preceding section that the purpose of any identification, which is to equate the properties of a pure culture with those of well-characterized and accepted species, is not possible with the traditional phenotypical tests aimed at identifying pseudomonads. The reason why identification keys for pseudomonads such as those presented in the familiar determinative system (STOLP and GADKARI, 1981) failed has been brought to light by the phylogenetic revolution of taxonomy described in the two introductory sections. A wide variety of physiological and biochemical reactions for identification of pseudomonads, such as sugar oxidations, nitrate reduction, gelatin liquefaction and others, which are the basis of the many commercially available multiple-test systems traditionally used in clinical microbiology, appear to be present at random in many genera and species of the dendrogram of aerobic Gram-negative bacteria now allocated to the class Proteobacteria (STACKEBRANDT et al., 1988). DE LEY (1992) presented the interesting idea that the traditional taxonomic studies of comparing physiological and biochemical reactions of bacteria might be focused too much on the active center of the enzymes involved, whereas the real difference will be hidden in other amino acid sequences.

However, there is a constant necessity for identification of newly isolated bacteria suspected to belong to the genus *Pseudomonas*. For example, the recent interest in exploiting the versatility of *Pseudomonas* species in biodegradation of xenobiotic compounds or in plant-growth promoting pseudomonads from the rhizosphere (DEFAGO and HAAS, 1990) has led to a great number of new isolates. It is obvious that many of them need to be characterized more specifically for the description of new species rather than merely to be identified. Some applications of nucleic acid (DNA, rRNA) techniques (STACKEBRANDT and GOODFELLOW, 1991) for the identification of pseudomonads are discussed in Sects. 5 and 7. If these techniques cannot be applied to the identification of purported pseudomonads because of the high costs, insufficient equipment or missing scientific knowledge, other chemical components of the bacterial cells that are of taxonomic interest (GOTTSCHALK, 1985; GOODFELLOW and MINNIKIN, 1985; TRÜPER and SCHLEIFER, 1992) provide the basis for an alternative rapid approach.

Any cellular component intended to be used as a chemotaxonomic marker should be ubiquitous within the microbial world. Its production must be stable and independent of media chosen for growth. Its determination should be simple, sensitive and rapid and preferably carried out through automated analysis. A characteristic distribution of such a marker is also required, i.e., variation between different taxons, but not within the same taxon. Among the cellular components, which comply with these requirements, quinones, fatty acids and polyamines have been proven to be especially useful for pseudomonads.

Different types of respiratory quinones can be recognized in the cytoplasmic membrane of prokaryotes (see Fig. 2). It is worth emphasiz-

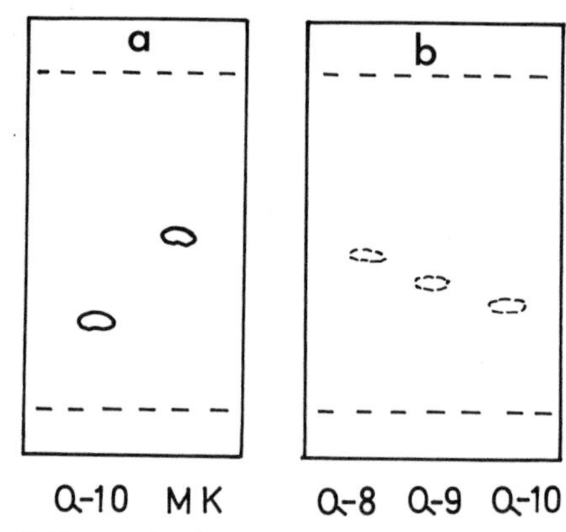

Fig. 3. Determination of the quinone system of *Pseudomonas* species by (a) preparative and (b) subsequent high-performance thin-layer chromatography according to KROPPENSTEDT (1982). Abbreviations: MK, menaquinone; Q-10, ubiquinone with a 10-carbon multiprenyl side chain.

ing that the inherent value of quinones as taxonomic criteria lies basically in the variation in the length of their multiprenyl side chains (COLLINS and JONES, 1981). The first systematic investigation on the distribution of quinones in pseudomonads (YAMADA et al., 1982) revealed that these bacteria possess ubiquinones, as opposed to Gram-positive bacteria which have menaquinones as the principal quinone. The same study also firmly established that species of the authentic genus *Pseudomonas* possess ubiquinone Q-9 (see Tab. 2), whereas the other misclassified pseudomonads which are phylogenetically distant from this nucleus, either have ubiquinone Q-8 or Q-10 (Tab. 3). The relative ease with which isoprenoid quinones can now be isolated even by simple thin-layer chromatography (Fig. 3) on ready-made high-performance reversed-phased plates, makes these molecules ideal chemotaxonomic markers (GOTTSCHALK, 1985; GOODFELLOW and MINNIKIN, 1985). The presence of ubiquinones in pseudomonads versus menaquinones in Gram-positive bacteria even suggests determination of the quinone class as a reliable alternative approach when Gram-staining gives conflicting results (AULING et al., 1986).

The fatty acids are the lipid components most widely used in chemotaxonomy. For identification of Gram-negative bacteria hydroxy fatty acids bound to lipopolysaccharides are very important. According to OYAIZU and KOMAGATA (1983) members of the authentic genus *Pseudomonas* possess 3-hydroxy fatty acids with ten or twelve carbon atoms (3-OH $C_{10:0}$, 3-OH $C_{12:0}$) of high diagnostic value (see

Tab. 3. Chemotaxonomic Characteristics and Phylogenetically Based Revision[a] of Misclassified Pseudomonads[b]

Previous Name	Status after Revision	Quinone	Polyamine[c]
	Gamma subclass		
P. maltophilia[d]	*Xanthomonas maltophilia*	Q-8	SPD
P. betle	*Xanthomonas maltophilia*	n.d.	SPD
P. gardneri	*Xanthomonas campestris*	n.d.	SPD
P. marina	*Deleya marina*	Q-9	SPD
	Beta subclass		
P. acidovorans	*Comamonas acidovorans*	Q-8	HPUT, PUT
P. facilis	*Acidovorax facilis*	Q-8	HPUT, PUT
P. flava	*Hydrogenophaga flava*	Q-8	HPUT, PUT
P. ruhlandii	*Alcaligenes xylosoxidans*	n.d.	HPUT, PUT
	Alpha subclass		
P. paucimobilis	*Sphingomonas paucimobilis*	Q-10	HSPD
P. carboxidovorans	*Oligotropha carboxidovorans*	Q-10	HSPD
P. aminovorans	*Aminobacter aminovorans*	Q-10	HSPD, PUT
P. mesophilica	*Methylobacterium mesophilicum*	Q-10	HSPD, PUT
Pseudomonas sp. ATCC 29600	*Chelatobacter heintzii*	Q-10	HSPD, PUT
P. comprансoris	*Zavarzinia compransoris*	Q-10	PUT, HSPD

[a] The misclassified species are grouped together in blocks according to their phylogenetic allocation into the gamma-, beta-, and alpha subclass of the Proteobacteria.

[b] Data have been collected from DE LEY, 1992; WOESE, 1987; YAMADA et al., 1982; OYAIZU and KOMAGATA, 1983; BOUSFIELD and GREEN, 1985; TAMAOKA et al., 1987; FRANZMANN et al., 1988; BUSSE and AULING, 1988; AKAGAWA and YAMASATO, 1989; DE VOS et al., 1989; WILLEMS et al., 1989, 1990; YABUUCHI et al., 1990; AULING et al., 1991 a; BUSSE and AULING, 1992; URAKAMI et al., 1992; MEYER et al., in press; AULING et al., in press.

[c] Characteristic polyamines are listed in decreasing order.

[d] The presence of branched fatty acids provides an additional argument for exclusion from the authentic genus *Pseudomonas*.

Protocol for Extraction and Analysis of Polyamines

steps

1. 40 mg lyophilized cells (log. growth phase)
2. add internal standard
3. 0.2 M $HClO_4$ (100° C, 30 min), adjust pH
4. "precolumn"-dansylation (60° C, 30 min)

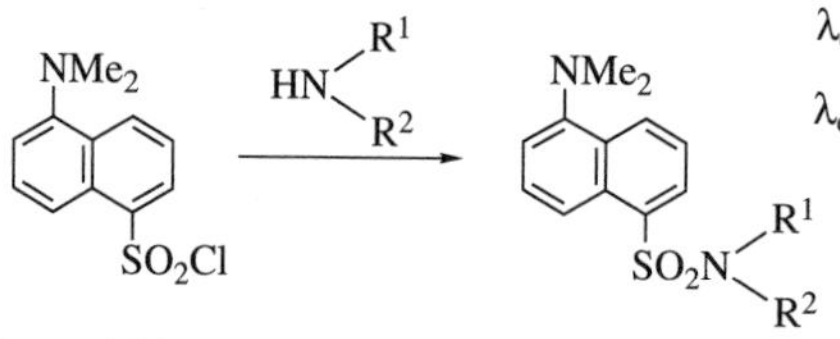

Reaction of dansylchloride with primary and secondary amines

5. extract and concentrate in 100 µL toluene
6. reversed phase HPLC, gradient elution (Acetonitrile/water : 40/60)

total time required per sample: 2 hours

Fig. 4. Extraction and analysis of polyamines (from AULING, 1992).

Tab. 2), whereas the even-numbered straight-chain fatty acids $C_{16:0}$, $C_{16:1}$, and $C_{18:1}$ were found as the major compounds. Although fatty acid compositions are now routinely used as analytical markers in diagnostic bacteriology and support the identification of authentic and other pseudomonads (automated identification is even possible), there are limitations which relate to the necessity of carefully controlled standardization of cultural and chemical techniques. The fatty acid pattern may be altered in response to growth conditions (TRÜPER and SCHLEIFER, 1992). Definitive identification of authentic pseudomonads by fatty acid profiles is impaired by the low ratio of the diagnostic hydroxy fatty acids (2 to 5% only) and by the observation that amide-bound hydroxy fatty acids may not be entirely liberated from cells by the hydrolytic method used (OYAIZU and KOMAGATA, 1983).

Our laboratory developed the use of polyamines as another useful chemotaxonomic marker which reflects the major branching of the phylogenetic dendrogram of the Proteobacteria (BUSSE and AULING, 1988). Polyamine profiles can be rapidly (Fig. 4) determined by gradient high-performance liquid chromatography (Fig. 5). Up to now, all members of the authentic genus *Pseudomonas* analyzed are characterized by putrescine and spermidine as the main polyamine compounds (see Tab. 2). On the other hand, the misclassified pseudomonads of the new genera *Comamonas*, *Acidovorax*, and *Hydrogenophaga*, now included into the new family Comamonadaceae (WILLEMS et al., 1991), can be easily allocated to the beta subclass of the Proteobacteria, solely by the presence of the beta subclass-specific polyamine 2-hydroxyputrescine (Tab. 3). Likewise, presence of *sym*-homospermidine allocates a number of diverse misclassified pseudomonads to the alpha subclass of the Proteobacteria, as does the presence of ubiquinone Q-10 (Tab. 3). Finally, ubiquinone analysis cannot distinguish between the misclassified yellow pseudomonads of the beta subclass of the Proteobacteria and members of the genus *Xanthomonas*, since a ubiquinone Q-8 is present in both, whereas polyamine profiles do (AULING et al., 1991a).

In conclusion, the phylogenetic approach has brought a solution to the old problem of defining an authentic genus *Pseudomonas*. Regarding the perplexing status of the traditionally defined genus we are currently facing a rapid reduction to its phylogenetic nucleus by stepwise exclusion of the distant species. Ubiquinones, fatty acids, and polyamines have

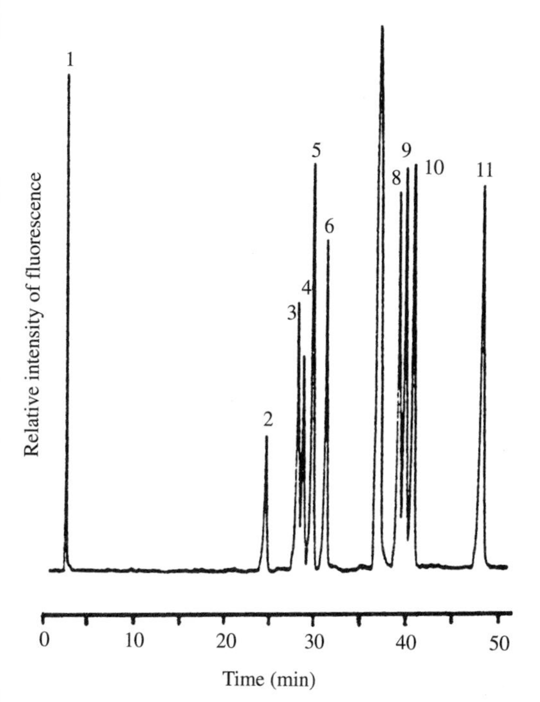

Fig. 5. HPLC-chromatogram of a mixture of dansylate polyamines. The numbers of peaks indicate: **1**, proline; **2**, 2-hydroxy-putrescine; **3**, 1,2-diaminoethane; **4**, 1,3-diaminopropane **5**, putrescine; **6**, cadaverine; **7**, 1,8-diaminoctane; **8**, *sym*-norspermidine; **9**, spermidine; **10**, *sym*-homospermidine; and **11**, spermine.

been evaluated as chemotaxonomic markers which reflect natural relationships indicated by phylogenetic dendrograms. This scientific progress is the basis for new efforts to develop improved identification schemes. The relative ease, the speed, and a possible option to automated analysis, evidently suggests that the determination of the markers discussed here should be included as the first obligatory step for both allocation to and exclusion from the authentic pseudomonads. In naming any freshly isolated Gram-negative, aerobic, motile rod *Pseudomonas*, there is no excuse not to examine at least its quinone system, because a determination can be done in any microbiological laboratory by simple reversed-phase partition thin-layer chromatography demonstrated in Fig. 3.

4 Genome Organization and Plasmids

Among the bacterial species for which physically constructed chromosome maps have been published (KRAWIEC and RILEY, 1990) *Pseudomonas aeruginosa* has a very large (5.9 Mb) genome (RÖMLING and TÜMMLER, 1991). Large genome sizes were also reported for misclassified hydrogen-oxidizing pseudomonads (AULING et al., 1980), now included into the new family Comamonadaceae (WILLEMS et al., 1991). More than 20 years after the initial proposal of the replicon model, an interesting hypothesis resulted from analysis of chromosomal origins of *Pseudomonas* and other bacteria. The term origin refers to the short sequence where chromosome replication initiates by binding and unwinding of a particular DNA region through the dnaA protein, and which can be cloned as autonomously replicating sequence. The origins of *P. aeruginosa* and *P. putida* rather than those of *Escherichia coli* or *Bacillus subtilis* have the most common features and are considered as the paradigm bacterial origin class (SMITH et al., 1991).

Based on intensive comparisons of the genomes of *P. aeruginosa* and *P. putida*, mainly by studies with traditional mapping techniques, HOLLOWAY et al. (1990) suggested that the present-day pseudomonads have evolved by integrating metabolic (preferentially catabolic) functions from plasmids into a historically smaller chromosome. Cooperation between genes from the plasmids and from the chromosome may indicate an intermediate evolutionary status. For example, the OCT plasmid originally described in *Pseudomonas putida* PpG6 codes for a number of proteins involved in growth on C_6 to C_{10} *n*-alkanes. Later studies with *P. aeruginosa* revealed a cooperation of plasmid- and chromosomally encoded genes for the catabolism of aliphatic hydrocarbons. The ancestral *Pseudomonas* chromosome may be represented by the one half of the present map of *P. aeruginosa* or *P. putida* where the housekeeping functions (anabolic and other central metabolic reactions) tend to be located. Interestingly, the linkage of genes for the β-ketoadipate pathway of aromatic compounds (cf. Sect. 6) known in *Pseudomonas* since the seventies is now also observed in *Acinetobacter*, another genus within the gamma subclass of the Proteobacteria (see Fig. 1), and referred to as supraoperonic clustering (ORNSTON and NEIDLE, 1991).

Another interesting hypothesis on the evolution of the pseudomonads was put forward by PALLERONI (1992a) and refers to the observation that *P. aeruginosa* is resistant to many antibiotics produced by streptomycetes. In soil pseudomonads coexist with streptomycetes which prefer similar growth conditions. Because pseudomonads are nutritionally very versatile with regard to growth on low-molecular weight compounds, they benefit from the capacity of streptomycetes to hydrolyze polymeric substrates. As a consequence of being associated in the same ecological niches intergeneric transfer in soil of antibiotic resistance factors which might have originated from the antibiotic-producing streptomycetes may be the reason why so many pseudomonads are notoriously refractory to antibiotics produced by members of the genus *Streptomyces*. The observation that rather few extrachromosomal determinants of resistance are present in *P. aeruginosa* compared with other species, has been explained as a result of active genetic exchange between resistance plasmids and the chromosome which led to the stable integration of the resistance genes in the genome (JACOBY, 1986).

The various plasmids of xenobiotic-degrading (FRANTZ and CHAKRABARTY, 1986), clinical (JACOBY, 1986), or plant-pathogenic (SHAW, 1987) pseudomonads can be classified on the basis of their compatibility (stable coexistence within the same host) into at least thirteen incompatibility groups named incP-1 to incP-13 (cf. Tab. 1 in JACOBY, 1986). Most of the degradative plasmids (FRANTZ and CHAKRABARTY, 1986), which contribute significantly to the metabolic versatility of some *Pseudomonas* species, belong to the incP-9 group. The plasmids TOL, NAH, and SAL, carrying genes involved in the degradation of toluene, naphthalene, and salicylate have been studied intensively and were recently reviewed (BURLAGE et al., 1989; ASSINDER and WILLIAMS, 1990). However, degradative plasmids are not restricted to the incP-9 incompatibility group. For example, plasmid pVI50, an incP-2 megaplasmid, confers efficient mineralization of phenol and methyl-substituted aromatics to *Pseudomonas* sp. CF600 (SHINGLER et al., 1992). Those plasmids whose host range exceeds the authentic genus *Pseudomonas* may also be classified according to incompatibility schemes developed for other bacterial groups, e.g., the Enterobacteriaceae. The *Pseudomonas* plasmids with a wide host range may participate in gene transfer from a gene pool (REANNEY et al., 1982) common to those Gram-negative bacteria now allocated to the class Proteobacteria (STACKEBRANDT et al., 1988). However, permanent transfer of *Pseudomonas* plasmids to other genera seems to be more restricted by their ability to replicate in a diversity of hosts and to ensure partitioning at the time of cell division than by their conjugative potential. Since the *Pseudomonas* plasmids of the RP4 type which belong to the incP-1 group, have an extraordinary wide host range (SCHUMANN, 1990) they must confer the ability to recognize a common structural feature of the surface of phylogenetically different bacteria. Consequently these plasmids have been exploited for the construction of delivery systems for transposon mutagenesis in Gram-negative bacteria (SIMON et al., 1983; MERMOD et al., 1986).

A matter of concern is that natural spreading of plasmid-encoded resistance genes is facilitated if they are located within transposon structures (JACOBY, 1986). Investigations for unravelling the evolution of such resistance determinants in pseudomonads and other bacteria by DNA sequencing are now very popular (see references cited in BISSONNETTE and ROY, 1992) and have led to the detection of a novel family of potentially mobile elements, the integrons. Recently the integron InO from the *Pseudomonas aeruginosa* plasmid pVS1 was characterized as an ancestor of more complex integrons (BISSONNETTE and ROY, 1992). From the practical point of view it is more important to understand the factors which control the expression range of the determinants for resistance to antibiotics or simple chemicals. For example, the well-studied integron in the *Tn21*-like transposons is made up of two conserved segments coding for an integrase-like protein and for sulfonamide resistance, between which discrete units which can be expressed at a high level are integrated. In this respect, the integrons are natural expression vectors.

The evolution of natural plasmids towards degradative plasmids adapted to recently released xenobiotic substrates is reminiscent of the spread of antibiotic resistance genes previously discussed. In the well-characterized degradative plasmid pWWO the TOL degradative pathway has been proven to reside on a 56 kb transposon which itself is part of a larger transposon of 70 kb (Fig. 6), giving rise to a whole family of TOL plasmids (ASSINDER and WILLIAMS, 1990). Although the archetypal TOL plasmid pWWO from *Pseudomonas putida* mt-2 has been referred to as having a broad host range, the expression range of its catabolic genes (cf. Fig. 6, lower part, and Sect. 6) is considered to be limited to *Pseudomonas* species (BURLAGE et al., 1989). However, upon conjugative transfer and transpositional insertion of the TOL catabolic genes into the degradative plasmid pSAH (JAHNKE et al., 1990) residing in *Alcaligenes* sp. O-1, a member of the beta subclass of the Proteobacteria, this strain acquired the ability to degrade methyl-substituted aromatic compounds (JAHNKE et al., in press).

Other catabolic transposons of *Pseudomonas* may be detected in the future as a result of continuing bacterial evolution. A transposon-like, possibly novel mobile element, encoding a dehalogenase function has recently been described for *P. putida* and may be present in further Gram-negative bacteria (THOMAS et al., 1992).

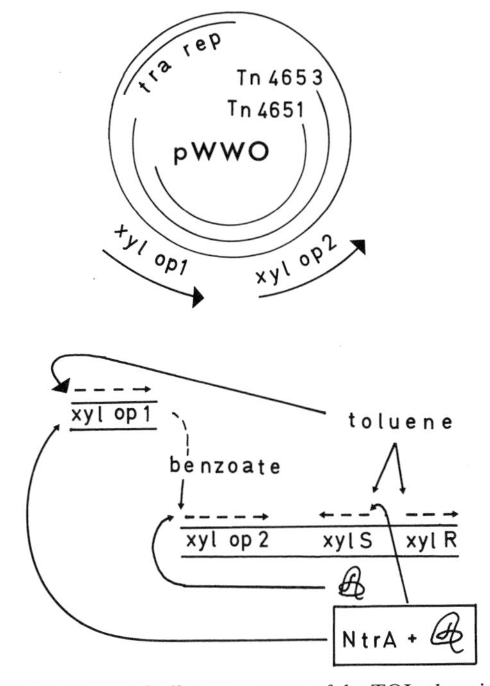

Fig. 6. The catabolic transposons of the TOL plasmid pWWO (above) and the regulation of its *xyl* genes (below). The transposable genes for xylene/toluene degradation (cf. Fig. 8) are organized in two independent operons, xyl operon 1 (xyl op 1), and xyl operon 2 (xyl op 2). The expression of both operons is induced by xylene/toluene. Two regulatory genes (*xylR*, *xylS*) and an alternative transcription factor (*NtrA* gene) are involved (modified from SCHUMANN, 1990, and BURLAGE et al., 1989).

5 Ecology of Pseudomonads

Members of the authentic genus *Pseudomonas* occur in many different habitats because their nutritional requirements are simple, whereas their metabolic versatility is extreme. However, their ability to occupy a certain habitat is limited by extremes of pH (both alkaline and acidic) and they are not thermophilic. Some psychrotrophic species are known (PALLERONI, 1992a, b).

Monitoring of environmental pseudomonads is nowadays enhanced by recombinant DNA technologies. However, the development of gene probes has begun with a focus on clinical and epidemiological applications (MACARIO and CONWAY DE MACARIO, 1990). Sometimes the DNA gene probes developed do not show the specificity provided by other approaches, e.g., a plasmid DNA probe for detection of xanthomonads was outcompeted by indirect immunofluorescence microscopy with a specific monoclonal antibody (WONG, 1991). It was expected that the sensitivity of hybridization methods might be greatly enhanced (JAIN et al., 1988). Future research will show whether gene probing becomes a feasible alternative to serological approaches in microbial ecology. Monitoring of biodegradative populations in polluted sites or waste water reactors by hybridization with extracted DNA, avoiding cultivation of microorganisms, has also been propagated. Colony hybridization with gene probes using TOL and NAH plasmid-DNA has been directly applied to detect the critical populations recovered from the environment in order to improve isolation techniques for bacteria capable of degrading polychlorinated biphenyls. However, when gene probes based on catabolic plasmids with a wide host range are used, they detect only sequences homologous to plasmid-encoded traits and do not reveal the identity of the corresponding bacteria. For this purpose chromosomally encoded probes are required.

Authentic pseudomonads (cf. Sect. 2) were identified within a large sample of characterized soil bacteria in a colony lift screen (STEWART-TULL, 1988) using a DNA hybridization probe developed from the 23S rRNA cistron of *P. aeruginosa* (FESTL et al., 1986). Recently the outer membrane protein gene (*oprI*) was successfully evaluated (SAINT-ONGE et al., 1992) for a similar purpose. Species identification of newly isolated pseudomonads, which has been proposed by using a sequence of increasingly specific probes will only be feasible following additional taxonomical studies for a redefinition of species in the authentic genus *Pseudomonas* (cf. Sect. 2). Meanwhile group-specific probes are available for differentiation between groups of species of the authentic genus *Pseudomonas.* For a "phylogenetic stain" (DELONG et al., 1989) fluorescently labeled oligonucleotides, based on ribosomal cistrons, have been proposed in combination with *in vitro* amplification (polymerase chain reaction) and flow cytometry for *in situ* detection of individual cells of pseudomonads. Whereas introduced *Pseudomonas aeruginosa* cells were detected, natural bacterial populations in soil did not result in significant hybridization signals without activation by adding nutrients (HAHN et al., 1992).

Clearly, ecological investigations of authentic pseudomonads benefit from serological methods introduced into soil biology by SCHMIDT (1974). Monoclonal antibodies specific for members of a phylogenetically defined family Pseudomonadaceae (see Tab. 1) are available (MUTHARIA et al., 1985). Polyclonal antibodies may also be used, provided they are specific enough, e.g., simple immunofluorescence testing permitted enumeration of nitrilotriacetic acid (NTA)-utilizing pseudomonads, now allocated to the new genus *Chelatobacter* (see Tab. 3). These special bacteria are ubiquituously distributed and were found to be tenfold enriched in activated sludge over aquatic ecosystems (WILBERG et al., in press). These immunological data correspond to data obtained from another investigation using a ^{14}C-NTA most probable number method (LARSON and VENTULLO, 1986). The special serological techniques, common for detection of plant-pathogenic pseudomonads, have been discussed by KLEMENT et al. (1990). Serotyping and various other typing methods practiced for *Pseudomonas aeruginosa* (PALLERONI, 1992b) and other pseudomonads (BUSSE et al., 1989) were recently completed by pulse field two-dimensional gel electrophoresis of DNA digested with rare cutting restriction enzymes.

Water is a common habitat of authentic pseudomonads where they can remain viable together with acinetobacters for long periods of time. They are not demanding, require no vitamins and many of them are able to grow under oligotrophic conditions, even in pure water. In a survey of mineral waters with an emphasis on *Pseudomonas*, the species *P. stutzeri*, *P. putida*, and *P. fluorescens* yielded the highest bacterial counts in either natural mineral water or in bottled water (ROSENBERG, 1990). Due to their ability to adhere to solid surfaces, pseudomonads may colonize tap-water distribution and cooling systems, power-plant condensor tubes or milk-transfer pipelines. Recently, the adhesion of *P. aeruginosa* to stainless steel, widely applied in the pharmaceutical industry, was demonstrated to be rapid and influenced by cell surface hydrophobicity rather than by cell surface charge (VANHAECKE et al., 1990, and references cited therein).

An anthropogenic habitat where pseudomonads, both authentic and those species now allocated to the beta- and alpha subclass of the Proteobacteria (see Tab. 3), may frequently occur is the activated sludge of sewage treatment plants. A simple chemotaxonomical approach, using either quinone profiles (HIRAISHI et al., 1991) or polyamine patterns (AULING et al., 1991 b) as biomarkers, which avoids isolation and cultivation of bacteria, has recently been introduced for a survey of pseudomonads and acinetobacters in activated sludge. In general, both techniques indicated that authentic pseudomonads were less dominant in sewage treatment plants, examined so far, than one might have expected. However, in certain plants, with high rates of enhanced biological phosphorus elimination, authentic pseudomonads and members of the beta-and alpha subclass of the Proteobacteria dominated the microbial flora, whereas acinetobacters were nearly absent in the cell aggregates of the sludge flocs.

This formation of cell aggregates, known as flocculation, is due to the production of microbial exopolysaccharides. Manipulation of the structure and the chemical and mechanical properties of these polysaccharides, which define their flocculation and bioadsorption characteristics, has gained recent interest in order to develop microbially based technologies for waste water treatment or metal reclamation. Alginate, which confers mucoidy to *Pseudomonas aeruginosa* and *Azotobacter vinelandii*, is a water-soluble heteropolymer of alternating oligomeric blocks consisting of either mannuronic acid or of mannuronic acid and guluronic acid (SUTHERLAND, 1990). Due to the absence of free guluronic acid, the alginate of *Pseudomonas* is not gelatinized in the presence of Ca^{2+} ions in contrast to alginate of higher algae which contains free guluronic acid. Other highly branched acid heteropolymers with different compositions are known, and in the case of *Zoogloea* strain 115 consist of glucose, galactose, and pyruvate (DUGAN et al., 1992). Whether *Pseudomonas* species really form cellulose (DEINEMA and ZEVENHUIZEN, 1971) will be answered by a taxonomic reinvestigation along the lines discussed in Sects. 2 and 3. The high self-affinity of cellulosic (ROSS et al., 1991) and other polysaccharide material is responsible for the formation of cell aggregates or microcolonies. These microcolonies appear to fulfill a structural role not only in activated sludge but also in natural habitats by conferring mechanical, chemical, and biological protection. Thus, microbial extracellular polysaccharides (EPS) have often been related to protection of microorganisms from desiccation. For example, a soil pseudomonad was shown to respond to desiccation by increased production of EPS which provide a microenvironment which holds water and dries more slowly than its surroundings (ROBERSON and FIRESTONE, 1992).

In an older numerical taxonomical study of petroleum-degrading bacteria in coastal water and marine sediment, members of the genus *Pseudomonas* were reported to be abundant. However, classification of marine pseudomonads has in the meantime been rearranged by allocation of many marine species to the genus *Deleya* (see Tab. 3) within the new family Halomonadaceae (DE LEY, 1992). Due to more recent reports on hydrocarbon-degrading microorganisms, members of the genus *Pseudomonas* appeared to be of minor importance in both marine environments and soil. There are early reports on increased degradation by seeding of contaminated aquatic environments with an oil-degrading *Pseudomonas* sp. However, the feasibility of this approach using either natural isolates or genetically engineered strains remains controversial (LEAHY and COLWELL, 1990). Although *Pseudomonas* is considered to be nutritionally versatile with regard to growth on low-molecular weight compounds only (PALLERONI, 1992a), certain members of the genus appear to attack high-molecular weight substrates as well. Recent molecular investigations on *Pseudomonas fluorescens* subsp. *cellulosa* revealed a cellulolytic system which differed in several respects from the cellulosomes of *Clostridium thermocellum* and other anaerobic bacteria which have a large surface-located complex mediating attachment of the cells to the insoluble polymeric carbohydrate substrates. The cellulolytic enzymes of *P. fluorescens* subsp. *cellulosa* appear to be post-transcriptionally modified, possibly by glycosylation, and have a cellulose-binding domain of 100 amino acids. Since they are secrected into the culture medium, the cells do not adhere to cellulose (HAZLEWOOD et al., 1992). *Pseudomonas* has also been reported as a major genus among the chitinolytic bacteria abundant in soil and the freshwater column (GOODAY, 1990). However, the author considered many of the reports in the literature on the distribution of certain physiological groups of bacteria in different ecosystems which claim to identify pseudomonads without any chemotaxonomic or molecular approach as meaningless with regard to identification.

The choice of suitable markers for monitoring the release of genetically engineered microorganisms is another challenge which was solved by introducing the luciferase genes of *Vibrio*, encoding bioluminescence, into *Pseudomonas* and other soil bacteria to study their fate after release. As another selectable, nonantibiotic marker, beta-galactosidase, has been introduced into a fluorescent pseudomonad in order to track the released organism during root colonization (JAIN et al., 1988). The field testing of genetically engineered microorganisms, including the "ice-minus" strain of *P. syringae* (cf. Sect. 7) has been reviewed (DRAHOS, 1991). In addition, numerous investigations on survival of released genetically engineered pseudomonads have been undertaken in different microcosms which simulate natural habitats in the laboratory. In general, the results obtained often indicate that the wild type outcompeted the modified strains (for further references see VAN ELSAS et al., 1991).

Natural gene transfer has become an area of interest with proposals for release of genetically altered organisms to the environment. *Pseudomonas stutzeri* undergoes natural transformation in soil and recently became a valuable model for studying this process in marine sediments as well, either in filter assays or with sediment columns (STEWART et al., 1991).

6 Metabolic Properties of Pseudomonads

The pseudomonads play a significant role in maintaining the global carbon cycle through complete mineralization of various organic molecules, noteworthy the aromatic compounds. Indeed, the benzene nucleus is one of the most abundant units of chemical structure in the biosphere. The pseudomonads participate in degrading the ever-increasing manmade chemicals (xenobiotic compounds) released into our environment. A wealth of information has accumulated regarding pathway chemistry, gene organization and regulation. Enzymology has been studied to a lesser extent, whereas taxonomy of the strains involved was completely ignored. The central role of catecholic intermediates in aerobic degradation of aromatic compounds by pseudomonads is well established. Various enzymes convert aromatics such as benzoate, phenol, salicylate and naphthalene (SMITH, 1990) to catechol (1,2-dihydroxybenzene), and substituted catechols are intermediates in the catabolism of methylated and chlorinated derivatives of these compounds (REINEKE and KNACKMUSS, 1988). The oxygenative ring fission occurs via the *ortho*- or the *meta*-cleavage pathway (Fig. 7). The first mode involves ring cleavage between the two hydroxyl groups, followed by a well-defined series of reactions leading to β-ketoadipate, and is usually chromosomally encoded. The alternative *meta*-pathway was originally described in *Pseudomonas* species and has also been detected in *Azotobacter* and *Alcaligenes*. It involves ring cleavage adjacent to-

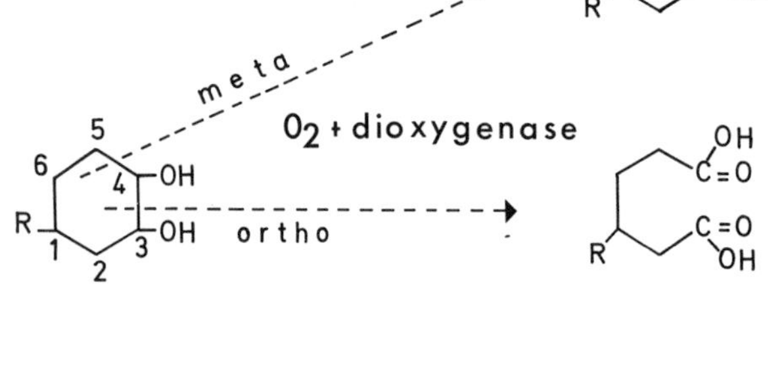

Fig. 7. Cleavage of catechol by *Pseudomonas* dioxygenases to *cis, cis*-muconic acid (*ortho*-fission) or to 3-hydroxymuconate semialdehyde (*meta*-fission).

the two catechol hydroxyls, followed by degradation of the ring cleavage product to pyruvate and a short-chain aldehyde. The biochemistry of the dioxygenases which are central in both pathways has been reviewed by SARIASLANI (1989). The most comprehensively studied *meta*-cleavage pathway is that of the incP-9 TOL plasmid pWWO which encodes toluene degradation organized in two operons with a common regulation (Fig. 6, lower part; Fig. 8). The TOL *meta*-cleavage pathway comprises 13 structural genes encoding the enzymes for conversion of benzoate, via catechol, to central metabolites, and a separate operon is required for conversion of toluene and xylene to benzoate. Many features of both gene organization and enzyme functions of the TOL pathway are preserved in other aromatic catabolic pathways in *Pseudomonas* species (ASSINDER and WILLIAMS, 1990). Recently it was shown that phenol degradation by the incP-2 plasmid pVS150 is encoded by only one operon with 15 genes and consists of a multicomponent phenol hydroxylase and a subsequent *meta*-pathway (SHINGLER et al., 1992). Various toluene degradation pathways following the *meta*-route are known in authentic pseudomonads, and a novel *ortho*-cleavage pathway was proposed as well (WHITED and GIBSON, 1992).

Reports on oxidative degradation of halogenated aliphatic hydrocarbons by pseudomonads are rare. These groundwater contaminants appear to be either the preferred substrate for methanotrophic bacteria or undergo reductive dechlorination by anaerobic methanogenic communities. On the other hand, chlorinated aromatic compounds have been reported to be degraded preferentially by four bacterial genera, *Pseudomonas*, *Alcaligenes*, *Flavobacterium*, and *Rhodococcus*. Pseudomonads have been reported to degrade chlorinated polycyclic hydrocarbons as well. However, none of the pseudomonads which attack either of the three classes of haloorganics has been thoroughly characterized with respect to its taxonomic status (CHAUDHRY and CHAPALAMADUGU, 1991). The recent progress regarding the biodegradation of the polychlorinated biphenyls (PCBs), commonly believed to be indestructible, has been reviewed by ABRAMOWICZ (1990). With respect to biochemistry of biphenyl degradation it appears that naphthalene, naphthalenesulfonic acids, and biphenyl can be degraded by a single set of enzymes, since naphthalene dioxygenase of *P. putida* 119 attacks biphenyl as well. Moreover, 1,2-dihydroxynaphthalene dioxygenase from strain BN6 converts 2,3-dihydroxybiphenyl, and the 2,3-dihydroxybiphenyl dioxygenase from *Pseudomonas paucimobilis* Q1 (cf. Tab. 3) metabolizes 1,2-dihydroxynaphthalene. Thus, the corresponding pathways might be evolutionarily related (KUHM et al., 1991).

The metabolism of monoterpenes and related compounds which are important renewable resources for the perfume and flavor industry has been investigated in the authentic pseudomonads *P. putida*, *P. mendocina*, and *P. citronellolis* (cf. Tab. 2), in *P. flava* (cf. Tab. 3) and in uncharacterized species of *Pseudomonas*. The enzymology of ring cleavage of monocyclic monoterpenes which possess an internal ether linkage like 1,8-cineole has been recently reviewed with a focus on ring hydroxylation strategy and biological Baeyer–Villiger oxygenation (TRUDGILL, 1990).

The presence of biosurfactants seems to be indispensable for the growth of microorganisms on water-immiscible substrates, and different types of rhamnolipids are synthesized by *Pseudomonas* species during growth on hydrocarbons. The enormous amount of work on regulatory principles of synthesis of rhamnolipids and other biosurfactants, since its original detection during growth of *P. aeruginosa* on *n*-paraffin, has been recently reviewed (HOMMEL, 1990). The conversion of *n*-alkane to the corresponding alkane-1-ol by means of a hydroxylase (monooxygenase) is linked to a rubredoxin electron carrier in authentic pseudomonads (SARIASLANI, 1989). The genetics and physiology of alkane utilization of *P. putida* and taxonomically uncharacterized *Pseudomonas* strains have been compared with that of other alkane-utilizing microorganisms in the recent review of WATKINSON and MORGAN (1990).

Some pseudomonads are also reported to attack the carbon-phosphorus bond by using alkyl phosphates as a phosphorus source for growth. The final taxonomic characterization of most of these isolates (cf. Tab. 16 in MACASKIE, 1991) remains to be elucidated. However, aside from the mainstream of neglecting

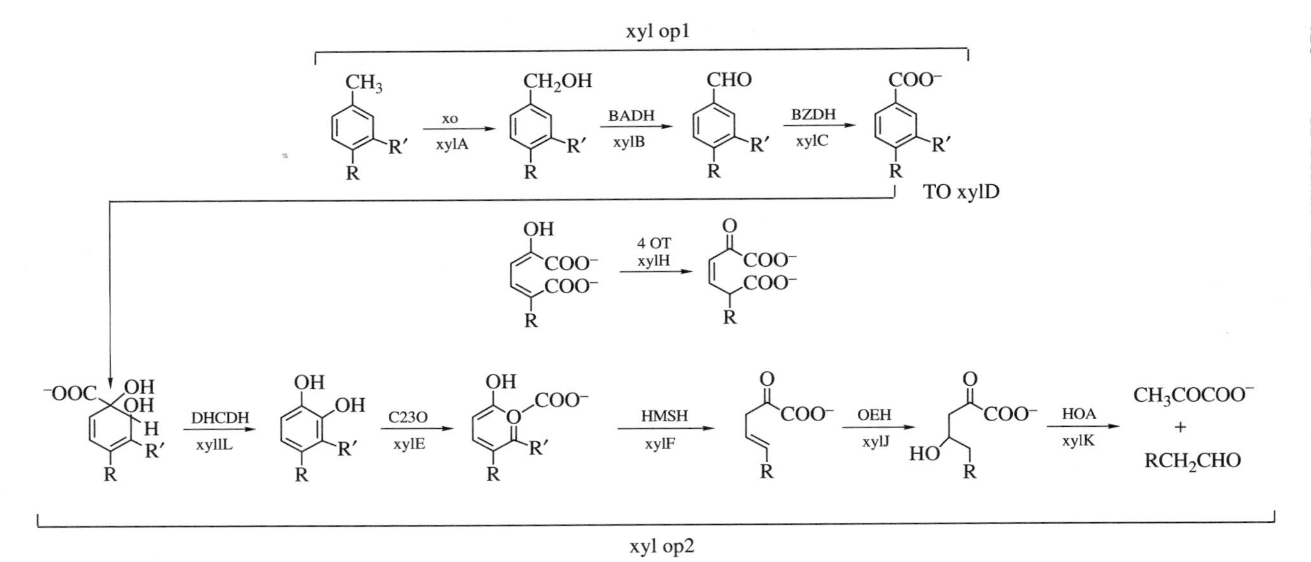

Fig. 8. Xylene/toluene degradation pathway encoded by the two operons (cf. Fig. 6) of the TOL plasmid. The abbreviated enzyme and gene designations are shown for every step with the modification (ASSINDER and WILLIAMS, 1990) that XO, the xylene oxygenase, is encoded by *xylMA*, and TO, the toluate dioxygenase, is encoded by *xylXYZ*. (Reprinted from FRANTZ and CHAKRABARTY, 1986, with permission.)

systematics of biodegradative isolates referred to as pseudomonads, there is an increasing number of investigations which were encouraged by the exciting progress towards a phylogenetic bacterial classification system (cf. Sects. 1–3). Pseudomonads degrading chloridazon, the active ingredient of the herbicide Pyramin, which has been used for more than 20 years for the control of weed in sugar beet- and beet-root culture, were, following thorough characterization, allocated to *Phenylobacterium immobile* (LINGENS et al., 1985). When a large collection of biodegradative pseudomonads attacking a variety of xenobiotic compounds was submitted to a polyphasic taxonomic characterization such as recommended in Sect. 3, only one-half were allocated to the authentic genus *Pseudomonas* (BUSSE et al., 1989). The simple chemotaxonomic approach based on the analysis of quinones and polyamines which identified the other half as members of the beta subclass of the Proteobacteria was confirmed by partial sequencing of 16S rRNA (BUSSE et al., 1992).

Degradation of the synthetic chelating agent nitrilotriacetate via iminodiacetate, glyoxalate, and glycine as intermediates (Fig. 9) has also been regarded for a long time as a metabolic capability which is confined to specialist pseudomonads (EGLI et al., 1990). However, considering the chemotaxonomic and phylogenetic data now available these bacteria are members of the alpha subclass of the Proteobacteria (AULING et al., in press) as are recently isolated EDTA-degrading bacteria (LAUFF et al., 1990; NÖRTEMANN, 1992; AULING, unpublished results). Likewise, no strain of the peculiar pseudomonads, which can detoxify carbon monoxide by oxidizing this trace gas in a branched electron transport chain by means of CO dehydrogenase and cytochrome b_{561} (MEYER et al., 1990), is an authentic pseudomonad (WILLEMS et al., 1989; AULING et al., 1988). On the other hand, detoxification/decomposition of cyanide (cyanate) as nitrogen source for growth has been recently studied in *Pseudomonas fluorescens*, and its cyanase gene resembles that of *Escherichia coli* (ANDERSON et al., 1990). Likewise, the availability of recombinant DNA technology led to the final elucidation of certain steps in the biosynthesis of vitamin B_{12} in *Pseudomonas denitrificans* (THIBAUT et al., 1992), previously not resolved in studies with *Propionibacterium shermanii.*

The cytochrome bc_1 complex which is the most common of the bacterial energy-transducing electron transfer complexes has been purified from *P. stutzeri*, a member of the authentic genus *Pseudomonas.* This bc_1 complex includes the Rieske iron-sulfur protein and is similar to the well-studied bc_1 complex of *Paracoccus denitrificans.* On the other hand, a Rieske-type iron-sulfur protein is also associated with the phthalate dioxygenase of *Pseudomonas cepacia*, a member of the beta subclass of the Proteobacteria. Under anaerobic, denitrifying conditions, ubiquinol is the low-potential electron donor for nitrate reductase (Fig. 10), bypassing the bc_1 complex and c-type cytochromes in an antimycin- and strobilurin-insensitive reaction. Two divergent electron transfer pathways were also found in the obligate aerobic plant pathogen *P. cichorii* which does not possess an aa_3-type oxidase, a situation reminiscent of *Rhodobacter capsulatus* (TRUMPOWER, 1990). Studies of azurin, the blue copper protein, which underwent a homologous evolution to cytochromes are another rapidly growing area of molecular research in *Pseudomonas aeruginosa* which combines site-directed mutagenesis, biochemical and physical approaches (KARLSON et al., 1991; NAR et al., 1991).

Intensive research on denitrification, the facultative respiration process that couples electron transport phosphorylation to the stepwise reduction of nitrogenous oxides, has been limited to a few genera, namely, *Paracoccus, Rhodobacter*, *Alcaligenes*, and *Pseudomonas*, and within *Pseudomonas* mainly to one species, *P. stutzeri* (formerly *P. perfectomarina*). Twenty eight *Pseudomonas* species were listed in a recent compilation of 130 denitrifying bacteria (cf. synopsis provided by ZUMFT, 1992). Comparing that compilation with Tabs. 2 and 3 of this chapter one immediately notes that merely one third of them belong to the authentic genus *Pseudomonas* which indeed harbors conspicuous denitrifiers. In general, the denitrification pathway is an assembly of three more or less independent processes: respiration of 1. nitrate, 2. nitrite, and 3. nitrous oxide. In *Pseudomonas aeruginosa* the genes for step 1 and 2 are separate and dispersed over the chromo-

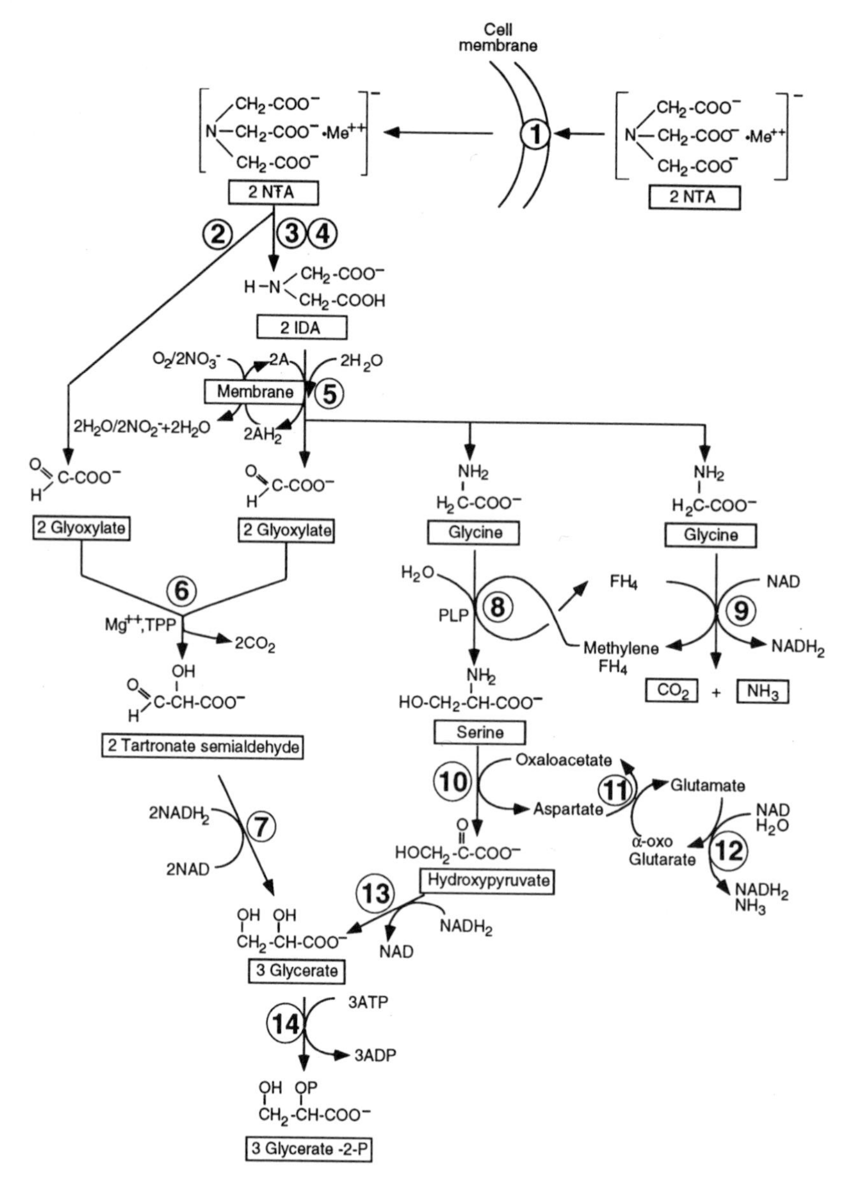

Fig. 9. Metabolic pathway proposed for nitrilotriacetic acid (NTA) in obligately aerobic and facultatively denitrifying Gram-negative bacteria. **1**, Transport across the membrane; **2**, NTA monooxygenase; **3**, NTA dehydrogenase; **4**, NTA dehydrogenase/nitrate reductase complex; **5**, iminodiacetate (IDA) dehydrogenase; **6**, glyoxylate carboligase; **7**, tartronate semialdehyde reductase; **8**, serine hydroxymethyl transferase; **9**, glycine synthase (decarboxylase); **10**, serine:oxaloacetate aminotransferase; **11**, transaminase; **12**, glutamate dehydrogenase; **13**, hydroxypyruvate reductase; **14**, glycerate kinase.
Methylene FH_4, (N^5,N^{10}-methylene) tetrahydrofolic acid; PLP, pyridoxal phosphate; TPP, thiamine pyrophosphate. (From Egli, 1992, with permission.)

some. In *P. stutzeri* the genes for steps 2 and 3 are closely linked. Interestingly, all four enzymes of denitrification have been identified as metalloproteins involving the transition metals molybdenum, iron, and copper (Fig. 10). A bactopterin containing a pterin moiety and an additional nucleotide is present in the respiratory nitrate reductase of *P. stutzeri.* As yet, it has not been investigated how the pseudomonads satisfy the increased metal ion demand for *de novo* synthesis of denitrification enzymes and their electron donors. The best studied denitrification enzyme is cytochrome cd_1, tetraheme nitrite reductase, which also serves as model for inter- and intramolecular electron transfer involving the presumed physiological electron donors cytochrome c_{551} and azurin. Since the copper binding center in the N_2O reductase of *P. stutzeri* (Fig. 10) is similar to that of subunit II in cytochrome c oxidase, it was speculated that it has preceded the evolution of the latter. Since truncated variants of denitrification (nitrate respiration being the most widely distributed one) occur in many bacterial species, the missing reactions might have been lost during evolution. Studies with *P. stutzeri* (and other bacteria as well) have shown that the products of denitrification can shift to exclusive liberation of NO and N_2O in response to changes in the oxygen concentration, thus reflecting differential sensitivities of individual reductases towards oxygen or even regulation of their biosynthesis by oxygen. In *P. stutzeri*, the synthesis of nitrate reductase is less sensitive towards oxygen than nitrite reductase, whereas N_2O reductase is constitutive. The awareness of increasing atmospheric contamination by nitrogenous oxides and the recognition that denitrification, although less bound to absence of oxygen than previously thought (ZUMFT, 1992), nevertheless is regulated individually in each bacterium by different oxygen tensions, requires further intensive research. The ecology of denitrifying bacterial communities and their regulation by oxygen has been reviewed (cf. ZUMFT, 1992).

Another new area of research comes from the recognition (EVANS and FUCHS, 1988) that denitrifiers may play an important role in the anaerobic degradation of aromatic compounds. The denitrifying bacteria involved are believed to be pseudomonads. However, since

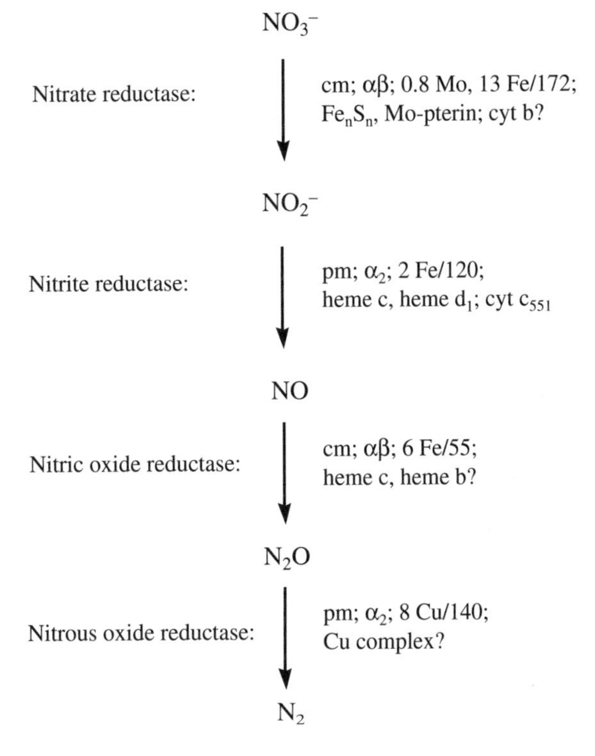

Fig. 10. Denitrification pathway of *Pseudomonas stutzeri* and properties of its enzymes based on the information provided by ZUMFT (1992). The data on nitrite reductase have been obtained with *Pseudomonas aeruginosa.* The properties of the enzymes involved are listed in the following order: localization, quaternary structure, number of metal atoms/M_r, prosthetic group, physiological acceptor. Abbreviations: cm, cytoplasmic; pm, periplasmic

the strains studied in the laboratories were either allocated to genera other than *Pseudomonas*, or have not been taxonomically characterized, it remains questionable whether denitrifying authentic pseudomonads are really able to grow by anaerobic degradation of aromatic compounds. A simple chemotaxonomic analysis (cf. Sect. 3) of the isolates under present investigation would answer this question.

A rather unexpected physiological response to oxygen limitation in the aerobic pseudomonads, studied intensively in organisms now excluded from the authentic genus *Pseudomonas*, is the derepression of dormant genes for synthesis of enzymes normally found only in anaerobic bacteria. Together with the appearance of lactate dehydrogenase, alcohol dehydrogenase, and butanediol dehydrogenase, typical fermentation products were observed, although the microorganims remained unable to grow anaerobically as fermentative organisms (VOLLBRECHT, 1987). However, *Pseudomonas aeruginosa* is capable of slow growth on rich media in the absence of oxygen and nitrate by the conversion of arginine to ornithine, CO_2, and NH_3 which is coupled to the production of ATP from ADP. This arginine deiminase pathway is thought to be transcriptionally activated by a regulatory protein homologous with the FNR protein of *Escherichia coli* (VERHOOGT et al., 1992).

Molecular studies on *Pseudomonas aeruginosa* provided new insights into the regulation of mucoidy by transcriptional control (osmolarity-induced activation) of alginate biosynthesis (MAY et al., 1991), the mechanism of protein secretion (STROM et al., 1991), and added another elastase to the zinc metalloprotease which appears to be remarkably similar to the *Bacillus* thermolysin (GALLOWAY, 1991). Studies on extracellular endoglucanase from *P. solanacearum*, phylogenetically a member of the beta subclass of Proteobacteria, indicated a two-part leader sequence of the gene. The first part directs the lipid modification and export of the endoglucanase across the inner membrane and is removed by signal peptidase II. The second part cooperates with other sequences of the protein (possibly directing the overall conformation) for export across the outer membrane (HUANG and SCHELL, 1992). The porins and structural outer membrane proteins of *Pseudomonas aeruginosa* have been intensively studied to unravel the high resistance of this bacterium to certain antibiotics which could be accounted for by the small number of large pores, and consequently less diffusion of antibiotics, in synergy with the action of periplasmic β-lactamases (HANCOCK et al., 1990).

7 Applications of Pseudomonads

The accumulation of the microbial reserve polyester poly(3-hydroxybutyrate) (PHB) has been described as an important property of the aerobic pseudomonads (STANIER et al., 1966). PHB-like inclusions containing a polyester of 3-hydroxyoctanoic acid (PHA) were found during growth of *Pseudomonas oleovorans* on short aliphatic hydrocarbons. In response to variation of the substrate from C_6 to C_{12} *n*-alkanes and C_8 to C_{10} *n*-alkenes, the bacterium can synthesize heteropolymers of varying composition and degree of unsaturation. This observation stimulated research on microbiological production for biologically derived plastics which are biodegradable. Although PHB and PHAs are partially crystalline polymers displaying similarity in structure to polypropylene, the granules *in vivo* are completely amorphous. The accumulation of PHAs containing medium-chain length (C_6 to C_{12}) 3-hydroxy acids, but not 3-hydroxybutyric acid, was reported to be characteristic of fluorescent (authentic) pseudomonads. However, since accumulation of PHB is widespread among different bacteria, it is to be expected that the ability to synthesize PHAs other than PHB is likewise not confined to a taxonomically limited group. Nevertheless, as yet the research on PHAs has been concentrated on two bacteria: *Pseudomonas oleovorans* (see Tab. 2) and *"Alcaligenes" eutrophus* which is not a member of a phylogenetically defined genus *Alcaligenes* (BUSSE and AULING, 1992). The state of the art of fundamental and applied research on PHAs has been reviewed by ANDERSON and DAWES (1990), and a first PHA consumer product was launched, consisting of biodegradable bottles from a 3-hydroxybutyrate-co-3-hydroxyvalerate-copolymer.

Interestingly, the psychrotrophic members of the authentic pseudomonads are able to produce thermally resistant proteases and/or lipases (GOUNOT, 1991), which may resist pasteurization and even ultra-high temperature treatment. The off-flavor of meat generated by the proteolytic action of pseudomonads is due to putrescine and cadaverine (DAINTY et al., 1986). The *Pseudomonas* species *P. aeruginosa*, *P. fluorescens*, and *P. fragi* have been repeatedly reported to produce extracellular lipases (triacylglycerol acylhydrolases; EC 3.1.1.3) which act at fat–water interfaces to catalyze the hydrolysis of long-chain triglycerides. The undesirable flavors of refrigerated milk generated by lipolytic action of pseudomonads are due to cleavage of even-carbon-number short-to-medium chain (C_4 to C_{12}) fatty acids from milk fat. On the other hand, microbial lipases are currently receiving considerable attention because of their potential applications in biotechnology (HARWOOD, 1989). This interest stems from their ability not only to hydrolyze ester bonds, transesterify triglycerides and resolve racemic mixtures, but also, in the reverse mode, to synthesize ester and peptide bonds. Lipase P from *Pseudomonas fluorescens* and Lipase CC from *Candida cylindrica* are the most widely used for lipase-catalyzed asymmetric transformation of synthetic chemicals (for further references cf. YAMAZAKI and HOSONO, 1990). Acylation of organometallic alcohols serves as a model to predict the different stereoselectivity of the two lipases. The recent success in the lipase-catalyzed enantioselective acylation of 2-hydroxyhexadecanoic acid prompted the use of this reaction to prepare (*R*)-2-hydroxy-4-phenylbutanoic acid which is an important intermediate for the synthesis of angiotensin-converting enzyme inhibitors (SUGAI and OHTA, 1991). The *Pseudomonas* lipases are probably related to each other, and are composed of a single type of subunit with a common size of M_r (relative molecular weight) between 29000 and 35000 Da. Non-terminal deletions or additions may explain certain variations of the average size, substantially larger enzymes are apparently due to a variable degree of aggregation. Improvement of lipases by genetic engineering has been patented for applications in detergent formulations. Recently a rather thermotolerant, EDTA-insensitive lipase was obtained from a strain of *Pseudomonas aeruginosa* by a more classic microbiological/biochemical approach (GILBERT et al., 1991). However, thermostability may not have the first priority considering the energy-saving trend to washing at lower temperature. Lipase is used for hydrolysis of fat stains which tend to remain on the fabric during washing at low temperatures. Since the products of hydrolysis are easily removed under alkaline washing conditions, only such lipases will be economically successful which are active under alkaline conditions and resist the bleaching agents and the protease of the detergent compositions (FALCH, 1991).

Another industrially important enzyme of authentic pseudomonads is amylase which has been used for production of maltooligosaccharides by enzymic digestion of starch. The α-amylases of pseudomonads hydrolyze the α-1,4-glycosidic bond by exo-glycolytic cleaving, opposed to the normal endo-glycolytic mode, forming α-anomeric oligosaccharides. Maltotetraose (G_4), maltopentaose (G_5), and maltohexaose (G_6) are produced using G_4-forming amylase from *Pseudomonas stutzeri* and *P. saccharophila* (a member of the beta subclass of the Proteobacteria). The novel G_5-forming amylase from a taxonomically uncharacterized *Pseudomonas* strain is so unusual that it produces selectively G_5 which is used as a reagent for measurement of serum amylase activity and as a nutrient for patients with renal failure. Remarkably, cloning and sequence comparison of the gene of the G_5-forming amylase with those of the G_4-forming amylases revealed almost no homology except for the COOH terminus, responsible for "starch binding". The molecular basis of the unique mode of exo-glycolytic cleavage by the G_5-forming amylase remains to be elucidated (for further references cf. SHIDA et al., 1992). The α-galactosidases, EC 3.2.1.22, are another interesting group of exo-type carbohydrases since they also catalyze transgalactosylation reactions. From *Pseudomonas fluorescens* an α-galactosidase with strong transfer activity and a wide acceptor specificity was purified making this enzyme a promising candidate for synthesis of various hetero-oligosaccharides containing α-galactosyl residues (HASHIMOTO et al., 1991).

Microbes or their enzymes are increasingly used as catalysts in organic synthesis for generation of chiral intermediates (DAVIES et al., 1990). The well-known metabolic activity of *Pseudomonas* species towards aromatic hydrocarbons suggests their recruitment for hydroxylation reactions involving aromatic compounds. Trifluoromethylbenzene is hydroxylated to the corresponding diol by *Pseudomonas putida*. The diol generated from toluene is used for the production of chiral prostaglandin intermediates. *cis*-Glycol dehydrogenase negative *Pseudomonas* mutants can generate optically active diols from quinoline and some of its derivatives (DAVIES et al., 1990). The conversion of various nitrile compounds to the corresponding amides is catalyzed by nitrile hydratase (EC 4.2.1.84) discovered in strains of *Rhodococcus* and in *Pseudomonas chlororaphis* B23. Some of them are successfully used for industrial production of acrylamide. The nitrile hydratase from *P. chlororaphis* B23 resembles the enzyme from *Rhodococcus* sp. strain N-774 and contains a nonheme iron and pyrroloquinoline quinone as prosthetic group (NISHIYAMA et al., 1992).

Pseudomonas putida is commercially exploited in the area concerning the formation of optically active alcohols from achiral or racemic substrates, i.e., in the conversion of isobutyric acid into (*S*)-β-hydroxybutyric acid. Recently a two-step production of 2-oxybutyric acid from crotonic acid via dihydroxybutyric acid was developed using intact cells of *P. putida*, since 2-oxybutyric acid is another important starting material for the production of chiral 2-aminobutyric acid and 2-hydroxybutyric acids. L-Tartrate as a cheap by-product of the wine industry is considered a good starting material for biotransformation to D-glyceric acid useful as a chiral synthon in synthetic organic chemistry and for enzymatic serine production. Production by an NAD^+-dependent conversion of D-glycerate from L-tartrate with a novel enzyme, L-tartrate decarboxylase, or simply with intact cells of *Pseudomonas* has been described (FURUYOSHI et al., 1991). The adaptive growth on "unnatural" D-(–)-tartrate is a property of authentic pseudomonads (AULING et al., 1986) which might be exploited in order to use the enzymes in optical assays. However, the absolute specificity and oxygen sensitivity of the *Pseudomonas* D-(–)-tartrate dehydratase involved in the initial attack of the substrate is inferior to the properties of a similar enzyme of *Rhodopseudomonas sphaeroides* which favors the application of the latter to enzymatic determination of D-(–)-tartrate in the presence of other tartaric acid isomers (RODE and GIFFHORN, 1982). Determination of amino acids is another important field where specific enzymes can be used. A novel L-phenylalanine oxidase of an uncharacterized *Pseudomonas* strain was immobilized on a nylon membrane and could be used as a sensor for aromatic amino acids by performing the measurement at pH 4.0 (NAKAJIMA et al., 1991).

The potential application of the pseudomonads in the up-coming field of remediation due to their well-known capability to degrade xenobiotic compounds, especially the halogenated organics, used as herbicides, plastics, solvents and degreasers, is not covered here since it will be the subject of Volume 11 of this series. The different types of halogenases of taxonomically uncharacterized pseudomonads (cf. Sect. 6), reported to attack these haloorganics, and their potential applications have been reviewed by HARDMAN (1991). Biotechnological treatment of wastes from the nuclear fuel cycle using biodegradation (cf. Sect. 6) of the radionuclide chelating agents (EDTA, NTA) has been evaluated (MACASKIE, 1991). The high radioactivity also present in some decontamination wastes may limit the use of living cells, therefore, it was considered to harness only the central catalytic enzymes (see Fig. 9).

The growing interest in beneficial interactions between soil pseudomonads and plants is driven by the motive of increased crop productivity while reducing the chemical inputs in farming based on the new insights obtained by analysis of the tripartite ecosystem plant-soil-microorganism, called rhizosphere. Optimization of fertilization (not maximization) may also reduce noxious gaseous evolution (cf. Sect. 6, Fig. 10) which might contribute to the "greenhouse effect". Whereas the market for crop protection and enhancement chemicals has increased on the average by about 10% in the last two decades, the success rate of discovery of new chemicals has decreased greatly, prompting the development of biological approaches. For example, there are no chemical procedures for the control of the plant disease, called take-all of cereals and induced by the fungus *Gaeumannomyces graminis*, which causes large crop losses worldwide (LYNCH, 1990). A breakthrough was the finding that antibiotics could be involved in biological control of take-all of wheat, because mutants of *P. fluorescens* deficient in synthesis of phenazine-1-carboxylic acid were less effective in biocontrol. The antibiotics and the antibiotic properties of pseudomonads which may play an important role for biocontrol of plant diseases have been reviewed (SMIRNOW and KIPRIANOVA, 1990). One of the effective strains, *Pseudomonas fluorescens* CHA0, which suppresses a variety of root diseases caused by soil-borne fungal pathogens, produces several secondary metabolites, notably HCN, 2,4-diacetylphloroglucinol, pyoluteorin, and indole-3-acetic acid. Both HCN and 2,4-diacetylphloroglucinol have been shown to be important factors in disease suppression (DEFAGO and HAAS, 1990).

The production of siderophores involved in iron uptake by pseudomonads is also suspected to play a role in suppression of soil-borne fungal pathogens (LEONG, 1986). Iron transport-mediated antagonism between plant-growth promoting and plant-deleterious pseudomonads was also reported. The species *P. fluorescens*, *P. aeruginosa*, *P. putida*, and *P. syringae* produce water-soluble fluorescent yellow-green pigments called pyoverdines. Due to their excellent iron-complexing abilities, the pyoverdines have the function of siderophores. The term pseudobactins is also used for some pyoverdines. The pyoverdines have a 2,3-diamino-6,7-dihydroxyquinoline derivative as chromophore, linked to a small peptide which differs among strains by the number and composition of amino acids (WINKELMANN, 1991).

Specificities of the siderophore receptors of *P. putida* strains involved in biocontrol have been described as well (see references cited in JURKEVITCH et al., 1992). Genes coding for siderophore receptors have been cloned from several plant-growth promoting pseudomonads, and recent studies suggest that the genes for siderophore biosynthesis of the opportunistic animal pathogen *P. aeruginosa* and the plant-associated species *P. syringae* and *P. putida* have evolved from a common ancestor (ROMBEL and LAMONT, 1992, and references cited therein). Pyochelin is another siderophore which has a thiazol derivative as phenolic ligand. *P. stutzeri* produces ferrioxamine, a hydroxamate-type siderophore (MEYER et al., 1987). The antagonistic action of pseudomonads towards soil-borne plant pathogens in the rhizophere has been discussed (DEFAGO and HAAS, 1990). Co-inoculation of fluorescent pseudomonads and non-pathogenic fungi in soilless cropping systems of heated green houses has also been developed for biocontrol of soil-borne fungal pathogens (EPARVIER et al., 1991). In addition to siderophores, many plant-beneficial *Pseudomonas* strains from the rhizosphere produce the phytohormone indole-3-acetic acid. However, a role for phytohormones is also suspected when bacterial pathogens induce growth abnormalities in the host. Different pathways for the biosynthesis of indole-3-acetic acid were found in *P. fluorescens* and *P. syringae* (OBERHÄNSLI et al., 1991).

Future application of siderophores from pseudomonads has also been proposed for control of lactic acid fermentations based on experiments which demonstrated that non-axenic fermentations could be directed towards lactic acid fermentations by complexation of iron through addition of the chemical chelator 2,2-dipyridyl. Certain strains of *P. fluorescens*, which may be misclassified, are able to secrete proferrorosamine, 2(2′-pyridyl)-1-pyrroline-5-carboxylic acid, which forms a pink complex with Fe^{2+} with an apparent stability constant of 10^{23} and even solubilizes iron from stainless steel. This water-soluble iron chelator is considered promising for general use in feed or food biotechnology, provided it is less toxic than its hydrophobic chemical analog 2,2′-dipyridyl (VANDE WOESTYNE et al., 1992).

Microbial herbicide detoxicification enzymes have been exploited for genetic engineering of herbicide-resistant plants (QUINN, 1990). For this purpose taxonomically uncharacterized *Pseudomonas* species which degrade the herbicide glyphosate, currently the most widely used non-selective weed killer, have been studied with respect to the degradation pathway of this herbicide. However, the growing international concern for environmental protection favors alternative strategies, i.e., reduction of chemical inputs in farming by exploiting the plant-growth promoting soil pseudomonads. There are first reports on successful biocontrol in field tests with cereals by inoculation with weed antagonistic pseudomonads and other bacteria from the rhizosphere.

Genetic engineering also provides the means to exploit traditional biocontrol agents, e.g., the insect pathogen *Bacillus thuringiensis* and others, because of their potential for biological control of insect infestations in plants. Consequently the delta-endotoxin gene from *Bacillus thuringiensis* was introduced into the chromosome of root-colonizing pseudomonads (Obukowicz et al., 1986). Meanwhile the first transgenic plants showing a certain resistance towards insects have also been created by introducing the delta-endotoxin gene into the plant genome. However, the final breakthrough for biocontrol technologies requires more substantial research budgets for ecophysiology and not only for molecular biology which clearly provided the technology to clone genes and identify plant-beneficial genes of microorganisms.

Furthermore, interesting new possibilities for applications of pseudomonads are raised by the rather unique phenomenon of bacterial ice nucleation. The epiphytic plant pathogen, *P. syringae*, possesses a membrane protein which confers the ability to nucleate crystallization in supercooled water, i.e., at a relatively warm temperature of −2 °C. The bacterium is responsible for frost damage on fruit trees and vines in spring. Furthermore, frost is known to be a predisposing factor to infection. The yearly frost damage by *P. syringae* to plants and associated agricultural crops is estimated to be several billion dollars worldwide (Margaritis and Bassi, 1991). For biological control by replacing the wild type, "ice-minus"-strains have been constructed for frost protection which are devoid of the ice-nucleating membrane protein. Currently, ice-nucleating bacteria are being used in snow making. They may replace silver iodide in cloud seeding and have potential applications in the production and texturing of frozen foods.

8 References

Abramowicz, D. A. (1990), Aerobic and anaerobic biodegradation of PCBs: a review, *Crit. Rev. Biotechnol.* **10,** 241–251.

Akagawa, M., Yamasato, K. (1989), Synonomy of *Alcaligenes aquamarinus*, *Alcaligenes faecalis* subsp. *homari* and *Deleya aesta: Deleya aquamarina* comb. nov. as the type species of the genus *Deleya*, *Int. J. Syst. Bacteriol.* **39,** 462–466.

Ambler, R. P. (1973), Bacterial cytochromes c and molecular evolution, *Syst. Zool.* **22,** 554–565.

Anderson, A. J., Dawes, E. D. (1990), Occurrence, metabolism, metabolic role, and industrial uses of bacterial polyhydroxyalkanoates, *Microbiol. Rev.* **54,** 450–472.

Anderson, P. M., Sung, Y. C., Fuchs, J. A. (1990), The cyanase operon and cyanate metabolism, *FEMS Microbiol Rev.* **87,** 247–252.

Assinder, S. J., Williams, P. A. (1990), The TOL plasmids: determinants of the catabolism of toluene and xylenes, *Adv. Microb. Physiol.* **31,** 1–69.

Auling, G. (1992), Polyamines, biomarker for taxonomy and ecology of phytopathogenic and other bacteria, *Belg. J. Bot.*, in press. (Symposium on Macromolecular Identification and Classification of Organisms, Antwerpen, 1991)

Auling, G., Dittbrenner, M., Maarzahl, M., Nokhal, T., Reh, M. (1980), Deoxyribonucleic acid relationships among hydrogen-oxidizing strains of the genera *Pseudomonas*, *Alcaligenes*, and *Paracoccus*, *Int. J. Syst. Bacteriol.* **30,** 123–128.

Auling, G., Probst, A., Kroppenstedt, R. M. (1986), Chemo- and molecular taxonomy of D(−)-tartrate-utilizing pseudomonads, *Syst. Appl. Microbiol.* **8,** 114–120.

Auling, G., Busse, J., Hahn, M., Hennecke, H., Kroppenstedt, R. M., Probst, A., Stackebrandt, E. (1988), Phylogenetic heterogeneity and chemotaxonomic properties of certain Gram-negative aerobic carboxydobacteria, *Syst. Appl. Microbiol.* **10,** 264–272.

Auling, G., Busse, H.-J., Pilz, F., Webb, L., Kneifel, H., Claus, D. (1991 a), Rapid differentiation by polyamine analysis of *Xanthomonas* from phytopathogenic pseudomonads and other members of the class Proteobacteria interacting with plants, *Int. J. Syst. Bacteriol.* **41,** 223–228.

Auling, G., Pilz, F., Busse, H.-J., Karrasch, S., Streichan, M., Schön, G. (1991 b), Analysis of the polyphosphate-accumulating microflora in phosphorus-eliminating, anaerobic-aerobic activated sludge systems by using diaminopropane as a biomarker for rapid estimation of *Acinetobacter* spp., *Appl. Environ. Microbiol.* **57,** 3585–3592.

Auling, G., Busse, H.-J., Egli, T., El-Banna, T., Stackebrandt, E., *Chelatobacter*, gen. nov., and

Chelatococcus, gen. nov., two novel genera of the alpha subclass of the *Proteobacteria* to accomodate the Gram-negative, obligately aerobic, nitrilotriacetate(NTA)-utilizing bacteria *Chelatobacter heintzii*, sp. nov., and *Chelatococcus asaccharovorans*, sp. nov., *Syst. Appl. Microbiol.*, in press.

BISSONNETTE, L., ROY, P. H. (1992), Characterization of InO of *Pseudomonas aeruginosa* plasmid pVS1, an ancestor of integrons of multiresistance plasmids and transposons of Gram-negative bacteria, *J. Bacteriol.* **174,** 1248–1257.

BOUSFIELD, I. J., GREEN, P. N. (1985), Reclassification of bacteria of the genus *Protomonas* Urakami and Komagata 1984 in the genus *Methylobacterium* (Patt, Cole, and Hanson) emend. Green and Bousfield 1983, *Int. J. Syst. Bacteriol.* **35,** 209.

BURLAGE, R. S., HOOPER, S. W., SAYLER, G. S. (1989), The TOL (pWWO) catabolic plasmid, *Appl. Environ. Microbiol.* **55,** 1323–1328.

BUSSE, H.-J., AULING, G. (1992), The genera *Alcaligenes* and "*Achromobacter*", in: *The Prokaryotes – A Handbook on the Biology of Bacteria: Ecophysiology, Isolation, Identification, Applications* (BALOWS, A., TRÜPER, H.-G., DWORKIN, M., HARDER, W., SCHLEIFER, K.-H., Eds.), 2nd Ed., pp. 2544–2555, New York: Springer-Verlag.

BUSSE, J., AULING, G. (1988), Polyamine pattern as a chemotaxonomic marker within the *Proteobacteria*, *Syst. Appl. Microbiol.* **11,** 1–8.

BUSSE, H.-J., EL-BANNA, T., AULING, G. (1989), Evaluation of different approaches for identification of xenobiotic-degrading pseudomonads, *Appl. Environ. Microbiol.* **55,** 1578–1583.

BUSSE, H.-J., EL-BANNA, T., OYAIZU, H., AULING, G. (1992), Identification of xenobiotic-degrading isolates from the beta subclass of the *Proteobacteria* by a polyphasic approach including 16S rRNA partial sequencing, *Int. J. Syst. Bacteriol.* **42,** 19–26.

CHAUDHRY, G. R., CHAPALAMADUGU, S. (1991), Biodegradation of halogenated organic compounds, *Microbiol. Rev.* **55,** 59–79.

COLLINS, M. D., JONES, D. (1981), Distribution of isoprenoid quinone structural types in bacteria and their taxonomic implifications, *Microbiol. Rev.* **45,** 316–354.

DAINTY, R. H., EDWARDS, R. A., HIBBARD, C. M., RAMANTANIS, S. V. (1986), Bacterial sources of putrescine and cadaverine in chill-stored vacuum-packaged beef, *J. Appl. Bacteriol.* **61,** 117–123.

DAVIES, H., G., GREEN, R. H., KELLY, D. R., ROBERTS, S. M. (1990), Recent advances in the generation of chiral intermediates using enzymes, *Crit. Rev. Biotechnol.* **10,** 129–152.

DEFAGO, G., HAAS, D. (1990), Pseudomonads as antagonists of soil borne plant pathogens: modes of action and genetic analysis, in: *Soil Biochemistry* (BOLLAG, J. M., STOTZKY, G., Eds.), Vol. 6., pp. 249-291, New York: Marcel Dekker.

DEINEMA, M. H., ZEVENHUIZEN, L. P. T. M. (1971), Formation of cellulose fibrils by Gram-negative bacteria and their role in bacterial flocculation, *Arch. Mikrobiol.* **78,** 42–57.

DE LEY, J. (1992), The proteobacteria: ribosomal RNA cistron similarities and bacterial taxonomy, in: *The Prokaryotes – A Handbook on the Biology of Bacteria: Ecophysiology, Isolation, Identification, Applications* (BALOWS, A., TRÜPER, H.-G., DWORKIN, M., HARDER, W., SCHLEIFER, K.-H., Eds.) 2nd Ed., pp. 2111–2140, New York: Springer.

DELONG, E. F., WICKHAM, G. S., PACE, N. R. (1989), Phylogenetic stains: ribosomal RNA-based probes for the identification of single cells, *Science* **243,** 1360–1363.

DE VOS, P., DE LEY, J. (1983), Intra- and intergeneric similarities of *Pseudomonas* and *Xanthomonas* ribosomal ribonucleic acid cistrons, *Int. J. Syst. Bacteriol.* **33,** 487–509.

DE VOS, P., VAN LANDSHOOT, A., SEGERS, P., TYTGAT, R., GILLIS, M., BAUWENS, M., ROSSAU, R., GOOR, M., POT, B., KERSTERS, K., LIZZARAGA, P., DE LEY, J. (1989), Genotypic relationships and taxonomic localization of unclassified *Pseudomonas* and *Pseudomonas*-like strains by deoxyribonucleic acid-ribosomal ribonucleic acid hybridizations, *Int. J. Syst. Bacteriol.* **39,** 35–49.

DOUDOROFF, M., PALLERONI, N, J. (1974), Family I. Pseudomonadaceae Winslow, Broadhurst, Buchanan, Krumwiede, Rogers and Smith 1917, in: *Bergey's Manual of Determinative Bacteriology* (BUCHANAN, R. E., GIBBONS, N. E., Eds.), 8th Ed., pp. 217-253, Baltimore: Williams & Wilkins.

DRAHOS, D. J. (1991), Field testing of genetically engineered microorganisms, *Biotechnol. Adv.* **9,** 157–171.

DUGAN, P. R, STONER, D. L., PICKRUM, H. M. (1992), The genus *Zoogloea*, in: *The Prokaryotes – A Handbook on the Biology of Bacteria: Ecophysiology, Isolation, Identification, Applications* (BALOWS, A., TRÜPER, H.-G., DWORKIN, M., HARDER, W., SCHLEIFER, K.-H., Eds.) 2nd Ed., pp. 3952–3964, New York: Springer.

EGLI, T. (1992), Biochemistry and physiology of the degradation of nitrilotriacetic acid and other metal complexing agents, *Biodegradation*, in press.

EGLI, T., BALLY, M., UETZ, T. (1990), Microbial degradation of chelating agents used in detergents with special reference to nitrilotriacetic acid (NTA), *Biodegradation* **1,** 121–132.

EPARVIER, A., LEMANCEAU, P., ALABOUVETTE, C. (1991), Population dynamics of non-pathogenic *Fusarium* and fluorescent *Pseudomonas* strains in rockwool, a substratum for soilless culture, *FEMS Microbiol. Ecol.* **86,** 177–184.

EVANS, W. C., FUCHS, G. (1988), Anaerobic degradation of aromatic compounds, *Annu. Rev. Microbiol.* **42,** 289–317.

FALCH, E. A. (1991), Industrial enzymes – developments in production and application, *Biotechnol. Adv.* **9,** 643–658,

FESTL, H., LUDWIG, W., SCHLEIFER, K. H. (1986), DNA hybridization probe for the *P. fluorescens* group, *Appl. Environ. Microbiol.* **52,** 1190–1194.

FIALHO, A. M., ZIELINSKI, N. A., FETT, W. F., CHAKRABARTY, A. M., BERRY, A. (1990), Distribution of alginate gene sequences in the *Pseudomonas* rRNA homology group I-*Azomonas-Azotobacter* lineage of superfamily B procaryotes, *Appl. Environ. Microbiol.* **56,** 436–443.

FRANTZ, B., CHAKRABARTY, A. M. (1986), Degradative plasmids in *Pseudomonas*, in: *The Bacteria*, Vol. 10: *The Biology of Pseudomonas* (SOKATCH, J. R., Ed.) pp. 295-323, Orlando: Academic Press.

FRANZMANN, P. D., WEHMEYER, U., STACKEBRANDT, E. (1988), Halomonadaceae fam. nov., a new family of the class Proteobacteria to accomodate the genera *Halomonas* and *Deleya*, *Syst. Appl. Microbiol.* **11,** 16–19.

FURUYOSHI, S., NISHIGOURI, J., KAWABATA, N., TANAKA, H., SODA, K. (1991), D-Glycerate production from L-tartrate by cells of *Pseudomonas* sp. with high content of L-tartrate decarboxylase, *Agric. Biol. Chem.* **55,** 1515–1519.

GALLOWAY, D. R. (1991), *Pseudomonas aeruginosa* elastase and elastolysis revisited: recent developments, *Mol. Microbiol.* **5,** 2315–2321.

GILBERT, E. J., CORNISH, A., JONES, C. W. (1991), Purification and properties of extracellular lipase from *Pseudomonas aeruginosa* EF2, *J. Gen. Microbiol.* **137,** 2223–2229.

GOODAY, G. W. (1990), The ecology of chitin degradation, in: *Advances in Microbial Ecology* (MARSHALL, K. C., Ed.), Vol. 11, pp. 387–430, New York: Plenum Press.

GOODFELLOW, M., MINNIKIN, D. E. (1985), *Chemical Methods in Bacterial Systematics*, London: Academic Press.

GOTTSCHALK, G. (1985), *Methods in Microbiology*, Vol. 18, London: Academic Press.

GOUNOT, A. M. (1991), Bacterial life at low temperature: physiological aspects and biotechnological applications, *J. Appl. Bacteriol.* **71,** 386–397.

HAHN, D., AMANN, R. I., LUDWIG, W., AKKERMANS, A. D. L., SCHLEIFER, K. H. (1992), Detection of micro-organisms in soil after *in situ* hybridization with rRNA-targeted, fluorescently labelled oligonucleotides, *J. Gen. Microbiol.* **138,** 879–887.

HANCOCK, R. E. W., SIEHNEL, R., MARTIN, N. (1990), Outer membrane proteins of *Pseudomonas*, *Mol. Microbiol.* **4,** 1069–1075.

HARDMAN, D. J. (1991), Biotransformation of halogenated compounds, *Crit. Rev. Biotechnol.* **11,** 1–40.

HARWOOD, J. (1989), The versatility of lipases for industrial uses, *Trends Biochem. Sci.* **14,** 125–126.

HASHIMOTO, H., GOTO, M., KATAYAMA, C., KITAHATA, S. (1991), Purification and some properties of α-galactosidase from *Pseudomonas fluorescens* H-601, *Agric. Biol. Chem.* **55,** 2831–2838.

HAZLEWOOD, G. P., LAURIE, J. I., FERREIRA, L. M. A., GILBERT, H. J. (1992), *Pseudomonas fluorescens* subsp. *celullulosa:* an alternative model for bacterial cellulase, *J. Appl. Bacteriol.* **72,** 244–251.

HIRAISHI, A., MORISHIMA, Y., TAKEUCHI, J.-I. (1991), Numerical analysis of lipoquinone patterns in monitoring bacterial community dynamics in wastewater treatment systems, *J. Gen. Appl. Microbiol.* **37,** 57–70.

HIRAISHI, A., SHIN, Y.-K., SUGIYAMA, J., KOMAGATA, K. (1992), Isoprenoid quinones and fatty acids of *Zoogloea*, *Antonie van Leeuwenhoek* **61,** 231–236.

HOLLOWAY, B. W., DHARMSTHITI, S., JOHNSON, C., KEARNEY, A., KRISHNAPILLAI, V., MORGAN, A. F., RATNANINGSIH, E., SAFFERY, R., SINCLAIR, M., STROM, D., ZHANG, C. (1990), Chromosome organisation in *Pseudomonas aeruginosa* and *Pseudomonas putida*, in: *Pseudomonas: Biotransformations, Pathogenesis and Evolving Biotechnology* (SILVER, S., CHAKRABARTY, A., IGLEWSKY, B., KAPLAN, S., Eds.), pp. 269–278, Washington, DC: American Society for Microbiology.

HOMMEL, R. K. (1990), Formation and physiological role of biosurfactants produced by hydrocarbon-utilizing microorganisms, *Biodegradation* **1,** 107–119.

HUANG, J., SCHELL, M. A. (1992), Role of the two-component leader sequence and mature amino acid sequences in extrcellular export of endoglucanase EGL from *Pseudomonas solanacearum*, *J. Bacteriol.* **174,** 1314–1323.

JACOBY, G. A. (1986), Resistance plasmids of *Pseudomonas*, in: *The Bacteria*, Vol. X: *The Biology of Pseudomonas* (SOKATCH, J. R., Ed.), pp. 265–293, Orlando: Academic Press.

JAHNKE, M., EL-BANNA, T., KLINTWORTH, R., AULING, G. (1990), Mineralization of orthanilic acid is a plasmid-associated trait in *Alcaligenes* sp. O-1, *J. Gen. Microbiol.* **136,** 2231–2249.

JAHNKE, M., LEHMANN, F., SCHOEBEL, A., AULING, G, Transposition of transposon *Tn4651* into plasmid pSAH confers sequential mineralization of sulpho- and methylsubtituted aromatics to *Alcaligenes* sp. O-1, *J. Gen. Microbiol.*, in press.

JAIN, R. K., BURLAGE, R. S., SAYLER, G. S. (1988), Methods for detecting recombinant DNA in the environment, *Crit. Rev. Biotechnol.* **8,** 33–84.

JURKEVITCH, E., HADAR, Y., CHEN, Y., LIBMAN, J.,

SHANZER, A. (1992), Iron uptake and molecular recognition in *Pseudomonas putida:* receptor mapping with ferrichrome and its biomimetic analogs, *J. Bacteriol.* **174,** 78–83.

KARLSON, B. G., NORDLING, M., PASCHER, T., TSAI, L.-C., SJÖLIN, L., LUNDBERG, L. G. (1991), Cassette mutagenesis of Meth121 in azurin from *Pseudomonas aeruginosa*, *Protein Eng.* **4,** 343–349.

KLEMENT, Z., RUDOLPH, K., SANDS, D. C. (1990), *Methods in Phytobacteriology*, Budapest: Akademiai Kiado.

KRAWIEC, S., RILEY, M. (1990), Organization of the bacterial chromosome, *Microbiol. Rev.* **54,** 502–539.

KROPPENSTEDT, R. M. (1982), Anwendung chromatographischer HP-Verfahren (HPTLC and HPLC) in der Bakterien-Taxonomie, *GIT-Labor Medizin* **5,** 266–275.

KUHM, A. E., STOLZ, A, KNACKMUSS, H.-J. (1991), Metabolism of naphthalene by the biphenyl-degrading bacterium *Pseudomonas paucimobilis* Q1, *Biodegradation* **2,** 115–120.

LARSON, R. J., VENTULLO, R. M. (1986), Kinetics of biodegradation of nitrilotriacetic acid (NTA) in an estuarine environment, *Ecotox. Environ. Saf.* **12,** 166–179.

LAUFF, J. L., STEELE, D. B., COOGAN, L.A., BREITFELLER, J. M. (1990), Degradation of the ferric chelate of EDTA by a pure culture of an *Agrobacterium* sp., *Appl. Environ. Microbiol.* **56,** 3346–3353.

LEAHY, J. G., COLWELL, R. R. (1990), Microbial degradation of hydrocarbons in the environment, *Microbiol. Rev.* **54,** 305–315.

LEONG, J. (1986), Siderophores: Their biochemistry and possible role in the biocontrol of plant pathogens, *Annu. Rev. Phytopathol.* **24,** 187–209.

LINGENS, F., BLECHER, R., BLECHER, H., BLOBEL, F., EBERSPÄCHER, J., FRÖHNER, C., GÖRISCH H., GÖRISCH, H., LAYH, G. (1985), *Phenylobacterium immobile* gen. nov., spec. nov., a Gram-negative bacterium that degrades the herbicide chloridazon, *Int. J. Syst. Bacteriol.* **35,** 26–39.

LYNCH, J. M. (1990), Beneficial interactions between micro-organisms and roots, *Biotechnol. Adv.* **8,** 335–346.

MACARIO, A. J. L., CONWAY DE MACARIO, E. (1990), *Gene Probes for Bacteria*, San Diego: Academic Press.

MACASKIE, L. E. (1991), The application of biotechnology to the treatment of wastes produced from the nuclear fuel cycle: Biodegradation and bioaccumulation as a means of treating radionuclide containig streams, *Crit. Rev. Biotechnol.* **11,** 41–112.

MARGARITIS, A., BASSI, A. S. (1991), Principles and biotechnological application of bacterial ice nucleation, *Crit. Rev. Biotechnol.* **11,** 277–295.

MAY, T. B., SHINABARGER, D., MAHARAJ, R., KATO, J., CHU, L., DEVAULT, J. D., ROYCHOUDURY, S., ZIELINSKI, N. A., BERRY, A., ROTHMEL, R. K., MISRA, T. K., CHAKRABARTY, A. M. (1991), Alginate synthesis by *Pseudomonas aeruginosa:* A key pathogenic factor in chronic pulmonary infections of cystic fibrosis, *Clin. Microbiol. Rev.* **4,** 191–206.

MERMOD, N., LEHRBACH, P. R., DON, R. H., TIMMIS, K. N. (1986), Gene cloning and manipulation in *Pseudomonas*, in: *The Bacteria*, Vol. X: *The Biology of Pseudomonas* (SOKATCH, J. R., Ed.), pp. 325–355, Orlando: Academic Press.

MEYER, J.-M., HALLE, F., HOHNADEL, D., LEMANCEAU, P., RATEFIARIVELO, H. (1987), Siderophores of *Pseudomonas*, in: *Iron Transport in Microbes, Plants and Animals* (WINKELMANN, G., VAN DER HELM, D., NEILANDS, J. B., Eds.), pp. 188–205, Weinheim: VCH.

MEYER, O., FRUNZKE, K., GADKARI, D., JACOBITZ, S., HUGENDIECK, I., KRAUT, M. (1990), Utilization of carbon monoxide by aerobes: recent advances, *FEMS Microbiol. Rev.* **87,** 253–260.

MEYER, O., STACKEBRANDT, E., AULING, G., Reclassification of ubiquinone Q-10 containing carboxidotrophic bacteria: Transfer of "[*Pseudomonas*] *carboxydovorans*" OM5^T to *Oligotropha*, gen. nov., as *Oligotropha carboxidovorans*, comb. nov., and transfer of "[*Alcaligenes*] *carboxydus*" DSM 1086^T to *Carbophilus*, gen. nov., as *Carbophilus carboxidus*, comb. nov.; transfer of "[*Pseudomonas*] *compransoris*" DSM 1231^T to *Zavarzinia*, gen. nov., as *Zavarzinia compransoris*, comb. nov., and amended descriptions of the new genera. *Syst. Appl. Microbiol.*, in press.

MUTHARIA, L. M., LAM, J. S., HANCOCK, R. E. W. (1985), The use of monoclonal antibodies in the study of common antigens of Gram-negative bacteria, in: *Monoclonal Antibodies Aqainst Bacteria* (MACARIO, A. J. L., CONWAY DE MACARIO, E., Eds.), Vol. II, pp. 131–142, Orlando: Academic Press.

NAKAJIMA, H., KOYAMA, H., SUZUKI, H. (1991), Immobilization of *Pseudomonas* L-phe oxidase on a nylon membrane for possible use as an amino acid sensor, *Agric. Biol. Chem.* **55,** 3117–3118.

NAR, H., MESSERSCHMIDT, A., HUBER, R., VAN DE KAMP, M. CANTERS, G. W. (1991), Crystal structure analysis of oxidized *Pseudomonas aeruginosa* azurin at pH 5.5 and pH 9.0. A pH-induced conformational transition involves a peptide bond flip, *J. Mol. Biol.* **221,** 765–772.

NISHIYAMA, M., HORINOUCHI, S., KOBAYASHI, M., NAGASAWA, T., YAMADA, H., BEPPU, T. (1992), Cloning and characterization of genes responsible for metabolism of nitrile compounds from *Pseudomonas chlororaphis* B23, *J. Bacteriol.* **173,** 2465–2472.

NORTEMANN, B. (1992), Total degradation of EDTA by mixed cultures and a bacterial isolate, *Appl. Environ. Microbiol.* **58,** 671–676.

OBERHÄNSLI, T., DEFAGO, G., HAAS, D. (1991), Indole-3-acetic acid (IAA) synthesis in the biocontrol strain CHAO of *Pseudomonas fluorescens:* role of tryptophan side chain oxidase, *J. Gen. Microbiol.* **137,** 2273–2279.

OBUKOWICZ, M. G., PERLAK, F. J, KUSANO-KRETZMER, K., MAYER, E. J. BOLTON, S. L, WATRUD, L. S. (1986), *Tn5*-mediated integration of the delta-endotoxin gene from *Bacillus thuringiensis* into the chromosome of root-colonizing pseudomonads, *J. Bacteriol.* **168,** 982–989.

ORNSTON, L. N., NEIDLE, E. L. (1991), Evolution of genes for the β-ketoadipate pathway in *Acinetobacter calcoaceticus*, in: *The Biology of Acinetobacter* (TOWNER, K., BERGOGNE-BEREZIN, E., FEWSON, C. A., Eds.), pp. 201–237, New York: Plenum Press.

OYAIZU, H., KOMAGATA, K. (1983), Grouping of *Pseudomonas* species on the basis of cellular fatty acid composition and the ubiquinone system with special reference to the existence of 3-hydroxy fatty acids, *J. Gen. Appl. Microbiol.* **29,** 17–40.

PALLERONI, N. J. (1984), Genus I *Pseudomonas* Migula 1894, in: *Bergey's Manual of Systematic Bacteriology* (KRIEG, N. R., HOLT, J. G., Eds.), Vol. 1, pp.141–199, Baltimore: Williams & Wilkins.

PALLERONI, N. J. (1992 a), Introduction to the family *Pseudomonadaceae*, in: *The Prokaryotes – A Handbook on the Biology of Bacteria: Ecophysiology, Isolation, Identification, Applications* (BALOWS, A., TRÜPER, H.-G., DWORKIN, M., HARDER, W., SCHLEIFER, K.-H., Eds.) 2nd Ed., pp. 3071–3085, New York: Springer.

PALLERONI, N. J. (1992 b), Human- and animal-pathogenic pseudomonads, in: *The Prokaryotes – A Handbook on the Biology of Bacteria: Ecophysiology, Isolation, Identification, Applications* (BALOWS, A., TRÜPER, H.-G., DWORKIN, M., HARDER, W., SCHLEIFER, K.-H., Eds.), 2nd Ed., pp. 3086–3103, New York: Springer.

PALLERONI, N. J., BALLARD, R. W., RALSTON, E., DOUDOROFF, M. (1972), Deoxyribonucleic acid homologies among some *Pseudomonas* species, *J. Bacteriol.* **110,** 1–11.

PALLERONI, N. J., KUNISAWA, R., CONTOPOULOU, R., DOUDOROFF, M. (1973), Nucleic acid homologies in the genus *Pseudomonas*, *Int. J. Syst. Bacteriol.* **23,** 333–339.

QUINN, J. P. (1990), Evolving strategies for the genetic engineering of herbicide resistance in plants, *Biotechnol. Adv.* **8,** 321–333.

REANNEY, D. C., ROBERTS, W. P., KELLY, W. J. (1982), Genetic interactions among microbial communities, in: *Microbial Interactions and Communities* (BULL, A. T., SLATER, J. H., Eds.), Vol. 1, pp. 287–322, London: Academic Press.

REINEKE, W., KNACKMUSS, H.-J. (1988), Microbial degradation of haloaromatics, *Annu. Rev. Microbiol.* **42,** 263–287.

ROBERSON, E. B., FIRESTONE, M. K. (1992), Relationship between desiccation and exopolysaccharide production in a soil *Pseudomonas* sp. *Appl. Environ. Microbiol.* **58,** 1284–1291.

RODE, H., GIFFHORN, F. (1982), D-(–)-Tartrate dehydratase of *Rhodopseudomonas sphaeroides:* purification, characterization, and application to enzymatic determination of D-(–)-tartrate, *J. Bacteriol.* **150,** 1061–1068.

ROMBEL, I. T., LAMONT, I. L. (1992), DNA homology between siderophore genes from fluorescent pseudomonads, *J. Gen. Microbiol.* **138,** 181–187.

RÖMLING, U., TÜMMLER, B. (1991), The impact of two-dimensional pulsed-field gel electrophoresis techniques for the consistent and complete mapping of bacterial genomes: refined physical map of *Pseudomonas aeruginosa* PAO, *Nucleic Acids Res.* **19,** 3199–3206,

ROSENBERG, F. A. (1990), The bacterial flora of natural mineral waters and potential problems associated with its digestion, *Riv. Ital. Ig.* **50,** 303–310.

ROSS, P., MAYER, R., BENZIMAN, M. (1991), Cellulose biosynthesis and function in bacteria, *Microbiol. Rev.* **55,** 35–58.

ROSSELLO, R., GARCIA-VALDES, E., LALUCAT, J., URSING, J. (1991), Genotypic and phenotypic diversity of *Pseudomonas stutzeri*, *Syst. Appl. Microbiol.* **14,** 150–157.

SAINT-ONGE, A., ROMEYER, F., LEBEL, P., MASSON, L., BROUSSEAU, R. (1992), Specificity of the *Pseudomonas aeruginosa* PAO1 lipoprotein gene as a DNA probe and PCR target region within the Pseudomonadaceae, *J. Gen. Microbiol.* **138,** 733–741.

SARIASLANI, F. S. (1989), Microbial enzymes for oxidation of organic molecules, *Crit. Rev. Biotechnol.* **9,** 171–257.

SCHMIDT, E. L. (1974), Quantitative autoecological study of microorganisms in soil by immunofluorescence, *Soil Sci.* **119,** 141–149.

SCHROTH, M. N., HILDEBRAND, D. C., PANOPOULOS, N. (1992), Phytopathogenic pseudomonads and related plant-associated pseudomonads, in: *The Prokaryotes – A Handbook on the Biology of Bacteria: Ecophysiology, Isolation, Identification, Applications* (BALOWS, A., TRÜPER, H.-G., DWORKIN, M., HARDER, W., SCHLEIFER, K.-H., Eds.), 2nd Ed., pp. 3104–3131, New York: Springer.

SCHUMANN, W. (1990), *Biologie bakterieller Plasmide*, Braunschweig: Vieweg.

SHAW, P. D. (1987), Plasmid ecology, in: *Plant-Mi-*

crobe Interactions: Molecular and Genetic Perspectives (KOSUGE, T., NESTER, E. W., Eds.), Vol. 2, pp. 3–39, New York: Macmillan.

SHIDA, O., TAKANO, T., TAKAGI, H., KADOWAKI, K., KOBAYASHI, S. (1992), Cloning and nucleotide sequence of the maltopentaose-forming amylase gene from *Pseudomonas* sp. KO-8940, *Biosci. Biotech. Biochem.* **56,** 76–80.

SHINGLER, V., POWLOWSKI, J., MARKLUND, U. (1992), Nucleotide sequence and functional analysis of the complete phenol/3,4-dimethylphenol catabolic pathway of *Pseudomonas* sp. strain CF600, *J. Bacteriol.* **174,** 711–724.

SILVER, S., CHAKRABARTY, A. M., IGLEWSKI, B. KAPLAN, S. (1990), *Pseudomonas – Biotransformations, Pathogenesis, and Evolving Biotechnology*, Washington, DC: American Society for Microbiology.

SIMON, R., PRIEFER, V., PÜHLER, A. (1983), A broad host range mobilization system for *in vivo* genetic engineering: Transposon mutagenesis in Gram-negative bacteria, *Bio/Technology* **1,** 784–790.

SMIRNOW, V. V., KIPRIANOVA, E. A. (1990), *Bacteria of the Pseudomonas Genus,* Kiev: Nauk Dumka.

SMITH, D. W., YEE, T. W., BAIRD, C., KRISHNAPILLAI, V. (1991), Pseudomonads replication origins: a paradigm of bacterial origins? *Mol. Microbiol.* **5,** 2581–2587.

SMITH, M. R. (1990), The biodegradation of aromatic hydrocarbons by bacteria, *Biodegradation* **1,** 191–206.

STACKEBRANDT, E., GOODFELLOW, M. (1991), *Nucleic Acid Techniques in Bacterial Systematics*, Chichester: John Wileys & Sons.

STACKEBRANDT, E., MURRAY, R. G. E., TRÜPER, H. G. (1988), *Proteobacteria* classis nov. a name for the phylogenetic taxon including the "purple bacteria and their relatives", *Int. J. Syst. Bacteriol.* **38,** 321–325.

STANIER, R. Y., PALLERONI, N. J., DOUDOROFF, M. (1966), The aerobic pseudomonads: A taxonomic study, *J. Gen. Microbiol.* **43,** 159–271.

STEAD, D. E. (1992), Grouping of plant-pathogenic and some other *Pseudomonas* spp. by using cellular fatty acid profiles, *Int. J. Syst. Bacteriol.* **42,** 281–295.

STEWART, G. J., SINIGALLIANO, C. D., GARKO, K. A. (1991), Binding of exogenous DNA to marine sediments and the effect of DNA/sediment binding on natural transformation of *Pseudomonas stutzeri* strain ZoBell in sediment columns, *FEMS Microbiol. Ecol.* **85,** 1–8.

STEWART-TULL, D. E. (1988), Round table 2: Detection methods including sequencing and probes, in: *Proc. First Int. Conf. on the Release of Genetically-Engineered Microorganisms* (SUSSMAN, M., COLLINS, C. H., SKINNER, F. A., STEWART-TULL, D. E., Eds.), pp. 207–230, London: Academic Press.

STOLP, H., GADKARI, D. (1981), Nonpathogenic members of the genus *Pseudomonas*, in: *The Prokaryotes – A Handbook on Habitats, Isolation and Identification of Bacteria* (STARR, M. R., STOLP, H., TRÜPER, H. G., BALOWS, A., SCHLEGEL, H. G., Eds.) pp. 719–741, Berlin: Springer-Verlag.

STROM, M. S., NUNN, D., LORY, S. (1991), Multiple roles of the pilus biogenesis protein PilD: Involvement of PilD in excretion of enzymes from *Pseudomonas aeruginosa*, *J. Bacteriol.* **173,** 1175–1180.

SUGAI, T., OHTA, H. (1991), A simple preparation of (*R*)-2-hydroxy-4-phenylbutanoic acid, *Agric. Biol. Chem.* **55,** 293–294.

SUTHERLAND, I. W. (1990), *Biotechnology of Microbial Exopolysaccharides*, Cambridge: Cambridge University Press.

SWINGS, J. P., DE VOS, P., VAN DEN MOOTER, M., DE LEY, J. (1983), Transfer of *Pseudomonas maltophilia* Hugh 1981 to the genus *Xanthomonas* as *Xanthomonas maltophilia* (Hugh 1981) comb. nov., *Int. J. Syst. Bacteriol.* **33,** 409–413.

TAMAOKA, J., HA, D.-M., KOMAGATA, K. (1987), Reclassification of *Pseudomonas acidovorans* den Dooren de Jong 1926 and *Pseudomonas testosteroni* Marcus and Talalay 1956 as *Comamonas acidovorans* comb. nov. and *Comamonas testosteroni* comb. nov., with emended description of the genus *Comamonas*, *Int. J. Syst. Bacteriol.* **37,** 52–59.

TCHAN, Y.-T. (1984), Family II. Azotobacteriaceae Pribram 1933, in: *Bergey's Manual of Systematic Bacteriology* (KRIEG, N. R., HOLT, J. G., Eds.), Vol. 1, pp. 219–220, Baltimore: Williams & Wilkins.

THIBAUT, D., COUDER, M., FAMECHON, A., DEBUSSCHE, L., CAMERON, B., CROUZOT, J., BLANCHE, F. (1992), The final step in the biosynthesis of hydrogenobyrinic acid is catalyzed by the *cobH* gene product with precorrin-8 × as the substrate, *J. Bacteriol.* **174,** 1043–1049.

THOMAS, A. W., SLATER, J. H., WEIGHTMAN, A. J. (1992), The dehalogenase gene *dehI* from *Pseudomonas putida* PP3 is carried on an unusual mobile genetic element designated *DEH*, *J. Bacteriol.* **174,** 1932–1940.

TRUDGILL, P. W. (1990), Microbial metabolism of monoterpenes - recent developments, *Biodegradation* **1,** 93–105.

TRUMPOWER, B. L. (1990), Cytochrome bc_1-complexes of microorganisms, *Microbiol. Rev.* **54,** 101–129.

TRÜPER, H.-G., SCHLEIFER, K. H. (1992), Prokaryote characterization and identification, in: *The Prokaryotes – A Handbook on the Biology of Bacteria: Ecophysiology, Isolation, Identification, Applications* (BALOWS, A., TRÜPER, H.-G., DWOR-

KIN, M., HARDER, W., SCHLEIFER, K.-H., Eds.), 2nd Ed., pp. 126–148, New York: Springer.

URAKAMI, T., ARAKI, H., OYANAGI, H., SUZUKI, K.-I., KOMAGATA, K. (1992), Transfer of *Pseudomonas aminovorans* (den Dooren de Jong 1926) to *Aminobacter* gen. nov. as *Aminobacter aminovorans* comb. nov. and description of *Aminobacter aganoensis* sp. nov. and *Aminobacter niigataensis* sp. nov., *Int. J. Syst. Bacteriol.* **42,** 84–92.

VANDE WOESTYNE, M., BRUYNEEL, B., VERSTRAETE, W. (1992), Physicochemical characterization of the microbial Fe^{2+} chelator proferrorosamine from *Pseudomonas roseus fluorescens*, *J. Appl. Bacteriol.* **72,** 44–50.

VAN ELSAS, J. D., VAN OVERBEEK, L. S., FELDMANN, A. M., DULLEMANS, A. M., DE LEEUW, O. (1991), Survival of genetically engineered *Pseudomonas fluorescens* in soil in competition with the parent strain, *FEMS Microbiol. Ecol.* **85,** 53–64.

VANHAECKE, E., REMON, J.-P., MOORS, M., RAES, F., DE RUDDER, D., VAN PETEGHEM, A. (1990), Kinetics of *Pseudomonas aeruginosa* adhesion to 304 and 316-L stainless steel: role of cell surface hydrophobicity, *Appl. Environ. Microbiol.* **56,** 788–795.

VAN ZYL, E., STEYN, P. L. (1992), Reinterpretation of the taxonomic position of *Xanthomonas maltophilia* and taxonomic criteria in this genus, *Int. J. Syst. Bacteriol.* **42,** 193–198.

VERHOOGT, H. J. C., SMIT, H., ABEE, T., GAMPER, M., DRIESSEN, A. J. M., HAAS, D., KONINGS, W. N. (1992), *arcD*, the first gene of the arc operon for anaerobic arginine catabolism in *Pseudomonas aeruginosa*, encodes an arginine-ornithine exchanger, *J. Bacteriol.* **174,** 1568–1573.

VOLLBRECHT, D. (1987), Der Einfluß von Elektronenakzeptoren auf den Stoffwechsel von Mikroorganismen, *Forum Mikrobiol.* **10,** 376–385.

WATKINSON, R. J., MORGAN, P. (1990), Physiology of aliphatic hydrocarbon-degrading microorganisms, *Biodegradation* **1,** 79–92.

WAYNE, L. G., BRENNER, D. J., COLWELL, R. R., GRIMONT, P. A. D., KANDLER, O., KRICHEVSKY, M. I., MOORE, L. H., MOORE, W. E., MURRAY, R. G. E., STACKEBRANDT, E., STARR, M. P., TRÜPER, H.-G. (1987), Report of the *ad hoc* committee on reconciliation of approaches to bacterial systematics, *Int. J. Syst. Bacteriol.* **33,** 215–239.

WHITED, G. M., GIBSON, D. T. (1992), Separation and partial purification of the enzymes of the toluene-4-monooxygenase catabolic pathway in *Pseudomonas mendocina* KR1, *J. Bacteriol.* **173,** 3017–3020.

WILBERG, E., EL-BANNA, T., AULING, G., EGLI, T., Serological studies on nitrilotriacetic acid (NTA)-utilizing bacteria: Distribution of *Chelatobacter heintzii* and *Chelatococcus asacharovorans* in sewage treatment plants and aquatic ecosystems, *Syst. Appl. Microbiol.*, in press.

WILLEMS, A., J. BUSSE, M., GOOR, B., POT, E., FALSEN, E., JANTZEN, B., HOSTE, M., GILLIS, K., KERSTERS, G., AULING, DE LEY, J. (1989), *Hydrogenophaga*, a new genus of hydrogen-oxidizing bacteria that includes *Hydrogenophaga flava* comb. nov. (formerly *Pseudomonas flava*), *Hydrogenophaga palleronii* (formerly *Pseudomonas palleronii*), *Hydrogenophaga pseudoflava* (formerly *Pseudomonas pseudoflava* and *"Pseudomonas carboxydoflava"*), and *Hydrogenophaga taeniospiralis* (formerly *Pseudomonas taeniospiralis*), *Int. J. Syst. Bacteriol.* **39,** 319–333.

WILLEMS, A., FALSEN, E., POT, B., JANTZEN, E., HOSTE, B, VANDAMME, P., GILLIS, M., KERSTER, K., DE LEY, J. (1990), *Acidovorax*, a new genus for *Pseudomonas facilis*, *Pseudomonas delafieldii*, EF group 13, EF group 16, and several clinical isolates, with the species *Acidovorax facilis* comb. nov., *Acidovorax delafieldii* comb. nov., and *Acidovorax temperans* sp. nov., *Int. J. Syst. Bacteriol.* **40,** 384–398.

WILLEMS, A., DE LEY, J., GILLIS, M., KERSTERS, K. (1991), Comamonadaceae, a new family encompassing the *acidovorans* rRNA complex, including *Variovorax paradoxus* gen. nov., comb. nov., for *Alcaligenes paradoxus* (Davis 1969), *Int. J. Syst. Bacteriol.* **41,** 445–450.

WINKELMANN, G. (1991), *Handbook of Microbial Iron Chelates*, Boca Raton: CRC Press.

WOESE, C. R. (1987), Bacterial evolution, *Microbiol. Rev.* **51,** 221–271.

WOESE, C. R. (1992), Prokaryote systematics: The evolution of a science, in: *The Prokaryotes – A Handbook on the Biology of Bacteria: Ecophysiology, Isolation, Identification, Applications* (BALOWS, A., TRÜPER H.-G., DWORKIN, M., HARDER, W., SCHLEIFER, K.-H, Eds.), 2nd Ed., pp. 3–18, New York: Springer.

WOESE, C. R., BLANZ, P., HESPELL, R. B., HAHN, C. M. (1982), Phylogenetic relationships among various helical bacteria, *Curr. Microbiol.* **7,** 119–124.

WOESE, C. R., BLANZ, P., HAHN, C. M. (1985), What isn't a pseudomonad: The importance of nomenclature in bacterial classification, *Syst. Appl. Microbiol.* **5,** 179–195.

WONG, W. C. (1991), Methods for recovery and immunodetection of *Xanthomonas campestris* pv. *phaseoli* in navy bean seed, *J. Appl. Bacteriol.* **71,** 124–129.

YABUUCHI, E., YANO, I., OYAIZU, H., HASHIMOTO, Y., EZAKI, T., YAMAMOTO, H. (1990), Proposals of *Sphingomonas paucimobilis* gen. nov. and comb. nov., *Sphingomonas parapaucimobilis* sp. nov., *Sphingomonas yanoikuyae* sp. nov., *Sphingomonas adhaesiva* sp. nov., *Sphingomonas cap-*

sulata comb. nov., and the two genospecies of the genus *Sphingomonas*, *Microbiol. Immunol.* **34,** 99–119.

YAMADA, Y., TAKINAMI-NAKAMURA, H. TAHARA, Y., OYAIZU, H. KOMAGATA, K. (1982), The ubiquinone systems in the strains of *Pseudomonas* species, *J. Gen. Appl. Microbiol.* **28,** 7–12.

YAMAZAKI, Y., HOSONO, K. (1990), Diametric stereoselectivity of *Pseudomonas fluorescens* lipase and *Candida cylindrica* lipase in the acylation of organometallic alcohols, *Agric. Biol. Chem.* **54,** 3357–3361.

ZUMFT, W. G. (1992), The denitrifying prokaryotes, in: *The Prokaryotes – A Handbook on the Biology of Bacteria: Ecophysiology, Isolation, Identification, Applications* (BALOWS, A., TRÜPER, H.-G., DWORKIN, M., HARDER, W., SCHLEIFER, K.-H., Eds.), 2nd Ed., pp. 554–582, New York: Springer.

13 Streptomycetes and Corynebacteria

WOLFGANG PIEPERSBERG
Wuppertal, Federal Republic of Germany

1 Introduction 434
2 Taxonomy and General Properties 434
3 Streptomycetes and Related Genera 436
3.1 Ecology, Isolation, and Culturing of Streptomycetes 436
3.2 Genetics and Genetic Manipulation of Streptomycetes 438
3.3 Primary Metabolism of Streptomycetes 442
3.4 Secondary Metabolism – Individualization Metabolism of Streptomycetes 443
3.5 Streptomycetes in Biotechnology 447
4 Corynebacteria and Related Genera 453
4.1 Ecology, Isolation, and Culturing of Corynebacteria 453
4.2 Genetics and Genetic Manipulation of Corynebacteria 453
4.3 Primary Metabolism of Corynebacteria 454
4.4 Corynebacteria in Biotechnology 456
5 Conclusions and Perspectives 458
6 References 459

1 Introduction

Historically, both streptomycetes and corynebacteria were the first microorganisms found in screening programs designed to identify the most efficient excreters for certain low-molecular weight natural compounds, secondary and primary metabolites as end products of biosynthetic pathways. The first group showed up when naturally occurring substances with antibiotic activity were searched during the first extensive screening programs in the early 1940s. Among members of the second group the first high-yielding amino acid producer *Corynebacterium glutamicum* was found in 1957 (KINOSHITA et al., 1957). A second, biological trait which unifies these two groups of eubacteria is their taxonomic relatedness (see below). However, in their general properties both genera are quite different in several respects. For the main purpose of this chapter it seemed justified to include some other groups like the nocardioforms, micromonosporae, actinoplanetes, and some coryneforms beyond the genus *Corynebacterium*.

The scope of this chapter is to describe the basics of biology and biotechnology of streptomycetes, corynebacteria, and some related genera which are under similar industrial use.

2 Taxonomy and General Properties

It is important to have at first a brief and comparative look at the taxonomical relationships and the basic properties of both groups of microorganisms in question. Also in view of the many production strains described under conflicting genus or species names and the patent situation with industrial strains this issue required additional emphasis in the past two or three decades. The genera *Streptomyces* and *Corynebacterium* belong to the high-G+C branch of the Gram-positive eubacteria comprising the actinomycetes in a wider sense (KORN-WENDISCH and KUTZNER, 1991; LIEBL, 1991). In Tab. 1 some of the characteristic and distinguishing properties of these and some other related genera are summarized. Today the taxonomical re-evaluation of both groups has not yet been settled. The methods for genus and species identification in the family Streptomycetaceae currently regarded as adequate are mainly chemotaxonomical methods in combination with micromorphological and some physiological criteria (WILLIAMS et al., 1989). However, an increasingly simplified set of genetic diagnostic methods will probably revolutionize this important field in the near future (STACKEBRANDT et al., 1991). This will lead soon to an unequivocal natural tree of evolutionary relationships among all the genera within the actinomycete branch of the Gram-positive bacteria. For species differentiation similar problems exist, especially for the streptomycetes, which are currently solved by grouping them into species clusters with a numerical classification based on physiological tests (WILLIAMS et al., 1989; KÄMPFER et al., 1991). Another sensitivity level could be reached by phage-typing (KORN-WENDISCH and SCHNEIDER, 1992), but also here the use of methods based on sequence relationships of nucleic acids will probably help to solve ambiguities in the future.

Both streptomycetes and corynebacteria are obligate chemoorganotrophs and use aerobic conditions for final oxidation of their carbon sources. However, in contrast to the obligately aerobic streptomycetes, most strains in the genus *Corynebacterium* are facultative anaerobes, fermenting carbon sources (glucose and some other sugars) under anaerobic conditions to acids (acetic, lactic, and succinic) in peptone media.

The corynebacteria form a group of irregularly shaped single-celled eubacteria lacking a differentiation cycle. None other than vegetative non-movable cell types are observed. They are characterized by their mycolic acid content in the cell walls. As a group of so-called "pro-actinomycetes" they are regarded as descending from an older evolutionary stage of the actinomycete line leading to the streptomycetes and other higher developed and differentiating "euactinomycetes" via intermediate stages which could have resembled the mycobacteria, rhodococci, and nocardias.

In the 16th edition of *Bergey's Manual of Systematic Bacteriology* (WILLIAMS et al., 1989)

Tab. 1. Taxonomical Differentiation of Corynebacteria, Streptomycetes and Some Other Euactinomycetes of Biotechnological Importance

Distinction or Properties[a]	*Corynebacterium*[b]	*Nocardia*	*Streptomyces*	*Micromonospora*	*Actinoplanes*	*Actinomadura*
Section in *Bergey's Manual*	15	26 (17)	29	28	28	30
Vegetative cells	Irregular rods	Irregular rods or fragm. hyphae	Branching hyphae	Branching hyphae	Branching hyphae	Branching hyphae
Substrate mycelium	No	No	Yes	Yes	Yes	Yes (chains)
Aerial mycelium	No	Yes	Yes	No	No	Yes
Spores	No	Conidia (short cains)	Chains of arthrospores	Single spores	Motile sporangio-spores	Arthrospores
Cell-wall type (diamino acid)	I (m-DAP)	IV (m-DAP)	I (LL-DAP)	II (m-DAP)	II (m-DAP)	III (m-DAP)
Mycolic acid (C atoms in chain)	Yes (20–36)	Yes (40–60)	No	No	No	No
Predominant menaquinones	MK-9 (H_2) or MK-8 (H_2)	MK-8 (H_4) or MK-9 (H_2)	MK-9 (H_6) or MK-9 (H_8)	MK-9/10 (H_6) or MK-10/12 (H_4)	MK-9 (H_4, H_6, H_8)	MK-9 (H_4, H_6, H_8)
Phospholipid type (predominating)	I (PI, PIM, DPG)	II (PE)	II (PE)	II (PE)	II (PE, PI)	I (IV) (PG, PIM, PI)
Fatty-acid type (composition)	1 a (S, U)	1 b (S, U, T)	2 c (S, A, I)	3 b (A, I, U)	2 d (A, I)	3 c (S, A, I)
G+C content of the DNA (%)	51–68	64–72	69–78	71–73	72–73	64–69 (77)
Genome size (Mbp)	2.5–2.8	n.d.	7.5–8.0	n.d.	n.d.	n.d.

[a] compiled from *Bergey's Manual for Determinative Bacteriology*, 8th Ed., Vol. 2 and 4 (1986, 1989); GOODFELLOW (1989); KORN-WENDISCH and KUTZNER (1991), LIEBL (1991). The abbreviations and figures mean: cell-wall types I, L-diaminopimelic acid (L-DAP) plus glycine; II, *meso*-DAP plus glycine; III, *m*-DAP; IV, *m*-DAP, arabinose plus galactose; menaquinones MK-9 (H_2) have two of the nine isoprenoid units hydrogenated; phospholipids are: PG, phosphatidylglycerol (variable); DPG, di-PG; PE, phosphatidylethanolamine; PI, phosphatidylinositol; PIM, PI mannosides; fatty acid patterns are classified according to KROPPENSTEDT (unpublished data; cited in WILLIAMS et al., 1989); S, saturated; U, unsaturated; A, *anteiso*-; I, *iso*-; T, methyl-branched acids; Mbp, mega base pairs; n.d., not determined

[b] including the genera *Arthrobacter* and *Brevibacterium*

the streptomycetes were entered as *section 29* listing two biotechnologically relevant genera, *Streptomyces* and *Streptoverticillium*. However, in this chapter we have to include some further actinomycete groups represented by the genera *Nocardia*, *Micromonospora*, *Actinoplanes*, and *Actinomadura*, which are also of importance for the production of bioactive metabolites, and for which similar basic conditions exist for cultivation, strain improvement, genetic engineering, and for industrial use (see Tab. 1). All these genera belong to the highly differentiating branch of the euactinomycetes. For example, the streptomycetes are characterized by the formation of a highly branched and septated substrate mycelium, which can differentiate into an aerial mycelium and further develop into variably shaped chains of arthrospores (conidia) by fragmentation of the aerial hyphae (Fig. 1). The pigmentation and morphology of spore chains and surfaces are stable and, therefore, valuable species characters. The general habitats of both groups of organisms are quite similar, such as soil, sewage, watery ecosystems, animal and plant sources (see Sects. 3.1 and 4.1). On the other hand, both groups do colonize very different ecological or physiological niches within their biotopes and fulfill different roles in the food chain during mineralization of organic material, even if they sometimes live in close neighborhood or even in contact to each other.

3 Streptomycetes and Related Genera

3.1 Ecology, Isolation, and Culturing of Streptomycetes

The genus *Streptomyces* is the euactinomycete with the highest incidence of isolation from any type of natural habitat including soils, sediments of lakes, rivers, estuaries, and marine environments. However, the well aerated sphere of soils rich in organic debris seem to be the preferred ecosystem (WILLIAMS et al., 1984; LOCCI, 1989; KORN-WENDISCH and KUTZNER, 1991). In the soil environment euactinomycetes seem to grow in microcolonies on or in the neigborhood of substrate-rich, but relatively inert rotten plant or animal material, where they carry out important steps in the hydrolytic phase of the aerobic mineralization processes. They are able to degrade most of the more stable biopolymers, e.g., lignocellulose, polyphenols, chitosans, keratins, pectins, agar agar, and other polysaccharides, proteins, and even synthetic polymers. For this purpose they excrete a large number of different hydrolases (PECZYNSKA-CZOCH and MORDARSKI, 1988; see Sect. 3.5). No requirements for elaborate growth factors, such as vitamins or aromatic amino acids, are observed. In some habitats,

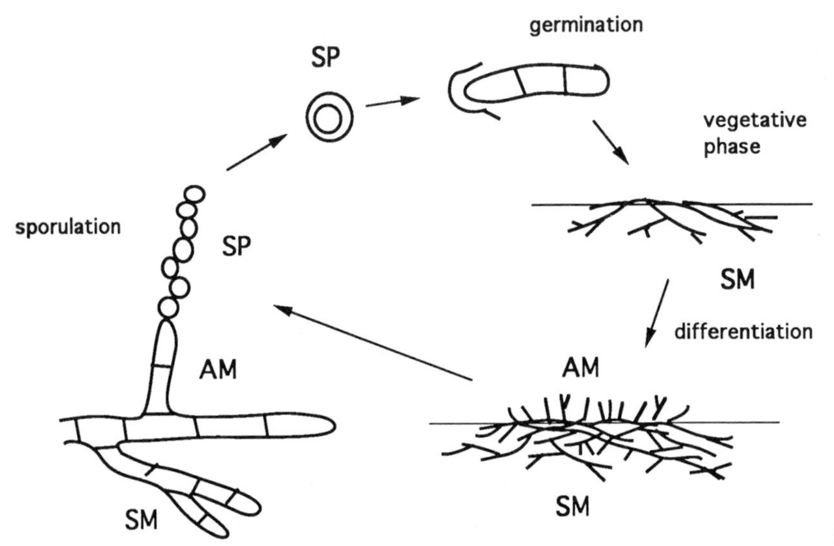

Fig. 1A

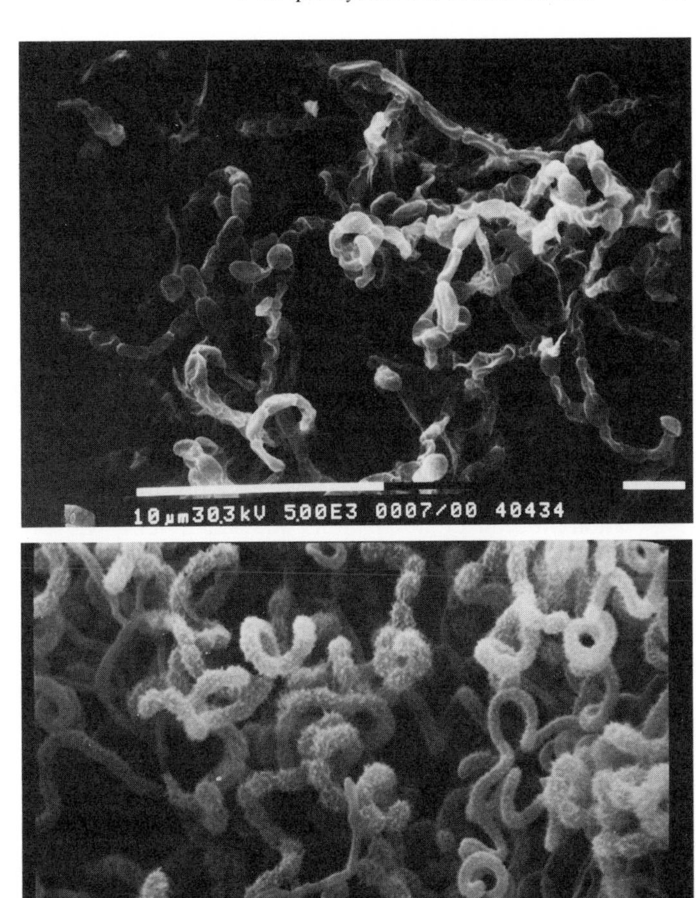

Fig. 1B

Fig. 1C

Fig. 1. Morphology and differentiation of streptomycetes. (A) Life cycle of a streptomycete. SP, arthrospores or conidia; SM, substrate mycelium; AM, aerial mycelium; the thin horizontal line symbolizes a substrate surface. (B, C) Scanning electron micrographs of sporulating streptomycete aerial hyphae forming chains of arthrospores. *Streptomyces lividans* 66 (DSM 40434) shows smooth spore surface (B), *Streptomyces* sp. (K 4007, soil from Crete) exhibits a spiked spore surface (C). The photographs were kindly contributed by F. KORN-WENDISCH, Technische Hochschule Darmstadt, FRG.

e.g., compost, manure, and hey, thermophilic streptomycetes (optimal growth at 50–60 °C) and other actinomycetes are frequent.

Although direct evidence is rare, a complex interaction of the differentiated cell types of streptomycetes among each other and with other microorganisms, e.g., single-celled eubacteria, protozoa and filamentous fungi, as well as with higher eukaryotes, such as plant roots (mycorrhiza), or animals, could occur in the soil biotope. Antibiotic-like compounds could play an important role in this interplay (see Sect. 3.4). Genetic cross-talk via plasmid transfer (MAZODIER and DAVIES, 1991; WEL-

LINGTON et al., 1992) or bacteriophage mediated interactions (CHATER, 1986) in natural habitats is likely to be frequent.

Very few human-pathogenic strains are known, like *Streptomyces somaliensis,* some *Nocardia* sp., and *Actinomadura* sp. inducing actinomycetomas (SCHAAL, 1988). Some plant-pathogenic varieties are also known, e.g., *Streptomyces scabies*, the etiologic agent of scabs on potato tubers or on sugar beet. Symbionts are only found in the taxonomically related genus *Frankia*, the members of which are all specialized in their aibility for nitrogen fixation in root nodules of a large variety of tree-forming angiosperms, the so-called actinorhizal plants (see WILLIAMS et al., 1989, for further reading).

Isolation procedures are manifold and simple, but rarely specific, whenever the full range of streptomycete-related organisms is the scope (WILLIAMS et al., 1989; KORN-WENDISCH and KUTZNER, 1991). Streptomycetes are enriched and isolated by various means, primarily based on their sporulation and antibiotic resistance properties. Sporulation can easily be induced by drying the samples, and viability of other microorganisms can be suppressed by mild heat treatment. The basic requirements for growth are simple carbon and nitrogen sources (DIETZ, 1986; NOLAN and CROSS, 1988). Optimal culturing in submersed liquid media requires high aeration rates and a means for preventing pellet formation. Physicochemical growth conditions are usually those of bacteria with mesophilic (ca. 15–40 °C) and neutral pH requirements. However, psychrophilic, thermophilic, or alkalophilic streptomycete strains can also be isolated.

Media for the general growth, antibiotic production and testing (WILLIAMS et al., 1989; KORN-WENDISCH and KUTZNER, 1991), and the genetic manipulation of streptomycetes (HOPWOOD et al., 1985) have been described in many variations and reflect the physiological flexibility of this group of bacteria. Antibiotic production can be either in the stationary phase or constitutively during vegetative growth for one and the same strain, depending on the medium composition, especially the combination of carbon and nitrogen source. The more rational design of selective and/or broad actinomycetes media is still a widely discussed problem to generate maximal chemical diversity through microbial diversity (WILLIAMS and VICKERS, 1988; NISBET and PORTER, 1989). Strategies involving the screening of less explored biotopic niches as sources and a direct feedback of measurable data into the next screening phase are discussed (WEYLAND and HELMKE, 1988; GOODFELLOW and O'DONNELL, 1989). Surprisingly, the increase of tested sample numbers from the same habitat and location can largely widen the overall number of different isolates (K. A. BOSTIAN, personal communication). Cultural instability, even under preservation conditions, is frequent and has to be taken into account in all strain histories (WILLIAMS et al., 1989).

Because of their unique differentiation pattern, the primary identification is generally easy by colony morphology and microscopical methods, such as the detection of spore surfaces and shapes of spore chains. More precise grouping, however, is complicated and requires the testing of many chemotaxonomical markers, for which microassays are now available (KÄMPFER et al., 1991).

3.2 Genetics and Genetic Manipulation of Streptomycetes

In streptomycetes the basic genetic material, the genome, is organized in a single circular chromosome as in other typical eubacteria: however, it has several specific and interesting features (CHATER and HOPWOOD, 1984; HOPWOOD and KIESER, 1990). It has a size of about 6 to 9 mega base pairs (Mbp). The chromosome of *Streptomyces coelicolor* A3(2) is estimated to be in the range of 7.5 to 8.0 Mbp by restriction mapping, which is almost double that of *Escherichia coli* and three times that of corynebacteria (cf. Tab. 1 and Sect. 4.2), and thereby it was also physically proven to be of circular nature (KIESER et al., 1992). The hyphal septae contain mostly several chromosomes physically organized in so-called nucleoids.

Genetic mapping by measurement of crossover frequencies in partial diploids, which are multiply allelic for several genetic loci, so-called heteroclones, allowed to establish a circular linkage map in several streptomycetes (CHATER and HOPWOOD, 1984; HOPWOOD and KIESER, 1990). One of the predominant basic characteristics of the genome is the functionally bipartite nature of the chromosome. The bulk of the essential genetic loci is assembled in two segments separated by large "silent" regions, which could make up almost half of the genome (Fig. 2). Frequently, parts of the genetic material needed for a common metabolic pathway are distributed, to both essential segments, and they are located in opposing positions (cf. example given in Fig. 2). This has led to the speculation that the chromosomal architecture is a result of a duplication event of a primordial streptomycete chromosome, which has been stabilized by unknown factors (HOPWOOD and KIESER, 1990). In later stages of development this doubled genetic material could have been functionally adapted by divergent evolution of genes, deletion events, and other rearrangements, such that two gene pools were created, an essential and more stable one and a more flexible one which could constitute the basis for adaptive variability.

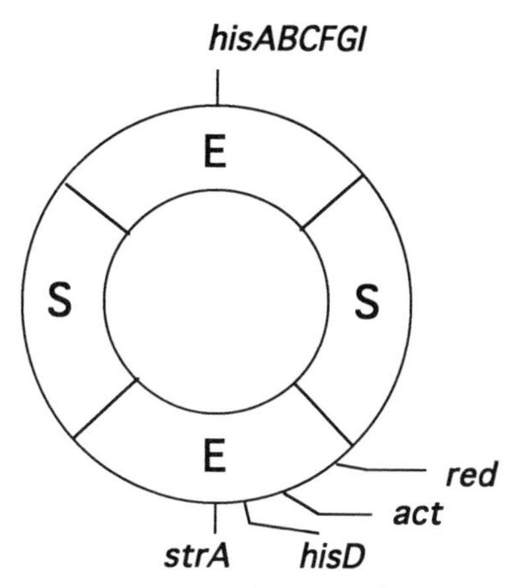

Fig. 2. The principal architecture of a streptomycete genome. The circular chromosome is functionally segmented into essential (E) gene assemblies and "silent" (S) regions. Some key genetic loci from the genetic map of *Streptomyces coelicolor* A3(2) are indicated. *his*, histidine biosynthetic genes; *str*, gene for ribosomal protein S12; *act*, actinorhodin biosynthetic gene cluster; *red*, prodigiosin biosynthetic gene cluster. For a detailed overview see HOPWOOD and KIESER (1990).

For many of the genes for secondary or "individualization" (see Sect. 3.4) metabolism, evidence has been presented that they are located in the silent sectors. Examples are the gene clusters for tetracycline production in *Streptomyces rimosus* (HUNTER and BAUMBERG, 1989), for methylenomycin in *S. coelicolor* A3(2) (HOPWOOD and KIESER, 1990), for chloramphenicol in *S. venezuelae* (VATS et al., 1987), or by circumstantial evidence also for 5′-hydroxystreptomycin in *S. glaucescens* (HÄUSLER et al., 1989). Some others, however, could be localized in the essential quadrants of the chromosome (see Fig. 2), as was shown, e.g., for the actinorhodin (*act* cluster) and prodigiosin (*red* cluster) production genes in *S. coelicolor* A3(2).

Laboratory-induced genetic instability via large deletions, up to more than 1 Mbp and/or large-scale tandem amplifications up to several hundred copies per chromosome of smaller DNA segments, seems to be another specific property of the genomes in many streptomycetes (CULLUM et al., 1989; HÄUSLER et al., 1989; ALTENBUCHNER and EICHENSEER, 1991; SCHREMPF, 1991; SIMONET et al., 1992). The mechanism(s) by which these deletions and amplifications arise have not been clarified as yet. Also, it is unknown whether they occur coordinately or independently and what exactly the inducing metabolic or other factors are. These properties can strongly affect the genetic stability of many strains, especially for differentiation, pigment and antibiotic production, or antibiotic resistance phenotypes. But primary metabolic traits can also be lost, such as the gene for the argininosuccinate synthase. Even considerable parts of such morphological and physiological markers used in species differentiation in *Streptomyces* can be lost or altered and, therefore, make the taxonomical relevance of these criteria questionable (SCHREMPF, 1991). The large and obviously contiguous deletions and other circumstantial evidence suggest that these large-scale alterations mainly occur in the silent quadrants of the genetic map (cf. Fig. 2; HOPWOOD and KIESER, 1990). In biotechnological terms these phenomena can be advantageous, e.g., the amplifiable units for the overproduction of gene products (cf. KOLLER and RIESS, 1987) and the deletions for avoiding unwanted physiological traits such as the production of melanin (HÄUSLER et al., 1989), or disadvantageous, e.g., because of the loss of valuable production properties by deletions. No generally applicable and straight-forward means, other than empirical media control or stable mutant selection for the individual strain, are currently available for avoiding genetic instability.

Another typical property of the genomic DNA of streptomycetes is the maximization of the G+C content which is constantly kept at values around 72 to 75 percent. Genes have the typically eubacterial characteristics and sizes (SENO and BALTZ, 1989). The high G+C content has led to an extreme codon bias. This in turn, formulated as a general rule in codon usage, which in G+C content is of intermediate concentration (ca. 70%) in the first codon position, lowest in the second (ca. 50%), and highest (ca. 90%) in the third codon position, can be used in gene identification. Other particular features, such as nearest neighbor preferences of bases in the context of a reading frame, the nearly complete avoidance of some codons, so-called rare codons (see below), and others, have also been observed (SENO and BALTZ, 1989). The eminent value of these findings for the detection and prediction of open reading frames in the DNA of streptomycetes has been proven in many cases.

Even unique characteristics of codon usage obviously result from the high G+C content, such as the possibility to use a rarely used codon (UUA) as a regulatory element in the control of differentiation and secondary or individualization metabolism (CHATER, 1992; see Sect. 3.4). Also, a gradient of enhanced and thereafter rapidly ceasing A+T content at the amino-terminal end of the first reading frames in transcription units was detected. This could mean that initiation of translation requires a less high G+C level (PISSOWOTZKI et al., 1991).

Beyond the chromosomal DNA, several other genetic elements were detected in streptomycetes, some of which such as plasmids and bacteriophages seem to be exceptionally frequent and wide-spread in nature for this particular bacterial group (CHATER and HOPWOOD, 1984; CHATER, 1986; HOPWOOD et al., 1986). Plasmids are mostly cryptic or exhibit a phenotype called "lethal zygosis" coupled with self-transmission via conjugation to plasmid-less receptor strains. These conjugative plasmids are frequent, use probably other transfer mechanisms than self-transmissible plasmids in Gram-negative bacteria, and mostly have broad host ranges within the family Streptomycetaceae. Several plasmids have now been studied in their molecular genetic properties; an example is pIJ101 (KENDALL et al., 1988). Giant linear plasmids have also been found, sometimes reversibly integrating into the chromosome, and in some cases could be concerned with antibiotic production phenotypes (KINASHI et al., 1992; HERSHBERGER et al., 1989).

The actinophages are mostly virulent and occur in each soil. Though all have been found to contain double-stranded DNA as nucleic acid, they are of very different structure, and some have the property to site-specifically integrate or to carry out generalized transduction (CHATER, 1986; VATS et al., 1987). A variety of transposons and insertion elements have been identified, which do not principally deviate in structure from those found in other bacterial groups (CHUNG and CROSE, 1990; KIESER and HOPWOOD, 1991). All these phenomena could support the view that genetic flexibility and the means of horizontal gene transfer are basic requirements for the life style of these soil organisms (see Sect. 3.4). No generally applicable protocols can be given to avoid or control actinophages in large-scale fermentations, other than the selection of phage-resistant mutants whenever a lytic virus occurs.

The transcriptional regulation of genes in streptomycetes seems to use the same general tools as in other bacteria (HÜTTER and ECKHARDT, 1988; SENO and BALTZ, 1989; LIRAS et al., 1990). However, there seems to be a larger variety of different promoter structures recognized by differential sigma factors (BUTTNER, 1989; TAKAHASHI et al., 1990). A recent compilation of 139 published streptomycete promoters lists only 29 of those with RpoD-like recognition sequences (STROHL, 1992; RpoD is the principal promoter-recognizing sigma factor of the RNA polymerase of *E. coli*). Special sigma factors together with several other regulatory mechanisms seem also to be engaged in the control of cell differentiation at the onset of sporulation and in antibiotic production (Fig. 3;

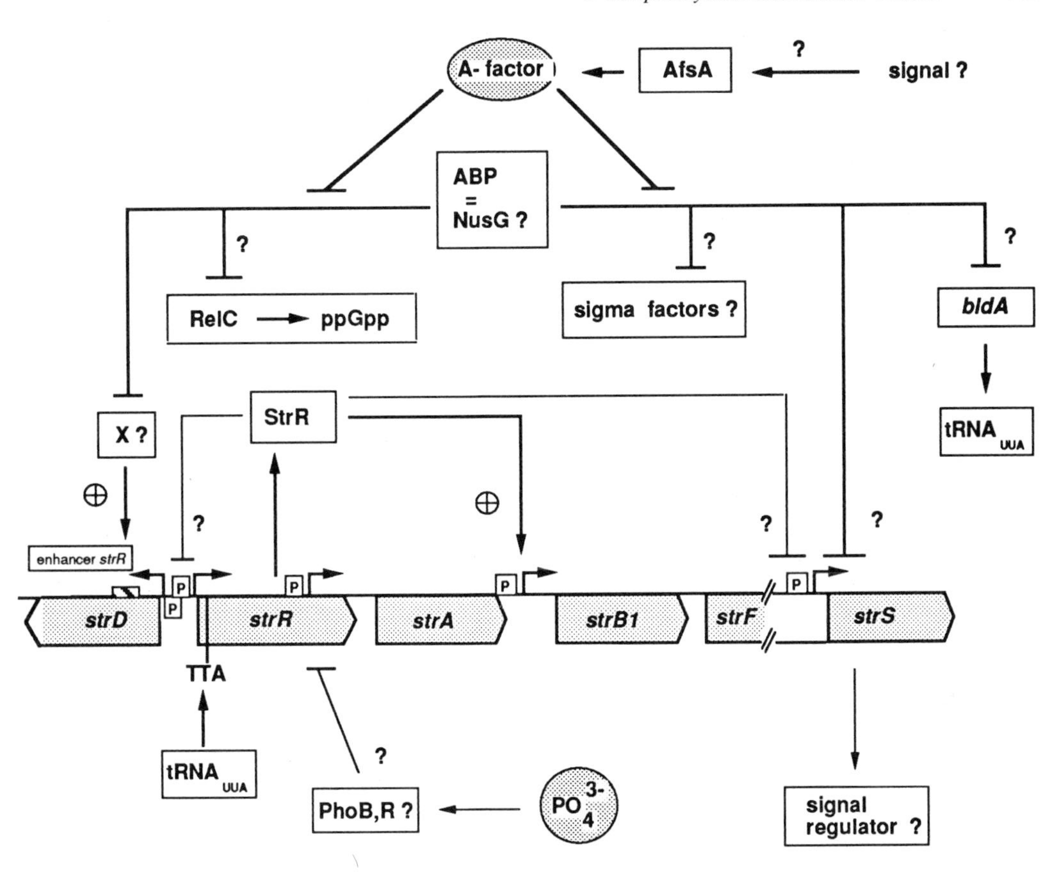

Fig. 3. Hypothetical regulatory network in a typical individualite (antibiotic) biosynthetic gene cluster. The central part of the gene cluster for streptomycin biosynthesis in *Streptomyces griseus* is shown, the regulation of which is subjected to the presence and concentrations of the autoregulator A-factor (AF; the generally distributed transcription factor NusG seems to be the AF-binding protein) and phosphate as globally regulating molecules, as well as to specific regulatory proteins encoded by the *strR* and *strS* genes. Other transcriptional-translational regulatory circuits seem to be dependent on either the presence of specific RNA polymerase sigma factors or on the rare TTA codon in N-terminal positions of crucial reading frames in connection with the growth phase-dependent activity of a leucyl-$tRNA_{UUA}$. The genes encode the following products: streptomycin 6-phosphotransferase (*strA*; resistance gene), scyllo-inosamine-4-phosphate amidinotransferase (*strB1*), dTDP-glucose synthase (*strD*), activator of *strB1* (*strR*), putative pleiotropic regulator (*strS*). The other symbols or abbreviations mean: AfsA, A-factor synthase; AfbP, A-factor binding protein; X, postulated A-factor and AfbP-dependent regulator of *strR*; *bldA*, $tRNA_{UUA}$ gene; PhoB,R, putative dual regulator of phosphate-controlled genes; RelA,C, stringent control factors involved in production of guanosine-tetraphosphate (ppGpp). For details, see DISTLER et al. (1992) and PIEPERSBERG (1992a).

CHATER, 1989; DISTLER et al., 1992; MCCUE et al., 1992). Further tools of regulation involve repressor/activator types of regulatory proteins, and dual-component sensor kinase/response regulator systems (HORINOUCHI and BEPPU, 1992). Diffusible autoregulators such as the A-factor and other compounds of the butyrolide class of bacterial hormones are typically observed in streptomycetes (KHOKHLOV, 1988; GRÄFE, 1989; HORINOUCHI and BEPPU, 1990). They are a special means to regulate differentiation and/or antibiotic production by governing switch phenomena during a decision phase in vegetative growth (DISTLER et al., 1990, 1992; MCCUE et al., 1992) (cf. Fig. 3; see also Sect. 3.4). Recently, the probable target site for the A-factor-related autoregulator virginia butyrolide C, a putative transcription control factor, was identified in *Streptomyces virginiae* to be related to the *E. coli* NusG protein (YAMADA et al., 1992). The homolog of NusG could also be effective in the streptomycin-producing and A-factor-dependent *S. griseus* strains (cf. Fig. 3) and other streptomycetes. Further substances, including proteins, with autoregulator-like or morphogenetic activities have been detected (KHOKHLOV, 1988; SZESZAK et al., 1991).

Another typical streptomycete regulatory component could be the rarely used codon (UUA), since a $tRNA_{UUA}$ (the product of the developmentally regulated *bldA* gene (CHATER, 1989; LESKIW et al., 1991; cf. Fig. 3) is needed only for developmentally regulated or antibiotic production genes. The BldA-tRNA or some factor coupled to it seems only to be active in *Streptomyces lividans* 66 in a certain growth phase, namely during the switch from vegetative growth to cell differentiation and secondary metabolism. Such genes with known vegetative expression patterns have never been found to contain TTA triplets in their respective reading frames. Taken together this might indicate that regulation takes place on both the transcriptional and translational levels, or even in steps which stringently couple these two expression levels both physically and via diffusible extra- and intracellular effectors (OCHI, 1990; DISTLER et al., 1992; BEPPU, 1992).

Genetic manipulation is well worked out now for streptomycetes, mainly due to the basic work of HOPWOOD and coworkers (HOPWOOD et al., 1985). These methods include mutant induction (BALTZ, 1986), protoplast formation and regeneration, construction, transfection, and transformation of phage and plasmid vectors (HOPWOOD et al., 1986; CHATER, 1986). Furthermore, more specific tools and more sophisticated methods are available, such as (1) promoter-probe and expression vectors (HOPWOOD et al., 1987), (2) vectors for integrative cloning and gene disruption or replacement (KIESER and HOPWOOD, 1991), (3) temperature-sensitive suicide vectors (MUTH et al., 1989), (4) transposon mutagenesis (CHUNG and CROSE, 1990; BALTZ et al., 1992), highly sensitive indicator systems for gene expression (SOBASKEY et al., 1992), or plasmid transfers avoiding protoplast transformation such as co-transduction (MCHENNEY and BALTZ, 1990) and mobilization in interspecific conjugation systems (MAZODIER and DAVIES, 1991).

Less generally developed, when compared to *Escherichia coli*, are the tools for homologous and heterologous expression systems, such as strong or inducible promoters (MURAKAMI et al., 1988; BRAWNER et al., 1990). Also, the now largely improving accessibility of signals for protein secretion from cloned genes encoding extracellular proteins, which differ in some minor respects from those of other bacterial groups (MANSOURI and PIEPERSBERG, 1991), constitutes a basis on which excretional expression systems can be developed (see Sect. 3.5).

Cloning systems for other industrially relevant euactinomycetes have also been reported, e.g., for *Micromonospora* sp. (HASEGAWA, 1992), or *Nocardia* sp. (HÜTTER and ECKHARDT, 1988).

3.3 Primary Metabolism of Streptomycetes

The primary metabolism, including the basic intermediary metabolism and macromolecular metabolism of replication, gene expression and cell wall formation, as well as its regulation are poorly studied in streptomycetes (VANEK et al., 1988). This probably results from the extremely one-sided interest in the so-called secondary metabolites (see Sect. 3.4). The resultant lack of knowledge about the metabolism during the vegetative phase (trophophase) was only filled when a direct and practical or general purpose initiated an investigation. This was mainly true for the acquisition of useful data for taxonomical identification (see Sect. 2), or for the control and yield improvement of fermentation processes (HOSTALEK et al., 1982), but also in the context of the investigation of cell differentiation processes (CHATER, 1989). However, a future understanding of the interactions between central and "individualization" (secondary) metabolism will focus research on this field. The main requirement for intensive investigations will be in two directions: (1) the unsolved question of precursor pool formation for a metabolite overproduced and excreted via a very complicated multi-step pathway, and (2) the metabolic and other control circuits governing the dynamics and switch phenomena between both metabolic compartments.

There is sufficient evidence that the streptomycetes and other euactinomycetes exhibit a typical eubacteria-like intermediary metabolism, especially related to other strictly oxygen-requiring chemoorganotrophs. The range of mono- and disaccharidic sugars and glycerol accepted as carbon sources suggests that their glycolytic (Embden–Meyerhof- and pentose phosphate pathways) and tricarboxylic acid pathways do not have anything in particular. Simple organic acids are usually not used as sole carbon sources (WILLIAMS et al., 1989). However, the presence of PTS sugar-uptake systems, which are generally used in facultative (e.g., *E. coli*) and anaerobic bacteria, has not yet been clearly identified, since conflicting data have been reported (HOSTALEK et al., 1982). Carbon catabolite regulation has been observed in many cases, and seems to involve regulation via intracellular cAMP-levels as in other bacteria (DEMAIN, 1989; VINING and DOULL, 1988). Under vigorous carbon catabolism α-ketoacids are excreted by many streptomycetes, causing decreased pH in the media during the vegetative phase (SHAPIRO, 1989).

Single enzymes, genes, or other components (e.g., components of the transcriptional and translational apparatus) of primary or central metabolism which were studied generally have significantly related counterparts in other bacteria. Some examples are: proline and tryptophan biosynthesis (HOOD et al., 1992); the histidine biosynthetic genes and enzymes (LIMAURO et al., 1990); the arginine biosynthetic genes and enzymes (PADILLA et al., 1991); the galactose operon and its regulation (BRAWNER et al., 1988); the glycerol utilization system (SMITH and CHATER, 1988).

Nitrogen is usually derived from ammonium or nitrate, or from amino acids such as arginine or asparagine. The nitrogen cycle seems to be also comparable to other bacteria (FISHER, 1992; SHAPIRO, 1989). Two glutamine synthetases, one of "eukaryotic type", seem to be common in streptomycetes and other actinomycetes. While the glutamine synthetase activity seems to be unaffected or repressed by ammonium, the glutamate synthetase and dehydrogenase activities vary largely with nitrogen source and/or concentration, e.g., they are high in presence of ammonium. The general regulatory mechanisms for the key enzymes of the nitrogen cycle, such as transcriptional control and regulation of enzyme activity, e.g., by reversible adenylylation of the glutamine synthetase, have also been demonstrated in streptomycetes, though species- or strain-specific differences may exist. Other nitrogen-regulated enzymes studied include alanine dehydrogenase which, however, does seem to play a major role in ammonium assimilation only in a limited number of strains, and nitrate reductase. Knowledge of amino acid uptake and catabolism is even less complete. Only the histidine utilization system has been studied in more detail which even can become the preferred route of entry for nitrogen in particular strains (WU et al., 1992).

Phosphate regulation is similarly observed as a widely used control element in streptomycetes, and it also seems to be involved in the switch phenomena between vegetative and "individualite" production phases (MARTÍN, 1989 a). The obvious occurrence of phosphate-control boxes in the DNA preceding phosphate-regulated genes caused the speculation that a similar dual component sensor kinase/response regulator system (PhoB/PhoR) exists in streptomycetes as found in *Escherichia coli* (LIRAS et al., 1990; DISTLER et al., 1990). Protein phosphorylation via protein kinases was also demonstrated in streptomycetes (STOWE et al., 1989).

3.4 Secondary Metabolism – Individualization Metabolism of Streptomycetes

The earlier used terms secondary metabolism/secondary metabolites, also "idiolites" or "natural products", are unfortunate, since they disregard or even deny a particular and essential function of the underlying biosynthetic pathways and their products (DAVIES et al., 1992). They were created primarily for the rare, genus- or species-specific, low-molecular-weight compounds of unknown function observed in many plants and microorganisms. Therefore, I here would like to introduce another term, "individual(ization) metabolism" ("individualites"), which should indicate that the products of this compartment of the overall metabolism are designed to individualize cell lines for multiple purposes, especially to protect them against their environment. This, in turn, is just one facet of the ubiquitous individual metabolism, as opposed to the central metabolism (encompassing primary, intermediary, and essential macromolecular metabolism), which is immanent to and distinguishing all living cells, e.g., the differentiated cells in the human body.

Therefore, we have to postulate two gene pools in all living beings, a stable one for the central metabolism, and a variable one for the individualization metabolism (PIEPERSBERG, 1992c). Also, we have to expect that there are strongly regulated links between the two metabolic compartments. The particular functions of specific traits or products of the individualization metabolism are manifold. Basically they all are designed for the cross-talk of the hosting cell with its environment, whether in form of intra- or interspecific cell-to-cell communication, as a means of participation in the competition for specific nutrients, or to colonize specific ecological niches. Two different categories of extracellularly made compounds, cell surface-attached (type A) or secreted and diffusible substances (type B individualites), could participate in these complex interactions. It seems likely that organisms exhibiting a sessile life style, differentiat-

Scheme 1. Possible Functions of Individualites (Secondary Metabolites)

Function(s)	Selective Advantage	Target(s) of Product
A. Early Evolution (molecular fossils)		
(1) **Ancient catalysts,** precursors of proteins, coenzymes, or else	Extinct (essential)	Some still existent? general ligands (substrate or else)
B. Late Evolution (recent inventions)		
(2) "**Selfish** metabolism", or products of "selfish DNA" activity	None	None to anything, specificity by chance
(3) **Hormones** (e.g., autoregulators, modulators of macromolecular synthesis, alarmones or pheromones, etc.)	Yes (reduced energy consumption via intraspecific cell-cell communication) for the producer itself via product	Receptors of signal transduction, for the intraspecific signaling
(4) **Repellants or attractants** (e.g., antibiotics, nodulation factors, etc.)	Yes (exclusion of competition versus symbiosis), for the producer itself via product	Essential complex/ receptors, for the interspecific signaling
(5) **Detoxification,** or **replacement** metabolism	Yes, for the producer itself; via metabolic shunt	None
(6) **Genetic reserve material,** "playground" of evolution	Yes, but not directly; for the offspring of all cell lines sharing a gene pool	None to anything under changing environmental conditions

ing into various cell types or organs, lacking an immune system, and exposing a large surface array to their environment, such as the filamentous and sporulating actinomycetes, fungi and plants, should have a more specific need for the production of type B substances for protection and hormone-like communication systems than movable or higher organisms with low surface exposure and a specific immune system. Since they cannot evade any situation of environmental confrontation, the former group of organisms should, therefore, also hold a larger gene pool for individualization metabolism, especially for type B individualites, than single-celled movable microorganisms. The latter organisms should be able to reach and contact their targets via the cell surface (type A individualites), or to evade environments with antagonistic effects. The evolution and the possible targets and meaning of low-molecular weight individualites in nature have often been discussed (PIEPERSBERG et al., 1988b; PIEPERSBERG, 1992c; VINING, 1992; DAVIES et al., 1992, and various contributions therein). A brief overview about six proposed solutions is given in Scheme 1. On the basis of our current rudimentary knowledge of microbial ecology in complex populations, the answers (3) and (4) in Scheme 1 are most likely to be in accord with the natural situation.

If we adopt the above view for the explanation of the highly variable and mostly strain-

Tab. 2. Molecular Biology of Gene Clusters for Actinomycete Bioactive Metabolites (Individualites) of Biotechnological Interest[a]

Antibiotic (Individualite)	Producer[b]	Genes in Cluster[c]	References
Actinorhodin	*S. coelicolor* A3(2)	B, R, P, T	HOPWOOD and SHERMAN (1990)
Granaticin	*S. violaceoruber*	B	HOPWOOD and SHERMAN (1990)
Oxytetracycline	*S. rimosus*	B, P, T	MCDOWALL et al. (1991)
Chlorotetracycline	*S. aureofaciens*	B, P, T	LOMOVSKAYA et al. (1991)
Tetracenomycin	*S. glaucascens*	B, R	HUTCHINSON (1992)
Daunorubicin	*S. peuceticus*	B, R	HUTCHINSON (1992)
Avermectin	*S. avermitilis*	B	MACNEIL et al. (1992)
Erythromycin	*Sacch. erythrea*	B, R, P, T	KATZ and DONADIO (1992)
Spiramycin	*S. ambofaciens*	B, R, P, T	KATZ and DONADIO (1992)
Tylosin	*S. fradiae*	B, R, P, T	SENO and BALTZ (1989)
Carbomycin	*S. thermotolerans*	B, R, P, T	EPP et al. (1989)
Nonactin	*S. griseus*	B, P	PLATER and ROBINSON (1992)
Streptomycin	*S. griseus*	B, R, P, E	DISTLER et al. (1992); Piepersberg (1992a)
5′-Hydroxy-streptomycin	*S. glaucescens*	B, R, P, E	DISTLER et al. (1992); PIEPERSBERG (1992a)
Fortimicin	*M. olivastereospora*	B, P	HASEGAWA et al. (1992)
Chloramphenicol	*S. venezuelae*	B	VATS et al. (1987)
Cephamycin C	*S. clavuligerus*, *N. lactamdurans*	B, R, P	QUEENER (1990), J. F. MARTÍN, personal communication
Phosphinothricin (bialaphos)	*S. hygroscopicus*,	B, R, P	THOMPSON et al. (1990)
	S. viridochromogenes	B, R, P	WOHLLEBEN et al. (1992)
Nosiheptide	*S. actuosus*	B, R, (P)	STROHL and CONNORS (1992)
Puromycin	*S. alboniger*	B, R, P	LACALLE et al. (1992)
Methylenomycin	*S. coelicolor*	B, R, T	CHATER and HOPWOOD (1989)
Lincomycin	*S. lincolnensis*	B, P, T	CHUNG and CROSE (1990); ZHANG et al. (1992)

[a] for an overview and further details see VINING, L. and STUTTARD, C. (Eds.), *Biochemistry and Genetics of Antibiotic Biosynthesis*, Stoneham: Butterworth-Heinemann; in preparation

[b] *S., Streptomyces; M., Micromonospora; N., Nocardia; Sacch., Saccharopolyspora*

[c] B, biosynthesis; E, extracellular processing; P, protection (resistance); R, regulation; T, transport (export)

specific formation of antibiotic-like individualites in streptomycetes, we possibly could find a basis for understanding the many pleiotropic phenomena governing their production, such as growth phase dependence (VINING and DOULL, 1988), nutrition and stress dependence (DEMAIN, 1989; SHAPIRO, 1989; VOTRUBA and VANEK, 1989), diffusible autoregulators (GRÄFE, 1989; HORINOUCHI and BEPPU, 1990; BEPPU, 1992), and internal control circuits (MARTÍN, 1989a; HUTCHINSON et al., 1989; DISTLER et al., 1990).

Currently the basics of individualization metabolism are being worked out, predominantly in the field of molecular biology of the biosynthetic pathways. This will lead to a collection of tools useful in the future design of pathways engineered by use of more sophisticated strategies (see Sect. 3.5). The individualites at present most intensively studied by means of molecular genetics and biochemistry are listed in Tab. 2. It turned out that in all cases the biosynthetic genes, even of mixed-type pathways, are clustered in gene assemblies of up to more than 100 kilobase pairs compact information (HUNTER and BAUMBERG, 1989; MARTÍN and LIRAS, 1989; SENO and BALTZ, 1989). These clusters not only include the biosynthetic genes, but also genes for regulation, resistance, export, and extracellular processing (see Tab. 2). Studies on protein and DNA relationships (hybridization) have shown that many of these genes, or parts thereof, are used in a modular fashion. That means that related tools can be employed to build up very different molecular structures. Some examples for modularly used enzymes are collected in Tab. 3. The first and best known of these are the polyketide synthases and their genes, which are highly conserved and even show DNA hybridization in many instances (MALPARTIDA et al., 1987). A further nice and even more wide-spread example might be the enzymes for the first steps in 6-deoxyhexose pathways occurring as branching parts in the biosyntheses of products from all chemical groups of actinomycete individualites (STOCKMANN and PIEPERSBERG, 1992). In the future this will largely facilitate the identification and speed up the analyses of other gene sets for individualites and their manipulation.

Another interesting phenomenon in connection with the individualization metabolism is its highly regulated supra-structure and

Tab. 3. Examples for Modular Design of Gene Products Used in Individualization (Secondary) Metabolism in Streptomycetes and Other Microorganisms[a]

Gene Product (Module)	Used in Pathway for (End Products)
Polyketide synthases, type II	Actinorhodin, granaticin, anthracyclines, tetracyclines, spore pigments
Polyketide synthases, type I	Avermectin, erythromycin, tylosin, carbomycin, spiramycin
Peptide synthases	β-Lactams, gramicidin, tyrocidin, actinomycin, phosphinothricin, lincomycin, nosiheptide
dTDP-glucose synthase (pyrophosphorylases) and dTDP-glucose dehydratase	Streptomycin, macrolides, anthracyclines, polyene macrolides, granaticin
dTDP-4-keto-6-deoxyglucose 3,5-epimerase	Streptomycin, avermectin, anthracyclines, polyene macrolides
PABA-synthases	Actinomycin, candicidin
Phosphine "synthases"	Phosphiniothricin
Export enzyme (Mmr-like, proton gradient-dependent)	Methylenomycin, tetracyclines, chloramphenicol, lincomycin, pristinamycin
Regulator (DegT/EryC1-like)	Extracellular hydrolases, erythromycin, streptomycin, puromycin, tylosin, daunorubicin, lipopolysaccharide O-chains
Antibiotic phosphotransferase	Aminoglycosides, viomycin
Antibiotic acetyltransferase	Aminoglycosides, puromycin
23S rRNA methyltransferase	Macrolides, lincosamides, thiopeptins
16S rRNA methyltransferase	Aminoglycosides

[a] for references see Tab. 2

built-in into the cell cycle and differentiation pathways. Complicated regulatory networks and the occurrence of extracellular and diffusible autoregulators (see Fig. 3) seem to govern individualite production and its coupling to certain switch phenomena in special growth phases (PIEPERSBERG et al., 1988a; PIEPERSBERG, 1992a; GRÄFE et al., 1989; MARTÍN, 1989a; BEPPU, 1992; DISTLER et al., 1990, 1992; YAMADA et al., 1992).

The antibiotic individualites are usually self-toxic for their producers. This requires the existence of resistance genes and the development of actively produced means of specific protection against these products, which occur in various types and mechanisms (PIEPERSBERG et al., 1988b; CUNDLIFFE, 1989). In many streptomycetes chromosomally encoded resistance determinants against not self-produced antibiotics, which are frequently cryptic or inducible, are resident, too. The original suggestion (BENVENISTE and DAVIES, 1973) that the plasmid-encoded antibiotic resistance genes, which are now wide-spread in clinically relevant groups of non-producing bacteria, originated from the actinomycete individualite producers can now be regarded as generally accepted.

Among the gene products inducing resistance two obviously large and wide-spread groups of (1) transmembrane-organized and proton gradient-driven (ZHANG et al., 1992), and of (2) ATP-dependent and membrane-bound (SCHONER et al., 1992) export enzymes are found. These export especially individualites with amphoteric chemistry such as (1) tetracyclines, lincosamides, methylenomycin, chloramphenicol, and pristinamycins, or (2) isochromanechinones (e.g., actinorhodin) and macrolides (e.g, erythromycin). The exporters causing resistance phenotypes are most likely designed to transport such compounds out of the cell, where they probably serve their intrinsic functions (see above), rather than being primarily resistance enzymes. In this context it is interesting to note that besides these export proteins, actinomycete individualite producers frequently express further resistance mechanisms against the same toxic compound (CUNDLIFFE, 1989; ZHANG et al., 1992).

3.5 Streptomycetes in Biotechnology

The predominant application of streptomycetes and related actinomycetes in industrial use is – and will in future be – the production of bioactive metabolites. They still account for about 2/3 to 3/4 of the newly described producers of natural products under the microorganisms. The numbers of published actinomycete "individualites" (secondary metabolites) per year is still rising and they total almost 10000 by now (Fig. 4; OMURA, 1992). Also the relative numbers, ca. 20% at present, of newly isolated bioactive microbial substances for other targets than the anti-infectives (see below) is rapidly increasing. Among the actinomycete genera found as producers, the following represent the most frequent, in decreasing order of instances mentioned: *Streptomyces*, *Micromonospora*, *Nocardia*, *Actinomadura*, *Actinoplanes*. The plants, which are the only competitors of bacteria for the first rank in overall product variation within the field of low-molecular-weight natural products, hold even higher numbers of individually described compounds (more than 40000), but these correspond to a more restricted number of chemical classes. The advantages of producers easily accessible to fermentation procedures give further weight to microorganisms, especially to the streptomycetes.

Recently most aspects and problems of the biotechnology of bioactive individualites produced by this group of organisms has been addressed in many reviews and books. Especially the screening strategies for both new producers (ZÄHNER et al., 1988; NOLAN and CROSS, 1988; GOODFELLOW and O'DONNELL, 1989) and products (ŌMURA, 1992), target-oriented testing (NISBET and PORTER, 1989), strain improvement by mutation (BALTZ, 1986; NORMANSELL, 1986; ELANDER and VOURNAKIS, 1986), the nutritional and physicochemical factors of control in fermentation processes (KELL et al., 1989; VOTRUBA and VANEK, 1989), and the possible impacts of molecular biology on most or all of these aspects (NISBET and PORTER, 1989) should be mentioned as currently deserving the highest attention.

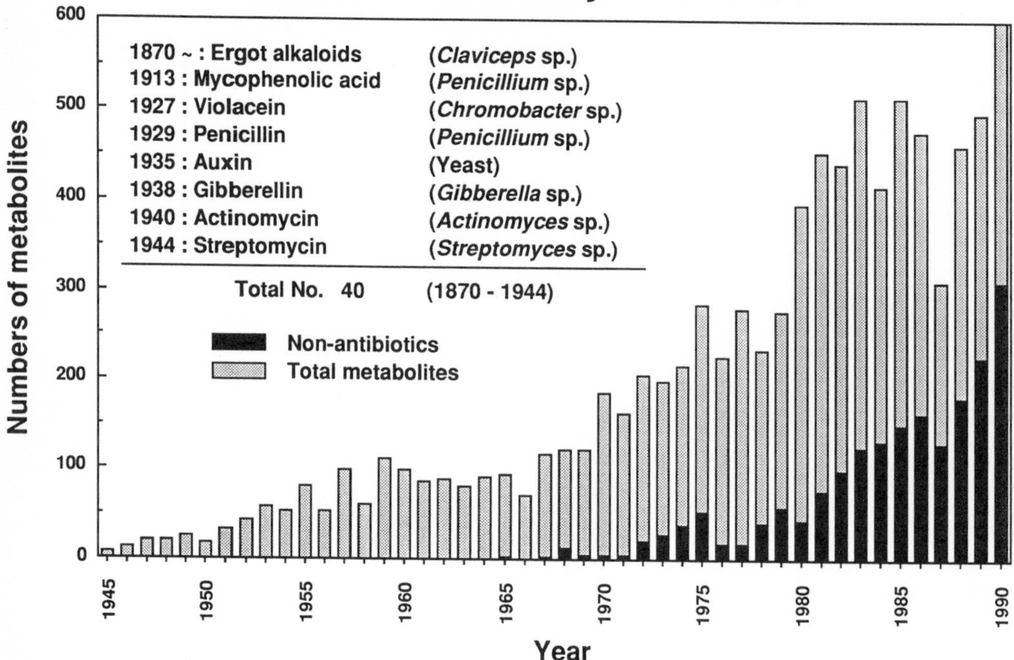

Fig. 4. The discovery of microbial bioactive compounds (individualites). This figure was kindly contributed by S. ŌMURA, The Kitasato Institute, Tokyo, Japan.

The basic philosophy underlying screening programs for biologically active individualites from actinomycetes currently employed by industrial companies and their targeting have considerably changed in the past two and a half decades since about 1965 (ŌMURA, 1992). The classical screening primarily selected antibiotic-like activities, including anti-microorganismal, antifungal, insecticidal, herbicidal, and cytostatic substances, which frequently were found in a common chemical family of compounds (Tab. 4). The modern screening programs are either strictly target-oriented (NISBET and PORTER, 1989), or non-oriented at all, as is the so-called chemical screening which looks primarily for new chemical structures (ZÄHNER et al., 1988), or a combination of both. An outline of several targets which are currently of major interest is given in Tab. 4. Further to be mentioned are the antiviral screens against human immunodeficiency virus (HIV) which, however, did not yet result in successful discoveries. With the increase in numbers of known individualites (see above), the problems of rational design and organization of screening programs increase. Examples for changing parameters in screening are the search for rare organismal groups, in extreme biotopes, or systematic trials on species richness as a function of sample number taken in the same biotope. Especially the need for identification of chemical structures, of target specificity, and for some basic toxicological data as early as possible in the process of searching for new bioactive molecules places a heavy burden on such developments. These will only to be coped with by improvement of specificity and automation of test systems, together with refinements of microchemical analyses for structure elucidation.

The strain improvement programs underwent similar, but less dramatic changes in the

Tab. 4. Old and New Screening Targets for Actinomycete Metabolites

Targets	Application as [a]	Selected References
Classical:		
Bacteria	Antiinfectant (P, A, O)	ŌMURA (1988); OKAMI and HOTTA (1988); NISBET and PORTER (1989); ŌMURA (1992)
Fungi	Antiinfectant (P, A, O)	NISBET and PORTER (1989); ŌMURA (1992)
Protozoa	Antiinfectant (P, A, O)	NISBET and PORTER (1989); ŌMURA (1992)
Viruses	Antiinfectant (P, O)	GHAZZOULI and HITCHCOCK (1988)
Mammalian cells	Antitumor (P, O)	
New:		
Immune cells	Immunomodulators (P)	NISBET and PORTER (1989); ASANO et al. (1989)
Immune cells	Antiinflammatory (P)	INOUYE and KONDO (1989)
Steroid production	Hormone control (P)	BUCKLAND et al. (1989); TOMODA and ŌMURA (1989)
Oncogene functions	Antitumor (P, O)	HORI (1988); UMEZAWA et al. (1989)
Cell cycle	Antitumor (P, O)	BEPPU and YOSHIDA (1989)
Protein kinases	Antitumor (P, O)	KASE (1988)
Tubulin assembly	Antifungal, -tumor (A, P)	IWASAKI (1989)
Cardiovascular (hypotension)	Vasoconstrictor (P)	BURG et al. (1988)
Cardiovascular (hypertension)	Vasodilator (P)	ANDOH and YOSHIDA (1989)
Reverse transcriptase	Antiviral, anti-HIV (P)	NAKAMURA and INOUYE (1989)
Alpha-amylase	Antidiabetes (P, O)	MÜLLER (1989)
Aldose reductase	Antidiabetes (P, O)	YAGINUMA et al. (1989)
Fatty acid oxidation	Antidiabetes (P, O)	KANAMURA and OKAZAKI (1989)
Lipoxygenase	Antiinflammatory (P)	KITAMURA et al. (1989)
Peptidases	Against many diseases (P, O)	AOYAGI and TAKEUCHI (1989)
Plant metabolism	Herbicide (A)	KIDA (1989)
Other enzymes	Enzyme inhibitors (P, A, O)	NISBET and PORTER (1989)

[a] (P), pharmaceuticals for human and animal use; (A), agricultural use (growth promoters in animal feed, herbicides, pest control); (O), other fields of application (research, diagnostics, etc.)

past by the introduction of more directed selection techniques in addition to the classical random mutation (BALTZ, 1986; ELANDER and VOURNAKIS, 1986). Examples for new selection methods are overlay bioassays, selection of resistance to self-toxic products or to analogs of rate-limiting intermediates, prototrophic reversion of auxotrophs, or protoplast fusions. This is exemplified by the following cases (reviewed in ELANDER and VOURNAKIS, 1986; NORMANSELL, 1986):

(1) Macrolide- and lincosamide-producing streptomycetes were found to enhance their production abilities significantly with relieving inducibility of resistance to constitutivity.

(2) A carbapenem-producing strain of *Streptomyces griseus* ssp. *cryophilus*, which originally produced a mixture of sulfated and non-sulfated end products, was modified subsequently by selection for selenate resistance and S-2-aminoethyl-L-cysteine resistance, which resulted in sulfate transport-deficient strains overproducing the non-sulfate compound C-19393 H_2 ten times. Reintroduction of the sulfate transport system into this strain by protoplast fusion resulted in an overproducer for the sulfated carbapenem C-19393 S_2.

(3) The selected relief of repressing regulatory factors should be mentioned which

was, e.g., achieved in the candicidin producer *S. griseus* by arsenate selection for phosphate-deregulated mutants. In the same strain resistance to analogs of the biosynthetic intermediate tryptophan also improved production significantly.

The impact of gene technology and the currently collected data on genetic organization, enzymology, and regulation of the secondary metabolic pathways (see Sect. 3.4) on the biotechnology of streptomycetes and related actinomycetes will soon become visible. An overview is given in Scheme 2, and first results indicate that:

Scheme 2. Biotechnological Perspectives in Individualite Biosynthesis

Strategies	Goals
1. "Genetic screening"	Diagnosis with distinct gene probes
2. Pathway engineering	Hybrid end products, biotransformations, newly designed products
3. Mutasynthesis	Planned pathway blocks and incorporation of modified intermediates
4. Overproduction	• Recognition of inducing factors • Enhanced gene dosage, especially of rate-limiting bottleneck enzymes • Manipulation of regulators (enhancement of activators; deletion of repressors) • Constitutive expression (?) • Enhancement of resistance?

(1) A "genetic screening" with DNA probes derived from indicator genes facilitates the isolation of gene clusters of related pathways and could help in future predictions of production potentials (MALPARTIDA et al., 1987; STOCKMANN and PIEPERSBERG, 1992).

(2) The formation of hybrid end products can be achieved by pathway fusions or single gene transfers (HOPWOOD et al., 1990; HUTCHINSON et al., 1989; STROHL and CONNORS, 1992). For instance, the transfer of the gene for an acyltransferase (*carE*) from the carbomycin producer *Streptomyces thermotolerans* to a spiramycin-producing strain of *S. ambofaciens* resulted in the formation of isovaleryl-spiramycin (EPP et al., 1989).

(3) Expression of latent genes after recombination, e.g., via protoplast fusions, might be another way of forcing strains to produce new metabolites (HUTCHINSON et al., 1989).

(4) Overproduction could be induced in a designed fashion by various strategies, such as the manipulation of regulators (CHATER, 1989; HUTCHINSON, 1992). An example is given by enhancing the gene copies for a putative two-component regulatory system via self-cloning in the daunorubicin producer *S. peuceticus* resulting in higher yields of intermediates or end products in various transformants (STUTZMAN-ENGWALL et al., 1992). Another way is key enzyme overproduction via enhanced gene dosage to relieve rate-limiting "bottlenecks" (HUTCHINSON et al., 1991). This strategy has already been very successfully used in the improvement of fungal cephalosporin producers (QUEENER, 1990): The expandase, the first enzyme after the common beta-lactam pathway branches off towards the cephalosporins and cephamycins from that leading to the penicillins, is such a rate-limiting enzyme. Its self-cloning in highly cephalosporin-producing strains of *Cephalosporium acremonium*, only doubling the gene copy number, improved the antibiotic production significantly by 15 to 40%. Enhancing the copy number of the isopenicillin N synthetase gene, which is in the common beta-lactam pathway, did not result in enhanced production rates. The genes/enzymes for engineering the beta-lactam pathways, in both fungi and bacteria, will certainly be taken from actinomycete sources (e.g., the producers of clavulanic acid and cephamycin C, *Streptomyces clavuligerus* and *Nocardia lactamdurans*), since they are non-spliced and

offer a much more varied pool of biochemical functions than those from the fungal producers. Third, the transfer of the whole genetic complement for biosynthesis to another production organism with advantageous properties could be a future perspective for the same purpose, whenever a strong physiological barrier hinders classical strain improvement techniques.

(5) Special future interest in pathway engineering should focus on the routes and the regulation of precursor pool formation (PIEPERSBERG, 1992b). Almost nothing is known about the regulatory networks connecting C- and energy source conversion with individualite biosynthesis during the production phase. Even specific genes (also coupled to the individualite biosynthetic gene clusters?) or enzymes could be involved. This is suggested, e.g., by the observations (a) that even the genes for nucleotide activation of glucose are frequently part of biosynthetic gene clusters where modified sugars are involved (PISSOWOTZKI et al., 1991; STOCKMANN and PIEPERSBERG, 1992), and (b) that certain branched-chain amino acids predominantly become degraded for precursor pool formation in macrolide-producing streptomycetes (ŌMURA and TANAKA, 1986).

(6) The ultimate goal of biotechnological application of molecular biology in the field of the development of production strains for low-molecular weight bioactive molecules would be the planned drug design via pathway and protein engineering techniques, as it is now generally achieved by chemical synthesis or modification (PIEPERSBERG, 1992b). An interesting basis for such a targeted redesign of the biosynthetic pathways are the complex polyketide synthases, e.g., involved in the formation of erythromycin or avermectin, or the highly variable patterns of 6-deoxyhexose modifying enzymes occurring in many antibiotic pathways (KATZ and DONADIO, 1992; MACNEIL et al., 1992; STOCKMANN and PIEPERSBERG, 1992). For instance, the 5,6-dideoxy-5-oxo or the 3-deoxy-3-oxo forms of erythronolide B could become first examples of such tailor-made products (KATZ and DONADIO, 1992). Other forms of even more advanced pathway engineering might follow.

Some problems of genetically engineered production strains of streptomycetes could result from structural instabilities of cloning vectors, especially of bifunctional vectors (PIGAC, 1991) or during elongated or continuous fermentation processes (ROTH et al., 1991). Integrational cloning and expression systems might therefore be more important in the future (KIESER and HOPWOOD, 1991).

Other fields of application of streptomycetes are envisaged or already explored. Among these, in the first place, the production of various technical enzymes and of enzymes for use in medicine and research has to be mentioned (PECZYNSKA-CZOCH and MORDARSKI, 1988). A list of cloned genes for streptomycete enzymes with a potential for use, and which are excreted into the medium, is given in Tab. 5. Many of these genes have been heterologously expressed in other streptomycete hosts, mostly derivatives of *S. lividans* 66 (for references see Tab. 5). Increased yield of the products after self-cloning into the original overproducer, from which the gene had been cloned, was reported for a lysozyme-producing *S. coelicolor* "Müller" (BRÄU et al., 1991). Heterologous excretional gene expression of human proteins for pharmaceutical application has also been achieved for the CD4 receptor (BRAWNER et al., 1990), human proinsulin (KOLLER et al., 1991), and interleukin-2 (BENDER et al., 1990). The only technical bulk enzymes of streptomycete origin marketed at present are the xylose (glucose) isomerases from various strains, which are used in large quantities in syrup (fructose) production, and the "pronase" complex from *Streptomyces griseus*. Small-scale production of enzymes and other proteins (e.g., proteinaceous enzyme inhibitors) have been reported for many more cases, such as restriction enzymes, other nucleases, proteinase and amylase inhibitors. The application of these is mainly for diagnostic and research purposes. The actual trends of

Tab. 5. Cloned Extracellularly Secreted Enzymes and Other Proteins from Streptomycetes with a Potential for Biotechnological Application[a]

Catalytic Activity[b]	Producer	References
Agarase	*S. coelicolor* A3(2)	KENDALL and CULLUM (1984); BUTTNER et al. (1987)
β-Galactosidase	*S. lividans 66*	BURNETT et al. (1985)
Protease A, serine-p.	*S. griseus*	HENDERSON et al. (1987)
Protease B, serine-p.	*S. griseus*	HENDERSON et al. (1987)
Trypsin, serine-p.	*S. griseus*	A[c]
Metalloprotease	*S. lividans 66*	LICHTENSTEIN et al. (1992
Metalloprotease	*S. cacoi*	CHANG et al. (1990)
Aminopeptidase	*S. griseus*	A
Carboxypeptidase, metallo-c.	*S. griseus*	A
α-Amylase	*S. hygroscopicus*	MCKILLOP et al. (1986)
α-Amylase	*S. limosus*	LONG et al. (1987)
Pullulanase	*S. lividans 66*	HÄNEL et al. (1991)
Xylanase	*S. lividans 66*	KLUEPFEL et al. (1991)
Cellulase (endoglucanase)	*S. lividans 66*	THEBERGE et al. (1992)
Chitinase	*S. lividans 66*	MIYASHITA et al. (1991)
Chitinase	*S. plicatus*	ROBBINS et al. (1992)
Alkaline phosphatase (StrK)	*S. griseus*	MANSOURI and PIEPERSBERG (1991)
Endo-β-N-acetylglucosaminidase H	*S. plicatus*	ROBBINS et al. (1984)
Lysozyme	*S. coelicolor "Müller"*	BIRR et al. (1989); BRÄU et al. (1991)
β-Lactamase	*S. badius, S. cacoi, S. fradiae, S. lavendulae*	FORSMAN et al. (1990)
Cholesterol oxidase	*S.* sp. SA-COO	ISHIZAKI et al. (1989)
Esterase	*S. scabies*	RAYMER et al. (1990)
α-Amylase inhibitor, tendamistat	*S. tendae*	KOLLER et al. (1991)
α-Amylase inhibitor, Haim-II	*S. griseosporeus*	NAGASO et al. (1988)
Proteinase inhibitor LEP-10	*S. lividans 66*	BRAWNER et al. (1990)
Proteinase inhibitor LTI	*S. longisporus*	BRAWNER et al. (1990)
β-Lactamase inhibitor	*S. clavuligerus*	DORAN et al. (1990)

[a] for an overview on actinomycete extracellular enzymes see PECZYNSKA-CZOCH and MORDARSKI (1988), on peptide enzyme inhibitors AOYAGI and TAKEUCHI (1989), UMEZAWA (1988)
[b] p, proteinase; c, carboxypeptidase
[c] A, C. ROESSLER and W. PIEPERSBERG, unpublished

development in these areas are generally based on testing a much broader organismal spectrum than in the production of bioactive molecules. Therefore, streptomycetes have to compete with other unrelated groups among the Gram-positive eubacteria (cf. Sect. 4.4), and the outcome of current and future evaluations, which hosts prove to be the best choice for extracellular protein production, is mostly open at present.

Biotransformation systems for antibiotics and other compounds (SEBEK, 1986) are not a major domain of the streptomycetes. However, their large and variable potential for different metabolic traits, especially in coping with many of the environmentally appearing organic substrates, could make them also of interest for such applications as steroid transformations. An example that this could be the case is the extracellular cholesterol oxidase produced by *Streptomyces* sp. SA-COO, which is used for diagnostic detection of cholesterol levels in serum (ISHIZAKI et al., 1989). A related enzyme was also detected in the coryneform *Brevibacterium sterolicum* (OHTA et al., 1991). Production of primary metabolites was rarely developed in streptomycetes, however, the industrial-scale production of vitamin B 12 in *S. olivaceus* should be mentioned as an exceptional case (BUSHELL and NISBET, 1981). Further applications of streptomycetes might come up in the future in agriculture, forestry, waste decomposition, and pollution control (LECHEVALIER, 1988; CRAWFORD, 1988; LACEY, 1988).

4 Corynebacteria and Related Genera

4.1 Ecology, Isolation, and Culturing of Corynebacteria

Although there are no systematic searches and estimates of cell numbers in different habitats available for this group of eubacteria, a wide-spread or even ubiquitous occurrence of *Corynebacterium* and related genera has to be assumed from the available data (LIEBL, 1991). Corynebacteria have been found in soil or aquatic habitats, on plants, in feces, on animal or human skin or in dairy products, e.g., on cheese surfaces. The non-specific growth requirements and physiology makes it difficult or impossible to design selective media for primary isolation, cultivation, and storage of representative members of this group. Rich to very rich media, in some cases containing growth factors such as biotin, thiamine, or *p*-aminobenzoic acid have been suggested. Preservation at –70 °C in the presence of glycerol (20%) or dimethylsulfoxide (7%) is suitable for long-term storage.

Taxonomical identification and differentiation are mainly based on the criteria listed in Tab. 1 (see above), where the coryneform morphology, the facultatively anaerobic growth, and the mycolic acid, *meso*-diaminopimelic acid (mDAP) and arabinogalactan content in the cell walls certainly are the dominant factors for classification (LIEBL, 1991). It is worth while to note that the biotechnologically relevant species have stable properties, and that the same species frequently can be re-isolated from natural samples, such as *C. glutamicum*, all isolates of which naturally produce large amounts of glutamate when grown aerobically on glucose.

4.2 Genetics and Genetic Manipulation of Corynebacteria

Genetic analysis and manipulation by biochemical methods of genetic engineering in corynebacteria have been started only recently. However, all the necessary tools, plasmids, bacteriophages, and transposons, have been detected and developed for use by now (MARTÍN, 1989b; SONNEN et al., 1992). Difficulties in these processes are envisaged in the rigidity of the mycolic acid-containing cell walls, which are highly resistant to the attack of lysozyme. Therefore, the conditions of formation and regeneration of protoplasts, an inevitable prerequisite for the isolation of plasmid and genomic DNA and of transformation, has to be worked out anew for each strain. Plasmid vectors for the expression of genes, excretion of proteins, and for insertional mutagenesis have been constructed and tested (MARTÍN et al., 1990; LEBLON et al., 1990; SCHWARZER and PÜHLER, 1991). Plasmid transfer can also be achieved by mobilization from *Escherichia coli*, avoiding protoplasting and transformation (MAZODIER and DAVIES, 1991).

Analysis of essential genes, especially of those encoding enzymes of the amino acid biosynthetic pathways (see Sect. 4.4), is easiest in *Escherichia coli* K12. This is because most or all of these genes seem to be expressed from their own promoters in *E. coli* (MARTÍN, 1989b; LIEBL, 1991) and since *E. coli* mutants for almost all steps of amino acid biosynthesis are available.

4.3 Primary Metabolism of Corynebacteria

In accord with the use of *Corynebacterium glutamicum*, *C. ammoniagenes*, *Arthrobacter* sp., and *Brevibacterium* sp. in the production of primary metabolites (see Sect. 4.4), the main current focus in biochemical and molecular genetic work is their primary metabolism. Here especially the topics of organization and regulation of amino acid and nucleotide biosynthesis, as well as the transport phenomena (uptake versus export), associated with the passing of essential constituents of cellular material through the cytoplasmic membrane, will have to be treated mainly.

A general view on the corynebacterial metabolism is difficult, since none of the strains in use were investigated to a similar extent as other single-celled eubacteria, e.g., *Escherichia coli*, *Bacillus subtilis*, *Pseudomonas putida*, or *P. aeruginosa* (see LIEBL, 1991). The predominant use of the glycolytic pathway (Embden-Meyerhof pathway) and tricarboxylic acid with built-in glyoxylic acid cycles as routes of the primary and energy-yielding carbohydrate consumption have been established. The use of acetate, various other organic acids, and ethanol as sole carbon source has also been observed. Use of ribose and gluconate via the pentose phosphate pathway is also possible. Sugars are probably taken up via the usual phosphoenolpyruvate-dependent sugar phosphotransferase (PTS) systems (MORI and SHIIO, 1987). Almost no data have been reported on the particular routes enzymes, and conditions of the fermentative energy metabolism under anaerobic growth conditions in amino acid-producing corynebacteria.

The organization and regulation of amino acid biosynthetic pathways attracted the largest interest of biochemical and genetic research in this bacterial group in the past two decades (literature reviewed in LIEBL, 1991). In this context it turned out that their components and controlling circuits are much more simple than in other bacteria having larger genomes and a more complicated physiology, such as *E. coli*. For example, the structure and regulation of the biosynthetic pathway leading to the aspartate family of amino acids (L-lysine, L-threonine, L-methionine, and L-isoleucine), which has been studied extensively in *Corynebacterium glutamicum* and *Brevibacterium flavum* (Fig. 5). In comparison to the respective pathway in *E. coli*, it is more simple in various respects:

(1) only one aspartate kinase gene/enzyme (*E. coli* has three isozymes with different allosteric properties),
(2) lack of transcriptional control by most end products (only L-methionine regulates on this level the homoserine branch of the pathway),
(3) the aspartate kinase is only feedback-regulated by L-lysine and L-threonine,
(4) the enzymes in the L-lysine pathway, which in contrast to *E. coli* and most other bacteria mainly proceeds via the diaminopimelate dehydrogenase reaction (saving three further enzymes relative to *E. coli*), are not regulated at all.

However, *C. glutamicum* additionally has the complement of genes/enzymes for the succinylase variant of diaminopimelate formation such as *E. coli* (SCHRUMPF et al., 1991). Similar simplifications are also found in the pathways of the aromatic amino acids (ITO, 1990) and the L-isoleucine/L-leucine/L-valine family of amino acids (CORDES et al., 1990).

The study of the secretory pathways and their underlying transport mechanism will be of equally high importance in our future understanding of the bio-processes leading to the excretion of amino acids and other primary metabolites from corynebacterial cells. In those cases studied in corynebacteria, efflux does not seem to be mediated by passive diffusion driven by an outward-directed gradient nor by the inverse of existing energy-dependent uptake systems. Rather, it was shown for the amino acids L-glutamate (HOISCHEN and KRÄMER, 1990), L-lysine (BRÖER and KRÄMER, 1991a, b), and L-isoleucine (EBBIGHAUSEN et al., 1989) that strictly substrate-specific and energy source-coupled export systems seem to be involved. In contrast, the active uptake system for L-glutamate is specifically inducible by L-glutamate, it is only expressed under conditions where the compound is employed as sole energy and carbon source, and is catabolite-repressible by glucose (KRÄMER et al., 1990).

4.4 Corynebacteria in Biotechnology

The corynebacteria became the classical amino acid producers after the detection in the late fifties that some strains of *Corynebacterium glutamicum* naturally excreted large amounts of L-glutamic acid when fed with glucose (KINOSHITA et al., 1957). This led to an

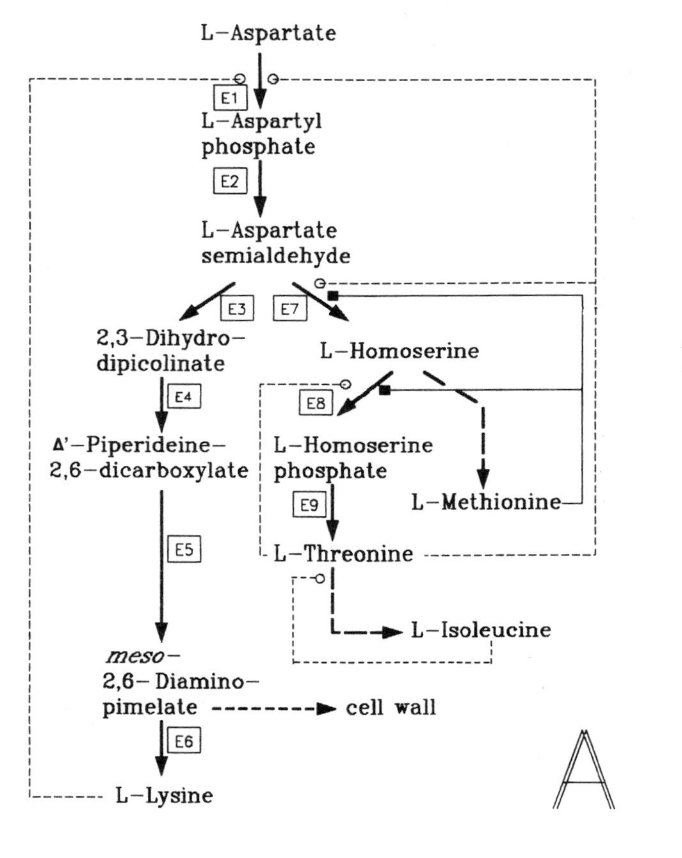

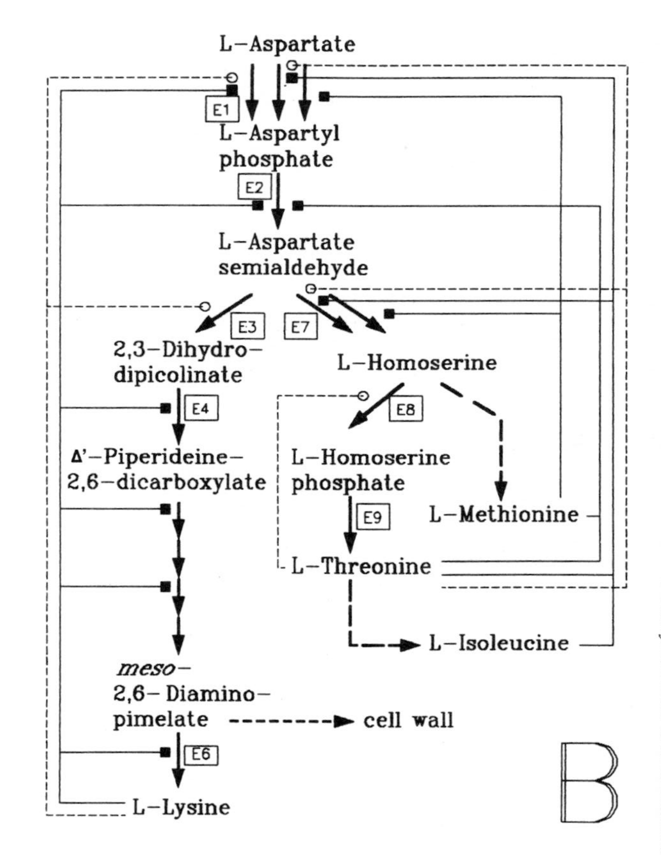

Fig. 5. Comparison of the pathways for the aspartate family of amino acids and their regulation in corynebacteria (A) and *Escherichia coli* (B) (taken from LIEBL, 1991, with permission). Regulation for the methionine and isoleucine branches of the pathway are not shown. Dashed lines ending in circles indicate enzyme inhibition by feedback control; thin lines ending in filled squares mean repression of gene expression. The indicated enzymes (E) are: E1, aspartate kinase; E2, aspartate semialdehyde dehydrogenase; E3, dihydrodipicolinate synthase; E4, dihydrodipicolinate reductase; E5, dioaminopimelate dehydrogenase; E6, dioaminopimelate decarboxylase; E7, homoserine dehydrogenase; E8, homoserine kinase; E9, threonine synthase.

extensive screening and breeding program for developing strains for the production of other L-amino acids, later also of vitamins and nucleotides, in this and other coryneform species (KINOSHITA, 1985). It turned out that the natural excretion of L-glutamic acid is an exceptional phenomenon, also in its physiology (see below). All other amino acids can only be produced in mutants with distinct alterations in particular biosynthetic pathways or in their regulation. The currently applied organisms of this group and their products are listed in Tab. 6. More and more it is found that competitive mutant producers of amino acids and other anabolic primary metabolites can also be selected in completely different bacterial and fungal (yeasts) groups (NIEDERBERGER, 1989). This can even be the enterobacterium *Escherichia coli.* However, the advantage of the corynebacteria is the generally simpler way of controlling their amino acid biosynthetic pathways relative to other bacteria (see Sect. 4.3).

The glutamic acid production in *C. glutamicum* has several unusual features and prerequisites which affect its biotechnological application. Firstly, the strong dependence on the limitation of biotin, an obligate growth factor for this species, in culture fluids is a long-discussed issue, but still a matter poorly understood (KINOSHITA, 1985). Oleate, C_{16}–C_{18} saturated fatty acids, penicillin, or surfactants can replace biotin in its effect. This suggests that alterations of membrane permeability due to changes in lipid composition, together with low activity of the α-ketoglutarate dehydrogenase complex, were the factors responsible for high-level L-glutamate excretion ("leak model"). Another hypothesis favored the view that the specific uptake system for L-glutamate in *C. glutamicum* functioned in the opposite direction under certain conditions ("inversion model"). These hypotheses were questioned recently, and new findings support the view that a glutamate-specific export system, non-identical with the uptake carrier, is present in this species (HOISCHEN and KRÄMER, 1990; KRÄMER et al., 1990).

Earlier mutant selection was mainly based on selection of blocks in unwanted pathways and feedback-deregulated alterations in the allosteric forward enzymes via suitable amino acid analogs (KINOSHITA, 1985). Breeding programs today largely involve genetic engineering for the detection and manipulation of the key-step biosynthetic enzymes, a large number of genes for which have already been cloned and analyzed (KATSUMATA et al., 1987; MARTÍN, 1989b; FOLLETTIE et al., 1990; ITO, 1990; CORDES et al., 1990). Recent strain constructions indicate that the pathway design via recombinant DNA technology is promising, and further enhancement of production levels in high-producing mutants can be achieved (see below).

The current strategies in this process of a strain improvement program by use of combined classical and genetic engineering methods will be exemplified here for the production

Tab. 6. Amino Acids Produced in *Corynebacterium glutamicum* and Related Coryneform Bacteria (for an overview see KINOSHITA, 1985)

Strain Derivation	Products	Organisms Involved
Wild type	L-Glutamic acid	*Corynebacterium glutamicum*
Mutants	L-Arginine L-Isoleucine L-Histidine L-Leucine L-Lysine L-Ornithine L-Phenylalanine L-Proline L-Threonine L-Tryptophan L-Tyrosine	*C. glutamicum, C. lilium, C. caleunae, C. acetoacidopohilum, Brevibacterium flavum, B. lactofermentum, B. divaricatum, B. immariophilum, B. thiogenitalis, B. reseum, Microbacterium ammoniaphilum*

of aromatic amino acids (Fig. 6; ITO, 1990; IKEDA and KATSUMATA, 1992). The classical procedure was, until recently, to enhance and channel the metabolic flux through the pathway via combinations of the following approaches:

(1) exclusion of feedback control by all three end products by mutation in the enzymes DS, AS, PRT, or CM (see legend to Fig. 6)
(2) block(s) by mutation in the branching pathway(s) leading to the unwanted end product(s)
(3) relief of the repression of expression of the enzymes encoded by the tryptophan operon by its end product, or
(4) removal of the repression of the enzymes of the upper aromatic pathway by tyrosine, or
(5) of the PD by phenylalanine.

An example of a strain of *Corynebacterium glutamicum* with similar alterations producing 18.1 g tryptophan per liter was reported (IKEDA and KATSUMATA, 1992; cf. Fig. 6). It has been successfully tried to superimpose further metabolite flow-enhancing properties onto such

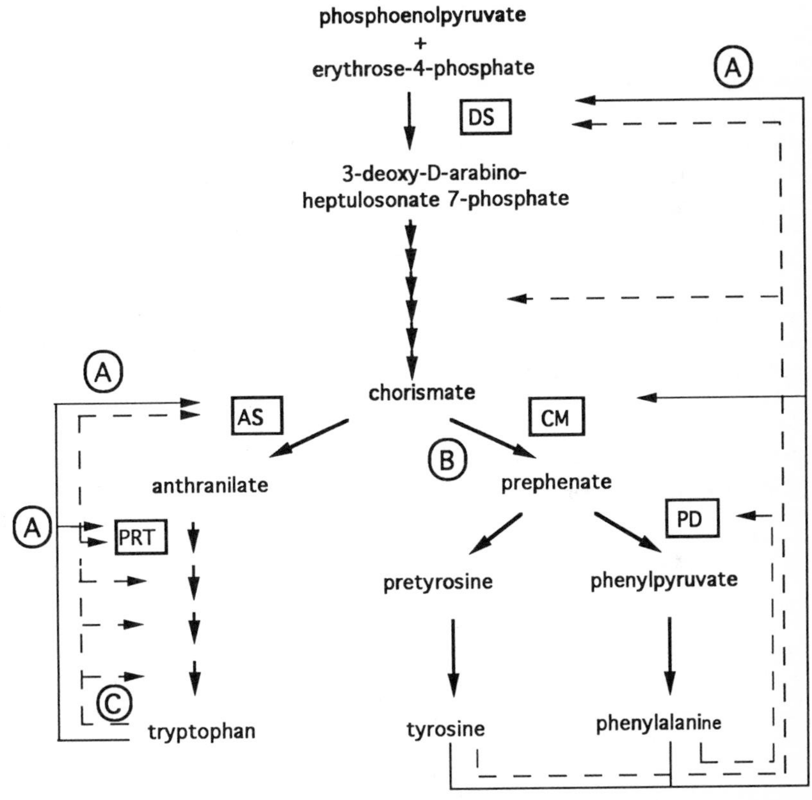

Fig. 6. The biosynthesis and regulation of aromatic amino acids in *Corynebacterium glutamicum* (modified from IKEDA and KATSUMATA, 1992). Negative regulation by feedback inhibition (full-lined arrows) and on the level of transcription (broken-lined arrows) is indicated. Regulated key enzymes are DS, 3-deoxy-D-arabino-heptulosonate 7-phosphate synthase; AS, anthranilate synthase; PRT, anthranilate phosphoribosyltransferase; CM, chorismate mutase; PD, pre-phenate dehydratase. As an example the encircled letters mark the positions where an optimal, mutationally bred overproducer of L-tryptophan must be (A) desensitized from product inhibition by mutation, (B) concomitantly blocked in the phenylalanine and tyrosine branch of the pathway, and (C) relieved from repression of gene expression.

strains by gene technology. The prerequisites and strategies for this are (a) cloning of the mutant feedback-deregulated key enzymes, and (b) enhancement of "bottleneck" (e.g., the DS, AS, PRT, or CM, and PD in our example; cf. Fig. 6) or of all enzyme activities by raising the gene dosage for the respective gene(s) after recloning of these genes into the high-producing strains.

When the genes for the end product-desensitized DS and the branch-point genes encoding the CM and PD (both also feedback-insensitive against L-tyrosine and L-phenylalanine) were assembled on a multicopy vector and reintroduced into the above mentioned L-tryptophan overproducer, this strain was switched to produce large amounts of phenylalanine (28 g/L). However, when the same strain was transformed with a plasmid containing only the desensitized DS and CM encoding genes, it was altered to overproduce L-tyrosine (26 g/L) and some phenylalanine (IKEDA and KATSUMATA, 1992). The cloned enzymes were simultaneously overexpressed about 7-fold in both strains. This high activity of the branch-point enzyme CM, deficient in the tryptophan-producing receptor strain, rechanneled the common intermediate chorismate into the L-tyrosine and/or L-phenylalanine pathways.

A similar example has been reported for the redesign of a L-threonine production strain on the basis of an L-lysine overproducer in *C. glutamicum* (KATSUMATA et al., 1987). In the future, further and advanced improvement strategies for metabolic flow will probably involve protein engineering for design of desired properties of biosynthetic enzymes, stable amplification of relevant genes in the chromosomes of producer strains, and the engineering of export systems (cf. Sect. 4.3) for amino acids and other suitable primary metabolites.

A second field of primary metabolite synthesis in corynebacteria is the production of the nucleotides IMP, XMP, and GMP as flavor enhancers in mutants of *Corynebacterium glutamicum* and *C.* (prior *Brevibacterium*) *ammoniagenes* by direct fermentation or by salvage synthesis from fed precursors (KUNINAKA, 1986). These strains are bred by very similar means as the amino acid overproducers, namely by the successive introduction of combinations of auxotrophic, analog-resistant, cofactor-insensitive, and leaky mutations into suitable wild-type strains.

Other applicable potentials or industrial uses of coryneform bacteria have been reported (cf. also MARTÍN, 1989b), such as their use in the cheese industry, or in biotransformations of steroids and terpenoids, or the production of emulsifiers. In some corynebacteria even the production of antibiotics of the chloramphenicol group was observed. The degradation of haloalkanes has also been reported (YOKOTA et al., 1987). Recently, the secretional production of extracellular enzymes in corynebacteria was suggested, since Gram-positive bacterial hosts are considered to be the better and simpler protein excreters due to the lack of an outer membrane, a barrier hindering this process in Gram-negatives, and cloning vectors suitable for protein excretion have been constructed (MARTÍN, 1989b; LIEBL, 1991; LIEBL et al., 1992).

5 Conclusions and Perspectives

Both bacterial groups treated in this chapter have been used in classical biotechnology since the introduction of sterile large-scale fermentation technology in the 40s and 50s of this century, and they will also in the future be of high impact for further developments in this field. It is clear that the still existing or even increasing need for specifically targeted pharmaceuticals, for food and feed additives, for diagnostics and technical enzymes makes streptomycetes, corynebacteria and related bacterial genera inevitable means in biotechnology. The record of long-term experience in GILSP methodology and GRAS standards existing for many different strains in both groups should also not be neglected in this context. However, the highest future value will primarily lie in the metabolic tools these bacteria provide to biotechnologists, the streptomycetes in the field of individualite (secondary metabolite) production, and the corynebacteria for primary metabolite formation. The basics, biochemistry and genetics of pathways and their regulation, are currently worked out with increasing speed. Their application already started with more advanced technology in pathway engineering in sharp focus. Other fields of application, such as heterologous protein production and biotransformation systems, have already shown promising results, and further developments can be expected.

6 References

ALTENBUCHNER, J., EICHENSEER, C. H. (1991), A new system to study DNA amplification in *Streptomyces lividans*, in: *Genetics and Product Formation in Streptomyces* (BAUMBERG, S., KRÜGEL, H., NOVACK, D., Eds.), pp. 253–263, New York: Plenum Press.

ANDOH, T., YOSHIDA, K. (1989), Screening of cardiovascular agents, in: *Novel Microbial Products for Medicine and Agriculture* (DEMAIN, A. L., SOMKUTI, G. A., HUNTER-CREVA, J. C., ROSSMOORE, H. W., Eds.), pp. 33–43 Amsterdam: Elsevier Science Publishers.

AOYAGI, T., TAKEUCHI, T. (1989), Low molecular weight enzyme inhibitors produced by microorganisms, in: *Novel Microbial Products for Medicine and Agriculture* (DEMAIN, A. L., SOMKUTI, G. A., HUNTER-CREVA, J. C., ROSSMOORE, H. W., Eds.), pp. 101–107, Amsterdam: Elsevier Science Publishers.

ASANO, M., KOHSAKA, M., AOKI, H., IMANAKA, H. (1989), Screening of immuno modulating agents, in: *Novel Microbial Products for Medicine and Agriculture* (DEMAIN, A. L., SOMKUTI, G. A., HUNTER-CREVA, J. C., ROSSMOORE, H. W., Eds.), pp. 45–48, Amsterdam: Elsevier Science Publishers.

BALTZ, R. H. (1986), Mutation in Streptomyces, in: *The Bacteria*, Vol. 9: *Antibiotic-Producing Streptomyces* (SOKATCH, J. R., ORNSTON, L. N., QUEENER, S. W., DAY, L. E., Eds.), pp. 61–94, Orlando: Academic Press.

BALTZ, R. H., HAHN, D. R., MCHENNY, M. A., SOLENBERG, P. J. (1992), Transposition of Tn5096 and related transposons in *Streptomyces* species, *Gene* **115,** 61–65.

BENDER, E., KOLLER, K. P., ENGELS, J. W. (1990), Secretory synthesis of human interleukin-2 by *Streptomyces lividans*, *Gene* **86,** 227–232.

BENVENISTE, R., DAVIES, J. (1973), Aminoglycoside antibiotic-inactivating enzymes in actinomycetes similar to those present in clinical isolates of antibiotic-resistant bacteria, *Proc. Natl. Acad. Sci. USA* **70,** 2276–2280.

BEPPU, T. (1992), Secondary metabolites as chemical signals for cellular differentiation, *Gene* **115,** 159–165.

BEPPU, T., YOSHIDA, M. (1989), Trichostatin and leptomycin: specific inhibitors of the G1 and G2 phases of eukaryotic cell cycle, in: *Novel Microbial Products for Medicine and Agriculture* (DEMAIN, A. L., SOMKUTI, G. A., HUNTER-CREVA, J. C., ROSSMOORE, H. W., Eds.), pp. 73–78, Amsterdam: Elsevier Science Publishers.

BIRR, E., WOHLLEBEN, W., AUFDERHEIDE, K., SCHNEIDER, T., PÜHLER, A., BRÄU, B., MARQUARDT, R., WÖHNER, G., PRÄVE, P., SCHLINGMANN, M. (1989), Isolation and complementation of mutants of *Streptomyces coelicolor* "Müller" DSM3030 deficient in lysozyme production, *Appl. Microbiol. Biotechnol.* **30,** 358–363.

BRÄU, B., HILGENFELD, R., SCHLINGMANN, M., MARQUARDT, R., BIRR, E., WOHLLEBEN, W., AUFDERHEIDE, K., PÜHLER, A. (1991), Increased yield of a lysozyme after self-cloning of the gene in *Streptomyces coelicolor* "Müller", *Appl. Microbiol. Biotechnol.* **34,** 481–487.

BRAWNER, M., INGRAM, C., FORNWALD, J., LICHENSTEIN, H., WESTPHELING, J. (1988), Structure of the *Streptomyces* galactose operon's catabolite controlled promoter, in: *Biology of Actinomycetes '88* (OKAMI, Y., BEPPU, T., OGAWARA, H., Eds.), pp. 35–40, Tokyo: Japan Scientific Societies Press.

BRAWNER, M., FORNWALD, J., ROSENBERG, M., POSTE, G., WESTPHELING, J. (1990), Heterologous gene expression in *Streptomyces,* in: *Proc. 6th Int. Symp. Genetics of Industrial Microorganisms* (HESLOT, H., DAVIES, J., FLORENT, J. BOBICHON, L., DURAND, G., PENEASSE, L., Eds.), pp. 85–93, Paris: Societé Française de Microbiologie.

BRÖER, S., KRÄMER, R. (1991 a), Lysine excretion by *Corynebacterium glutamicum* – Identification of a specific secretion carrier system, *Eur. J. Biochem.* **202,** 131–135.

BRÖER, S., KRÄMER, R. (1991 b), Lysine excretion by *Corynebacterium glutamicum* – Energetics and mechanism of the transport system, *Eur. J. Biochem.* **202,** 137–143.

BUCKLAND, B., GBEWONYO, T., HALLADA, T., KAPLAN, L., MASUREKAR, P. (1989), Production of lovastatin, an inhibitor of cholesterol accumulation in humans, in: *Novel Microbial Products for Medicine and Agriculture* (DEMAIN, A. L., SOMKUTI, G. A., HUNTER-CREVA, J. C., ROSSMOORE, H. W., Eds.), pp. 161–169, Amsterdam: Elsevier Science Publishers.

BURG, R. W., HUANG, L., MONAGHAN, R. L., STAPLEY, E. O. (1988), Cholecystokinin antagonists and other natural products from microorganisms, in: *Biology of Actinomycetes '88* (OKAMI, Y., BEPPU, T., OGAWARA, H., Eds.), pp. 153–158, Tokyo: Japan Scientific Societies Press.

BURNETT, W. V., BRAWNER, M., TAYLOR, D. P., FARE, L. R., HENNER, J., ECKHARDT, T. (1985), Cloning and analysis of an exported beta-galactosidase and other proteins from *Streptomyces lividans*, in: *Microbiology – 1985* (LEIVE, L., Ed.), pp. 441–444, Washington, DC: American Society for Microbiology.

BUSHELL, M. E., NISBET, L. J. (1981), A technique for eliminating recurring producers of known metabolites in antibiotic screens. *Zentralbl. Bakteriol. Mikrobiol. Hyg. Abt. 1 Suppl.* **11,** 507–514.

BUTTNER, M. J. (1989), RNA polymerase heterogeneity in *Streptomyces coelicor* A3(2), *Mol. Microbiol.* **3,** 1653–1659.

BUTTNER, M. J., FEARNLEY, I. M., BIBB, M. J. (1987), Two promoters from *Streptomyces lividans* pIJ101 and their expression in *Escherichia coli*, *Gene* **51,** 179–186.

CHANG, P C., KUO, T. C., TSUGITA, A., WU LEE, Y. H (1990), Extracellular metalloprotease gene of *Streptomyces cacaoi:* structure, nucleotide sequence and characterization of cloned gene product, *Gene* **88,** 87–95.

CHATER, K. F. (1986), *Streptomyces* phages and their application to *Streptomyces* genetics, in: *The Bacteria*, Vol. 9: *Antibiotic-Producing Streptomyces* (SOKATCH, J. R., ORNSTON, L. N., QUEENER, S. W., DAY, L. E., Eds.), pp., 119–158, Orlando: Academic Press.

CHATER, K. F. (1989), Multilevel regulation of *Streptomyces* differentiation, *Trends Genet.* **5,** 372–377.

CHATER, K. F. (1992), Genetic regulation of secondary metabolic pathways in microbes, in: *Secondary Metabolites: Their Function and Evolution, Ciba Foundation Symposium* No. **171** (DAVIES, J., CHADWICK, D., WHELAN, J., WIDDOWS, K., Eds.), Chichester: John Wiley & Sons, in press.

CHATER, K. F., BRAIN, P., DAVIS, N. K., LESKIW, B. K., PLASKITT, K. A., SOLIVERI, J., TAN, H. (1990), Developmental pathways in *Streptomyces:* a comparision with endospore formation in *Bacillus*, in: *Proc. 6th Int. Symp. Genetics of Industrial Microorganisms* (HESLOT, H., DAVIES, J., FLORENT, J., BOBICHON, L., DURAND, G., PENEASSE, L., Eds.), pp. 373–378, Paris: Societé Française de Microbiologie.

CHATER, K. F., HOPWOOD, D. A. (1984), *Streptomyces* genetics, in: *The Biology of Actinomycetes* (GOODFELLOW, M., MORDARSKI, M., WILLIAMS, S. T., Eds.), pp. 229–286, London: Academic Press.

CHATER, K. F., HOPWOOD, D. A. (1989), Antibiotic biosynthesis in *Streptomyces*, in: *Genetics of Bacterial Diversity* (HOPWOOD, D. A., CHATER, K. F., Eds.), pp. 129–150, London: Academic Press.

CHUNG, S. T., CROSE, L. L. (1990), Transposon Tn4556 mediated DNA insertion and site-directed mutagenesis, in: *Proc. 6th Int. Symp. Genetics of Industrial Microorganisms* (HESLOT, H., DAVIES, J., FLORENT, J., BOBICHON, L., DURAND, G., PENEASSE, L., Eds.), pp. 207–218, Paris: Societé Française de Microbiologie.

CORDES, C., EGGELING, L., SAHM, H. (1990), Production of L-isoleucine by recombinant *Corynebacterium glutamicum* strains after cloning, analysis and amplification of genes involved in isoleucine biosynthesis, in: *Proc. 6th Int. Symp.Genetics of Industrial Microorganisms* (HESLOT, H., DAVIES, J., FLORENT, J., BOBICHON, L., DURAND, G., PENEASSE, L., Eds.), pp. 339–351, Paris: Societé Française de Microbiologie.

CRAWFORD, A. L. (1988), Biodegradation of agriculture and urban wastes, in: *Actinomycetes Biotechnology* (GOODFELLOW, M., WILLIAMS, S. T., MORDARSKI, M., Eds.), pp. 433–459, London: Academic Press.

CULLUM, J., FLETT, F., PIENDL, W. (1989), Genetic instability, deletions, and DNA amplification in *Streptomyces* species, in: *Genetics and Molecular Biology of Industrial Microorganisms* (HERSHBERGER, C. L., QUEENER, S. W., HEGEMAN, G., Eds.), pp. 127–132, Washington, DC: American Society for Microbiology.

CUNDLIFFE, E. (1989), How antibiotic-producing organisms avoid suicide. *Annu. Rev. Microbiol.* **43,** 207–233.

DAVIES, J., CHADWICK, D., WHELAN, J., WIDDOWS, K. (1992), *Secondary Metabolites: Their Function and Evolution, Ciba Foundation Symposium* No. **171,** Chichester: John Wiley & Sons, in press.

DEMAIN, A. L. (1989), Carbon source regulation of idiolite biosynthesis in actinomycetes, in: *Regulation of Secondary Metabolism in Actinomycetes* (SHAPIRO, S., Ed.), pp. 127–134, Boca Raton: CRC Press, Inc.

DIETZ, A. (1986), Structure and taxonomy of *Streptomyces*, in: *The Bacteria*, Vol. 9: *Antibiotic-Producing Streptomyces* (SOKATCH, J. R., ORNSTON, L. N., QUEENER, S. W., DAY, L. E., Eds.), pp. 1–25, Orlando: Academic Press.

DISTLER, J., MANSOURI, K., MAYER, G., PIEPERSBERG, W. (1990), Regulation of biosynthesis of streptomycin, in: *Proc. 6th Int. Symp. Genetics of Industrial Microorganisms* (HESLOT, H., DAVIES, J., FLORENT, J., BOBICHON, L., DURAND, G., PENEASSE, L., Eds.), pp. 379–392, Paris: Societé Française de Microbiologie.

DISTLER, J., MANSOURI, K., MAYER, G., STOCKMANN, M., PIEPERSBERG, W. (1992), Streptomycin biosynthesis and its regulation in Streptomycetes, *Gene* **115,** 105–111.

DORAN, L. J., LESKIW, B. K., AIPPERSBACH, S., JENSE, S. E. (1990), Isolation and characterization of a β-

lactamase-inhibitor protein from *Streptomyces clavuligerus* and cloning and analysis of the corresponding gene, *J. Bacteriol.* **172,** 4909–4918.

EBBIGHAUSEN, H., WEIL, B., KRÄMER, R. (1989), Isoleucine excretion in *Corynebacterium glutamicum:* evidence for a specific efflux carrier system, *Appl. Microbiol. Biotechnol.* **31,** 184–190.

ELANDER, R. P., VOURNAKIS, J. N. (1986), Genetic aspects of overproduction of antibiotics and other secondary metabolites, in: *Overproduction of Microbial Metabolites* (VANEK, Z., HOSTALEK, Z., Eds.), pp. 63–79, Boston: Butterworths.

EPP, J. K., HUBBER, M. L., TURNER, J. R., SCHONER, B.E. (1989), Molecular cloning and expression of carbomycin biosynthesis and resistance genes from *Streptomyces thermotolerans*, in: *Genetics and Molecular Biology of Industrial Microorganisms* (HERSHBERGER, C. L., QUEENER, S. W., HEGEMAN, G., Eds.), pp. 35–39, Washington, DC: American Society for Microbiology.

FISHER, S. H. (1992), Glutamine synthesis in *Streptomyces* – a review, *Gene* **115,** 13–17.

FOLLETTIE, M. T., PEOPLES, O. P., ARCHER, J. A. C., SINSKEY, A. J. (1990), Metabolic engineering in corynebacteria, in: *Proc. 6th Int. Symp. Genetics of Industrial Microorganisms* (HESLOT, H., DAVIES, J., FLORENT, J., BOBICHON, L., DURAND, G., PENEASSE, L., Eds.), pp. 315–325, Paris: Societé Française de Microbiologie.

FORSMAN, M., HÄGGSTRÖM, B., LINDGREN, L., JAURIN, B. (1990), Molecular analysis of β-lactamases from four species of *Streptomyces:* comparison of amino acid sequences with those of other β-lactamases, *J. Gen. Microbiol.* **136,** 589–598.

GHAZZOULI, I., HITCHCOCK, M. J. M. (1988), Antiviral screening: An overview, in: *Biology of Actinomycetes '88* (OKAMI, Y., BEPPU, T., OGAWARA, H., Eds.), pp. 178–183, Tokyo: Japan Scientific Societies Press.

GOODFELLOW, M., O'DONNELL, A. G. (1989), Search and discovery of industrially significant actinomycetes, in: *Microbial Products: New Approaches* (BAUMBERG, S., HUNTER, L., RHODES, M., Eds.), pp. 343–383, Cambridge: Cambridge University Press.

GRÄFE, U. (1989), Autoregulatory secondary metabolites from actinomycetes, in: *Regulation of Secondary Metabolism in Actinomycetes* (SHAPIRO, S., Ed.), pp. 75–126, Boca Raton: CRC Press, Inc.

HÄNEL, F., KRÜGEL, H., PESCHKE, T. (1991), Pullulan-hydrolyzing enzymes of streptomycetes: effect of carbon source on their production and characterization of a pullulanase-negative mutant, in: *Genetics and Product Formation in Streptomyces* (BAUMBERG, S., KRÜGEL, H., NOVACK, D., Eds.), pp. 215–226, New York: Plenum Press.

HASEGAWA, M. (1992), A novel, highly efficient gene-cloning system in *Micromonospora* applied to the genetic analysis of fortimicin biosynthesis, *Gene* **115,** 85–91.

HÄUSLER, A., BIRCH, A., KREK, W., PIRET, J., HÜTTER, R. (1989), Heterogeneous genomic amplification in *Streptomyces glaucescens*, structure, location and junction sequence analysis, *Mol. Gen. Genet.* **217,** 437–446.

HENDERSON, G., KRYGSMAN, P., LIU, C. J., DAVEY, C. C., MALEK, L. T. (1987), Characterization and structure of genes for protease A and B from *Streptomyces griseus*, *J. Bacteriol.* **169,** 3779–3784.

HERSHBERGER, C. L., ARNOLD, B., LARSON, J., SKATRUD, P., REYNOLDS, P. S., ROSTECK, P. R., Jr., SWARTLING, J., MCGILVRAY, D. (1989), Role of giant linear plasmids in the biosynthesis of macrolide and polyketide antibiotics, in: *Genetics and Molecular Biology of Industrial Microorganisms* (HERSHBERGER, C. L., QUEENER, S. W., HEGEMAN, G., Eds.), pp. 147–155, Washington, DC: American Society for Microbiology.

HOISCHEN, C., KRÄMER, R. (1990), Membrane alteration is necessary but not sufficient for effective glutamate secretion in *Corynebaterium glutamicum*, *J. Bacteriol.* **172,** 3409–3416.

HOOD, D. W., HEIDSTRA, R., SWOBODA, U., K., HODGSON, D. A. (1992), Molecular genetic analysis of proline and tryptophan biosynthesis in *Streptomyces coelicolor A3(2):* interaction between primary and secondary metabolism – a review, *Gene* **115,** 5–12.

HOPWOOD, D. A., KIESER, T. (1990), The *Streptomyces* genome, in: *The Bacterial Chromosome* (DRLICA, K., RILEY, M., Eds.), pp. 147–162, Washington, DC: American Society for Microbiology.

HOPWOOD,. D. A., SHERMAN, D. H. (1990), Molecular genetics of polyketides and its comparison to fatty acid biosynthesis, *Annu. Rev. Genet.* **24,** 37–66.

HOPWOOD, D. A., BIBB, M. J., CHATER, K. F., KIESER, T., BRUTON, C. J., KIESER, H. M., LYDIATE, D. J., SMITH, C. P., WARD, J. M., SCHREMPF, H. (1985), *Genetic Manipulation of Streptomyces: A Laboratory Manual*, Norwich: The John Innes Foundation.

HOPWOOD, D. A., KIESER, T., LYDIAT E, D., BIBB, M. J. (1986), *Streptomyces* plasmids: their biology and use as cloning vectors, in: *The Bacteria*, Vol. 9: *Antibiotic-Producing Streptomyces* (SOKATCH, J. R., ORNSTON, L. N., QUEENER, S. W., DAY, L. E., Eds.), pp. 159–229, Orlando: Academic Press.

HOPWOOD, D. A., BIBB, M. J., CHATER, K. F., KIESER, T. (1987), Plasmid and phage vectors for gene cloning and analysis in *Streptomyces*, *Methods Enzymol.* **153,** 116–166.

HOPWOOD, D. A., SHERMAN, D. H., KHOSLA, C., BIBB, M. J., SIMPSON, T. J., FERNANDEZ-MORENO, M. A.,

MARTINEZ, E., MALPARTIDA, F. (1990), "Hybrid" pathways for the production of secondary metabolites, in: *Proc. 6th Int. Symp. Genetics of Industrial Microorganisms* (HESLOT, H., DAVIES, J., FLORENT, J., BOBICHON, L., DURAND, G., PENEASSE, L., Eds.), pp. 259–270, Paris: Societé Française de Microbiologie.

HORINOUCHI, S., BEPPU, T. (1990), Autoregulatory factors of secondary metabolism and morphogenesis in actinomycetes, *CRC Crit. Rev. Biotechnol.* **10,** 191–204.

HORINOUCHI, S., BEPPU T. (1992), Regulation of secondary metabolism and cell differentiation in *Streptomyces:* A-factor as a microbial hormone and the AfsR protein as a component of a two-component regulator system, *Gene* **115,** 167–172.

HOSTALEK, Z., BEHAL, V., NOVOTNA, J., ERBAN, V., CURDOVA, E., JECHOVA, V. (1982), Regulation of expression of chlortetracycline synthesis in *Streptomyces aureofaciens*, in: *Overproduction of Microbial Products* (KRUMPHANZL, V., SIKYTA, B., VANEK, Z., Eds.), pp. 47–61, London: Academic Press.

HUNTER, I. S., BAUMBERG, S. (1989), Molecular genetics of antibiotic formation, in: *Microbial Products: New Approaches* (BAUMBERG, S., HUNTER, I., RHODES, M., Eds.), pp. 121–162, Cambridge: Cambridge University Press.

HUTCHINSON, C. R. (1992), Anthracycline antibiotics, in: *Biochemistry and Genetics of Antibiotic Biosynthesis* (VINING, L., STUTTARD, C., Eds.), Stoneham: Butterworth-Heinemann, in press.

HUTCHINSON, C. R., BORELL, C. W., OTTEN, S. L., STUTZMAN-ENGWALL, K. J., WANG, Y. G. (1989), A perspective on drug discovery and development through the genetic engineering of antibiotic-producing microorganisms, *J. Med. Chem.* **32,** 929–937.

HUTCHINSON, C. R., BORELL, C. W., DONOVAN, M. J., KATO, F., MOTAMEDI, H., NAKAYAMA, H., RUBIN, R. L., STREICHER, S. L., SUMMERS, R. G., WENDT-PIENKOWSKI, E., WESSEL, W. L. (1991), The genetic and biochemical basis of polyketide metabolism in microorganisms and its role in drug discovery and development, *Planta Med.* **57,** 36–43.

HÜTTER, R., ECKHARDT, T. (1988), Genetic manipulation, in: *Actinomycetes in Biotechnology* (GOODFELLOW, M., WILLIAMS, S. T., MORDARSKI, M., Eds.), pp. 89–184, London: Academic Press.

IKEDA, M., KATSUMATA, R. (1992), Metabolic engineering to produce tyrosine or phenylalanine in a tryptophan-producing *Corynebacterium glutamicum* strain, *Appl. Environ. Microbiol.* **58,** 781–785.

INOUYE, S., KONDO, S. (1989), Amicoumacin and SF-2370, pharmacologically active agents of microbial origin, in: *Novel Microbial Products of Medicine and Agriculture* (DEMAIN, A. L., SOMKUTI, G. A., HUNTER-CREVA, J. C., ROSSMOORE, H. W., Eds.), pp. 179–193, Amsterdam: Elsevier Science Publishers.

ISHIZAKI, T., HIRAYAMA, N., SHINAKAWA, H., NIMI, O., MUROOKA, Y. (1989), Nucleotide sequence of the gene for cholesterol oxidase from a *Streptomyces* sp., *J. Bacteriol.* **171,** 596–601.

ITO, H. (1990), Molecular breeding of aromatic amino acid-producing strains of *Brevibacterium lactofermentum*, in: *Proc. 6th Int. Sym. Genetics of Industrial Microorganisms* (HESLOT, H., DAVIES, J., FLORENT, J., BOBICHON, L., DURAND, G., PENEASSE, L., Eds.), pp. 327–337, Paris: Societé Française de Microbiologie.

IWASAKI, S. (1989), Rhizoxin, an inhibitor of tubulin assembly, in: *Novel Microbial Products for Medicine and Agriculture* (DEMAIN, A. L., SOMKUTI, G. A., HUNTER-CREVA, J. C., ROSSMOORE, H. W., Eds.), pp. 79–89, Amsterdam: Elsevier Science Publishers.

KÄMPFER, P., KROPPENSTEDT, R. M., DOTT, W. (1991), A numerical classification of the genera *Streptomyces* and *Streptoverticillium* using miniaturized physiological tests, *J. Gen. Microbiol.* **137,** 1831–1891.

KANAMURA, T., OKAZAKI, H. (1989), Emriamine: a new inhibitor of long chain fatty acid oxidation and its antidiabetic activity, in: *Novel Microbial Products for Medicine and Agriculture* (DEMAIN, A. L., SOMKUTI, G. A., HUNTER-CREVA, J. C., ROSSMOORE, H. W., Eds.), pp. 135–144, Amsterdam: Elsevier Science Publishers.

KASE, H. (1988), New inhibitors of protein kinases from microbial source, in: *Biology of Actinomycetes '88* (OKAMI, Y., BEPPU, T., OGAWARA, H., Eds.), pp. 159–164, Tokyo: Japan Scientific Societies Press.

KATSUMATA, R., MIZUKAMI, T., OZAKI, A., KIKUCHI, Y., KINO, K., OKA, T., FURUYA, A. (1987), Gene cloning in glutamic acid bacteria: the system and its applications, in: *Proc. 4th. Eur. Congr. Biotechnology*, Vol. 4, pp. 767–776, Amsterdam: Elsevier.

KATZ, L., DONADIO, S. (1992), Macrolides, in: *Biochemistry and Genetics of Antibiotic Biosynthesis* (VINING, L., STUTTARD, C., Eds.), Stoneham: Butterworth-Heinemann, in preparation.

KELL, D. B., VAN DAM, K., WESTERHOFF, H. V. (1989), Control analysis of microbial growth and productivicity, in: *Microbial Products: New Approaches* (BAUMBERG, S., HUNTER, I., RHODES, M., Eds.), pp. 61–93, Cambridge: Cambridge University Press.

KENDALL, K. J., CULLUM, J. (1984), Cloning and expression of an extracellular agarase gene from *Streptomyces coelicolor* A3(2) in *Streptomyces lividans* 66, *Gene* **29,** 315–321.

KENDALL, K. J., COHEN, S. N. (1988), Complete nucleotide sequence of the *Streptomyces lividans* plasmid pIJ101 and correlation of the sequence with genetic properties, *J. Bacteriol.* **170,** 4634–4651.

KHOKHLOV, A. S. (1988), Results and perspectives of actinomycete autoregulators studies, in: *Biology of Actinomycetes '88* (OKAMI, Y., BEPPU, T., OGAWARA, H., Eds.), pp. 338–345, Tokyo: Japan Scientific Societies Press.

KIDA, T. (1989), Screening of microbial products affecting plant metabolism, in: *Novel Microbial Products for Medicine and Agriculture* (DEMAIN, A. L., SOMKUTI, G. A., HUNTER-CREVA, J. C., ROSSMOORE, H. W., Eds.), pp. 195–202, Amsterdam: Elsevier Science Publishers.

KIESER, T., HOPWOOD, D. A. (1991), Genetic manipulation of *Streptomyces:* Integrating vectors and gene replacement, *Methods Enzymol.* **204,** 430–458.

KIESER, H. M., KIESER, T., HOPWOOD, D. A. (1992), A combined genetic map of the *Streptomyces coelicolor* A3(2) chromosome, *J. Bacteriol.* **174,** 5496–5507.

KINASHI, H., SHIMAJI-MURAYAMA, M., HANAFUSA, T. (1992), Integration of SCP1, a giant linear plasmid, into the *Streptomyces coelicolor* chromosome, *Gene* **115,** 35–41.

KINOSHITA, S. (1985), Glutamic acid bacteria, *Biol. Ind. Microorg.* **6,** 115–142.

KINOSHITA, S., UDAKA, S., SHIMONO, M. (1957), The production of amino acids by fermentation process, *J. Gen. Appl. Microbiol.* **3,** 193–205.

KITAMURA, S., HASHIZUME, K., LIDA, T., OHMORI, K., KASE, H. (1989), Lipoxygenase inhibitors, in: *Novel Microbial Products for Medicine and Agriculture* (DEMAIN, A. L., SOMKUTI, G. A., HUNTER-CREVA, J. C., ROSSMOORE, H. W., Eds.), pp. 145–150, Amsterdam: Elsevier Science Publishers.

KLUEPFEL, D., MOROSOLI, R., SHARECK, F. (1991), Homologous cloning of the xylanase genes and their expression in *Streptomyces lividans* 66, in: *Genetics and Product Formation in Streptomyces* (BAUMBERG, S., KRÜGEL, H., NOVACK, D., Eds.), pp. 207–214, New York: Plenum Press.

KOLLER, K. P., RIESS, G. (1987), Heterologous expression of the α-amylase inhibitor gene cloned from an amplified genomic sequence of *Streptomyces tendae*, *J. Bacteriol.* **171,** 4953–4957.

KOLLER, K. P., RIESS, G., SAUBER, K., VERTESY, L., UHLMANN, E., WALLMEIER, H. (1991), The tendamistat expression-secretion system: synthesis of proinsulin fusion proteins with *Streptomyces lividans*, in: *Genetics and Product Formation in Streptomyces* (BAUMBERG, S., KRÜGEL, H., NOVACK, D., Eds.), pp. 227–233, New York: Plenum Press.

KORN-WENDISCH, F., KUTZNER, H. J. (1991), The family Streptomycetaceae, in: *The Prokaryotes* (BALOWS, A, TRÜPER, H. G., DWORKIN, M., HARDER, W., SCHLEIFER, K.-H. Eds.), 2nd. Ed., pp. 921– 995, New York: Springer Verlag.

KORN-WENDISCH, F., SCHNEIDER, J. (1992), Phage typing – A useful tool in actinomycetes systematics, *Gene* **115,** 243–247.

KRÄMER, R., LAMBERT, C., HOISCHEN, C., EBBIGHAUSEN, H. (1990), Uptake of glutamate in *Corynebacterium glutamicum* – evidence for a primary active transport system, *Eur. J. Biochem.* **194,** 929–935.

KUNINAKA, A. (1986), Nucleic acids, nucleotides, and related compounds, in: *Biotechnology* (REHM, H.-J., REED, G., Eds.), Vol. 4: *Microbial Products II*, pp. 71–117, Weinheim: VCH Verlagsgesellschaft.

LACALLE, R. A., TERCERO, J. A., JIMENEZ, A. (1992), Cloning of the complete biosynthetic gene cluster for an aminonucleoside antibiotic, puromycin, and its regulated expression in heterologous hosts, *EMBO J.* **11,** 785–792.

LACEY, J. (1988), Actinomycetes as biodeteriogens and pollutants of the environment, in: *Actinomycetes in Biotechnology* (GOODFELLOW, M., WILLIAMS, S. T., MORDARSKI, M., Eds.), pp. 359–432, London: Academic Press.

LEBLON, G., REYES, O., GUYONVARCH, A., BONAMY, C., LABARRE, J., WOJCIK, F. (1990), Construction and utilization of integrative vectors for Corynebacteria, in: *Proc. 6th Int. Symp. Genetics of Industrial Microorganisms* (HESLOT, H., DAVIES, J., FLORENT, J., BOBICHON, L., DURAND, G., PENEASSE, L., Eds.), pp. 293–314, Paris: Societé Française de Microbiologie.

LECHEVALIER, M. P. (1988), Actinomycetes in agriculture and forestry, in: *Actinomycetes in Biotechnology* (GOODFELLOW, M., WILLIAMS, S. T., MORDARSKI, M., Eds.), pp. 327–358, London: Academic Press.

LESKIW, B. K., BIBB, M. J., CHATER, K. F. (1991), The use of a rare codon specifically during development? *Mol. Microbiol.* **5,** 2861–2867.

LIRAS, P., ASTURIAS, J. A., MARTÍN, J. F. (1990), Phosphate control sequences involved in transcriptional regulation of antibiotic biosynthesis, *TIBTECH.* **8,** 184–189.

LICHTENSTEIN, H. S., BUSSE, L. A., SMITH, G. A., NAHRI, L. O., MCGINLEY, M. O., ROHDE, M. F., KATZOWITZ, J. L., ZUKOWSKI, M. M. (1992), Cloning and characterization of a gene encoding extracellular metalloprotease from *Streptomyces lividans*, *Gene* **111,** 125–130.

LIEBL, W. (1991), The genus *Corynebacterium*-Nonmedical, in: *The Prokaryotes* (BALOWS, A, TRÜPER, H. G., DWORKIN, M., HARDER, W., SCHLEIFER, K.-H. Eds.), 2nd. Ed., pp. 1157–1171, New York: Springer Verlag.

LIEBL, W., SINSKEY, A., SCHLEIFER, K. H. (1992), Expression, secretion, and processing of staphylococcal nuclease by *Corynebacterium glutamicum*, *J. Bacteriol.* **174,** 1854–1861.

LIMAURO, D., AVITABILE, A., CAPPELLANO, M., PUGLIA, A. M., BRUNI, C. B. (1990), Cloning and characterization of the histidine biosynthetic gene cluster of *Streptomyces coelicolor* A3(2), *Gene* **90,** 31–40.

LOCCI, R. (1989), Streptomycetes and related genera, in: *Bergey's Manual of Systematic Bacteriology* (WILLIAMS, S. T., SHARPE, M. E., HOLT, J. G, MURRAY, R. G. E., BRENNER, D. J., KRIEG, N. R., MOULDER, J. M., PFENNIG, N., SNEATH, P. H. A., STALEY, J. T., Eds.), Vol. 4, pp. 2451–2452, Baltimore: Williams & Wilkins.

LOMOVSKAYA, N., SEZONOV, G., ISAYEWA, L., CHINENOVA, T. (1991), Genetic characterization and cloning of genes for chlorotetracycline resistance in *Streptomyces aureofaciens* strains, in: *Genetics and Product Formation in Streptomyces* (BAUMBERG, S., KRÜGEL, H., NOVACK, D., Eds.), pp. 117–127, New York: Plenum Press.

LONG, C. M., VIROLLE, M. J., CHANG, S. Y., CHANG, S., BIBB, M. J. (1987), α-Amylase gene of *Streptomyces limosus:* nucleotide sequence, expression motifs, and amino acid sequence homology to mammalian and invertebrate α-amylases, *J. Bacteriol.* **169,** 5745–5754.

MACNEIL, D. J., OCCI, J. L., GEWAIN, K. M., MACNEIL, T., GIBBONS, P. H., RUBY, C. L., DANIS, S. J. (1992), Complex organization of the *Streptomyces avermitilis* genes encoding the avermectin polyketide synthase, *Gene* **115,** 119–125.

MALPARTIDA, F., HALLAM, S. E., KIESER, H. M., MOTAMEDI, H., HUTCHINSON, C. R., BUTLER, M. J., SUGDEN, D. A., WARREN, M., MCKILLOP, M., BAILEY, C. R., HUMPHREYS, G. O., HOPWOOD, D. A. (1987), Homology between *Streptomyces* genes coding for synthesis of different polyketides used to clone antibiotic biosynthesis genes, *Nature* **325,** 818–821.

MANSOURI, K., PIEPERSBERG, W. (1991), Genetics of streptomycin production in *Streptomyces griseus:* nucleotide sequence of five genes, *str FGHIK*, including a phosphatase gene, *Mol. Gen. Genet.* **228,** 459–469.

MARTÍN, J. F. (1989 a), Molecular mechanisms for the control by phosphate of the biosynthesis of antibiotics and other secondary metabolites, in: *Regulation of Secondary Metabolism in Actinomycetes* (SHAPIRO, S., Ed.), pp. 213–237, Boca Raton: CRC Press, Inc.

MARTÍN, J. F. (1989 b), Molecular genetics of aminoacid producing corynebacteria, in: *Microbial Products: New Approaches* (BAUMBERG, S., HUNTER, I., RHODES, M., Eds.), pp. 25–59, Cambridge: Cambridge University Press.

MARTÍN, J. F., LIRAS, P. (1989), Organization and expression of genes involved in the biosynthesis of antibiotics and other secondary metabolites, *Ann. Rev. Microbiol.* **43,** 173–206.

MARTÍN, J. F., CADENAS, R. F., MALUMBRES, M., MATEOS, L. M., GUERRERO, C., GIL, J. A. (1990), Construction and utilization of promoter probe and expression vectors in corynebacteria. Characterization of corynebacterial promoters, in: *Proc. 6th Int. Symp. Genetics of Industrial Microorganisms* (HESLOT, H., DAVIES, J., FLORENT, J., BOBICHON, L., DURAND, G., PENEASSE, L., Eds.), pp. 283–292, Paris: Societé Française de Microbiologie.

MAZODIER, P., DAVIES, J. (1991), Gene transfer between distantly related bacteria, *Ann. Rev Genet.* **25,** 147–171.

MCCUE, L. A., KWAK, J., BABCOCK, M. J., KENDRICK, K.F. (1992), Molecular analysis of sporulation in *Streptomyces griseus*, *Gene* **115,** 173–179.

MCDOWALL, K. J., DOYLE, D., BUTLER, M. J., BINNIE, C., WARREN, M., HUNTER, I. S. (1991), Molecular genetics of oxytetracycline production by *Streptomyces rimosus*, in: *Genetics and Product Formation in Streptomyces* (BAUMBERG, S., KRÜGEL, H., NOVACK, D., Eds.), pp. 105–116, New York: Plenum Press.

MCHENNEY, M., BALTZ, R. H. (1990), Transduction of plasmid DNA containing the *erm* E gene and expression of erythromycin-resistance in streptomycetes, *J. Antibiot.* **44,** 1267–1269.

MCKILLOP, C., ELVIN, P., KENTEN, J. (1986), Cloning and expression of an extracellular α-amylase gene from *Streptomyces hygroscopicus* in *Streptomyces lividans* 66, *FEMS Microbiol. Lett.* **36,** 3–7.

MIYASHITA, K., FUJII, T., SAWADA, Y. (1991), Molecular cloning and characterization of chitinase genes from *Streptomyces lividans* 66, *J. Gen. Microbiol.* **137,** 2065–2072.

MORI, M., SHIO, I. (1987), Phosphoenolpyruvate: sugar phosphotransferase systems and sugar metabolism in *Brevibacterium flavum*, *Agric. Biol. Chem.* **51,** 2671–2678.

MÜLLER, L. (1989), Chemistry, biochemistry and therapeutic potential of microbial α-glucosidase inhibitors, in: *Novel Microbial Products for Medicine and Agriculture* (DEMAIN, A. L., SOMKUTI, G. A., HUNTER-CREVA, J. C., ROSSMOORE, H. W., Eds.), pp. 109–116, Amsterdam: Elsevier Science Publishers.

MURAKAMI, T., HOLT, T., THOMPSON, C. J. (1988), The regulation of a thiostrepton-induced promoter in *Streptomyces lividans,* in: *Biology of Actinomycetes '88* (OKAMI, Y., BEPPU, T., OGAWARA, H., Eds.), pp. 371–373, Tokyo: Japan Scientific Societies Press.

MUTH, G., NUSSBAUMER, B., WOHLLEBEN, W., PÜHLER, A. (1989), A vector system with temperature-sensitive replication for gene disruption and mutational cloning in streptomycetes, *Mol. Gen. Genet.* **219,** 341–348.

NAGASO, H., SAITO, S., SAITO, H., TAKAHASHI, H. (1988), Nucleotide sequence and expression of a *Streptomyces griseosporeus* proteinaceous α-amylase inhibitor (Haim II) gene, *J. Bacteriol.* **170,** 4451–4457.

NAKAMURA, S., INOUYE, Y. (1989), Reverse transcriptase inhibitors, in: *Novel Microbial Products for Medicine and Agriculture* (DEMAIN, A. L., SOMKUTI, G. A., HUNTER-CREVA, J. C., ROSS-MOORE, H. W., Eds.), pp. 91–99, Amsterdam: Elsevier Science Publishers.

NIEDERBERGER, P. (1989), Amino acid production in microbial eukaryotes and prokaryotes other than coryneforms, in: *Microbial Products: New Approaches* (BAUMBERG, S., HUNTER, I., RHODES, M., Eds.), pp. 1–24, Cambridge: Cambridge University Press.

NISBET, L. J., PORTER, N. (1989), The impact of pharmacology and molecular biology on the exploitation of microbial products, in: *Microbial Products: New Approaches* (BAUMBERG, S., HUNTER, I., RHODES, M., Eds.), pp. 309–342, Cambridge: Cambridge University Press.

NOLAN, R. D., CROSS, T. (1988), Isolation and screening of actinomycetes, in: *Actinomycetes in Biotechnology* (GOODFELLOW, M., WILLIAMS, S. T., MORDARSKI, M., Eds.), pp. 1–32, London: Academic Press.

NORMANSELL, I. D. (1986), Isolation of *Streptomyces* mutants improved for antibiotic production, in: *The Bacteria*, Vol. 9: *Antibiotic-Producing Streptomyces* (SOKATCH, J. R., ORNSTON, L. N., QUEENER, S. W., DAY, L. E., Eds.), pp. 95–118, Orlando: Academic Press.

OCHI, K. (1990), *Streptomyces griseus,* as an excellent object for studying microbial differentiation, *Actinomycetologica* **4,** 23–30.

OHTA, T., FUJISHIRO, K., YAMAGUCHI, K., TAMURA, Y., AISAKA, K. UWAJIMA, T., HASEGAWA, M. (1991), Sequence of gene *choB* encoding cholesterol oxidase of *Brevibacterium sterolicum:* comparison with *choA* of *Streptomyces* sp. SA-COO, *Gene* **103,** 93–96.

OKAMI, Y., HOTTA, K. (1988), Search and discovery of new antibiotics, in: *Actinomycetes in Biotechnology* (GOODFELLOW, M., WILLIAMS, S. T., MORDARSKI, M., Eds.), pp. 33–67, London: Academic Press.

ŌMURA, S. (1988), Search for bioactive compounds from microorganisms – strategies and methods, in: *Biology of Actinomycetes '88* (OKAMI, Y., BEPPU, T., OGAWARA, H., Eds.), pp. 26–34, Tokyo: Japan Scientific Societies Press.

ŌMURA, S. (1992), The expanded horizon for microbial metabolites – a review, *Gene* **115,** 141–149.

ŌMURA, S., TANAKA, Y. (1986), Macrolide antibiotics. in: *Biotechnology* (REHM, H.-J., REED, G., Eds.), Vol. 4: *Microbial Products II*, pp. 359–391, Weinheim: VCH Verlagsgesellschaft.

PADILLA, G., HINDLE, Z., CALLIS, R., CORNER, A., LUDOVICE, M., LIRAS, P., BAUMBERG, S. (1991), The relationship between primary and secondary metabolism in streptomycetes, in: *Genetics and Product Formation in Streptomyces* (BAUMBERG, S., KRÜGEL, H., NOVACK, D., Eds.), pp. 35–45, New York: Plenum Press.

PECZYNSKA-CZOCH, W., MORDARSKI, M. (1988), Actinomycete enzymes, in: *Actinomycetes in Biotechnology* (GOODFELLOW, M., WILLIAMS, S. T., MORDARSKI, M., Eds.), pp. 219–283, London: Academic Press.

PIEPERSBERG, W. (1992a), Streptomycin and related aminoglycoside antibiotics, Chapter 17, in: *Biochemistry and Genetics of Antibiotic Biosynthesis* (VINING, L., STUTTARD, C., Eds.), Stoneham: Butterworth-Heinemann, in press.

PIEPERSBERG, W. (1992b), Pathway engineering in actinomycetes, in: *Biotechnology Focus V* (BUCHHOLZ, R., BUCKEL, P., ESSER, K., LEMKE, P. A., PRÄVE, P., SCHLINGMANN, M., THAUER, R. K., WAGNER, F., Eds.), München: Carl Hanser Verlag, in preparation.

PIEPERSBERG, W. (1992c), Metabolism and cell individualization, in: *Secondary Metabolites: Their Function and Evolution, Ciba Foundation Symp.* **171** (DAVIES, J., CHADWICK, D., WHELAN, J., WIDDOWS, K., Eds.), pp. 294–299. Chichester: John Wiley & Sons.

PIEPERSBERG, W., DISTLER, J., EBERT, A., HEINZEL, P., MANSOURI, K., MAYER, G., PISSOWOTZKI, K. (1988a), Expression of genes for streptomycin biosynthesis, in: *Biology of Actinomycetes '88* (OKAMI, Y., BEPPU, T., OGAWARA, H., Eds.), pp. 86–91, Tokyo: Japan Scientific Societies Press.

PIEPERSBERG, W., DISTLER, J., HEINZEL, P., PEREZ-GONZALEZ, J.-A. (1988b), Antibiotic resistance by modification: Many resistance genes could be derived from cellular control genes in actinomycetes. – A hypothesis. *Actinomycetology* **2,** 83–98.

PIGAC, J. (1991), Structural instability of bifunctional vectors in *Streptomyces*, in: *Genetics and Product Formation in Streptomyces* (BAUMBERG, S., KRÜGEL, H., NOVACK, D., Eds.), pp. 287–294, New York: Plenum Press.

PISSOWOTZKI, K., MANSOURI, K., PIEPERSBERG, W. (1991), Genetics of streptomycin production in *Streptomyces griseus:* molecular structure and putative function of genes *strELMB2N*, *Mol. Gen. Genet.* **231,** 113–123.

PLATER, R., ROBINSON, J. A. (1992), Cloning and sequence of a gene encoding macrotetrolide antibiotic resistance from *Streptomyces griseus*, *Gene* **112,** 117–122.

QUEENER, S. W. (1990), Molecular biology of penicillin and cephalosporin biosynthesis, *Antimicrob. Agents Chemother.* **34,** 943–948.

RAYMER, G., WILLARD, J. M. A., SCHOTTEL, J. L. (1990), Cloning, sequencing, and regulation of expression of an extracellular esterase gene from the plant pathogen *Streptomyces scabies*, *J. Bacteriol.* **172,** 7020–7026.

ROBBINS, P. W., TRIMBLE, R. B., WIRTH, D. F., HERING, C., MALEY, F., MALEY, G. F., DAS, R., GIBSON, B. W., ROYAL, N., BIEMANN, K. (1984), Primary structure of the *Streptomyces* enzyme endo-b-N-acetylglucosaminidase H, *J. Biol. Chem.* **259,** 7577–7583.

ROBBINS, P. W., OVERBYE, K., ALBRIGHT, C., BENFIELD, B., PERO, J. (1992), Cloning and high-level expression of chitinase-encoding gene of *Streptomyces plicatus*, *Gene* **111,** 69 –76.

ROTH, M., MÜLLER, G., NEIGENFIND, M., HOFFMEIER, C., GEUTHER, R. (1991), Partitioning of plasmid in *Streptomyces:* Segregation in continuous culture of a vector with temperature-sensitive replication, in: *Genetics and Product Formation in Streptomyces* (BAUMBERG, S., KRÜGEL, H., NOVACK, D., Eds.), pp. 305–313, New York: Plenum Press.

SCHAAL, K. P. (1988), Actinomycetes as human pathogens, in: *Biology of Actinomycetes '88* (OKAMI, Y., BEPPU, T., OGAWARA, H., Eds.), pp. 277–282, Tokyo: Japan Scientific Societies Press.

SCHONER, B., GEISTLICH, M., RAO, R., SENO, E., REYNOLDS, P., COX, K., BURGETT, S., HERSHBERGER, C. (1992), Sequence similarity between macrolide-resistance determinants and ATP-binding transport proteins, *Gene* **115,** 93–96.

SCHREMPF, H. (1991), Genetic instability in *Streptomyces*, in: *Genetics and Product Formation in Streptomyces* (BAUMBERG, S., KRÜGEL, H., NOVACK, D., Eds.), pp. 245–252, New York: Plenum Press.

SCHRUMPF, B., SCHWARZER, A., KALINOWSKI, J., PÜHLER, A., EGGELING, L., SAHM, H. (1991), A functionally split pathway for lysine synthesis in *Corynebacterium glutamicum*, *J. Bacteriol.* **173,** 4510–4516.

SCHWARZER, A., PÜHLER, A. (1991), Manipulation of *Corynebacterium glutamicum* by gene disruption and replacement, *Biotechnology* **9,** 84–87.

SEBEK, O. K. (1986), Antibiotics, in: *Biotechnology* (REHM, H.-J., REED, G., Eds.), Vol. 6 a: *Biotransformations*, pp. 239–276, Weinheim: VCH Verlagsgesellschaft.

SENO, E. T., BALTZ, R. H. (1989), Structural organization and regulation of antibiotic biosynthesis and resistance genes in actinomycetes, in: *Regulation of Secondary Metabolism in Actinomycetes* (SHAPIRO, S., Ed.), pp. 1–48, Boca Raton: CRC Press, Inc.

SHAPIRO, S. (1989), Nitrogen assimilation in actinomycetes and the influence of nitrogen nutrition on actinomycete secondary metabolism, in: *Regulation of Secondary Metabolism in Actinomycetes* (SHAPIRO, S., Ed.), pp. 135–211, Boca Raton: CRC Press, Inc.

SIMONET, J. M., SCHNEIDER, D., VOLFF, J.-N., DARY, A., DECARIS, B. (1992), Genetic instability in *Streptomyces ambofaciens:* inducibility and associated genome plasticity, *Gene* **115,** 49–54.

SMITH, C. (1992), Molecular genetics of glycerol catabolism in *Streptomyces*, *Gene*, in press.

SMITH, C. P., CHATER, K. F. (1988), Structure and regulation of controlling sequences for the *Streptomyces coelicolor* glycerol operon, *J. Mol. Biol.* **204,** 569–580.

SOBASKEY, C. D., IM, H., NELSON, A. D., SCHAUER, A. T. (1992), Tn4566 and luciferase: synergistic tools for visualizing transcription in *Streptomyces*, *Gene* **115,** 67–71.

SONNEN, H., THIERBACH, G., KAUTZ, S., KALINOWSKI, J., SCHNEIDER, J., PÜHLER, A. KUTZNER, H. J. (1992), Characterization of pGA1, a new plasmid from *Corynebacterium glutamicum* LP-6, *Gene*, in press.

STACKEBRANDT, E., LIESACK, W., WEBB, R., WITT, D. (1991), Towards a molecular identification of *Streptomyces* species in pure culture and in environmental samples, *Actinomycetologica* **5,** 38–44.

STOCKMANN, M., PIEPERSBERG, W. (1992), Gene probes for the detection of 6-deoxyhexose metabolism in secondary metabolite-producing streptomycetes, *FEMS Microbiol. Lett.* **90,** 185–190.

STOWE, D. J., ATKINSON, T. H., MANN, N., (1989), Protein kinase activities in cell-free extracts of *Streptomyces coelicolor* A3(2), *Biochimie* **71,** 1101–1105.

STROHL, W. R. (1992), Compilation and analysis of DNA sequences associated with apparent streptomycete promoters, *Nucleic Acids Res.* **20,** 961–974.

STROHL, W. R., CONNERS, N. C. (1992), Significance of anthraquinone formation resulting from the cloning of actinorhodin genes in heterologous streptomycetes, *Mol. Microbiol.* **6** (2), 147–152.

STROHL, W. R., BARTEL, P. L., CONNORS, N. C., ZHU, C. B., DOSCH, D. C., BEALE, J. M., Jr., FLOSS, H. G., STUTZMAN-ENGWALL, K., OTTEN, S. L., HUTCHINSON, C. R. (1989), Biosynthesis of natural and hybrid polyketides by anthracycline-producing streptomycetes, in: *Genetics and Molecular Biology of Industrial Microorganisms* (HERSH

BERGER, C. L., QUEENER, S. W., HEGEMAN, G., Eds.), pp. 68–84, Washington, DC: American Society for Microbiology.

STUTZMAN-ENGWALL, K., OTTEN, S., HUTCHINSON, C. R. (1992), Regulation of secondary metabolism in *Streptomyces* spp. and overproduction of daunorubicin in *Streptomyces peucetius*, *J. Bacteriol.* **174,** 144–154.

SZESZAK, F., VITALIS, S., BEKESI, I., SZABO, G. (1991), Presence of factor C in streptomycetes and other bacteria, in: *Genetics and Product Formation in Streptomyces* (BAUMBERG, S., KRÜGEL, H., NOVACK, D., Eds.), pp. 11–18, New York: Plenum Press.

TAKAHASHI, H., TANAKA, K., SHIINA, T., MASUDA, S. (1990), Biological significance of multiple principal sigma factors in eubacteria, in: *Proc. 6th Int. Symp. Genetics of Industrial Microorganisms* (HESLOT, H., DAVIES, J., FLORENT, J., BOBICHON, L., DURAND, G., PENEASSE, L., Eds.), pp. 363–372, Paris: Societé Française de Microbiologie.

THEBERGE, M., LACAZE, P., SHARECK, F., MOROSOLI, R., KLUEPFEL, D. (1992), Purification and characterization of an endoglucanase from *Streptomyces lividans* 66 and DNA sequence of the gene, *Appl. Environ. Microbiol.* **58,** 815–820.

THOMPSON, C., HOLT, T. RAIBAUD, A., LAURENT, C., CHANG, C., MEYERS, P., GARRELS, J., MURAKAMI, T., DAVIS, J. (1990), Regulation and synthesis of the peptide antibiotic bialaphos in *Streptomyces hygroscopicus* in: *Proc. 6th Int. Symp. Genetics of Industrial Microorganisms* (HESLOT, H., DAVIES, J., FLORENT, J., BOBICHON, L., DURAND, G., PENEASSE, L., Eds.), pp. 393–402, Paris: Societé Française de Microbiologie.

TOMODA, H., ŌMURA, S. (1989), Triacsins, acyl-CoA synthase inhibitors and F244, a hydroxymethylglutaryl-CoA synthase inhibitor, in: *Novel Microbial Products for Medicine and Agriculture* (DEMAIN, A. L., SOMKUTI, G. A., HUNTER-CREVA, J. C., ROSSMOORE, H. W., Eds.), pp. 161–169, Amsterdam: Elsevier Science Publishers.

UMEZAWA, H. (1988), Low-molecular weight enzyme inhibitors and immunomodifiers, in: *Actinomycetes in Biotechnbology* (GOODFELLOW, M., WILLIAMS, S. T., MORDARSKI, M., Eds.), pp. 285–325, London: Academic Press.

UMEZAWA, K., HORI, M., TAKEUCHI, T. (1989), Microbial secondary metabolites inhibiting oncogene functions, in: *Novel Microbial Products for Medicine and Agriculture* (DEMAIN, A. L., SOMKUTI, G. A., HUNTER-CREVA, J. C., ROSSMOORE, H. W., Eds.), pp. 57–62, Amsterdam: Elsevier Science Publishers.

VANEK, Z., NOVAK, J., JECHOVA, V. (1988), Primary and secondary metabolism, in: *Biology of Actinomycetes '88* (OKAMI, Y., BEPPU, T., OGAWARA, H., Eds.), pp. 389–394, Tokyo: Japan Scientific Societies Press.

VATS, S., STUTTARD, C., VINING, L. C. (1987), Transductional analysis of chloramphenicol biosynthesis genes in *Streptomyces venezuelae*, *J. Bacteriol.* **169,** 3809–3813.

VINING, L. C. (1992), Secondary metabolism, inventive evolution and biochemical diversity – a review, *Gene* **115,** 135–140.

VINING, L. C., DOULL, J. L. (1988), Catabolite repression of secondary metabolism in actinomycetes, in: *Biology of Actinomycetes '88* (OKAMI, Y., BEPPU, T., OGAWARA, H., Eds.), pp. 406–411, Tokyo: Japan Scientific Societies Press.

VOTRUBA, J., VANEK, Z. (1989), Physicochemical factors affecting actinomycete growth and secondary metabolism, in: *Regulation of Secondary Metabolism in Actinomycetes* (SHAPIRO, S., Ed.), pp. 263–281, Boca Raton: CRC Press, Inc.

WELLINGTON, E. M. H., CRESSWELL, N., HERRON, P. R. (1992), Gene transfer between Streptomycetes in soil, *Gene* **115,** 193–198.

WEYLAND, H., HELMKE, E. (1988), Actinomycetes in the marine environment, in: *Biology of Actinomycetes '88* (OKAMI, Y., BEPPU, T., OGAWARA, H., Eds.), pp. 294–299, Tokyo: Japan Scientific Societies Press.

WILLIAMS, S. T., VICKERS, J.C. (1988), Detection of actinomycetes in the natural environment – problems and perspectives, in: *Biology of Actinomycetes '88* (OKAMI, Y., BEPPU, T., OGAWARA, H., Eds.), pp. 265–270, Tokyo: Japan Scientific Societies Press.

WILLIAMS, S. T., LANNING, S., WELLINGTON, E. M. H. (1984), Ecology of actinomycetes, in: *The Biology of Actinomycetes* (GOODFELLOW, M., MORDARSKI, M., WILLIAMS, S. T. Eds.), pp. 481–528, London: Academic Press.

WILLIAMS, S. T., GOODFELLOW, M., ALDERSON, G. (1989), Genus *Streptomyces* Waksman and Henrici 1943, 339AL, in: *Bergey's Manual of Systematic Bacteriology* (WILLIAMS, S. T., SHARPE, M. E., HOLT, J. G, MURRAY, R. G. E., BRENNER, D. J., KRIEG, N. R., MOULDER, J. M., PFENNIG, N., SNEATH, P. H. A., STALEY, J. T., Eds.), Vol. 4, pp. 2452–2492, Baltimore: Williams & Wilkins.

WOHLLEBEN, W., ANJAH, R., DORENDORF, J., HILLEMANN, D., NUSSBAUMER, B., PELZER, S. (1992), Identification and characterization of phosphinothricintripeptide biosynthetic genes in *Streptomyces viridochromogenes*, *Gene* **115,** 127–132.

WU, P. C., KROENING, T. A., WHITE, P. J., KENDRICK, K. F. (1992), Histidine ammonia-lyase from *Streptomyces griseus*, *Gene* **115,** 19–25.

YAGINUMA, S., ASAHI, A., TAKADA, M., HAYASHI, M., TEUJINO, M., MIZUNO, K. (1989), Aldostatin, a novel aldose reductase inhibitor, in: *Novel Mi-*

crobial Products for Medicine and Agriculture (DEMAIN, A. L., SOMKUTI, G. A., HUNTER-CREVA, J. C., ROSSMOORE, H. W., Eds.), pp. 127–133, Amsterdam: Elsevier Science Publishers.

YAMADA, Y., NIHIRA, T., SAKUDA, S. (1992), Biosynthesis and receptor protein of butyrolactone autoregulator of *Streptomyces virginiae*, *Actinomycetologica* **6,** 1–8.

YOKOTA, T., OMORI, T., KODAMA, T. (1987), Purification and properties of haloalkane dehalogenase from *Corynebacterium* sp. strain m15-3, *J. Bacteriol.* **169,** 4049–4054.

ZÄHNER, I., DRAUTZ, H., FIEDLER, P., GROTE, R., KELLER-SCHIERLEIN, W., KÖNIG, W. A., ZEECK, A. (1988), Ways to new metabolites from actinomycetes, in: *Biology of Actinomycetes '88* (OKAMI, Y., BEPPU, T., OGAWARA, H., Eds.), pp. 171–177, Tokyo: Japan Scientific Societies Press.

ZHANG, H. Z., SCHMIDT, H., PIEPERSBERG, W. (1992), Molecular cloning and characterization of two lincomycin-resistance genes, *lmrA* and *lmrB*, from *Streptomyces lincolnensis* 78-11, *Mol. Microbiol.*, in press.

14 Yeasts

Jürgen J. Heinisch
Cornelis P. Hollenberg
Düsseldorf, Federal Republic of Germany

1 Introduction 470
2 Species of Biotechnological Importance 470
2.1 Natural Occurrence, Isolation, and Cultivation 471
2.2 Life Cycles 472
2.3 Organization of the Genetic Material and Methods of Manipulation 474
2.3.1 Chromosomal Organization 474
2.3.2 Extrachromosomal Inheritance 476
2.3.3 Manipulation of the Genetic Material 476
3 Metabolism of *Saccharomyces cerevisiae* 478
3.1 Carbohydrate Metabolism 478
3.1.1 Glycolysis and Pentose Phosphate Pathway 480
3.1.2 Gluconeogenesis 484
3.1.3 TCA Cycle, Glyoxylate Cycle, and Respiration 484
3.1.4 Storage Carbohydrates 485
3.1.5 Regulatory Mechanisms in Carbohydrate Metabolism 486
3.2 Amino Acid Metabolism 489
3.3 Lipids, Vitamins, and Inorganic Phosphate 490
4 Regulatory Mechanisms of Gene Expression 491
4.1 Transcriptional Regulation 491
4.1.1 Transcription of Glycolytic Genes 492
4.1.2 Transcription of Galactose-Inducible Genes 493
4.2 Post-Transcriptional Regulation 494
5 Industrial Applications 495
5.1 "Classical" Applications 495
5.2 Use of Yeasts for Heterologous Gene Expression 496
5.2.1 Expression Vectors 499
5.2.2 Secretion 500
6 Perspectives 501
7 References 501

1 Introduction

Yeasts are unicellular eukaryotes which with few exceptions at least at some stages in their life cycle divide by budding (PHAFF, 1990). Although a large number of species belong to this group (BARNETT et al., 1983; KREGER-VAN RIJ, 1984), *Saccharomyces cerevisiae* is the organism commonly referred to, when the term "yeast" is not further specified. This species has been used for thousands of years by mankind in the preparation of wine, beer, and bread and can thus be regarded as a GRAS organism (generally regarded as safe). For scientific purposes, yeasts combine the advantages of prokaryotes (fast growth, relatively small genomes) with the opportunity to study the functions of a eukaryotic cell with regard to gene expression, post-translational protein modifications, and secretion. In fact, yeast has attracted that much attention in the last decade that it has been called the "*E. coli* of eukaryotic cells" (WATSON et al., 1987). For industrial applications this resulted in yeasts being more and more exploited as hosts for the expression and/or isolation of heterologous genes (see Sect. 5.2).

From a genetic point of view, yeast had been well characterized by a large set of mutants even before the onset of gene technology, providing the basis for the development of transformation techniques. These methods also gained profit from the fact that yeast can be stably maintained both in the haploid and the diploid state, making it possible to study mutants in essential genes to some extent. The ease with which genetic material can be isolated, manipulated *in vitro*, and introduced into the yeast cell led to the characterization of more than 700 genes by 1989 (MORTIMER et al., 1989), a number likely to be in excess of 1000, today. In addition, with the combined efforts of the European Community and a worldwide project, the sequence of the complete yeast genome will be available in the foreseeable future (GRIVELL and PLANTA, 1990; VASSAROTTI and GOFFEAU, 1992). This in turn will make every part of each chromosome available for genetic manipulation and lead to new findings in the metabolic relationships, making yeast the ideal production organism. The following sections will give an impression of different yeast species, their genetics and physiology and of their possible use.

A number of monographs have been published on different aspects of yeast taxonomy, metabolism, and genetics. Thus, the second edition of *The Yeasts* (ROSE and HARRISON, 1989) provides an extensive overview on different yeast species and their metabolism. Genetic aspects on *S. cerevisiae* are covered in two excellent volumes (STRATHERN et al., 1981, 1982), whose second edition in three volumes is being published (BROACH et al., 1991). A recent volume of *Methods in Enzymology* (GUTHRIE and FINK, 1991) was dedicated to the techniques used in yeast molecular genetics and a large part of another volume (GOEDDEL, 1990) concerns yeast secretion systems. More concise information on these subjects is also available (CAMPBELL and DUFFUS, 1988; SPENCER et al., 1983; HICKS, 1986; BARR et al., 1989). In addition, numerous review articles have been published on specialized aspects of yeast biology, which will be cited in the following wherever appropriate. Due to the vast amount of publications available on the physiology, biochemistry, and genetics of yeasts, we will largely refrain from citing original work. Rather, recent reviews and research papers on specific topics will be mentioned, from which the interested reader should easily find all relevant literature.

2 Species of Biotechnological Importance

As mentioned above, *Saccharomyces cerevisiae* has long been used in the production of wine, beer, and bread. It was also the first yeast studied to be used for the production of heterologous proteins such as human interferon (HITZEMAN et al., 1981) and hepatitis B surface antigen (VALENZUELA et al., 1982). Furthermore, *Hansenula polymorpha*, *Pichia pastoris*, *Yarrowia lipolytica*, and *Kluyveromyces lactis* have been used successfully for heterologous protein production (Tab. 1; REISER et al., 1990;

Tab. 1. Yeast Species Used for Heterologous Protein Production

Species	Proteins
Saccharomyces cerevisiae	see Tab. 5 for extensive listings
Kluyveromyces lactis	Chymosin, IL-1β, human serum albumin
Pichia pastoris	Human IGF-1, human lysozyme, aprotinin, HIV gp120, tetanus C
Hansenula polymorpha	Human serum albumin, human lipase, hepatitis B antigens
Schizosaccharomyces pombe	Human α-1 antitrypsin, human antithrombin III Human blood coagulation factor XIII a
Yarrowia lipolytica	Human tPA, porcine α-1 interferon, calf and bovine prochymosin

Only some examples of proteins produced in non-*Saccharomyces* yeasts are given. See BURKE et al. (1989), BUCKHOLZ and GLEESON (1991), REISER et al. (1990) for more extensive listings and references.

BUCKHOLZ and GLEESON, 1991; GELLISSEN et al., 1992).

Yeasts of the genera *Hansenula*, *Candida*, and *Pichia* have been developed because of their capacity to use cheap compounds such as methanol as a carbon source (GLEESON and SUDBERY, 1988). *Candida boidinii* and some species of *Torulopsis* have been found to utilize C1-compounds as the sole carbon source and some methylated amines as a nitrogen source (HARDER and VEENHUIS, 1989; SAHM, 1977). Apparently such methylotrophic yeasts have adapted their metabolism to the utilization of decomposing plant material. Among these yeasts, *H. polymorpha* seems unusual, as it grows well at temperatures around 40 °C. Utilization of methanol starts with oxidation to formaldehyde, catalyzed by methanoloxidase (MOX; also generally referred to as alcohol oxidase), a process in which hydrogen peroxide is formed, that is immediately converted to water and oxygen by the action of catalase. Formaldehyde dehydrogenase and formate dehydrogenase reactions then lead to the formation of carbon dioxide. Whereas the first two reactions take place in peroxisomes, the latter two enzymes are located in the cytoplasm. For the assimilation of methanol, a special xylulose monophosphate pathway, with the key dihydroxyacetone synthase (DAS) reaction, has been detected in yeast, being different from the ribulose monophosphate pathway found in methylotrophic bacteria (HARDER and VEENHUIS, 1989). Methanol metabolism is regulated by carbon catabolite repression by readily available carbon sources like glucose or ethanol (EGGELING and SAHM, 1978).

Other yeast species have been selected for their ability to utilize *n*-alkanes as a carbon source. Thus, *Candida tropicalis*, *Yarrowia lipolytica*, and species of *Torulopsis* were favored for the production of single cell protein, but also to produce intermediates of the tricarboxylic acid cycle, vitamins and coenzymes (TANAKA and FUKUI, 1989). *n*-Alkanes in such yeasts are metabolized through their oxidation to fatty acids and conversion through acyl-CoA-intermediates. The metabolic routes of β-oxidation and the glyoxylate pathway are localized in peroxisomes in these yeasts.

A few species of *Pichia* and *Candida* are also capable of utilizing pentoses such as xylose for aerobic growth. As pentoses are the principal components of hemicelluloses, interest has been focused upon their bioconversion (JEFFRIES, 1990). However, no useful products such as ethanol are produced in adequate quantities by such yeasts. Thus, attempts are made to express the genes encoding the key enzymes for xylose utilization in *S. cerevisiae* (see Sect. 5.1).

2.1 Natural Occurrence, Isolation, and Cultivation

Yeasts belong to the fungi. They lack photosynthetic abilities and are saprophytic or parasitic, depending on organic carbon sources provided by other organisms (PHAFF, 1990). Thus, yeasts were first identified in sugar-containing natural substrates such as fruits and berries. They are also found associated with other parts of plants and have been shown to occur in soil and aquatic environments, wherever organic carbon sources are available (PHAFF and STARMER, 1987). This depends on their capability for using available carbon sources and their sensitivity to substances like antibiotics and phenolic compounds. Several yeast species are found in special habitats. Yeasts have also been frequently found on insects, providing a means for their distribution. Only a few opportunistic yeast pathogens of humans (*Candida*, *Cryptococcus*) are known (ERNST, 1989).

The fact that different yeast species are found in different habitats, can be used for their isolation. For these isolations one can also take advantage of the fact that yeasts can grow in a broad range of pH values preferring acidic media (pH 3.5 to 5.0), whereas bacteria generally grow better in more alkaline environments. Thus, yeast extract (0.3%), malt extract (0.3%), peptone (0.5%), and glucose (1%) are used at pH 3.7–3.8, occasionally in the presence of antibiotics like chloramphenicol, to avoid bacterial contamination (PHAFF, 1990).

For the laboratory, defined media for the growth of wild-type strains and their mutant derivatives as well as general guidelines for their handling have been published (NAKAJIMA, et al., 1991; GUTHRIE and FINK, 1991). For larger-scale production, cheaper substrates such as molasses (for *S. cerevisiae*) and whey (for *K. lactis*) can be used. It is common practice in baking, breweries, and wineries to add yeast for a fast fermentation (ODA and OUCHI, 1989a; BOULTON, 1991; HENICK-KLING, 1988). Although different species names are still used for such strains defining their specific performances (e.g., *S. carlsbergensis*, *S. uvarum*), taxonomically in fact all belong to *S. cerevisiae* (BARNETT, 1992).

2.2 Life Cycles

The life cycles of only a few yeast species have been worked out in detail. The different stages for *Saccharomyces cerevisiae*, the yeast best studied so far, are shown in Fig. 1. One of the great advantages for geneticists is the ability of this yeast to grow both in the haploid and in the diploid state (HERSKOWITZ, 1989). Haploid laboratory strains are defined by one of two possible mating types (*MAT*a or *MAT*α). Mating at high frequency will only occur between two yeast cells of opposite mating type, and karyogamy will immediately follow zygote formation. Using strong selection procedures, one can construct a diploid being *a* or α in mating type. Such a diploid can then be crossed with another strain of the opposite mating type, thus creating tri- or even tetraploids for genetic studies (MAYER and AGUILERA, 1990).

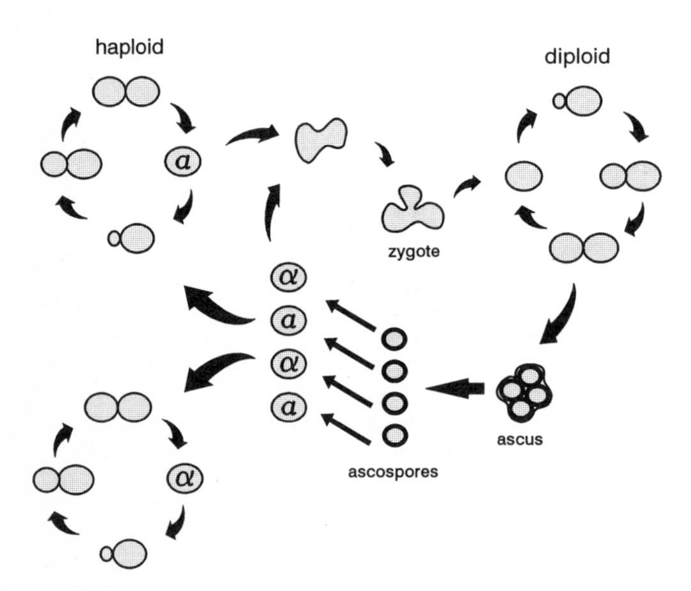

Fig. 1. Cell cycle of the budding yeast *Saccharomyces cerevisiae.*

However, this method is only used for special applications. The process of mating involves the production of pheromones by the haploid cells, called *a*- and α-factor, respectively (FIELDS, 1990), which are not produced by diploid cells heterozygous for the *MAT* locus. After binding to a receptor in cells of opposite mating type, the pheromones elicit a signal transduction pathway causing severe metabolic changes (ANDREWS and HERSKOWITZ, 1990; MARSH et al., 1991). Thus, cells are arrested in the G1 phase of their cell cycle and start to form "shmoos", indicating their ability to mate. A zygote is formed which initiates the diploid life cycle by budding. Diploid cells are slightly larger, lack the ability to mate but are otherwise phenotypically indistinguishable from haploid yeast cells. By adjusting the growth conditions, i.e., transfer to a medium lacking a nitrogen source and providing only a poor carbon source (1% potassium acetate), diploid cells can be forced to undergo meiosis. The four haploid spores (a "tetrad" in genetic terminology) produced in this process are contained in an ascus. Upon germination, the spores will mate, but in case of separation they can resume the haploid life cycle. It should be mentioned at this point that although the spores are slightly more resistant to heat than vegetative cells they do not survive temperatures above 60 °C, unlike most bacterial spores.

Standard genetic procedures (GUTHRIE and FINK, 1991) can be used for crosses and dissection of the tetrads, to obtain recombined haploid yeast strains. Tetrad analysis not only provides a means of combining any set of genetic markers from two yeast strains, but has been invaluable for the mapping of the respective genes (MORTIMER et al., 1989). Following the segregation of two markers in a cross between two haploid strains, one can determine the number of parental ditypes, non-parental ditypes, and tetratypes in the progeny (MORTIMER and HAWTHORNE, 1966). A simple equation, Eq. (1), gives a fairly accurate estimation of how closely the two markers are linked. The equation is used up to about 35 centiMorgan (cM) and adjustments for larger distances can be made (MORTIMER et al., 1989). On the average, 1 cM equals a physical distance of about 3 kb, with significant variations for specific chromosomal loci (KABACK et al., 1989).

$$\mathrm{cM} = \frac{\mathrm{T} + 6\ \mathrm{NPD}}{2\ (\mathrm{PD} + \mathrm{NPD} + \mathrm{T})} \times 100 \quad (1)$$

where PD is the parental ditype, NPD the non-parental ditype, and T the tetratype.

The natural life cycle is that of a homothallic yeast strain, which is able to switch the mating type due to a gene called *HO*. However, heterothallic strains have been isolated early in the development of yeast genetics being defective in the *HO* function (MARSH et al., 1991). The gene encodes an endonuclease initiating a special form of homologous recombination (i.e., gene conversion) leading to replacement of the mating type allele with that of the opposite mating type present as silent copies on the same chromosome as *MAT* (*HML*α and *HMR*a are located on chromosome III on the left and right arm, respectively).

As yeast strains commonly employed for baking, brewing, and wine making are polyploid or aneuploid (BAKALINSKY and SNOW, 1990; BOULTON, 1991; HENICK-KLING, 1988), genetic analysis as described above is usually not carried out. It is also made difficult by a low spore viability, if diploid industrial strains can be put through meiosis (ODA and OUCHI, 1989a). However, methods are being developed that will allow strain improvement by crossing with laboratory strains (THORNTON, 1991; ODA and OUCHI, 1989a).

A similar life cycle as in Fig. 1 can be drawn for other yeasts, including *Kluyveromyces lactis*. In contrast to *S. cerevisiae* this yeast does normally not propagate in the diploid phase, mating being followed rapidly by sporulation upon entry into the stationary growth phase. This behavior somewhat limits the possibilities to work with genes whose products are essential for cell survival. Nevertheless, *K. lactis* strains of opposite mating types are available (JOHANNSEN, 1984). However, the mode of inheritance for the mating type in this yeast is not as clear-cut as in *S. cerevisiae*.

The *HO* gene has been isolated from *Schizosaccharomyces pombe*, the fission yeast, that usually grows only in the haploid state as well and undergoes meiosis immediately after mating. With the help of mutants, stable diploid cell cycles could be established there (FORSBURG and NURSE, 1991). This yeast differs from *S. cerevisiae* in that the cell cycle seems to be controlled predominantly at the G2/M transition point, rather than at the G1/S transition.

For *Hansenula polymorpha*, too, mating has been shown to be possible, and four-spored asci can be found (GLEESON and SUDBERY, 1988). However, as they are smaller than the ones of *S. cerevisiae*, micromanipulation has proved to be extremely difficult, and has rarely been carried out successfully. Little is known about the mating type system, but it is generally assumed that no stable diploid phase exists in this yeast. Thus, in culture it propagates as a mixture of haploid and diploid cells (R. ROGGENKAMP, personal communication).

Even less is known at present about species like *Candida albicans* and *Schwanniomyces occidentalis*. The first draws attention because of its opportunistic nature as a pathogenic yeast (SCHERER and MAGEE, 1990; ERNST, 1989), whereas the main biotechnological interest in *S. occidentalis* is its ability to degrade starch by the production of α-amylase and glucoamylase (STRASSER et al., 1989).

2.3 Organization of the Genetic Material and Methods of Manipulation

The size of the yeast genome (about 15 Mbp, depending on the species) is only about 4 times that of *Escherichia coli*. Thus, it is reasonably small to be studied and, as mentioned above, projects for complete sequencing of the yeast nuclear genome are in progress. Nevertheless, the organization of the genetic material is substantially different from that of bacteria, and gene expression has to be brought about differently, as well. Whereas the DNA transcriptional and translational machinery is located in the same compartment in bacteria, yeasts as eukaryotes contain them separated by the nuclear membrane. This also implies a completely different mode of regulation of gene expression for the two groups (see Sect. 4), with yeast being used as a simple model whose data provide insights for gene regulation in higher eukaryotes.

2.3.1 Chromosomal Organization

The genes of yeast are organized in linear chromosomes of different sizes, the numbers of which vary between different species (Tab. 2; SOR and FUKUHARA, 1989; STEENSMA et al., 1988). The chromosomes can be separated by pulse-field or orthogonal-field alternating gel electrophoresis (PFGE, OFAGE), visualized with ethidium bromide and/or transferred to DNA binding matrices for Southern analysis (JOHNSTON et al., 1988). This method greatly accelerated the speed by which new genes can be mapped in the yeast genome, in conjunction with conventional tetrad analysis. Each chromosome apparently consists of a single double-stranded DNA molecule (FANGMAN and BREWER, 1991), associated with histones H2,

Tab. 2. Number of Chromosomes and Genome Sizes in Different Yeasts

Species	Number of Chromosomes	Chromosome Size (Mbp)	Haploid Genome Size (Mbp)
Saccharomyces cerevisiae	16	0.2–1.6	15
Kluyveromyces lactis	6	1.0–3.0	12
Schizosaccharomyces pombe	3	3.7–5.7	14
Candida albicans	8	1.0–3.0	14–18
Pichia (Hansenula) canadiensis	6–8	>1.0	n.d.
Hansenula polymorpha	4–6	>1.0	n.d.
Schwanniomyces occidentalis	5	>1.0	n.d.

Data for *S. cerevisiae* are from CARLE and OLSON (1985), for *K. lactis* from SOR and FUKUHARA (1989), for *S. pombe* from SMITH et al. (1987), for *C. albicans* from LASKER et al. (1989), for *H. polymorpha* from G. GELLISSEN (personal communication) and for the other species from JOHNSTON et al. (1988).
n. d., not determined

H3, and H4 (PEREZ-ORTIN et al., 1989), to form nucleosomes as in higher eukaryotes. Although other proteins are associated with DNA in yeast, at least for *S. cerevisiae* histone H1 could not be detected.

Replication origins are located at an average distance of about 40 kb on yeast chromosomes and are presumably identical with so-called ARS elements (autonomously replicating sequence; FANGMAN and BREWER, 1991). The latter have first been identified by their ability to confer high frequency transformation to plasmids in yeast (STRUHL et al., 1979). A centromeric region, with specific sequence requirements (see Fig. 2; CLARKE, 1990) and devoid of nucleosomes, enhances stability of such plasmids. Finally, telomeres are localized at the ends of each chromosome. They, too, have a special structure (Fig. 2; WRIGHT et al., 1992), and special enzymes are needed to ensure their replication (D'MELLO and JAZWINSKI, 1991). The three functional elements of a chromosome (CEN, ARS, and TEL) have been manipulated *in vitro* to construct yeast vectors (YAC = yeast artificial chromosome) able to accomodate large fragments of foreign DNA (see Sect. 5.2.1).

Proper segregation of chromosomes during mitosis and meiosis is crucial for eukaryotic organisms. Thus, in tests for mutagenicity of certain chemicals, bacterial systems have only been useful, as far as these substances act directly on the DNA. However, malsegregation of chromosomes due to interaction of drugs with other components, such as tubulin, can only be detected in eukaryotic organisms. Again, yeast has served as a simple model system, and tests for mutagenic (and possibly cancerogenic) effects have been developed for this special purpose (ALBERTINI and ZIMMERMANN, 1991). For basic research, yeast systems have also been used to study the mechanisms of spontaneous and induced mutations (LEE et al., 1988).

Molecular analysis of isolated DNA fragments gives the impression that the nuclear genome of yeast is tightly packed with protein coding information and only a few repetitive sequences can be detected. The largest of these is located on chromosome XII and contains about 100 copies of tandemly repeated rRNA genes (WARNER, 1989). Besides this gene cluster, there are three classes of repetitive DNA located in the chromosomes of *S. cerevisiae.* They belong to the group of transposable elements and are called Ty (transposon yeast; EIGEL and FELDMANN, 1982), sigma (=Ty3; CHALKER and SANDMEYER, 1990), and tau (CHISHOLM et al., 1984). They are frequently found in the promoter regions of genes activating or repressing transcription (FARABAUGH et al., 1989). By far the best studied group are the Ty elements, being about 6 kb in size, flanked by direct repeats of so-called delta elements. Ty elements can be lost through recombination, leaving solo-delta elements at their original position. The complete Ty element is transcribed as one unit encoding two polypeptide

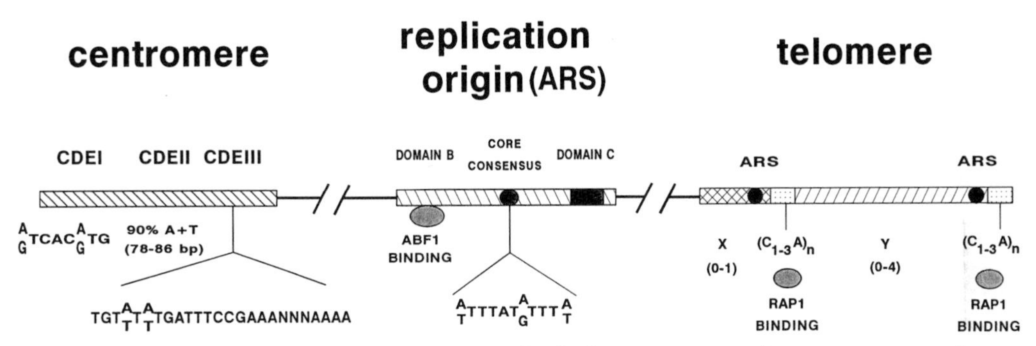

Fig. 2. Functional elements of a yeast chromosome. Each chromosome contains a centromere, telomere sequences at the two ends, and several replication origins (5–50, depending on the chromosome length). Consensus sequences derived from several chromosomes for each element are given. See Fig. 6 for consensus sequences of ABF1 and RAP1 binding sites. X and Y regions are dispensable for function and found at most, but not all, yeast telomeres.

chains that are produced by a translational frameshift mechanism (BELCOURT and FARABAUGH, 1990). One of the proteins has a reverse transcriptase activity (MÜLLER et al., 1991), and it has been shown that transposition of Ty elements occurs in a retrovirus-like manner (FINK et al., 1986). Virus-like particles could also be detected in yeast, but apparently are not set free by living cells and are not infectious. They have been proposed as possible carriers for heterologously produced proteins.

A major difference in the organization of the genomes of most yeasts, such as *S. cerevisiae*, and *Schizosaccharomyces pombe* is the frequency of introns found within their genes. Whereas the latter yeast resembles higher eukaryotes in that many genes contain intervening sequences (NASIM et al., 1989), in *S. cerevisiae* only few examples of nuclear genes containing introns have been found (RUBY and ABELSON, 1991).

2.3.2 Extrachromosomal Inheritance

Within the nucleus *Saccharomyces cerevisiae* contains 60–100 copies of the 2 μm plasmid (HOLLENBERG et al., 1970; VOLKERT et al., 1989). The function of this plasmid for yeast is not clear, as cir^0 strains lacking these extrachromosomal elements are indistinguishable from wild-type strains carrying them (ERHART and HOLLENBERG, 1981). This high copy number plasmid is extremely useful for the overproduction of proteins in yeast, as will be explained in Sect. 5.2.1. Its special structure, which consists of two halves divided by two inverted repeated sequences, seems to be closely related to its mode of replication (HOLLENBERG et al., 1976). Homologous recombination between these sequences leads to the natural occurrence of two forms of the plasmid, named A and B. If the recombination event occurs shortly after the onset of the bidirectional replication, the orientation of both forks becomes the same, leading to the production of concatamers of 2 μm DNA, which are resolved by another recombination event. Thus, with only a single start of replication during each cell cycle, multiple rounds of replication can be achieved. The copy number is controlled by an autocatalytic mechanism, not yet completely understood (VOLKERT et al., 1989). Plasmids with similar structural features, but no apparent sequence similarities have been detected in other yeasts, e.g., *K. lactis* (CHEN et al., 1986) and some species of *Zygosaccharomyces* (TOH-E et al., 1982).

Another nuclear determinant that does not follow the rules of Mendelian inheritance is the so-called killer phenotype. Some strains of *S. cerevisiae* and *K. lactis* are able to produce a killer toxin, inhibiting growth of sensitive strains of the same species (STARK et al., 1990; DUJON, 1981). The genetic information for the toxin is carried on linear, double-stranded RNA or DNA molecules, respectively. The signal sequence for secretion of the toxin has been exploited for directing secretion of foreign proteins from yeast (see Sect. 5.2.2). The killer phenomenon has been applied in clinical trials to biotyping of pathogenic yeasts (POLONELLI et al., 1991) and can play an important role for beer and wine yeasts.

Other extrachromosomal DNA is found in mitochondria, whose circular DNA is about 75 kb in length (HOLLENBERG et al., 1970) and is present in about 40–50 copies per cell (WICKNER, 1981). For industrial purposes, mitochondrial genetics has not yet been employed.

2.3.3 Manipulation of the Genetic Material

In this section, we will concentrate on methods for genetic manipulation mainly of yeast chromosomes. The features of vectors used for special purposes are to be discussed later (Sect. 5.2).

Today, common forms of genetic manipulation depend on the possibility of introducing DNA modified *in vitro* back into the living cell. For yeast, several methods of transformation are in use (Tab. 3). For long-term storage of competent yeast cells and fast transformation of self-replicating plasmids, the freeze method (KLEBE et al., 1983) has proved to be most useful in our hands (DOHMEN et al., 1991). It can also be applied to yeasts other than *S. cerevisiae*, and seems to be fairly independent of the

Tab. 3. Methods for Yeast Transformation

Method	Transformants per μg DNA[a]	Comments
Spheroplasts	10^5–10^6	Used for transforming genomic libraries and yeast artificial chromosomes, complex procedure
Lithium acetate	10^3–10^4	Whole cell transformation, used in everyday experiments
Improved lithium acetate	10^6–10^7	Used for construction of genomic libraries directly in yeast, variations in transformation frequencies are strain-dependent
Freeze method	10^3–10^4	Competent cells can be used after long-term storage; no strain dependencies in transformation frequency observed; can be used for other yeast species
Electroporation	10^6–10^7	Transformation procedures should be optimized for each strain
E. coli conjugation	low [b]	Not yet used routinely

see text for references
[a] Transformation rates are given for 2 μm-based vectors.
[b] As no free DNA is involved, transformation frequencies per μg of DNA cannot be determined.

genetic background of the strains used. The lithium acetate transformation procedure (ITO et al., 1983), on the other hand, does not give significantly higher transformation frequencies (an improved method has been proposed recently by GIETZ and SCHIESTL, 1991), and varies from strain to strain. However, as it is not very laborious and does not require temperatures of –70 °C, the method has found widespread use. The production of spheroplasts for yeast transformation (BURGERS and PERCIVAL, 1987) efficiently introduces extragenic DNA, but is quite complicated and needs special care, e.g., in taking cells at the right optical density and not leaving them for too long with the cell-wall degrading enzymes. Therefore, although it was the first method developed for yeast transformation (HINNEN et al., 1978; BEGGS, 1978), it is now only used for special purposes, such as transformation with gene libraries and the introduction of very large YAC-based plasmids into the cell. Another high-efficiency transformation procedure has been developed lately, using electroporation (BECKER and GUARENTE, 1990). Here, too, the method has to be optimized for each strain used, and one needs special equipment. However, as the procedure also yields extremely high transformation frequencies for *E. coli* (>10^9 transformants/μg DNA) it might soon belong to the standard procedures in laboratories working on molecular genetics. Most transformation procedures have been recorded to result in cells growing more slowly in the absence of 2 μm DNA by causing a dominant mutation in a single gene (DANHASH et al., 1991). This heritable damage does not seem to be of great practical importance, but might be avoided when plasmids are introduced by promiscuous conjugation, where DNA is directly transferred from *E. coli* to yeast (HEINEMANN and SPRAGUE, 1989).

Once inside the cell, maintenance of DNA can be assured by either of two pathways:

(1) The introduced DNA contains an origin of replication and a selectable marker and can thus be inherited as a plasmid. This has the disadvantage that plasmids are unstable and lost when the selective pressure is removed.
(2) The newly introduced DNA becomes an integral, and thus stable, part of a yeast chromosome and is inherited along with it. The latter possibility makes use of the very efficient system for homologous recombination in yeast.

The relatively small genome, the information gathered by classical genetics, the availibility of vectors, transformation systems, and an efficient system for homologous recombination have made yeast both an ideal system for studying molecular mechanisms in eukaryotes, as well as for the production of foreign proteins.

3 Metabolism of *Saccharomyces cerevisiae*

The metabolism of any eukaryotic cell is a complex subject exceeding the scope of this chapter. Thus, the following will be limited to *S. cerevisiae* and only special aspects of other yeasts will be mentioned. Furthermore, we will concentrate on the description of some general and well studied aspects of yeast metabolism. Again, the reader with special interest in yeast is referred to more extensive monographs (ROSE and HARRISON, 1989; STRATHERN et al., 1982).

Metabolism is not only a function of the set of enzymes that any given organism provides for the conversion of substrates, but also of their compartmentation within a cell. A yeast cell, as a simple eukaryote, harbors the organelles known from higher eukaryotic cells. The bulk of its DNA is localized within the nucleus. Connected to the nuclear membrane is the endoplasmic reticulum (ER) that in the rough ER is covered at the cytoplasmic side of the membrane by polyribosomes. Non-cytoplasmic proteins that are destined for secretion or transport to the vacuole, usually have to pass through the ER, where they are core-glycosylated. Proteins destined to mitochondria, peroxisomes, or the nucleus are synthesized in the cytoplasm with special targeting sequences (SCHEKMAN, 1985; HERMAN and EMR, 1990). As an exception for secretory proteins, the yeast pheromone *a* factor apparently not passes the ER, but uses a strikingly different secretory pathway (KUCHLER et al., 1989). Other secreted proteins are transferred from the ER to the Golgi apparatus and from there to the plasma membrane, involving clathrin-coated vesicles (PAYNE, 1990).

Whereas glycolysis, as the main route of carbohydrate utilization in yeast, is carried out in the cytoplasm, the Krebs cycle and respiration are confined to the mitochondria. In contrast to cells from vertebrates, yeast cells carry a vacuole, being the major compartment for proteolytic degradation. Yeast peroxisomes are the exclusive sites of fatty acid β-oxidation and possibly also contain the enzymes of the glyoxylate cycle (OSUMI and FUJIKI, 1990). Whereas in *S. cerevisiae* peroxisomes are found in moderate numbers upon induction, they can produce giant organelles in methylotrophic yeasts like *Hansenula polymorpha* (ROGGENKAMP et al., 1989).

The interior of a yeast cell is stabilized by a cytoskeleton, whose main components are actin and tubulin fibers (SOLOMON, 1991). At the outside, a rigid cell wall made up of glucans, mannans, and glycoproteins serves for stabilization and as a first barrier for macromolecules (DE NOBEL and BARNETT, 1991). Enzymes secreted by yeast are primarily found in the periplasmic space between cell wall and cytoplasmic membrane (DE NOBEL and BARNETT, 1991). The significance of the compartments just described for yeast metabolism and its utilization as a production organism will be treated in the following sections.

3.1 Carbohydrate Metabolism

The broad use of yeast by mankind was brought about by its unique ability to ferment sugars, yielding ethanol and CO_2 as the end products for use in the brewery and the prepa-

Fig. 3. Schematic representation of carbohydrate metabolism in *Saccharomyces cerevisiae*. Boxed reactions belong to the pentose phosphate pathway. Enzymes designated by numbers are: **1–12**, glycolytic pathway enzymes (see Tab. 4); **13**, fructose-1,6-bisphosphatase; **14**, glycerol-3-phosphate dehydrogenase; **15**, glycerol-1-phosphatase; **16**, glycerolkinase; **17**, phosphoenolpyruvate carboxykinase; **18**, lactate dehydrogenase; **19**, pyruvate dehydrogenase; **20, 23**, citrate synthase; **21**, aldehyde dehydrogenase; **22**, acetyl-CoA synthetase; **24**, pyruvate carboxylase; **25**, 6-phosphogluconate dehydrogenase; **26**, ribose phosphate isomerase; **27**, ribose phosphate epimerase; **28**, transketolase; **29**, transaldolase; **30**, glucose-6-phosphate dehydrogenase. Different steps catalyzed by alcohol dehydrogenase and citrate synthase are performed by different isoenzymes (see text for detailed descriptions).

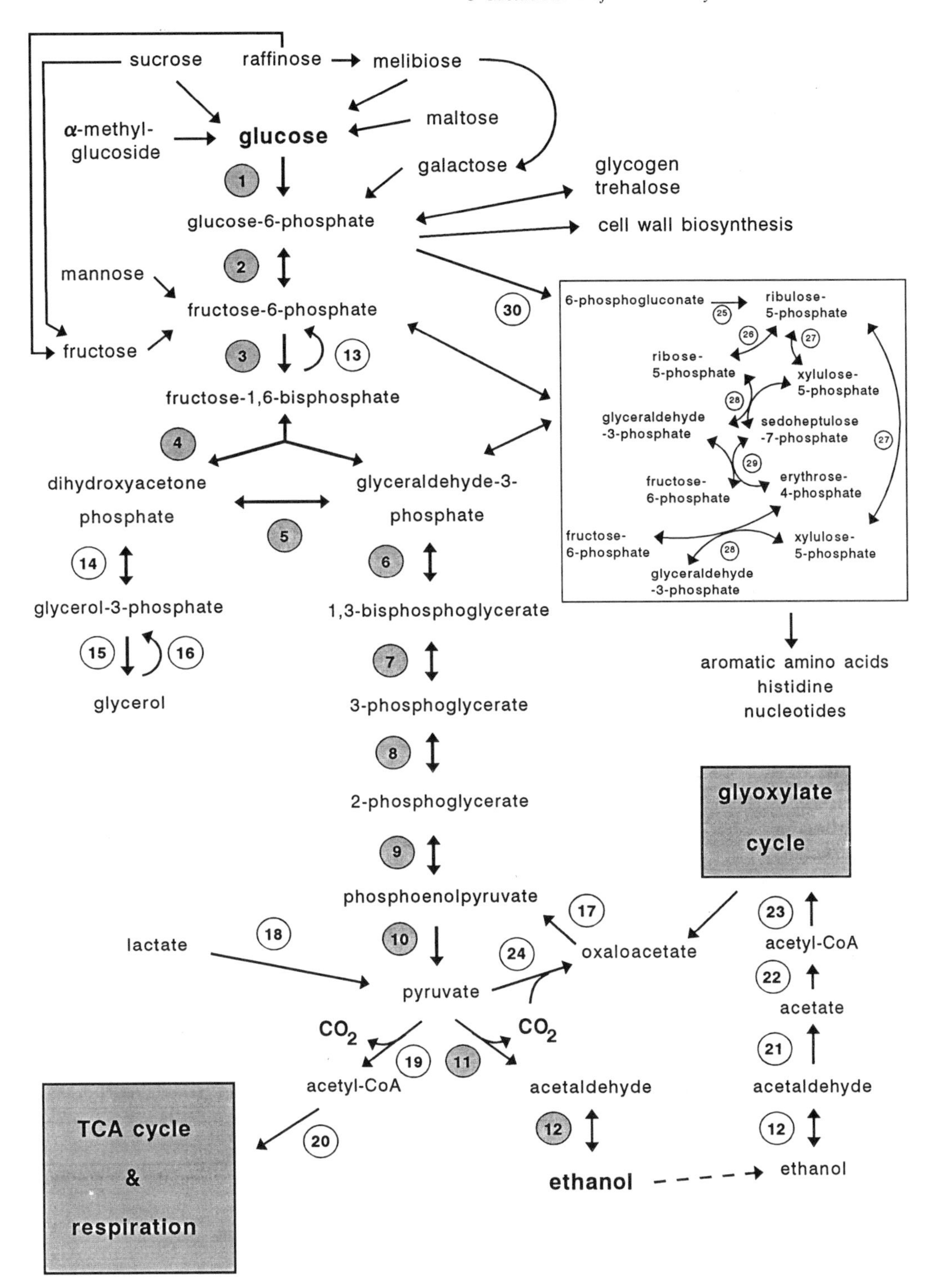

sucrose
raffinose
melibiose
maltose
α-methyl-glucoside
glucose
galactose
glycogen
trehalose
glucose-6-phosphate
cell wall biosynthesis
mannose
fructose-6-phosphate
fructose
fructose-1,6-bisphosphate
dihydroxyacetone phosphate
glyceraldehyde-3-phosphate
glycerol-3-phosphate
1,3-bisphosphoglycerate
glycerol
3-phosphoglycerate
2-phosphoglycerate
phosphoenolpyruvate
lactate
oxaloacetate
pyruvate
CO_2
CO_2
acetyl-CoA
acetaldehyde
TCA cycle & respiration
ethanol
6-phosphogluconate
ribulose-5-phosphate
ribose-5-phosphate
xylulose-5-phosphate
glyceraldehyde-3-phosphate
sedoheptulose-7-phosphate
fructose-6-phosphate
erythrose-4-phosphate
fructose-6-phosphate
xylulose-5-phosphate
glyceraldehyde-3-phosphate
aromatic amino acids
histidine
nucleotides
glyoxylate cycle
acetyl-CoA
acetate
acetaldehyde
ethanol

ration of bread. No yeast other than *Saccharomyces cerevisiae* and only a few other microorganisms have a metabolism so specialized for alcoholic fermentation of mono- and some disaccharides. A schematic representation of the entry of different carbohydrates into the yeast metabolism is given in Fig. 3 (also reviewed in GANCEDO and SERRANO, 1989, and WILLS, 1990).

3.1.1 Glycolysis and Pentose Phosphate Pathway

Whereas all genes encoding the glycolytic enzymes of *S. cerevisiae* have been cloned and sequenced (see Tab. 4), molecular data on its pentose phosphate pathway have emerged only recently. Thus the genes encoding glucose-6-phosphate dehydrogenase (*ZWF1*, THOMAS et al., 1991; NOGAE and JOHNSTON, 1990), transaldolase (*TAL1*, SCHAAFF et al., 1990) and transketolase (*TKT1*, FLETCHER et al., 1992) have been isolated and sequenced. For the first two, null mutants have been constructed, showing that the gene products are not essential for cell viability. Deletion mutants of *TKT1* are also viable (I. SCHAAFF-GERSTENSCHLÄGER, personal communication). It seems noteworthy that several enzymes of the glycolytic pathway are encoded by genes present at least in duplicate. Thus, glucose can be phosphorylated in the first step by either of two hexokinases, encoded by the genes *HXK1* and *HXK2* (KOPETZKI et al., 1985; FRÖHLICH et al., 1985), or a gluco-

Tab. 4. Glycolytic Enzymes and Encoding Genes in *Saccharomyces cerevisiae*

No.[a]	Enzyme (Gene)	E.C. Number	Size[b]	Subunits[c]	Reference[d]
1	Hexokinase PI (*HXK1*)	2.7.1.1	485	2	KOPETZKI et al. (1985)
	Hexokinase PII (*HXK2*)	2.7.1.1	486	2	FRÖHLICH et al. (1985)
	Glucokinase (*GLK1*)	2.7.1.2	500	2	ALBIG and ENTIAN (1988)
2	Phosphoglucose isomerase (*PGI1*)	5.3.1.9	554	2	TEKAMP-OLSON et al. (1988); GREEN et al. (1988)
3	Phosphofructokinase α (*PFK1*)	2.7.1.11	987	4	HEINISCH et al. (1989)
	Phosphofructokinase β (*PFK2*)		959	4	HEINISCH et al. (1989)
4	Aldolase (*FBA1*)	4.1.2.13	359	2	SCHWELBERGER et al. (1989)
5	Triosephosphate isomerase (*TPI1*)	5.3.1.1	248	2	ALBER and KAWASAKI (1982)
6	Glyceraldehyde-3-phosphate dehydrogenase (*TDH1*)	1.2.1.12	322	4	HOLLAND et al. (1983)
	(*TDH2*)		332	4	
	(*TDH3*)		332	4	
7	Phosphoglycerate kinase (*PGK1*)	2.7.2.3	416	1	HITZEMAN et al. (1982)
8	Phosphoglycerate mutase (*GPM1*)	2.7.5.3	247	4	WHITE and FOTHERGILL-GILMORE (1988); HEINISCH et al. (1991b)
9	Enolase (*ENO1*)	4.2.1.11	437	2	MCALISTER and HOLLAND (1985)
	(*ENO2*)		437	2	
10	Pyruvate kinase (*PYK1*)	2.7.1.40	499	4	BURKE et al. (1983)
11	Pyruvate decarboxylase (*PDC1*)	4.1.1.1	563	4	KELLERMANN et al. (1986)
	(*PDC5*)		563		SEEBOTH et al. (1990)
	(*PDC6*)		563		HOHMANN (1991 a)
12	Alcohol dehydrogenase (*ADH1*)	1.1.1.1	348	4	BENNETZEN and HALL (1982)

[a] Numbers refer to the step in glycolysis catalyzed and are the same as in Fig. 3.
[b] Deduced number of amino acids for each subunit encoded by the open reading frames.
[c] Number of subunits an active enzyme is composed of; in the case of phosphofructokinase only a heterooctamer of 4 α- and 4 β-subunits is active.
[d] References given refer to sequence data; see text for additional references on genetic and physiological studies.

kinase, encoded by *GLK1* (ALBIG and ENTIAN, 1988). The first two enzymes can also phosphorylate fructose as a substrate and apparently have a function in the regulation of carbohydrate metabolism (ROSE et al., 1991, see also Sect. 3.1.5). Triple mutants with defects in all three genes fail to grow on media containing glucose, whereas double mutants in the two hexokinase genes do not grow on fructose anymore. Isomerization between glucose-6-phosphate and fructose-6-phosphate is catalyzed by phosphoglucose isomerase. The encoding *PGI1* gene has been isolated and used for the construction of deletion mutants (AGUILERA and ZIMMERMANN, 1986; AGUILERA, 1986). Such mutants have been shown to require trace amounts of glucose when grown on non-fermentable carbon sources, presumably because of the requirement of glucose-6-phosphate as a precursor in cell wall biosynthesis. On the other hand, higher amounts of glucose (above 0.2%) will inhibit growth of such mutants. This can be attributed to repression of gluconeogenesis and respiration under such conditions (see below), blocking the utilization of the non-fermentable carbon sources. This result also indicates that the capacity of the oxidative part of the pentose phosphate pathway in *S. cerevisiae* does not allow for fermentation of the sugar. The existence of a second structural gene for phosphoglucose isomerase, *CDC30* (DICKINSON, 1991), has been suggested. However, the data provided still await further molecular analysis.

The first irreversible reaction specific for glycolysis is catalyzed by phosphofructokinase. Genetic analysis produced some puzzling results. The enzyme is composed of 4 α- and 4 β-subunits, encoded by the genes *PFK1* and *PFK2*, respectively (KOPPERSCHLÄGER et al., 1977; HEINISCH, 1986; HEINISCH et al., 1989). Mutants carrying disruptions in any one of these genes render the cells devoid of a detectable PFK activity, but still able to ferment glucose (BREITENBACH-SCHMITT et al., 1984) and indistinguishable from wild type in their growth characteristics. Only complete deletions of the *pfk2* coding sequence lead to cells growing more slowly on fermentable carbon sources (HEINISCH, unpublished results). Double mutants in both genes do not grow on glucose-containing media. These results can be interpreted as each subunit still having catalytic activity undetectable by *in vitro* assays, or by the involvement of the subunits in a separate pathway of glucose fermentation (HEINISCH and ZIMMERMANN, 1985). Further analysis to distinguish between these possibilities is in progress.

Disruption mutants of *FBA1*, encoding aldolase, constructed after cloning and sequencing of the gene (SCHWELBERGER et al., 1989), indicate that this enzyme is essential for glycolysis and gluconeogenesis. The mutants fail to grow on media containing glucose and only produce small colonies, when grown on rich medium with non-fermentable carbon sources. Their growth behavior upon addition of trace amounts of glucose, as tested for *pgi1* mutants, has not been reported.

The interconversion of triosephosphates by the triosephosphate isomerase (TPI) reaction naturally has its equilibrium at the side of dihydroxyacetone phosphate. However, due to the rapid flux of glyceraldehyde-3-phosphate through the lower part of glycolysis, glucose consumption is effectively channeled into this direction under normal growth conditions. Again, the enzyme is essential for yeast glycolysis, as mutants grow extremely slowly in the presence of glucose (CIRIACY and BREITENBACH, 1979). They also do not grow with either ethanol or glycerol as the sole carbon source. This is presumably due to the inability of such mutants to complete either glycolysis or gluconeogenesis, as the enzymatic activity is needed for both pathways (LAM and MARMUR, 1977). Second site revertants of *tpi1* deletion mutants isolated by their ability to do so, fall into different complementation groups, surprisingly also containing mutants in the gene encoding phosphoglycerate kinase (*PGK1*; M. CIRIACY, personal communication).

Conversion of glyceraldehyde-3-phosphate to 1,3-bisphosphoglycerate is mediated by either of three isoenzymes of glyceraldehyde-3-phosphate dehydrogenases, encoded by the genes *TDH1*, *TDH2*, and *TDH3* (HOLLAND et al., 1983). The isoenzymes are present in a 1:3:6 ratio within the cell, independent of the carbon source used for growth (MCALISTER and HOLLAND, 1985). However, total GAPDH activity is about twofold higher on glucose than on ethanol media (BITTER et al., 1991). The phenotype of triple mutants devoid of detectable GAPDH activity has not been reported.

PGK1, encoding phosphoglycerate kinase, is one of the best expressed genes of a yeast cell, making up about 5% of the total soluble protein (FRAENKEL, 1982). Therefore, it was one of the first glycolytic genes to be sequenced (HITZEMAN et al., 1982), and much attention has been paid to its transcriptional regulation (CHAMBERS et al., 1988, 1990; MELLOR et al., 1987), as well as to the characterization of the protein structure (MISSIAKAS et al., 1990). Mutational analysis has revealed amino acid residues in its catalytic domain, that have been conserved among phosphoglycerate kinase enzymes from a variety of organisms (JOHNSON et al., 1991). *pgk1* deletion mutants do not grow on glucose media. As is true for *tpi1* mutants and mutants in genes whose products catalyze reversible reactions in the lower glycolytic pathway, they fail to grow on ethanol or glycerol as the sole carbon source (LAM and MARMUR, 1977).

The *GPM1* gene, encoding phosphoglycerate mutase, has been cloned, sequenced, and deleted from the haploid yeast genome (RODICIO and HEINISCH, 1987; WHITE and FOTHERGILL-GILMORE, 1988; HEINISCH et al., 1991b). Deletion mutants show a phenotype reminiscent of the ones described for *tpi1* and *pgk1*. However, a residual GPM activity has been detected in such deletion mutants, although Southern analysis does not show other homologous sequences (C. GANCEDO, personal communication, and J. HEINISCH, unpublished data). A sequence similarity to the phosphatase domain of mammalian phosphofructo-2-kinase (WHITE and FOTHERGILL-GILMORE, 1988), the enzyme producing the regulator fructose-2,6-bisphosphate, might not be sufficient to explain this residual activity. Attempts to isolate second site revertants from such deletion mutants that would be able to grow on glucose have also failed so far.

There are two isoenzymes of enolase in yeast, encoded by the genes *ENO1* and *ENO2*. Here, there is a clear functional distinction. Whereas *ENO1* is expressed constitutively (COHEN et al., 1987), about 20 times higher levels of the isoenzyme encoded by *ENO2* are found when cells are grown on glucose as opposed to cells grown on non-fermentable carbon sources (COHEN et al., 1986). A deletion of *ENO1* does not have an apparent phenotype (MCALISTER and HOLLAND, 1982). Again, double deletion mutants have not been described.

The last step of glycolysis is the basically irreversible conversion of phosphoenolpyruvate to pyruvate. This step is catalyzed by pyruvate kinase, encoded by *PYK1* (BURKE et al., 1983). The overall specific activities of most glycolytic enzymes do not vary to a great extent between cells grown on fermentable or non-fermentable carbon sources in *S. cerevisiae* (FRAENKEL, 1982). Seemingly contradictory data are frequently cited but were in fact obtained with a hybrid yeast, being a fusion product of *S. cerevisiae* and *K. fragilis* (MAITRA and LOBO, 1971; SHUSTER, 1989). PYK activities measured in crude extracts of *S. cerevisiae* are three- to four times higher on glucose than on ethanol media, regulation being exerted at the transcriptional level (NISHIZAWA et al., 1989).

In yeast, the term glycolysis is often used in a broader sense, including the two following enzymatic conversions that lead to the final end products of fermentation. At the branching point between fermentation and respiration there is the pyruvate decarboxylase reaction. The first structural gene, *PDC1*, has been isolated by complementation of a mutant unable to grow on glucose as the sole carbon source (SCHMITT and ZIMMERMANN, 1982). Transcription of the gene, in concert with specific enzymatic activities, was shown to be much higher on glucose than on non-fermentable carbon sources (factors vary between 5 and 20, depending on the system used to monitor transcription (SCHMITT et al., 1983; BUTLER and MCCONNELL, 1988; KELLERMANN and HOLLENBERG, 1988)). Surprisingly, a deletion mutant turned out to ferment glucose, in contrast to the originally isolated *pdc1* mutant. The puzzle was solved when a second structural gene, *PDC5*, was found, encoding an isoenzyme (SCHAAFF et al., 1989a; SEEBOTH et al., 1990). Apparently, *PDC5* is only transcribed, when *PDC1* is deleted, and the original *pdc1* mutant did not allow for this autoregulation. Recently, a third structural gene has been reported, being silent and only activated upon a recombination event providing a promoter function (HOHMANN, 1991a, b).

Finally, alcohol dehydrogenase (ADH) interconverts acetaldehyde and ethanol. Again, a small gene family encodes ADH isoenzymes. Thus, *ADH1* encodes a constitutively expressed enzyme, primarily used in the fermentative pathway (CIRIACY, 1975). Accordingly, mutants defective for this enzyme grow slowly on glucose media. In contrast, *ADH2* transcription is repressed on glucose media, indicating that this isoenzyme serves a function in gluconeogenesis (CIRIACY, 1979). Third, *ADH3* encodes an isoenzyme located in mitochondria, whose function might be primarily the consumption of ethanol for energy production through Krebs cycle and respiration (CIRIACY, 1975). A fourth gene, *ADH4*, was detected in revertants of an *adh1* mutant, but its physiological significance is unclear (WALTON et al., 1986). Mutants devoid of all four enzymes have been obtained and shown to produce reduced but significant amounts of ethanol (DREWKE et al., 1990). This has been attributed to other mitochondrial functions (THIELEN and CIRIACY, 1991).

Other glucose-negative mutants have been isolated in the gene encoding 6-phosphogluconate dehydrogenase (LOBO and MAITRA, 1982), where inhibition by glucose has been speculated to be brought about by a toxic accumulation of the substrate 6-phosphogluconate. This conclusion was reached after the isolation of glucose-positive second site revertants with reduced glucose-6-phosphate dehydrogenase activity.

It should be stressed again that *S. cerevisiae* is an exception among yeasts in that it is extremely well adapted for fermentation. Thus, even in the presence of oxygen, about 80% of the glucose consumed is channeled to be fermented, and less than 5% contribute to respiration (see GANCEDO and SERRANO, 1989). The regulatory mechanisms underlying this behavior will be discussed in Sect. 3.1.5. The so-called Pasteur effect postulates that in the presence of oxygen, due to higher energy yield by using respiration, microorganisms will reduce their fermentation rates and thus the consumption of sugars during growth. Unfortunately, *S. cerevisiae* has frequently been cited as an organism showing this effect. However, this yeast does not show a Pasteur effect under normal growth conditions (LAGUNAS et al., 1982). Such an effect can only be observed under special circumstances, e.g., in continuous or fed batch culture using glucose limitation (FIECHTER et al., 1981). This indicates that the intracellular concentration of pyruvate is the critical parameter for the switch between respiratory and fermentative metabolism, with low concentrations being channeled into TCA cycle and respiration. In contrast, other yeasts, such as species of *Kluyveromyces*, *Hansenula*, and *Candida*, have been reported to ferment less than 30% of the glucose consumed under aerobic conditions (GANCEDO and SERRANO, 1989). This indicates that basically different modes of regulation of carbohydrate metabolism are operating in such yeasts, meriting investigation.

In principle, glucose can also be metabolized through the pentose phosphate pathway (see Fig. 3). That this pathway can account for a major part of carbohydrate utilization is indicated by the fact that some yeast species, e.g., *Pichia stipitis*, can use pentoses as sole carbon sources (BARNETT, 1976). Although this is not the case for *S. cerevisiae*, it could be shown that this yeast can be brought to grow on riboses aerobically. Thus, when the *Pichia stipitis* genes encoding xylose-reductase (*XYL1*) and xylitol-dehydrogenase (*XYL2*) are expressed in a special strain of *S. cerevisiae*, cells grow on xylose as the sole carbon source (KÖTTER et al., 1990; AMORE et al., 1991). However, they do not efficiently convert the pentose to ethanol which has been attributed to an imbalance of redox equivalents. The pentose phosphate pathway can be divided into an oxidative part, producing NADPH and a non-oxidative part, yielding precursors for aromatic amino acid biosynthesis as well as phosphoriboses. The latter part of the pathway can be interconnected with glycolysis through glyceraldehyde-3-phosphate and fructose-6-phosphate. In the oxidative part of this pathway, decarboxylation of the C1 carbon atom of glucose leads to CO_2 production. Using radioactive labeling studies, it has been concluded that this part of the pentose phosphate pathway contributes to a maximum of 2.5% to glucose consumption (LAGUNAS and GANCEDO, 1973). This value can be further reduced by using rich media for growth, obliviating the need for amino acid biosynthesis and thus for production of

NADPH. The existence of an L-type pentose phosphate pathway postulated for liver tissues (WILLIAMS, 1980) cannot be ruled out in yeast, especially as one of its intermediates, octulose phosphate, can be produced by yeast extracts (DATTA and RACKER, 1961).

3.1.2 Gluconeogenesis

Yeast shows a diphasic growth behavior when incubated in glucose media under aerobic conditions. This is attributed to the production of ethanol in the first stage, that can be used as a carbon source in the second stage. Enzymatically, this does not constitute a major change, as most of the glycolytic reactions are reversible (see Fig. 3), with the exception of the phosphofructokinase and the pyruvate kinase steps. The reactions catalyzed by these enzymes can be reversed by fructose-1,6-bisphosphatase (FBPase) and the actions of pyruvate carboxylase (PYC) and phosphoenolpyruvate carboxykinase (PEP-CK), respectively. The genes encoding these enzymes have been cloned and sequenced (SEDIVY and FRAENKEL, 1985; ENTIAN et al., 1988; ROGERS et al., 1988; STUCKA et al., 1988, 1991). Their expression is repressed by glucose, avoiding futile cycling (HOLZER, 1976). Again, a functional redundancy is observed in the PYC reaction, as two genes have been found (STUCKA et al., 1991). Interestingly, overproduction of FBPase in cells growing on glucose does not have a marked effect (DE LA GUERRA et al., 1988). Under these as well as under gluconeogenic growth conditions catabolic and anabolic enzymes are present simultaneously, indicating the need for allosteric mechanisms to avoid waste of energy (see Sect. 3.1.5).

Ethanol, glycerol, and lactate are frequently used as gluconeogenic carbon sources. Their entry into metabolism is brought about by specific enzymes, usually repressed on glucose. Thus, ethanol is converted by the ADHII or ADHIII isoenzymes to acetaldehyde (see above). This in turn is oxidized to acetic acid, a reaction mediated by aldehyde dehydrogenases (TAMAKI and HAMA, 1982; DICKINSON and HAYWOOD, 1987). Acetyl-CoA synthetase then produces acetyl-CoA that by the action of the glyoxylate cycle (see next section) with its key enzymes isocitrate lyase and malate synthase, forms oxaloacetate, the substrate for PEP-CK. From there, gluconeogenesis can proceed mainly through the reversible reactions catalyzed by glycolytic enzymes as described above. It should be noted that ethanol crosses the yeast plasma membrane quite rapidly by the way of passive diffusion (LOURIERO and FERRIERA, 1983). Glycerol also diffuses through the yeast plasma membrane and enters yeast metabolism through the action of glycerol kinase and a mitochondrial glycerol-phosphate-ubiquinone reductase. In contrast, lactate seems to be taken up by yeast cells through an active transport system, that is also repressed by glucose. It is then oxidized to pyruvate by a mitochondrial lactate-dehydrogenase. For a discussion of the enzymes involved in glycerol and lactate utilization and discrepancies in their nomenclature see WILLS (1990) and GANCEDO and SERRANO (1989).

3.1.3 TCA Cycle, Glyoxylate Cycle, and Respiration

When grown on non-fermentable carbon sources, yeast has to produce glucose-6-phosphate as a major precursor for polymer biosynthesis through gluconeogenesis as described above. In addition, energy is provided by reactions of the TCA cycle coupled with respiration. Both are localized in the mitochondria. A special feature of *Saccharomyces* and *Kluyveromyces* among yeasts is that they contain only the coenzyme Q6 of ubiquinone in their respiratory chain (YAMADA et al., 1976) and lack a complex I type NADH:Q reductase (GRIVELL, 1989). As is true for other organisms, mitochondrial DNA in yeast encodes only a few proteins. A special kind of self-splicing RNA is observed in mitochondrial transcripts containing introns (GRIVELL and SCHWEYEN, 1989). One can deduce from the limited coding capacity of the mitochondrial genome that more than 95% of the proteins needed to produce functional mitochondria are encoded by nuclear genes (GRIVELL, 1989). Therefore, complex interactions in the regulation of nuclear and mitochondrial genes exist (COSTANZO and

FOX, 1990). Thus, most enzymes needed for the TCA cycle are partially repressed by glucose (PERLMAN and MAHLER, 1974). Respiratory functions are also glucose-repressed and in addition are regulated by the availability of oxygen and heme (ZITOMER and LOWRY, 1992). It should be noted that yeast as a facultative anaerobe is not dependent on mitochondrial functions and a large collection of petite mutants are available (TZAGOLOFF and DIECKMANN, 1990). We will not describe the components of the respiratory chain and their regulation in any detail, as these have been covered extensively by the reviews cited and by DE VRIES and MARRES (1987). As will be evident from the following, the enzymes of the TCA and the glyoxylate cycle are quite redundant. Whereas the TCA cycle is located in the mitochondria, the glyoxylate cycle enzymes are probably located in peroxisomes. Metabolites are interchanged between these pathways across the mitochondrial membrane by different mechanisms, such as the malate-succinate shuttle or the acetyl-carnitine shuttle (WILLS et al., 1986).

The glyoxylate pathway bypasses the two decarboxylation steps in the TCA cycle and thus can lead to the synthesis of C_4 compounds from C_2 precursors. Reactions common to TCA and glyoxylate cycle are catalyzed by different isoenzymes. Thus, two genes encoding citrate synthase (*CIT1* and *CIT2*) have been isolated and sequenced (LEWIN et al., 1990). *CIT1* encodes the mitochondrial enzyme, whereas the protein encoded by *CIT2* is likely to be directed to peroxisomes. The gene encoding isocitrate lyase, *ICL1*, has been isolated recently, and disruption mutants confirmed that it is essential for utilization of ethanol as the sole carbon source (FERNÁNDEZ et al., 1992). As the deduced amino acid sequence does not contain a peroxisomal targeting sequence at its C-terminus, again the subcellular localization of the enzyme is not clear. Two genes encoding malate synthase can be detected by Southern analysis in *S. cerevisiae* (R. RODICIO, personal communication). There appear to exist three structural genes encoding malate dehydrogenases, with *MDH1* directing the synthesis of the mitochondrial isoenzyme (MINARD and MCALISTER-HENN, 1991) involved in the TCA cycle. *MDH2* encodes the major non-mitochondrial isoenzyme whose localization (cytoplasm or peroxisomes) is not yet clear. The role of *MDH3* also remains to be elucidated. Aconitase is encoded by only one gene (*ACO1*) which is synergistically repressed by glucose and glutamate (GANGLOFF et al., 1990). *KGD1* and *KGD2* are the structural genes for two of the three subunits of the α-ketoglutarate dehydrogenase complex in yeast. The genes have been cloned, and their regulation and disruption have been reported (REPETTO and TZAGOLOFF, 1982, 1990). They share the third subunit, a lipoamide dehydrogenase, with the pyruvate dehydrogenase complex (REPETTO and TZAGOLOFF, 1991). Its gene has also been isolated and shown to be subject to carbon catabolite repression (BOWMAN et al., 1992). Genes encoding subunits of the succinate dehydrogenase complex have also been cloned and disrupted (LOMBARDO and SCHEFFLER, 1989; LOMBARDO et al., 1990; ROBINSON et al., 1991).

3.1.4 Storage Carbohydrates

Both the polysaccharide glycogen (α-1,4- and α-1,6-linked glucose monomers) and the disaccharide trehalose (α-1,1-linked glucose monomers) have been reported to accumulate upon entry of yeast into stationary phase and to be degraded when growth is resumed (LILLIE and PRINGLE, 1980). However, different kinetics in the two processes indicate that they serve a different function in yeast metabolism (WINKLER et al., 1991). From the data available, it seems that glycogen is the actual storage compound and provides the energy upon depletion of other carbon sources. On the other hand, trehalose has been implicated to serve a crucial function in conferring resistance to heat shock to cells (HOTTIGER et al., 1989). Accumulation and degradation of both glycogen and trehalose are regulated through a cAMP-dependent mechanism; the synthetic enzymes, glycogen synthase and trehalase synthase, are inactivated by phosphorylation, whereas the enzymes used for degradation, glycogen phosphorylase and trehalase, are active in their phosphorylated form (PANEK and PANEK, 1990).

3.1.5 Regulatory Mechanisms in Carbohydrate Metabolism

In nature, the composition and concentration of carbohydrates vary between different microenvironments or, within a given environment, in time, e.g., by the action of microorganisms metabolizing them. Therefore, yeasts, as any other (micro-)organism, have developed ways to regulate the set of enzymes they need at any given point or time. This is exemplified by the different utilization of carbon sources by different yeast species when grown under optimal laboratory conditions (see GANCEDO and SERRANO, 1989). Thus, the *Pasteur effect*, discussed above as normally not applicable to *S. cerevisiae*, can be observed in other yeast species with a predominantly respiratory metabolism (e.g., *Candida*). The *Kluyver effect*, meaning that some yeasts are able to ferment glucose aerobically, but some other sugars only anaerobically, has been interpreted at its biochemical basis as a specialized Pasteur effect (GANCEDO and SERRANO, 1989). Whereas the Pasteur effect postulates that under aerobic conditions respiration would increase and the rate of sugar consumption decreases, some yeasts show a contradictory behavior, called *Crabtree effect* (VAN URK et al., 1989, 1990). There, respiration is inhibited during aerobic growth, if sufficient sugar is provided and fermentation prevails (e.g., species of *Saccharomyces*, *Schizosaccharomyces*, and *Brettanomyces*). In other yeast species, producing significant amounts of acetate, a *Custers effect* can be observed, as they ferment a sugar more effectively under aerobic than under anaerobic conditions. This has been attributed at the physiological level to the need for balancing the intracellular NAD/NADH ratio (see GANCEDO and SERRANO, 1989, for a more detailed discussion of the effects described).

The phenomena just described are partly brought about by allosteric regulation of the activities of specific enzymes. As it stands, the common view is that more than 80% of the glucose consumed by *S. cerevisiae* are channeled through glycolysis and alcoholic fermentation. Given that glycolytic enzymes are synthesized more or less constitutively (FRAENKEL, 1982), this raises the question of how the flux of metabolites through glycolysis is controlled and what might be the rate-limiting steps. At present, two possible mechanisms are being discussed. First, the cell clearly cannot ferment more glucose than is actually provided as a glycolytic substrate, i.e., glucose transport is one of the crucial control points (RAMOS et al., 1988). It occurs by facilitated diffusion, and the proteins involved are themselves subject to glucose regulation (BISSON, 1988). Second, it has long been assumed that irreversible steps in a pathway can serve to regulate the flux through it. Although reason for the latter relies on formal biochemistry, data have been presented indicating that phosphofructokinase und pyruvate kinase might control the glycolytic flux (HESS and BOITEUX, 1971, reviewed in GANCEDO and SERRANO, 1989). However, overproduction of glycolytic enzymes did not lead to an increase in fermentation rates showing that the amount of enzyme is not limiting in wild-type yeast (SCHAAFF et al., 1989b). On the other hand, enzymes catalyzing irreversible reactions in a pathway are usually liable to allosteric regulation. Thus, increasing the amount of enzyme might not significantly affect its overall *in vivo* activity. Two points for such a regulation in glycolysis have been postulated, namely the phosphofructokinase and the pyruvate kinase reactions (see above). Phosphofructokinase activity is increased by ammonium ions, AMP, and fructose-2,6-bisphosphate as the most potent activator (HOFMANN and KOPPERSCHLÄGER, 1982; NISSLER et al., 1983; BANUELOS et al., 1977). It is inhibited by ATP, and altogether 21 allosteric effectors have been listed for the enzyme from different organisms before the discovery of fructose-2,6-bisphosphate (SOLS, 1981). Pyruvate kinase, on the other hand, has been reported to be activated by the product of phosphofructokinase, fructose-1,6-bisphosphate (GANCEDO and SERRANO, 1989). Thus, upon addition of glucose to cells grown on non-fermentable carbon sources, fructose-2,6-bisphosphate concentrations will rise. This will result in an activation of phosphofructokinase and glycolysis and an inactivation of fructose-1,6-bisphosphatase and gluconeogenesis (YUAN et al., 1990; VAN SCHAFTINGEN and HERS, 1981a, b). Recently, overproduction of PFK *in vivo* has been reported to decrease the intracellular fructose-2,6-bisphosphate concentration (DAVIES and BRINDLE, 1992). This has been attributed to a down-regulation of phosphofructo-2-kinase, encoded by *PFK26* (KRETSCHMER and FRAENKEL, 1991) producing this metabolite. One could thus speculate on the role of PFK in other regulatory networks. However, the observation that yeast cells devoid of detectable PFK activity can still ferment glucose (see Sect. 3.1.1) argues in favor of glucose uptake as the critical determinant of fermentation rates. Clearly, the flux through a pathway is dependent on the availability of its substrates. Thus, interest has been focused in the past years on the systems available for glucose uptake. Biochemically, one can define two uptake systems, namely high-affinity and low-affinity glucose uptake (RAMOS et al., 1988). The first is induced at low concentrations of glucose in the medium (<0.1%), whereas the latter is constitutive and operates at higher concentrations. As for the genetics, *SNF3* has been claimed to encode the high-affinity glucose transporter (MARSHALL-CARLSON et al., 1990), but recent data suggest that additional systems also exist (LEWIS and BISSON, 1991). As hexokinase mutants show a reduced level of glucose uptake, some interaction between glucose transport and phosphorylation has been deduced (BISSON and FRAENKEL, 1983). Although it seems clear that a phosphotransferase system as in bacteria is not operating, the exact mechanism of this interaction has not yet been elucidated.

Other means of specific carbon-source dependent regulation of carbohydrate metabolism have again mostly been provided for *S. cerevisiae* and will be discussed in the following.

As mentioned above, accumulation and degradation of storage carbohydrates are regulated by the state of phosphorylation of key enzymes. The RAS/cAMP signaling pathway by which the extracellular nutrient situation is monitored has drawn much attention in the past years (GIBBS and MARSHALL, 1989; BROACH and DESCHENES, 1990; BROACH, 1991). This cannot be attributed purely to growing interest in yeast metabolism, but mainly to the fact that *RAS* genes were first discovered as protooncogenes in humans. Yeast has been used as a simple eukaryotic model system to elucidate the details of this pathway. Although adenylate cyclase is not the target of the RAS proteins in humans, more and more factors modulating RAS activity common to both organisms are being discovered (MICHAELI et al., 1989; BALLESTER et al., 1989, 1990). A simplified version of the pathway in yeast is drawn in Fig. 4. Addition of glucose to the medium is sensed (directly or indirectly) by the membrane bound CDC25 protein (JONES et al., 1991; VAN AELST et al., 1990). This induces a substitution of GDP bound to the RAS1 and RAS2 proteins for GTP, activating the two proteins. An intrinsic GTPase activity of the RAS proteins can in turn be activated by the products of the genes *IRA1* and *IRA2*, whose homologs have also been detected in humans (BALLESTER et al., 1990; TANAKA et al., 1990). RAS-GTP activates adenylate cyclase (encoded by *CYR1*) which is associated with the SRV2 (or CAP) protein (FEDOR-CHAIKEN et al., 1990; GERST et al., 1991; VOJTEK et al., 1991). This activation can be counteracted by the products of the *PDE1* and *PDE2* genes, encoding phosphodiesterases that convert cAMP to AMP. cAMP produced by the activated adenylate cyclase promotes the dissociation of the regulatory subunits (encoded by the single gene *BCY1*) of three protein kinases from the catalytic subunits (TPK1–TPK3; also called protein kinase A). These can now, directly or indirectly, trigger the phosphorylation of cellular enzymes (GIBBS and MARSHALL, 1989). Thus, phosphorylation of the key enzymes in storage carbohydrate metabolism is brought about, as well as for some enzymes involved in gluconeogenesis and transcriptional regulators of nuclear gene expression (see below). Elucidation of this signaling pathway has shown the power of molecular genetics combined with formal classical yeast genetics. It should be stressed, however, that this is not the only mechanism by which a yeast cell can respond to nutrients in general or specifically to glucose (CAMERON et al., 1988). In fact, the phenomenon to be discussed in the following seems to be mostly independent of the system just described.

In the presence of glucose, yeast does not synthesize enzymes needed for the degradation of disaccharides such as maltose and sucrose or of galactose (MOEHLE and HINNEBUSCH, 1991).

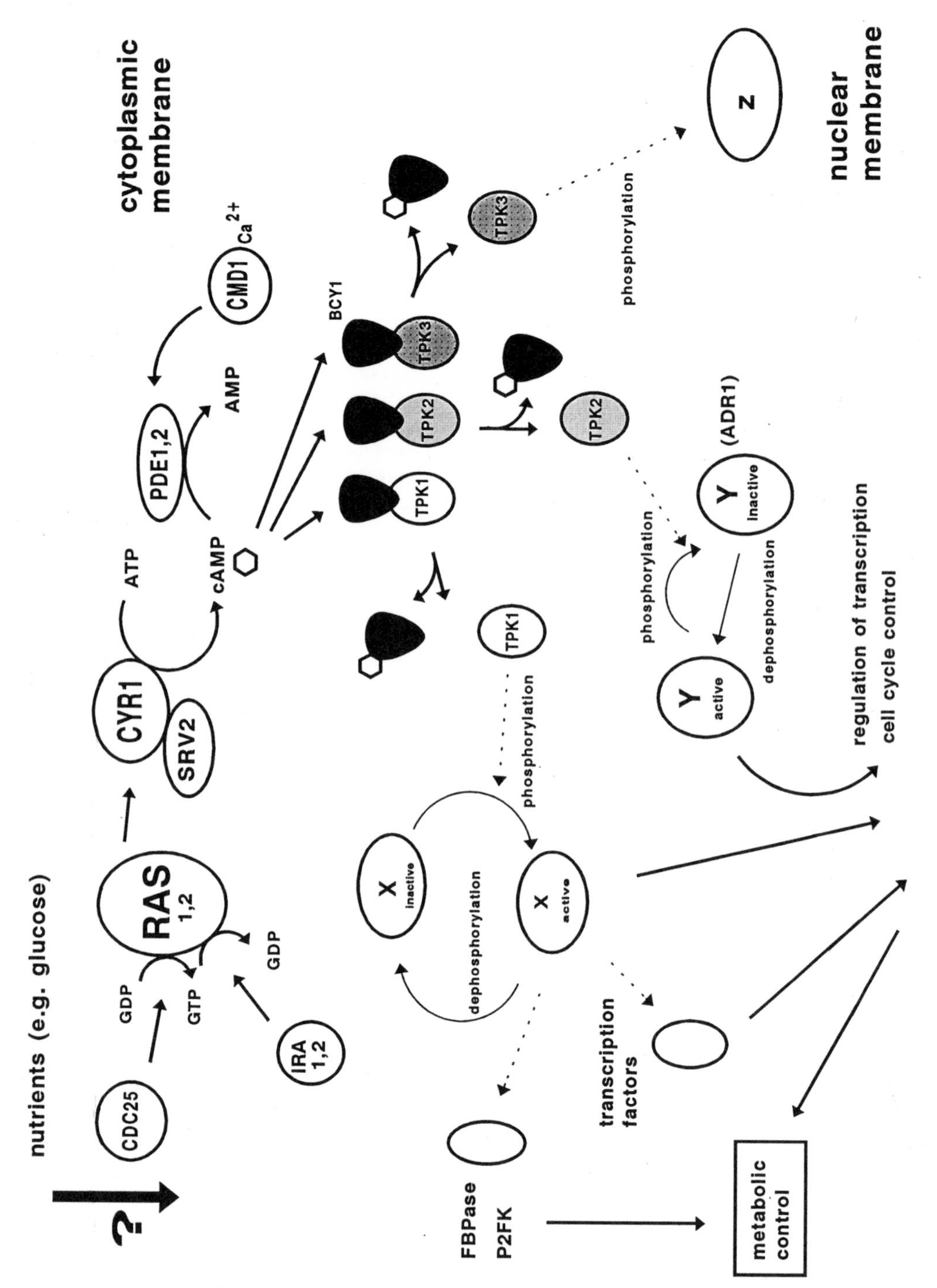

Fig. 4. The RAS/cAMP pathway of *Saccharomyces cerevisiae*. ? means: the mode by which the extracellular nutrient concentration is monitored and initiates the regulatory cascade is not yet clear. x, y, z are primary target proteins of protein kinase A.

Furthermore, expression of the genes encoding gluconeogenic enzymes is also blocked (GANCEDO and GANCEDO, 1979), and the expression of respiratory enzymes and a variety of other systems is reduced (PERLMAN and MAHLER, 1974; DE VRIES and MARRES, 1987). This phenomenon has been termed carbon catabolite repression, in analogy to bacterial systems (GANCEDO, 1992). Despite numerous efforts, the mechanism by which the extracellular signal "glucose" is transformed intracellularly to regulate the transcription of all these genes has not yet been elucidated. On the other hand, a large amount of data has been gathered on how transcription of individual genes is regulated by proteins binding to their promoters (to be discussed in detail in Sect. 4). What is still missing, are the links between these proteins and the primary signal "glucose".

If glucose is added to yeast cells grown under gluconeogenic conditions, not only is the transcription of the respective enzymes being repressed, but the specific activities of the enzymes drop rapidly, and they are proteolytically degraded. This mechanism has been termed carbon catabolite inactivation (HOLZER, 1976). The enzyme best studied in this regard is fructose-1,6-bisphosphatase (MAZON et al., 1982). Phosphorylation of this enzyme is brought about by the RAS/cAMP pathway and accounts for a loss of about half of the enzymatic activity within the first few minutes after glucose addition. Whereas the first step of inactivation of fructose-1,6-bisphosphatase is reversible by dephosphorylation, the enzyme is being degraded in the following two hours after glucose addition. This degradation takes place in the vacuole (CHIANG and SCHEKMAN, 1991), and contrary to earlier speculations is not triggered by the previous phosphorylation (ROSE et al., 1988). *In vitro*, the enzyme is allosterically inhibited by fructose-2,6-bisphosphate (GANCEDO et al., 1983) that simultaneously acts as a potent activator of phosphofructokinase (AVIGAD, 1981). Levels of this effector increase dramatically upon glucose addition, as the enzyme producing it, phosphofructo-2-kinase (P2FK), is being activated by a cAMP-dependent phosphorylation. A gene encoding P2FK has been isolated and sequenced (KRETSCHMER and FRAENKEL, 1991). It has been concluded from this work that at least one more structural gene for P2FK is present in yeast. In contrast to the mammalian enzyme that carries both kinase and phosphatase activities, the fructose-2,6-bisphosphatase activity in yeast is exerted by different enzymes, one of which is encoded by *PHO8* (PLANKERT et al., 1991).

Regulatory mechanisms in the carbohydrate metabolism of other yeasts have been studied only in isolated cases. Lactose metabolism in *Kluyveromyces lactis* has been shown to use similar strategies in glucose repression as *S. cerevisiae* for its galactose metabolism (BREUNIG and KUGER, 1987; MEYER ET al., 1991; WEBSTER and DICKSON, 1988). Glucose repression has also been observed in peroxisome proliferation in *Hansenula polymorpha*, a yeast capable of growing on methanol as the sole carbon source (GLEESON and SUDBERY, 1988). When grown on glucose media, the genes encoding the enzymes needed to utilize methanol are completely repressed. Upon depletion of glucose and addition of methanol, however, these genes are vigorously expressed, and the cells can be packed with giant peroxisomes, harboring the enzymes methanol oxidase (MOX), dihydroxyacetone synthase (DAS), and catalase (ROGGENKAMP et al., 1989).

3.2 Amino Acid Metabolism

A description of amino acid metabolism in yeast, prepared more than ten years ago, has taken 119 pages (JONES and FINK, 1982). Since then, several reviews on selected topics have appeared, e.g., on regulation and compartmentalization of arginine biosynthesis (DAVIS, 1986) and biosynthesis of aromatic amino acids (BRAUS, 1991). As has been shown for carbohydrate metabolism, in this research area the interest has also shifted from the formal description of the biochemical reactions to the isolation of genes encoding the respective enzymes, the construction of mutants, and the study of the underlying regulatory mechanisms. Transcription of a large number of genes involved is controlled by promoters functioning according to the standard scheme (Fig. 5), whose main regulatory component, the *GCN4* gene product, will be discussed in Sect. 4. Apart from the general control, the different

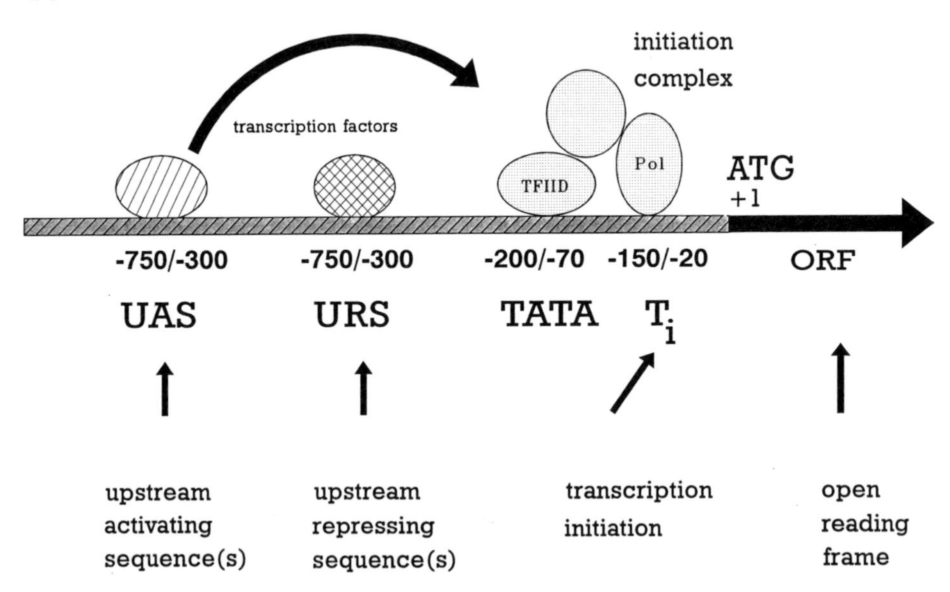

Fig. 5. Schematic representation of a typical yeast promoter. Pol is RNA polymerase II, ATG marks the translation start codon, numbers refer to the distance in base pairs relative to it. As distances may vary considerably, the position range where the majority of the elements are found, is indicated.

routes of amino acid synthesis are regulated both at the level of gene expression and by allosteric mechanisms. Thus, a 5- to 10-fold increase in the level of a specific mRNA (*ARG3*) upon a shift to starvation conditions, can be accompanied by a more than 150-fold increase in the enzymatic activity (JONES and FINK, 1982). However, among the different biosynthetic routes, activation of enzyme activities under such conditions shows a large variation, with some being only 3-fold.

It should also be noted that amino acid metabolism in yeast takes advantage of the compartmentalization within the eukaryotic cells. Thus, a major part of the amino acid pools is located in the vacuole, whereas synthesis is partly carried out in the mitochondria, due to the need for acetyl-CoA as a precursor, and partly in the cytoplasm (DAVIS, 1986).

3.3 Lipids, Vitamins, and Inorganic Phosphate

In general, lipids are major components of all cellular membranes. However, large variations in both lipid content and composition are found between different yeast species (with lipids making up 1.5% to more than 50% of the cells' dry weight) and within a given yeast in dependence of temperature and growth conditions (RATLEDGE and EVANS, 1989). Phospholipids and sterols, as the principal lipid components, are synthesized by yeast under aerobic conditions from precursors such as glycerol, fatty acids, and inositol (HENRY, 1982). For anaerobic growth, sterols and fatty acids have to be supplied externally, as their synthesis requires molecular oxygen (HENRY et al., 1975). Of the precursor molecules, glycerol can be supplied by the upper part of the glycolytic pathway. Fatty acids are synthesized by the action of fatty acid synthetase, an enzyme composed of two large subunits with different functional domains. The subunits are encoded by the genes *FAS*1 and *FAS*2 whose genetics and regulation have been studied in detail (SCHÜLLER et al., 1992). In yeast, mainly unsaturated fatty acids with a length of 16–18C are produced by this enzyme (OKUYAMA et al., 1979). Unsaturated fatty acids, not produced in bacteria, are abundant in yeasts as in higher eukaryotes (RATLEDGE and EVANS, 1989). Inositol-1-phosphate is synthesized from glucose-6-phosphate by the action of inositol-1-phosphate synthase. From the precursors mentioned, phospholipids such as phosphatidylinositol, phosphatidylserine, phosphatidylethanolamine, phosphatidylcholine, and phosphatidylglycerines (e.g., cardiolipin) are synthesized to make up the main fraction of membrane lipids (HENRY et al., 1975). Different phosphorylated forms of phosphatidylinositol act as second messengers in higher eukaryotes. A similar role for yeast and an interrelation with the RAS/cAMP pathway (see Sect. 3.1.5) have been proposed, but await further analysis (WHITE et al., 1991). Mutants in lipid metabolism have first been isolated by their requirement for inositol and ethanolamine/choline (HENRY, 1982). In turn, the insights gained by studying phospholipid biosynthesis have allowed the development of a powerful enrichment procedure called inositol starvation (HENRY et al., 1975). In principle, growing cells being deprived of inositol will die, whereas non-growing mutants (e.g., in genes encoding enzymes of amino acid synthesis after deprivation for the respective amino acid) will not. These can then be grown by addition of the amino acid and inositol.

Catabolic enzymes of lipids, such as lipases and phospholipases, are of potential commercial interest (RATLEDGE and EVANS, 1989).

Inositol has been classified among vitamins such as thiamine, biotin, riboflavin, and choline. Although yeast is capable of synthesizing all these substances (UMEZAWA and KISHI, 1989), they are taken up by active transport systems when supplied in the medium (COOPER, 1982).

Yeast has also developed systems to supply the phosphate needed for metabolic processes. The cell contains two types of phosphatases distinguished by their pH optima: Alkaline phosphatase is an intracellular enzyme, whereas two isoenzymes of acid phosphatase have been reported to be secreted into the periplasmic space (VOGEL and HINNEN, 1990). The transcription of genes in this pathway is strongly controlled by the availability of inorganic phosphate in the medium. Thus, it is completely repressed when phosphate is abundant and is only turned on upon phosphate depletion (VOGEL and HINNEN, 1990). Especially the secreted acid phosphatase encoded by *PHO5* and the regulation of its gene expression have been extensively studied. Transcription is activated by two DNA binding proteins, encoded by *PHO2* and *PHO4* (BÜRGLIN, 1988; OGAWA and OSHIMA, 1990), and negatively controlled by the products of the genes *PHO80* and *PHO85* (GILLIQUET et al., 1990). The latter has been shown to encode a protein with strong similarities to a protein kinase (TOH-E et al., 1988). Both the secretory signal sequence of *PHO5* and its tight regulation have been employed for heterologous gene expression in yeast (see Sect. 5.2).

4 Regulatory Mechanisms of Gene Expression

It has been shown above that metabolic pathways in yeast, especially those of glycolysis, gluconeogenesis, and amino acid biosynthesis, can be effectively controlled by allosteric regulations. In addition, protein kinases exert regulatory functions in a variety of cellular processes (HOEKSTRA, et al., 1991a, b), some of them being regulated by the cAMP/RAS pathway (GIBBS and MARSHALL, 1989; HUBBARD et al., 1992). General control mechanisms found in higher eukaryotes are also working in yeast. Thus, calcium-dependent regulation mediated by calmodulin has been observed (OHYA et al., 1991; JENNISSEN et al., 1992). In addition to regulation at the level of enzyme activity there are two other possibilities for control of gene expression, i.e., transcription of the genes in question and post-transcriptional mechanisms (splicing, polyadenylation, translation). Whereas the first type of control has been investigated for a large number of genes in a variety of pathways, the latter is also beginning to draw attention, and first interesting data are becoming available.

4.1 Transcriptional Regulation

Transcriptional regulation of yeast genes is mediated mainly by specific transcription factors, binding to promoter regions, which, like transcriptional terminators, are generally AT-rich (about 70% AT). A growing number of such factors are being characterized (GUARENTE and BERMINGHAM-MCDONOGH, 1992; STRUHL, 1989). In this section, we will describe the general principles underlying transcriptional regulation and then describe only two examples in further detail. The basis for regulatory effects on gene expression from sequences located within the open reading frame in some glycolytic genes (*PGK1*, MELLOR et al., 1987; *PYK1*, PURVIS et al., 1987; *PFK1*, HEINISCH et al., 1991 a) and in the gene-encoding lipoamide dehydrogenase (*LPD1*, ZAMAN et al., 1992) is not yet clear. It has been speculated that specific sequences ("DAS", downstream activating sites) function as transcriptional enhancers in these cases (ZAMAN et al., 1992).

Three types of RNA polymerases are found in yeast, as in other eukaryotes, where RNA polymerase I (Pol I) transcribes ribosomal gene clusters, RNA polymerase II (Pol II) transcribes the bulk of protein encoding genes, and RNA polymerase III (Pol III) is needed for transcription of tRNA genes. We will focus on Pol II mediated gene transcription which is controlled by the formation of a transcriptional complex with the factor TFIID binding at the TATA-box as a primary event (Fig. 5; HAHN et al., 1989; PUGH and TJIAN, 1992). As in the case of RAS, the regulatory proteins controlling gene expression in yeast and humans are functionally interchangeable (KELLEHER et al., 1992). The efficiency of transcription initiation is dependent on binding of transcription factors in the 5′ non-transcribed parts of the gene, the promoter (Fig. 5). A similarity between yeast promoters and those of higher eukaryotes has been noted quite early, in that DNA sequences enhancing transcription (called UAS, upstream activating sites in yeast) can be found at a distance of several hundred base pairs from the point of initiation. Whereas in yeast the UASs have to be located 5′ to this point (with the possible exceptions of elements located in the open reading frame as noted above), enhancers of mammalian gene transcription are position independent and can be found at even larger distances (GUARENTE and BERMINGHAM-MCDONOGH, 1992). It is likely that DNA bending (MATTHEWS, 1992) and nucleosome positioning (THOMA, 1992) are involved in transcriptional activation.

In practice, promoter elements controlling transcription have been tested by fusions to a reporter gene, in the majority of cases the *lacZ* gene of *E. coli* encoding β-galactosidase (GUARENTE, 1983). Such experiments have shown that in addition to binding sites for transcriptional activators (UASs), URSs, upstream repressing site), are frequently located in yeast promoters (GUARENTE and BERMINGHAM-MCDONOGH, 1992). Deletion of such sequences will lead to an increased transcription compared to the wild type.

4.1.1 Transcription of Glycolytic Genes

Consistent with the enormous activity of glycolysis, glycolytic gene promoters are among the strongest promoters found in yeast (SHUSTER, 1989). For this reason, these promoters have been studied extensively and are used to direct heterologous gene expression (see Sect. 5.2.1). Originally, sequence homologies were identified between different glycolytic gene promoters, and several consensus elements were postulated to play a role in gene expression (DOBSON et al., 1982). As the methodology for gel retardation assays and DNAse footprinting developed, such elements could be shown to be the binding sites for regulatory proteins. A model glycolytic gene promoter is presented in Fig. 6.

The activities of most glycolytic enzymes (with the exceptions of hexokinase and phosphofructokinase) are strongly affected by the *GCR1* ("glycolysis regulation") gene product (BAKER, 1986; FRAENKEL, 1982). In all cases examined, this regulation could be shown to be exerted at the transcriptional level (SCOTT et al., 1990). Recently, strong evidence has been provided that the GCR1 protein binds to a promoter element containing a CTTCC box (BAKER, 1991), previously reported to enhance transcription (CHAMBERS et al., 1988). However, the sequence does not show an activating effect on transcription by itself, when placed into a heterologous promoter. It seems that it potentiates the activation effects obtained from hinding of the RAP1 protein at its consensus sequence, also found in most glycolytic

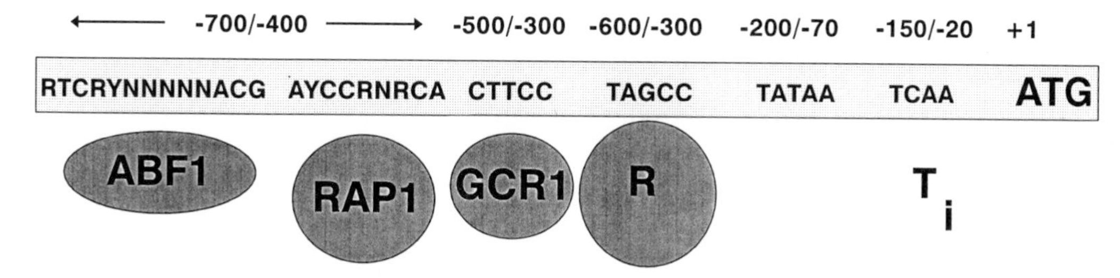

Fig. 6. Regulatory elements in a model glycolytic gene promoter. T_i is the point of transcription initiation, R the postulated transcriptional repressor protein.

gene promoters (TORNOW and SANTANGELO, 1990). The RAP1 binding sites as well as the actual binding of the protein were first discovered in the promoters of genes encoding ribosomal proteins ("RPG box", ribosomal protein gene box). Binding is also found at yeast telomeres (CONRAD et al., 1990) and in regions necessary for silencing of mating type information (KURTZ and SHORE, 1991). From this variety of affected loci, the picture emerges that RAP1 binding does not have a specific effect on transcription, but facilitates the binding of specific regulatory proteins interacting with the transcriptional apparatus. Many glycolytic gene promoters contain also a consensus binding site for ABF1, a protein shown to bind at yeast replication origins (RHODE et al., 1992). RAP1 and ABF1 have been shown to share some sequence similarities (DIFFLEY and STILLMAN, 1989), with the possibility of both having a similar function in transcriptional regulation. Both have been shown to activate or repress transcription depending on their promoter context. It is likely that some form of interaction between both proteins takes place in glycolytic gene promoters where their consensus binding sites are found in close proximity to each other (CHAMBERS et al., 1990; HEINISCH et al., 1991a; HOLLAND et al., 1990). Whereas the RAP1 binding site has been shown to confer UAS functions in a heterologous gene promoter (NISHIZAWA et al., 1989), some promoters of glycolytic genes also contain URS elements. A consensus binding site already detected in URS elements of promoters of amino acid biosynthetic genes has been proposed (LUCHE et al., 1990; RODICIO et al., *Gene*, in press). Thus, although the transcription of most glycolytic genes does not vary to a great extent in relation to the carbon source used for growth, they still contain a subset of binding sites for regulatory proteins in their promoters, indicating the possibility for subtle transcriptional regulations.

Other genes involved in carbohydrate metabolism have been shown to be controlled by other or additional DNA binding proteins, such as general transcriptional suppressors encoded by *MIG1* (NEHLIN and RONNE, 1990) and *TUP1* (WILLIAMS et al., 1991). The latter has recently been shown to act in conjunction with the *SSN6* gene product to repress transcription of haploid-specific and glucose repressible genes (KELEHER et al., 1992). A model for its interaction with MCM1 and MIG1 is presented by the authors. The first case where phosphorylation of a transcription factor is involved in regulation was reported for ADR1, which is inactivated by phosphorylation leading to a decrease in *ADH2* gene expression (CHERRY et al., 1989). This factor has been shown to bind DNA (THUKRAL et al., 1992) and to have a more general function in regulating transcription of genes other than *ADH2*, such as those encoding peroxisomal proteins (SIMON et al., 1991).

4.1.2 Transcription of Galactose-Inducible Genes

In contrast to glycolysis, the enzymes needed for galactose metabolism are strongly regulated (up to 1000-fold in galactose as opposed to glucose medium; JOHNSTON and DAVIS, 1984). Transcription from the encoding genes is repressed by glucose, derepressed by a non-fermentable carbon source, and induced by galactose (JOHNSTON, 1987). This is achieved by the interaction of three components: (1) GAL4 is the general activator protein, containing a DNA binding domain and acidic domains mediating the activation (JOHNSTON et al., 1986). (2) Genes under the control of the GAL4 protein contain one or more 17 bp conserved palindromic sequences in their promoters, to which the protein can bind (GINIGER et al., 1985; BRAM et al., 1986). (3) In the absence of galactose, the protein encoded by *GAL80* binds to the C-terminal domain of the GAL4 protein and thus blocks activation (SALMERON et al., 1990). Recently it could be shown that relatively small variations (less than a factor of two) in the amount of GAL4 protein can account for the regulation of the entire pathway (GRIGGS and JOHNSTON, 1991). The L4 protein appears to be phosphorylated under inducing conditions, but the *in vivo* significance of this phenomenon is not yet clear (SADOWSKI et al., 1991). The GAL system is the best studied case of regulatory mechanisms in eukaryotes. The presence of a DNA binding domain and an activation domain is a common feature in transcriptional activators, and the yeast elements are functional even in mammalian systems (GILL et al., 1990). *Vice versa*, mammalian regulatory DNA elements responsive to hormones have been shown to function in yeast (PURVIS et al., 1991). The power of regulation achieved with promoters susceptible to this control is demonstrated by the expression of lethal enzymes, e.g., *EcoRI*, after transfer to galactose medium, when the cells have been pregrown without any harm in the presence of glucose (BARNES and RINE, 1985).

The regulatory components just described have been conserved in other yeasts. Thus, lactose metabolism in *Kluyveromyces lactis* is controlled by the LAC9 protein sharing some amino acid similarities with GAL4 (SALMERON and JOHNSTON, 1986; WRAY et al., 1987). Glucose repression observed in some strains (BREUNIG, 1989; KUGER et al., 1990) is mediated by slight variations in the intracellular concentration of this GAL4 homolog (KUZHANDAIVELU et al., 1992), reminiscent of the regulation in *Saccharomyces cerevisiae*. Interestingly, the *GAL*1 gene product, galactokinase, is also a regulatory protein required for the induction of the lactose/galactose regulon in *K. lactis* and can functionally replace the GAL3 protein in *gal*3 mutants of *S. cerevisiae* (MEYER et al., 1991).

In summary, transcriptional regulation of yeast genes is mediated by DNA binding proteins acting at specific sites in the promoters. Regulation is rarely, if at all, achieved by a single activator or repressor, but by the interaction of several elements. This general principle applies to transcriptional regulation in higher eukaryotes, too. In fact, transcriptional activators have been shown to function in both yeast and mammalian cells and more and more homologs to such factors found in yeast are detected in other systems (GUARENTE and BERMINGHAM-MCDONOGH, 1992). This includes proteins involved in oncogenic transformation, like fus and jun (STRUHL, 1989). Although we are beginning to understand the mechanisms involved in transcriptional regulation, in most systems knowledge about the transduction of a (extracellular) signal onto the factors regulating gene transcription is still lacking.

4.2 Post-Transcriptional Regulation

Once a gene has been transcribed, several factors can influence its actual expression. Thus, sequences in the 3′-untranslated region of the mRNA can determine its stability (DEMOLDER et al., 1992). In addition, consensus sequences surrounding the translation start codon (CIGAN and DONAHUE, 1987; MÜLLER and TRACHSEL, 1990) can influence the efficiency of translation. The latter can also be influenced by the codon usage (EGEL-MITANI et al., 1988; BENNETZEN and HALL, 1982; ERNST, 1988), in a few cases leading to a weaker expression of heterologous genes with unbiased codons (KOTULA and CURTIS, 1991). mRNA stability and thereby gene expression are also dependent on an efficient transcription termination. Consensus sequences for termination and polyadenylation have been proposed (ZARET and SHERMAN, 1982; OSBORNE and GUARENTE, 1989). A second type of transcriptional terminator not following these rules has also been described in yeast (IRNIGER et al., 1991). Still, efficient heterologous gene expression is very much a matter of trial and error, and optimization of all the steps mentioned frequently leads only to marginal improvements.

In higher eukaryotes, alternative splicing is used in a number of cases for regulation of gene expression (SMITH et al., 1989). As mentioned above, *S. cerevisiae* does not contain very many introns and alternative splicing has not been reported for this yeast. The mechanism of mRNA splicing in yeast involves an intron-specific "TACTAAC"-box (LANGFORD et al., 1984) and a lariat formation between the 5′-splice junction and the A at the fifth nucleotide of this element (RUBY and ABELSON, 1991; WOOLFORD, 1989). Changing this nucleotide by *in vitro* mutagenesis allows assessment of the physiological role of an intron after substitution for the wild-type copy.

A well studied case of translational control is provided for a gene encoding a transcriptional activator binding to the promoters of genes whose expression is regulated by general amino acid control, *GCN4*. Translation of its message is negatively controlled by a number of proteins encoded by *GCD* genes first identified genetically (see HINNEBUSCH, 1988, for a review). These in turn are negatively controlled by the products of *GCN1*, *GCN2*, and *GCN3*, which are activated under conditions of amino acid starvation. Under these conditions the GCN proteins are activated, leading to an inactivation of the GCD system and, therefore, an efficient translation of *GCN4* mRNA. This in turn leads to a transcriptional activation of genes encoding a number of amino acid biosynthetic enzymes. *GCN2* has been shown to encode a protein kinase that is associated with ribosomes, where it phosphorylates initiation factor 2α (RAMIREZ et al., 1991; HINNEBUSCH, 1990; DEVER et al., 1992). Translation of *GCN4* mRNA is controlled by the presence of 4 short open reading frames preceding the actual translation start codon. Generally, translation in yeast starts at the first AUG present at the 5′ end of the mRNA (KOZAK, 1983). Apparently, in *GCN4* the first of the short open reading frames is translated, and reinitiation of translation can take place at the fourth of these elements, preventing translation of the protein coding region. However, under conditions of amino acid starvation, reinitiation does not take place at this element, and translation starts at the AUG of the main open reading frame (MILLER and HINNEBUSCH, 1989; HINNEBUSCH et al., 1988).

More indirect evidence for translational control has come from studies on the distribution of polyribosomes along specific mRNAs. Thus, it could be shown that overproduction of the mRNA for the glycolytic enzyme pyruvate kinase (PYK) in yeast leads to a shift from the mRNA found predominantly associated with polysomes in a wild-type strain to association with few ribosomes in the overproducing strain. This presumably causes a less effective translation, and gene expression is not directly correlated to gene copy number (MOORE et al., 1990). Interestingly, overproduction of one specific mRNA might also influence the translation of the mRNA encoded by a different gene. It has been observed that overproduction of *PYK1* mRNA led to a reduction of polysomes associated with *PFK2* mRNA, (MOORE et al., 1990). Clearly, such mechanisms will be of growing importance for future investigations of the regulation of gene expression.

5 Industrial Applications

As mentioned in the introduction, yeasts belong to a unique group of microorganisms that have been commercially exploited by man for thousands of years, due to their ability to efficiently convert sugars into ethanol and carbon dioxide. Thus, predominantly strains of *Saccharomyces cerevisiae* (and synonyms like *S. uvarum* and *S. carlsbergiensis*) are used in the preparation of wine, beer and bread.

5.1 "Classical" Applications

Despite their historical use in breweries and wineries, the yeast strains as well as the techniques used are continuously being improved (BOULTON, 1991; JAVELOT et al., 1991; ODA and OUCHI, 1989b). Although some strains used in wine production are amenable to techniques of genetic manipulation described in

Sect. 2.2, others as well as most of the brewing strains are not (BOULTON, 1991). They are usually polyploid or aneuploid for some chromosomes (BAKALINSKY and SNOW, 1990), do not mate and exhibit low sporulation frequencies and spore viabilities (ODA and OUCHI, 1989a). Nevertheless, methods for their genetic analysis have been devised (KIELLAND-BRANDT et al., 1983), and crossing with haploid laboratory strains using dominant selection markers has been achieved (ODA and OUCHI, 1989a; THORNTON, 1991). One might predict that an increased use of such techniques and the possibility of manipulations at the laboratory scale will be of growing interest, to avoid variabilities in properties of strains used in production cultures (BOULTON, 1991). Another possibility for classical genetic manipulations is exemplified by a recent paper (JAVELOT et al., 1991) showing that specific mutations in sterol biosynthesis can be introduced into wine yeast for flavor adjustments. In addition to changes in the yeasts themselves, new techniques are being introduced in the fermentation processes. Both fermentation conditions and strains used are adjusted to allow control of flocculation, which when occurring too early impairs fermentation capacity, but is desirable in later stages for the separation of biomass. The genetic control of yeast flocculation has been reviewed recently (STRATFORD, 1992). Addition of vitamins, sterols and unsaturated fatty acids is also used to improve fermentation processes (HENICK-KLING, 1988). A crucial factor in alcoholic fermentation is the sensitivity of the organism to toxic effects of the end product, namely ethanol (D'AMORE et al., 1990; JONES and GREENFIELD, 1985; ROSA and SÁ-CORREIA, 1992). Strains of yeast with high ethanol tolerance have been used for fuel production in Brazil (LALUCE, 1991).

Aside from the predominant utilization of *S. cerevisiae* for fermentation processes, other yeasts are of potential industrial use, e.g., in the production of lipases for highly specific modifications of oil components (BJÖRKLING et al., 1991; DEL RIO et al., 1990), in the conversion of pentoses, the principal components of hemicelluloses (JEFFRIES, 1990), and in the use of killer strains in biotyping of pathogenic yeasts (POLONELLI et al., 1991). The use of baker's yeast as a reagent in the synthesis of various organic compounds has been reviewed recently (SERVI, 1990).

Recombinant DNA technology already yielded significant improvements of yeast strains used for classical purposes. The introduction of genes encoding amylolytic enzymes from another yeast, *Schwanniomyces castelli*, makes starch saccharification prior to alcoholic fermentation possible to prepare low-calorie beer and to ferment non-malted substrates (STRASSER et al., 1989). The introduction of the *XYL*1 and *XYL*2 genes in *S. cerevisiae* described above is the first step on the way to a xylose-fermenting yeast strain (KÖTTER et al., 1990; AMORE et al., 1991).

5.2 Use of Yeasts for Heterologous Gene Expression

The advantages in using yeasts as a "factory" for the production of heterologous proteins are manifold. To name a few, yeast is a non-pathogenic organism obviating the need for extensive purifications of most products, e.g., from pyrogens often associated with bacterial production systems. The genetic basis available as outlined above provides a variety of selection systems, such as amino acid requirements, but also dominant selection markers are available. Furthermore, numerous expression systems, both for high-level gene expression and secretion of the products have been developed.

Several reviews deal with heterologous protein production in *S. cerevisiae* (KINGSMAN et al., 1985, 1987, 1990) and other yeasts (REISER et al., 1990; BUCKHOLZ and GLEESON, 1991; SHUSTER, 1989; GELLISSEN et al., 1992). During the preparation of this manuscript, an extensive review on the subject appeared (ROMANOS et al., 1992). Tab. 5 gives a list of some proteins heterologously expressed in *S. cerevisiae* in the last few years. In the following we will concentrate on a description of expression strategies used in yeast. Once the proteins are produced, other problems might have to be dealt with, e.g., degradation by the ubiquitin system (JOHNSON et al., 1992) and proteases (HEE LEE et al., 1992; SUAREZ RENDUELEZ and WOLF,

Tab. 5. Heterologous Gene Expression in *Saccharomyces cerevisiae*

Protein	Reference
Human Proteins	
Interferon-gamma	BITTER, G. A., and EGAN, K. M. (1988), *Gene* **69,** 193
Interleukin-1 alpha	LIVI, G. P. et al. (1990), *Gene* **88,** 297
Interleukin-1 beta	LIVI, G. P. et al. (1991), *J. Biol. Chem.* **266,** 15348
Interleukin-2	ERNST, J. F., and RICHMAN, L. H. (1989), *Bio/Technology,* **7,** 716
Interleukin 5	INGLEY, E. et al., (1991), *Eur. J. Biochem.* **196,** 623
Interleukin 6	GUISEZ, Y. et al. (1991), *Eur. J. Biochem.* **198,** 217
DNA topoisomerase 1	BJORNSTI, M.-A. et al. (1989), *Cancer Res.* **49,** 6318–6323.
alpha-Amylases	SHIOSAKI, K. et al. (1990), *Gene* **89,** 253
cAMP phosphodiesterase	MCHALE, M. M. et al. (1991), *Mol. Pharmacol.* **39,** 109
Glutathione S-transferases	BLACK, S. M. et al. (1990), *Biochem. J.* **268,** 309
Placental aromatase	POMPON, D. et al. (1989), *Mol. Endocrinol.* **3,** 1477–1487
Lysosomal β-hexosaminidase	PREZANT, T.R. (1990), *Biochem. Biophys. Res. Commun.* **170,** 383
Monoamine oxidases	URBAN, P. et al. (1991), *FEBS Lett.* **286,** 142
Asparagine synthetase	VAN HEEKE, G., and SCHUSTER, S. M. (1990), *Protein Eng.* **3,** 739
5-Lipoxygenase	NAKAMURA, M. et al. (1990), *Gene* **89,** 231
Cytochrome P450	WU, D.-A. et al. (1991), *DNA Cell Biol.* **10,** 201
Multiple drug resistance	KUCHLER, K., and THORNER, J. (1992), *Proc. Natl. Acad. Sci. USA* **89,** 2302
Insulin	EGEL-MITANI, M. et al. (1988), *Gene* **73,** 113
Insulin-like growth factor I	STEUBE, K. et al. (1991), *Eur. J. Biochem.* **198,** 651
Fibroblast growth factors	BARR, P. J. et al. (1988), *J. Biol. Chem.* **263,** 16471
Granulocyte-macrophage colony-stimulating factor	ERNST, J. F. et al. (1987), *Bio/Technology* **5,** 831
Nerve growth factor	SAKAI, A. et al. (1991), *Bio/Technology* **9,** 1382
Epidermal growth factor	CLEMENTS, J. M . et al. (1991), *Gene* **106,** 267
Tumor necrosis factor	SREEKRISHNA, K. et al. (1989), *Biochemistry* **28,** 4117
Factor XIII	BISHOP, P. D. et al. (1990), *J. Biol. Chem.* **265,** 13888
Chromosomal proteins	SRIKANTHA, T. et al. (1990), *Exp. Cell. Res.* **191,** 71
Vitamin D receptor	SONE, T. et al. (1990), *J. Biol. Chem.* **265,** 21997
M1 muscarinic receptor	PAYETTE, P. et al. (1990), *FEBS Lett.* **266,** 21
Leukocyte common antigen CD45	FERNANDEZ-LUNA, J. L. et al. (1991), *Genomics* **10,** 756
P-glycoprotein	SAEKI, T. et al. (1991), *Agric. Biol. Chem.* **55,** 1859
Erythropoietin	ELLIOTT, S., et al. (1989), *Gene* **79,** 167
Profilin	ASPENSTRÖM, P. et al. (1991), *J. Muscle Res. Cell Motility* **12,** 201
alpha-Fetoprotein	YAMAMOTO, R. et al. (1990), *Life Sci.* **46,** 1679
Lysozyme	SUZUKI, K. et al., (1989), *Mol. Gen. Genet.* **219,** 58
Apolipoprotein E	STURLEY, S. L. et al. (1991), *J. Biol. Chem.* **266,** 16273
Ribophorin I	SANDERSON, C. M. et al. (1990), *J. Cell. Biol.* **111,** 2861
reg Protein	ITOH, T. et al. (1990), *FEBS Lett.* **272,** 85
alpha-1-Antitrypsin	GEORGE, P. M. et al. (1989), *Blood* **73,** 490
Mouse/human antibody Fab fragment	HORWITZ, A.H. et al. (1988), *Proc. Natl. Acad. Sci. USA* **85,** 8678
Other Mammalian Proteins	
Catalytic antibody (against chorismate mutase)	BOWDISH, K. et al. (1991), *J. Biol. Chem.* **266,** 11901
Mouse IG kappa chain	KOTULA, L., and CURTIS, P. J. (1991), *Bio/Technology* **9,** 1386
Cytochrome P450	NISHIKAWA, M. et al. (1992), *Curr. Genet.* **21,** 101
Avian gag-myc oncogene	DURRENS, P. et al. (1990), *Curr. Genet.* **18,** 7
Bovine β-lactoglobulin	TOTSUKA, M. et al. (1990), *Agric. Biol. Chem.* **54,** 3111
Bovine myoglobin	SHIMADA, H. et al. (1989), *J. Biochem.* **105,** 417
Bovine β-casein	JIMENEZ-FLORES, R. et al. (1990), *J. Agric. Food Chem.* **38,** 1134
Goat α-lactalbumin	TAKEDA, S. et al. (1990), *Biochem. Biophys. Res. Commun.* **173,** 741

Tab. 5. Continued

Protein	Reference
Rat cytochrome c	CLEMENTS, J. M. et al. (1989), *Gene* **83,** 1
Rat c-erbA β thyroid hormone receptor	LU, C. et al. (1990), *Biochem. Biophys. Res. Commun.* **171,** 138
Rat α-fetoprotein	NISHI, S. et al. (1988), *J. Biochem.* **104,** 968
Rat C1-tetrahydrofolate synthase	THIGPEN, A. E. et al. (1990), *J. Biol. Chem.* **265,** 7907
Viral Proteins	
Hepatitis B surface antigens	KURODA, S. et al. (1992), *J. Biol. Chem.* **267,** 1953
Hepatitis B virion-like particles	SHIOSAKI, K. et al. (1991), *Gene* **106,** 143
HIV-1 reverse transcriptase	BATHURST, I. C. et al. (1990), *Biochem. Biophys. Res. Commun.* **171,** 589
HIV-1 gag-polyprotein precursor	BATHURST, I. C. et al, (1989), *J. Virol.* **63,** 3176
HIV-1 p55 precursor core protein	EMINI, E. A. et al. (1990), *AIDS Res. Hum. Retroviruses* **6,** 1247
HTLV-1 tax transactivator	KRAMER, R. A. et al. (1990), *AIDS Res. Hum. Retroviruses* **6,** 1305
Human papillomavirus proteins	CARTER, J. J. et al. (1991), *Virology* **182,** 513
Rous sarcoma virus	BONNET, D., and SPAHR, P.-F. (1990), *J. Virol.* **64,** 5628
Bluetongue virus proteins VP7 and NS3	MARTYN, J. C. et al. (1991), *Virus Res.* **18,** 165
Moloney murine leukemia virus integration protein	STRAUSS, W. M., and JAENISCH, R. (1992), *EMBO J.* **11,** 417
Bursal disease virus proteins	JAGADISH, M. N. et al. (1990), *Gene* **95,** 179
Other Proteins of Higher Eukaryotes	
Echistatin	CARTY, C. E. et al. (1990), *Biotechnol. Lett.* **12,** 879
Hirudin	ACHSTETTER, T. et al. (1992), *Gene* **110,** 25
Chicken calmodulin	OHYA, Y., and ANRAKU, Y. (1989), *Biochem. Biophys. Res. Commun.* **158,** 541
Chicken β-actin	KARLSSON, R. (1988), *Gene* **68,** 249
Chicken oviduct progesterone receptor	MAK, P. et al. (1989), *J. Biol. Chem.* **264,** 21613
Nicotinic acetylcholine receptor	JANSEN, K. U. et al. (1989), *J. Biol. Chem.* **264,** 15022
Somatostatin	BOURBONNAIS, Y. et al. (1991), *J. Biol. Chem.* **266,** 13203
Drosophila DNA topoisomerase II	WYCKOFF, E., and HSIEH, T.-S. (1988), *Proc. Natl. Acad. Sci. USA* **85,** 6272
Insect immune peptides	REICHHART, J.-M., and ACHSTETTER, T. (1990), *Res. Immunol.* **141,** 943
Silkworm eclosion hormone	HAYASHI, H. et al. (1990), *Biochem. Biophys. Res. Commun.* **173,** 1065
(Barley) α-amylase	KOTYLAK, Z., and EL-GEWELY, M. R. (1991), *Curr. Genet.* 20, 181
Proteins from Other Fungi	
α-Amylase (*Schwanniomyces occidentalis*)	STRASSER, A. W. M. et al. (1989), *Eur. J. Biochem.* **184,** 699
Glucoamylase	SUNTSOV, N. I. et al. (1991), *Yeast* **7,** 119
Cellulase (*Trichoderma reesei*)	ZURBRIGGEN, B. D. et al. (1990), *J. Biotechnol.* **13,** 267
Endo-β-glucanase (*Trichoderma reesei*)	ZURBRIGGEN, B. D. et al. (1991), *J. Biotechnol.* **17,** 133
Alkaline protease (*Aspergillus oryzae*)	TATSUMI, H. et al. (1989), *Mol. Gen. Genet.* **219,** 33
Neutral protease II (*Aspergillus oryzae*)	TATSUMI, H. et al. (1991), *Mol. Gen. Genet.* **228,** 97
β-Galactosidase (*Aspergillus niger*)	KUMAR, V. et al. (1992), *Bio/Technology* **10,** 82
Rennin (*Mucor*)	AIKAWA, J. et al. (1990), *J. Biol. Chem.* **265,** 13955
Bacterial Proteins	
Tetanus toxin fragment	ROMANOS, M. A. et al. (1991), *Nucleic Acids Res.* **19,** 1461
Bacterial enterotoxoid	SCHONBERGER, O. et al. (1991), *Mol. Microbiol.* **5,** 2663
α-Amylase	RUOHONEN, L. et al. (1991), *Yeast* **7,** 337
β-Galactosidase	MORACCI, M. et al. (1992), *J. Bacteriol.* **174,** 873
Staphylococcal nuclease	PINES, O., and LONDON, A. (1991), *J. Gen. Microbiol.* **137,** 771

1988), and the recovery of proteins from the yeast cell (KING et al., 1988). The reader is referred to the works cited for detailed descriptions of these mechanisms.

5.2.1 Expression Vectors

The vectors used to express homologous and heterologous proteins in yeast are almost exclusively yeast/*Escherichia coli* shuttle vectors. Thus, they contain selectable markers for the two organisms as well as a bacterial replication origin as minimal requirements. Extensive listings of suitable vectors have been given in several recent reviews (TUITE, 1992; CHRISTIANSON et al., 1992; GUTHRIE and FINK, 1991; see also Tab. 6). For *S. cerevisiae*, genes complementing auxotrophic phenotypes are usually used as selection markers contained in the vector, e.g., requirements for amino acids (*TRP1*, *HIS3*, *LEU2*, etc.), or nucleotides (*URA3*). For the use in industrial strains, where nutritional markers are rarely available, positive selection markers are available, e.g., copper resistance, and resistance to antibiotics such as cycloheximide, tunicamycin, G418, and methotraxate (REISER et al., 1990). According to their mode of replication in yeast, two basic types of vectors exist for *S. cerevisiae.* Vectors containing no yeast origin of replication (YIp = yeast integrative plasmid) are dependent on integration into the genome for their stable inheritance. In contrast, plasmids containing either a chromosomal origin of replication, called ARS (autonomously replicating sequence; YRp = yeast replicating plasmid) or the replication origin and stability locus of the yeast 2 μm plasmid (YEp = yeast episomal plasmid) are inherited as independent replication units. The choice of vectors to be used for a special purpose, depends on the requirements one has for stability, insert size, and copy number of the gene to be introduced. Thus, integrative vectors show a stable inheritance, but suffer from low copy number and laborious methods when the vector is to be recovered from the yeast cell. Nevertheless, based on repetitive DNA sequences (LOPES et al., 1991) or sequences that can be amplified (MACREADIE et al., 1991), higher copy numbers of integrative vectors can be achieved. A fairly stable mode of inheritance is obtained by including a centromeric region (YCp = yeast centromere plasmid) and telomeres (YLp = yeast linear plasmid) in the vector in conjunction with a replication origin and long inserts. Still, stability is at least two orders of magnitude below that of natural chromosomes. Such vectors are maintained in low copy numbers. A specially developed set of such vectors is called YAC (yeast artificial chromosomes; SCHLESSINGER, 1990). They accomodate extremely large DNA fragments (up to 2 Mbp; COFFEY et al., 1992) and are used for mapping of mammalian DNA as well as isolation and propagation of intron-containing genes encoded by large DNA fragments (ANAND, 1992). In several cases, expression of genes carried on a YAC vector in mammalian cells has been reported (e.g., GNIRKE and HUXLEY, 1991; STRAUSS and JAENISCH, 1992). Finally, if high copy numbers are needed for maximal expression, YEp vectors carrying a 2 μm origin of replication plus the *STB* region have been largely employed (VOLKERT et al., 1989; CHRISTIANSON et al., 1992). They are inherited with a moderate stability in cir^+ strains.

Heterologous genes are rarely expressed from their own promoters in yeast. Therefore, expression vectors have been constructed containing two principal types of yeast promoters Originally, strong constitutive promoters of glycolytic genes (*TDH3*, *PGK1*) have been used, yielding up to 50% of the soluble cell protein in homologous, but "only" 5% in heterologous expression (HITZEMAN et al., 1981). Although most proteins can be produced in yeast without interfering with metabolism or viability, some proteins have deleterious effects, especially when overproduced. In addition, for large-scale productions it might be necessary to first grow a population under optimal conditions for a maximum yield of biomass, before the onset of expression of the heterologous protein. For this purpose, vectors with inducible promoters are available. These might have been obtained from natural yeast systems (*ADH2*, *GAL1–10*, *PHO5*) or are of a hybrid nature (*TDH3/GAL1–10*). Transcription termination signals frequently have to be used to obtain stable transcripts. As both promoter and terminator elements can be varied considerably in their distance to the translation start and termination codons, "shotgun" experiments using suitable restriction sites in the heterologous DNA fragment are usually sufficient (SCHNEIDER and GUARENTE, 1990).

5.2.2 Secretion

A special type of expression vectors for yeast makes use of the fact that some yeast proteins are naturally secreted into the periplasmic space. Clearly, proteins thus localized outside the plasma membrane are much more easily purified and may even diffuse into the medium, allowing removal of cells by simple centrifugation. Proteins secreted by yeast generally pass through the endoplasmic reticulum (ER) where they are core-glycosylated and then transferred to the Golgi apparatus, where outer chain glycosylation occurs. They are directed into this pathway by signal sequences located at the amino terminal end (SCHEKMAN, 1985). Such sequences are placed in frame to the protein coding sequence and direct secretion of heterologous proteins (GOEDDEL, 1990). They contain either only a pre-sequence that is cleaved off during the entry into the ER (SCHEKMAN, 1985) or an additional pro-sequence cleaved off further on in the secretion process by a specific KEX2 protease with self-processing activity (GERMAIN et al., 1992).

Although some heterologous secretion signals work in yeast, for most practical purposes signals from homologous yeast proteins are used (HITZEMAN et al., 1990). The most commonly used leader sequences are those of the mating pheromone α-factor, invertase (*SUC2* product), an acid phosphatase (*PHO5* product), and the killer factor (HITZEMAN et al., 1990). It should be noted that not every protein can be manipulated at the genetic level to be secreted. Although rarely reported, genes are frequently encountered that will not be properly expressed at different stages of the process, i.e., their transcripts might either not get translated or the proteins are being rapidly degraded. The latter might be produced without having enzymatic activity in yeast cells, or they might get stuck in the secretory pathway (BIELEFELD and HOLLENBERG, 1989). Thus, predictions of the fate of the expression of a given gene in yeast are difficult to make, and most work has to rely on trial and error. However, one can assume that proteins naturally secret-

Tab. 6. Features of Yeast Vectors

Vector	Inheritance	Maintenance	Copy Number per Cell	Yeast Sequences
YIp	integrative	stable	1–few	Yeast selectable marker, no origin of replication
YRp	episomal	unstable	5–10[a]	Chromosomal origin of replication, yeast selectable marker
YEp	episomal	moderate	60–100	2 μm origin of replication, yeast selectable marker
YCp	episomal	stable	1–2	Chromosomal or 2 μm origin of replication, yeast selectable marker, centromere sequences, circular
YLp	episomal	stable	1–2	Chromosomal or 2 μm origin of replication, yeast selectable marker, centromere and telomere sequences, more stable with larger inserts
YAC	episomal	stable	1–2	Chromosomal origin of replication, at least two yeast selectable markers, centromere and telomere sequences, accomodate large DNA inserts

[a] The average copy number per cell is given. Due to their mode of replication, more than 50% of cells grown under selective conditions do not contain any plasmid, whereas the others contain it in higher copy numbers.

ed by the donor organism have a better chance of being secreted by yeast, too. A problem that may occur with mammalian proteins being secreted by yeast is their overglycosylation. Mutants affected in the glycosylation enzymes leading to shorter mannan chains are available (BALLOU, 1990). Another obstacle in foreign gene expression is that protein processing can occur at unexpected sites in the protein leader sequences or in the foreign protein. This may lead to undesired 5′-termini or product degradation.

Although these problems can be dealt with by the sophisticated genetic methods available for *S. cerevisiae* as described above, other yeast species have been investigated as alternative hosts. Thus, efficient protein secretion has been observed in strains of *Pichia pastoris*, *K. lactis*, and *H. polymorpha* (Tab. 1; ROMANOS et al., 1992), e.g., for bovine lysozyme (DIGAN et al., 1989), prochymosin (VAN DEN BERG et al., 1990) and glucoamylase (GELLISSEN et al., 1991), respectively. Although the genetic and molecular basis of secretion in these yeasts is yet poorly understood, they may be the organisms of choice for future applications.

6 Perspectives

In this chapter, we tried to give an overview on the current state of the art in working with yeasts. Although we mainly concentrated on strains of *Saccharomyces cerevisiae*, it should be obvious that the special metabolic capacities of other yeasts are only beginning to be explored. In certain instances the use of yeasts other than *S. cerevisiae* has already shown their advantages in production processes. Thus, the use of *Hansenula polymorpha* as a host for foreign gene expression is likely to be increased due to its high production capacity. Similarly, we are only beginning to explore the possibilities of yeasts like *Kluyveromyces lactis*, which has the advantage of being almost as amenable to genetic manipulations as *S. cerevisiae*. The great number of genes already isolated from *S. cerevisiae* and its close evolutionary relatedness to *K. lactis* will ensure a rapid progress in our knowledge in this yeast.

Acknowledgement

We like to thank our colleagues ROSAURA RODICIO (Oviedo), MICHAEL CIRIACY (Düsseldorf) and KARIN BREUNIG (Düsseldorf) for critical reading of the manuscript and many useful suggestions.

Work in our laboratory has been supported by grants from BMFT (to CPH) and DFG (to JJH and CPH).

7 References

AGUILERA, A. (1986), Deletion of the phosphoglucose isomerase structural gene makes growth and sporulation glucose dependent in *Saccharomyces cerevisiae*, *Mol. Gen. Genet.* **204**, 310–316.

AGUILERA, A., ZIMMERMANN, F. K. (1986), Isolation and molecular analysis of the phosphoglucose isomerase structural gene of *Saccharomyces cerevisiae*, *Mol. Gen. Genet.* **202**, 83–89.

ALBER, T., KAWASAKI, G. (1982), Nucleotide sequence of the triose phosphate isomerase gene of *Saccharomyces cerevisiae*, *J. Mol. Appl. Genet.* **1**, 419–434.

ALBERTINI, S., ZIMMERMANN, F. K. (1991), The detection of chemically induced chromosomal malsegregation in *Saccharomyces cerevisiae* D61.M: A literature survey (1984–1990), *Mutat. Res. Rev. Genet. Toxicol.* **258**, 237–258.

ALBIG, W., ENTIAN, K.-D. (1988), Structure of yeast glucokinase, a strongly diverged specific aldohexose-phosphorylating isoenzyme, *Gene* **73**, 141–152.

AMORE, R., KÖTTER, P., KÜSTER, C., CIRIACY, M., HOLLENBERG, C. P. (1991), Cloning and expression in *Saccharomyces cerevisiae* of the NAD(P)H-dependent xylose reductase-encoding gene (*XYL1*) from the xylose-assimilating yeast *Pichia stipitis*, *Gene* **109**, 89–97.

ANAND, R. (1992), Yeast artificial chromosomes (YACs) and the analysis of complex genomes, *Trends Biotechnol.* **10**, 35–40.

ANDREWS, B. J., HERSKOWITZ, I. (1990), Regulation of cell cycle-dependent gene expression in yeast, *J. Biol. Chem.* **265**, 14057–14060.

AVIGAD, G. (1981), Stimulation of yeast phosphofructokinase activity by fructose 2, 6-bis-phosphate, *Biochem. Biophys. Res. Commun.* **102**, 985–991.

BAKALINSKY, A. T., SNOW, R. (1990), The chromosomal constitution of wine strains of *Saccharomyces cerevisiae*, *Yeast* **6**, 367–382.

BAKER, H. V. (1986), Glycolytic gene expression in

Saccharomyces cerevisiae: Nucleotide sequence of GCR1, null mutants, and evidence for expression, *Mol. Cell. Biol.* **6,** 3774–3784.

BAKER, H. V. (1991), *GCR1* of *Saccharomyces cerevisiae* encodes a DNA binding protein whose binding is abolished by mutations in the CTTCC sequence motif, *Proc. Natl. Acad. Sci. USA* **88,** 9443–9447.

BALLESTER, R., MICHAELI, T., FERGUSON, K., XU, H.-P., MCCORMICK, F., WIGLER, M. (1989), Genetic analysis of mammalian GAP expressed in yeast, *Cell* **59,** 681–686.

BALLESTER, R., MARCHUK, D., BOGUSKI, M., SAULINO, A., LETCHER, R., WIGLER, M., COLLINS, F. (1990), The *NF1* locus encodes a protein functionally related to mammalian GAP and yeast IRA proteins, *Cell* **63,** 851–859.

BALLOU, C. E. (1990), Isolation, characterization, and properties of *Saccharomyces cerevisiae mnn* mutants with nonconditional protein glycosylation defects, *Methods Enzymol.* **185,** 440–470.

BANUELOS, M., GANCEDO, C., GANCEDO, J. M. (1977), Activation by phosphate of yeast phosphofructokinase, *J. Biol. Chem.* **252,** 6394–6398.

BARNES, G., RINE, J. (1985), Regulated expression of endonuclease EcoRI in *Saccharomyces cerevisiae:* nuclear entry and biological consequences, *Proc. Natl. Acad. Sci. USA* **82,** 1354–1358.

BARNETT, J. (1976), The utilization of sugars by yeast, *Adv. Carbohydr. Chem. Biochem.* **32,** 126–234.

BARNETT, J. A. (1992), The taxonomy of the genus *Saccharomyces* Meyen *ex* Reese: A short review for non-taxonomists, *Yeast* **8,** 1–23.

BARNETT, J. A., PAYNE, R. W., YARROW, D. (1983), *Yeasts: Characteristics and Identification*, Cambridge, UK: Cambridge University Press.

BARR, P. J., BRAKE, A. J., VALENZUELA, P. (1989), *Yeast Genetics Engineering*, Boston: Butterworth.

BECKER, D. M., GUARENTE, L. (1990), High-efficiency transformation of yeast by electroporation, *Methods Enzymol.* **194,** 182–186.

BEGGS, J. (1978), Transformation of yeast by a replicating hybrid plasmid, *Nature* **275,** 104–108.

BELCOURT, M. F., FARABAUGH, P. J. (1990), Ribosomal frameshifting in the yeast retrotransposon Ty: tRNAs induce slippage on a 7 nucleotide minimal site, *Cell* **62,** 339–352.

BENNETZEN, J. L, HALL, B. D. (1982), Codon selection in yeast, *J. Biol. Chem.* **257,** 3026–3031.

BIELEFELD, M., HOLLENBERG, C. P. (1989), Mutant invertase proteins accumulate in the yeast endoplasmic reticulum, *Mol. Gen. Genet.* **215,** 401–406.

BISSON, L. F. (1988), High-affinity glucose transport in *Saccharomyces cerevisiae* is under general glucose repression control, *J. Bactceriol.* **170,** 4838–4845.

BISSON, L. F., FRAENKEL, D. G. (1983), Involvement of kinases in glucose and fructose uptake by *Saccharomyces cerevisiae*, *Proc. Natl. Acad. Sci. USA* **80,** 1730–1734.

BITTER, G. A., CHANG, K. K. H., EGAN, K. M. (1991), A multi-component upstream activation sequence of the *Saccharomyces cerevisiae* glyceraldehyde-3-phosphate dehydrogenase gene promoter, *Mol. Gen. Genet.* **231,** 22–32.

BJÖRKLING, F., GODTFREDSEN, S. E., KIRK, O. (1991), The future impact of industrial lipases, *TIBTECH* **9,** 360–363.

BOULTON, C. A. (1991), Developments in brewery fermentation, *Biotechnol. Genet. Eng. Rev.* **9,** 127–181.

BOWMAN, S. B., ZAMAN, Z., COLLINSON, L. P., BROWN, A. J. P., DAWES, I. W. (1992), Positive regulation of the *LPD1* gene of *Saccharomyces cerevisiae* by the HAP2/HAP3/HAP4 activation system, *Mol. Gen. Genet.* **231,** 296–303.

BRAM, R. J., LUE, N. F., KORNBERG, R. D. (1986), A GAL family of upstream activating sequences in yeast: Roles in both induction and repression of transcription, *EMBO J.* **5,** 603–608.

BRAUS, G. H. (1991), Aromatic amino acid biosynthesis in the yeast *Saccharomyces cerevisiae:* a model system for the regulation of a eukaryotic biosynthetic pathway, *Microbiol. Rev.* **55,** 349–370.

BREITENBACH-SCHMITT, I., HEINISCH, J., SCHMITT, H. D., ZIMMERMANN, F. K. (1984), Yeast mutants without phosphofructokinase activity can still perform glycolysis and alcoholic fermentation, *Mol. Gen. Genet.* **195,** 530–535.

BREUNIG, K. D. (1989), Glucose repression of *LAC* gene expression in yeast is mediated by the transcriptional activator LAC9, MGG, *Mol. Gen. Genet.* **216,** 422–427.

BREUNIG, K. D., KUGER, P. (1987), Functional homology between the yeast regulatory proteins GAL4 and LAC9: LAC9-mediated transcriptional activation in *Kluyveromyces lactis* involves protein binding to a regulatory sequence homologous to the GAL4 protein-binding site, *Mol. Cell. Biol.* **7,** 4400–4406.

BROACH, J. R. (1991), *Ras* genes in *Saccharomyces cerevisiae:* Signal transduction in search of a pathway, *TIG* **7,** 28–33.

BROACH, J. R., DESCHENES, R. J. (1990), The function of *RAS* genes in *Saccharomyces cerevisiae*, *Adv. Cancer Res.* **54,** 79–140.

BROACH, J. R., JONES, E. W., PRINGLE, J. R. (1991), *The Molecular Biology of the Yeast Saccharomyces*, 2nd Ed., Cold Spring Harbor: Cold Spring Harbor Laboratory Press.

BUCKHOLZ, R. G., GLEESON, M. A. G. (1991), Yeast systems for the commercial production of heterologous proteins, *Bio/Technology* **9,** 1067–1072.

BURGERS, P. J. M., PERCIVAL, K. J. (1987), Transformation of yeast spheroplasts without cell fusion, *Anal. Biochem.* **163,** 391–397.

BÜRGLIN, T. R. (1988), The yeast regulatory gene *PHO2* encodes a homeo box, *Cell* **53,** 339–340.

BURKE, D., GASDASKA, P., HARTWELL, L. (1989), Dominant effects of tubulin overexpression in *Saccharomyces cerevisiae*, *Mol. Cell. Biol.* **9,** 1049–1059.

BURKE, R. L., TEKAMP-OLSON, P., NAJARAN, R. C. (1983), The isolation, characterization, and sequence of the pyruvate kinase gene of *Saccharomyces cerevisiae*, *J. Biol. Chem.* **258,** 2193–2201.

BUTLER, G., MCCONNELL, D. J. (1988), Identification of an upstream activation site in the pyruvate decarboxylase structural gene (*PDC1*) of *Saccharomyces cerevisiae*, *Curr. Genet.* **14,** 405–412.

CAMERON, S., LEVIN, L., ZOLLER, M., WIGLER, M. (1988), cAMP-independent control of sporulation, glycogen metabolism, and heat shock resistance in *S. cerevisiae*, *Cell* **53,** 555–566.

CAMPBELL, I., DUFFUS, J. H. (1988), *Yeast – A Practical Approach*, Oxford–Washington: IRL press.

CARLE, G. F., OLSON, M. V. (1985), An electrophoretic karyotype for yeast, *Proc. Natl. Acad. Sci. USA* **82,** 3756–3760.

CHALKER, D. L., SANDMEYER, S. B. (1990), Transfer RNA genes are genomic targets for *de novo* transposition of the yeast retrotransposon Ty3, *Genetics* **126,** 837–850.

CHAMBERS, A., STANWAY, C., KINGSMAN, A. J., KINGSMAN, S. M. (1988), The UAS of the yeast *PGK* gene is composed of multiple functional elements, *Nucleic Acids Res.* **16,** 8245–8260.

CHAMBERS, A., STANWAY, C., TSANG, J. S. H., HENRY, Y., KINGSMAN, A. J., KINGSMAN, S. M. (1990), ARS binding factor 1 binds adjacent to RAP1 at the UASs of the yeast glycolytic genes *PGK* and *PYK1*, *Nucleic Acids Res.* **18,** 5393–5399.

CHEN, X. J., SALIOLA, M., FALCONE, C., BIANCHI, M. M., FUKUHARA, H. (1986), Sequence organization of the circular plasmid pKD1 from the yeast *Kluyveromyces drosophilarum*, *Nucleic Acids Res.* **14,** 4471–4481.

CHERRY, J. R., JOHNSON, T. R., DOLLARD, C., SHUSTER, J. R., DENIS, C. L. (1989), Cyclic AMP-dependent protein kinase phosphorylates and inactivates the yeast transcriptional activator ADR1, *Cell* **56,** 409–419.

CHIANG, H.-L., SCHEKMAN, R. (1991), Regulated import and degradation of a cytosolic protein in the yeast vacuole, *Nature* **350,** 313–318.

CHISHOLM, G. E., GENBAUFFE, F. S., COOPER, T. G. (1984), *tau*, a repeated DNA sequence in yeast, *Proc. Natl. Acad. Sci. USA* **81,** 2965–2969.

CHRISTIANSON, T. W., SIKORSKI, R. S., DANTE, M., SHERO, J. H., HIETER, P. (1992), Multifunctional yeast high-copy-number shuttle vectors, *Gene* **110,** 119–122.

CIGAN, A. M., DONAHUE, T. F. (1987), Sequence and structural features associated with translational initiator regions in yeast – a review, *Gene* **59,** 1–18.

CIRIACY, M. (1975), Genetics of alcohol dehydrogenase in *Saccharomyces cerevisiae*, *Mutat. Res.* **29,** 315–326.

CIRIACY, M. (1979), Isolation and characterization of further *cis*- and *trans*-acting regulatory elements involved in the synthesis of glucose-repressible alcohol dehydrogenase (ADHII) in *Saccharomyces cerevisiae*, *Mol. Gen. Genet.* **176,** 427–431.

CIRIACY, M., BREITENBACH, I. (1979), Physiological effects of seven different blocks in glycolysis in *Saccharomyces cerevisiae*, *J. Bacteriol.* **139,** 152–160.

CLARKE, L. (1990), Centromeres of budding and fission yeasts, *TIG* **6,** 150–154.

COFFEY, A. J., ROBERTS, R. G., GREEN, E. D., COLE, C. G., BUTLER, R., ANAND, R., GIANNELLI, F., BENTLEY, D. R. (1992), Construction of a 2.6-Mb contig in yeast artificial chromosomes spanning the human dystrophin gene using an STS-based approach, *Genomics* **12,** 474–484.

COHEN, R., HOLLAND, J. P., YOKOI, T., HOLLAND, M. J. (1986), Identification of a regulatory region that mediates glucose-dependent induction of the *Saccharomyces cerevisiae* enolase gene *ENO2*, *Mol. Cell. Biol.* **6,** 2287–2297.

COHEN, R., YOKOI, T., HOLLAND, J. P., PEPPER, A. E., HOLLAND, M. J. (1987), Transcription of the constitutively expressed yeast enolase gene *ENO1* is mediated by positive and negative *cis*-acting regulatory sequences, *Mol. Cell. Biol.* **7,** 2753–2761.

CONRAD, M. N., WRIGHT, J. H., WOLF, A. J., ZAKIAN, V. A. (1990), RAP1 protein interacts with yeast telomeres *in vivo:* Overproduction alters telomere structure and decreases chromosome stability, *Cell* **63,** 739–750.

COOPER, T. G. (1982) Transport in *Saccharomyces cerevisiae*, in: *The Molecular Biology of the Yeast Saccharomyces: Metabolism and Gene Expression* (STRATHERN, J. N., JONES, E. W., BROACH, J. R., Eds.), pp. 399–461,Cold Spring Harbor, NY: Cold Spring Harbor Laboratory Press.

COSTANZO, M. C., FOX, T. D. (1990), Control of mitochondrial gene expression in *Saccharomyces cerevisiae*, *Annu. Rev. Gen.* **24,** 91–113.

D'AMORE, T., PANCHAL, C. J., RUSSEL, I., STEWART, G. G. (1990), A study of ethanol tolerance in yeast, *Crit. Rev. Biotechnol.* **9,** 287–304.

DANHASH, N., GARDNER, D. C. J., OLIVER, S. G. (1991), Heritable damage to yeast caused by transformation, *Bio/Technology* **9,** 179–182.

DATTA, A. G., RACKER, E. (1961), Mechanism of action of transketolase, *J. Biol. Chem.* **236,** 617–623.

DAVIES, S. E. C., BRINDLE, K. M. (1992), Effects of overexpression of phosphofructokinase on glycolysis in the yeast *Saccharomyces cerevisiae*, *Eur. J. Biochem.* **31,** 4729–4735.

DAVIS, R. H. (1986), Compartmental and regulatory mechanisms in the arginine pathways of *Neurospora crassa* and *Saccharomyces cerevisiae*, *Microbiol. Rev.* **50,** 280–313.

DE LA GUERRA, R., VALDÉS-HEVIA, M. D., GANCEDO, J. M. (1988), Regulation of yeast fructose-1,6-bisphosphatase in strains containing multicopy plasmids coding for this enzyme, *FEBS Lett.* **242,** 149–152.

DEL RIO, J. L., SERRA, P., VALERO, F., POCH, M., SOLÁ, C. (1990), Reaction scheme of lipase production by *Candida rugosa* growing on olive oil, *Biotechnol. Lett.* **12,** 835–838.

DEMOLDER, K., FIERS, W., CONTRERAS, R. (1992), Efficient synthesis of secreted murine interleukine-2 by *Saccharomyces cerevisiae:* influence of 3′-untranslated regions and codon usage, *Gene* **111,** 207–213.

DE NOBEL, J. G., BARNETT, J. A. (1991), Passage of molecules through yeast cell walls: a brief essay-review, *Yeast* **7,** 313–323.

DEVER, T. E., FENG, L., WEK, R. C., CIGAN, A. M., DONAHUE, T. F., HINNEBUSCH, A. G. (1992), Phosphorylation of initiation factor 2 α by protein kinase GCN2 mediates gene-specific translation control of *GCN4* in yeast, *Cell* **68,** 585–596.

DE VRIES, S., MARRES, C. A. M. (1987), The mitochondrial respiratory chain of yeast. Structure and biosynthesis and the role in cellular metabolism, *Biochim. Biophys. Acta* **895,** 205–239.

DICKINSON, F. M., HAYWOOD, G. W. (1987), The role of metal ion in the mechanism of the K^+ activated aldehyde dehydrogenase of *Saccharomyces cerevisiae*, *Biochem. J.* **247,** 377–383.

DICKINSON, J. R. (1991), Biochemical and genetic studies on the function of, and relationship between, the *PGI1*- and *CDC30*-encoded phosphoglucose isomerases in *Saccharomyces cerevisiae, J. Gen. Microbiol.* **137,** 765–770.

DIFFLEY, J. F. X., STILLMAN, B (1989), Similarity between the transcriptional silencer binding proteins ABF1 and RAP1, *Science* **246,** 1034–1038.

DIGAN, M. E., LAIR, S. V., BRIERLY, R. A., SIEGEL, R. S., WILLIAMS, M. E., ELLIS, S. B., KELLARIS, P. A., PROVOW, S. A., CRAIG, W. S., VELICELEBI, G. et al. (1989), Continuous production of a novel lysozyme via secretion from the yeast *Pichia pastoris, Bio/Technology* **7,** 160–164.

D'MELLO, N. P., JAZWINSKI, S. M. (1991), Telomere length constancy during aging of *Saccharomyces cerevisiae*, *J. Bacteriol.* **173,** 6709–6713.

DOBSON, M. J., TUITE, M. F., ROBERTS, N. A., KINGSMAN, A. J., KINGSMAN, S. M. (1982), Conservation of high efficiency promoter sequences in *Saccharomyces cerevisiae*, *Nucleic Acids Res.* **10,** 2625–2637.

DOHMEN, R. J., STRASSER, A. W. M., HÖNER, C. B., HOLLENBERG, C. P. (1991), An efficient transformation procedure enabling long-term storage of competent cells of various yeast genera, *Yeast* **7,** 691–692.

DREWKE, C., THIELEN, J., CIRIACY, M. (1990), Ethanol formation in adh^0 mutants reveals the existence of a novel acetaldehyde-reducing activity in *Saccharomyces cerevisiae, J. Bacteriol.* **172,** 3909–3917.

DUJON, B. (1981), Mitochondrial genetics and functions, in: *The Molecular Biology of the Yeast Saccharomyces: Life Cycle and Inheritance* (STRATHERN, J. N., JONES, E. W., BROACH, J. R., Eds.), pp. 505–635, Cold Spring Harbor, NY: Cold Spring Harbor Laboratory Press.

EGEL-MITANI, M., HANSEN, M. T., NORRIS, K., SNEL, L., FIIL, N. P. (1988), Competitive expression of two heterologous genes inserted into one plasmid in *Saccharomyces cerevisiae*, *Gene* **73,** 113–120.

EGGELING, L., SAHM, H. (1978), Derepression and partial insensitivity to carbon catabolite repression in the methanol dissimilating enzymes in *Hansenula polymorpha*, *Eur. J. Appl Microbiol. Biotechnol.* **5,** 197–202

EIGEL, A., FELDMANN, H. (1982), Ty1 and delta elements occur adjacent to several tRNA genes in yeast, *EMBO J.* **1,** 1245–1250.

ENTIAN, K.-D., VOGEL, R. P., ROSE, M., HOFMANN, L., MECKE, D. (1988), Isolation and primary structure of the gene encoding fructose-1,6-bisphosphatase from *Saccharomyces cerevisiae*, *FEBS Lett.* **236,** 195–200.

ERHART, E., HOLLENBERG, C. P. (1981), Curing of *Saccharomyces cerevisiae* 2-μm DNA by transformation, *Curr.Genet.* **3,** 83–89.

ERNST, J. F. (1988), Codon usage and gene expression, *TIBTECH* **6,** 196–199.

ERNST, J. F. (1989), Molecular genetics of pathogenic fungi: some recent developments and perspectives, *Mycoses* **33,** 225–229.

FANGMAN, W. L., BREWER, B. J. (1991), Activation of replication origins within yeast chromosomes, *Annu. Rev. Cell. Biol.* **7,** 375–402.

FARABAUGH, P., LIAO, X.-B., BELCOURT, M., ZHAO, H., KAPAKOS, J., CLARE, J. (1989), Enhancer and silencerlike sites within the transcribed portion of a Ty2 transposable element of *Saccharomyces cerevisiae*, *Mol. Cell. Biol.* **9,** 4824–4834.

FEDOR-CHAIKEN, M., DESCHENES, R. J., BROACH, J. R. (1990), *SRV2*, a gene required for *RAS* activation of adenylate cyclase in yeast, *Cell* **61,** 329–340.

FERNÁNDEZ, E., MORENO, F., RODICIO, R. (1992),

The *ICL*1 gene from *Saccharomyces cerevisiae*, *Eur. J. Biochem.* **204,** 983–990.

FIECHTER, A., FUHRMANN, G. F., KÄPPELI, O. (1981), Regulation of glucose metabolism in growing yeast cells, *Adv. Microb. Physiol.* **22,** 123–183.

FIELDS, S. (1990), Pheromone response in yeast, *TIBS* **15,** 270–273.

FINK, G. R., BOEKE, J. D., GARFINKEL, D. J. (1986), The mechanism and consequences of retrotransposition, *TIG* **2,** 118–123.

FLETCHER, T. S., KWEE, I. L., NAKADA, T., LARGMAN, C., MARTIN, B. M. (1992), DNA sequence of the yeast transketolase gene, *Biochemistry* **31,** 1892–1896.

FORSBURG, S.L., NURSE, P. (1991), Cell cycle regulation in the yeasts *Saccharomyces cerevisiae* and *Schizosaccharomyces pombe*, *Annu. Rev. Cell. Biol.* **7,** 227–256.

FRAENKEL, D. G. (1982), Carbohydrate metabolism, in: *The Molecular Biology of the Yeast Saccharomyces* (STRATHERN, J., JONES, E. W., BROACH. J. R., Eds.), pp. 1–37, Cold Spring Harbor, NY: Cold Spring Harbor Laboratory Press.

FRÖHLICH, K. U., ENTIAN, K. D., MECKE, D. (1985), The primary structure of the yeast hexokinase PII gene (*HXK2*) which is responsible for glucose repression, *Gene* **36,** 105–111.

GANCEDO, C., SERRANO, R. (1989), Energy-yielding metabolism, in: *The Yeasts,* (ROSE, A. H., HARRISON, J. S., Eds.), pp. 205-259, London: Academic Press.

GANCEDO, J. M. (1992), Carbon catabolite repression in yeast, *Eur. J. Biochem.* **206,** 297–313.

GANCEDO, J. M., GANCEDO, C. (1979), Inactivation of gluconeogenic enzymes in glycolytic mutants of *Saccharomyces cerevisiae*, *Eur. J. Biochem.* **101,** 455–460.

GANCEDO, J. M., MAZON, M. J., GANCEDO, C. (1983), Fructose-2,6-bisphosphatase activates the cAMP-dependent phosphorylation of yeast fructose-1,6-bisphosphatase *in vitro*, *J. Biol. Chem.* **258,** 5998–6007.

GANGLOFF, S. P., MARGUET, D., LAUQUIN, G. J. M. (1990), Molecular cloning of the yeast mitochondrial aconitase gene (*ACO1*) and evidence of a synergistic regulation of expression by glucose plus glutamate, *Mol. Cell. Biol.* **10,** 3551–3561.

GELLISSEN, G., JANOWICZ, Z. J., MERCKELBACH, A., PIONTEK, M., KEUP, P., WEYDEMANN, U., HOLLENBERG, C. P., STRASSER, A. W. (1991), Heterologous gene expression in *Hansenula polymorpha:* Efficient secretion of glucoamylase, *Bio/Technology* **9,** 291–295.

GELLISSEN, G., WEYDEMANN, U., STRASSER, A. W. M., PIONTEK, M., HOLLENBERG, C. P., JANOWICZ, Z. A. (1992), Progress with using methylotrophic yeasts as expression systems, *Antonie van Leeuwenhoek*, in press.

GERMAIN, D., DUMAS, F., VERNET, T., BOURBONNAIS, Y., THOMAS, D. Y., BOILEAU, G. (1992), The proregion of the Kex2 endoprotease of *Saccharomyces cerevisiae* is removed by self-processing, *FEBS Lett.* **299,** 283–286.

GERST, J. E., FERGUSON, K., VOJTEK, A., WIGLER, M., FIELD, J. (1991), CAP is a bifunctional component of the *Saccharomyces cerevisiae* adenylyl cyclase complex, *Mol. Cell. Biol.* **11,** 1248–1257.

GIBBS, J. B., MARSHALL, M. S. (1989), The ras oncogene – An important regulatory element in lower eukaryotic organisms, *Microbiol. Rev.* **53,** 171–185.

GIETZ, R. D., SCHIESTL, R. H. (1991), Applications of high efficiency lithium acetate transformation of intact yeast cells using single-stranded nucleic acids as carrier, *Yeast* **7,** 253–263.

GILL, G., SADOWSKI, I., PTASHNE, M. (1990), Mutations that increase the activity of a transcriptional activator in yeast and mammalian cells, *Proc. Natl. Acad. Sci. USA* **87,** 2127–2131.

GILLIQUET, V., LEGRAIN, M., BERBEN, G., HILGER, F. (1990), Negative regulatory elements of the *Saccharomyces cerevisiae PNO* system: Interaction between PHO80 and PHO85 proteins, *Gene* **96,** 181–188.

GINIGER, E., VARNUM, S., PTASHNE, M. (1985), Specific DNA binding of GAL4, a positive regulatory protein of yeast, *Cell* **40,** 767–774.

GLEESON, M. A., SUDBERY, P. E. (1988), The methylotrophic yeasts, *Yeast* **4,** 1–15.

GNIRKE, A., HUXLEY, C. (1991), Transfer of the human *HPRT* and *GART* genes from yeast to mammalian cells by microinjection of YAC DNA, *Somat. Cell Mol. Genet.* **17,** 573–580.

GOEDDEL, D. V. (1990), Gene expression technology, *Methods Enzymol.* **185,** 231–482.

GREEN, J. B. A., WRIGHT, A. P. H., CHEUNG, W. Y., LANCASHIRE, W. E., HARTLEY, B. S. (1988), The structure and regulation of phosphoglucose isomerase in *Saccharomyces cerevisiae*, *Mol. Gen. Genet.* **215,** 100–106.

GRIGGS, D. W., JOHNSTON, M. (1991), Regulated expression of the *GAL4* activator gene in yeast provides a sensitive genetic switch for glucose repression, *Proc. Natl. Acad. Sci. USA* **88,** 8597–8601.

GRIVELL, L. A. (1989), Nucleo-mitochondrial interactions in yeast mitochondrial biogenesis, *Eur. J. Biochem.* **182,** 477–493.

GRIVELL, L. A., PLANTA, R. J. (1990), Yeast: the model 'eurokaryote'? *TIBTECH* **8,** 241–243.

GRIVELL, L. A., SCHWEYEN, R. J. (1989), RNA splicing in yeast mitochondria: Taking out the twists, *TIG* **5,** 39–41.

GUARENTE, L. (1983), Yeast promoters and *lacZ* fusions designed to study expression of cloned genes in yeast, *Methods Enzymol.* **101,** 181–191.

GUARENTE, L., BERMINGHAM-MCDONOGH, O. (1992), Conservation and evolution of transcriptional mechanisms in eukaryotes, *TIG* **8,** 27–32.

GUTHRIE, C., FINK, G. R. (1991), *Methods in Enzymology*, Vol. 194: *Guide to Yeast Genetics and Molecular Biology,* San Diego–New York–Boston–London: Academic Press.

HAHN, S., BURATOWSKI, S., SHARP, P. A., GUARENTE, L. (1989), Yeast TATA-binding protein TFIID binds to TATA elements with both consensus and nonconsensus DNA sequences, *Proc. Natl. Acad. Sci. USA* **86,** 5718–5722.

HARDER, W., VEENHUIS, M. (1989), Metabolism of one-carbon compounds, in: *The Yeasts* (ROSE, A. H., HARRISON, J. S., Eds.), Vol. 3, pp. 289–316, London: Academic Press.

HEE LEE, D., TANAKA, K., TAMURA, T., HA CHUNG, C., ICHIHARA, A. (1992), *PRS3* encoding an essential subunit of yeast proteasomes homologous to mammalian proteasome subunit C5, *Biochem. Biophys. Res Commun.* **182,** 452–460.

HEINEMANN, J. A., SPRAGUE, G. P., Jr. (1989), Bacterial conjugative plasmids mobilize DNA transfer between bacteria and yeast, *Nature* **340,** 205–209.

HEINISCH, J. (1986), Isolation and characterization of the two structural genes coding for phosphofructokinase in yeast, *Mol. Gen.Genet.* **202,** 75–82.

HEINISCH, J., ZIMMERMANN, F. K. (1985), Is the phosphofructokinase reaction obligatory for glucose fermentation by *Saccharomyces cerevisiae*? *Yeast* **1,** 173–175.

HEINISCH, J., RITZEL, R. G., VON BORSTEL, R. C., AGUILERA, A., RODICIO, R., ZIMMERMANN, F. K. (1989), The phosphofructokinase genes of yeast evolved from two duplication events, *Gene* **78,** 309–321.

HEINISCH, J., VOGELSANG, K., HOLLENBERG, C. P. (1991 a), Transcriptional control of yeast phosphofructokinase gene expression, *FEBS Lett.* **289,** 77–82.

HEINISCH, J., VON BORSTEL, R. C., RODICIO, R. (1991 b), Sequence and localization of the gene encoding yeast phosphoglycerate mutase, *Curr. Genet.* **20,** 167–171.

HENICK-KLING, T. (1988), Yeast and bacterial control in wine making, in: *Wine Analysis* (LINSKENS, H. F., JACKSON, J. F., Eds.), Heidelberg: Springer-Verlag,

HENRY, S. A. (1982), Membrane lipids of yeast: Biochemical and genetic studies, in: *The Molecular Biology of the Yeast Saccharomyces – Metabolism and Gene Expression* (STRATHERN, J. N., JONES, E. W., BROACH, J. R., Eds.), pp. 101–158, Cold Spring Harbor, NY: Cold Spring Harbor Laboratory Press.

HENRY, S. A., DONAHUE, T., CULBERTSON, M. (1975), Selection of spontaneous mutants by inositol starvation in *Saccharomyces cerevisiae*, *Mol. Gen. Genet.* **143,** 5–12.

HERMAN, P. K., EMR, S. D. (1990), Characterization of *VPS34*, a gene required for vacuolar protein sorting and vacuole segregation in *Saccharomyces cerevisiae*, *Mol. Cell. Biol.* **10,** 6742–6754.

HERSKOWITZ, I. (1989), A regulatory hierarchy for cell specialization in yeast, *Nature* **342,** 749–757.

HESS, B., BOITEUX, A. (1971), Oscillatory phenomena in biochemistry, *Annu. Rev. Biochem.* **4,** 237–258.

HICKS, J. (1986), *Yeast Cell Biology*, New York: Allan R. Liss.

HINNEBUSCH, A. G. (1988), Mechanisms of regulation in the general control of amino acid biosynthesis in *Saccharomyces cerevisiae*, *Microbiol. Rev.* **52,** 248–273.

HINNEBUSCH, A. G. (1990), Involvement of an initiation factor and protein phosphorylation in translational control of *GCN4* mRNA, *TIBS* **15,** 148–152.

HINNEBUSCH, A. G., JACKSON, B. M., MUELLER, P. P. (1988), Evidence for regulation of reinitiation in translational control of *GCN4* mRNA, *Proc. Natl. Acad. Sci. USA* **85,** 7279–7283.

HINNEN, A., HICKS, J. B., FINK, J. R. (1978), Transformation of yeast, *Proc. Natl. Acad. Sci. USA* **75,** 1929–1933.

HITZEMAN, R. A., HAGIE, F. F., LEVINE, H. L., GOEDDEL, D. V., AMMERER, G., HALL, B. D. (1981), Expression of a human gene for interferon in yeast, *Nature* **293,** 717–722.

HITZEMAN, R. A., HAGIE, F. E., HAYFLICK, J. S., CHEN, C. Y., SEEBURG, P. H., DERYNCK, R. (1982), The primary structure of the *Saccharomyces cerevisiae* gene for 3-phosphoglycerate kinase, *Nucleic Acids Res.* **10,** 7791–7808.

HITZEMAN, R. A., CHEN, C. Y., DOWBENKO, D. J., RENZ, M. E., LIU, C., PAI, R., SIMPSON, N. J., KOHR, W. J., SINGH, A., CHISHOLM, V., et al. (1990), Use of heterologous and homologous signal sequences for secretion of heterologous proteins from yeast, *Methods Enzymol.* **185,** 421–439.

HOEKSTRA, M. F., DEMAGGIO, A. J., DHILLON, N. (1991 a), Genetically identified protein kinases in yeast I: transcription, translation, transport and mating, *TIG* **7,** 256–260.

HOEKSTRA, M. F., DEMAGGIO, A. J., DHILLON, N. (1991 b), Genetically identified protein kinases in yeast II: DNA metabolism and meiosis, *TIG* **7,** 293–297.

HOFMANN, E., KOPPERSCHLÄGER, G. (1982), Phosphofructokinase from yeast, *Methods Enzymol.* **90,** 49–60.

HOHMANN, S. (1991 a), Characterization of *PDC6*, a third structural gene for pyruvate decarboxylase in *Saccharomyces cerevisiae*, *J. Bacteriol.* **173,** 7963–7969.

HOHMANN, S. (1991 b), *PDC6*, a weakly expressed pyruvate decarboxylase gene from yeast, is activated when fused spontaneously under the control of the *PDC1* promoter, *Curr. Genet.* **20,** 373–378.

HOLLAND, J. P., LABIENIEC, L., SWIMMER, C., HOLLAND, M. J. (1983), Homologous nucleotide sequences at the 5′ termini of messenger RNAs synthesized from the yeast enolase and glyceraldehyde-3-phosphate dehydrogenase gene families: The primary structure of a third yeast glyceraldehyde-3-phosphate dehydrogenase gene, *J. Biol. Chem.* **258,** 5291–5299.

HOLLAND, J. P., BRINDLE, P. K., HOLLAND, M. J. (1990), Sequences within an upstream activation site in the yeast enolase gene *ENO2* modulate repression of *ENO2* expression in strains carrying a null mutation in the positive regulatory gene *GCR1*, *Mol. Cell. Biol.* **10,** 4863–4871.

HOLLENBERG, C. P., BORST, P., VAN BRUGGEN, E. F. J. (1970), A 25-μ closed circular duplex DNA molecule in wild-type yeast mitochondria. Structure and genetic complexity, *Biochim. Biophys. Acta* **209,** 1–15.

HOLLENBERG, C. P., KUSTERMANN-KUHN, B., and ROYER, H. D. (1976), Characterization of 2-μm DNA of *Saccharomyces cerevisiae* by restriction fragment analysis and integration in an *Escherichia coli* plasmid, *Proc. Natl. Acad. Sci. USA* **73,** 2072–2076.

HOLZER, H. (1976), Catabolite inactivation in yeast, *TIBS* **1,** 178–181.

HOTTIGER, T., BOLLER, T., WIEMKEN, A. (1989), Correlation of trehalose content and heat resistance in yeast mutants altered in the RAS/adenylate cyclase pathway: Is trehalose a thermoprotectant, *FEBS Lett.* **255,** 431–434.

HUBBARD, E. J. A., YANG, X., CARLSON, M. (1992), Relationship of the cAMP-dependent protein kinase pathway to the SNF1 protein kinase and invertase expression in *Saccharomyces cerevisiae*, *Genetics* **130,** 71–80.

IRNIGER, S., EGLI, C. M., BRAUS, G. H. (1991), Different classes of polyadenylation sites in the yeast *Saccharomyces cerevisiae*, *Mol. Cell. Biol.* **11,** 3060–3069.

ITO, H., FUKUDA, Y., MURATA, K., KIMURA, A. (1983), Transformation of intact yeast cells treated with alkali cations, *J. Bacteriol.* **153,** 163–168.

JAVELOT, C., GIRARD, P., COLONNA-CECCALDI, B., VLADESCU, B. (1991), Introduction of terpene-producing ability in a wine strain of *Saccharomyces cerevisiae*, *J. Biotechnol.* **21,** 239–252.

JEFFRIES, T. W. (1990), Fermentation of D-xylose and cellobiose, in: *Yeast Biotechnology and Biocatalysis* (VERACHTERT, H., DE MOT, R., Eds.), pp. 349–394, New York: Marcel Dekker.

JENNISSEN, H. P., BOTZET, G., MAJETSCHAK, M., LAUB, M., ZIEGENHAGEN, R., DEMIROGLOU, A. (1992), Ca^{2+}-dependent ubiquitination of calmodulin in yeast, *FEBS Lett.* **296,** 51–56.

JOHANNSEN, E. (1984), *Antonie van Leeuwenhoek* **46,** 177–189.

JOHNSON, C. M., COOPER, A., BROWN, A. J. P. (1991), A comparison of the reactivity and stability of wild type and His388 → Gln mutant phosphoglycerate kinase from yeast, *Eur. J. Biochem.* **202,** 1157–1164.

JOHNSON, E. S., BARTEL, B., SEUFERT, W., VARSHAVSKY, A. (1992), Ubiquitin as a degradation signal, *EMBO J.* **11,** 497–505.

JOHNSTON, J. R., CONTOPOULOU, R., MORTIMER, R. K. (1988), Karyotyping of yeast strains of several genera by field inversion gel electrophoresis, *Yeast* **4,** 191–198.

JOHNSTON, M. (1987), A model fungal gene regulatory mechanism: the *GAL* genes of *Saccharomyces cerevisiae*, *Microbiol. Rev.* **51,** 458–476.

JOHNSTON, M., DAVIS, R. W. (1984), Sequences that regulate the divergent GAL1–GAL10 promoter in *Saccharomyces cerevisiae*, *Mol. Cell. Biol.* **4,** 1440–1448.

JOHNSTON, S. A., ZAVORTINK, M. J., DEBOUCK, C., HOPPER, J. E. (1986), Functional domains of the yeast regulatory protein GAL4, *Proc. Natl. Acad. Sci. USA* **83,** 6553–6557.

JONES, E. W., FINK, G. R. (1982), Regulation of amino acid and nucleotide biosynthesis in yeast, in: *The Molecular Biology of the Yeast Saccharomyces: Metabolism and Gene Expression* (STRATHERN, J. N., JONES, E. W., BROACH, J. R., Eds.), pp. 181–299, Cold Spring Harbor, NY: Cold Spring Harbor Laboratory Press.

JONES, R. P., GREENFIELD, P. F. (1985), Replicative inactivation and metabolic inhibition in yeast ethanol fermentations, *Biotechnol. Lett.* **7,** 223–228.

JONES, S., VIGNAIS, M.-L., BROACH, J. R. (1991), The CDC25 protein of *Saccharomyces cerevisiae* promotes exchange of guanine nucleotides bound to ras, *Mol. Cell. Biol.* **11,** 2641–2646.

KABACK, D. B., STEENSMA, H. Y., DE JONGE, P. (1989), Enhanced meiotic recombination on the smallest chromosome of *Saccharomyces cerevisiae*, *Proc. Natl. Acad. Sci. USA* **86,** 3694–3698.

KELEHER, C. A., REDD, M. J., SCHULTZ, J., CARLSON, M., JOHNSON, A. D. (1992), Ssn6-Tup1 is a general repressor of transcription in yeast, *Cell* **68,** 709–719.

KELLEHER, R. J., III, FLANAGAN, P. M., CHASMAN, D. I., PONTICELLI, A. S., STRUHL, K., KORNBERG, R. D. (1992), Yeast and human TFIIDs are interchangeable for the response to acidic transcriptional activators *in vitro*, *Genes Dev.* **6,** 296–303.

KELLERMANN, E., HOLLENBERG, C. P. (1988), The glucose- and ethanol-dependent regulation of PDC1 from *Saccharomyces cerevisiae* are controlled by two distinct promoter regions, *Curr. Genet.* **14,** 337–344.

KELLERMANN, E., SEEBOTH, P. G., HOLLENBERG, C. P. (1986), Analysis of the primary structure and promoter function of a pyruvate decarboxylase gene (*PDC1*) from *Saccharomyces cerevisiae*, *Nucleic Acids Res.* **14,** 8963–8977.

KIELLAND-BRANDT, M. C., NILLSON-TILLGREN, T., PETERSEN, J. G.L., HOLMBERG, S., GJERMANSEN, C. (1983), Approaches to the genetic analysis and breeding of brewer's yeast, in: *Yeast Genetics. Fundamental and Applied Aspects* (SPENCER, J. F. T., SPENCER, D. M., SMITH, A. R. W., Eds.), pp. 421–437, New York: Springer Verlag.

KING, D. J., WALTON, F., SMITH, B. W., DUNN, M., YARRANTON, G. T. (1988), Recovery of recombinant proteins from yeast, *Biochem. Soc. Trans.* **16,** 1083–1086.

KINGSMAN, S. M., KINGSMAN, A. J., DOBSON, M. J., MELLOR, J., ROBERTS, N. A. (1985), Heterologous gene expression in *Saccharomyces cerevisiae*, *Biotechnol. Genet. Eng. Rev.* **3,** 377–416.

KINGSMAN, S. M., KINGSMAN, A. J., MELLOR, J. (1987), The production of mammalian proteins in *Saccharomyces cerevisiae*, *TIBTECH* **5,** 53–57.

KINGSMAN, S. M., COUSENS, D., STANWAY, C. A., CHAMBERS, A., WILSON, M., KINGSMAN, A. J. (1990), High-efficiency yeast expression vectors based on the promoter of the phosphoglycerate kinase gene, *Methods Enzymol.* **185,** 329–340.

KLEBE, R. J., HARRIS, J. V., SHARP, Z. D., DOUGLAS, M. G. (1983), A general method for polyethylene glycol induced genetic transformation of bacteria and yeast, *Gene* **25,** 333–341.

KOPETZKI, E., ENTIAN, K. D., MECKE, D. (1985), Complete nucleotide sequence of the hexokinase PI gene (*HXK1*) of *Saccharomyces cerevisiae*, *Gene* **39,** 95–102.

KOPPERSCHLÄGER, G., BÄR, J., NISSLER, K., HOFMANN, E. (1977), Physicochemical parameters and subunit composition of yeast phosphofructokinase, *Eur. J. Biochem.* **81,** 317–325.

KÖTTER, P., AMORE, R., HOLLENBERG, C. P., CIRIACY, M. (1990), Isolation and characterization of the *Pichia stipitis* xylitol dehydrogenase gene, *XYL2*, and construction of a xylose utilizing *Saccharomyces cerevisiae* transformant, *Curr. Genet.* **18,** 493–500.

KOTULA, L., CURTIS, P. J. (1991), Evaluation of foreign gene codon optimization in yeast: Expression of a mouse IG kappa chain, *Bio/Technology* **9,** 1386–1389.

KOZAK, M. (1983), Translation of insulin-related polypeptides from messenger RNAs with tandemly reiterated copies of the ribosome binding site, *Cell* **34,** 971–978.

KREGER-VAN RIJ, N. J. W. (1984), *The Yeasts – A Taxonomic Study*, 3rd Ed., Amsterdam: Elsevier Science.

KRETSCHMER, M., FRAENKEL, D. G. (1991), Yeast 6-phosphofructo-2-kinase: Sequence and mutant, *Biochemistry* **30,** 10663-10672.

KUCHLER, K., STERNE, R. E. THORNER, J. (1989), STE6 gene product: a novel pathway for protein export in eukaryotic cells, *EMBO J.* **8,** 3973–3984.

KUGER, P., GOEDECKE, A., BREUNIG, K. D. (1990), A mutation in the Zn-finger of the GAL4 homolog LAC9 results in glucose repression of its target gene, *Nucleic Acids Res.* **18,** 745-751.

KURTZ, S., SHORE, D. (1991), RAP1 protein activates and silences transcription of mating-type genes in yeast, *Genes Dev.* **5,** 616–628.

KUZHANDAIVELU, N., JONES, W. K., MARTIN, A. K., DICKSON, R. C. (1992), The signal for glucose repression of the lactose-galactose regulon is amplified through subtle modulation of transcription of the *Kluyveromyces lactis Kl-GAL4* activator gene, *Mol. Cell. Biol.* **12,** 1924–1931.

LAGUNAS, R., GANCEDO, J. M. (1973), Reduced pyridine nucleotide balance in glucose-growing *Saccharomyces cerevisiae*, *Eur. J. Biochem.* **37,** 90–96.

LAGUNAS, R., DOMINGUEZ, C., BUSTURIA, A., SÁEZ, M. J. (1982), Mechanisms of appearance of the Pasteur effect in *Saccharomyces cerevisiae:* Inactivation of sugar transport systems, *J. Bacteriol.* **152,** 19–25.

LALUCE, C. (1991), Current aspects of fuel ethanol production in Brazil, *Crit. Rev. Biotechnol.* **11,** 149–161.

LAM, K. B., MARMUR, J. (1977), Isolation and characterization of *Saccharomyces cerevisiae* glycolytic pathway mutants, *J. Bacteriol.* **130,** 746–749.

LANGFORD, C. J., KLINZ, F.-J., DONATH, C., GALLWITZ, D. (1984), Point mutations identify the conserved, intron-contained TACTAAC box as an essential splicing signal sequence in yeast, *Cell* **36,** 645–653.

LASKER, B. A., CARLE, G. F., KOBAYASHI, G. S., MEDOFF, G. (1989), Comparison of the separation of *Candida albicans* chromosome-sized DNA by pulsed-field gel electrophoresis techniques, *Nucleic Acids Res.* **17,** 3783–3793.

LEE, G. S., SAVAGE, E. A., RITZEL, R. G., VON BORSTEL, R. C. (1988), The base-alteration spectrum of spontaneous and ultraviolet radiation-induced forward mutations in the *URA3* locus of *Saccharomyces cerevisiae*, *Mol. Gen. Genet.* **214,** 396–404.

LEWIN, A. S., HINES, V., SMALL, G. M. (1990), Citrate synthase encoded by the *CIT2* gene of *Saccharo-*

myces cerevisiae is peroxisomal, *Mol. Cell. Biol.* **10,** 1399–1405.

LEWIS, D. A., BISSON, L. F. (1991), The *HXT1* gene product of *Saccharomyces cerevisiae* is a new member of the family of hexose transporters, *Mol. Cell. Biol.* **11,** 3804–3813.

LILLIE, S. H., PRINGLE, J. R. (1980), Reserve carbohydrate metabolism in *Saccharomyces cerevisiae:* Response to nutrient limitation, *J. Bacteriol.* **143,** 1384–1394.

LOBO, Z., MAITRA, P. K. (1982), Pentose phosphate pathway mutants of yeast, *Mol. Gen. Genet.* **185,** 367–368.

LOMBARDO, A., SCHEFFLER, I. E. (1989), Isolation and characterization of a *Saccharomyces cerevisiae* mutant with a disrupted gene for the IP subunit of succinate dehydrogenase, *J. Biol. Chem.* **264,** 18874–18877.

LOMBARDO, A., CARINE, K., SCHEFFLER, I. E. (1990), Cloning and characterization of the iron-sulfur subunit gene of succinate dehydrogenase from *Saccharomyces cerevisiae, J. Biol. Chem.* **265,** 10419–10423.

LOPES, T. S., HAKKAART, G.-J. A. J., KOERTS, B. L., RAUÉ, H. A., PLANTA, R. J. (1991), Mechanism of high-copy-number integration of pMIRY-type vectors into the ribosomal DNA of *Saccharomyces cerevisiae, Gene* **105,** 83–90.

LOURIERO, V., FERRIERA, H. G. (1983), On the intracellular accumulation of ethanol in yeast, *Biotechnol. Bioeng.* **25,** 2263–2267.

LUCHE, R. M., SUMRADA, R., COOPER, T. G. (1990), A *cis*-acting element present in multiple genes serves as a repressor protein binding site for the yeast *CAR1* gene, *Mol. Cell. Biol.* **10**, 3884–3895.

MACREADIE, I. G., HORAITIS, O., VERKUYLEN, A. J., SAVIN, K. W. (1991), Improved shuttle vectors for cloning and high-level Cu^{2+}-mediated expression of foreign genes in yeast, *Gene* **104,** 107–111.

MAITRA, P. K., LOBO, Z. (1971), A kinetic study of glycolytic enzyme synthesis in yeast, *J. Biol. Chem.* **246,** 475–488.

MARSH, A. M., NEIMAN, A. M., HERSKOWITZ, I. (1991), Signal transduction during pheromone response in yeast, *Annu. Rev. Cell. Biol.* **7,** 699–728.

MARSHALL-CARLSON, L., CELENZA, J. L., LAURENT, B. C., CARLSON, M. (1990), Mutational analysis of the SNF3 glucose transporter of *Saccharomyces cerevisiae, Mol. Cell. Biol.* **10,** 1105–1115.

MATTHEWS, K. S. (1992), DNA looping, *Microbiol. Rev.* **56,** 123–136.

MAYER, V. W., AGUILERA, A. (1990), High levels of chromosome instability in polyploids of *Saccharomyces cerevisiae, Mutat. Res. Fund. Mol. Mech. Mutagenesis* **231,** 177–186.

MAZON, M. J., GANCEDO, J. M., GANCEDO, C. (1982), Phosphorylation and inactivation of yeast fructose-bisphosphatase *in vivo* by glucose and by proton ionophores. A possible role for cAMP, *Eur. J. Biochem.* **127,** 605–608.

MCALISTER, L., HOLLAND, M. J. (1985), Differential expression of the three yeast glyceraldehyde-3-phosphate dehydrogenase genes, *J. Biol. Chem.* **260,** 15019–15027.

MELLOR, J., DOBSON, M. J., KINGSMAN, A. J., KINGSMAN, S. M. (1987), A transcriptional activator is located in the coding region of the yeast *PGK* gene, *Nucleic Acids Res.* **15,** 6243–6259.

MEYER, J., WALKER-JONAH, A., HOLLENBERG, C. P. (1991), Galactokinase encoded by *GAL1* is a bifunctional protein required for the induction of the *GAL* genes in *Kluyveromyces lactis* and able to suppress the gal3 phenotype in *Saccharomyces cerevisiae, Mol. Cell. Biol.* **11,** 5454–5461.

MICHAELI, T., FIELD, J., BALLESTER, R., O'NEILL, K., WIGLER, M. (1989), Mutants of H-*ras* that interfere with *RAS* effector function in *Saccharomyces cerevisiae, EMBO J.* **8,** 3039–3044.

MILLER, P. F., HINNEBUSCH, A. G. (1989), Sequences that surround the stop codons of upstream open reading frames in *GCN4* mRNA determine their distinct functions in translational control, *Genes Dev.* **3,** 1217–1225.

MINARD, K. I., MCALISTER-HENN, L. (1991), Isolation, nucleotide sequence analysis, and disruption of the *MDH2* gene from *Saccharomyces cerevisiae:* Evidence for three isozymes of yeast malate dehydrogenase, *Mol. Cell. Biol.* **11,** 370–380.

MISSIAKAS, D., BETTON, J.-M., MINARD, P., YON, J. M. (1990), Unfolding-refolding of the domains in yeast phosphoglycerate kinase: Comparison with the isolated engineered domains, *Biochemistry* **29,** 8683–8689.

MOEHLE, C. M., HINNEBUSCH, A. G. (1991), Association of RAP1 binding sites with stringent control of ribosomal protein gene transcription in *Saccharomyces cerevisiae, Mol. Cell. Biol.* **11,** 2723–2735.

MOORE, P. A., BETTANY, A. J. E., BROWN, A. J. P. (1990), Expression of a yeast glycolytic gene is subject to dosage limitation, *Gene* **89,** 85–92.

MORTIMER, R. K., HAWTHORNE, D. C. (1966), Genetic mapping in *Saccharomyces, Genetics* **53,** 165–173.

MORTIMER, R. K., SCHILD, D., CONTOPOULOU, C. R., KANS, J. A. (1989), Genetic and physical maps of *Saccharomyces cerevisiae, Yeast* **5,** 321–403.

MÜLLER, F., LAUFER, W., POTT, U., CIRIACY, M. (1991), Characterization of products of TY1-mediated reverse transcription in *Saccharomyces cerevisiae, Mol. Gen. Genet.* **226,** 145–153.

MÜLLER, P. P., TRACHSEL, H. (1990), Translation and regulation of translation in the yeast *Saccharomyces cerevisiae, Eur. J. Biochem.* **191,** 257–261.

NAKAJIMA, H., KONO, N., YAMASAKI, T., HOTTA, K.,

KAWACHI, M., HAMAGUCHI, T., NISHIMURA, T., MINEO, I., KUWAJIMA, M., NOGUCHI, T., et al. (1991), A genetic defect in muscle phosphofructokinase deficiency, a typical clinical entity presenting myogenic hyperuricemia, *Adv. Exp. Med. Biol.* **309B,** 141–144.

NASIM, A., JOHNSON, B. F., YOUNG, P. (1989) *Molecular Biology of the Fission Yeast*, New York: Academic Press.

NEHLIN, J. O., RONNE, H. (1990), Yeast MIG1 repressor is related to the mammalian early growth response and Wilms' tumour finger proteins, EMBO J. **9,** 2891-2898.

NISHIZAWA, M., ARAKI, R., TERANISHI, Y. (1989), Identification of an upstream activating sequence and an upstream repressible sequence of the pyruvate kinase gene of the yeast *Saccharomyces cerevisiae*, *Mol. Cell. Biol.* **9,** 442–451.

NISSLER, K., OTTO, A., SCHELLENBERGER, W., HOFMANN, E. (1983), Similarity of activation of yeast phosphofructokinase by AMP and fructose-2,6-bisphosphate, *Biochem. Biophys. Res. Commun.* **111,** 294–300.

NOGAE, I., JOHNSTON, M. (1990), Isolation and characterization of the ZWF1 gene of *Saccharomyces cerevisiae*, encoding glucose-6-phosphate dehydrogenase, *Gene* **96,** 161–169.

ODA, Y., OUCHI, K. (1989 a), Genetic analysis of haploids from industrial strains of baker's yeast, *Appl. Environ. Microbiol.* **55,** 1742–1747.

ODA, Y., OUCHI, K. (1989 b), Principal-component analysis of the characteristics desirable in baker's yeasts, *Appl. Environ. Microbiol.* **55,** 1495–1499.

OGAWA, N., OSHIMA, Y. (1990), Functional domains of a positive regulatory protein, PHO4, for transcriptional control of the phosphatase regulon in *Saccharomyces cerevisiae*, *Mol. Cell. Biol.* **10,** 2224–2236.

OHYA, Y., KAWASAKI, H., SUZUKI, K., LONDESBOROUGH, J., ANRAKU, Y. (1991), Two yeast genes encoding calmodulin-dependent protein kinases. Isolation, sequencing, and bacterial expressions of *CMK1* and *CMK2*, *J. Biol. Chem.* **266,** 12784–12794.

OKUYAMA, H., SAITO, M., JOSKI, V., GUNSBERG, S., WAKIL, S. (1979), Regulation by temperature of chain length of fatty acids in yeast, *J. Biol. Chem.* **254,** 12281–12289.

OSBORNE, B. I., GUARENTE, L. (1989), Mutational analysis of a yeast transcriptional terminator, *Proc. Natl. Acad. Sci. USA* **86,** 4097–4101.

OSUMI, T., FUJIKI, Y. (1990), Topogenesis of peroxysomal proteins, *BioEssays* **12,** 217–222.

PANEK, A. D., PANEK, A. C. (1990), Metabolism and thermotolerance function of trehalose in *Saccharomyces:* A current perspective, *J. Biotechnol.* **14,** 229–238.

PAYNE, G. S. (1990), Genetic analysis of clathrin function in yeast, *J. Membr. Biol.* **116,** 93–105.

PEREZ-ORTIN, J. E., MATALLANA, E., FRANCO, L. (1989), Chromatin structure of yeast genes, *Yeast* **5,** 219–238.

PERLMAN, P. S., MAHLER, H. R. (1974), Derepression of mitochondria and their enzymes in yeast: regulatory aspects, *Arch. Biochem. Biophys.* **162,** 248–271.

PHAFF, H. J. (1990), Isolation of yeasts from natural sources, in: *Isolation of Biotechnological Organisms from Nature* (LABEDA, D. P., Ed.), pp. 53–79, New York: McGraw-Hill.

PHAFF, H. J., STARMER, W. T. (1987), Yeasts associated with plants, insects and soil, in: *The Yeasts* (ROSE, A. H., HARRISON, J. S., Eds.), Vol. 1, pp. 123–180, London: Academic Press.

PLANKERT, U., PURWIN, C., HOLZER, H. (1991), Yeast fructose-2,6-bisphosphate 6-phosphatase is encoded by *PHO8*, the gene for nonspecific repressible alkaline phosphatase, *Eur. J. Biochem.* **196,** 191–196.

POLONELLI, L., CONTI, S., GERLONI, M., MAGLIANI, W., CHEZZI, C., MORACE, G. (1991), Interfaces of the yeast killer phenomenon, *Crit. Rev. Microbiol.* **18,** 47–87.

PUGH, B. F., TJIAN, R. (1992), Diverse transcriptional functions of the multisubunit eukaryotic TFIID complex, *J. Biol. Chem.* **267,** 679–682.

PURVIS, I. J., LOUGHLIN, L., BETTANY, A. J. E., BROWN, A. J. P. (1987), Translation and stability of an *E. coli* β-galactosidase mRNA expressed under the control of pyruvate kinase sequences in *Saccharomyces cerevisiae*, *Nucleic Acids. Res.* **15,** 7963–7973.

PURVIS, I. J., CHOTAI, D., DYKES, C. W., LUBAHN, D. B., FRENCH, F. S., WILSON, E. M., HOBDEN, A. N. (1991), An androgen-inducible expression system for *Saccharomyces cerevisiae*, *Gene* **106,** 35–42.

RAMIREZ, M., WEK, R. C., HINNEBUSCH, A. G. (1991), Ribosome association of GCN2 protein kinase, a translational activator of the *GCN4* gene of *Saccharomyces cerevisiae*, *Mol. Cell. Biol.* **11,** 3027–3036.

RAMOS, J., SZKUTNICKA, K., CIRILLO, V. P. (1988), Relationship between low- and high affinity glucose transport systems of *Saccharomyces cerevisiae*, *J. Bacteriol.* **170,** 5375–5377.

RATLEDGE, C., EVANS, C. T. (1989), Lipids and their metabolism, in: *The Yeasts* (ROSE, A. H., HARRISON, J. S., Eds.), pp. 367–455, London: Academic Press.

REISER, J., GLUMOFF, V., KÄLIN, M., OCHSNER, U. (1990), Transfer and expression of heterologous genes in yeasts other than *Saccharomyces cerevisiae*, *Adv. Biochem, Eng. Biotechnol.* **43,** 75–102.

REPETTO, B., TZAGOLOFF, A. (1989), Structure and

regulation of *KGD1*, the structural gene for yeast alpha-ketoglutarate dehydrogenase, *Mol. Cell. Biol.* **9,** 2695–2705.

REPETTO, B., TZAGOLOFF, A. (1990), Structure and regulation of *KGD2*, the structural gene for yeast dihydrolipoyl transsuccinylase, *Mol. Cell. Biol.* **10,** 4221–4232.

REPETTO, B., TZAGOLOFF, A. (1991), *In vivo* assembly of yeast mitochondrial α-ketoglutarate dehydrogenase complex, *Mol. Cell. Biol.* **11,** 3931–3939.

RHODE, P. R., ELSASSER, S., CAMPBELL, J. L. (1992), Role of multifunctional autonomously replicating sequence binding factor 1 in the initiation of DNA replication and transcriptional control in *Saccharomyces cerevisiae*, *Mol. Cell. Biol.* **12,** 1064–1077.

ROBINSON, K. M., VON KIECKEBUSCH-GÜCK, A., LEMIRE, B. D. (1991), Isolation and characterization of a *Saccharomyces cerevisiae* mutant disrupted for the succinate dehydrogenase flavoprotein subunit, *J. Biol. Chem.* **266,** 21347–21350.

RODICIO, R., HEINISCH, J. (1987), Isolation of the yeast phosphoglyceromutase gene and construction of deletion mutants, *Mol. Gen. Genet.* **206,** 133–140.

ROGERS, D. T., HILLER, E., MITSOCK, L., ORR, E. (1988), Characterization of the gene for fructose-1,6-bisphosphatase from *Saccharomyces cerevisiae* and *Schizosaccharomyces pombe*, *J. Biol. Chem.* **263,** 6051–6057.

ROGGENKAMP, R., DIDION, T., KOWALLIK, K. V. (1989), Formation of irregular giant peroxisomes by overproduction of the crystalloid core protein methanol oxidase in the methylotrophic yeast *Hansenula polymorpha*, *Mol. Cell. Biol.* **9,** 988–994.

ROMANOS, M. A., SCORER, C. A., CLARE, J. J. (1992), Foreign gene expression in yeast: A review, *Yeast* **8,** 423–488.

ROSA, M. F., SÁ-CORREIA, I. (1992), Ethanol tolerance and activity of plasma membrane ATPase in *Kluyveromyces marxianus* and *Saccharomyces cerevisiae*, *Enzyme Microb. Technol.* **14,** 23–27.

ROSE, A. H., HARRISON, J. S. (1989), *The Yeasts*, 2nd Ed., Vol. 3, London– San Diego–New York–Boston–Toronto: Academic Press.

ROSE, M., ENTIAN, K.-D., HOFMANN, L., VOGEL, R. F., MECKE, D. (1988), Irreversible inactivation of *Saccharomyces cerevisiae* fructose-1,6-bisphosphatase independent of protein phosphorylation at Ser^{11}, *FEBS Lett.* **241,** 55–59.

ROSE, M., ALBIG, W., ENTIAN, K.-D. (1991), Glucose repression in *Saccharomyces cerevisiae* is directly associated with hexose phosphorylation by hexokinases PI and PII, *Eur. J. Biochem.* **199,** 511–518.

RUBY, S. W., ABELSON, J. (1991), Pre-mRNA splicing in yeast, *TIG* **7,** 79–85.

SADOWSKI, I., NIEDBALA, D., WOOD, K., PTASHNE, M. (1991), GAL4 is phosphorylated as a consequence of transcriptional activation, *Proc. Natl. Acad. Sci. USA* **88,** 10510–10514.

SAHM, H. (1977), Metabolism of methanol by yeast, *Adv. Biochem. Eng.* **6,** 77–103.

SALMERON, J. M., JOHNSTON, S. A. (1986), Analysis of the *Kluyveromyces lactis* positive regulatory gene *LAC9* reveals functional homology to, but sequence divergence from, the *Saccharomyces cerevisiae* GAL4 gene, *Nucleic Acids Res.* **14,** 7767–7781.

SALMERON, JR., J. M., LEUTHER, K. K., JOHNSTON, S. A. (1990), *GAL4* mutations that separate the transcriptional activation and GAL80-interactive functions of the yeast GAL4 protein, *Genetics* **125,** 21–27.

SCHAAFF, I., GREEN, J. B. A., GOZALBO, D., HOHMANN, S. (1989a), A deletion of the *PDC1* gene for pyruvate decarboxylase of yeast causes a different phenotype than previously isolated point mutations, *Curr. Genet.* **15,** 75–81.

SCHAAFF, I., HEINISCH, J., ZIMMERMANN, F. K. (1989b), Overproduction of glycolytic enzymes in yeast, *Yeast* **5,** 285-290.

SCHAAFF, I., HOHMANN, S., ZIMMERMANN, F. K. (1990), Molecular analysis of the structural gene for yeast transaldolase, *Eur. J. Biochem.* **188,** 597–603.

SCHEKMAN, R. (1985), Protein localization and membrane traffic in yeast, *Annu. Rev. Cell. Biol.* **1,** 115–143.

SCHERER, S., MAGEE, P. T. (1990), Genetics of *Candida albicans*, *Microbiol. Rev.* **54,** 226–241.

SCHLESSINGER, D. (1990), Yeast artificial chromosomes: Tools for mapping and analysis of complex genomes, *TIG* **6,** 248–258.

SCHMITT, H. D., ZIMMERMANN, F. K. (1982), Genetic analysis of the pyruvate decarboxylase reaction in yeast glycolysis, *J. Bacteriol.* **151,** 1146–1152.

SCHMITT, H. D., CIRIACY, M., ZIMMERMANN, F. K. (1983), The synthesis of yeast pyruvate decarboxylase is regulated by large variations in the messenger RNA level, *Mol. Gen. Genet.* **192,** 247–252.

SCHNEIDER, J. C, GUARENTE, L. (1990), Vectors for expression of cloned genes in yeast: Regulation, overproduction, and underproduction, *Methods Enzymol.* **194,** 373–388.

SCHÜLLER, H.-J., HAHN, A., TRÖSTER, F., SCHÜTZ, A., SCHWEIZER, E. (1992), Coordinate genetic control of yeast fatty acid synthase genes *FAS1* and *FAS2* by an upstream activation site common to genes involved in membrane lipid biosynthesis, *EMBO J.* **11,** 107–114.

SCHWELBERGER, H. G., KOHLWEIN, S. D., PALTAUF, F. (1989), Molecular cloning, primary structure and disruption of the structural gene of aldolase from *Saccharomyces cerevisiae*, *Eur. J. Biochem.* **180,** 301–308.

SCOTT, E. W., ALLISON, H. E., BAKER, H. V. (1990), Characterization of *TPI* gene expression in isogenic wild-type and *gcr1*-deletion mutant strains of *Saccharomyces cerevisiae*, *Nucleic Acids Res.* **18,** 7099–7107.

SEDIVY, J. M., FRAENKEL, D. G. (1985), Fructose bisphosphatase of *Saccharomyces cerevisiae.* Cloning, disruption and regulation of the *FBP1* structural gene, *J. Mol. Biol.* **186,** 307–319.

SEEBOTH, P. G., BOHNSACK, K., HOLLENBERG, C. P. (1990), *pdc1*^{0} mutants of *Saccharomyces cerevisiae* give evidence for an additional structural *PDC* gene: Cloning of *PDC5*, a gene homologous to *PDC1*, *J. Bacteriol.* **172,** 678–685.

SERVI, S. (1990), Baker's yeast as a reagent in organic synthesis, *Synthesis*, 1–25.

SHUSTER, J. R. (1989), Regulated transcriptional systems for the production of proteins in yeast: Regulation by carbon source, in: *Yeast Genetic Engineering* (BARR, P. J., BRAKE, A.J., VALENZUELA, P., Eds.), pp. 83–108, Stoneham: Butterworth Publishers.

SIMON, M., ADAM, G., RAPATZ, W., SPEVAK, W., RUIS, H. (1991), The *Saccharomyces cerevisiae* ADR1 gene is a positive regulator of transcription of genes encoding peroxisomal proteins, *Mol. Cell. Biol.* **11,** 699–704.

SMITH, C. L., MATSUMOTO, T., NIWA, O., KLEO, S., FAN, J. B., YANAGIDA, M., CANTOR, C. R. (1987), An electrophoretic karyotype for *S. pombe* by pulsed field gel electrophoresis, *Nucleic Acids Res.* **15,** 4481–4489.

SMITH, C. W. J., PATTON, J. G., NADAL-GINARD, B. (1989), Alternative splicing in the control of gene expression, *Annu. Rev. Genet.* **23,** 527–577.

SOLOMON, F. (1991), Analyses of the cytoskeleton in *Saccharomyces cerevisiae*, *Annu. Rev. Cell. Biol.* **7,** 633–662.

SOLS, A. (1981), Multimodulation of enzyme activity, *Curr. Top. Cell. Regul.* **19,** 77–101.

SOR, F., FUKUHARA, H. (1989), Analysis of chromosomal DNA patterns of the genus *Kluyveromyces*, *Yeast* **5,** 1–10.

SPENCER, J. F. T., SPENCER, D. M., SMITH, A. R. W. (1983), *Yeast Genetics, Fundamental and Applied Aspects*, New York: Springer-Verlag.

STARK, M. J. R., BOYD, A., MILEHAM, A. J., ROMANOS, M. A. (1990), The Plasmid-encoded killer system of *Kluyveromyces lactis*: A review, *Yeast* **6,** 1–29.

STEENSMA, H. Y., DE JONGH, F. C. M., LINNEKAMP, M. (1988), The use of electrophoretic karyotypes in the classification of yeasts: *Kluyveromyces marxianus* and *K. lactis*, *Curr. Genet.* **14,** 311–317.

STRASSER, A. W. M., SELK, R., DOHMEN, R. J., NIERMANN, T., BIELEFELD, M., SEEBOTH, P., TU, G., HOLLENBERG, C. P. (1989), Analysis of the α-amylase gene of *Schwanniomyces occidentalis* and the secretion of its gene product in transformants of different yeast genera, *Eur. J. Biochem.* **184,** 699–706.

STRATFORD, M. (1992), Yeast flocculation: Reconciliation of physiological and genetic viewpoints, *Yeast* **8,** 25–38.

STRATHERN, J. N., JONES, E. W., BROACH, J. R. (1981), *The Molecular Biology of the Yeast Saccharomyces – Life Cycle and Inheritance*, Cold Spring, Harbor, NY: Cold Spring Harbor Laboratory Press.

STRATHERN, J. N., JONES, E. W., BROACH, J. R. (1982), *The Molecular Biology of the Yeast Saccharomyces – Metabolism and Gene Expression*, Cold Spring, Harbor, NY: Cold Spring Harbor Laboratory Press.

STRAUSS, W. M., JAENISCH, R. (1992), Molecular complementation of a collagen mutation in mammalian cells using yeast artificial chromosomes, *EMBO J.* **11,** 417–422.

STRUHL, K. (1989), Molecular mechanisms of transcriptional regulation in yeast, *Annu. Rev. Biochem.* **58,** 1051–1077.

STRUHL, K., STINCHCOMB, D. T., SCHERER, S., DAVIS, R. W. (1979), High-frequency transformation of yeast: Autonomous replication of hybrid DNA molecules, *Proc. Natl. Acad. Sci. USA* **76,** 1035–1039.

STUCKA, R., VALDÉS-HEVIA, M. D., GANCEDO, C., SCHWARZLOSE, C., FELDMANN, H. (1988), Nucleotide sequence of the phosphoenolpyruvate carboxykinase gene from *Saccharomyces cerevisiae*, *Nucleic Acids Res.* **16,** 10926.

STUCKA, R., DEQUIN, S., SALMON, J. M., GANCEDO, C. (1991), DNA sequences in chromosomes II and VII code for pyruvate carboxylase isoenzymes in *Saccharomyces cerevisiae:* analysis of pyruvate carboxylase-deficient strains, *Mol. Gen. Genet.* **229,** 307–315.

SUAREZ RENDUELEZ, P., WOLF, D. H. (1988), Proteinase function in yeast: biochemical and genetic approaches to a central mechanism of post-translational control in the eukaryote cell, *FEMS Microbiol. Rev.* **54,** 17–46.

TAMAKI, N., HAMA, T. (1982), *Methods Enzymol.* **89,** 469–473.

TANAKA, A., FUKUI, S. (1989), Metabolism of *n*-alkanes, in: *The Yeasts* (ROSE, A. H., HARRISON, J. S., Eds.), Vol. 3, pp. 261–288, London: Academic Press.

TANAKA, K., NAKAFUKU, M., SATOH, T., MARSHALL, M. S., GIBBS, J. B., MATSUMOTO, K., KAZIRO, Y., TOH-E, A. (1990), *S. cerevisiae* genes IRA1 and IRA2 encode proteins that may be functionally equivalent to mammalian ras GTPase activating protein, *Cell* **60,** 803–807.

TEKAMP-OLSON, P., NAJARIAN, R., BURKE, R. L. (1988), The isolation, characterization and nucleotide sequence of the phosphoglucoisomerase gene of *Saccharomyces cerevisiae*, *Gene* **73,** 153–161.

THIELEN, J., CIRIACY, M. (1991), Biochemical basis of mitochondrial acetaldehyde dismutation in *Saccharomyces cerevisiae*, *J. Bacteriol.* **173,** 7012–7017.

THOMA, F. (1992), Nucleosome positioning, *Biochim. Biophys. Acta* **1130,** 1–19.

THOMAS, D., CHEREST, H., SURDIN-KERJAN, Y. (1991), Identification of the structural gene for glucose-6-phosphate dehydrogenase in yeast. Inactivation leads to a nutritional requirement for organic sulfur, *EMBO J.* **10,** 547–553.

THORNTON, R. J. (1991), Wine yeast research in New Zealand and Australia, *Crit. Rev. Biotechnol.* **11,** 327–345.

THUKRAL, S. K., MORRISON, M. L., YOUNG, E. T. (1992), Mutations in the zinc fingers of ADR1 that change specificity of DNA binding and transactivation, *Mol. Cell. Biol.* **12,** 2784–2792.

TOH-E, A., TADA, S., OSHIMA, Y. (1982), 2-μm DNA-like plasmid in the osmophilic haploid yeast *Saccharomyces rouxii*, *J. Bacteriol.* **151,** 1380–1390.

TOH-E, A., TANAKA, K., UESONO, Y., WICKNER, R. B. (1988), *PHO85*, a negative regulator of the PHO system, is a homolog of the protein kinase gene, *CDC28*, of *Saccharomyces cerevisiae*, *Mol. Gen. Genet.* **214,** 162–164.

TORNOW, J., SANTANGELO, G. M. (1990), Efficient expression of the *Saccharomyces cerevisiae* glycolytic gene *ADH1* is dependent upon a *cis*-acting regulatory element (UAS_{RPG}) found initially in genes encoding ribosomal proteins, *Gene* **90,** 79–85.

TUITE, M. F. (1992), Strategies for the genetic manipulation of *Saccharomyces cerevisiae*, *Crit. Rev. Biotechnol.* **12,** 157–188.

TZAGOLOFF, A., DIECKMANN, C. L. (1990), *PET* genes of *Saccharomyces cerevisiae*, *Microbiol. Rev.* **54,** 211–225.

UMEZAWA, C., KISHI, T. (1989), Vitamin metabolism, in: *The Yeasts* (ROSE, A. H., HARRISON, J. S., Eds.), pp. 457-488, London: Academic Press.

VALENZUELA, P., MEDINA, A., RUTTER, W. J., AMMERER, G., HALL, B. D. (1982), Synthesis and assembly of hepatitis B virus surface antigen particles in yeast, *Nature* **298,** 347–350.

VAN AELST, L., BOY-MARCOTTE, E., CAMONIS, J. H., THEVELEIN, J. M., JACQUET, M. (1990), The C-terminal part of the *CDC25* gene product plays a key role in signal transduction in the glucose-induced modulation of cAMP level in *Saccharomyces cerevisiae*, *Eur. J. Biochem.* **193,** 675–680.

VAN DEN BERG, J. A., VAN DER LAKEN, K. J., VAN OOYEN, A. J. J., RENNIERS, T. C. H. M., REITVELD, K., SCHAAP, A., BRAKE, A. J., BISHOP, R. J., SCHULTZ, K., MOYER, D., et al (1990), *Kluyveromyces* as a host for heterologous gene expression: Expression and secretion of prochymosin, *Bio/Technology* **8,** 135–139.

VAN SCHAFTINGEN, E., HERS, H. G. (1981 a), Inhibition of fructose-1,6-bisphosphatase by fructose 2,6-bisphosphate, *Proc. Natl. Acad. Sci. USA* **78,** 2861–2863.

VAN SCHAFTINGEN, E., HERS, H. G. (1981 b), Phosphofructokinase 2 – The enzyme that forms fructose 2,6-bisphosphate from fructose-6-phosphate and ATP, *Biochem. Biophys. Res. Commun.* **101,** 1078–1084.

VAN URK, H., POSTMA, E., SCHEFFERS, W. A., VAN DIJKEN, J. P. (1989), Glucose transport in Crabtree-positive and Crabtree-negative yeasts, *J. Gen. Microbiol.* **135,** 2399–2406.

VAN URK, H., VOLL, W. S. L., SCHEFFERS, W. A., VAN DIJKEN, J. P. (1990), Transient-state analysis of metabolic fluxes in Crabtree-positive and Crabtree-negative yeasts, *Appl. Environ. Microbiol.* **56,** 281–287.

VASSAROTTI, A., GOFFEAU, A. (1992), Sequencing the yeast genome: The European effort, *Trends Biotechnol.* **10,** 15–18.

VOGEL, K., HINNEN, A. (1990), The yeast phosphatase system, *Mol. Microbiol.* **4,** 2013–2017.

VOJTEK, A., HAARER, B., FIELD, J., GERST, J., POLLARD, T. D., BROWN, S., WIGLER, M. (1991), Evidence for a functional link between profilin and CAP in the yeast *S. cerevisiae*, *Cell* **66,** 497–505.

VOLKERT, F. C., WILSON, D. W., BROACH, J. R. (1989), Deoxyribonucleic acid plasmids in yeasts, *Microbiol. Rev.* **53,** 299–317.

WALTON, J. D., PAQUIN, C. E., KANEKO, K., WILLIAMSON, V. M. (1986), Resistance to antimycin A in yeast by amplification of ADH4 on a linear, 42 kb palindromic plasmid, *Cell* **46,** 857–863.

WARNER, J. R. (1989), Synthesis of ribosomes in *Saccharomyces cerevisiae*, *Microbiol. Rev.* **53,** 256–271.

WATSON, J. D., HOPKINS, N. H., ROBERTS, J. W., STEITZ, J. A., WEINER, A. M. (1987), *Molecular Biology of the Gene*, 4th Ed., Menlo Park, CA: Benjamin Cummings Publishing Company.

WEBSTER, T. D., DICKSON, R. C. (1988), The organization and transcription of the galactose gene cluster of *Kluyveromyces lactis*, *Nucleic Acids Res.* **16,** 8011–8028.

WHITE, M. F., FOTHERGILL-GILMORE, L. A. (1988), Sequence of the gene encoding phosphoglycerate mutase from *Saccharomyces cerevisiae*, *FEBS Lett.* **229,** 383–387.

WHITE, M. J., LOPES, J. M., HENRY, S. A. (1991), Inositol metabolism in yeasts, *Adv. Microb. Physiol.* **32,** 2–51.

WICKNER, R. (1981), Killer systems in *Saccharomyces cerevisiae*, in: *The Molecular Biology of the Yeast Saccharomyces: Life Cycle and Inheritance* (STRATHERN, J. N., JONES, E. W., BROACH, J. R., Eds.), pp. 415–444, Cold Spring Harbor, NY: Cold Spring Harbor Laboratory Press.

WILLIAMS, F. E., VARANASI, U., TRUMBLY, R. J. (1991), The CYC8 and TUP1 proteins involved in glucose repression in *Saccharomyces cerevisiae* are associated in a protein complex, *Mol. Cell. Biol.* **11,** 3307–3316.

WILLIAMS, J. F. (1980), A critical examination of the evidence for the reactions of the pentose pathway in animal tissues, *TIBS* **5,** 315–320.

WILLS, C. (1990), Regulation of sugar and ethanol metabolism in *Saccharomyces cerevisiae*, *CRC Crit. Rev. Biochem. Mol. Biol.* **25,** 245–280.

WILLS, C. T., MARTIN, T., MELHAM, T. (1986), Effect on gluconeogenesis of mutants blocking two mitochondrial transport systems in the yeast *Saccharomyces cerevisiae*, *Arch. Biochem. Biophys.* **246,** 306–313.

WINKLER, K., KIENLE, I., BURGERT, M., WAGNER, J.-C., HOLZER, H. (1991), Metabolic regulation of the trehalose content of vegetative yeast, *FEBS Lett.* **291,** 269–272.

WOOLFORD, J. L. (1989), Nuclear pre-mRNA splicing in yeast, *Yeast* **5,** 439–457.

WRAY, L. V., JR., WITTE, M. M., DICKSON, R. C., RILEY, M. I. (1987), Characterization of a positive regulatory gene, *LAC9*, that controls induction of the lactose–galactose regulon of *Kluyveromyces lactis:* Structural and functional relationships to *GAL4* of *Saccharomyces cerevisiae*, *Mol. Cell. Biol.* **7,** 1111–1121.

WRIGHT, J. H., GOTTSCHLING, D. E., ZAKIAN, V. A. (1992), *Saccharomyces* telomeres assume a non-nucleosomal chromatin structure, *Genes Dev.* **6,** 197–210.

YAMADA, Y., NOJIRI, M., MATSUYAMA, M., KONDO, K. (1976), Coenzyme Q system in the classification of the ascosporogenous yeast genera *Debaryomyces*, *Saccharomyces*, *Kluyveromyces*, and *Endomycopsis*, *J. Gen. Appl. Microbiol.* **22,** 325–337.

YUAN, Z., MEDINA, M. A., BOITEUX, A., MÜLLER, S. C., HESS, B. (1990), The role of fructose 2,6-bisphosphate in glycolytic oscillations in extracts and cells of *Saccharomyces cerevisiae*, *Eur. J. Biochem.* **192,** 791–795.

ZAMAN, Z., BROWN, A. J. P., DAWES, I. W. (1992), A 3′ transcriptional enhancer within the coding sequence of a yeast gene encoding the common subunit of two multi-enzyme complexes, *Mol. Microbiol.* **6,** 239–246.

ZARET, K. S., SHERMAN, F. (1982), DNA sequence required for efficient transcription termination in yeast, *Cell* **28,** 563–573.

ZITOMER, R. S., LOWRY, C. V. (1992), Regulation of gene expression by oxygen in *Saccharomyces cerevisiae*, *Microbiol. Rev.* **56,** 1–11.

15 Filamentous Fungi

FRIEDHELM MEINHARDT
Münster, Federal Republic of Germany

KARL ESSER
Bochum, Federal Republic of Germany

1 Introduction 516
2 Classification 516
3 Morphology and Propagation 517
4 Procurement and Cultivation 520
  4.1 Procurement 520
  4.2 Cultivation 521
    4.2.1 Surface Culture 521
    4.2.2 Submerged Culture 521
    4.2.3 Solid-State Fermentation (SSF) 521
    4.2.4 Immobilization 522
5 Production of Useful Metabolites 522
  5.1 Phycomycetes 522
  5.2 Ascomycetes 525
  5.3 Basidiomycetes 532
6 Perspectives 536
7 References 537

1 Introduction

The fungi are eukaryotes, most probably derived from colorless representatives of unicellular algae. They may, however, have a polyphyletic origin, i.e., their course of development may have differed in evolutionary terms. Irrespective of phylogenetic relations, fungi share similarities concerning organization, proliferation, and physiology.

Since fungi are unable to grow in the absence of organic substances, and since no fungus possesses photosynthetic pigments, they are heterotrophic. Most of the fungi are saprophytic, feeding on dead, organic matter only. Others are parasites of man, domestic animals, insects, plants, soil eelworms, or protozoa and obtain their nutrients from living host organisms. Regardless of their saprophytic or parasitic mode of living, many fungal species can be grown on artificial culture media.

The physiological properties of fungi can be used and exploited for production purposes. Under special conditions, mainly depending on the nutrients supplied, fungi can produce a great variety of complex organic compounds, such as proteins, antibiotics, alcohols, enzymes, organic acids, vitamins, drugs, pigments, and fats.

Vegetative cells and spores of almost all fungi are surrounded by a cell wall consisting of chitin, a polymer of *N*-acetylglucosamine. Cellulose, the main component of the cell walls of plants, is only formed in a few taxa.

This chapter is in part based on a previous publication (ESSER and MEINHARDT, 1987); definitions are derived from ESSER (1986). It necessarily contains selected topics. The reader who desires more information on different aspects of fungi is referred to the following literature:

Applied mycology: BÖTTICHER (1974), CHANG (1972), FLEGG et al. (1985), GROSSBARD (1979), LELLEY and SCHMAUS (1976), REHACEK and SAJDL (1990), SMITH (1969), ARORA et al. (1992)

General mycology: AINSWORTH and SUSSMAN (1965–1968), BURNETT (1968), GÄUMANN (1964), INGOLD (1973), MÜLLER and LÖFFLER (1971), STEVENS (1974), WEBSTER (1980)

Genetics: BURNETT (1975), ESSER and KUENEN (1967), DAY (1974), FINCHAM and DAY (1971), RAPER (1966), TIMBERLAKE (1985)

Physiology: COCHRANE (1963), LILLY and BARNETT (1951), SMITH and BERRY (1974), TURNER (1971)

Practicals: ALEXOPOULOS and BENEKE (1964), BOTH (1971), DADE and GUNNEL (1969), ESSER (1986), HUDSON (1986), KOCH (1972), STEVENS (1974), STEVENSON (1970), KING (1974)

Special groups of fungi: BRESINSKY and BESL (1985), CHANG and HAYES (1978), COLE and SAMSON (1979), FREY et al. (1979), GRAY (1970), LANGE (1988), LINCOFF and MITCHELL (1977), MOORE et al. (1985), REISS (1986), VAN GRIENSVEN (1991)

Taxonomy: KREISEL (1969), BESSEY (1968)

2 Classification

Modern molecular biological methods and techniques are being applied to basic research concerned with the taxonomy of filamentous fungi. It is therefore not surprising that the historical fungal taxa, created mainly on artificial criteria and aimed to rapidly and properly identify fungal species, undergo severe changes when phylogenesis becomes the basis of taxonomy.

Although a so-called phylogenetic system has a great value in fundamental mycology, this does not necessarily hold true for biotechnology. On the one hand, there is no evident correlation between metabolic abilities exploited in biotechnology. On the other hand, the "modern scientific system" based on sequence similarities of rRNA genes, G+C values, DNA complementarity, and restriction enzyme analysis does not familiarize biotechnologists with the morphology and life cycles of fungi, both of which are necessary in the handling of fungi for biotechnological purposes.

Taxonomic investigations on fungi are countless, but at present, there is no satisfying systematic approach with a durable outlook. Therefore, we subdivide the fungi here exclusively according to their practicability in biotechnology.

Tab. 1. Subdivision of Filamentous Fungi (biotechnologically relevant taxa are given in bold face)

Class	Subclass	Order
Phycomycetes		1. Chytridiales
		2. Blastocladiales
		3. Monoblepharidales
		4. Hyphotrichiales
		5. Oomycetales
		6. **Zygomycetales**
Ascomycetes	Protoascomycetidae	1. **Endomycetales**
		2. Taphrinales
	Plectomycetidae	1. **Plectascales**
		2. Erysiphales
	Loculomycetidae	1. Myriangiales
		2. Microthyriales
		3. Hysteriales
		4. Pleosporales
		5. Dothideales
	Pyrenomycetidae	1. **Sphaeriales**
		2. **Clavicipitales**
		3. Laboulbeniales
	Discomycetidae	1. Pezizales
		2. Helotiales
		3. Phacidiales
		4. **Tuberales**
Basidiomycetes	Phragmobasidiomycetidae	1. Uredinales
		2. Ustilaginales
		3. **Auriculariales**
		4. **Tremellales**
	Holobasidiomycetidae	1. Exobasidiales
		2. **Poriales**
		3. **Agaricales**
		4. **Gastromycetales**

Therefore, we follow the general classification used by ESSER (1986), well aware of the fact that taxonomists may criticize this approach.

The Myxomycota are no longer considered to be a class of fungi, because they are evidently more closely related to protozoa. Since there are no biotechnological representatives among this group, we shall not deal with these organisms.

We will only deal with fungi belonging to the Eumycota, a group to which nearly all of the fungi belong. To facilitate an overview, the fungi treated here are divided into classes, subclasses, and orders, as presented in Tab. 1. Since there is a separate chapter on yeasts (Chapter 14), we do not detail the Endomycetales here. Only those taxa of the filamentous fungi which are of biotechnological importance will be thoroughly described (bold face, Tab. 1).

3 Morphology and Propagation

The vegetative body of fungi consists of filamentous branched structures, the hyphae. These are either non-septate (Phycomycetes) or septate (Ascomycetes and Basidiomycetes). The septa are usually perforated, ensuring cytoplasmic continuity. Therefore, the fungi are coenocytic organisms. The entire hyphae complex is defined as the mycelium. As may be evident from this rather short description of fungal vegetative structures, to understand the mycological literature and — more importantly — to understand the biology of fungi, one must be familiar with some of the terminology used by mycologists. The minimum vocabulary is presented in Tab. 2, a glossary of mycological terms.

Having spread to nearly every substrate, including all kinds of soil, stored food, organic debris, and many other habitats, fungi belong to a group of the most widely distributed living forms. As a consequence, they can be found nearly everywhere all over the world, very often forming extremely dense populations. In order to facilitate distribution and to maintain the density of the populations, fungi have evolved rather efficient propagation systems, i.e., they produce large amounts of spores. A fruiting body of the edible giant puff ball *Calvatia gigantea* contains more than 7×10^{12} spores, the large bracket fungus *Ganoderma applanatum* discharges approximately 3×10^{10} spores per day, and the members of the genus *Penicillium* (blue molds) bear about 4×10^{8} co-

Tab. 2. Glossary of Mycological Terms

Hyphae	Branched septate or non-septate filaments
Mycelium	The mass of hyphae. If the hyphae are septate, the mycelium constitutes a coenocyte, since exchange of cytoplasm and all cell organelles is possible
Plectenchyma	Felt or rope-like aggregation of hyphae forming a tissue-like composition mediated by swelling of the cell walls
Rhizomorphs	Bundles of hyphae visible with the naked eye
Chlamydospores/ Gemmae	Thick-walled resting structures
Sclerotia	Multicellular, plectenchymatous organs rendering possible survival over a long period of time
Stromata	Sclerotia-like complexes eventually giving rise to fruiting bodies
Planospores	Also called zoospores, mobile spores of aquatic and some parasitic fungi
Aplanospores	Also called conidiospores, non-mobile spores of terrestrial fungi
Ascus	Characteristic tubular or sac-like meiosporangium of the Ascomycetes, usually containing 8 spores
Basidium	Characteristic club-shaped meiosporangium of the Basidiomycetes, on which usually four spores are exogenously generated
Ascocarp	Fruiting body of the Ascomycetes
Basidiocarp	Fruiting body of the Basidiomycetes
Hymenium	The meiosporangium-bearing layer (asci or basidia)

nidiospores on a colony of 2.5 cm in diameter (INGOLD, 1971). Although the majority of fungi are spread by spores, some employ other means of dispersal, e.g., by forming thick-walled resting structures called gemmae or chlamydospores which allow survival even in hostile environments. Even the sclerotia of the well known ergot fungus *Claviceps purpurea* may be involved in dispersal. However, distribution of this fungus is mainly by wind-borne ascospores and insect-borne conidiospores (see also the description of the life cycle, given later).

Genetic variation is necessary to render an organism successful in different environments or habitats. Genetic diversity makes it possible for each species to find its ecological niche. Genetic diversity is the basis of evolution and is brought about by mutation or recombination. Recombination is strongly enhanced in sexually propagating organisms, as most of the fungi are classified.

Sexuality requires karyogamy, resulting in the formation of a zygote from which, as a consequence of meiosis, gametes are generated. In diploid (2n) organisms, a vegetative body is formed which produces the gametes in specific sex organs. In the haplonts, however, the vegetative body is haploid, since meiosis takes place immediately after karyogamy. A combination of cycles may occur when the organism functions in both the haploid (n) and the diploid (2n) manner.

In this context, two different life cycles can characterize fungi, a quality exclusively observed in the fungi. This is explained by the fact that fusion of the gametes (plasmogamy) is not immediately followed by fusion of the gamete nuclei (karyogamy), i.e., between the haploid and the diploid phase a dikaryotic phase is inserted in which both haploid nuclei multiply mitotically. As these nuclear divisions are usually conjugate (i.e., synchronous), it is safe to say that at the end of the dikaryotic phase the zygote is derived from the fusion of descendants of the two gamete nuclei.

Two cycles can be distinguished:

1. *Haplo-dikaryotes:* The vegetative body is haploid, only the zygote is diploid, and vegetative spores can be formed in both phases. A typical life cycle is represented by that of the Ascomycetes in Fig. 1.

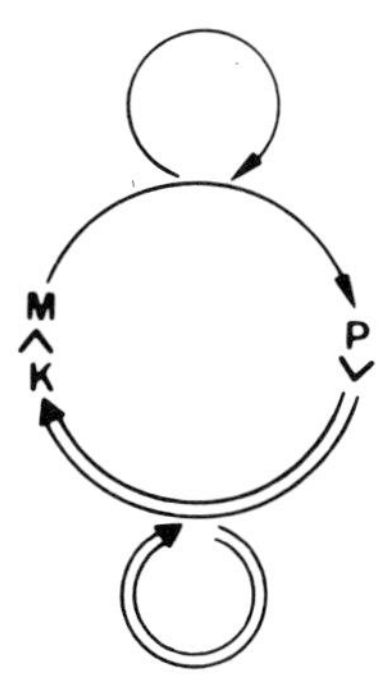

Fig. 1. A generalized scheme of the life cycle of a haplo-dikaryote. The haploid phase is represented by a single line, the diploid phase corresponds to the double line. M, meiosis; K, karyogamy; P, plasmogamy. Small circles on the top and below represent vegetative cycles.

2. *Dikaryotes:* The vegetative body is dikaryotic, because meiotic spores immediately fuse with each other. Only the zygote is diploid. Asexual reproduction is possible through dikaryotic spores. A typical life cycle is represented by that of the Basidiomycetes in Fig. 2.

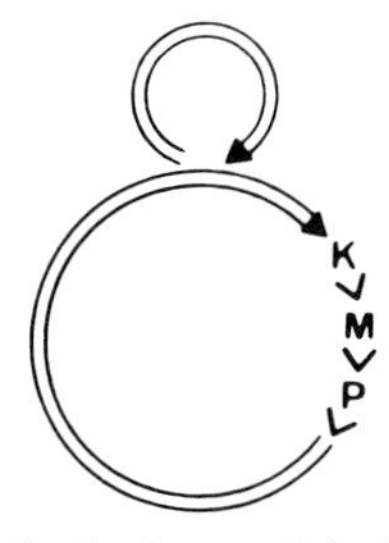

Fig. 2. A generalized scheme of the life cycle of a dikaryote. For explanation see Fig. 1.

Parasexual cycle: For completeness, we must mention another phenomenon of reproduction which cannot be classified in the life cycles mentioned above. This is the parasexual cycle, which has to date been reported only in the fungi. As its name suggests, this cycle comprises a process similar to a sexual cycle which, indeed, involves the recombination of genetic material. Recombination occurs in the parasexual cycle during a mitotic division and not during meiosis, which is expressed by the prefix "para".

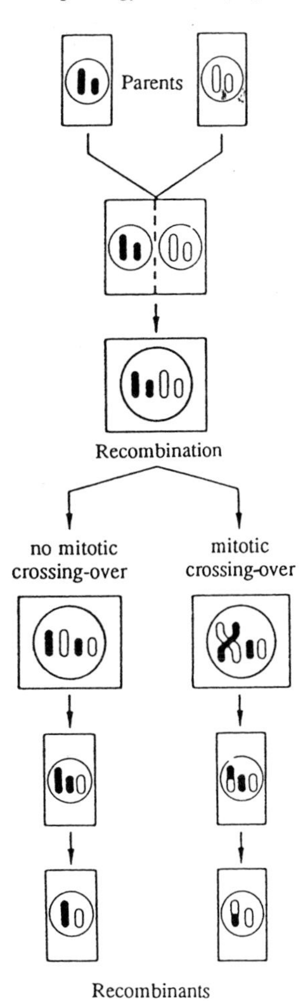

Fig. 3. Schematic representation of the parasexual cycle. For details see text.

The individual steps of the parasexual cycle (Fig. 3), first discovered in *Aspergillus*, are as follows:

- heterokaryon formation: fusion of haploid cells with genetically different nuclei
- production of diploid nuclei after the fusion of haploid nuclei of the heterokaryon
- mitotic crossing-over: chromosomal exchanges during mitosis of the diploid nuclei in the heterokaryon
- haploidization: change of diploid nuclei to haploid during the course of numerous mitotic divisions (e.g., by loss of chromosomes).

4 Procurement and Cultivation

4.1 Procurement

Many fungi can be collected from nature. Apart from the fruiting bodies of the higher Basidiomycetes, which are visible with the naked eye, a magnifying lens is necessary to locate the fruiting bodies of the remaining taxa. Specific substrates for saprotrophs or parasites should also be carefully considered.

Crude cultures may be prepared by removing small pieces of plectenchyma from the inside of a large fruiting body by using a sterile scalpel and placing it onto the agar in a Petri dish. Spore smears may be made from smaller fruit bodies, or the fruit bodies may be transferred as a whole onto the agar. A widely used method to obtain fungi from nature is by *baiting*, used principally for aquatic fungi, but also for terrestrial forms. An optimal substrate is normally used as the bait, e.g., protein-rich plant seeds or animal protein for the aquatic Oomycetales and carbohydrate-rich media for the terrestrial representatives of Zygomycetales. In addition to the many fungi which can be collected from nature, numerous species can be obtained from different culture collections located throughout the world. Most of the strains from the culture collections are available to the public, i.e., to individuals operating in a professional environment with suitable facilities for handling the material. Most of the culture collections offer an identification and characterization service for unknown specimens. Within a relatively short period of time, freshly collected material is treated and correctly named. Some well known organizations are summarized below:

ATCC	American Type Culture Collection, 12301 Parklawn Drive, Rockville, MD 20852, USA
CBS	Centraalbureau voor Schimmelcultures, Oosterstraat 1, Baarn, The Netherlands
COM	Commonwealth Mycological Institute, Ferry Lane, Kew, Surrey, UK
DSM	Deutsche Sammlung für Mikroorganismen, Mascheroder Weg 1, D-3300 Braunschweig-Stöckheim, FRG
IHEM	Institute of Hygiene and Epidemiology/Mycology, IHEM Culture Collection, Rue J Witsman 14, B-1050 Bruxelles, Belgium
MUCL	Mycotheque de l'Université Catholique de Louvain, MUCL Culture Collection, Place Croix du Sud 3, B-1348 Louvain la Neuve, Belgium
CCBAS	Culture Collection of Basidiomycetes, Czech Academy of Sciences, Institute of Microbiology, Department of Experimental Mycology, Videnska 1083, 142 20 Prague 4 Krc, Czech Republic
FGSC	Fungal Genetics Stock Center, Department of Microbiology, University of Kansas Medical Center, Kansas City, Kansas 66106, USA
LGMACC	Laboratoire de Génétique Moleculaire et Amélioration des Champignons Cultivés, INRA Département de Génétique Moleculaire et d'Amélioration des Plantes, Université de Bordeaux II, Domaine de la Grande Ferrade, F-33140 Pont de la Maye, France
CBSC	Biology/Science Materials, Carolina Biological Supply Company, Burlington, NC 27215, USA

Isolation procedures are required for fungi collected from nature, as well as for those strains obtained from culture collections should infection occur. There are several possible methods, and sometimes a combination is necessary to obtain a pure culture.

Manual isolation under the dissecting microscope is the most convenient method. However, it requires cells or spores which are at least 10 μm in size. All manipulations under the dissecting microscope are usually carried out in a Petri dish containing hard agar, i.e., 5% water agar. This method is especially suitable for fungi that actively eject their spores. A Petri dish containing the hard agar is inverted over the fungal culture and the edges sealed with Parafilm. After ejection, the spores stick to the agar surface and can be isolated and transferred to fresh medium.

A *plating method* is required for smaller organisms. A suspension of cells or spores is plated on nutrient agar and, after germination, single cultures can be transferred to fresh medium.

Bacterial infections may cause problems, but they can be easily avoided by adding appropriate amounts of antibiotics. Fast growing filamentous fungi can also be cleaned up by dissecting hyphal tips which are very often free from contaminating bacteria. Fungal cultures can be easily maintained at room temperature without illumination. Stock cultures of fungi are usually stored at 4 °C.

4.2 Cultivation

In nature filamentous fungi mainly colonize solid substrates. In biotechnology, however, growth of molds in liquid culture media is of great practical importance. In general four different methods for cultivation of filamentous fungi can be applied in industrial processes.

4.2.1 Surface Culture

The fungus is grown in liquid medium without stirring or agitating the flask or vessel. This is a classical procedure for the production of citric acid with *Aspergillus niger*, or penicillin with *Penicillium* spp. Huge numbers of containers must be supplied with relatively small volumes of nutrient media. Inoculation is laborious and time-consuming, incubation periods usually exceed 6 days.

4.2.2 Submerged Culture

In many respects this method is more economical than surface culture. Liquid medium is inoculated with the organism of choice and then stirred or shaken. When growing filamentous fungi, two morphologies can be distinguished: the filamentous and the pellet form. In the pellet form the mycelium shows a clear morphological differentiation, because it develops stable spherical aggregates consisting of a dense, branched network of hyphae. Pellets can reach several millimeters in diameter, and pellet suspensions are usually less viscous than suspensions in which filamentous growth occurs.

Penicillin was previously produced in surface cultures for many years. Today, however, it is produced exclusively in submerged cultures. The reduction in incubation time, as well as minor technical efforts concomitant with an increase in yield, were only possible because of the enormous efforts dedicated to generate high-producing strains.

4.2.3 Solid-State Fermentation (SSF)

Since this type of fermentation takes place without an aqueous phase (i.e., no water is exudated from the microorganism/substrate system), it is extremely useful for fermentation of lignocellulose-containing material. SSF of lignocellulose is mainly accomplished with wood-decomposing white- and brown rotting Basidiomycetes. White rotting fungi degrade lignin and cellulose, whereas the brown rotting species only utilize cellulose.

Applications of SSF include fruiting body production of edible fungi (HÜTTERMANN, 1989), increasing the digestibility of lignocellulose-containing material for animal feeds (CZAIJKOWSKA and ILNICKA-OLEIJNICZAK, 1985), delignification for production of adhesives and gums in the wood industry (ERIKSON, 1990), cleaning of contaminated soil and air (HÜTTERMANN et al., 1988, 1989) and production of enzymes such as cellulases and phenoloxidases.

During the growth of filamentous fungi on solid substrates, morphological differentiation is likely to occur, as well as limited exhaustion of nutrients. A mathematical description of growth is, therefore, hampered. However, there are promising approaches to model the fermentation of highly porous material by filamentous fungi (KÖRNER et al., 1991; PROSSER and TOUGH, 1991).

4.2.4 Immobilization

Immobilized cells have been defined as cells that are fixed within or associated with an insoluble matrix. The various methods of immobilization include (1) covalent crosslinking to supports (e.g., silanization to silica support, coupling by glutaraldehyde or isocyanate, metal hydroxide precipitation), (2) adsorption onto organic or inorganic high-surface material (polysaccharides, Kieselgur, collagen, diatomaceous earth, activated carbon, sintered glass, ceramic, wood), (3) entrapment in a three-dimensional polymer network (agar, calcium alginate, polyurethane, polyacrylamide, κ-carrageenan), (4) confinement in a liquid emulsion (e.g., encapsulation in oil drops), (5) entrapment in semipermeable membranes (polyvinylidene fluoride, polyether imide). Attachment of miroorganisms is brought about by electrostatic interactions between cells and supports, simple entrapment, covalent-bond formation, and free radical polymerization with covalent bonding (KOLOT, 1988).

5 Production of Useful Metabolites

5.1 Phycomycetes

For simplicity, it seems appropriate to treat the Phycomycetes as an independent class. This affords a clear arrangement for those who are working with fungi for the first time, and allows the distinction between the Phycomycetes (coenocytic forms without transverse walls), the Ascomycetes (sporangia = asci) and the Basidiomycetes (sporangia = basidia).

The only order which is relevant for biotechnology are the Zygomycetales. They comprise terrestrial fungi propagating vegetatively either via endogenously emerging sporangiospores or via exogenous conidiospores. Sexual reproduction is accompanied by the fusion of gametangia (conjugum formation). The order consists of three families, of which only the Mucoraceae, which comprise about thirty genera, are important with respect to biotechnology. Biotechnologically exploited members of the Mucoraceae are listed in Tab. 3, together with the processes and/or products involved.

Enzymes derived from representatives of the Mucoraceae are being used for food fermentation in Asia (tempeh, ragi, arak, tapé)

Tab. 3. Biotechnologically Important Representatives of the Mucoraceae

Species	Product/Process
Mucor miheii	Rennet, casein hydrolyzation, cheese making
Mucor pusillus	Proteases, tenderizer, deodorant
Mucor rouxii	Rennet, proteases, tempeh, Asian food fermentation
Mucor racemosus	Sufu, Asian food fermentation
Phycomyces blakesleeanus	Linolic acid
Rhizopus arrhizus	Steroid conversions, organic acids
Rhizopus nigricans	Steroid conversions, organic acids
Rhizopus delemar	Glucose oxidase
Rhizopus oligosporus	Tempeh, Asian food fermentation
Rhizopus oryzae	Tempeh, Asian food fermentation
Rhizopus stolonifer	Tempeh, Asian food fermentation
Chlamydomucor oryzae	Ragi, arak, tapé, Asian food
Blakeeslea trispora	Trisporic acids, β-carotene
Choanephora cucurbitarum	Trisporic acids, β-carotene
Entomophtera thaxteriana	Fly killer, biological insecticide

and for cheese making in European countries. The biology of *Phycomyces blakesleeanus* has been thoroughly investigated, and the life cycle, presented in Fig. 4, may serve as an example of the Mucoraceae, since the mode of propagation for the other family members differs only slightly from that of *Phycomyces.*

Although fungi produce a panoply of organic acids, only fumaric and lactic acid are produced with representatives of the Mucoraceae. Fumaric acid, produced with *Rhizopus* species, is used as a food additive and also as an industrial chemical, e.g., in fruit juices, baked goods and as a salt substitute (FOSTER, 1954). Fermentation of fumaric acid has been extensively studied: the most important enzyme is pyruvate carboxylase; carboxylation of pyruvate yields oxaloacetic acid, which is further metabolized by reductive reactions of the tricarboxylic acid cycle to malic and fumaric acid (ROMANO et al., 1967; BUCHTA, 1983).

Lactic acid is mainly used as a food preservative or alcidulant, but is also added to cosmetics and pharmaceuticals. Esterification provides useful polymers. Fermentation of lactic acid is usually carried out with homofermentative lactic acid bacteria. A 90% yield can be achieved in bacterial fermentation with glucose as the carbon source. *Rhizopus* fermentations, however, have simpler nutritional requirements than bacterial cultures, while reaching the same production level by direct fermentation of starch to lactic acid (HANG, 1989). Since fermentation with immobilized fungal strains is also efficient (HANG et al., 1989), these organisms have become important in lactic acid fermentation.

Many representatives of the Mucoraceae have the ability to accumulate lipids, which could be used as food additives. However, industrial use of microbially produced fat is more likely to occur. Microorganisms producing more than 20% lipids are called oleaginous. A comprehensive list of oleaginous molds can be found in WEETE (1980). The biochemistry and physiology of fatty acid synthesis by oleaginous yeasts and molds is well documented (WEETE, 1984; RATLEDGE, 1986) (Fig. 5).

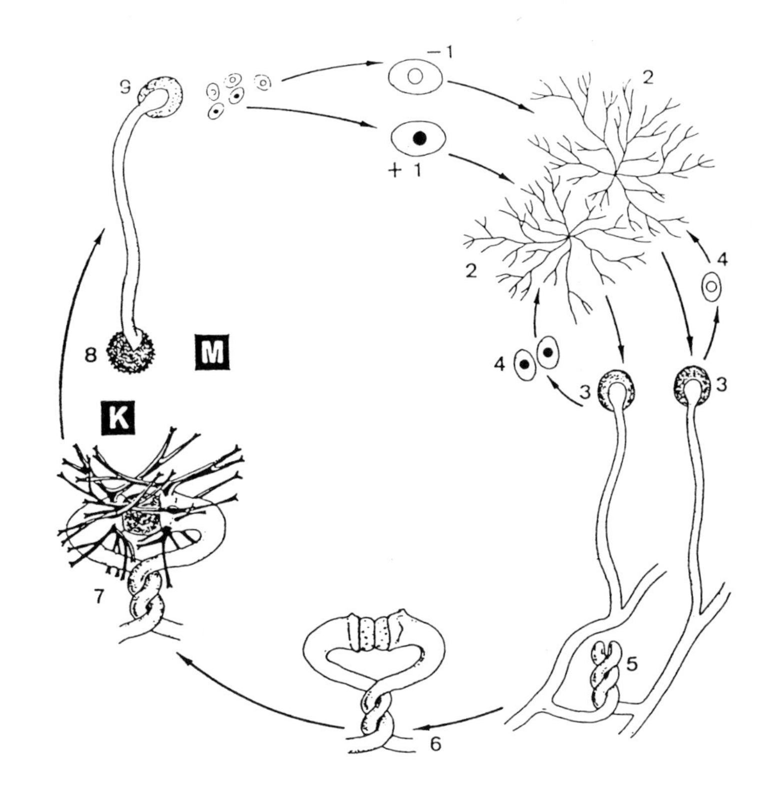

Fig. 4. Life cycle of *Phycomyces blakesleeanus.* (**1**) meiospores of opposite mating types (+/−) form a mycelium (**2**) which can individually propagate in a vegetative fashion through conidiospores (**3** and **4**). Fusion of morphologically alike gametangia (**6** + **7**) is preceded by conjugium formation (**5**). In the zygospore (**7**), karyogamy (K) and meiosis (M) occur (**8**). After mitotic divisions, numerous sexually formed spores are liberated (from ESSER, 1986, with permission).

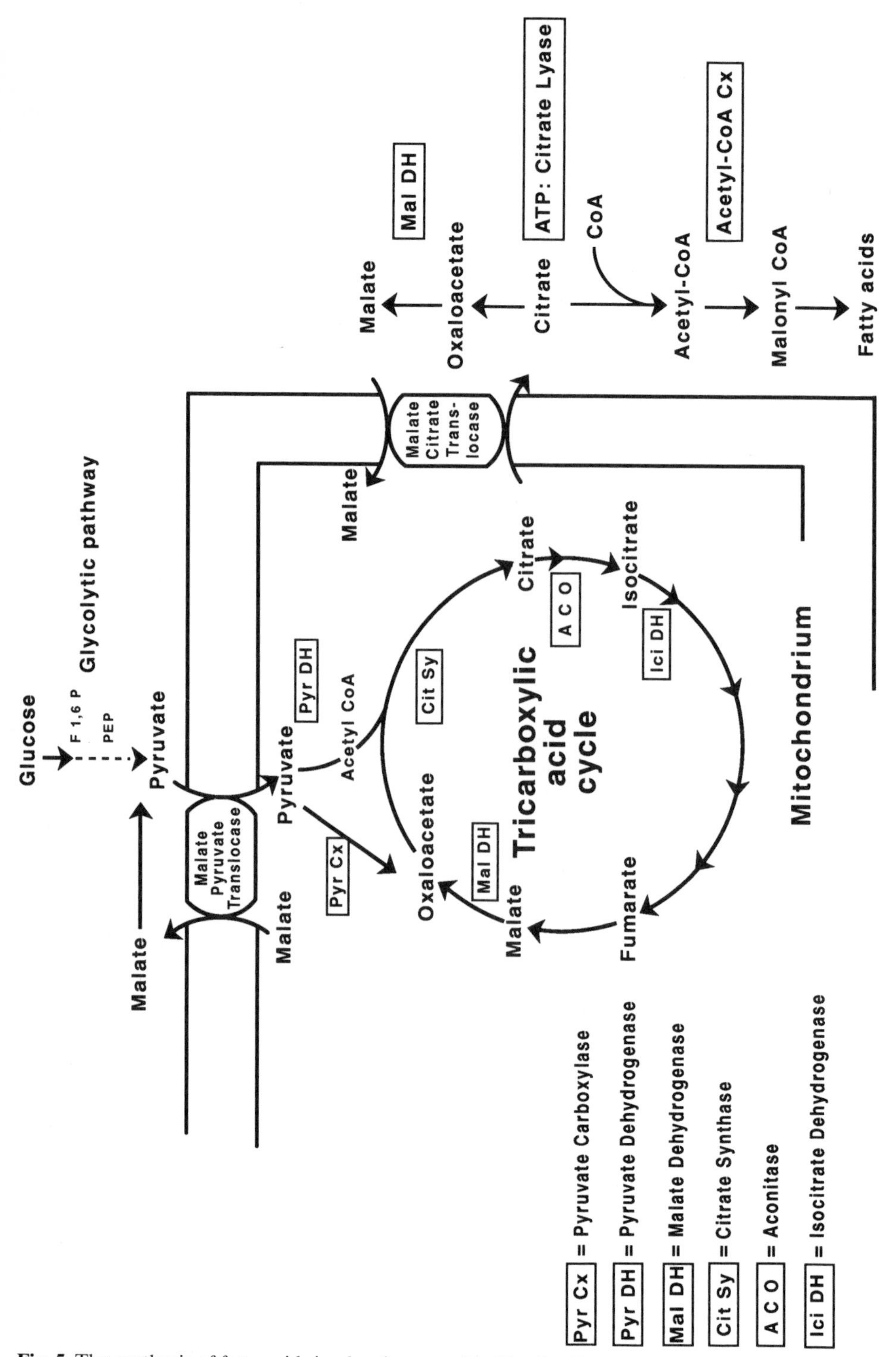

Fig. 5. The synthesis of fatty acids in oleaginous molds. For details see text.

Citrate drastically enhances lipid synthesis by activating acetyl-CoA carboxylase, the pacemaker reaction in fatty acid synthesis. An excess of carbon source, such as glucose, coincidental with a limited supply of nitrogen or other important growth factors, eventually leads to fat accumulation. Since depletion of nitrogen results in decreasing levels of AMP, which itself is necessary for the activity of the isocitrate dehydrogenase, the mitochondrially localized tricarboxylic acid cycle is arrested. As a consequence, isocitrate and citrate accumulate and are transported into the cytoplasm via the malate/citrate translocase system. In the cytoplasm malate is then split by the ATP: citrate lyase into acetyl-CoA, for fatty acid synthesis and oxaloacetate. Conversion of oxaloacetate to malate by malate dehydrogenase keeps the malate/citrate translocase operating. Since acetyl-CoA cannot be transported across the mitochondrial membrane, ATP: citrate lyase, which is found only in oleaginous yeasts and filamentous fungi, is the key enzyme for fat-accumulating microorganisms.

There are many reports on the production of fat by filamentous fungi (for complete lists see REHM, 1980; RATLEDGE, 1984; WEETE, 1984). The fat content of the mycelia depends on the substrate used for cultivation (HOFFMANN and REHM, 1976) and varies between 10 and 25%. However, more than 50% was also reported for *Mucor ramannis*. In large-scale fermentation, nitrogen and phosphorus must be at low levels to ensure that the synthesis of proteins and fat is inversely correlated. In this manner, fermentation will result in an excess amount of lipids and less protein from a given amount of carbohydrate.

5.2 Ascomycetes

Containing about 20000 species the Ascomycetes embrace approximately 30% of all fungi. They are saprotrophs or parasites, and are predominantly terrestrial. Many of the saprotrophs are coprophilous (i.e., they grow on animal feces, from which they can easily be isolated). Among the parasites, numerous agents of plant diseases are found.

Asexual reproduction takes place by means of conidiospores, for which different modes of development are known, or by means of oidia.

Sexual reproduction: In the Ascomycetes, there is a higher degree of uniformity in the sexual process than that found in the Phycomycetes. The class receives its name from the organ associated with an important point in the reproduction cycle, the meiosporangium, which is also termed the ascus. Within this structure, with only a few exceptions, karyogamy occurs, immediately followed by meiosis. In the higher Ascomycetes, the asci arise in fruiting bodies (ascocarps).

Tab. 4. Biotechnologically Relevant Ascomycetes

Protoascomycetidae
No fruiting body is produced

Saccharomyces cerevisiae: Ethanol
Eremothecium ashbyii: Riboflavin

Plectoascomycetidae
Closed spherical fruiting bodies

Aspergillus niger: Organic acids
A. oryzae: Enzymes
A. terreus: Enzymes
Penicillium chrysogenum: Antibiotics
P. notatum: Antibiotics
P. stoloniferum: Cheese making
P. camemberti: Cheese making
P. roquefortii: Cheese making
Cephalosporium: Antibiotics

Loculomycetidae
Fruiting bodies in cavities

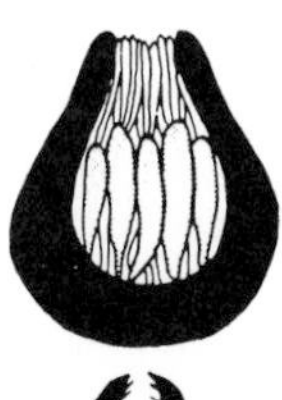

Cochliobolus lunatus:
Steroid transformation

Pyrenomycetidae
Closed flask-like fruiting bodies

Claviceps purpurea:
Alkaloids

Discomycetidae
Disc-shaped to cup-shaped fruiting bodies

Morchella esculenta: Edible mushroom
Tuber melanosporum: Edible mushroom

One of the best known and biotechnologically important members of the Ascomycetes (Tab. 4) is the ergot fungus *Claviceps purpurea*. Its life cycle is illustrated in Fig. 6 and may serve as an example for the entire class.

The filamentous ascospores are dispersed in the spring by wind. Upon reaching the inflorescence of a grass, they germinate, and infect the ovary with their hyphae. After the disintegration of the ovarian tissue, the fungus forms a white, loosely packed mycelium over its surface, and then gives rise to numerous conidia. In the process of conidium formation, a nectar-like product (honey dew) is produced which attracts many insects, and in this way the conidia are carried to other uninfected inflorescences.

As grain formation occurs in the infected inflorescences, the mycelia harden to form reddish, and later black sclerotia (ergots), which replace the rye grain, but are generally larger in size. The sclerotia overwinter in the soil and germinate the next spring, giving rise to several stalked stromata, 1–2 cm tall, with head-like swellings at their tips. Just below the surface of these swellings, the sex organs are produced, which develop into the sunken perithecia after fertilization. The ascospores are actively ejected. The economic importance of *Claviceps* lies not only in its character as a cereal pathogen, but as a producer of alkaloids within the sclerotia.

In the following we shall use a classification scheme based upon the structure of the fruiting body because it is easy to employ, and the different types of fruiting bodies can generally be identified with the naked eye. According to the different types of fruit bodies, the ascomycetes can be subdivided into five subclasses. These are listed in Tab. 4 along with biotechnologically important representatives, as well as products and processes in which they are involved.

Since yeasts are treated separately in Chapter 14, this subclass is only listed for the sake of completeness. Representatives of the Plectomycetiadae, however, are the classical organisms which have been exploited in biotechnology for centuries.

Aspergillus niger accumulates high concentrations of citric acid from glucose, depending on a set of conditions. Citric acid is one of the most important chemicals produced by industrial fermentation using a filamentous fungus. Worldwide, more than 350000 tons are pro-

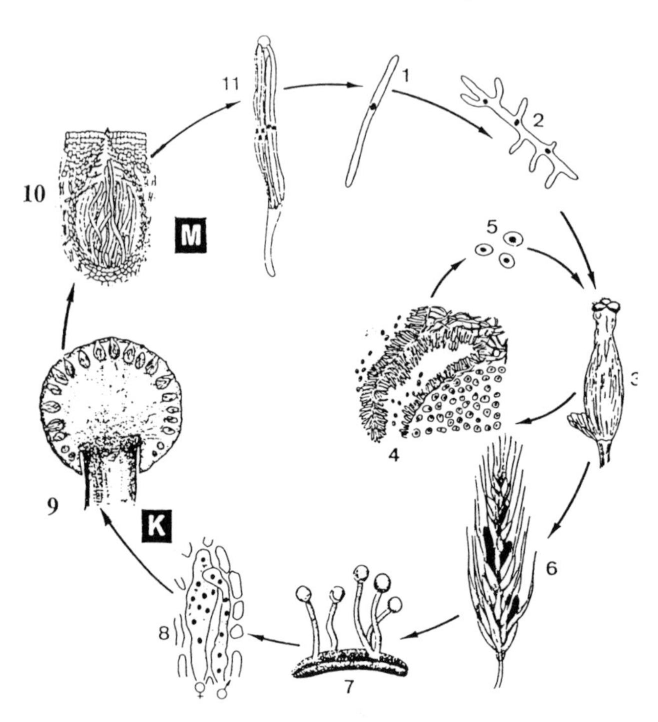

Fig. 6. Life cycle of *Claviceps purpurea.* Ascospores (**1**) germinate (**2**) and infect ovaries of host plants (**3**). Disintegration of infected plant tissue (**4**) is accompanied by the formation of conidia (**5**) and honeydew. Conidiospores are dispersed by insects. Sclerotia (**7**) (ergots) are formed instead of grain (**6**). After the winter, these give rise to several stalked stromata, in which the sexual cycle of the fungi is completed (**8**, fusion of gametangia; **9**, karyogamy in perithecia; **10** and **11**, ripe perithecia and asci). From ESSER (1986) with permission.

duced per year (KUBICEK and RÖHR, 1986). It is used as an alcidulant, flavor enhancer, preservative, chelating agent, antioxidant, stabilizer, buffer, emulsifier, and cleaning agent. There are many review articles providing detailed information on the subject, a recent one by BIGELIS and ARORA (1992) is highly recommended. *Aspergillus niger* is an imperfect fungus, i.e., the sexual cycle is lacking. That is why most of the work aimed at the development of high production strains was based on mutagenesis and screening. In the fermentation process, liquid medium is inoculated with conidia which were collected from a surface culture. The fungus grows, submerged in the form of small pellets which are less than 1 mm in diameter. Growth strictly depends on low levels of manganese, which must be in the ppb range. Manganese deficiency is essential for citric acid production, as is a low pH level (pH 1.5–3.0). A growth phase, during which the sugar is assimilated, is followed by a production phase, in which almost no growth is observed and in which the sugar is converted to citric acid (RÖHR et al., 1983; KUBICEK and RÖHR, 1986). The yield, which can reach 70–90 kg per 100 kg sugar, is strain-dependent and also depends on the fermentation conditions, which must be individually developed for each strain (BIGELIS and ARORA, 1992).

Citric acid synthesis begins with the cleavage of carbon compounds via glycolysis, yielding pyruvate. The action of pyruvate carboxylase gives rise to the formation of oxaloacetate, from which citric acid is formed by reactions of the tricarboxylic acid cycle (see also Fig. 5). It is, thus, evident that for industrial citric acid production, a high glycolytic flux is essential in order to provide the high concentrations of precursors, pyruvate and oxaloacetate (XU et al., 1989). The role of fructose-2,6-bisphosphate as the key regulator of glycolysis and gluconeogenesis has been well documented for several eukaryotes. However, the phosphofructokinase 2 from *A. niger*, which catalyzes its formation, has been studied only recently (HARMSEN et al., 1992). Whereas most of the other eukaryotic enzymes are regulated by cyclic AMP-dependent phosphorylation/dephosphorylation, the *A. niger* enzyme is evidently regulated directly by fructose-6-phosphate (F-6-P), in a way that high concentrations of F-6-P result in activation of the phosphofructokinase 2. This also explains why citric acid accumulation is only possible from certain sugars, such as glucose and sucrose, but not from mannose or galactose.

Members of the genera *Penicillium* and *Cephalosporium* are being exploited for the production of antibiotics of the β-lactam type, still the most important antimicrobial agents used in the therapy of bacterial infections. Penicillins are *N*-acyl derivatives of 6-aminopenicillanic acid. They are synthesized from three amino acids (L-α-aminoadipic acid, L-cysteine, and L-valine) (Fig. 7). All three amino acids are subsequently linked together resulting in the formation of the tripeptide L-α-aminoadipyl-L-cysteinyl-D-valine, which is converted to isopenicillin N by the action of isopenicillin synthase. If no side-chain precursor is provided, this is deacylated to 6-aminopenicillanic acid (6-APA). However, benzylpenicillin is produced if phenylacetic acid is available as a side-chain precursor. In a similar manner, many different penicillins can be obtained depending upon the type of side-chain precursor that is provided (for detailed information see EL-SAYED, 1992). Chemical or enzymatic acylation of 6-APA with specific side chains is another widely used method to obtain a wide range of different penicillins.

Some members of the penicillia, however, do not form 6-APA or any of its derivatives. They can epimerize the L-α-aminoadipyl side chain to the D-form, leading to isopenicillin N. The enzymatic conversion of the thiazolidine ring of penicillin N to a dihydrothiazine ring results in production of diacetoxycephalosporin C (see Fig. 7). A further reaction, catalyzed by a dioxygenase, converts the product to deacetylcephalosporin C, which is acetylated to give cephalosporin C. Thus, all cephalosporin producers also form isopenicillin N at least as an intermediate.

Penicillin production on an industrial scale exploits high-producing strains of *Penicillium chrysogenum*, whereas cephalosporins are produced with high-yielding strains of *C. acremonium.* There are numerous publications on the penicillin production process, covering nearly all aspects of batch- and fed-batch fermentations, including productivity, nutrient supply,

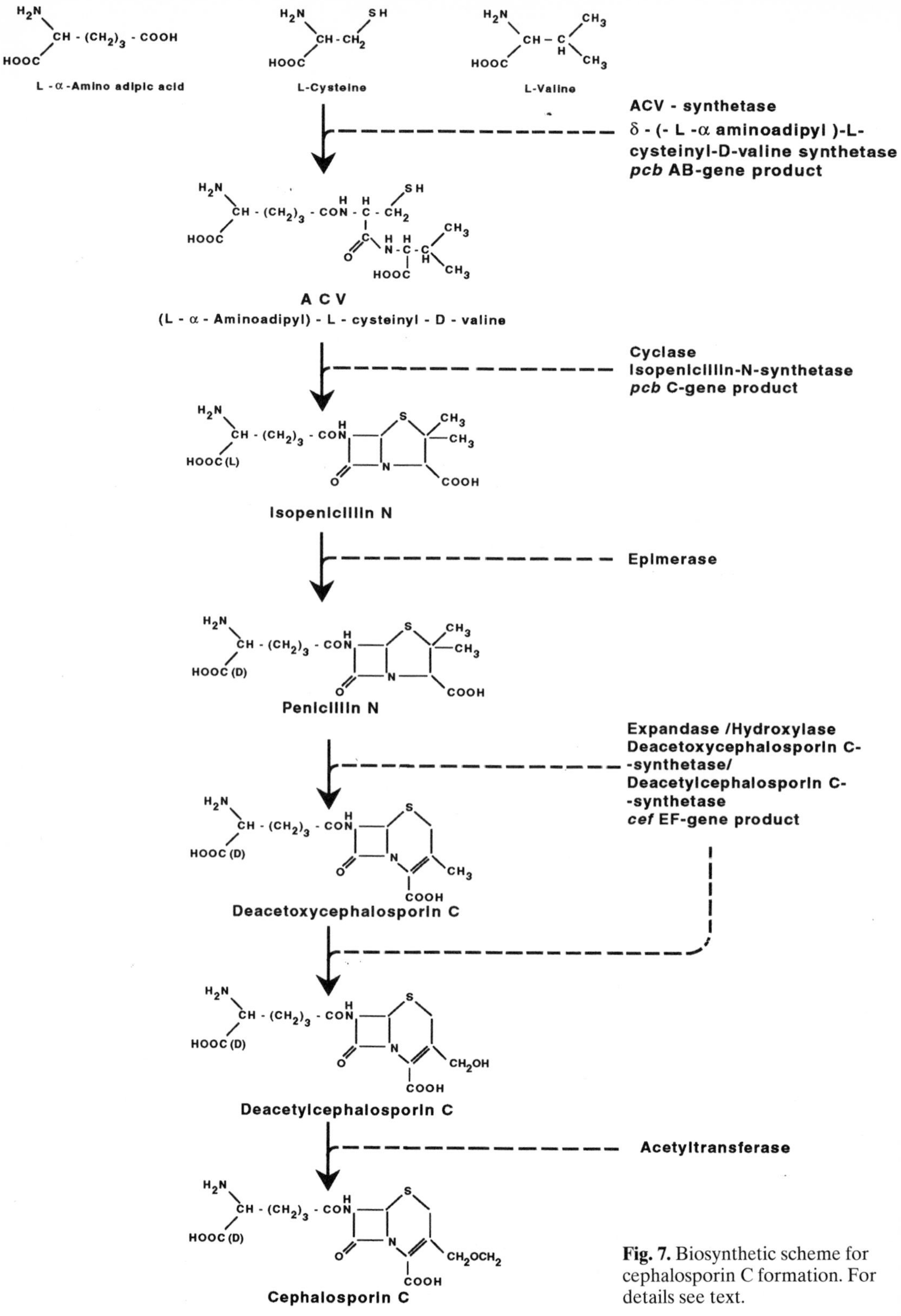

L-α-Amino adipic acid

L-Cysteine

L-Valine

ACV - synthetase
δ - (- L -α aminoadipyl)-L-cysteinyl-D-valine synthetase
pcb AB-gene product

A C V
(L - α - Aminoadipyl) - L - cysteinyl - D - valine

Cyclase
Isopenicillin-N-synthetase
pcb C-gene product

Isopenicillin N

Epimerase

Penicillin N

Expandase /Hydroxylase
Deacetoxycephalosporin C-
-synthetase/
Deacetylcephalosporin C-
-synthetase
cef EF-gene product

Deacetoxycephalosporin C

Deacetylcephalosporin C

Acetyltransferase

Cephalosporin C

Fig. 7. Biosynthetic scheme for cephalosporin C formation. For details see text.

feeding, oxygen transfer, yield and modeling (see reviews by HEERSBACH et al., 1984, and EL-SAYED, 1992). Since its discovery, the production yield of cephalosporin C has dramatically increased with the aid of process and strain improvement (NEWTON and ABRAHAM, 1955). Cephalosporin C is used for treatment of human infections caused by penicillinase-producing bacteria. The currently exploited high producers were derived by multiple rounds of mutagenic treatment, screening, and selection. Further improvement using these methods, however, is associated with enormous efforts, making it economically unattractive. Recently, the *pcb* C gene, encoding isopenicillin N synthetase (cyclase), and the *cef* EF gene, encoding the bifunctional enzyme deacetoxy-cephalosporin C synthetase/deacetylcephalosporin C synthetase (expandase/hydroxylase), have been cloned and their nucleotide sequences determined (SAMSON et al., 1985, 1987). The gene coding for the multifunctional enzyme δ-L-α-aminoadipyl-L-cysteinyl-D-valine synthetase (ACV-synthetase) was also cloned (HOSKINS et al., 1990), and its DNA sequence is available (GUTIERREZ et al., 1991). Using these genes as probes in hybridization experiments, it could be shown that a high-producing strain possesses multiple copies of these genes (SMITH et al., 1989). The rapid progress made in understanding the biosynthetic genes, as well as techniques to transfer DNA into fungi, makes recombinant DNA technology the method of choice to further improve the yield. In fact, by cloning the *cef* EF gene into an industrial cephalosporin C producer, the yield was enhanced by 15% (SKATRUD et al., 1989).

Irrespective of these successful examples of the use of recombinant DNA techniques, in general, enhancement of productivity by genetic manipulation requires the knowledge of the rate-limiting steps in the biosyntheis of β-lactam antibiotics. Based on published data, kinetic analysis of cephalosporin biosynthesis led to a mathematical model which judged ACV-synthetase to be the rate-limiting enzyme (MALMBERG and HU, 1992). Since the corresponding gene is cloned (see above), further improvement of the yield is to be expected by providing additional copies of this gene to producer strains. For comprehensive information concerning ACV-synthetase, see ZHANG and DEMAIN (1992). When it is also taken into account that transformants of *Penicillium chrysogenum* were stable in a fermentor for 312 h, the potential of recombinant DNA technology in this field is obvious (REMIO et al., 1990).

Among the Loculomycetidae, only *Cochliobolus lunatus* (*Curvularia lunata*) is to be mentioned with respect to biotechnology, because of its capability for steroid transformation. Unlike the microbial 11α-hydroxylation of progesterone, which can be performed by a variety of fungi (e.g., *Mucor*, *Rhizopus*, *Fusarium*), the introduction of the hydroxyl group into position 11β can only be performed by a few fungi, the most important of which is *Cochliobolus lunatus* (*Curvularia lunata*) (Fig. 8). The accumulation of by-products, mainly of other hydroxy derivatives, significantly decreases the economic efficiency of the process. Changes in media composition and variation of fermentation parameters could not improve the cortisol yield (SMITH, 1984). However, mutagenesis of protoplasts generated strains producing only small amounts of by-products (WILMANSKA et al., 1992). The establishment of gene transfer techniques may also help to overcome these obstacles (OSIEWACZ and WEBER, 1989; DERMASTIA et al., 1991).

The Pyrenomycetidae include organisms which are well known in basic research, such as *Neurospora*, *Podospora*, and *Sordaria*. In bio-

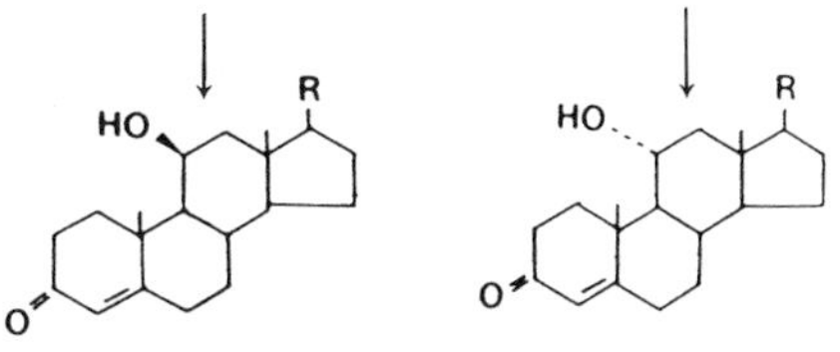

Fig. 8. Hydroxylation of steroid hormones by fungi.

technology, however, only some *Gibberella* species, which produce the plant hormone gibberellin, some imperfect *Fusarium* species, and the ergot fungus *Claviceps purpurea* are important. The ergot alkaloids accelerate the peristaltic movement of the bowels and uterus, serving to quicken birth or act as abortifacients. If sclerotia are mixed with the rye in bread, serious consequences may result, often causing death after poisoning symptoms. This disease, known as ergotism or St. Antony's fire, was widespread in earlier centuries, be-

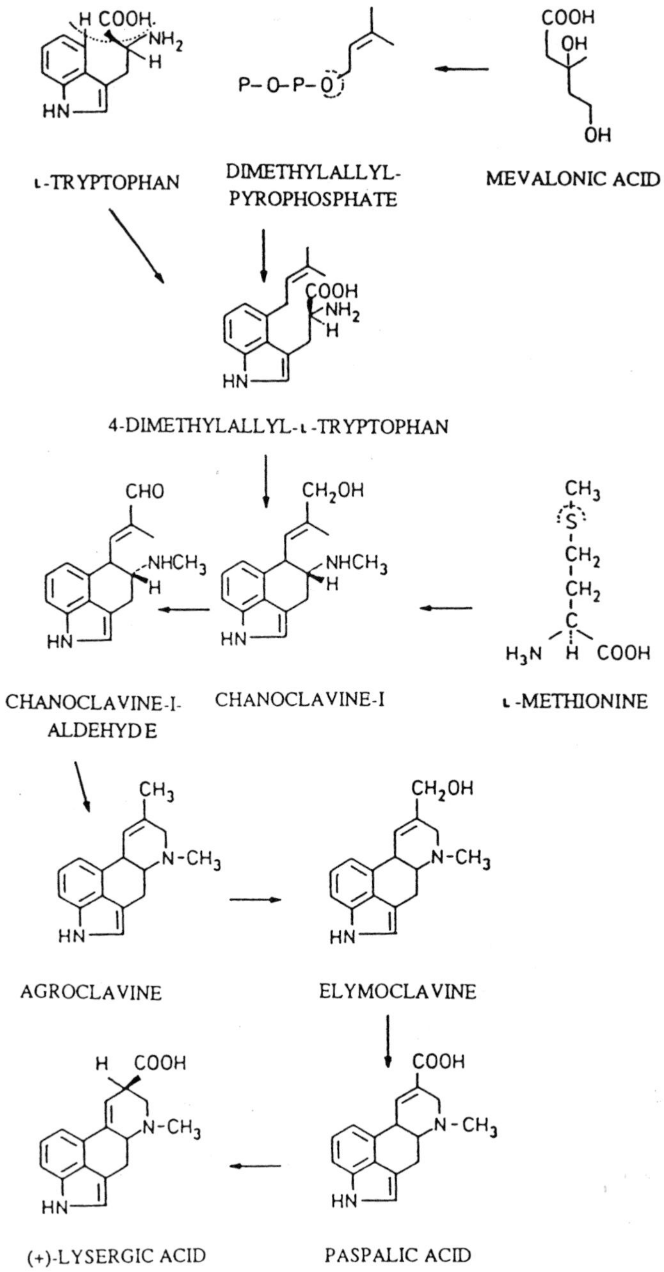

L-TRYPTOPHAN

DIMETHYLALLYL-PYROPHOSPHATE

MEVALONIC ACID

4-DIMETHYLALLYL-L-TRYPTOPHAN

CHANOCLAVINE-I-ALDEHYDE

CHANOCLAVINE-I

L-METHIONINE

AGROCLAVINE

ELYMOCLAVINE

(+)-LYSERGIC ACID

PASPALIC ACID

Fig. 9. Biosynthesis of lysergic and paspalic acid by *Claviceps purpurea.*

fore the significance of the ergots was known. Because the sclerotium alkaloids and their derivatives are used today in medicine, they are commercially important.

During the last three decades, numerous investigations concerned the biosynthesis and physiology of the ergot alkaloids. Authorative reviews provide all the available information on precursors, intermediates, and regulation (KOBEL and SANGLIER, 1986; SOČIČ and GABERC-POREKAR, 1992). Despite the fact that some details remain to be investigated, the biosynthetic pathway is proposed as given in Fig. 9. Tryptophan, mevalonic acid, and *S*-adenosyl-methionine are the precursors from which lysergic acid is synthesized. Clavines, which are the main alkaloids secreted by certain *Claviceps* strains are produced as intermediates. Many derivatives, however, can be formed from the lysergic acid. The most important of these are the peptide alkaloids, especially those carrying a cyclic tripeptide (cyclols). Only ergotamine and ergometrine are used as therapeutic agents without any further chemical modifications, all other medically relevant ergot alkaloids are chemically changed (Fig. 10).

According to the diversity of their chemical structure, no other group of substances isolated from nature exhibits such a spectrum of diverse biological effects, i.e., influence on blood pressure, hypothermia induction, uterotonic action, and control of hormone release.

Structural similarities to several neurotransmitters (serotonin, noradrenaline, dopamine) are considered to be the reason for the different effects caused by the ergot alkaloids. In their most frequent use, ergot alkaloids are therapeutics in the treatment of migraine, uterus atonia, cerebral insufficiency, and hypertension (FANCHAMPS, 1979).

The biosynthesis of ergot alkaloids is certainly a genetically regulated process, but due to their rather complicated life cycle, genetic studies in *Claviceps* are rather scant (ESSER and TUDZYNSKI, 1978; ESSER and DÜVELL, 1984). Genetic approaches to improve the yield were performed mainly by classical procedures; i.e., by mutation and selection. Different mutagens, such as UV light, ethylmethane sulfonate, nitrous acid, and methyl-nitro-nitrosoguanidine were successfully employed (KOBEL and SANGLIER, 1986; SOČIČ and GABERC-POREKAR, 1992). Meiotic recombination via the

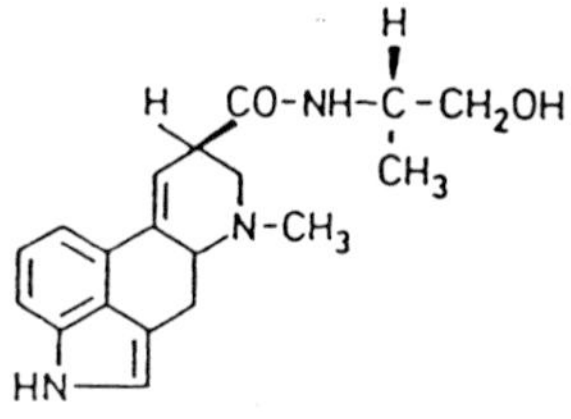

ERGOMETRINE

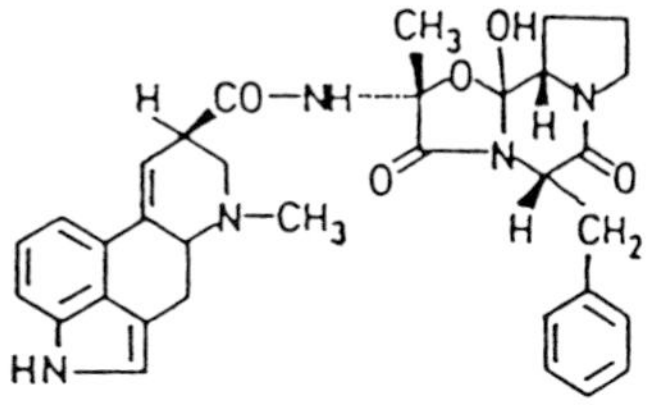

ERGOTAMINE

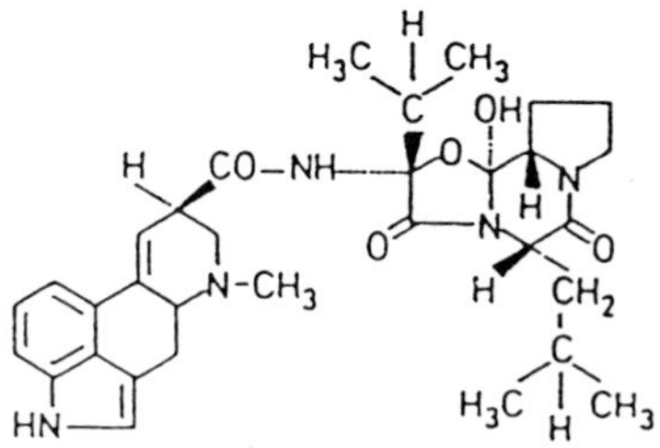

α-ERGOKRYPTINE

Fig. 10. Structure of medically important ergot alkaloids.

sexual cycle can only be achieved in nature. TUDZYNSKI et al. (1982) performed crosses between two mutant strains and found among the recombinant progeny some strains exhibiting an alkaloid content higher than either parent. Since the parasexual cycle also exists in *Claviceps* (BRAUER and ROBBERS, 1987), mitotic recombination may also be exploited for strain improvement. The fusion of protoplasts, generated by the action of lytic enzymes, can be performed in intra- and interspecific matings (ROBBERS, 1984; SOČIČ and GABERC-PORЕКAR, 1992). Most of the fusants obtained in these experiments could only be maintained on minimal medium, because segregation is frequently observed on complete media. Thus, it is desirable to have a system facilitating the addition and stable maintenance of specific genes in high-producing strains. First attempts used mitochondrial genetic material (ESSER et al., 1983; TUDZYNSKI and ESSER, 1986). The first successful molecular transformation of *Claviceps*, however, was achieved using a bacterial phleomycin resistance gene, governed by a nuclear promoter, which was derived from *Aspergillus nidulans*. Multiple copies of the transforming DNA were found to be integrated at multiple sites in the genome (VAN ENGELENBURG et al., 1989). One of the main problems in those experiments were the very low transformation rates obtained. That is why a homologous system was developed. For this purpose an orotidine-5′-monophosphate decarboxylase (OMPD)-deficient mutant was generated by UV mutagenesis and the corresponding gene was isolated from a lambda-genomic library of *Claviceps*. Based on this gene, a homologous transformation system proved to be efficient, also in cotransformation experiments, in which the bacterial β-glucuronidase was not only expressed in axenic culture, but also in the parasitic cycle during honey-dew formation (SMIT and TUDZYNSKI, 1992).

Since the OMPD encoding gene is the first genetic trait cloned from *Claviceps* chromosomal DNA, it becomes obvious that much work remains to be done along this line to improve alkaloid formation.

Among the Discomycetes only the edible morels and truffles are economically important. These fungi are saprophytic species whose mycelia live in the soil. Some are mycorrhizal fungi, i.e., they form a symbiosis with roots of higher land plants. The edible truffles are mostly associated with Mediterranian species of *Quercus*. Fruiting body formation has not been observed in laboratory cultures; however, spores can be germinated on nutrient agar media. However, the mycelia grow very poorly in axenic cultures (1–2 mm per month) and that is why they are not available from culture collections. Truffle farms are fairly common in the Mediterranean region. For the production of truffles, the roots of oaks are inoculated with mycelia. This laborious operation leads to the first flash of fruit bodies only after about seven years. In Southern France and Italy, however, experiments are in progress with laboratory cultures as an attempt to create a truffle industry comparable to the mushroom industry using *Agaricus* species (see later). Economically important edible fungi include *Tuber melanosporum* (Périgord truffle), *Tuber magnatum* (Piedmont truffle), *Tuber brumale* (winter truffle), and *Tuber aestivum* (summer truffle).

5.3 Basidiomycetes

The Basidiomycetes contain about 12000 species, i.e., about 20% of all fungi. They are almost exclusively terrestrial, and live as saprotrophs and parasites. Among the saprotrophs, the edible fungi are the best known, and of the parasitic fungi the rusts and smuts (Uredinales, Ustilaginales) are the most important, causing many plant diseases. Wood-rotting fungi (mostly saprotrophs) are of practical importance in their decomposition of cellulose and lignin. The fruit bodies of some of them are edible. Many Basidiomycetes live as symbionts in roots (sheathing mycorrhiza).

Like the Ascomycetes, the Basidiomycetes form richly branching septate mycelia, whose cell walls contain chitin. Both taxa are classified together as the higher fungi, as opposed to the lower fungi (Phycomycetes). One essential difference between the mycelia of the Ascomycetes and the Basidiomycetes is in the ultrastructure of the transverse septa. While the former show a circular opening in the septa (simple septum), the pores of the Basidiomycetes, about 0.1–0-2 μm in diameter, are

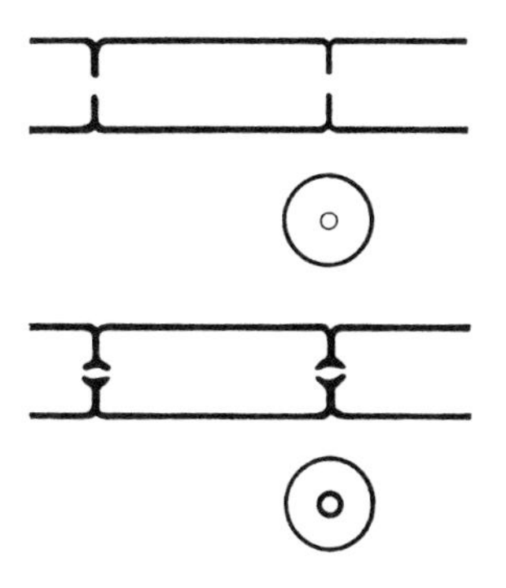

Fig. 11. Schematic representation of (above) simple septa (Ascomycetes) and (below) dolipore septa (Basidiomycetes).

surrounded by a flag-like projection. Both openings of this "dolipore" are covered with a perforated cap (parenthosome), which is an extrusion of the endoplasmic reticulum (Fig. 11).

Like the Ascomycetes, the Basidiomycetes form no motile reproductive cells.

Asexual reproduction proceeds by conidiospores, chlamydospores, or oidia, although specialized conidiophores, such as those occurring in the Ascomycetes, are only rarely formed.

Sexual reproduction takes place with the production of basidiospores, which are meiospores, arising on typical club-shaped or sac-like structures, the basidia. The development of the basidia is comparable with that of asci; however, the spores are produced exogenously rather than endogenously. Asci and basidia are thus homologous organs.

The life cycle of *Agrocybe aegerita* as a representative species of the Basidiomycetes is shown in Fig. 12. We have chosen *A. aegerita* for several reasons: first, it is genetically well-known (MEINHARDT and ESSER, 1981; ESSER and MEINHARDT, 1977; MEINHARDT, 1980), second, it is easily cultured in the laboratory and on straw (ZADRAZIL, 1977), and finally, antibiotics may be produced with this fungus (MEINHARDT, 1980). As it becomes evident from Fig. 12, *A. aegerita* is a typical dikaryotic fungus, whose breeding system is controlled by complex genetic traits (A and B). Fruiting body formation is controlled by a series of genes, as it is in other Basidiomycetes (ESSER and MEINHARDT, 1986).

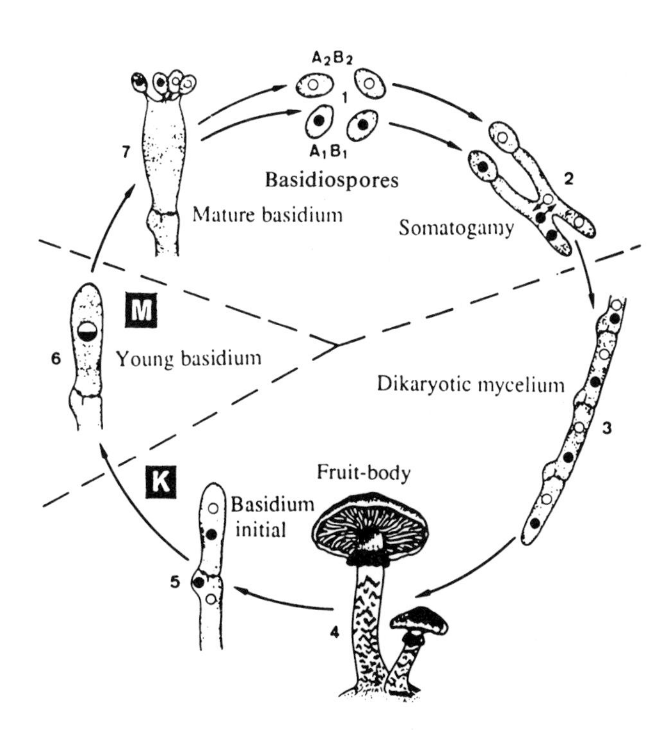

Fig. 12. Life cycle of *Agrocybe aegerita.* Immediately after germination, genetically different basidiospores fuse to give a dikaryotic mycelium upon which the fruiting bodies are formed. Basidiospores are released from the basidia, carried in the fruiting bodies.

The systematic subdivision of the Basidiomycetes is not based, as in many other cryptogam taxa, on phylogenetic concepts. Phyletic connections within the individual orders are only limited, therefore, the subdivision of the Basidiomycetes is, in principle, artificial. The mode of formation and the final form of the basidia serve as the main criteria. The differentiation into the two sub-classes of Phragmo- and Holobasidiomycetidae is based on whether or not the basidia are septate. The further subdivision into orders is based on further characteristics of the basidia, the form of the hymenium, and the habitat of the fruiting body.

Sub-class Phragmobasidiomycetidae

1st Order: Uredinales. Rust fungi, which are obligate endoparasites of vascular plants.

2nd Order: Ustilaginales. Smut fungi, endoparasites of endosperm plants.

3rd Order: Auriculariales. Ear fungi, predominantly saprotrophs, only few live as plant parasites.

4th Order: Tremellales. Jelly fungi, saprophytic fungi living on dead wood.

Sub-class Holobasidiomycetidae

The basidia are non-septate. Further subdivision is based on the development of the hymenium and the way in which the hymenium is arranged on the fruiting body. In the Poriales the hymenium occurs in tubes or teeth-like projections, in the Agaricales on lamellae, and in the Gastromycetales in tube-like, interconnected cavities. The economic importance of the Poriales, also called Aphyllophorales, lies principally in the fact that they make up the bulk of the wood-rotting fungi. Apart from saprotrophic forms on tree stumps and fallen branches, several species are parasites on living wood. Two types are distinguished.

The *white-rot fungi* can decompose both cellulose and lignin. This so-called corrosion-rot results in a bleaching and a fibrous disintegration. Decomposition of the wood is correlated with the secretion of oxidoreductases (phenol oxidases) in a high proportion of such fungi, but the function of these enzymes is not fully understood. Representatives: *Fomes fomentarius*, *Polyporus brumalis*, *Polyporus versicolor*, *Trametes confragosa.*

The *brown-rot fungi*, on the other hand, are only able to decompose cellulose. They cause a destruction-rot, which reduces wood to brown cubic structures. Representatives: *Laetiporus sulphureus*, *Serpula lacrymans*, *Polyporus* (*Piptoporus*) *betulinus*.

Apart from the Poriales, representatives of the Tremellales, Agaricales and a few Ascomycetes are also involved in the decomposition of wood. The edible fungi among the Poriales include not only the well-known chanterelle, *Cantharellus cibarius*, which lives saprotrophically on the forest floor, but also several fungal wood decomposers. Under the name Agaricales (Phyllophorales) or the leaf fungi, those fungi are grouped having a typically stalked, umbrella-like fruit body, known also as mushrooms or toadstoals.

In recent years, the classification has been considerably modified, and today, up to 16 families are distinguished, which is too many to deal with in this chapter. One of the most important taxonomic criteria is the anatomy of the trama (lamella-plectenchyma), which is the basis of the classification used by ALEXOPOULOS and MIMS (1979). The continual rearrangement of the classification of the Agaricales has resulted in numerous synonyms (200 genera, 350 synonyms; 3250 specific names).

The edible fungi have a special economic importance, e.g., cultivated mushroom (*Agaricus bisporus*), cepe (*Boletus edulis*), oyster mushroom (*Pleurotus ostreatus*), *Kuehneromyces mutabilis*, honey fungus (*Armillaria mellea*). In Eastern Asia the shii-take fungus (*Lentinus edodes*) and the paddy-straw fungus (*Volvariella volvacea*) are cultivated. Economically important Basidiomycetes are listed in Tab. 5.

As stated above, the white-rot fungi are capable of degrading lignin. Some of the organisms involved have been reported to produce one or several extracellular enzymes called ligninases or lignin peroxidases. These enzymes are involved in radical-mediated attack of lignin (HAMMEL et al., 1986; BUMPUS and AUST, 1987a). Some white-rot fungi can also cleave a variety of organic chemicals, and thus they can be used for detoxification of xenobiotics, such as polycyclic hydrocarbons (BUMPUS et al., 1989; HAEMMERLI et al., 1986; SANGLARD et al., 1986), DDT (KÖHLER et al., 1988; BUMPUS and AUST, 1987b), polychlorinated biphenyls

Tab. 5. Economically Important Basidiomycetes

Phragmobasidiomycetidae
Septate basidia

Rusts and smuts, economically important as parasites of wheat, rye, corn and other cereals

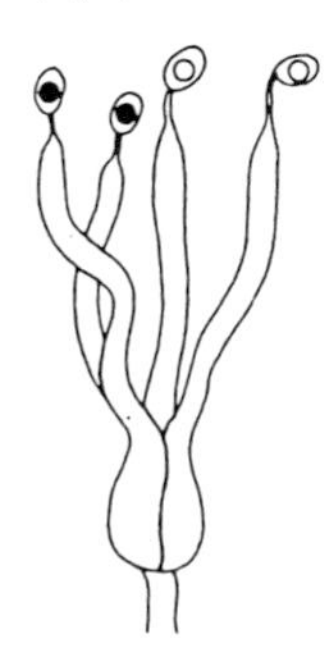

Jelly fungi, ear fungi (*Auricularia auricula*, *Tremmella fuciformis*), edible mushrooms predominantly cultivated in Asia

Holobasidiomycetidae
Non-septate basidia

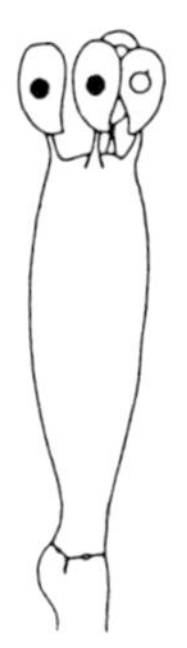

Agaricus bisporus, *A. bitorquis*	Edible mushroom
Coprinus fimetarius, *C. atramentarius*	Coprin, an antabus-like substance
Flammulina velutipes	Edible mushroom
Kuehneromyces mutabilis	Edible mushroom
Lentinus edodes	Edible mushroom
Oudemansiella mucida	Mucidermin, antibiotics
Phanerochaete chrysosporium	Lignin degradation
Pholiota nameko	Edible mushroom
Pleurotus ostreatus	Edible mushroom
Stropharia rugoso-annulata	Edible mushroom
Trichoderma matsutake	Edible mushroom
Volvaria volvacea	Edible mushroom

(EATON, 1985), hexochlorocyclohexane (BUMPUS et al., 1985), or dichloroaniline (HALLINGER et al., 1988), and pentachlorophenol (MILESKI et al., 1988). Most of the studies concerning degradation of xenobiotic compounds have employed *Phanerochaete chrysosporium* (see also a review by HAMMEL, 1992), but other wood-rotting fungi, such as *Trametes* species, also can degrade xenobiotics (MORGAN et al., 1991).

The purification and characterization of two extracellular peroxidases operating in lignin degradation (lignin peroxidase and manganese peroxidase), and the study of their catalytic mechanism have significantly increased our knowledge of the biochemistry of lignin degradation (KIRK and FARRELL, 1987; GOLD et al., 1989; WARIISHII et al., 1989). The sequences of both genes are known, and they evidently belong to related gene families (ASADA et al., 1988; DE BOER et al., 1987; TIEN and TU, 1987). DNA-mediated genetic transformation of *Phanerochaete chrysosporium* (ALIC et al., 1990, 1991) makes possible study of gene regulation, as well as new approaches to investigate structure–function relations.

The production of edible mushrooms is, however, the economically most important role of Basidiomycetes. *Agaricus bisporus* is without doubt the most abundantly cultivated fungus throughout the world. In Europe it has been grown since ca. 1700 (GAYLEY, 1938); more than 700000 tons are being produced each year (DELCAIRE, 1978). In Asia *Lentinus edodes*, the shii-take, is cultivated in Japan, Taiwan, China, and also in South Korea.

Annual average production is, however, only about one third of the value for *Agaricus bisporus*.

Cultivation of *A. bisporus* is strictly dependent on complex organic compounds synthesized by other microorganisms, and that is why mushrooms are grown on a prefermented compost which, after colonization with the mycelium, is cased with a layer of soil to facilitate the final stages of the life cycle, i.e., fruit body formation. Since microorganisms involved in composting and in the casing layer directly influence the yield of biomass obtained, much effort is necessary to preserve the desired properties and to eliminate or inactivate harmful organisms by pasteurization. The wood-destroying fungi, such as shii-take, are grown on logs of dead wood; no pasteurization is necessary for the cultivation of these fungi. The biology and cultivation of the twelve most popular edible mushrooms has been comprehensively treated by several expert authors in a book edited by CHANG and HAYES (1978), which still can be recommended because it provides extensive information not only on the biology and cultivation, but also on the economics of different fungi.

While many species of fungi have already been transformed, relatively few of the higher Basidiomycetes have been investigated in this respect. Only recently genetic transformation of mycorrhiza-forming Basidiomycetes has been reported (BARRETT et al., 1990; MARMEISSE et al., 1992), in context of improving the mycorrhizal abilities. The development of transformation systems for the edible *Pleurotus ostreatus*, the oyster mushroom, may prove to be useful in genetically engineering this fungus to produce mushrooms of improved nutritional value (PENG et al., 1992).

6 Perspectives

In recent years, the number of projects is growing in which recombinant DNA technology is successfully applied to strain improvement of microorganisms involved in biotechnology. For genetic engineering of fungi, the availability of a transformation system is a prerequisite (ESSER and MOHR, 1986; WÖSTEMEYER and BURMESTER, 1987; RUTTKOWSKI and KHAN, 1991). Most transformation systems developed for fungi use auxotrophic mutant strains as recipients, and prototrophic strains are searched on minimal media after DNA-mediated transformation (RAMBOSEK and LEACH, 1987). For microorganisms used in industrial fermentation processes, normally auxotrophic mutants are not available, and that is why dominant selectable marker genes must be used. The *amd*S gene codes for the enzyme acetamidase, which converts acetamide to acetate and ammonia, permitting wild-type *Aspergillus nidulans* to use acetamide as the sole carbon source. Other fungi, such as *Aspergillus niger*, and *amd*S mutant strains of *A. nidulans* and *Cochliobolus heterostrophus*, do not grow or only poorly grow on acetamide containing media. Transformants can be selected by searching for abundant growth on these media. Transformation on the basis of this system could be obtained for *A. nidulans* (TILBURN et al., 1983), *A. niger* (KELLY and HYNES, 1985), or *Cochliobolus heterostrophus* (TURGEON et al., 1985).

The gene conferring resistance to the antibiotic hygromycin B (PUNT et al., 1987) is another widely used dominant selectable marker in transformation experiments performed with filamentous fungi, as is also the phleomycin resistance gene (VAN ENGELENBURG et al., 1989). Transformation efficiency is greatly enhanced when these resistance genes, which were derived from prokaryotic species, are governed by fungal regulatory sequences (see, e.g., KÜCK et al., 1989).

When fungi are used as hosts for the production of homologous and heterologous proteins, high expression levels can be reached (FINKELSTEIN, 1987). Correctly processed, biologically active mammalian proteins are secreted into the medium in relatively large amounts, compared to other systems (SAUNDERS et al., 1989).

Findings reported by HANEGRAAF et al. (1991) are of highest biotechnological importance. They constructed *A. nidulans* strains carrying several copies of the glyceraldehyde-3-phosphate dehydrogenase (GPD) gene (*gpd*A). Production of GPD was proportional to the gene dosage and reached up to 22% of the total soluble cell protein. GPD overproduction seemed to resolve some "bottle neck" in the growth of *A. nidulans*, resulting in more efficient growth. These results may be of general importance, since the effects of overproducing glycolytic enzymes may be similar in different fungi. For example, *A. niger* strains exhibiting increased glycolytic capacity grow faster and have increased citric acid production (SCHREFERL-KUNAR et al., 1989).

At present, transformation systems for lower fungi exist as well as for Asco- and Basidiomycetes. Application to industrially important strains will certainly improve already existing procedures and will also lead to microbial production of novel compounds and processes.

As outlined, fungi exhibit an enormous array of diverse forms, ranging from simple unicellar types to complex macroscopic Asco- and Basidiomycetes. They have the ability to invade solid substrates, secrete enzymes, rot, transform, and spot organic material. Unique secondary metabolites are present in many species. Thus, fungi play a major role in nature, in commerce their abilities are exploited. Fungi are biotechnologically very important organisms; they are used in the food as well as the pharmaceutical industry. There is also no doubt that fungi produce many more secondary metabolites than we know today, some of which could certainly be used in medicine. Target-directed screening, e.g., for heart drugs, psychopharmaceuticals, and anticancer compounds, will discover new and useful compounds from fungi. Cyclosporin may serve as an example, originally isolated as an antimycotic agent from *Tolypocladium polysporum*, it was soon discovered that this cyclic undecapeptide also selectively inhibits the activation of T-lymphocytes. Today it is used to prevent rejection of transplanted organs (AGATHOS et al., 1986). Thus, the application of new screening methods (FRANCO and COUTINHO, 1991) and gene transfer techniques will greatly enhance exploitation of fungi in the near future.

7 References

AGATHOS, S. N., MARSHALL, J. W., MORAITI, C., PAREKH, R., RADHOSINGH, D. (1986), Physiological and genetic factors for progress development of cyclosporin fermentations, *J. Ind. Microbiol.* **1,** 39–48.

AINSWORTH, G. C., SUSSMAN, A. S. (Eds.) (1965–1968), *The Fungi*, Vol. I–III, New York–London: Academic Press.

ALEXOPOULOS, C. J., BENEKE, E. S. (1964), *Laboratory Manual for Introductory Mycology*, Minneapolis: Burgess.

ALEXOPOULOS, C. J., MIMS, C. W. (1979), *Introductory Mycology*, 3rd Ed., New York–Chichester–Brisbane–Toronto: John Wiley & Sons.

ALIC, M., CLARK, E. K., KORNEGAY, J. R., GOLD, M. H. (1990), Transformation of *Phanerochaete chrysosporium* and *Neurospora crassa* with adenine biosynthetic genes from *Schizophyllum commune*, *Curr. Genet.* **17,** 305–311.

ALIC, M., MAYFIELD, M. B., AKILESWARAN, L., GOLD, M. H. (1991), Homologous transformation of the lignin-degrading basidiomycete *Phanerochaete chrysosporium*, *Curr. Genet.* **19,** 491–494.

ARORA, D. K., ELANDER, R. P., MUKERJI, K. G. (Eds.) (1992), *Handbook of Applied Mycology*, Vol. 4: *Fungal Biotechnology*, New York: Marcel Dekker.

ASADA, Y., KIMURA, Y., KUWUHARA, M., TSUKAMOTO, A., KOIDE, K., OKA, A., TAKANAMI, M. (1988), Cloning and sequencing of a ligninase gene from a lignin-degrading basidiomycete *Phanerochaete chrysosporium*, *Appl. Microbol. Biotechnol.* **29,** 469–473.

BARRETT, V., DIXON, R. K., LEMKE, P. A. (1990), Genetic transformation of a mycorrhizal fungus, *Appl. Microbiol. Biotechnol.* **33,** 313–316.

BESSEY, E. A. (1968), *Morphology and Taxonomy of Fungi*, New York–London: Hafner.

BIGELIS, R., ARORA, R. K. (1992), Organic acids of fungi, in: *Handbook of Applied Mycology*, Vol. 4: *Fungal Biotechnology* (ARORA, R. K., ELANDER, R. P., MUKERJI, K. G., Eds.), pp. 357–376, New York: Marcel Dekker.

BOTH, C. (Ed.) (1971), *Methods in Microbiology*, Vol. 4, London: Academic Press.

BÖTTICHER, W. (1974), *Technologie der Pilzverwertung. Biologie, Chemie, Kultur, Verwertung, Untersuchung*, Stuttgart: Eugen Ulmer.

BRAUER, K. L., ROBBERS, J. E. (1987), Induced parasexual processes in *Claviceps* strain SD-58. *Appl. Environ. Microbiol.* **53,** 70–73.

BRESINSKY, A., BESL, H. (Eds.) (1985), *Giftpilze mit einer Einführung in die Pilzbestimmung. Ein Handbuch für Apotheker, Ärzte und Biologen,*

Stuttgart: Wissenschaftliche Verlagsgesellschaft mbH.

BUCHTA, K. (1983), Organic acids of minor importance, in: *Biotechnology* (REHM, H.-J., REED, G., Eds.), Vol. 3, *Microbial Products, Biomass and Primary Products*, pp. 407–478, Weinheim–Deerfield Beach/Florida–Basel: Verlag Chemie.

BUMPUS, J. A. (1985), Biodegradation of natural and synthetic humic acids by the white rot fungus *Phanerochaete chrysosporium*, *Appl. Environ. Microbiol.* **55,** 1282–1285.

BUMPUS, J. A., AUST, S. D. (1987a), Biodegradation of environmental pollutants by the white rot fungus *Phanerochaete chrysosporium:* involvement of the lignin degrading system, *BioEssays* **6,** 116–120.

BUMPUS, J. A., AUST, S. D. (1987b), Biodegradation of DDT [1,1,1-trichloro-2,2-bis(4-chlorophenyl)-ethane] by the white rot fungus *Phanerochaete chrysosporium*, *Appl. Environ. Microbiol.* **53,** 2001–2008.

BUMPUS, J. A., TIM, M., WRIGHT, D., AUST, S. D. (1989), Oxidation of persistent environmental pollutants by a white rot fungus, *Science* **228,** 1434–1436.

BURNETT, J. H. (Ed.) (1968), *Fundamentals of Mycology*, London: Edward Arnold Publishers Ltd.

BURNETT, J. H. (1975), *Mycogenetics. An Introduction to the General Genetics of Fungi*, London–New York–Sydney–Toronto: John Wiley & Sons.

CHANG, S. T. (Ed.) (1972), *The Chinese Mushroom (Volvariella volvacea) — Morphology, Cytology, Genetics, Nutrition and Cultivation*, Hong Kong: The Chinese University.

CHANG, S. T., HAYES, W. A. (Eds.) (1978), *The Biology and Cultivation of Edible Mushrooms*, New York–San Francisco–London: Academic Press.

COCHRANE, V. W. (1963), *Physiology of Fungi*, New York: John Wiley & Sons.

COLE, G. T., SAMSON, R. A. (Eds.) (1979), *Patterns of Development in Conidial Fungi*, London–San Francisco–Melbourne: Pitman.

CZAIJKOWSKA, D., ILNICKA-OLEIJNICZAK, O. (1985), Biosynthesis of protein by microscopic fungi in solid state fermentation, *Acta Biotechnol.* **8,** 407–413.

DADE, H. A., GUNNEL, J. (1969), *Class Work with Fungi*, 2nd Ed., Kew: Commonwealth Mycological Institute.

DAY, P. R. (Ed.) (1974), *Genetics of Host–Parasite Interaction*, San Francisco: Freeman Company.

DE BOER, H. A., ZHANG, Y. Z., COLLINS, C., REDDY, C. A. (1987), Analysis of nucleotide sequences of two ligninase cDNAs from a white rot filamentous fungus, *Phanerochaete chrysosporium*, *Gene* **60,** 93–102.

DELCAIRE, J. R. (1978), Economics of cultivated mushrooms, in: *Biology of Edible Mushrooms* (CHANG, S. T., HAYES, W. A., Eds.), pp. 728–793, New York: Academic Press.

DERMASTIA, M., ROZMAN, D., KOMEL, R. (1991), Heterologous transformation of *Cochliobolus lunatus*, *FEMS Microbiol. Lett.* **77,** 145–150.

EATON, D. C. (1985), Mineralization of polychlorinated biphenyls by *Phanerochaete chrysosporium:* a lignolytic fungus, *Enzyme Microb. Technol.* **8,** 209–212.

EL-SAYED, A. H. M. M. (1992), Production of penicillins and cephalosporins by fungi, in: *Handbook of Applied Mycology*, Vol. 4: *Fungal Biotechnology* (ARORA, D. K., ELANDER, R. P., MUKERJI, K. G., Eds.), pp. 517–564, New York: Marcel Dekker.

ERIKSON, K. E. (1990), Biotechnology in the pulp and paper industry, *Wood Sci. Technol.* **24,** 79–101.

ESSER, K. (1986), *Kryptogamen — Blaualgen, Algen, Pilze, Flechten*, 2nd Ed., Berlin–Heidelberg–New York: Springer.

ESSER, K., DÜVELL, A. (1984), Biotechnological exploitation of the ergot fungus (*Claviceps purpurea*), *Process Biochem.* **19,** 143–149.

ESSER, K., KUENEN, R. (1967), *Genetik der Pilze*, Berlin–Heidelberg–New York: Springer.

ESSER, K., MEINHARDT, F. (1977), A common genetic control of dikaryotic und monokaryotic fruiting in the basidiomycete *Agrocybe aegerita*, *Mol. Gen. Genet.* **155,** 113–115.

ESSER, K., MEINHARDT, F. (1986), Ectomycorrhizal fungi: state of art, application and perspectives for research under consideration of molecular biology, *Symbiosis* **2,** 125–137.

ESSER, K., MEINHARDT, F. (1987), Microorganisms — biology and genetic procedures for strain improvement, in: *Fundamentals of Biotechnology* (PRÄVE, P., FAUST, U., SITTIG, W., SUKATSCH, D. A., Eds.), pp. 17–65, Weinheim–Deerfield Beach/Florida–Basel: VCH.

ESSER, K., MOHR, G. (1986), Integrative transformation of filamentous fungi with respect to biotechnological application, *Process Biochem.* **21,** 153–159.

ESSER, K., TUDZYNSKI, P. (1978), Genetics of the ergot fungus *Claviceps purpurea.* I. Proof of a monoecious life cycle and segregation patterns for morphology and alkaloid production, *Theor. Appl. Genet.* **53,** 145–149.

ESSER, K., KÜCK, U., STAHL, U., TUDZYNSKI, P. (1983), Cloning vectors of mitochondrial origin for eukaryotes, a new concept in genetic engineering, *Curr. Genet.* **7,** 239–243.

FANCHAMPS, A. (1979), *J. Pharmacol.* **10,** 567–587.

FINCHAM, J. R. S., DAY, P. R. (1971), *Fungal Genetics*, 3rd Ed., Oxford–Edinburgh: Blackwell.

FINKELSTEIN, D. B. (1987), Improvement of enzyme production in *Aspergillus*, *A. v. Leeuwenhoek* **53,** 349–352.

FLEGG, P. B., SPENCER, D. M., WOOD, D. A. (Eds.) (1985), *The Biology and Technology of the Cultivated Mushrooms*, Chichester–New York–Brisbane–Toronto–Singapore: John Wiley & Sons.

FOSTER, J. W. (1954), Fumaric acid, in: *Industrial Fermentations*, Vol. 1 (UNDERKOFLER, L. A., HIDLEY, R. J., Eds.), pp. 470–487, New York: Chemical Publishing Company.

FRANCO, C. M. M., COUTINHO, L. E. L. (1991), Detection of novel secondary metabolites, *CRC Crit. Rev. Biotechnol.* **11,** 193–276.

FREY, D., OLDFIELD, R. J., BRIDGER, R. C. (Eds.) (1979), *A Colour Atlas of Pathogenic Fungi*, London: Wolfe Medical Publications Ltd.

GÄUMANN, E. (1964), *Die Pilze*, 2nd Ed., Basel–Stuttgart: Birkhäuser.

GAYLEY, D. (1938), Experimental spawn and mushroom culture, *Ann. Appl. Biol.* **25,** 322–340.

GOLD, M. H., KUWUHARA, M., CHIN, A. A., GLENN, J. K. (1989), Purification and characterization of an extracellular H_2O_2-requiring diaryl-propane oxygenase from the white rot basidiomycete *Phanerochaete chrysosporium*, *Arch. Biochem. Biophys.* **234,** 353–362.

GRAY, W. D. (1970), *The Use of Fungi as Food and in Food Processing*, Part II, London: Butterworth.

GROSSBARD, E. (Ed.) (1979), *Straw Decay and its Effect on Disposal and Utilization, Proc. Symp. Straw Decay Workshop Assessment Techniques*, Hatfield Polytechnic, Chichester–New York–Brisbane–Toronto: Wiley & Sons.

GUTIERREZ, S., DIEZ, B., MONTENEGRO, E., MARTÍN, J. F. (1990), Characterization of the *Cephalosporium acremonium pcp* AB gene, a large multidomain peptide synthetase: linkage to the *pcb* C gene as a cluster of early cephalosporin biosynthetic genes and evidence of multiple functional domains, *J. Bacteriol.* **173,** 2354–2365.

HAEMMERLI, S. D., LEISOLA, M. S. A., SANGLARD, D., FIECHTER, A. (1986), Oxidation of benzo(a)pyrene by extracellular ligninases of *Phanerochaete chrysosporium*, *J. Biol. Chem.* **261,** 6900–6903.

HALLINGER, S., ZIEGLER, W., WALLNÖFER, P. R., ENGELHARDT, G. (1988), Verhalten von 3,4-Dichloranilin in wachsenden Pilzkulturen, *Chemosphere* **17,** 543–550.

HAMMEL, K. E. (1992), Oxidation of aromatic pollutants by lignin degrading fungi and their extracellular peroxidases, in: *Metal Ions in Biological Systems*, Vol. 28, *Degradation of Environment Pollutants by Microorganisms and their Metalloenzymes* (SIGEL, H., SIGEL, A., Eds.), New York: Marcel Dekker.

HAMMEL, K. E., KALYANARAMAN, B., KIRK, T. (1986), Oxidation of polycyclic aromatic hydrocarbons and dibenzo-*p*-dioxins by *Phanerochaete chrysosporium* ligninase, *J. Biol. Chem.* **261,** 16948–16952.

HANEGRAAF, P. P. F., PUNT, P. J., VAN DEN HONDEL, C. A. M. J. J., DEKKER, J., YAP, W., VAN VERSEVELD, H. W., STOUTHAMER, A. H. (1991), Construction and physiological characterization of glyceraldehyde-3-phosphate dehydrogenase overproducing transformants of *Aspergillus nidulans*, *Appl. Microbiol. Biotechnol.* **34,** 765–771.

HANG, Y. D. (1989), Direct fermentation of corn to L(+)-lactic acid by *Rhizopus oryzae*, *Biotechnol. Lett.* **11,** 299–300.

HANG, Y. D., HAMMACI, H., WOODANS, E. E. (1989), Production of L(+)-lactic acid by *Rhizopus oryzae* immobilized in calcium alginate gels, *Biotechnol. Lett.* **11,** 119–120.

HARMSEN, H. J. M., KUBICEK-PRANZ, E. M., RÖHR, M., VISSER, J., KUBICEK, C. P. (1992), Regulation of 6-phosphofructo-2-kinase from the citric acid accumulating fungus *Aspergillus niger*, *Appl. Microbiol. Biotechnol.* **37,** 784–788.

HEERSBACH, G. J. M., VAN DER BEEK, C. P., VAN DIJEK, P. W. M. (1984), The penicillins: properties, biosynthesis and fermentation, in: *Biotechnology of Industrial Antibiotics* (VANDAMME, E. J., Ed.), Vol. 22, pp. 45–100, New York: Marcel Dekker.

HOFFMANN, B., REHM, H.-J. (1976), Degradation of long chain *n*-alkanes by Mucorales, I., *Eur. J. Appl. Microbiol.* **3,** 19–30.

HOSKINS, J. A., O'CALLAGHAN, N., QUEENER, S. W., CANTWELL, C. A., WOOD, J. S., CHEN, V. J., SKATRUD, P. L. (1990), Gene disruption of the *pcb* AB gene encoding ACV synthetase in *Chephalosporium acremonium*, *Curr. Genet.* **18,** 523–530.

HUDSON, H. J. (Ed.) (1986), *Fungal Biology — A Series of Student Texts in Contemporary Biology*, London: Edward Arnold Ltd.

HÜTTERMANN, A. (1989), Verwendung von Weißfäulepilzen in der Biotechnologie, *GIT Fachz. Lab. Biotechnol.* **10,** 943–945.

HÜTTERMANN, A., TROJANOWSKI, A. J., LOSKE, D. (1988), Verfahren zum Abbau schwer abbaubarer Aromaten mit ligninabbauenden Pilzen in festen Substraten, z.B. Erdboden, *Ger. Patent* 3731816.

HÜTTERMANN, A., ZADRAZIL, F., MAJCHERCZYK, (1989), Verfahren zum Dekontaminieren von sauerstoffhaltigen Gasen, insbesondere Abgasen, *Ger. Patent* 3807033.

INGOLD, C. T. (1971), *Fungal Spores*, Oxford: Clarendon.

INGOLD, C. T. (Ed.) (1973), *The Biology of Fungi*, 2nd Ed., London: Hutchinson Educational Ltd.

KELLY, J. M., HYNES, M. J. (1985), Transformation of *Aspergillus niger* by the *amd*S gene of *Aspergillus nidulans*, *EMBO J.* **4,** 475–479.

KING, R. C. (Ed.) (1974), *Handbook of Genetics*, Vol. 1, New York–London: Plenum Press.

KIRK, T. K., FARRELL, R. L. (1987), Enzymatic "combustion": the microbial degradation of lignin, *Annu. Rev. Microbiol.* **41,** 465–505.

KOBEL, H., SANGLIER, J.-J. (1986), Ergot alkaloids, in: *Biotechnology*, Vol. 4 (REHM, H.-J., REED, G., Eds.), pp. 569–609, Weinheim–Deerfield Beach/Florida–Basel: VCH.

KOCH, W. J. (1972), *Fungi in the Laboratory, a Manual and Text*, 2nd Ed., Carolina: Student Stores.

KÖHLER, A., JÄGER, A., WILLERSHAUSEN, H., GRAF, H. (1988), Extracellular ligninase of *Phanerochaete chrysosporium* Burdsall has no role in the degradation of DDT, *Appl. Microbiol. Biotechnol.* **29,** 618–620.

KOLOT, F. B. (1988), *Immobilized Microbial Systems: Principles, Techniques and Industrial Applications*, Melbournen/Florida: R. E. Krieger Publishing Company.

KÖRNER, S., WOLF, K. H., PECINA, H. (1991), Ein Beitrag zur Modellierung des Wachstums von Basidiomyceten bei der Solid-State-Fermentation, *BIOforum* **14,** 346–349.

KREISEL, H. (1969), *Grundzüge eines natürlichen Systems der Pilze*, Vaduz: J. Cramer.

KUBICEK, C. P., RÖHR, M. (1986), Citric acid fermentation, *CRC Crit. Rev. Biotechnol.* **3,** 331–373.

KÜCK, U., MOHR, G., MRACEK, M. (1989), The 5′-sequence of the isopenicillin N-synthase gene (*pcbc*) from *Cephalosporium acremonium* directs the expression of the prokaryotic hygromycin B phosphotransferase gene (*hph*) in *Aspergillus niger*, *Appl. Microbiol. Biotechnol.* **31,** 358–365.

LANGE, L. (Ed.) (1988), Zoosporic Plant Pathogens — with Main Emphasis on the Obligate Species, *PhD Thesis*, University of Copenhagen.

LELLEY, J., SCHMAUS, F. (Eds.) (1976), *Pilzanbau*, Vol. 12, *Handbuch des Erwerbsgärtners*, Stuttgart: Eugen Ulmer.

LILLY, V. G., BARNETT, H. L. (1951), *Physiology of the Fungi*, New York: McGraw-Hill.

LINCOFF, G., MITCHELL, D. H. (1977), *Toxic and Hallucinogenic Mushroom Poisoning — A Handbook for Physicians and Mushroom Hunters*, New York–Cincinnati–Atlanta–Dallas–San Francisco–London–Toronto–Melbourne: van Nostrand Reinhold Company.

MALMBERG, L. H., HU, W. S. (1992), Identification of rate limiting steps in cephalosporin C biosynthesis in *Cephalosporium acremonium:* a theoretical analysis, *Appl. Microbiol. Biotechnol.* **38,** 122–128.

MARMEISSE, R., GAY, G., DEBAUD, J. C., CASSELTON, L. A. (1992), Genetic transformation of the symbiotic basidiomycete fungus *Hebeloma cylindrosporum*, *Curr. Genet.* **22,** 41–46.

MEINHARDT, F. (1980), Untersuchungen zur Genetik des Fortpflanzungsverhaltens und der Fruchtkörper- und Antibiotikabildung des Basidiomyceten *Agrocybe aegerita*, *Bibliotheca Mycologica* **56,** Vaduz: J. Cramer.

MEINHARDT, F., ESSER, K. (1981), Genetic studies of the basidiomycete *Agrocybe aegerita*. II. Genetic control of fruit body formation and its practical implication, *Theor. Appl. Genet.* **60,** 265–268.

MILESKI, G. J., BUMPUS, J. A., JUREK, M. A., AUST, S. D. (1988), Biodegradation of pentachlorophenol by the white-rot fungus *Phanerochaete chrysosporium*, *Appl. Environ. Microbiol.* **54,** 2885–2889.

MOORE, D., CASSELTON, L. A., WOOD, D. A., FRANKLAND, J. C. (Eds.) (1985), *Developmental Biology of Higher Fungi. Symposium British Mycological Society*, Manchester, 1984, Cambridge–London–New York–New Rochelle–Melbourne–Sydney: Cambridge University Press.

MORGAN, P., LEWIS, S. T., WATKINSON, R. J. (1991), Comparison of abilities of white-rot fungi to mineralize selected xenobiotic compounds, *Appl. Microbiol. Biotechnol.* **34,** 693–696.

MÜLLER, E., LÖFFLER, W. (1971), *Mykologie*, 4th Ed., Stuttgart: Georg Thieme.

NEWTON, G. G. F., ABRAHAM, E. P. (1955), Cephalosporin C, a new antibiotic containing sulfur and D-α-aminoadipic acid, *Nature* **175,** 548–556.

OSIEWACZ, H. D., WEBER, A. (1989), DNA mediated transformation of the filamentous fungus *Curvularia lunata* using a dominant selectable marker, *Appl. Microbiol. Biotechnol.* **30,** 375–380.

PENG, M., SINGH, N. K., LEMKE, P. A. (1992), Recovery of recombinant plasmids from *Pleurotus ostreatus* transformants, *Curr. Genet.* **22,** 53–59.

PROSSER, J. I., TOUGH, A. J. (1991), Growth mechanisms and growth kinetics of filamentous microorganisms, *CRC Crit. Rev. Biotechnol.* **10,** 253–274.

PUNT, P., OLIVER, R. P., DINGEMANSE, M. A., POUWELS, P. H., VAN DEN HONDEL, C. A. M. J. J. (1987), Transformation of *Aspergillus* based on the hygromycin B resistance marker from *Escherichia coli*, *Gene* **56,** 117–124.

RAMBOSEK, J., LEACH, J. (1987), Recombinant DNA in filamentous fungi: progress and prospects, *CRC Crit. Rev. Biotechnol.* **6,** 357–393.

RAPER, J. R. (1966), *Genetics of Sexuality in Higher Fungi*, New York: Ronald Press.

RATLEDGE, C. (1986), Lipids, in: *Biotechnology*, Vol. 4: *Microbial Products II* (REHM, H.-J., REED, G., Eds.), pp. 185–213, Weinheim–New York–Basel–Cambridge: VCH Verlagsgesellschaft.

REHÁCEK, Z., SAJDL, P. (Eds.) (1990), *Ergot Alkaloids — Chemistry, Biological Effects, Biotechnology*, Amsterdam: Elsevier.

REHM, H.-J. (1980), *Industrielle Mikrobiologie*, 2nd Ed., Berlin–Heidelberg–New York: Springer-Verlag.

REISS, J. (Ed.) (1986), *Schimmelpilze — Lebensweise, Nutzen, Schaden, Bekämpfung*, Berlin–Heidelberg–New York–Tokyo: Springer-Verlag.

REMIO, D. V., SAUNDERS, G., BULL, A. T., HOLT, G. (1990), The genetic stability of *Penicillium chrysogenum* transformants in a fermentor, *Appl. Microbiol. Biotechnol.* **34,** 364–367.

ROBBERS, J. E. (1984), The fermentative production of ergot alkaloids, in: *Advances in Biotechnological Processes*, Vol. 3, pp. 197–239, New York: A. R. Liss.

RÖHR, M., KUBICEK, C. P., KOMINEK, J. (1983), Citric acid, in: *Biotechnology*, Vol. 3: *Microbial Products, Biomass and Primary Products* (REHM, H.-J., REED, G., Eds.), pp. 419–454, Weinheim–Deerfield Beach/Florida–Basel: Verlag Chemie.

ROMANO, A. H., BRIGHT, M. M., SCOTT, W. E. (1967), Mechanism of fumaric acid accumulation in *Rhizopus nigricans*, *J. Bacteriol.* **93,** 600–604.

RUTTKOWSKI, E., KHAN, N. Q. (1991), Filamentöse Pilze als Expressionssysteme für technisch nutzbare Enzyme, *GIT Fachz. Lab.* **12,** 1309–1315.

SAMSON, S. M., BELAGAJE, R., BLANKENSHIP, D. T., CHAPMAN, J. L., PERRY, D., SKATRUD, P. L., VAN FRANK, R. M., ABRAHAM, E. P., BALDWIN, J. E., QUEENER, S. W. (1985), Isolation, sequence determination and expression in *Escherichia coli* of the penicillin N synthetase gene from *Cephalosporium acremonium*, *Nature* **318,** 191–194.

SAMSON, S. M., DOTZLAF, J. F., SLISZ, M. L., BEKKER, G. W., VAN FRANK, R. M., VEAL, L. E., YEH, W. K., MILLER, J. R., QUEENER, S. W., INGOLIA, T. D. (1987), Cloning and expression of the fungal expandase/hydroxylase gene involved in cephalosporin biosynthesis, *Bio/Technology* **5,** 1207–1214.

SANGLARD, D., LEISOLA, M. S. A., FIECHTER, A. (1986), Role of extracellular ligninases in biodegradation of benzo(a)pyrene by *Phanerochaete chrysosporium*, *Enzyme Microb. Technol.* **8,** 209–212.

SCHREFERL-KUNAR, G., GROTZ, M., RÖHR, M., KUBICEK, C. P. (1989), Increased citric acid production by mutants of *Aspergillus niger* with increased glycolytic capacity, *FEMS Microbiol. Lett.* **59,** 297–300.

SKATRUD, P. L., TIETZ, A. J., INGOLIA, T. D., CANTWELL, C. A., FISHER, D. L., CHAPMAN, J. L., QUEENER, S. W. (1989), Use of recombinant DNA to improve production of cephalosporin C by *Cephalosporium acremonium*, *Bio/Technology* **7,** 477–485.

SMIT, R., TUDZYNSKI, P. (1992), Efficient transformation of *Claviceps purpurea* using pyrimidine auxotrophic mutants: cloning of the OMP decarboxylase gene, *Mol. Gen. Genet.* **234,** 297–305.

SMITH, G. (Ed.) (1969), *An Introduction to Industrial Mycology*, 6th Ed., London: Edward Arnold Publishers.

SMITH, L. L. (1984), Steroids, in: *Biotechnology*, Vol. 6a: *Biotransformations* (REHM, H.-J., REED, G., Eds.), pp. 31–78, Weinheim–Deerfield Beach/Florida–Basel: Verlag Chemie.

SMITH, J. E., BERRY, D. R. (Eds.) (1974), *Biochemistry of Fungal Development — An Introduction*, London–New York: Academic Press.

SMITH, J. S., BULL, J. H., EDWARDS, J., TURNER, G. (1989), Amplification of the isopenicillin N synthetase gene in a strain of *Penicillium chrysogenum* producing high levels of penicillin, *Mol. Gen. Genet.* **216,** 492–497.

SOČIČ, H., GABERC-POREKAR, V. (1992), Biosynthesis and physiology of ergot alkaloids, in: *Handbook of Applied Mycology*, Vol. 4, *Fungal Biotechnology* (ARORA, D. K., ELANDER, R. P., MUKHERJI, H. G., Eds.), pp. 475–516, New York: Marcel Dekker.

STEVENS, R. B. (Ed.) (1974), *Mycology Guidebook*, Seattle–London: University of Washington Press.

STEVENSON, G. (Ed.) (1970), *The Biology of Fungi, Bacteria and Viruses — A Series of Student Texts in Contemporary Biology*, 2nd Ed., London: Edward Arnold Ltd.

TIEN, M., TU, C. P. D. (1987), Cloning and sequencing of a cDNA for a ligninase from *Phanerochaete chrysosporium*, *Nature* **362,** 520–523.

TILBURN, J., SCAZZOCHIO, T., TAYLOR, G. G., ZABICKY-ZISSMAN, J. H., LOCKINGTON, R. A., DAVIES, R. W. (1983), Transformation by integration in *Aspergillus nidulans*, *Gene* **26,** 205–221.

TIMBERLAKE, W. E. (Ed.) (1985), Molecular genetics of filamentous fungi. *Proc. UCLA Symp.*, Keystone, Colorado–New York: Alan R. Liss Inc.

TUDZYNSKI, P., ESSER, K. (1986), Extrachromosomal genetics of *Claviceps purpurea*. II. Plasmids in various wild strains and integrated sequences in mitochondrial genome DNA, *Curr. Genet.* **10,** 463–467.

TUDZYNSKI, P., ESSER, K., GRÖSCHEL, H. (1982), Genetics of the ergot fungus. II. Exchange of genetic material via meiotic recombination, *Theor. Appl. Genet.* **61,** 97–100.

TURGEON, B. G., GARBER, R. C., YODER, O. C. (1985), Transformation of the fungal maize pathogen *Cochliobolus heterostrophus* using the *Aspergillus nidulans amd*S gene, *Mol. Gen. Genet.* **201,** 450–453.

TURNER, W. B. (1971), *Fungal Metabolites*, London–New York: Academic Press.

VAN ENGELENBURG, F., SMIT, R., GOOSEN, T., VAN DEN BROEK, H., TUDZYNSKI, P. (1989), Transformation of *Claviceps purpurea* using a bleomycin resistance gene, *Appl. Microbiol. Biotechnol.* **30,** 364–370.

VAN GRIENSVEN, L. J. L. D. (Ed.) (1991), Genetics and breeding of *Agaricus*, *Proc. 1st Int. Seminar on Mushroom Science*, Wageningen, The Netherlands: Pudoc.

WARIISHII, H., VALLI, K., GOLD, M. H. (1989), Oxidative cleavage of a phenolic diaryl propane lignin model dimer by manganese peroxidase from *Phanerochaete chrysosporium*, *Biochemistry* **28,** 6017–6023.

WEBSTER, J. (1980), *Introduction to Fungi*, 2nd Ed., Cambridge: University Press.

WEETE, J. D. (1980), *Lipid Biochemistry of Fungi and Other Organisms*, New York: Plenum Press.

WEETE, J. D. (1984), *Lipid Biochemistry*, New York: Plenum Press.

WILMANSKA, D., MILCZAREK, K., RUMIJOWSKA, A., BARTNICKA, K., SEDLACZEK, L. (1992), Elimination of by-products in 11β-hydroxylation of substance S using *Curvularia lunata* clones regenerated from NTG-treated protoplasts, *Appl. Microbiol. Biotechnol.* **37,** 626–630.

WÖSTEMEYER, J., BURMESTER, A. (1987), Genetische Manipulation von Pilzen, *Biol. Unserer Zeit* **17,** 114–121.

XU, D. B., MADRID, C. P., RÖHR, M., KUBICEK, C. P. (1989), The influence of type and concentration of the carbon source on production of citric acid by *Aspergillus niger*, *Appl. Microbiol. Biotechnol.* **30,** 553–558.

ZADRAZIL, F. (1977), The conversion of straw into feed by basidiomycetes, *Eur. J. Appl. Microbiol.* **4,** 273–281.

ZHANGH, J., DEMAIN, A. L. (1992), ACV synthetase, *CRC Crit. Rev. Biotechnol.* **12,** 245–260.

16 Bacteriophages

HEINRICH SANDMEIER
JÜRG MEYER
Basel, Switzerland

1 Introduction 545
2 Fundamentals 546
2.1 What is a Bacteriophage? 546
2.2 Morphology, Structure of Nucleic Acids and Classification 547
2.3 Host Receptors 552
3 Physiology of Phage Reproduction 552
3.1 Phage Adsorption and Injection of Nucleic Acid 552
3.2 Impact of Phage Infection on Host Cells 553
3.3 Replication of Phage Genomes 554
3.4 Expression of Phage Genes 554
3.5 Morphogenesis and DNA Packaging 557
3.6 Release of Progeny Phage 557
4 Lysogeny 558
5 Horizontal Gene Transfer and Recombination 559
5.1 Generalized and Specialized Transduction 559
5.2 Bacterial Defense Mechanisms and Counteracting Phage Functions 559
5.3 Evolution of Phage Genomes 560
6 Practical Aspects 560
6.1 Methods for Detection of Phages 560
6.2 Characterization of Phages 561
6.3 Phage Control 561
7 Description of Specific Phages 562
7.1 *E. coli* Phages 562
7.1.1 The Temperate dsDNA Phage Lambda 562
7.1.2 The Temperate dsDNA Phage Mu 562
7.1.3 The Temperate dsDNA Phage P1 563
7.1.4 The Virulent dsDNA Phage T4 563
7.1.5 The Filamentous ssDNA Phages M13, fd, and f1 564
7.1.6 The ssRNA Phages 565

7.2 Phages of Methanogens 565
7.3 Phages of Methylotrophs 566
7.4 Phages of Clostridia 566
7.5 Phages of Lactic Acid Bacteria 566
7.6 Phages of Bacilli 567
7.6.1 Temperate dsDNA Phages of *Bacillus subtilis* 567
7.6.2 Virulent dsDNA Phages of *Bacillus subtilis* 567
7.6.3 Other *Bacillus* Phages 568
7.7 Phages of Pseudomonads 568
7.8 Actinophages 568
7.9 Phages of Coryneform Bacteria 568
8 References 569

1 Introduction

About 20 years after the discovery of plant and animal viruses as a new class of infectious agents, viruses infecting bacteria were discovered by TWORT (1915) and D'HÉRELLE (1917), probably independently (DUCKWORTH, 1976). While TWORT determined a "glassy transformation" of colonies of micrococci to be a consequence of viral infection, D'HÉRELLE observed lysis of *Shigella* cells. He coined the name bacteriophage and proposed to use phages as therapeutic agents against bacterial infections. This stimulated phage research in a wide range of bacterial taxa, particularly in those of medical importance. But in spite of a vast literature – RAETTIG (1958) cited over 5600 publications for the period 1917 to 1956 – progress towards therapeutic applications and a better understanding of the nature and life cycle of phages was rather slow.

A new period started, when in the early nineteen forties a group of scientists animated by M. DELBRÜCK, known as "the phage group", restricted their cooperative research to a small number of phages (T1 to T7) of a single host organism, *Escherichia coli*, and annually organized a course on bacteriophages in Cold Spring Harbor. This cooperation revealed the potential of these simple model systems and catalyzed directly the emergence of molecular biology. Many of the key discoveries in molecular biology were made by exploiting phages and their hosts (CAIRNS et al., 1966; LEDERBERG, 1987). Phages and functional genetic units derived from them have played an important role at the origin of genetic engineering and still represent work horses for genetic analyses. Phage genomes were among the first genetic units to be entirely sequenced.

An old observation by BURNET and LUSH (1936) linked a phenotypic trait, pigment change of staphylococci, to phage lysogenization. Related phenomena are also of medical relevance, as they concern pathogenic bacteria. The production of toxins is an important virulence factor by which bacteria cause human disease. Some toxin genes are located on phage genomes and are transferred into nontoxigenic bacteria upon infection or lysogenization (see Sect. 4): Toxin production by *Corynebacterium diphtheriae* results from lysogenic conversion (FREEMAN, 1951) by the β-family of corynephages or the unrelated phage δ. The erythrogenic toxin of *Streptococcus pyogenes*, the etiologic agent of scarlet fever, is phage-encoded. Similarly, the production of the shigalike toxin by enteropathogenic *E. coli*, a staphylococcal enterotoxin and a botulinus toxin are all phage-associated. In neither of these cases is the toxin gene essential for phage reproduction (BISHAI and MURPHY, 1988).

Another aspect of medical importance is the transfer of drug resistance determinants by phages (transduction, Sect. 5) among clinically important, particularly Gram-positive bacteria. Furthermore, phages with a narrow host range have been used in diagnostic laboratories for the identification of bacterial strains. This phage typing has been very helpful for epidemiological purposes, and it is complemented today by gene technology procedures. D'HÉRELLE's idea to use phages in the therapy of bacterial infections has not been substantiated satisfactorily due to the rapid emergence of phage-resistant bacterial strains. It was almost totally abandoned when a multitude of antibiotics became available. Only recently the efficacy of phage therapy in gastrointestinal infections was carefully re-evaluated and its use advocated as an efficient means to reduce the number of bacteria which carry multiple drug resistance (SMITH et al., 1987).

Another motivation for research on phages was their often devastating role in the manufacture of fermented dairy products (WHITEHEAD and COX, 1935; WHITEHEAD, 1953) and in industrial fermentation processes for solvents or antibiotic production (MCCOY et al., 1944; RUDOLPH, 1978). A number of strategies and procedures have been developed to avoid phage infections in industrial fermentations, which will be summarized in Sect. 6. While initial research had focused on the negative effects of phages, recent efforts try to exploit phages for the improvement of bacterial production strains.

Progress in phage research has been extremely rapid in the last 20 years and cannot be presented comprehensively in this chapter. Instead, the reader is referred to recent monographs (ACKERMANN and DUBOW, 1987; GOYAL et al., 1987; CALENDAR, 1988; KLAUS et al.,

1992; WEBSTER and GRANOFF, 1993) which contain detailed descriptions of a large number of bacteriophages. Here, we will first briefly outline the general characteristics of bacteriophages. A discussion of some practical aspects is then followed by a description of specific phages of *E. coli* and the bacterial groups important in biotechnology. These host bacteria are presented in other chapters of this volume, where additional information may be found.

2 Fundamentals

2.1 What is a Bacteriophage?

Viruses are obligate cellular parasites which for replication depend on the metabolic activities of their host cell. Virus particles, called virions, contain either a DNA or an RNA genome packaged into a protein shell, called the capsid. Upon infection of a suitable host cell, the viral genome directs the host's biosynthetic machinery to produce progeny virions identical to the parental virion which are then liberated and are able to propagate in new hosts.

Bacterial viruses, commonly called bacteriophages or phages, fulfill the criteria of this virus definition. They are distinguished from other viruses by their host range: they infect only bacteria and propagate within them. Infection starts with the specific interaction, called adsorption, between parts of the viral capsid and a particular component of the bacterial cell surface, called the receptor (Fig. 1). The next step involves the transfer of the viral genome across the bacterial cell wall leaving all or most of the capsid outside. This separation of nucleic acid from the capsid is a prerequisite for a successful infection. The components of the progenote virus particles are synthesized by the host metabolic machinery, which is to a variable degree brought under the genetic control of the virus genome. The components are assembled in an ordered fashion into complete particles which are released altogether by lysis of the host or, in a few cases, one by one from intact cells.

Usually, a phage is specific for a bacterial genus or species which represent the host range of this phage. Phages with a narrow host range are able to replicate in only a limited number of strains within the taxon.

Bacteriocins are biologically active molecules which resemble bacteriophages in some aspects (KONISKY, 1982; LURIA and SUIT,

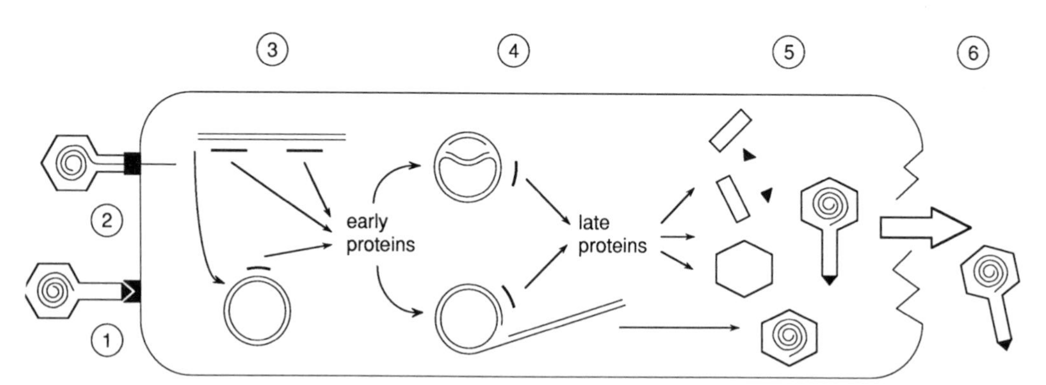

Fig. 1. Typical stages of the replication cycle of large dsDNA phages. The specific interaction of a viral component with the receptor on the bacterial surface (1) precedes irreversible adsorption and subsequent transfer of the viral DNA into the cell (2). Early proteins are synthesized from mRNA (thick lines) transcribed from both linear dsDNA and circularized dsDNA (3). Then replication of viral DNA is initiated resulting in θ-form and σ-form intermediates (4). Concomitant late transcription yields mRNA (thick lines) for late proteins. Tail and head structures are assembled separately from late proteins, and viral DNA is packaged into heads (5). After assembly of complete virions, the cell is lysed and progeny particles are released (6).

1987). Bacteriocins are proteins produced by bacteria which kill or inhibit other strains of the same or closely related bacterial species. Bacteria producing a bacteriocin are specifically immune to this bacteriocin, but not to others. Bacteriocins are distinct from antibiotics, which have a wider spectrum of activity. Bacteriocins are produced by a large variety of bacterial cells and are named according to the species producing them, e.g., *Escherichia coli* produce colicins, *Bacillus subtilis* produce subtilisins and so forth. Bacteriocin-sensitive bacteria carry receptors with which the bacteriocin has to interact before it can exert its action. The killing activity of bacteriocins is due to disruption of an essential cellular function: some cause membrane depolarization resulting in leakage, others act as a DNA endonuclease or inactivate ribosomes. Often the genes for bacteriocins are located on plasmids.

2.2 Morphology, Structure of Nucleic Acids and Classification

Phages are widely distributed among the bacterial phyla, but most of them have been isolated from a few well studied groups of bacteria. About 3000 phage isolates have been analyzed by electron microscopy (ACKERMANN, 1987). The simplest virions are composed of a single or very few capsid protein(s) arranged in an icosahedron or a helical fila-

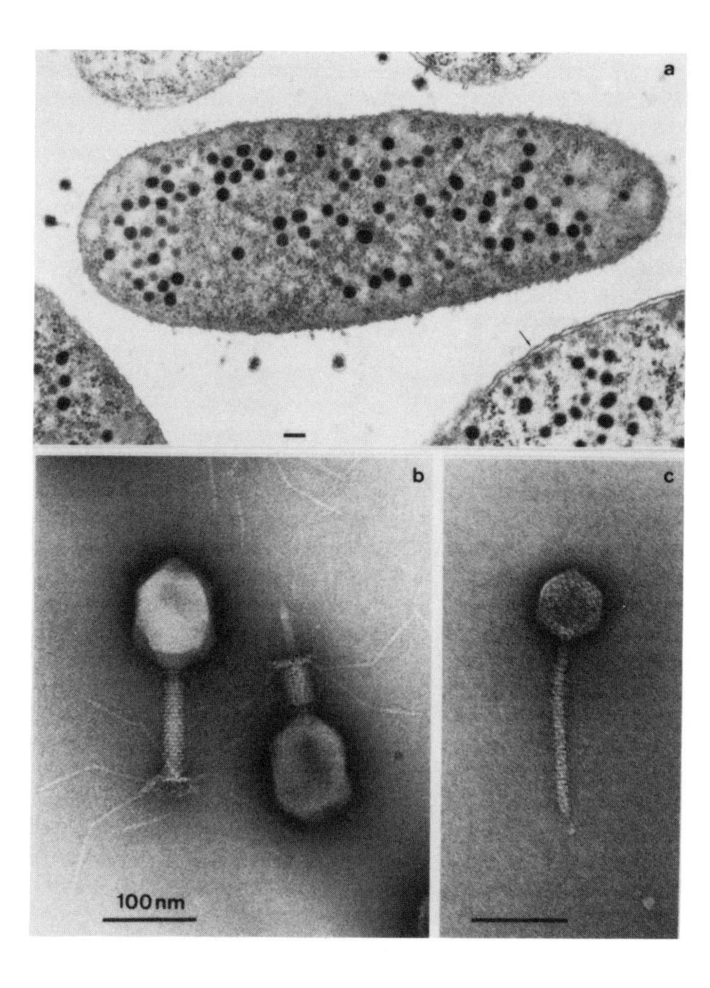

Fig. 2. Electron micrographs of representative phages: (a) Section through *Escherichia coli* cells at late stages of phage T4 infection. Infecting virions are attached to the cell surface, and the cytoplasm contains particles representing various stages of morphogenesis. The arrow points to a scaffold. Courtesy of M. MAEDER and M. WURTZ. (b) Phage T4 particles, one with a contracted tail. (c) Phage lambda. Bars represent 100 nm. Courtesy of M. WURTZ.

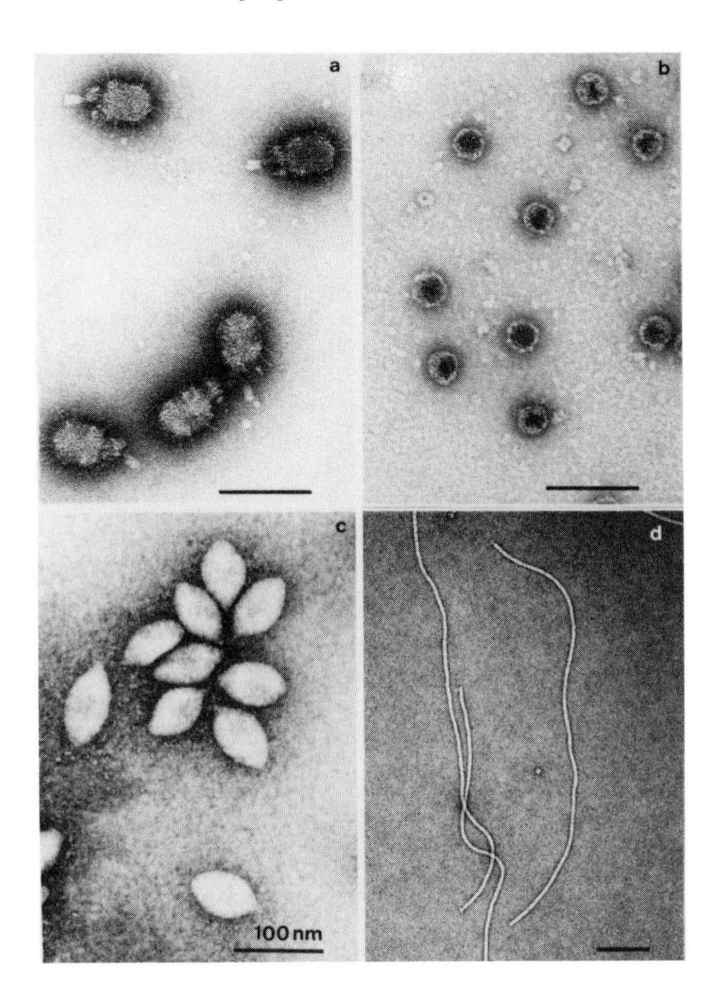

Fig. 3. Electron micrographs of representative phages: (a) Phage ø29 of *Bacillus subtilis*. (b) The ssRNA phage Qβ. (c) *Sulfolobus shibatae* phage SSV1. Courtesy of P. JANEKOVIC and W. ZILLIG. (d) The filamentous ssDNA phage fd. Bars represent 100 nm. Courtesy of M. WURTZ.

ment. However, the most typical phage structure is that of a tailed phage, composed of a regular or elongated icosahedral head, containing double-stranded DNA (dsDNA) of between 19 kb and 200 kb, and a tail which may have appendages like fibers or whiskers. The tail is either contractile or non-contractile and may be flexible. These complex virions are composed of many (perhaps 50) different head and tail proteins. High-resolution electron microscopy combined with image processing has revealed the molecular organization of many model phages (WURTZ, 1992).

A lipid envelope is a rather unusual structural component among phages (MINDICH and BAMFORD, 1988). Several novel morphotypes have been discovered among bacteriophages and phage-like particles from archaebacteria (ZILLIG et al., 1986, 1988). A few prominent representatives of some of these phage groups are shown in Figs. 2 and 3. Filamentous and cubic phages are rarely isolated and are restricted to a few bacterial groups. The majority of isolates are tailed phages from diverse bacterial taxa (Tab. 1) .

A classification scheme developed by BRADLEY (1967) and refined by ACKERMANN and EISENSTARK (1974) is primarily based on morphological criteria of the virion and properties of the nucleic acid (Fig. 4 and Tab. 2). The International Committee on Taxonomy of Viruses (ICTV) has defined criteria for the classification of phages and approved twelve families and 16 genera including several hun-

Tab. 1. Distribution of Tailed Phages by Host Groups (from ACKERMANN, 1987)

Host Genus or Group	Phages Surveyed	Phages Classified	Phage "Species"
Actinomycetes	308	149	35
Aeromonas	34	22	9
Agrobacterium	39	37	5
Bacillus	306	153	24
Brucella	45	45	1
Clostridium	152	36	5
Cyanobacteria	18	18	9
Enterobacteria	569	342	30
Gram-positive cocci	465	239	31
Lactobacillus	69	39	9
Listeria	49	45	5
Mycoplasmas	7	4	3
Pasteurella	25	23	4
Pseudomonads	175	96	22
Rhizobium	101	50	14
Vibrio	89	62	16
Total	2451	1360	222

dred species (FRANCKI et al., 1991). Problems arise because (a) there is no agreement on what a species is, and (b) phages of different morphotypes may share genetic homology and recombine to form viable hybrids.

A more general classification scheme for viruses of eukaryotes and prokaryotes by BALTIMORE (1971) considers the strategy of viral genomes, i.e., the structure and replication of the viral nucleic acid and the formation of translatable mRNA = (+)ssRNA (Fig. 5): Viruses of *class I* carry dsDNA, and mRNA can be transcribed from both strands (all dsDNA phages are included).

The single-stranded DNA (ssDNA) genome of *class II* viruses has to be converted to a dsDNA intermediate for transcription of mRNA. Only phages with ssDNA, that is noncomplementary to the mRNA, i.e., (+)ssDNA, are known (filamentous phages). Viruses with a dsRNA genome are grouped in *class III*. Only one phage of this class is known (ø6), and its genome is segmented. The mRNA is transcribed from only one strand of each segment by an endogenous RNA polymerase.

The (+)ssRNA of the *class IV* viruses functions directly as mRNA and is replicated by a virus-encoded replicase. Phages MS2, Qβ and Sp can synthesize additional mRNA species by transcription of the (–)strand.

Phages corresponding to *class V* viruses which contain (–)ssRNA have not been found so far. The retroviruses of *class VI* carry a (+)ssRNA genome which does not function as mRNA upon infection. A virion-associated reverse transcriptase converts this RNA into a dsDNA molecule from which mRNA is synthesized. Recently, retron elements encoding reverse transcriptase activity were discovered in prokaryotes, and one was located on a phage genome in *E. coli* (INOUYE et al., 1991; INOUYE and INOUYE, 1991). However, its function, particularly with respect to phage replication, awaits further elucidation.

Phages of the same morphotype which share the same host species, a common life style, and the same genome organization have been considered phylogenetically close relatives. However, CAMPBELL and BOTSTEIN (1983) defined a phage family exclusively by the ability to exchange genetic information among its members by homologous recombination. According to this classification, phages with grossly different architecture become relatives, e.g., P22 and lambda, or kappa and theta (DOSKOČIL et al., 1988).

Tab. 2. Characteristics of the Main Phage Groups

Shape	Family[a]	Nucleic Acid[b]	Morphological and Functional Characteristics	n[c]	Representative Phage	Morphotypes[d]
Tailed	Myoviridae	dsDNA, L	Contractile tail	811	T4	A1, A2, A3 (A)
	Siphoviridae	dsDNA, L	Long, non-contractile tail	1469	λ	B1, B2, B3 (B)
	Podoviridae	dsDNA, L	Short tail	441	T7	C1, C2, C3 (C)
Cubic	Microviridae	ssDNA, C	Appendices	30	ØX174	D1 (D)
	Leviviridae	ssRNA, L		41	MS2	D2
	Corticoviridae	dsDNA, C, S	Complex structure, lipids	2	PM2	D3
	Tectiviridae	dsDNA, L	Double capsid, lipids, pseudotail	14	PRD1	D4
Cubic enveloped	Cystoviridae	dsRNA, L, seg	Protein capsid, lipid envelope, RNA polymerase	1	$\phi 6$	E
Filamentous	Inoviridae, Inovirus	ssDNA, C	Long filaments	25	fd	F1 (F)
	Inoviridae, Plectovirus	ssDNA, C	Short rods	15	L51	F2
Enveloped	Plasmaviridae	dsDNA, C, S	Lipid envelope, no detectable capsid, pleomorphic	2	L2	G
	SSV1 group		Lemon-shaped	1+[e]	SSV1	
	Lipothrixviridae	dsDNA, L	Lipid envelope, protrusions at both ends	3	TTV1	

[a] Viral families and genera of Inoviridae are as recognized by the International Committee on Taxonomy of Viruses (Francki et al., 1991).
[b] ds, double-stranded; ss, single-stranded; C, circular; L, linear; S, superhelical; seg, segmented.
[c] Number of phages computed by Ackermann (1987).
[d] Designation of morphotypes is according to Ackermann (1987), and those in parentheses are according to Bradley (1967).
[e] Additional possible members.

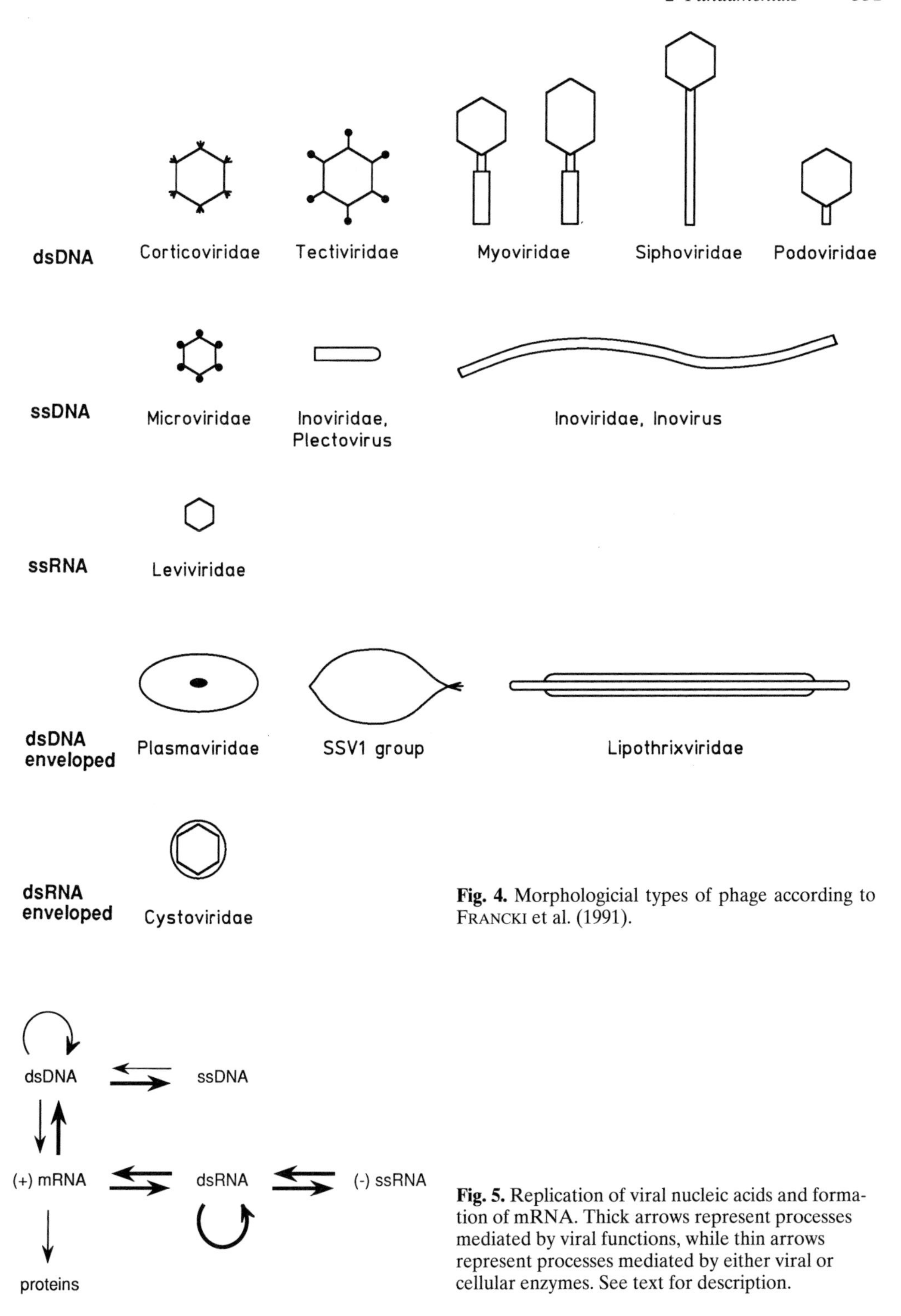

Fig. 4. Morphologicial types of phage according to FRANCKI et al. (1991).

Fig. 5. Replication of viral nucleic acids and formation of mRNA. Thick arrows represent processes mediated by viral functions, while thin arrows represent processes mediated by either viral or cellular enzymes. See text for description.

2.3 Host Receptors

The external structures of the bacterial cell are complex. They include parts of the envelope (peptidoglycan, outer membrane of Gram-negatives, capsule) and appendages (flagella, pili) and are composed of different kinds of macromolecules which serve diverse functions and are often multifunctional. Phages have evolved to specifically interact with one or the other of these surface molecules (BEUMER et al., 1984; Fig. 6; Tab. 3). This interaction is the first step in the recognition of a potential host and retains the virus at or close to the host cell surface.

Receptors in Gram-positive bacteria have not been studied as extensively as those of Gram-negative bacteria. Components of the cell wall of *Bacillus subtilis* and *Lactobacillus* sp. constitute adsorption partners for phages as well as the flagellum of the bacilli. In the outer membrane of the Gram-negative bacteria, both the lipopolysaccharides and the outer membrane proteins which form pores or specific transport systems, can serve as phage receptors (e.g., the A protein of the *E. coli* outer membrane functions for the uptake of vitamin B12, for the adsorption of colicin E and of phage BF23). Molecules of a capsule which may surround the bacterial cell, the pili adhesion organelles and flagella can also be used as receptors. Mutations altering the receptor may abolish the specific interaction, thus leading to phage resistance.

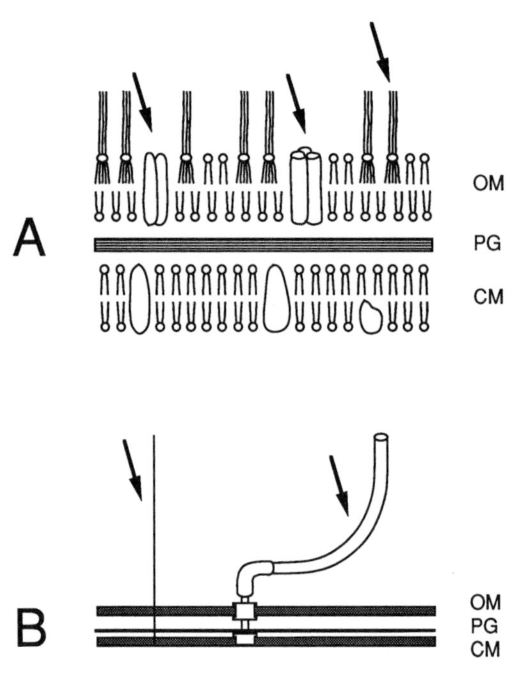

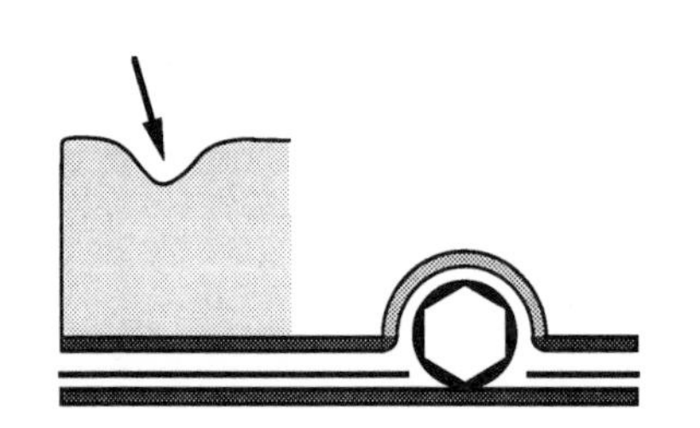

Fig. 6. Phage receptors (arrows) of the Gram-negative cell envelope and of bacterial appendages. (A) Gram-negative bacteria contain only a thin peptidoglycan layer (PG) between the cytoplasmatic membrane (CM) and the outer membrane (OM). Membrane proteins are embedded in the phospholipid-bilayer of the CM as well as in the phospholipid/lipopolysaccharide bilayer of the OM. Phages adsorb also to pili (thin bar), to flagella (B) and to capsules (C). The latter can be degraded by viral enzymes. Phages with a lipid envelope may enter the cell by fusion of their envelope with the OM (C).

3 Physiology of Phage Reproduction

3.1 Phage Adsorption and Injection of Nucleic Acid

Binding of host receptor and the complementary phage structure, which is often a specialized tail fiber in large dsDNA phages, represents a very specific interaction. Generally, a reversible step precedes the irreversible adsorption step. Environmental conditions, such as pH, ionic strength and the presence of divalent cations, influence the kinetics of this reaction. Some phages require adsorption cofactors (e.g., L-tryptophan). After irreversible binding, the phage nucleic acid is transferred to the cytoplasm. In phages with a contractile

Tab. 3. Bacterial Structures Serving as Phage Receptors (Modied from BEUMER et al., 1984)

Phage Receptor	Phage	Host
Gram-Negative Bacteria		
Outer membrane		
Protein OmpC	T4	*Escherichia coli*
Protein MalB (LamB)	λ	*E. coli*
Lipopolysaccharide	P1, Mu	*E. coli*
Pili		
F pili	M13, fd, f1,	*E. coli*
	MS2, Qβ, SP	*E. coli*
Polar pili	M6	*Pseudomonas aeruginosa*
Plasmid encoded pili	PR4	*E. coli, P. aeruginosa*
Flagellum	7-7-1	*Rhizobium lupini*
	(chi)	*E. coli*
Outer polysaccharide capsule		
Exopolysaccharide	M-1	*Rhizobium japonicum*
Glycolipoprotein	8	*P. aeruginosa*
Gram-Positive Bacteria		
Cell wall		
Teichoic acid + peptidoglycan + D-glucose moiety	SPO1, SPO2, SP50, SP3	*Bacillus subtilis*
L-Rhamnose moiety in outer polysaccharide layer	PL-1	*Lactobacillus* sp.
Flagellum	PBS1	*B. subtilis, Bacillus pumilus*

tail, interactions of the base plate triggers contraction of the tail sheath and penetration of the central tail tube.

The host cell contributes to the uptake of phage DNA, e.g., by maintaining the electrochemical potential across the cytoplasmic membrane. In some phages, proteins that are linked to the DNA or RNA, pilot the nucleic acid into the cell. In coliphage T5, only the terminal 8% of the linear dsDNA is transferred to the cytoplasm, unless genes required for transferring the rest of the DNA are expressed from this genome segment. These first steps of infection are generally poorly understood and have been investigated for only a few phages (e.g., T4, see Sect. 7.1.4).

Bacteriophages containing a lipid envelope may enter a host cell by a mechanism common among enveloped animal viruses: by membrane fusion. Phage ø6 was shown to mediate fusion between its envelope and the outer membrane of its *Pseudomonas syringae* host (BAMFORD et al., 1987; MINDICH and BAMFORD, 1988).

Viral DNA genomes can enter cells by other means: During conjugation not only plasmids and the chromosome, but also phage DNA can be transferred from one cell to another. Populations of certain bacterial taxa (particularly Gram-positive species) contain a high proportion of cells competent for DNA uptake, others (e.g., *E. coli*) can be treated to increase this proportion. The transfer of naked viral DNA by transformation is called transfection and is rather inefficient, particularly for large genomes. Once inside the cell, the viral genome can exert its activities.

3.2 Impact of Phage Infection on Host Cells

The metabolic activities of the host are not significantly altered upon infection by a filamentous ssDNA phage. These phages are reproduced continuously and released through the intact membranes. Other phages alter host metabolism drastically and direct synthesis of macromolecules largely to their own needs. Large dsDNA phages with complex genomes (e.g., T4, see Sect. 7.1.4) can replace many host enzymes by phage-encoded analogs. They may even enzymatically degrade the host genome, thus blocking host transcription and translation.

An infecting phage often prevents a succeeding infection by a phage of the same type or, rarely, by a phage of a different type. This phenomenon, called mutual exclusion or superinfection exclusion, may be accomplished by a change in the phage receptor molecule (e.g., T5 receptor) or by blocking the superinfecting phage DNA in the periplasmic space (e.g., T4 and other T-even phages). This phenomenon is different from superinfection immunity of lysogenic cells (see Sect. 4). In addition, bacteria can acquire new phenotypic traits by lysogenic conversion (Bishai and Murphy, 1988).

3.3 Replication of Phage Genomes

Most of the dsDNA phages carry a linear molecule in their head. A few of these genomes are covalently linked at their 5′ ends to a terminal protein which plays an essential role in the initiation of DNA replication (e.g., ø29, see Sect. 7.6.2). Thus, these linear phage DNAs are replicated from their ends (Salas, 1988a, b). Replication of ds phage-DNA usually starts at a specific position within the molecule, the origin of replication (Keppel et al., 1988; Wickner, 1992). From this origin, either a single replication fork proceeds in one direction (unidirectional replication), or two forks move in both directions (bidirectional). A round of replication is completed, when the replication forks reach the ends of the linear dsDNA or when the forks meet in a circular dsDNA. Several linear phage DNAs become circularized after injection by recombination between terminal repeats or by annealing of complementary single-stranded (cohesive) ends.

In later stages of replication of circular DNA molecules, a switch from a bidirectional replication with typical "θ"-form intermediates to a so-called rolling circle replication with typical "σ"-form intermediates is observed (Fig. 7). The latter replication mode is initiated by a specific cut of one DNA strand at a replication origin which is distinct from that for "θ"-replication. The free 3′ end serves then as a primer, and the displaced 5′ sequences serve as the template for dsDNA synthesis resulting in a molecule, often comprising several genomes (concatamer). Phage Mu and some phages of pseudomonads are unique among the class I viruses in that they replicate by transposition (Symonds et al., 1987; Harshey, 1988; Pato, 1989; see Sects. 7.1.2 and 7.7).

The replication of many ssDNA phages of class II (e.g., M13, see Sect. 7.1.5) proceeds in two steps (Baas and Jansz, 1988). First, the circular (+)ssDNA is converted to the circular dsDNA replicative form which is multiplied by the rolling circle mode. The (–)strand serves then as the template for synthesis of the (+)strand which is packaged into virions.

RNA phages of classes III and IV must encode their own RNA replicase, since cellular enzymes do not accept RNA as a template for the formation of new RNA strands. The phage with a dsRNA genome, ø6 (class III), carries a virion-associated replicase and cotransfers this enzyme with the nucleic acid (Mindich and Bamford, 1988). From viral (+)ssRNA genomes (see Sect. 7.1.6), the replicase gene has first to be translated, before viral RNA replication can start. One fraction of the replicated (+)ssRNA serves as mRNA, while another fraction is packaged into virions (Fiers, 1979; van Duin, 1988).

3.4 Expression of Phage Genes

Many phages have evolved sophisticated means to optimize gene expression for formation of progeny and to use the host's resources economically. As is true for prokaryotic gene expression in general, regulation at the transcriptional level is of particular importance. Many phages achieve an optimal order in gene expression by organizing genes of related functions in the same operon (i.e., the transcription unit). Most large dsDNA phages show a biphasic gene expression: an early phase before DNA replication and a late phase (see Figs. 1

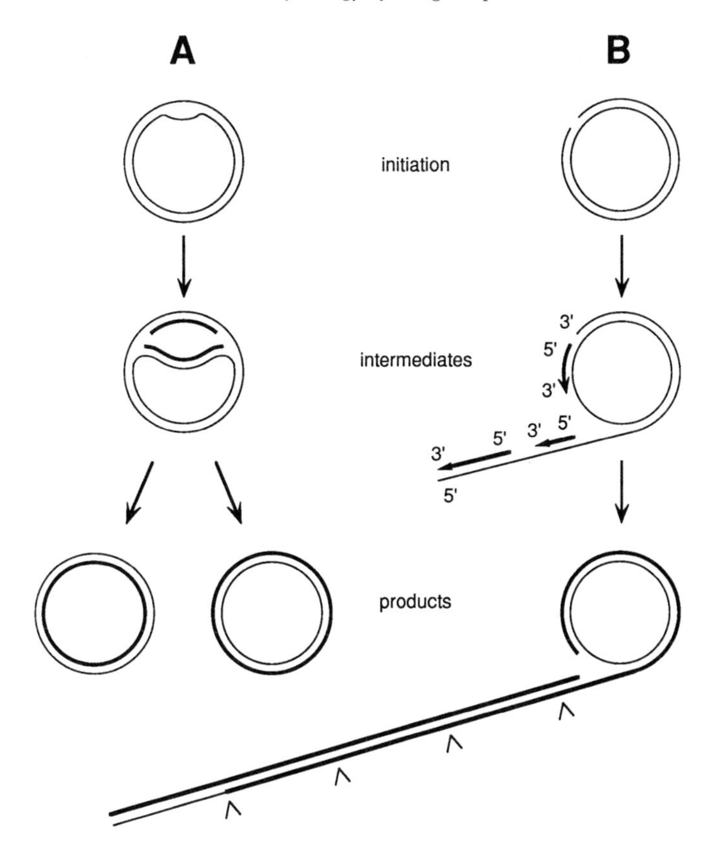

Fig. 7. Modes of replication of circular dsDNA (simplified from WATSON et al., 1987). Bidirectional replication via a θ-form intermediate results in two identical circular products (A) and rolling-circle replication with σ-form intermediates produces linear concatameric DNA (B). Newly synthesized strands are drawn as thick lines. DNA synthesis proceeding in the $5'$ to $3'$ direction is indicated in (B). Marks along the concatamers indicate one genome length (not to scale).

and 8). During the early phase, only a few genes (early genes) are activated. Often they are under control of one or a few strong promoters which have to compete for the host's RNA polymerase with a great number of cellular promoters. Early gene products are typically involved in the initiation of phage DNA replication, the regulation of late transcription, the modulation of the host metabolism and, for all temperate phages, in the decision of the lytic vs. lysogenic pathway. The late genes code for morphogenetic functions, structural proteins of the phage particle and host cell lysis. Continuing DNA replication helps to increase synthesis of late proteins by means of a gene dosage effect.

A major strategy of large dsDNA phages for regulating gene expression involves the mod-

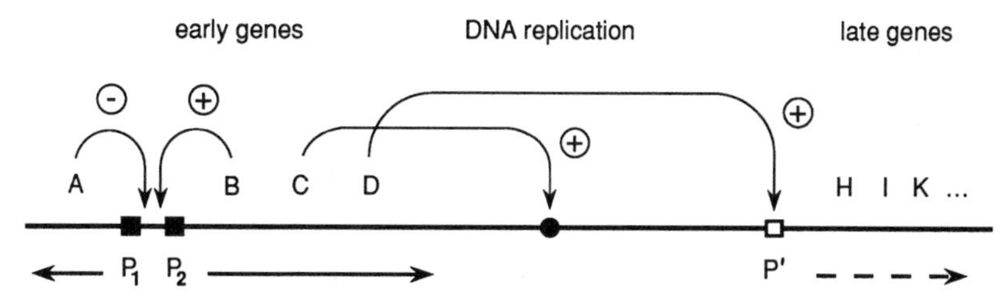

Fig. 8. Simplified model for the gene organization and regulation of gene expression of large dsDNA phages. Early genes (A, B, C, D) are transcribed from two promoters (P_1 and P_2) which are negatively and positively controlled by products of early genes. The products of other early genes activate replication of viral DNA and a promoter (P') for the transcription of late genes (H, I, K ...).

ification of the host RNA polymerase by one or several phage-encoded sigma-factor(s) (GEIDUSCHEK and KASSAVETIS, 1988). The modified form of the enzyme specifically recognizes phage promoters which differ in their sequence from host promoters. Other controls include the negative regulation by repression of transcription by a phage-encoded repressor (PTASHNE, 1987) and by extension of transcription past a termination signal mediated by a phage-encoded antiterminator (FRIEDMAN, 1988; e.g., in phage lambda, Sect. 7.1.1). The strength of a promoter, termination signals and secondary structure of phage DNA are important factors as well. Differential stability of mRNA species and processing of transcripts have also been observed (GUARNEROS, 1988).

Similar to the regulation of transcription, translation is efficiently controlled at the initiation step. Critical factors are ribosome-binding

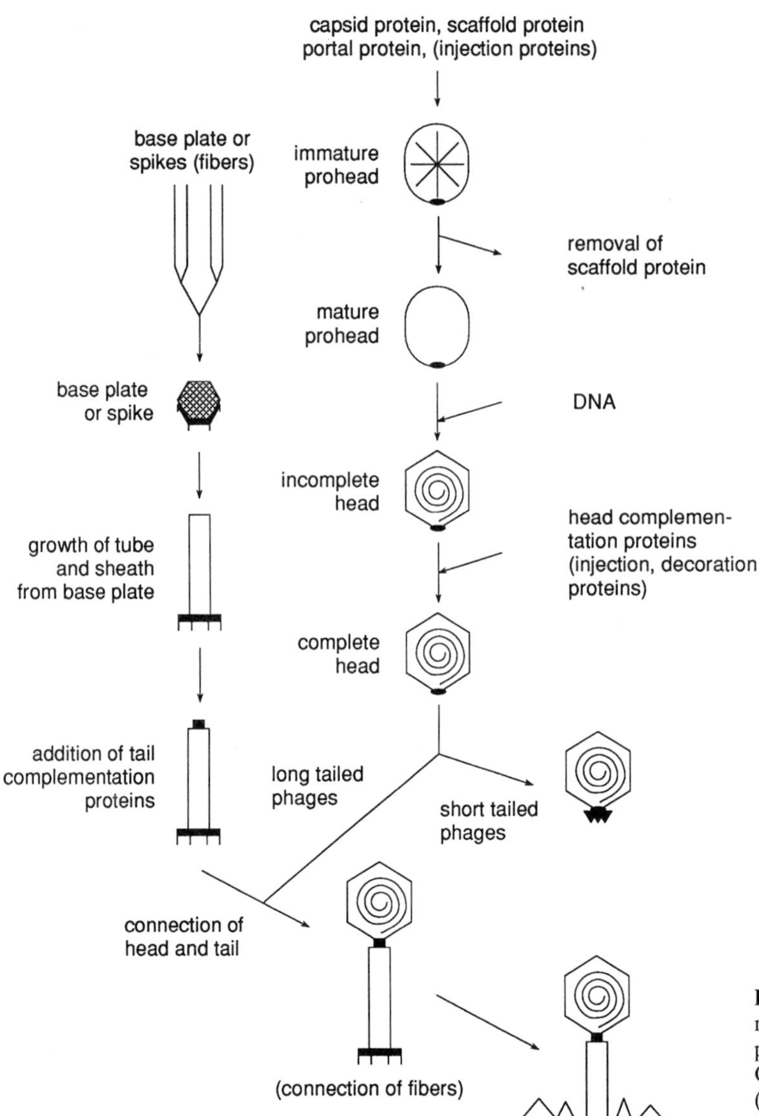

Fig. 9. General scheme for morphogenesis of dsDNA phage with tail according to CASJENS and HENDRIX (1988). Structures indicated in parentheses need not be present in all phages.

to mRNA at the Shine-Dalgarno sequence, the sequence around the initiation codon and the secondary structure of mRNA which may mask initiation signals.

3.5 Morphogenesis and DNA Packaging

A number of genetically well characterized dsDNA phages served as model systems for the study of assembly of complex structures from macromolecular units (CASJENS, 1985; CASJENS and HENDRIX, 1988). The potential use of the insights into self-assembling structures for *in vitro* packaging of recombinant DNA has propelled such studies. Fig. 9 shows a general scheme for the morphogenesis of tailed dsDNA phages. Although host-encoded proteins may play a role in phage morphogenesis, they are not incorporated into the virion as a structural entity. Hundreds of protein units encoded by a number of genes (6 in ø29 and >40 in T4) form these complex virions.

Capsid morphogenesis is initiated at the future head-tail connector or portal complex by interactions of the major capsid protein(s) around a protein scaffold. In subsequent steps, proteins are removed from the scaffold, and capsid proteins may become modified. These processes determine the size and symmetry parameters of the icosahedral head.

The empty prohead is then filled with DNA, and the stability of the capsid is increased by so-called decoration proteins which are in some cases dispensable. The DNA is packaged into the prohead intermediate in a highly condensed form (20- to 50-fold higher than the cellular nucleoid). Energy supplied by the hydrolysis of ATP is required to channel the DNA into the capsid through the portal complex (BAZINET and KING, 1985; BLACK, 1988, 1989). Newly replicated DNA of some phages has a circular or linear form of one genome length and can be packaged directly. DNA resulting from rolling circle replication is concatameric and needs to be cut into packageable pieces. This is achieved in two ways: Concatameric phage DNA is cut at specific sites to generate DNA of exactly one genome length with cohesive ends (e.g., *cos* in phage lambda, see Sect. 7.1.1). Alternatively, concatamers may be cut first at a single site (a specific *pac*-site in phage P1, see Sect. 7.1.3, or an undetermined site in phage T4, see Sect. 7.1.4), from where packaging proceeds in one direction, until the phage head is full. Then a second cut is made. By this so-called headful mechanism several heads can be filled sequentially from one concatamer. Since phage heads can take up DNA corresponding to more than one genome (112% of a genome length for P1 and 102% for T4), the genomes packaged represent a population of circularly permuted, terminally redundant DNA molecules.

Phage appendages (tails, base plates, fibers, collars) are more variable in their architecture than phage heads. They are assembled separately and then connected to the mature head. Assembly of the tail tube and the contractile or non-contractile sheath starts at the base plate. Their length is determined by a ruler or tape-measure protein (HENDRIX, 1988). The base plate, which often possesses a 6-fold symmetry, provides the attachment points for tail fibers. dsDNA phages may contain between 1 (phage lambda) and 12 (phage ø29) fibers composed of one to a few protein subunits. The tail fibers may be very short spikes or long extended filaments of up to 160 nm. Proteins of both structures, base plates and tail fibers, may be enzymatically active on host surface macromolecules.

3.6 Release of Progeny Phage

Except for filamentous phages which are assembled at the host cell membrane, phage assembly occurs intracellularly (see Figs. 1 and 2). Complete phage particles are finally released by lysis of the host cell which is directly or indirectly accomplished by phage-encoded proteins (YOUNG, 1992). Practical use of cellular autolysins and phage lysis systems has been made for disruption of *E. coli* cells (DABORA and COONEY, 1990). The latency period and number of progeny virions (burst size) may vary considerably and depend on the phage-host combination and the physiological conditions.

4 Lysogeny

After infecting a cell, virulent phages readily produce progeny phages. Temperate large dsDNA phages can also undergo a lytic infection with ultimate vegetative phage production and cell lysis. Alternatively, the functions for phage reproduction can be repressed by a phage-encoded repressor and the cell survives the infection (Fig. 10). The repressed phage genome, called prophage, is stably maintained in such lysogenic cells for many generations. The prophage can enter the lytic cycle (induction) spontaneously, or induction can be stimulated by DNA damaging agents, such as UV-light irradiation, X-rays or mitomycin C treatment. A prophage which is transferred to a new host cell by plasmid-mediated conjugation can be induced in the recipient cell (zygotic induction) due to the lack of repressor molecules.

Most prophages (e.g., phage lambda, see Sect. 7.1.1) integrate into the host chromosome and are coreplicated with it. This recombination usually occurs between a specific *att*P site on the phage genome and a preferred *att*B site on the host chromosome, and it is mediated by one or a few phage proteins in concert with several host functions. Other prophages (e.g., phage P1, see Sect. 7.1.3) assure replication and proper segregation to daughter cells by replicating autonomously as a plasmid. In a lysogen only few regulatory viral genes are expressed which are required for maintaining the prophage state. Usually repressor molecules act on operator sites to block transcription of vegetative functions (PTASHNE, 1992). Since the repressor protein also acts on operators of a superinfecting phage of the same kind or a similar phage with closely related operators, they block expression of the second infecting phage. Thus, the lysogen is immune to superinfection (homoimmunity). However, related phages with altered operators (heteroimmune phages) or totally dissimilar phages can still infect a lysogen and hence, cells can carry several prophages at the same time.

A bacterium can acquire new phenotypes by phage-encoded genes (lysogenic conversion). Examples are the *Bacillus subtilis* phage SPβ which is responsible for betacin production, and *Corynebacterium diphtheriae* phages which encode the diphtheria toxin (BISHAI and MURPHY, 1988). Prophages are subjected to

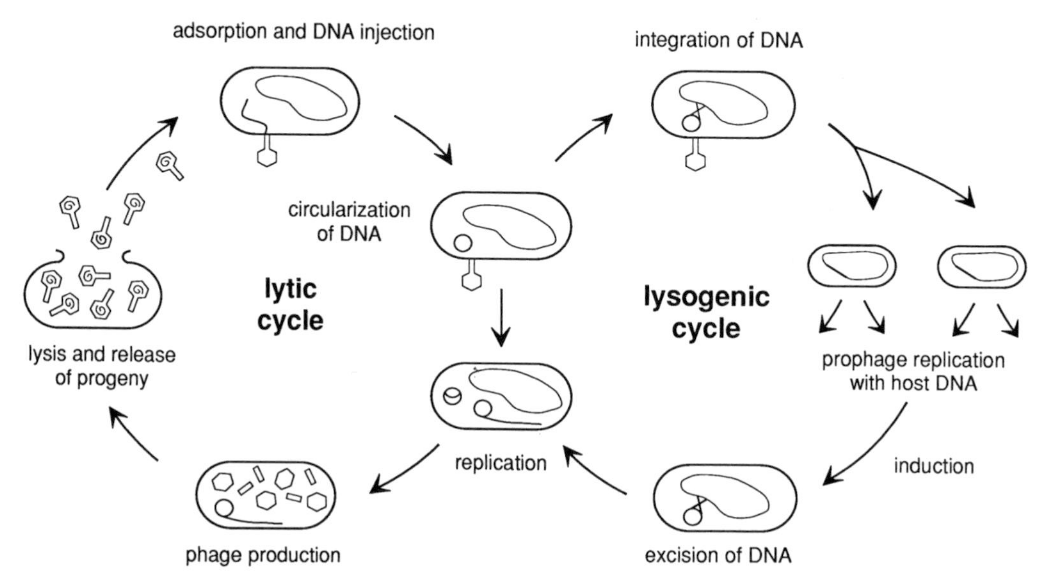

Fig. 10. Cycles of a temperate phage. After infection, the phage genomes enter the lytic cycle in one portion of the bacterial population, and in another portion the viral DNA is maintained as a prophage and inherited to the cellular progeny. Prophage induction results in lytic replication.

mutational events, some of which may distort or inactivate their functional integrity. Such defective prophages may still have phenotypic effects and participate in recombinational events promoting phage evolution (CAMPBELL, 1988). Some may be induced resulting in the production of some phage proteins, perhaps even incomplete phage particles and/or cell lysis.

A culture of lysogenic bacteria always contains free phages due to spontaneous prophage induction in a small fraction of cells. The presence of free phages is used as an indication of lysogenic bacteria by plating on a lawn of sensitive indicator bacteria (see Sect. 6.1). The spontaneous loss of a prophage occurs at a low frequency (e.g., 10^{-6}/cell and generation for phage lambda), but can be stimulated by UV-light irradiation (curing). From such cured cells, a culture free of phages can be obtained.

Pseudolysogenic bacteria carry phages for many generations without lysing. This so-called carrier state does not confer immunity and requires repetitive reinfection. A culture of pseudolysogens can be freed of phages by treatment with antiphage serum.

5 Horizontal Gene Transfer and Recombination

5.1 Generalized and Specialized Transduction

During phage reproduction, bacterial DNA is packaged into the head of some phages at a low frequency and is subsequently injected into a host bacterium. The process by which phages transfer a genetic trait to a recipient cell is called transduction. Two kinds of transducing phages – generalized and specialized transducers – are distinguished. Specialized transducing phages contain bacterial DNA covalently linked to viral DNA (WEISBERG, 1987). These phages are viable and replicate either autonomously or in mixed infection with a helper phage. Usually, only a few genes of the host chromosome can be transduced. Such phages generally integrate into the chromosome of the lysogen and mobilize the genes flanking the integration site (e.g., phage lambda, see Sect. 7.1.1).

In generalized transduction (MARGOLIN, 1987; STERNBERG and MAURER, 1991), phage heads usually contain only bacterial DNA randomly packaged from any chromosomal location. As these particles do not contain any phage DNA, they are non-infectious, i.e., they inject the DNA but it is not replicated. The prototype of a generalized transducing phage is P1 (see Sect. 7.1.3) which can package up to 2% of the *Escherichia coli* chromosome.

Generalized transducing phages can be used as a tool for fine mapping studies of bacterial chromosomes as well as for *in vivo* engineering of bacterial strains. If the transducing DNA fragment is not integrated into chromosomes of the recipient, it is lost, but may be expressed transiently (abortive transduction). In a phenomenon called plasmid transduction, plasmid DNA is packaged into transducing phages.

5.2 Bacterial Defense Mechanisms and Counteracting Phage Functions

Phage infection may be abortive due to mutual exclusion or homoimmunity against superinfection as well as for a number of other reasons. Cell wall components can mask phage receptors, or a mutation of the host's receptor gene may result in structurally altered receptors. Such point mutations can be readily counterbalanced by natural selection of mutations in the phage adhesion genes.

Another frequent cause for low efficiency of infection are host-encoded restriction and modification systems (BICKLE, 1987; WILSON and MURRAY, 1991). The restriction enzyme can cleave DNA endonucleolytically after recognition of a short specific sequence, while the modification function enzymatically methylates single residues within the same recognition sequence. The host DNA is fully modified and, therefore, protected from restriction. However, infecting phage DNA, which has been replicated in a host lacking the same restriction and modification system as the infected cell, is degraded with high efficiency. Hundreds of enzymes of one class of restriction enzymes, the type II endonucleases, have been isolated from diverse bacteria and are indispensable tools for *in vitro* genetic engineering (ROBERTS, 1990).

Phages have evolved several diverse means by which they evade host restriction (KRÜGER and BICKLE, 1983). Some phages' DNA is modified by virus-encoded methylases (e.g., *Bacillus* phages, see Sect. 7.6), and other phage DNA includes unusual nucleotides which prevent restriction (e.g., *Bacillus* phages, T4 and relatives, see Sect. 7.1.4). Recognition sites for several endonucleases common among their hosts are rare in many phage genomes, since their presence has been counterselected during evolution. Furthermore, specialized phage proteins may directly interfere with the restriction enzymes (e.g., P1, T3 and T7 of *E. coli*, *B. subtilis* phages), or phage functions may degrade cofactors required by some endonucleases (e.g., T3 of *E. coli*). In lactococci a number of phage-insensitive plasmids have been described (see Sect. 7.5).

5.3 Evolution of Phage Genomes

There is an obvious resemblance between the genomes of several large DNA phages and parts of bacterial chromosomes, plasmids and transposons. Among the members of a closely related phage group, the borders of homologous and non-homologous genome segments often coincide with borders of functional units, called modules (CAMPBELL and BOTSTEIN, 1983; MEYER et al., 1986; BAAS and JANSZ, 1988; DOSKOČIL et al., 1988; STASSEN et al., 1992). This suggests that these genomes represent mosaics of functional units assembled by recombinational mechanisms. The recombination events which may explain the genesis of such assemblies and which reshape phage genomes under experimental conditions, include homologous and site-specific recombination, transposition and other recombination mechanisms not requiring sequence homology (CAMPBELL, 1984; SYVANEN, 1984; IIDA et al., 1987; CRAIG, 1988; SMITH, 1988; KUCHARLAPATI and SMITH, 1988; BERG and HOWE, 1989; HIGHTON et al., 1990; SANDMEIER et al., 1992). Existing data have recently been reviewed for the best studied group of phages, the lambdoid phages of *E. coli*, by CAMPBELL (1988). He documented that the natural hosts provide a gene pool from which individual phage genomes are built by recombination. Since superinfection and multiple lysogeny are common phenomena in nature, the opportunity for production of new phages must be numerous. However, it is striking that the overall genome organization (e.g., grouping of early functions, related morphogenetic functions, etc.) appears conserved, while the sequences of functionally related genes are rather dissimilar between phage groups. This seems to indicate that these evolutionarily successful pastiches are of ancient origin (CAMPBELL, 1988).

6 Practical Aspects

6.1 Methods for Detection of Phages

The presence of a phage can become apparent by lysis of a growing culture. A definite proof is the electron microscopic visualization of a phage particle by the negative staining technique. This technique is, however, unsuitable for monitoring routine processes. More sensitive and simpler methods are used to visualize the presence of phages indirectly (PEITERSEN, 1991). A suspension to be tested for the presence of phages is mixed with about 10^7 to 10^8 phage-sensitive bacteria and poured in a thin layer of soft agar (0.7% agar) onto a nutrient medium. During subsequent incubation a confluent turbid bacterial layer is formed in the soft agar. Infected bacteria will lyse or slow down growth and release progeny phages which can infect neighboring sensitive bacteria. During several succeeding propagation cycles round lysis zones, called plaques, develop around those bacteria which have been infected prior to pouring (Fig. 11). Since viruses depend on the host metabolic activities, their propagation stops when growth of the bacteria stops due to exhaustion of nutrients or due to

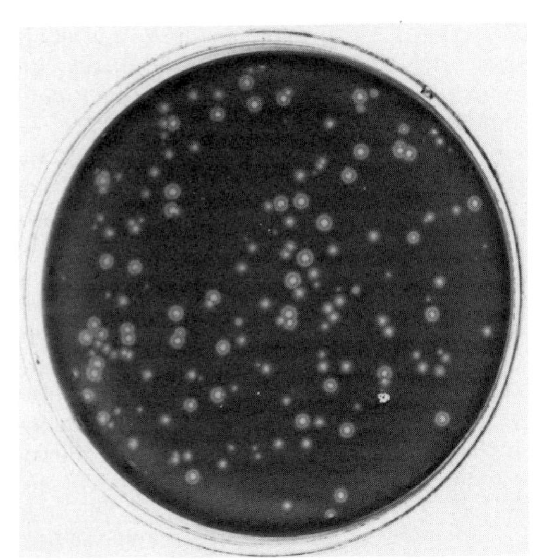

Fig. 11. Plaques of different phage T4 mutants on an *E. coli* indicator strain. Courtesy of E. KELLENBERGER.

accumulation of toxic metabolites. The phage titer (number of infectious particles per mL) can be accurately determined by counting plaques on plates prepared with serially diluted suspension.

If only qualitative results are required, this plating technique can be modified. An aliquot of an indicator strain is poured onto an agar plate together with soft agar, and a small amount (5 μL) of a suspension to be tested for phage presence is deposited on the hardened soft agar layer. If a phage is present, a clear spot in the developing bacterial lawn will remain at the site at which the phage suspension was applied. If the phage suspension is diluted enough, single plaques may be apparent. These plating methods can also be used to detect lysogens, since prophages are induced spontaneously at a low frequency. Instead of a phage suspension, a small amount of a bacterial culture is applied, and a lysis zone in the lawn of indicator bacteria will develop around lysogenic cultures.

Immunological methods using antibodies directed against phage surface proteins or DNA hybridization techniques involving phage genome or gene probes can be used, if the phage to be searched for or related phages are already available.

6.2 Characterization of Phages

Each phage, depending on the host strain and the growth conditions, will display plaques with characteristic form, size and turbidity. The plaque morphology can be altered by minor mutation of phage genes, e.g., a virulent mutant of a temperate phage exhibits a clear rather than a turbid plaque. Electron microscopic analysis of particles leads to the classification according to morphological types (see Figs. 2 to 4). In one-step growth experiments the latent period, i.e., the time required for release of progeny phages after adsorption, and the burst size, i.e., the average number of progeny particles per infected cell, are determined. These parameters depend on the host cell and the growth conditions. Reference strains from the same bacterial species, genus or perhaps even related genera are screened to determine the host range of the phage. Molecular methods, e.g., the analysis of major capsid protein profiles by polyacrylamide gel electrophoresis or restriction enzyme analysis of the phage genome, are laborious methods used only for detailed studies. Close relationships among phages can be traced by immunological or DNA hybridization methods.

6.3 Phage Control

It is appropriate to distinguish between phages that do not disturb a biotechnological process and are, hence, considered as part of the normal biological system, and phages that do cause problems. The major step to avoid interfering phages is, of course, the prevention of access of the phage to the system. This might not be achieved readily in processes which are not run under sterile conditions or if the substrates cannot be sterilized (e.g., in the cheese industry; PEITERSEN, 1991). It is generally very difficult to isolate phage-free bacteria from a lysogenic culture, although UV-light irradiation stimulates this process. Pseudolysogens (or carrier state bacteria, see Sect. 4) can be successfully treated with antisera against the phage.

7 Description of Specific Phages

7.1 *E. coli* Phages

7.1.1 The Temperate dsDNA Phage Lambda

In terms of its physiology, biochemistry and genetics, bacteriophage lambda together with its host, *E. coli* K-12, is the best understood biological system (HENDRIX et al., 1983). It is composed of an icosahedral head of 54 nm diameter to which a flexible tail of 150 nm length is attached (see Fig. 2). Adsorption to an *E. coli* host is initiated when the single tail fiber residing at the end of the tube interacts with the MalB outer membrane protein. After interactions of components of the tail tube with the mannose-specific permease of the inner membrane, the linear DNA is injected. Inside the host cell, the 48502 bp-long lambda DNA (SANGER et al., 1982; DANIELS et al., 1983) is circularized by annealing of the 12 nucleotide-long cohesive single-stranded ends and becomes supercoiled. Expression of early operons is initiated from two strong divergent promoters. Each operon contains about a dozen genes and several transcription termination signals. The decision between lysogeny and lytic growth is influenced by the physiological state of the cell and depends mainly on the relative levels of early gene products (HO and ROSENBERG, 1988). Either the lambda repressor *c*I and the site-specific *int* (=integrase) recombinase genes are expressed leading to integration of lambda DNA into the chromosome in the absence of vegetative functions (LANDY, 1989), or DNA replication and late transcription are initiated by other early gene products.

A lysogen is induced when the lambda CI repressor is inactivated resulting in transcription of early genes (PTASHNE, 1992). Inducing signals like UV-light irradiation or chemical inhibitors of DNA synthesis act indirectly via the host's SOS-response (WALKER, 1987). Subsequently, the prophage is excised by two phage proteins and the lytic program is switched on.

During lytic growth, gene expression is controlled at several levels (FRIEDMAN, 1988; GUARNEROS, 1988). Lambda DNA replication proceeds bidirectionally from the origin and requires two phage proteins in addition to several host functions. Fifteen minutes after infection or induction, about 50 progeny circles have been produced by θ-DNA replication. DNA replication then continues predominantly by the rolling circle mechanism, resulting in concatamers of 2–8 genome lengths. For head morphogenesis, ten phage genes and two host genes are required. Concatameric DNA is cut by the lambda terminase at the *cos*-sites and hence, exactly one genome is packaged into the heads (MURIALDO, 1991). The tail tube, the length of which is determined by the "ruler" protein, is attached to the head (KATSURA, 1987; HENDRIX, 1988).

Derivatives of lambda were among the first vectors and are still today widely used for molecular cloning (MURRAY, 1991). Since large amounts of foreign DNA (up to 25 kb) can be easily produced by infection with recombinant, virulent phages and a large number of phage suspensions can easily be stored for long periods, lambda derivatives are used to construct gene libraries of whole bacterial and eukaryotic genomes. Up to 45 kb can be ligated into cosmids which are plasmids that carry the *cos*-site for DNA packaging. DNA containing *cos*-sites can be packaged into lambda virions *in vitro* (COLLINS and HOHN, 1978) to circumvent inefficient transfection of large recombinant DNA molecules into *E. coli. In vitro* packaging is achieved by mixing the DNA with two complementing lysates of bacteria that each carry a heat-inducible defective lambda derivative and ATP as an energy source. The heads assemble spontaneously in the tube, DNA of between 75% and 105% of the length of lambda genome carrying two *cos*-sites is packaged, and the complete tails attach properly to produce infective particles.

7.1.2 The Temperate dsDNA Phage Mu

The 37 kb-long linear genome of Mu is peculiar in that it is always linked to host DNA and it behaves as a transposable element (SYMONDS et al., 1987; HARSHEY, 1988; PATO,

1989). As Mu DNA can be inserted at many sites on the bacterial chromosome, it causes mutations – hence the name Mu (for "mutator"). Another peculiarity of Mu is that its host range is altered by a programmed DNA inversion (KOCH et al., 1987; GLASGOW et al., 1989).

Phage Mu particles look like other tailed phages: they possess an icosahedral head (54 nm in diameter), a contractile tail (18 nm thick and 100 nm long) and a base plate to which six tail fibers are attached. Lipopolysaccharide serves as the phage receptor. The viral DNA is packaged by a headful mechanism from Mu prophages integrated individually in the host DNA. The process requires a *pac*-site and results in linear dsDNA of 39 kb length: 37 kb of Mu DNA and 2 kb host DNA linked to the ends. After 45 to 60 min, 50 to 200 progeny phages are released upon cell lysis.

Since Mu transposition occurs at much higher levels than that of other transposons and the insertion points are almost randomly selected along the target DNA, derivatives of Mu are very versatile tools to carry out *in vivo* genetic engineering in *E. coli* and other bacteria (GROISMAN, 1991). A large number of mini-Mu constructions are available for cloning, strain construction by generalized transduction, mutagenesis, gene fusions, DNA sequencing and other purposes.

7.1.3 The Temperate dsDNA Phage P1

Phage P1 has considerable practical importance as a generalized transducer (YARMOLINSKY and STERNBERG, 1988; STERNBERG and MAURER, 1991; MEYER, 1993). P1 lysates contain particles with a contractile tail, 18 nm wide and 210 nm long, and an icosahedral head of either 85, 65, or 47 nm in diameter which contain differently sized DNA molecules. The virions with big heads represent the majority of particles. Only these are infectious, because they contain 100 kb linear dsDNA, thus encompassing the 90 kb genome. Probably six tail fibers are attached to the base plate, and they recognize lipopolysaccharide as receptor. The viral DNA is circularized shortly after injection by means of homologous recombination of the 10 kb-long redundant ends or by one of the two phage-encoded site-specific recombination systems.

Unlike most other temperate phages, the P1 prophage does not integrate into the host chromosome. Rather P1 adopts the life style of a plasmid and encodes functions and sites for plasmid replication, partitioning and incompatibility. Lytic functions are, as in most other phages, repressed by a phage-encoded repressor protein. The genetic control of this repressor is complex and involves at least four other P1 genes. In the lytic mode DNA replication is initiated at another origin and proceeds by the rolling circle mechanism. DNA packaging starts at a *pac*-site, ends when the head is full and can successively process 2–3 heads of either size from the same concatamer. One in about 3000 to 100000 P1 particles contains up to 100 kb host DNA packaged from any segment of the host chromosome or from plasmids; these are the transducing phage particles.

P1 encodes a switch for host range which is largely homologous to that of phage Mu (GLASGOW et al., 1989). The P1 host range has been expanded and exploited for the gene transfer into a variety of host genera, even in the absense of phage replication (KAISER and DWORKIN, 1975; MUROOKA and HARADA, 1979). The studies of the P1 restriction and modification system by W. ARBER (ARBER and DUSSOIX, 1962) led the way to *in vitro* genetic engineering, although the P1 restriction endonuclease has not been useful. Recently, P1 derivatives, in which foreign DNA fragments up to 100 kb can be cloned, have been constructed (STERNBERG, 1990).

7.1.4 The Virulent dsDNA Phage T4

Together with its close relatives T2 and T6, T4 has been intensively studied as a model system for a large virulent phage (MATHEWS et al., 1983; MOSIG and EISERLING, 1988). The virions contain 171 kb of linear dsDNA (166 kb representing the T4 genome and 5 kb terminally redundant DNA) in a head of 85 × 115 nm size, to which a contractile tail of 110 nm is attached. In addition to six long-tail fibers attached to the base plate, shorter fibers, so-called whiskers, extrude from the neck between head and tail. These appendages and the short-tail spikes located at the bottom of the base plate are required for infection. Depending on the host strain, either lipopolysaccharide or the OmpC outer membrane protein is recognized by the long fibers. Enzymatic functions, e.g., a T4 lysozyme, are associated with the virion. After irreversible binding, the base plate undergoes a conformational change and comes in contact with the receptor. Concomitant with contraction of the tail sheath, the inner tail tube penetrates the outer membrane and the DNA is injected.

The complex T4 genome contains 130 identified and 70 suspected genes and encodes a battery of enzymes which replace host functions. Early after infection, host translation is inhibited and transcription is directed to phage genes. While the expression of very early genes initiates at promoters resembling *E. coli* promoters, delayed early and late genes are sequentially transcribed from different promoter sequences which are specifically recognized by the host RNA polymerase modified by one of the T4-encoded sigma factors (GEIDUSCHEK, 1991). DNA replication is initiated at multiple origins and involves the products of 30 phage genes (SELICK et al., 1987). Transcription, DNA replication and packaging are intimately linked to DNA recombination and repair (MOSIG, 1987). Morphogenesis of heads involves at least 24 phage proteins, but only ten protein species are found in the head. DNA is packaged by a "headful" mechanism, but it is not cut at a specific site. The tail and the tail fibers require 26 structural proteins. After a latent period of about 23 min, up to several hundred particles are released upon lysis of the host.

T4 DNA is protected from attack by cellular nucleases and by T4 enzymes that degrade host DNA during infection, since it contains hydroxymethylcytosine (HMC) instead of cytosine. The phage itself has not been used in biotechnology or gene technology, but a variety of T4 enzymes (e.g., T4 DNA ligase, T4 RNA ligase, T4 DNA polymerase, T4 polynucleotide kinase) are indispensable tools for the generation and manipulation of recombinant DNA *in vitro*. In the future the product of the T4 *pin* gene might be useful to protect exogenous protein from proteolysis in bacterial expression systems.

7.1.5 The Filamentous ssDNA Phages M13, fd, and f1

With the exception of the F3 group of archaebacteriophages, which contain dsDNA and a lipid envelope (Tab. 2, ZILLIG et al., 1986), all filamentous phages carry ssDNA. The small rod-shaped phages of the Plectovirus group are found in mycoplasms. The long and flexible particles of the Inovirus group are closely related and infect Gram-negative bacteria by attaching to either F-pili or polar pili (MODEL and RUSSELL, 1988; BAAS and JANSZ, 1988). The *Escherichia coli* phages M13, fd, and f1, also called the Ff-group (RASCHED and OBERER, 1986), are male-specific, i.e., they infect only cells carrying the F-plasmid or related plasmids (i.e., male bacteria) which express the F-pili receptor. Filamentous phages are small viruses and have a simple morphology. The filaments are 900 nm long, 6–10 nm thick and the capsid is composed of about 2700 copies of one protein species in a helical array. One end is formed by five copies each of two proteins and the other end by five copies of another two proteins. After adsorption of one end to a pilus, the phage gets to the cell surface, where the circular DNA is transported into the cytoplasm and the capsid protein monomers associate with the inner membrane and can be reused for progeny phages.

The 6.4 kb-long genome encodes three genes for replication functions, five structural genes for the phage particles and two accessory genes required in morphogenesis. The information density of these genomes is extremely high showing multifunctional sequences (overlap among coding and regulatory sequences) (ZINDER and HORIUCHI, 1985; MODEL and RUSSELL, 1988). Gene expression is regulated at the transcriptional and translational level and does not impair the expression machinery of the host. Replication of the phage DNA involves a dsDNA intermediate (replicative form) which is synthesized exclusively from the infecting (+)strand by host enzymes. One of the viral replication functions cuts the replicative form at the *ori*-site to initiate replication by a rolling circle mechanism producing concatamers of (+)ssDNA. Another phage protein binds the newly synthesized (+)ssDNA concatamers to prevent generation of a (–)strand and at the same time forms a precursor for DNA packaging. Assembly of filamentous particles is unique in that it occurs at the cytoplasmic membrane and involves replacement of the phage protein bound to the (+)ssDNA by the major capsid protein previously inserted into the membrane.

Progeny phages are released without damage to the cell as early as ten minutes after infection. Phage release stops at the transition of the cell to the stationary phase.

Derivatives of the Ff phage family provide several advantages when used as cloning vehicles (GEIDER, 1986): Large quantities of very pure ssDNA can be isolated from cultures, thus providing a template for DNA sequencing using the Sanger method or a substrate for site-directed mutagenesis. From the same culture, the dsDNA form can be obtained for *in vitro* manipulation of recombinant DNA; recombinant phages can be easily transferred to other host strains; the length of the phage particles is determined by the length of the genome. Signal sequences for DNA replication and DNA packaging have also been transferred from Ff phages to plasmid vectors. Such vectors containing foreign DNA show a higher stability than the phage genomes and retain the capability of ssDNA production upon co-infection with a helper phage.

7.1.6 The ssRNA Phages

The genomes of ssRNA phages, 3.5 to 4.3 kb in length, are among the smallest molecules which suffice to produce infectious virus particles (FIERS, 1979; VAN DUIN, 1988). The morphologically and genetically very similar ssRNA phages of *E. coli* (MS2, Qβ, SP and others) are very simple structures, built by only three macromolecular species. An icosahedral head, about 25 nm in diameter, consists of 180 copies of the capsid protein and one copy of the adsorption protein, and encapsidates the highly folded ssRNA. (Some phages carry also 3 to 14 molecules of an internal protein which is required for adsorption and gene expression.)

Like the ssDNA phages, ssRNA phages recognize F-pili of *E. coli* or polar pili in *Pseudomonas* spp. and *Caulobacter* spp. as the receptor, but attach along the side of pili. As a consequence of binding to the receptor, the adsorption protein is split and the RNA released from the capsid. The infecting (+)ssRNA functions both as mRNA for translation of viral proteins and as template for RNA replication. The synthesis of viral proteins which are required in various amounts and at different times is finely regulated by RNA secondary structure and by binding of the capsid protein within translation initiation regions. The viral replicase assembles with four pre-existing host proteins into an active complex to produce complementary (–)strands which do not base-pair extensively with (+)strands. Then (+)strands are synthesized by a complex of the viral replicase and only three of the four host proteins. The progeny (+)strands serve as mRNA or are encapsidated. Some ssRNA phages produce a lysis protein whose gene overlaps those of the capsid protein and the replicase subunit, while cell lysis is accomplished by the overproduction of the capsid maturation protein by other ssRNA phages. About 10000 progeny particles are produced by one cell after 30 to 60 minutes.

7.2 Phages of Methanogens

While several bacteriophages have been isolated from the genus *Halobacterium* (ZILLIG et al., 1986, 1988), there are only few reports on bacteriophages of methanogens. The virulent phage ΨM1 of *Methanobacterium thermoautotrophicum* is well characterized (JORDAN et al., 1989; MEILE et al., 1989). It carries linear dsDNA of about 30 kb which is circularly permuted and exhibits a terminal redundancy of about 10%. ΨM1 has been shown to transfer chromosomal markers by generalized transduction and to encapsidate plasmid DNA (LEISINGER and MEILE, 1990). A phage-like particle has also been isolated from *Methanococcus voltae* (WOOD et al., 1989). There are preliminary reports on a transduction-like gene transfer in *M. voltae* and on bacteriophage of *Methanobrevibacter smithii* (see LEISINGER and MEILE, 1990).

7.3 Phages of Methylotrophs

So far only a few phages of methylotrophic bacteria have been described. They originated mainly from fermentation cultures in which irregularities were noticed. A recent report describes a short-tailed phage with a dsDNA genome of 46 kb growing on strains of the genus *Methylophilus* (ALMUMIN et al., 1990). Earlier reports described phages of *Methanomonas methylovora* (OKI et al., 1972), of *Methylomonas methanolis* and *M. ceredia* (ICHIKAWA et al., 1978), and of *Acetobacter methanolicus* (WÜNSCHE et al., 1983a, b; KIESEL et al., 1989).

Bacteriophages of methanotrophs, particularly of *Methylosinus* and *Methylocystis* species, have been isolated from several different environments (TYUTIKOV et al., 1980, 1983; BESPALOVA et al., 1982; KRETOVA et al., 1987). Transduction has not been reported.

7.4 Phages of Clostridia

Infections of solvent-producing clostridia by phages during industrial fermentation have caused serious problems (JONES and WOODS, 1986). Many phages derived from these clostridia have been characterized (OGATA and HONGO, 1979; NIEVES et al., 1981), and mainly included tailed dsDNA phages. Furthermore, bacteriocins and defective phage-like particles have been demonstrated for *Clostridium saccharoperbutylacetonicum*, *C. acetobutylicum*, and other solvent-producing strains. A few phages have been used for transfection of *C. acetobutylicum* protoplasts (REID et al., 1983). Recently, a 6.6 kb-long ssDNA complexed with protein in a filamentous form was found to be continuously released from a *C. acetobutylicum* culture. The intracellular presence of a double-stranded form of the same DNA which has homology to the ssDNA genome of filamentous *E. coli* phages suggests that this strain contains a defective phage (KIM and BLASCHEK, 1991). These findings represent the first example of a filamentous phage in Gram-positive bacteria.

Phages of pathogenic *Clostridium* spp., particularly the lysogenic conversion to toxin production and modulation of sporulation have been studied extensively (EKLUND and POYSKY, 1974; EKLUND et al., 1974; GRANT and RIEMANN, 1976; STEWART and JOHNSON, 1977; MAHONY, 1979; SCHALLEHN and EKLUND, 1980; SELL et al., 1983; BISHAI and MURPHY, 1988; CANARD and COLE, 1990; ROOD and COLE, 1991).

7.5 Phages of Lactic Acid Bacteria

In the manufacture of fermented dairy products, such as cheese, involving lactococci, phages are common and represent a serious problem (PEITERSEN, 1991). All known phages of lactococci are tailed and contain dsDNA (JARVIS and MEYER, 1986; TEUBER and LOOF, 1987; BRAUN et al., 1989; NEVE and TEUBER, 1991; TEUBER et al., 1992).

Virulent phages isolated from cheese factories often showed small isometric heads or prolate heads, whereas large isometric headed phages and others with a peculiar morphology occurred occasionally. Their genomes measured between 18 and 54 kb, and most of them had cohesive ends. They contain few restriction sites in their DNA, a property that most probably has been counterselected during evolution by the hosts' DNA restriction and modification systems. The temperate phages have been less well studied, often because indicator strains are lacking (DAVIDSON et al., 1990). Most carry linear dsDNA of between 38 and 46 kb in isometric heads. These genomes are frequently terminally redundant and circularly permuted.

Many undefined starter cultures used in cheese factories are lysogenic for one or several prophages, but often prophages do not impair acid production. Defined starter cultures are often selected for higher phage resistance. Some lactococcal strains are naturally resistant to phage infection. The resistance is encoded on 5 to 100 kb-long self-transmissible plasmids (SING and KLAENHAMMER, 1990) and prevent infection by virulent phages by blocking successful adsorption or replication. The molecular genetic analysis of the responsible genes from some of these plasmids is currently being investigated (KLAENHAMMER et al., 1991; COFFEY et al., 1991).

7.6 Phages of Bacilli

Since bacilli and particularly *Bacillus subtilis* are the most intensively studied Gram-positive bacteria, it is not surprising that a large number of their phages have been isolated and characterized (HEMPHILL and WHITELEY, 1975). So far, only large dsDNA phages, but no ssDNA, RNA or transposon-like phages have been found.

7.6.1 Temperate dsDNA Phages of *Bacillus subtilis*

Based on serology, immunity, host range and site of adsorption, the temperate phages were classified in five groups (RUTBERG, 1982; ZAHLER, 1988). In lysogens the prophage is integrated into the chromosome.

Phage ø105 (BIRDSELL et al., 1969), as a representative of *group I*, has an icosahedral head of 52 nm in diameter, a flexible non-contractile tail, 10 nm wide and 220 nm long, and contains a 40 kb-long linear dsDNA. When a lysogen is induced, the prophage is replicated together with adjacent bacterial sequences, before it is excised from the chromosome. After a latency period of about 40 minutes, 100 to 200 progeny phages are released. Lysogenic cells, competent for DNA uptake, can be induced by adding DNA, homologous to the chromosome. DNA transformation efficiency drops when the ø105 prophage is induced (e.g., by means of mitomycin C treatment). ø105 is capable of specialized transduction of chromosomal genes flanking the integration site and of plasmids, carrying homologies to the phage DNA. Cloning vectors which can take up 4 kb of foreign DNA have been derived from ø105.

Phage SPO2 is a member of *group II,* but is very similar to ø105; it is also capable of transduction. 5.7 kb of its 40 kb long genome are homologous to ø105 DNA.

Group III contains large phages. During lytic growth, they protect their DNA from endonucleolytic attack by the host restriction systems by expressing multispecific DNA methyltransferases. These modify cytosines at several recognition sites (GÜNTHERT and TRAUTNER, 1984). Phage SPβ exerts lysogenic conversion on its host, i.e., it can induce the production of betacin, a substance that kills neighboring, non-lysogenic cells. *Group III* phages are also used as cloning vehicles.

Phages of *group IV* (e.g., SP16 which can also infect *B. amyloliquefaciens* and *B. licheniformis*) and of *group V* (containing only defective prophages) have not been studied extensively.

7.6.2 Virulent dsDNA Phages of *Bacillus subtilis*

Phage ø29 (SALAS, 1988a, b) is one of the smallest dsDNA phages found to date. The genome is only about 19 kb long and contains 17 genes. The head measures 42×32 nm and the tail 6×32 nm. ø29 replication shows two peculiarities: A protein is covalently bound to the 5' ends of the linear dsDNA and can prime DNA replication. The packaging of viral DNA requires a 120 nucleotide-long phage-encoded RNA molecule, which is not found in the mature virion. About 570 progeny phages are released after 45 min. Derivatives of ø29 are used as gene cloning or expression vectors, and an *in vitro* packaging system is also available.

Phage SPO1 and relatives (STEWART, 1988; HOET et al., 1992) are very big dsDNA phages with a head of 90×90 nm, a tail of 200 nm length, and linear DNA of about 150 kb (including a terminal redundancy of 10%). The DNA contains 5-hydroxymethyluracil instead of thymine. Like other phages with large genomes (e.g., T4), SPO1 can block cellular macromolecular synthesis efficiently. To regulate its gene expression, the virus encodes different sigma factors which modify the host RNA polymerase to specifically recognize different classes of phage promoters. DNA replication starts from primary origins and is followed by extensive recombination. The major part of replicative forms are concatamers of up to 20 genomes in length. The latency period is 50 minutes and the burst size 100–200 particles.

Phage SPP1 is a medium-sized generalized transducer which transduces plasmids carrying homologies to phage DNA with high frequency. Transducing particles contain exclusively plasmid DNA in long concatameric forms.

7.6.3 Other *Bacillus* Phages

For genetic mapping analysis in *Bacillus subtilis*, the generalized transducing phage PBS1 has been used, since it can package 150 to 200 kb long chromosomal fragments (DEDONDER et al., 1977; PIGGOT and HOCH, 1985; HOCH, 1991). However, PBS1 is a very poor plaque former and has not been thoroughly studied.

From *B. licheniformis* strains used in the production of bacitracin, temperate kappa phages (THORNE and KOWALSKI, 1976) and usually virulent theta phages have been isolated. Although the kappa and theta phages are morphologically different, their genomes share extensive sequence homologies and can form natural, infectious recombinants (DOSKOČIL et al., 1988).

7.7 Phages of Pseudomonads

Phages are quite common among *Pseudomonas aeruginosa* strains, and a large body of data has been accumulated (HOLLOWAY and KRISHNAPILLAI, 1975; HOLLOWAY et al., 1979). Most of the known phages carry dsDNA and a tail, but filamentous and spherical phages (some containing RNA) have been isolated as well. A large number of lytic phages have been used for strain typing (BERGAN, 1978; ACKERMANN et al., 1988). Transducing phages represented invaluable tools for genetic studies. In recent years, M3112 and its derivatives have proven useful (ROTHMEL et al., 1991). M3112 is one of many Mu-like bacteriophage isolates from *P. aeruginosa* which use transposition for their replication and consequently offer the possibility of *in vivo* genetic manipulation. Analogous to mini-Mu vectors (see Sect. 7.1.2), mini-M3112 elements for cloning, insertion mutagenesis or transduction have been constructed.

Phages seem to be less common among other *Pseudomonas* species and very few have been studied (HOLLOWAY and KRISHNAPILLAI, 1975). The *P. syringae* phage ø6 is unusual in that it contains dsRNA in a segmented form, and its icosahedral head is surrounded by a membraneous envelope containing phospholipids and glycoproteins (MINDICH and BAMFORD, 1988). Another lipid-containing phage is PM2 from a marine *Pseudomonas* (reclassified as *Alteromonas espejiiani*) which carries a circular dsDNA genome (FRANKLIN, 1974).

7.8 Actinophages

A fair number of both virulent and temperate *Streptomyces* phages have been described. All are tailed and contain dsDNA (LOMOVSKAYA et al., 1980; CHATER, 1986). While some phage infections had devastating effects on antibiotics production, others resulted in a significant yield increase during strain improvement, as documented for the production of rifamycins (GHISALBA et al., 1984). SV1 of *Streptomyces venezuelae* is a generalized transducing phage and has been used for genetic mapping (VATS et al., 1987). Another generalized transducer, øSF1, functions as a conjugative plasmid in the prophage form (CHUNG, 1982). Phage øC31, a temperate phage with a genome of 41 kb, can infect a wide range of *Streptomyces* species. Vectors have been derived from øC31 for cloning and for *in vivo* gene disruption (CHATER, 1986; KIESER and HOPWOOD, 1991) .

7.9 Phages of Coryneform Bacteria

The existence of phages which interfere with the industrial production of amino acids by coryneform bacteria has been recognized for many years (HONGO et al., 1972), and procedures for practical phage control have been described (YAMANAKA et al., 1975; KASHIMA et al., 1976). Recently, virulent tailed dsDNA phages infecting *Corynebacterium glutamicum* and *C. lilium* (PATEK et al., 1985; TRAUTWETTER et al., 1987a, b) or *Brevibacterium* or

Arthrobacter strains (TRAUTWETTER and BLANCO, 1988) have been described in more detail. Two partly related phages of *Brevibacterium flavum* and *C. glutamicum* have been explored for their use as cloning vehicles (SONNEN et al., 1990a, b; SONNEN, 1991)

Acknowledgements

We thank T. A. BICKLE, M. DWORKIN and O. GHISALBA for comments on the manuscript, H. BAHL, D. HAAS, H. KÖNIG, L. MEILE, F. G. PRIEST, H. SONNEN and M. TEUBER for advice on specific phage groups, M. WURTZ and M. ZOLLER for photographic work and E. VITZTHUM for patiently typing the various versions of the manuscript. Work in the authors' laboratory is supported by grants from the Swiss National Science Foundation (grants No. 31-25680.88 and 31-25680.88/2).

8 References

ACKERMANN, H. W. (1987), Bacteriophage taxonomy in 1987, *Microbiol. Sci.* **4,** 214–218.

ACKERMANN, H. W., DUBOW, M. S. (1987), Viruses of Prokaryotes, Boca Raton, Florida: CRC Press Inc.

ACKERMANN, H. W., EISENSTARK, A. (1974), The present state of phage taxonomy, *Intervirology* **3,** 201–219.

ACKERMANN, H. W., CARTIER, C., SLOPEK, S., VIEU, J. F. (1988), Morphology of *Pseudomonas aeruginosa* typing phages of the Lindberg set, *Ann. Inst. Pasteur/Virol.* **139,** 389–404.

ALMUMIN, S., KADRI, M., MAMAT, U., ENGEL, J. (1990), Isolation and primary characterization of a new bacteriophage of obligate methylotrophic bacteria, *J. Basic Microbiol.* **30,** 627–632.

ARBER, W., DUSSOIX, D. (1962), Host specificity of DNA produced by *Escherichia coli*. I. Host controlled modification of bacteriophage lambda, *J. Mol. Biol.* **5,** 18–36.

BAAS, P. D., JANSZ, H. S. (1988), Single-stranded DNA phage origins, *Curr. Top. Microbiol. Immunol.* **136,** 31–70.

BALTIMORE, D. (1971), Expression of animal virus genomes, *Bacteriol. Rev.* **35,** 235–241.

BAMFORD, D. H., RAMANTSCHUK, M., SOMERHARJU, P. J. (1987), Membrane fusion in prokaryotes: bacteriophage ø6 membrane fuses with the *Pseudomonas syringae* outer membrane, *EMBO J.* **6,** 1467–1473.

BAZINET, C., KING, J. (1985), The DNA translocation vertex of dsDNA bacteriophage, *Annu. Rev. Microbiol.* **39,** 109–129.

BERG, D. E., HOWE, M. M. (Eds.) (1989), *Mobile DNA,* Washington, DC: American Society for Microbiology.

BERGAN, T. (1978), Phage typing of *Pseudonomas aeruginosa, Methods Microbiol.* **10,** 169–196.

BESPALOVA, I. A., TYUTIKOV, F. M., MARTYNKINA, L. P., GALCHENKO, V. K., KRIVISKY, A. S. (1982), Lysogeny in methanotrophic bacteria, *Mikrobiologia* **51,** 403–408.

BEUMER, J., HANNECART-POKORNI, E., GODARD, C. (1984), Bacteriophage receptors, *Bull. Inst. Pasteur* **82,** 173–253.

BICKLE, T. A. (1987), DNA restriction and modification systems, in: *Escherichia coli and Salmonella typhimurium: Cellular and Molecular Biology* (NEIDHARDT, F. C., Ed.), pp. 692–696, Washington, DC: American Society for Microbiology.

BIRDSELL, D. C., HATHAWAY, G. M., RUTBERG, L. (1969), Characterization of temperate *Bacillus* bacteriophage ø105, *J. Virol.* **4,** 264–270.

BISHAI, W. R., MURPHY, J. R. (1988), Bacteriophage gene products that cause human disease, in: *The Bacteriophages* (CALENDAR, R., Ed.), Vol. 2, pp. 683–723, New York: Plenum Press.

BLACK, L. W. (1988), DNA packaging in dsDNA bacteriophages, in: *The Bacteriophages* (CALENDAR, R., Ed.), Vol. 2, pp. 321–373, New York: Plenum Press.

BLACK, L. W. (1989), DNA packaging in dsDNA bacteriophages, *Annu. Rev. Microbiol.* **43,** 267–292.

BRADLEY, D. E. (1967), Ultrastructure of bacteriophages and bacteriocins, *Bacteriol. Rev.* **31,** 230–314.

BRAUN, Jr., V., HERTWIG, S., NEVE, H., GEIS, A., TEUBER, M. (1989), Taxonomic differentiation of bacteriophages of *Lactococcus lactis* by electron microscopy, DNA-DNA hybridization, and protein profiles, *J. Gen. Microbiol.* **135,** 2551–2560.

BURNET, F. M., LUSH, D. (1936), Induced lysogenity and mutation of bacteriophage within lysogenic bacteria, *Austr. J. Exp. Biol. Med. Sci.* **14,** 27–38.

CAIRNS, J., STENT, G. S., WATSON, J. D. (1966), Phages and the origins of molecular biology, Cold Spring Harbor, NY: Cold Spring Harbor Laboratory Press.

CALENDAR, R. (Ed.) (1988), *The Bacteriophages,* New York: Plenum Press.

CAMPBELL, A. (1984), Types of recombination: Common problems and common strategies, *Cold Spring Harbor Symp. Quant. Biol.* **49,** 834–844.

CAMPBELL, A. (1988), Phage evolution and speciation, in: *The Bacteriophages* (Calendar, R., Ed.), Vol. 1, pp. 1–14, New York: Plenum Press.

Campbell, A., Botstein, D. (1983), Evolution of the lambdoid phages, in: *Lambda II* (Hendrix, R. W., Roberts, J. W., Stahl, F. W., Weisberg, R. A., Eds.), pp. 365–380, Cold Spring Harbor, NY: Cold Spring Harbor Laboratory Press.

Canard, B., Cole, S. T. (1990), Lysogenic phages of *Clostridium perfringens:* mapping of the chromosomal attachment sites, *FEMS Microbiol. Lett.* **66,** 323–326.

Casjens, S. (Ed.) (1985), *Virus Structure and Assembly*, Boston: Jones & Bartlett Publ. Inc.

Casjens, S., Hendrix, R. W. (1988), Control mechanisms in dsDNA bacteriophage assembly, in: *The Bacteriophages* (Calendar, R., Ed.), Vol. 1, pp. 15–92, New York: Plenum Press.

Chater, K. F. (1986), *Streptomyces* phages and their application to *Streptomyces* genetics, in: *The Bacteria.* Vol. 9: *Antibiotic-Producing Streptomyces* (Sokatch, J. R., Ornston, L. N., Queener, S. W., Day, L E., Eds.), pp. 119–158, Orlando: Academic Press.

Chung, S. T. (1982), Isolation and characterization of *Streptomyces fradiae* plasmids which are prophages of the actinophage øSF1, *Gene* **17,** 239–246.

Coffey, A., Costello, V., Daly, C., Fitzgerald, G. (1991), Plasmid-encoded bateriophage insensitivity in members of the genus *Lactococcus*, with special reference to pC1829, in: *Genetic and Molecular Biology of Streptococci, Lactococci, and Enterococci* (Dunny, G. M., Cleary, P. P., McKay L. L., Eds.), pp. 131–135, Washington, DC: American Society for Microbiology.

Collins, J., Hohn, B. (1978), Cosmids: A type of plasmid gene-cloning vector that is packageable *in vitro* in bacteriophage lambda heads, *Proc. Natl. Acad. Sci. USA* **75,** 4242–4246.

Craig, N. L. (1988), The mechanism of conservative site-specific recombination, *Annu. Rev. Genet.* **22,** 77–105.

Dabora, R. L., Cooney, C. L. (1990), Intracellular lytic enzyme systems and their use for disruption of *Escherichia coli*, *Adv. Biochem. Eng. Biotechnol.* **43,** 11–30.

Daniels, D. L., Schroeder, J. L., Szybalski, W., Sanger, F., Coulson, A. R., Hong, G. F., Hill, D. F., Petersen, G. B., Blattner, F. R. (1983), Complete annotated lambda sequence, in: *Lambda II* (Hendrix, R. W., Roberts, J. W., Stahl, F. W., Weisberg, R. A., Eds.), pp. 519–676, Cold Spring Harbor, NY: Cold Spring Harbor Laboratory Press.

Davidson, B. E., Powell, I. B., Hillier, A. J. (1990), Temperate bacteriophages and lysogeny in lactic acid bacteria, *FEMS Microbiol. Rev.* **87,** 79–90.

Dedonder, R. A., Lepesant-Kejzlarová, J. A., Billault, A., Steinmetz, M., Kunst, F. (1977), Construction of a kit of reference strains for rapid genetic mapping in *Bacillus subtilis* 168, *Appl. Environ. Microbiol.* **33,** 989–993.

d'Hérelle, F. (1917), Sur un microbe invisible antagoniste des bacilles dysenteriques, *C. R. Acad. Sci.* (Paris) **165,** 373–375.

Doskočil, J., Štokrová, J., Štorchová, H., Forstová, J., Meyer, J. (1988), Correlation of physical maps and some genetic functions in the genomes of the kappa-theta phage family of *Bacillus licheniformis*, *Mol. Gen. Genet.* **214,** 343–347.

Duckworth, D. H. (1976), Who discovered bacteriophage? *Bacteriol. Rev.* **40,** 793–802.

Eklund, M. W., Poysky, F. T. (1974) Interconversion of type C and D strains of *Clostridium botulinum* by specific bacteriophages, *Appl. Environ. Microbiol.* **27,** 251–258.

Eklund, M. W., Poysky, F. T., Meyers, J. A., Pelroy, G. A. (1974), Interspecies conversion of *Clostridium botulinum* type C to *Clostridium novyi* type A by bacteriophage, *Science* **186,** 456–458.

Fiers, W. (1979), RNA bacteriophages, *Compr. Virol.* **13,** 69–204.

Francki, R. I. B., Fauquet, C. M., Knudson, D. L., Brown, F. (Eds.) (1991), Classification and Nomenclature of Viruses (Fifth Report of the International Committee on Taxonomy of Viruses), *Arch. Virol. Suppl.* 2.

Franklin, R. M. (1974), Structure and synthesis of bacteriophage PM2 with particular emphasis on the viral lipid bilayer, *Curr. Top. Microbiol. Immunol.* **68,** 107–159.

Freeman, V. J. (1951), Studies on the virulence of bacteriophage-infected strains of *Corynebacterium diphtheriae*, *J. Bacteriol.* **61,** 675–688.

Friedman, D. I. (1988), Regulation of phage gene expression by termination and antitermination of transcription, in: *The Bacteriophages* (Calendar, R., Ed.), Vol. 2, pp. 263–319, New York: Plenum Press.

Geider, K. (1986), DNA cloning vectors utilizing replication functions of the filamentous phages of *Escherichia coli*, *J. Gen. Virol.* **67,** 2287–2303.

Geiduschek, E. P. (1991), Regulation of expression of the late genes of bacteriophage T4, *Annu. Rev. Genet.* **25,** 437–460.

Geiduschek, E. P., Kassavetis, G. A. (1988), Changes in RNA polymerase, in: *The Bacteriophages* (Calendar, R., Ed.), Vol. 1, pp. 93–115, New York: Plenum Press.

Ghisalba, O., Auden, J. A., Schupp, T., Nüesch, J. (1984), The rifamycins: Properties, biosynthesis, and fermentation, in: *Biotechnology of Industrial Antibiotics* (Vandamme, E. J., Ed.), pp. 281–327, New York: Marcel Dekker.

Glasgow, A. C., Hughes, K. T., Simon, M. I. (1989),

Bacterial DNA inversion systems, in: *Mobile DNA* (BERG, D. E., HOWE, M. M., Eds.), pp. 637–659, Washington, DC: American Society for Microbiology.

GOYAL, S. M., GERBA, S P., BITTON, G. (1987), *Phage Ecology*, New York: Wiley Interscience Publ.

GRANT, R. B., RIEMANN, H. P. (1976), Temperate phages of *Clostridium perfringens* type C[1], *Can. J. Microbiol.* **22,** 603–610.

GROISMAN, E. A. (1991), *In vivo* genetic engineering with bacteriophage Mu, *Methods Enzymol.* **204,** 180–212.

GUARNEROS, G. (1988), Retroregulation of bacteriophage lambda *int* gene expression, *Curr. Top. Microbiol. Immunol.* **136,** 1–19.

GÜNTHERT, U., TRAUTNER, T. A. (1984), DNA methyltransferases of *Bacillus subtilis* and its bacteriophages, *Curr. Top. Microbiol. Immunol.* **108,** 11–22.

HARSHEY, R. M. (1988), Phage Mu, in: *The Bacteriophages* (CALENDAR, R., Ed.), Vol. 2, pp. 193–234, New York: Plenum Press.

HEMPHILL, H. E., WHITELEY, H. R. (1975), Bacteriophages of *Bacillus subtilis*, *Bacteriol. Rev.* **39,** 257–315.

HENDRIX, R. W. (1988), Tail length determination in double-stranded DNA bacteriophages, *Curr. Top. Microbiol. Immunol.* **136**, 21–29.

HENDRIX, R. W., ROBERTS, J. W., STAHL, F. W., WEISBERG, R. A. (1983), *Lambda II*, Cold Spring Harbor Laboratory, NY: Cold Spring Harbor Laboratory Press.

HIGHTON, P. J., CHANG, Y., MYERS, R. J. (1990), Evidence for the exchange of segments between genomes during the evolution of lambdoid bacteriophages, *Mol. Microbiol.* **4,** 1329–1340.

HO, Y. S., ROSENBERG, M. (1988), The structure and function of the transcription activator protein *c*II and its regulatory signals, in: *The Bacteriophages* (CALENDAR, R., Ed.), Vol. 2, pp. 725–756, New York: Plenum Press.

HOCH, J. A. (1991), Genetic analysis in *Bacillus subtilis*, *Methods Enzymol.* **204,** 305–320.

HOET, P. P., COENE, M. M., COCITO, C. G. (1992), Replication cycle of *Bacillus subtilis* hydroxymethyluracil-containing phages, *Annu. Rev. Microbiol.* **46,** 95–116.

HOLLOWAY, B. W., KRISHNAPILLAI, V. (1975), Bacteriophages and bacteriocins, in: *Genetics and Biochemistry of Pseudomonas* (CLARKE, P. H., RICHMOND, M. H., Eds.), pp. 99–132, London: John Wiley & Sons.

HOLLOWAY, B. W., KRISHNAPILLAI, V., MORGAN, A. F. (1979), Chromosomal genetics of *Pseudomonas*, *Microbiol. Rev.* **43,** 73–103.

HONGO, M., OKI, T., OGATA, S. (1972), Phage contamination and control, in: *The Microbial Production of Amino Acids* (YANADA, K., KINOSHITA, S., TSUNIDA, T., AIDA, K., Eds.), Tokyo: Tokyo Kodansha.

ICHIKAWA, T., TAHARA, T., HOSHINO, J. (1978), Isolation and characterization of bacteriophages active against methanol-assimilating bacteria, in: *Abstracts of the II. International Symposium on Microbial Growth on C_1 Compounds, Pushchino, USSR,* p. 47, Moscow: Nauka.

IIDA, S., MEYER, J., ARBER, W. (1987), Mechanisms involved in the formation of plaque-forming derivatives from over-sized hybrid phages between bacteriophage P1 and the R plasmid NR1, *FEMS Microbiol. Lett.* **43,** 117–120.

INOUYE, M., INOUYE, S. (1991), Retroelements in bacteria, *Trends Biochem. Sci.* **16,** 18–21.

INOUYE, S., SUNSHINE, M. G., SIX, E. W., INOUYE, M. (1991), Retronphage ϕR73: An *E. coli* phage that contains a retroelement and integrates into a tRNA gene, *Science* **252,** 969–971.

JARVIS, A. W., MEYER J. (1986), Electron microscopic heteroduplex study and restriction endonuclease cleavage analysis of the DNA genomes of three lactic streptococcal bacteriophages, *Appl. Environ. Microbiol.* **51,** 566–571.

JONES, D. T., WOODS, D. R. (1986), Acetone-butanol fermentation revisited, *Microbiol. Rev.* **50,** 484–524.

JORDAN, M., MEILE, L., LEISINGER, T. (1989), Organization of *Methanobacterium thermoautotrophicum* bacteriophage ΨM1 DNA, *Mol. Gen. Genet.* **220,** 161–164.

KAISER, D., DWORKIN, M. (1975), Gene transfer to a myxobacterium by *Escherichia coli* phage P1, *Science* **187,** 653–654.

KASHIMA, N., YAMANAKA, S., MITSUGI, K., HIROSE, Y. (1976), Inhibition of bacteriophages of amino acid-producing bacteria by N-acylamino acids, *Agric. Biol. Chem.* **40,** 41–47.

KATSURA, I. (1987), Determination of bacteriophage lambda tail length by a protein ruler, *Nature* **327,** 73–75.

KEPPEL, F., FAYET, O., GEORGOPOULOS, C. (1988), Strategies of bacteriophage DNA replication, in: *The Bacteriophages* (CALENDAR, R., Ed.), Vol. 2, pp. 145–161, New York: Plenum Press.

KIESEL, B., MAMAT, U., WÜNSCHE, L. (1989), Phagen methylotropher Bakterien, *Math.-nat. wiss. Reihe* **38,** 287, Leipzig: Karl-Marx-Universität.

KIESER, T., HOPWOOD, D. A. (1991), Genetic manipulation of *Streptomyces:* integrating vectors and gene replacement, *Methods Enzymol.* **204,** 430–458.

KIM, A. Y., BLASCHEK, H. P. (1991), Isolation and characterization of a filamentous viruslike particle from *Clostridium acetobutylicum* NCIB 6444, *J. Bacteriol.* **173,** 530–535.

KLAENHAMMER, D. R., ROMERO, D., SING, W., HILL, C. (1991), Molecular analysis of pTR2030 gene systems that confer bacteriophage resistance to lactococci, in: *Genetics and Molecular Biology of Streptococci, Lactococci, and Enterococci* (DUNNY, G. M., CLEARY, P. P., MCKAY L. L., Eds.), pp. 124–130, Washington, DC: American Society for Microbiology.

KLAUS, S., KRÜGER, D. H., MEYER, J. (1992), *Bakterienviren*, Jena: Gustav Fischer Verlag.

KOCH, C., MERTENS, G., RUDT, F., KAHMANN, R., KANAAR, R., PLASTERK, R. H. A., VAN DE PUTTE, P., SANDULACHE, R., KAMP, D., TOUSSAINT, A., HOWE, M. M. (1987), The invertible G segment, in: *Phage Mu* (SYMONDS, N., TOUSSAINT, A., VAN DE PUTTE, P., HOWE, M. M., Eds.) pp. 75–91, Cold Spring Harbor, NY: Cold Spring Harbor Laboratory Press.

KONISKY, J. (1982), Colicins and other bacteriocins with established modes of action, *Annu. Rev. Microbiol.* **36,** 125–144.

KRETOVA, A. F., BESPALOVA, I. A., TYUTIKOV, F. M., NOVIKOVA, E. G., TIKHONENKO, A. S. (1987), Electron microscopic study of the structure and antigenic relationship of methanotrophous bacterial phages isolated from different natural sources, *Mikrobiologia* **56,** 742–745.

KRÜGER, D. H., BICKLE T. A. (1983), Bacteriophage survival: Multiple mechanisms for avoiding the deoxyribonucleic acid restriction systems of their hosts, *Microbiol. Rev.* **47,** 345–360.

KUCHERLAPATI, R., SMITH, G. R. (Eds.) (1988), *Genetic Recombination*, Washington, DC: American Society for Microbiology.

LANDY, A. (1989), Dynamic, structural, and replicatory aspects of lambda site-specific recombination, *Annu. Rev. Biochem.* **58,** 913–949.

LEDERBERG, J. (1987), Genetic recombination in bacteria: a discovery account, *Annu. Rev. Genet.* **21,** 23–46.

LEISINGER, T., MEILE, L. (1990), Approaches to gene transfer in methanogenic bacteria, in: *Microbiology and Biochemistry of Strict Anaerobes Involved in Interspecies Hydrogen Transfer* (BÉLAICH, J. P., BRUSCHI, M., GARCIA, J. L., Eds.), pp. 11–23, New York: Plenum Press.

LOMOVSKAYA, N. D., CHATER, K. F., MKRTUMIAN, N. M. (1980), Genetics and molecular biology of *Streptomyces* bacteriophages, *Microbiol. Rev.* **44,** 206–229.

LURIA, S. E., SUIT, J. L. (1987), Colicins and col plasmids, in: *Escherichia coli and Salmonella typhimurium: Cellular and Molecular Biology* (NEIDHARDT, F. C., Ed.), pp. 1616–1624. Washington, DC: American Society for Microbiology.

MAHONY, D. E. (1979), Bacteriocin, bacteriophage and other epidemiological typing methods for the genus *Clostridium*, in: *Methods in Microbiology* (BERGAN, T., NORRIS, J. R., Eds.), pp. 1–30, New York: Academic Press.

MARGOLIN, P. (1987), Generalized transduction, in: *Escherichia coli and Salmonella typhimurium: Cellular and Molecular Biology* (NEIDHARDT, F. C., Ed)., pp. 1154–1168, Washington, DC: American Society for Microbiology.

MATHEWS, C. K., KUTTER, E. M., MOSIG, G., BERGET, P. B. (Eds.) (1983), *Bacteriophage T4*, Washington, DC: American Society for Microbiology.

MCCOY, E., MCDANIEL, L. E., SYLVESTER, J. C. (1944), Bacteriophage in a butyl fermentation plant, *J. Bacteriol.* **47,** 433.

MEILE, L., JENAL, U., STUDER, D., JORDAN, M., LEISINGER, T. (1989), Characterization of ΨM1, a virulent phage of *Methanobacterium thermoautotrophicum* Marburg, *Arch. Microbiol.* **152,** 105–110.

MEYER, J. (1993), Bacteriophage P1, in: *Encyclopedia of Virology: Bacteriophages* (WEBSTER, R. G., GRANOFF, A., Eds.). London: Academic Press, in press.

MEYER, J., STÅLHAMMAR-CARLEMALM, M., STREIFF, M., IIDA, S., ARBER, W. (1986), Sequence relations among the IncY plasmid p15B, P1, and P7 prophages, *Plasmid* **16,** 81–89.

MINDICH, L., BAMFORD, D. H. (1988), Lipid-containing bacteriophages, in: *The Bacteriophages* (CALENDAR, R., Ed.), Vol. 2, pp. 475–519, New York: Plenum Press.

MODEL, P., RUSSELL, M. (1988), Filamentous phages, in: *The Bacteriophages* (CALENDAR, R., Ed.), Vol. 2, pp. 375–456, New York: Plenum Press.

MOSIG, G. (1987), The essential role of recombination in phage T4 growth, *Annu. Rev. Genet.* **21,** 347–371.

MOSIG, G., EISERLING, F. (1988), Phage T4 structure and metabolism, in: *The Bacteriophages* (CALENDAR, R., Ed.), Vol. 2, pp. 521–605, New York: Plenum Press.

MURIALDO, H. (1991), Bacteriophage lambda DNA maturation and packaging, *Annu. Rev. Biochem.* **60,** 125–153.

MUROOKA, Y., HARADA, T. (1979), Expansion of the host range of coliphage P1 and gene transfer from enteric bacteria to other gram-negative bacteria, *Appl. Environ. Microbiol.* **38,** 754–757.

MURRAY, N. E. (1991), Special uses of lambda phage for molecular cloning, *Methods Enzymol.* **204,** 280–304.

NEVE, H., TEUBER, M. (1991), Basic microbiology, and molecular biology of bacteriophage of lactic acid bacteria in dairies, *Bull. Int. Dairy Fed.* **263,** 3–15.

NIEVES, B. M., GIL, F., CASTILLO, F. J. (1981), Growth inhibition activity and bacteriophage and bacteriocin-like particles associated with different species of *Clostridium*, *Can. J. Microbiol.* **27,** 216–225.

Ogata, S., Hongo, M. (1979), Bacteriophages of the genus *Clostridium*, *Adv. Appl. Microbiol.* **25,** 241–273.

Oki, T., Nishida, H., Ozaki, A. (1972), Deoxyribonucleic acid bacteriophage of *Methanomonas methylovora*, *J. Virol.* **9,** 544–546.

Patek, M., Ludvik, J., Benada, O., Hochmannova, J., Krumphanzl, V., Bucko, M. (1985), New bacteriophage-like particles in *Corynebacterium glutamicum*, *Virology* **140,** 360–363.

Pato, M. L. (1989), Bacteriophage Mu, in: *Mobile DNA* (Berg, D. E., Howe, M. M., Eds.), pp. 23–52. Washington, DC: American Society for Microbiology.

Peitersen, N. (Ed.) (1991), Practical phage control, *Bull. Int. Dairy Fed.* **263,** 1–43.

Piggot, P. J., Hoch, J. A. (1985), Revised genetic linkage map of *Bacillus subtilis*, *Microbiol. Rev.* **49,** 158–179.

Ptashne, M. (1992), *A Genetic Switch*, Palo Alto: Blackwell Sci. Publ. & Cell Press.

Raettig, H. (1958), *Bacteriophagie 1917–1956*, Stuttgart: Fischer Verlag.

Rasched, I., Oberer, E. (1986), Ff coliphages: structural and functional relationship, *Microbiol. Rev.* **50,** 401–427.

Reid, S. J., Allcock, E. R., Jones, T. D., Woods, D. R. (1983), Transformation of *Clostridium acetobutylicum* protoplasts with bacteriophage DNA, *Appl. Environ. Microbiol.* **45,** 305–307.

Roberts, R. J. (1990), Restriction enzymes and their isoschizomers , *Nucleic Acids Res.* **18,** 2331–2365.

Rood, J. I., Cole, S. T. (1991), Molecular genetics and pathogenesis of *Clostridium perfringens*, *Microbiol. Rev.* **55,** 621–648.

Rothmel, R. K., Chakrabarty, A. M., Berry, A., Darzins, A. (1991), Genetic systems in *Pseudomonas*, *Methods Enzymol.* **204,** 485–514.

Rudolph, V. (1978), Bacteriophages in fermentation, *Process Biochem.* **13,** 16–26.

Rutberg, L. (1982), Temperate bacteriophages of *Bacillus subtilis*, in: *Molecular Biology of the Bacilli* (Dubnau, D. A., Ed.), pp. 247–268, New York: Academic Press.

Salas, M. (1988a), Phages with proteins attached to the DNA ends, in: *The Bacteriophages* (Calendar, R., Ed.), Vol. 1, pp. 169–191, New York: Plenum Press.

Salas, M. (1988b), Initiation of DNA replication by primer proteins: Bacteriophage ϕ29 and its relatives, *Curr. Top. Microbiol. Immunol.* **136,** 71–88.

Sandmeier, H., Iida, S., Arber, W. (1992), DNA inversion regions Min of plasmid p15B and Cin of bacteriophage P1: Evolution of bacteriophage tail fiber genes, *J. Bacteriol.* **174,** 3936–3944.

Sanger, F., Coulson, A. R., Hong, G. F., Hill, D. F., Petersen, G. B. (1982), Nucleotide sequence of bacteriophage lambda DNA, *J. Mol. Biol.* **162,** 729–773.

Schallehn, G., Eklund, M. W. (1980), Conversion of *Clostridium novyi* type D (*C. haemolyticum*) to alpha toxin production by phages of *C. novyi* type A, *FEMS Microbiol. Lett.* **7,** 83–86.

Selick, H. E., Barry, J., Cha, T.A., Munn, M., Nakanishi, M., Wong, M. L. (1987), Studies on the T4 bacteriophage DNA replication system, in: *Mechanisms of DNA Replication and Recombination* (Kelly, T., McMacken, R., Eds.), pp. 183–214, New York: Alan R. Liss Publ.

Sell, T. L., Schaberg, D. R., Fekety, D. R. (1983), Bacteriophage and bacteriocin typing scheme for *Clostridium difficile*, *J. Clin. Microbiol.* **54,** 69–73.

Sing, W. D., Klaenhammer, T. R. (1990), Plasmid-induced abortive infection in lactococci: a review, *J. Dairy Sci.* **73,** 2239–2251.

Smith, G. R. (1988), Homologous recombination in prokaryotes, *Microbiol. Rev.* **52,** 1–28.

Smith, H. W., Huggins, M. B., Shaw, K. M. (1987), The control of experimental *Escherichia coli* diarrhoea in calves by means of bacteriophages, *J. Gen. Microbiol.* **133,** 1111–1126.

Sonnen, H. (1991), Molekulargenetische Charakterisierung von Phagen-Wirt-Beziehungen bei coryneformen Aminosäure-Produzenten, *Dissertation*, Technische Hochschule Darmstadt, FRG.

Sonnen, H., Schneider, J., Kutzner, H. J. (1990a), Characterization of ϕGA1, an inducible phage particle from *Brevibacterium flavum*, *J. Gen. Microbiol.* **136,** 567–571.

Sonnen, H., Schneider, J., Kutzner, H. J. (1990b), Corynephage Cog, a virulent bacteriophage of *Corynebacterium glutamicum*, and its relations to ϕGA1, an inducible phage particle from *Brevibacterium flavum*, *J. Gen. Virol.* **71,** 1629–1633.

Stassen, A. P. M., Schoenmakers, E. F. P. M., Yu, M., Schoenmakers, J. G. G., Konings, R. N. H. (1992), Nucleotide sequence of the genome of the filamentous bacteriophage-I2-2: module evolution of the filamentous phage genome, *J. Mol. Evol.* **34,** 141–152.

Sternberg, N. (1990), Bacteriophage P1 cloning system for the isolation, amplification, and recovery of DNA fragments as large as 100 kilobase pairs, *Proc. Natl. Acad. Sci. USA* **87,** 103–107.

Sternberg, N. L., Maurer, R. (1991), Bacteriophage-mediated generalized transduction in *Escherichia coli* and *Salmonella typhimurium*, *Methods Enzymol.* **204,** 18–42.

Stewart, A. W., Johnson, M. G. (1977), Increased numbers of heat-resistant spores produced by two strains of *Clostridium perfringens* bearing temperate phage s9, *J. Gen. Microbiol.* **103,** 45–50.

STEWART, C. (1988), Bacteriophage SPO1, in: *The Bacteriophages* (CALENDAR, R., Ed.), Vol. 1, pp. 477–516, New York: Plenum Press.

SYMONDS, N., TOUSSAINT, A., VAN DE PUTTE, P., HOWE, M. M. (Eds.) (1987), *Phage Mu*, Cold Spring Harbor, NY: Cold Spring Harbor Laboratory Press.

SYVANEN, M. (1984), The evolutionary implications of mobile genetic elements, *Annu. Rev. Genet.* **18,** 271–293.

TEUBER, M., LOOF, M. (1987), Genetic characterization of lactic streptococcal bacteriophages, in: *Streptococcal Genetics* (FERRETTI, J. J., CURTIS III, R., Eds.), pp. 250–258, Washington, DC: American Society for Microbiology.

TEUBER, M., GEIS, A., NEVE, H. (1992), The genus *Lactococcus*, in: *The Prokaryotes*, 2nd Ed. (BALOWS, A., TRÜPER, H. G., DWORKIN, M., HARDER, W., SCHLEIFER, K.-H., Eds.), pp. 1482–1501, New York: Springer Verlag.

THORNE, C. B., KOWALSKI, J. B. (1976), Temperate bacteriophages for *Bacillus licheniformis*, in: *Microbiology – 1976* (SCHLESINGER, D., Ed.), pp. 303–314, Washington, DC: American Society for Microbiology.

TRAUTWETTER, A., BLANCO, C. (1988), Isolation and preliminary characterization of twenty bacteriophages infecting either *Brevibacterium* or *Arthrobacter* strains, *Appl. Environ. Microbiol.* **54,** 1466–1471.

TRAUTWETTER, A., BLANCO, C., SICARD, M. (1987a) Structural characteristics of the *Corynebacterium lilium* bacteriophage CL31, *J. Virol.* **61,** 1540–1545.

TRAUTWETTER, A., BLANCO, C., BONNASSIE, S. (1987b), Characterization of the corynebacteriophage CG33, *J. Gen. Microbiol.* **133,** 2945–2952.

TYUTIKOV, F. M., BESPALOVA, I. A., REBENTISH, B. A., ALEXANDRUSHKINA, N. N., KRIVISKY, A. S. (1980), Bacteriophages of methanotrophic bacteria, *J. Bacteriol.* **144,** 375–382.

TYUTIKOV, F. M., YESIPOVA, V. V., REBENTISH, B. A., BESPALOVA, I. A., ALEXANDRUSHKINA, N. I., GALCHENKO, V. V., TIKHONENKO, A. S. (1983), Bacteriophages of methanotrophs isolated from fish, *Appl. Environ. Microbiol.* **46,** 917–924.

TWORT, F. W. (1915), An investigation on the nature of ultra microscopic viruses, *Lancet* **II,** 1241–1243.

VAN DUIN, J. (1988), The single-stranded RNA bacteriophages, in: *The Bacteriophages* (CALENDAR, R., Ed.), Vol. 2, pp. 117–168, New York: Plenum Press.

VATS, S., STUTTARD, C., VINING, L. C. (1987), Transductional analysis of chloramphenicol biosynthesis genes in *Streptomyces venezuelae*, *J. Bacteriol.* **169,** 3809–3813.

WALKER, G. C. (1987), The SOS response of *Escherichia coli*, in: *Escherichia coli and Salmonella typhimurium: Cellular and Molecular Biology* (NEIDHARDT, F. C., Ed.), pp. 1346–1357, Washington, DC: American Society for Microbiology.

WATSON, J. D., HOPKINS, N. H., ROBERTS, J. W., STEITZ, J. A., WEINER, A. M. (1987), *Molecular Biology of the Gene*, Menlo Park: The Benjamin/Cummings Publ. Co., Inc.

WEBSTER, R. G., GRANOFF, A. (Eds.) (1993), *Encyclopedia of Virology: Bacteriophages*, London: Academic Press, in press.

WEISBERG, R. A. (1987), Specialized transduction, in: *Escherichia coli and Salmonella typhimurium: Cellular and Molecular Biology* (NEIDHARDT, F. C., Ed.), pp. 1169–1176, Washington, DC: American Society for Microbiology.

WHITEHEAD, H. R. (1953), Bacteriophage in cheese manufacture, *Bacteriol. Rev.* **17,** 109–123.

WHITEHEAD, H. R., COX, G. A. (1935), The occurrence of bacteriophages in starter cultures of lactic streptococci, *New Zealand J. Sci. Technol.* **16,** 319–320.

WICKNER, S. (1992), DNA replication of plasmids and phages, *Annu. Rev. Genet.* **26,** in press.

WILSON, G. G., MURRAY, N. E. (1991), Restriction and modification systems, *Annu. Rev. Genet.* **25,** 585–627.

WOOD, A. G., WHITMAN, W. B., KONISKY, J. (1989), Isolation and characterization of an archaebacterial viruslike particle form *Methanococcus voltae* A3, *J. Bacteriol.* **171,** 93–98.

WÜNSCHE, L., FISCHER, H., KIESEL, B. (1983a), Lysogenie und lysogene Konversion bei methylotrophen Bakterien. I. Nachweis des lysogenen Zustandes des fakultativ methanolassimilierenden Stammes *Acetobacter* MB58/1 und Charakterisierung seines temperenten Phagen MO1, *Z. Allg. Mikrobiol.* **23,** 81–94.

WÜNSCHE, L., KIESEL, B., FISCHER, H. (1983b), Lysogenie und lysogene Konversion bei methylotrophen Bakterien. II. Lysogene Konversion bei fakultativ methanolassimilierenden *Acetobacter*-Stämmen, *Z. Allg. Mikrobiol.* **23,** 189.

WURTZ, M. (1992), Bacteriophage structure, *Electron. Microscop. Rev.* **5,** 283–309.

YAMANAKA, S., KASHIMA, N., MITSUGI, K. (1975), Method for limiting damage due to bacteriophages in fermentation media, *U.S. Patent* [19]. 3.880.718 [11]. [45] Apr. 29, 1975.

YARMOLINSKY, M. B., STERNBERG, N. (1988), Bacteriophage P1, in: *The Bacteriophages* (CALENDAR, R., Ed.), Vol. 1, pp. 291–438, New York: Plenum Press.

YOUNG, R. (1992), Bacteriophage lysis: mechanism and regulation, *Microbiol. Rev.* **56,** 430–481.

ZAHLER, S. A. (1988), Temperate bacteriophages of *Bacillus subtilis*, in: *The Bacteriophages* (CALEN-

DAR, R., Ed.), Vol. 1, pp. 559–593, New York: Plenum Press.

ZILLIG, W., GROPP, F., HENSCHEN, A., NEUMANN, H., PALM, P., REITER, W. D., RETTENBERGER, M., SCHNABEL, H., YEAT, S. (1986), Archaebacterial virus-host systems, *Syst. Appl. Microbiol.* **7**, 58–66.

ZILLIG, W., REITER, W. D., PALM, P., GROPP, F., NEUMANN, H., RETTENBERGER, M. (1988), Viruses of Archaebacteria, in: *The Bacteriophages* (CALENDAR, R., Ed.), Vol. 1, pp. 517–558, New York: Plenum Press.

ZINDER, N. D., HORIUCHI, K. (1985), Multiregulatory element of filamentous bacteriophages, *Microbiol. Rev.* **49**, 101–106.

17 Plant Cell Cultures

MAIKE PETERSEN
AUGUST WILHELM ALFERMANN
Düsseldorf, Federal Republic of Germany

1 The Organization of Higher Plants and Plant Cells and Their Impact on Plant Cell Biotechnology 578
1.1 The Organization of the Higher Plant Cell 578
1.2 The Organization of the Higher Plant Cormus 579
1.3 Cell Types in Higher Plants and Plant Cell, Tissue and Organ Cultures 579
1.4 Metabolic Features of Plants and Plant Cell, Tissue and Organ Cultures 580
2 *In vitro* Cultures of Plants and Plant Cells 581
2.1 Types of Plant Cell, Tissue and Organ Cultures and Their Historic Development 581
2.2 Callus Cultures 582
2.3 Suspension Cultures 583
2.3.1 Characteristics of Suspension Cultures 583
2.3.2 Cultivation in Bioreactors 586
2.3.3 Assessment of Growth in Suspension Cultures 588
3 Plant Cell Cultures and the Production of Secondary Compounds 589
3.1 Optimization of the Culture Media 590
3.2 Selection of High-Producing Cell Lines 592
3.3 Stress: Elicitation, Induction and Stimulation 593
3.4 Design of Culture Vessels and Physical Conditions 595
3.5 Feeding of Metabolic Precursors and Biotransformation Reactions 595
3.6 Immobilization of Plant Cells 598
3.7 Permeabilization and Two-Phase Cultures 598
3.8 Differentiated Cultures 600
3.9 Genetic Engineering of Plant Cells 601
4 Application of Plant Tissue Culture Techniques in Agriculture 603
5 Outlook 604
6 References 606

1 The Organization of Higher Plants and Plant Cells and Their Impact on Plant Cell Biotechnology

1.1 The Organization of the Higher Plant Cell

The cells of higher plants (Fig. 1) are typical eukaryotic cells. They have a nucleus which harbors the nuclear DNA organized in chromosomes together with histones and other nuclear proteins and one or more nucleoli. The nucleus is surrounded by the nuclear membrane with pores which mediate the contact with the cytoplasm. In the cytoplasm, there are membraneous organelles like the endoplasmic reticulum (ER), the Golgi apparatus, microbodies and vesicles. Mitochondria, surrounded by a double unit membrane and containing mitochondrial DNA, are genetically semi-autonomous organelles. Their internal membranes contain the respiratory electron transport chain. Ribosomes as the protein synthesizing machinery are found in the cytoplasm as well as in mitochondria and plastids. Plant cells further contain microtubules and microfilaments. The cell is surrounded by the plasmalemma, a unit membrane. Special to plant cells are the rigid cell wall, plastids as another type of DNA-containing, semi-autonomous organelles and the large central vacuole of the differentiated cell. The cell wall is secreted by the protoplast on the outside of the plasma membrane and consists of the middle lamella, which holds adjoining cells together, the primary and the secondary walls. Chemically, the cell wall is formed from pectins, hemicellulose and cellulose together with wall proteins. The rigid cell wall imposes problems on certain tissue culture techniques such as isolation of single cells, cell fusion and genetic manipulation of plant cells. Therefore, the cell wall and the middle lamella are often removed by treatment with cellulases and pectinases giving rise to protoplasts. These protoplasts are able to regenerate their cell walls after removal of the hydrolytic enzymes. The vacuole is surrounded by a single unit membrane, the tonoplast. Meristematic cells contain a number of small vacuoles, whereas differentiated cells have one large central vacuole which can occupy more than 90% of the cell volume. In the vacuole water, ions, enzymes, storage products and other compounds, which have to be isolated from the metabolism of the cytoplasm, are accumulated. The water-storing vacuole in cooperation with the rigid cell wall maintains the turgor of the cell which is necessary for land plants to keep their body upright. The plastids are typical organelles of

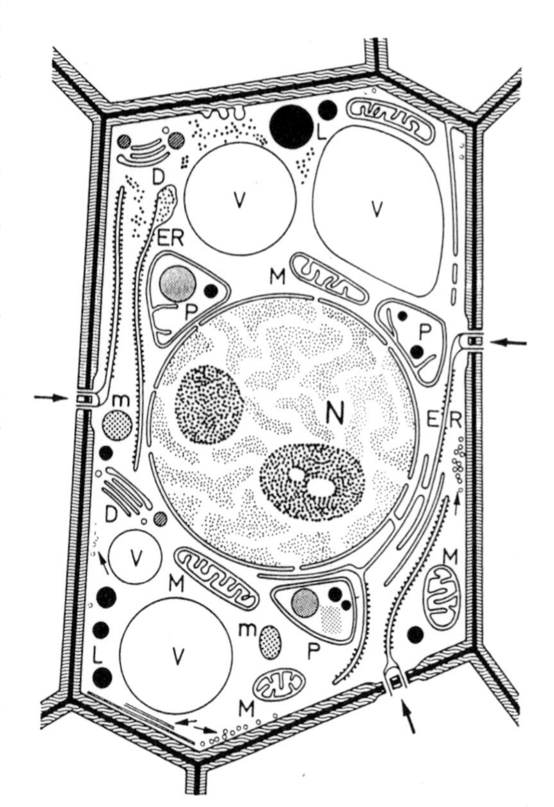

Fig. 1. Scheme of a meristematic plant cell. N, nucleus with chromatin and two nucleoli; ER, endoplasmic reticulum with and without ribosomes; D, dictyosomes (components of the Golgi apparatus); V, vacuoles; M, mitochondria; P, plastids (here: proplastids); m, microbodies; L, lipid bodies (oleosomes). The cell is surrounded by the plasma membrane (plasmalemma) and a primary cell wall. The arrows outside of the cell indicate plasmodesmata. The arrows inside the cell point to microtubules (from MOHR and SCHOPFER, 1992, with kind permission of the authors and Springer Verlag).

the plant. They are surrounded by two unit membranes and contain DNA. Photosynthesizing plastids (chloroplasts) have a developed intramembrane system carrying the photosynthetic pigments, chlorophylls and carotenoids and the protein complexes responsible for the photosynthetic electron transport. The accumulation of assimilation products can take place in form of starch grains in the plastids. When plant cells are kept away from light, the chloroplasts degenerate to small colorless proplastids without photosynthetic capacity. This is the case in a great number of plant cell or tissue cultures which are kept as heterotrophic cultures. In plant cell and tissue cultures fully active chloroplasts can only be found in photosynthetically active cell cultures cultivated in light.

For further details about the structure of plant cells, the interested reader is referred to the respective textbooks (e.g., WEIER et al., 1974; RAVEN et al., 1976; VON SENGBUSCH, 1988; STRASBURGER et al., 1991).

1.2 The Organization of the Higher Plant Cormus

Higher plants as the most developed cormophytes show a typical differentiation into roots, stems and leaves. The stem supports leaves and flowers and contains the elements for water and nutrient transport into the above ground parts of the plant as well as for the translocation of photosynthetic products. The shoot is anchored in the ground by the root system which is also responsible for the uptake of mineral nutrients from the ground. The leaves are attached to the shoot and its branches usually in such a way that they can capture the light they need for photosynthesis in an optimal manner. Typical leaves are flat organs where the cells are arranged in layers: upper epidermis, the parenchymateous mesophyll layers and the lower epidermis. Typically shoots and roots have meristematic centers which allow them to grow. Thus plants are – in contrast to animal systems – open systems with generally unlimited growth. A unique feature of plants as open systems is, besides the existence of primary meristematic centers, the capability of almost every single living plant cell – irrespective of its state of differentiation – to dedifferentiate and to become a meristematic cell again. Thus plant cells can be considered to be omnipotent. This process is the basis for all plant tissue culture techniques, since dedifferentiated cells can be manipulated to grow further in this dedifferentiated state or to redifferentiate to all possible cell types and organs of a higher plant as well as to a complete plant again.

1.3 Cell Types in Higher Plants and Plant Cell, Tissue and Organ Cultures

The major structural features of plant cells in culture are the same as of organized plant cells. However, the most common cell types of higher plants which have some relevance for plant tissue culture techniques will be discussed here. For a more detailed list of plant cell types see ALBERTS et al. (1989).

Meristematic cells have retained their dividing capacity and their essentially undifferentiated state. They are small, approximately isodiametric and contain only small vacuoles. Their cell walls are comparatively thin and have a low degree of lignification, thus making them suitable for protoplast isolation. Meristematic cells show a high metabolic activity and a high cell division rate. Since they have a low degree of differentiation, meristematic cells are well suited for the initiation of plant cell cultures and morphogenic cultures. Meristematic cells and cell layers in all types of plant tissues, e. g., the cambium layers of the vascular bundles, can give rise to the proliferation of organized or unorganized cells as the basis of all types of plant cell, tissue and organ cultures.

Layers of epidermal cells form the boundary of the plant cormus to the environment. These cells have rigid cell walls and incrusted or acrusted layers of cutin, suberin or waxes on their outside thus forming a real barrier against the loss of water or the penetration of chemicals or microorganisms. In plant cell cultures this cell type does not occur normally. In organ cultures – root or shoot cultures – or *in vitro* regenerated plants epidermal cells occur, but cannot serve their purpose properly due to a low degree of lignification and insufficient wax layers. However, such plants can be acclimatized to normal environmental conditions.

Vascular cells, the elongated cells of phloem and xylem, have the task of translocation of water, inorganic nutrients and metabolic products in intact plants. Phloem cells still have a living cytoplasm, but have lost their nucleus or vacuoles. Serving their purpose of product translocation phloem cells fuse to a huge syncytium with porous transverse cell walls. In contrast, xylem cells serve their purpose as translocation elements, mainly for water and inorganic nutrients, as dead cells. Longitudinal cell walls are highly lignified, whereas the transversal cell walls are greatly reduced. Vascular cells are present in regenerated plants or organ cultures. However, highly elongated cells with typical wall reinforcements can also be found in suspension and callus cultures. This differentiation process is considered to be influenced by the phytohormone balance and the stage of the growth cycle.

Mesophyll cells form the main photosynthetic tissue in organized leaves. Consequently they contain the major part of the chloroplasts. These are located mainly in the thin cytoplasmic plasma layer surrounding the large central vacuole. In leaves of the "normal" C_3-plants two types of cells form the mesophyll: the more cylindrical cells of the palisade parenchyma and the irregularly shaped cells of the spongy parenchyma with lower numbers of plastids. Cells equivalent to photosynthesizing mesophyll cells in suspension and callus cultures can only be found in mixotrophic and photoautotrophic cell cultures. In organized shoot cultures as well as regenerated *in vitro* cultured plants this cell type occurs as in normal plants. In mixo- and heterotrophic cell cultures a frequent cell type, besides the small plasma-rich cells similar to meristematic cells, are large cells with a large central vacuole but without fully developed chloroplasts due to the cultivation with sugar as a carbon source and the low or missing irradiation. These cells can be compared to mesophyll cells in organized tissues.

The cell type and the structural differentiation in plant cell, tissue and organ cultures is mostly determined by the culture conditions, e. g., the phytohormone balance. The origin of a plant tissue used for the establishment of an *in vitro* culture is of less importance especially after longer periods of cultivation.

1.4 Metabolic Features of Plants and Plant Cell, Tissue and Organ Cultures

The major metabolic features of plant cell, tissue and organ cultures are widely similar to organized plants, and the reader is referred to plant physiology textbooks (e. g., BIDWELL, 1974; LEVITT, 1974; MOHR and SCHOPFER, 1992; RICHTER, 1988; STUMPF and CONN, 1980–1988). Concerning primary metabolic processes cultured plant cells show essentially the same properties as naturally organized plant cells as far as their metabolism is not influenced by the culture process itself. A substantial difference of suspension-cultured plant cells is the fact that they are submersed in a liquid containing all necessary nutrients. These nutrients can enter the cells either by diffusion or by active uptake. The nutrients can enter the cells against a concentration gradient only by active uptake, and this process therefore requires energy. Investigations on the uptake of nutrients into suspension cells have shown that active transport systems occur for all kinds of nutrients (for a review see DOUGALL, 1980).

Carbohydrates, usually sugars, from the culture media are either metabolized via glycolysis, the tricarboxylic acid cycle and the pentose phosphate pathway in order to yield energy and metabolic intermediates, or they are used for the synthesis of polysaccharides such as plant cell wall components or starch. These metabolic processes have been shown to be active in a number of cell cultures (for reviews see MARETZKI et al., 1974; DOUGALL, 1980). Even heterotrophically grown cell cultures are able to fix CO_2 as shown by NESIUS and FLETCHER (1975) and GATHERCOLE et al. (1976). This process becomes more important in photoheterotrophically cultivated cell cultures.

Chlorophyll-containing cells grown with addition of a carbohydrate source can contribute a considerable amount of their accumulated dry matter by photosynthetic CO_2 fixation (NATO et al., 1977). A number of cell cultures have been reported to grow photoautotrophically. Usually these cultures require elevated CO_2 levels but no exogenously added organic carbon source (for reviews see HÜSEMANN, 1988; HÜSEMANN et al., 1989). In these cultures the common photosynthetic reactions provide the cells with carbon compounds, although CO_2 fixation by PEP carboxylase is more active in cell cultures than in common C_3-plants (HÜSEMANN et al., 1989; U. FISCHER, personal communication). Another peculiarity is the blue-light dependent chloroplast differentiation (HÜSEMANN et al., 1989).

Cell cultures can be fed with different kinds of nitrogen sources: nitrate, ammonium and amino acids are the most common ones. When different nitrogen compounds are given simultaneously, amino acids are usually depleted first followed by ammonium and nitrate (DOUGALL, 1980). Therefore, the most reduced nitrogen source is consumed first. Nitrate and nitrite reductase have been shown to be active in cell cultures. The incorporation of ammonium into organic compounds is the same as in whole plants: glutamic dehydrogenase, glutamate synthase (GOGAT) and glutamine synthase are active in cell cultures (for reviews see DOUGALL, 1977; MIFLIN and LEA, 1977).

Sulfate is the most common sulfur source in plant cell cultures besides the amino acids cysteine and cystine. Since this is the oxidation state required for further metabolization, the cells only couple the sulfate with ATP by ATP-sulfurylase giving rise to adenosine-5'-phosphosulfate (APS) and pyrophosphate (REUVENY and FILNER, 1976).

Phytohormones are supplied exogenously to most plant cell cultures which leads to a certain metabolic and differentiation state of the cultured cells. In the plant itself not all cells synthesize phytohormones although the genetic information for hormone synthesis is present in all the cells. Growth and development are regulated by phytohormone gradients across the plant organs. Therefore, cell cultures mostly require exogenous phytohormones. However, cell cultures are known which synthesize their own phytohormones and therefore are called "habituated". The state of hormone autotrophy can also be accomplished by transformation with *Agrobacterium* species as shown in Sect. 3.8.

Plant secondary metabolism shows much more variation in organized plants, plant organ cultures or undifferentiated cultures, since it is often coupled to a certain degree of differentiation. The special features of plant secondary metabolism in cultured plant cells or organs will be discussed in more detail in Sect. 3.

2 *In vitro* Cultures of Plants and Plant Cells

2.1 Types of Plant Cell, Tissue and Organ Cultures and Their Historic Development

In vitro cultures of plants are nearly as old as the century. A short overview of the development of cultivation techniques for plant cells and organs is given in Tab. 1.

In vitro cultures of plants or plant cells can have various degrees of differentiation. Whole plants or seedlings can be grown aseptically on defined culture media. Organ cultures, such as roots and shoots, can be propagated with the help of phytohormone-containing media or as transformed organs (see Sects. 3.8 and 4). Embryos can be isolated from seeds and can be further cultivated, or somatic embryos can be formed from dedifferentiated cells under a suitable phytohormone regime. Undifferentiated tissues are propagated as callus cultures on solid or liquid media (see Sect. 2.2). Cells and cell aggregates of up to 200 undifferentiated cells are cultivated as suspension cultures in liquid media (see Sect. 2.3) These cultures can be grown as heterotrophic, mixotrophic or photoautotrophic cultures. After removal of the cell wall protoplast cultures are obtained which can regenerate the cell wall and further grow as suspension or callus or regenerate organs or whole plants. The different types of *in vitro* cultures can be converted into each other by appropriate techniques, mostly by the phytohormone regime. This interrelationship also makes *in vitro* cultures of plants a tool for other scientific interests, e. g., developmental biology.

Tab. 1. Short Historical Overview of *in vitro* Culture of Plants, Organs and Plant Cells

1902	HABERLANDT	Maintenance of single plant cells in simple nutrient media without cell division
1904	HANNIG	Embryo culture
1934	WHITE	Establishment of an actively growing clone of tomato roots
1937	WHITE	Discovery of the importance of B vitamins for the growth of roots
1937	WENT and THIMANN	Discovery of the role of auxin in the control of plant growth
1939	WHITE; GAUTHERET; NOBÉCOURT	Establishment of callus cultures from dicotyledonous plants
1947	LEVINE	Detection of somatic embryogenesis
1950	MOREL	Callus cultures from monocotyledonous plants
1954	MUIR, HILDEBRANDT and RIKER	Establishment of suspension cultures
1955	SKOOG and coworkers	Detection of kinetin
1960	BERGMANN	Single cell cloning
1960	COCKING	Isolation of protoplasts
1964	MOREL	Micropropagation
1964	GUHA and MAHESHWARI	Anther cultures, regeneration of haploid plants
1966	LUTZ	Regeneration of whole plants from single cells
1972	CARLSON and coworkers	Protoplast fusion and regeneration of somatic hybrid plants
1978	MELCHERS and coworkers	Intergeneric fusion of tomato and potato protoplasts and hybrid plant regeneration

2.2 Callus Cultures

Callus cultures are larger aggregates of undifferentiated plant cells usually grown on solidified nutrient media (Fig. 2). The state of undifferentiated growth is maintained by the phytohormone balance, mainly auxins and cytokinins, added to the medium. Other components of the medium are mineral salts, carbohydrate sources and supplements such as vitamins and amino acids. The media are solidified with agar-agar or other gelling agents, mostly polysaccharides. In normal plant life, callus tissue is formed after wounding, the so-

Fig. 2. Callus culture (right) and suspension culture (left) of *Coleus blumei.*

called wound callus. This cell mass helps to close the wound rapidly. For the establishment of sterile *in vitro* cultures this process is also exploited: a tissue is wounded and the induced callus is further cultivated on nutrient media. Almost every living plant tisssue can give rise to callus cultures. However, young tissues (e. g., from seedlings), which are not highly lignified, and tissues with meristematic or promeristematic centers are preferred, but many fully differentiated plant cells can dedifferentiate and thus form callus as well. The most severe problem during the establishment of callus cultures is the high degree of microbial contamination of wild-grown plants. The plant tissue therefore must be sterilized prior to transferring it to the nutrient medium, since the media used for plant tissue cultures are also good media for microbial growth. For sterilization ethanol, hypochlorite solutions, hydrogen peroxide, mercuric chloride and other agents are used alone or in combination. Sterilized explants are transferred to the solid media in Petri dishes or other suitable containers which should be sealed in order to minimize evaporation. The containers are incubated in the light or dark at approximately 25 °C. Callus formation can be observed after a few days up to several weeks. Contamination is usually detected earlier. Some cultures develop better in the light, on the other hand, "browning" caused by polyphenols is also higher in the light. Therefore, the optimal conditions for callus formation as well as the suitable sterilizing procedures and nutrient media have to be determined empirically. Suitable media for a number of plants have been compiled by GEORGE et al. (1987, 1988). Callus formation is commonly observed at the cut surface of the explant, but sometimes callus formation also starts in the interior of the explant and breaks through the surface of the tissue. After formation of a fairsized callus, the callus is cut from the explant and transferred to fresh medium. If suitable hormone concentrations are used, the callus grows further in this undifferentiated state and does not regenerate organs or embryoids.

Although it is possible to get callus cultures from nearly every species, some plants may prove to be recalcitrant. Attaining sterility is often a problem, other problems may be browning or missing callus formation. For "tricks" for these recalcitrant plants the reader is referred to textbooks dealing with theory and practice of tissue culture (see, e.g., BHOJWANI and RAZDAN, 1983; GEORGE and SHERRINGTON, 1984; DIXON, 1985; SEITZ et al., 1985).

A rather promising source for the establishment of cultures are seeds of the respective plants. Usually seeds are easily sterilized, since they are much more resistant to the sterilizing agents. Sterile germinated seedlings may then be used as an explant source for the establishment of callus cultures as well as for organ cultures. Here also problems may arise, e.g., for seeds which have to pass through the digestive tract of animals prior to germination.

The texture of callus cultures can be very variable: hard, compact callus clumps occur as well as friable callus or more slimy cell masses. After a number of subcultivations, which are usually performed every ten days to six weeks, the appearance of the culture may change until a stable callus line is established. Selection of the desired parts from heterogeneous cultures (e. g., for pigmentation) enhances the establishment of a cell line with the desired traits.

Callus cultures are not commonly used for production of chemicals, for investigations of biosynthetic pathways or for developmental or physiological investigations. However, they are the source for the establishment of suspension cultures (see Sect. 2.3) and usually form the stock of a tissue culture laboratory. Physiological traits, such as the production of secondary compounds (see Sect. 3) are more stable in callus cultures than in the respective suspensions, and the callus stock therefore provides the material for the establishment of new suspensions after the loss of their production capacities. Moreover, callus cultures are more suitable as stock, since they have longer culture periods.

2.3 Suspension Cultures

2.3.1 Characteristics of Suspension Cultures

Callus cultures can be transferred into liquid medium to establish a suspension culture (Figs.

2 and 3), which is placed on a shaker to supply the cells with sufficient oxygen. See DIXON (1985), SEITZ et al. (1985), STREET (1977), and MORRIS and FOWLER (1981) for practical performance and isolation of fine cell suspension cultures. Some species can turn out to be recalcitrant to give fine suspensions. Parallel to a change in aggregate size a change in physiological behavior or even of the genetics of the cells may occur during several subcultivations. The latter can be seen in changing chromosome numbers of many plant cell cultures (BAYLISS, 1973; SUNDERLAND, 1977). Plant suspension cultures are known to be relatively instable, e.g., with respect to production of natural products (DEUS-NEUMANN and ZENK, 1984), a phenomenon which is known for microorganisms as well. Callus cultures, obviously, are more stable in this respect. Therefore, after a possible loss of productivity, new producing suspension cultures can be initiated in most cases from the stock callus cultures. To overcome these problems with genetic instability of plant callus and cell suspension cultures protocols were elaborated in many laboratories to cultivate cells at reduced temperature or to store them at –178 °C in liquid nitrogen and to preserve by this treatment a given genotype over a long time. There is an increasing list of reports confirming the success of this concept (WITHERS, 1986; KARTHA, 1987; SEITZ, 1987; KAIMORI and TAKAHASHI, 1989).

Normally, cell division in suspensions is not synchronized. Synchronized cell cultures are of importance, if cell-cycle dependent reactions are investigated (KOMAMINE et al., 1990). Several culture regimes have been described making use mainly of cell starvation by reducing certain nutrients or by a temporary blocking of cell division by inhibitors of DNA replication or treatment with hormones, to achieve at least partial synchrony of cell division for several cell cycles (CONSTABEL et al., 1977; ERIKSSON, 1966; BAYLISS and GOULD, 1974; KING and STREET, 1977; OKAMURA et al., 1973; AMINO et al., 1985; for a review see KING, 1980).

To establish a well growing cell suspension culture it is necessary, to inoculate at least 10^4 cells per mL, otherwise the cells may not divide. This value depends, however, also on the aggregate size. The aggregate size has an influence on natural product formation as well (SHULER, 1981; HULST et al., 1989). The reason why plant cells need such relatively high cell densities for undergoing division surely is that they lose hormones and/or vitamins and nutrients to the surrounding medium. This could be demonstrated by the so-called feeder culture experiments in culturing protoplasts at low cell

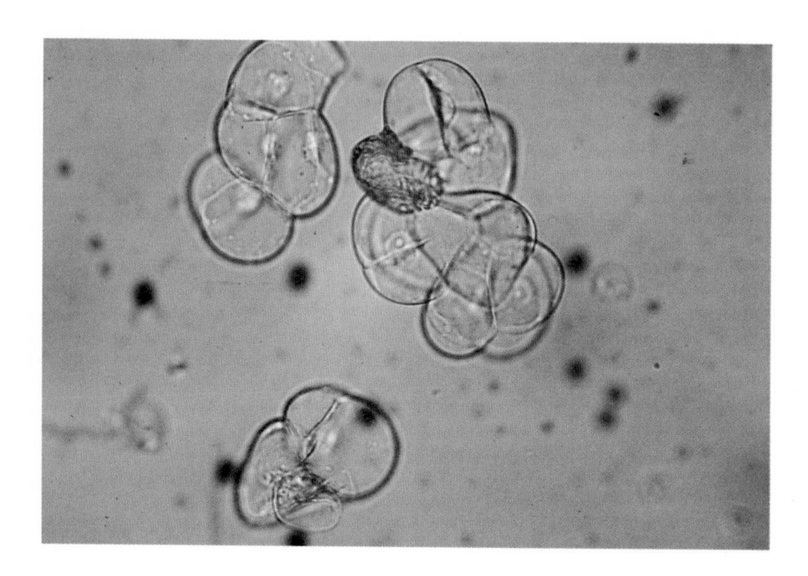

Fig. 3. Cell aggregates of a *Digitalis* cell suspension culture.

densities (MUIR et al., 1958; WEBER and LARK, 1979) or by culturing single cells or protoplasts in extremely small volumes (KOOP et al., 1983). In contrast to microorganisms, plant cells have a long doubling time (T_D), normally between 2 and 5 days. During the exponential growth phase, however, a doubling time of 15 h has been found (NOGUCHI et al., 1977). One normally speaks of a good growing suspension culture, if the cells divide on the average every second day. This means that during a subculture time of 7 days only 3–4 cell divisions occur. Tab. 2 compares typical features of bacterial and plant cells.

In a batch culture of plant cells the same different phases of growth can be observed as with microorganisms (Fig. 4). After a lag phase an exponential phase follows. Normally, the exponential phase is very short and is followed by a linear growth phase indicating that growth is already hampered by one or more cultural factors. A deceleration phase can be observed before a stationary phase, which may be followed by a death or decline phase. Fig. 4 shows a batch culture of carrot cells in Erlenmeyer flasks over a growth cycle of 15 days. The cells multiply from about 500000 cells to about 4 million cells per mL, which means that the cells divide about 3 times. Fig. 4 shows as well that the fresh weight of the cells increases for a somewhat longer time than the dry weight indicating that – typical for plant cells – the vacuole is still enlarging. Inversely to the growth of the cells, the nutrients of the medium decrease, shown here by the decrease of the nitrate content and conductivity, sucrose concentration and osmolarity.

Transfer of the cells to new medium is usually performed using a pipette with an enlarged opening; in some laboratories a transfer is achieved with a spatulum or a spoon with small holes. With the first method, which is very well applicable to fine suspension cultures, a considerable amount of spent medium is transferred during inoculation besides the cells. This medium – sometimes referred to as "conditioned medium" – may have a certain influence on cell growth, as besides the unused nutrients it may contain substances excreted by the cells. With the latter technique, mainly cells and only minute amounts of spent medium are inoculated. As stated above, in both cases an inoculation of a sufficient amount of cells into the new medium is important. Although 10^4 cells per mL normally are sufficient to achieve cell division, it is advisable to inocu-

Tab. 2. Comparison of Some Typical Features of Bacteria and Plant Cells in Suspension (modified after HESS, 1992)

Character	Bacteria	Plant Cells
Inoculum	Low	High (<10^4 cells · mL^{-1}, 5–10 vol. % of suspension)
Size	Small	Large
Diameter (µm)	1 – 10	40 – 200
Surface ($µm^2$)	7 – 70	4500[a]
Volume ($µm^3$)	1 – 40	900000[a]
Growth	Quick	Slow
Doubling time (h)	0.33[b]	15 – 180
Spec. growth rate (1/h)	2.10[b]	0.010 – 0.046
Behavior of growth	Single cells	Single cells, but mainly aggregates of few to several hundred cells
Sensitivity to shear	Low	Relatively high
Demand for oxygen	High	Much lower than bacteria
Genetic instability	Low	Relatively high
Localization of secondary products	Extracellular	Intracellular, sometimes extracellular

[a] *Coleus blumei*
[b] *Escherichia coli*

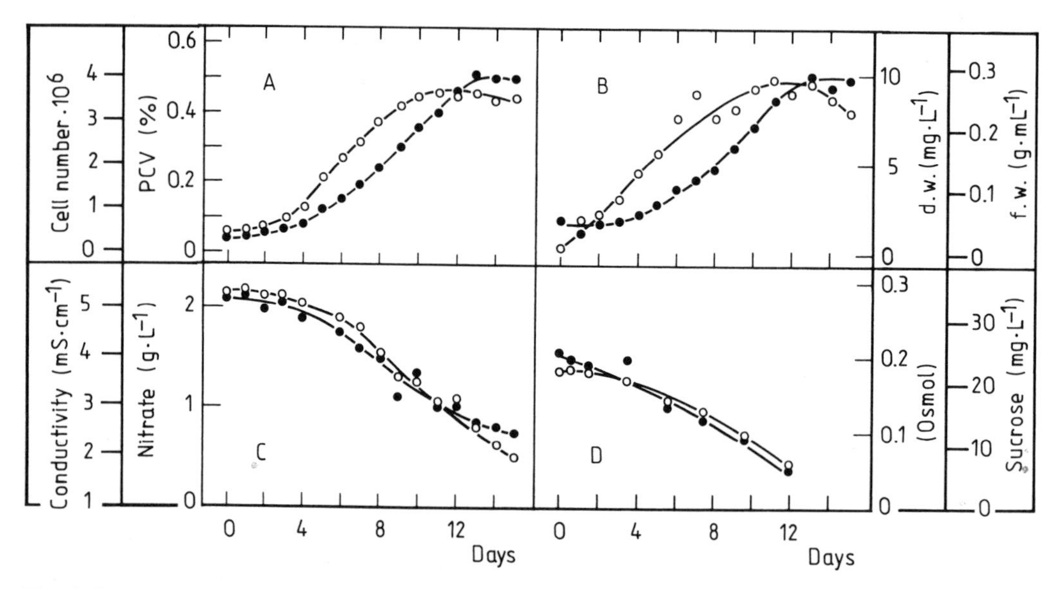

Fig. 4. Characterization of a suspension culture of *Daucus carota* over a culture period of 15 days. (A) Cell number (o–o), packed cell volume (PCV, ●–●); (B) dry weight (d. w., o–o), fresh weight (f. w., ●–●); (C) conductivity (o–o), nitrate content of the medium (●–●); (D) osmolarity (o–o), sucrose content of the medium (●–●) (modified from SEITZ et al., 1985, with kind permission of Gustav Fischer Verlag, Stuttgart).

late higher cell densities. Normally, cell densities of about $5 \cdot 10^5$ cells/mL, which is equivalent to about 1 g dry weight/L or 20–50 g fresh weight/L, or an inoculum of 5 to 25% of the total volume, are practical values.

2.3.2 Cultivation in Bioreactors

Plant cell suspensions can be handled in Erlenmeyer flasks of up to 5 liters containing about 2 L culture volume. A commercial production of natural products by plant cell cultures, however, is only achievable, if the cells can be grown in large bioreactor volumes to reduce the production costs. Therefore, already in 1959/1960 TULECKE and NICKEL cultivated plant cell cultures in simple bioreactor systems such as aerated carboys of 9 L volume, but also in steel tanks up to 134 L working volume. In the meantime, plant cell cultures were cultivated semi-continuously in bioreactors of 75 m^3 with a working volume of 60 m^3 for several months without problems with respect to sterility (WESTPHAL, 1990; WESTPHAL and RITTERSHAUS, 1991; for a recent review see, e.g., SCRAGG, 1992).

After the first experiments of TULECKE and NICKEL in 1959, in several laboratories simple, lab-made reactors of limited volume were used such as the system described by VELIKY and MARTIN (1970), later on used by WILSON for continuous culture (WILSON, 1980), or the culture device developed in the laboratory of STREET (STREET, 1977; KING and STREET, 1977). More recently normal commercial reactors developed for cultivation of microorganisms were used as such or in slightly modified version for cultivation of plant cells as well.

As documented in Tab. 2, plant cells show some important differences compared with microorganisms. They grow much more slowly, which results in longer cultivation times and therefore in higher demands for sterility of the reactors. Plant cells form quite large aggregates; therefore, the pipes for transfer as well as for sampling have to be enlarged. Due to the rigid cellulose cell wall, plant cells are relatively sensitive to shear stress, although after an appropriate selection one may find cell lines with a higher shear resistance (SCRAGG, 1990).

In 1977, WAGNER and VOGELMANN reported on the impact of shear stress on secondary product formation in plant cells cultivated in bioreactors using cell cultures of *Morinda citrifolia*. In these experiments it turned out that accumulation of anthraquinones in the *Morinda* cells was substantially enhanced, when the cells were cultivated in an airlift reactor, as compared to a reactor equipped with a flat turbine impeller and battle plates, or a reactor with a draft tube and a Kaplan turbine (type b50, Giovanola Frères, Monthey, Switzerland), a perforated disk impeller or even with the cells cultivated in shake flasks. It was concluded by the authors that this higher productivity was due to the lower shear stress imposed by the airlift reactor. *Morinda* cells are especially sensitive to shear stress at the end of the growth phase and during the stationary phase, when the cells synthesize the anthraquinones, because they consist of long chains of cells. These results prompted many laboratories to use airlift reactors for cultivation of plant cells. Indeed, airlift reactors turned out to be a very useful and simple instrument for cultivation of plant cells. However, they have also some disadvantages, especially if cells have to be cultivated at high densities to get as high product yields as possible in terms of weight per volume. Optimal mixing becomes difficult, and oxygen transfer rates decrease decisively (TANAKA, 1982a,b).

Although the oxygen demand of plant cells is much lower than that of microorganisms, in cases where a high oxygen tension is favorable for increased product formation, a sufficient oxygenation requires high ventilation rates. This can result in problems with foaming and this may cause problems with using necessary antifoams or even problems with sterility if the culture foams over. Furthermore, high ventilation rates favor the adhesion of a crust or "meringue" consisting of cells, foam and excreted polysaccharides at the top of the vessel (FOWLER, 1982). Additionally, these high ventilation rates may result in stripping off gaseous compounds, namely CO_2, which turned out to be benificial for the growth of the plant cells, although these are not photosynthetic under the conditions used (HEGARTY et al., 1986).

Furthermore, another gaseous compound, ethylene, may influence product formation. KOBAYASHI et al. (1991) showed for *Thalictrum* cells cultivated in Erlenmeyer flasks that ethylene influences berberine formation. SCRAGG et al. (1988) showed that not all plant cell cultures were as sensitive against shear as *Morinda* cells. Additionally, SCRAGG (1990) found that it is possible to select cell lines with a reduced shear sensitivity from a mother culture with high shear sensitivity. As a result, several research groups have tested other types of bioreactors and especially stirred tank reactors with modified stirrer types again. In summary, it can be said that plant cells can be cultivated not only in airlift reactors of different configurations, but also in stirred tank reactors.

An interesting new reactor is the so-called rotated drum reactor which was used by TANAKA et al. (1983) and FUJITA and TABATA (1987) for cultivating *Catharanthus roseus* and *Lithospermum erythrorhizon* at very high cell densities. Higher product yields than with stirred tank reactors were achieved. When using a normal stirred tank reactor, it is advisable to work at a relatively low stirrer speed. WAHL (1985) reported that *Digitalis* cells could be cultivated in a 1 m^3 reactor only, when the stirrer velocity was not higher than 60 rpm which is equivalent to an energy input of less than 0.2 $kW \cdot m^3$. Due to problems with insufficient mixing at such low stirrer speeds, several groups have tested other stirrer types. The so-called spin stirrer (Chemap AG, Volketswil/Switzerland; SPIELER et al., 1985), the paddle stirrer (TANAKA, 1981, 1987), the helical stirrer or "modul spiral stirrer" (ULBRICH et al., 1985) and an "Intermig stirrer" (RITTERSHAUS et al., 1989; K. WESTPHAL, personal communication) turned out to be better suited for cultivation of plant cells. However, it should not be forgotten that NOGUCHI et al. (1977) successfully cultivated tobacco cells in a reactor of a working volume of 15500 L equipped with a normal blade stirrer. As mentioned above, these tobacco cells showed the shortest doubling time (15 h) of all plant cell suspension cultures studied until today. Tab. 3 summarizes some reports on the use of bioreactors of various configurations.

Although, in general, it is stated that plant cell cultures need a lower aeration rate for growth and product formation than microorganisms, it turned out in recent years that there exist certain optimal conditions of oxygen supply on growth and product formation. Already

Tab. 3. Some Examples of Various Bioreactor Types Used for Cultivation of Plant Cells

Type of Reactor	Working Volume (L)	Plant Species Cultivated	Reference
Airlift, outer draught tube	85	*Catharanthus roseus*	SMART and FOWLER, 1984
Airlift, inner draught tube	10	*Morinda citrifolia*	WAGNER and VOGELMANN,
		C. roseus	1977
	200	*Digitalis lanata*	ALFERMANN et al., 1983
	1000	*D. lanata*	WAHL, 1985
Stirred tank reactor			
flat bladed stirrer	1000	*D. lanata*	WAHL, 1985
	750	*Lithospermum erythrorhizon*	FUJITA et al., 1982
	15500	*Nicotiana tabacum*	NOGUCHI et al., 1977
Paddle stirrer	10	*Cudrania tricuspidata*	TANAKA, 1981
	27	*D. lanata*	SPIELER et al., 1985
	3	*N. tabacum*	HOOKER et al., 1990
	20000	*Panax ginseng*	USHIYAMA, 1991
Modul spiral stirrer	32	*Coleus blumei*	ULBRICH et al., 1985
	390	*Podophyllum versipelle*	ULBRICH et al., 1988
Intermig stirrer	60000	*Echinacea purpurea*	WESTPHAL, 1990; and pers. commun.
Rotating drum reactor	4	*Catharanthus roseus*	TANAKA et al., 1983
	1000	*L. erythrorhizon*	FUJITA and TABATA, 1987
Membrane stirrer reactor	20	*Thalictrum rugosum*	PIEL et al., 1988

in 1975 KATO et al. reported on the influence of different oxygen transfer rates (k_La) on the growth of tobacco cells. SPIELER et al. (1985) and BREULING et al. (1985) demonstrated that product formation (biotransformation of β-methyldigitoxin and protoberberine alkaloid formation, respectively) can be increased by cultivating the cell cultures at constant dissolved oxygen concentrations optimal for growth and production. In a recent study, LECKIE et al. (1991) demonstrated that a higher oxygen supply improves alkaloid formation in *Catharanthus roseus*. On the other hand, SCRAGG et al. (1987) found that serpentine formation in *C. roseus* is not much affected, if the initial dissolved oxygen concentration is no lower than 20% of saturation. As already stated before, it can be very important not to use too high ventilation rates at the beginning of a batch culture cycle, when the biomass concentration is still low. This can lead to the stripping off of necessary gases from the medium or an overoxygenation of the cells resulting in long lag phases (HEGARTY et al., 1986).

2.3.3 Assessment of Growth in Suspension Cultures

Growth in plant cell suspension cultures is commonly measured by cell counting, determination of total cell volume (packed cell volume, pcv) or fresh (fw) and dry weight (dw) increase of the cell mass. Additionally, DNA or protein content of the cells is determined. Measuring different medium parameters, such as sugar and ion content, osmolarity and conductivity, the growth of the cells can be followed indirectly. It is important that not all cells of a given culture are viable during the whole cultivation time. Therefore, measuring the cell viability with different vital stains or physiological tests, e.g., using 2, 3, 5-triphenyl-tetrazolium chloride can be important. See BHOJWANI and RAZDAN (1983), BLOM et al. (1992), DIXON (1985), and SEITZ et al. (1985) for practical instructions.

3 Plant Cell Cultures and the Production of Secondary Compounds

Besides scientific interest, one of the great expectations regarding plant cell cultures is their use for the production of plant secondary compounds. Plants synthesize an enormous array of so-called secondary compounds which are not important in primary metabolism but rather in the survival of a plant in its environment (HARTMANN, 1985), e.g., as protecting compounds against fungal or microbial attack or herbivores, as flower colors or aromatic attractants. More than 20000 different chemical structures have been isolated from plants up to now, and many of them are used as pharmaceuticals, dyes, gums, fragrances or flavors. The advantages of a production of these compounds by plant cell cultures would be manyfold: High yields could be achieved by the optimization of the production conditions. The production would be independent of political, economical, climatic and geographical circumstances. A loss of plant material by herbivores or pathogens would be excluded. A standardized production would result in standardized contents of the active compounds. Very toxic compounds such as narcotics could be produced under controlled conditions. And finally, the adaptation to changes in the market would be faster and easier (HAHLBROCK, 1986).

Besides the production of secondary compounds, the isolation of enzymes from plant cell cultures as tools for chemical reactions or as diagnostics has also been seen as a future aspect for plant cell biotechnology (STAFFORD and FOWLER, 1991). Especially glycosylating enzymes which occur abundantly in plants have attracted the attention of chemists, since site- and stereo-specific glycosylation reactions are not easy to perform chemically and microbial enzymes for this purpose are not as abundant.

The hope placed into plant cell cultures for the production of natural compounds has not been fulfilled up to now, although a great number of secondary products have been described (STABA, 1980; BERLIN, 1986). Only a limited number of processes have come to commercial application (Tab. 4).

One of the major drawbacks is the fact that many of the valuable natural products which can be isolated from the whole plant are not synthesized by cell cultures at all or in only very low amounts. On the other hand, secondary products hence unknown to a special plant can also be found in the respective cell cultures, and absolutely new compounds have been isolated from plant cell cultures (RUYTER and STÖCKIGT, 1989), although it is doubted that all these compounds really do not occur in the respective plant (BERLIN, 1986). Since plant cells can be regarded as totipotent, the question arises as to why a biosynthetic pathway expressed in a plant is not expressed in the cell culture. The genes encoding for this pathway are surely present, but the physiological state of the cell which is determined by the neighboring cells or environmental or nutritional factors does not allow an expression of these genes (BERLIN, 1986). We therefore must learn to establish this special physiological

Tab. 4. Economical Processes for the Production of Secondary Compounds by Plant Cell Cultures

Product	Species	Company	Reference
Shikonin	*Lithospermum erythrorhizon*	Mitsui Petrochemical Ind. Ltd.	FUJITA et al., 1982
Ginsenosides	*Panax ginseng*	Nitto Denko Corp.	USHIYAMA, 1991
Purpurin	*Rubia akane*	Mitsui Petrochemical Ind. Ltd.	personal communication

state even in our artificial culture systems, if we want to achieve secondary product formation. For this purpose we have to keep in mind that the accumulation of so-called secondary products always is an equilibrium process of synthesis, transformation, secretion and degradation and that secondary products are not side products or waste products accumulated in the cell, but serve a specific function and may be metabolized.

Cell suspension cultures are easy to handle with respect to industrial production processes. They can be cultivated in bioreactors similarly to bacteria (see Sect. 2.3.2). However, the production of some natural compounds, e.g., cardiac glycosides, some alkaloids, terpenoids, is correlated to some degree of differentiation. Cardiac glycosides, for example, are usually not found in cell cultures of *Digitalis lanata* any more. But as soon as a differentiation in direction of embryogenic or shoot cultures is observed, the synthesis of these compounds is also restored though only low levels of cardiac glycosides were found (KUBERSKI et al., 1984; LUCKNER and DIETTRICH, 1985; SEIDEL and REINHARD, 1987; GREIDZIAK et al., 1990). This shows that the missing product synthesis is not due to a loss of the respective genes or a complete inactivation of those genes, but due to a regulatory process not yet understood. Therefore, more success in the production of plant secondary compounds will need more knowledge of the biosynthetic pathways and their regulation in plants.

Even though the commercial production of secondary compounds by plant cell cultures has not come to the success we hoped for, cell cultures were found to be "a pot of gold" (ZENK, 1991) for another purpose, the elucidation of biosynthetic pathways for secondary compounds. Cell culture material is available all over the year in the same quality and in almost unlimited amounts. Moreover, the metabolic activity of cell cultures is much higher, and a considerable synthesis of secondary products can be achieved in one or a few weeks of cultivation, whereas the development of plants takes several weeks to months. Furthermore, plants contain high levels of phenols and tannins which can inhibit enzyme activities to zero, when cells are broken in order to isolate enzymes. Sometimes even the secondary product levels are higher in cell cultures than in plants. This leads to high activities of the enzymes involved in the biosynthetic pathways of the respective secondary compounds and, therefore, facilitates the isolation and characterization of the enzymes and the elucidation of the metabolic pathways. A number of important biosynthetic pathways have been completely clarified with the help of plant cell cultures, e.g., those leading to flavonoids and several classes of alkaloids (STÖCKIGT et al., 1989; ZENK, 1991).

Up to now, most approaches to optimize the production of secondary compounds by plant cell cultures have been empirical – but not without any success. In the following we will give an overview and some examples for the strategies used for the optimization of product formation.

3.1 Optimization of the Culture Media

Most of the cell culture processes are based on the limited number of so-called classical culture media (MURASHIGE and SKOOG, 1962; WHITE, 1963; LINSMAIER and SKOOG, 1965; GAMBORG et al., 1968). However, the effect of different nutrient mixtures on the production of secondary compounds must be taken into consideration. The "classical" culture media were mainly optimized for growth. Growth and secondary product synthesis by cell cultures, however, are in many cases not correlated and therefore have different nutrient requirements. Growth and secondary product formation behave according to the so-called trophophase-idiophase development described for plant cells by LUCKNER et al. (1977). As an example, the accumulation of rosmarinic acid in cell cultures of *Coleus blumei* has been tested in a number of those "classical" media, and pronounced differences in their effects on growth and rosmarinic acid formation were observed (ULBRICH et al., 1985). This often leads to an effort to optimize every single component of the medium. However, a combination of the singly-tested optimal concentrations of the nutrients may or may not result in an optimal medium (FUJITA et al., 1981a, b; DE-EKNAMKUL and ELLIS, 1985).

Since up to now only a limited number of cell cultures are known to grow photoautotrophically (HÜSEMANN, 1988), and these cell cultures usually have slow growth rates and require a high energy input (irradiation), the choice is to use heterotrophic or mixotrophic cultures which require a carbohydrate source. The most common carbon sources added to the media are glucose and sucrose. Of the macronutrients normally applied to the culture media, generally nitrate, potassium, ammonium and phosphate are regarded to have the highest influence on cell growth. In many cases depletion of these nutrients leads to an enhancement of secondary product formation. Addition of phytohormones and their balance is a necessary and crucial point for successful secondary product formation (exceptions are habituated, oncogenic or transformed cultures). Usually kinetin or benzylaminopurine (BAP) as cytokinins and indole acetic acid (IAA), 2,4-dichlorophenoxyacetic acid (2,4-D) or naphthaleneacetic acid (NAA) as auxins are added to the culture media. Especially the auxins have differing effects on secondary product synthesis. As an example, the effect of different auxins on the production of berberine by cell cultures of *Thalictrum minus* (NAKAGAWA et al., 1986) is shown in Fig. 5. In many cases the addition of 2,4-D represses the respective synthesis as reported for berberine (NAKAGAWA et al., 1986), anthraquinones (ZENK et al., 1975) or indole alkaloids (ZENK et al., 1977b). The natural auxin IAA, on the other hand, is not light- and thermostable, and there-

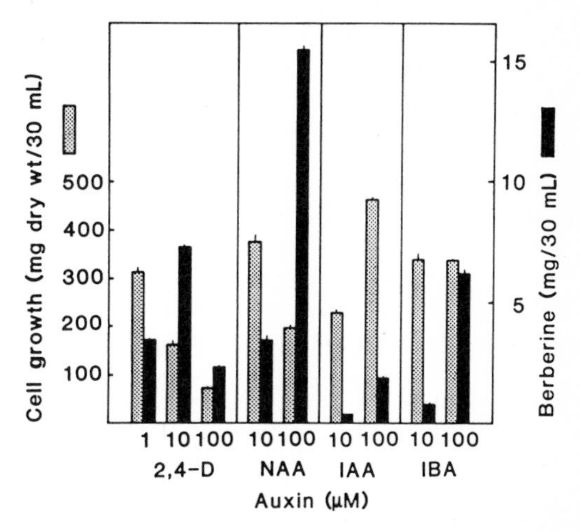

Fig. 5. Effects of various auxins in combination with benzylaminopurine (BAP) (10 µM) on cell growth and berberine production in suspension cultures of *Thalictrum minus* during a culture period of 16 days (IBA, indole-butyric acid) (from NAKAGAWA et al., 1986, with kind permission of the authors and Springer Verlag).

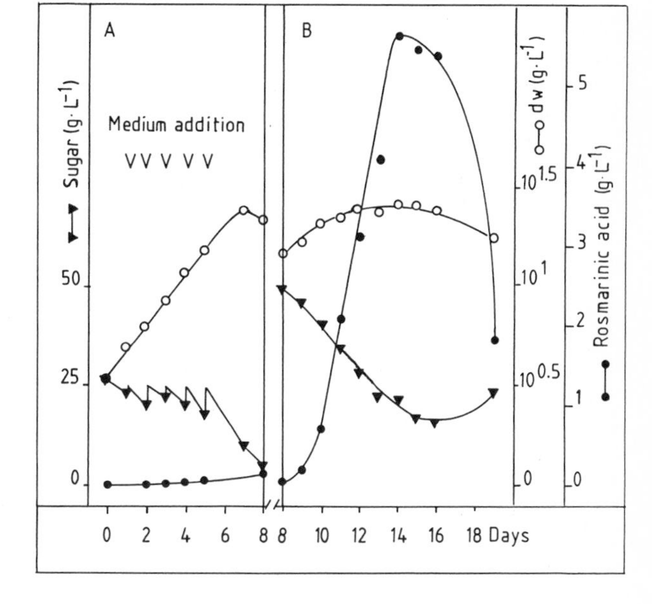

Fig. 6. Two-stage process of rosmarinic acid production by *Coleus blumei* cell line CBL 2B. Cells were first grown in growth medium in an airlift reactor (A) and then transferred to production medium in a stirred tank reactor (B). The working volume of both reactors was 32 L (from ULBRICH et al., 1985, with kind permission of the authors and Springer Verlag).

fore the actual concentration in the medium is doubtful. Often NAA is the auxin of choice, especially concerning secondary product formation. A result of optimization processes is often the formulation of two different media, one optimal for growth, the other one for secondary product formation. Two-stage culture processes have been successfully established for the production of shikonin by cell cultures of *Lithospermum erythrorhizon* (TABATA and FUJITA, 1985), rosmarinic acid by cell cultures of *Coleus blumei* (ULBRICH et al., 1985; Fig. 6) and berberine from cell cultures of *Coptis japonica* (FUJITA and TABATA, 1987).

3.2 Selection of High-Producing Cell Lines

Different cell culture strains derived from the same (high-producing) plant can show different rates of secondary product formation. Even in one callus strain differently producing cells can be found. However, extensive investigations have shown that the probability to get high-producing cell lines is higher when explants from high-producing plants are used (ZENK et al., 1977b; KINNERSLEY and DOUGALL, 1981). The type of tissue used as an explant should not play a role in the ability of the culture to produce secondary compounds, since plant cells are regarded to be totipotent. Though, examples have been reported where tissues from high-producing organs were the best explants for achieving high-producing cultures (STABA, 1982). A high product level in a plant tissue, however, does not refer to a high biosynthetic capacity because of the translocation and accumulation characteristics of secondary products in plants. The heterogeneity of cell cultures is especially obvious when colored products, such as anthocyanins, shikonin or berberine, are accumulated. In this case the selection of high-producing cells can be done by eye. Other methods such as microspectrophotometry (CHAPRIN and ELLIS, 1984), the use of a cell sorter (AFONSO et al., 1985; ALEXANDER et al., 1985; BARIAUD-FONTANEL et al., 1988), ELISA or RIA techniques (ZENK et al., 1977b; OKSMAN-CALDENTEY and STRAUSS, 1986) or other sophisticated analytical methods have been used for selection processes.

Small cell clusters or single cells obtained by sieving or by plating of protoplasts are the basis for the selection of high-producing cell lines (TABATA et al., 1987). In the case of *Lithospermum erythrorhizon* (MIZUKAMI et al., 1978; FUJITA et al., 1985; Tab. 5 or *Coptis japonica* (SATO and YAMADA, 1984) this selection leads to the establishment of a rather stable cell line with better production than the parent cell line. On the other hand, clonally propagated cell lines of *Anchusa officinalis* did not show a correlation between the production capacity for rosmarinic acid of the mother cell and the established cell lines (ELLIS, 1985). It has to be kept in mind that the capacity of a cell strain to keep its high-producing qualities varies considerably and cannot be predicted. A loss of the production capacity for secondary products after numerous subcultures is often observed and is explained by somaclonal variation (LAR-

Tab. 5. Growth Rate, Shikonin Content and Shikonin Productivity of Selected Cell Lines Obtained by Protoplast Culture (FUJITA et al., 1984)

Cell Line	Growth Rate (g/g inoculum/23 days)	Shikonin Content (%)	Productivity (g shikonin/g inoculum/23 days)
Original Cell Line	23.9	17.6	4.20
A	27.8	23.2	6.45
B	26.4	22.6	5.97
C	31.2	18.5	5.78
D	29.1	19.5	5.68
E	24.6	21.6	5.32

KIN and SCOWCROFT, 1981) and epigenetic variability (MEINS and BINNS, 1978; DOUGALL, 1980).

3.3 Stress: Elicitation, Induction and Stimulation

Various stress factors exerted on plant cell cultures can influence the accumulation of secondary compounds (BRODELIUS, 1988):

- medium stress (e.g., sugar, nitrogen, phosphate, phytohormones)
- physical stress (e.g., light, UV, aeration, osmolarity, pH)
- chemical stress (e.g., heavy metals, abiotic elicitors)
- infectional stress (e.g., pathogens, biotic elicitors)

Some of these stress factors will be discussed in further detail.

Elicitation in the strict sense of its meaning is the induction of the *de novo* synthesis of secondary compounds by the addition of a natural or denatured preparation of a fungus or microorganism, either complex or sometimes reduced to a defined polysaccharide preparation. The secondary compound thus induced is called a phytoalexin and is regarded to be a defense compound directed against an invading pathogen. The synthesis of those phytoalexins can in some cases also be induced by abiotic factors (e.g., UV light, alkaline pH, osmotic pressure, heavy metal ions). On the other hand, the synthesis of compounds already present in the plant cell can also be stimulated by stress factors, which – in the strict sense – is not elicitation. An example is the stimulation of rosmarinic acid accumulation after addition of yeast extract to cell cultures of *Orthosiphon aristatus* (SUMARYONO et al., 1991). Therefore, the lines between elicitation, induction and stimulation should not be drawn too strictly. Usually, the reaction of plant cells to these kinds of stimulation is very fast, several hours to a few days, and makes the application of such easy stimulation procedures enticing. Moreover, many of the secondary products which are formed after stimulation are released to the culture medium which makes the product recovery easy. This excretion is a natural and "logical" side effect, since defense compounds are most useful on the outside of the cell.

For a successful elicitation process using microbial or fungal preparations, several factors have to be taken into consideration: sensitivity of the cell culture strain, elicitor specificity, concentration of the elicitor and timing and duration of the elicitor exposure. Each of these factors has to be optimized independently for each process. Our understanding of the molecular events leading from the contact between the elicitor and the plant cell to gene activation and phytoalexin synthesis is far from completeness, but more and more steps of the signal transduction pathways are being elucidated.

A successful large-scale application of elicitation effects is the elicitation of cell cultures of *Papaver somniferum*, the opium poppy, with a sterile *Botrytis* preparation which resulted in a production of sanguinarine, a benzophenanthridine alkaloid usually not found in this plant

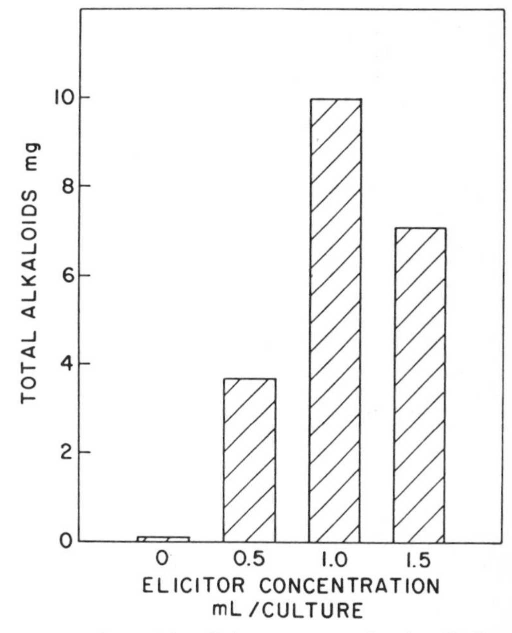

Fig. 7. Effect of the elicitor concentration (sterile *Botrytis* culture homogenate) on the total alkaloid yield (sanguinarine and dihydrosanguinarine) from 100 mL *Papaver somniferum* suspension culture (from TYLER et al., 1989, with kind permission of the authors and Springer Verlag).

(EILERT et al., 1985; TYLER et al., 1989; Fig. 7). Twelve hours after elicitor addition the alkaloid accumulation started and, also unusual for this product, 40–60% of the total alkaloids were excreted to the medium. Since the cells remained viable after a medium exchange, a semi-continuous process was established with successive elicitation/medium exchange cycles and isolation of the alkaloids from the harvested cells and the medium (KURZ et al., 1988). Sanguinarine has antibiotic activity against bacteria responsible for plaque formation on teeth and is used as an active ingredient for dental care products (CONSTABEL, 1990). For more information on the elicitation of secondary compounds in plant cell cultures see, e.g., HEINSTEIN (1985), KURZ et al. (1988), BARZ et al. (1988), BRODELIUS (1988).

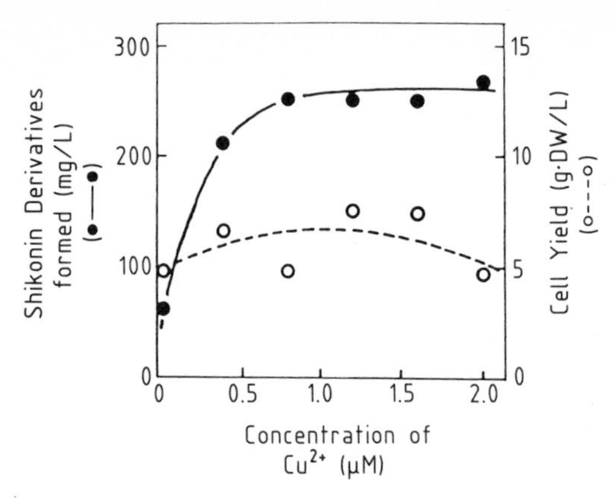

Fig. 8. Effect of the concentration of copper on growth and shikonin production in suspension cultures of *Lithospermum erythrorhizon* (from FUJITA et al., 1981b, with kind permission of the authors and Springer Verlag).

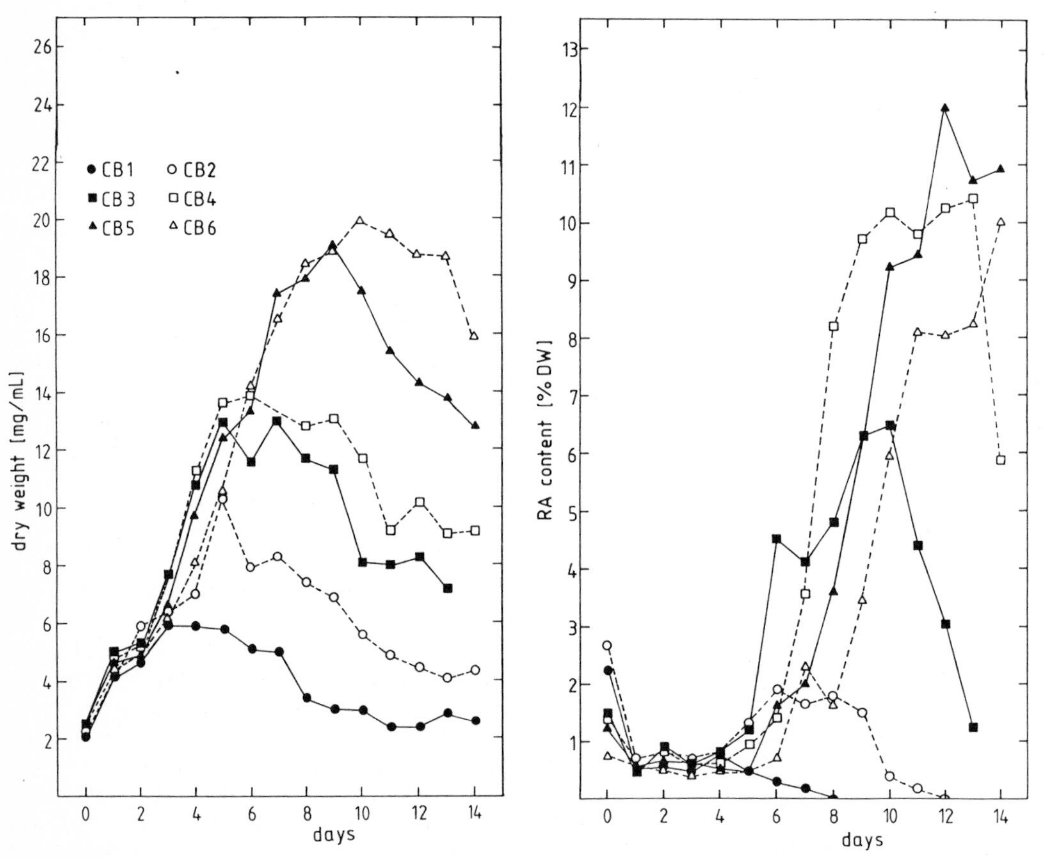

Fig. 9. Effect of the sucrose (1 – 6%, CB1–CB6) concentration in the medium on growth (dry weight; left) and rosmarinic acid (RA) production (right) in suspension cultures of *Coleus blumei.*

The formation of secondary compounds can also be induced or stimulated by abiotic factors. These observations often are made by chance, and the determination of abiotic stimulators or inducers has to be made empirically. Medium components can be among these factors, e.g., polysaccharides from agar (FUKUI et al., 1983) or copper ions (FUJITA et al., 1981b; Fig. 8) which both stimulated shikonin production in cell cultures of *Lithospermum erythrorhizon*. Environmental factors such as osmotic stress or UV light also influence product formation in some cell cultures. Osmotic stress by high mannitol concentrations, for instance, stimulated indole alkaloid production in periwinkle cell cultures (RUDGE and MORRIS, 1986). UV irradiation resulted in an enhanced anthocyanin formation in carrot suspensions (GLEITZ and SEITZ, 1989) or induced flavonoid synthesis in parsley cell cultures (SCHEEL et al., 1986). Stimulation often directly meets the optimization of the culture medium, e.g., when extremely high sucrose levels are used for the enhancement of secondary product formation. High sucrose levels, for example, stimulate rosmarinic acid accumulation in cell cultures of *Coleus blumei* (ZENK et al., 1977a; GERTLOWSKI and PETERSEN, 1993; Fig. 9) or anthraquinone formation in cultures of *Morinda citrifolia* (ZENK et al., 1975). In the case of *Coleus blumei*, the high osmotic pressure is not the stimulating factor; for other cultures this has not been investigated.

3.4 Design of Culture Vessels and Physical Conditions

Basic investigations on growth and secondary product formation are most commonly performed in small culture vessels, e.g., Erlenmeyer flasks, on a laboratory scale. The physical conditions adjustable at this stage are temperature, mode and vigor of shaking and the light regime. The decision of what optimal conditions have to be applied depends on the individual culture. Agitation of the culture flasks is most commonly performed on a rotary shaker at 80–150 revolutions per minute. However, the speed of shaking greatly influences metabolite and oxygen transfer to the cells as well as the shear forces applied to the cells. Light is not necessary for a number of cultures synthesizing secondary products, e.g., shikonin (*Lithospermum erythrorhizon*) or rosmarinic acid (*Coleus blumei*). The formation of other substances can be enhanced by light, e.g., the production of flavonoid glycosides by parsley cultures (GRISEBACH and HAHLBROCK, 1974), or it strictly requires irradiation (SEIBERT and KADKADE, 1980; MANTELL and SMITH, 1983). Temperature dependence of growth and secondary metabolism of plant cell cultures has not been investigated to a large extent; usually temperatures around 25 °C are applied (MANTELL and SMITH, 1983). For industrial production, upscaling to large reaction volumes is necessary. Many plant cell cultures do not change their growth and production rates significantly during this process, others can change positively (ULBRICH et al., 1985; FUJITA and TABATA, 1987) or negatively (SCRAGG, 1990). Since plant cell cultures have a rather slow growth rate and many secondary compounds are synthesized only at the end of the growth phase or in the stationary phase, the problem of comparatively long culture cycles arises. Except for the productivity problems, problems relating to sterility during these long fermentation runs can be solved nowadays. The problems of bioreactor design, aeration and agitation concerning plants cell cultures have been discussed in Sect. 2.3.2 and in several other references (MARTIN, 1980; SCRAGG and FOWLER, 1985; PAREILLEUX, 1988; SHULER, 1988; KREIS and REINHARD, 1989; FOWLER and STAFFORD, 1992; BUITELAAR and TRAMPER, 1992).

3.5 Feeding of Metabolic Precursors and Biotransformation Reactions

Pathways leading to secondary plant products are necessarily linked to primary metabolism. The first enzymes linking primary to secondary metabolism are often thought to be regulatory enzymes regulating the throughput of the primary metabolite to the secondary product. This, however, is not always the case, as can be seen in the biosynthetic pathway of berberine in cells of *Thalictrum glaucum* where the activities of three intermediary methyl

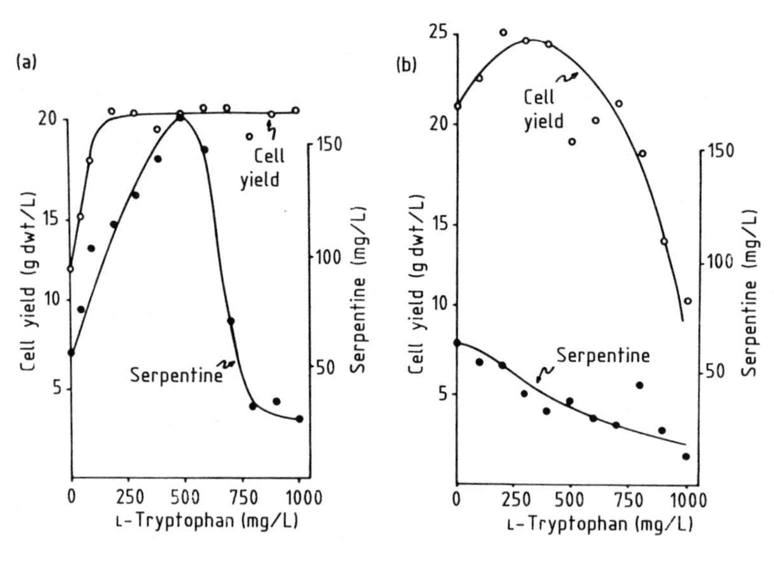

Fig. 10. Effect of L-tryptophan concentration on cell yield and serpentine production in two cell strains of *Catharanthus roseus* derived from the same parent culture. (a) Stimulation of alkaloid and cell yield; (b) inhibition of alkaloid and cell yield (from ZENK et al., 1977b, with kind permission of the authors and Springer Verlag).

transferases seem to determine the synthesis of the end product and not the introductory enzyme of the pathway (GALNEDER and ZENK, 1990). Feeding of limiting precursors, either primary metabolites or intermediary metabolites of biosynthetic pathways, can result in enhanced product formation. One of the first to show this was the group of ZENK who fed tryptophan as precursor for indole alkaloids to cell cultures of *Catharanthus roseus* (ZENK et al., 1977b; Fig. 10) or phenylalanine to cell cultures of *Coleus blumei* resulting in enhanced rosmarinic acid biosynthesis (ZENK et al., 1977a). Precursors from primary metabolism as well as intermediary precursors were fed to cell cultures of *Capsicum frutescens* producing capsaicin and led to increased secondary product formation (YEOMAN et al., 1980). Intermediary precursors are especially effective, when the activity of one of the intermediary enzymes of the pathway is reduced. In anthocyanin-producing cell cultures of *Daucus carota* gibberellic acid addition to the culture medium blocks anthocyanin synthesis at the level of chalcone synthase. Feeding of anthocyanin precursors located after the block can restore anthocyanin formation (HINDERER et al., 1984). This leads to the special case of bio-

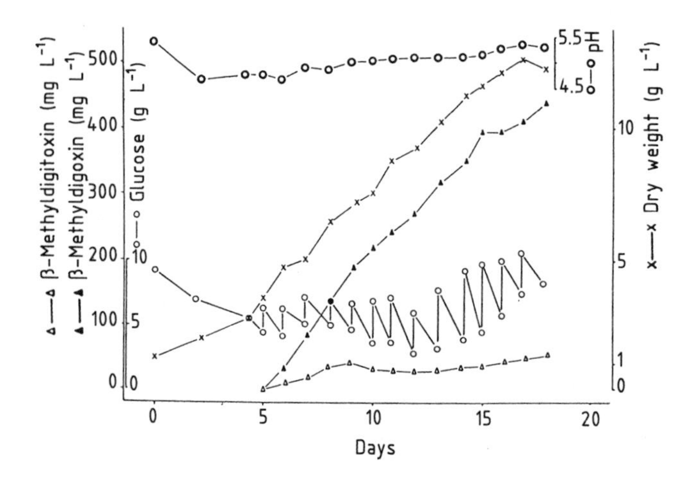

Fig. 11. Biotransformation of β-methyldigitoxin (Mdt) to β-methyldigoxin (Mdg) in a 200 L reactor. On day 0 the reactor was inoculated with 35 L cell suspension of the preculture. From day 5 of the culture time the reactor is incubated with an appropriate amount of Mdt in methanol and the concentration of Mdt is maintained at around 50 mg $\cdot$ L^{-1} throughout the reaction time. On day 18 a product yield (Mdg) of about 0.5 g $\cdot$ L^{-1} is achieved. Due to the uptake of glucose by the cells, this has to be fed additionally from day 5. Modified from ALFERMANN (1983), with kind permission of Metzler Verlag, Stuttgart.

transformation reactions, where sometimes only one single transformation reaction is effectively performed by the cell culture after feeding of a compound. Cell cultures which are not able to synthesize secondary products *de novo* have, in some cases, retained the ability to perform special reactions to compounds fed to them. This can be exploited to transform cheap substances into rare and expensive products. The reactions catalyzed by the plant cell cultures are regio- and stereoselective hydroxylations, oxidation and reduction reactions, coupling with glycosyl residues, hydrolysis, isomerization, epoxidation and various other reactions (ALFERMANN and REINHARD, 1988; SUGA and HIRATA, 1990).

Biotransformation reactions have been coupled with the application of immobilized plant cells making the process more feasible for economical exploitation (see also Sect. 3.6). One of the best known biotransformation examples is the hydroxylation of β-methyldigitoxin to β-methyldigoxin (Fig. 11) by cell cultures of *Digitalis lanata* which otherwise are no longer able to synthesize cardiac glycosides (ALFERMANN et al., 1983). Fig. 11 shows a typical time schedule of the hydroxylation reaction using β-methyldigitoxin as the substrate in a 200 L airlift reactor. This process could be scaled up to a 1 m^3 reactor (WAHL, 1985). More recently, KREIS und REINHARD (1988) could select special *Digitalis* cell lines, which are able to hydroxylate digitoxin at C-12 and produce deacetyllanatoside C (cf. Fig. 12), which is an additional interesting starting material for the production of cardioactive drugs. Biotransformation experiments using hydroquinone as substrate have shown that plant cells are able to accumulate secondary products up to more than 40% of their dry weight (SUZUKI et al., 1987). This process of arbutine formation by plant cell cultures has been explored by the Shiseido Company with the goal of practical application (YOKOYAMA and YANAGI, 1991, and personal communication).

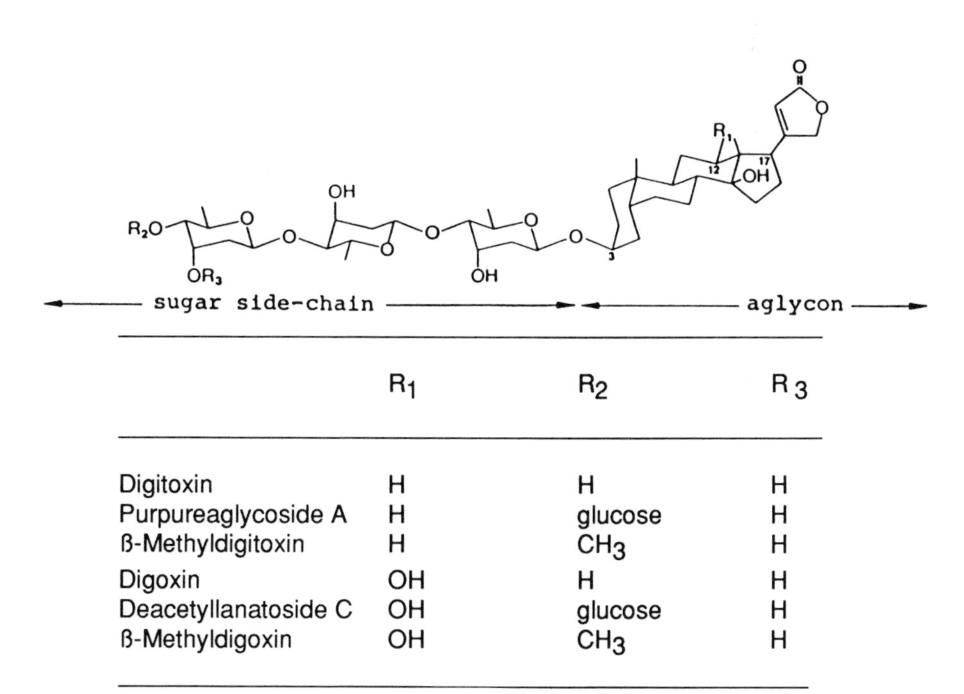

	R_1	R_2	R_3
Digitoxin	H	H	H
Purpureaglycoside A	H	glucose	H
ß-Methyldigitoxin	H	CH_3	H
Digoxin	OH	H	H
Deacetyllanatoside C	OH	glucose	H
ß-Methyldigoxin	OH	CH_3	H

Fig. 12. Structural formulas of some cardenolides. The main difference between digitoxin, β-methyldigitoxin and purpureaglycoside A on one side and digoxin, β-methyldigoxin and deacetyllanatoside C on the other side is the additional hydroxyl function at C-12.

3.6 Immobilization of Plant Cells

The immobilization of plant cells for the production of secondary compounds has been investigated extensively because of the obvious advantages for use in biotechnological processes. One advantage would be the possible re-use of the cells after harvesting of the medium, but an absolute prerequisite for this is the secretion of the secondary compounds into the medium. Unfortunately, most natural compounds are stored within the plant cell, and procedures for permeabilization (see Sect. 3.7) mostly reduce the viability of the plant cells. In most cases the immobilization process had positive effects on the production of secondary compounds, partially due to higher synthetic rates, partially due to longer production periods. Also the protection of the plant cells against shear forces and an enforced contact between the plant cells have beneficial effects on product formation. Positive effects were reported for, e.g., berberine production by *Thalictrum minus* cells (KOBAYASHI et al., 1988; Fig. 13), capsaicin synthesis by *Capsicum frutescens* (LINDSEY et al., 1983) or shikonin production by immobilized cells of *Lithospermum erythrorhizon* (KIM and CHANG, 1990). Negative effects of the immobilization on product formation were reported as well, e.g., due to the diffusion barriers for oxygen and nutrients formed by the immobilization matrix. Immobilized cells proved to be useful for the biotransformation reaction of β-methyldigitoxin to β-methyldigoxin by cell cultures of *Digitalis lanata* (Fig. 14) where high conversion rates (<90%) were combined with the release of the product into the medium (ALFERMANN et al., 1983). The predominant method for immobilization is still the entrapment of the cells in calcium alginate beads (Fig. 15) or other gelling agents (BRODELIUS and NILSSON, 1980). Other methods are immobilization in polyurethane foam or the use of fiber or membrane reactors. For reviews of immobilization methods and their effects on secondary product formation see LINDSEY and YEOMAN, 1983, NOVAIS, 1988; BUITELAAR and TRAMPER (1992) and the literature cited therein.

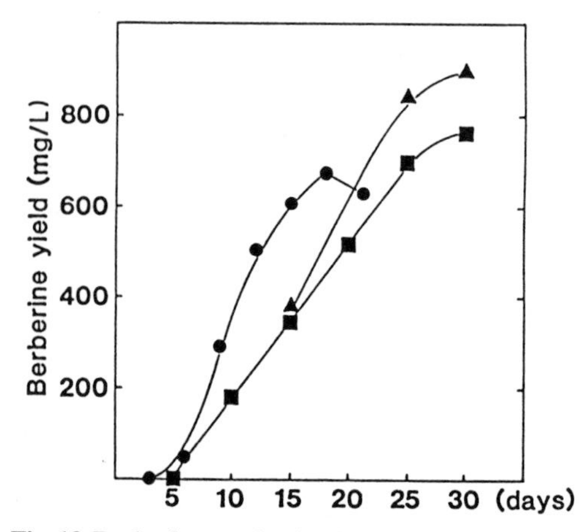

Fig. 13. Berberine production in free and alginate-immobilized cells of *Thalictrum minus:* free cells in shake flask (●), immobilized cells in shake flask (■) and immobilized cells in a bioreactor (▲) (from KOBAYASHI et al., 1988, with kind permission of the authors and Springer Verlag).

3.7 Permeabilization and Two-Phase Cultures

An obstacle to the economical use of plant cell cultures to the production of secondary compounds is that the majority of secondary compounds are stored intracellularly which renders product recovery very expensive. The place where plant cells store their secondary compounds is a matter of their physiological function and is therefore genetically encoded and not easy to manipulate. Only a limited number of secondary compounds is deliberately released to the medium by plant cells (GUERN et al., 1987), e.g., capsaicin from *Capsicum frutescens* (YEOMAN et al., 1980) or a number of alkaloids (TABATA, 1988). In one case it was proven that the secretion of the secondary compound was strain-specific: a specific cell strain of *Thalictrum minus* secretes berberine into the medium where it crystallizes (NAKAGAWA et al., 1984), whereas in other strains of *T. minus* or in other species the alkaloids are stored in the vacuole (TABATA, 1988). This, however, is the exception. Efforts have been made to permeabilize cell cultures without lowering their viability too much in order to isolate the secondary compounds from the culture medium. This is a prerequisite when

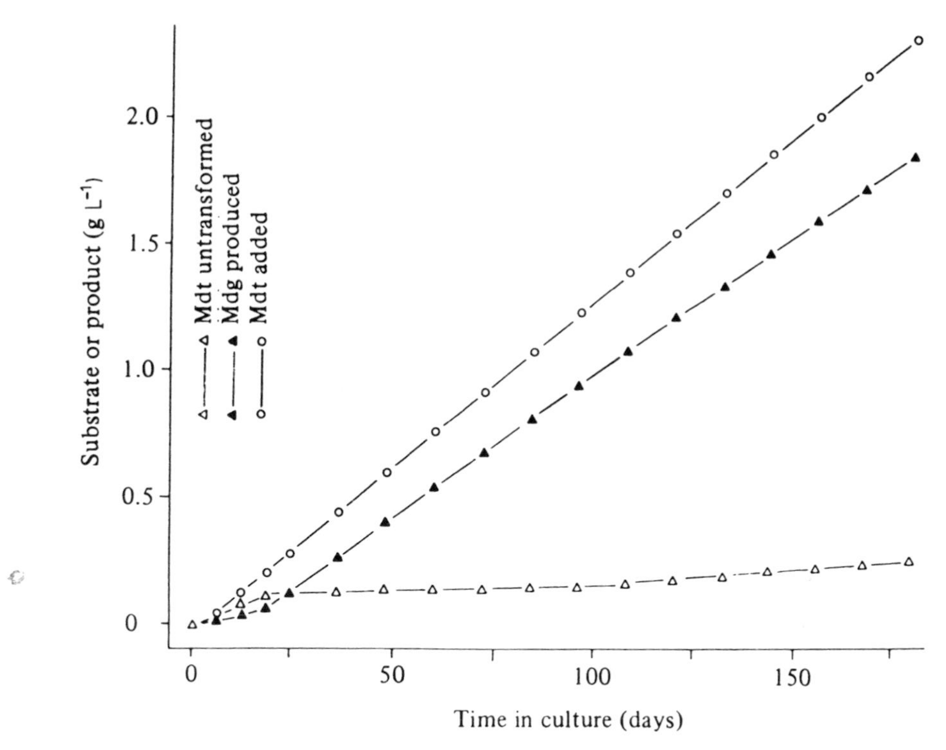

Fig. 14. Biotransformation of β-methyldigitoxin by *Digitalis lanata* cells immobilized in alginate gel during a cultivation time of 170 days. Most of the product β-methyldigoxin is excreted into the medium. Calculating the untransformed substrate the product yield is about 90% (from ALFERMANN et al. (1983) with kind permission of Cambridge University Press).

Fig. 15. Alginate beads with immobilized *Digitalis lanata* cells.

immobilized cell systems (see Sect. 3.6) are used for product formation. Physical and chemical methods have been applied to the permeabilization of plant cells which will be discussed briefly (for reviews see FONSECA et al., 1988; BUITELAAR and TRAMPER, 1992).

Addition of dimethylsulfoxide (DMSO), a chemical that renders membranes more permeable, has been tried successfully in only a few systems, e.g., *Catharanthus roseus* (BRODELIUS and NILSSON, 1983) or *Coleus blumei* (PARK and MARTINEZ, 1992). Mostly cells do not survive a treatment with DMSO or other organic chemicals (chloroform, propanol) probably due to a loss of internal compartmentation.

Water-immiscible organic solvents were used more successfully forming two phases with the medium. Lipophilic secondary compounds are trapped within the organic layer and extracted from the culture medium. Inhibitory effects as well as solubility problems can thus be overcome. The organic solvent must meet some important prerequisites, e.g., it must be sterilizable and non-toxic or not too toxic to the cells and should not interfere with the medium or the oxygen transfer. Hexadecane is often used as the organic solvent of choice due to its very low toxicity, e.g., for the extraction of anthraquinones from the medium of *Morinda citrifolia* cultures (BUITELAAR and TRAMPER, 1992). Miglyol is able to trap terpenoid compounds produced by cell cultures in low amounts, thus enhancing the overall yield (BERLIN et al., 1984; CORMIER and AMBID, 1987). The Miglyol phase probably functions as a substitute for accumulating cells for these compounds in the differentiated plant. The producing cells which are not able to store higher amounts of terpenoids are therefore able to dipose of these compounds. Solid adsorbents have also been used as the second phase in two-phase culture systems. Amberlite XAD-4 and XAD-7 resins proved to be the most effective adsorbents which do not only adsorb the secondary products from the medium but also enhance the productivity (BUITELAAR and TRAMPER, 1992). These resins are able to shift the equilibrium concentrations of the secondary compounds in the medium by adsorbing them and therefore increase the overall production.

In some cases the composition of the medium affects the secretion of secondary compounds. A low phosphate concentration in the medium enhanced the secretion of protoberberine alkaloids from *Thalictrum rugosum* cell cultures (BERLIN et al., 1988). A special case has already been mentioned in Sect. 3.3: elicitation often leads to secretion of the respective phytoalexins.

The physical methods used for permeabilization are higher temperatures, electrical permeabilization and ultrasonication. Mild heat treatment enhances the product release from plant cells; however, the viability also decreases with higher temperatures (WEATHERS et al., 1990). The same effect was found for electroporation, which is able to trigger a substantial release of secondary products to the medium. The voltage applied and the viability of the cells are negatively correlated (BRODELIUS et al., 1988; FONSECA et al., 1988). Short intervals of ultrasonication led to a release of betalains from cell cultures of *Beta vulgaris* (KILBY and HUNTER, 1990).

3.8 Differentiated Cultures

The synthesis of a number of plant secondary compounds is supposed to be coupled with a specific degree of differentiation of the plant tissue. This would explain the inability to get certain compounds from suspension cultures. As a rule, with the onset of differentiation, the formation of somatic embryoids, roots or shoots (Figs. 16 and 17), these compounds are detectable again in the plant tissue (CHARLWOOD et al., 1990). However, organized cultures can no longer be termed "cell cultures" and therefore do not fall within the scope of this chapter. Furthermore, there are still some problems regarding the mass cultivation of organ cultures, especially shoot cultures for the production of secondary compounds.

The initiation and propagation of differentiated cultures can be controlled by the phytohormone regime. Moreover, root and shoot cultures can be attained by transformation with *Agrobacterium tumefaciens* or *A. rhizogenes*. As a result of the transformation, bacterial genes encoding for enzymes, which catalyze the plant phytohormone biosynthesis or inter-

Fig. 16. Shoot cultures of *Apocynum cannabinum.*

Fig. 17. Root cultures of *Hyoscyamus albus.*

fere with it, are stably integrated into the plant genome (reviews: ZAMBRYSKI et al., 1989; HOOYKAAS and SCHILPEROORT, 1992). The transformed roots ("hairy roots") and shoots, therefore, grow independently of exogenous phytohormones.

Transformed and untransformed root and shoot cultures synthesize more or less the same spectrum of secondary compounds as does the respective organ of the differentiated plant (HAMILL et al., 1987; CHARLWOOD et al., 1990; SPENCER et al., 1990; WOERDENBAG et al., 1992). The quantities range from rather low levels to concentrations exceeding those of normal plant organs (Tab. 6). An advantage of transformed cultures is their fast growth (Tab. 6); however, their applicability for the commercial production of substances has to be proven. Special types of bioreactors have been designed for the large-scale culture of root and shoot cultures. Especially "hairy root" cultures have been under extensive investigation for the biosynthesis of secondary compounds, mainly alkaloids (FLORES et al., 1987; RHODES et al., 1990). Due to their high growth rate and their stable production of secondary compounds, they are suitable systems for the investigation of the respective biosynthetic pathways and may be used for production in the future.

3.9 Genetic Engineering of Plant Cells

During the last years a number of enzymes involved in the biosynthetic pathways for secondary plant products have been identified and isolated, and the corresponding genes have been isolated and cloned (MOL et al., 1988), e.g., phenylalanine ammonia-lyase (KUHN et al., 1984), 4-coumarate:CoA ligase (DOUGLAS et al., 1987), chalcone synthase (SOMMER and SAEDLER, 1986; WIENAND et al., 1986), tryptophan decarboxylase (DE LUCA et al., 1989), strictosidine synthase (KUTCHAN et al., 1988) or hyoscyamine 6β-hydroxylase (HASHIMOTO et al., 1990). First attempts to exploit gene technology for secondary product formation were based on the assumption that the biosynthetic capacity of some pathways leading to secondary products may be limited by the en-

Tab. 6. Natural Product Formation in Untransformed and Transformed Root Cultures (from RHODES et al., 1990)

Species		Hormone	Growth fresh wt (g/L/day)	Growth dry wt (g/L/day)	Product Accumulation (% dry wt)
Atropa belladonna (1)	N	---	---	0.1	0.8
	T	---	---	0.4	1.32
Calystegia sepium (1)	N	---	---	0.02	0.25
	T	---	---	0.44	0.30
Chaenactis douglasii (2)	N	NAA	---	0.016*	0.05
	T	---	---	0.021*	0.1
Hyoscyamus muticus (3)	N	---	0.51	---	0.61
	T	---	4.95	---	0.52
Hyoscyamus niger (3)	N	---	0.45	---	0.55
	T	---	4.75	---	0.43
Panax ginseng (4)	N	---	---	0.03	0.38
	N	IBA/K	---	0.31	0.91
	T	---	---	0.4	0.36
	T	IBA/K	---	1.05	0.95
Nicotiana tabacum (3)	N	---	0.3	---	2
	T	---	2.05	---	3

NAA, naphthylacetic acid; IBA, indolebutyric acid; K, kinetin; N, normal roots; T, transformed roots
* g/flask/day
(1) JUNG and TEPFER (1987); (2) CONSTABEL and TOWERS (1988); (3) FLORES and FILNER (1985); (4) YOSHIKAWA and FURUYA (1987)

zyme activities connecting primary and secondary metabolism. Introduction of genes for these rate-limiting enzymes under the control of constitutively expressed promoters might enhance the throughput through these pathways (BERLIN, 1984).

However, our knowledge of secondary product pathways and their regulation as well as the impact of such genetically altered biosynthetic capacities is rather limited. In a very simple biosynthetic pathway – the biosynthesis of serotonin from tryptophan by cell cultures of *Peganum harmala* which is catalyzed by only two enzymes (tryptophan decarboxylase, tryptamin 5-hydroxylase) – introduction of a plant gene for tryptophan decarboxylase under the control of the 35S-promoter of cauliflower mosaic virus enhanced the synthesis of serotonin (BERLIN et al., 1992). This example shows that in principle genetic manipulation of secondary pathways is possible. The more complex, however, those pathways are, the less probable is a successful manipulation in the plant. The same is true for the expression of single plant proteins or enzymes or complex biosynthetic pathways in bacteria. Single plant proteins or enzymes can be successfully expressed in microbial systems, such as strictosidine synthase which was used for the production of strictosidine (KUTCHAN, 1989). The probability of expressing complex biosynthetic pathways in microbial systems is very low, and up to now we are far from understanding their regulation in their natural environment. Moreover, expression of foreign proteins in *E. coli* is impaired by the fact that recombinant proteins often are packed into dense inclusion bodies.

Antisense DNA technology may be successful in at least partially blocking biosynthetic pathways which are not desired in the respective plants (MOL et al., 1990). Anthocyanin formation in flowers of *Petunia* has been blocked by introducing an antisense gene for chalcone synthase (VAN DER KROL et al., 1988). The inhibition of metabolic side branches restricting the biosynthesis of desired secondary compounds may be possible in the future. Manipulation of regulatory genes may lead to an expression of tissue-specific or developmentally regulated pathways in undifferentiated cells. Here again the missing knowledge about the regulatory principles of biosynthetic pathways in plants renders a successful application of this technique in the near future unlikely.

4 Application of Plant Tissue Culture Techniques in Agriculture

As stated above, plant cells are omnipotent. This was demonstrated very convincingly by SKOOG and MILLER (1957), who were able to regenerate plants and roots or to induce callus formation from undifferentiated cells by varying the concentration of an auxin and a cytokinin in the culture medium. REINERT (1959) found that undifferentiated cells cannot only regenerate organs, but also form embryoids. This phenomenon of development of embryoids, in contrast to formation of embryos from a zygote, often is called "somatic embryogenesis". In many cases, these embryoids develop from single cells, as it was shown for carrot by BACKS-HÜSEMANN and REINERT (1970). A certain problem in using these techniques is that plants regenerated from a callus may show genetic diversity due to genetic changes during callus subcultivation. Although this so-called somaclonal variation (LARKIN and SCOWCROFT, 1981; SCOWCROFT et al., 1987) may be used for selection of plants with new interesting genotypes (EVANS et al., 1984; WENZEL, 1992), the multiplication of an interesting genotype into a great number of identical plantlets is not easily possible. This, however, can be achieved at least with dicotyledonous plants using meristems as explants under conditions, where regeneration of plantlets is achieved more or less directly. In the case of monocots, most often somatic embryogenesis must still be used for micropropagation. These regenerants are multiplied afterwards under *in vitro* conditions. An additional advantage over conventional propagation methods is that under appropriate conditions plants free of pathogens (viruses, fungi, bacteria, etc.) can be produced resulting in increased vigor and productivity of the plants. This phenomenon was observed already in 1952 by MOREL and MARTIN who regenerated virus-free dahlia plants from meristems. Later on, MOREL (1960) demonstrated that this technique can also be used for propagation of pathogen-free orchids. These key experiments led to a world-wide application of this technique not only for the propagation of ornamental plants (e.g., orchids, roses, *Pelargonium, Ficus*), but also of many crop plants (potato, cassava, sugar cane and sugar beet, oil and date palms, and rubber trees, among many others), fruit plants and trees such as strawberries, bananas, apple, plum, cherry and *Citrus* trees or many forest trees (e.g., oak, *Eucalyptus,* pines). Fig. 18 shows the flow diagram for the isolation of virus-free potato plants by this technique as an example.

In 1988 more than 170 million plants were propagated by tissue culture techniques in Europe alone (DEBERGH and ZIMMERMAN, 1991). In addition to micropropagation, plant tissue culture methods are used for plant breeding. A possible use of somaclonal variation in selecting plants of a desired new genotype has already been mentioned. Such a selection can be achieved, e.g., by cultivating tissue cultures in the presence of a toxin and by regenerating resistant mutants. Haploid plants produced by anther culture (GUHA and MAHESHWARI, 1964) or as in barley by ovary culture (KASHA and KAO, 1970) are important for improving and accelerating breeding programs. This is also true for the use of plant protoplasts, e.g., in protoplast fusion but especially in their use for gene transfer with molecular biological techniques. Today, special, selected genes can be transferred into plant cells using *Agrobacterium tumefaciens* and *A. rhizogenes* as vectors or by direct gene transfer by DNA uptake by protoplasts or bombarding plant tissues with DNA coated on gold or tungsten particles (for more details see BERNATZKY and TANKSLEY, 1987; BHOJWANI, 1990; DAY and LICHTENSTEIN, 1992; HORSCH et al., 1987; KUCKUCK et al., 1991; LÖRZ et al., 1987; OOMS, 1992; POTRYKUS et al., 1987; VASIL, 1988; VASIL et al., 1991; WENZEL, 1992).

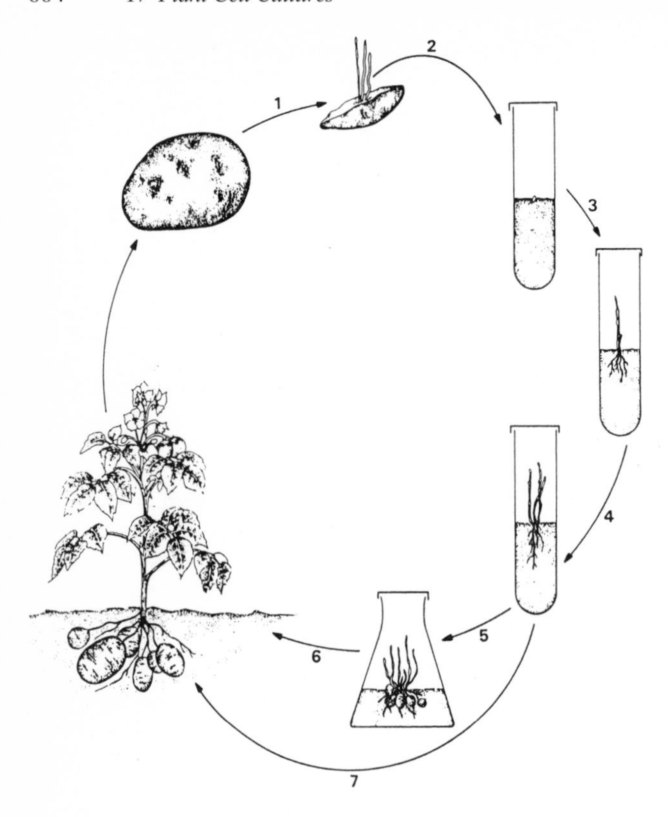

Fig. 18. Micropropagation of potato via meristem culture. Potato tubers are obtained from plants grown in the field. The tuber is sectioned in 2.5 cm pieces and cultivated on moist vermiculite after treatment with 0.03 M gibberellic acid for 1 h to break dormancy and to induce shooting (**1**). Meristems from the newly formed shoots are excised and planted on medium without growth hormones (**2**). Small shoots with roots develop (**3**), after transfer to medium with 0.05 μM naphthalene acetic acid (**4, 5**) axillary buds grow out. Plants with a well developed root system can be planted into soil (**6**). Additionally, the small plantlets (stage 4) can be subcultivated on medium with 22–44 μM benzylaminopurine to induce formation of microtubers, which can be transferred to soil for plant formation (**7**). From DIETER HESS: *Biotechnologie der Pflanzen* (1992), with kind permission of Eugen Ulmer Verlag, Stuttgart.

5 Outlook

Nowadays, the use of plant tissue culture techniques in agriculture is very well established. Every year several hundred million plants are produced world-wide. In recent years it became possible to propagate not only herbaceous but also many woody plants (DEBERGH and ZIMMERMAN, 1991). Especially in the latter area many species are still recalcitrant, but this is also true for many herbaceous species, e.g., many Fabaceae. Very recently, a breakthrough has been achieved with regeneration of important monocot species via embryogenesis (VASIL, 1988; VASIL et al., 1991). Of course, much effort is still necessary to elaborate appropriate procedures for further plant species and varieties. A series of transgenic plants has been produced with molecular biological methods, so that the commercial use of new varieties can be expected in the near future (GASSER and FRALEY, 1992; HUTTNER et al., 1992; WRUBEL et al., 1992). Much scientific effort, however, will be needed for studying the different factors involved in the integration of the transferred genes in the plant genome as well as its organ- and/or time specific expression. Especially in this area, more effort is still needed to ensure that the biotechnological techniques are safe for our environment and to convince the public of the necessity and safety of the methods used.

Compared with recent progress in plant transformation, the practical use of plant cell culture techniques for the production of phytochemicals is still in its infancy. Up to now, only *Lithospermum, Rubia* and ginseng cultures are used for commercial production (cf. Tab. 4). Shikonin produced from *Lithospermum* cell cultures is only used for cosmetic products and as a dye for silk products, since the Japanese authorities have not yet allowed its use for me-

Tab. 7. Plant-Derived Drugs Widely Used in Western Medicine (after FARNSWORTH, 1985)

Acetyldigoxin	Ephedrine[a]	Pseudoephedrine[a]
Aescin	Hyoscyamine	Quinidine
Ajmalicine	Khellin	Quinine
Allantoin[a]	Lanatoside C	Rescinnamine
Altropine	Leurocristine	Reserpine
Bromelain	α-Lobeline	Scillarens
Caffeine[a]	Morphine	Scopolamine
Codeine	Narcotine	Sennosides
Colchicine	Ouabain	Sparteine
Danthron[a]	Papain	Strychnine
Deserpidine	Papaverine[a]	Tetrahydrocannabinol
Digitoxin	Physostigmine	Theobromine[a]
Digoxin	Picrotoxin	Theophylline[a]
L-DOPA	Pilocarpine	Tubocurarine
Emetine	Protoveratrines	Vincaleukoblastine
		Xanthotoxin

[a] industrially synthesized

dicinal purposes. Therefore, the yearly production of this compound by tissue culture methods is still relatively small. In contrast, every month ginseng cells are cultivated in a 20 m^3 reactor, and extracts of these cells are used as additives in the food industry (Nitto Denko Comp., personal communication). Several other systems are being tested world-wide by companies, so that some other species may be included in the list of Tab. 4 in the near future.

The main reason for the limited success in this field is that many plant cell cultures do not produce the products of interest of the differentiated plant at all, or that the product yields remain insufficiently low despite strong efforts to increase them by various medium or culture technique modifications. Tab. 7 shows a list of plant-derived drugs widely used in western medicine published by FARNSWORTH (1985). With some exceptions (caffeine: FRISCHKNECHT et al., 1977; L-DOPA: WICHERS et al., 1983) none of these compounds can be produced by undifferentiated tissue cultures in such amounts that commercial production could be thought of at all.

This drawback is caused by the fact that many of these products are only synthesized in specialized organs or cells of the differentiated plant. In nature, they may be transported to specialized storage cells in other organs. Obviously, very often such specialized cells are not present in "undifferentiated" cell suspension cultures, and the appropriate enzymes are not expressed. This opinion is sustained by the fact that organ cultures such as hairy roots of the Solanaceae or shoot cultures, e.g., of mint are able to produce and accumulate the appropriate target compounds. Mass cultivation of such organ cultures, however, is still in its infancy. An economic use of such *in vitro* organ cultures surely is much more difficult than the use of cell suspension cultures. Due to the long generation times of the plant cells, the costs for mass cultivation of cell suspension cultures will be very high. For organ cultures, the problems with regard to transfer of the organs to larger bioreactors, simple harvesting of the cell mass, etc. have still to be solved. At least in some cases shoot organ cultures will have to be illuminated to induce product formation which, of course, would increase the production costs. However, in every case it has to be tested whether illumination is necessary. Dark-grown shoot cultures of *Hypericum perforatum,* for example, are able to accumulate substantial amounts of hypericine (ZDUNEK and ALFERMANN, unpublished results), whereas shoot cultures of *Pelargonium graveolens* accumulate essential oils only after transfer into light (ALFERMANN et al., unpublished).

A concept to overcome the above mentioned problems in our opinion is to study in more detail, how in the plant this organ-specific expression of the genes responsible for synthesis as well as of those involved in translocation, storage and/or degradation is performed. When we know more about these factors, especially on the molecular and gene level, then perhaps we will be able to influence the undifferentiated plant cell suspension culture by physiological or genetic methods to produce and accumulate the products of interest in such amounts that a biotechnological production on a commercial scale becomes feasible.

6 References

AFONSO, C. L., HARKINS, K. R., THOMAS-COMPTON, M. A., KREJCI, A. E., GALBRAITH, D. W. (1985), Selection of somatic hybrid plants in *Nicotiana* through fluorescence-activated sorting of protoplasts, *Bio/Technology* **3**, 811–816.

ALBERTS, A., BRAY, D., LEWIS, J., RAFF, M., ROBERTS, K., WATSON, J. D. (Eds.) (1989), *Molecular Biology of the Cell*, 2nd Ed., New York: Garland.

ALEXANDER, R. G., COCKING, E. C., JACKSON, P. J., JETT, J. H. (1985), The characterization and isolation of plant heterokaryons by flow cytometry, *Protoplasma* **128**, 52–58.

ALFERMANN, A. W. (1983), Gewinnung von Arzneistoffen durch pflanzliche Zellkulturen, in: *Biotechnologie* (DOHMEN, K., Ed.), pp. 47–62, Stuttgart: Metzler.

ALFERMANN, A. W., REINHARD, E. (1988), Biotransformation of synthetic and natural compounds by plant cell cultures, in: *Plant Cell Biotechnology, NATO ASI Ser.* Vol. H 18 (PAIS, M. S. S., MAVITUNA, F., NOVAIS, J. M., Eds.), pp. 275–283, Berlin - Heidelberg: Springer.

ALFERMANN, A. W., BERGMANN, W., FIGUR, C., HELMBOLD, U., SCHWANTAG, D., SCHULLER, I., REINHARD, E. (1983), Biotransformation of β-methyldigitoxin to β-methyldigoxin by cell cultures of *Digitalis lanata*, in: *Plant Biotechnology* (MANTELL, S. H., SMITH, H., Eds.), pp. 67–74, Cambridge: Cambridge University Press.

AMINO, S., TAKEUCHI, Y., KOMAMINE, A. (1985), Changes in synthetic activity of cell walls during the cell cycle in a synchronous culture of *Catharanthus roseus, Physiol. Plant.* **64**, 202–206.

BACKS-HÜSEMANN, D., REINERT, J. (1970), Embryoidbildung durch isolierte Einzelzellen aus Gewebekulturen von *Daucus carota, Protoplasma* **70**, 49–60.

BARIAUD-FONTANEL, A., JULIEN, M., COUTOS-THEVENOT, P., BROWN, S., COURTOIS, D., PETIARD, V. (1988), Cloning and cell sorter, in: *Plant Cell Biotechnology, NATO ASI Ser.* Vol. H 18 (PAIS, M. S. S., MAVITUNA, F., NOVAIS, J. M., Eds.), pp. 403–419, Berlin-Heidelberg: Springer.

BARZ, W., DANIEL, S., HINDERER, W., JAQUES, U., KESSMANN, H., KÖSTER, J., TIEMANN, K. (1988), Elicitation and metabolism of phytoalexins in plant cell cultures, in: *Plant Cell Biotechnology, NATO ASI Ser.* Vol. H 18 (PAIS, M. S. S., MAVITUNA, F., NOVAIS, J. M., Eds.), pp. 211–230, Berlin-Heidelberg: Springer.

BAYLISS, M. W. (1973), Origin of chromosome number variation in cultured plant cells, *Nature* **246**, 529–530.

BAYLISS, M. W., GOULD, A. R. (1974), Studies on the growth in culture of plant cells, *J. Exp. Bot.* **25**, 772–783.

BERLIN, J. (1984), Plant cell cultures – a future source of natural products?, *Endeavour* **8**, 5–8.

BERLIN, J. (1986), Secondary products form plant cell cultures, in: *Biotechnology* (REHM, H. J., REED, G., Eds.) Vol. 4, pp. 629–658, Weinheim-Deerfield Beach, FL-Basel: VCH.

BERLIN, J., WITTE, L., SCHUBERT, W., WRAY, V. (1984), Determination and quantification of monoterpenoids secreted into the medium of cell cultures of *Thuja occidentalis, Phytochemistry* **23**, 1277–1279.

BERLIN, J., MOLLENSCHOTT, C., WRAY, V. (1988), Triggered efflux of protoberberine alkaloids from cell suspension cultures of *Thalictrum rugosum, Biotechnol. Lett.* **10**, 193–198.

BERLIN, J. FECKER, L., HERMINGHAUS, S., RÜGENHAGEN, C., THOLL, D. (1992), Overproduction of amines by genetic transformation, *Planta Med.* **58**, *Suppl.* **1**, A 575.

BERNATZKY, R., TANKSLEY, S. D. (1987), Using molecular markers to analyze genome organization and evolution, in: *Plant Tissue and Cell Culture* (GREEN, C. E., SOMERS, D. A., HACKETT, W. P., BIESBOER, D. D., Eds.) pp. 331–339, New York: Alan R. Liss.

BHOJWANI, S. S. (Ed.) (1990), *Plant Tissue Culture: Applications and Limitations*, Amsterdam: Elsevier.

BHOJWANI, S. S., RAZDAN, M. K. (1983), *Plant Tissue Culture: Theory and Practice*, Amsterdam: Elsevier.

BIDWELL, R. G. S. (1974), *Plant Physiology*, New York: Macmillan.

BLOM, T. J. M., KREIS, W., VAN IREN, F., LIBBENGA, K. R. (1992), A non-invasive method for the routine-estimation of fresh weight of cells grown in batch suspension cultures, *Plant Cell Rep.* **11,** 146–149.

BREULING, M., ALFERMANN, A. W., REINHARD, E. (1985), Cultivation of cell cultures of *Berberis wilsoniae* in 20-l airlift reactors, *Plant Cell Rep.* **4,** 220–223.

BRODELIUS, P. (1988), Stress-induced secondary metabolism in plant cell cultures, in: *Plant Cell Biotechnology, NATO ASI Ser.* Vol. H 18 (PAIS, M. S. S., MAVITUNA, F., NOVAIS, J. M., Eds.), pp. 195–209, Berlin-Heidelberg: Springer.

BRODELIUS, P., NILSSON, K. (1980), Entrapment of plant cells in different matrices. A comparative study, *FEBS Lett.* **122,** 312–326.

BRODELIUS, P., NILSSON, K. (1983), Permeabilization of immobilized plant cells resulting in a release of intracellularly stored products with preserved viability, *Eur. J. Appl. Microbiol. Biotechnol.* **17,** 275–280.

BRODELIUS, P. E., FUNK, C., SHILLITO, R. D. (1988), Permeabilization of cultivated plant cells by electroporation for release of intracellularly stored secondary products, *Plant Cell Rep.* **7,** 186–188.

BUITELAAR, R. M., TRAMPER, J. (1992), Strategies to improve the production of secondary metabolites with plant cell cultures: a literature review, *J. Biotechnol.* **23,** 111–141.

CHAPRIN, N., ELLIS, B. E. (1984), Microspectrophotometric evaluation of rosmarinic acid accumulation in single cultured plant cells, *Can. J. Bot.* **62,** 2278–2282.

CHARLWOOD, B. V., CHARLWOOD, K. A., MOLINA-TORRES, J. (1990), Accumulation of secondary compounds by organized plant cultures, in: *Secondary Products from Plant Tissue Culture* (CHARLWOOD, B. V., RHODES, M. J. C., Eds.), pp. 167–200, Oxford: Clarendon Press.

CONSTABEL, F., (1990), Medicinal plant biotechnology, *Planta Med.* **56,** 421–425.

CONSTABEL, F., TOWERS, G. H. N. (1988), Thiarubrine accumulation in hairy root cultures of *Chaenactis douglasii, J. Plant Physiol.* **133,** 67–72.

CONSTABEL, F., KURZ, W. G. W., CHATSON, K. B., KIRKPATRICK, J. W. (1977), Partial synchrony in soybean cell suspension cultures induced by ethylene, *Exp. Cell Res.* **105,** 263–268.

CORMIER, F., AMBID, C. (1987), Extractive bioconversion of geraniol by a *Vitis vinifera* cell suspension employing a two-phase system, *Plant Cell Rep.* **6,** 427–430.

DAY, A. G., LICHTENSTEIN, C. P. (1992), Plant genetic transformation, in: *Plant Biotechnology* (FOWLER, M. W., WARREN, G. S., Eds.), pp. 151–182, Oxford: Pergamon Press.

DEBERGH, P. C., ZIMMERMAN, R. H. (Eds.) (1991), *Micropropagation, Technology and Application,* Dordrecht: Kluwer Academic Publishers.

DE-EKNAMKUL, W., ELLIS, B. E. (1985), Effects of macronutrients on growth and rosmarinic acid formation in cell suspension cultures of *Anchusa officinalis, Plant Cell Rep.* **4,** 46–49.

DE LUCA, V., MARINEAU, C., BRISSON, N. (1989), Molecular cloning and analysis of cDNA encoding a plant tryptophan decarboxylase: comparison with animal DOPA decarboxylase, *Proc. Natl. Acad. Sci. USA* **86,** 2582–2586.

DEUS-NEUMANN, B., ZENK, M. H. (1984), Instability of indole alkaloid production in *Catharanthus roseus* cell suspension cultures, *Planta Med.* **50,** 427–431.

DIXON, R. A. (Ed.) (1985), *Plant Cell Culture – a Practical Approach,* Oxford: IRL Press.

DOUGALL, D. K. (1977), Current problems in the regulation of nitrogen metabolism in plant cell cultures, in: *Plant Tissue Culture and Its Bio-Technological Application* (BARZ, W., REINHARD, E., ZENK, M. H., Eds.), pp. 76–84, Berlin: Springer.

DOUGALL, D. K. (1980), Nutrition and metabolism, in: *Plant Tissue Culture as a Source of Biochemicals* (STABA, E. J., Ed.), pp. 21–58, Boca Raton: CRC Press.

DOUGLAS, D., HOFFMANN, H., SCHULZ, W., HAHLBROCK, K. (1987), Structure and elicitor or UV light stimulated expression of two 4-coumarate:CoA ligase genes in parsley, *EMBO J.* **6,** 1189–1195.

EILERT, U. F. K., KURZ, W. G. W., CONSTABEL, F. (1985), Stimulation of sanguinarine accumulation in *Papaver somniferum* cell cultures by fungal elicitors, *J. Plant Physiol.* **119,** 65–76.

Ellis, B. E. (1985), Characterization of clonal cultures of *Anchusa officinalis* derived from single cells of known productivity, *J. Plant Physiol.* **119,** 149–158.

ERIKSSON, T. (1966), Partial synchronization of cell division in suspension cultures of *Haplopappus gracilis, Physiol. Plant.* **19,** 900–910.

EVANS, D. A., SHARP, W. R., MEDINA-FILHO, H. P. (1984), Somaclonal and gametoclonal variation, *Am. J. Bot.* **71,** 759–774.

FARNSWORTH, N. R. (1985), The role of medicinal plants in drug development, in: *Natural Products and Drug Development* (KROGSGAARD-LARSEN, P., BROGGER CHRISTENSEN, S., KOFOD, H., Eds.), pp. 17–30, Copenhagen: Munksgaard.

FLORES, H. E., FILNER, P. (1985), Metabolic relationships of putrescine, GABA and alkaloids in cell and root cultures of Solanaceae, in: *Primary and Secondary Metabolism of Plant Cell Cultures* (NEUMANN, K. H., BARZ, W., REINHARD, E., Eds.), pp. 174–185, Berlin: Springer Verlag.

FLORES, H. E., HOY, M. W., PICKARD, J. J. (1987), Secondary metabolites from root cultures, *TIBTECH* **5,** 64–69.

FONSECA, M. M. R., MAVITUNA, F., BRODELIUS, P. (1988), Engineering aspects of plant cell culture, in: *Plant Cell Biotechnology, NATO ASI Ser.* Vol. H 18 (PAIS, M. S. S., MAVITUNA, F., NOVAIS, J. M., Eds.), pp. 389–401, Berlin-Heidelberg: Springer.

FOWLER, M.W. (1982), The large scale cultivation of plant cells, *Prog. Ind. Microbiol.* **17,** 209–220.

FOWLER, M. W., STAFFORD, A. M. (1992), Plant cell culture, process systems and product synthesis, in: *Plant Biotechnology, Comprehensive Biotechnology,* 2nd. Supplement (FOWLER, M. W., WARREN, G. S., Eds.), pp. 79–98, Oxford: Pergamon Press.

FRISCHKNECHT, P. M., BAUMANN, T. W., WANNER, H. (1977), Tissue culture of *Coffea arabica*. Growth and caffeine formation, *Planta Med.* **31,** 344–350.

FUJITA, Y., TABATA, M. (1987), Secondary metabolites from plant cells – pharmaceutical applications and progress in commercial production, in: *Plant Tissue and Cell Culture* (GREEN, C. E., SOMERS, D. A., HACKETT, W. P., BIESBOER, D. D., Eds.), pp. 169–185, New York: Alan R. Liss.

FUJITA, Y., HARA, Y., OGINO, T., SUGA, C. (1981a), Production of shikonin derivatives by cell suspension cultures of *Lithospermum erythrorhizon* I. Effects of nitrogen sources on the production of shikonin derivatives, *Plant Cell Rep.* **1,** 59–60.

FUJITA, Y., HARA, Y., SUGA, C., MORIMOTO, T. (1981b), Production of shikonin derivatives by cell suspension cultures of *Lithospermum erythrorhizon* II. New medium for the production of shikonin derivatives, *Plant Cell Rep.* **1,** 61–63.

FUJITA, Y., TABATA, M., NISHI, A., YAMADA, Y. (1982), New medium and production of secondary compounds with the two-staged culture method, in: *Plant Tissue Culture 1982* (FUJIWARA, A., Ed.), pp. 399–400, Tokyo: Maruzen Co.

FUJITA, Y., TAKAHASHI, S., YAMADA, Y. (1984), Selection of cell lines with high productivity of shikonin derivatives through protoplast of *Lithospermum erythrorhizon,* in: *Third European Congress on Biotechnology, München,* pp. I-161-I-166, Weinheim-Deerfield Beach, FL-Basel: Verlag Chemie.

FUJITA, Y., TAKAHASHI, S., YAMADA, Y. (1985), Selection of cell lines with high productivity of shikonin derivatives by protoplast culture of *Lithospermum erythrorhizon* cells, *Agric. Biol. Chem.* **49,** 1755–1760.

FUKUI, H., YOSHIKAWA, N., TABATA, M. (1983), Induction of shikonin formation by agar in *Lithospermum erythrorhizon* cell suspension cultures, *Phytochemistry* **22,** 2451–2453.

GALNEDER, E., ZENK, M. H. (1990), Enzymology of alkaloid production in plant cell cultures, in: *Progress in Plant Cellular and Molecular Biology* (NIJKAMP, H. J. J., VAN DER PLAS, L. H. W., VAN AARTRIJK, J., Eds.), pp. 754–762, Dordrecht-Boston-London: Kluwer Academic Publishers.

GAMBORG, O. L., MILLER, R. A., OJIMA, K. (1968), Nutrient requirements of suspension cultures of soybean root cells, *Exp. Cell Res.* **50,** 151–158.

GASSER, C. S., FRALEY, R. T. (1992), Transgenic crops, *Sci. Am.*, June, 34–39.

GATHERCOLE, R. W. E., MANSFIELD, K. J., STREET, H. E. (1976), Carbon dioxide as an essential requirement for cultured sycamore cells, *Physiol. Plant.* **37,** 213–217.

GEORGE, E. F., SHERRINGTON, P. D. (1984), *Plant Propagation by Tissue Culture.* Handbook and Directory of Commercial Laboratories, Basingstoke: Exegetics Ltd.

GEORGE, E. F., PUTTOCK, D. J. M., GEORGE, H. J. (1987), *Plant Culture Media,* Vol. 1: *Formulations and Uses,* Basingstoke: Exegetics Ltd.

GEORGE, E. F., PUTTOCK, D. J. M., GEORGE, H. J. (1988), *Plant Culture Media,* Vol. 2: *Commentary and Analysis,* Basingstoke: Exegetics Ltd.

GERTLOWSKI, C., PETERSEN, M. (1993), Influence of the carbon source on growth and rosmarinic acid production in suspension cultures of *Coleus blumei, Plant Cell Tissue Organ Cult.*, in press.

GLEITZ, J., SEITZ, H. U. (1989), Induction of chalcone synthase in cell suspension cultures of carrot (*Daucus carota* L. ssp. *sativus*) by ultraviolet light: evidence for two different forms of chalcone synthase, *Planta* **179,** 323–330.

GREIDZIAK, N., DIETTRICH, B., LUCKNER, M. (1990), Batch cultures of somatic embryos of *Digitalis lanata* in gaslight fermenters. Development and cardenolide accumulation, *Planta Med.* **56,** 175–178.

GRISEBACH, H., HAHLBROCK, K. (1974), Enzymology and regulation of flavonoid and lignin biosynthesis in plants and plant cell suspension cultures, in: *Metabolism and Regulation of Secondary Plant Products* (RUNECKLES, V. C., CONN, E. E., Eds.), pp. 21–52, New York: Academic Press.

GUERN, J., RENAUDIN, J. P., BROWN, S. C. (1987), The compartmentation of secondary metabolites in plant cell cultures, in: *Cell Culture and Somatic Cell Genetics of Plants*, Vol. 4 (CONSTABEL, F., VASIL, I. K., Eds.), pp. 43–76, San Diego: Academic Press.

GUHA, S., MAHESHWARI, S. C. (1964), *In vitro* production of embryos from anthers of *Datura, Nature* **204,** 497.

HAHLBROCK, K. (1986), Secondary products, in: *Biotechnology: Potentials and Limitations, Dahlem Konferenzen 1986* (SILVER, S., Ed.), pp. 241–257, Berlin: Springer.

HAMILL, J. D., PARR, A. J., RHODES, M. J. C., ROBINS, R. J., WALTON, N. J. (1987), New routes to plant secondary products, *Biotechnology* **5,** 800–804.

HARTMANN, T. (1985), Principles of plant secondary metabolism, *Plant Syst. Evol.* **150,** 15–34.

HASHIMOTO, T., MATSUDA, J., OKABE, S., AMANO, Y., YUN, D. J., HAYASHI, A., YAMADA, Y. (1990), Molecular cloning and tissue- and cell-specific expression of hyoscyamine 6β-hydroxylase, in: *Progress in Plant Cellular and Molecular Biology* (NIJKAMP, H. J. J., VAN DER PLAS, L. H. W., VAN AARTRIJK, J., Eds.), pp. 775–780, Dordrecht-Boston-London: Kluwer Academic Publishers.

HEGARTY, P. K., SMART, N. J., SCRAGG, A. H., FOWLER, M. W. (1986), The aeration of *Catharanthus roseus* L. G. Don suspension cultures in airlift bioreactors: The inhibition effect at high aeration rates on culture growth, *J. Exp. Bot.* **37,** 1911–1920.

HEINSTEIN, P. F. (1985), Future approaches to the formation of secondary natural products in plant cell suspension cultures, *J. Nat. Prod.* **48,** 1–9.

HESS, D. (1992), *Biotechnologie der Pflanzen*, Stuttgart: Eugen Ulmer Verlag.

HINDERER, W., PETERSEN, M., SEITZ, H. U. (1984), Inhibition of flavonoid biosynthesis by gibberellic acid in cell suspension cultures of *Daucus carota* L., *Planta* **160,** 544–549.

HOOKER, B. S., LEE, J. M., AN, G. (1990), Cultivation of plant cells in a stirred vessel: Effect of impeller design, *Biotechnol. Bioeng.* **35,** 296–304.

HOOYKAAS, P. J. J., SCHILPEROORT, R. A. (1992), *Agrobacterium* and plant genetic engineering, *Plant Mol. Biol.* **19,** 15–38.

HORSCH, R., FRALEY, R., ROGERS, S., FRY, J., KLEE, H., SHAH, D., MCCORMICK, S., NIEDERMEYER, J., HOFFMANN, N. (1987), *Agrobacterium*-mediated transformation of plants, in: *Plant Tissue and Cell Culture* (GREEN, C. E., SOMERS, D. A., HACKETT, W. P., BIESBOER, D. D., Eds.), pp. 317–330, New York: Alan R. Liss.

HULST, A. C., MEYER, M. T., BRETELER, H., TRAMPER, J. (1989), Effect of aggregate size in cell cultures of *Tagetes patula* on thiophene production and cell growth, *Appl. Microbiol. Biotechnol.* **30,** 18–25.

HÜSEMANN, W. (1988), Physiological and biochemical characteristics of photoautotrophic plant cell cultures, in: *Plant Cell Biotechnology, NATO ASI Ser.* Vol. H 18 (PAIS, M. S. S., MAVITUNA, F., NOVAIS, J. M., Eds.), pp. 179–193, Berlin-Heidelberg: Springer.

HÜSEMANN, W., FISCHER, K., MITTELBACH, I., HÜBNER, S., RICHTER, G., BARZ, W. (1989), Photoautotrophic plant cell cultures for studies on primary and secondary metabolism, in: *Primary and Secondary Metabolism of Plant Cell Cultures II* (KURZ, W. G. W., Ed.), pp. 35–46. Berlin-Heidelberg: Springer Verlag.

HUTTER, S. L., ARNTZEN, C., BEACHY, R., BREUNING, G., NESTER, E., QUALSET, C., VIDAVER, A. (1992), Revising oversight of genetically modified plants, *Bio/Technology* **10,** 967–971.

JUNG, G., TEPFER, D. (1987), Use of genetic transformation by the Ri T-DNA of *Agrobacterium rhizogenes* to stimulate biomass and tropane alkaloid production in *Altropa belladonna* and *Calystegia sepium* roots grown *in vitro*, *Plant Sci.* **50,** 145–151.

KAIMORI, N., TAKAHASHI, N. (1989), Plant regeneration from dried callus of carrot (*Daucus carota* L.), *Jpn. J. Breed.* **39,** 379–382.

KARTHA, K. (1987), Advances in the cryopreservation technology of plant cells and organs, in: *Plant Tissue and Cell Culture* (GREEN, C. E., SOMERS, D. A., HACKETT, W. P., BIESBOER, D. D., Eds.), pp. 447–458. New York: Alan R. Liss.

KASHA, K. J., KAO, K. N. (1970), High frequency haploid production in barley (*Hordeum vulgare* L.), *Nature* **225,** 874–876.

KATO, A., SHIMIZU, Y., NAGAI, S. (1975), Effect of initial k_La on the growth of tobacco cells in batch culture, *J. Ferment. Technol.* **53,** 744–751.

KILBY, N. J., HUNTER, C. S. (1990), Repeated harvest of vacuole-located secondary product from *in vitro* grown cells using 1.02 MHz ultrasound, *Appl. Microbiol. Biotechnol.* **33,** 448–451.

KIM, D. J., CHANG, H. N. (1990), Enhanced shikonin production from *Lithospermum erythrorhizon* by *in-situ* extraction and calcium alginate immobilization, *Biotechnol. Bioeng.* **36,** 460–466.

KING, P. J. (1980), Plant tissue cultures and cell cycle, *Adv. Biochem. Eng.* **18,** 1–38.

KING, P. J., STREET, H. E. (1977), Growth patterns in cell cultures, in: *Plant Tissue and Cell Culture* (STREET, H. E., Ed.), pp. 307–387, Oxford: Blackwell Scientific Publishers.

KINNERSLEY, A. M., DOUGALL, D. K. (1981), Correlation between nicotine content of tobacco plants and callus cultures, in: *W. Alton Jones Cell Science Center Annual Report,* pp. 7–8, Lake Placid: W. Alton Jones Cell Science Center.

KOBAYASHI, Y., FUKUI, H., TABATA, M. (1988), Berberine production by batch and semi-continuous cultures of immobilized *Thalictrum* cells in an improved bioreactor, *Plant Cell Rep.* **7,** 249–252.

KOBAYASHI, Y., HARA, M., FUKUI, H., TABATA, M. (1991), The role of ethylene in berberine production by *Thalictrum minus* cell suspension cultures, *Phytochemistry* **30,** 3605–3608.

KOMAMINE, A., MATSUMOTO, M., TSUKAHARA, M., FUJIWARA, A., KAWAHARA, M., ITO, M., SMITH, J., NOMURA, K., FUJIMURA, T. (1990), Mechanisms of somatic embryogenesis in cell cultures – physiology, biochemistry and molecular biology, in: *Progress in Plant Cellular and Molecular Biology* (NIJKAMP, H. J. J., VAN DER PLAS, L. H. W., VAN AARTRIJK, J., Eds.), pp. 307–313, Dordrecht-Boston-London: Kluwer.

KOOP, H.-U., WEBER, G., SCHWEIGER, H.-G. (1983), Individual culture of selected single cells and protoplasts of higher plants in microdroplets of defined media, *Z. Pflanzenphysiol.* **112,** 21–34.

KREIS, W., REINHARD, E. (1988), 12β-Hydroxylation of digitoxin by suspension-cultured *Digitalis lanata* cells. Production of deacetyllanatoside C using a two-stage culture method, *Planta Med.* **54,** 143–148.

KREIS, W., REINHARD, E. (1989), The production of secondary metabolites by plant cells cultivated in bioreactors, *Planta Med.* **55,** 409–416.

KUBERSKI, C., SCHEIBNER, H., STEUP, C., DIETTRICH, B., LUCKNER, M. (1984), Embryogenesis and cardenolide formation in tissue cultures of *Digitalis lanata, Phytochemistry* **23,** 1407–1412.

KUCKUCK, H., KOBABE, G., WENZEL, G. (1991), *Fundamentals of Plant Breeding*, Berlin: Springer Verlag.

KUHN, D., CHAPPELL, J., BOUDET, A., HAHLBROCK, K. (1984), Induction of phenylalanine ammonialyase and 4-coumarate: CoA ligase mRNAs in cultured plant cells by UV light or fungal elicitor, *Proc. Natl. Acad. Sci. USA* **81,** 1102–1106.

KURZ, W. G. W., TYLER, R. T., ROEWER, I. A. (1988), Elicitation – A method to induce metabolite production by plant cell cultures, in: *Proc. 8th Int. Biotechnol. Symp.,* Vol. 1 (DURAND, G., BOBICHON, L., FLORENT, J., Eds.), pp. 193–204, Paris: Société Française de Microbiologie.

KUTCHAN, T. M. (1989), Expression of enzymically active cloned strictosidine synthase from the higher plant *Rauwolfia serpentina* in *Escherichia coli, FEBS Lett.* **257,** 127–130.

KUTCHAN, T. M., HAMPP, N., LOTTSPEICH, F., BEYREUTHER, K., ZENK, M. H. (1988), The cDNA clone for strictosidine synthase from *Rauwolfia serpentina.* DNA sequence determination and expression in *E. coli, FEBS Lett.* **237,** 40–44.

LARKIN, P. J., SCOWCROFT, W. R. (1981), Somaclonal variation – a novel source of variability from cell cultures for plant improvement, *Theor. Appl. Genet.* **60,** 197–214.

LECKIE, F., SCRAGG, A. H., CLIFFE, K. C. (1991), An investigation into the role of initial $k_{L}a$ on the growth and alkaloid accumulation by cultures of *Catharanthus roseus, Biotechnol. Bioeng.* **37,** 364–370.

LEVITT, J. (1974), *Introduction to Plant Physiology,* 2nd. Ed., Saint Louis: The C. V. Mosby Comp.

LINDSEY, K., YEOMAN, M. M. (1983), Novel experimental systems for studying the production of secondary metabolites by plant tissue cultures, in: *Plant Biotechnology* (MANTELL, S. H., SMITH, H., Eds.), pp. 39–65, Cambridge-New York-New Rochelle-Melbourne-Sydney: Cambridge University Press.

LINDSEY, K., YEOMAN, M. M., BLACK, G. M., MAVITUNA, F. (1983), A novel method for the immobilization and culture of plant cells, *FEBS Lett.* **155,** 143–149.

LINSMAIER, E. F., SKOOG, F. (1965), Organic growth factor requirements of tobacco tissue cultures, *Physiol. Plant.* **18,** 100–127.

LÖRZ, H., JUNKER, B., SCHELL, J., DE LA PENA, A. (1987), Gene transfer in cereals, in: *Plant Tissue and Cell Culture* (GREEN, C. E., SOMERS, D. A., HACKETT, W. P., BIESBOER, D. D., Eds.), pp. 303–316, New York: Alan R. Liss.

LUCKNER, M., DIETTRICH, B. (1985), Formation of cardenolides in cell and organ cultures of *Digitalis lanata,* in: *Primary and Secondary Metabolism of Plant Cell Cultures* (NEUMANN, K. H., BARZ, W., REINHARD, E., Eds.), pp. 154–163, Berlin: Springer.

LUCKNER, M., NOVER, L., BÖHM, H. (1977), *Secondary Metabolism and Cell Differentiation,* Berlin-Heidelberg-New York: Springer.

MANTELL, S. H., SMITH, H. (1983), Cultural factors that influence secondary metabolite accumulations in plant cell and tissue cultures, in: *Plant Biotechnology* (MANTELL, S. H., SMITH, H., Eds.), pp. 75–108, Cambridge-New York-New Rochelle-Melbourne-Sydney: Cambridge University Press.

MARETZKI, A., THOM, M., NICKELL, L. G. (1974), Utilization and metabolism of carbohydrates in cell and callus cultures, in: *Tissue Culture and Plant Science 1974* (STREET, H. E., Ed.), pp. 329–361, London-New York: Academic Press.

MARTIN, S. M. (1980), Mass culture systems for plant cell suspension, in: *Plant Tissue Culture as a Source of Biochemicals* (STABA, E. J., Ed.), pp. 149–166, Boca Raton: CRC Press.

MEINS, F., JR., BINNS, A. R. (1978), Epigenetic clonal variation and the requirement of plant cells for cytokinins, in: *The Clonal Basis for Development* (SUBTELNY, S., SUSSEX, I. M., Eds.), pp. 185–201, New York: Academic Press.

MIFLIN, B. J., LEA, P. J. (1977), Amino acid metabolism, *Annu. Rev. Plant Physiol.* **28,** 299–329.

MIZUKAMI, H., KONOSHIMA, M., TABATA, M. (1978), Variation in pigment production in *Lithospermum erythrorhizon* callus cultures, *Phytochemistry* **17,** 95–97.

MOHR, H., SCHOPFER, P. (1992), *Pflanzenphysiologie,* 4th Ed., Berlin-Heidelberg-New York: Springer Verlag.

MOL, J. N. M., STUITJE, T. R., GERATS, A. G. M., KOES, R. E. (1988), Cloned genes of phenylpropanoid metabolism in plants, *Plant Mol. Biol. Rep.* **6,** 274–279.

MOL, J. N. M., DE LANGE, P., OOSTDAM, A., VAN DER PLAS, L. H. W. (1990), Use of genetic engineering to improve yields in cell cultures, e. g. (anti)sense DNA technology, in: *Progress in Plant Cellular and Molecular Biology* (NIJKAMP, H. J. J., VAN DER

PLAS, L. H. W., VAN AARTRIJK, J., Eds.), pp. 712–716, Dordrecht-Boston-London: Kluwer Academic Publishers.

MOREL, G. (1960), Producing virus-free *Cymbidium. Am. Orchid. Soc. Bull.* **29,** 495–497.

MOREL, G., MARTIN, C. (1952), Guérison de dahlias atteint d'une maladie a virus. *C. R. Acad. Sci. Paris* **235,** 1324–1325.

MORRIS, P., FOWLER, M. W. (1981), A new method for the production of fine cell suspension cultures, *Plant Cell Tissue Organ Cult.* **1,** 15–24.

MUIR, W. H., HILDEBRANDT, A. C., RIKER, A. J. (1958), The preparation, isolation and growth in culture of single cells from higher plants, *Am. J. Bot.* **45,** 589–597.

MURASHIGE, T., SKOOG, F. (1962), A revised medium for rapid growth and bio assays with tobacco tissue cultures, *Physiol. Plant.* **15,** 473–497.

NAKAGAWA, K., KONAGAI, A., FUKUI, H., TABATA, M. (1984), Release and crystallization of berberine in the liquid medium of *Thalictrum minus* cell suspension cultures, *Plant Cell Rep.* **3,** 254–257.

NAKAGAWA, K., FUKUI, H., TABATA, M. (1986), Hormonal regulation of berberine production in cell suspension cultures of *Thalictrum minus, Plant Cell Rep.* **5,** 69–71.

NATO, A., BAZETOUX, S., MATHIE, Y. (1977), Photosynthetic capacities and growth characteristics of *Nicotiana tabacum* (cv. Xanthi) cell suspension cultures, *Physiol. Plant.* **41,** 116–123.

NESIUS, K. K., FLETCHER, J. S. (1975), Contribution of nonautotrophic carbon dioxide fixation to protein synthesis in suspension cultures of Paul's Scarlett rose, *Plant Physiol.* **55,** 643–645.

NOGUCHI, M., MATSUMOTO, T., HIRATA, Y., YAMAMOTO, K., KATSUYAMA, A., KATO, A., AZECHI, S., KATO, K. (1977), Improvement of growth rates of plant cell cultures, in: *Plant Tissue Culture and its Bio-technological Application* (BARZ, W., REINHARD, E., ZENK, M. H., Eds.), pp. 85–94, Berlin: Springer Verlag.

NOVAIS, J. M. (1988), Methods of immobilization of plant cells, in: *Plant Cell Biotechnology, NATO ASI Ser.* Vol. H 18 (PAIS, M. S. S., MAVITUNA, F., NOVAIS, J. M., Eds.), pp. 353–363, Berlin-Heidelberg: Springer.

OKAMURA, S., MIYASAKA, K., NISHI, A. (1973), Synchronization of carrot cell cultures by starvation and cold treatment, *Exp. Cell Res.* **78,** 467–470.

OKSMAN-CALDENTEY, K. M., STRAUSS, A. (1986), Somaclonal variation of scopolamine content in protoplast derived cell culture clones of *Hyoscyamus muticus, Planta Med.* **52,** 6–12.

OOMS, G. (1992), Genetic engineering of plants and cultures, in: *Plant Biotechnology* (FOWLER, M. W., WARREN, G. S., Eds.), pp. 223–258, Oxford: Pergamon Press.

PAREILLEUX, A. (1988), The large-scale cultivation of plant cells, in: *Plant Cell Biotechnology, NATO ASI Ser.* Vol. H 18 (PAIS, M. S. S., MAVITUNA, F., NOVAIS, J. M., Eds.), pp. 313–328, Berlin-Heidelberg: Springer.

PARK, C. H., MARTINEZ, B. C. (1992), Enhanced release of rosmarinic acid from *Coleus blumei* permeabilized by dimethylsulfoxide (DMSO) while preserving viability and growth, *Biotechnol. Bioeng.* **40,** 459–464.

PIEL, G.-W., BERLIN, J., MOLLENSCHOTT, C., LEHMANN, J. (1988), Growth and alkaloid production of a cell suspension culture of *Thalictrum rugosum* in shake flasks and membrane-stirrer reactors with bubble free aeration, *Appl. Microbiol. Biotechnol.* **29,** 456–461.

POTRYKUS, I., PASZKOWSKI, J., SAUL, M. W., NEGRUTIO, I., SHILITO, R. R. (1987), Direct gene transfer to plants: Facts and future, in: *Plant Tissue and Cell Culture* (GREEN, C. E., SOMERS, D. A., HACKETT, W. P., BIESBOER, D. D., Eds.) pp. 289–302, New York: Alan R. Liss.

RAVEN, P. H., EVERT, R. F., CURTIS, H. (1976), *Biology of Plants,* 2nd Ed., New York: Worth Publishers Inc.

REINERT, J. (1959), Über die Kontrolle der Morphogenese und die Induktion von Adventivembryonen an Gewebekulturen aus Karotten, *Planta* **53,** 318–333.

REUVENY, Z., FILNER, P. (1976), A new assay for ATP sulfurylase based on differential solubility of the sodium salts of adenosine 5'-phosphate and sulfate, *Anal. Biochem.* **75,** 410–428.

RHODES, M. J. C., ROBINS, R. J., HAMILL, J. D., PARR, A. J., HILTON, M. G., WALTON, N. J. (1990), Properties of transformed root cultures, in: *Secondary Products from Plant Tissue Culture* (CHARLWOOD, B. V., RHODES, M. J. C., Eds.), pp. 201–225, Oxford: Clarendon Press.

RICHTER, G. (1988), *Stoffwechselphysiologie der Pflanzen,* 5th Ed., Stuttgart: Thieme.

RITTERSHAUS, E., ULRICH, J., WEISS, A., WESTPHAL, K. (1989), Großtechnische Fermentation von pflanzlichen Zellkulturen. Planung, Bau und Inbetriebnahme einer Fermentationsanlage (75.000 Liter) für pflanzliche Zellkulturen, *BioEngineering* **5,** 8–10.

RUDGE, K., MORRIS, P. (1986), The effect of osmotic stress on growth and alkaloid accumulation in *Catharanthus roseus,* in: *Secondary Metabolism in Plant Cell Cultures* (MORRIS, P., SCRAGG, A. H., STAFFORD, A., FOWLER, M. W., Eds.), pp. 75–81, Cambridge: Cambridge University Press.

RUYTER, C. M., STÖCKIGT, J. (1989), Novel natural products from plant cell and tissue culture – an update. *GIT Fachz. Lab.* **4,** 283–293.

SATO, F., YAMADA, Y. (1984), High berberine-producing cultures of *Coptis japonica* cells, *Phytochemistry* **23,** 281–285.

SCHEEL, D., DANGL, J. L., DOUGLAS, C., HAUFFE, K. D., HERRMANN , A., HOFFMANN, H., LOZOYA, E., SCHULZ, W., HAHLBROCK, K. (1986), Stimulation of phenylpropanoid pathways by environmental factors, in: *Plant Molecular Biology* (VON WETTSTEIN, D., CHUA, N. H., Eds.), pp. 315–326, New York: Plenum Press.

SCOWCROFT, W. R., BRETTELL, R. I. S., RYAN, S. A., DAVIES, P. A., PALLOTTA, M. A. (1987), Somaclonal variation and genomic flux, in: *Plant Tissue and Cell Culture* (GREEN, C. E., SOMERS, D. A., HACKETT, W. P., BIESBOER, D. D., Eds.), pp. 275–286, New York: Alan R. Liss.

SCRAGG, A. H. (1990), Fermentation systems for plant cells, in: *Secondary Products from Plant Tissue Culture* (CHARLWOOD, B. V., RHODES, M. J. C., Eds.), pp. 243–263, Oxford: Clarendon Press.

SCRAGG, A. H. (1992), Bioreactors for mass cultivation of plant cells, in: *Plant Biotechnology* (FOWLER, M. W., WARREN, G. S., Eds.), pp. 45–62, Oxford: Pergamon Press.

SCRAGG, A. H., FOWLER, M. W. (1985), The mass culture of plant cells, in: *Cell Culture and Somatic Cell Genetics of Plants,* Vol. 2, pp. 103–128, New York: Academic Press.

SCRAGG, A. H., MORRIS, P., ALLAN, E. J., BOND, P., FOWLER, M. W. (1987), Effect of scale-up on serpentine formation by *Catharanthus roseus* suspension cultures, *Enzyme Microb. Technol.* **9,** 619–624.

SCRAGG, A. H., ALLAN, E. J., LECKIE, F. (1988), Effect of shear on the viability of plant cell suspensions, *Enzyme Microb. Technol.* **10,** 361–367.

SEIBERT, M., KADKADE, P. G. (1980), Environmental factors: A. Light, in: *Plant Tissue Cultures as a Source of Biochemicals* (STABA, E. J., Ed.), pp. 123–142, Boca Raton: CRC Press.

SEIDEL, S., REINHARD, E. (1987), Major cardenolide glycosides in embryogenic suspension cultures of *Digitalis lanata, Planta Med.* **53,** 308–309.

SEITZ, U. (1987), Cryopreservation of plant cell cultures, *Planta Med.* **52,** 311–314.

SEITZ, H. U., SEITZ, U., ALFERMANN, A. W. (1985), *Pflanzliche Gewebekultur, ein Praktikum*, Stuttgart: Fischer Verlag.

SHULER, M. L. (1981), Production of secondary metabolites from plant tissue cultures, *Ann. N. Y. Acad. Sci.* **369,** 65–69.

SHULER, M. L., (1988), Bioreactor for plant cell culture, in: *Plant Cell Biotechnology, NATO ASI Ser.* Vol. H 18 (PAIS, M. S. S., MAVITUNA, F., NOVAIS, J. M., Eds.), pp. 329–342, Berlin-Heidelberg: Springer.

SKOOG, F., MILLER C. O. (1957), Chemical regulation of growth and organ formation in plant tissues cultured *in vitro*, *Symp. Soc. Exp. Biol.* **11,** 118–130.

SMART, N. J., FOWLER, M. W. (1984), An airlift column reactor suitable for large scale cultivation of plant cell suspensions, *J. Exp. Bot.* **35,** 531–537.

SOMMER, H., SAEDLER, H. (1986), Structure of the chalcone synthase gene of *Antirrhinum majus, Mol. Gen. Genet.* **202,** 429–434.

SPENCER, A., HAMILL, J. D., RHODES, M. J. C. (1990), Production of terpenes by differentiated shoot cultures of *Mentha citrata* transformed with *Agrobacterium tumefaciens* T37, *Plant Cell Rep.* **8,** 601–604.

SPIELER, H., ALFERMANN, A. W., REINHARD, E. (1985), Biotransformation of β-methyldigitoxin by cell cultures of *Digitalis lanata* in airlift and stirred tank reactors, *Appl. Microbiol. Biotechnol.* **23,** 1–4.

STABA, E. J. (Ed.) (1980), *Plant Tissue Culture as a Source of Biochemicals,* Boca Raton: CRC Press.

STABA, E. J. (1982), Production of useful compounds from plant tissue cultures, in: *Plant Tissue Culture 1982* (FUJIWARA, A., Ed.), pp. 25–30, Tokyo: Maruzen Co.

STAFFORD, A., FOWLER, M. W. (1991), Plant cell culture and product opportunities, *Agro-Industry Hi-Tech* **2,** 19–23.

STÖCKIGT, J., SCHMIDT, D., RUYTER, C. M. (1989), Pflanzliche Zellkulturen – Biosynthese, Enzyme und Enzymprodukte, *Dtsch. Apoth. Ztg.* **129,** 2767–2772.

STRASBURGER E., SITTE, P., ZIEGLER, H., EHRENDORFER, F., BRESINSKY, A. (1991), *Lehrbuch der Botanik,* 33rd Ed., Stuttgart–New York: Gustav Fischer.

STREET, H. E. (1977), Cell (suspension) cultures – techniques, in: *Plant Tissue and Cell Culture* (STREET, H. E., Ed.), pp. 61–102, Oxford: Blackwell Scientific Publ.

STUMPF, P. K., CONN, E. E. (Eds.) (1980–1988), *The Biochemistry of Plants. A Comprehensive Treatise,* Vols. 1–14, London: Academic Press.

SUGA, T., HIRATA, T. (1990), Biotransformation of exogenous substrates by plant cell cultures, *Phytochemistry* **29,** 2393–2406.

SUMARYONO, W., PROKSCH, P., HARTMANN, T., NIMTZ, M., WRAY, V. (1991), Induction of rosmarinic acid accumulation in cell suspension cultures of *Orthosiphon aristatus* after treatment with yeast extract, *Phytochemistry* **30,** 3267–3271.

SUNDERLAND, N. (1977), Nuclear cytology, in: *Plant Tissue and Cell Culture* (STREET, H. E., Ed.), pp. 177–205, Oxford: Blackwell Scientific Publ.

SUZUKI, T., YOSHIOKA, T., TABATA, M., FUJITA, Y. (1987), Potential of *Datura innoxia* cell suspension cultures for glucosylating hydroquinone, *Plant Cell Rep.* **6,** 275–278.

TABATA, M. (1988), Secretion of secondary products by plant cell cultures, in: *Proc. 8th Int. Biotechnol-*

ogy Symp., Paris, Vol. 1 (DURAND, G., BOBICHON, L., FLORENT, J., Eds.), pp. 167–178, Paris: Société Française de Microbiologie.

TABATA, M., FUJITA, Y. (1985), Production of shikonin by plant cell cultures, in: *Biotechnology in Plant Science* (DAY, P., ZAITLIN, M., HOLLÄNDER, A., Eds.), pp. 207–218, Orlando: Academic Press.

TABATA, M., OGINO, T., YOSHIOKA, K., YOSHIKAWA, N., HIRAOKA, N. (1978), Selection of cell lines with high yield of secondary products, in: *Frontiers of Plant Tissue Culture* (THORPE, T. A., Ed.), pp. 313–322, Calgary: University of Calgary.

TANAKA, H. (1981), Technological problems in cultivation of plant cells at high density, *Biotechnol. Bioeng.* **23,** 1203–1218.

TANAKA, H. (1982a), Some properties of pseudocells of plant cells, *Biotechnol. Bioeng.* **24,** 2591–2596.

TANAKA, H. (1982b), Oxygen transfer in broths of plant cells at high density, *Biotechnol. Bioeng.* **24,** 425–442.

TANAKA, H. (1987), Large scale cultivation of plant cells at high density: A review. *Process Biochem.,* August, 106–113.

TANAKA, H., NISHIJIMA, F., SUWA, M., IWAMOTO, T. (1983), Rotating drum fermenter for plant cell suspension cultures, *Biotechnol. Bioeng.* **25,** 2359–2365.

TISSERAT, B. (1985), Embryogenesis, organogenesis and plant regeneration, in: *Plant Cell Culture, a Practical Approach* (DIXON, R. A., Ed.), pp. 79–105, Oxford: IRL Press.

TULECKE, W., NICKEL, L. G. (1959), Production of large amounts of plant tissue by submerged culture, *Science* **130,** 863–864.

TULECKE, W., NICKEL, L. G. (1960), Methods, problems, and results of growing plant cells under submerged conditions, *Trans. N. Y. Acad. Sci.* **22,** 196–206.

TYLER, R. T., EILERT, U., RIJNDERS, C. O. M., ROEWER, I. A., MCNABB, C. K., KURZ, W. G. W. (1989), Studies on benzophenanthridine alkaloid production in elicited cell cultures of *Papaver somniferum,* in: *Primary and Secondary Metabolism of Plant Cell Cultures II* (KURZ, W. G. W., Ed.), pp. 200–207, Berlin–Heidelberg: Springer Verlag.

ULBRICH, B., WIESNER, W., ARENS, H. (1985), Large scale production of rosmarinic acid from plant cell cultures of *Coleus blumei,* in: *Primary and Secondary Metabolism of Plant Cell Cultures* (NEUMANN, K. H., BARZ, W., REINHARD, E., Eds.), pp. 293–303, Berlin: Springer.

ULBRICH, B., OSTHOFF, H., WIESNER, W. (1988), Aspects of screening plant cell cultures for new pharmacologically active compounds, in: *Plant Biotechnology* (PAIS, M. S. S., MAVITUNA, F., NOVAIS, J. M., Eds.), pp. 461–474, Berlin: Springer.

USHIYAMA, K. (1991), Large scale cultivation of ginseng, in: *Plant Cell Culture in Japan* (KOMAMINE, A., MISAWA, M., DICOSMO, F., Eds.), pp 92–98, Tokyo: CMC Co.

VAN DER KROL, A. R., LENTING, P. E., VEENSTRA, J., VAN DER MEER, I. M., KOES, R. M., GERATS, A. G. M., MOL, J. N. M., STUITJE, A. R. (1988), An antisense chalcone synthase gene in transgenic plants inhibits flower pigmentation, *Nature* **333,** 866–869.

VASIL, I. K. (1988), Progress in the regeneration and genetic manipulation of cereal crops. *Bio/Technology* **6,** 397–402.

VASIL, V., BROWN, S. M., RE, D., FROMM, M. E., VASIL, I. K. (1991), Stably transformed callus lines from microprojectile bombardment of cell suspension cultures of wheat. *Bio/Technology* **9,** 743–747.

VELIKY, I. A., MARTIN, S. M. (1970), A fermenter for plant cell suspension cultures, *Can. J. Microbiol.* **16,** 223–226.

VON SENGBUSCH, P. (1988), *Botanik,* Hamburg: McGraw Hill Book Company.

WAGNER, F., VOGELMANN, H. (1977), Cultivation of plant tissue cultures in bioreactors and formation of secondary products, in: *Plant Tissue Culture and its Bio-Technological Application* (BARZ, W., REINHARD, E., ZENK, M. H., Eds.), pp. 245–252, Berlin: Springer.

WAHL, J. (1985), Adaption konventioneller Fermenter zur Züchtung von Pflanzenzellen zum Zwecke der Gewinnung von Naturstoffen, in: *Pflanzliche Zellkulturen* (Federal Minister for Research and Technology, Ed.) pp. 35–43, Jülich: Projektträger Biotechnologie, KFA Jülich.

WEATHERS, P. J., DIIORIO, A., CHEETHAM, R., O'LEARY, M. (1990), Recovery of secondary metabolites with minimal loss of cell viability, in: *Progress in Plant Cellular and Molecular Biology* (NIJKAMP, H. J. J., VAN DER PLAS, L. H. W., VAN AARTRIJK, J., Eds.), pp. 582–586, Dordrecht-Boston-London: Kluwer Academic Publishers.

WEBER, G., LARK, K. G. (1979), An efficient plating system for rapid isolation of mutants from plant cell suspensions, *Theor. Appl. Genet.* **55,** 81–86.

WEIER, T. E., STOCKING, C. R., BARBOUR, M. G. (1974), *Botany – An Introduction to Plant Biology,* 5th Ed., New York-London-Sydney-Toronto: John Wiley & Sons.

WENZEL, G. (1992), Application of unconventional techniques in classical plant breeding, in: *Plant Biotechnology* (FOWLER, M. W., WARREN, G. S., Eds.), pp. 259–281, Oxford: Pergamon Press.

WESTPHAL, K. (1990), Large scale production of new biologically active compounds in plant cell cultures, in: *Progress in Plant Cellular and Molecular Biology* (NIJKAMP, H. J. J., VAN DER PLAS, L. H. W., VAN AARTRIJK, J., Eds.), pp. 601–608, Dordrecht-Boston-London: Kluwer Academic Publ.

WESTPHAL, K., RITTERSHAUS, E. (1991), Großtechnische Fermentation von Pflanzenzellkulturen. Erfolge, Möglichkeiten und Probleme bei der Herstellung von Produkten, *Paper presented at the Meeting of the German IAPTC Section,* Hamburg, September 18–19, 1991.

WHITE, P. R. (1963), *The Cultivation of Animal and Plant Cells,* New York: Ronald Press.

WICHERS, H. J., MALINGRÉ, T. M., HUIZING, H. J. (1983), The effect of some environmental factors on the production of L-DOPA by alginate-entrapped cells of *Mucuna pruriens, Planta* **158,** 482–486.

WIENAND, U., WEYDEMANN, U., NIESBACH-KLÖSGEN, U., PETERSON, P. A., SAEDLER, H. (1986), Molecular cloning of the c2 locus of *Zea mays,* the gene coding for chalcone synthase, *Mol. Gen. Genet.* **203,** 202–207.

WILSON, G. (1980), Continuous culture of plant cells using the chemostat principle, *Adv. Biochem. Eng.* **16,** 1–25.

WITHERS, L. A. (1986), Cryopreservation and genebanks, in: *Plant Cell Culture Technology* (YEOMAN, M. M., Ed.), pp. 96–140, Oxford: Blackwell.

WOERDENBAG, H. J., LÜERS, J. F. J., VAN UDEN, W., PRAS, N., MALINGRÉ, T. M., ALFERMANN, A. W. (1992), Production of the new antimalarial drug artemisinin in shoot cultures of *Artemisia annua* L., *Plant Cell Tissue Organ Cult.* **32,** 247–257.

WRUBEL, R. P., KRIMSKY, S., WETZLER, R. E. (1992), Field testing transgenic plants. An analysis of the US Department of Agriculture's environmental assessments. *BioScience* **42,** 280–289.

YEOMAN, M. M., MIEDZYBRODZKA, M. B., LINDSEY, K., MCLAUGHLAN, W. R. (1980), The synthetic potential of cultured plant cells, in: *Plant Cell Cultures: Results and Perspectives* (SALA, F., PARISI, B., CELLA, R., CIFERRI, O., Eds.), pp. 327–343, Amsterdam-New York-Oxford: Elsevier.

YOKOYAMA, M., YANAGI, M. (1991), High-level production of arbutin by biotransformation, in: *Plant Cell Culture in Japan* (KOMAMINE, A., MISAWA, M., DICOSMO, F., Eds.), pp. 79–91, Tokyo: CMC Co.

YOSHIKAWA, T., FURUYA, T. (1987), Saponin production by cultures of *Panax ginseng* transformed with *Agrobacterium rhizogenes, Plant Cell Rep.* **6,** 449–453.

ZAMBRYSKI, P., TEMPE, J., SCHELL, J. (1989), Transfer and function of T-DNA genes from *Agrobacterium* Ti and Ri plasmids in plants, *Cell* **56,** 193–201.

ZENK, M. H. (1991), Chasing the enzymes of secondary metabolism: Plant cell cultures as a pot of gold, *Phytochemistry* **30,** 3861–3863.

ZENK, M. H., EL-SHAGI, H., SCHULTE, U. (1975), Anthraquinone production by cell suspension cultures of *Morinda citrifolia, Planta Med. Suppl.,* 79–101.

ZENK, M. H., EL-SHAGI, H., ULBRICH, B. (1977a), Production of rosmarinic acid by cell suspension cultures of *Coleus blumei, Naturwissenschaften* **64,** 585–586.

ZENK, M. H., EL-SHAGI, H., ARENS, H., STÖCKIGT, J., WEILER, E. W., DEUS, B. (1977b), Formation of indole alkaloids serpentine and ajmalicine in cell suspension cultures of *Catharanthus roseus,* in: *Plant Tissue Culture and its Bio-Technological Application* (BARZ, W., REINHARD, E., ZENK, M. H., Eds.), pp. 27–43, Berlin-Heidelberg-New York: Springer.

Index

A

acetamidase, gene (*amdS*) 536
acetate production, by clostridia 309
Acetogenium kivui, growth kinetics 133
acetone-butanol fermentation, in clostridia 309
acetyl-CoA 72
acid phosphatase, production 239
acidic tomato broth (ATB), for *Leuconostoc oenos* isolation 348
Acidomonas 270
acidophilic bacilli *see* bacilli
Actinomadura, chemotaxonomic marker 435
– G+C content 435
– morphology 435
actinomycetes, product applications 449
– production of antibiotics 445
actinophage, of *Streptomyces* 568
Actinoplanes, chemotaxonomic marker 435
– G+C content 435
– morphology 435
actinorhodin 215
activation energy 66
ACV-synthetase, gene 529
adenylate system 54
adhesin 18
Aerobacter aerogenes see Klebsiella pneumoniae
Agaricus bisporus 534f
– cultivation 536
agglutinins, in budding yeast 33
Agrobacterium radiobacter, exopolysaccharide 172
Agrobacterium rhizogenes, effect in plant organ culture 600
Agrobacterium tumefaciens, effect in plant organ culture 600
agroclavine production, by immobilized *Claviceps* cells 240
Agrocybe aegerita, life cycle 533
L-alanine production 210
Alcaligenes eutrophus, heterologous gene expression 26
– poly-β-hydroxybutyrate (PHB) production 212
– substrate spectrum 196
alcohol dehydrogenase, activity, in cross-linked yeast cells 238
– gene (*ADH*) 483
alcohol fermentation, in clostridia 300f
alcohol oxidase, in methylotrophic yeasts 271, 273, 280
alcohol oxidoreductase, in methylotrophic bacteria 273
aldolase, gene (*FBA1*) 481
alginate, in pseudomonads 413
alginate bead *see* cell immobilization
Alicyclobacillus 371
Alicyclobacillus acidocaldarius 378
alkaliphilic bacilli *see* bacilli
alkaloid production, by plant cell culture 596
– by *Claviceps purpurea* 240
– calcium alginate effect 240
alkanols, effect on microbial growth 135
amine oxidase 274
amino acid, deamination 91f
– degradation 91
– – in clostridia 304f
– estimation methods 126
– transport in *Lactococcus* 358
amino acid biosynthesis 92
– aromatic amino acids 93
– aspartate family 454f
– by corynebacteria 454ff
– – regulation 454, 457
– by methylotrophs 279
– feedback inhibition 101
– oxaloacetate family 93
– oxoglutarate family 93
– phosphoglycerate family 93
– pyruvate family 93
– strain improvement 457
amino acid metabolism 91f
– in lactic acid bacteria 354, 358
– in yeast 489

– transamination 91
amino acid production 164
– by coryneforms 454
– by methylotrophic microorganisms 279
– strain construction 206ff
6-aminopenicillanic acid (6-APA) 527
ammonia 174f
– effect on microbial growth 133
– estimation methods 126
– futile cycle 178
– membrane transport 175
ammonium assimilation 92f, 174f, 387
amphibolic pathway 52
Amycolatopsis methanolica, amino acid production 279
amylase 27
– from pseudomonads 421
– secretion 27
– thermostable amylolytic enzyme 28
α-amylase, extracellular 384
– gene (*AMY1*), *Schwanniomyces occidentalis* 201
– gene regulation 386
– in clostridia 310
– mutants 386
– operon 386
– synthesis, regulation in *Bacillus subtilis* 384
Anabaena azollae 237
– hydrogen production 237
anabolism 50
anaerobic respiration 382
anaplerotic reaction 85
animal cell 38f
– culture 40f
– envelope 38
– requirements 41
animal tissue 40
– cell-cell connection 40
antibiotic resistance 447
– as selectable marker 536
– genes, in pseudomonads 410
– – in streptomycetes 447
– types 447
antibiotics, gene cluster 445
– history of discovery 448
– producers 445
– production 214, 441ff
– – by filamentous fungi 527
– – novel synthetic antibiotics 214
– – regulation of 441
– streptomycetes genes 439
antibody-specific marker 11
antiport, over membranes 74ff
6-APA *see* 6-aminopenicillanic acid
Aphyllophorales *see* Poriales
Apocynum cannabinum, shoot culture 601
apoenzyme 70
apple cider 334
archaea 252f
archaebacteria, cell wall 20
arginine deiminase pathway, in pseudomonads 420
arginine dihydrolase pathway, in lactic acid bacteria 358
aromatic hydrocarbons, hydroxylation by pseudomonads 422
aromatics, degradation, by clostridia 300ff
– – by denitrifiers 419
– – by pseudomonads 414
– – of chloro-aromatics 202f
– – *ortho-/meta*-cleavage pathway 414
– – strain improvement 202f
– demethylation, in clostridia 302
Arrhenius equation 128
arthrospores, of streptomycetes 437
Ascomycetes 517, 525ff
– reproduction 525
L-ascorbate 211
Aspergillus 525
Aspergillus amstelodami, specific growth rate 130
Aspergillus nidulans, strain improvement 536f
Aspergillus niger, citric acid production 526f
– morphology 116, 233f
– oxygen consumption 228
– strain improvement 536f
ATP hydrolysis 54
ATPase 65
– ABC-family 77
– export ATPase 76
– types 74
autonomously replicating sequence (*ARS*), in yeast chromosomes 475
autotrophy 50
auxin, effect in plant cell culture 591
Azomonas, phylogenetic status 403
Azotobacter, phylogenetic status 403

B

bacilli 367ff
– acidophilic 381
– – mesophilic 381
– – thermophilic 381
– alkaliphilic 379f
– – adaptation 380
– – bioenergetics 380
– – biotechnological application 380
– – ecology 380
– – proteases 380
– α-amylase 384f
– bacteriophages 567
– carbohydrate uptake 385f
– carbon metabolism 383ff
– – genes 385f

– cellulase 28
– DegS/DegU system 385
– DNA homology 372
– extracellular enzymes 384ff
– food fermentation 368
– genetic diversity 369
– genetic engineering 394ff
– glutamine synthetase 387
– habitats 374f
– heterologous gene products 396
– hexose catabolism 385
– insect toxins 397
– insertional mutagenesis 395
– interleucin production 396
– morphological groups 368
– nitrate reduction 383
– nitrogen metabolism 387f
– oxygen requirements 381
– pentose catabolism 385
– phylogenetic analysis 372
– physiological diversity 369
– physiology 381ff
– plasmid vectors 395
– – stability 395
– polysaccharide-degrading enzymes 384f
– psychrophilic 379
– soil bacilli 374
– sporulation 390ff
– – *cot* (spore coat protein) genes 391
– – gene expression timing 393
– – genetic regulation 392
– – *ger* (germination) genes 391
– – initiation 391f
– – morphological stages 390f
– – mutations 391
– – *out* (outgrowth) genes 391
– – RNA polymerase regulation 393
– – *spo* (sporulation) genes 391
– – *ssp* (SASP) genes 391
– taxonomy 368ff
– thermophilic 378ff
– – ecology 378
– – enzyme production 379
– – genetic engineering 379
– – product characteristics 378
– transformation 394f
– – of competent cells 395
– – plasmid introduction 395
– – protoplast transformation 395
Bacillus, chemotaxonomy 369ff
– classification 368ff
– – thermophilic bacilli 370f
– ecology 373ff
– endospores 388ff
– evolution 372
– identification by 368, 372ff
– – computerized systems 373
– – DNA probes 373
– – fatty acid composition 373
– – pyrolysis mass spectrometry 373
– numerical phenetics 369f
Bacillus-animal association 376ff
Bacillus-plant association 375f
Bacillus alcalophilus 380
Bacillus anthracis 376
Bacillus brevis, relatives 369f
Bacillus cereus, food poisoning 375
Bacillus circulans 372
Bacillus coagulans 378
Bacillus fastidiosus 374
Bacillus firmus 374
Bacillus lentimorbus 376
Bacillus lentus 374
Bacillus licheniformis 368, 374, 382
Bacillus macerans 382f
Bacillus megaterium, endospore germination 394
Bacillus methanolicus 270
– alcohol oxidoreductase 273
Bacillus natto 368
Bacillus polymyxa 375, 382f
– relatives 369f
Bacillus popiliae 376
Bacillus sphaericus 376
– relatives 369f
Bacillus stearothermophilus 379
Bacillus subtilis 372
– phage φ29 548
– relatives 369f
– temperate phages 567
– virulent phages 567
Bacillus thuringiensis 368, 377f
– delta-endotoxin gene 377, 424
– – transfer to plants 424
– inclusion bodies 26
– toxin characteristics 377
bacteria *see also* microorganisms
– anti-phage defense 559
– cell surface layer 22
– cell wall synthesis 113
– fimbriae 17
– gliding bacteria 114f
– growth 113 *see also* microbial growth
– infection by phages 546
– lysogenic conversion 558
– methanol oxidation 271
– methylotrophic 269
– morphology 114
– phage receptor 552f
– phage species 548ff
– pili 17
– pseudolysogenic 559
– reproduction 113
– suspension characteristics 585
bacterial capsule 16

– electron microscopy 17
– Gram-negative bacteria 17
– Gram-positive bacteria 17
bacterial cell *see* prokaryotic cell
bacterial L-form, cell wall 20
bacterial protein production, by *Saccharomyces cerevisiae* 498
bacterial slime 16
– electron microscopy 17
– exopolysaccharide 16
bacteriocin 358ff, 546
bacteriophage(s) 543ff
– actinophage 440, 568
– adsorption 546, 552f
– bacterial defense mechanisms 559
– characteristics 546f, 561
– classification 547
– control 561
– detection methods 560
– DNA packaging 556f
– – headful mechanism 557
– DNA protection 559
– dsDNA phage, morphogenesis 556f
– families 550
– Ff-group 564
– gene expression 554f
– gene organization 555
– host range 548
– host receptor 552f
– impact on host cells 553
– infection efficiency 559
– infection mechanism 546
– lysogeny 558f
– main groups 550
– morphogenesis 556f
– morphology 547
– mutual exclusion 554
– nucleic acid injection 552f
– nucleic acid structure 547
– of bacilli 567
– of clostridia 566
– of coryneform bacteria 568
– of *Escherichia coli* 562ff
– of lactic acid bacteria 566
– of methanogenic bacteria 565
– of methylotrophic bacteria 566
– of *Pseudomonas* 568
– progeny release 557
– prophage 558f
– receptor 546
– replication cycle 546
– routine monitoring 560
– species diversity 548
– ssRNA phages 565
– SSV1 group 550
– superinfection 554, 558
– temperate 558, 562
– – life cycle 558, 565
– therapeutic applications 545
– toxin gene transfer 545
– virulent 558, 563
bacteriophage φ29 548, 567
bacteriophage φ105 567
bacteriophage fd 548
bacteriophage lambda 562f
bacteriophage M13 564
bacteriophage Mu 562
bacteriophage P1 563
bacteriophage "Psi"M1 565
bacteriophage Qβ 548
bacteriophage SPβ 567
bacteriophage SPO1 567
bacteriophage SPO2 567
bacteriophage SSV1 548
bacteriophage T1 545ff
bacteriophage T4 547, 563f
– detection 561
bacteriophage genome 549ff
– replication, σ-form 554f
– – φ-form 554f
– chromosome integration 558
– concatamers 557
– evolution 560
– recombination 560
– replication 546, 551, 554
– – rolling circle mechanism 554
bagasse 150
Basidiomycetes 517, 532 ff
– characteristics 532
– life cycle 533
– morphology 532
– reproduction 533
– taxonomy 534
batch culture 150
– model simulation 153
beer production 334
– beer sarcina 339
– light beer 200
– process improvement 495
– *Saccharomyces* strains 496
benzoate degradation 202
berberine production, by *Thalictrum minus* cells 598f
Betabacteria 329
Bifidobacterium 335, 346
– as starter culture 346
– habitats 346
bile acid conversion, by clostridia 311f
binary fission 113f
biochemical reaction, relaxation time 138
bioenergetics, of lactic acid bacteria 350f
biofilm, poison resistance 243
– substrate profile 227
biogas production 263

– overall quantity 263
biomass estimation analyte, estimation methods 126 *see also* cell mass estimation
biomembrane *see* membrane
biopolymer production *see* polymer production
bioreactor, continuous culture, dilution rate 156
– continuous flow, for NMR 193
– design 595
– for plant cell cultures 586ff
– microbial growth 129
– oxygen supply 135
– oxygen transfer 226ff
– parameter values 151
– stirred tank reactor, continuous - 150
– – oxygen transfer rate 125
– stirrer types 587
– types 587f
biosurfactant 415
biotin 72
biphytanyl tetraether lipid, of methanogens 259
Blakeslea 522
botulinum toxin 311
bread production, history 327
brown-rot fungi 534
budding, in yeast 115f

C

C_1 metabolism, of methylotrophic microorganisms 266ff
callus culture 582f
– media 582f
– sterilization 583
– texture 583
– tissue sources 583
Calvin cycle 84, 276
cAMP, in yeast 487f
Candida 269
Candida albicans 474
Candida tropicalis 32
capsid 546
– morphogenesis 556f
carbohydrate metabolism, biochemistry 352
– energetics 351
– enzyme coding genes 480ff
– in microorganisms 81
– mutants 480ff
– of bacilli 383ff
– of clostridia 297ff
– of lactic acid bacteria 350ff
– of methylotrophs 266, 271ff
– of *Saccharomyces cerevisiae* 478ff
– – regulation 486ff
carbohydrates, estimation methods 126
– transport *see* saccharide transport
carbon assimilation 276
carbon catabolite inactivation 489
carbon catabolite repression, in yeasts 489
carbon dioxide, effect on microbial growth 135
carbon dioxide fixation, in methanogens 261
– in plant cell cultures 580
carbon flow 51
carbon limitation, metabolic effect 172
carbon metabolism, of bacilli 383
– of methanogens 261
carbon source 80 *see also* substrates
– effect on microbial growth 132
– for growth media 143
– microbial utilization 80
carboxylic acid degradation, in clostridia 303
cardenolide 597
Carnobacterium 335
β-carotene production 211
carrier transport *see* membrane transport
β-casein digestion 354ff
cassava (*Manihot esculenta*) 149
catabolism 50
catabolite repression 386
– carbon catabolites in yeast 489
catalysis *see* enzyme catalysis
catechol degradation, by pseudomonads 414
Catharanthus roseus, serpentine production 596
cell compartment 30f
cell component labeling 11
– nucleic acid preparation 11
– protein preparation 11
cell culture 581 *see also* plant cell culture
– eukaryotic cells 40f
– requirements 41
– substrates 195ff
cell differentiation, effect in plant cell culture 600
cell energy status, measurement by NMR 193
cell flocculation 129
cell flotation 129
cell immobilization 223ff, 598 *see also* immobilized microorganisms
– by adsorption 224ff
– – materials 224
– by aggregate formation 225
– by covalent coupling 225
– – reaction mechanism 225
– by microorganism entrapment 225
– *Comamonas acidovorans* 129
– for process improvement 226
– in alginate beads 598
– methods 224ff
– plant cells 598
– suitable polymers 225
cell mass estimation 121ff
– calorimetric estimation 126
– capacitance 127
– conductivity 127
– dry cell weight 121f

– elemental balance 124
– image analysis 123
– mass balance 124
– model-based estimation 124
– kinetic model 126
– off-line 122
– on-line 122
– oxygen consumption 125
– packed cell volume 123
– real-life measurement 122
– turbidity 123
– viscosity 126
cell metabolism *see* metabolism
cell number estimation 119ff
– Coulter counter 121
– direct microscopic count 119
– flow cytometer 121
– membrane filtration 120
– viable plate count 120
cell organelle 31
cell pellet 116
– physical structure 117f
cell structure 5ff
cell surface layer, bacterial 22
– in plants 34
cell suspension, ^{31}P NMR spectrum 192f
cell topology 9
cell wall 578
– bacterial 19ff
– cellulose synthase complex 35
– lignin 36
– murein 19f
– of archaebacteria 21
– of bacterial L-forms 20
– of halobacteria 22
– of methanogens 21, 257
– of mycoplasms 20
– of plants 35, 578
– synthesis in bacteria 113
cellulase 28
– carboxymethylcellulase 28
– *Clostridium thermocellum* 28
– – action mechanism 29
cellulose degradation, by clostridia 297
cellulosome 297
– function 29
– structure 29
centrifugal filtration 191
cephalosporin *C*, synthesis, biosynthetic pathway 528
– production 214
Cephalosporium 525ff
Cephalosporium acremonium, cephalosporin production 214, 527
chanoclavine production, by immobilized *Claviceps* cells 240
cheese production *see* milk products
cheese whey 149
chemical reaction energetics 67
– transition state 66
chemiosmotic theory 63
chemostat 156
– dilution rate 156
chemostat pulse method 146
chemotaxonomical markers 404ff
– of euactinomycetic genera 435
chemotaxonomy *see* taxonomy
chemotrophy 50
Chlamydomucor 522
chloro-aromatics *see* aromatics
4-chloro-2-nitrophenol degradation 229, 235
chlorophyll 59
chloroplast 578
Choanephora 522
citrate metabolism, in clostridia 306
citric acid production, by *Aspergillus niger* 526f
– conditions 527
cladistic classification 372
Claviceps purpurea 525ff
– alkaloid formation 240
– genetic recombination 532
– immobilization effect 240
– life cycle 526
– morphology 234
– strain improvement 532
clostridia 285ff
– acetate production 309
– acetyl-CoA 299
– alcohol fermentation 300
– amino acid fermentation 304f
– – products 304f
– – reaction mechanisms 304f
– bacteriophages 566
– citrate metabolization 303
– complex medium 295
– degradation of, aromatics 300
– – cellulose 297
– – flavonoids 303
– – nitrogenous compounds 304f
– – organic acids 300
– – pectin 297
– – purines 306f
– – pyrimidines 308
– – starch 297
– – xylan 297
– enrichment culture 296
– extracellular enzymes 310
– G+C content 286ff
– glucose fermentation 299
– growth characteristics 293ff
– growth temperature 287ff
– handling 295
– homoacetogenic metabolism 300
– industrial application 308ff

– isolation 295
– media, complex - 295
– – group selective - 293f, 296
– – inoculation 296
– – redox potential 293f, 296
– – supplements 294
– – synthetic 295
– metabolism 297ff
– – reduction reactions 311f
– molecular genetics 293
– nitrogen fixation 308
– nutritional requirements 294ff
– oxygen sensitivity 293
– pathogenicity 286ff
– pH requirements 294
– physiology 297ff
– product stereospecificity 311
– product tolerance 309
– product yield 309
– products 308ff
– properties 286ff
– protein production 310
– proteolysis 287ff
– saccharolysis 287ff
– species diversity 286
– sporulation 292f, 297
– starch-degrading enzymes 310
– storage 295f
– substrates 287ff
– – carbon dioxide 308
– – carbon monoxide 308
– – gaseous substrates 308
– – hydrogen 308
– – metabolization of 297ff
– succinate metabolism 303f
– temperature requirements 294
– transport systems 298
Clostridium 285ff, 368
Clostridium aceticum 300
Clostridium acetobutylicum 286
– acetone-butanol fermentation 309
– butyrate toxicity 181
– growth medium 295
– molecular genetics 293
Clostridium bifermentans, growth medium 295
Clostridium botulinum 286f
– toxin 311
– – medical application 311
Clostridium butyricum, growth kinetics 133
Clostridium formicoaceticum 292f
– spore formation 292f
Clostridium histolyticum 293
Clostridium perfringens 293
– molecular genetics 293
Clostridium sphenoides 300f
Clostridium tetani 293
Clostridium thermocellum 297, 310
– cellulase 28
– cellulosome 297
CO_2 fixation *see* carbon dioxide fixation
Cochliobolus lunatus 525, 529
cocoa fermentation 375
coenzymes 70ff, 259
– distribution of 71
– F_{420} 259
– F_{430} 259
– function of 71
– methanofuran 259
– of methanogenic bacteria 72, 259
Coleus blumei, callus culture 582
– rosmarinic acid production 590ff
– – sucrose effect 594
– suspension culture 582
colony-forming unit (CFU) 120
Comamonas acidovorans, immobilized cell 129
Condromyces crocatus, fruiting body 115
conidia, of *Aspergillus* 116
– of streptomycetes 437
conjugation, in *Lactococcus* 361
continuous culture 155
– model simulation 157
copper, effect on plant cell culture 594
corn steep liquor (CSL) 149
Corticoviridae 550
corynebacteria 433, 453ff
– amino acid production 456ff
– amino acid biosynthesis 454ff
– bacteriophages 568
– biotechnological application 434, 456ff
– chemotaxonomic marker 435
– culture 453
– ecology 453
– genetic engineering 453
– genetics 453
– G+C content 435
– isolation 453
– morphology 435
– nucleotide production 458
– primary metabolism 454ff
– product alteration 458
– secretory pathways 454
– strain improvement 457f
– substrates 454
– taxonomy 434, 453
Corynebacterium glutamicum 206, 434
– aromatic amino acid production 457ff
– – strain improvement 457f
– glutamic acid production 456
– – biotin effect 456
– lactose metabolization 195
– lysine export 209
cosmid 562
Coulter counter 121
counting chamber 119

Crabtree effect, in yeast 133f, 154, 486
crenarchaeota 252
Cryptococcus elinovii, immobilization effect 242f
– phenol degradation 241f
– phenol resistance 243
culture media *see* media
Curvularia lunata see Cochliobolus lunatus
Custers effect, in yeast 486
cyclodextrin 381
cyclodextrin glucanotransferase (CGT) 381
Cystoviridae 550
cytochrome bc_1 complex, in pseudomonads 417
cytoplasm 15
cytoplasmic membrane *see* membrane
cytoskeleton 30, 38
cytosol 14

D

dairy industry *see* milk products
Daucus carota, suspension culture characteristics 586
DDM degradation 229
DDT degradation 229, 235
deacetyllanatoside *C* 597
defective mutants, glycolysis-defective yeasts 480
dehalogenation, by clostridia 313
– – tetrachloromethane degradation 313
denitrification, degradation of aromatics 419
– enzyme properties 417ff
– in pseudomonads 417, 419ff
dental caries 344
desmosome 40
D,L-diaminopimelate biosynthetic pathway 208
dictyosome 40
diffusion 72
– facilitated - 77
– over membranes 73
Digitalis, suspension culture 584
Digitalis lanata, cells in alginate bead 599
– β-methyldigitoxin hydroxylation 596f
digitoxin 597
digoxin 597
dimethylsulfide (DMS) oxidation 274
2,4-dinitrophenol (DNP), effect on microbial metabolism 172
– membrane proton transport 181
diol fermentation, in clostridia 301
dipicolinic acid (DPA), in endospores 388
DNA binding protein 106
– activator 106
– effector 106
– in yeast 492f
– repressor 106
DNA homology, in bacilli 372
DNA repetitive sequence, Ty (transposon yeast) 475
– in yeasts 475
DNA replication 95ff
DNA uptake competence, of bacilli 395
DNA-rRNA hybridization 403
– $T_{m(e)}$ dendrogram 403
doubling time, of cells 118, 139, 585
drugs, plant-derived 605

E

Eadie-Hofstee kinetics 68
electrochemical potential 55, 62
– standard electrode potential 55
electron microscopy 9ff
– electron spectroscopic imaging (ESI) 13
– element analysis 13
– sample visualization 12f
electron transport, in pseudomonads 422
– photosynthetic 60
– respiratory 58f
electron transport phosphorylation 57ff
– chemo-lithotrophic 59
elicitation, of plant secondary compound synthesis 593f
– signal transduction 593
elymoclavine production, by immobilized *Claviceps* cells 240
Embden-Meyerhof-Parnas (EMP) pathway 82
endergonic process 53
endoplasmatic reticulum (ER) 39
endospore(s) 368, 388ff *see also* sporulation
– activation 394
– coat protein 391
– DNA conformation 388
– formation, in clostridia 292
– germinants 394
– germination 394f
– – genes 394
– – mutants 394
– internal structure 389
– of bacilli 388ff
– resistance properties 388ff
– small acid-soluble spore protein (SASP) 388f
– spore cortex 391
– sporulation 390ff
– structure 388ff
– thermoresistance 388
– UV resistance 388
– water content 388
endothermic reaction 52
energy balance, in microorganisms 142
energy charge 55
energy coupling 62ff
– chemical - 62
– electrochemical - 62
– – in halobacteria 64
– proton gradient 64

– uncouplers 64
energy flow 51
energy metabolism 50ff
– energy yielding reactions 260
– of lactic acid bacteria 350ff
– of methanogens 260
– of microorganisms 170f
energy status *see* cell energy status
energy surplus 174
enolase gene (*ENO*) 482
Enterococcus 346
– habitats 346
– isolation media 349
– pathogenicity 346
Enterococcus faecalis 171, 346
– pyruvate fermentation 182
Enterococcus faecium 346
enthalpy 52
Entner-Doudoroff (KDPG) pathway 82f
Entomophtera 522
entrapment *see* cell immobilization
entropy 52
enzyme(s) 66, 165f, 492 *see also* extracellular enzymes, proteins
– active site 67
– classification 66
– control coefficient 165f
– – negative 166
– glycolytic 480
– – *GCR1* (glycolysis regulation) gene 492
– – gene promoters 492
– – gene transcription 492
– – regulation 486
– housekeeping 105
– induced fit 67
– mitochondrial 485
– of *Saccharomyces cerevisiae* 480
– oxygen detoxifying 352
– starch-degrading 310
enzyme activity, in immobilized microorganisms 236
– regulation 70, 101
– – allosteric control 101
– – binding site 101
– – covalent modification 101
– – effector 101
– – feedback inhibition 101
enzyme catalysis 65ff
– energetics 66
enzyme-catalyzed reaction 67
– kinetics 67
– ping-pong mechanism 69
– sequential mechanism 69
enzyme cofactor, of methanogens 259 *see also* coenzymes
enzyme construction 216
enzyme inhibition, competitive 70
– non-competitive 70
enzyme production *see also* protein production, extracellular enzymes
– *Bacillus* protease 380
– by filamentous fungi 522
– by clostridia 310
– by immobilized microorganisms 239
– by plant cell cultures 589
– by pseudomonads 421
– by streptomycetes 451
– by white-rot fungi 534f
– kinetics 239
– methane monooxygenase 280
– thermostable amylolytic - 28
enzyme synthesis regulation 105 *see also* translation regulation, gene expression
– – glycolytic enzymes 480ff
– – two-component system 106
epidermis, of plants 579
ER *see* endoplasmatic reticulum
Eremothecium ashbyii 525
α-ergocryptine 531
ergometrine 531
– production, by immobilized *Claviceps* cells 240
ergot alkaloids 530f
– biological effect 530
– biosynthesis 531
ergotamine 531
Erwinia, vitamin precursor production 211
Escherichia coli 168
– bacteriophages 545ff, 562ff
– basal media supplements 143
– ethanol production 205
– F pilus 18
– growth kinetics 133
– lactose operon transfer 195f
– membrane lipid-protein ratio 243
– nutrient limitation effect 173
– phage f1 564
– phage fd 564
– phage lambda 562f
– phage M13 564
– phage MS2 565
– phage Mu 562f
– phage P1 563
– phage Qβ 565
– phage SP 565
– phage T4 563
– phage-infected cell 547
– potassium efflux systems 177
– potassium uptake systems 176
– 1,3-propanediol production 206
– specific growth rate 128
ethanol, effect, on membranes 136
– – on microbial growth 136
– estimation methods 126
– fermentation in clostridia 300

– from pentoses 204
– from xylose 197
ethanol production 204f
– artificial operon construction 205
– genetic improvement 204f
– in immobilized *Saccharomyces cerevisiae* 236
– spoilage by lactobacilli 334
4-ethylbenzoate degradation 202
ethylene, effect on plant cell cultures 587
euactinomycetes 435
eukaryotic cell 30ff
– structural organization 30ff
eukaryotic protein production, by *Saccharomyces cerevisiae* 498
Eumycota 517ff
euryarchaeota 252
evolution, of bacilli 372
evolution tree, *Clostridium* branch 328f
exergonic process 53
exon 97
exopolysaccharides *see also* polysaccharides
– in pseudomonads 413
– in bacteria 16
exothermic reaction 52
expression vector, in yeast 499 *see also* vectors
– – integrative vectors 499
extracellular enzymes *see also* enzymes
– bacterial 27f
– of bacilli 384
– of clostridia 310
– of streptomycetes 451f

F

fatty acids 87f
– biosynthesis 88
– – in oleaginous molds 523f
– degradation of 87
– in chemotaxonomy 405, 407
flavine adenine dinucleotide (FAD) 71
fed-batch culture 154
– model simulation 155
– penicillin production 156
fermentation 56, 86, 381
– end products 86
– heat evolution 142
– processes in clostridia 297ff
filamentous fungi 515ff *see also* fungi
– characteristics 516
– classification 516
– cultivation 520ff
– culture collections 520
– dikaryotes 519
– distribution 518
– fatty acid synthesis 523f
– genetic diversity 518
– genetic engineering 536
– haplo-dikaryotes 518
– immobilization 522
– isolation 520f
– lactic acid production 523
– life cycle 518ff
– lipid production 523f
– metabolite production 522
– morphology 517
– oleaginous 523
– procurement 520f
– production of, fumaric acid 523
– – lactic acid 523
– – lipids 523f
– – metabolites 522f
– propagation 517
– solid-state fermentation 521
– spore 517
– submerged culture 521
– submerged pellet 116
– surface culture 521
– taxa 517
– white-rot fungi 534f
fimbriae 17, 32
– structure 18
flavin coenzyme 71
flavonoid degradation, in clostridia 303
flow cytometer 121
fluid mosaic model 23
fluorescence microscopy 8
flux control analysis, metabolic flux 165
food fermentation, by filamentous fungi 522
food production, with lactic acid bacteria 327, 362f
formaldehyde dehydrogenase 275
formaldehyde fixation 275
– dissimilatory RuMP cycle 275
– linear oxidation pathway 275
formate oxidation 275
Frateuria, phylogenetic status 404
fructose-1,6-bisphosphatase, carbon catabolite inactivation 489
fumaric acid production, by filamentous fungi 523
fungal growth *see* mycelial growth
fungal protein production, by *Saccharomyces cerevisiae* 498
fungi 515ff *see also* filamentous fungi, yeasts
– biocontrol of plant pathogenic - 423
– edible - 534
– parasexual cycle 519
– selectable marker 536
futile cycle 180ff
– proton leakage 181

G

Gaeumannomyces graminis, biocontrol 423
galactose metabolization, in lactic acid bacteria 354
– in yeasts 493f
galactose-inducible gene 493
– regulation 493
α-galactosidase, from pseudomonads 417, 421
gap junction 40
G + C content, of euactinomycetic genera 435
– of clostridia 286ff
gene expression 97ff, 496ff *see also* heterologous gene expression, transcription regulation, translation regulation
– bacteriophage genes 554f
– in bacteria 97ff
– in eukaryotes 107f
– in *Lactococcus* 361f
– in yeast 491ff
– post-transcriptional gene regulation 494
– regulation by DNA binding proteins 492
– regulatory sites 104
gene(s), amylase - 200f
– antibiotic - 439ff
– for glycolytic enzymes 480ff
– in bacilli, sporulation - 390ff
– heterologous expression *see* heterologous gene expression
– – in yeasts 491f
– probing 411f
– *RAS* - 487f
– transfer 195f, 545
– – horizontal 559f
genetic engineering *see also* strain improvement
– galactose operon transfer 195f
– of bacilli 394ff
– of filamentous fungi 536
– of lactic acid bacteria 360ff
– of plant cells 601f
– of yeasts 476f
genetic recombination, in fungal parasexual cycle 519
ginsenosides, production, by cell cultures 589
glucoamylase, gene (*GAM1*), *Schwanniomyces occidentalis* 201
– in clostridia 310
glucokinase, gene (*GLK*) 481
gluconate production 178
gluconeogenesis 85, 484
glucose, degradation of, heterofermentative 170f
– – homofermentative 170f
– energy yield 141
– fermentation 182
– – in clostridia 299f
– – nutrient limitation effect 173f
– – pathways 170ff
– – pH dependence 182
– – product quantity 182
– uptake, in yeast 487
glucose dehydrogenase 173ff, 178f, 182
glucose-6-phosphate dehydrogenase, gene (*ZWF*) 480
α-glucosidase, in clostridia 310
– production 239
β-glucosidase, production 239
glutamate dehydrogenase 93
glutamate synthase 93
glutamic acid production 206
– in *Corynebacterium glutamicum* 456
glutamine synthetase, in streptomycetes 443
– operon 387
– regulation 387
glyceraldehyde-3-phosphate dehydrogenase, gene (*TDH*) 481
glycerol metabolization, in clostridia 298, 301
glycocalyx 38
glycogen polymerization 95
glycogen synthesis 174, 485
glycolysis 102, 479ff
– allosteric control 102
– enzyme coding genes 480
glycolytic enzymes *see* enzymes
glyoxisome 33
– enzyme set 33
glyoxylate cycle 85, 484
– genetics in *Saccharomyces cerevisiae* 484
glyphosate degradation, by pseudomonads 423
GOGAT *see* glutamate synthase
Golgi apparatus 39
Gram-negative bacteria 19ff
– capsule 17
– envelope 19
– facultative methylotrophs 270
– methanogens 257
– molecular envelope composition 19
– nitrilotriacetic acid metabolization 417f
– obligate methylotrophs 269
– outer membrane 19
– periplasmic space 19
– phage receptors 552f
Gram-positive bacteria 20
– capsule 17
– cell wall assembly 21
– envelope 20
– methanogens 256
– methylotrophs 270
– peptidoglycan layer 20
– phage receptors 552f
group transfer potential 54
growth, of microorganisms 111ff *see also* microbial growth, mycelial growth
– doubling time 118, 139, 585
– of plant cell cultures 585f
growth energetics, of methylotrophs 278

growth media *see* media, microbial growth media
growth rate, medium osmolality effect 130
– specific - 118, 585
– temperature dependence 128

H
habitats
– of bacilli 374
– of lactic acid bacteria 328ff, 341, 344
– of methanogens 255f
– of methylotrophs 268
– of pseudomonads 411
– of streptomycetes 436
– of yeasts 471f
– soil 244
halobacteria, cell wall 22
halogenated hydrocarbons, degradation, by pseudomonads 415
– detoxification 313f
Hansenula 269
Hansenula polymorpha 272, 280, 470ff
– life cycle 474
– peroxisomes 34
heat evolution, microbial 141
herbicides, degradation 201
– detoxification, by pseudomonads 423
Herpetosiphon auranticus 114
heterologous gene expression *see also* enzyme production, gene expression, protein production
– in bacilli 396
– in streptomycetes 450f
– in yeast 496ff
– protein inclusion bodies 26f
heterotrophy 50
hexokinase, activity 236
– gene (*HXK1*) 480
β-hexosaminidase, production 239
hexose fermentation, in bacilli 385
– in clostridia 300
– in lactic acid bacteria 353
hexose phosphate metabolization, in clostridia 298
Homohiochii medium 348
HP medium 348
HPLC chromatography, of polyamines 409
human protein production, by *Saccharomyces cerevisiae* 496
hydrocarbon, degradation, by pseudomonads 413
– estimation methods 126
hydrocarbon metabolism 89
– aliphatic 89
– aromatic 89
hydrogen production, by *Anabaena azollae* 237
hydrolase 66
Hyoscyamus albus, root culture 601
hypha 116, 517
– branching 116
– chain elongation 116
Hyphomicrobium 267, 270

I
immobilization *see* cell immobilization
immobilized microorganisms 223ff *see also* cell immobilization
– alterations of, macromolecular composition 236
– – morphology 233
– – physiology 234ff
– – product formation 239ff
– colonies, formation 230
– – growth 230
– – morphology 231ff
– *Cryptococcus elinovii* 242f
– development 226
– enzyme activity 236
– for process improvement 226
– fungi 522
– growth 235
– in natural habitats 244
– interface effect 235
– membrane composition 243
– microfilms 231
– oxygen availability 228
– oxygen limitation effect 234
– oxygen supply 226ff
– phenol effect 242
– poison resistance 242
– – protection mechanism 243
– productivity prolongation 242
– *Pseudomonas putida* 242f
– viability 239ff
– water activity effect 230
immunofluorescence microscopy 7
inclusion bodies 26
– heterologous protein 27
– of *Bacillus thuringiensis* 26
– protein inclusion 26
indole alkaloid, production by *Catharanthus roseus* 596
Inoviridae 550
insect toxin, from *Bacillus thuringiensis* 377
integron, in pseudomonads 410
interleucin production 396
intestinal flora 330f
– *Enterococcus* 346
intron 97
iron limitation, metabolic effect 179
irreversible thermodynamics 52ff
IS element, in *Lactococcus* 361
isolation media, for lactic acid bacteria 347ff
isomerase 66

J

Jacob-Monod kinetics *see* Monod kinetics

K

KDPG pathway *see* Entner-Doudoroff pathway
kefir 332
– microflora 332
2-ketogluconate production 178
2-keto-L-gulonic acid biosynthesis 211
Klebsiella oxytoca, ethanol production 206
Klebsiella planticola, ethanol production 204
- strain improvement 204
Klebsiella pneumoniae 168ff
– *dha* regulon 206
– glucose conversion 172
– glucose dehydrogenase 178
– nutrient limitation effect 172
Kluyver effect, in yeast 486
Kluyveromyces lactis 470ff
– galactose metabolism, regulation 494
– life cycle 473
Krebs cycle *see* tricarboxylic acid (TCA) cycle
Kuehneromyces mutabilis 534f

L

lactate, metabolization, in *Lactobacillus* 352
– export, in *Lactococcus* 350
lactic acid bacteria 325ff
– arginine dihydrolase pathway 358
– bacteriocins 358ff
– bacteriophages 566
– bioenergetics 350f
– biotopes 328ff
– cultivation 347ff
– evolution 328
– for vitamin determination 364
– functional classification 328
– galactose metabolization 353f
– genetic engineering 360ff
– genetics 360ff
– growth requirements 330
– hexose fermentation 353
– in animals 330
– in biological research 328
– in historical food preparations 327f
– in man 326
– in modern food 327, 362f
– in plants 330
– isolation 347ff
– lactose metabolization 354
– metabolism 350ff
– – amino acid - 354
– – carbohydrate - 350ff
– – oxygen - 352
– occurrence 328ff
– phage resistance 566
– specific isolation media 347ff
– taxonomy 328ff
lactic acid production, by filamentous fungi 523
lactic fermentation 327ff
– history of exploration 328
– starter culture 328
Lactobacillus 329ff
– acid tolerant - 335
– facultative heterofermentative - 335ff
– food preserve spoilage 335
– growth conditions 336ff
– habitats 330ff
– in beer 334
– in diet 331
– in meat 331
– in milk products 332
– in the human vagina 331
– in the intestinal tract 330
– in the oral cavity 330
– in the stomach 331
– isolation media 347ff
– obligately heterofermentative - 335ff
– obligately homofermentative - 335f
– pathogenicity 331
– species characteristics 336ff
– substrates 336ff
– wine production 334
Lactobacillus brevis 333, 335
Lactobacillus delbrueckii ssp. *bulgaricus* 332, 335
Lactobacillus fermentum 333
Lactobacillus helveticus 333, 335
Lactobacillus plantarum 333
lactococcin A, nucleotide sequence 359f
Lactococcus 329, 343
– DNA transfer 361
– energy consumption 351
– energy generation 351
– gene expression 361f
– habitats 343
– isolation media 348
– lactate gradient 351
– membrane potential 351
– peptidase 355f
– pH gradient 351
– plasmid coded functions 361
– plasmid profiles 361
– proton motive force 351
– species characteristics 343
– substrates 343
Lactococcus lactis 328, 339, 343
– cell wall proteinase, genes 355
– – properties 355
– – specificity 354f
– molecular genetics 360
lactose, metabolization, by improved cultures 195

– – in lactic acid bacteria 354
lactose operon, transfer from *Escherichia coli* 195f
lantibiotic 359
laser scanning microscopy 8
lectin 35
Lentinus edodes 534f
Leuconostoc 329, 340ff
– genus differentiation 342
– growth conditions 342
– habitats 341
– isolation media 348
– pathogenicity 341
– psychrophilic 341
– species characteristics 342
– substrates 342
Leuconostoc mesenteroides 339, 341
Leuconostoc oenos 341
Leviviridae 550
life cycle, of bacteriophages 558
– of *Claviceps purpurea* 526
– of filamentous fungi 518f, 523, 526, 533
– of streptomycetes 436
– of yeasts 472ff
ligase 66
light microscopy 7ff
– confocal *see* laser scanning microscopy
– conventional 7
– Nomarski interference contrast 7
– phase contrast 7
– sample preparation 7
– sample visualization 7ff
lignin degradation, by filamentous fungi 534f
lignin peroxidase, gene regulation 534f
lindane dechlorination 313
Lineweaver-Burk kinetics 69
lipase, from pseudomonads 421
– in detergent formulations 421
– stereoselectivity 421
lipid synthesis, by filamentous fungi 523f
– in yeast 490f
Lipothrixviridae 550
lithotrophy 50
Lithospermum erythrorhizon, commercial shikonin production 589, 592f, 604
lyase 66
lysergic acid, biosynthesis 530
lysine, biosynthetic pathway 207
– secretion 209
L-lysine production, pathway control 206
lysogen *see* prophage

M
magnesium limitation, effect of calcium 179
– metabolic effect 179
magnesium transport system 180
maintenance energy 141
maintenance energy requirement 168
mammalian protein production, by *Saccharomyces cerevisiae* 497
manganese peroxidase, gene regulation 535
manioc *see* cassava (*Manihot esculenta*)
meat, microflora 331
– spoilage 341
medermycin 215
mederrhodin 215
media 143ff, 170ff *see also* microbial growth media
– for callus cultures 582
– for clostridia 293ff
– for *Escherichia coli* 143, 173
– for lactic acid bacteria isolation 347ff
– for plant cell cultures 590, 594
– M17 348
– optimization 145, 590
– pH value 180
membrane 23
– bacterial plasmalemma, fluidity 23
– – membrane protein 23
– – molecular composition 23
– composition, in immobilized *Escherichia coli* 243
– – lipid-protein ratio 243
– – of methanogens 259
– – phenol effect 243
– ethanol effect 136
– intracellular 24
– – dinitrogen fixer 25
– – dynamic changes 24
– – non-phototrophic bacteria 25
– – phototrophic bacteria 24
membrane channel 73
membrane flow 31
membrane gradient, in *Lactococcus* 351
membrane permeabilization, in plant cells 598
– physical methods 600
membrane skeleton 38
membrane transport 72ff
– energetics 73
– group translocation 75
– in lactic acid bacteria 350
– kinetics 76
– macromolecules 79
– mechanism 73
– models 73ff
– primary - 73f
– proton leakage 181
– rate 78
– secondary - 73ff
– studies by NMR 193
– substrate-carrier interaction 73
membrane transport system 74ff
– binding protein dependent 75

- classification 74
- PEP-dependent sugar transport 75
menaquinone 404
meristeme 34, 579
mesophilic microorganisms, specific growth rate 128
mesophyll cell 580
mesosome 25
messenger RNA (mRNA) 97
- half-life 97
- splicing 97
metabolic branchpoint 167
metabolic design 189ff
metabolic energy turnover, in *Lactococcus* 350
metabolic engineering 164, 190ff
metabolic flux control 104
- modification 216
metabolic flux control analysis 165f
- summation theorem 165f
metabolic flux manipulation 183
metabolic pathway 165
- rate-limiting step 166
metabolism 15, 47ff, 442ff *see also* carbohydrate metabolism
- central 81
- compartmentation 478
- concept 50
- coordination 100
- flux control 104
- of lactic acid bacteria 350ff
- of methanogens 255ff
- of methylotrophs 266ff
- of plant cells 580
- of pseudomonads 414f
- of *Saccharomyces cerevisiae* 478
- of streptomycetes 442ff
- pathway redesign 214
- regulation 100
metabolite, overproduction 163ff, 450
- production improvement 204
- quantification 190ff
- screening 447f
- - target 449
methane formation reactions 260f
- autotrophic CO_2 fixation 261
- cofactors 260f
methane monooxygenase (MMO) 271, 280
methane oxidation, in methanotrophic organisms 271
Methanobacteriales 253
Methanobacterium thermoautotrophicum 261
- genome 262
- methane formation 261
Methanobacterium uliginosum, electron micrograph 256
Methanobrevibacter smithii, electron micrograph 256
methanochondroitin 258
Methanococcales 253
Methanococcus janaschii, electron micrograph 257
Methanocorpusculaceae 255
methanogenic bacteria 251ff
- bacteriophages 565
- carbon metabolism 261
- cell envelope 21, 257
- coenzymes 72
- culture 259ff
- - handling 260
- - maintenance techniques 260
- DNA-associated proteins 262
- ecology 255
- energy metabolism 260
- environmental distribution 255
- enzyme cofactors 259
- extreme habitats 256
- Gram-negative 257
- Gram-positive 256
- in animals 255
- medium optimization 260
- membrane 259
- metabolic pathways 255
- molecular biology 262
- morphology 256
- nutrient requirements 260
- restriction enzymes 263
- ribosome structure 262
- substrates 255
- taxonomy 252
- temperature requirements 260
methanol dehydrogenase (MDH), in methylotrophic bacteria 273
methanol fermentation, in clostridia 300
methanol oxidation, in microorganisms 271ff
- in yeasts 271, 471
Methanolobus vulcani, electron micrograph 257
Methanomicrobiaceae 254
Methanomicrobium mobile, electron micrograph 257
Methanoplanaceae 254
Methanoplanus limicola, electron micrograph 257
Methanopyrus 255
Methanosarcinaceae 254
Methanosphaera stadtmanae, electron micrograph 256
Methanospirillum hungatei, electron micrograph 257
Methanothermus fervidus, electron micrograph 256
methanotrophic microorganisms 268f *see also* methylotrophic microorganisms
- characteristics 269
- methane oxidation 271

methyl aromatics, degradation of 202f
methylamine dehydrogenase 274
methylamine oxidation 274
β-methyldigitoxin (Mdt) 597
– biotransformation 596f, 599
β-methyldigoxin (Mdg) 597
Methylobacillus 267, 269
Methylobacter 267, 269
Methylobacterium 267, 270
Methylobacterium extorquens, methylamine dehydrogenase 274
Methylococcus 267, 269
Methylocystis 267, 269
Methylomonas 267, 269
Methylophaga 267, 269
Methylophilus 267, 269
Methylosinus 267, 269
methylotrophic bacteria 269
– bacteriophages 566
methylotrophic microorganisms 265ff
– amino acid synthesis 279
– definitions 268
– ecology 268
– growth energetics 278
– industrial applications 278
– isolation 268
– metabolism 266, 271
– methylamine oxidation 274
– non-methane-utilizing 269
– physiology 271
– products 278
– single cell protein (SCP) production 278
– species characteristics 266f
– substrate pathways 266
– substrates 266f
– symbioses 268
– taxonomy 268ff
– rRNA sequence data 271
methylotrophic yeasts 269
Methylovorus 267, 269
Michaelis constant 69
Michaelis-Menten equation 68
Michaelis-Menten kinetics 132
– plot 68
microbial extracellular polysaccharide (EPS) *see* exopolysaccharides
microbial growth 111ff *see also* mycelial growth, bacteria
– acceleration phase 152
– bacterial - 113
– balanced - 139
– characterization 118ff
– clostridia 293ff
– conditions, for lactic acid bacteria 336ff
– – for streptomycetes 438
– continuous culture, steady state level 156f
– death phase 153
– diauxic growth 154
– doubling time 118, 139, 585
– effects of, alkanol 135
– – ammonia 140
– – ammonium 177
– – carbon dioxide 135
– – carbon source 132
– – energy supply 140
– – environment 127ff
– – mechanical stress 129
– – medium pH 131
– – nitrogen source 133
– – nutrients 137, 146
– – organic acids 136
– – oxygen 134
– – sodium ion concentration 131
– – temperature 127f
– – water activity 130
– effects on macromolecular composition 138
– energetics 168
– – cell integrity maintenance 169
– – cell material synthesis 169
– – energy drain manipulation 169
– – energy-spilling reaction 169
– energy balance 142
– energy demand 141
– energy requirements 141
– exponential growth 152
– immobilized microorganisms 235
– kinetics 152f
– lag phase 152
– limiting steps 139
– measurement 118ff *see also* cell number estimation, cell mass estimation
– modes 113ff
– physical constraints 138
– rates in batch culture 153
– relaxation time 138
– retardation phase 152
– stationary phase 152
– substrate uptake 139
– variability 138
– yeasts 115f
microbial growth media 143ff
– bagasse 150
– carbon compounds 143
– carbon limitation 170f
– cassava 149
– cheese whey 149
– chemostat pulse method 146
– complex natural media 148
– computer-assisted optimization 147f
– corn steep liquor 149
– energy limitation 170f
– factorial design 146
– for clostridia 294f
– for methanogens 260

– for streptomycetes 436f
– growth factors 145
– macro-elements 145
– material balances 145
– medium optimization 145
– molasses 148
– nitrogen compounds 144
– nutrient limitation 172ff
– optimization, by neural network 147
– Placket-Burman two-factorial design 147
– trace elements 145
microbial physiology 164
microbial product, modification 204ff
– – spectrum 204ff
– – yield 204ff
microbial product formation 170ff
– effects of, carbon limitation 170f
– – carbon source 133
– – medium composition 170ff
– – microbial growth 169
– – nutrient limitation 170ff
– – potassium limitation 178
– energetics 167f, 169
– energy limitation 170f
– primary metabolites 170
microbodies 33, 36
microfilaments 30
microfilm, of microorganisms 231 *see also* biofilm
Micromonospora, chemotaxonomic marker 435
– G+C content 435
– morphology 435
– production of antibiotics 445
microorganisms, batch culture 150
– cell mass estimation 121ff
– cell number estimation 119ff
– continuous culture 155
– culture techniques 150ff
– effects of, ethanol 136
– – mechanical stress 129
– – medium pH 131
– – temperature 128
– – water activity 130
– energy demand 141
– fed-batch culture 154
– growth kinetics 133
– growth *see* microbial growth
– heat evolution 141
– nutrient requirements 133
– phage host groups 549
– specific growth rate 118
– specific substrate uptake rate 132
– substrate uptake 139
microscopy 7ff
microtubules 30
Microviridae 550
milk fermentation, history 327f
milk products, cheese production 332
– *Enterococcus* 346
– *Lactobacillus* 332
– lactococcal starter culture 349
– *Lactococcus* 343
– *Leuconostoc* 341
– yoghurt 332
minerals, estimation methods 126
mitochondrial enzymes, genetics 485
mitosis spindle 30
molasses 148
– constituents 148
molds *see* fungi
molecular genetics, of clostridia 293
Monod kinetics 132
monoterpene metabolism, of pseudomonads 415
Morchella 525
morel 532
Morinda citrifolia, shear stress effect 587
morphology, of bacilli 368
– of bacteria 13ff, 114
– of euactinomycetic genera 435
– of eukaryotic cells 36f, 578
– of filamentous fungi 517f
– of immobilized microorganisms 231ff
– of methanogens 256f
– of phages 547
– of plant cells 36f, 578
– of streptomycetes 435, 437
mouth flora 330
mRNA *see* messenger RNA
MRS agar 347
Mucor 522
murein 19f
mushrooms, edible 534
mycelial growth 116 *see also* microbial growth
– model based simulation 117
mycelium 116, 517
mycoplasm, cell wall 20
Myoviridae 550
myxobacterium, sporangiophore 115

N

natural gene transfer, markers 414
neural network, use in media design 147
nicotinamide adenine dinucleotide (NAD) 71
nicotinamide adenine dinucleotide phosphate (NADP) 71
nisin 358
– gene 358ff
– – nucleotide sequence 360
nitrate, effect on microbial growth 134
– estimation methods 126
nitrate reductase, in pseudomonads 419
nitrile hydratase 422
nitrilotriacetic acid (NTA) degradation, in Gram-negative bacteria 417f

nitrite reductase, in pseudomonads 419
nitrogen, effect on microbial growth 133
nitrogen cycle 90
– in streptomycetes 443
nitrogen limitation, metabolic effect 172ff
nitrogen metabolism 90f
– of bacilli 387f
nitrogen supply, in plant cell culture 581
N_2O reductase, in pseudomonads 419
Nocardia, chemotaxonomic marker 435
– G+C content 435
– morphology 435
– production of antibiotics 445
nuclear magnetic resonance (NMR) spectroscopy 192ff
– bioreactor 193
– cell spectrum 192f
– *in vivo* application 192
– membrane transport studies 193
– ^{31}P 192
nucleic acid, electron microscopy 11
nucleotide production, by corynebacteria 458
nutrient availability, effect on microbial growth 131ff *see also* media
nutrient limitation, metabolic effect 170ff
– pH effect 176
– polysaccharide production 172
nutrient requirements, of microorganisms 143
nutrient supply, plant cell culture 591

O
oleaginous microorganisms, filamentous fungi 523
organic acids, conversion, by *Pseudomonas putida* 422
– effect on microbial growth 136
– metabolization, in clostridia 300
– production, by filamentous fungi 523
organotrophy 50
osmolality, effect on microbial growth 130
oxidation, thermodynamics 55
β-oxidation 87
oxidoreductase 66
oxygen, effect on microbial growth 134
– estimation methods 126
– exclusion techniques 259, 296
– profile, in alginate beads 229
– requirements, of bacilli 381
– supply, of plant cell cultures 587f
– toxicity 135, 293
oxygen consumption, carbon-limited 168
– energy balance 142
– fungal 228
– microbial 168
– nitrogen-limited 168
– of pseudomonads 420
– phosphate-limited 168
– sulfate-limited 168
oxygen detoxification 293
– enzymes 352
– in clostridia 293
– in lactic acid bacteria 352
oxygen metabolism, energetics 352
oxygen transfer, in bioreactors, effectiveness factor 228f
– – in alginate 228
– – in gels 227
– – influence of microbial growth 227f
– – into microfilms 227
– – to immobilized cells 226ff

P
Panax ginseng, commercial ginsenoside production 589, 604
Papaver somniferum, alkaloid yield 593
– elicitor effect 593
parasexual cycle, in fungi 519
paspalic acid, biosynthesis 530
Pasteur effect, in yeast 483, 486
pathogenic microorganisms, bacilli 376
– – for insects 376
– – for man and mammals 376
– clostridia 286ff
– lactic acid bacteria 331, 344ff
pectin degradation, in clostridia 297
Pediococci, isolation media 348
Pediococcus 329, 339f
– growth conditions 340
– habitats 339
– in food and feed 339
– species characteristics 340
– substrates 340
Pediococcus damnosus (*cerevisiae*) 339
– beer spoilage 339
Pediococcus dextrinicus 339
penicillin production, amount 156
– by filamentous fungi 156, 527
Penicillium 525ff
– penicillin production 156
Penicillium chrysogenum, growth kinetics 133
– pellet 118
– penicillin production 527
Penicillium frequentans, immobilization effect 231
pentose, fermentation pathway 204
– metabolization, in bacilli 385
– – in clostridia 298
pentose phosphate pathway (PPP) 83, 479ff, 483
– reductive 84
PEP:carbohydrate phosphotransferase, PTS system 385
peptidase, in *Lactococcus* 355f

peptide, metabolism in lactic acid bacteria 354
– transport in *Lactococcus* 358
peptidoglycan 19f
– in *Bacillus* spores 391
peroxisome 33
Petroff-Hausser slide 119
pH tolerance, of pediococci 340
pH value, of media 180f
– optima, for microorganisms 131
PHA *see* polyhydroxyalkanoic acid
Phaenerochaete chrysosporium 535
phage *see* bacteriophage
phage reproduction, physiology 552ff
PHB *see* poly-β-hydroxybutyrate
phenol degradation, by immobilized microorganisms 241f
phloem, cells 580
phosphatase, in yeast 490f
phosphate, estimation methods 126
phosphate bond, energy-rich 54
phosphate limitation, metabolic effect 179
phosphate uptake system 179
phosphofructokinase, activity, in immobilized microorganisms 237f
– gene 481
– regulation 486
phosphoglucose isomerase, genes, (*CDC30*) 481
– – (*PGI1*) 481
phosphoglycerate kinase, gene (*PGK1*) 482
phosphoglycerate mutase, gene (*GPM1*) 482
phosphor limitation, effect on *Escherichia coli* 173
– – metabolism 172ff
photophosphorylation 59f
photosynthesis 59f
– electron transport chain 59
– – cyclic electron flow 60
– – Z-scheme 60
– energy transduction 61ff
– photosystems 59ff
– reaction center 59
– reducing equivalent 61
– reversed electron-flow 62
photosynthetic membrane 23
photosystems, in plants 59ff
phototrophic bacteria, intracytoplasmic membranes 24
– plasmalemma 24
phototrophy 50
Phycomyces blakesleeanus 522
– life cycle 523
Phycomycetes 517, 522ff
phylogenetic system, rRNA superfamily 403
– tree of organisms 252
– tripartite 402
phytanyl diether lipid, of methanogens 259
phytoalexin 593
– induction of synthesis 593
phytochemical production, commercial 604
phytohemagglutinin 35
phytohormones, in plant cell culture 581, 591
Pichia 269
Pichia pastoris 280, 470ff
pickle fermentation 333
pili 17
– structure 18
plant, as *Bacillus* habitat 375
– biocontrol of pathogens 423
– cormus organization 579
– metabolism 580
– propagation 603
– – pathogen-free plants 603
– protoplast 36
plant breeding, by plant tissue culture 603
plant cell 34f, 577ff
– antisense DNA technology 602
– compartments 578
– differentiation 579
– epidermal 579
– genetic engineering 601f
– membrane permeabilization 598
– meristematic 579
– mesophyll 580
– morphology 578
– organization 578ff
– shear sensitivity 586
– surface layer 34
– suspension characteristics 585
– types 579
– vascular 580
– wall 35, 578
plant cell culture 40f, 579ff
– biomass kinetics 586
– biosynthetic pathway elucidation 590
– biotransformation reactions 595f, 599
– cell density 584
– cell differentiation 590
– cell division synchrony 584
– cell immobilization 598
– cell line selection 592
– characteristics 583f
– CO_2 fixation 580
– differentiated - 600f
– doubling time 585
– enzyme production 589
– genetic stability 584
– growth, assessment 588
– – phases 585f
– – rate 592
– in bioreactors 586f
– *in vitro* 579, 581ff
– macronutrients 591
– medium optimization 590
– metabolism 580

– nutrient requirements 580f
– oxygen supply 587f
– – effects on product formation 587
– physical conditions 595
– physiological stability 584
– phytohormones 581, 591
– productivity 592
– requirements 41
– secondary compound production 589ff
– stirrer types 587
– substrate kinetics 586
– suspension - 583ff
– – ^{31}P-NMR spectrum 192f
– two-phase - 598
– types 581
– ventilation 587
– vessel design 595
plant material fermentation 333
– *Lactobacillus mesenteroides* 341
– microbial succession 333
plant-microorganism interaction 422
plant organ culture 579, 605
– metabolism 580
– transformation 600
– types 581
plant secondary metabolites, abiotic induction/stimulation factors 595
– elicitation by stress 593f
– extraction with solid adsorbents 600
– industrial production 589ff
– isolation 598f
– precursor feeding 595
– production induction/stimulation 593f
– secretion 598
– two-phase extraction 600
plant tissue 37
plant tissue culture 579, 582f
– application in agriculture 603
– for plant breeding 603
– metabolism 580
– types 581
plasmalemma of bacteria *see* membrane
Plasmaviridae 550
plasmids, in bacilli 395
– in *Lactococcus* 361
– in pseudomonads 409f
– – host range 410
– in streptomycetes 440
– incompatibility group (incP) 410
– pAMβ1 362
– pWWO 201, 411
– – transposon 411
– – *xyl* gene regulation 411
– segregational instability 119
– TOL 201, 411
– types in yeast 475, 499 *see also* yeast plasmids
plasmodesmata 34
plastics, production by pseudomonads 420
Plectovirus 550
Pleurotus ostreatus 534f
Podoviridae 550
polyhydroxyalkanoic acid (PHA) 26
– production by pseudomonads 420
poly-β-hydroxybutyrate (PHB) 26
– biosynthetic pathway 212
– heterologous gene expression 26
– microbial production 420
polyamines, analysis of 408
– extraction of 408
– HPLC chromatography of 409
– in chemotaxonomy 408
– in pseudomonads 405ff
polychlorinated biphenyl (PCB), degradation of, by pseudomonads 415
polyhydroxyalkanoate production 212
polymer production 174, 212
polysaccharides 413 *see also* exopolysaccharides
– degradation, by bacilli 384
– – by clostridia 297
– production, by *Agrobacterium radiobacter* 173
– – nutrient limitation effect 172
polysome 14
Poriales 534
porin 76
potassium efflux system, *Escherichia coli* 177
potassium limitation, effect of ammonium 177
– effect on product formation 178
– energetic effects 177
– metabolic effect 172, 176
potassium uptake system, *Escherichia coli* 176
potato, micropropagation 604
precursor metabolite 81
primary metabolism 15, 442ff
prokaryotic cell 13ff *see also* microorganisms
– adhesion 18
– macromolecular composition 15
– metabolism 15
– structural organization 13ff
promoter *see* transcription promoter
1,3-propanediol production 206
prophage 558f
– detection 561
– homoimmunity 558
prosthetic group *see* coenzymes
proteases, cell wall proteinase, of *Lactococcus lactis* 354f
– from pseudomonads 421
proteins, electron microscopy 11
– import 79
– metabolism, in lactic acid bacteria 354ff
– overproduction, effect on polysomes 495
– – in *Aspergillus* 536f
– translocation 79

protein production 380ff *see also* enzyme production, heterologous gene expression
– by clostridia 310
– by plant cell cultures 589
– by pseudomonads 421
– heterologous -, by bacilli 396ff
– – human interleucin 396
– – in yeast 496ff
– single cell protein 278
protein secretion 79
– in yeast 500
– leader sequence 500
protein synthesis 99
– elongation 99
– initiation 99
– termination 99
proteinase, of *Lactococcus lactis* 354f
Proteobacteria 402
proton-motive force 64
– in *Lactococcus* 351
Pseudomonadaceae 401ff
– aromatic hydrocarbon hydroxylation 422
– as biocontrol agents 423
– catabolic transposons 411
– characteristics 402
– chemotaxonomical identification 406ff
– degradation of, aromatics 414
– – halogenated hydrocarbons 415
– – hydrocarbons 413
– – polychlorinated biphenyls (PCBS) 415
– – xenobiotics 414
– DNA hybridization 412
– ecology 410ff
– electron transfer pathway 417
– flocculation 413
– gene probing 411f
– genome organization 409f
– – integrons 410
– habitats 411
– herbicide detoxification 423
– identification techniques 411f
– *in situ* detection 412
– interaction with plants 422
– markers for monitoring 414
– metabolic properties 414
– – genetic organization 414
– monoterpene metabolism 415
– natural classification 404
– oxygen limitation response 420
– phylogenetic status 402
– plasmids 409f
– production of, amylase 421
– – α-galactosidases 421
– – lipases 421
– – proteases 421
– – reserve polyester (PHB) 420
– – siderophores 423
– resistance to antibiotics 410
– serological identification 412
– α subclass 407
– β subclass 407
– γ subclass 407
– tartrate metabolization 422
Pseudomonas 401ff
– authentic genus 403f
– bacteriophages 568
– degradation of xenobiotics 201
– diagnostic fatty acid content 405
– genus structure 402
– *ortho*-pathway for 3-chlorobenzoate 203
– – expansions 203
– polyamine content 405
– species 405
– substrate spectrum expansion 201
– TOL-plasmid pWWO 201
– – gene regulation 202
Pseudomonas aeruginosa, arginine deiminase pathway 420
– genome 409
Pseudomonas capacia, growth kinetics 133
Pseudomonas fluorescens ssp. *cellulosa* 405
– cellulolytic system 413
Pseudomonas putida 405
– genome 409
– immobilization effect 242f
– organic acid conversion 422
– phenol degradation 242
Pseudomonas stutzeri, denitrification pathway 419
Pseudomonas syringae, frost damage in plants 424
pseudomurein 258
– of methanogenic bacteria 21
pseudomycelia, of yeasts 115
psychrophilic bacilli *see* bacilli
psychrophilic microorganisms, specific growth rate 128
pullulanase 27
– in clostridia 310
purinolysis 306f
– biotechnological application 306
purpureaglycoside A 597
purpurin production, by cell cultures 589
putrescine, in pseudomonads 405ff
pyoverdine 423
Pyrenomycetidae, biotechnologically important species 529
pyridine nucleotide coenzymes 71
pyridoxalphosphate 72
pyrimidine degradation, in clostridia 308
pyrroloquinoline quinone (PPQ) 71, 173ff, 178f, 274
– metabolic function 173
pyruvate, anaerobic dissimilation, in lactic acid bacteria 355

– fermentation 182
– – pH dependence 182
– – product quantity 182
pyruvate decarboxylase, gene (*PDC1*) 482
pyruvate kinase, gene (*PYK*) 482
– regulation 486
pyruvate metabolism, in clostridia 298

Q
quinone coenzyme 71
quinones, in chemotaxonomy 406
– in pseudomonads 404f
– isolation 406

R
RAS gene, in yeast 487f
RAS/cAMP signaling pathway 487f
reaction equilibrium 53
– constant 53
recombination 559
reduction reaction, thermodynamics 55
relaxation time, of processes in microorganisms 138
respiration 484
– anaerobic 382
– energy transduction 64
– heat evolution 142
respiratory chain 57
– branched, in *Escherichia coli* 59
– electron flow 58
– redox potential 58
restriction enzyme 559
Rhizopus 522
rhizosphere, plant-pseudomonad interaction 422
Rhodopseudomonas palustris 23
rhodoquinone 404
riboflavine 5'-phosphate (FMN) 71
ribosomal RNA (rRNA) 98
– superfamily 403
ribosome 99
– subunits 99
ribulose monophosphate (RuMP) cycle 84, 276
– dissimilatory 275
RNA polymerase 105, 393
– σ-factor 105, 393
– in yeast 492
RNA polymerization 97ff
RNA replicase 554
root culture, hairy root culture 601
– of *Hyoscyamus albus* 601
– secondary metabolite production 601f
– transformation effect 601f
rosmarinic acid, production by *Coleus blumei* cultures 590f
– – sucrose effect 594
rotated drum reactor 587f
Rubia akane, commercial purpurin production 589, 604

S
saccharide metabolization, in clostridia 298
saccharide transport 75, 385f
– in bacilli 385f
saccharide uptake system, in clostridia 298
Saccharomyces cerevisiae 115, 197ff, 470ff, 478ff, 525 *see also* yeast
– alcohol dehydrogenase activity 238
– budding 115f
– carbohydrate metabolism 478ff
– – regulation 486ff
– covalent binding effect 238
– diauxic growth 154
– galactose metabolism, regulation 493
– glycolytic enzymes, genes 480
– glyoxylate cycle, genes 484
– growth kinetics 133
– hexokinase activity 236
– immobilization effect 230ff
– – metabolic changes 236
– life cycle 472
– 2m plasmid 476
– macromolecular composition 235f
– metabolism 478ff
– oxygen consumption 228
– pentose phosphate pathway 483
– phosphofructokinase activity 237f
– production of
– – bacterial proteins 498
– – eukaryotic proteins 498
– – fungal proteins 498
– – human proteins 496f
– – mammalian proteins 497
– – viral proteins 498
– respiration 484
– starch degradation 200
– strain improvement 197ff
– strains for wine/beer production 496
– TCA cycle 484
– xylose metabolization 198f
Saccharopolyspora, production of antibiotics 445
Salmonella, binary fission 114
Salmonella newport, specific growth rate 130
sanguinarine, production in cell culture 593
SASP *see* small acid soluble protein
sauerkraut fermentation 333
sausage fermentation 332
scanning electron microscopy 9
– conventional - 12
– principle 12
– sample preparation 9
– transmission - 12

Schizosaccharomyces pombe 473
Schwanniomyces occidentalis 474
– α-amylase gene (*AMY1*) 201
– glucoamylase gene (*GAM1*) 201
secondary metabolism, functions 444
– of plant cell cultures 589ff, 595ff
– pathway redesign 214
secondary metabolites, function 165
secretion 76
– in corynebacteria 454
– metabolite transport system 76
selectable marker, in fungi 536
serine pathway 277
serine production, by methylotrophic microorganisms 279
Serpens, phylogenetic status 404
shear stress, effect on plant cells 586
shii-take *see Lentinus edodes*
shikonin production, by cell cultures 589, 592
– copper effect 594
shoot culture, of *Apocynum cannabinum* 601
siderophore 180
– production by pseudomonads 423
– receptors in pseudomonads 423
– synthesis regulation 180
silage 333
– fermentation 341
silicone oil centrifugation 191
single cell protein (SCP) production, by methylotrophic microorganisms 278
Siphoviridae 550
SL medium, selective for *Lactobacillus* 347
small acid soluble protein (SASP), in *Bacillus* endospores 388f
sodium, effect on microbial growth 131
soil microorganisms, microenvironments 244
solid-state fermentation, with Basidiomycetes 521
solute transport systems, in *Lactococcus* 351 *see also* membrane transport
somaclonal variation 603
somatic embryogenesis 603
sour dough 333
spermidine, in pseudomonads 405ff
spores *see also* endospores
– fungal 517
Sporosarcina 368
sporulation, genes 390
– of bacilli 390ff
– of clostridia 292
– of streptomycetes 437
staining 10
– Gram- 10
– negative - 11
– ruthenium red - 10
Staphylococcus aureus, specific growth rate 130
starch degradation, by improved cultures 200
– in clostridia 297
– – enzymes 310
– in *Saccharomyces cerevisiae* 200
starch-degrading enzymes, in clostridia 310
starter culture, basic medium 349
– *Bifidobacterium* 346
– for food production 349
– for lactic fermentation 328
– preparation 349
– preservation 349
– propagation 349
– storage 349
sterilization 129
– of callus cultures 583
steroid conversion, by clostridia 311f
steroid hormone, hydroxylation by fungi 529
stirred tank reactor 587f
– oxygen transfer rate 125
storage polymer production 174
strain improvement, L-amino acid production 206ff
– bacilli 396
– corynebacteria 456
– degradation of, aromatics 202f
– – starch 200
– – xenobiotics 210ff
– ethanol production 204f
– filamentous fungi 532, 536f
– lactose metabolization 195
– microbial product modification 204ff
– plant cell culture 592
– streptomycetes 449ff
– xylose metabolization 196f
Streptobacteria 329
streptococci 170f
– homolactic fermentation 172
Streptococcus 329, 344f
– binary fission 114
– habitats 344
– hemolytic - 344f
– – isolation of 349
– oral -, isolation media 349
– – species 345
– pathogenicity 344f
Streptococcus faecalis see Enterococcus faecalis
Streptococcus mutans 344
Streptococcus thermophilus 332, 335, 344f
– habitats 344
– isolation media 348
Streptomyces, actinophages 568
– chemotaxonomic marker 435
– chromosome 438f
– – architecture 439
– – silent region 439
– G+C content 435, 440
– morphology 435ff
– production of antibiotics 445

Streptomyces aureofaciens, microcolonies 232
Streptomyces coelicolor 438
Streptomyces griseus, streptomycin synthesis regulation 441
Streptomyces lividans 66 451
streptomycetes 433ff
– actinophages 440
– antibiotic production, genes 439
– antibiotic resistance, genes 447
– arthrospores 437
– attractant synthesis 444
– biotechnological application 434, 447ff
– biotransformation of antibiotics 452
– cell-cell interaction 437
– conidia 437
– conjugative plasmid 440
– culture 436f
– – genetic stability 439
– differentiation 437
– diffusible autoregulator 441
– – A-factor 441
– drug design 451
– ecology 436f
– extracellular enzyme production 451f
– gene regulation 440
– genetic engineering 438, 442, 450
– genetic screening 450
– genetics 438ff
– growth conditions 438
– hormone synthesis 444
– human protein production 451
– industrially applied species 447
– isolation 436ff
– β-lactam pathway, improvement 450
– life cycle 436
– metabolite overproduction 450
– modular gene product 446
– mutasynthesis 450
– mycelium 437
– nitrogen cycle 443
– pathway engineering 450
– phosphate regulation 443
– precursor formation improvement 451
– primary metabolism 442
– products 447
– regulatory codon 442
– repellant synthesis 444
– RNA polymerase, σ-factor 440
– secondary metabolism 443ff
– – regulation of 447
– secondary metabolites 443ff
– strain improvement 449
– substrates 436f
– taxonomy 434
– transcription promoter 440
– – RpoD-like sequence 440
– vectors 442
streptomycin, biosynthesis, gene cluster 441
– – regulation 441
Streptoverticillium 436
substrate concentration, profiles in biofilms 227
substrate inhibition 132f
substrate level phosphorylation 56
substrate uptake, effect on microbial growth 139
– in plant cell cultures 586
– specific rate 132
substrates, for cell cultures 195ff
– for clostridia 287ff, 308
– for corynebacteria 454
– for lactic acid bacteria 334ff
– for methanogens 255
– for methylotrophs 266f
– for pediococci 340
– for streptomycetes 436f
– spectrum expansion 195, 201
succinate metabolization, in clostridia 303
sucrose, effect on plant cell culture 594
sugar acid metabolization, in clostridia 298
sugar processing, microflora 335
sugar transport *see* saccharide transport
Sulfolobus shibatae, phage SSV1 548
sulfur, physiological function 175
– supply, in plant cell culture 581
sulfur cycle 93f
sulfur limitation, effect, on *Escherichia coli* 173
– – on metabolism 172ff
– – on pyruvate production 176
sulfur metabolism 93
– assimilatory sulfate reduction 94
– desulfurylation 94
– dissimilatory sulfate reduction 94
– sulfur oxidation 95
– sulfur respiration 95
superoxide dismutase activity, in clostridia 293
suspension culture *see* plant cell culture
symport, over membranes 74
synchrony, in plant cell cultures 584
synthase *see* ligase

T

tagatose-6-P pathway 353
tapioca *see* cassava (*Manihot esculenta*)
tartrate metabolization, by pseudomonads 422
taxonomy, of bacilli 368ff
– of bacteriophages 548ff
– of corynebacteria 434, 453
– of lactic acid bacteria 328ff
– of methanogens 255ff
– of methylotrophs 268ff
– of pseudomonads 402, 404ff
– of streptomycetes 434f
TCA cycle *see* tricarboxylic acid (TCA) cycle
Tectiviridae 550

tetrachloromethane degradation 313
tetrad analysis, of yeast spores 473
Tetragenococcus 340
tetrahydrofolic acid 72
Tetrahymena thermophila, enzyme production 239
Thalictrum minus, auxin effect in suspensions 591
– berberine production 598f
Thermoactinomyces 368
Thermobacteria 329
thermodynamics 52ff
– first law 52
– free energy 53
– internal energy 52
– irreversible 52ff
– of microbial growth 141
– second law 52
– standard free energy of formation 53
thermophilic bacilli *see* bacilli
thermophilic microorganisms, specific growth rate 128
thiamine diphosphate (TPP) 72
threonine production 209f
tight junction 40
tissue organization 7
toluene degradation, by pseudomonads 416
– gene regulation 411
tonoplast 36
topoquinone (TPQ) 274
Torulopsis 269
toxin, of *Bacillus thuringiensis* 377
– gene transfer, by phages 545
transaldolase, gene (*TAL*) 480
transcription promoter 97, 106
– in streptomycetes 440
– in yeast 489, 499
transcription regulation, bacteriophage genes 554f
– yeast genes 491f
transduction 545
– generalized - 559
– specialized - 559
transfer RNA (tRNA) 98
– aminoacyl-tRNA 100
transferase 66
transformation, methods 477
– of bacilli 395
– of plant organ cultures 601
– of yeasts 476f
transketolase, gene (*TKT*) 480
translation 97
translation regulation 106 *see also* enzyme synthesis regulation
– attenuator model 106
– GCD/GCN system 495
– in yeast 494
– of phage genes 556
transmission electron microscopy 10, 12
– conventional 12
– cytochemical sample preparation 10
– immunocytochemical sample preparation 11
– low temperature 12
– – sample preparation 10
– shock freezing 10
– ultrathin sectioning 10
– – chemical fixation 10
transport systems, in clostridia 298
transposons, in pseudomonads 410f
– pWWO plasmid 411
trehalose, in yeast 485
tricarboxylic acid (TCA) cycle 85, 484
2,4,5-trichlorophenoxyacetic acid degradation 201
Trichoderma reesei, growth kinetics 133
Trichosporon cutaneum, growth kinetics 133
triosephosphate isomerase, gene (*TPI*) 481
truffle 532
trypticase-yeast-extract-cystine (TYC) agar 349
tryptophan, effect on plant cell culture 596
tryptophan tryptophyl quinone (TTQ) 274
Tuber 532
Tuber melanosporum 525

U

ubiquinone 71, 404
ultramicrotomy 10
– cryo ultramicrotomy 10
– freeze etching 10
– freeze fracturing 10
uniport, over membranes 74ff
upstream activating site (*UAS*), in yeast 492f
upstream repressing site (*URS*), in yeast 492f
urea, estimation methods 126

V

vacuole 36, 578
vaginal flora 331
vectors, phage lambda 562
– phage Mu 562
– for streptomycetes 442
– in *Lactococcus* 362
– in yeast 499
viable plate count 120
viral protein production, by *Saccharomyces cerevisiae* 498
virion 546
virus 546 *see also* bacteriophage
– classification scheme 548ff
– genome structure 549
– mRNA formation 551
– nucleic acid replication 551

vitamin, determination, by lactic acid bacteria 364
– precursor production 211
– synthesis, in yeast 490f
vitamin C *see* L-ascorbate
Volvariella volvacea 534f

W

water activity, effect on microbial growth 130
weak organic acids, membrane proton transport 181
whisky production, microflora 334
white-rot fungi 534f *see also* filamentous fungi
– enzyme production 534
wine production 334
– process improvement 496
– red wine 341
– *Saccharomyces* strains 495

X

xanthan gum production, modification 213
Xanthomonas, phylogenetic status 404
Xanthomonas campestris, xanthan gum production 213
xenobiotics 201ff
– degradation 235
– – by clostridia 311f
– – by pseudomonads 414ff
– – pathway construction 201
– – plasmid compatibility 410
– – strain improvement 201ff
– detoxification, by white-rot fungi 534f
xylan degradation, in clostridia 297
xylem, cells 580
xylene degradation, by pseudomonads 416
– gene regulation 411
xylose isomerase 196
xylose metabolization 205
– by improved cultures 196f
– cofactor regeneration 198f
xylulose monophosphate (XuMP) cycle 84, 276

Y

Yarrowia lipolytica 470ff
yeast artificial chromosome (YAC) 499f
yeast centromer plasmid (YCp) 499f
yeast chromosomes 474f
– repetitive DNA sequence 475
– replication origins (*ARS*) 475
yeast episomal plasmid (YEp) 499f
yeast growth *see* microbial growth
yeast integrative plasmid (YIp) 499
yeast linear plasmid (YLp) 499f
yeast plasmids 475
– 2m plasmid 476
yeast replicative plasmid (YRp) 499
yeasts 32, 469ff
– amino acid metabolism 489
– basal media supplements 143
– biotechnologically important species 470
– budding 115f
– carbohydrate metabolism 478ff
– – defective mutants 480
– – regulation 486ff
– cell compartmentation 478
– cell envelope organization 32
– cell structure 32
– characteristics 470
– Crabtree effect 133, 154
– cultivation 471
– DNA binding proteins 492
– expression vector 499
– extrachromosomal inheritance 476
– galactose-inducible gene 493
– GCD/GCN system 495
– gene expression regulation 491ff
– genetic manipulation 476
– genetic organization 474
– gluconeogenesis 484
– haploid strain production 473
– heterologous gene expression 496ff
– industrial applications 495
– isolation 471
– killer phenotype 476
– life cycle 472ff
– lipid synthesis 490f
– mating types 472
– – *HO* genes 473
– metabolic compartmentation 478
– methanol metabolization 271, 471
– methylotrophic 269
– occurrence 471
– phosphatases 490f
– – genes 491
– pseudomycelia 115
– Pol II mediated gene transcription 492
– post-transcriptional gene regulation 494
– protein secretion 500
– *RAS* genes 487f
– RAS/cAMP signaling pathway 487f
– RNA polymerase 492
– selective marker 499
– storage carbohydrates 485
– tetrad analysis 473
– transcription, promoter 489, 499
– – regulation 491f
– transformation methods 476f
– transposable element 475
– vitamin synthesis 490f
– xylose metabolization 196
yoghurt *see* milk products

Z

Zoogloea, phylogenetic status 404
Zygomycetales, metabolite production 522
Zymomonas mobilis, L-alanine production 210
– continuous cultivation kinetics 158
– elemental fluxes 125
– ethanol production 136
– growth kinetics 133
– lactose metabolization 195
– ^{31}P NMR spectrum 192f
– xylose metabolization 198f